Lecture Notes in Mech

Lecture Notes in Mechanical Engineering (LNME) publishes the latest developments in Mechanical Engineering - quickly, informally and with high quality. Original research reported in proceedings and post-proceedings represents the core of LNME. Volumes published in LNME embrace all aspects, subfields and new challenges of mechanical engineering. Topics in the series include:

- Engineering Design
- Machinery and Machine Elements
- Mechanical Structures and Stress Analysis
- Automotive Engineering
- Engine Technology
- Aerospace Technology and Astronautics
- Nanotechnology and Microengineering
- Control, Robotics, Mechatronics
- MEMS
- Theoretical and Applied Mechanics
- Dynamical Systems, Control
- Fluid Mechanics
- Engineering Thermodynamics, Heat and Mass Transfer
- Manufacturing
- Precision Engineering, Instrumentation, Measurement
- Materials Engineering
- Tribology and Surface Technology

To submit a proposal or request further information, please contact the Springer Editor in your country:

China: Li Shen at li.shen@springer.com
India: Priya Vyas at priya.vyas@springer.com
Rest of Asia, Australia, New Zealand: Swati Meherishi at swati.meherishi@springer.com
All other countries: Dr. Leontina Di Cecco at Leontina.dicecco@springer.com

To submit a proposal for a monograph, please check our Springer Tracts in Mechanical Engineering at http://www.springer.com/series/11693 or contact Leontina.dicecco@springer.com

Indexed by SCOPUS. The books of the series are submitted for indexing to Web of Science.

More information about this series at http://www.springer.com/series/11236

BBVL. Deepak · DRK Parhi · Pankaj C. Jena
Editors

Innovative Product Design and Intelligent Manufacturing Systems

Select Proceedings of ICIPDIMS 2019

Editors
BBVL. Deepak
Department of Industrial Design
National Institute of Technology
Rourkela, Odisha, India

DRK Parhi
Department of Mechanical Engineering
National Institute of Technology
Rourkela, Odisha, India

Pankaj C. Jena
Department of Production Engineering
Veer Surendra Sai University of Technology
Burla, Odisha, India

ISSN 2195-4356 ISSN 2195-4364 (electronic)
Lecture Notes in Mechanical Engineering
ISBN 978-981-15-2698-5 ISBN 978-981-15-2696-1 (eBook)
https://doi.org/10.1007/978-981-15-2696-1

This Springer imprint is published by the registered company Springer Nature Singapore Pte Ltd.
The registered company address is: 152 Beach Road, #21-01/04 Gateway East, Singapore 189721, Singapore

Preface

This volume contains research articles selected from the First International Conference on Innovative Product Design and Intelligent Manufacturing System (ICIPDIMS 2019) held at the National Institute of Technology, Rourkela, India. This conference provided professors, researchers and students an international forum to present and share knowledge in the areas of design and manufacturing technologies.

The chapters in this volume revolve around eight themes: (i) design in aesthetics, ergonomics, CAD and UX/UI; (ii) design forecast and sustainability; (iii) design for creativity and optimization; (iv) smart manufacturing, precision engineering and metrology; (v) artificial intelligence in manufacturing processes; (vi) optimization and simulation; (vii) virtual manufacturing; and (viii) robotics and automation, mechatronics.

We are grateful to our reviewers and the advisory board for performing the peer-review process in order to enhance the quality of the research. We are also thankful to the technical session chairs and institute administration for their help and contribution to make the conference a success. The quality of presentations delivered by the authors and their passion to communicate with the other participant made this conference series a grand success.

Rourkela, India — BBVL. Deepak
Rourkela, India — DRK Parhi
Burla, India — Pankaj C. Jena
(Co-ordinators, ICIPDIMS)

Contents

About the Editors

Dr. BBVL. Deepak is currently working in National Institute of Technology, Rourkela as head of the department of Industrial Design. He received his Master's and PhD from National Institute of Technology, Rourkela in 2010 and 2015 respectively. He has 8 years of research and teaching experience in robotics and product design fields. Currently, he is supervising 8 doctoral students and published more than 100 papers in various peer-reviewed journals and conferences. He is also currently handling two sponsored research projects in the field of robotics. He received the IEI Young Engineer Award in the field of Mechanical Engineering in 2017.

Dr. DRK Parhi is working in NIT Rourkela as a Professor. He is currently heading the department of Mechanical Engineering. He has received his PhD in Mobile Robotics from Cardiff School of Engineering, UK. He has 26 years of research and teaching experience in robotics and artificial intelligence fields. He has guided more than 20 doctoral students and published more than 300 papers in various journals and conferences. He has also completed and is currently handling several sponsored research projects in the field of robotics.

Dr. Pankaj C. Jena is currently an Associate Professor at the Department of Production Engineering, Veer Sirendra Sai University of Technology, Burla, Odisha, India. He obtained his B.E. (Mechanical Engineering) and M.Tech (Mechanical System Design) from Biju Patnaik University of Technology, Odisha, and Ph.D. from the Jadavpur University, Kolkata. His major areas of research include composite material structures, fault diagnosis, fuzzy technique, vibration engineering and use of waste materials in engineering applications. He has published 19 papers in reputed international journals, 15 papers in international conferences, 6 book chapters, as well as one book. Currently, he is an editorial board member of the International Journal of Materials Manufacturing Technology.

Design in Aesthetics, Ergonomics, CAD and UX/UI

A Strategy for Ergonomic Design of Amphibian Vehicle

Debashis Majumder and Anirban Chowdhury

Abstract Amphibian Vehicle (AV) is a vehicle that can run on land as well as in the water. Several attempts have been made to construct various models of these vehicles in the past. Because of the need for such a vehicle, it had gained popularity. Most of the models did not run for a long time because of the high cost and few operational problems. Attempts have been made in this paper, to look at the design aspects of AV and to develop a framework considering ergonomic and styling aspects of design. This framework can represent the AV design process in a better way considering ergonomics and styling aspects. This framework will help to customize AV for different areas of applications like military application, flood rescue operation, etc. This framework is useful to improve aesthetics, styling and ergonomics (comfort and safety) of AV, apart from engineering.

Keywords Amphibian Vehicle · Automobile · Concept · Design · Emotion

1 Introduction and Background

It has been observed that the human race started settling on land approximately 200,000 years back across the globe. The specific vehicle has been designed for moving on land and on water separately. Very few attempts have been made where the same vehicle used to move on the water as well as on land. The experience of bi-terrestrial requirement led engineers to create the same vehicle suitable for running in water as well as land. The kind of vehicles which rover on multiple territories (land and water) are called AV [1].

D. Majumder · A. Chowdhury (✉)
School of Design (SoD), University of Petroleum and Energy Studies (UPES), Dehradun, Uttarakhand 248007, India

BBVL. Deepak et al. (eds.), *Innovative Product Design and Intelligent Manufacturing Systems*, Lecture Notes in Mechanical Engineering,
https://doi.org/10.1007/978-981-15-2696-1_1

1.1 Amphibian Vehicle for Military Requirement

Several military vehicles are converted to AV for specific requirement of the field. The DUKW of Germany has produced many models of Amphibian Trucks during World War II. Several models of Amphibian Trucks with 6 × 6 wheels carrying guns and soldiers across coastal areas were deployed. Refer Fig. 1c for DUKW Amphibian Truck. In the UK, Alvis Stalwart produced Amphibian Trucks with 6 × 6 wheels for cargo and military transport [1]. Refer Fig. 1d for Alvis Stalwart Amphibian Truck. It looks like a truck on Hull [1].

1.2 Amphibian Bus

Amphibian Bus is another manifestation of recreational vehicle in history. The splash tour in Rotterdam, the Netherlands, is a popular tourist attraction. It looks like a normal bus with a high-level floor. The internal mechanisms are different on the water and on roads. The vehicle can carry 25 passengers and go at 8 knots on water [1. 2]. Refer Fig. 1e for Amphibious Bus used in Rotterdam, Netherland.

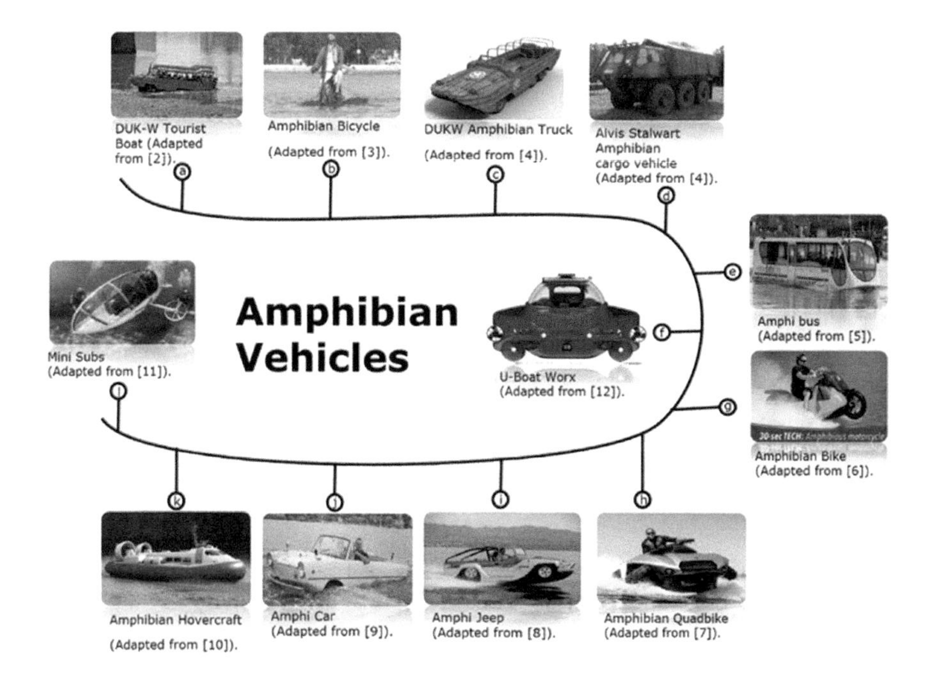

Fig. 1 Miscellaneous types of AV

1.3 Amphibian Car

This is a very popular version of Amphibian Vehicle found in history. One company in Germany produced and sold in the USA and Europe. It looked like a car but with 4 × 4 drive watertight chassis. It was with 43 HP engine, Porche 356 transmission and BMW control unit. It could sail on rivers and run on roads. People liked the model as it almost looked like a car [1, 2]. The Amphi Car had more ground clearance than a jeep and low first gear ratio. Power-assisted steering and oversize tire. Refer Fig. 1j for Amphi Car. Though there are mentions about different types of AV design in literature, the user-centric approach is not well established for AV design. Therefore, the aim of this study is to develop a user-centric framework for AV design and development-based critical review of the literature.

2 Possibility of Applications of AVs for Recreational Purpose

Probably, the biggest sector in AV is found in recreational activity. History has seen various types of vehicles in the 1950s. DUKW of Boston made various models of the ferry, which could run on roads [1]. Refer Fig. 1e for Amphi Bus.

2.1 Amphibian Cycle

This is a kind of bicycle that can run on land as well as on water. The first coverage about the Amphibian Cycle was done by BBC News and by Discovery Channel. It is known as Saidullah's Bicycle (please refer Fig. 1b). It used four rectangular air filled tanks for buoyancy. It used a tricycle frame and a fan blade for propelling. The manual peddled cycle could reach a speed of 1.12 m/s on water. This was appreciated by urban areas in applications like floods and in the leisure industry [2].

2.2 Amphibious Bike

The most known Amphibious Bike is made by Gibbs of the USA. It is having 55 HP twin-cylinder engine. This has 80 mph speed on land and 35 mph on water. The two-wheeled AV weighs 502 lb [3]. Refer Fig. 1g for Amphi Bike made by Gibbs.

2.3 Amphi Jeep

The most popular form of AV is Amphibian Jeep, which almost looks like a jeep but moves on the water at a speed of 44 mph. As of now, it is the fastest watercraft vehicle. The most popular company is 'Water Craft' in California, USA. It has a 3.7-lt V6 engine, sold under the brand name of 'panther' [4]. Refer Fig. 1h for Amphibian Quadbike and Fig. 1i For Amphi Jeep.

2.4 Amphi Submarine

Scubster is a model of Amphibian submarine used for recreation and adventure by Scubster team in the USA in 2011. It took part in the international naval race championship. It is an electric version vehicle with the two-seater arrangement. Two electric motors operate it and it has a maximum speed of 8 km/h inside water [5]. Refer Fig. 1l for model of mini Subs. Like the above, there are many more forms of recreation vehicles, which are mainly for water sports and adventure [6, 7].

3 Design Philosophy of Existing Amphibian Vehicle

If we observe the design philosophy of the existing Amphibian Vehicle, we find that all the vehicles are the need-based conversion of a land vehicle. The vehicle shapes are of either bike, jeep or car. If we do the task analysis of an AV that is different from the land vehicle, the shape and structure are not conducive to a water vehicle. Currently, user aspirations are not taken for the design of an AV. Since the vehicle has completely different features and functions, a user aspiration survey is done for the design of a recreational Amphibian Vehicle [8].

3.1 Current Design Strategies

Literature survey reveals that all the Amphibian Vehicle design is converting road vehicle into a water vehicle. For instance, in the case of Amphibian Jeep or Amphibian Car, the same look is maintained. The principles of Hull design and ergonomic consideration of ingress and outgress in water are different from land vehicles. Moreover, Amphibian Vehicle characteristics are unique in nature, so they deserve aesthetics not matching with boats or land vehicle. For example, Amphibian Bus looks like a bus and the structure is similar to a bus. While the surprise factors for the spectators (i.e. bus can run on water) are also counted, still it does not last long. Based on experiments, the different kinds of AVs were made at a different time in history for specific purposes. Mostly, AVs were sold for defence purpose; however, very limited and customized edition of AVs are available for public use. Therefore, there is no user-centric standard and process available for AV design. It is accepted that the full range of human feelings influences decisions to purchase. Sensory aspects of design should be congruent with the product's appearance. Boundaries can be difficult to establish relevant to human perception. Kansei Engineering process is a useful tool to match visual perception with the form of the vehicle. This process allows the styling of the vehicle as per the sensory need of the consumer [8].

4 A Framework for AV Design

A framework is a mapping of a sequence of actions and micro-actions taken while designing a product adopted by designers. The richness of the process depends on minute considerations of factors that may affect the design deliverable. It is also based on the current situation and probable factors for a holistic design solution [9].

The formal concepts of Amphibian Vehicles will be different in consideration of a land vehicle. The buoyancy and stability considerations are very specific to move on water. The propulsion requirement is also different, as it needs to float on water and travel on water. The drag calculation on water for high-speed boats is areas where the Amphibian Vehicle is different from land vehicle and boats (Fig. 2).

The user considerations of the Amphibian Vehicle for example ingress and outgoes from the vehicle are also different from a car or any other land vehicle. The speed characteristics and bouncing effect on water reveal that the seating arrangement can be customized to the appropriate arrangement required for AV for recreation [10].

4.1 Requirement Analysis

For the user requirement and their aspirations in detail, 'user survey' is performed. As per the Kansei Method, the aspirations of users and keyword surveys are done. This survey is generally conducted using a questionnaire, direct interview and secondary research method. Current scenario reveals two-seater vehicle, which can run on the road as well as on water and underwater [11].

4.2 Conceptualization

Sketching and brainstorming forms the basis of AV design. The key aspirations and mood board are formed 'safety' as per the user survey. The forms of the vehicle are

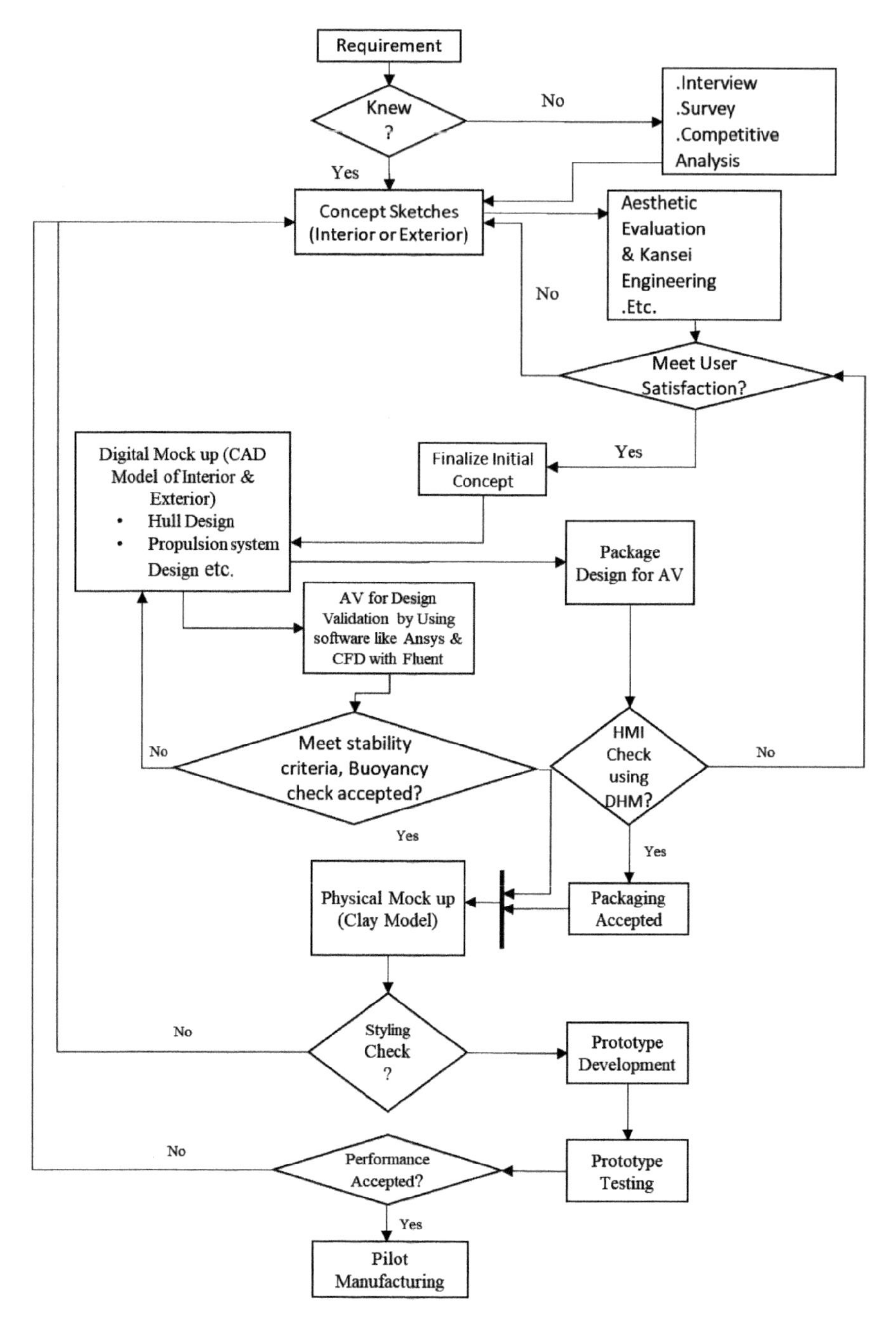

Fig. 2 A novel framework for Amphibian Vehicle (AV) design

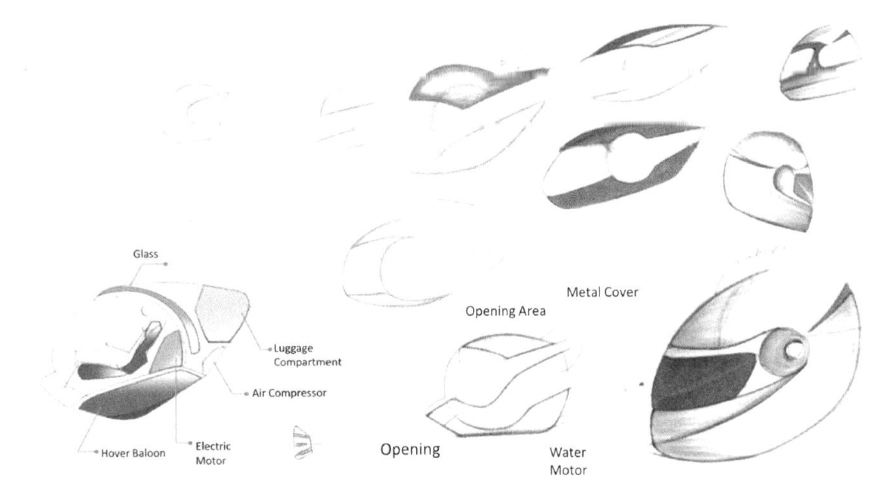

Fig. 3 Initial sketches of two-seater AV

completely different from the existing vehicle [12]. The initial inspiration of dynamic, safety and elegant was found in 'helmet'. The form generation started with helmet as basic form and transformed into Amphibian Vehicle (Fig. 3).

4.3 Interior and Exterior Sketching

The detail sketching takes into account the ergonomic considerations and packaging principles. Here, the Hull design, propulsion method selection, stability and buoyancy criteria are taken into account [13].

4.4 Design Evaluation

The evaluation of the final design is a complex process. With reference to the use of AV, emotional aspects are integrated into design processes, which also improve product identity. Amphibian Vehicle has a definite requirement of styling. It should not mix with boats or cars. The operation and comforts factors are not the same as cars or boats. Hence, the vehicle needs such a design strategy to relate to its application. In the framework, the Kansei Engineering (KE) method has been proposed which will help to design AV in a systematic manner considering all factors to arrive at an aesthetically meaningful Amphibian Vehicle. It is done through a user survey. The aesthetics and personality assessment is a key factor for the success of the product. The KE method is followed to capture and implement user opinions about vehicle aesthetics [14].

4.5 Preparation of Digital Mock-up

The vehicle concepts are converted into digital mock-ups by using CATIA/SolidWorks/Fusion 360, etc. software. The scaling of the vehicle and material allocation is also done simultaneously. In this phase, all the vehicle-engineering considerations are taken and based on vehicle design finite element analysis is also done for particular parts of the vehicle. The digital mock-up also takes care of modifications demanded as a result of FEA results and aesthetics. The assemblies are made as per available parts and some newly designed parts. This is the stage where engineering and aesthetics meet. The ergonomic considerations are also checked by using digital human models [15].

4.6 Finalizing CAD Design

Finalizing geometrical design involves design for stability and buoyancy. It also involves hydrodynamic characteristics of the floating body. At this point only, we do aggregate (e.g. engine, transmission system, etc.) selection for the vehicle [16].

The CAD design involves packaging of internal components of AV for recreation. In this case, of electric AV, the battery selection and motor selection are prime work. Once the two are selected, other items like a propeller, water jet motor, shaft, chassis and body dimensions are decided. The digital mock-up is an exact replica of actual AV, which is important for ergonomics check using digital human models [16]. While constructing the digital mock-up of AV, other considerations like materials and processes also can be considered. In the construction of AV the lightweight and volume of water, displacement by 'Hull' is considered for calculation of buoyancy. The frame material could be carbon steel, which is strong and having lightweight. For lightweight, expanded polyurethane can be considered as filling material in the body design of AV. There is a Hover fan, which is used for travel on water and on the beach also. Packaging of Hover fan and its frame attachment with the body is critical in the design of this AV. The styling and aesthetics are critical while constructing the AV design. The initial concept sketches act as a guideline while constructing the digital mock-up of AV design. For calculating efficiency, preliminary calculations can be done manually. Advanced design validation can be done by using the software. Package finalization is the first stage in design finalization. Optimization applied to materials and processes is another criterion for design finalization.

4.7 Human Machine Interface (HMI) Check Applying DHM

The design of AV is not complete without HMI check. In this case, the digital human model is used for checking a digital mock-up of vehicle concept to finalize the overall dimensions, panel dimensions, door dimensions, etc. The concepts are worked out in sketches and validated by digital mock-up [15]. The digital human models are supposed to be a replica of the human body where all attributes like tall, short, male, female, strong and lean criteria can be applied, based on the requirement. The dimensions based on postures like seating, standing and contact area in seating, grips, etc. are done with a software model. In the design of AV for recreation, DHM plays a critical role. The interior of the vehicle with seats and leg space is easily done using DHM. The visibility through the windscreen and head clearance with a ceiling of the vehicle is done using DHM easily [17–19].

4.8 Final Styling Check and Modification

Since the design grows with a particular theme in mind from the beginning, it remains even after a lot of alteration done for various reasons, for example, engineering packaging, etc. The styling check is done after every alteration in order to regain the essence of the visual design aspect. The CAD models have altered for styling considerations also. Finally, after many dummy models are done, 1:1 scale clay modelling is done for finalizing the very fine aspects of visual design. The criteria for modification were in terms of styling. The other modification was for the placement of rotor and wheels and fan [20].

4.9 Prototype Construction

The full-scale prototype is to be prepared for concept evaluation of the Amphibian Vehicle. Currently, it is not in the scope of this paper. Testing of the engineering parameters is done to check for proof of concept. After the successful satisfaction of test results, detail drawings are generated for pilot manufacturing [21].

New methods and materials that are lightweight and float on the water can be thought of using in the construction of the body of the vehicle. Plastic foam parts using rotomolding can be used for simple and easy to construct form, may be used.

4.10 Prototype Testing

Prototype testing under real condition is the ultimate stage for confirmation of result. Design modification still happens based on fine-tuning of results, at this stage. The prototype testing can be done after the construction of the prototype [22].

Generally, reliability of AV is gained from prototype testing. Prototype testing can be of three types: (1) Virtual prototype testing, (2) Actual prototype testing in laboratory, (3) Prototype testing in the field.

(1) *Virtual prototype testing* is generally done with a digital mock-up and various analysis software. For Chassis validation, static and dynamic stress analysis is done. There are several software like ANSYS, LS-DYNA, Abacus used for crash analysis. The drag on AV body can be calculated by using fluid analysis software like Fluent, Star-CD, etc.
(2) The *actual prototype test* of AV can be done by attaching instruments and measurement can be done by running the AV in simulated condition. The observed and measured reading can be used for further analysis of life prediction.
(3) The third type of testing is *field testing* of AV can be done by actually running on water and beach several times under changed load conditions. Observation can be made on all functional parameters.

5 Conclusion

Development of AV for recreation is a new area of application. The current study has developed a framework for the inclusion of important design features specific to the development of this type of vehicle. The process mentioned under the proposed framework will enable consideration of user needs and psychological aspects for AV development. If the designer considers the framework, a complete product meets the user requirement in totality. The proposed framework might be useful for another kind of vehicle design (for aesthetics and ergonomics check). Different concepts of AVs can be checked in the future using the proposed framework, which is currently not under the scope of the present study.

References

1. Shirsath PS, Hajare MS, Sonawane GD, Kuwar A, Gunjal SU (2015) A review on design and analysis of AV. Int J Sci Tech Manag 4(1):43–51
2. Alamy (2015) A bicycle that rides on water. Invented by this 60 year old man from Bihar, available at https://www.alamy.com/stock-photo/london-amphibious-tour.html. Accessed 30 Oct 2018

3. Amphi Jeep (1999) Youtube; Amphi Jeep, available at https://www.youtube.com/watch?v=m5xQy8MIuec. Accessed 30th Oct 2018
4. Gibbs Sports, Biski (2018) Sports Amphibian Bike, available at https://www.youtube.com/watch?v=6bvTlvjnGLI. Accessed 30th Oct 2018
5. Ridden P (2015) Electric Scubster personal sub dives into crowd funding pond. https://newatlas.com/scubster-nemo/38827/. Accessed on 12th April 2019
6. Hardegree M (2018) Amphicoach amphibious tourist bus: greyhound meets Dolphin, available at https://en.wikipedia.org/wiki/Alvis_Stalwart. Accessed 30th Oct 2018
7. Gibbs Sports Amphibian (2018) Quadski, available at https://www.gibbssports.com/about/. Accessed 30th Oct 2018
8. Tovey M (1997) Styling and design: intuition and analysis in industrial design. Des Studs 18(1):5–31
9. Nagamachi M, Lokman AM (2016) Innovations of Kansei engineering. CRC Press
10. Gawande AM, Mali AP (2016) Amphibious vehicle. Int Res J Eng Technol 3(10):137–141
11. Börekçi NA, Kaygan P, Hasdoğan G (2016) Concept development for vehicle design education projects carried out in collaboration with industry. Procedia CIRP 50:751–758
12. Tovey M, Porter S, Newman R (2003) Sketching, concept development and automotive design. Des Studs 24(2):135–153
13. Rieuf V, Bouchard C, Meyrueis V, Omhover JF (2017) Emotional activity in early immersive design: sketches and mood boards in virtual reality. Des Studs 48:43–75
14. Wycoff J (1991) Mindmapping: your personal guide to exploring creativity and problem-solving. Berkley Books, New York
15. Chaffin DB (2001) Digital human modeling for vehicle and workplace design. Society of Automotive Engineers Inc., Warrendale PA, USA
16. Tovey M, Owen J (2000) Sketching and direct CAD modeling in automotive design. Des Studs 21(6):569–588
17. Thaneswer P, Sanjog J, Chowdhury A, Karmakar S (2013) Applications of DHM in agricultural engineering: a review. Adv Eng Forum Trans Tech Publ 10:16–21
18. Sanjog J, Chowdhury A, Karmakar S (2012) Digital human modeling software in the secondary manufacturing sector: a review. In: Proceedings of international conference on recent trends in computer science and engineering (ICRTCSE-2012), 3–4
19. Sanjog J, Karmakar S, Patel T, Chowdhury A (2015) Towards virtual ergonomics: aviation and aerospace. aircraft engineering and aerospace technology. Int J 87(3):266–273
20. Ferreira J, Furini F, Silva N (2007) Framework for an advanced design vehicle process development. In: Proceedings of IASTED international conference on modelling, identification, and control (MIC 2007), 12–14
21. Huang T, Kong CW, Guo H, Baldwin A, Li H (2007) A virtual prototyping system for simulating construction processes. Automat Constr 16(5):576–585
22. First M, Kocabicak U (2004) Analytical durability modeling and evaluation—complementary techniques for physical testing of automotive components. Eng Fail Anal 11(4):655–674

Sustainable Design of Sprocket Through CAD and CAE: A Case Study

Anju Mathew, Arjun Santhosh, S. Vinodh and P. Ramesh

Abstract During the recent days, product development practices are to be incorporated with sustainability principles for minimization of environment impacts. This paper presents the sustainable product design of a sprocket. The existing design of sprocket is modelled using computer-aided design (CAD). The existing design is subjected to analysis using computer-aided engineering (CAE) and the weight has been reduced to derive two modified designs. The existing and modified designs are subjected to environmental impact analysis using sustainability analysis module. The environmental impacts are assessed under four categories. Based on the study, it has been found that weight of the sprocket got reduced by 14% and carbon-based impact got reduced by 2%.

Keywords Product design, sustainability · Sprocket · Environmental impacts · Carbon footprint

1 Introduction

Sustainability concepts in designs enable the development of products with minimal environmental impact. Automotive component manufactures adopt sustainability in product development practices. Sustainable designs that are environment-friendly and commercially viable are in popular demand. Moreover, reduced material usage, better material choices, increased ease of disassembly, and lower rates of hazardous materials are other distinct features of manufacturing industry that has massive potential in showcasing sustainable change. This article presents the sustainable design of sprocket. In this study, existing model of sprocket is developed. The sprocket design is further optimized. Environmental impacts were computed for existing and modified designs. The reduction in environmental impacts is

A. Mathew · A. Santhosh · S. Vinodh (✉) · P. Ramesh
Department of Production Engineering, National Institute of Technology, Tiruchirappalli 620 015, India

BBVL. Deepak et al. (eds.), *Innovative Product Design and Intelligent Manufacturing Systems*, Lecture Notes in Mechanical Engineering,
https://doi.org/10.1007/978-981-15-2696-1_2

computed. CAD modelling and simulation had been done with SolidWorks and ANSYS, respectively. The implications for design engineers are presented. The novelty of this paper is to integrate CAD, CAE, and sustainability analysis for developing an automotive component (sprocket).

2 Literature Review

Vinodh [1] carried out a case study in Indian sprocket manufacturing organization. Using computer-aided design (CAD), design optimization and sustainability analysis of an existing sprocket were done upon which assessment of environmental impact took place. The following results revealed that design optimization and CAD could aid in the development of sustainable designs having minimal environmental impacts.

Lim and Song [2] designed a highly efficient and lightweight cassette system specifically for a road bicycle. They conducted various analyses to impart sufficient mechanical strength and rigidity. Analysis under various boundary conditions was done to ensure the safety of the entire system. Branker and Jeswiet [3] proposed the usage of a new economic model for optimized selection of machining parameters taking milling example. The instance involves gathering input data for sprocket milling with regard to different feed rates and speeds. This requires both contour milling and ramping that utilize two distinct tools. The underlying approach for modification of the economic model to incorporate more complex parts was also stated. Wang et al. [4] presented issues arising from the feature of non-conjugated meshing. The author designed a new tooth profile for sprocket aimed at reducing meshing impact and polygonal action at high speeds. Analysis of the meshing and fluctuation impacts of the chain was done under various rotational speeds. The results claimed that newly designed sprocket profiles are capable of efficiently reducing the friction and meshing chain effect.

Hassan et al. [5] proposed a different approach for the application of Product Sustainability Index (ProdSI) in the selection of optimal product design configurations. They used the proposed approach for assessment of a product's sustainability performance. This approach was particularly helpful for product designers in the generation of many other similar configurations and in making the ideal choice of design possessing the most sustainable configuration. Yun et al. [6] analyzed the environmental effects of employing a cold extrusion process in the manufacture of helical gears. This analysis was done to overcome shortcomings including an excess of material loss, reduced productivity, increased consumption of energy, and considerable carbon consumption due to conventional processes. As per ISO 14000, life cycle assessment (LCA) method was followed to compare the environmental impacts of both the processes. Williams et al. [7] researched the basics of sprocket design and the construction of the reverse sprocket of Yamaha CY80 motorcycle. The research was conducted by utilizing concepts of reverse engineering. Hingve and Vanalkar [8] presented a study to eliminate faults related to sprocket failure through fault diagnosis, new tooth generation, and modelling approach of the sprockets. Suggestions and results associated with designing of such a sprocket tooth profile were briefly discussed. Gupta et al. [9] presented a paper reviewing the current technology available for sustainable processes of gear manufacturing. They also made recommendations for improving quality and productivity while promoting environmental sustainability. Moreover, certain sustainable manufacturing strategies were discussed taking into consideration environment-friendly lubrication techniques and advanced gear manufacturing processes.

Suresh et al. [10] presented an evaluation dealing with product sustainability by identifying changes in material and manufacturing processes. Four essential sources including the total consumption of energy, air acidification, eutrophication of water, and creation of carbon footprint in addition to the material financial figures for cost-effectiveness were considered. The analysis was done utilizing a sustainability module available in SolidWorks CAD software. The crankshaft in a four-cycle engine was used as an instance. Nikam and Tanpure [11] conducted analyses regarding static and fatigue aspects of the sprocket design harnessing ANSYS software and finite element analysis of chain sprocket for safety and reliability. The results obtained were utilized for optimization of the sprocket for weight reduction. Laxmikant et al. [12] optimized and experimentally validated the weight of sprocket for track hoe at various torque conditions. The author also reviewed existing geometry and redesigned the sprocket to adhere to drawbacks in the design by optimizing geometry, developing a modified CAD model, and conducting FEA on the obtained model. Suresh et al. [13] researched sustainable product design arising out of integration of Design for Environment (DFE) with Design for Manufacture and Assembly (DFMA) techniques. This research was conducted through a case study that took place within a charge alternator pulley. The study supported DFE and DFMA integration concept by highlighting the scope for sustainable design development with minimal environmental impact and a possible decline in product cost. Ambole and Pravin [14] presented a comparison between the sprocket of a Bajaj Pulsar 180 and the sprocket of carbon fibre material for stress analysis of the chain drive. They designed the sprocket as per standard procedures with input from Bajaj Pulsar 180, developed the CAD model and performed post-processing using ANSYS 13.

Han et al. [15] conducted a comparative analysis for sprocket heating processes through finite element numerical computation. Upon the analysis and comparison of the causes of several electromagnetic and geometric parameters along evaluation indicators, it was concluded that there was the possibility for a potential change in the heat transfer process taking place during sprocket induction heating. This change would be made possible by tweaking specific profile coil parameters and further performance improvement of gear. Suwannahong and Suvanjumrat [16] proposed new installed cases of the sprocket and roller chain modelling technique. They followed finite element method for the simulation of behaviour of those parts that are damageable upon the integration with the multi-body system consisting of rigid parts present in driving system. This integrating simulation would be capable of explaining any installed cases of sprocket and roller chain uniquely. Choi et al. [17] performed a study on the comparison between environmental threats imposed by the machining process and hot forging process for manufacture of sprocket wheels. They identified the critical design factors and observed energy consumption of the hot forging technique used for fabrication of sprocket wheels. Barua and Kar [18] conducted a study on the design optimization of sprockets utilizing different processes and techniques. This paper reviewed the designing of chain sprocket, analysis using FEA (whose results were used to research on how the optimization of sprocket for weight reduction was carried out). They also provided a brief overview of various techniques opted by researchers to improve sprocket efficiency through design optimization.

Naji and Kurt [19] presented a study to determine the load distribution on an elastic sprocket for an elastic roller chain. They have also discussed how the load distribution is affected by elastic properties, geometric variations and lubrication.

Shuaib et al. [20] proposed ProdSI, a product sustainability method. This method covers TBL, total life-cycle and 6R concept and put forward a set of product sustainability metrics.

3 Methodology

The methodology flow chart is shown in Fig. 1.

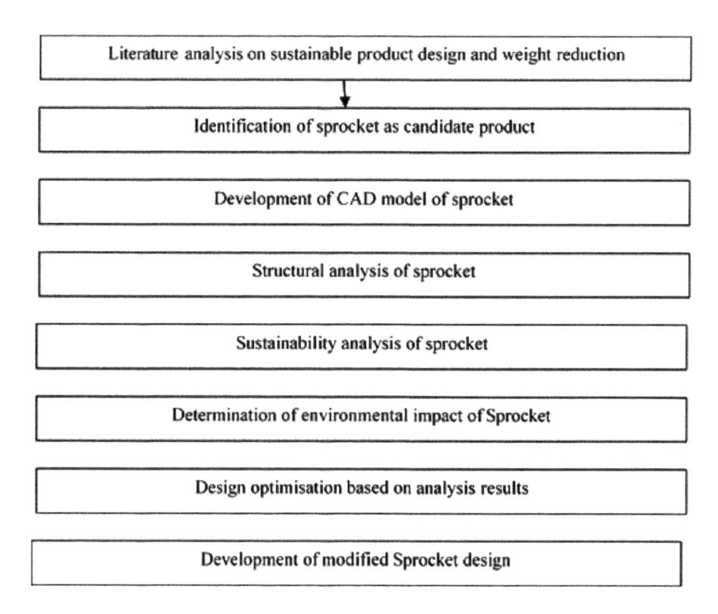

Fig. 1 Methodology flow chart

4 Development of CAD Model of Sprocket

CAD model of standard bicycle sprocket is taken as a case product. 3D model of sprocket is developed using SolidWorks platform. Figure 2 shows 3D model of sprocket and Fig. 3 presents detailed view of existing design.

Further, two more different designs were developed using same software platform. Figures 4 and 5 state 3D model of modified design and detailed view of modified design. Figure 4 shows simple holes were developed to reduce weight of sprocket and chamfer and fillets were added to deduce weight.

5 Finite Element Simulation of Sprocket

The force is applied tangentially by resolving the axis. Force magnitude is considered from the research article by Nikam and Tanpure [11]. The forces applied on respective sprocket tooth are shown in Table 1. Figure 6 shows boundary condition used for sprocket design. Material Considered is AISI 1045 Steel.

Fig. 2 CAD model of existing design

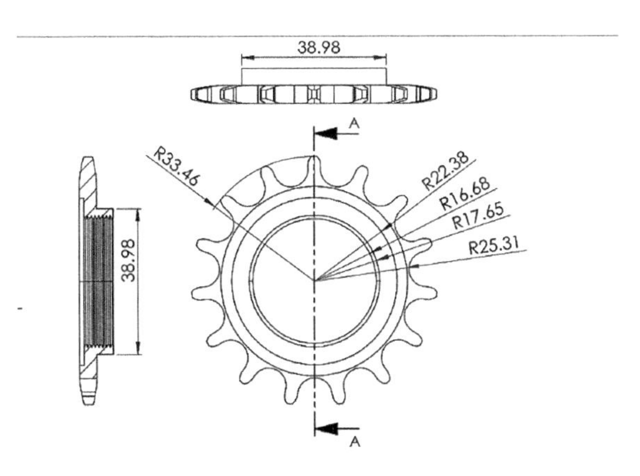

Fig. 3 Detailed view of existing design

Fig. 4 CAD model of modified design

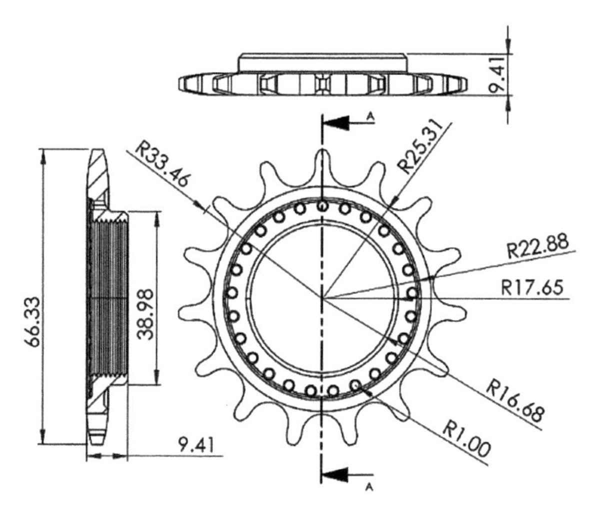

Fig. 5 Detailed view of modified design

Table 1 Force applied on respective sprocket tooth

Tooth number	Force in '*N*'
T1	1177.9
T2	1251.9
T3	846.16
T4	567.99
T5	381.27

Figure 7 shows the total displacement of existing design and the maximum displacement observed is 0.022637 mm. The same kind of boundary conditions is followed for all designs. Force is applied tangentially by resolving the axis. Force applied in T1, T2, T3, T4, T5 are 1177.9 N, 1251.9 N, 846.16 N, 567.99 N, 381.27 N, respectively.

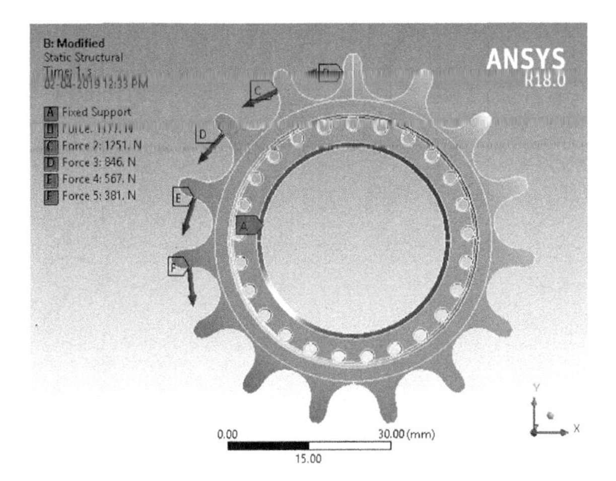

Fig. 6 Boundary condition of sprocket

Fig. 7 Total deformation of existing design

Figure 8 shows Von Mises stress of existing design and stress developed is 385.69 MPa. Same kind of boundary conditions is followed for all designs. Force is applied tangentially by resolving the axis.

Figure 9 shows the total displacement of modified design and the maximum displacement observed is 0.027444 mm. After analysis of present sprocket design, design optimization has been done to find the optimized design parameters. Same kind of boundary conditions is followed for all designs. Force is applied tangentially by resolving the axis. Figure 10 shows the Von Mises stress of modified design and the stress developed is 348.22 MPa. Same kind of boundary conditions is followed for all designs. Force is applied tangentially by resolving the axis.

Fig. 8 Von Mises stress of existing design

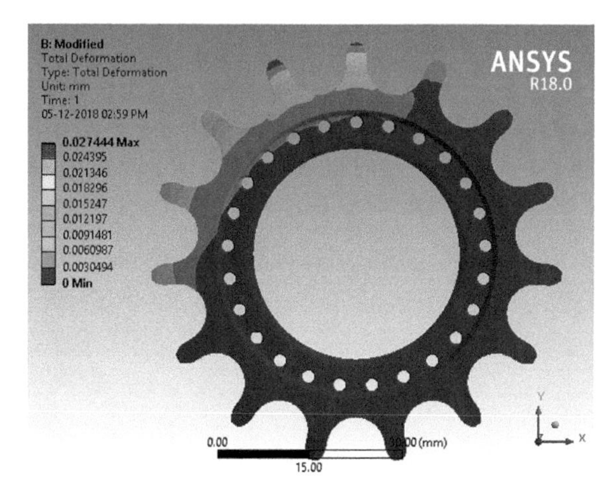

Fig. 9 Total deformation of modified design 1

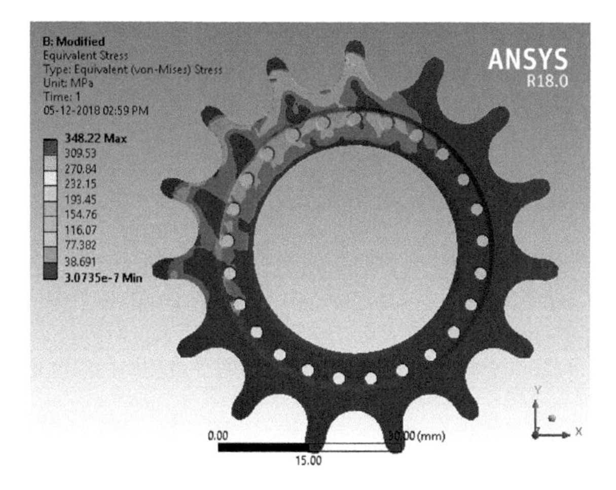

Fig. 10 Von Mises stress of modified design 1

6 Topology Analysis

Figure 11 shows results of topology analysis; after using the topology analysis module of ANSYS Workbench, a weight reduction of 7% was obtained by marginal removal of material near the fixed support.

6.1 Simulation Results of Modified Design 2

Figure 12 shows the total deformation of modified design 2 and Fig. 13 shows the Von Mises stress of modified design 2.

The comparison of design parameters obtained after structural analysis is represented in Table 2.

AISI 1010 steel is chosen as sprocket material for both existing design and modified design. Figure 14 expresses environmental impact of sprocket existing design. The environmental impacts are assessed using SolidWorks sustainability analysis module. Inputs for the analysis are CAD model of design, material, manufacturing process, and manufacture and user location.

Figure 15 shows the environmental impact comparison of existing and modified designs with same material. Table 3 compares the results of environmental impact and energy consumptions associated with different design strategies.

Nikam and Tanpure [11] were carried out CAE analysis for sprocket and found stress developed. Same boundary condition was followed for present study. The idea of present study is to develop a sustainable product through CAD, CAE without affecting physical performance of sprocket.

Fig. 11 Topology analysis

Fig. 12 Total deformation of modified design 2

Fig. 13 Von Mises stress of modified design 2

Table 2 Comparison of design parameters obtained after structural analysis

Design	Total volume of sprocket in mm^3	Total displacement in 'mm'	Von Mises stress in MPa
Existing design	8082.06	0.02263	385.69
Modified design 1	7296.52	0.02744	348.22
Modified design 2	6785.77	0.02901	328.21

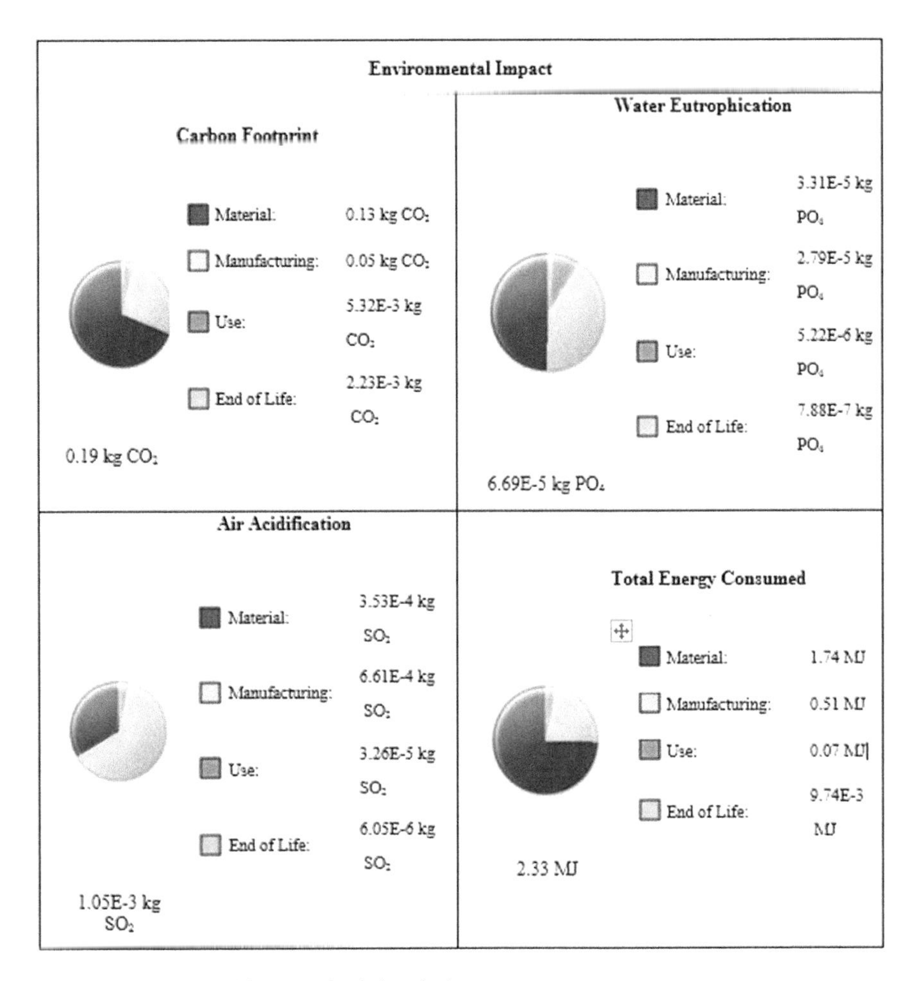

Fig. 14 Environmental impact of existing design

7 Conclusion

This article aims at sustainable development of sprocket. After the development of CAD model of sprocket design, engineering analysis is done using CAE. The design is optimized. Environmental impacts in terms of carbon foot print, acidification, eutrophication, and energy consumption are determined for both existing and modified designs. The conduct of the study facilitated the development of product with minimal environmental impacts. In the present study, CAD and CAE have been used for developing sustainable design. In future, optimization approaches also could be attempted towards developing sustainable design. Also, sustainable design of other automotive products could be attempted.

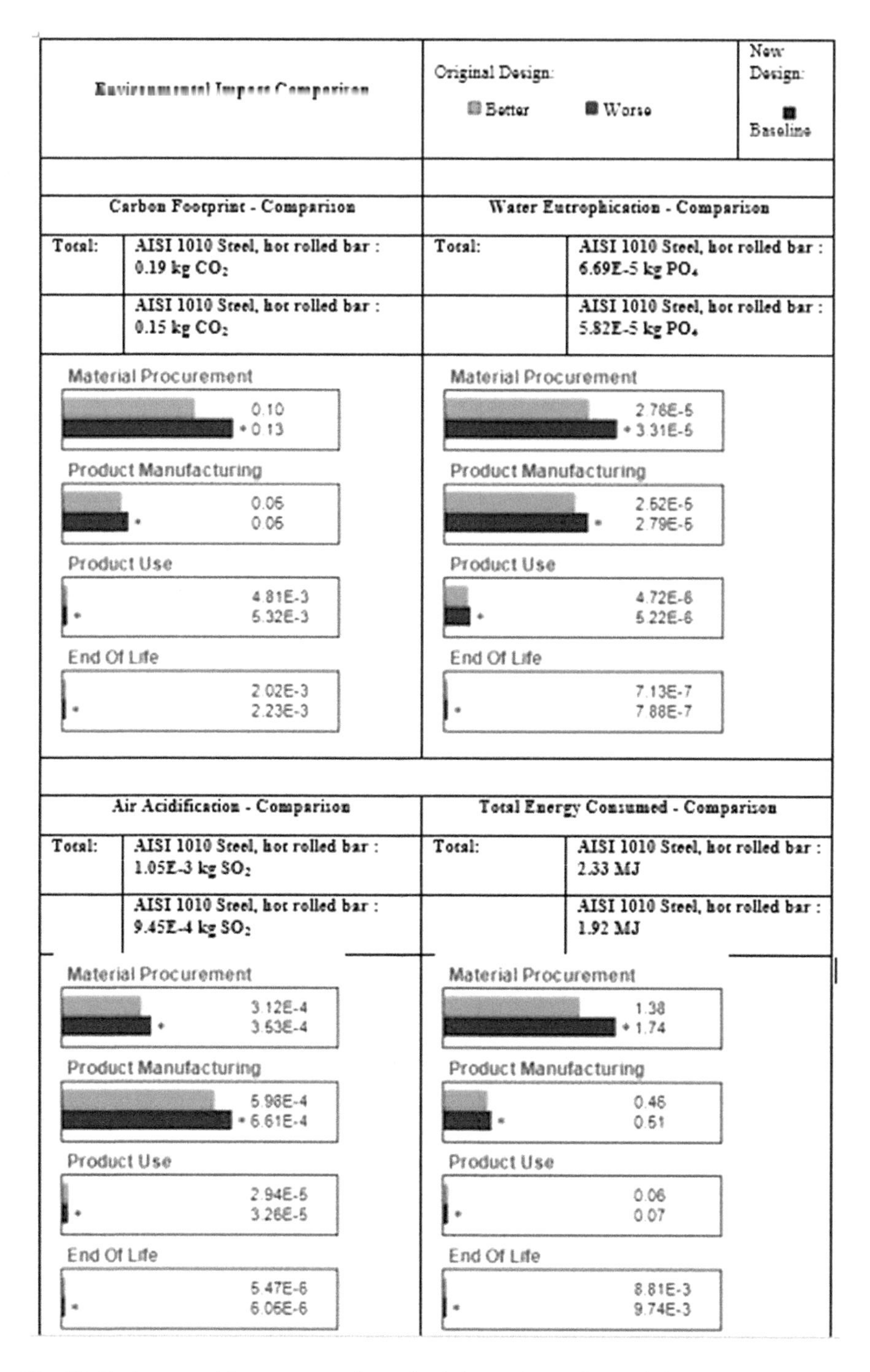

Fig. 15 Environmental impact comparison of existing design and modified design

Table 3 Comparison of results of different design strategies and existing design

Impacts design strategies	Carbon footprint in kg CO_2 (kg)	Water eutrophication in kg PO_4 (kg)	Air acidification in kg SO_2 (kg)	Total energy consumed in MJ
Existing design	0.19	6.69E-5	1.05E-3	2.33
Design strategy 1	0.15	5.82E-5	9.45E-4	1.92
Design strategy 2	0.13	4.84E-5	8.5E-4	1.72

References

1. Vinodh S (2011) Sustainable design of sprocket using CAD and Design Optimisation. Environ Dev Sustain 13(5):939–951
2. Lim YS, Song BU, Shin YW, Seo TI (2013) Design of lightweight and highly efficient cassette system for road bicycle by using finite element analysis. In: Applied mechanics and materials, vol 271 Trans Tech Publications, pp 720–726
3. Branker K, Jeswiet J (2012) Using a new economic model with LCA-based carbon emission inputs for process parameter selection in machining. In: In Leveraging technology for a sustainable world. Springer, Berlin, Heidelberg, pp 323–328
4. Wang Y, Ji D, Zhan K (2013) Modified sprocket tooth profile of roller chain drives. Mech Mach Theory 70:380–393
5. Hassan MF, Saman M, Sharif S, Omar B, Zhang X, Badurdeen F, Jawahir IS (2013) Selection of product design configuration for improved sustainability using the product sustainability index (ProdSI) scoring method. Appl Mech Mater 315:51–56
6. Yun JH, Jeong MS, Lee SK, Jeon JW, Park JY, Kim GM (2014) Sustainable production of helical pinion gears: environmental effects and product quality. Int J Precision Eng Manufact-Green Technol 1(1):37–41
7. Williams ES, Emmanuel A, Peter O (2014) Fundamentals of sprocket design and reverse engineering of rear sprocket of a yamaha CY80 motorcycle. Int J Eng Technol 4(4):170–179
8. Hingve MTS, VanalkarAV Faults diagnosis and design of the sprockets for dispenser in ginning industries
9. Gupta K, Laubscher R, Davim J, Jain N (2015) Recent developments in sustainable manufacturing of gears: a review. J Clean Prod 112:3320–3330
10. Suresh P, Ramabalan S, Natarajan U (2016) Cost effectiveness and sustainable analysis of crankshaft assembly vis-a-vis material and manufacturing process. Ecol Environ Conserv 22:257–266
11. Nikam P, Tanpure R (2016) Design optimization of chain sprocket using finite element analysis. Int. J Eng Res Appl ISSN 2248–9622
12. Laxmikant PS, Karajagi P, Kulkarni R (2016) Design and development of optimized sprocket for Track hoe. Int J Eng Trends Technol 181–185
13. Suresh P, Ramabalan S, Natarajan U (2016) Integration of DFE and DFMA for the sustainable development of an automotive component. Int J Sustain Eng 9(2):107–118
14. Ambole NP, Pravin RK (2016) Carbon fiber sprocket: finite element analysis and experimental validation
15. Han Y, Wen H, Yu E (2016) Study on electromagnetic heating process of heavy-duty sprockets with circular coils and profile coils. Appl Therm Eng 100:861–868

16. Suwannahong W, Suvanjumrat C (2017) An integrating finite element method and multi-body simulation for drive systems analysis. Eng J 21(1):221–234
17. Choi YJ, Lee SK, Lee IK, Hwang SK, Yoon JC, Choi CY, Lee YS, Jeong MS (2018) Hot forging process design of sprocket wheel and environmental effect analysis. J Mech Sci Technol 32(5):2219–2225
18. Barua A, Kar S (2018) Review on design optimization of sprocket wheel using different techniques. Int J Adv Mech Eng 8:55–62
19. Naji MR, Kurt MM (1983) Analysis of sprocket load distribution. Mech Mach Theory 18 (5):349–356
20. Shuaib M, Seevers D, Zhang X, Badurdeen F, Rouch KE, Jawahir IS (2014) Product sustainability index (ProdSI) a metrics based framework to evaluate the total life cycle sustainability of manufactured products. J Ind Ecol 18(4):491–507

Redesign of the Walking Stick for the Elderly Using Design Thinking in the Indian Context

Shivangi Pande, Akshay Kenjale, Aditya Mathur, P. Daniel Akhil Kumar and Biswajeet Mukherjee

Abstract This paper addresses the walking stick issues with the target demographic being Indian elderly. The study includes an understanding of the various types of walking aids, the requirements of the target demographic, and the duration of usage of pertaining walking aid by them. Methodologies both existing and constructed were deployed to further understand the root level needs and reform of current designs. Ideas were generated using Indian anthropometric dimensions and requirements.

Keywords Design thinking · Concept design · Product ideation · Design for the elderly

1 Introduction

A large segment of the elderly is affected by pain in the knees, osteoarthritis, stooping back, and other ailments usually associated with old age, which affect their daily lives by restricting them from walking without support or aids [1]. The purposes of this study was to identify the challenges associated with the current design of walking aids available and understand the perception of elderly people toward them through experimental behavior analysis and changes in their perception with changes in the functionality and aesthetics of the aids, and then propose a solution to overcome these challenges to reduce their discomfort.

S. Pande · A. Kenjale · A. Mathur · P. Daniel Akhil Kumar · B. Mukherjee (✉)
PDPM Indian Institute of Information Technology, Design and Manufacturing, Jabalpur 482005, India
e-mail: biswajeet.26@gmail.com

BBVL. Deepak et al. (eds.), *Innovative Product Design and Intelligent Manufacturing Systems*, Lecture Notes in Mechanical Engineering,
https://doi.org/10.1007/978-981-15-2696-1_3

2 Walking Stick

It is a product used to facilitate walking and for clarity of definition, a very personal non-medical usage has been referred to in the study we are referring to its usage for a very personal non-medical usage; used by elderly for support, relief and confidence while walking. The basic parts of a walking stick majorly are the hand grip, the lower stand, and the base grip [2]. The various designs in the market were available with different types of hand grips, and bases are shown in Figs. 1 and 2, respectively.

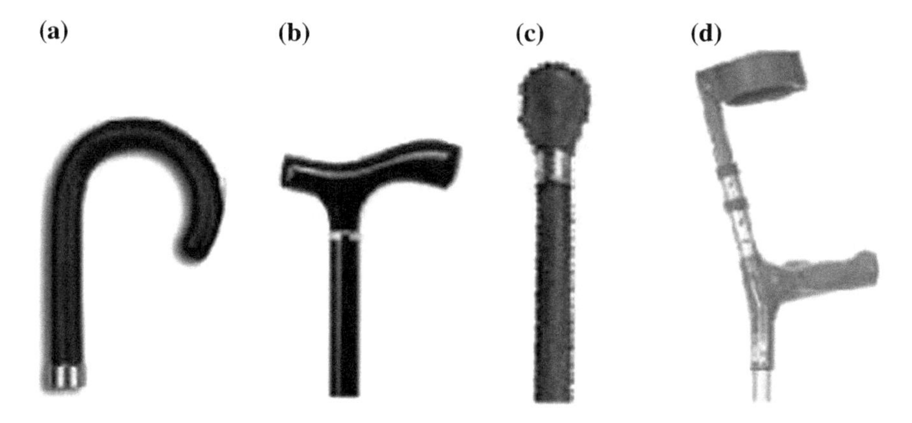

Fig. 1 Types of handles of a walking stick

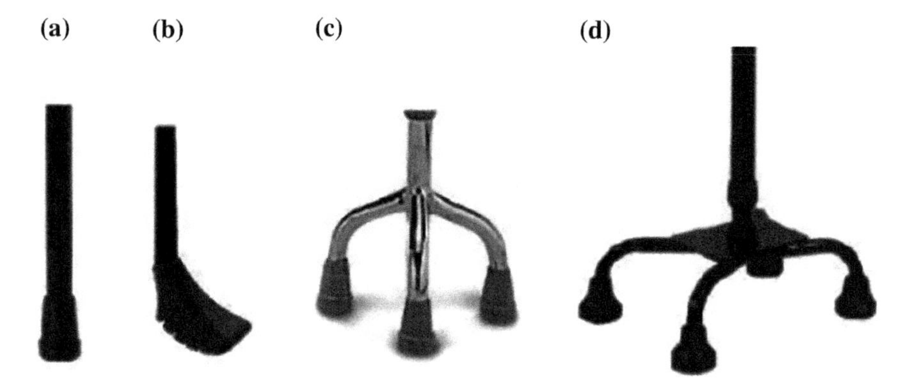

Fig. 2 Types of bases of a walking stick

3 Design Process

The design process was divided into four distinct phases Discover (Research), Define (Insights), Develop (Ideate) and Prototype. The double diamond [3] is a simple visual map of the design process. Throughout the process, a number of possible ideas are created (divergence) before refining and narrowing down to the best idea (convergence).

3.1 Discover (Research)

The design process according to the double diamond model begins with the discovery of user needs, current design and choice of various methods for the same, is depicted as Red color in Fig. 3.

Literature review—Existing patent of walking canes includes the 'Illuminating walking stick' which has a light source disposed within cane shaft adjacent to a translucent section for emitting flashes of light, a battery housed within the cane for supplying power to said light source, the generic 'walking stick' comprising an elongated hollow stick having upper and lower parts which are provided on its

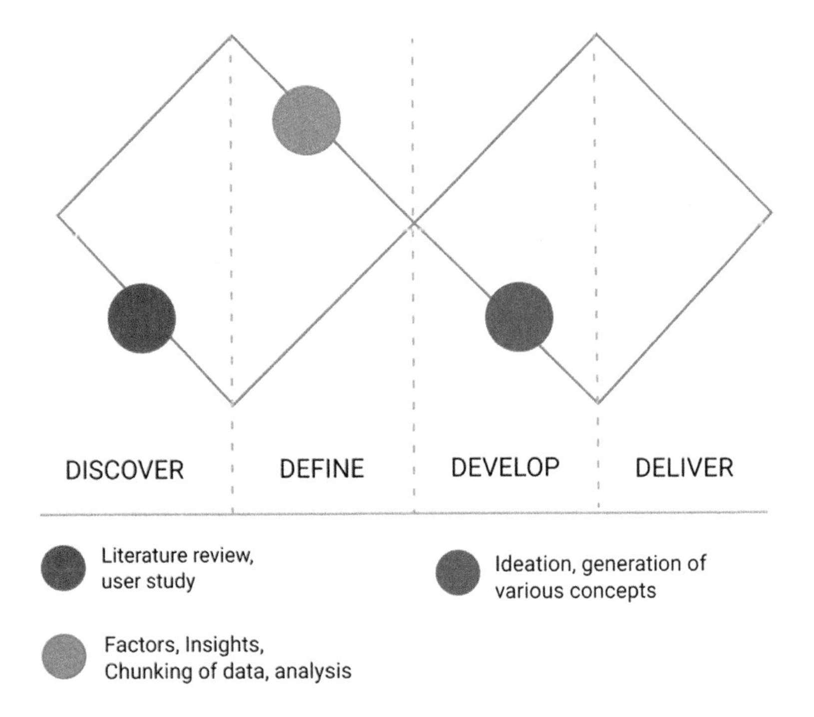

Fig. 3 Double diamond model

lower part with two supporting legs which are tiltable between a folded-in position and a folded-out, active position, and finally the 'multi-functional walking stick,' its electric torch can be adjusted by a finger at any time for pinpointing a small location or for shining a wide area. The shaft is adjustable by double-safe construction for keeping adjusted shaft firmly in position and is more solid and stable than any prior art after adjusting.

User study—The methods deployed for user study were:

Shadowing is the behavioral observation of a user in their natural environment that provides direction for further user research. Shadowing as a tool is much stronger than interviews and questionnaires as it refrains from bias in answers under the Hawthorne effect [4].

Interviews—Semi-structured interviews were conducted with all the participants of the study to collect qualitative and quantitative data among different categories: objective measures of the environment, residents' perceptions and attitudes about the environment and walking stick, walking behavior data, pain reception, and habit formation. In a lot of the cases, the interview was the correct alternative to the questionnaire and was narrated.

Questionnaires A set of questions for obtaining statistically useful or personal information from individuals. The questionnaire had both qualitative data and quantitative data collection.

Likert scale—Qualitative data describing the pain felt on various parts of the hand were also quantified using the Likert scale. It is a psychometric scale commonly involved in research that employs questionnaires. It is the most widely used approach to scaling responses in survey research. The kind we used quantified pain as 0 is not felt at all and 10 being extremely painful. The various pain points were then marked on the image of the palm by the users to describe the regions where they felt it.

3.2 Define (Insights)

Defining the areas of interest and development by analysis of data collected in the previous stage. Factors associated include storage, reason for use, parallel tasks, illumination, and the terrain. The define is depicted as green color in Fig. 3.

Analysis The analysis of various insights and user data [5].

Mapping the pain points of the palm—Users were asked various parts of their palm which pained while using their walking stick, users then marked various spots on the image of their palm. These images were then traced and converted into 10% opacity overlays and then all marked on one common hand frame to pinpoint areas which were common in the complaint about pain. The higher the opacity of the area the usual the spot was for pain or discomfort. This graphically explained and analyzed the common pain points with similar handles in Fig. 4.

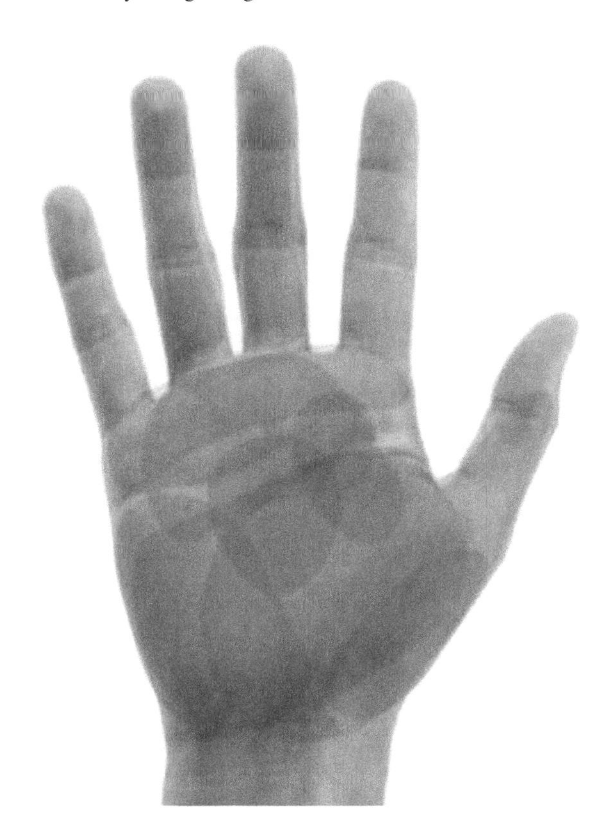

Fig. 4 Pain points on palm

Graphical chunking of insights—This step included identifying the quantum of users that fall under different categories depending on their usage and preference. The users were asked what was the reason they opted to start using a walking stick in their daily life and what purpose do they think it serves for them and what they did with the stick when not in use for the short and long duration of time (minutes while sitting to hours at night). A repetition was observed in the responses after interviewing less than twenty-five percent of the participants, and thus the data were separated into the major categories and plotted on a pie chart to show the similarity in user preference and behavior.

From the plotted data, we come to know that bedside and under the bed are the most preferred location of storage for a longer duration, which is shown in Fig. 5 and that most people believe that their walking stick helps them keep balance and gives them confidence which is shown in Fig. 6.

Insights The various pain points collected by user study are shown in Fig. 4.

The purpose of using a walking stick include The recommendation by a doctor, purpose for strengthening the gait, to provide the elderly with confidence and independence, reduces pain caused by knee or leg ailments, and reduces the fear of falling.

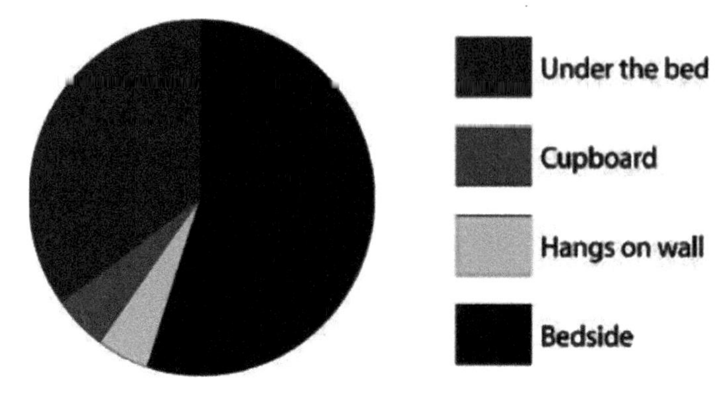

Fig. 5 Graphical data of storage

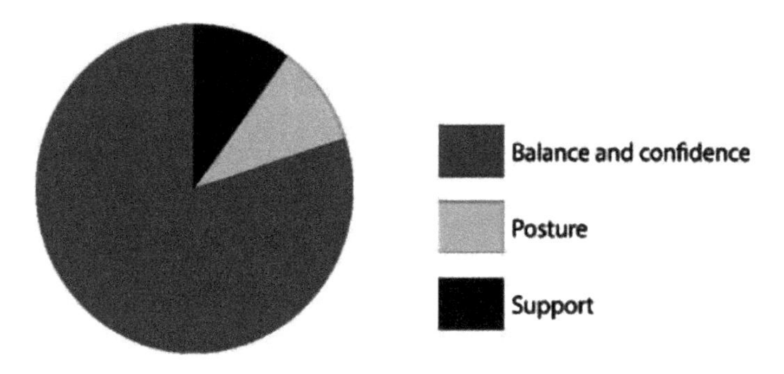

Fig. 6 Graphical data of utility

The various issues with the walking sticks in the market include incorrect or damaging usage by the elderly, the tendency to develop over-dependence and obsolescence by under usage.

Reasons for issues with a walking stick—The lack of independence or fear of over-dependence, poor or damaging usage due to lack of semantics, the lack of features required by the elderly, bulky, and restricting style of walking sticks and the general stigma associated with the walking stick that restricts either usage or positive association with the walking aid.

3.3 Ideation

The ideation process is depicted as purple color in Fig. 3, Various designs worked with different feasible solutions followed Indian anthropometric dimensions and had certain constraints in mind:

1. Great illumination
2. Ergonomic
3. Easy storage
4. More of a fashion accessory than a medical aid
5. Adjustable
6. In tandem to the Indian elderly's mental model.

The anthropometric constraints considered for ideation is tabulated below (Table 1) [6].

Concept 1

Proposed ideation of Concept 1 is shown in Fig. 7. *The features are as follows*:

1. Height adjustable to both extremes; the length is adjustable from 5th percentile gluteal furrow height to the 95th percentile; The furrow between the buttocks and the thigh muscles make up the gluteal furrow.
2. The base is the step height of 19.7 cm standard step size so while climbing stairs the elderly do not have to open or raise the height of foot more than 19.7 cm breaking down the step height into smaller steps.
3. The bearing attached to handle to trunk allows multi-angular movement, easy to rest against wall, or change grips; easy product departing.
4. The cavity also allows better storage by means of hanging on walls.
5. The illumination button is a simple slider taken from generic torches to the mental model of the elderly.
6. The loop-shaped handle reminds elderly of bus handles/support which is up to the mental model of an active youngster thus detaching the walking stick from the stigma of a medicinal/elderly aid.

Table 1 Anthropometry reference to Figs. 7, 8, and 9 [4]

Reference	Anthropometry	Dimension (mm)
Handle width	Handbreadth without the thumb, at the metacarpal	117
Grip diameter	Grip inside diameter	42
Cavity	finger-tip depth	18
Height min	Gluteal furrow min	682
Height max	Gluteal furrow max	923

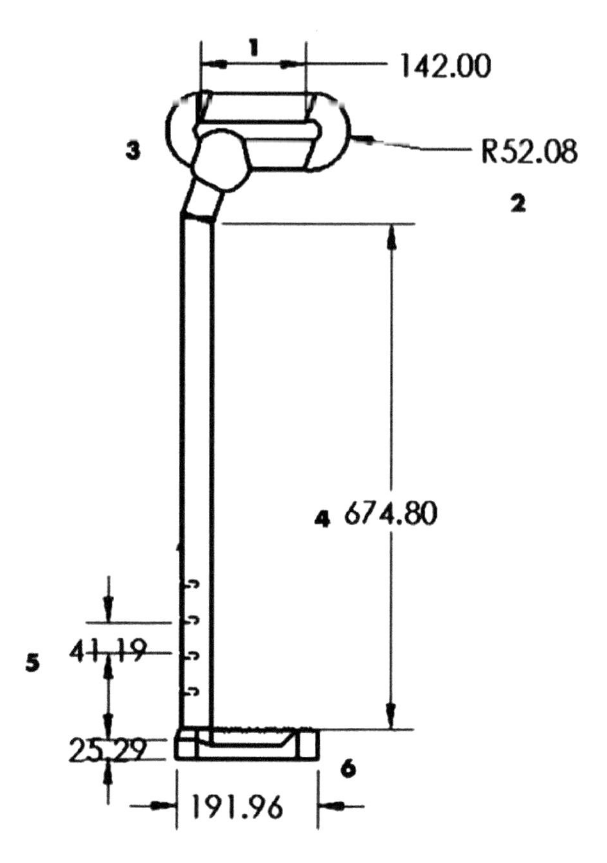

Fig. 7 Dimensions of Concept 1

Concept 2

Proposed ideation of Concept 2 is shown in Fig. 8. *The features are as follows*:

1. Two gripping postures are enabled by the handle. Depending on the comfortable wrist angle, the user can switch between the two postures.
2. An accessory like look is given by the triangular form of the stick. This is in contrary with the circular form which is associated with a walking stick used for medical aid.
3. As per the user's preference, the base can be switched from 1 leg to 3 legs depending on the surface.
4. Depending on the context of use, there are 4 combinations the user can adjust as per their comfort.
5. The body of the walking stick can be made of glass fiber reinforced polymer (GFRP), also known as fiberglass.
6. Mechanical advantages like high strength, water resistance, and low weight-to-strength ratio will be provided by GFRP.
7. The look of GFRP will suit the triangular form of the stick and give an overall premium aesthetic to the walking stick.

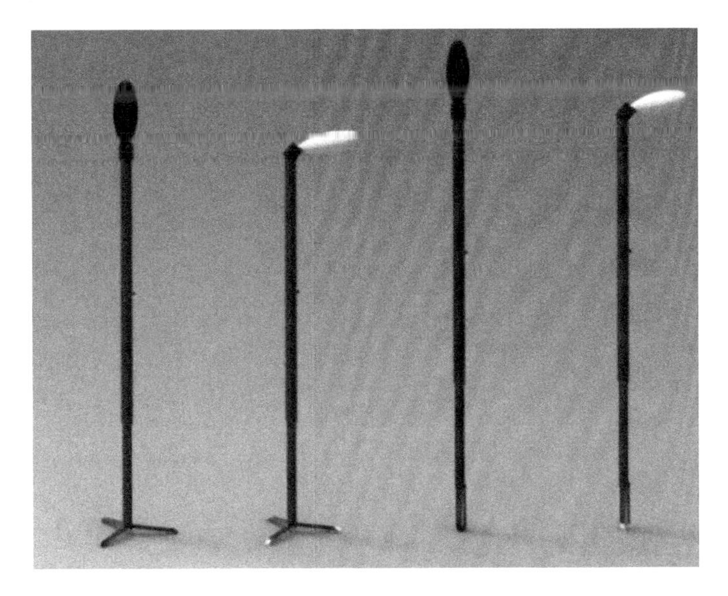

Fig. 8 Concept 2

4 Discussion

The study follows the design thinking methodology for creation of ideas that solve the issues with the current designs of walking aid for elderly, specifically to the Indian context. The comparison of ideas and current designs in the market is on the basis of various qualities. Figure 9 shows the existing and proposed designs of walking sticks [7–13].

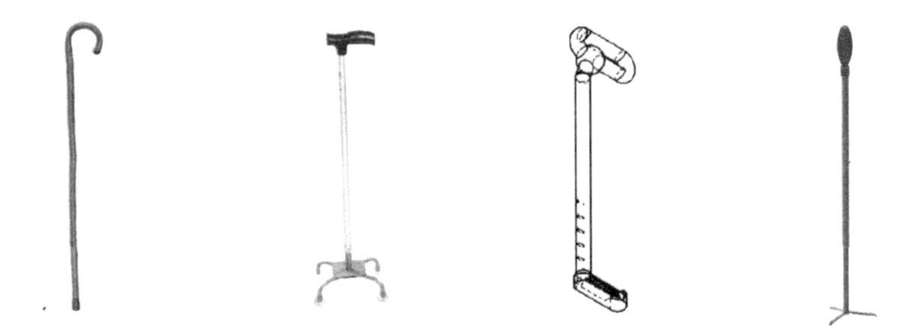

Fig. 9 Existing designs compared to proposed designs

Table 2 Design comparison of proposed with existing products

Designs	Balance	Grip	Ergonomic	Adjustability	Storage	Illumination	Aesthetic	Total
E.D 1	2	0	1	0	2	0	1	6
E.D 2	2	1	2	2	1	0	1	9
P.D 1	2	2	2	2	1	2	1	12
P.D 2	2	2	2	2	1	0	2	11

E.D. Existing Design, *P.D.* Proposed Design

Table 2 compares proposed designs with existing designs. The features of the designs are evaluated on 0–2 scale (minimized Likert scale).

0—No presence of the feature

1—Weak presence of the feature

2—Strong presence of the feature.

Proposed designs scored more score than the existing designs.

5 Conclusions

The study helped in creating walk aids that cater to the needs of the elderly. The survey was done in institute and no ethical clearance was needed. Also, the data are anonymized. First is the multipurpose usage (Concept 1) and second is the aesthetics and accessory like appearance (Concept 2). The target audience includes the urban elderly and ideally elderly.

With previous experience of using a walking aid, the aim of the walking aid is to not only fulfill the needs of the elderly but also extend the needs of people involved in their lives.

Acknowledgements The authors acknowledge the support of SEED, DST project no. SP/YO/002/2017.

References

1. Tutuncu Z, Kavanaugh A (2007) Rheumatic disease in the elderly: rheumatoid arthritis. Rheum Dis Clin North America 33(1):57–70
2. Van der Esch M, Heijmans M, Dekker J (2003) Factors contributing to possession and use of walking aids among persons with rheumatoid arthritis and osteoarthritis. Arthritis Care Res Official J Am Coll Rheumatology 49(6):838–842
3. Informa Healthcare, Karwowski W (2006) International encyclopedia of ergonomics and human factors, 2nd edn., vol 3. CRC Press
4. Norman AD (2013) Design of Everyday Things, expanded and revised. Basic Books Publishing
5. Moustakas, C (1990) Heuristic research: design, methodology, and applications. Sage Publications

6. Chakrabati D (1977) Indian anthropometric dimensions for ergonomic design practices. National Institute of Design
7. Hollday R (2018) Walking stick papers Al Haines
8. Kroemer KHE (2005) Extra-ordinary ergonomics. CRC Press
9. Williamson D (1988) Walking stick standards. Physiotherapy 74(3):121
10. Robinson JP (2000) Phases of the qualitative research interview with institutionalized elderly individuals. J Gerontological Nursing 26(11):17–23
11. Sainsbury R, Mulley GP (1982) Walking sticks used by the elderly 284(6331):1751–51
12. Kumar D, Shankar H (2018) Prevalence of chronic diseases and quality of life among elderly people of rural Varanasi IJCMR 5(7)
13. Fabbro PA (1965) Walking aid 45(1):34–34

Human-Guided Following Trolley Mechanism and Integrated Shopping Mechanism Using RFID

M. L. N. Vital, V. Hari Vamsi, T. Purnachandra Rao and K. Srikanth

Abstract A supermarket is actually a place where varieties of goods are available. The objective of these supermarkets is to provide all the products available and also to save the time of customers. But these customers get very frustrated waiting at the billing counters and randomly changing queues, get confused while comparing price. In this era of e-commerce shopping, the demand for these types of system is corroding day by day. To eliminate these problems, a trolley is designed that actually does not make customers wait for queues. These supermarkets can use this technique as a strategy to increase their customers and it included a human-guided following system that makes trolley to floor the user. This concept is a smaller prototype of automatic self-checkout system which makes customers do their payment for items they want to purchase before leaving the store. This is to release pressure at the counters during peak hours and increase malls efficiency.

Keywords Hand-guided movement · RFID · IR · Trolley

1 Introduction

This paper explores and demonstrates how the hand-guided trolley model was developed in the evaluation of a business opportunity from a technology idea proposed for commercialization purposes. The hand-guided trolley mainly concentrates on the objective of reducing pressure at tills in supermarkets. The customer's products are scanned automatically once they put in the trolley. So, they can easily their bills reducing the time consumed waiting in queues. The trolley is added with a new feature that makes it follow the users autonomously. It is

M. L. N. Vital (✉) · V. Hari Vamsi · T. Purnachandra Rao · K. Srikanth
Department of EEE, VR Siddhartha Engineering College, Vijayawada, AP 520007, India
e-mail: mlnvital@vrsiddhartha.ac.in

V. Hari Vamsi
e-mail: harivamsi@vrsiddhartha.ac.in

BBVL. Deepak et al. (eds.), *Innovative Product Design and Intelligent Manufacturing Systems*, Lecture Notes in Mechanical Engineering,
https://doi.org/10.1007/978-981-15-2696-1_4

equipped with infrared sensors so that users can control its movement from their hand-equipped sensors.

1.1 Literature Survey

The research conducted by Box Technologies and processor manufacturing giant Intel found that 38% of customers said that they have given up trying to make a purchase because of long queues, with nine minutes the average time consumers are prepared to wait before leaving without a product [1]. Around a quarter of people said five minutes is the maximum they are prepared to wait before departing a store. This results in a decrease in the mall's efficiency. Any robot that has human-guided functions can be developed to carry these heavy loads while shopping and also a feature to scan the products in the cart itself and also to process the payment. This improves the mall's efficiency and better profits.

2 System Design

The system design for hardware as well as software for the development of the automatic human-guided shopping trolley is described in Table 1.

2.1 Hardware Implementations

In the hardware design, Arduino Uno (microcontroller) is used for the portable robot. The hardware components comprise RFID reader, IR sensors, motor driver, LCD 16 × 2, and DC gear motors which are connected to the Arduino Uno.

Table 1 Trolley mechanism

Parameters	Specification
Features	1. The mechanism is attached to the trolley 2. The robot can follow user 3. The robot has 12 V DC motors supporting speed of 200 rpm
Platform	Arduino Uno, infrared sensors, RFID RC522
Actuator and payload capacity	12 V DC gear motors, L293D motor driver (can pull 3 kgs)
Sensor	IR sensor, RFID RC522 reader
Algorithm	1. Human-guided following system by using the IR 2. The polling method is by using RFID algorithm

The RFID reader is used to read RFID UID's data and send data to LCD display. IR sensors are used for human-guided movement of trolley [2]. The motor driver is used to drive the DC gear motor. There is a robot-based mechanism installed in the shopping trolley. The microcontroller, IR sensors, motor driver, and a 12 V acid or lithium-ion battery are put on the trolley base in order to control the shopping trolley. Figure 1 illustrates the robot mechanism with attached hardware and components. Table 1 summarizes the specifications of the developed robot (Fig. 2).

3 RFID Shopping Mechanism

The RFID reader is attached to the robot so that it can read the RFID tag cards which are attached to the products placed in the cart [3]. When the RFID reader reads a tag card and identifies that the tag's UID code is the same as the selected item, tag UID code is sent to the Arduino. The data of scanned card is processed and if matched with the coordinates of the item then the product data is displayed in LCD.

Fig. 1 Smart trolley

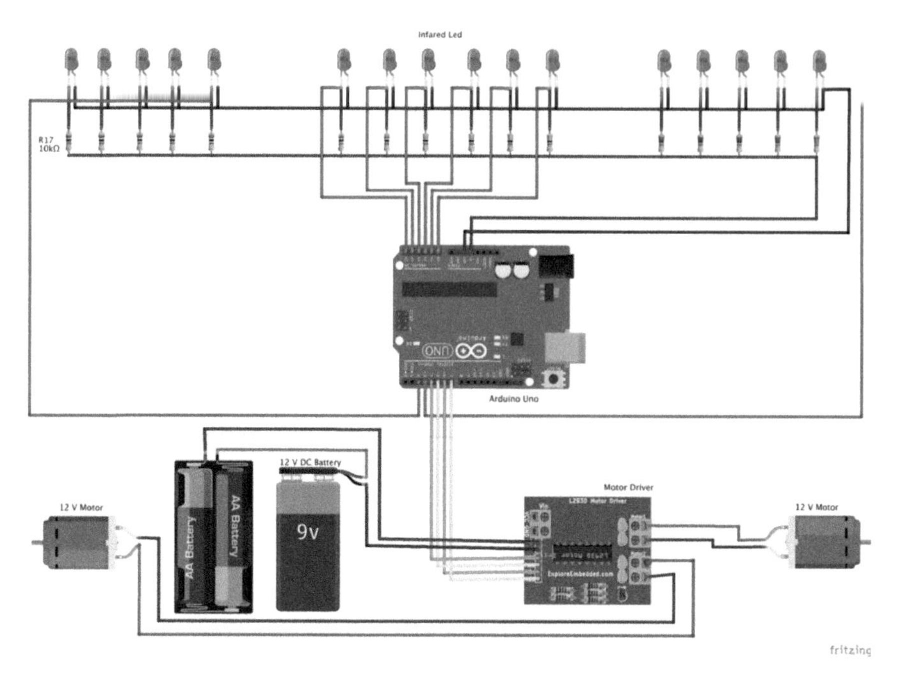

Fig. 2 Circuit diagram of hand-guided trolley movement

4 Software Design

The software design approach in this project involves C programming language-based Arduino operation. The flowchart for the human-guided trolley movement is shown in Fig. 3 and for RFID system in Fig. 4.

5 Development of Trolley

The configuration of the apparatus required for developing smart trolley is specified.

6 Hand-Guided Movement

The hand-guided following system is brought to life by using infrared sensors. The IR receivers equipped on the trolley play the key role in trolley's movement. The users are provided with an infrared transmitter powered using a 9 V battery source. The transmitter can be hand-held or can be attached to the user's back.

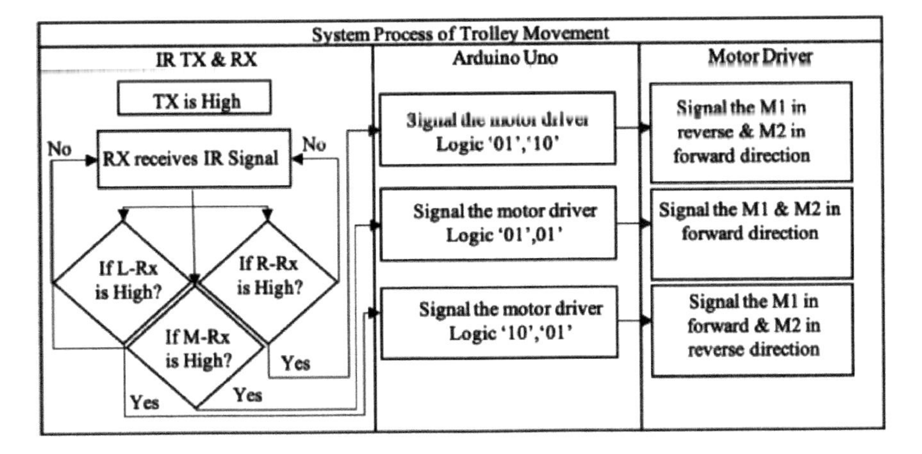

Fig. 3 Flowchart for trolley movement

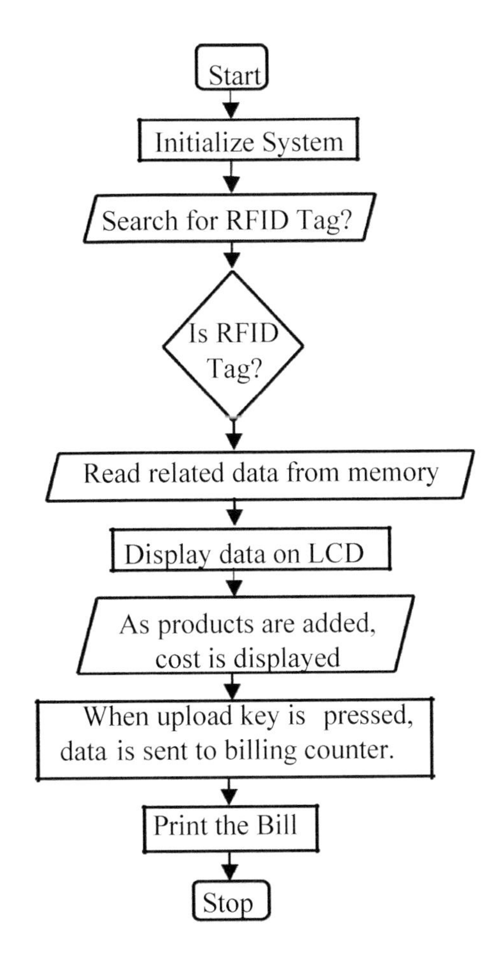

Fig. 4 Flowchart for RFID operation

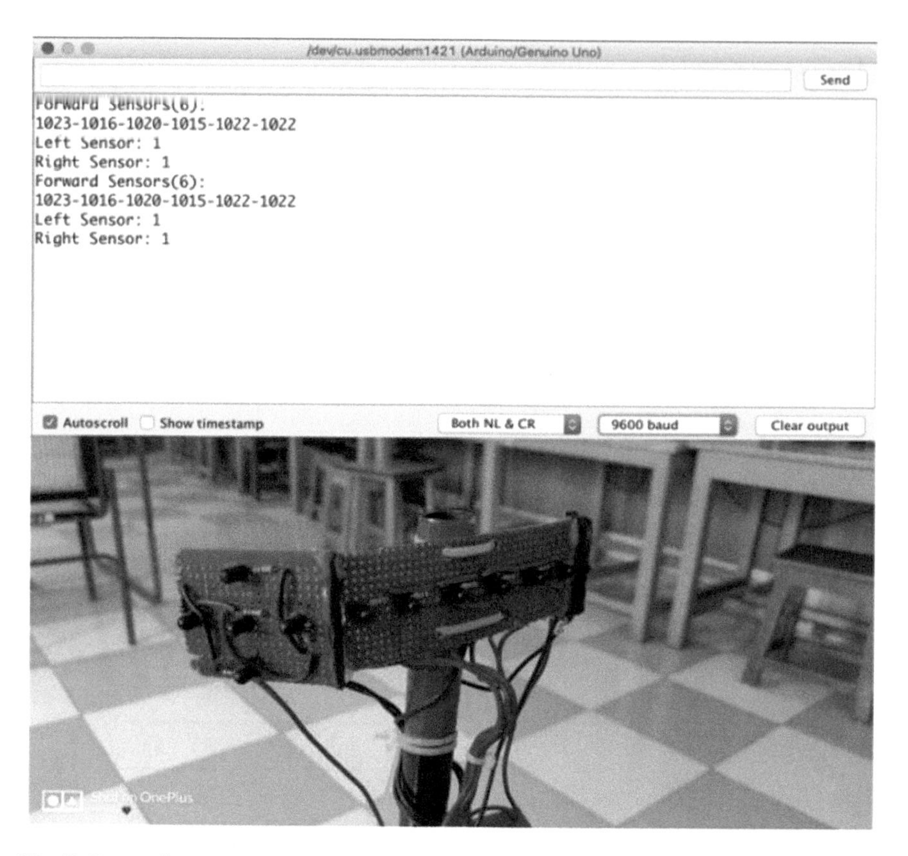

Fig. 5 Data collected from 6 IR sensors and 6 + 2 IR sensors ED's attached

Pointing this transmitter toward the trolley makes the trolley move toward the user. There are 6 + 2 IR sensors used in the project. Figure 5 shows a snapshot of the output of the 6 + 2 infrared sensors. The infrared sensor is tested for the following function. Figure 6 shows IR sensor that has detected an Infrared signal from the transmitter, which has resulted in the related DC gear motor to run and the result is that the trolley moves toward the user. Figure 5 shows data collected from trolley when no infrared rays are being received. Figure 6 shows data collected from trolley when guided by a transmitter. So, these results are calibrated for movement of the trolley in a particular direction. If the 6 infrared sensors receive values in the determined scale, the trolley moves forward until it receives specified values. The 2 infrared sensors placed on either side of 6 IR sensors move trolley in the left or right direction upon receiving infrared signal from transmitter. Figure 7 shows CRO receives 5 V (in analog 1024) at VA3 pin when the transmitter is OFF. Thus, the resistance across the IR will be high. When photodiode is exposed to infrared light, electrical resistance across the diode decreases thereby increasing the reverse current as shown in Fig. 7 below circuit.

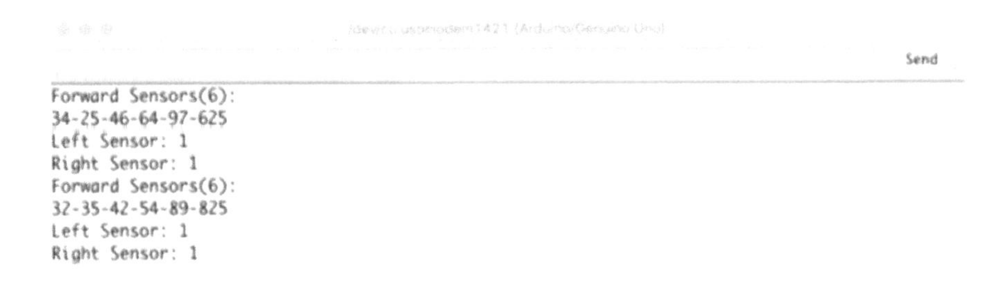

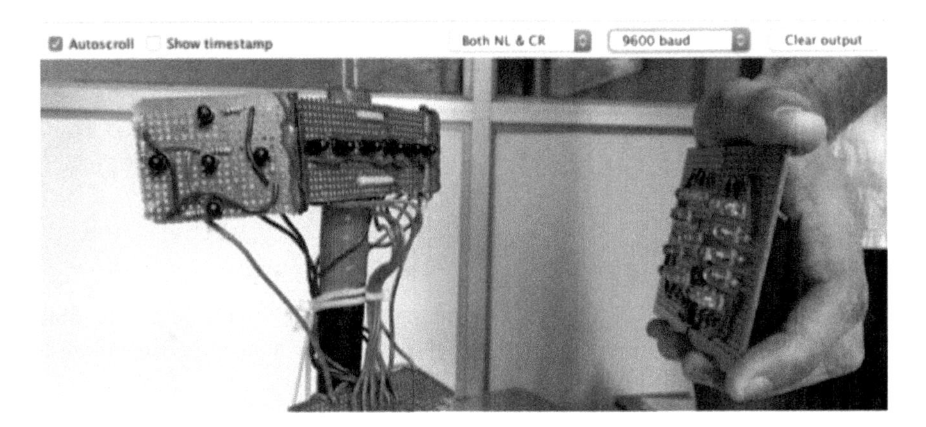

Fig. 6 Data collected from 6 IR sensors when transmitter is ON

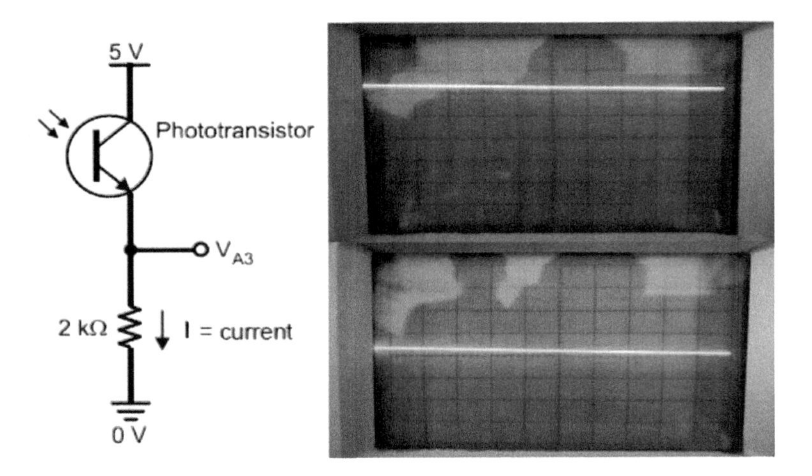

Fig. 7 IR circuit and calibration on CRO

6.1 RFID Shopping

The products in shopping malls are stickered with unique RFID tags based on the product type. The RFID sensor is attached to the trolley, so that whenever a product is dropped into the trolley, the product is scanned displaying the cost on LCD for the user. The user can add or remove the products and check the added products list by using CHECK and REMOVE buttons in the trolley as shown in Fig. 8. The following figures illustrate the shopping mechanism employed in smart trolley. The RFID is a feature where the products can be scanned at a faster rate or at a group. There is no specified code on the product to search for code and scanning it, and the RFID mechanism is simple where the product can just be wiped across the sensor. This feature can be upgraded to a higher extent where all the products are scanned at once by using moving the trolley across the sensor gate. This method is bit costly but rapid development in technology is reducing the price day by day and it would be cheap as a printout in the near future. This feature helps in having an exact count of products shipped and prevents theft of products.

7 Applications

The smart trolley can be employed in shopping malls with both its features enabled, i.e., human-guided movement and RFID shopping. The human-guided trolley can be employed in airport terminals.

8 Advantages and Limitations

8.1 Advantages

It reduces duplicate goods entering the market, saves time to customers rather than waiting in long queues, increases mall's efficiency, generates e-receipt, and saves paper. The human-guided following system saves customers' efforts from pulling heavy loads. Malls profits increase due to low workforce requirement. RFID is much faster than the traditional barcode system and is easy to scan even under obstacles. Theft of products from the store is not possible. IR sensors are cheaper than that of ultrasonic sensors and image processing.

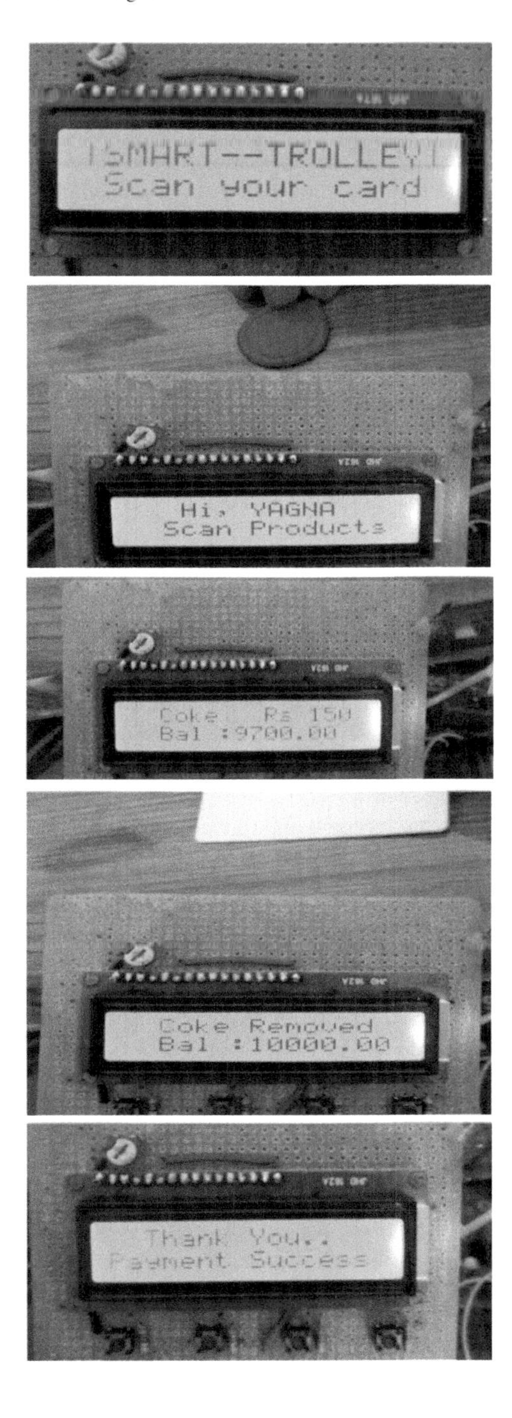

Fig. 8 RFID shopping

8.2 *Limitations*

Human-guided following system can malfunction under sunlight, and this is neglected since malls are constructed as enclosed structures. RFID can increase the cost of production but cheaper than traditional barcode system when compared to bulk production. Another method is image processing. This is a method to perform some operations on an image, in order to get an enhanced image or to extract some useful information from it. It is a type of signal processing in which input is an image and output may be image or characteristics/features associated with that image. But the cost is very high if we adopt this system.

9 Future Scope

From the above discussion, the smart trolley can be equipped with cloud storage capability to give customers a feature of revising previously purchased products and compare previously purchased costs with present costs. This benefits the malls with exact selling data and an exact estimation of future imports.

10 Conclusions

This prototype will pull weight of 3 Kgs. So, that a real product equipped with 36 V DC gear motor and a high count of IR sensors will improve its semi-autonomous capability and can pull higher loads. Each product in the shop or a mall will have an RFID tag on it. Each cart will have an RFID reader. If the product is removed, it must get deleted from bill too. So, by making use of this, the supermarket shopping system will become easier. It will also provide anti-theft system for a supermarket. It will enable online transaction procedure for billing, and it will also give suggestions to the user for buying products, display offers, etc.

References

1. Boamah D, Jalloh AB, Ali S, Tambi IT, Riaz M, Sun M, Fernado H (2015, May 15) Technology Evaluation and commercialization, Smart Trolley. London South bank University
2. Ghatol SD, Jahagirdar VS, Pratiksha DK (2015, May) Smart shopping using smart trolley. In: Proceedings of IRJET, vol 5, issue 5, pp 1532–1535
3. Mohit K, Jaspreet S, Anju, Varun S (2015) Smart trolley with instant billing to ease queues at shopping malls using ARM7 LPC2148: a review. In: Proceedings of IJARCCE, vol 4, issue 8, pp 39–42

Design Optimization of Innovative Foldable Iron Box

Alok R. Shivappagoudar, Amit S. Gali, Anirudh V. Kuber, Sadashivu I. Giraddi, Arshad N. Havaldar, Arun Y. Patil, Basavaraj B. Kotturshettar and R. Keshavamurthy

Abstract A foldable iron box is an optimized design concept unique from any current existing iron boxes in the world. This design focuses on the surface area of the base plate (sole plate) of the iron box. The alteration of the surface area of the sole plate is not available with current iron boxes. Hence, the same surface area is used to iron all types of clothing. This design focuses on alteration of the surface area of the sole plate depending on the field of cloth to be ironed. The new model has provision for the supply of heat to the required surface area, in turn, will deliver a cost-effective and energy-efficient solution. This design is built with the usage of surface modeling software CATIA. This new design mainly focuses on the industrial design of the iron box. The ergonomics and aesthetic are given at most consideration in this design.

Keywords Optimization · Sole plate · Ergonomics · Foldable

1 Introduction

A typical iron box has a triangular base plate or sole plate which when heated removes the creases on clothes. Ironing happens because of loosening the tie between the long chains of molecules in polymer fiber material. The heating effect of the sole plate ensures that the fibers are stretched and eventually the fabric maintains the new shape when cooled. In early days, irons were heated by the combustion

A. R. Shivappagoudar · A. S. Gali · A. V. Kuber · S. I. Giraddi · A. N. Havaldar · A. Y. Patil (✉) · B. B. Kotturshettar
School of Mechanical Engineering, KLE Technological University, Vidya Nagar, Hubli, Karnataka 580031, India

R. Keshavamurthy
Mechanical Engineering, Dayananda Sagar College of Engineering, Bangalore, India

BBVL. Deepak et al. (eds.), *Innovative Product Design and Intelligent Manufacturing Systems*, Lecture Notes in Mechanical Engineering,
https://doi.org/10.1007/978-981-15-2696-1_5

process either by fire or with internal arrangements. The electric flat iron was invented by US inventor Henry W Seeley and patented on June 6, 1882. The evolution of iron box started from charcoal iron and reached till electric iron. The problem with the existing iron box is that the same surface area of the sole plate is used to iron all areas of clothing. Hence, the heat generated by the iron box is not utilized thoroughly. The research on extracting the waste heat is in process. The new design has provision for altering the surface area of the sole plate which is proportional to the area of the cloth to be ironed. In addition to this, there is provision for the supply of heat to the respective regions of a sole plate in working. Hence, the heat supply is also used effectively. The handle of the iron box plays a vital role in muscular wrist pain of the user. Therefore, the main focus is on the various angles of the handle of the iron box and is analyzed using respective software [1]. The folding and unfolding of creases are also crucial while ironing and there are several methods to do so [2]. Muscular fatigue and body pain are one of the most dangerous problems faced by users; it differs from lightweight iron and heavyweight iron, also it differs from female to male candidates [3]. The ironing of clothes manually is a time-consuming process. Hence, an automatic ironing machine with innovations and motorized mechanisms, effective utilization of time is focused, and a machine that can iron the cloth both top and bottom simultaneously is built [4]. The user experience of the electrical iron box is also an important parameter [5]. The weight of the iron box is also considered a benchmark for user experience. Hence, the number of parts is lowered, and the iron box is made significantly lighter using design for manufacturing and assembly (DFMA) method [6]. The two kinds of iron boxes are steam iron and gas pressing iron. The gas pressing iron box is helpful when the power supply is not available [7]. The process of ironing involved lot of strain to home makers or old age people due to too much of heat dissipation while operation, this will lead to serious damage to intervertebral disc [8].

2 Concept Generation and Testing

2.1 *Material Details*

From Table 1, it can be inferred that the cover and knob are assigned with PLA material which is light in weight and biocompatible material.

In Fig. 1, the overall assembly was done combining eight 3D parts. Following figures represent the parts. From Fig. 1, it can be inferred that the whole assembly contains in whole eight parts. The components with their material and quantity were

Table 1 Description of components

Title	Quantity	Material
Center base plate	1	Aluminum alloy
Left base plate	1	Aluminum alloy
Right base plate	1	Aluminum alloy
Center coil	1	Copper alloy
Side coil	2	Copper alloy
Cover	1	PLA
Knob	1	PLA
ECU	1	–

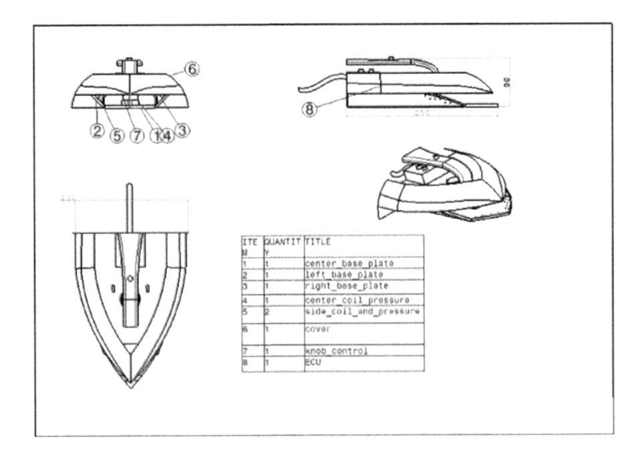

Fig. 1 Overall assembly of a novel iron model

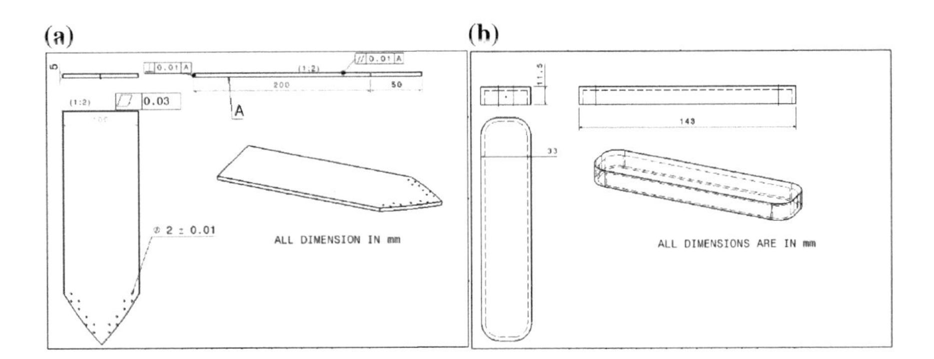

Fig. 2 **a** Base plate (sole plate); **b** center coil and pressure plate

also mentioned. From Fig. 2a, it is understood that the sole plate is a thick triangular-shaped slab that forms a base over which electric iron is built up. This is the part that gets heated and is utilized for ironing. The material used is aluminum

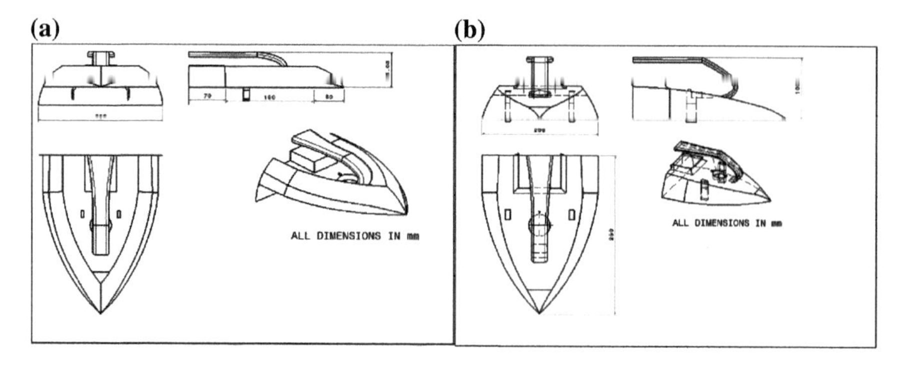

Fig. 3 **a** Cover (lifted); **b** cover (original position)

alloy. The sole plate holds pressure plate and cover plate in position. Figure 2b illustrates the coil makes use of thermostat which is heated to supply heat to the sole plate. The thermostat is made up of the bi-metallic strip to operate the switch is connected in series with a resistance. The bi-metallic piece is a simple element that converts temperature changes into mechanical displacement.

From Fig. 3a, we can confirm that cover which shields all the inner parts of the iron box and the material used here is polylactic acid. This is predominantly used the material in fast-moving consumer goods sector (FMCG). Figure 3b illustrates the cover (original position refers to the alignment of protection concerning all three sole plates) which includes an envelope of iron box for better aesthetic and ergonomic appeal.

Figure 4a depicts the knob used to set the amount of heat required based on the type of material of the cloth. Figure 4b demonstrates the function of the left side base plate is similar to the center base plate.

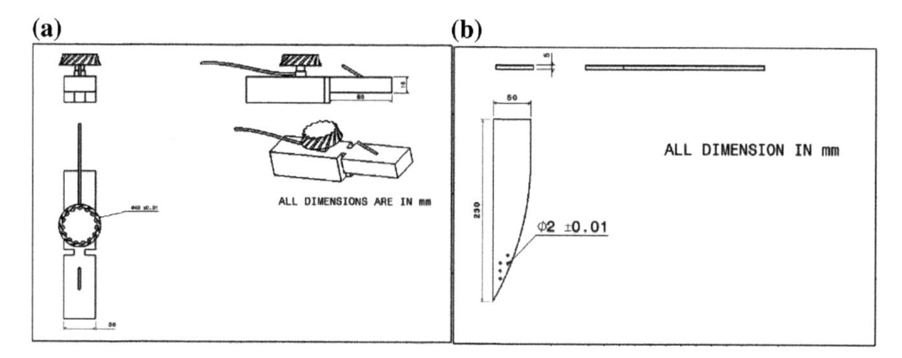

Fig. 4 **a** Knob control; **b** left side base plate

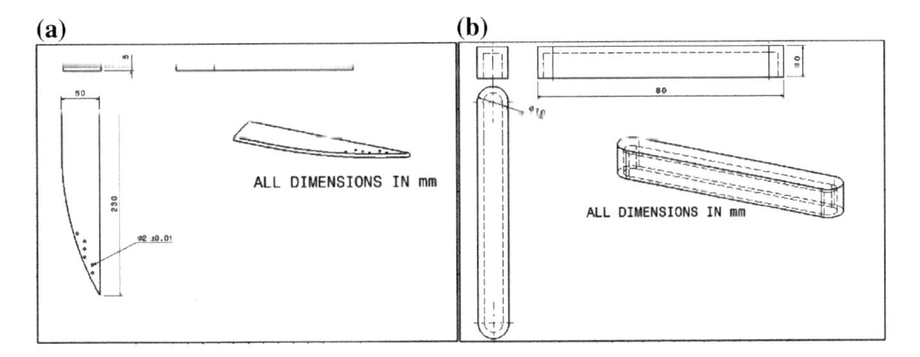

Fig. 5 **a** Right side base plate; **b** Pin for locking the right side and left side base plate

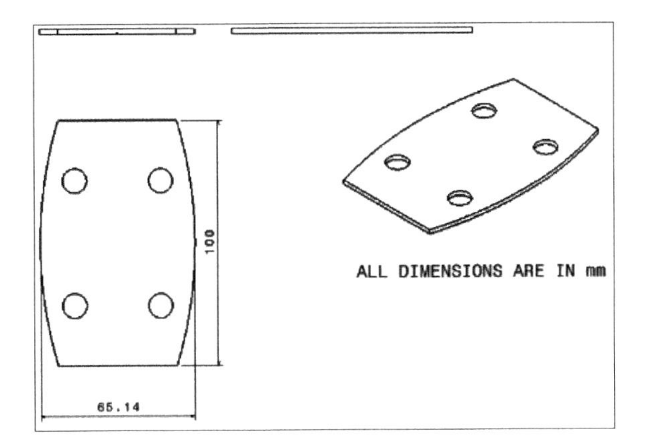

Fig. 6 Coil and pressure plate (side parts)

The function of the right side base plate is similar to the center base plate, as illustrated in Fig. 5a. Depending on the requirement of the surface area the right side or left side plate can be aligned with the center plate. Coil and pressure plate have been shown in Fig. 5b which focuses on adequate amount of load on the cloth for various types of material.

The additional plate connector functions as additional support are shown in Fig. 6.

This new plate is utilized in ironing the cloths of higher surface area as illustrated in Fig. 7a. For long size clothes, a unique feature has been embedded in the unit which is depicted in Fig. 7b.

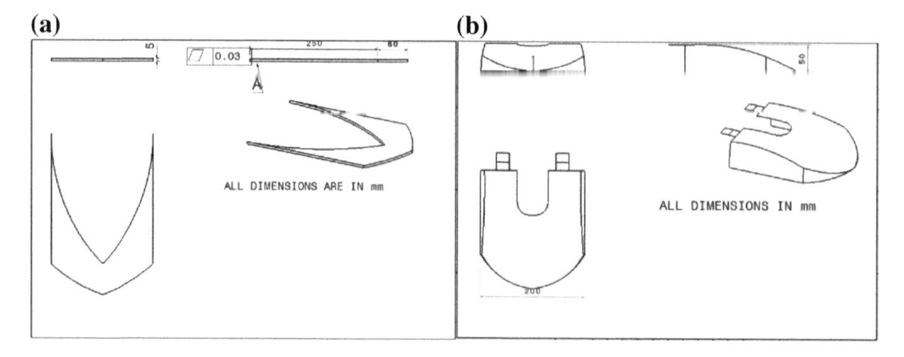

Fig. 7 **a** Base plate (additional); **b** cover for additional plate

3 Working of Product

This design provides the provision of altering the area of a base plate according to the area of cloth to be ironed. The supply of heat to the area of the base plate can be altered based on the area to be used. This design consists of an add on the cover, plate, additional plate connector, center base plate, center coil, and pressure plate, knob and water control, left side base plate, right side base plate, side coil, and pressure plate. This design is done about BAJAJ Dx7 lightweight iron. The model is validated for its optimal design based on the parameter of optimization [8]. From Table 2 a, comparison with sole plate, iron box, and the heating element was carried out from ancient days to the modern era to an innovative design. The comparative study exhibits a cost-effective material usage along with less depletion of the ozone layer by releasing lower CO_2 emission.

The current design as depicted in Table 3 has the edge over ancient and modern design, and perhaps biodegradability, recyclability, and CO_2 emission are the prominent parameters which will define the future trends.

4 Conclusion

This paper depicts the optimal usage of the iron box. The surface area of the sole plate is altered based on the requirement. Power and heat supply are cut off when the surface areas of single plates are not in use. This helps in less consumption of electricity. This provision also helps in less wastage of heat or supply during ironing. This design also focuses on the reduction of weight of the iron box, so the base plate is provided with the feature of lifting the left and right side sole plates and only required an amount of area that can be utilized with optimum power supply. The new single plate is detachable and thus helps in reduction of weight. The material used is also biodegradable. The design also focuses on the ergonomics

Table 2 Comparative study of iron box components from its inception to recent design

Components	Ancient design	Modern design	New design
Sole plate	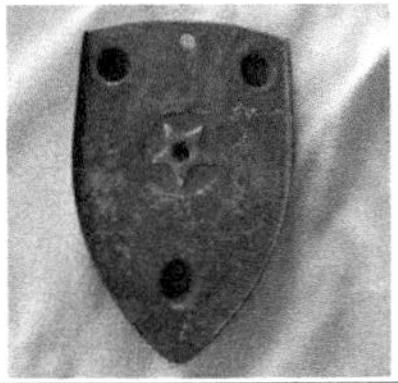	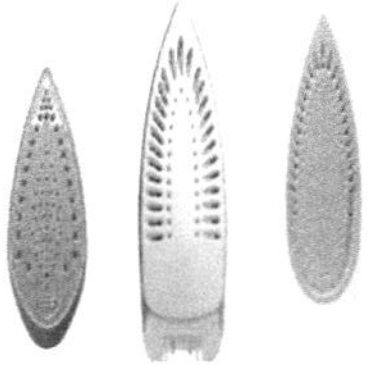	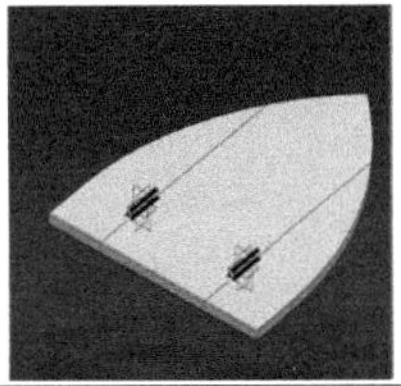
Iron box		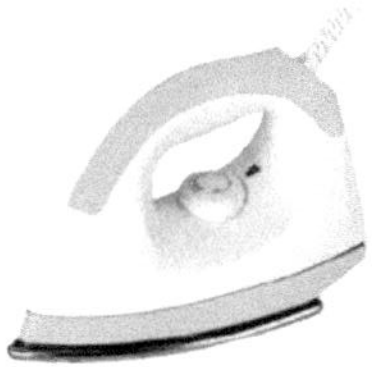	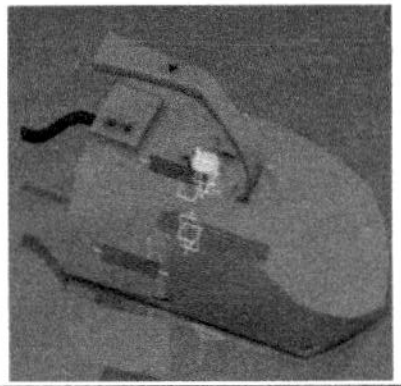
Heating element		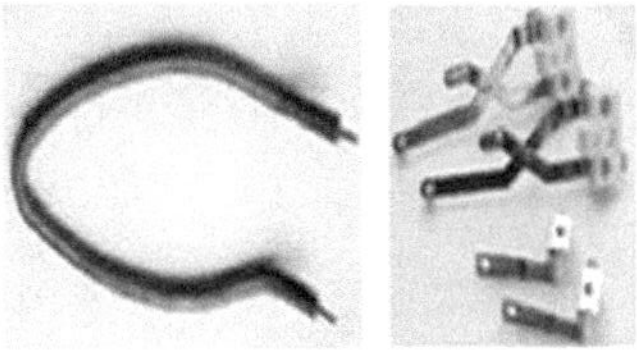	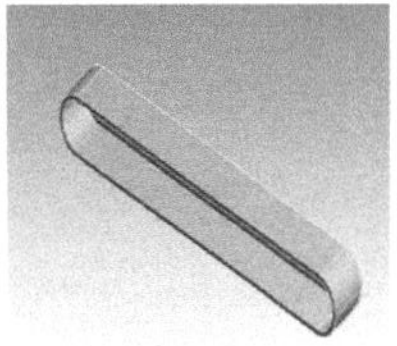

Table 3 Parametric study for optimization [9–11]

Parameters	Ancient design	Modern design (steam-based)	New design (steam-based)
Overall weight (in kg)	3–8	2.3–2.5	2.2–2.406
Cost (in Rs.)	3100–8500	700–8000	600–700
CO_2 emission per kg during material manufacturing per Kg	5.45	6.8	0.8
Material(cover)	Wood	Fiber/polymer	Polylactic acid (PLA)
Biodegradability	Non-biodegradable	Non-biodegradable	Partial biodegradable (except metallic parts)
Recyclability	Costly	Costly	Nominal price
Unique selling points (USP)	No power consumption	Weight reduction/ focus on industrial design	Effective utilization of power/optimal usage of surface area (base plate)

of the iron box. The handle of the iron box is also an optimized design, and fatigue life is also a considered parameter. The ECU controls the power supply to the heating element or coils. A knob is provided to adjust the heat supply to the sole plate. Henceforth, the design focuses on all parameters and is optimized in the best possible way.

Acknowledgements The support extended by KLE Technological University authorities and CAE team is highly appreciated and acknowledged with due respect. The author's team is thankful to the management of KLE Technological University.

References

1. Balaji B, Porchilamban S (2015) Conceptual design of iron box handles with ergonomic aspects pp 766–767. ISSN 1662-7482
2. Benusiglio A, Mansard V, Biance AL, Bocquet L (2018) The anatomy of a crease, from folding to ironing. 8:3342
3. Aujla P, Sindhu P, Kaur R (2008) An ergonomic study of muscular fatigue during ironing cloths with selected irons. J Hum Ecol 24(1):31–34
4. Kaushik A, Mishra A, Singh H, Hemalatha B (2014) Automatic ironing machine, vol 1. ISSN 2319–2801
5. Wee LK (2014) Umea, Enhancing the user experience using an electric iron. Umeå Institute of Design Umeå University, Sweden, SE901 87
6. Hazuan Bin M, Hawawi M (2009) Design and analysis of electric iron box using Boothroyd dewhursd dema methodology. Thesis of Bachelor of Mechanical Engineering, Universiti Teknikal Malaysia Melaka
7. Aghawn S (2014) Design and fabrication of gas pressing iron. J Multi Sci Technol 1(5). ISSN: 3159–0040

8. Poornakanta H, Kadam K, Pawar D, Medar K, Makandar I, Patil AY, Kottturshettar BB (2018) Optimization of sluice gate under fatigue life subjected for forced vibration by fluid flow. Sciendo J Mech Eng 68(3):129–142
9. Arun YP, Umbrajkar HN, Basavaraj GD, Gireesha RC, Krishnaraja GK (2018) Influence of bio-degradable natural fiber embedded in polymer matrix, materials. Today Proc 5:7532–7540
10. Yang (2018) Sustainable packaging bio composites from poly lactic acid and wheat straw: Enhanced physical performance by solid state shear milling process. Compos Sci Technol 158:34–42
11. Vilarinho et al (2018) Monitoring lipid oxidation in a processed meat product packaged withnanocomposite poly (lactic acid) film. Euro Polym J 98:362–367

Microcontroller-Based Office Digitization

Vijitaa Das and Sharmistha B. Pandey

Abstract The office Digitization system is an application of information technology. This system has been developed in order to minimize wasting time in a particular task. Usually, employees take long time to complete one task where it requires searching of paper, documents, and files. The main objective is to make paperless work environment where it will greatly eliminate paper usage. This will benefit employees to work efficiently, better, and quicker. This computer-based system ensures to perform functions in aspects like storing all inventory details, managing business documents in digital form, updating work order status, alerting with notification in mobile with latest notice, and many more. It stores scanned documents. Using an Arduino UNO interfaced with OV7670 camera module can capture the image of order number, convert it into text, and save it with accurate date and time. Computer processing becomes quite accurate if the task to be performed is properly prepared. So, office Digitization ensures better accuracy. Chances of error will be eliminated.

Keywords Digitization · Arduino UNO · OV7670 · Tesseract OCR

1 Introduction

The term "Office Digitization" is common since desktop computers were introduced in the business world. To optimize office environment, typewriter was initially mechanized to minimize manual task. Office Digitization is intended to simplify and improve basic office task, different activities, communication, and collaboration within multiple branches of a company or organization. It includes storing inventory details, exchanging of information, management of business documents, work

V. Das (✉)
Narula Institute of Technology, MAKAUT, West Bengal, Kolkata, India
e-mail: vijitaadas2@gmail.com

S. B. Pandey
Regional Remote Sensing Centre—East, ISRO, West Bengal, Kolkata, India

BBVL. Deepak et al. (eds.), *Innovative Product Design and Intelligent Manufacturing Systems*, Lecture Notes in Mechanical Engineering,
https://doi.org/10.1007/978-981-15-2696-1_6

order status, event schedules, etc. Earlier, experts were needed for typesetting, printing, and electronic recording which are integrated in this system, so that it eliminates requirements for large number of staffs.

LAN plays a major role in storing, transmitting, and receiving raw data information and helps to connect with other destination through connecting network. In our system, we have made a server. There will be two or three admin users whose name would be already stored in database. New user will have to register their details in order to enter their name in database. Registered employees have to fill the custodian details and then about their respective system details. If they face any sort of problem regarding the system, they can register any complain which will be visible to admin, so that an immediate step can be taken to solve the problem. A registered employee can only see their work order status, store and transfer documents, communicate with higher authorities, and will receive news and updates notification about any event schedules in the organization.

Office Management requires to manage business documents including images and graphics, business dates, appointments, meetings, and client information contacts which can be edited, stored, and retrieved if necessary. Business documents can be scanned and stored in a common file where all employees can use it. Using Arduino UNO microcontroller interfaced with OV7670 camera module, office order numbers will be captured and stored in the server connected with MySQL database. The image will be converted to text with the help of Tesseract OCR. In the display page of office order details, order number along with the description of the item, the date and time of delivery, and receipt will be displayed.

2 Literature Survey

Office Digitization is a broadly used term, and it represents a new profession, a new integration of technologies and a new perception of the potential of information tools available to man. It is primarily based on two factors—computers and communication technology. The computer is moving from being an independent system to a component embedded in a whole range of office devices. Communication technology integrates these devices and people. It provides an effective communications infrastructure. So, office Digitization is the use of various technologies (e.g., computer and telecommunication) to simplify and support routine office functions, improve communication, increase office productivity, and enhance the quality of clerical output.

In the recent past, we can view almost everywhere Digitization playing a crucial role. Digitization does not mean only to mechanize anything automatically, but it has the capability to function in such a way, so that the data or information can be recorded and transferred via some mode [1]. Digitization system has already been introduced in the business world. In any case, Digitization is something more than motorization since mechanization is a self-directed process in which the work is finished with the least human endeavors [2]. The self-controlled process goes for a consistent stream of data without least human mediation. So, in a word, the word Digitization signifies the specialty of recording, preparing, and controlling the data naturally by mechanical and electronic machine [3].

Nowadays, schools and colleges are working with the help of Digitization system. It automates the processes that take place in the institutional office. There are many information need to be managed through the online method with the help of some application. The activities, student details, and many matters like this can be better maintained sequentially by the applications which are capable of doing this task. The admission details, students and faculty's information, feedback reviews regarding faculties, the performance of the students, reviews from outside institutes, details regarding fees clearance, and suggestions which can be given in some matters needed for improvement can all be handled easily using the office Digitization system application [4]. Employees want less pressure and loads in their workplace. Paper-based office work is highly costly and makes huge wastage of time in search of an old file or documents. It also decreases their productivity. The office Digitization manager should assume an active role as a change agent, collaborate effectively with various staff groups, co-ordinate the skills of the office Digitization team, and understand the business requirements [5].

3 Methodology

The block diagram of this system has been shown in Fig. 1.

New user has to register with the username and password. After submitting, they can access the login page with their registered username and password. Administrators name are already stored in the database. So, these users can easily login with their credentials. If any of the field is null or found wrong during accessing login, there will appear a warning dialog box showing "Invalid Credentials." If the user put valid credentials, they can successfully login.

After logging in, there comes a home page. In the home page, there will appear different options to choose. Under office Digitization, the image of the order_id will be converted into text and will be saved with a description of the item along with date and time of delivery and receipt. Arduino UNO interfacing with OV7670 module will help to capture the image of any printed written number, and with the help of Tesseract OCR, it will convert the image into text.

Arduino UNO will supply 3.3 V power to OV7670 camera module to get activated along with relevant pin connections. The hardware components will work according to the instructions given in the program of Arduino_IDE software. After capturing and uploading the image successfully, Tesseract OCR will help to convert it into text. The text will immediately get inserted and saved into MySQL database. So, in the web server, the text of order number along with the description of the item delivered or received will be displayed.

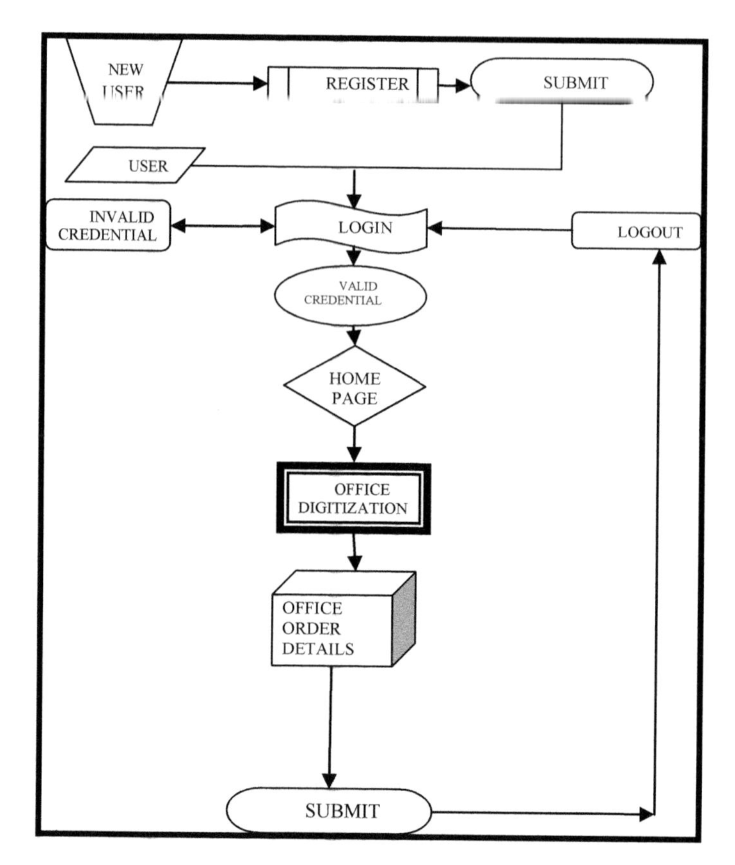

Fig. 1 Block diagram of office Digitization

4 Components

In this system, we have used two main hardware components—Arduino UNO and OV7670 camera module. For connection, we needed one breadboard, two resistors, each of 10 kΩ, and another two resistors of 4.7 kΩ. This paper outlines a cost-effective solution using the highly effective system. It costs around ₹680 (approx $9.78).

4.1 Arduino UNO

Arduino UNO is an open-source 8-bit microcontroller board based on ATmega328P. This single board has 14 digital input/output pins (D0–D13)—out of which 6 pins (D3, D5, D6, D9, D10, and D11) generate PWM output. It has 6

Fig. 2 Arduino UNO

analog input pins (A0–A5). It generates a 5 V supply to power the board, and its DC current is 40 mA. It also has a 3.3 V power supply which works as an on-board regulator and can draw a maximum of 50 mA current. It consists of a crystal oscillator of 16 MHz, an ICSP header. There are two serial pins—0(Rx—receiver) and 1(Tx—transmitter). There is an in-build LED on pin 13 which is ON when it is HIGH and OFF when it is LOW. Arduino_IDE software helps to upload and run the program onto the board via a USB cable. The image of Arduino UNO has been shown in Fig. 2.

4.2 OV7670

The OV7670 is an image sensor which has a CMOS processor and a VGA camera. It is of small size and a low-voltage camera module. Its resolution is 640 × 480 VGA which is equal to 0.3 Mpx. It is powered by 3.3 V supply (in our case, we have powered from Arduino UNO board). It has an input SCL and bi-directional SDATA. It has two output pins: VSYNC and HREF. It has a clock signal (XCLK) pin whose frequency range is from 10 to 48 MHz. By default, PCLK output pin also has a frequency similar as XCLK pin. The D7–D0 (pixel data output) pins must be sampled when HREF is high. These pins are YUV/RGB video component output pins. This image sensor helps in the detection and tracking of moving objects. The image of OV7670 has been shown in Fig. 3.

The image of the connected circuit has been shown in Fig. 4.

Fig. 3 OV7670

Fig. 4 Circuit connection

5 Software's

5.1 Eclipse Java EE

It is an integrated development environment (commonly known as Eclipse IDE). It supports in the development of server like Tomcat. The Eclipse Web Tools Platform (WTP) project is used for developing and designing Web using Java EE (EE stands for enterprise edition) applications. In Eclipse, two Web components are necessary for developing Web: Servlet and JSP (JavaServer Pages). In programming, we used "doPost" method which provides data security for applying encoding, encryption, and security algorithms. It helps to send data in form fields to the server along with the request.

5.2 Heidisql

It is an open-source tool for database systems such as MySQL, MariaDB, Microsoft SQL Server, and PostgreSQL. It is very useful for Web developers. It is such a platform where creating and editing tables and other functions can easily be performed. From Java, while sending any SQL query to the database, we used “Prepared Statement” command. It protects from any SQL injection attacks. So, it secures all the data.

5.3 *Arduino_IDE*

The Arduino Integrated Development Environment (IDE) provides a software library from the wiring project. The programs written in this software are C, C++, and Java. Programs written in Arduino IDE are called sketches. There is a serial monitor to display the serials sent from the Arduino board. It connects with the Arduino hardware (Arduino UNO, in our case) via a USB cable to upload the program and communicate with them.

5.4 *Tesseract OCR*

Tesseract is the software which is used as an open-source optical character recognition (OCR) engine. Its function is to convert any scanned images of handwritten, typewritten, or printed into a text. It supports a wide variety of languages.

6 Results

See Figs. 5, 6, and 7.

Fig. 5 Is the login page where user has to put valid credentials for logging in

Fig. 6 Shows the table of order details where the image of order number can be uploaded

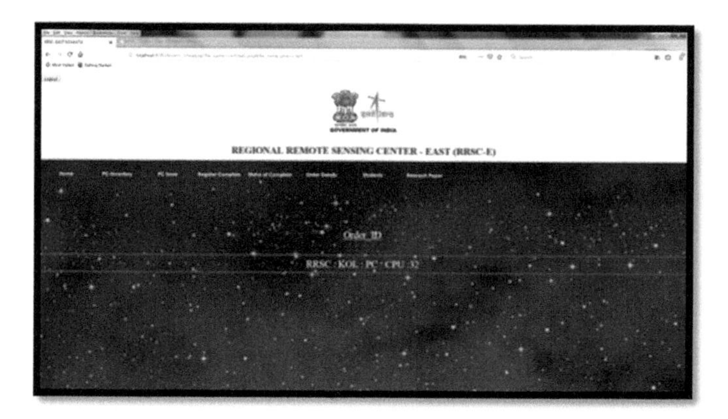

Fig. 7 Shows the uploaded image of order number which has been converted into text and displayed

7 Advantages

- Arduino UNO and OV7670 camera module is a low power consumption device
- Cost-effective components
- Enhanced productivity
- Optimum usage of power
- Big savings
- Monitoring and control of complete space
- Reduce the storage space
- Better communications
- Reduce wastage of time.

8 Conclusion

Automated office systems can provide a powerful mechanism for increasing productivity and improving the quality of work life by changing the fundamental nature of organizational information processing. The office automation manager should assume an active role as a change agent, collaborate effectively with various staff groups, co-ordinate the skills of the office automation team, and understand the business requirements. Therefore, the term office automation refers to the use of integrated computer and communication systems to support administrative procedures in an office environment.

9 Limitation and Future Scope

If there are multiple strings of alphanumeric characters in the header section, then it is difficult to differentiate office order no. from the other alphanumeric strings. In the future, we have a plan to build a barcode module in our system that would be beneficial to scan the documents with high speed and accuracy. It will also help to search for any scanned document with ease. And most importantly, it will help to keep all the scanned files and documents as online backup.

References

1. Canning RG (1978, September) The automated office: Part I. EDP Analyzer 16(9)
2. Canning RG (1978, October) The automated office: Part II. EDP Analyzer, 16(10)
3. Islam MS, Rashedul Alam MM (1999) Bangladesh journal of computer and information technology
4. Meyer ND (1999, March) Office automation. A progress report: office: technology and people
5. https://en.wikipedia.org/wiki/office.automation

Design and Evaluation of User Interface of a Mobile Application for Aiding Entrepreneurship

Anirban Chowdhury

Abstract Business incubators (BIs) are helping to grow start-up businesses, and thus, registration in BI is beneficial for budding entrepreneurs to make their carrier in the business world. Preferably, space for office, business mentoring support, technological support, design support, legal and IPR support, etc., are functions of a BI. However, synchronized support by BI for these functions is limited in current business incubation scenarios. The objectives of this study are (1) to design user interface (UI) of a mobile app and (2) human factors centric validation of it. In the current study, a design concept of a mobile app for BI is developed, and its prototype was created. Then, the heuristics evaluation of the prototype was conducted. Nielsen's ten heuristic principles were applied to assess the prototype of the mobile app. It was observed that the user interface of the mobile app is quiet promising to solve the synchronized management of BI activities as the flow of the app is good and users can access all important information in a single platform. The prototype of the mobile app also satisfied most of the heuristics. It can be concluded that the planned mobile app might be beneficial to provide a synchronized BI support to the budding entrepreneurs to pursue their business.

Keywords Business incubation · Design · Heuristics · Human–computer interaction (HCI) · Human factors · User interface (UI)

1 Introduction

The concept of BI becomes popularized since 1980s by many scholars and business practitioners. In a book entitled "New business incubator," Smilor (1986) had mentioned about a framework of a business incubator and its role in the growth of start-up businesses [1]. It has been unveiled in a comparative study that there is a

A. Chowdhury (✉)
School of Design (SoD), University of Petroleum and Energy Studies (UPES), Bidholi, Dehradun, Uttarakhand 248007, India

BBVL. Deepak et al. (eds.), *Innovative Product Design and Intelligent Manufacturing Systems*, Lecture Notes in Mechanical Engineering,
https://doi.org/10.1007/978-981-15-2696-1_7

positive impact of "*Business-to-Business Networks*" on innovation. In addition, institutional mechanisms (such as incubators, clusters, science parks) play an important role in entrepreneurship and the growth of start-up businesses [2]. Isabelle [3] reported about five key factors influencing technology entrepreneur's choice of incubator or accelerator. These factors include

(1) *The stage of their new venture*: Incubators are generally better suited to a very early stage venture, whereas an accelerator tends to focus on growing a firm quickly.
(2) *The fit between the entrepreneur's needs and incubator's mission, purpose, and sector focus*: With the proliferation of incubators and accelerators, technology entrepreneurs must pay close attention to their short and long-term needs to ensure an adequate fit with potential incubators or accelerators.
(3) *The selection and graduation policies*: Entrepreneurs should also consider the flexibility in how these policies are applied.
(4) *The nature and extent of services provided*: Here, technology entrepreneurs need to objectively assess their most urgent needs and the capacity of the incubator or accelerator to meet these needs in a timely fashion, and at a reasonable fee.
(5) *The network of partners*: Entrepreneurs should look for a variety of expertise to support firms (e.g., legal, regulatory, technical, intellectual property, finance).

The success of technology business incubators (TBIs) depends on management policies and practices of and incubation services offered by the TBIs. A recent study indicates that TBIs of China are better than TBIs of India in terms of TBI indicators [4]. These TBI indicators are venture capitals, incubation funds, employees, number of tenant firms, incomes of tenants and incubators, survival rates of tenant firms, and number of graduated tenants for which data are available, patents of tenants and national S and T projects undertaken by tenants (are used to assess the high-tech nature of tenants). The capability of TBI to pool funding resources is indicated by the number of venture capitals obtained, the effectiveness of TBI in creating jobs and contributing to local economic growth is demonstrated by the number of staffs in tenants, the number of tenants, and the incomes of tenants and incubators. This study also reported that services provided by Indian TBIs include access to venture capitals and seed funding, market research, public relations and press conferences, marketing services such as help for new ventures to establish networks with customers, suppliers, dealers, and other incubated firms, provide training related to tax, marketing and management skills, and intellectual property rights (IPR), effort to link up globally through networking with incubators, funding, advisory role, and involving experts.

Studies have been carried out on business incubators across the globe [4–8]; however, there is no clear indication about the use of computing system which assists for all the business incubation services. Therefore, there is a scope of design and development of a system which will assist the start-up owner to avail all the incubation services in a single platform. The interface of such computing system should be designed in such a way that it ensures user satisfaction. Nielsen [9] has proposed ten usability heuristics for user interface design to ensure user satisfaction. These heuristic principles might be helpful for ensuring the usability of the software. Based on these facts, the current study aimed to design a mobile app ensures its usability applying heuristic principles.

2 Methods

2.1 System Design

A system design workshop was conducted with one system architect (have 8 years of experience), one user experience designer (have 9 years of experience), and one incubation manager (20 years of managerial experience). Initially, all of them were introduced about the structure and functions of a business incubator. Then, they were introduced to TBI indicators which have been introduced by Tang et al. [4] which is followed by a brainstorming season for structuring an enterprise resource planning (ERP) system design using use case diagram for channelizing the BI-related activities. Figure 1 is indicating the use case diagram for the designed mobile app.

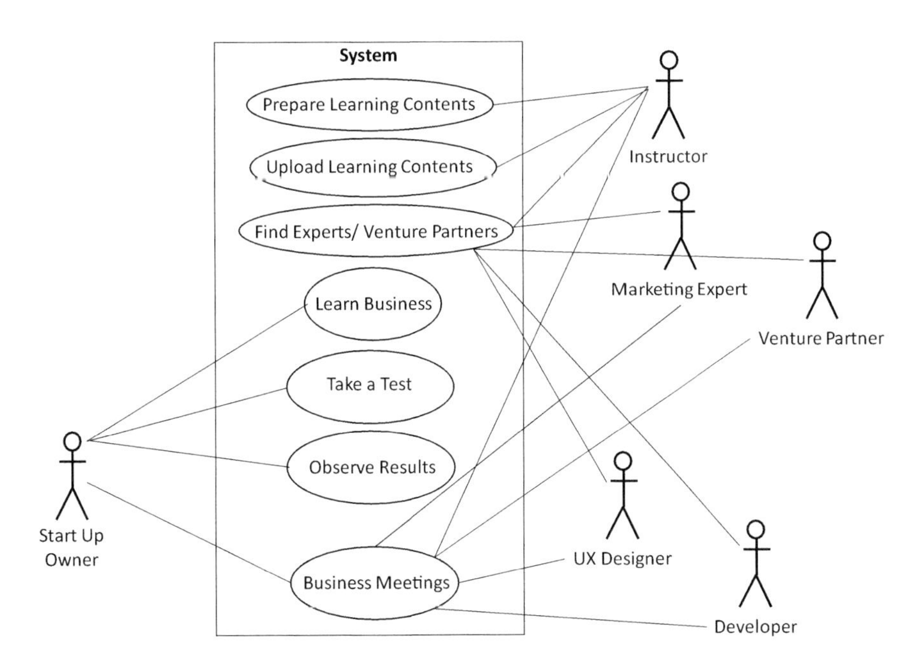

Fig. 1 Use case diagram of ERP system for mobile app

2.2 UI Design and Prototyping

The UI of the mobile app was designed based on the use case diagram. An IT business incubator was planned, and a name (Business Spector) was assigned to the mobile app. In addition, a dummy brand language was developed through a logo design (Please see Fig. 2). Following this hypothetical brand language, UI screens

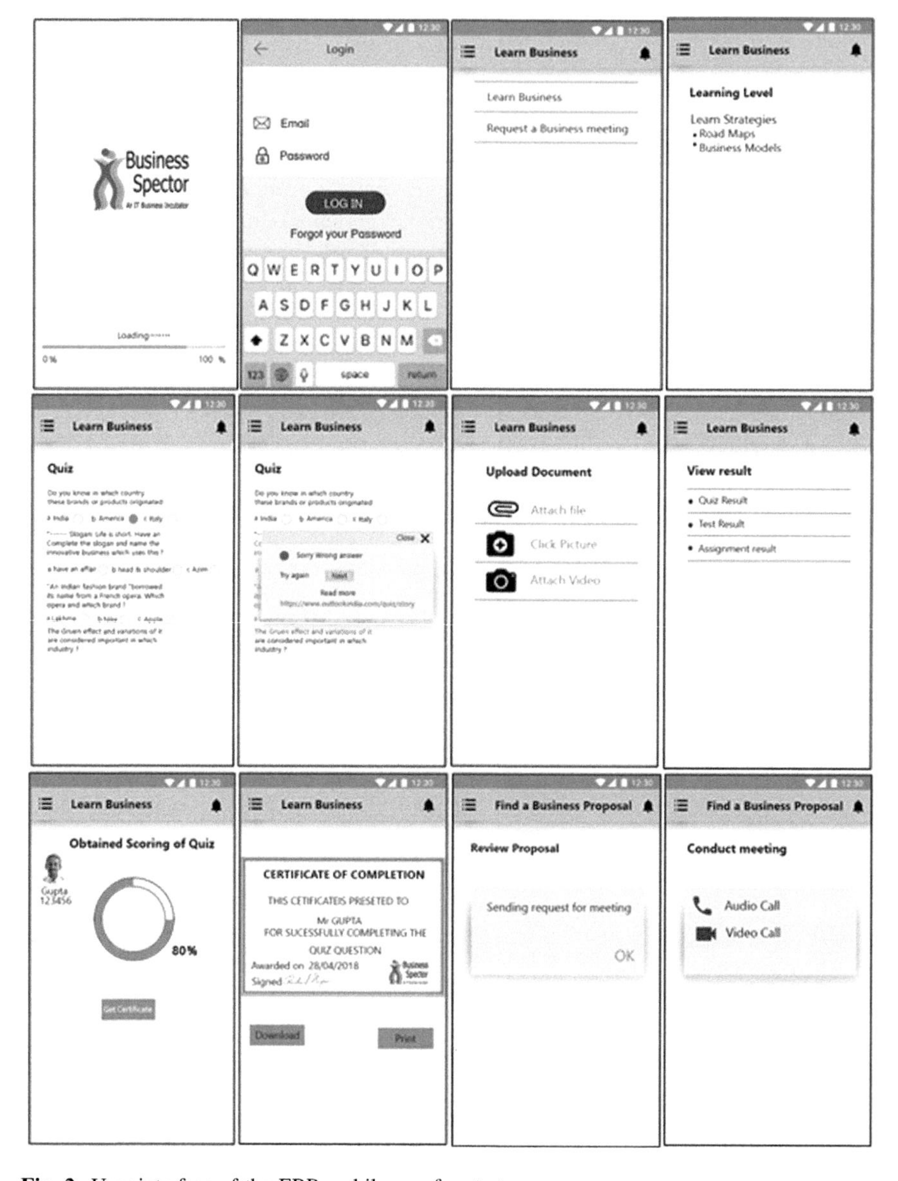

Fig. 2 User interface of the ERP mobile app for start-up owner

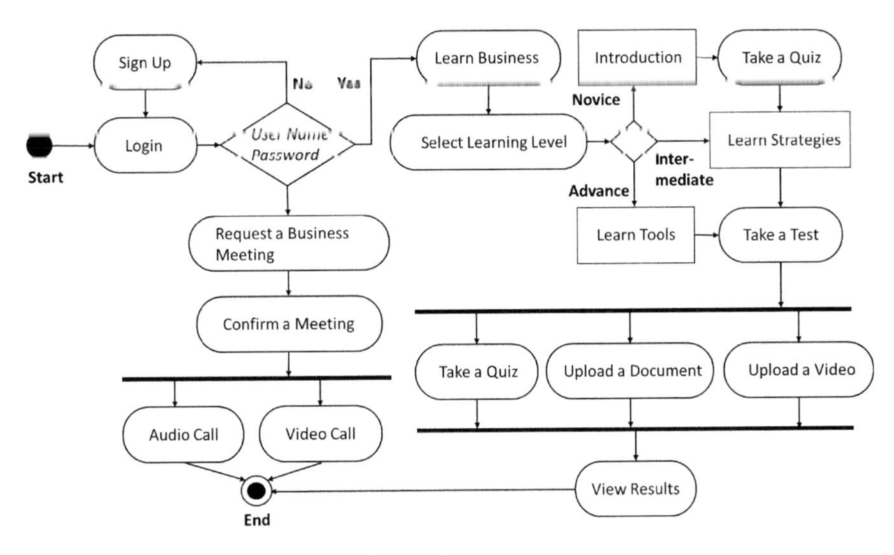

Fig. 3 User interface of the ERP mobile app for start-up owner

for the mobile app were designed using Adobe Photoshop CC, and interactive prototypes were developed using MS PowerPoint 2010, for all target user groups (start-up owner, instructor, marketing expert, venture partner, UX designer, and developer). UI screens of the mobile app for a start-up owner are depicted in Fig. 2. An activity diagram for a start-up owner is presented in Fig. 3. This diagram explained that a start-up business owner may learn the business (through online business learning kit line, a learning management system or LMS), and they can request different business meetings for different business supports. For instance, an entrepreneur might ask for design support from UX designer, business learning support from the instructor, marketing support from marketing expert, software development support from software developer, and financial support from a venture capitalist.

2.3 Heuristic Evaluation

To determine the usability of the proposed mobile application and to envisage the user satisfaction, Nielsen's ten heuristics principles [9] were applied. These heuristics are as follows:

(1) *Visibility of system status*: Time to time feedback shown to users through UI, depending on what is happening in the system
(2) *Match between system and the real world*: Use of familiar user-friendly terminologies instead of system terminologies

(3) *User control and freedom*: Autonomy to avoid the unwanted state of a system
(4) *Consistency and standards*: Same UI elements should not be represented through different ways
(5) *Error prevention*: Avoid error-prone situation or provide confirmation message before certain actions
(6) *Recognition rather than recall*: Reduction of short-term memory load through clearly visible objects, actions, and options
(7) *Flexibility and efficiency of use*: Provide step-by-step breakdowns for a task for novice users and shortcuts for expert users
(8) *Aesthetic and minimalist design*: Do not provide unwanted information and take care of visibility of the information
(9) *Help users recognize, diagnose, and recover from errors*: Clearly define the cause of an error in the error message to avoid the confusions among users
(10) *Help and documentation*: Provide instructions or guide for system use.

A total of three usability/human factors experts (average work experience: 8 years) participated in the heuristic evaluation of the designed prototype of the mobile app. They tried to identify the usability problems in it using Nielsen's ten heuristics principles.

Initially, experts ensured that all functionalities are existing in the mobile app as presented in the use case diagram (please find Fig. 1). Then, all experts have checked the activity flow of the users supposed to do using the designed interface, as per the activity diagram (please see Fig. 3). Thereafter, usability experts went through all UI screens of the prototype, and they tried to identify the usability-related problems exist (if any) in the designed UI of the prototype, using a checklist based on Nielsen's ten heuristics. They have identified and noted the user interface-related problems in the checklist. Please find the sample of the usability evaluation checklist in Fig. 4.

3 Results and Discussions

The interface features always play a crucial role in UI acceptance by target users [10, 11]. Furthermore, usability is the key predictor for user acceptance [12]. Literature also suggests that aesthetics is an important factor for UI or its prototype design and might contribute to the usability perception of the software interface [13–15].

In this study, all three usability experts agreed that visibility of system status, user control and freedom, consistency and standards, error prevention strategies, and flexibility and efficiency of use are fine. However, "*the visual language of the interface (UI color scheme) can be explored further*" as suggested by the one usability expert. Hence, the UI should be improved in terms of aesthetics and minimalistic design. In addition, "*navigation support, dialog, and notification system of the mobile app can be improved further to follow the heuristics principle*

Heuristics	Activity1	Activity2	Activity3	Activity (….n)
Visibility of system status	✓	✗	✓	-
Match between system and the real world	✗	✓	✓	-
User control and freedom	✗	✓	✓	-
Heuristic (……n)				-
Interpretation/ Note	There are issues in "Match between system and the real world" and "User control and freedom"	There is an issue in user feedback and thus, in "Visibility of system status"	All heuristics are well considered for this activity	-

Fig. 4 Sample checklist for heuristics analysis

recognition rather than recall," suggested by two experts. "*Current mobile application is average in terms of clarity of cause of an error in the error message,*" suggested by all usability experts. Therefore, there is a scope of improvement of the proposed mobile app to help users recognize, diagnose, and recover from errors [9]. Experts said that "*current prototype of the mobile app doesn't have any instruction manual or guide for usage of the mobile app.*" Thus, the help and documentation part requires improvement which is necessary for the novice users [16–18]. All three experts identified that "*menu tag terminology for the function of upload a business plan and request business meeting function were same.*" Therefore, a proper menu title should be assigned to ensure the match between system and the real world [9, 19].

4 Conclusion

Currently designed mobile app is promising in the context of providing a synchronized and various business supports to the budding entrepreneurs by business incubators through an online platform. However, the planned mobile app is a preliminary one with a requirement of few UI improvements in terms of usability factors such as aesthetics and minimalistic design and helps users recognize, diagnose, and recover from errors, help and documentation, and match between system and the real world. Therefore, user experience designer needs to ensure the quality of the UI design as per these heuristics, suggested by the usability/human factors experts. The detailed design of the mobile app has to be strategized in future, and a high-fidelity prototype with a database has to be tested before the implementation of the mobile app with the correction of heuristics-related issues.

Usability testing of the current mobile application was not conducted with users. However, a real-time usability testing can be conducted in near future after developing a high-fidelity prototype using Metrics for Usability Standards in Computing (MUSiC) method with the help of DRUM (diagnostic recorder for usability measurement) [20].

References

1. Smilor RW (1986) New business incubator. Lexington Books, New York
2. Pittaway L, Robertson M, Munir K, Denyer D, Neely A (2004) Networking and innovation: a systematic review of the evidence. Int J Manag Rev 5(3–4):137–168
3. Isabelle DA (2013) Key factors affecting a technology entrepreneur's choice of incubator or accelerator. Tech Inno Manag Rev 3(2):16–22
4. Tang M, Baskaran A, Pancholi J, Lu Y (2013) Technology business incubators in China and India: a comparative analysis. J Glob Inf Tech Man 16(2):33–58
5. Hutabarat Z, Pandin M (2014) Absorptive capacity of business incubator for SME's rural community located in indonesia's village. Proc Soc Behav Sci 115:373–377
6. Masutha M, Rogerson CM (2014) Small business incubators: an emerging phenomenon in South Africa's SMME economy. Urbani Izziv 25:S47–S62
7. Masutha M, Rogerson CM (2014) Small enterprise development in South Africa: the role of business incubators. Bull Geog Socio-eco Ser 26(26):141–155
8. Salem MI (2014) The role of business incubators in the economic development of Saudi Arabia. Int Bus Econ Res J 13(4):853
9. Nielsen J (1995) 10 Usability heuristics for user interface design. Available at: https://www.nngroup.com/articles/ten-usability-heuristics/. Accessed 28 Oct 2018
10. Koli A, Chowdhury A, Dhar D (2016) Requirement of new media features for enhancing online shopping experience of smartphone users. In: Berretti S, Thampi S, Dasgupta S (eds) Intelligent systems technologies and applications. advances in intelligent systems and computing, vol 385. Springer, Cham, pp 423–435
11. Chowdhury A, Karmakar S, Reddy SM, Ghosh S, Chakrabarti D (2013) Product personality rating style for satisfaction of tactile need of online buyers—A human factors issue in the context of e-retailers' web-design. In: International conference on human computer interactions (ICHCI). Chennai, pp 1–8
12. Chowdhury A, Karmakar S, Reddy SM, Ghosh S, Chakrabarti D (2014) Usability is more valuable predictor than product personality for product choice in human-product physical interaction. Int J Ind Ergonom 44(5):697–705
13. Chowdhury A (2019) Design and development of a stencil for mobile user interface (UI) design. In: Chakrabarti A (ed) Research into design for a connected world. smart innovation, systems and technologies, vol 135. Springer, Singapore, pp 629–639
14. Shirole D, Chowdhury A, Dhar D (2018) Identification of aesthetically favourable interface attributes for better user experience of social networking application. In: Ergonomics in caring for people. Springer, Singapore, pp 251–259

15. Karkun P, Dhar D, Chowdhury A (2014) Usability and Cognitive interference in Technology adoption for elderly—a critical review of literatures in HCI context. In: Proceedings of international ergonomics conference HWWE 2014, vol 1. Tata McGraw-Hill, India, pp 408–411
16. Benyon D (2014) Designing interactive systems: a comprehensive guide to HCI, UX and interaction design, 3rd edn. Pearson, United Kingdom
17. Bay S, Ziefle M (2003) Design for all: User characteristics to be considered for the design of devices with hierarchical menu structures. Hum Factors Organ Design Management 503–508
18. Medhi I, Sagar A, Toyama K (2006) Text-free user interfaces for illiterate and semi-literate users. In: ICTD'06: international conference on technologies and development, information and communication technologies and development. IEEE 72–82
19. Simon HA (1996) The sciences of the artificial. MIT Press, USA
20. Bevan N (1995) Measuring usability as quality of use. Softw Q J 4(2):115–130

Novel Dexterity Kit Concept Based on a Review of Hand Dexterity Literature

Gaurav Saraf and Dhananjay Singh Bisht

Abstract Dexterity is commonly defined by the quality of fine, voluntary movements used to manipulate objects during a specific task involving the movement of wrist, hands, arm, and fingers. Dexterity assessment kits are used to determine a person's skilled task abilities through performance parameters such as speed, accuracy, and precision. This study proposes that one parameter that is as critical as the traditionally measured parameters is finger strength which could be measured as the amount of force or effort that a human hand exerts during object manipulation through fingers. In this paper, a detailed literature review was conducted of the traditional dexterity assessment methods and their kits used in the past. Thereafter, a novel dexterity kit has been proposed which incorporates measurement of finger strength data in addition to the traditional dexterity parameters during hand dexterity assessment. An experiment suggested that a significantly greater finger force is required for peg manipulation in the new test kit than in the traditional one.

Keywords Dexterity · Precision · Finger strength · Hand function · Rehabilitation

1 Introduction

Dexterity is the skill in performing precision hand-based tasks. A dexterity test is used to figure out the manual ability of an individual. For a person to perform a precision activity, both dexterity and hand strength are necessary to produce manipulative actions. Some general examples of dexterous activities are picking up objects using thumbs and fingers, writing carefully using a pen, playing sports, playing finger instruments, etc. Dexterity can be broadly divided into two categories [1]:

G. Saraf (✉) · D. S. Bisht
Department of Industrial Design, National Institute of Technology, Rourkela, Odisha 769008, India
e-mail: gsaraf35@gmail.com

BBVL. Deepak et al. (eds.), *Innovative Product Design and Intelligent Manufacturing Systems*, Lecture Notes in Mechanical Engineering,
https://doi.org/10.1007/978-981-15-2696-1_8

1. ***Manual dexterity***: It refers to the overall gross movement of the hand and the ability of the hand to handle objects.
2. ***Fine motor dexterity***: It involves the fine movement exhibited by the different parts of hand like fingers, wrists with precision, and accuracy.

The two main fields in which the dexterity assessments play a major role are in:

1. ***Rehabilitation***: Dexterity is the best predictor of independence in activities of daily living (ADL) [2]. The gross and the fine motor skills can be impaired due to injury, illness, and stroke or development disabilities causing problems like lack of coordination between the hand, fingers, and eyes. An occupational therapist makes use of the suitable dexterity assessment tool to help re-develop, recover, and maintain the meaningful activities or occupation.
2. ***Industrial use***: The dexterity tools are very useful in evaluating employment interests and pursuits, employment seeking and acquisition, job performance, and retirement preparation [3].

Dexterity assessment helps to measure a person's speed, accuracy, and precision, i.e. the quality of movement as the hand manipulate objects and tools in the context of self-care, work, or any other activity of daily life in order to predict the abilities and the disabilities of a person [3].

In the past, a number of studies have been focused on the concerns of hand dexterity and strength that an individual should possess in his day to day working life. Bell et al. [20] demonstrate the importance of hand dexterity and strength in accessing packaging (plastic bottles, jars, and crisp packets) as the persons taking the test faced difficulty in accessing the package if they lacked either of the two parameters. The data obtained from this study were compared with the Purdue Pegboard Test data to highlight the distinction from pegboard test which focuses only on the dexterity of the person and not on finger strength. Therefore, a means to measure and improve both an individual's finger strength and dexterity shall be highly desirable. In the past, hand function skills like control precision have been measured using Grooved Pegboard Test [4]; manual dexterity has been measured by using kits like Minnesota Manual Dexterity test [5], Functional Dexterity Test [6], and Box and Block test [7]; finger tactility has been measured using O'Connor Finger Dexterity Test [8]; speed, i.e. the ability to make repeated hand movements, rapidly has been measured using tapping test [9]; fine motor dexterity has been measured using Moberg Pick-Up Test [10], Nine-Hole Peg Test [21], Jebsen-Taylor Test of Hand Function [11], and Purdue Pegboard Test [12]. Fleishman [13] identified the existence of at least two types of dexterity: arm-and-hand (gross) and wrist-and-finger (fine). Fleishman [13] later investigated the nature of factors responsible for the manipulative performance and identified five factors responsible for the effect on manual dexterity, namely finger dexterity, manual dexterity, wrist-finger speed, aiming, and positioning based on the various variables obtained by Purdue Pegboard, Tapping, and Punch board. The best measure of finger dexterity factors is the Purdue Pegboard Test. The three printed tests, namely square marking, marking accuracy, and tracing, were regarded as better measures of aiming than the two tapping tests [14] have presented a review on the innovative evaluation of dexterity for infants and children. The paper presented the measurement concepts being incorporated in assessment tools for pediatrics such as rate of completion, in-hand manipulation, and dynamic force control. Functional dexterity test (FDT) and strength dexterity test (SDT) were two novel assessment tools used in this paper [15] evaluated hand dexterity, grip, and pinch strength of the children. All these tests have been successfully measuring the skills for which they have been designed. In this study, the authors propose that along with all these "traditional" parameters of hand dexterity measurement, finger strength or force exerted by the human hand is also an important parameter which when measured along with the other "traditionally" measured dexterity skills can offer a more comprehensive hand dexterity test.

2 Methodology

2.1 Discussion on a Proposed Dexterity Kit

From the study of the various available dexterity kits, it is evident that these kits focus chiefly on the speed and accuracy of the user to complete the task. But in the real-world situations for which these tests are targeted, strength is also a very important factor to determine the user's performance and abilities. In this paper, a design of dexterity kit has been proposed to make the use of a layout similar to existing test kits and to introduce a mechanism to fulfil the finger strength criteria that is realized as missing during this review.

A CAD model of the proposed design has been created using SolidWorks 2015 ×64 edition [16] as shown in Fig. 1. The board consists of four columns of holes with 12 holes in each row (tentative proposal). The board has radial spring mechanisms within it underneath each hole. The hole diameters are kept decreasing along the width of the board to study the finger force variation with respect to pin diameter. The spring stiffness is kept increasing along the length of the board to study the variation of finger force with respect to the stiffness of the spring. The shape of the holes is circular with rectangular slits on diametrically opposite sides to provide for a locking option. Similar to the mating part, the pins are also cylindrically shaped with two rectangular pins on either side of the diameter. The pegboard is hollow inside where springs are placed beneath each hole. The spring is fixed at the bottom with a plate at the top which covers the hole from inside. A user inserting the pin into the hole will have to unlock by applying a suitable force (according to the stiffness of the spring used) to the correct orientation such that the extensions on the pin mates with the hole extensions. Also, to keep the pin inside the hole after the applied force is removed the user must turn the pin inside the hole to lock. Using such an arrangement, it becomes convenient to estimate the force exerted by the fingers in addition to the hand dexterity. An important parameter for the proposed design of the

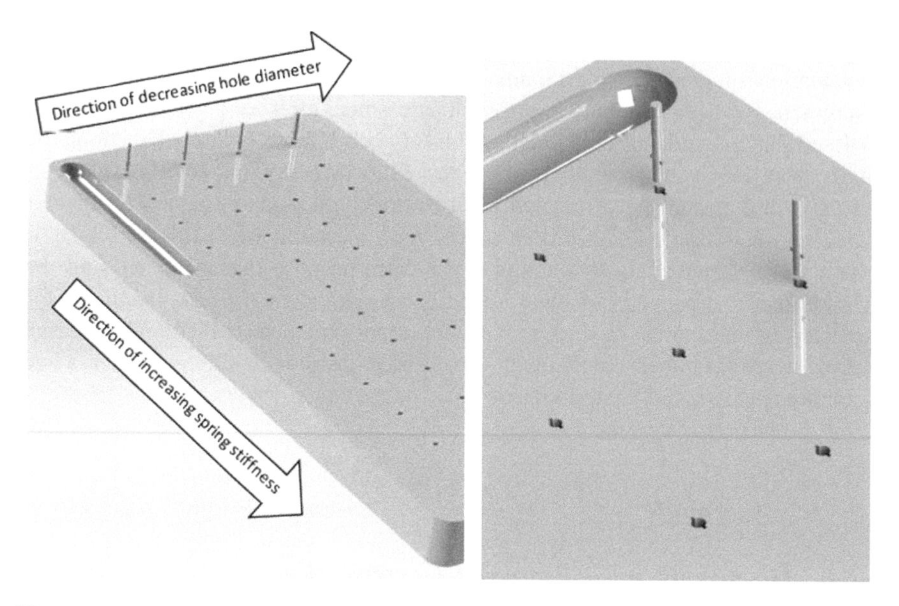

Fig. 1 Proposed kit design

kit is the diameter of the pin and step size with which it will vary. Some previous related works have been helpful in determining the size of the pins. Some of the useful data has been discussed below:

I. 60 Aluminium knobs for determination of torque capacity were used ranging from 12.7 mm (0.5 inch) to 25.4 mm (1 inch) with step size of 3.175 mm, 25.4 mm (1 inch) to 76.2 mm (3 inch) with step size of 6.35 mm, and 76.2 mm (3 inch) to 127 mm (5 inch) with step size of 12.7 mm [17].
II. Cylindrical handles were used for determination of grasping force with diameters ranging from 31 to 116 mm [18].
III. Cylindrical handles were used for determining grip force with diameters ranging from 25 to 50 mm [19].

From the study of these previous works, it can be seen that the diameter measure for which the power gripping or grasping forces has been determined is above 25 mm. Therefore, this test kit is focused on the analysis of pinch (precision) force exerted by a human finger in the smaller dimension range of 5–15 mm with a step size of 2 mm. To validate this range of diameters and the step sizes, a prototype model was developed on which the participants could perform the insertion, twisting, and locking tasks, and the traditional dexterity data along with the finger force data could be recorded.

2.2 Proposed Prototype Design

A CAD model of the prototype of the proposed design is shown in Fig. 2 made using SolidWorks 2015 ×64 edition. The prototype shows only a single hole spring arrangement. This prototype is proposed to manipulate/vary two important parameters for the kit hole diameter and the spring stiffness by trial and error method, for analysing the observed dexterity and finger force. The prototype was developed using MDF board, plywood board, acrylic fixtures, and an ABS rapid-prototyped peg for preliminary experimentation as can be seen in Fig. 2.

2.3 Working of the Prototype

The pin needs to be inserted in the hole with proper orientation, i.e. when the extensions on both the pin and hole are mating, then the pin is rotated inside the hole to fix it or lock it inside the hole by applying a turning moment.

2.4 Experiment

Eight participants, all post-graduate students of design, were first given a short briefing before the test to ensure that they understood the procedure completely. Participants were requested to comply with the following guidelines:

- Assume a sitting position.
- Ensure the elbow is flexed at a 90° angle.
- Ensure that the forearm is in a neutral position.
- Use the preferred hand to pinch the pin.

Fig. 2 Proposed prototype model

The following procedure was used for performing the task:

1. On the thumb and index finger of the participants, pressure sensors were attached.
2. The participants were requested to pinch the pin firmly and apply downward force.
3. Then the participants are asked to give a clockwise torque to lock the pin.
4. The pin was to be released.
5. Then a counter-clockwise torque was applied to unlock the pin.
6. Steps 1–5 were repeated for a duration of 60 s.

The fabricated model was used to carry out experiments to estimate the finger strength data from the available test kits. The individuals were asked to insert the pins in the slots, and their reading of force exertion was recorded using is Finger Tactile Pressure Sensing (FingerTPS) system developed by Pressure Profile Systems (PPS). At first, the spring was not loaded into the kit, and the individuals were asked to insert the pin (diameter 7 mm) into the slot. This step was performed to observe the amount of finger force exerted using traditional test kits. Then the spring was loaded into the kits, the individuals perform the same task again, and the readings were recorded. The individuals were made to perform the same task again on the spring-loaded test kit but with varying the pin diameter (11, 15 mm).

3 Results and Discussions

From Fig. 3, we can see that each of the eight participants during the traditional test condition, i.e. without spring-loaded pin, has to exert much lesser force than that in the spring-loaded condition. The reading after the spring was loaded into the kit

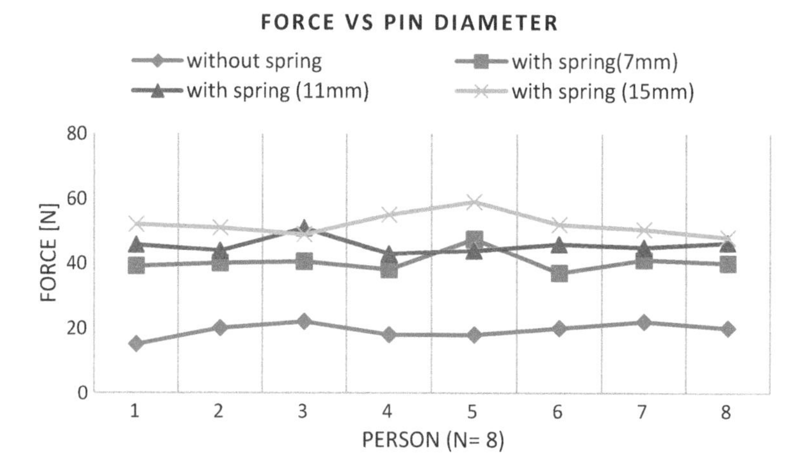

Fig. 3 Variation of finger force under traditional and spring-loaded conditions

Table 1 Comparisons of the various parameters measured using the dexterity kits

References	Control precision	Manual dexterity	Fine dexterity	Finger tactility	Speed	Finger force
Merker and Podell [4]	Measured		Measured			
Berger et al. [8]		Measured		Measured		
Aaron and Jansen [6]		Measured				
Desrosiers et al. [5]		Measured				
Mathiowetz et al. [7]		Measured				
Shimoyama et al. [9]					Measured	
Moberg [10]		Measured	Measured			
Kellor et al. (1971)			Measured			
Jebsen et al. [11]		Measured	Measured			
Tiffin et al. [12]		Measured	Measured			Measured
This study	Measured		Measured			Measured

shows that the finger force also varies with changing pin diameters. These results imply that a significant amount of finger force exertion was demanded in the newly developed test kit as compared to the non-spring-loaded kit. This result verified that the newly designed test kit fulfils the original aim of this work which was to additionally introduce a factor of finger force in the traditional dexterity kit architecture. Traditional ones tend to focus more on parameters such as accuracy, errors, and task time. Table 1 shows a comparative study of the various skill sets measured by the various available dexterity kits.

4 Conclusion

In this work, the design of a new dexterity measurement kit has been proposed which can incorporate concerns of hand dexterity and finger strength parameters during a dexterity test. On comparing the peak force measures obtained from spring-loaded pegs against regular pegs, it can be observed that significant differences do seem to exist in terms of the amount of force applied. Apparently, there exist a number of shortcomings in the work described in this paper; one of them is that the design proposed has only one port for performing the test, but in a professional kit, the test kit should include a battery of ports. An estimation of designing the professional kit could be based on the results obtained from the experiment conducted with a single port model. Also, greater breadth of experiments could be conducted on this single port model, e.g. to study the variation of finger strength with a change in pin diameters or the change in spring stiffness, as well as inputs regarding perceived finger fatigue could be collected. All such data will serve to provide inputs for scaling this test kit into a professional kind with a matrix of several ports to measure speed, accuracy, and precision and finger strength in a single test kit. Another area in which this model needs improvement is the accuracy of construction.

References

1. Backman C, Gibson SCD, Parsons J (1992) Assessment of hand function: the relationship between pegboard dexterity and applied dexterity. Can J Occup Ther 59(4):208–213
2. Williams ME, Hadler NM, Earp JAL (1982) Manual ability as a marker of dependency in geriatric women. J Chronic Dis 35(2):115–122
3. Gallus J, Mathiowetz Vl (2003) Test–retest reliability of the Purdue Pegboard for persons with multiple sclerosis. Am J Occupational Ther 57(1):108–111
4. Merker B, Podell K (2011) Grooved pegboard test. In: Encyclopedia of clinical neuropsychology. Springer, New York, pp 1176–1178
5. Desrosiers J, Rochette A, Hébert R, Bravo G (1997) The Minnesota manual dexterity test: reliability, validity and reference values studies with healthy elderly people. Can J Occupational Ther 64(5):270–276
6. Aaron DH, Jansen CWS (2003) Development of the Functional Dexterity Test (FDT): construction, validity, reliability, and normative data. J Hand Ther 16(1):12–21
7. Mathiowetz V, Volland G, Kashman N, Weber K (1985) Adult norms for the box and block Test of manual dexterity. Am J Occupational Ther 39(6):386–391
8. Berger MA, Krul AJ, Daanen HA (2009) Task specificity of finger dexterity tests. Appl Ergon 40(1):145–147
9. Shimoyama I, Ninchoji T, Uemura K (1990) The finger-tapping test: a quantitative analysis. Arch Neurol 47(6):681–684
10. Moberg E (1958) Objective methods for determining the functional value of sensibility in the hand. J bone Joint Surg. Br Volume 40(3):454–476
11. Jebsen RH, Taylor NEAL, Trieschmann RB, Trotter MJ, Howard LA (1969) An objective and standardized test of hand function. Arch Phys Med Rehabilitation 50(6):311–319
12. Tiffin J (1968) Purdue Pegboard examiner manual. Science Research Associates Inc., Chicago, Illinois
13. Fleishman EA (1953) Testing for psychomotor abilities by means of apparatus tests. Psychol Bull 50(4):241
14. Duff SV, Aaron DH, Gogola GR, Valero-Cuevas FJ (2015) Innovative evaluation of dexterity in pediatrics. J Hand Ther 28(2):144–150
15. Lee-Valkov PM, Aaron DH, Eladoumikdachi F, Thornby J, Netscher DT (2003) Measuring normal hand dexterity values in normal 3-, 4-, and 5-year-old children and their relationship with grip and pinch strength. J Hand Ther 16(1):22–28
16. Manual, SolidWorks–Users (2009) Dassault Systems Solid Works Corporation. SolidWorks Corp,Concord
17. Sharp ED (1962) Maximum torque exertable on knobs of various sizes and rim surfaces. No. MRL-TDR-62-17. Air Force Aerospace Medical Research Lab Wright-Patterson AFB OH
18. Amis AA (1987) Variation of finger forces in maximal isometric grasp tests on a range of cylinder diameters. J Biomed Eng 9(4):313–320

19. Kong Y-K, Lowe BD (2005) Optimal cylindrical handle diameter for grip force tasks. Int J Ind Ergon 35(6):495–507
20. Bell A, Walton K, Yoxall A (2017) Measure for measure: pack performance versus human dexterity and grip strength. Packag Technol Sci 30(4):117–126
21. Mathiowetz V, Weber K, Kashman N, Volland G (2016) Adult norms for the Nine Hole Peg Test of finger dexterity. Occup Ther J Res 5(1):24–38

Allocentric and Egocentric Behaviour of People While Wayfinding

Pranjali Pachpute, Shubham Johari and Wricha Mishra

Abstract Wayfinding is a daily task which is associated with efficient manoeuvrability within given space and time. Factors affecting wayfinding can be categorised as external and internal. Current study has been done to understand allocentric and egocentric behaviour of people while wayfinding. The study was conducted in two different phases on the participants of age group 25–45 years old. In the first phase, 180 participants were taken into consideration for asking a face-to-face questionnaire on wayfinding. A real-time experiment was conducted in Pune on another eight participants in the second phase. The area chosen for the experiment was unfamiliar to the participants. Results showed that ability to remember the landmarks differs significantly with the age and does not affect with the gender. Significant correlation found out between individual characteristics of a person and different wayfinding factors. The identified dominant factors which affect wayfinding were—modes of wayfinding, spatial anxiety and environmental factors.

Keywords Allocentric behaviour · Egocentric behaviour · Wayfinding strategy · Navigation · Factors affecting wayfinding

1 Introduction

Wayfinding is a daily task which involves understanding the spatial factors while navigating in an unknown area. Wayfinding involves the ability to navigate successfully across the territory. More specifically, wayfinding is the capability to recognise one's location and approach to the destinations in the environment, both cognitively and behaviourally [1].

P. Pachpute · S. Johari · W. Mishra (✉)
MIT Institute of Design, Loni-Kalbhor, Pune, India

BBVL. Deepak et al. (eds.), *Innovative Product Design and Intelligent Manufacturing Systems*, Lecture Notes in Mechanical Engineering,
https://doi.org/10.1007/978-981-15-2696-1_9

Kevin A. Lynch has discovered that during wayfinding cognitive maps of locations are processed by individuals [2] and defined a cognitive map as a mental representation made up of routes, paths and environmental relationships, such as cardinal directions, which an individual uses for making wayfinding decisions [3].

With the help of different methods in the experiment, wayfinding performance can be evaluated. Self-evaluation methods have included revealing one's ability to know directions [4] and describing the different methodologies one uses while wayfinding. Behavioural methods have included evaluating distances, drawing maps of an area, exploring familiar or unfamiliar area [5], giving verbal or written directions [6] and indicating unseen locations in the area [1]. Large number of factors can influence individual's ability to find their way and to realise they have reached to their destination.

1.1 Factors Affecting Wayfinding

Factors influencing wayfinding can be categorised as external and internal factors. External factors include environmental factors like landmarks, street structures, number of buildings in an area, etc. Internal factors include characteristics of people like their ability of wayfinding, their behaviour while wayfinding, and familiarity of the environment [7]. A change in wayfinding behaviour has observed when the area is familiar. Wayfinding behaviour is also influenced by the existence of population. There are four environmental factors that affect wayfinding behaviour: visual access, architectural differentiation, floor plan and signage [8]; Presence or absence of these factors affects the wayfinding performance of a person.

1.2 Allocentric and Egocentric Behaviour

Route and survey knowledge are two main types of spatial knowledge according to the theory of cognitive maps [3]. A sequence of memory about how to get from the starting location to the next place is represented by route knowledge and the gestalt like memory of interconnections between locations is represented by survey knowledge [9]. Allocentric/survey knowledge strategy includes global perspective while egocentric/route knowledge strategy involves local features like landmarks. We explored the behaviour of people while wayfinding, strategies they use while navigating in an unknown area and their spatial knowledge from these previous studies. In the current study, we have focused on these different types of behaviours and have conducted a real-time experiment to understand those, also we have investigated the influence of navigational support on the navigational performance. The following hypothesis was considered:

(1) Difference between allocentric and egocentric behaviour
(2) Difference in behaviour between finding a way under stress and without stress
(3) Correlation between individual characteristics and selection of landmarks of a person while wayfinding.

2 Methodology

Through self-report questionnaire investigation and a wayfinding navigation experiment, the differences between navigational (signs and map) and non-navigational assist, tasks showing wayfinding capabilities and wayfinding strategies (egocentric and allocentric) were explored.

The following study was approved from institutional ethics committee, MIT institute of design, Pune, India. Also consent was taken from all the participants involved in the experiment, while the data we gathered from questionnaire survey was anonymised.

Phase 1. One hundred eighty participants of 25–45 years old were taken into consideration for asking a face-to-face questionnaire on wayfinding. They were college students and professionals from Pune and Mumbai. Questions were based on their ability to remember the landmarks and directions, possibility of taking wrong turn in hurry, tendency to follow the crowd when lost, perceived feeling of confusion due to architectural layouts and behaviour of people while navigating. We also referred Lawton strategy scale (1995) for the questionnaire. A 5-point Likert scale was adopted to estimate the typical wayfinding strategies of the participants.

Phase 2. An experiment was conducted in Pune on another eight participants of 25–35 years old in an unfamiliar area. This age group was preferred because these individuals travel and navigate more due to their daily routine, and they are well versed with the use of wayfinding applications like Google maps and have the cognitive ability to find the quickest possible route. Initial survey of 180 participants proved that Google maps application is mostly used solution for wayfinding; hence, we used that in the second part of the experiment. Thus, Google maps helped to explore survey or allocentric strategy while wayfinding. Two tasks corresponding to the egocentric (route) and allocentric (survey) strategies were given for the experiment:

(1) The task was to find out the given location without using Google maps (using signage, landmarks and asking people), to return to the initial point and to draw the way travelled (map) on the paper. Area chosen was less crowded.
(2) The task was to find out the given location using Google maps (without asking people), to return to the initial point and to draw the way travelled (map) on the paper. Area chosen was of more traffic and turns.

The criteria of the experiment and tasks were first explained to the participants. There was no time limitation. Observer was monitoring the activities of the participants throughout the experiment.

Statistical Analysis. The Cronbach's alpha was calculated to check the reliability of the questionnaire which was found to be 0.690 which is good [9]. Statistical analysis of this quantitative research approach was conducted in SPSS version 17. To test the differentiation between the factors related to wayfinding and demographic data like age and gender Mann–Whitney test was executed. Spearman correlation test was performed to examine the association between the various factors related to wayfinding. To get help in information interpretation and to diminish the quantity of factors factor analysis was done on the correlation matrix of the observed variables. It was found that a factor analysis was appropriate when the variables were evaluated with KMO (0.744) and Bartlett's test ($p < 0.001$). The data provided by factor analysis was analysed and used for the further experiment.

3 Results

3.1 Analysis from Questionnaire

Table 1 depicts that there is significant difference in age for the ability to remember the landmarks, but no significant difference found in gender.

Table 1 Differentiation between the factors related to wayfinding with the age and gender

Demographics	Factors	Mann–Whitney *U*	*Z*	Asymp. Sig. (2-tailed)
Age	Frequency of visiting unknown places	449.000	−1.033	0.302**
	Ability to remember the landmarks	341.500	−2.613	0.009*
	Ability to remember the directions	384.000	−1.769	0.077**
	Frequency of using online maps	454.000	−0.911	0.362**
Genders	Frequency of visiting unknown places	832.000	−0.088	0.930**
	Ability to remember the landmarks	826.000	−0.161	0.872**
	Ability to remember the directions	778.500	−0.604	0.546**
	Frequency of using online maps	780.000	−0.590	0.555**

*Significant at level $p < 0.05$
**Not significant

Table 2 Correlation test results

Factors	Sig. value (p value)	Type of correlation (r value)
Ability to remember the landmarks—ability to find a way in the dark	0.000	0.390
Ability to remember the landmarks—difference in the experience of travelling a same path with different modes of transport	0.000	0.547
Ability to remember the directions—ability to find a way in the dark	0.000	0.272
Ability to remember the directions—difference in the experience of travelling a same path with different modes of transport	0.000	0.408
Possibility of taking wrong turn in hurry—tendency to follow the crowd when lost while navigating	0.000	0.533
The feeling of confusion due to architectural layouts while wayfinding—ability to remember the landmarks	0.046	−0.149
The feeling of confusion due to architectural layouts while wayfinding—tendency to follow the crowd when lost while navigating	0.034	−0.158

*Significant at $p < 0.05$

Table 2 depicts that the ability to remember the landmarks and directions was significantly correlated to ability to find a way in the dark. People who were able to remember the landmarks and directions agreed for the difference in the experience of travelling a same path with different modes of transport. Possibility of taking wrong turn in hurry was significantly correlated with the tendency to follow the crowd when lost while navigating. The perceived feeling of confusion due to architectural layouts while wayfinding is negatively correlated with the ability to remember the landmarks and tendency to follow the crowd when lost while navigating.

Table 3 represents the three clear patterns of response among the respondents—one pattern of modes of wayfinding (or not), one pattern of spatial anxiety while wayfinding (or not) and one pattern of environmental factors (or not). These independent three tendencies (i.e. not correlated) were considered for the further experiment.

3.2 Analysis of the Experiment

Pre-experiment and post-experiment questionnaires were asked to the participants to test their ability to recall the landmarks, directions and the path they travelled based on the Lawton Wayfinding Strategy Scale.

Table 3 Factor analysis table for wayfinding behaviour

	Factor 1: Modes of wayfinding	Factor 2: Spatial anxiety	Factor 3: Environmental factors
Ability to remember the landmarks	0.814		
Ability to remember the directions	0.808		
Mode of transport while wayfinding	0.740		
Follow crowd when lost		0.859	
Afraid of dark		0.758	
Wrong turn in hurry		0.618	
Contribution of architectural layouts for wayfinding			0.710
Contribution of colours for wayfinding			0.612

Table 4 Representing task analysis in percentage ($n = 8$)

Task	Reaching to the destination (%)	Return to the initial point (%)	Recall and draw (%)	Asked for help (%)
Finding location without using map (Phase 1)	100	67	67	100
Finding location with map (Phase 2)	100	100	33	0

Table 4 depicts task analysis in percentage to complete different tasks like reaching to the destination, return to the starting point, recall and draw the way they travelled. It was also observed that whether they are asking for help to find their way.

4 Discussion

Studies have revealed that the egocentric oriented support mode is usually used by females compared to males [3]. In the present study, Table 1 depicts no significant difference in gender for the ability to remember the landmarks, but it is found in age. Lawton (1996) demonstrated experimentally that participants suffering from spatial anxiety show incorrect landmark identification which is supported by this study. Table 2 depicts that the feeling of confusion due to architectural layouts while wayfinding is negatively correlated with the ability to remember the landmarks and tendency to follow the crowd when lost while navigating.

Studies have revealed that landmarks and sense of direction help to build cognitive model of the area which helps in wayfinding [10]. We must know where we are in relation to the streets and landmarks and we must update this information while moving through the area. Sometimes in the situation of anxiety, fear and stress, this task becomes difficult. The ability to do all these appears to be captured well in the present study of the perception of one's own sense of direction. We identified the comprehensive list of all the factors that affect wayfinding. Table 3 depicts the dominant factors identified—modes of wayfinding, spatial anxiety and environmental factors.

Studies have revealed that a valid component of human wayfinding experience is a sense of direction [11]. Different cognitive processes are involved in finding a way in an environment. The names of streets, blocks, where the streets are located and how they lay in relation to each other must be remembered. Studies have also shown that people in India rarely use maps for navigation. They rely primarily on asking around and navigate using landmarks [12]. In the present study, analysis of the experiment showed in Table 4 depicts that behaviour of people while finding a given location without map (egocentric) and using map (allocentric) was different. Recalling the landmarks, directions and the way they travelled were different in both allocentric and egocentric behaviour of the participants. Most participants reported a greater reliance on route strategy and a low to moderate use of map strategy.

There are few studies which have conducted real-time experiment of wayfinding. In the present study, with the survey and the experiment in real-time environment we observed and analysed wayfinding ability of the people.

5 Conclusion

In this paper, we discussed an egocentric (without using map) and allocentric (using map) wayfinding behaviour in the real physical world. The results showed that the ability to remember the landmarks differs significantly with the age and does not affect with the gender. Also there is difference between egocentric and allocentric behaviour of people while wayfinding. Subsequently, the correspondence between the various factors related to wayfinding was interpreted by the current study. It identified the comprehensive list of all the factors that affect wayfinding. The dominant factors identified were—modes of wayfinding, spatial anxiety and environmental factors.

Large sample size of the participants can be used for real-time experiment in future. This study indicates that technologies supporting egocentric wayfinding behaviour can be useful and can reduce the cognitive load while finding the destination. In future study, other kinds of navigational interface systems can be used. There is a scope of designing and developing a system which will make daily task of wayfinding simpler and which will help to perceive this cognitive process of wayfinding.

References

1. Prestopnik JL, Roskos-Ewoldsen B (2000) The relations among wayfinding strategy use, sense of direction, sex, familiarity, and wayfinding ability. J Environ Psychol 20(2):177–191
2. Nadel L (2013) Cognitive maps
3. Chen CH, Chang WC, Chang WT (2009) Gender differences in relation to wayfinding strategies, navigational support design, and wayfinding task difficulty. J Environ Psychol 29(2):220–226
4. Hegarty M, Richardson AE, Montello DR, Lovelace K, Subbiah I (2002) Development of a self-report measure of environmental spatial ability. Intelligence 30(5):425–447
5. Hund AM, Nazarczuk SN (2009) The effects of sense of direction and training experience on wayfinding efficiency. J Environ Psychol 29(1):151–159
6. Iachini T, Sergi I, Ruggiero G, Gnisci A (2005) Gender differences in object location memory in a real three-dimensional environment. Brain Cogn 59(1):52–59
7. Calori C, Vanden-Eynden D (2015).Signage and wayfinding design: a complete guide to creating environmental graphic design systems. Wiley
8. Emo B, Hoelscher C, Wiener J, Dalton R (2012) Wayfinding and spatial configuration: evidence from street corners
9. Sadeghian P, Kantardzic M, Lozitskiy O, Sheta W (2006) The frequent wayfinding-sequence (FWS) methodology: finding preferred routes in complex virtual environments. Int J Hum-Comput Stud 64(4):356–374
10. NHS Estates (2005) Wayfinding: effective wayfinding and signing systems; guidance for healthcare facilities. The Stationery Office
11. Cornell EH, Sorenson A, Mio T (2003) Human sense of direction and wayfinding. Ann Assoc Am Geogr 93(2):399–425
12. Patil A Aagey se right: exploring wayfinding in the Indian context

Design for Cost and Sustainability

Dynamic Stability Analysis of an Asymmetric Sandwich Beam on a Sinusoidal Pasternak Foundation

Dipesh Kumar Nayak, Madhusmita Pradhan, Prabir Kumar Jena and Pusparaj Dash

Abstract The dynamic stability of an asymmetric sandwich beam with viscoelastic core resting on a sinusoidal varying Pasternak foundation subjected to parametric vibration is observed. The effects of different parameters such as temperature gradient of each elastic layer, the ratio of modulus of the shear layer of Pasternak foundation to Young's modulus of the elastic layer, core loss factor, stiffness of Pasternak foundation and elastic foundation parameter on the dynamic stability are investigated. Hamilton's principle, generalized Galerkin's method and Hill's equations are utilized, followed by Saito–Otomi conditions to obtain the results.

Keywords Asymmetric sandwich beam · Temperature gradient · Sinusoidal Pasternak foundation · Dynamic stability

Nomenclature

$A_i(i = 1, 2, 3)$	Cross-sectional Area of *i*th layer
B	Beam width
$E_i(i = 1, 3)$	Young's modulus of *i*th elastic layer
g	Shear parameter
G_s	Foundation's shear layer modulus
G_2^*	Complex shear modulus of core
$h_i(i = 1, 2, 3)$	*I*th layer's thickness at '*x*'
$I_i(i = 1, 2, 3)$	Second moment of inertia about relevant axis
l	Length of beam
l_{h1}	l/h_{10}
d	Shear layer thickness of foundation
m	Mass per unit length of beam
ρ_i	*I*th layer's density
$\overline{\omega}$	Nondimensional forcing frequency

D. K. Nayak · M. Pradhan · P. K. Jena · P. Dash (✉)
Veer Surendra Sai University of Technology, Burla, Sambalpur 768018, India

BBVL. Deepak et al. (eds.), *Innovative Product Design and Intelligent Manufacturing Systems*, Lecture Notes in Mechanical Engineering,
https://doi.org/10.1007/978-981-15-2696-1_10

$\delta_i(i = 1,3)$	Constant temperature gradient of ith layer
t	Time
$\bar{t}$	Nondimensional time
$w(x,t)$	Lateral deflection of beam at 'x'
p	Number of functions

1 Introduction

The problem of beams on elastic foundations has an important place in modern structural and foundation engineering. Studies were done on asymmetric sandwich beam to reduce the weight of the system by Dash et al. [1], without compromising the stability of the system. Many research works focused on beams' stability and its vibration when the beams are on springs. The extensive static research work is in Hetenyi's book [2]. Wang and Stephan [3] studied the natural frequencies for Timoshenko beam and found that shear deformation, rotary inertia and foundation constants for various boundary conditions were affecting it. Kar and Sujata [4] studied the influence of constant temperature gradient on a Timoshenko beam and found that the stability was exaggerated. The temperature effect on Young's modulus of a nonuniform beam in parametric vibration was illustrated by the same authors [5]. Nayak et al. [6] examined the temperature effect on the sandwich beam, considering two different boundary conditions along with temperature gradient. Pradhan et al. [7] considered the effect of temperature gradient in case of a sandwich beam and observed that the stability was affected by it. It was exposed that the temperature gradient was affecting the stability of a tapered asymmetric sandwich beam by Pradhan et al. [8]. Pradhan et al. [9] studied the effect of the temperature gradient on a sandwich beam, asymmetric in nature, placed on Pasternak foundation and observed that stability was affected due to the stiffness of the foundation and temperature gradient. Ray and Kar [10] also investigated parametric instability of sandwich beams for a number of end conditions. Kerwin [11] observed the damping characteristic of the constrained layer. Then, Rao and Stuhler [12] considered symmetric sandwich beam and studied about loss factor and frequency. Saito and Otomi [13] investigated the parametric response of a beam that is viscoelastically supported. Dash and Nayak [14] determined the profile of an asymmetric rotating sandwich beam which has maximum stability and will be economical.

From the accessible literature, it has been discovered that till now, most of the work has been done for linear and parabolic variation of Pasternak foundation with no work done to study the effect of sinusoidal Pasternak foundation.

So, the present analysis explains the dynamic stability of an asymmetric sandwich beam resting on a sinusoidal varying Pasternak foundation under pinned—pinned and clamped—free conditions at the ends. The objective of using sinusoidal foundation is that any type of deflected shape, whether linear or parabolic, can be expressed in sinusoidal form. The effect of various parameters on the regions of parametric instability is examined by the computational method, and the results are graphically presented.

2 Problem Design

A generalized sandwich beam of length 'l' resting on a sinusoidal Pasternak foundation is shown in Fig. 1. It consists of two elastic layer and middle viscoelastic layer. Here, G_2 is the in-phase shear modulus of the viscoelastic core, and G_2^* is the complex shear modulus. Hence, $G_2^* = G_2(1 + j\eta)$ where 'η' is the core loss factor of the core and $j = \sqrt{-1}$.

Since the system is assumed to be only in the pure bending condition in one plane with a bending plane in the x-z plane, the axial load $P(t) = P_0 + P_1 \cos \omega t$ is applied at the C.G. of the transverse cross section. In the above expression, 'ω' is the frequency of the excitation and 'P_1' and 'P_0' are dynamic and static load amplitudes, and 't' is time. The system is resting on a foundation, which is made of close and equally placed springs having variable stiffness $k(x)$ for the springs, added with a shear layer of modulus of rigidity 'G_S'.

The assumptions considered to obtain the governing equations of motion are as [7].

According to Kerwin's [11] assumption, $E_1A_1(x)U_{1,x} + E_3A_3(x)U_{3,x} = 0$

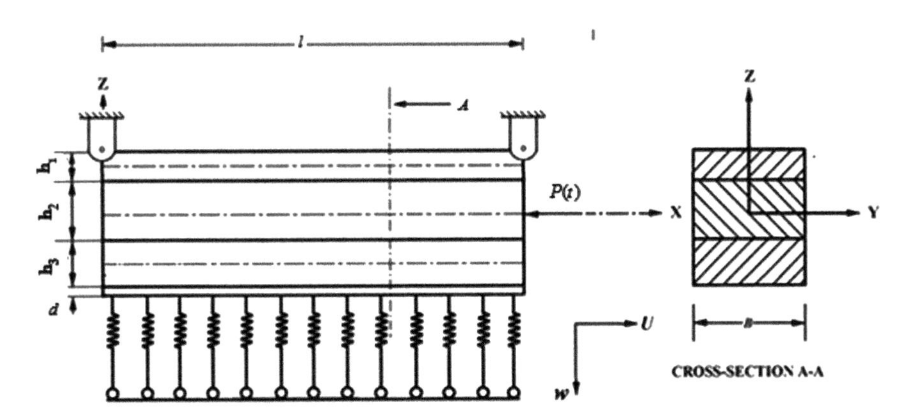

Fig. 1 Configuration of the system

The kinetic energy (T), potential energy (V) and work done (W_P) expressions are given by

$$V = \frac{1}{2}\int_0^l E_1(x)u_{1,x}^2 dx + \frac{1}{2}A_3\int_0^l E_3(x)u_{3,x}^2 dx + \frac{1}{2}I_1\int_0^l E_1(x)w_x^2 dx + \frac{1}{2}I_3\int_0^l E_3(x)w_x^2 dx + \frac{1}{2}G_2^* A_2\int_0^l \gamma_2^2 dx + \frac{1}{2}G_S Bd\int_0^l w_{,x}^2 dx + \frac{B}{2}\int_0^l k(x)w^2 dx \tag{1}$$

$$T = \frac{1}{2}m\int_0^l w_{,t}^2 dx \tag{2}$$

and

$$W_P = \frac{1}{2}\int_0^l P(t)w_x^2 dx \tag{3}$$

The shear strain in the middle layer, $\gamma_2 = \frac{u_1 - u_3}{2h_2} - \frac{cw_{,x}}{2h_2}$.

By Kerwin's assumption [11], u_3 is eliminated. The expression for nondimensional equations of motion is obtained by applying the extended Hamilton's energy principle.

$$\bar{w},_{\bar{t}\bar{t}} + (1+Y)\bar{w},_{\bar{x}\bar{x}\bar{x}\bar{x}} - \left[\frac{3}{2}\frac{G_S}{E_1(x)}\frac{d}{l}\frac{l_{h_1}^3}{1+E_{31}h_{31}^3} - \bar{P}(\bar{t})\right]\bar{w},_{\bar{x}\bar{x}} + \frac{3}{2}\frac{k(x)l}{E_1(x)}\frac{l_{h_1}^3}{1+E_{31}h_{31}^3}\bar{w} + Y\frac{2h_2}{c}\gamma_{2,\bar{x}\bar{x}\bar{x}} = 0 \tag{4}$$

$$\frac{2E_1(x)A_1h_2c}{(1+\alpha)l}\frac{2h_2}{c}\gamma_{2,\bar{x}\bar{x}} - \frac{G_2^* A_2 lc}{2h_2}\frac{2h_2}{c}\gamma_2 + \frac{2E_1(x)A_1h_2c}{(1+\alpha)l}\bar{w},_{\bar{x}\bar{x}\bar{x}} = 0 \tag{5}$$

where $\bar{w},_{\bar{x}\bar{x}\bar{x}\bar{x}} = \frac{\partial^4 \bar{w}}{\partial \bar{x}^4}, \bar{w},_{\bar{x}\bar{x}} = \frac{\partial^2 \bar{w}}{\partial \bar{x}^2}, \gamma_2, \bar{x}\bar{x}\bar{x} = \frac{\partial^3 \gamma_2}{\partial \bar{x}^3}, \gamma_2, \bar{x}\bar{x} = \frac{\partial^2 \gamma_2}{\partial \bar{x}^2}$.

The geometric parameter, $Y = \frac{E_1(x)A_1c^2}{D(1+\alpha)}$ where $\alpha = (E_1A_1)/(E_3A_3)$, $D = E_1(x)I_1 + E_3(x)I_3$ and $\overline{m} = 1 + \left(\frac{\rho_2}{\rho_1}\right)\left(\frac{h_2}{h_1}\right) + \left(\frac{\rho_3}{\rho_1}\right)\left(\frac{h_3}{h_1}\right)$

Equation (5) can be simplified as

$$\frac{2h_2Y}{c}\gamma_{2,\bar{x}\bar{x}} - \frac{2g^* Yh_2\gamma_2}{c} + Y\bar{w},_{\bar{x}\bar{x}\bar{x}} = 0 \tag{6}$$

The nondimensional end conditions at $\bar{x} = 0$ and $\bar{x} = 1$ are given by

$$(1 + Y)\bar{w}_{,\bar{x}\bar{x}\bar{x}} + Y\frac{2h_2}{c}\gamma_{2,\bar{x}\bar{x}} + \left[\bar{P}(\bar{t}) - \frac{3G_S d l_{h_1}^3}{2E_1(x)l(1 + E_{31}h_{31}^3)}\right]\bar{w}_{,\bar{x}} = 0 \tag{7}$$

or

$$\bar{w} = 0 \tag{8}$$

$$(1 + Y)\bar{w}_{,\bar{x}\bar{x}} + Y\frac{2h_2}{c}\gamma_{2,\bar{x}} = 0 \tag{9}$$

or

$$\bar{w}_{,\bar{x}} = 0 \tag{10}$$

$$Y\frac{2h_2}{c}\gamma_{2,\bar{x}} + \bar{w}_{,\bar{x}\bar{x}} = 0 \tag{11}$$

or

$$\gamma_2 = 0 \tag{12}$$

In the above, $\bar{x} = x/l$, $h_{31} = h_{30}/h_{10}$, $h_{21} = h_{20}/h_{10}$, $\bar{P}(\bar{t}) = \bar{P}_0 + \bar{P}_1\cos(\bar{\omega}\bar{t})$, $\bar{P}_0 = P_0 l^2/(E_1(x)I_1 + E_3(x)I_3)$, $\bar{P}_1 = P_1 l^2/(E_1(x)I_1 + E_3(x)I_3)$, $\bar{w}_{,\bar{x}} = \frac{\partial\bar{w}}{\partial\bar{x}}$ and $\bar{w}_{,\bar{t}} = \frac{\partial\bar{w}}{\partial\bar{t}}$, etc.

$g^* = \frac{G_2^* l_{h1}^2(1 + E_{31}h_{31})}{4E_3(x)h_{21}h_{31}}$ is the complex shear parameter and also $g^* = g(1 + j\eta)$

2.1 *Approximate Series of Solution*

For Eqs. (4) and (6), the series of approximate solutions are assumed in the form

$$\bar{w}(\bar{x},\bar{t}) = \sum_{i=1}^{i=p} w_i(\bar{x})f_i(\bar{t}) \tag{13}$$

$$\bar{\gamma}_2(\bar{x},\bar{t}) = \sum_{k=p+1}^{k=2p} \gamma_k(\bar{x})f_k(\bar{t}) \tag{14}$$

Here, the shape functions are w_i and γ_k . f_i and f_k are generalized coordinates. w_i and γ_k are chosen so that they satisfy the equations for motion and maximum number of possible end conditions [12]. The shape functions given in Ray and Kar [10] used for the following end conditions.

1. Pinned-Pinned (P–P): $w_i(\bar{x}) = \sin(i\pi\bar{x})$, $\gamma_k(\bar{x}) = \cos(k\pi\bar{x})$
2. Clamped-Free (C–F): $w_i(\bar{x}) = (i+2)(i+3)\bar{x}^{i+1} - 2i(i+3)\bar{x}^{i+2} + i(i+1)\bar{x}^{i+3}$
   $\gamma_k(\bar{x}) = \bar{x}^k - [k/(k+1)]\bar{x}^{k+1}$

where $i = 1, 2, 3, \ldots p$, $\bar{k} = p+1, p+2, \ldots 2p$.

Substituting the above-mentioned shape functions in Eqs. (4) and (6), the following equations of motion in matrix form are obtained by means of Galerkin's method.

$$[m]\{\ddot{Q}_1\} + [k_{11}]\{Q_1\} + [k_{12}]\{Q_2\} = \{0\} \tag{15}$$

$$[k_{21}]\{Q_1\} + [k_{22}]\{Q_2\} = \{0\} \tag{16}$$

where

$$\{Q_1\} = \{f_1, \ldots, f_p\}^T \tag{17}$$

$$\{Q_2\} = \{f_{p+1}, \ldots, f_{2p}\}^T \tag{18}$$

The various matrix elements are

$$M_{ij} = \int_0^1 w_i w_j d\bar{x} \tag{19}$$

$$k_{11ij} = \int_0^1 (1+Y)w_i''w_j''d\bar{x} + \phi\int_0^1 w_i w_j d\bar{x} + \int_0^1 [\psi - \bar{P}(\bar{t})]w_i'w_j'd\bar{x} \tag{20}$$

$$k_{12jk} = \int_0^1 Yw_i'u_k'd\bar{x} \tag{21}$$

$$k_{22kl} = \int_0^1 Yu_k'u_l'd\bar{x} + \int_0^1 g^* Yu_k u_l d\bar{x} \tag{22}$$

In the above, $u_k = \frac{2h_2}{c}\gamma_k$, $u_l = \frac{2h_2}{c}\gamma_l$ and $w_i' = \frac{\partial w_i}{\partial x}$, $\lambda_s = \left(\frac{kl}{E_1}\right)$, $\phi = \frac{3\lambda_s l_{h_1}^3}{2\left(1+E_{31}h_{31}^3\right)}$,

$$\psi - \frac{3G_S dl_{h_1}^3}{2E_1(x)l(1 + E_{31}h_{31}^3)}$$

$$[k_{21}] = [k_{12}]^T \tag{23}$$

Equations (15) and (16) further simplified to

$$[m]\{\ddot{Q}_1\} + [[k] - \bar{P}_0[H]]\{Q_1\} - \bar{P}_1 \cos(\bar{\omega}\bar{t})[H]\{Q_1\} = \{0\} \tag{24}$$

where

$$[k] = [\bar{k}_3] - [k_{12}][k_{22}]^{-1}[k_{12}]^T \tag{25}$$

$$H_{ij} = \int_0^1 w_i{}' w_j' d\bar{x} \tag{26}$$

and

$$[\bar{k}_3]_{ij} = \int_0^1 (1 + Y)w_i'' w_j'' d\bar{x} + \int_0^1 \phi w_i w_j d\bar{x} + \int_0^1 \psi w_i' w_j' d\bar{x} \tag{27}$$

2.2 Instability Regions

Let for $[m]^{-1}\{k\}$,$[L]$ be a modal matrix. Introducing the linear transformation $\{Q_1\} = [L]\{u\}$, where a set of new generalized coordinates is $\{u\}$ and a set of Hill's equation with complex coefficient is obtained from (24).

$$\ddot{u}_N + \omega_N^{*2} u_N + 2\varepsilon \cos \bar{\omega}\bar{t}[B]u_N = 0 \tag{28}$$

where the distinct eigenvalues for $[m]^{-1}\{k\}$ are ω_N^{*2} and are given by $\varepsilon = \frac{\bar{P}_1}{2} < 1$

$$[B] = -[m]^{-1}[L]^{-1}[H][L]$$

Equation (28) can be rewritten as

$$\ddot{u}_N + \omega_N^{*2} u_N + 2\varepsilon \cos \bar{\omega}\bar{t} \sum_{M=1}^{M=p} b_{NM} u_N = 0 \tag{29}$$

$$N = 1, 2 \ldots p$$

where N varies from $1, 2...p$; b_{NM} are the elements of $[B]$; the complex quantities ω_N^* and b_{NM} are given by

$$\omega_N^* = \omega_{N,R} + j\omega_{N,I} \text{ and } b_{NM} = b_{NM,R} + jb_{NM,I}$$

The main and combination resonances are obtained by conditions of Saito and Otomi [13].

3 Discussion and Results

In this case, the stiffness of the springs is of sinusoidal type and is given as $K(\bar{x}) = K_0(1 - \gamma_e \sin(\pi\xi))$ where γ_e denote the foundation parameter and K_0 is the constant spring stiffness. Unless stated, the following parameter values have been considered for sandwich beam.

$$\eta = 0.1, g = 0.01, \delta_1 = 0.1, \delta_2 = 0.4, P_0 = 0.05, G_2/E_1 = 0.003, G_s/E_1 = 0.001, \rho_2/\rho_1 = 0.05, \rho_3/\rho_1 = 0.5, \gamma_e = 0.2, K_0 = 500, p = 3$$

The temperature referring to reference temperature, anywhere ξ, is $\psi = \psi_0(1 - \xi)$. Choosing $\psi = \psi_0$ as the reference temperature, the variation of Young's modulus of the beam, $E(\xi) = E[1 - \lambda\psi_1(1 - \xi)] = E_1T(\xi), 0 \leq \lambda\psi_1 < 1$ where λ is the coefficient of thermal expansion of the beam material, $\delta = \lambda\psi_1$ and $T(\xi) = [1 - \delta(1 - \xi)]$. Here,

$$\alpha = \frac{E_1A_1}{E_3A_3} = \frac{E_1T(\xi)A_1}{E_3T(\xi)A_3} = \frac{E_1A_1[1 - \delta_1(1 - \xi)]}{E_3A_3[1 - \delta_2(1 - \xi)]}$$

where δ_1 and δ_2 are thermal gradient in the top and bottom elastic layer, respectively (Figs. 2, 3 and 4).

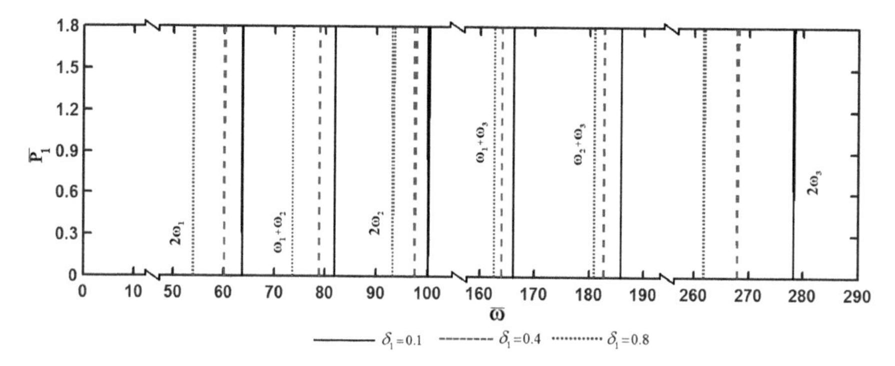

Fig. 2 Effect of δ_1 in case of P–P

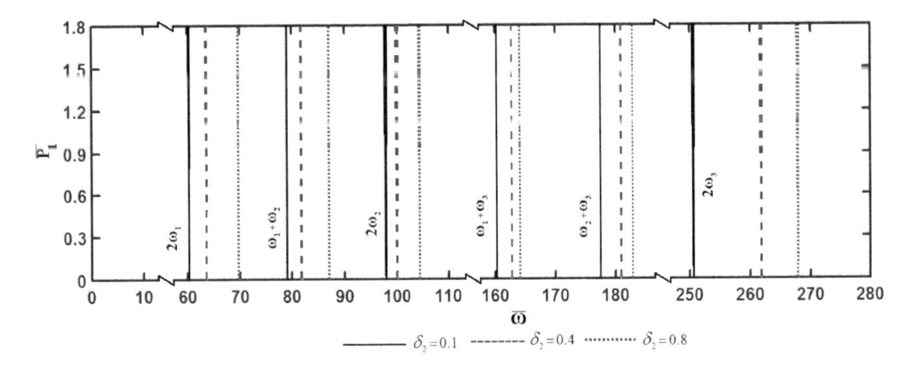

Fig. 3 Effect of δ_2 in case of P–P

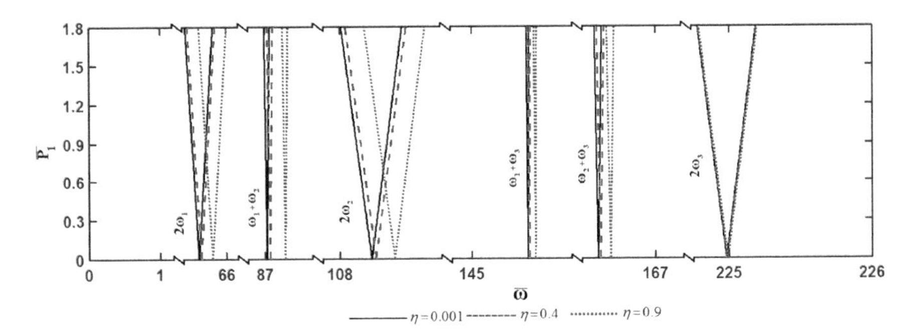

Fig. 4 Effect of η in case of P–P

The increase of δ_2 and decrease of δ_1 shift the instability zones away from the origin as in Figs. 2 and 3. Increase in δ_2 and decrease in δ_1 improve the rigidity of the system. Increase of η is due to the increase of intermolecular friction. This increase results in loss of energy possessed by the system which ultimately makes the system more stable. This nature has been established in Fig. 4. The increase of K_0(constant spring stiffness) and decrease of γ_e are both responsible for the decrease in maximum deflection of the system. Because of this reason, the effect of the increase in K_0 and decrease in γ_e is making the system to have better stability as in Figs. 5 and 6. All the results are developed with pinned-pinned boundary conditions. To limit the number of figures, the zones of instability for clamped-free conditions are developed for δ_1 only. The nature of graphs for clamped-free boundary condition is the same as that of pinned-pinned boundary condition. However, from all the above figures, it can be inferred that sandwich beams with pinned-pinned boundary conditions will have greater stability as compared to clamped-free boundary conditions (Fig. 7).

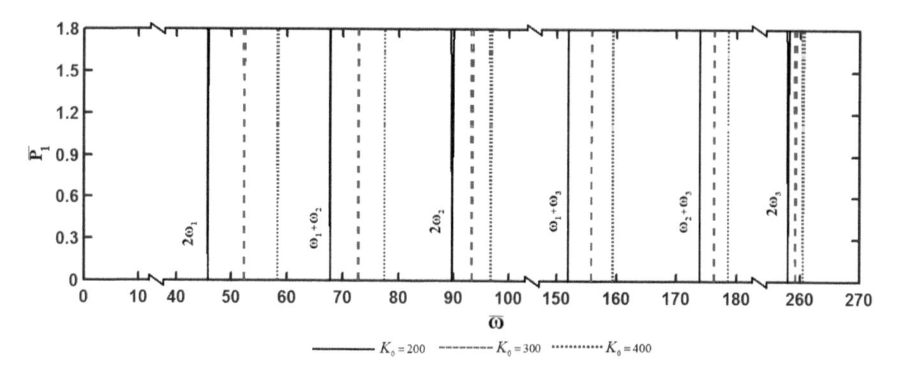

Fig. 5 Effect of K_0 in case of P–P

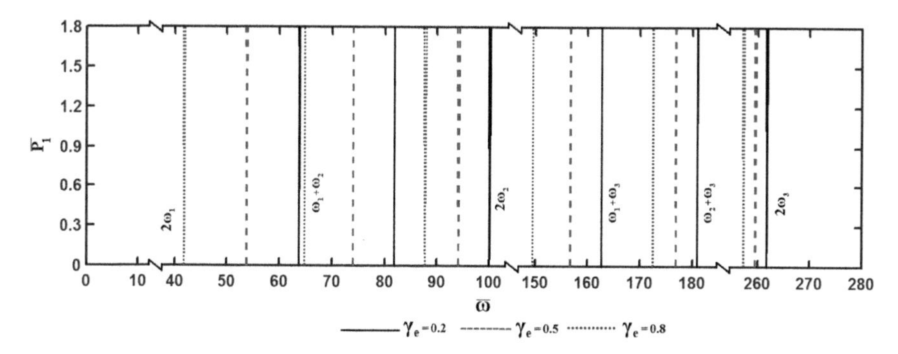

Fig. 6 Effect of γ_e in case of P–P

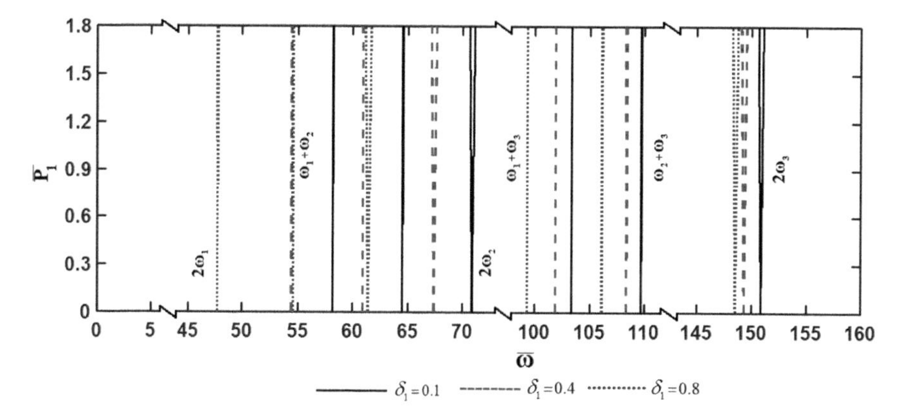

Fig. 7 Effect of δ_1 in case of C–F

4 Conclusion

Two types of boundary conditions have been studied, and these are pinned-pinned and clamped-free. It is found that increase of δ_2, K_0, η and decrease of γ_e, δ_1 is responsible for improvising dynamic stability for the above two boundary conditions. In all the cases, the systems with the pinned-pinned condition in comparison with the clamped-free condition are found to have better stability. This is obvious because the earlier one is more rigid than the later one.

References

1. Dash PR, Maharathi BB, Ray K (2010) Dynamic stability of an asymmetric sandwich beam resting on a pasternak foundation. J Aerosp Sc Technol 62(1):66
2. Hetényi M (1962) Beams on elastic foundation: theory with applications in the fields of civil and mechanical engineering. University of Michigan Press
3. Wang TM, Stephens JE (1977) Natural frequencies of Timoshenko beams on Pasternak foundations. J Sound Vibr 51:149–155
4. Kar RC, Sujata T (1990) Parametric instability of Timoshenko beam with thermal gradient resting on a variable Pasternak foundation. Comput Struct 36 (4):659–665
5. Kar RC, Sujata T (1988) Parametric instability of a non-uniform beam with thermal gradient resting on a Pasternak foundation. Comput Struct 29(4):591–599
6. Nayak S, Bisoi A, Dash PR, Pradhan PK (2014) Static stability of a viscoelastically supported asymmetric sandwich beam with thermal gradient. Int J Adv Struct Eng 6(3):65
7. Pradhan M, Dash PR (2016) Stability of an asymmetric tapered sandwich beam resting on a variable Pasternak foundation subjected to a pulsating axial load with thermal gradient. Compos Struct 140:816–834
8. Pradhan M, Mishra MK, Dash PR (2016) Stability analysis of an asymmetric tapered sandwich beam with thermal gradient. Procedia Eng 144:908–916
9. Pradhan M, Dash PR, Pradhan PK (2016) Static and dynamic stability analysis of an asymmetric sandwich beam resting on a variable Pasternak foundation subjected to thermal gradient. Meccanica 51(3):725–739
10. Ray K, Kar RC (1995) Parametric instability of a sandwich beam under various boundary conditions. Comput Struct 55(5):857–870
11. Kerwin EM Jr (1959) Damping of flexural waves by a constrained viscoelastic layer. J Acoust Soc Am 31(7):952–962
12. Rao DK, Stuhler W (1977) Frequency and loss factors of tapered symmetric sandwich beams. J Appl Mech 44(3):511–513
13. Saito H, Otomi K (1979) Parametric response of viscoelastically supported beams. J Sound Vib 63(2):169–178
14. Dash P, Nayak DK (2019) Determination of economical and stable rotating tapered sandwich beam experiencing parametric vibration and temperature gradient. Journal of the Institution of Engineers (India): Series C, 100(6):879–890

A Parametric Study of Functionally Graded Variable Thickness Longitudinal Fin Under Fully Wet Condition

Upendra Bajpai, Vivek Kumar Gaba and Shubhankar Bhowmick

Abstract Thermal analysis and comparison of the functionally graded longitudinal fin, having a different profile with an insulated tip in fully wet condition, are reported in the present work. In many air-conditioning and refrigeration equipments, the performance of the cooling coil is affected due to vapor condensation on its surface. In this work, the thermal conductivity of the longitudinal fin is varied with exponential law. For analysis and comparison of a fin having different profiles, their weight is assumed to be constant. With the help of the psychometric chart, a nonlinear cubic polynomial relationship is established between specific humidity and corresponding fin surface temperature. Considering a volume element of fin under steady state, energy balance concept is used to derive nonlinear differential heat transfer equation. This differential equation is solved using ***bvp4c*** command in MATLAB®. This technique is very useful to solve boundary value problems by collocation method. Further for a different combination of grading parameters, geometry parameters, and relative humidity, a differential equation is solved and results are shown in graphical form. The formulation is verified with standard results, and relative error obtained between these two results is negligible. These results give a better understanding of the thermal performance of functionally graded longitudinal wet fin, and generated data can be used for design purpose.

Keywords Functionally graded · Longitudinal fin · Fully wet

U. Bajpai · V. K. Gaba (✉) · S. Bhowmick
Department of Mechanical Engineering, National Institute of Technology, Raipur, Raipur 492010, India
e-mail: vgaba.mech@nitrr.ac.in

S. Bhowmick
e-mail: sbhowmick.mech@nitrr.ac.in

BBVL. Deepak et al. (eds.), *Innovative Product Design and Intelligent Manufacturing Systems*, Lecture Notes in Mechanical Engineering,
https://doi.org/10.1007/978-981-15-2696-1_11

1 Introduction

To increase the heat transfer rate in the heat exchanger, finned tubes are used. By attaching fin at tube surface, overall surface area increases and consequently heat transfer rate also increases. In many refrigeration, air-conditioning, and chemical processing industrial applications, generally the cooling coil surface temperature is below the dew point temperature of surrounding air. Moisture present in the air condenses on the tube, due to this performance of cooling coil gets affected. Many researchers have taken an interest in this problem and presented their work. To improve the thermal performance along with optimizing the design parameters, the variation in material properties can also be done. In functionally graded material, properties vary gradually with dimension. This property of functionally graded material is used here to improve the thermal performance of fin.

A comprehensive study of the literature on thermal performance analysis of fully wet fin has been performed. Kuehn et al. [1] assumed enthalpy difference affects the simultaneous heat and mass transfer and derived an analytical expression to find overall efficiency. McQuiston [2] derived a governing equation and solved it for efficiency under simultaneous heat and mass transfer conditions. It was considered that the driving force for mass transfer process was the difference between the specific humidity of surrounding moist air and saturated moist air on the surface. Chilton and Colburn [3] proposed a relationship between heat transfer coefficient and mass transfer coefficient, called Chilton–Colburn analogy. Elmahdy and Biggs [4] proposed an algorithm to calculate the efficiency of a longitudinal circular fin having uniform thickness under simultaneous heat and mass transfer. Provided results showed a decrease in efficiency with an increase in relative efficiency. Convey et al. [5] investigated the performance of vertical longitudinal fin in moist airflow and concluded that condensate film thickness and temperature gradient along the fin length depend on airflow and fin effectiveness which decreases due to the thermal resistance of condensate layer. Second-degree polynomial relationship was assumed between specific humidity of saturated air and fin surface temperature. Wu and Bong [6] took a linear relationship between the specific humidity at the fin surface and fin surface temperature. Both partially and fully wet condition fin efficiency were examined. Lin et al. [7] experimentally investigated the thermal performance of rectangular fin in wet condition. For fully wet condition, there was a very slight decrement in efficiency with an increase in relative efficiency. Xu et al. [8] proposed a modified McQuiston model to calculate fin efficiency. Effect of motion of condensate film on fin surface was also considered. Assuming linear relationship between humidity ratio on fin surface and corresponding surface temperature of the fin, Kundu [9] analyzed the performance and optimum design of longitudinal and pin fins. A very small change in efficiency with an increase in relative efficiency was found. Later Kundu [10] considered the polynomial relationship between humidity ratio and dry bulb temperature. With the help of ADM method, an approximate analytical solution for thermal performance of fin was proposed. Sharqawya and Zubair [11] mainly focused on rectangular fin in dry, partially and full wet condition assuming a linear relationship between specific humidity and corresponding fin base temperature. The overall fin efficiency depends on the condition of the fin surface, whether it is dry, partially wet, or fully wet, and also on atmospheric pressure. It increased with an increase in atmospheric pressure. Gaba et al. [12] examine the performance of functionally graded annular parabolic fins. For different grading and geometry parameters, the thermal performance was analyzed and their comparison was done. Udupa et al. [13] represented an overview of basic concepts, properties, classification, preparation methods, and applications of functionally graded composite materials. A case study on CNT reinforced aluminum matrix was discussed. By varying the weight percentage of CNT composition, its effect on grain boundaries and structure was reported. Bhavar et al. [14] provided brief information about functionally graded material. Its properties, classifications, method of manufacturing, and application have been reported. Subramaniam et al. [15] proposed the temperature distribution in functionally graded longitudinal fins of varying geometry for insulated tip condition. The temperature distribution over the fin length of constant thickness longitudinal fin was plotted for different grading parameters. It was concluded that the fin performance could be enhanced either by varying geometry or by varying the thermal conductivity.

From the above-mentioned literature review, it can be said that in wet fins, considering the polynomial relationship between fin surface temperature and corresponding specific humidity, the effect of variation of thermal conductivity with dimensions is not reported by any researchers. This gap gives the scope for further research in this area, and hence, an analytical treatment of functionally graded variable thickness longitudinal fin under fully wet condition has been reported in the present work.

2 Mathematical Formulation

A longitudinal fin of length $\boldsymbol{L}$, base thickness $\boldsymbol{\delta_o}$, width $\boldsymbol{w}$, and constant base temperature $\boldsymbol{T_b}$ are considered as shown in Fig. 1. During the thermal analysis, wet fin is assumed to be steady, one dimensional, with moist air near the condensate film being saturated and effect of temperature variation on thermal conductivity is neglected as the fin is of functionally graded material with spatial variation of thermal properties only as it is supposed to work in narrow temperature range. A cubic relationship between specific humidity and corresponding fin temperature is assumed.

$$w = A + B * T + C * T^2 + D * T^3. \tag{1}$$

A, B, C, D are constants whose values are given in Table 1, calculated with the help of psychrometric chart and regression analysis for the temperature range 0–30 °C [10].

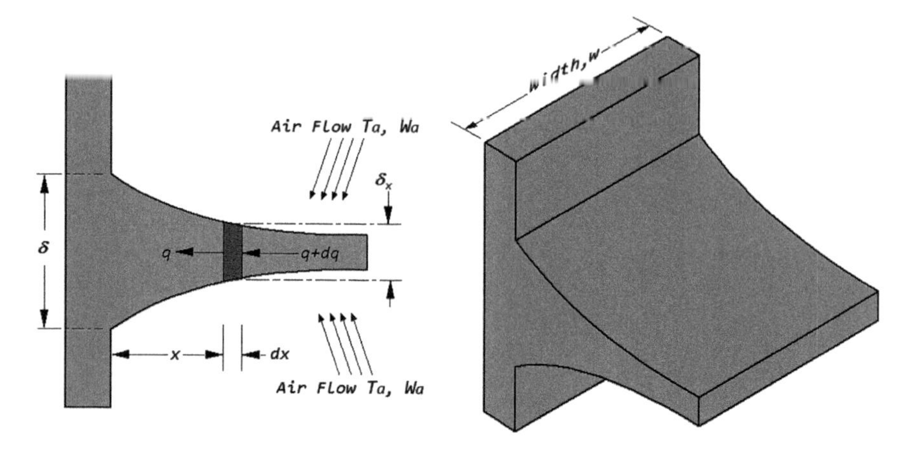

Fig. 1 Schematic diagram of a completely wet straight longitudinal fin

Table 1 Values of constant *A, B, C, D*

A	*B*	*C*	*D*
0.0037444	0.0003078	0.0000046	0.0000004

Thermal conductivity ***K*** is taken as the exponential function of ***X*** as

$$K = ae^{(-bX)}. \quad \text{Where } X = \frac{x}{L}. \tag{2}$$

where ***a*** is the conductivity coefficient and has the dimension of W/m K.

The thickness of fin is assumed the parabolic function of length co-ordinate given as

$$\delta = \delta_o(1 - n * X^m). \tag{3}$$

where ***m*** and ***n*** are dimensionless constants.

Applying the energy balance on a small element dx, governing equation of wet fin is derived as follows:

$$(q + \mathrm{d}q) + h * P * \mathrm{d}x * (T_a - T) + h_m * P * \mathrm{d}x * (w_a - w) - q = 0. \tag{4}$$

Equation (4) is a boundary value problem and is given for following boundary conditions:

$$T(x)_{x=0} = T_b. \tag{5}$$

$$\frac{\mathrm{d}T}{\mathrm{d}x_{x=L}} = 0. \tag{6}$$

Relationship between heat transfer $\boldsymbol{h}$ and mass transfer $\boldsymbol{h_m}$ is taken from [3] and is given by

$$\frac{h}{h_m} = C_p * Le^{\frac{1}{3}}. \tag{7}$$

where $\boldsymbol{Le}$ is the Lewis coefficient and $\boldsymbol{C_p}$ specific heat of moist air.

Solving Eq. (4) with the help of Eqs. (1)–(3) and (5)–(7) and reducing it into normalized form with the help of Eqs. (8) and (9) given below.

$$\text{Normalized Length } X = \frac{x}{L}. \tag{8}$$

$$\text{Normalized Temperature } \theta = \frac{T_a - T}{T_a - T_b}. \tag{9}$$

Differential equation becomes,

$$\frac{d^2\theta}{dX^2} + \left[-b + \frac{-n * m * X^{(m-1)}}{1 - n * X^m}\right]\frac{d\theta}{dX}\frac{2 * h * L^2 * (\delta + W)}{K * W * \delta o * (1 - n * X^m)}\left[K_1 + K_2\theta - K_3\theta^2 + K_4\theta^3\right]. \tag{10}$$

where

$$K_1 = \xi\left[\frac{w_a - A - B * T_a - C * T_a^2 - D * T_a^3}{T_a - T_b}\right]. \tag{11}$$

$$K_2 = 1 + \xi\left[B + 2 * C * T_a + 3 * D * T_a^2\right]. \tag{12}$$

$$K_3 = \xi[(C + 3 * D * T_a)(T_a - T_b)]. \tag{13}$$

$$K_4 = D * \xi(T_a - T_b)^2. \tag{14}$$

$$\xi = \frac{h_{fg}}{C_p * Le^{\left(\frac{2}{3}\right)}}. \tag{15}$$

where $\boldsymbol{K_1}$, $\boldsymbol{K_2}$, $\boldsymbol{K_3}$, $\boldsymbol{K_4}$ given in Eqs. (11)–(14) are constants, and ξ is a dimensionless latent heat parameter. The above differential Eq. (10) has been solved using ***bvp4c*** function in MATLAB® software.

3 Results and Discussion

The temperature distribution of functionally graded longitudinal fin having a different profile with an insulated tip in fully wet condition is obtained by solving second-order partial differential equation, using subroutine ***bvp4c*** in MATLAB® [16]. From the literature survey, it is found that study related to thermal performance analysis of longitudinal fin under simultaneous heat and mass transfer, considering functionally grading material as fin material is not reported yet. The formulation is validated for rectangular longitudinal fin under simultaneous heat and mass transfer with Sharqawy and Zubair [11] in Fig. 2, and a good agreement between these two results are obtained for various relative humidity conditions.

Aluminum has been used as a fin base material. Fin base temperature 7 °C, surrounding air temperature 27 °C, fin length 0.2 m, air heat transfer coefficient 36 W/m^2K, fin parameter ($\boldsymbol{mL}$) 0.8, and different relative humidity 60, 80, 100% are considered to solve Eq. (10), using ***bvp4c*** in MATLAB®, for different grading and geometry parameters. These numerical results and a variation of dimensionless temperature along with dimensionless co-ordinates are plotted and shown in Figs. 3 and 4.

For geometry parameters $\boldsymbol{m}$ = 0.5, $\boldsymbol{n}$ = 0.5 and relative humidity 60%, the effect of grading parameter $\boldsymbol{b}$ is shown in Fig. 3a. With a decrease in value of $\boldsymbol{b}$, normalized temperature increases at a particular position of the fin. Consequently, fin surface temperature decreases at that particular position, which in turn results in greater temperature difference with decreasing value of $\boldsymbol{b}$, causing more heat transfer. For the small length of the fin, the effect of grading parameter $\boldsymbol{b}$ can be neglected.

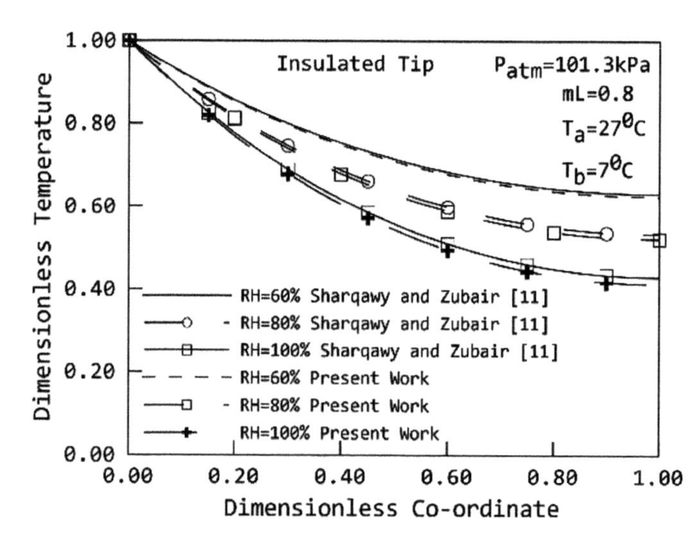

Fig. 2 Comparison of normalized temperature distribution over fin length with that of obtained by Sharqawy and Zubair [11]

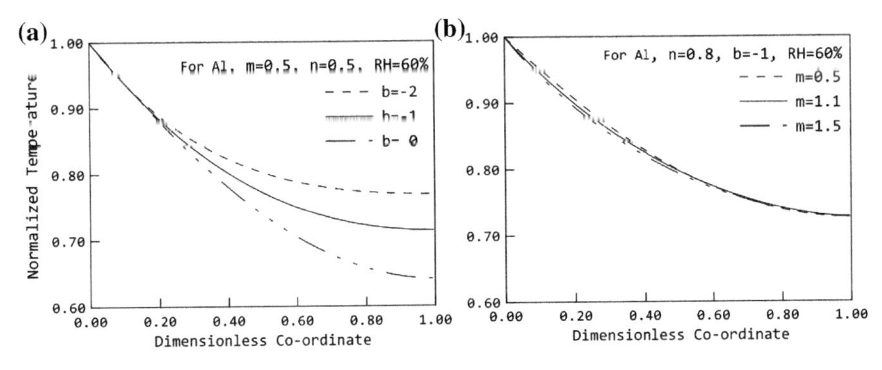

Fig. 3 Normalized temperature distribution along with dimensionless co-ordinate for different values of grading parameter

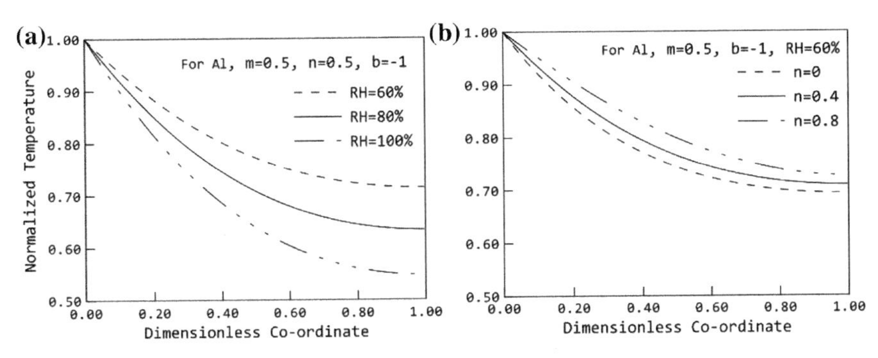

Fig. 4 Normalized temperature distribution along with dimensionless co-ordinate for different values of RH

Effect of variation of geometry parameter $\boldsymbol{m}$, for $\boldsymbol{n} = 0.8$, $\boldsymbol{b} = -1$, and relative humidity 60% is shown in Fig. 3b. Keeping $\boldsymbol{n}$ constant, the effect of variation of $\boldsymbol{m}$ on fin surface temperature is very less.

Figure 4a shows that an increase in relative humidity increases fin surface temperature at a particular location, which can be justified as follows. Due to condensation on the fin surface, fin surface temperature increases. As relative humidity increases, vapor present in air increases, so more vapor condenses on fin surface. As such, more heat of condensation liberates on the fin, surface causes an increase in its temperature.

Considering the effect of increment in the value of $\boldsymbol{n}$, taking $\boldsymbol{m} = 0.5$, relative humidity 60%, $\boldsymbol{b} = -1$, it is shown in Fig. 4b that with an increase in $\boldsymbol{n}$, the difference between fin surface temperature and air temperature increases at a particular location. For rectangular fin ($\boldsymbol{n} = 0$), this temperature difference is lower resulting in lesser heat transfer. Varying the value of m gives scope to enhance the heat transfer.

4 Conclusion

Parametric study of functionally graded variable geometry longitudinal wet fin is presented. Temperature distributions along the fin length for different values of relative humidity ***RH***, grading parameter ***b***, and geometric parameters ***m*** and ***n*** are shown in Figs. 3 and 4. Increase in relative humidity reduces the temperature difference between saturated wet film temperature and corresponding fin surface temperature at a particular location, which results in a reduction of heat transfer. Fin performance decreases as relative humidity increases. It is remarked that a higher value of ***n*** and lower value of ***m*** gives better results for the same fin parameter. Fin performance can be increased by using negative grading parameter ***b***. Overall, by varying the material property along the fin length or varying the fin profile or combination of both, the performance of the wet fin can be enhanced.

References

1. Kuehn TH, Ramsey JW, Threlkeld JL (1998) Thermal environmental engineering, 3rd edn. Prentice-Hall Inc., Upper Saddle River, New Jersey
2. McQuiston FC (1975) Fin efficiency with combined heat and mass transfer. ASHRAE Trans 81(1):350–355
3. Chilton TH, Colburn AP (1934) Mass transfer (absorption) coefficients. Ind Eng Chem 26:1183–1187
4. Elmahdy AH, Biggs RC (1983) Efficiency of extended surfaces with simultaneous heat and mass transfer. ASHRAE Trans 89:135–143
5. Coney JE, Sheppard CG, El-Shafei EA (1989) Fin performance with condensation from humid air: a numerical investigation. Int J Heat Fluid Flow 10:224–231
6. Wu G, Bong TY (1994) Overall efficiency of a straight fin with combined heat and mass transfer. ASHRAE Trans 100(1):367–374
7. Lin YT, Hsu KC, Chang YJ (2001) Performance of rectangular fin in wet conditions: visualization and wet fin efficiency. ASME J Heat Transf 123:827–836
8. Xu X, Xia L, Chan M, Deng S (2008) A modified McQuiston model for evaluating efficiency of wet fin considering effect of condensate film moving on fin surface. Energy Convers Manag 49:2403–2408
9. Kundu B (2007) Performance and optimum design analysis of longitudinal and pin fin with simultaneous heat and mass transfer: unified and comparative investigation. Appl Therm Eng 27:976–987
10. Kundu B (2009) Approximate analytic solution for performances of wet fin with a polynomial relationship between humidity ratio and temperature. Int J Therm Sci 48:2108–2118
11. Sharqawy MH, Zubair SM (2011) Efficiency and optimization of straight rectangular fin with combined heat and mass transfer. Heat Transf Eng 29:1018–1026
12. Gaba VK, Bhowmick S, Tiwari AK (2014) Thermal performance of functionally graded parabolic annular fins having constant weight. J Mech Sci Technol 28:4309–4318
13. Udupa G, Rao SS, Gangadharan KV (2014) Functionally graded composite materials: an overview. Proc Mater Sci 5:1291–1299
14. Bhavar V, Kattire P, Thakare S, Patil S, Singh RKP (2017) A review on functionally gradient materials (FGMs) and their applications. Mater Sci Eng 229

15. Subramaniam SK, Gaba VK, Bhowmick S (2018) Temperature distribution in functionally graded longitudinal fins of varying geometry. In: Hierarchical composite materials, 1st edn, pp 90–97
16. Shampine LF, Reichelt MW, Kierzenka J (2010) Solving boundary value problems for ordinary differential equations in Matlab with bvp4c. www.mathworks.com/bvp_tutorial

A DEMATEL Approach to Evaluate the Enablers for Effective Implementation of Ecodesign in Sustainable Product Development: A Case of MSMEs

Singh Prashant Kumar and Sarkar Prabir

Abstract Increasing pressure of producing environmentally friendly products and services has forced companies to adopt ecodesign practices in their production process, especially in micro, small and medium enterprises (MSMEs). These practices help the designers to mitigate the environmental issues such as climate change and depleting natural resources by designing sustainable products. There are certain enablers that need to be realized by companies for effective implementation of eco-friendly practices. The main focus of this research is to identify and evaluate the key enablers for sustainable product development. A Decision Making Trial and Evaluation Laboratory (DEMATEL) approach is used in this study to evaluate the identified enablers. A case study based on an Indian manufacturing MSME is carried out to present the real-life applicability of the proposed study. The findings of this study show that training of designers to use various available ecodesign methods and tools is the most important enabler that can have a significant effect in the implementation of ecodesign practices for sustainable product development.

Keywords Ecodesign · Sustainable development · MSMEs · Enablers · DEMATEL

1 Introduction

Currently, the world is facing various environmental issues such as increasing amount of greenhouse gases in the atmosphere, melting glaciers, increasing occurrences of flood and drought and continuously depleting natural resources. One of the main reasons which is responsible for these issues is the current irresponsible production and consumption patterns. The current pattern of production and consumption may lead the world toward an unavoidable collapse by the end of

S. P. Kumar (✉) · S. Prabir
Department of Mechanical Engineering, IIT Ropar, Rupnagar, Punjab, India
e-mail: pksiitrpr@gmail.com

BBVL. Deepak et al. (eds.), *Innovative Product Design and Intelligent Manufacturing Systems*, Lecture Notes in Mechanical Engineering,
https://doi.org/10.1007/978-981-15-2696-1_12

twenty-first century [1]. An active participation is required by all stakeholders of the society (i.e., government, industries and academia) to mitigate these issues and to achieve sustainable development. The term 'sustainable development' is a philosophy which says that we should consume our resources to fulfill our current need in such a way that the ability of our future generations to meet their need is not compromised. Ecodesign can be a potential approach to deal with these issues through a responsible production and consumption of various products and services. Ecodesign is defined as 'an approach that considers the integration of environmental criteria in the development of products throughout the entire life cycle of products' [2]. The Biofore Company defines ecodesign as 'design and produce products responsibly to be safe and sustainable, and ensure that, whenever possible, the end of products' life cycle is the birth of something new.' It is encouraged to consider ecodesign principles in the development of a product in the earlier stage of the design, i.e., in conceptual design because almost 80% of environmental load of a product is determined in this stage [3–5].

Adopting eco-friendly practices is a challenge for micro, small and medium enterprises (MSMEs) because they have limited resources of finance and personnel [6]. Generally, MSMEs are considered as the backbone of the developing countries because they have a significant contribution to the growth of a developing nation. MSMEs are defined on the basis of various criteria such as a number of employees, investment in plant and equipment, sales turnover and production capacity [7]. In India, MSMEs are defined on the basis of investment in plant and machinery, as given in Table 1.

The objective of this research is to identify and evaluate various enablers to improve the implementation of ecodesign practices in sustainable product development in MSMEs using DEMATEL approach. The enablers are identified through a literature survey. A case study is conducted by applying the proposed idea in an Indian manufacturing MSME.

The remaining work is structured as follows: Sect. 2 provides a brief description of the enablers in ecodesign implementation which is identified through the literature search. The methodology of this work is explained in Sect. 3 which is followed by Sect. 4 in which a case study is carried out based on the proposed methodology in an Indian MSME and some managerial implications are provided based on the results of the case study. The conclusion of the study is provided in Sect. 5.

Table 1 Definitions of micro, small and medium enterprises (MSMEs) in India

Enterprises	Investment in plant and machinery/equipment	
	Manufacturing sector	Service sector
Micro	< INR 25 lakh	< INR 10 lakh
Small	> INR 25 lakh but < INR 5 crore	> INR 10 lakh but < INR 2 crore
Medium	> INR 5 crore but < INR 10 crore	> INR 2 crore but < INR 5 crore

2 Literature Review

A literature survey is carried out using the online sources such as Web of Science, Scopus and Google Scholar to find out the enablers for effective implementation of ecodesign in sustainable product development in MSMEs. Only those enablers which are mentioned in multiple studies are considered in this work. Variously identified enablers are presented in Table 2 with a short description and references.

Table 2 Enablers for effective implementation of ecodesign practices in sustainable product development in MSMEs

Code	Enablers	Short description	Refs.
E1	Support from top management	Strong and continuous support is required from the top management so that there is no lack of resources such as finance and personnel, especially in MSMEs	[8, 9]
E2	Using a life cycle approach to design products	Almost all products consume resources and cause emission during each phase of their life cycle beginning from raw material to the end of life. Therefore, the life cycle approach should be used to design products	[10]
E3	Training of designers	Training of the designers should be conducted through green activities such as seminars and workshops. It provides an opportunity for the designers to understand the environmental aspects of a product	[11, 12]
E4	Collaboration between design domains	Ecodesign is a multidisciplinary activity and requires collaboration not only among internal but also among external stakeholders	[13, 14]
E5	Building an information system to exchange data	There should be an information system in place so that the important data related to ecodesign of products can be exchanged between different departments of the firm	[15–17]
E6	Considering ecodesign strategies as essential practices	Most of the MSMEs focus only on earning profit and consider environmental issues as a short-term goal. This attitude should be changed and ecodesign strategies should be considered as essential practices for long term	[15, 18]
E7	Including an environmental expert in the product development team	Generally, MSMEs are reluctant to recruit personnel for a specific task. But, the inclusion of an environmental expert can improve the development of eco-friendly products	[19, 20]

(continued)

Table 2 (continued)

Code	Enablers	Short description	Refs.
E8	Considering ecodesign approaches in the initial design phase	Almost 80% of the environmental load of a product is determined in the conceptual design phase. Therefore, ecodesign approaches should be considered in the initial design phase	[3–5]
E9	Making the customers aware about environmental issues	The companies which are based on consumer products and services can make their customers aware of the environmental issues so that the products are used in an eco-friendly manner	[21, 22]
E10	Environmental certifications	Environmental certifications provide a way not only to adopt green practices but also to develop an environmentally friendly infrastructure	[23, 24]

3 Methodology

The methodology of this work is based on the DEMATEL approach. This approach was introduced by Battelle Memorial Institute of Geneva in 1976. This approach has been used in various multi-criteria decision problems to obtain the interrelationship among different factors [25, 26]. DEMATEL divides the factors into cause and effect groups and provides their interrelationship with the help of a causal diagram.

A flow diagram of the research methodology is shown in Fig. 1. Various steps involved in this study are as follows:

Step 1: In this step, various enablers for effective implementation of ecodesign are identified and finalized through an extensive literature survey and opinion of the experts.

Step 2: Form a direct assessment matrix using the judgment of each expert with the help of a linguistic scale. A score is provided to each factor with respect to the linguistic judgment of the experts. The scale is given as:

0—No influence; 1—Low influence; 2—Medium influence; 3—High influence

Step 3: Construct the average matrix after receiving the inputs of all experts. For each expert, a non-negative matrix of the order $n \times n$ is constructed as $X^k = \left[x_{ij}^k\right]$ where k is the number of experts with $1 \leq k \leq P$ and n indicates the number of factors. $X^1, X^2, \ldots, X^P$.

Are the matrices obtained through P experts? The average matrix $A = \left[a_{ij}\right]$ can be established as:

$$a_{ij} = \frac{1}{P}\sum_{k=1}^{P} x_{ij}^k \quad (1)$$

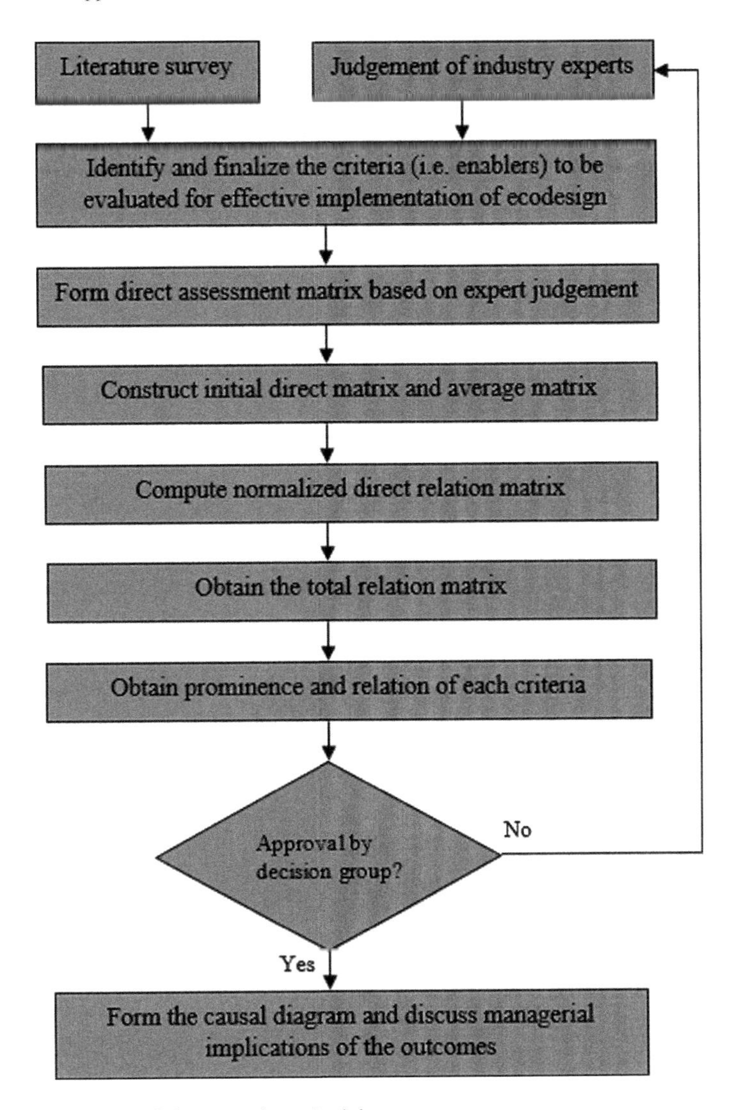

Fig. 1 Flow diagram of the research methodology

Step 4: Compute the normalized direct relation matrix D as $D = A \times B$, where B is given as

$$B = \frac{1}{\max\limits_{1 \leq i \leq n} \sum_{j=1}^{n} a_{ij}}. \tag{2}$$

Each element of matrix D lies within 0 and 1.

Step 5: Calculate total relation matrix T as $T = D\,(I - D)^{-1}$ where I represents identity matrix.

Step 6: Obtain prominence and relation for each factor which is represented by $(r + c)$ and $(r - c)$, respectively. Here, r and c denote the sum of rows and columns of total relation matrix for individual factors, respectively. If r_i be the sum of ith row of matrix T, then it shows the total effect of factor i on other factors. If c_j be the sum of jth column of matrix T, then it shows the total effect received by factor j from the other factors. The sum $(r + c)$ of a factor shows the degree of importance of that factor on the entire system. If $(r - c)$ of a factor is positive, then it is known as 'cause factor' and is known as a receiver if $(r - c)$ is negative.

4 Case Study and Managerial Implications

A case study based on the proposed methodology is carried out in an Indian manufacturing MSME situated in the northern part of India. This company manufactures automotive components such as rings, pistons and connecting rods. This company falls under the medium category of MSME and has an ISO 14001:2015 certification. The company is doing business from the last 40 years and has an eco-friendly state-of-the-art infrastructure spread over 20,000 m^2. The officials showed a keen interest to be a part of this study because of their commitment to adopt and improve the environmentally friendly practices in their production process.

Detailed information of the current study was shared with a team of five experts out of which two are environmental engineers, two senior designers and one production manager. Each of them has experience of at least 10 years with more than 3 years in the case company. All the 10 identified enablers for effective ecodesign implementation were discussed with the experts, and eight enablers (E1 − E8) were finalized. Two enablers (E9 and E10), i.e., 'Making the customers aware about environmental issues' and 'Environmental certifications' were removed because the company does not produce consumer goods and is already registered with ISO 14001:2015 certification. Each expert was asked to provide his linguistic judgment about the relative importance of each enabler on the basic of a linguistic scale as mentioned in Sect. 3. On the basis of the input provided by the experts, the initial direct matrix for each expert is constructed as follows:

$$X^1 = \begin{bmatrix} 0 & 1 & 0 & 3 & 2 & 2 & 2 & 2 \\ 0 & 0 & 1 & 2 & 1 & 2 & 2 & 0 \\ 1 & 2 & 0 & 2 & 1 & 3 & 2 & 3 \\ 3 & 1 & 0 & 0 & 1 & 0 & 2 & 2 \\ 2 & 2 & 3 & 0 & 0 & 1 & 1 & 2 \\ 2 & 2 & 3 & 2 & 2 & 0 & 2 & 3 \\ 2 & 1 & 2 & 3 & 1 & 2 & 0 & 1 \\ 2 & 1 & 1 & 0 & 2 & 1 & 2 & 0 \end{bmatrix} \quad X^2 = \begin{bmatrix} 0 & 0 & 1 & 2 & 2 & 1 & 2 & 0 \\ 1 & 0 & 2 & 3 & 3 & 1 & 2 & 1 \\ 0 & 1 & 0 & 2 & 3 & 3 & 1 & 2 \\ 1 & 3 & 1 & 0 & 1 & 1 & 3 & 1 \\ 3 & 0 & 2 & 1 & 0 & 0 & 1 & 0 \\ 2 & 1 & 3 & 1 & 1 & 0 & 1 & 2 \\ 3 & 2 & 1 & 1 & 1 & 2 & 0 & 3 \\ 1 & 1 & 3 & 1 & 1 & 2 & 0 & 0 \end{bmatrix}$$

$$X^3 = \begin{bmatrix} 0 & 2 & 1 & 1 & 1 & 0 & 2 & 1 \\ 1 & 0 & 2 & 3 & 1 & 1 & 1 & 1 \\ 1 & 3 & 0 & 1 & 1 & 2 & 2 & 1 \\ 1 & 2 & 1 & 0 & 0 & 1 & 1 & 1 \\ 1 & 2 & 2 & 1 & 0 & 1 & 0 & 1 \\ 1 & 0 & 2 & 1 & 1 & 0 & 2 & 1 \\ 1 & 2 & 2 & 1 & 0 & 1 & 0 & 1 \\ 1 & 0 & 1 & 1 & 1 & 2 & 2 & 0 \end{bmatrix} \quad X^4 = \begin{bmatrix} 0 & 1 & 2 & 2 & 1 & 1 & 1 & 1 \\ 2 & 0 & 3 & 2 & 2 & 0 & 1 & 2 \\ 2 & 3 & 0 & 1 & 3 & 3 & 1 & 2 \\ 3 & 3 & 2 & 0 & 2 & 2 & 1 & 0 \\ 1 & 1 & 3 & 2 & 0 & 2 & 2 & 1 \\ 1 & 1 & 3 & 0 & 0 & 0 & 1 & 3 \\ 2 & 1 & 1 & 3 & 2 & 1 & 0 & 2 \\ 0 & 2 & 3 & 2 & 2 & 3 & 1 & 0 \end{bmatrix}$$

$$X^5 = \begin{bmatrix} 0 & 2 & 1 & 1 & 2 & 2 & 1 & 2 \\ 1 & 0 & 3 & 3 & 1 & 1 & 0 & 1 \\ 1 & 2 & 0 & 0 & 3 & 3 & 1 & 2 \\ 1 & 3 & 1 & 0 & 1 & 1 & 1 & 1 \\ 1 & 2 & 2 & 1 & 0 & 1 & 1 & 2 \\ 0 & 2 & 3 & 1 & 1 & 0 & 1 & 1 \\ 1 & 1 & 1 & 1 & 1 & 0 & 0 & 1 \\ 1 & 1 & 3 & 1 & 1 & 3 & 2 & 0 \end{bmatrix}$$

The average matrix A is constructed using Eq. (1) and normalized matrix D using Eq. (2) as follows:

$$A = \begin{bmatrix} 0.0 & 1.2 & 1.0 & 1.8 & 1.6 & 1.2 & 1.6 & 1.2 \\ 1.0 & 0.0 & 2.2 & 2.6 & 1.6 & 1.0 & 1.2 & 1.0 \\ 1.0 & 2.2 & 0.0 & 1.2 & 2.2 & 2.8 & 1.4 & 2.0 \\ 1.8 & 2.4 & 1.0 & 0.0 & 1.0 & 1.0 & 1.6 & 1.0 \\ 1.6 & 1.4 & 2.4 & 1.0 & 0.0 & 1.0 & 1 & 1.2 \\ 1.2 & 1.2 & 2.8 & 1.0 & 1.0 & 0.0 & 1.4 & 2.0 \\ 1.8 & 1.4 & 1.4 & 1.8 & 1.0 & 1.2 & 0.0 & 1.6 \\ 1.0 & 1.0 & 2.2 & 1.0 & 1.4 & 2.2 & 1.4 & 0.0 \end{bmatrix}$$

$$D = A \times \frac{1}{\max\limits_{1 \le i \le 8} \sum_{j=1}^{8} a_{ij}}$$

$$= \begin{bmatrix} 0.00 & 0.09 & 0.08 & 0.14 & 0.12 & 0.09 & 0.12 & 0.09 \\ 0.08 & 0.00 & 0.17 & 0.20 & 0.12 & 0.08 & 0.09 & 0.08 \\ 0.08 & 0.17 & 0.00 & 0.09 & 0.17 & 0.22 & 0.11 & 0.15 \\ 0.14 & 0.18 & 0.08 & 0.00 & 0.08 & 0.08 & 0.12 & 0.08 \\ 0.12 & 0.11 & 0.18 & 0.08 & 0.00 & 0.08 & 0.08 & 0.09 \\ 0.09 & 0.09 & 0.22 & 0.08 & 0.08 & 0.00 & 0.11 & 0.15 \\ 0.14 & 0.11 & 0.11 & 0.14 & 0.08 & 0.09 & 0.00 & 0.12 \\ 0.08 & 0.08 & 0.17 & 0.08 & 0.11 & 0.17 & 0.11 & 0.00 \end{bmatrix}$$

Finally, the total relation matrix T is computed as:

$$T = D(I - D)^{-1} = \begin{bmatrix} 0.36 & 0.50 & 0.56 & 0.53 & 0.49 & 0.49 & 0.48 & 0.46 \\ 0.49 & 0.48 & 0.70 & 0.62 & 0.54 & 0.54 & 0.50 & 0.51 \\ 0.56 & 0.70 & 0.67 & 0.62 & 0.66 & 0.74 & 0.59 & 0.65 \\ 0.50 & 0.59 & 0.58 & 0.43 & 0.48 & 0.50 & 0.49 & 0.47 \\ 0.48 & 0.53 & 0.67 & 0.49 & 0.40 & 0.51 & 0.46 & 0.48 \\ 0.50 & 0.56 & 0.76 & 0.53 & 0.52 & 0.49 & 0.53 & 0.58 \\ 0.52 & 0.55 & 0.63 & 0.56 & 0.49 & 0.53 & 0.40 & 0.52 \\ 0.48 & 0.54 & 0.70 & 0.52 & 0.53 & 0.61 & 0.51 & 0.43 \end{bmatrix}$$

Prominence and relation of the enablers calculated using matrix T are given in Table 3. Based on $(r + c)$ values, the degree of importance of the enablers in effective implementation of ecodesign practices is obtained in the order as E3 > E6 > E2 > E8 > E4 > E7 > E5 > E1. Training of designers (E3) comes out to be the most important enabler with a value of 10.46 followed by E6, i.e., considering ecodesign strategies as essential practices with 8.88. Support from the top management (E1) is the least important enabler with a value of 7.76. Considering ecodesign strategies as essential practices (E6), including an environmental expert in product development team (E7) and considering ecodesign approaches in the initial design phase (E8) together form the cause group as they are having positive values of $(r - c)$, whereas support from top management (E1), using life cycle approach to design products (E2), training of designers (E3), collaboration between design domains (E4) and building an information system to exchange data

Table 3 Prominence and relation of the enablers

Enablers	r	c	$r + c$	$r - c$
E1	3.87	3.89	7.76	−0.02
E2	4.38	4.45	8.83	−0.07
E3	5.19	5.27	10.46	−0.08
E4	4.04	4.30	8.34	−0.26
E5	4.02	4.11	8.13	−0.09
E6	4.47	4.41	8.88	0.06
E7	4.2	3.96	8.16	0.24
E8	4.32	4.10	8.42	0.22

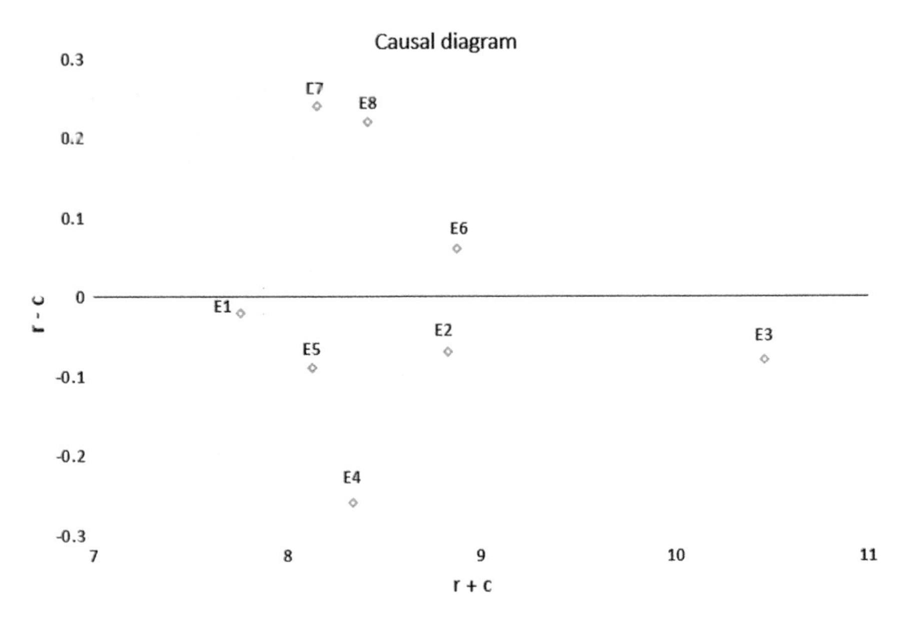

Fig. 2 Causal diagram

(E5) are the enablers that form effect group and fall under receivers category with negative values of $(r - c)$. Cause and effect groups are also shown with the help of a causal diagram in Fig. 2. Support from the top management (E1) is the most independent enabler among all because it neither affects nor gets affected by the other criteria. It happens because many individuals in the case company are self-motivated to learn about eco-friendly practices.

Some important managerial implications are suggested to the case company on the basis of results obtained through the DEMATEL approach. The case company should pay much attention to the causes (E6, E7, E8) in comparison with the receivers (E1, E2, E3, E4, E5). Considering ecodesign strategies as essential practices (E6) is not only the second most important enabler but also a cause factor. Therefore, the company needs to develop a culture in which environmentally friendly practices are treated as a necessity. Including an environmental expert in product development team (E7) is also crucial because it is a cause factor and does not get affected by the other factors. However, the appointment of an environmental expert in the product development team will further enhance two enablers, i.e., training of designers (E3) and collaboration between design domains (E4), i.e., collaboration between environmental and engineering design domains.

5 Conclusion

This study proposes a methodology based on the DEMATEL approach to evaluate the enablers for effective implementation of ecodesign practices in sustainable product development in MSMEs. The proposed methodology is applied in an Indian manufacturing MSME for evaluating the identified enablers. Results show that training of designers is the most important enabler. It provides an opportunity for the designers to learn about various challenges to develop environmentally friendly products and different ecodesign methods and tools to deal with these challenges. Support from the top management is the least important enabler among all identified enablers. Apart from that, the appointment of an environmental expert in the product development team is another essential enabler because an environmental expert can not only assist a designer about the requirement of an eco-friendly product but also act as a link between environmental and engineering design domains.

References

1. Meadows DH, Meadows DL, Randers J, Behrens W (1972) Los límites del crecimiento: informe al Club de Roma sobre el predicamento de la Humanidad. Universidad Politécnica de Madrid, Madrid, Spain
2. ISO I, ISO (2011) Environmental management systems—guidelines for incorporating ecodesign. ISO 14006. ISO, Geneva, Switzerland. https://www.iso.org/standard/43241.html
3. Bakker (1995) Environmental information for industrial designers
4. McAloone T, Holloway L (1996) From product designer to environmentally conscious product designer. In: Proceedings of applied concurrent engineering conference. Seattle, WA
5. Frei M (1998) Eco-effective product design: the contribution of environmental management in designing sustainable products. J Sustain Prod Des 16–25
6. Hillary R (2004) Environmental management systems and the smaller enterprise. J Cleaner Production 12:561–569
7. Banik S (2018) Small scale industries in India: opportunities and challenges. Int J Creative Res Thoughts 6:337–341
8. Lee S (2008) Drivers for the participation of small and medium-sized suppliers in green supply chain initiatives. Supply Chain Manage Int J 13:185–198
9. Pigosso DCA, McAloone TC, Rozenfeld H (2014) Systematization of best practices for ecodesign implementation. In: Design organization and management.Dubrovnik-Croatia pp 1651–1662
10. Ritzén S (2000) Akademisk avhandling som, med tillstånd av Kungliga Tekniska Högskolan i Stockholm, framläggs till offentlig granskning för avläggande av Teknologie doktorsexamen, tisdagen den 21 mars 2000, klockan 10.00 i sal M3, Brinellvägen 64, Kungliga Tekniska Högskolan. 67
11. Collado-Ruiz D, Ostad-Ahmad-Ghorabi H (2010) Influence of environmental information on creativity. Design Stud 31:479–498
12. Bovea MD, Pérez-Belis V (2012) A taxonomy of ecodesign tools for integrating environmental requirements into the product design process. J Cleaner Prod 20:61–71
13. Boks C (2006) The soft side of ecodesign. J Cleaner Prod 14:1346–1356

14. Wolf J (2013) Improving the sustainable development of firms: the role of employees: the role of employees. Bus Strategy Environ 22:92–108
15. Reyes T, Millet D (2013) An exploratory study for the long-term integration of ecodesign in SMEs: the environmental Trojan horse strategy. Progress Industrial Ecol Int J 8:67
16. Rio M, Reyes T, Roucoules L (2013) Toward proactive (eco)design process: modeling information transformations among designers activities. J Cleaner Prod 39:105–116
17. Zhang F, Rio M, Allais R, Zwolinski P, Carrillo TR, Roucoules L, Mercier-Laurent E, Buclet N (2013) Toward an systemic navigation framework to integrate sustainable development into the company. J Cleaner Prod 54:199–214
18. Rochlin S, Bliss R, Jordan S, Kiser C.Y (2015) Project ROI—report defining the competitive and financial advantages of corporate responsibility and sustainability. Babson College
19. Millet D, Bistagnino L, Lanzavecchia C, Camous R, Poldma T (2007) Does the potential of the use of LCA match the design team needs? J Cleaner Prod 15:335–346
20. Kozemjakin da Silva M, Guyot E, Remy S, Reyes T (2013) A product model to capture and reuse ecodesign knowledge. In: Bernard A, Rivest L, Dutta D (eds.) Product lifecycle management for society. Springer, Berlin, Heidelberg, pp 220–228
21. Noci G, Verganti R (1999) Managing “green” product innovation in small firms. R D Manag 29:3–15
22. Cloquell-Ballester, V-A, Monterde-Díaz R, Cloquell-Ballester V-A, Torres-Sibille A. del C (2008) Environmental education for small- and medium-sized enterprises: Methodology and e-learning experience in the Valencian region. J Environ Manag 87:507–520
23. Diabat A, Govindan K (2011) An analysis of the drivers affecting the implementation of green supply chain management. Res Conserv Recycling 55:659–667
24. Rao P, Holt D (2005) Do green supply chains lead to competitiveness and economic performance? Int J Operations Prod Manag 25:898–916
25. Mangla S, Kumar P, Barua MK (2014) An evaluation of attribute for improving the green supply chain performance via DEMATEL method. Int J Mech Eng Robot Res 1:30–35
26. Chang A-Y (2011) Analysing critical factors of introducing RFID into an enterprise—an Application of AHP and DEMATEL method. Int J Industr Eng Theory Application Practice 18

Experimental Investigation of Paddy Grain Drying Mechanism

Mummina Vinod, M. Raghuraman and V. Mahesh Chakravarthi

Abstract The main purpose of this paper is to solve the problems that occur during reaping of rice grains from paddy fields and it involves high labor cost. The traditional way of drying grains takes a time of 40–50 h when the grains kept on the paddy field but with grain drying mechanism lot of time can be saved. For the drying process, the semi-dried grains having 22–25% of moisture content are taken off from the paddy fields for the drying to happen manually for 40 h in sunlight. But this process has some drawbacks which can make the grains wet again by untimely rains, causes fungal infections, pest, and it increases the labor costs. To avoid the problems of drying associated with mechanical dryers, a drying mechanism is proposed where we can bring down the level of moisture to a safe level of 5–6% from 25% moisture content in the semi-dried grains.

Keywords Paddy · Moisture · Harvesting · Drying · Fields

1 Introduction

For harvesting of paddy we use different methods, they are mainly classified into traditional and modern harvesting techniques. Orido et al. [1] state that modern harvesting techniques are the best method to reduce human effort, cost, and time. Combined harvesting machines are used in which the processes of reaping, threshing, cleaning, and hauling are performed but drying is not integrated into the modern harvesting machines. In order to dry the grains after the hauling stage, a grain drying mechanism is employed where a heat transfer mechanism is used to reduce the moisture content of the grain to the possible extent (5–6%). The sudden removal of moisture in the grain, i.e., more than 5–6% by rapid drying at high temperatures creates internal tension, and hence, it produces tiny cold cracks which reduce the quality of the grain. So removing of moisture to a safe level of 5–6% is recommended. Harvesting is the process of collecting the rice crop from the field.

M. Vinod (✉) · M. Raghuraman · V. M. Chakravarthi
Vishnu Institute of Technology, Bhimavaram 534202, India

BBVL. Deepak et al. (eds.), *Innovative Product Design and Intelligent Manufacturing Systems*, Lecture Notes in Mechanical Engineering,
https://doi.org/10.1007/978-981-15-2696-1_13

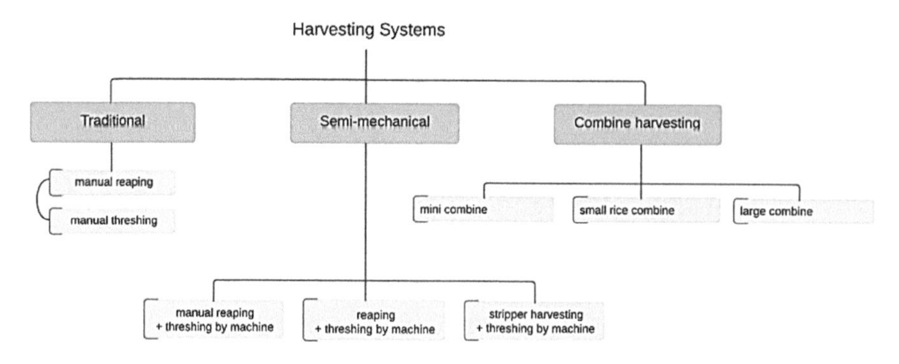

Fig. 1 Harvesting system

These processes will be carried out in three different techniques which are traditional, semi-mechanical, and combined harvesting as shown in Fig. 1.

1.1 Combined Harvesting Systems

The combined harvesting machine Dumitru [2] is designed for the effective utility of machinery to accomplish the harvesting of a wide variety of grain crops. Nowadays majority of the farmers are using combined harvesting machine to harvest. Most of the modern combined harvesting machine was not equipped with drying systems. Since there is no drying process equipped in the harvesting machines, farmers have been using traditional and mechanical systems with varying technologies and capacities for either farm or commercial level as shown in Fig. 2.

1.2 Grain Drying

The process of drying removes the moisture content present in the grains. During reaping of paddy by using harvesting machine, paddy contains approximately $\sim$ 25% moisture Janier and Maidin [3]. Due to this moisture natural respiration occurs and deteriorates paddy by forming fungal molds and pests. Even for a short duration of storage with high moisture in the grain may reduce the quality of grain. To eradicate this problem after harvesting, the drying process should begin within 24 h after harvesting. Majority of the farmers are using traditional and mechanical systems with different technologies and capacities for either farm or commercial level as shown in Fig. 3.

Fig. 2 Combined harvesting machine

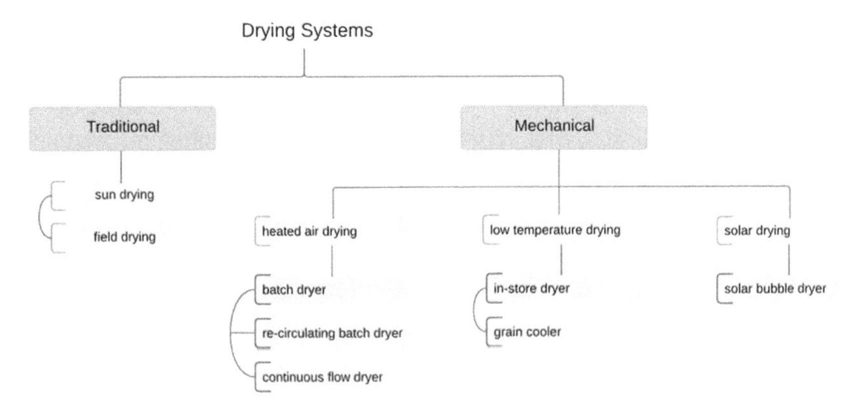

Fig. 3 Classification of drying systems

1.3 Traditional Method

In this method, floor drying is one of the oldest agricultural methods for removal of moisture up to the safe-moisture level for food preservation, for drying the paddy completely by conventional method takes more than 168–240 h. Due to improper drying of paddy, most of the yielded grains gets infected with fungal and micro-bacterial attacks. Most of the farmers are still using floor drying method, and it has many disadvantages like untimely rains, fungal infections, pests, molds, require a large area to dry Jambhulkar et al. [4] which reduces the quality of the rice as shown in Fig. 4.

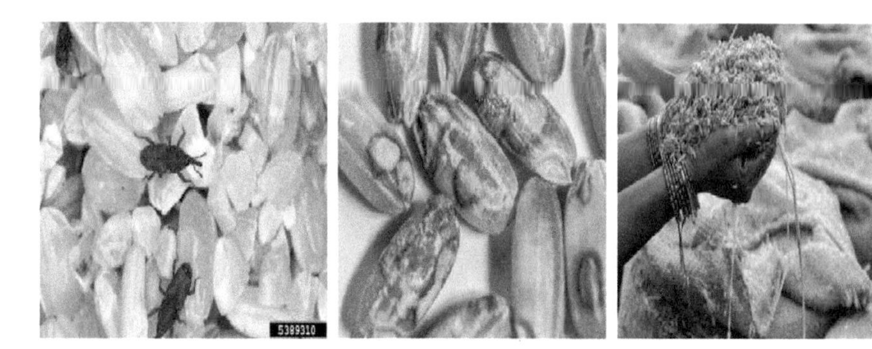

Fig. 4 Problems in traditional drying

1.4 Mechanical Method

In mechanical drying system, different methods like heated air drying, in-store drying, solar drying, etc. are used. By using a mechanical drying process, the moisture content in the grain can be removed quickly in comparison with the traditional method. Although mechanical method of drying is best compared to traditional drying, this process contains many problems such as storage of paddy, manpower, and maintenance cost.

Since this mechanical drying system is placed away from the farm for implementation farmer needs to transport the paddy to the process station. A survey states that buyer purchases the moisture paddy for a low price as the farmer needs to pay heavily to transport to mills. To overcome these problems, farmers keep all their efforts to dry the paddy and to gain profit by using floor drying method. The quality of rice gets reduced by considering the above disadvantages and to avoid these disadvantages a mechanism is created to solve the problems under feasibility.

1.5 Solar Bubble Dryer (SBD)

The energy from the sun is useful to SBD De Bruin et al. [5] in two different ways as shown in Fig. 5. SBD receives sun rays and converts that solar energy into temperature to enhance the temperature of drying air. SBD also equipped with battery and controller to generate electricity. This generator drives a small blower to push air in drying tunnels. This air evaporates water in the grains placed in the tunnel as shown in Fig. 6. The drying time in SBD is depending on the weather conditions and initial moisture content in the grain. SBD takes nearly 20 h/ton time under sunlight to dry the paddy from initial moisture of 24–14%. SBD takes more time to dry the paddy during rainy seasons, cloudy days, night time, and when the relative humidity is high.

Table 1 shows the moisture content for safe storage provided by Gummert et al. [6].

Fig. 5 Solar bubble dryer

Fig. 6 Grains inside drying tunnel

Table 1 Moisture content for safe storage

Required moisture content for safe storage (%)	Safe storage period	Potential problems
≤ 14	≤ 8 months	Molds, discoloration, insect damage, moisture adsorption, etc.
≤ 13	8–12 months	Insect damage
≤ 12	Storage of farmer's seeds	Loss of germination
≤ 9	<1 year	Loss of germination

2 Experiment Setup for Drying Mechanism

In this drying mechanism, a special attachment has been developed by using an air compressor and heating coils. Paddy is passed through hollow steel tube with the help of compressed air and heating coils are placed outside of the tubes. The temperature of the coil can be controlled by the supplying of electricity. Now the compressed air absorbs the heat from the coil and becomes warm, due to this warm air the moisture in the paddy evaporates continuously. Initially, the paddy having approximately 25% moisture content can be reduced more than 8% but the sudden removal of moisture in the grain, i.e., more than 8% by rapid drying at high temperatures creates internal tension and produces tiny cold cracks which reduce the quality of the grain. Removing of moisture up to 5–6% is done within the temperature range of 45–55 °C, but increasing of temperature more than 55 °C will tend to decrease the quality of grains.

2.1 Design and Fabrication of Drying Mechanism

The design of the drying mechanism is done by using CATIA V5 software as shown in Fig. 7 and the fabricated setup as shown in Fig. 8. The moisture content in the paddy can be measured with the help of digital moisture meter as shown in Fig. 8 (Table 2).

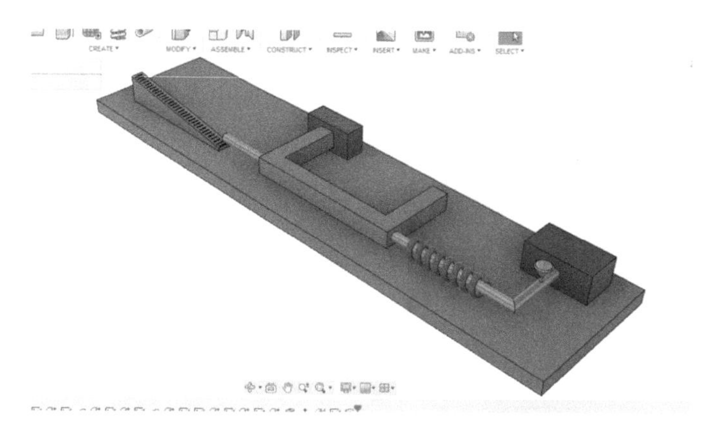

Fig. 7 Isometric view

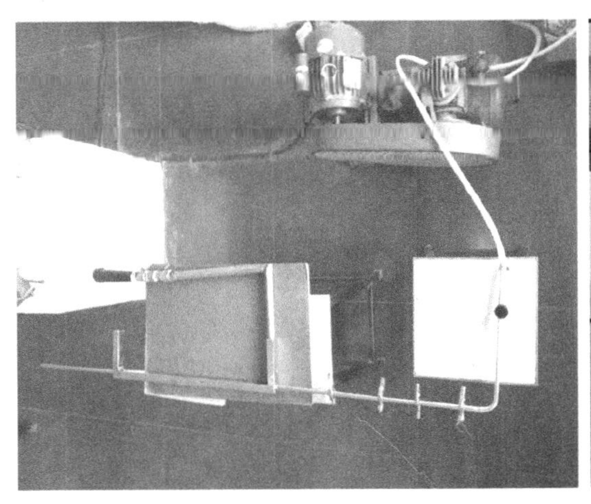

Fig. 8 Experiment setup and moisture meter

Table 2 Specifications of components

Components	Specifications
Circular section stainless steel pipe	Length = 1800 mm
	Diameter = 15 mm
Heating coil	4 × 10 W heating elements
Air compressor	0.5 HP, 0.37 KW

3 Results and Discussion

The readings from the experiment are tabulated as shown in Table 3.

Table 3 Experimental Readings

Pipe temperature	Air compressor pressure	Moisture content (in %)		Moisture shrink
(in °C)	(in kg/cm^2)	Before heating	After heating	(in %)
35	1	24.0	22.1	2.43
40	1	23.5	21.6	2.42
48	1	23.5	19.5	4.49
52	1	23.5	19.2	5.32

Table 4 Comparison of sun-D, SBD, and NPD

Items	Unit	Sun-D	SBD	PPD
Investment cost	RS	7400	140,000	13,000
Life span	years	–	5	10
Paddy initial MC (wet basis)	%	24	24	24
Paddy final MC	%	14	14	19
Capacity	ton/batch	1	1	1
Land area used for drying operations	m^2/ton	40	28	15
Drying time	h/ton	40	20	5
Operating labor	h/ton	16	4	1

3.1 Calculations

Grain moisture shrink can be calculated using the formula: Moisture shrink (%)

$$= \frac{\text{IMC} - \text{FMC}}{100 - \text{FMC}} X100 \quad (1)$$

where IMC = Initial moisture content.

FMC = Final moisture content.

For reading 1 at 35 °C pipe temperature and other respective readings from Table 4,

Moisture shrink $= \frac{23.9-22.1}{100-22.1} = 2.43\%$.

Sun-D = Sun drying; SBD = Solar bubble dryer; PPD = Proposed paddy dryer.

Table 4 shows the comparison of sun-D, SBD, and NPD values. Nguyen-Van-Hung et al. [7], provided the values for solar bubble dryer in which the initial moisture content for all methods are same but the final moisture content of NPD is different from the other two methods. The reason for different moisture content is due to the sudden removal of moisture content from 24 to 14% within a short time that will affect the quality of grain.

4 Conclusion

To overcome the problems in traditional drying method and SBD incorporated with combined harvesting machine (CHM), a drying mechanism is developed which can give a perfect solution for reduction in area required for drying grains, labor cost, drying time, fungal infections, pests, and it also helps to prevent grains getting wet from untimely rains. In this experiment setup, it has been observed that there is shrinkage of moisture with an increase of coil temperature at constant pressure. By using conventional drying method, the complete drying of paddy takes 40 h in sunlight when the moisture percentage is about 25% but after adopting this drying method, drying of paddy takes up to 5 h when the moisture percentage has been brought down around 4–5%. If you adopt this mechanism in CHM, then there is no need to use any traditional or mechanical drying methods it means all process like reaping, threshing, cleaning, hauling, and drying done in CHM. This method can be extended by using exhaust gases as a primary source to dry the paddy in CHM.

References

1. George O, Musa N, Ngunjiri G (2017) Grain drying simulation in a GT-380 dryer using energy recovered from ICE exhaust. IOSR J Agricu Vet Sci 10:01–06. https://doi.org/10.9790/2380-1006020106
2. Dumitru M (2009) Researches on grain harvesting machines. 66:1843–5386
3. Janier JB, Maidin MB (2011) Paddy moisture content detector. J Appl Sci 11(7):1476–1478
4. Jambhulkar AC, Pawar VB, Pawar SB, Dharwadkar AS Solar drying techniques and performance analysis: a review. J Mech Civil Eng (IOSR-JMCE) 35–39
5. De Bruin T, Villers P, Navarro S (201) Worldwide developments in ultra hermetic™ storage and solar drying technologies. In: 11th international working conference on stored product protection, 24–28 November 2014. Chiang Mai, Thailand
6. Gummert M, Aldas R, Barredo IR, Muehlbauer W, Quick GR (1993) Low-temperature in-store drying system. Project report, IRRI-GTZ Project Postharvest Technologies in the Humid Tropics
7. Nguyen-Van-Hung, Tran-Van-Tuan, Meas P, Tado CJM, Kyaw MA, Gummert M (2019) Best practices for paddy drying: case studies in Vietnam, Cambodia, Philippines, and Myanmar. Plant Prod Sci 22(1):107–118

Analytic Hierarchy Process for Prioritization of Design Requirements for Domestic Plumbing Services

Sachin Shivaji Jadhav, Pratul Ch. Kalita and Amarendra Kumar Das

Abstract Sustainable trends of servitization have progressively increased over the last couple of decades. So developing product–service–system design is important for academic researchers and leading manufacturers. Design principles and customer needs typically cast the requirements at the early stages of design development. The prime objective of this study is to prioritize design requirements for domestic plumbing services. Design requirements of domestic plumbing are taken as an example to demonstrate the application of the analytic hierarchy process and rough group. We conducted in-depth interviews including exploratory surveys. We applied the analytic hierarchy process and rough group method to prioritize the design requirements and product–service components of plumbing. Altogether 34 design requirements were identified from previous studies and interactions with stakeholders for domestic plumbing. These design requirements are categorized into a hierarchical structure of product, service, and system. The results show that the most important product-related design requirements are efficiency and flexibility. Service-related design requirements are response/delivery and availability. System-related design requirements are skills and communication of plumber. The study provides a design management insight for PSS implementation in plumbing services in the domestic sector. This study includes ranking and prioritization of design requirements, which may help design managers and designers to make effective and efficient decisions on the design of product–service systems.

Keywords Product–service–system · Analytic hierarchy process · Rough group · Design management · Design methods · Domestic plumbing

---

S. S. Jadhav (✉) · P. Ch. Kalita · A. K. Das
Department of Design, Indian Institute of Technology Guwahati, Guwahati 781039, Assam, India
e-mail: sjsachin@iitg.ac.in

P. Ch. Kalita
e-mail: pratulkalita@iitg.ac.in

A. K. Das
e-mail: dasak@iitg.ac.in

BBVL. Deepak et al. (eds.), *Innovative Product Design and Intelligent Manufacturing Systems*, Lecture Notes in Mechanical Engineering,
https://doi.org/10.1007/978-981-15-2696-1_14

1 Introduction

In today's competitive market, customer requirements determine what companies supply. Developers do not always know which requirements are most important to the customers, and customers cannot judge the cost and technical difficulty associated with specific requirements [1]. Designers and users should team up on requirements prioritization. Requirement elicitation and prioritization have decisive roles in the development process [2]. Most development process models in product engineering and service engineering consider requirements engineering only during the first phases of the overall development process. The goal of the requirements engineering's activities is to elicit and specify the requirements [3]. However, Müller et al. [4] presented a guideline to elicit and analyze requirements of PSS properties and quality. They developed a checklist of criteria in terms of lifecycle activities, values, contracts, business and operation models, structure, behavior, technical artefacts, service, information, communication, and actors. Berkovich [5] proposed a requirements data model to facilitate an integrated requirements engineering approach for a PSS described at different levels of abstraction. Lindström [6] reviewed the literature to identify potential through lifecycle aspects that needed to be considered during the development and operation of functional products. With regard to product–service–system (PSS) transformation drivers, the key contributors related to provider, joint operation, sacrifice, and configuration for contract incentives and performance indicators were identified by Priya Datta and Roy [7]. There is a lack of research in identifying customers' needs and prioritizing design requirements in the context of a product–service–system. The concept of product–service–system, also named as 'functional sales,' was proposed by the United Nations Environment Program (UNEP) in the late 1990s [8]. Its core idea is to provide solutions to customers by integration of 'products' and 'services,' meeting customers' requirements while reducing resource consumption and environmental impact at the same time. Several authors have defined a product–service–system in terms of economic and environmental context. As per the definition of Mark J. Goedkoop, 'a marketable set of products and services capable of jointly fulfilling a user's needs. The product–service ratio in this set can vary, either in terms of function fulfillment or economic value' [9]. Surveys by authors Baines et al. [10] and Beuren et al. [11] showed the benefits of PSS to the consumer, provider, environment, and society. Benefits result from a higher level of satisfaction, increased competitiveness, decreased environmental impact and materials savings.

1.1 Aim and Objectives

The aim of the research is to identify product–service–system aspects of domestic plumbing and prioritize design requirements. The objectives of the research are

- To identify the interrelationships of product–service–system aspects in domestic plumbing.
- To identify and prioritize design requirements for the domestic plumbing sector.

2 Methodology

We conducted in-depth interviews including exploratory surveys. The prime focus of the survey was to study the ‘plumbing tools’ and ‘service aspect’ in domestic plumbing. A structured interview with various stakeholders revealed plumbing tools, viz. adjustable wrench, pliers, metal file, hacksaw, lubricants, and replacement parts, and also aspects of customer service requirements, viz. corrective maintenance, preventive maintenance, operation time, service frequency, replacement of spare parts, consumables, fittings. Further, insights from the interview were utilized in structuring the hierarchy of product-, service-, and system-related design requirements. We applied the analytic hierarchy process and rough group method to prioritize the design requirements and product–service components of plumbing. A total of 34 design requirements were identified from previous studies and interactions with stakeholders for domestic plumbing [5]. These design requirements are categorized into a hierarchical structure of product, service, and system. Product-related design requirements comprise technical function, economic, and quality. Service-related design requirements comprise process, interaction, timing, and reliability. System-related design requirements comprise human resources, facility, material, information, and capital. Pair-wise comparison between designs requirements is conducted in each hierarchy. Experts were chosen from three sectors, viz. designer, maintenance engineer and technician. Expert’s judgment on the importance between each requirement was checked for consistency.

3 Results and Discussion

AHP is a basic theory of subjective measurement and is a popular tool for assigning weights to compare certain criteria or alternatives [12]. The analytic hierarchy process method provides valuable aid in organizing, assessing requirements, ranking, and incorporating the judgments of multiple experts. Comparison values may be taken from surveys or measurements from the respondents using fundamental scales. AHP was deployed in the following four steps.

STEP 1: Identify and form a hierarchy of design criteria related to product, service, and system. Develop separately a group of pair-wise comparison matrix for product, service, and system. A group of ‘k’ experts is formed to rate importance. Where $k = 1, 2, 3, \ldots$.

STEP 2: To make a pair-wise comparison, each expert from group 'k' is invited. Then, obtain priority weights of the data matrix. The 'k' expert's pair-wise comparison matrix A^k is as follows:

$$A^k = \begin{bmatrix} 1 & r_{12}^k & \cdots & r_{1n}^k \\ r_{21}^k & 1 & \cdots & r_{2n}^k \\ \vdots & \vdots & \ddots & \vdots \\ r_{n1}^k & r_{n2}^k & \cdots & 1 \end{bmatrix}_{n \times n}$$

where r_{ij}^k is the kth expert's judgment for the ith design requirement importance compared with jth design requirement and n is the number of design requirements.

STEP 3: Check for consistency of pair-wise comparison matrix. Consistency test is conducted by the following equations

$$\mathrm{CI} = \frac{\lambda_{\max} - n}{n - 1} \tag{1}$$

$$\mathrm{CR} = \left(\frac{\mathrm{CI}}{\mathrm{RI}(n)}\right) \tag{2}$$

where CI is consistency index, $\lambda_{\max}$ is the largest eigenvalue of matrix A^k, and n is the dimension of the matrix A^k. CR is the consistency ratio. RI is the random index which depends on the dimension of the matrix as shown in Table 1 [13].

Consistency test pair-wise comparison matrix is acceptable, When CR is less than 0.1, if CR is greater than 0.1, experts need to adjust pair-wise comparison until clearing the consistency test.

STEP 4: After combining all pair-wise matrixes from expert's opinions, develop group evaluation matrix B of design requirements and sub-requirements.

$$B = \begin{bmatrix} 1 & r_{12} & \cdots & r_{1n} \\ r_{21} & 1 & \cdots & r_{2n} \\ \vdots & \vdots & \ddots & \vdots \\ r_{n1} & r_{n2} & \cdots & 1 \end{bmatrix}_{n \times n} \quad \text{where } r_{ij} = \left[r_{ij}^1, r_{ij}^2, r_{ij}^3, \ldots, r_{ij}^k\right]$$

Rough numbers: Assume that there is a set of m classes of human judgments. $J = \left\{r_{ij}^1, r_{ij}^2 \ldots r_{ij}^k \ldots r_{ij}^m\right\}$ ordered in the manner of $r_{ij}^1 \prec r_{ij}^2 \prec \cdots \prec r_{ij}^k \cdots \prec r_{ij}^m$. U is

Table 1 Random index

Order	1	2	3	4	5	6	7	8	9	10
RI(n)	0	0	0.52	0.89	1.11	1.25	1.35	1.40	1.45	1.49

the universe including all the objects, and Y is an arbitrary object of U. Then, lower and upper approximations of r_{ij}^k can be defined as [14],

$$\text{Lower approximation} \quad : \quad \underline{\text{Apr}}\left(\boldsymbol{r}_{ij}^k\right) = \cup\left\{Y \in U/J(Y) \leq \boldsymbol{r}_{ij}^k\right\}$$
$$\text{Upper approximation} \quad : \quad \overline{\text{Apr}}\left(\boldsymbol{r}_{ij}^k\right) = \cup\left\{Y \in U/J(Y) \geq \boldsymbol{r}_{ij}^k\right\}$$

STEP 5: Convert the element r_{ij} in group decision matrix B into $RN\left(r_{ij}^k\right)$ of r_{ij} as:

$$RN\left(r_{ij}^k\right) = \left[r_{ij}^{kL}, r_{ij}^{kU}\right] \tag{3}$$

where r_{ij}^{kL} is the lower limit and r_{ij}^{kU} is the upper limit of rough number $RN\left(r_{ij}^k\right)$ in kth pair-wise comparison matrix, respectively,

$$r_{ij}^k = \underline{\text{Lim}}\left(r_{ij}^k\right) = \left(\prod_{m=1}^{N_L} x_{ij}\right)^{1/N_L}, \quad r_{ij}^k = \overline{\text{Lim}}\left(r_{ij}^k\right) = \left(\prod_{m=1}^{N_U} y_{ij}\right)^{1/N_U},$$

where x_{ij} and y_{ij} are the elements of lower and upper approximations for r_{ij}^k. N_L and N_U are the number of objects included in the lower and upper approximations of r_{ij}^k, respectively.

STEP 6: Then, we obtain rough sequence number as

$$RN(r_{ij}) = \left\{\left[r_{ij}^{1L}, r_{ij}^{1U}\right], \left[r_{ij}^{2L}, r_{ij}^{2U}\right], \ldots, \left[r_{ij}^{kL}, r_{ij}^{kU}\right]\right\}$$

The average rough interval $\overline{RN(r_{ij})}$ is obtained by using equation

$$\overline{RN(r_{ij})} = \left[r_{ij}^L, r_{ij}^U\right] \tag{4}$$

where $r_{ij}^L = \sqrt[k]{r_{ij}^{1L} \times r_{ij}^{2L} \times \cdots r_{ij}^{kL}}$ and $r_{ij}^U = \sqrt[k]{r_{ij}^{1U} \times r_{ij}^{2U} \times \cdots r_{ij}^{kU}}$.

Then, rough group decision matrix M is formed as

$$M = \begin{pmatrix} [1,1] & \left[r_{12}^L, r_{12}^U\right] & \cdots & \left[r_{1n}^L, r_{1n}^U\right] \\ \left[r_{21}^L, r_{21}^U\right] & [1,1] & \cdots & \left[r_{2n}^L, r_{2n}^U\right] \\ \vdots & \vdots & \ddots & \vdots \\ \left[r_{n1}^L, r_{n1}^U\right] & \left[r_{n2}^L, r_{n2}^U\right] & \cdots & [1,1] \end{pmatrix}$$

STEP 7: Calculate rough-based weight and its normalized counterparts as follows:

$$W_i = \left(W_i^L, W_i^U\right) = \left[\left(\prod_{i=1}^{n} r_{ij}^L\right)^{1/n}, \left(\prod_{i=1}^{n} r_{ij}^U\right)^{1/n}\right] \quad (5)$$

$$NW_i = \left(NW_i^L, NW_i^U\right) = \left[\frac{W_i^L}{\max\left(W_i^U\right)}, \frac{W_i^U}{\max\left(W_i^U\right)}\right] \quad \text{where} : i = 1, 2, 3\ldots \quad (6)$$

3.1 *Case Study: Prioritization of Design Requirements for Domestic Plumbing Services*

STEPS 1 and 2: Form a hierarchy of design criteria related to product–service–system. Develop separately a group of pair-wise comparison matrix for product, service, and system as shown in Fig. 1.

STEPS 3 and 4: Pair-wise comparison between design requirements is conducted in each hierarchy until each comparison matrix gets through consistency test. To illustrate the computation process, below matrixes show expert's judgments on the first level of product-related design requirements of domestic plumbing, viz. technical functions, economic and quality. According to Eqs. (1) and (2), the consistency test and consistency ratio are calculated.

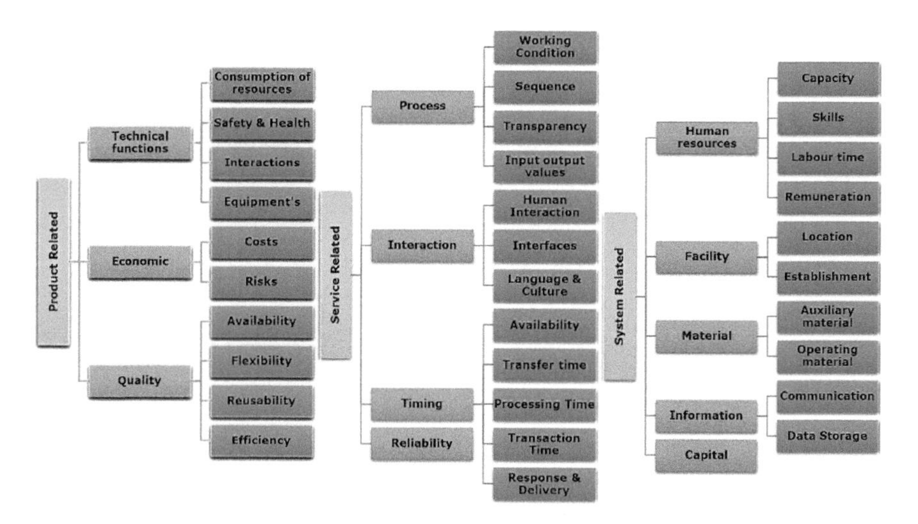

Fig. 1 Hierarchy of product-, service-, system-related requirements for domestic plumbing

$$A^1 = \begin{bmatrix} 1 & 5 & 1/3 \\ 1/5 & 1 & 1/7 \\ 3 & 7 & 1 \end{bmatrix} \quad A^2 = \begin{bmatrix} 1 & 2 & 1 \\ 1/2 & 1 & 1/2 \\ 1 & 2 & 1 \end{bmatrix} \quad A^3 = \begin{bmatrix} 1 & 1 & 1/7 \\ 1 & 1 & 1/3 \\ 7 & 3 & 1 \end{bmatrix}$$

$$\lambda = 3.066, \mathrm{CR} = 0.063 < 0.1 \quad \lambda = 3, \mathrm{CR} = 0.00 < 0.1 \quad \lambda = 3.082, \mathrm{CR} = 0.078 < 0.1$$

$$A^4 = \begin{bmatrix} 1 & 1 & 1/5 \\ 1 & 1 & 1/6 \\ 5 & 6 & 1 \end{bmatrix} \quad A^5 = \begin{bmatrix} 1 & 4 & 1/3 \\ 1/4 & 1 & 1/6 \\ 3 & 6 & 1 \end{bmatrix}$$

$$\lambda = 3.004, \mathrm{CR} = 0.004 < 0.1 \quad \lambda = 3.054, \mathrm{CR} = 0.052 < 0.1$$

Then, the rough group evaluation matrix B of first-level product-related design requirements can be obtained by combining the above five pair-wise matrixes together.

$$B = \begin{bmatrix} 1,1,1,1,1 & 5,2,1,1,4 & 1/3,1,1/7,1/5,1/3 \\ 1/5,1/2,1,1,1/4 & 1,1,1,1,1 & 1/7,1/2,1/3,1/6,1/6 \\ 3,1,7,5,3 & 7,2,3,6,6 & 1,1,1,1,1 \end{bmatrix}$$

The same procedure can be conducted to other level of hierarchical structure to get their comparison matrixes.

STEPS 5 and 6: To get the rough form of group comparison matrix, transfer the elements of B matrix into rough number form according to Eq. (3). Now, to find lower and upper approximations, consider matrix B element $C_{12} = (5, 2, 1, 1, 4)$. The rough number conversion process is as follows and is shown in Table 2:

$$\underline{\mathrm{Lim}}(5) = (5 \times 4 \times 2 \times 1 \times 1)^{1/5} = 2.091, \overline{\mathrm{Lim}}(5) = 5$$
$$\underline{\mathrm{Lim}}(2) = (2 \times 1 \times 1)^{1/3} = 1.259, \overline{\mathrm{Lim}}(2) = (2 \times 4 \times 5)^{1/3} = 3.419$$
$$\underline{\mathrm{Lim}}(1) = 1, \overline{\mathrm{Lim}}(1) = (1 \times 1 \times 2 \times 4 \times 5)^{1/5} = 2.091$$
$$\underline{\mathrm{Lim}}(4) = (4 \times 2 \times 1 \times 1)^{1/4} = 1.681, \overline{\mathrm{Lim}}(4) = (4 \times 5)^{1/2} = 4.472$$

The average rough interval $\overline{RN(r_{12})}$ is obtained by using Eq. (4). Similarly, other elements of the rough sequence table are obtained. Then, rough group decision matrix M is formed as

$$\overline{RN(r_{12})} = (1.34, 3.19)$$

$r_{12}^L = \sqrt[5]{2.09 \times 1.25 \times 1 \times 1 \times 1.68} = 1.34$. And $r_{12}^U = \sqrt[5]{5 \times 3.41 \times 2.09 \times 2.09 \times 4.47} = 3.19$

Table 2 Rough number conversion for matrix *B*

Experts	C_{11}	Lower limit	Upper limit	C_{12}	Lower limit	Upper limit	C_{13}	Lower limit	Upper limit
1	1	1	1	5	2.09	5	1/3	0.23	0.48
2	1	1	1	2	1.25	3.41	1	0.31	1
3	1	1	1	1	1	2.09	1/7	0.14	0.31
4	1	1	1	1	1	2.09	1/5	0.16	0.38
5	1	1	1	4	1.68	4.47	1/3	0.23	0.48
Experts	C_{21}	Lower limit	Upper limit	C_{22}	Lower limit	Upper limit	C_{23}	Lower limit	Upper limit
1	1/2	0.2	0.47	1	1	1	1/7	0.23	0.23
2	1/2	0.29	0.79	1	1	1	1/2	0.23	0.5
3	1	0.47	1	1	1	1	1/3	0.19	0.19
4	1	0.47	1	1	1	1	1/6	0.15	0.26
5	1/4	0.22	0.59	1	1	1	1/6	0.15	0.26
Experts	C_{31}	Lower limit	Upper limit	C_{32}	Lower limit	Upper limit	C_{33}	Lower limit	Upper limit
1	3	2.08	4.21	7	4.32	7	1	1	1
2	1	1	3.15	2	2	4.32	1	1	1
3	7	3.15	7	3	2.44	5.24	1	1	1
4	5	2.59	5.916	6	3.83	6.31	1	1	1
5	3	2.08	4.21	6	3.83	6.31	1	1	1

$$M = \begin{bmatrix} (1.00, 1.00) & (1.34, 3.19) & (0.21, 0.48) \\ (0.31, 0.74) & (1.00, 1.00) & (0.19, 0.27) \\ (2.04, 4.71) & (3.15, 5.75) & (1.00, 1.00) \end{bmatrix}$$

STEP 7: The rough-based weight is calculated using Eq. (5) as

$$W_1 = \left(W_1^L, W_1^U\right), W_1^L = (1 \times 1.34 \times 0.21)^{1/3} = 0.66, W_1^U = (1 \times 3.19 \times 0.48)^{1/3} = 1.16$$

$$W = \begin{bmatrix} (0.66, 1.16) \\ (0.39, 0.59) \\ (1.86, 3.00) \end{bmatrix}$$

The above matrix W gives the rough weight of first-level hierarchical structure for product-related design requirements of domestic plumbing, i.e., technical functions, economic and quality. Similarly, the rough weights are calculated for other levels of hierarchical structure and shown in Tables 3, 4, and 5. The final overall weights are calculated using multiplication synthesis method from top level to bottom level. Normalized rough weights are calculated using Eq. (6).

Table 3 Overall weights and normalized rough weights for product-related design requirements

First-level requirements			Second-level requirements			Overall weights		Normalized rough weights	
	Lower limit	Upper limit		Lower limit	Upper limit	Lower limit	Upper limit	Lower limit	Upper limit
Technical functions	0.66	1.16	Consumption of resources	1.19	2.26	0.79	2.63	0.10	0.32
			Safety and health	1.71	3.40	1.13	3.95	0.14	0.48
			Interaction	0.36	0.72	0.24	0.84	0.03	0.10
			Equipment's	0.35	0.70	0.23	0.82	0.03	0.10
Economic	0.39	0.59	Costs	0.58	1.68	0.23	0.99	0.03	0.12
			Risks	0.59	1.73	0.23	1.02	0.03	0.12
Quality	1.86	3.00	Availability	0.39	1.25	0.73	3.76	0.09	0.46
			Flexibility	0.54	1.59	1.00	4.78	0.12	0.59
			Reusability	0.42	1.45	0.79	4.36	0.10	0.53
			Efficiency	1.42	2.72	2.65	8.17	0.32	1.00

Table 4 Overall weights and normalized rough weights for service-related design requirements

First-level requirements			Second-level requirements			Overall weights		Normalized rough weights	
	Lower limit	Upper limit		Lower limit	Upper limit	Lower limit	Upper limit	Lower limit	Upper limit
Process	0.30	0.58	Working conditions	0.53	0.98	0.16	0.57	0.03	0.11
			Sequence	0.85	1.73	0.25	1.01	0.05	0.20
			Transparency	0.59	0.89	0.18	0.52	0.03	0.10
			Input and output values	1.12	2.24	0.34	1.30	0.07	0.25
Interaction	0.64	1.20	Human interaction	1.24	2.60	0.80	3.11	0.16	0.61
			Interfaces	0.44	1.24	0.28	1.49	0.05	0.29
			Language and culture	0.49	1.16	0.32	1.39	0.06	0.27
Timing	0.86	2.12	Availability	1.12	2.19	0.96	4.63	0.19	0.90
			Transfer time	0.27	0.48	0.23	1.02	0.05	0.20
			Processing time	0.86	2.13	0.74	4.51	0.14	0.88
			Transaction time	0.55	1.43	0.48	3.03	0.09	0.59
			Response and delivery	0.89	2.42	0.76	5.12	0.15	1.00
Reliability	1.50	2.73				1.50	2.73	0.29	0.53

Table 5 Overall weights and normalized rough weights for system-related design requirements

First-level requirements			Second-level requirements			Overall weights		Normalized rough weights	
	Lower limit	Upper limit		Lower limit	Upper limit	Lower limit	Upper limit	Lower limit	Upper limit
Human resources	0.53	1.32	Capacity	0.31	0.48	0.17	0.63	0.03	0.13
			Skills	2.22	3.82	1.18	5.04	0.23	.00
			Labor time	0.66	1.15	0.35	1.51	0.07	0.30
			Remuneration	0.75	1.40	0.40	1.85	0.08	0.37
Facility	0.77	1.59	Location	0.86	1.59	0.66	2.53	0.13	0.50
			Establishment	0.63	1.16	0.48	1.85	0.10	0.37
Material	0.57	1.01	Auxiliary material	0.53	0.81	0.30	0.82	0.06	0.16
			Operating material	1.23	1.89	0.71	1.91	0.14	0.38
Information	0.80	1.73	Communication	0.86	1.87	0.69	3.24	0.14	0.64
			Data storage	0.53	1.16	0.43	2.01	0.09	0.40
Capital	0.91	1.55				0.91	1.55	0.18	0.31

Table 6 Crisp weight and ranking for product-, service-, and system-related design requirements

Product			Service			System		
Criteria	Crisp weight	Rank	Criteria	Crisp weight	Rank	Criteria	Crisp weight	Rank
Consumption of resources	0.209	6	Working conditions	0.071	12	Capacity	0.079	11
Safety and health	0.311	4	Sequence	0.123	11	Skills	0.616	1
Interaction	0.066	9	Transparency	0.067	13	Labor time	0.184	9
Equipment's	0.064	10	Input and output values	0.159	9	Remuneration	0.223	8
Costs	0.074	8	Human interaction	0.381	5	Location	0.316	3
Risks	0.077	7	Interfaces	0.172	7	Establishment	0.231	7
Availability	0.275	5	Language and culture	0.166	8	Auxiliary material	0.111	10
Flexibility	0.354	2	Availability	0.545	2	Operating material	0.259	4
Reusability	0.315	3	Transfer time	0.122	10	Communication	0.389	2
Efficiency	0.662	1	Processing time	0.512	3	Data storage	0.241	6
			Transaction time	0.342	6	Capital	0.244	5
			Response and delivery	0.574	1			
			Reliability	0.413	4			

Rough weights' prioritization and ranking are given to crisp value. To convert rough weights into crisp value, authors Song et al. [15] have introduced the optimistic indicator $\lambda(0 \le \lambda \le 1)$. If decision-makers are more optimistic about their judgments, then λ can be selected greater than 0.5. If decision-makers are more pessimistic about their judgments, then λ can be selected lesser than 0.5. If decision-makers are more moderate about their judgments, then λ can be selected 0.5. Using equation $= (1 - \lambda)NW_i^L + \lambda NW_i^U$, crisp weight and ranking for product-, service-, and system-related priority of requirements are shown in Table 6.

4 Conclusions

The paper starts by outlining the need for design managers to correctly prioritize design requirements of the product, services, and systems (PSS). This would help to manage resources better and result in higher levels of customer satisfaction, increased competitiveness, decreased environmental impact, and material savings. A structured interview with various stakeholders revealed the aspects of customer service requirements and related components, viz. corrective maintenance, preventive maintenance, operation time, service frequency, replacement of spare parts, consumables, fittings, and pricing. Further, insights from the interview were utilized in structuring the hierarchy of product-, service-, and system-related design requirements. In this study, a methodology to determine how to prioritize the various aspects of PSS is described and demonstrated the technique to domestic plumbing services. The results show that the most important product-related design requirements are efficiency and flexibility. Service-related design requirements are response/delivery and availability. System-related design requirements are skills and communication of plumber. Future research may be conducted to develop stakeholder's activities and function modeling for plumbing services in the domestic sector. It would finally contribute to design for development.

References

1. Wiegers KE (1999) First things first: prioritizing requirements 1, no. September, pp 1–6
2. Song W (2017) Computers in industry requirement management for product-service systems: Status review and future trends. Comput Ind 85:11–22
3. Berkovich M, Esch S, Leimeister JM, Krcmar H (2009) Requirements engineering for hybrid products as bundles of hardware, software and service elements—a literature review product as bundles of hardware, software and service elements—a literature review
4. Müller P, Schulz F, Stark R (2010) Guideline to elicit requirements on industrial product-service systems. In: Proceedings of the 2nd CIRP IPS2 conference, pp 109–116
5. Berkovich M, Marco J, Hoffmann A, Krcmar H (2014) A requirements data model for product service systems, pp 161–186
6. Lindström J (2015) Through-lifecycle aspects for functional products to consider during development and operation: a literature review. In: Redding L, Roy R (eds) Through-life engineering services: motivation, theory, and practice. Springer International Publishing, Cham, pp 187–207
7. Priya Datta P, Roy R (2011) Operations strategy for the effective delivery of integrated industrial product-service offerings. Int J Oper Prod Manag 31(5):579–603
8. Qu M, Yu S, Chen D, Chu J, Tian B (2016) State-of-the-art of design, evaluation, and operation methodologies in product service systems. Comput Ind 77(127):1–14
9. Mont O (2002) Clarifying the concept of product–service system. J Clean Prod 10(3):237–245
10. Baines TS et al (2007) State-of-the-art in product-service systems. Proc Inst Mech Eng Part B J Eng Manuf 221(10):1543–1552
11. Beuren FH, Gomes Ferreira MG, Cauchick Miguel PA (2013) Product-service systems: a literature review on integrated products and services. J Clean Prod 47:222–231
12. Prasad L, Kim Y (2018) An analysis on barriers to renewable energy development in the context of Nepal using AHP. Renew Energy 129:446–456
13. Saaty TL, The analytic network process
14. Yang Q, Du PA, Wang Y, Liang B (2017) A rough set approach for determining weights of decision makers in group decision making. PLoS ONE 12(2):1–16
15. Song W, Ming X, Han Y, Wu Z (2013) A rough set approach for evaluating vague customer requirement of industrial product-service system, 7543

Issues with Indian SMEs: A Sustainability-Oriented Approach for Finding Potential Barriers

Sudeep Kumar Singh and A. M. Mohanty

Abstract Manufacturing had never been more focused on environmental issues as of today.> The path toward sustainability is not deprived of barriers for the successful realization of economic, environmental and social goals of manufacturing. Sustainability has become a major issue today for the better life of our future generations. The large industries are continually equipping themselves with increasing green manufacturing practices. But in practice, the scenario is quite different for small and medium enterprises (SMEs). SMEs in developing countries like India face a different set of challenges in the post-globalization era, such as stiff competition and reduced market domination from their multi-national counterparts, mainly due to their failure to embrace modernization and innovation. The present work tries to fill the gap by identifying the potential barriers for successful implementation of sustainability for Indian SMEs, after analyzing the issues faced by the manufacturing sector since 1991, till date, i.e., the post-globalization period in the Indian manufacturing industry.

Keywords Sustainability · SME · India

1 Introduction

Manufacturing is often regarded as the backbone of the modern economy [1]. In developing countries like India, it plays a crucial role in both job creation and economic growth. These objectives will disperse in thin air if policymakers and entrepreneurs wont devise a sustainable strategy acceptable by the industry [2–4]. The Intergovernmental Panel on Climate Change in a special report titled "Global Warming of 1.5 °C" has stated that impacts of global warming have attained 1.5 °C above pre-industrial levels. The manufacturing sector is the primary environmental polluter, in the form of emission from greenhouse gases, in both direct and indirect

S. K. Singh (✉) · A. M. Mohanty
Centurion University of Technology and Management, Gajapati, Odisha, India

BBVL. Deepak et al. (eds.), *Innovative Product Design and Intelligent Manufacturing Systems*, Lecture Notes in Mechanical Engineering,
https://doi.org/10.1007/978-981-15-2696-1_15

forms. An immediate step toward the implementation of sustainable practices in industries is the need of the hour.

SMEs are more disorganized and poorly regulated than larger industries. They does not seem to be aware of many effective tools and techniques for quality and productivity improvement available and fail to implement largely due to the lack of knowledge and resources to implement them [5]. In spite of all these problems, the development of small-scale industries (SSIs) had been one of the major contributors to the economic development of India since its independence. Since post-globalization in 1991, SSI in India has been facing intensely growing competitive environment mainly due to the liberalization of investments in the 1990s and the formation of the World Trade Organization (WTO) in 1995 forcing its member countries including India to reduce restrictions on imports [6, 7]. The last decade witnessed an increased number of studies conducted, exploring the innovation pattern in small firms, compared to the same for large firms [8].

A combined approach for understanding the effect of globalization, WTO regulations, along with the environmental, social and economic sustainability is necessary to successfully analyze the current scenario of Indian SMEs. The authors did not find any work dealing with the sustainability-oriented post-globalization scenario to find the potential barriers in the context of Indian SMEs. This has been the motivation of the present research.

The objective of this research paper is to collect empirical data for analyzing the functioning of SMEs in this era of globalization and modernization and identify potential barriers, which shall help device a workable methodology for promoting sustainable growth of SMEs in India. This survey was conducted through a set of interviews with SME entrepreneurs situated in the Indian state of Odisha and relevant literature review. Considering the importance of the engineering professionals for building a sustainable industrial ecology, engineering students of Centurion University, Odisha, were interviewed to analyze their views for finding the gaps in the prevalent education system to cope with the industrial needs.

The distribution of the paper is as follows: Sect. 2 elaborates the literature survey, Sect. 3 explains the research methodology adopted, and Sect. 4 presents the outcome of the research along with the identification of most hazardous barriers with respect to Indian SMEs.

2 Literature Survey

Recent years have witnessed increasing attention from the research fraternity, toward the difficulties faced by SMEs. The following subsections describe the current situation regarding SME's current status and sustainability issues faced with respect to Indian scenario reported by different researchers.

2.1 *SME in India*

In India, the involvement of the micro-, small and medium enterprise (MSME) sector toward socioeconomic development of the country is much prominent. This is evident from the fact that the SME sector provides employment to over 80 million people spread across 36 million MSME units in India. These units contribute about 8% to the Indian GDP producing more than 6000 products, share 45% of the total manufacturing output and also are responsible for 40% of the country's exports [9]. Still, many units are either sick or forced to close due to various reasons such as stiff market competition and inability to modernize operations leading to reduced market share. In India, manufacturing industries are struggling mainly in two areas, namely quality and process technology [10, 11]. But the majority of the manufacturing strategies reported in the published literature remains limited to either one of the two factors. Few researchers attempted to analyze the difficulties faced by small manufacturing units in India. A framework was even developed to identify and analyze critical success factors for SME by [12].

2.2 *Sustainability in SMEs*

Sustainability in the context of manufacturing organizations aims at increasing the proportion of SMEs around the world, promising profitability and resilience, positive social and environmental impacts [13]. Adapting to environmental-friendly operations is the inevitable solution. Implementation of this depends on various factors, and SMEs are often late to respond to the change [14]. As a result, there remains inadequate penetration of green technologies (GTs) in India [15]. Sustainable manufacturing has gained higher attention recently for its benefits directed at the triple bottom line (TBL) factors (social, environmental, financial). Characteristics of SMEs have been studied by [16]. Some authors have also proposed strategies for implementing GT in the Indian manufacturing sector [17–19]. Still, no reliable guideline exists to guarantee the successful implementation of sustainable manufacturing [20].

3 Research Methodology

The approaches followed in the present survey to identify the potential barriers for the successful implementation of sustainable practices in order to achieve all the three TBL factors in Indian SME are as follows:

(a) In-depth personal interviews were conducted at the factory site with 25 SME entrepreneurs in the state of Odisha, India. These included units functioning only in the machining and welding businesses. A questionnaire was framed, which focused mainly on the current market scenario, business competition, internal factors like skilled manpower availability, adoption to automated technology and external factors like government regulations, transport facility, raw material availability. The questions in this section were of the polar type having yes/no as available options. Few questions were included inviting suggestions for the improvement of the SME sector, which were descriptive in nature.

(b) A survey was conducted in order to gather the opinion of pre-final year engineering students of Centurion University, Bhubaneswar, Odisha, India. A 30-min-long lecture was delivered to introduce the students about the key aspects like sustainability and the present scenario of SME in India. A number of students successfully bonded with the concepts and started contributing naturally to the discussion as most were belonging to different industrial areas or their relatives working in factories in and around Bhubaneswar. This followed by a test which included questions mainly directed at finding out their viewpoint on sustainability, Indian SMEs and issues faced by them.

(c) A literature review was conducted by searching articles from online databases using keywords like SME, India, sustainability and barriers.

The collected data were analyzed in order to find the root cause of problems faced by SMEs. The complete process of identifying the barriers for practical and successful implementation of sustainability approach for Indian SME was brought down to four major issues. All the issues found their root to the year 1991, the post-globalization period. It had become clear by then that SMEs in India have started showing signs of sickness mostly ignored or treated with futility way back in the 1990s. The four major issues in focus are as follows:

(i) Failure to **modernization**
(ii) Failure to **competition** (with overseas players)
(iii) Failure to (market) **domination**
(iv) Failure to **innovation** (product and process).

Table 1 Major issues with Indian SMEs and their reason

S. No.	Major issues with Indian SMEs		Reason for failure		
			Why happened	What happened	How happened
1	Failure to	Modernization	Focus on short term return	Business goals not met	General economic slowdown
2		Competition	Poor on-time delivery	Lost brand value	Poor process and quality management
3		Market domination	Slow productivity	Lost market share	Lack of modern infrastructure
4		Innovation	Lack of proper planning	Focus only on growing market demand	Brain drain Rudimentary education system

These four pillars of failure were then expanded with the help of three "wh-words," namely why, what and how.

Table 1 depicts that during the pre-globalization period, the Indian manufacturing sector was undergoing a huge increase in demand in goods and services across different sectors like consumer goods, electronics, automobile to name a few. The domestic manufacturers remained busy feeding their products to the market and remained content with the gains from it. They paid less interest in coping with the latest technological developments, modern processes and methods of manufacturing, innovation at the plant level to reduce waste, processing cost, and allied expenditure, which continued even after globalization. The international players slowly gained control over the Indian market, which the native players were not ready for. This resulted in stiff competition among the two classes; domestic players struggled for their own survival and international players for increasing their market share. Slowly, the domestic players lost their market domination as they were not equipped with modern tools in manufacturing and supply chain management.

The information collected from the survey were divided into fifteen barriers and distributed among the four principal failure criteria. Figure 1 shows the four failure criteria, five broad gaps and fifteen identified barriers.

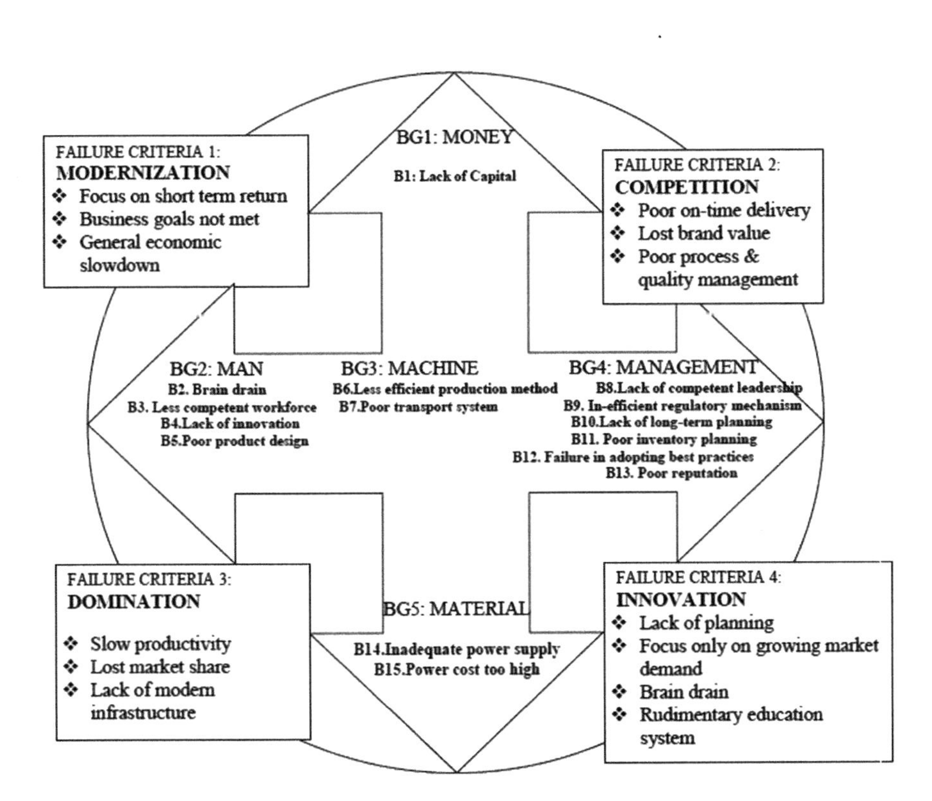

Fig. 1 Distribution of barriers among failure criteria and broad gaps

It is evident from Fig. 1 that the lack of modernization and innovation triggered competition among domestic and foreign players and resulted in market domination of the latter. Thus, in order to regain the lost market share, the Indian SMEs need to promote and implement the concepts of modernization and innovation at plant and supply chain level. Simultaneously, the policymakers should take steps to resolve the key issues like capital infusion, creation of infrastructure for strengthening transport system, regulating power failure and subsidizing the cost of power to a level that would be affordable by the SMEs.

Finally, the fifteen identified barriers were distributed among five broad gaps as shown in Fig. 2. Bringing out a sustainable methodology to sort out the barriers is expected to result in multifold progress in the Indian SMEs.

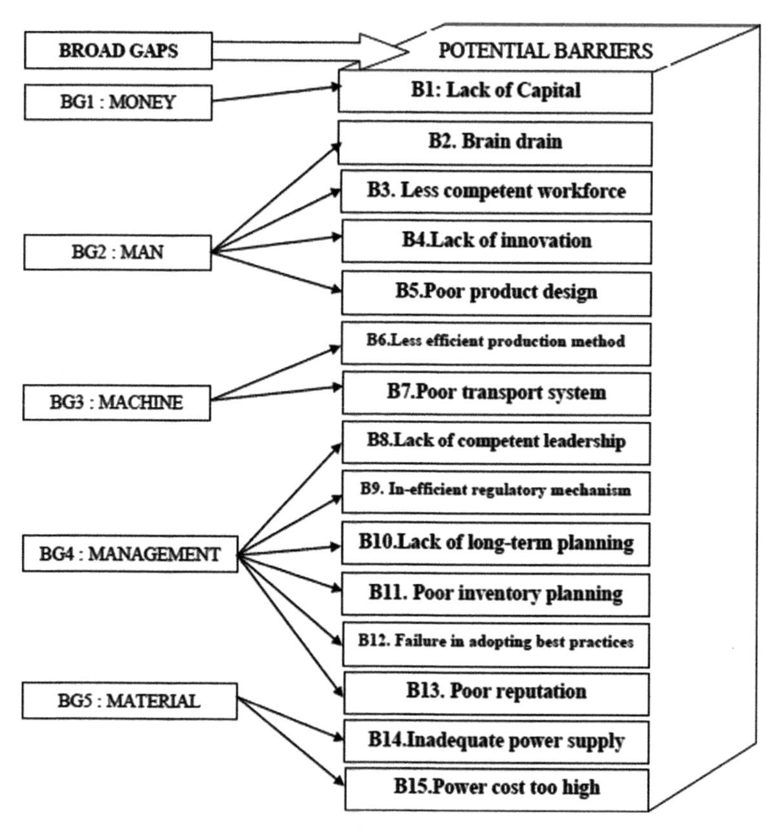

Fig. 2 Interrelation among the broad gaps and the potential barriers

4 Conclusion and Future Work

In the past decade, there has been increased interest from the research community pertaining to sustainability practices in the manufacturing sector. SMEs have also attracted increased attention from the research fraternity. Still, Indian SMEs continue to face very tough competition simultaneously from large industries and their multi-national counterparts. Although the Indian government has made several policy initiatives to safeguard the interests of SMEs, their successful implementation needs proper coordination and implementation. The recent "Make in India" campaign needs to be transmuted from mere slogan to successful practice. Negative effects of modernization and globalization need to be reduced and converted into opportunities keeping in mind the sustainability aspects and without sacrificing profitability.

The collected empirical data revealed lack of capital, intermittent power failure, poor Indian transport system, rudimentary education system as the prominent barriers, which need to be addressed with higher priority. SME entrepreneurs need to be educated through seminars and workshops. Training of workmen should be carried out at regular intervals, for developing sustainable practices as culture, at the plant level. Incentivizing and recognizing best contributors, in order to felicitate the innovative best practices, suggestions for improvement can help to trigger consistent growth in the performance of the organization. The existing laws governing the flow of capital to SMEs should be broadened to ensure adequate availability of resources. Establishment of more technology-oriented and internationally competitive SME needs to be encouraged.

Future studies can use sophisticated statistical techniques to gain further clearer insight into individual barriers. This will help to develop a robust strategical methodology to resolve the barriers and find out a practically implementable roadmap and devise a feasible plan of action. This would provide a new direction for attaining sustainability in Indian SMEs without sacrificing productivity, quality and also a significant reduction in the cost of production.

5 Compliance with Ethical Standards

The approval for the presented study was obtained from Head of the Department, Mechanical Engineering, Centurion University of Technology and Management, Odisha, India. All the questions involved in the study were clearly explained to the individuals participated in the survey. The data collected from the survey were anonymized based on the requests of the subjects involved.

References

1. Thurner TW, Roud V (2016) Greening strategies in Russia's manufacturing—from compliance to opportunity. J Cleaner Prod 112:2851–2860
2. Wang Z, Sarkis J (2017) Corporate social responsibility governance, outcomes, and financial performance. J Clean Prod 162:1607–1616
3. Moktadir MA et al (2018) Drivers to sustainable manufacturing practices and circular economy: a perspective of leather industries in Bangladesh. J Cleaner Prod 174:1366–1380
4. Bhanot N, Venkateswara Rao P, Deshmukh SG (2017) An integrated approach for analysing the enablers and barriers of sustainable manufacturing. J Cleaner Prod 142:4412–4439
5. Desai DA (2008) Cost of quality in small-and medium-sized enterprises: case of an Indian engineering company. Prod Plan Control 19(1):25–34
6. Subrahmanya MHB (2005) Small-scale industries in India in the globalisation era: performance and prospects. Int J Manage Enterp Dev 2(1):122–139
7. Lahiri R (2012) Problems and prospects of micro, small and medium enterprises (MSMEs) in India in the era of globalization. In: International conference on the interplay of economics, politics, and society for inclusive growth
8. Singh D, Khamba JS, Nanda T (2017) Influence of technological innovation on performance of small manufacturing companies. Int J Prod Performance Manage 66(7):838–856
9. Government of India, Ministry of Micro, Small and Medium enterprises, "MSME AT A GLANCE 2016", New Delhi (2016)
10. Thanki SJ, Thakkar J (2014) Status of lean manufacturing practices in Indian industries and government initiatives: a pilot study. J Manuf Technol Manage 25(5):655–675
11. Khatri Jk, Metri B (2016) SWOT-AHP approach for sustainable manufacturing strategy selection: a case of Indian SME. Glob Bus Rev 17(5):1211–1226
12. Thanki SJ, Thakkar J (2018) Interdependence analysis of lean-green implementation challenges: a case of Indian SMEs. J Manuf Technol Manag 29(2):295–328
13. Caldera HTS, Desha C, Dawes L (2019) Evaluating the enablers and barriers for successful implementation of sustainable business practice in 'lean' SMEs. J Clean Prod 218:575–590
14. Natarajan GS, Wyrick DA (2011) Framework for implementing sustainable practices in SMEs in the United States. In: Proceedings of the world congress on engineering, vol 1
15. Rehman AHA, Shrivastava RR, Shrivastava RL (2013) Validating green manufacturing (GM) framework for sustainable development in an Indian steel industry. Univers J Mech Eng 1(2):49–61
16. Calder HTS, Desha C, Dawes L (2018) Exploring the characteristics of sustainable business practice in small and medium-sized enterprises: experiences from the Australian manufacturing industry. J Cleaner Prod 177:338–349
17. Nulkar G (2014) SMEs and environmental performance–a framework for green business strategies. Procedia-Soc Behav Sci 133:130–140
18. Maruthi GD, Rashmi R (2015) Green manufacturing: it's tools and techniques that can be implemented in manufacturing sectors. Mater Today Proc 2(4–5):3350–3355
19. Gupta H, Barua MK (2018) A framework to overcome barriers to green innovation in SMEs using BWM and Fuzzy TOPSIS. Sci Total Env 633:122–139
20. Shankar KM, Kannan D, Udhaya Kumar P (2017) Analyzing sustainable manufacturing practices—a case study in Indian context. J Cleaner Prod 164:1332–1343

Design and Development of Technology-Enabled Biomass Stoves

Bogala Konda Reddy, Korla Mohanakrishna Chowdary and Chintalapudi Pallavi

Abstract The gradual depletion of fossil fuels leads to gaining prominence on biomass fuels in rural and urban sectors. Biomass is a carbon-neutral fuel, and utilization of wood waste ensures sustainability. In today's scenario, biomass is the most efficient cooking fuel for rural households to attain sustainable development by reducing the usage of firewood. Improved version of biomass cooking stoves can reduce greenhouse emissions through enhancing the combustion process by providing means of sufficient air for combustion. The present work concentrates mainly on the design and development of biomass stoves with automation by using technology-enabled tools for controlling and operating the stove for better performance and also discusses various features of the designed and developed models.

Keywords Design · Development · Biomass stoves · Automation · Technology-enabled tools

1 Introduction

Cookstoves are commonly used for cooking and heating of food in households using solid fuels such as firewood, crop residue, dung cake and coal. As per the 2011 census of India, the fuels used for cooking in rural and urban areas are shown in Fig. 1. The traditional method of cooking on three-stone stoves gives the following disadvantages: (i) more emissions which cause health problems to women and children [1, 2], (ii) more fuel consumption which causes deforestation and (iii) more heat losses which cause less performance of stoves. In order to control

B. Konda Reddy (✉) · K. Mohanakrishna Chowdary (✉) · C. Pallavi
Department of Mechanical Engineering, RGUKT-AP, IIIT RK Valley, Vempalli, Kadapa, Andhra Pradesh 516330, India
e-mail: bogala.kondareddy@rguktrkv.ac.in

K. Mohanakrishna Chowdary
e-mail: mohanakrishna.k@rguktrkv.ac.in

BBVL. Deepak et al. (eds.), *Innovative Product Design and Intelligent Manufacturing Systems*, Lecture Notes in Mechanical Engineering,
https://doi.org/10.1007/978-981-15-2696-1_16

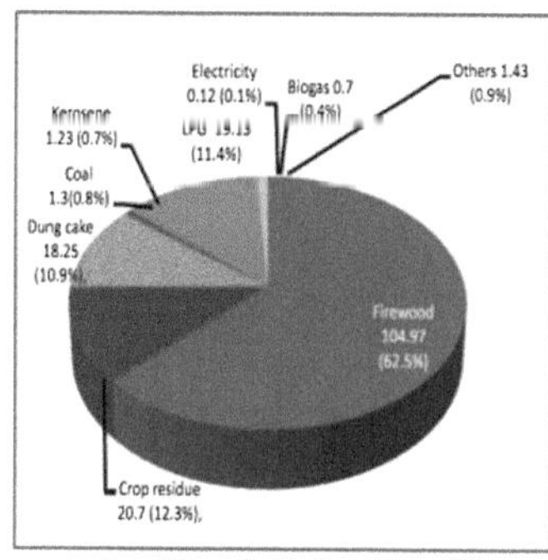

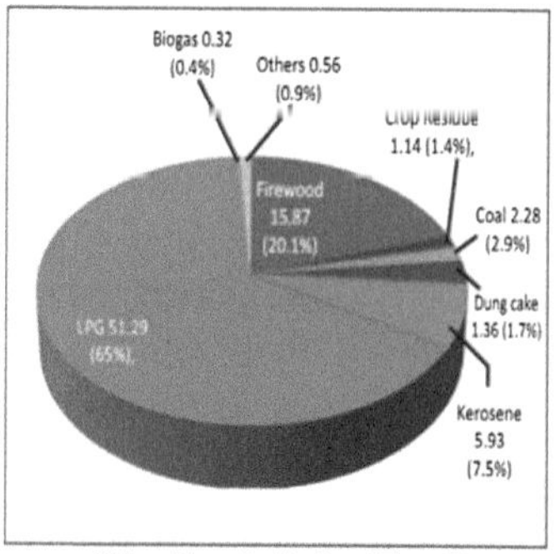

Rural Households in million (%) **Urban Households in million (%)**

Fig. 1 Indian household energy scenario in rural and urban areas as per 2011 census of India

these disadvantages of traditional stoves, improved cookstoves (ICSs) came into picture.

Biomass is an organic matter that is derived from plant residues (agriculture and forest). Biomass fuel may be in the form of pellets or briquettes as shown in Fig. 2, generally a combination of agriculture (sawdust, wheat straw, corn stalk, rice husk, coconut shells, cotton stalks, maize straw, etc.) and forest wastes (firewood, wood chips, wood chunks, etc.). With gradual depletion of fossil fuels, once again biomass fuels are gaining prominence in the rural and urban sectors. Biomass is a carbon-neutral fuel, and utilization of wood waste ensures sustainability [3] as shown in Fig. 3.

The Indian standard for portable solid biomass cookstoves, IS 13152 (Part 1), was first published in 1991 [4] and revised in 2013 [5]. This standard covers requirements of different designs and types of solid biomass portable cookstoves for domestic and community/commercial applications. Natural and forced draught models of cookstoves were developed by various manufacturers for domestic and community applications, and those models were approved by Ministry of New and Renewable Energy (MNRE) on the basis of their performance testing as per IS

Pellets Briquettes

Fig. 2 Pellets and briquettes

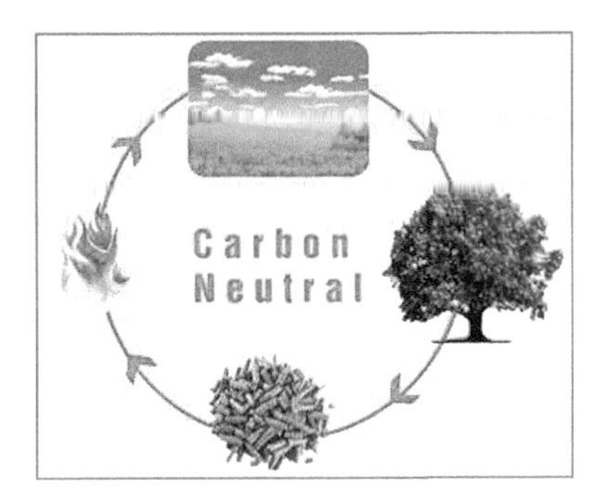

Fig. 3 Carbon neutral cycle with biomass fuel

13152 (Part 1) 2013 specification. The approved models supplied by respective manufacturers/developers with stipulated performance parameters are given in Ref. [6] and also the models developed by various manufacturers are given in Ref. [7].

Out of the given models in Ref. [6], the automated models of biomass stoves were not developed. In view of developing automated biomass stoves, the present study concentrated on the design and development of biomass stoves with technology-enabled tools for controlling and operating the stoves for better performance.

2 Design of Biomass Stoves

Based on the instructions of the user and signals from sensors, the automatic control panel will give instructions to operating devices to control the flow rate of air, fuel feeding and opening and closing of the aperture of the combustion chamber for obtaining better performance. The flow chart for the automation of biomass stove is shown in Fig. 4.

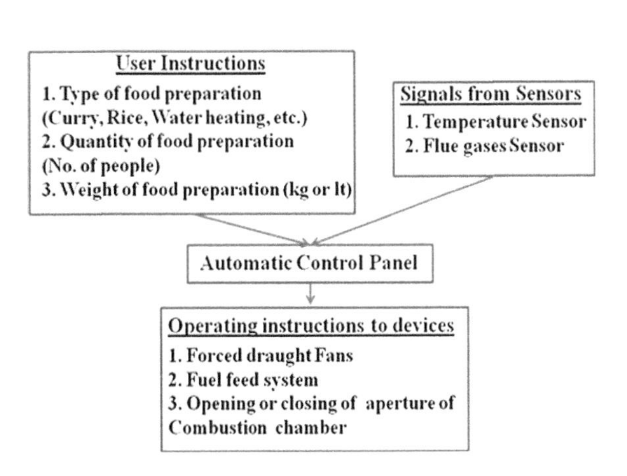

Fig. 4 Flow chart for automation of biomass stove

Two models of biomass stoves are designed in the modelling software as shown in Figs. 5 and 6. The proportionate dimensions of stove are taken as per specifications of IS 13152. The stove is designed in circular section and generally consists of combustion chamber, grate, ashtray at bottom, support at top for keeping cooking vessel and provision for circulation of air in combustion chamber. Apart from these basic parts of the stove, there are provisions designed in Model-I for automatic fuel feed system, iris mechanism for opening and closing of aperture of the combustion chamber and primary and secondary air circulation for combustion. Model-I is designed with some more provisions to provide sensors and to tap heat losses. In Model-II, more number of air circulation holes on combustion chamber is provided to increase the combustion performance.

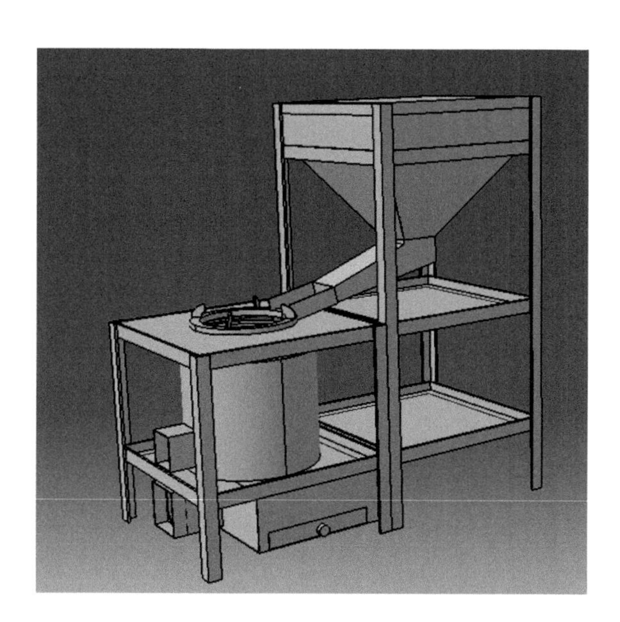

Fig. 5 Model-I of biomass stove (designed model)

Fig. 6 Model-II of biomass stove (designed model)

In the design, measures are taken care to integrate all these components of stove with an automatic control panel. Based on the requirement of air for combustion, respective number of holes is provided on combustion chamber and the respective flow rate of fans is chosen for primary and secondary air supply.

3 Development of Biomass Stoves

Model-I: As per the design, the model is constructed with respective materials as shown in Fig. 7. The components of Model-I are shown in Fig. 8. The functions of these technology-enabled tools of biomass stove are as follows: (i) forced draught fans to supply primary and secondary air for better combustion and regulate the supply of combustion gases for cooking, (ii) TEG modules with heat sink to convert thermal energy into electrical energy by consuming waste heat loss from the stove, and the electrical energy will be stored in the battery for running variable speed forced draught fans based on requirement, (iii) automatic feeding mechanism to supply the required quantity of biomass for combustion, (iv) iris mechanism for variable opening and closing of aperture of combustion chamber based on the requirement of fire during cooking and also used to shut the fire after cooking and (v) an automatic control system to control or to operate the devices based on the instructions of the user and also to improve the performance of the stove.

Fig. 7 Model-I (constructed) of biomass stove

Fig. 8 Components of Model-I (from top left—combustion chamber, grate, thermal insulation and temperature sensor, rack and pinion mechanism, battery, primary and secondary fans, iris mechanism for closing and opening of aperture of combustion chamber, control panel, flue gas sensor, thermoelectric generator (TEG) module with heat sink, hopper and ashtray)

Model-II: As per the design, this model is constructed with respective materials as shown in Fig. 9. The components of Model-II are same as Model-I except the combustion chamber to which more number of holes are provided for more circulation of air for better combustion.

Fig. 9 Model-II (constructed) of biomass stove

The programme was developed and embedded in the control panel board. Based on the instructions of the user for the type of food preparation, the control panel will give instructions to devices to supply a respective quantity of pellets into the combustion chamber and respective speed of air through primary and secondary fans. Hence, the control of combustion takes place effectively. Based on water boiling test, these two models have recorded the thermal efficiency of 35–40%.

4 Conclusions

In the present work, two models of automated biomass stoves were developed with technology-enabled tools to improve the performance of stoves. The following components were used as technology-enabled tools: (i) forced draught fans, (ii) automatic pellets feeding mechanism, (iii) iris mechanism and (iv) an automatic control system. The designed and developed models of stoves have been recorded 35–40% thermal efficiency.

Acknowledgements We wish to express our sincere thanks, deep sense of gratitude and indebtedness to the Design Innovation Centre (DIC) of IIIT RK Valley and JNTU Kakinada for the support in developing working models of biomass stoves.

References

1. Bates MN, Chandyo RK, Valentiner-Branth P, Pokhrel AK, Mathisen M, Basnet S (2013) Acute lower respiratory infection in childhood and household fuel use in Bhaktapur, Nepal. Environ Health Perspect 121:637–642
2. Smith KR, Samet JM, Romieu I, Bruce N (2000) Indoor air pollution in developing countries and acute lower respiratory infections in children. Thorax 55:518–532
3. http://biomassproject.blogspot.com/2014/01/carbon-positive-carbon-carbon-neutral.html
4. https://archive.org/details/gov.in.is.13152.1.1991/page/n5
5. Indian Standard for Portable Solid Bio-mass Cookstove (Chulha), IS 13152 (Part 1): 2013 (First Revision)
6. https://mnre.gov.in/file-manager/UserFiles/approved-models-of-portable-improved-biomass-cookstove-manufactures.pdf
7. Daniel MK (2011) Household cookstoves, environment, health, and climate change: a new look at an old problem. The Environment Department (Climate Change) The World Bank

Damage Assessment of Beam Structure Using Dynamic Parameters

Basna Bidisha Bal, Sarada P. Parida and Pankaj C. Jena

Abstract The occurrences of damages in the engineering structures are very common in practicality. In the current eve of time of emergence of composite materials, they are used widely. Hence, it is very important to study the reliability and in crack mechanical properties. So, damage diagnosis becomes crucial and necessary in dynamic members. This work depicts the dynamic behavior of epoxy–glass fibers composite beams taking the effect of bending and torsion coupling to consideration in 1-D model. A Timoshenko beam is assumed for the purpose that includes shear force and rotatory inertia. For the study, cracked beams are modeled as two beams attached by a massless spring with one end fixed boundary condition by FEA. The bending natural frequency is calculated and compared with the natural frequency of the uncracked beam and expressed by its non-dimensional form as relative natural frequency. The effect of depth of crack and location on the dynamic characteristics of the beams was studied and presented.

Keywords Composite · Beam · Crack · Assessment

1 Introduction

The presence of damage as fracture, cracks, and surface unevenness in a beam structure for extended time intensifies the possibility of the system failure that causes loss of properties and life. The classic presentation of structural beam assembly or components of machinery is hindered by the presence of damage. Presently, various mechanical structures encounter the failure because of material

B. B. Bal
Department of Mechanical Engineering, Gandhi Institute of Technology and Management, Bhubaneswar, Odisha, India

S. P. Parida · P. C. Jena (✉)
Department of Production Engineering, Veer Surendra Sai University of Technology, Burla, Odisha 768018, India

BBVL. Deepak et al. (eds.), *Innovative Product Design and Intelligent Manufacturing Systems*, Lecture Notes in Mechanical Engineering,
https://doi.org/10.1007/978-981-15-2696-1_17

fatigue that causes the development of various forms of damage such as fracture, cracks, or any unevenness. So that to detect and localize is the major subject of the study for different research workers worldwide. The stiffness of the structural element is altered with the presence of a crack. Due to the vibration of the whole structure, the dynamic characteristics like natural frequencies and mode shapes are affected.

Douka et al. [1] methodologically presented the effect of crack location and its depth in the dual cracked beam. Natural frequency and anti-resonance are used as diagnosing parameters of crack. Huh et al. [2] used vibration energy of cracked beams based on the accelerations to identify the local damage. To validate the study, an open cracked uniform beam is analyzed by a numerical and experiment method. Nahvi and Jabbari [3] used natural frequencies and mode shapes of the beam structure as input parameters for the analytical and finite element method to identify the crack in the cantilever beam. Darpe et al. [4] analyzed the response of centrally cracked rotor to axial forces using electrodynamic exciter both in rotating and idle conditions. Tada et al. [5] formulated compliance matrices for the damaged structure to identify the crack location and depth. Dado [6] mathematically observed and presented the crack position and severity for various end conditioned beams assuming them as Euler–Bernoulli beam. He made a conclusion that though analytical results are useful for determination of crack properties, however, the assumptions made do not have practicality. Gounaris and Papadopoulos [7] related modal response of an open crack beam with crack parameter and presented them by functional relationships which were validated by comparing the Eigen frequencies of damaged and undamaged beams. Ravi and Liew [8] made a modal study of aluminum sheet with microcracks due to compression and the deformation is traced by the acoustic emission technique. Wang et al. [9] used Castigliano's theorem and CLT to determine the effect of surface edge crack for fiber-reinforced composite beam and presented as a frequency spectrum with aspect ratio. Sekhar [10] summarized the effect of multiple cracks on the composite structures. Liu et al. [11] used a crack detection method to determine the crack location and depth for fiber-reinforced composite material. Yan et al. [12] studied the dynamic response of open edge cracked FGM beam with a transverse moving load and presented as a function of natural frequency. Jena et al. [13–17] have studied intact and cracked beam dynamics by taking different materials of beam structures. Zhu et al. [18] introduced one new damage index using wavelet coefficient for assessment of natural frequency and mode shape of edge open crack FGM beams. Kim et al. [19] studied vibration behavior of the cracked composite beam using Jacobi–Ritz method and FSDT; the results obtained by numerical and FEA study are compared and presented.

Engineering structures are often exposed to different loading environments depending upon their uses. Hence, the development of damage is a usual process. Sometimes, the loading and environmental conditions affect the performance of structures and life span. Analytical models can be used to investigate the effect of the damage on the vibration characteristics of the engineering structures. Generally, to incorporate a mathematical equation, a stiffness spring model is used to model the damaged structure with the boundary conditions to express the vibration characteristics. In this study, polyester–glass fiber-reinforced composite beam with cracks is taken and the modal study is carried and compared with the uncracked beam.

2 Theoretical Formulation

In general, classical beam theory known as *Bernoulli–Euler* theory is used in most of the cases. The deflection of the beam according to this can be written as

$$\frac{\mathrm{d}^2 W}{\mathrm{d}x^2} = \frac{M}{EI_0} \tag{1}$$

The strain energy per unit length is given by

$$U = \frac{1}{2} EI_0 \left(\frac{\mathrm{d}^2 W}{\mathrm{d}x^2} \right)^2 \tag{2}$$

$$\Delta U = C_q k C_a \left(\frac{\partial U(x,t)}{\partial x} - V(x,t) \right)_{x=e} \tag{3}$$

Considering the cracked beam as two discontinued beams attached by a spring (shown in Fig. 1), the change in vertical displacement is given in Eq. (3). The generalized frequency equation for the cracked beam is given by

$$f^2 = \frac{\psi^2 - \frac{w}{l}\frac{s^2}{r^2} q(x)(U'(e) - V(e))^2 + r^2 \varphi(\beta)(V'(e))^2}{r^2 \int_0^1 U(\mu) + s^2 V(\mu)^2 \mathrm{d}\mu} \tag{4}$$

where $\boldsymbol{C_q}$ and $\boldsymbol{C_a}$ are flexibility constants of the spring.

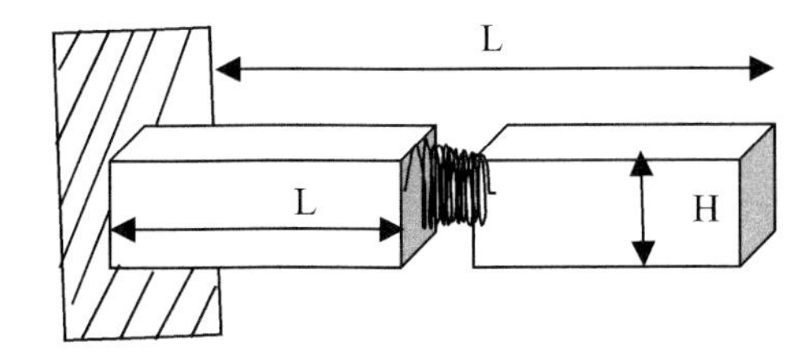

Fig. 1 A theoretical cantilever cracked beam spring model

Fig. 2 Discretized FEA model RCP 0.2 and RCD of 0.1

3 FEA Analysis

Finite element analysis (FEA) is a commonly used a numerical method to study the vibration of structures. ANSYS is an FEA software tool. Here, we used this for free vibration analysis of a composite beam. A composite beam with 600 mm × 15 mm × 15 mm dimension with fifteen layers of reinforcement glass fibers layer with polyester base is taken for determining the natural frequencies and mode shapes with and without crack taking the end condition of the beam as a cantilever. The crack position and depth were varied. For each experiment, the crack location was changed ranging from 120 to 480 mm with 120 mm steps along the length from the fixed end. The depth of crack was also varied for a specific crack position from 1.5 to 7.5 mm. The first three bending natural frequencies obtained for cracked beams are compared with the natural frequency of uncracked beam model (shown in Figs. 3, 4, and 5) and expressed in a non-dimensional form as relative natural frequency obtained

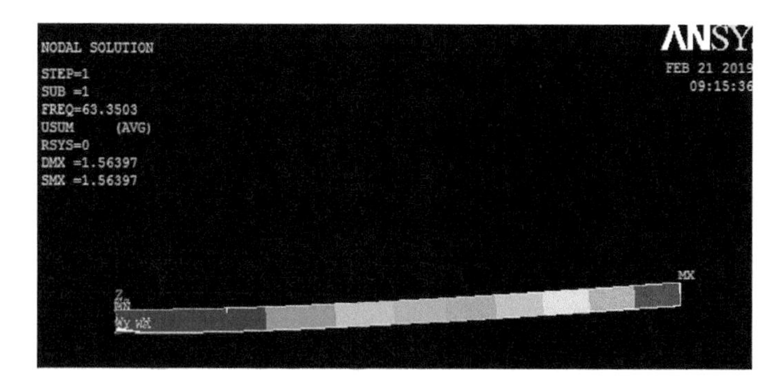

Fig. 3 First mode vibration of the beam

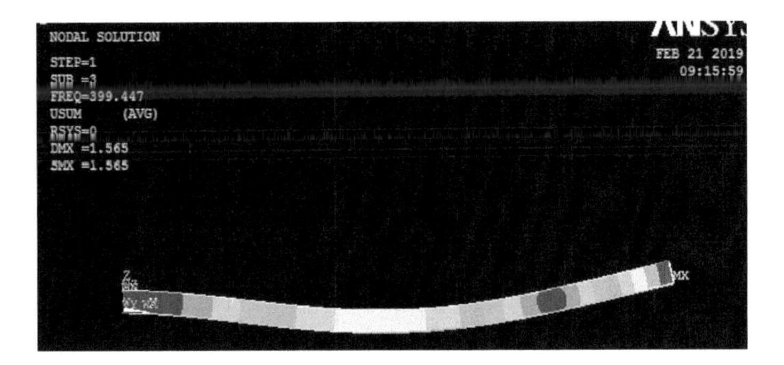

Fig. 4 Second mode vibration of the beam

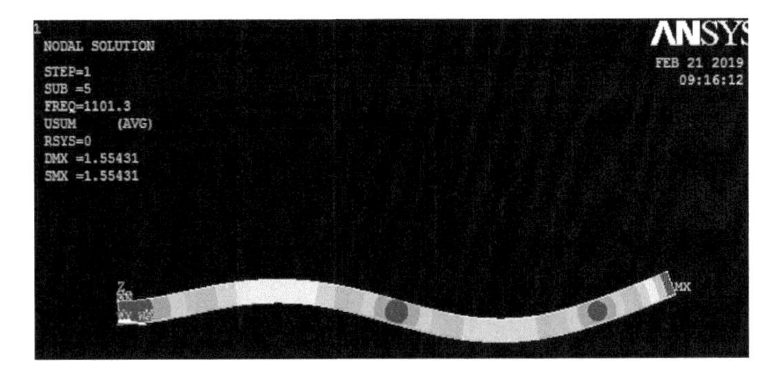

Fig. 5 Third mode vibration of a cracked beam

$$\text{Relative frequency ratio (RFR)} = \frac{\text{frequency of cracked beam}}{\text{frequency of uncracked beam}}$$

The results obtained are analyzed and the effects of crack on the dynamic behavior of the beam are expressed subsequently. The changes in relative natural frequency with relative crack depth and location are plotted and discussed (Fig. 2).

4 Result and Discussions

In this study, the cracks are given at the positions of 120, 240, 360, and 480 mm from the fixed end. At each position, the depth of the crack is also varied. The depths of the cracks taken are 1.5, 3, 6, and 7.5 mm. When the depth of crack is increased from 10 to 50% with 10% increment, it is found that there is a significant drop in natural frequency from 1.3 to 15% as depicted in Fig. 5. The same natures of variation of natural frequencies are observed when cracks of different depths are taken at other locations. Figures 6, 7, 8, and 9 present the variation of first three

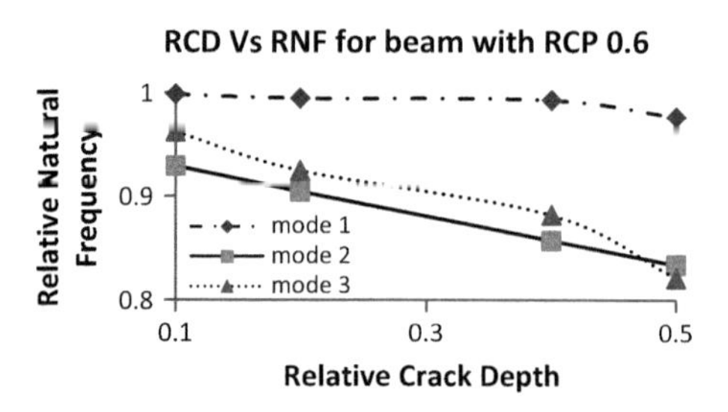

Fig. 6 Variation of RNF with RCD for cantilever beam with cracks at 120 mm from the fixed end

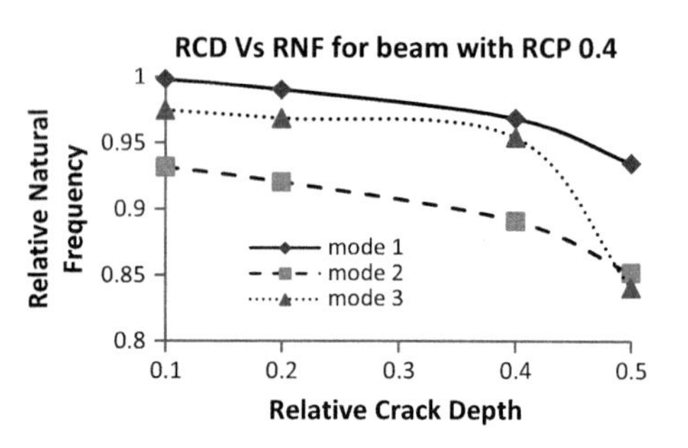

Fig. 7 Variation of RNF with RCD for cantilever beam with cracks at 240 mm from the fixed end

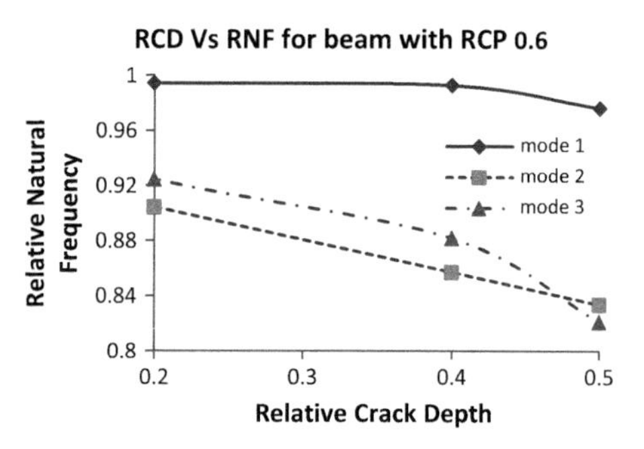

Fig. 8 Variation of RNF with RCD for cantilever beam with cracks at 360 mm from the fixed end

relative modal natural frequencies of vibrations of the polyester–glass composite beam at the crack location of 240 mm, 360 mm, and 480 mm from the fixed end, respectively. From all the cases, it can be observed that the natural frequencies start to decrease when the depth of crack increases as depicted through the graph in Fig. 9.

The variance of the relative natural frequency with the crack position for fixed crack depth is studied and presented through Figs. 10, 11, 12, and 13. For the relative crack of 10% crack depth, the variation of RNF is of negligible importance as depicted in Fig. 10. It is assumed that the crack of 10% depth has the least effect at any position and does have significant variation.

The effect of relative crack position for other crack depths has a remarkable signature on relative natural frequency. The second mode of vibration and the third mode of vibration have higher variation than the first mode of vibration as depicted

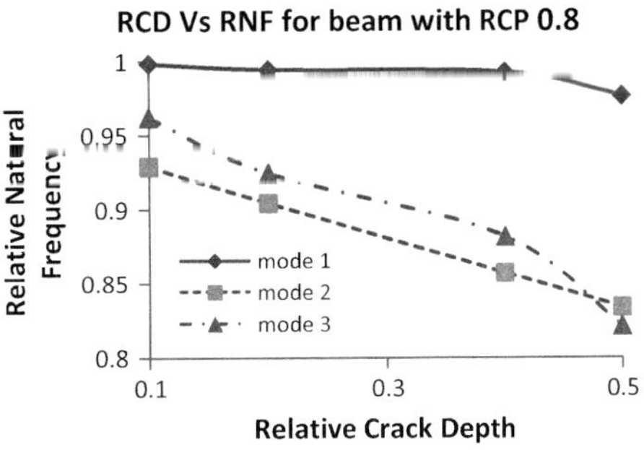

Fig. 9 Variation of RNF with RCD for cantilever beam with cracks at 480 mm from the fixed end

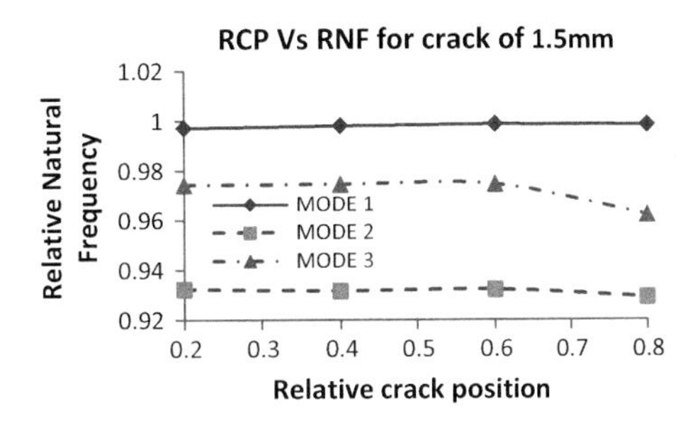

Fig. 10 Variation of RNF with RCP for cantilever beam with 10% crack depth

through Figs. 10, 11, 12 and 13. It is found out that by increasing the RCP, the RNF increases and then decreases subsequently.

The slope of the curves for cracked beams with 40 and 50% crack depth is more abrupt than the cracked beams of 10 and 20% crack depth. Another finding of observation is of consequent increase and decrease of RNF with the relative crack position for the beams of all kinds.

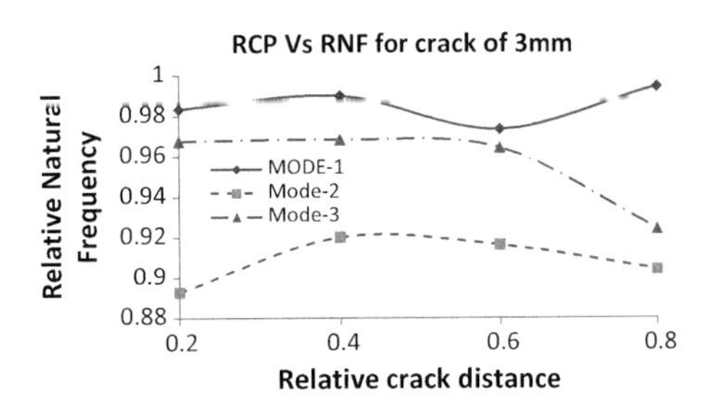

Fig. 11 Variation of RNF with RCP for cantilever beam with 20% crack depth

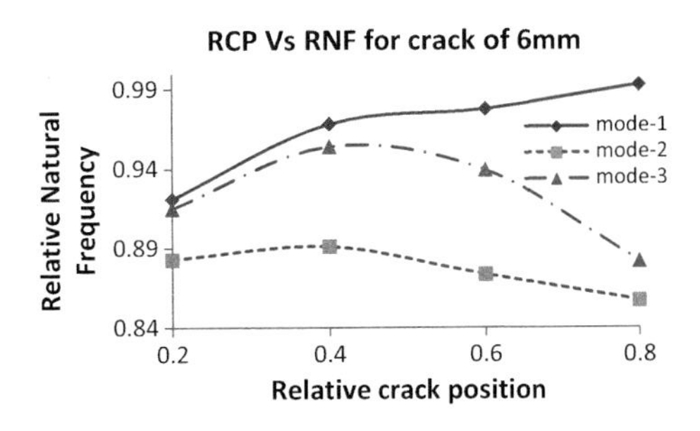

Fig. 12 Variation of RNF with RCP for cantilever beam with 40% crack depth

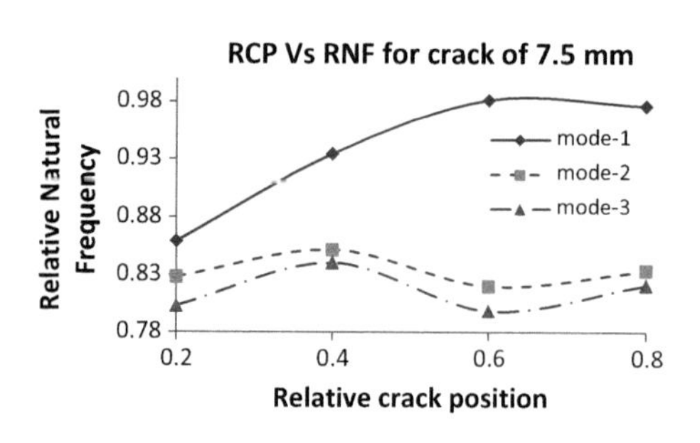

Fig. 13 Variation of RNF with RCP for cantilever beam with 50% crack depth

5 Conclusion

The natural frequencies start to decrease when the depth of crack increases. The slope of the curves for cracked beams with 40 and 50% crack depth than the cracked beams of 10 and 20% crack depth is more abrupt signifying the prominent loss of stiffness to weight ratio. The relative natural frequency increases and then decreases subsequently with respect to relative crack distance.

References

1. Douka E, Bamnios, Trochidis G (2004) A method for determining the location and depth of crack in a double cracked beam. Appl Acoust 65:997–1008
2. Huh YC, Chung TY, Moon SJ, Kil HG, Kim JK (2007) Damage detection in beams using vibratory power estimated from the measured accelerations. J Sound Vib 330(15):3645–3665
3. Nahvi H, Jabbari M (2005) Crack detection in beams using experimental modal data and finite element model. Int J Mech Sci 47:1477–1497
4. Darpe AK, Gupta K, Chawla A (2003) Experimental investigations of the response of a cracked rotor to periodic axial excitation. J Sound Vib 260:265–286
5. Tada H, Paris PC, Irwin GR (1973) The stress analysis of cracks handbook. Del Research Corporation, Hellertown, Pennsylvania
6. Dado MH (1997) A comprehensive crack identification algorithm for beams under different end conditions. Appl Acoust 51:381–398
7. Gounaris GD, Papadopoulos CA (1997) Analytical and experimental crack identification of beam structures in the air or in fluid. Comput Struct 65(5):633–639
8. Ravi D, Liew KM (2000) A study of the effect of micro-crack on the vibration mode shape. Eng Struct 22:1097–1102
9. Wang K, Inman DJ, Farrar CR (2005) Modeling and analysis of a cracked composite cantilever beam vibrating in coupled bending and torsion. J Sound Vib 284(1–2):23–49
10. Sekhar AS (2008) Multiple cracks effects and identification. Mech Syst Signal Process 22 (4):845–878
11. Liu Y, Li Z, Zhang W (2010) Crack detection of fiber reinforced composite beams based on continuous wavelet transform. J Nondestr Test Eval 25
12. Yan TY, Kitinporchai S, Yang J, He SQ (2011) Dynamic behavior of edge-cracked shear deformable functionally graded beams on an elastic foundation under a moving load. Compos Struct 93(11):2992–3001

13. Jena PC, Parhi DR, Pohit G (2014) Theoretical, numerical (FEM) and experimental analysis of composite cracked beams of different boundary conditions using vibration mode shape curvatures. Int J Eng Technol 6(2):309–318
14. Jena PC, Parhi DR, Pohit G, Samal BP (2015) Crack assessment by FEM of AMMC beam produced by modified stir casting method. J Mater Today: Proc, 2267–2276
15. Jena PC, Parhi DR, Pohit G (2016) Dynamic study of composite cracked beam by changing the angle of bi-directional fibers. Iran J Sci Technol Trans A 40(1):27–37
16. Jena PC, Pohit G, Parhi DR (2017) Fault measurement in composite structure by fuzzy-neuro hybrid technique from the natural frequency and fiber orientation. J Vib Eng Technol 5 (2):123–138
17. Jena PC (2018) Free vibration analysis of short bamboo fiber based polymer composite beam structure. J Mater Today: Proc 5(2.1):5870–5875
18. Zhu LF, Ke LL, Zhu XQ, Xiang Y, Wang YS (2019) Crack identification of functionally graded beams using continuous wavelet transform. Compos Struct 210(15):473–485
19. Kim K, Choe K, Kim S, Wang Q (2019) A modeling method for vibration analysis of cracked laminated composite beam of uniform rectangular cross-section with arbitrary boundary condition. Compos Struct 208:127–140

Bending Stress Analysis of PM Composite Beam

Ch. Siva Ramakrishna, K. V. Subbarao, Saineelkamal Arji and B. Harisankar

Abstract In the present work, the stress behavior of laminated composite plate under compressive loading using a four-node element with six degrees of freedom at each node and translations in the *x* and *y* directions is done. In the present study, the modeling is done in Abaqus. Investigations were carried on plates starting with three layers of the top location of 0° angle-ply laminated composite plates at clamped boundary condition. Similarly, with three layers of top location, 0°, 30° and −45° angle ply are laminated. By changing the location of ply orientations the bending stress may be improved. The effect of changing the ply orientation is to increase or decrease the stresses. The composite plate has been analyzed for various orientations and their effects on stresses so as to find the optimized conditions.

Keywords Laminated composite plate · Numerical method · Ply orientation · Abaqus

1 Introduction

Composite materials are made from two or more constituent materials which are having different physical and chemical properties when compared to individual components. The composite material can be lighter, stronger and also can be lightweight. The reinforcement and matrix give us composites. The finite element

Ch. S. Ramakrishna (✉) · K. V. Subbarao · S. Arji · B. Harisankar
Department of Mechanical Engineering, Vignan's Institute of Information Technology, Visakhapatnam 530049, India

BBVL. Deepak et al. (eds.), *Innovative Product Design and Intelligent Manufacturing Systems*, Lecture Notes in Mechanical Engineering,
https://doi.org/10.1007/978-981-15-2696-1_18

method (FEM) is a numerical method for solving problems of engineering and mathematical physics. It is also referred to as finite element analysis (FEA). The analytical solution of this problem generally requires the solution to boundary value problems for partial differential equations. In present work, calculate the bending stress values by changing the location of the ply orientations, i.e., 0° at the top, 30° at top and similarly −45° angle ply at top of the laminated composite

2 Literature Review

Sino and Baranger [1] worked on the dynamic instability of an internally damped rotating composite shaft. A homogenized finite element beam model, which takes into account internal damping, is introduced and then used to evaluate natural frequencies and instability thresholds. The results are compared to obtained by using equivalent modulus beam theory (EMBT), modified EMBT and layerwise beam theory (LBT). Topal [2] presented a multi-objective optimization of laminated cylindrical shells to maximize a weighted sum of the frequency and buckling load under external load. The layer fiber orientation is used as the design variable and the multi-objective optimization is formulated as the weighted combinations of the frequency and buckling under external load. Topal and Uzman [3] proposed a multi-objective optimization of symmetrically angle-ply square laminated plates subjected to biaxial compressive and uniform thermal loads. The design objective is the maximization of the buckling load for a weighted sum of the biaxial compressive and thermal loads. The first-order shear deformation theory (FSDT) is used in the mathematical formulation of buckling analysis of laminated plates. Roos and Bakis [4] analyzed the flexible matrix composites which consist of low modulus elastomers such as polyurethanes which are reinforced with high-stiffness continuous fibers such as carbon. Kayikci and Sonmez [5] studied and optimized the natural frequency response of symmetrically laminated composite plates. The optimal frequency response of laminates subjected to static loads was also investigated. Abadi and Daneshmehr [6] developed the buckling analysis of composite laminated beams based on modified coupled stress theory. By applying the principle of minimum potential energy and considering two different beam theories and analysis is conducted with Abaqus for investigations

3 Methodology

The engineering elastic constants of the unidirectional graphite/epoxy lamina are considered to solve the problem with the formulae below. Compliance matrix elements are,

$$S_{11} = \frac{1}{E_1}$$

$$S_{12} = \frac{-\vartheta 12}{E_1}$$

$$S_{22} = \frac{1}{E_2}$$

$$S_{66} = \frac{1}{G_{12}}$$

and the reciprocal relationship is

$$\frac{\vartheta_{12}}{E_1} = \frac{\vartheta_{21}}{E_2}$$

$$\vartheta_{21} = \frac{\vartheta_{12}}{E_1} \times E_2$$

The reduced stiffness matrix $[Q]$ elements are

$$Q_{11} = \frac{E_1}{1 - \vartheta_{21}\vartheta_{12}}$$

$$Q_{12} = \frac{\vartheta_{12}E_2}{1 - \vartheta_{21}\vartheta_{12}}$$

$$Q_{22} = \frac{E_2}{1 - \vartheta_{21}\vartheta_{12}}$$

The compliance matrix for an orthotropic plane stress problem can be written as,

$$\begin{bmatrix} \varepsilon_1 \\ \varepsilon_2 \\ \gamma_{12} \end{bmatrix} = \begin{pmatrix} S_{11} & S_{12} & 0 \\ S_{12} & S_{22} & 0 \\ 0 & 0 & S_{66} \end{pmatrix} \begin{bmatrix} \sigma_1 \\ \sigma_2 \\ \tau_{12} \end{bmatrix}$$

The reduced stiffness matrix for 0° graphite/epoxy ply is

$$[Q] = \begin{pmatrix} Q_{11} & Q_{12} & 0 \\ Q_{12} & Q_{22} & 0 \\ 0 & 0 & Q_{66} \end{pmatrix}$$

The transformed reduced stiffness matrix $\left[\overline{Q}\right]$ for each of the three plies is

$$\left[\overline{Q}\right]=\begin{pmatrix}\overline{Q}_{11} & \overline{Q}_{12} & \overline{Q}16\\ \overline{Q}_{12} & \overline{Q}_{22} & \overline{Q}26\\ \overline{Q}16 & \overline{Q}26 & \overline{Q}_{66}\end{pmatrix}$$

The transformed reduced stiffness matrix $\left[\overline{Q}\right]$ for 0°, 30° and −45° ply is developed. The total thickness of the laminate is $h = 0.005 \times 3 = 0.015$ m. The midplane is 0.0075 m from the top and the bottom of the laminate. The locations of play surfaces are placed evenly from the surface of midplane such that each ply is placed is oriented as shown in Fig. 1.

Coupling stiffness matrix $[B]$

$$B_{ij}=\frac{1}{2}\sum_{k=1}^{3}\left[\overline{Q}_{ij}\right]_k\left(h_k^2-h_{k-1}^2\right);\ i=1,2,6;\ j=1,2,6;$$

The bending stiffness matrix $[D]$

$$D_{ij}=\frac{1}{3}\sum_{k=1}^{3}\left[\overline{Q}_{ij}\right]_k\left(h_k^3-h_{k-1}^3\right);i=1,2,6;j=1,2,6;$$

$$\begin{Bmatrix}k_x\\ k_y\\ k_{xy}\end{Bmatrix}=[C']\begin{Bmatrix}N_x\\ N_y\\ N_{xy}\end{Bmatrix}$$

By using the following relations to find strains and curvatures of each ply with different orientations,

The laminate strains can be written as,

$$\begin{Bmatrix}\varepsilon_x\\ \varepsilon_y\\ \gamma_{xy}\end{Bmatrix}=\begin{Bmatrix}\varepsilon_x^0\\ \varepsilon_y^0\\ \gamma_{xy}^0\end{Bmatrix}+Z\begin{Bmatrix}K_X\\ K_Y\\ K_{XY}\end{Bmatrix}$$

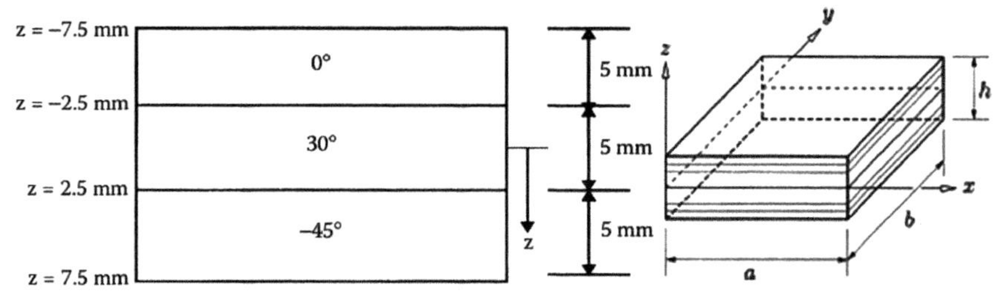

Fig. 1 Schematic diagrams of plies

Find the strains where $Z = -0.0025$ mm for 30°, 60° and −45° angle ply is calculated. Similarly, using stress–strain relation for top location of 30° ply is

$$\begin{bmatrix} \sigma_x \\ \sigma_y \\ \tau_{xy} \end{bmatrix} = \begin{pmatrix} \bar{Q}_{11} & \bar{Q}_{12} & \bar{Q}_{16} \\ \bar{Q}_{12} & \bar{Q}_{22} & \bar{Q}_{26} \\ \bar{Q}_{16} & \bar{Q}_{26} & \bar{Q}_{66} \end{pmatrix} \begin{Bmatrix} \varepsilon_x \\ \varepsilon_y \\ \gamma_{xy} \end{Bmatrix}$$

Following the above procedure and placing the ply's at the orientation of −45°/0°/30° calculated bending stress and then calculate

$$A^* = A^{-1}$$
$$D^* = [D] - \left[[B][A]^{-1}\right][B]$$
$$[A]^1 = [A]^* - [B^*][D^*]^{-1}[C^*]$$
$$[C]^1 = -[D^*]^{-1}[C^*]$$

$$\begin{bmatrix} \varepsilon_X^0 \\ \varepsilon_Y^0 \\ \gamma_{XY}^0 \end{bmatrix} = A^1 \begin{bmatrix} N_X \\ N_Y \\ N_{XY} \end{bmatrix}, \begin{bmatrix} K_X \\ K_Y \\ K_{XY} \end{bmatrix} = C^1 \begin{bmatrix} N_X \\ N_Y \\ N_{XY} \end{bmatrix}$$

$$\begin{bmatrix} \varepsilon_X \\ \varepsilon_Y \\ \gamma_{XY} \end{bmatrix} = \begin{bmatrix} \varepsilon_X^0 \\ \varepsilon_Y^0 \\ \gamma_{XY}^0 \end{bmatrix} + Z \begin{bmatrix} K_X \\ K_Y \\ K_{XY} \end{bmatrix} (\because Z = -0.0025)$$

Similarly, using stress–strain relation for top location of 0° ply

$$\sigma = [Q]_0 \begin{bmatrix} \varepsilon_X \\ \varepsilon_Y \\ \gamma_{XY} \end{bmatrix}$$

Following the above procedure and placing the plies at the orientation of 30°/45°/0°, calculated the bending stress is as follows

$$[A^1] = [A^*] - [B^*][D^*]^{-1}[C^*]$$

$$\begin{bmatrix} \varepsilon_X^0 \\ \varepsilon_Y^0 \\ \gamma_{XY}^0 \end{bmatrix} = [A^1] \begin{bmatrix} 1000 \\ 1000 \\ 0 \end{bmatrix}, \begin{bmatrix} K_X \\ K_Y \\ K_{XY} \end{bmatrix} = C^1 \begin{bmatrix} N_X \\ N_Y \\ N_{XY} \end{bmatrix}$$

$$\begin{bmatrix} \varepsilon_X \\ \varepsilon_Y \\ \gamma_{XY} \end{bmatrix} = \begin{bmatrix} \varepsilon_X^0 \\ \varepsilon_Y^0 \\ \gamma_{XY}^0 \end{bmatrix} + (-0.0025) = \begin{bmatrix} \varepsilon_X^0 \\ \varepsilon_Y^0 \\ \gamma_{XY}^0 \end{bmatrix} + \begin{bmatrix} 2.105 \times 10^{-4} \\ -4.885 \times 10^{-6} \\ -4.0705 \times 10^{-4} \end{bmatrix}$$

$$\begin{bmatrix} \sigma_X \\ \sigma_Y \\ \gamma_{XY} \end{bmatrix} = \begin{bmatrix} \varepsilon_X \\ \varepsilon_Y \\ \gamma_{XY} \end{bmatrix} [Q]_{-45}$$

Similarly, using stress–strain relation for top location of −45° ply is

$$\begin{bmatrix} \sigma_X \\ \sigma_Y \\ \gamma_{XY} \end{bmatrix} = \begin{bmatrix} 6.27 \times 10^7 \\ 1.11 \times 10^6 \\ -8.02 \times 10^7 \end{bmatrix}$$

4 Results and Discussions

The Numerical results are compared with simulated results to estimate the percentage of error is less being obtained hence the simulation with Abaqus is validated. The defined problem with different locations was solved using numerical method; the results are tabulated and the same was analyzed using FEM-based Abaqus Software and the results are shown in the figure below.

4.1 *Stress Analysis of Composite Ply at the Orientation of [0°/30°/−45°]*

The ply at [0°/30°/−45°] orientation is analyzed in *x*-direction, *y*-direction and in *xy*-direction (shear), results obtained are shown in Table 1. The results obtained through Abaqus on solving are shown in Fig. 2.

4.2 *Stress Analysis of Composite Ply at the Orientation of [−45°/0°/30°]*

The ply at −45°/0°/30° orientation is analyzed in *x*-direction, *y*-direction and in *xy*-direction (shear), results obtained are shown in Table 2. The results obtained through Abaqus on solving are shown in Fig. 3.

Table 1 Stresses in the orientation of [0°/30°/−45°]

x-direction	σ_x	6.93×10^7 Pa
y-direction	σ_y	7.495×10^5 Pa
xy-direction	τ_{xy}	3.676×10^7 Pa

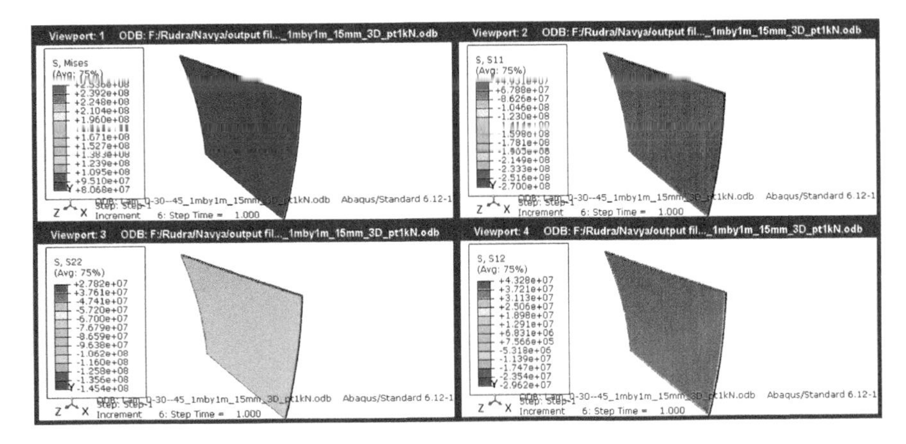

Fig. 2 Stress on [0°/30°/−45°] composite

Table 2 Stresses in the orientation of [−45°/0°/30°]

x-direction	σ_x	-1.85×10^8 Pa
y-direction	σ_y	8.56×10^5 Pa
xy-direction	τ_{xy}	-3.05×10^7 Pa

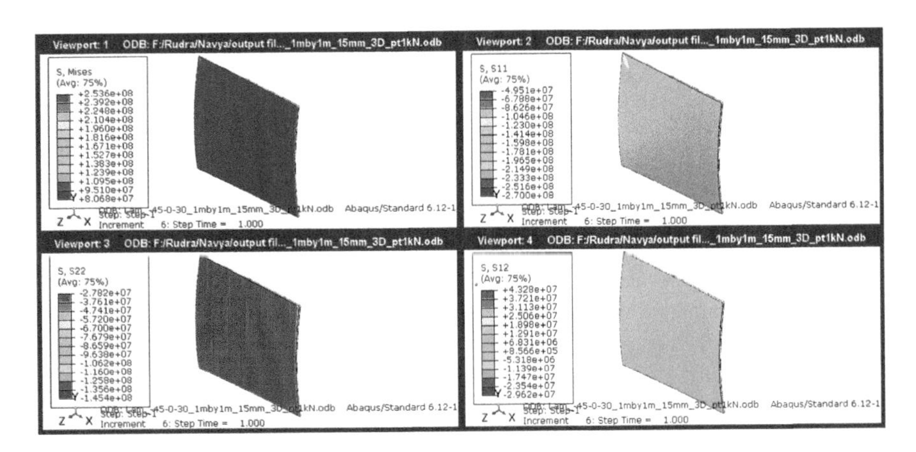

Fig. 3 Stress on [−45°/0°/30°] composite

4.3 Stress Analysis of Composite Ply at the Orientation of [30°/−45°/0°]

The ply at [0°/30°/−45°] orientation is analyzed in x-direction, y-direction and in xy-direction (shear), results obtained are shown in Table 3.

The results obtained through Abaqus on solving are shown in Fig. 4

Table 3 Stresses in the orientation of [30°/−45°/0°]

x-direction	σ_x	6.27×10^7 Pa
y-direction	σ_y	1.11×10^6 Pa
xy-direction	τ_{xy}	-8.02×10^7 Pa

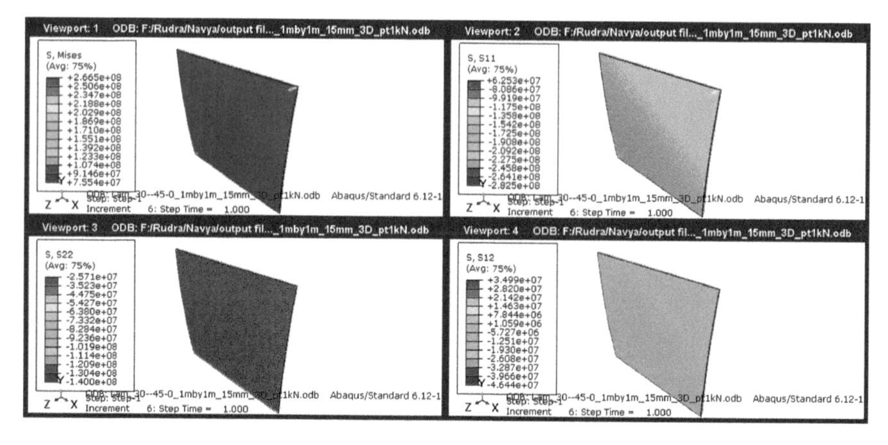

Fig. 4 Stress on [30°/−45°/0°] composite

4.4 *Stress Analysis Plots*

Figure 5 describes the variation of stresses, i.e., bending stress in x-direction, with respect to ply orientation.

Similarly, Fig. 6 develops the variation of stresses, i.e., bending stress in y-direction, with respect to ply orientation and Fig. 7 gives the variation of stresses, i.e., bending stress in xy-direction, with respect to ply orientation. Figure 5 estimates that at ply orientation of [0°/30°/−45°] maximum bending stress in x-direction is obtained while minimum bending stress is obtained at the arrangement of [−45°/0°/30°]. From Fig. 6 the stresses at ply orientation of [30°/−45°/0°] maximum bending stress in the y direction is obtained while minimum bending stress is obtained at the arrangement of [0°/30°/−45°]. With reference to Fig. 7, it

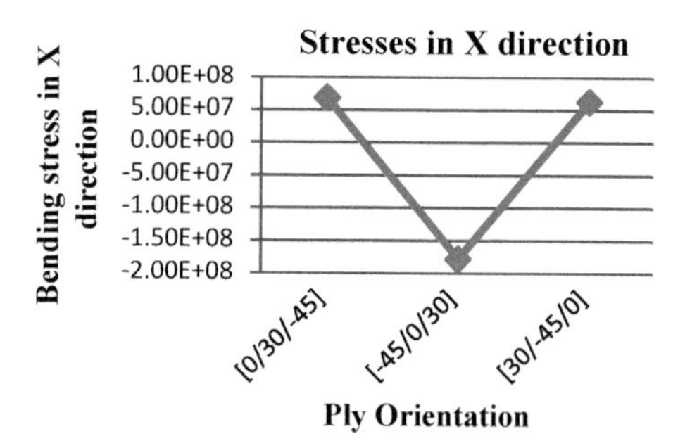

Fig. 5 Bending stress in x-direction versus ply orientation

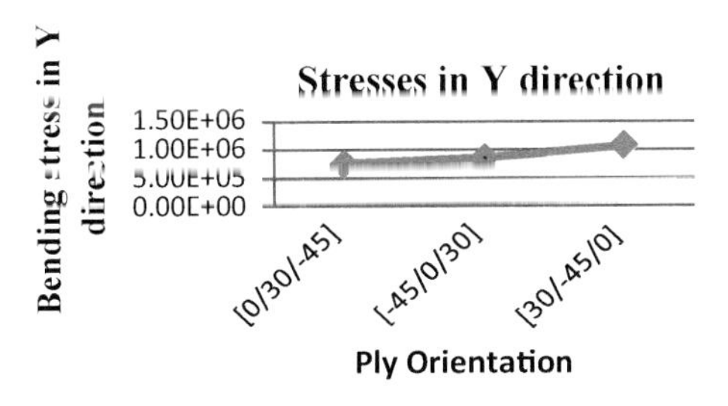

Fig. 6 Bending stress in y-direction versus ply orientation

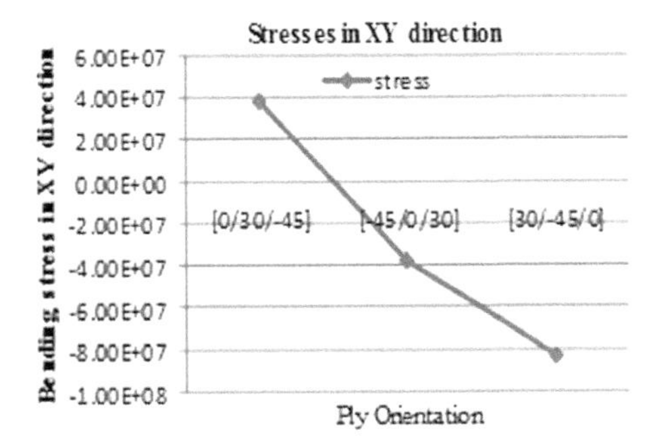

Fig. 7 Bending stresses in xy-direction versus ply orientation

can be concluded that at ply orientation of [0°/30°/−45°] maximum bending stress in xy-direction is obtained while minimum bending stress is obtained at the arrangement of [−30°/−45°/0°].

5 Conclusions

Ply orientation at [0°/30°/−45°] shows maximum bending stress values in both x, y and shear directions. It is also observed that the values of shear stress are less when compared to bending stresses in x and y directions for different plies. The lesser values of bending stresses in x-direction are observed when the ply orientation is at [−45°/0°/30°]. It is also observed that the values of stresses in y-direction are less when the ply orientation is at [30°/−45°/0°]. It is also observed that the values of shear stresses in xy-direction are less when the ply orientation is at [30°/−45°/0°]. The future scope of work use of the same concept on sandwich composites.

References

1. Sino R, Baranger TN (2008) Dynamic analysis of a rotating composite shaft. Compos Sci Technol 68:337–345
2. Topal U (2009) Multiobjective optimization of laminated composite cylindrical shells for maximum frequency and buckling load. Mater Des 30:2584–2594
3. Topal U, Uzman U (2010) Multiobjective optimization of angle-ply laminated plates for maximum buckling load. Finite Elem Anal Des, 273–279
4. Roos C, Bakis CE (2011) Multi-physics design and optimization of flexible matrix composite driveshafts. Compos Struct 93(9):2231–2240
5. Kayikci R, Sonmez FO (2012) Design of composite laminates for optimum frequency response. J Sound Vib 331:1759–1776
6. Abadi M, Daneshmehr AR (2014) An investigation of modified couple stress theory in buckling analysis of micro-composite laminated Euler-Bernoulli and Timoshenko beams. Int J Eng Sci 75:40–53

An Intelligent Drone for Agriculture Applications with the Aid of the MAVlink Protocol

Umamaheswara Rao Mogili and BBVL. Deepak

Abstract In the present, agriculture fields in India are facing the problems not only with their wages, but also with labor because of the change in climate and crop product losses due to the pests. This could challenge the additional requirements in precision agriculture technologies. The electronic and mechanic fields are made a revolution in monitoring the agriculture fields introducing unmanned aerial vehicles (UAV). The proposed work introduces a multi-rotor drone system which comes with a solution that can exterminate the perilous situations faced by our farmers. The current study, introduced a low cost, lightweight drone system which can operate at low altitude operating conditions during the flight over the crop field. This drone system has combined the implementation of various mechatronic components like flight controller (FC), brushless motors (BLDC), electronic speed control (ESC), global positioning system (GPS), telemetry radio link, and radio-controlled transmitter and receiver. The motion of the drone system is done over the crop filed as per the one planned path using "APM Planner" application. It is controlled with the help of a MAVlink protocol which is an open-source, point-to-point networking protocol which carries telemetry. The communication between multi-rotor and ground control station (GCS) is done using MAVLink networking protocol. The drone can be controlled remotely by sending and receiving data through the MAVlink protocol to the drone control modules: flight controller (ATmega2560 processor) autopilot and the APM Planner application. With the availability of above mentioned new technologies, it is possible to observe the water stress management, crop management, crop harvesting, and crop protection. These drones have great potential due to its flexibility in handling to increase crop productivity. This model developed that brings technological and economic support to small farmers.

Keywords Quadcopter · Brushless motors (BLDC) · Propellers · ESC · GPS · Precision agriculture

---

U. R. Mogili (✉) · BBVL. Deepak
National Institute of Technology, Rourkela, Odisha 769008, India

BBVL. Deepak et al. (eds.), *Innovative Product Design and Intelligent Manufacturing Systems*, Lecture Notes in Mechanical Engineering,
https://doi.org/10.1007/978-981-15-2696-1_19

1 Introduction

An unmanned aerial vehicle (UAV) simple called drone system is an autonomous aircraft which can fly without an onboard pilot. Since the 1990s, the development of this type of unmanned aircraft is increased rapidly and has become a research topic. The main advantage of the quadcopter drone is that it has a capability of vertical takeoff and landing (VTOL), which enables the drone to launch and landing within less space, high hover capability, and maneuverability. It has four symmetry equivalent sizes of the propellers to create thrust. Because of its configuration, drone possesses more advantages in terms of small size, efficiency, and safety. At the same time, the drone has a possible capability to fly in indoor and outdoor environments [1]. Due to these reasons, drones are playing a good role in military and surveillance applications [2]. Later on, it is adapted to precision agriculture applications [3]. In the present scenario, UAVs are playing a major role in the development of agriculture productivity with low cost. Because controlling functions of the drone are included in the firmware rather than to develop an expensive hardware system [4]. Development of a drone test bed plays a key role to verify our method in a real-time environment. Some basic and key components which are used in the construction of a fully autonomous drone include hardware platform configuration and its dynamics, firmware system design and its integration, and different parameters tuning and implementations. After all, using this test bed, any developed new control strategy can be implemented and then verify its performance in the real system. This drone system developed using open-source project platforms, which are very flexible in both software and hardware, makes easier to modify the specific requirements of the users [5]. The attached sensors are generally used to detect the changes in the basic functionality of the drone system. Also, these are the ability to maintain stability for better maneuver [6]. The proportional–integral–derivative (PID) is a controlling method that could be used to stabilize the drone in various movements [7]. The tuning of the PID controller senses the controlling and stability of the drone system in open fields [8]. Furthermore, these drones are to be used only for specific purpose experiments in indoor and outdoor environments. Commonly, the signal of the GPS is not good in indoor environments for localization, which is very useful in positioning system to realize autonomous navigation and path planning. Because of that, some of the drone systems use vision-based tracking system in order to track, instead of using GPS [9]. In outdoor environments, the GPS data is captured by the GPS sensor, and it captures the location of the field and sends to ground control station using MAVlink protocol [10].

In this study, a procedure to build a quadcopter drone is explained, and the required components are specified in detail as shown in Fig. 1.

The applications of the drone systems in precision agriculture are explained. The paper as three sections as follows: Sect. 2 presents the mandatory components required to build a drone system for agriculture applications. Section 3 represents the software and hardware setup of the proposed set up. Finally, Sect. 3 describes

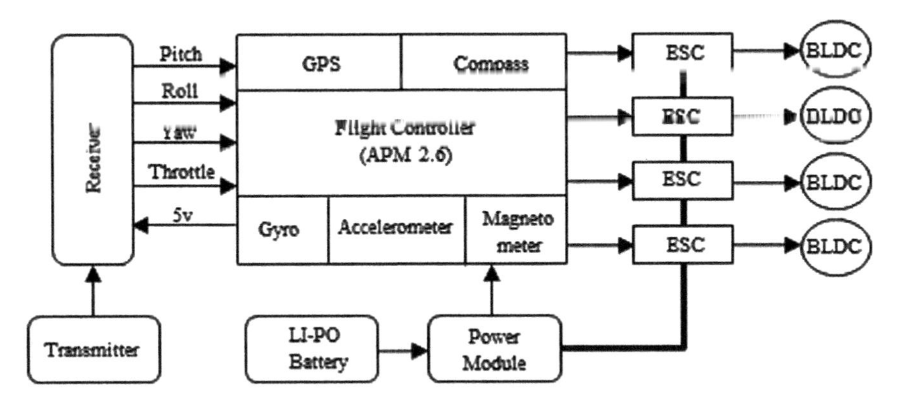

Fig. 1 Block diagram of multi-rotor drone system

the drone systems in precision agriculture applications. Therefore, the main objective of the paper is to design a quadcopter with various hardware components and loaded a proper firmware for its calibrations. Using these devices, a prominent advantage explores the current issues of precision agriculture in India.

2 Mandatory Materials

The design of the drone system is depended upon the total weight of the mandatory components used in. This prototype includes various components such as APM 2.6 flight controller board, DJI 920 kV motor, receiving the module, 2 pair propeller (clockwise and anticlockwise), global positioning system (GPS) module, and telemetry

2.1 Drone System

There are many important peripherals used to operate the UAV system as listed below.

2.1.1 APM 2.6 Flight Controller

It is a complete open-source autopilot system and is capable of performing programmed GPS missions with waypoints. It controls the speed of the motor according to the signal that is getting from the RC transmitter which is controlled manually. This board contains Atmel's ATMEGA 2560 and ATMEGA323U-2 chips for processing and USB functions, respectively. It includes the 3-axis gyro, 6 DoF accelerometer along with highly performed barometer. But, there is no onboard compass to get rid of magnetic interference between power and motor sources. Moreover, it allows off-board uBlox NEO6 GPS to identify waypoints.

2.1.2 Electronic Speed Controller (ESC)

It is a brushless ESC, takes commands directly from the FC, and provides power width modulation (PWM) signal to the brushless motor. It has 16 kHz of motor frequency, giving the fastest response of the motor and no cutoff of voltage and over temperature. It has the onboard battery eliminator circuit (BEC), which provides 5 V power and its input range from 7.4 to 14.8 V and gives 30 A continuous current.

2.1.3 DJI 2212 920 KV Brushless Motor (BLDC)

It is a brushless DC motor that works on 12 V and gives 1200 grams of max thrust so that it can lift up the whole weight of UAV and fly. Its output speed is 920 kV, and it consumes standard current at 15–25 A. It runs by 3S or 4S lipo battery with a strongly attached 8.0 mm shaft which cannot be easily bent by jerks.

2.1.4 RC Transmitter and Receiver

There is a fly sky FS-i6AFHDS 2.4GHZ 6 channel radio-controlled transmitter module and FS-iA6B receiver pair which works in the radio frequency of 2.40–2.48 GHz channel. By this device, the user sends the signal to the multi-rotor in the purpose of controlling within the radio frequency. This system uses low power electronic components and sensitive receiver chip. The RF modulation uses intermittent signal, thus reducing even more power consumption. It is powered up in the range of 4.0–6.5 V with 500 kHz bandwidth with the transmitting power to less than 20 dBm.

2.2 *GPS Module*

It is a Ublox 7 M GPS module attached with an HMC5883L digital compass which gets a signal from the satellite so that it can perform automatically course reversal, provides accurate positioning and smart directional control function. Using these characteristics, it can lock altitude, longitude, and latitude use for accurate hovering. This series is high sensitivity and low power GPS modules with a working voltage of 5 V, I2C EEPROM storage, and position accuracy is 0.9 m. It connects with GPS 5 pin or compass 4 pins for the APM 2.6 flight controller.

2.3 *Telemetry*

It's a ground control station (GCS) component which is used to get the live data from the UAV during its flight with frequency band 915 MHz. It uses MAVLink protocol framing and status reporting for this wireless communication. It uses open-source firmware with a bidirectional amplifier for even more range with auto-correction capacity up to 25% data bit error. Telemetry transmit power upto 20 and 117 dBm and receives sensitivity with air data rates up to 250 kbps. The firmware upgrades and fully supported for APM Mission Planner application.

2.3.1 MAVLink

The Micro Air Vehicle Link communication protocol (MAVLink) is a protocol to allow bidirectional wireless communication between drone system and GCS. This real-time digital communication is done in the form of messages which are sent as data packets. The GCS sends the commands to the drone system fly properly, and drone system sends telemetry status information.

This protocol specially designs for sharing the information such as radio control channels, Euler angles (pitch, yaw, and roll), drone speed, altitude, drone climb rate, airspeed, IMU details, and collected GPS raw data. MAVLink messages are sent by byte-wise to the GCS with a checksum field for packet error detection. The length of the MAVLink message is 256 bytes, and it has different fields with its frame format as shown in Fig. 2.

2.3.2 Ground Control Station

The GCS is wirelessly communicating either telemetry or RC transmitter device equipped with software which is installed in Android app or laptop. The GCS is used to configure the vehicle parameter, controlling and continuously monitoring

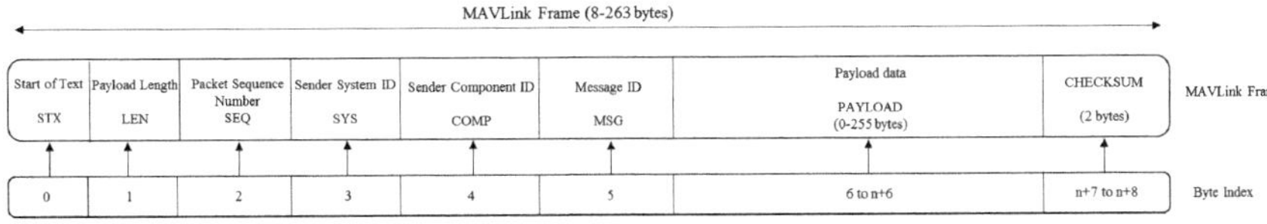

MAVLink Frame (8-263 bytes)

Start of Text STX	Payload Length LEN	Packet Sequence Number SEQ	Sender System ID SYS	Sender Component ID COMP	Message ID MSG	Payload data PAYLOAD (0-255 bytes)	CHECKSUM (2 bytes)	MAVLink Frame
0	1	2	3	4	5	6 to n+6	n+7 to n+8	Byte Index

Fig. 2 MAVLink frame format

the drone system in a real-time environment. There are a number of control stations: They are Mission Planner, APM Planner, AndroPilot, and MavProxy. All these software interfaces are supported by the MAVLink communication protocol, and they are creative commons license anyone can use. In this prototype, Mission Planner version 1.3.62 was selected as a ground control station running on any operating system. Its features include point-and-click waypoint entry, provide mission commands from drop-down menus, able to download mission log files and analyze them directly, and easy to configure flight controller settings for any air-frame design. All the flight modes (Stabilize, Alt hold, RTL, Loiter, and Auto) are set using this firmware. The tuning of PID controllers is done using the firmware.

3 Hardware and Software Setup

3.1 Hardware Setup

A multi-rotor drone system structure airframe consists of two plates (baseplate and top plate) and four arms. These plates and arms are placed in the center of gravity frame and in the correct direction. One plate is made with electronic control board (ECB) to connect the four electronic speed control (ESC) and LIPO battery. The top plate is allocated for assembling the APM 2.8 FC, sensors, GPS, receiver, and a telemetry device. The four arms support to the propulsion system of the drone system using X configuration frame to mount all electrical and mechanical mandatory components. The propulsion system is composed of four ESC, BLDC, and propellers connect with each other shown in Fig. 3. Moreover, the speed of the BLDC motors creates some vibrations in the arms and plates. Because of this behavior, flight controller does not receive signal frequencies coming from the radio transmitter properly. To rectify this problem, FC board is supported by a bumper system. The bumper system stabilize the motion of the drone system otherwise it may crash. The communication between the drone system and GCS is done in two ways, either a radio-controlled transmitter nor the telemetry system. These two are connected within FC, and then, the telemetry radio device was directly connected to the USB port of ground control station. Basically, both communications are used for the initial flight configuration, real-time flight monitoring, and to manually control quadcopter movement and orientation. The movement and orientation are depended upon the thrust created by the system. The relationship between thrust, power, and induced velocity of a UAV is proposed in momentum theory [11]. Based on momentum considerations, thrust and maximum power can be expressed in Table 1.

$$T = \left(\frac{\pi}{2} * D^2 * \rho * P_{\text{in}}^2\right)^{0.33} \tag{1}$$

Fig. 3 Designed a drone system for agriculture purpose

Table 1 Specifications of the typical UAV system

BLDC	Requirement
KV	920
Weight	53 g
Max thrust	1200 g
Max power	370 W
Length	49 mm
Battery	3S LiPO
ESC (recommended)	30 A

3.2 *Software Setup*

The software setup of the FC APM 2.8 includes downloading and installing the APM Mission Planner application as a GCS. Figure 4 shows a flow chart describing the software setup of the FC APM 2.8 with APM mission planner firmware.

This one is the open-source software, free to customize, and easy to install. Load this firmware into APM 2.8 via USB cable and then perform initial configuration and calibration. The calibration includes the type of frame, radio calibration, sensor and ESC calibration, and flight mode setups. The tuning of parameters roll, pitch, and yaw are important in order to get the best response and optimal flying performance. There are two ways to tune these parameters by using manual or autotune feature. Compared to manual, autotune mode is more critical that is why manual tuning is preferable. From the GCS, adjust these basic roll, pitch, yaw, and throttle tuning using RC transmitter. The mandatory controller configuration of PID autopilot parameter adjustment is made to the drone system fly and gives reliable performance in the wind. There are many PID parameters can be tuned in the drone systems to get optimal performance.

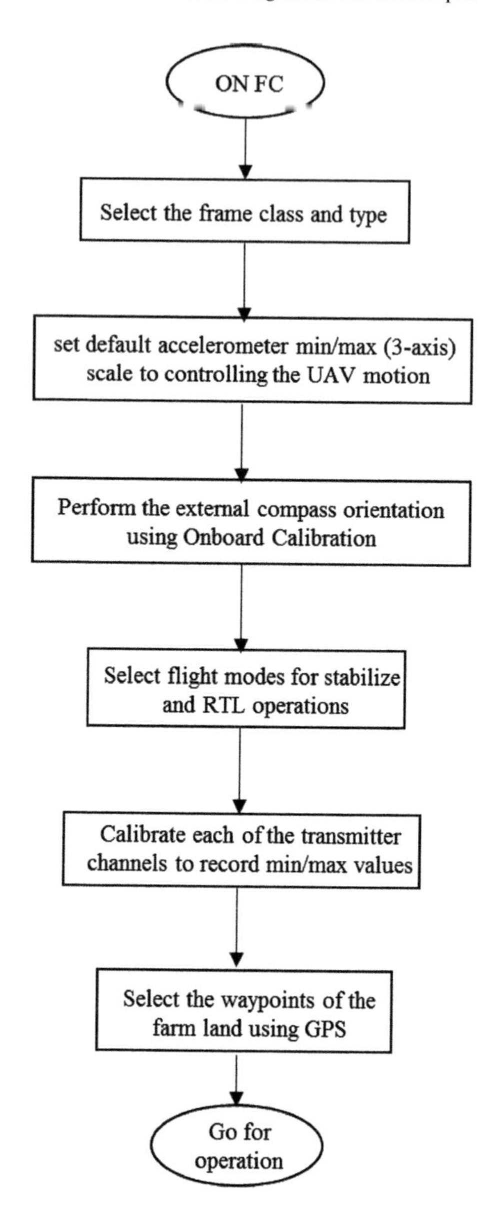

Fig. 4 Flowchart of firmware calibration

4 Agriculture Applications Related to the Drone Systems

The above-configured drone systems give technical support to small farmers that could help to increase farm productivity. In the precision agriculture following innovative developments, technology, and new methodologies useful for improvement in small size field farming.

4.1 GPS and Laser for an Estimate the Properties of Land

To plan and design soil properties of the land, farmers can obtain the two and three-dimensional models using laser and GPS waypoints mounted with the drone system. According to that, the amount of water required is estimated, and the liquid is taken to each corner, and unnecessary accumulations are avoided.

4.2 Drone Systems for Crop Monitoring

The GPS mounted to the drone system allows to identify the above models that are problematic in the grounds. The drones are equipped with different types of camera sensors (multispectral, hyper, and thermal) that allow identifying infected areas of low growth and changes in the color of the plants. This could be done using vegetation and color indices with the different wavelengths of the sensor rays. Early care can prevent the spread of disease and reduce losses by various problems that affect crop productivity.

4.3 Drones for Pesticide Spraying

The recent developments in the drones involve the incorporation of atomizers of high-precision sprinkler systems in drones. The combination of drone system and pesticide sprinkler show a high impact favorably in productivity of the crops with precision speeds. Which leads to reduction of pesticides and fertilizers used in the field. An increase of beneficial species (spider) which are native and biological control of some of the pests, they are quite efficient to controlling pests of rice crops.

4.4 Weed Identification

Drone-mounted multispectral sensors create weed map using the captured blue, green, near-infrared light wavelengths data and employing normalized difference vegetation index (NDVI) using image processing techniques. Seeing this map, farmers can easily identify the weed regions in the crop filed. Also, developed automated weed pressure software is available in the market to analyze the high-intensity weed growth. This firmware provides a highly accurate analysis of the weed growth in the healthy crop plants.

5 Conclusion

As a conclusion, the development of the drone system for precision agriculture applications are explored in detail that includes the general standards of the framework, different sensor techniques, controllers, and distinguishing components to build. The scope of this proposed system is to develop a user-friendly drone system, which can be executed and its stable flying in the agriculture fields to monitor the esteemed crop. For smooth flying, the parameters tuning and trimming phases of this system are done using the PID controller of the APM Planner. The drone technology is suitable for achieving greater production per acre and receive a higher income. Still, these technologies are in its early stages in countries like India that has agriculture as its main economy. However, several studies have included that small increas in the productivity of local agriculture can allow significant reductions in poverty. But, there are still some limitations related to drones including high costs, platform reliability, sensor capability, and weather considerations. After all, in many areas, strict aviation regulations and knowledge over the technology from the farmers may impede using drones.

6 Future Scope

However, it is expected that new developments in the drone technologies, image processing techniques, and sensor designs are giving more prominent advantages over the manual equipments in the agriculture field.

Acknowledgements This work is supported by SERB, Govt. of India, with the Sanction order No. ECR/2017/000140 on Dt. 05th July 2017.

References

1. Toji J, Iwata Y, Ichihara H (2015) Building quadrotors with Arduino for indoor environments. In: 2015 10th Asian control conference (ASCC). IEEE, pp 1–6
2. Cai G, Chen BM, Lee TH (2010) An overview of the development of miniature unmanned rotorcraft systems. Front Electr Electron Eng China 5(1):1–14
3. Mogili UR, Deepak BBVL (2018) Review on application of drone systems in precision agriculture. Procedia Comput Sci 133:502–509
4. Zhang C, Kovacs JM (2012) The application of small unmanned aerial systems for precision agriculture: a review. Precision Agric
5. Lim H, Park J, Lee D, Kim HJ (2012) Build your own quadrotor: open-source projects on unmanned aerial vehicles. IEEE Robot Autom Mag 19(3):33–45
6. Gupte S, Mohandas PIT, Conrad JM (2012) A survey of quadrotor unmanned aerial vehicles. In: 2012 proceedings of IEEE Southeastcon. IEEE, pp 1–6

7. Qasim M, Susanto E, Wibowo AS (2017) PID control for attitude stabilization of an unmanned aerial vehicle quad-copter. In: 2017 5th international conference on instrumentation, control, and automation (ICA). IEEE, pp 109–114
8. Sohail S, Nasim S, Khan NH (2017) Modeling, controlling and stability of UAV quad copter. In: 2017 international conference on innovations in electrical engineering and computational technologies (ICIEECT). IEEE, pp 1–8
9. Dai B, He Y, Gu F, Yang L, Han J, Xu W (2017) A vision-based autonomous aerial spray system for precision agriculture. In: 2017 IEEE international conference on robotics and biomimetics (ROBIO). IEEE, pp 507–513
10. Śmigielski P, Raczyński M, Gosek Ł (2017). Visual simulator for MavLink-protocol-based UAV applied for search and analyze task. In: 2017 federated conference on computer science and information systems (FedCSIS). IEEE, pp 1177–1185
11. http://ardupilot.org/planner/docs/mission-planner-overview.html © 2016 by ArduPilot.ORG, 201904120950

Development of Novel Cost-Effective Automatic Packing System for Small-Scale Industries

Vaibhav Gaunkar and A. P. Sudheer

Abstract The paper mainly deals with novel design and fabrication of an automated packing system which incorporates boxes of a different shape for getting custom shaped boxes for small-scale industries. Fully automated packing systems are being used in large-scale industries extensively. As large packaging systems are unaffordable and space-consuming, small industries tend to do their packaging manually which led to low production. The aim of the project is to design and fabricate automated packing system which will be incorporating boxes of different shapes for getting custom shaped boxes for small-scale industries. The kinematic modeling and dynamic analysis of subassemblies are done. The static structural analysis is done for the critical part of the mechanism for analyzing stress and deformation in the system. All subsystems are fabricated and assembled together. Coordination between all the subsystems is tested, and the whole mechanism is also tested for the packing operation.

Keywords Automated packing · Computer-aided design · Kinematic modeling · Dynamic modeling

1 Introduction

Current packaging methodology for varying sized boxes is a very convoluted process while considering small-scale industries. It involves various types of mechanisms for cutting, folding, and wrapping the corrugated boxes. All these mechanisms need to be tuned according to the size of the boxes before the packing operation. Also, all these mechanisms are assisted by transportation systems which include conveyor and suction grippers. All these add to the cost of the packaging system significantly. The average cost of a packaging system would be around 1.5–

V. Gaunkar · A. P. Sudheer (✉)
Mechatronics/Robotics Laboratory, Department of Mechanical Engineering, National Institute of Technology Calicut, Kozhikode 673601, India
e-mail: apsudheer@nitc.ac.in

BBVL. Deepak et al. (eds.), *Innovative Product Design and Intelligent Manufacturing Systems*, Lecture Notes in Mechanical Engineering,
https://doi.org/10.1007/978-981-15-2696-1_20

2 lakhs. Along with this cost, the whole setup occupies a large area. Both conditions are unsuitable for the small-scale industries. Various types of existing methodologies were studied [1]. It was observed that existing methods were suitable for batch production with less room for flexibility in one particular batch. Different types of mechanisms were considered for this methodology. Teo et al. [3] have dynamically modeled and designed adaptive control of an *H*-type gantry stage. Giam et al. [4] presented a survey of control schemes and also the development of enhanced schemes for the coordinated motion control of moving gantry stages. Palomares et al. [8] proposed a straightforward model to accurately predict force–displacement behavior using as a basic experimental observation for several pressures and harmonic displacements of the rod. He et al. [10] presented a three-step method that is used to determine a proper way to accelerate a speed-controlled belt conveyor during transient operation.

For the packing operation, corrugated box of varying size has been chosen to incorporate different sized packages. Model of packing mechanism is based on the gantry system. For the folding operation of boxes, various subsystems are integrated which fold the cardboard boxes according to the size of the package and, in turn, reduce the space acquired by the entire workstation. Combination of actuators such as pneumatic cylinder and lead screws are used. More workspace with less of total space is achieved by using a combination of linear actuators in the form of 2P and 2PR mechanism. For the cost reduction, most of the materials used were either easily available or recycled. The size of the entire automaton was kept to a minimum.

2 Modeling of the Robotic Mechanism

As the different parts of the robotic mechanism perform different operations, the whole operation, as well as the mechanism, can be divided into three subsystems for simplicity of control as well as a fabrication. These are 1. suction transfer mechanism, 2. box folding mechanism, and 3. box transfer mechanism. The suction transfer system is used to transfer corrugated boxes from storage area to the folding area. Cardboard transfer takes place through the rotary motion from the storage area to box folding area which is placed right across it. Box folding mechanism uses a combination of linear rails, pneumatic cylinders, and belt drive for folding the corrugated cardboard. It is having various subassemblies with a combination of linear rail and pneumatics. The box transfer mechanism is basically a conveyor system which is used for transferring boxes after the folding operation. The cad model is designed in SOLIDWORKS 2016 which is shown in Fig. 1.

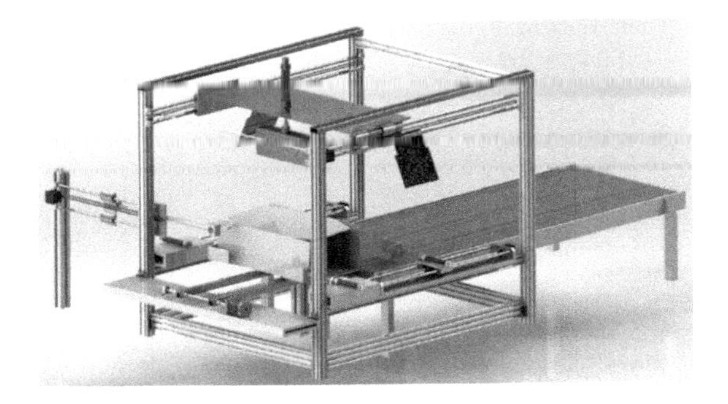

Fig. 1 CAD model

3 Kinematic Modeling

Kinematic modeling is formulated for the following two mechanisms which are inherent for packing mechanism: (1) 2P mechanism and (2) 2PR mechanism.

3.1 2P Mechanism

In this mechanism, two prismatic joints are placed at the lower portion of the mechanism as shown in Fig. 2. DH parameters for this mechanism are shown in Table 1.

Final transformation matrix is given in Eq. (1).

$$T_2^0 = \begin{bmatrix} 1 & 0 & 0 & 0 \\ 0 & 0 & 1 & d_2 \\ 0 & -1 & 0 & d_1 \\ 0 & 0 & 0 & 1 \end{bmatrix} \quad (1)$$

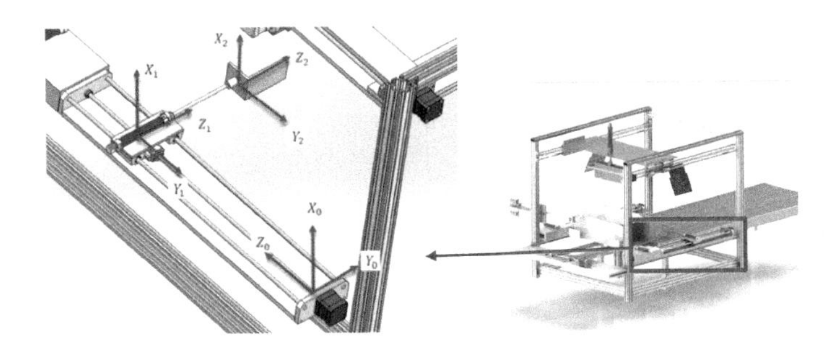

Fig. 2 2P mechanism

Table 1 Denavit–Hartenberg (DH) parameters

Joint	θ_i	d_i	a_i	α_i
1	0	d_1	0	−90
2	0	d_2	0	0

3.2 2PR Mechanism

This mechanism consists of two prismatic joints and one revolute joint and placed at the upper portion of the mechanism as shown in Fig. 3. DH parameters are shown in Table 2.

Final transformation matrix is given in Eq. (2).

$$T_3^0 = \begin{bmatrix} 0 & 0 & 1 & 0 \\ -S_3 & C_3 & 0 & d_5 \\ C_3 & -S_3 & 0 & d_4 \\ 0 & 0 & 0 & 1 \end{bmatrix} \tag{2}$$

Inverse kinematic modeling helps to get the joint variable to implement the motion for the automation. Equations (1) and (2) are equated with required poses for inverse kinematic solutions.

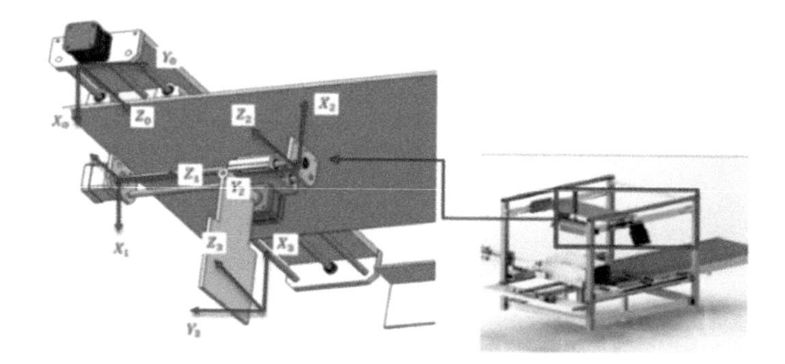

Fig. 3 2PR mechanism

Table 2 Denavit–Hartenberg (DH) parameters

Joint	θ_i	d_i	a_i	α_i
1	0	d_4	0	−90
2	−90	d_5	0	−90
3	θ_3	0	0	0

4 Static Structural Analysis

The static analysis for a critical part of the mechanism is performed in ANSYS software.

Trial 1: First the analysis is done for the upper moving platform as it is a heavily loaded mechanism, using aluminum alloy material for the upper moving platform. The material is selected as it is easily available. Hence, after analysis, it is found that deformation is unacceptable as shown in Fig. 4. Even though stress for the system is within allowable limits as shown in Fig. 5, the material is needed to be replaced.

Trial 2: The analysis is repeated using lighter material without compromising the integrity of the structure. The material selected was multi-wood which was separately bought for the purpose of fabrication as the analysis results were satisfactory. The total deformation is very small as compared to aluminum as the weight of multi-wood is comparably less. Total deformation and von Mises stress are shown in Figs. 6 and 7, respectively.

5 Dynamic Modeling

The dynamic model is developed for the 2P and 2PR mechanisms by using Lagrangian–Euler method.

Case (i): 2P mechanism: As per the Lagrangian–Euler formulation, torque of *i*th joint is:

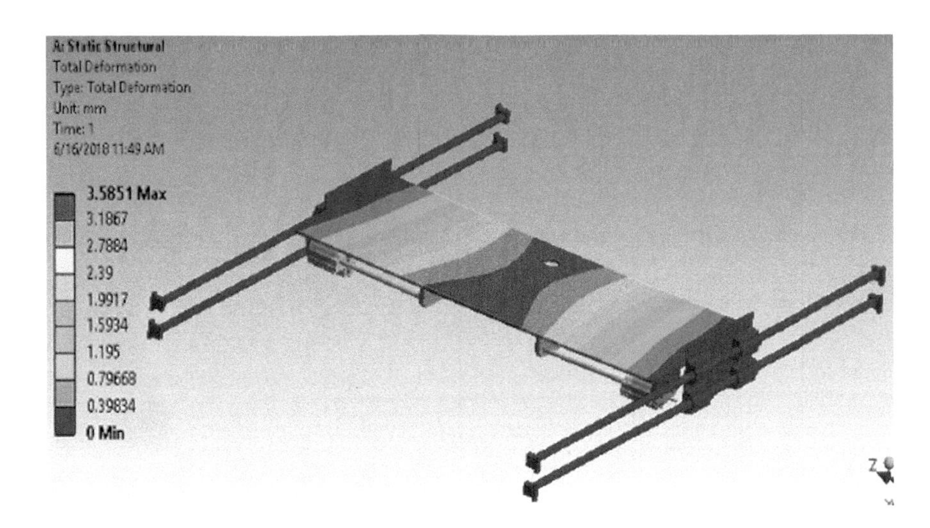

Fig. 4 Total deformation of upper platform

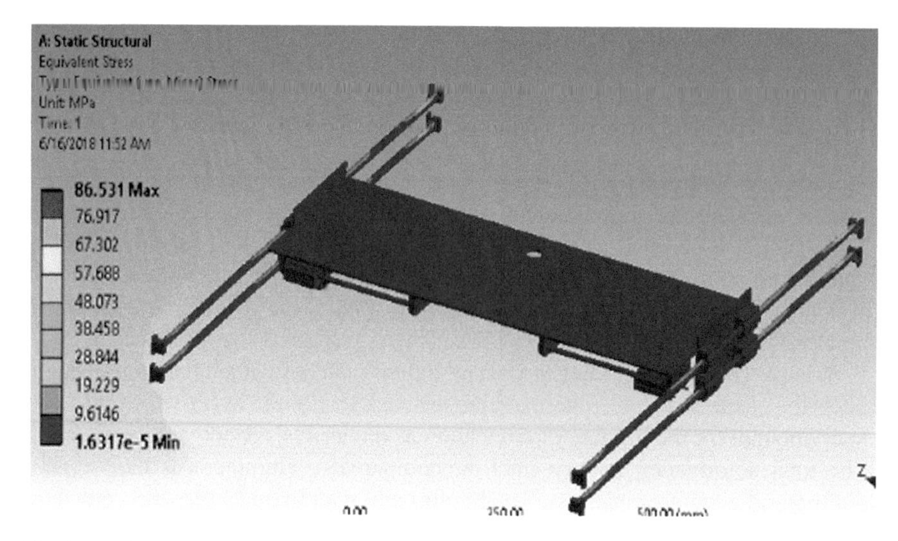

Fig. 5 Equivalent von Mises stress

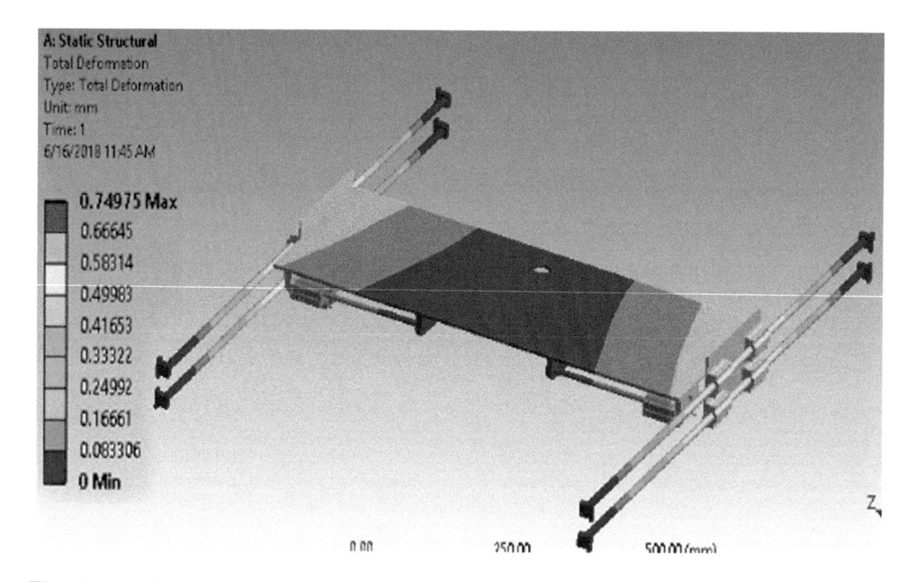

Fig. 6 Total deformation

$$\tau_i = \frac{d}{\mathrm{d}t}\left(\frac{\partial L}{\partial \dot{d}_i}\right) - \left(\frac{\partial L}{\partial d_i}\right)$$

Kinetic energy

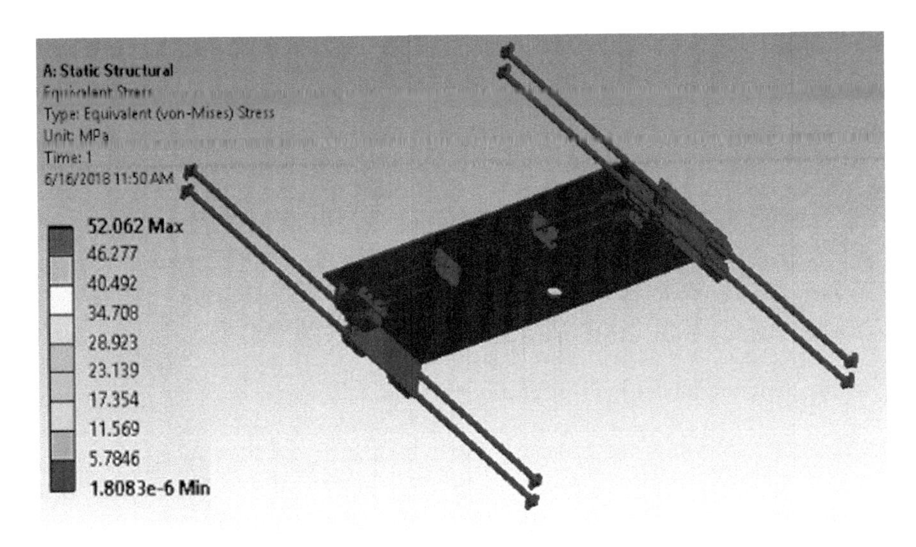

Fig. 7 Equivalent von Mises stress

$$K_1 = \frac{1}{2} m_1 \dot{d}_1^2 \tag{3}$$

$$K_2 = \frac{1}{2} m_2 \left(\dot{d}_2^2 + \dot{d}_1^2 \right) \tag{4}$$

Potential energy $P = 0$

$$\text{Lagrangian} : L = K - P$$

$$L = \frac{1}{2} m_1 \dot{d}_1^2 + \frac{1}{2} m_2 \left(\dot{d}_2^2 + \dot{d}_1^2 \right) - 0 \tag{5}$$

$$F_1 = m_1 \ddot{d}_1 \tag{6}$$

$$F_2 = m_2 \ddot{d}_2$$

$$\begin{bmatrix} F_1 \\ F_2 \end{bmatrix} = \begin{bmatrix} m_1 + m_2 & 0 \\ 0 & m_2 \end{bmatrix} \begin{bmatrix} \ddot{d}_1 \\ \ddot{d}_2 \end{bmatrix} \tag{7}$$

Case (ii): 2PR mechanism: Similarly, the Lagrangian function can be determined as explained above.

$$L = \frac{1}{2} m_1 \dot{d}_1^2 + \frac{1}{2} m_2 \left(\dot{d}_2^2 + \dot{d}_1^2 \right) + \frac{1}{2} m_3 \left(L_1 \dot{\theta} \right)^2 + \frac{1}{2} I_{zz} \dot{\theta}^2 - (-m_3 g L_3 S_3) \tag{8}$$

$$\tau_i - \frac{d}{dt}\left(\frac{\partial L}{\partial \dot{d}_i}\right) - \left(\frac{\partial L}{\partial d_i}\right)$$

$$F_1 = [m_1 + m_2]\ddot{d}_1 \tag{9}$$

$$F_2 = m_2\ddot{d}_2 \tag{10}$$

$$\tau_3 = \frac{1}{4}m_3L_2^2\ddot{\theta} + I_{zz}\ddot{\theta} - m_3gLCos\theta_3 \tag{11}$$

The dynamic model is given in Eq. (12).

$$\begin{bmatrix} F_1 \\ F_2 \\ \tau_3 \end{bmatrix} = \begin{bmatrix} m_1 + m_2 & 0 & 0 \\ 0 & m_2 & 0 \\ 0 & 0 & \frac{1}{4}m_3L_2^2 + I_{zz} \end{bmatrix} \begin{bmatrix} \ddot{d}_1 \\ \ddot{d}_2 \\ \ddot{\theta} \end{bmatrix} - \begin{bmatrix} 0 \\ 0 \\ m_3gLCos\theta_3 \end{bmatrix} \tag{12}$$

The dynamic model is used for selection of the other actuators. This can also be used for controller design and power or energy calculation which is not within the purview of this paper.

6 Fabrication and Automation of Robotic Mechanism

After detailed designing of the mechanism, the fabrication started as per the CAD model. For fabrication, the complete model is disintegrated into small submechanisms such as suction transfer mechanism, box folding mechanism, and box transfer mechanism (conveyor). After fabricating each submechanism, everything is integrated and assembled to automate the entire mechanism to obtain the desired output. The material for fabrication is selected so that it is economical, easy to machine, and also based on the availability. Figure 8 shows the complete assembly of the automatic packing mechanism.

Controlling is done using Arduino mega 2560 microcontroller by interfacing it with RAMPS 1.4. Arduino IDE software is used to write control logic code to control Arduino mega. Motors are connected to Arduino through motor drivers which amplify the signal received from the microcontroller. For pneumatic components, such as pneumatic cylinders and suction cups, 5/2 solenoid valve is used to operate them. Relay board is used to control this circuit by getting signal from Arduino. Control block diagrams are shown in Fig. 9.

Fig. 8 Automatic packing mechanism

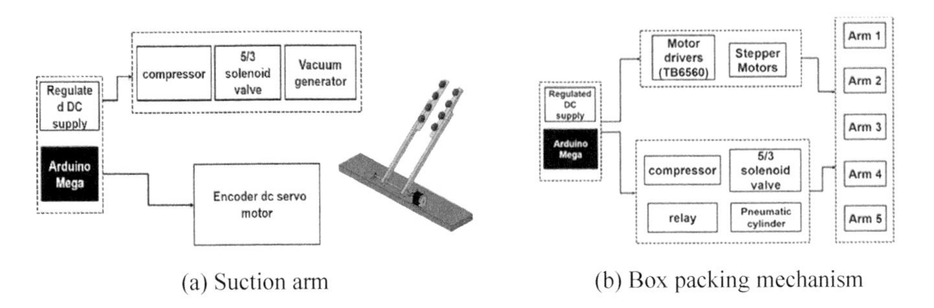

(a) Suction arm (b) Box packing mechanism

Fig. 9 Control block diagram

7 Experimentation of Packing Mechanism

Each subsystem is tested and the experimental verification is carried out for folding by using various parts of the packing system.

Testing of suction transfer arm is shown with snapshots as shown in Fig. 10. Experimentation of box packing process is shown in Fig. 11.

Fig. 10 Testing suction transfer arms

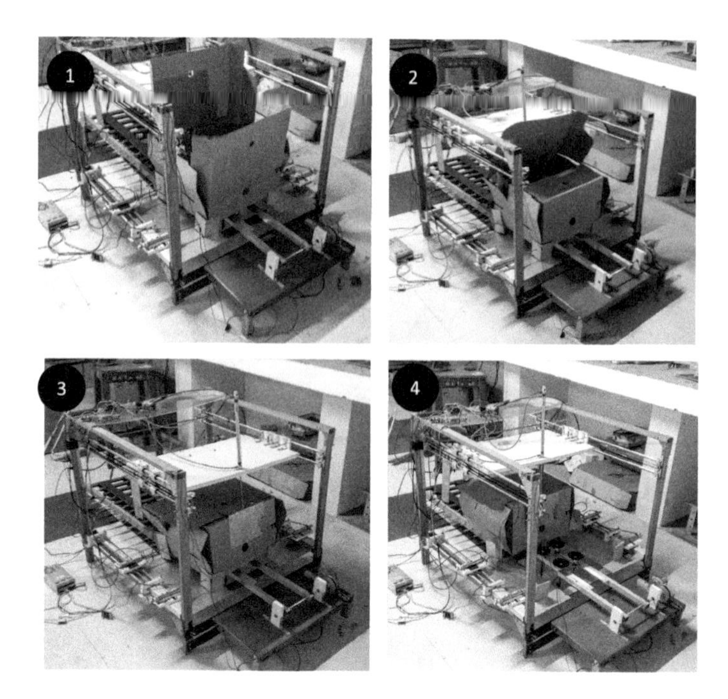

Fig. 11 Snap shots of testing of packing mechanism

8 Results and Discussion

Conceptual modeling of the packing mechanism is done using SOLIDWORKS software. Motion analysis is also carried out using the same software to depict the working of the mechanism. Various numerical calculations are carried out for finding dimensions of various components. The diameter of the shaft of transfer arm and diameter of guide rods for the upper moving platform is selected as 8 and 12 mm, respectively. Motor for driving upper moving platform through belt drive is selected as 0.43 Nm with safety factor. Structural analysis is done using ANSYS workbench to find out stresses and deformation in the system. Dynamic analysis is carried out for mechanism for the actuator selection. Trials of analysis are conducted for the upper moving platform for different materials. For aluminum, the deflection in the system is unallowable. So, the material for the platform is chosen as multi-wood. The total deformation is almost negligible, and the stress induced is very less as compared to the failure strength of the material used.

Automation and control are done using Arduino mega 2650 microcontroller for all the subsystems. Different pneumatic and electronic actuators are used in the system for packing boxes of maximum size 600 × 500 × 600 mm and minimum size of 400 × 250 × 200 mm. The components used in control of the mechanism are RAMPS 1.4, DRV8825 motor driver, vacuum generator, 5/2 pneumatic valve, and relay board. Coordination between various subsystems is tested. Packing process is implemented which is depicted in Fig. 11.

9 Conclusions

This robotic mechanism is small in size as well as very economical as compared to the machinery from the large-scale industries. The components which are used in fabrication are affordable as well as easily available. Most of the actuators selected are able to perform the required task in the packing process. The lead screw mechanism is precise for positioning as compared to others. As for the pneumatic cylinder, controlling of the stroke is a difficult task which is needed for the varying box sizes. 5/3 solenoid valve is used to attain three stroke positions of the pneumatic cylinder. Hence, this packing process is restricted to boxes of three different sizes which can be remedied by using step cylinders.

The robotic mechanism for automatic packing for varying sized boxes is designed, fabricated, and tested for which is economical for small-scale industries. The cost of fabrication was around 60 thousand as compared to lakhs of rupees for conventional methodologies. This work can be extended as follows: Packaging system can be integrated with an actual production system and packing can be done online. Appropriate controllers can be designed for checking the efficiency of controllers through implementation. Artificial intelligence and IoT can be implemented for increasing the productivity, quality, and efficiency of this system.

References

1. Tallian RB, Weinstein MA (1998) Flexible automation solutions for today's bakeries. Ind Rob Int J 25(1):20–26
2. Hong JW, Kim YO, Ha IJ (2008) Simplified time-optimal path planning of XY gantry systems along circular paths. Automatica, 44(1):149–156
3. Teo CS, Tan KK, Lim SY, Huang S, Tay EB (2007) Dynamic modeling and adaptive control of a H-type gantry stage. Mechatronics 17:361–367
4. Giam TS, Tan KK, Huang S (2007) Precision coordinated control of multi-axis gantry stages. ISA Trans 46:399–409
5. García-Herreros I, Kestelyn X, Gomand J, Coleman R, Barre PJ (2013) Model-based decoupling control method for dual-drive gantry stages: a case study with experimental validations. Control Eng Practice 21(3):298–307
6. Li K, Zhang Y, Hu Q (2019) Dynamic modelling and control of a Tendon-Actuated Lightweight Space Manipulator. Aerosp Sci Technol 84:1150–1163
7. Yu YQ, Zhang N (2019) Dynamic modeling and performance of compliant mechanisms with inflection beams. Mech Mach Theor 134:455–475
8. Palomares E, Nieto AJ, Morales Al, Chicharro JM, Pintado P (2017) Dynamic behaviour of pneumatic linear actuators. Mechatronics 45:37–48

9. Li J, Kawashima K, Fujita T, Kagawa T (2013) Control design of a pneumatic cylinder with distributed model of pipelines. Precision Eng 37(4):880–887
10. He D, Pang T, Lodewijks G (2016) Speed control of belt conveyors during transient operation. Powder Technol 301:622–636

Comparison of Mechanical Behavior for Cow- and Goat-Fiber-Reinforced Epoxy Composites

K. Ch Sekhar, Srinivas Kona, V. V. Rama Reddy and A. Lakshumu Naidu

Abstract These days' researchers are fascinated to develop environment-friendly materials. Natural fiber composites are environment-friendly, renewable, and biodegradable materials. These organic composites have more advantages like low cost, lower density, and easily available in nature than inorganic fibers. Manufacturing of the natural fiber composite is easy compared to the conventional methods. Also, environmental awareness and growing concern with the greenhouse effect have initiated by various industries to pay attention to eco-friendly materials for replacing the hazardous materials. Epoxy-based composites are prepared by using animal fibers, and short fiber is obtained from the animals (goat and cow). These fibers are treated with the chemicals. Groundnut shells ash is used as the filler material. Mechanical properties such as impact strength, compressive strength, surface roughness, and hardness numbers are measured. It is found that interfacial bonding between the two phases enhances the mechanical behavior of the composites. Among all, C3 composites exhibit better mechanical behavior than other composites. The microstructure of the composites is observed by the scanning electron microscopy. It is exposed that incorporation of groundnut shells ash acts as filler material, which enhances the interfacial bonding inside the composite.

Keywords Animal fibers · Mechanical properties · Reinforced polymer composites

K. Ch Sekhar (✉) · S. Kona · V. V. Rama Reddy
Department of Mechanical Engineering, Lendi Institute of Engineering and Technology, Vizianagaram, India

A. L. Naidu
Department of Mechanical Engineering, GMR Institute of Technology, Rajam, India

BBVL. Deepak et al. (eds.), *Innovative Product Design and Intelligent Manufacturing Systems*, Lecture Notes in Mechanical Engineering,
https://doi.org/10.1007/978-981-15-2696-1_21

1 Introduction

Natural fiber composites are growing rapidly due to their ease of availability in nature. These fiber composites have the potential of replacing the synthetic fiber composites with the lower cost along with improved sustainability. These fibers are used for reinforcement of polymers because they are precise of lower density than inorganic fibers, environmentally friendly, and relatively easy to obtain [1]. These fibers are extracted from the different sources available in nature. Sources of natural fibers are plants, animals, and minerals. Plant fibers are extracted from the parts of plants like stems/stacks, leaves, and fruits. A plant fiber mainly contains cellulose that supports the stalk of the plant. Plant fibers include flax, banana, jute, sisal, and coir. Mineral fibers commonly adopted fibers acquire from minerals. Anthophyllite, amphiboles, asbestos, and serpentine are the classification of the mineral fibers. These fibers are now avoided for causing health issues. Animal fibers are extracted from the hair of the animals. These fibers are comprised of proteins. Animal fibers include mohair, fleece, silk, alpaca, and goat. Cow and goat hairs are the type of human hair which are composed of proteins. Goat hair is used in textile applications. They are considered as good thermal insulators. Cow and goat fibers are waste produced from the animals, and they are largely available in India [2, 3].

Jayaseelan studied the effect of the mechanical behavior of composites when goat hair, graphene and goat fiber reinforced with the epoxy polymer. Jayaseelan concluded that mechanical properties of the composites are increasing with the addition of the goat hair [4]. Oladele investigates the effect of fiber treatment on the mechanical behavior of composites. Oladele manufactured the CFF/LDPE and cow hair fiber/LDPE composites with different volume fractions (2, 4, 6, 8, and 10%). Oladele stated that the addition of CFF and cow hair fiber increases the flexural strength of the composites. The chemically (NaOH)-treated animal fibers obtained a better performance than the untreated fibers [5]. Oladele used the 10 mm length of the cow fibers. Oladele concluded that flexural and Young's modulus of the composites increase with the addition of the cow fiber up to 15% volume fraction. The composites reinforced with 20% cow fiber obtain the lower flexural and Young's modulus compared to the 15% cow-fiber-reinforced composites. When these fibers were reinforced with LDPE, composites got the tensile modulus of 232 MPa. When it comes to polyester matrix, reinforced cow fiber composites obtain the 756.16 MPa of tensile modulus. The mechanical properties of the composites vary with the matrix material. In this study, we utilize the epoxy resin (XIN 100 IN) as the matrix material. Oladele also studied the microstructure of the composite. These composites have voids which are present in the composite. Cow fibers are constituted of keratin and amino acids. Keratin and amino acids are basically hydrophobic nature, which tends to absorb the moisture from the atmosphere. This phenomenon results in the dimensional changes of the composite [6–9]. Bishnu studied the effect of the mechanical and microstructural behavior of composites when short human fiber reinforced with the epoxy composites. Bishnu manufactured the human hair fiber/epoxy composites with different volume fractions (2, 4, 6, and 8%). He observed that the addition of fiber to polymers influences mechanical behavior. The mechanical properties of the composites are increasing with the addition of the human hair fiber [10]. Lee studied the thermal behavior of the plant fiber composites. Lee stated that composites contain high voids which absorb more moisture from the atmosphere by the hydrogen bonds between the two phases. To reduce the void percentage, we utilize the groundnut shells ash. Groundnut shells are waste materials produced from the agriculture. In this research, we utilized groundnut shells ash as the filler material [11].

In the present study, cow hair fiber and goat hair fibers are used as the reinforced materials. Cow and goat fibers are waste produced from the animals, and groundnut shells are waste produced from the plants, dumping of these products into the environment causes environmental pollution. Utilization of these waste products into the polymer reduces the impact on the environment. Fibers are treated with NaOH and H_2SO_4 to reduce hydrophobic nature. Epoxy resin (XIN 100IN) was used as the matrix material. In this study, we study the effect of short fiber-reinforced composites on mechanical properties. We performed surface roughness, hardness, impact, and compressive tests to analyze the behavior of the composites.

2 Experimental

2.1 Materials

Epoxy resin (XIN 100 IN) and hardener (XIN 900 IN) are purchased from Araldite Pvt. Ltd. Groundnut shell is obtained from the local sources in Rajam, Srikakulam, India. Ash is obtained from burning of groundnut shell. Groundnut shells are burnt in the furnace at 1000 °C for an hour. These shells are then converted into ash, and the ash is dried in a muffle furnace at 150 °C for 4 h. Goatskin and cow hair are purchased from the local sources in Rajam, Srikakulam, India. NaOH and H_2SO_4 with concentration of 98% were purchased from Home Projects Pvt. Ltd., Maharashtra, India.

2.2 Extraction of Fiber and Composite Preparation

Goatskin is dried in a normal atmosphere for 30 h, and then the fiber is extracted from the skin. The fiber is washed with normal water for the removal of external impurities, and then the fiber is dried at room temperature in a closed atmosphere. Cow hair is washed with normal water, and then dried under the sunlight for 72 h. A uniform fixed amount of fiber is cut and washed with distilled water. Goat hair fiber is soaked in 0.25 M NaOH solution for half an hour. Then the soaked fiber is

Table 1 Material compositions and codes

Codes	Polymer matrix composite materials composition
C1	Pure epoxy
C2	95% epoxy + 5% groundnut shells ash
C3	80% epoxy + 5% ash + 15% goat hair fiber
C4	80% epoxy + 5% ash + 15% cow hair fiber
C5	80% epoxy + 5% Ash + 7.5% Goat and 7.5% cow hair fibers

dried in the open atmosphere for 24–48 h till the skiving of the moisture. Cow hair fiber is treated with H_2SO_4 solution and is soaked in the 0.04 M H_2SO_4 solution for half an hour. Then this fiber is dried in the open atmosphere for 24–48 h till the skiving of the moisture. Neatly separated goat and cow hair fibers are cut to a uniform length of 40 mm. The epoxy is mixed with a hardener in the ratio of 10:1. The groundnut shell ash is poured into the matrix material without allowing the formation the air bubbles. The composite slabs are made by the conventional hand-lay-up method. Five different composites (pure epoxy, epoxy + 5% ash, epoxy + 5% ash + 15% goat hair fiber, epoxy + 5% ash + 15% cow hair fiber, epoxy + 5% ash + 7.5% goat and 7.5% cow hair fiber) with different weight ratios are prepared. The specimens of composites are cut as per ASTM standards. Manufactured specimens are listed in Table 1.

2.3 *Composite Testing*

Surface Roughness Talysurf was used for measuring the surface roughness for the specimens. Average surface roughness value (R_a) and ten-point mean surface roughness (R_z) are investigated for these specimens using Taylor Hobson precision surtronic 3 + talysurf.

Hardness Test Hardness is the measure of the plastic deformation of the object against the compression load. Saroj hardness tester was used for hardness measurement. The tester had a one-fourth ball indenter which is forced into the material under a load of 60 Kgf.

Impact Strength The impact test specimens are prepared according to the required dimensions and testing following the ASTM-D256-10 standard. Both Izod and Charpy tests are conducted to the five specimens. The energy required to break the material will be scrutinized by using the impact test. The effect of Izod and Charpy tests is analyzed in this paper. In Izod, the test material is tested in the vertical direction, and in the Charpy test, it is tested in a horizontal direction.

Compression Test Compression test specimens are prepared according to the required dimensions and testing following ASTM-D695 standard. The compressive strength is usually obtained experimentally by means of a compressive test using UTM.

3 Results and Discussion

The surface roughness of the epoxy-based fiber composites is shown in Figs. 1 and 2. Both top and bottom surface roughness values are examined. The composite C1, i.e., pure epoxy composite, attained 4.885 μm and 11.85 μm of surface roughness at the bottom and top surface. The epoxy/groundnut shells ash (C2) and goat hair fiber and groundnut shells fiery remain/epoxy composite (C3) show the lower surface harshness than the composite. The dairy animal's fiber-fortified epoxy composites, for example, C4 composite, displays higher surface unpleasantness than alternate composites. It was discovered that the surface unpleasantness of the composite C3, for example, goat-hair-fiber-strengthened groundnut shells fiery remains/epoxy mixed composite, achieved less surface unpleasantness among alternate composites. In the event that the surface unpleasantness increments for a material, the wear rate of that material likewise builds and results in a decline in its lifetime. Expansion of the goat fiber to the composite gets low surface harshness. Goat fiber has the better interfacial holding with the epoxy and groundnut shells powder contrasted with the cow fiber. The composite C3 achieves the surface harshness of 4.84 μm and 8.16 μm at the base and best surface. The surface unpleasantness of the base side of these composites shows slightly higher when contrasted and the plant fiber composites. The surface unpleasantness of the best side of these composites displays bring down when contrasted and the plant fiber composites. Andrea [12] revealed that wood/plastic composites are achieving 4.48 μm at the base surface and 33.94 μm at the best surface. The hand-lay-up strategy creates low unpleasantness esteems one side as it were.

The hardness number of the epoxy-based fiber composites is introduced in Fig. 3. The composite C1, for example, unadulterated epoxy composite accomplished 45.5 hardness number. The epoxy/groundnut shells ash (C2) composite accomplished 40.5 hardness number. C2 composite has a lower hardness number than the C1 composite. The goat hair fiber and groundnut shells cinder/epoxy composite (C3) show the higher hardness number than the other composite. The cow-fiber-strengthened epoxy composites, for example, C4 composite, show higher hardness number than the C1 composite and lower hardness number than the C3 composite. It was discovered that the hardness number of the composite C3, for example, goat hair fiber and groundnut shells fiery debris/epoxy composite, achieves the higher hardness number which is 46.6, which demonstrates that the

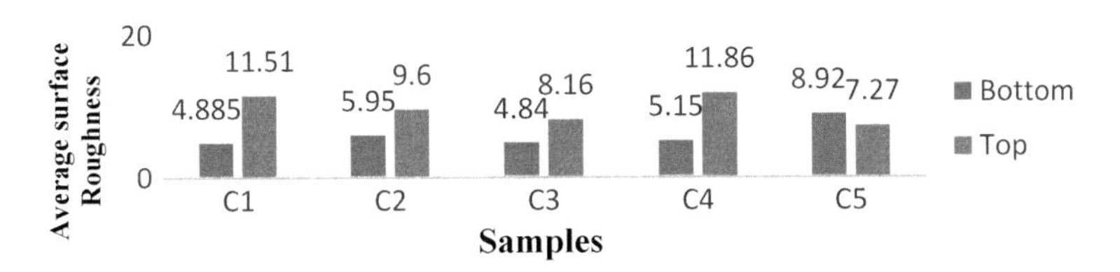

Fig. 1 Comparison of average surface roughness of top and bottom surfaces of composites

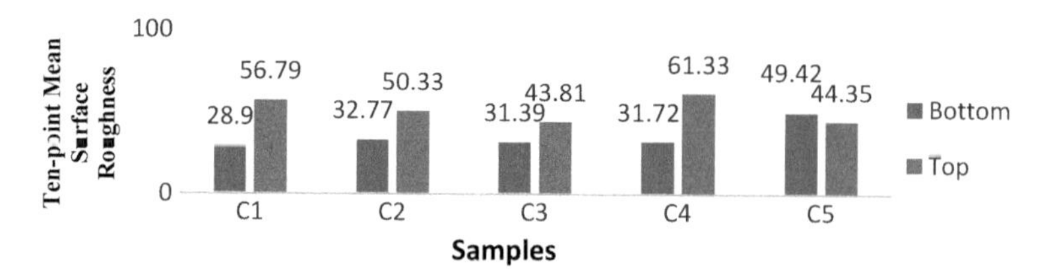

Fig. 2 Comparison of ten-point mean surface roughness of the top and bottom surfaces of composites

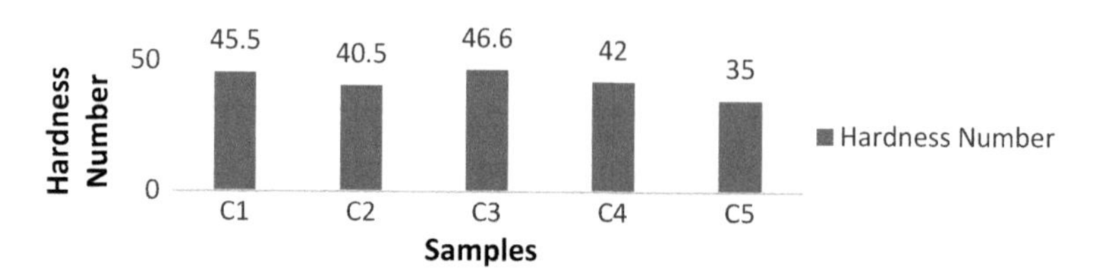

Fig. 3 Comparison of hardness numbers of composites

communication between the framework and the material is greatly contrasted with alternate materials. In this investigation, we saw that surface harshness impacts the hardness number of the composites. Among the all C3 composites accomplished low surface harshness, it achieved the higher hardness number. In this investigation, we saw that the communication between the two stages has a bigger number of impacts on hardness number than the surface harshness of the composites. Naidu [3] hardness number of plant-based strands achieved 37.3 lower than the goat hair fiber and groundnut shells fiery debris/epoxy composite.

The impact strength of the epoxy-based fiber composites is shown in Fig. 4. In the impact test, both Izod and Charpy tests were conducted and compared below. Impact test determines the amount of energy absorbed by a material during the fraction. The absorbed energy is the measurement of the toughness of a material and acts as a tool to study temperature reliant on ductile–brittle adhesion. In Izod, the test material is tested in the vertical direction, and in the Charpy test, it is tested in the horizontal direction. The composite C1, i.e., pure epoxy composite, attained 1.007 J/mm^2 in Izod and 1.123 J/mm^2 in Charpy impact strength test. The epoxy/ groundnut shells ash (C2) composite attained lower impact strength than the C1 composite. The goat hair fiber and groundnut shell ash/epoxy composite (C3) exhibit a higher Charpy strength than the other composite. The cow-fiber-reinforced epoxy composites, i.e., C4 composite, exhibit higher hardness number than the C1, C2, and C3 composites. It was found out that the hardness number of the composite C3, i.e., goat hair fiber, and groundnut shells ash/epoxy composite attains the higher Charpy strength (i.e., 1.235 J/mm^2) when compared with the cow-fiber-reinforced

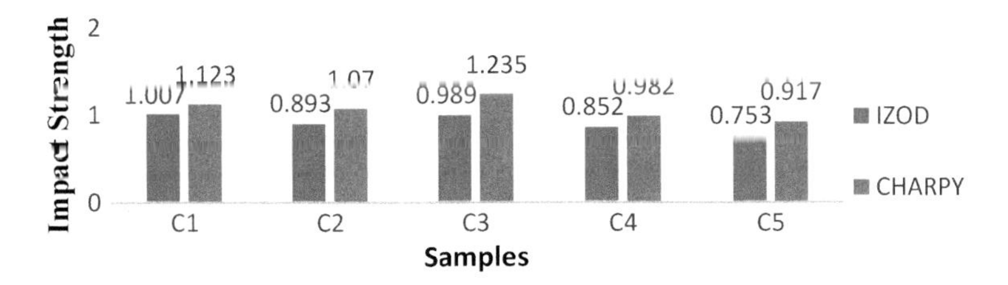

Fig. 4 Comparison of Izod and Charpy values of composites

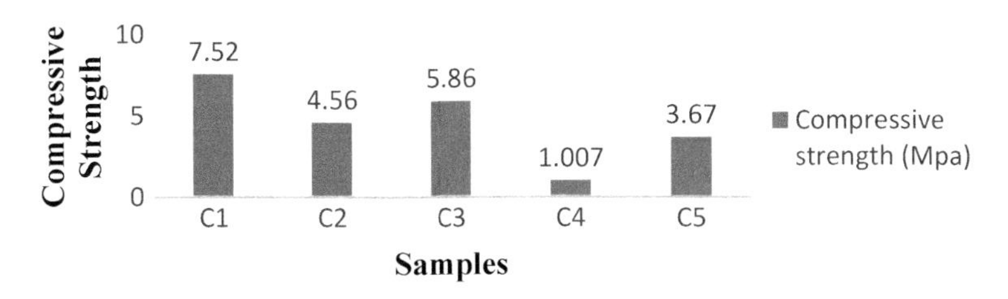

Fig. 5 Comparison of compressive strength of composites

epoxy composites, i.e., C4. Among all cow- and goat-hair-fiber-reinforced groundnut shells ash/epoxy-blended composite, C5 attained lower impact strength than the other composites. Naidu [3] impact strength of plant-based fibers attained 0.650 J/mm^2 lower than that of the goat hair fiber and groundnut shells ash/ epoxy composite.

The compressive strength of the epoxy-based fiber composites is shown in Fig. 5. Among all, pure epoxy sample, C1, attained the higher compressive strength after the goat-hair-fiber-reinforced groundnut shells ash/epoxy-blended composite C3 attained the 5.86 MPa strength. Compressive strength depends on the matrix material, reinforcement, and also on the bonding between the matrix and reinforcement. Cow-hair-fiber-reinforced groundnut shells ash/epoxy-blended composite has the lower compressive strength than that of the other composites. C2, C4, and C5 composites attain lower compressive strength than other composites. Naidu [3] compressive strength of plant-based fibers attained 24.4 MPa higher than the goat hair fiber and groundnut shells ash/epoxy composite. Human hair/epoxy composite exhibits higher compressive strength than the goat hair fiber and groundnut shells ash/epoxy composite [10].

SEM micrographs (X 1000) images of five composites (impact velocity 50–70 m/s, impingement angle 90° and erodent size 500 μm) are shown in Fig. 6. SEM images exposed that groundnut shells' ash acts as filler material. Addition of the filler material helps to enhance the interfacial bonding between the two phases.

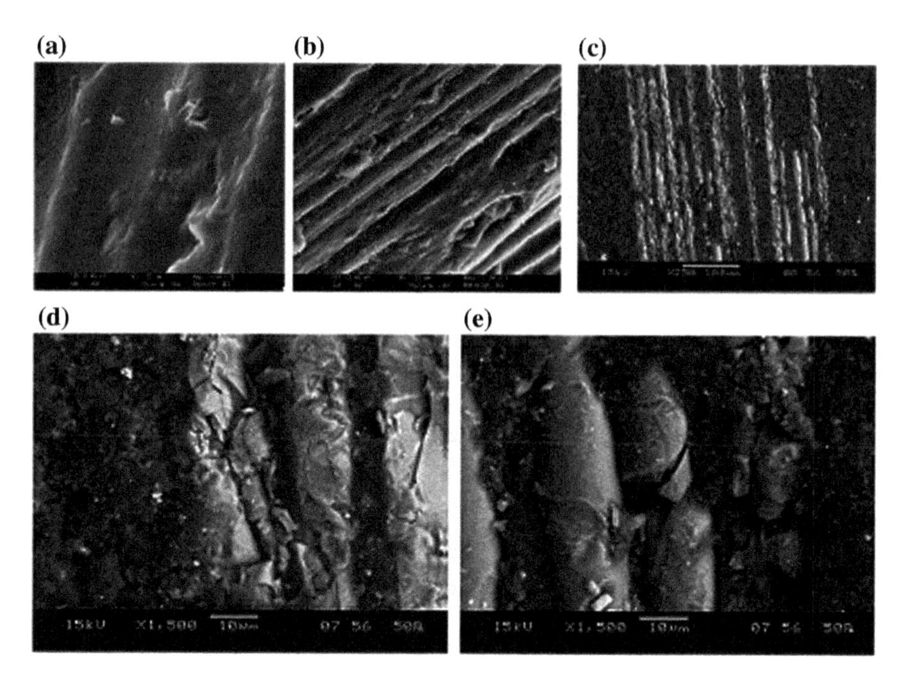

Fig. 6 Microstructure of composites **a** C1, **b** C2, **c** C3, **d** C4, and **e** C5

Mechanical behavior of the composites depends on the interfacial bonding of polymer and fiber, and better interfacial bonded composites obtain better mechanical behavior.

4 Conclusion

In this paper, five different composites by the hand-lay-up process are prepared and simultaneously tested. Compared to cow fiber composite (C4), goat fiber composite (C3) has superior mechanical properties. Interfacial bonding between the fiber and matrix is the major impact factor in the polymer matrix composite. Among all composites, goat fiber composite (C3) has better interfacial bonding. Also, observed that Hand-lay-up process produces better surface finish only on one side. Cow fibers are constituted of keratin and amino acids. Keratin and amino acids are basically hydrophobic nature, which tends to absorb the moisture from the atmosphere. This phenomenon results in the dimensional changes of the composite. Chemical treatment of the fibers reduces the moisture absorption of composites. These composites can be used in various automobile applications.

References

1. Pickering KL, Efendy MA, Le TM (2016) A review of recent developments in natural fibre composites and their mechanical performance. Compos A Appl Sci Manuf 83:98–112
2. Srinivas K, Naidu AL, Raju Bahubalendruni MVA (2017) A review on chemical and mechanical properties of natural fiber reinforced polymer composites. Int J Performability Eng 13(2):189–200
3. Naidu AL, Srinivas K (2018) Experimental study of the mechanical properties of banana fiber and groundnut shell ash Reinforced epoxy hybrid composite. Int J Eng (IJE), IJE Trans A: Basics 31(4):659–665
4. Jayaseelan J, Vijayakumar KR, Ethiraj N, Sivabalan T (2017) The effect of fibre loading and graphene on the mechanical properties of goat hair fibre epoxy composite. IOP Conf Ser: Mater Sci Eng 282(1):01
5. Oladele IO, Olajide JL, Ogunbadejo AS (2015) The influence of chemical treatment on the mechanical behavior of animal fibre-reinforced high density polyethylene composites. Am J Eng Res 4(2):19–26
6. Oladele IO (2018) Effects of fiber fraction on the mechanical and abrasion properties of treated cow hair fiber reinforced polyester composites. Tribol Ind 40(2)
7. Verma A, Singh VK, Verma SK, Sharma A (2016) Human hair: a biodegradable composite fiber–a review. Int J Waste Resour 6(206):2
8. Barone JR, Schmidt WF (2005) Polyethylene reinforced with keratin fibers obtained from chicken feathers. Compos Sci Technol 65(2):173–181
9. Cheng S, Lau KT, Liu T, Zhao Y, Lam PM, Yin Y (2009) Mechanical and thermal properties of chicken feather fiber/PLA green composites. Compos B Eng 40(7):650–654
10. Nanda BP, Satapathy A (2017, February) Processing and characterization of epoxy composites reinforced with short human hair. IOP Conf Ser: Mater Sci Eng 178(1):01
11. Lee SM, Cho D, Park WH, Lee SG, Han SO, Drzal LT (2005) Novel silk/poly (butylene succinate) biocomposites: the effect of short fibre content on their mechanical and thermal properties. Compos Sci Technol 65(3):647–657
12. Andrea W, Hiziroglu S (2007) Some of the properties of wood–plastic composites. Build Environ 42(7):2637–2644

A Systematic Approach to Identify the Critical Parameters of Two-Wheeler E-Vehicles

Deepak Singh, Golak Bihari Mahanta and BBVL. Deepak

Abstract A developing country like India is facing increasing challenges to make urban transportation sustainable and to manage the continuously growing air pollution. As in India, most people using the conventional vehicle, it is better to use an electric vehicle for saving our environment. Based on data publicly available from the official Web site of several well-known manufacturers, this paper aims at presenting a statistical analysis of standard specification and finding the critical parameter that will influence the customer to switch to EV. For detecting the critical parameter while purchasing an e-motorbike/e-scooter, a small survey was conducted. It will help the designer while designing an electric vehicle to improve critical parameters and make electric vehicle more practical to market.

Keywords Electric vehicle (EVs) · Speed · Range · Motor power · Weight · Statistical analysis

1 Introduction

Air pollution is a major issue in developing country like India. It is very challenging to make urban transportation sustainable and to manage the continuously rising air pollution. As air pollution is increasing with a very high rate, one of the primary sources is the use of the vehicle that almost contributes 11% of overall carbon emission [1]. As electric vehicles are used in European and China, this report is focused on finding out the main critical parameter that creates hindrance in purchasing of electric vehicle (EVs) in India. An electric vehicle uses electrical energy as the source of power as petrol and diesel use in the conventional vehicle [1]. As petrol or diesel use for running the internal combustion engine (IC Engine), when the piston moves from the top dead center (TDC) to bottom dead center (BDC) crankshaft rotate and that crankshaft rotation uses as the mechanical source of energy, and this energy is used for propelling the vehicle [2]. In the process of

D. Singh (✉) · G. B. Mahanta · BBVL. Deepak
Industrial Design Department, NIT Rourlela, Orissa 769008, India

BBVL. Deepak et al. (eds.), *Innovative Product Design and Intelligent Manufacturing Systems*, Lecture Notes in Mechanical Engineering,
https://doi.org/10.1007/978-981-15-2696-1_22

converting chemical energy into mechanical energy, some harmful gases are releases in the environment like carbon dioxide, carbon monoxide etc. As we are using the IC engine from a very long time, it creates the very worst scenario for the next generation. An electric vehicle uses electric energy to run the motor, and rotation of the shaft of the motor is used as a mechanical source of energy [1]. During the conversion of chemical energy into electrical energy, no waste gas is produced, so that is why it is called as green energy or clean energy [3]. As we are suffering from various diseases due to air pollution, we have to switch to the electric vehicle to save our environment. In India, 11 states are included in the top 20 most polluted states in the world [4]. If this scenario is continued, there will be a very adverse effect on future generations. One of the primary sources of air pollution is the transportation system. The government is also planning to increase market penetration of electric vehicle by giving incentives to buyer and manufacturer for that government launches faster adoption and manufacturing of electric vehicle in India (FAME-2) scheme [5]. This scheme will help the manufacturer and buyer to switch from a conventional vehicle. In FAME-2, government increases the tax for manufacturers and buyer of the convention vehicle and money, which collected by this will be used in giving incentive to an electric buyer and electric manufacturers [5]. Globally, adoption of an electric vehicle is increasing at a high rate. China, UK and in the USA—these countries are giving an incentive to the buyer, free or reduced the price for parking this increase the rate of adoption for an electric vehicle [6]. Renewable sources of energy can be used to recharge the battery of EVs, and in a country like India, where the third fourth of year solar energy available in plenty amount manufacturer should focus on the improvement of our transportation system and decreases our dependency on fossil fuel. The main critical parameter a buyer sees in any of the electric vehicles is speed, the range on per charge, charging time, charging location, and price of the electric vehicle.

2 Literature Review

Air pollution causes many diseases as it is showed in study on Nature of air pollution, emission sources, and management in the Indian cities by Guttikunda et al. [7, 8] find that if we proceed with current pace, then by 2030 we will face number of diseases like premature death, ischemic heart disease, chronic obstructive pulmonary diseases, lower respiratory infections, and cancers (in trachea, lungs, and bronchitis). They also suggest using an electric vehicle to save our environment. A case study on an electric scooter by Bishop et al. [9] had done the investigating the technical, economic, and environmental performance of electric vehicles in the real world by using the electric scooter and tracking with the help of global positioning satellite (GPS). The key finding in this study was that if we use electric vehicle, then we can save our environment from GHG by many tons, and if we use electric energy producing by other means, then it will be more helpful for our environment. A study on the usage of the electric scooter in urban condition is by Hardt and Bogenberger [10] in Munich, Germany. In this paper, they try to use e-scooter as an alternative to the car. The key finding of this study is that, parking of e-scooter was easy as compared to car but when we compared e-scooter to the car in other feature like weather dependent, storage capacity, and subjective safety then e-scooter somewhat having disadvantages over the car segment. Zhang et al. [11] design a regenerative shock absorber to generate electricity and store it for extension of range. Regenerative shock absorber mainly has four component suspension vibration input modules, transmission system, generator, and power storage capacitor. While the vehicle is running on the rough road, then vibration input to the shock absorber and transmission system passes this vibration to the generator with the conversion of motion. Generator is producing electric energy, and this electric energy is stored in a capacitor which will be used for range extension. Electric energy is a source of energy for propelling the vehicle, and Qiu and Wang [12] introduce the regenerative braking system. The key finding of this study is that regeneration braking system can increase the range of EVs by 24.63%.

3 Methodology

To find out the primary objective of this paper i.e. find out critical parameter that will help the designer of e-scooter/e-motorbike to satisfy the customer need and will help to increase the market penetration of the electric vehicle. For the fulfillment of the above objective, the process is described in Fig. 1.

3.1 Data Collection

Speed and range—these two parameters are most important for an electric scooter or electric motorbike. To know the dependence of these parameters on other parameters, data were collected from the public source of well-known manufacturers. Speed, range, motor power, weight, and price are the basic parameters which were to be taken from the specification of model for the fulfillment of the study.

3.2 Statistical Analysis and Data Interpretation

Once all the specification is compiled, a statistical analysis of data can be conducted. Accordingly, an overview of the resulting analyzed data is shown in Table 1 which lists mean, median, mode, and standard deviation of all the selected specification of the e-motorbike/e-scooter (Tables 2 and 3).

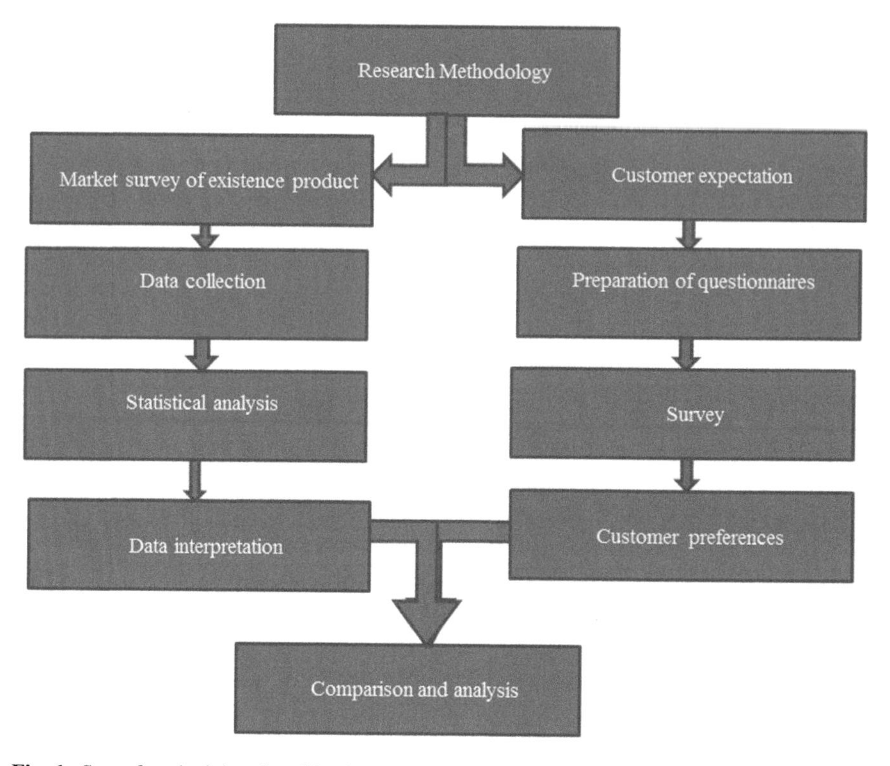

Fig. 1 Steps for obtaining the objective

3.3 *Preparation of Questionnaire*

A detailed questionnaire has been structured remembering the goals of the investigation. A multiple-choice question has been utilized relying upon the idea of data looked for from the shopper. The survey was conducted in conformance to the protocols of the Institute Ethics Committee (IEC), NIT Rourkela (India). All the data collected was anonymized to ensure the protection of participant identities. The poll has been intended to begin with a couple of warm-up inquiries in the initial and afterward advance into increasingly pointed inquiries that will help comprehend the buyer inclinations and decision making. The aim is to hold the interaction under 10 min and will establish between 15 and 20 inquiries taking all things together. The questionnaire is divided into five sections:

1. Age and gender variation: this section captures the age and gender of participants. It will help to understand the choice of different ages.
2. Awareness about the electric vehicle: this section captures awareness about an electric vehicle that includes awareness about incentive given by government and government scheme about an electric vehicle.

Table 1 Statistical values of the standard specification

	Mean		Median		Mode		Standard deviation	
	E-motorbike	E-scooter	E-motorbike	E-scooter	E-motorbike	E-scooter	E-motorbike	E scooter
Speed (km/hr)	153	49	158	45	137	25	8.8105	3.9
Range (km)	151	83.85	150	70	200	70	11.6028	[illegible]4.34
Power (W)	48,102	3243	40,000	1000	45,000	250	5587.894	8[illegible]1.16
Charging time (h)	6	6.35	5	7	4	8	3.022	0.31[illegible]0

Table 2 Statistical values of the standard specification

	E-motorbike		E-scooter		Range	
	Max	Min	Max	Mini	E-motorbike	E-scooter
Speed (km/h)	350	50	160	18	300	142
Range (km)	337	43	280	35	294	245
Power (W)	149,140	1000	35,000	137	14,840	34,863
Charging time (h)	12	2	12	2	10	10

Table 3 Distribution of different parameter and comparison between them

	E-motorbike	E-scooter
Average battery backup time (h)	1–1.5	0.5–1
Carrying load capacity (kg)	150	130

3. Participants preference about the electric vehicle: To increase the market penetration of electric vehicle, it is essential to understand the behavior of customer about an electric vehicle. The behavior of customer is captured in this section.
4. The expectation of participants: this section captures expectation about speed, range, charging time, and price. It is also essential to know expectation about critical parameter so that designer will design according to expectation.
5. Expectation about charging location: this area captures expectation about charging location, facility and if they want to be charging location at public, then the expectation of distance between two charging locations.

4 Result and Discussion

A survey was conducted in which 109 people have participated. With the help of questionnaire, try to include all the critical data that will help the manufacturers try to understand why they lack with a comparison of the internal combustion engine and the overviews of the survey are summarized below:

Gender variation: In the survey, 88.1% male and 11.9% female have participated. In Fig. 2, distribution of gender variation is shown with the help of a chart.

Age variation: In Fig. 3, age variation of the participant has been shown. As it can be observed from histogram that most of the participants are from young ages which vary from 18 years to 30 years, it can be concluded that the result is close to real scenario because this age group is more prone to buy any type of vehicle in the near future.

Percentage of participants has an electric vehicle: From Fig. 4, it can be observed 7.3% of participants have their electric vehicle (EV) and 92.7% of do not have their EV.

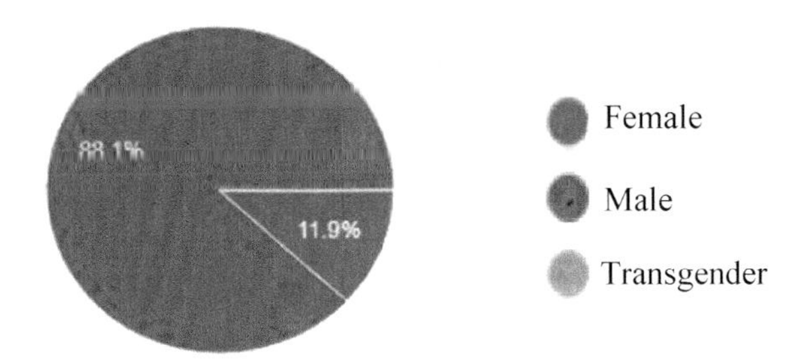

Fig. 2 Distribution of gender

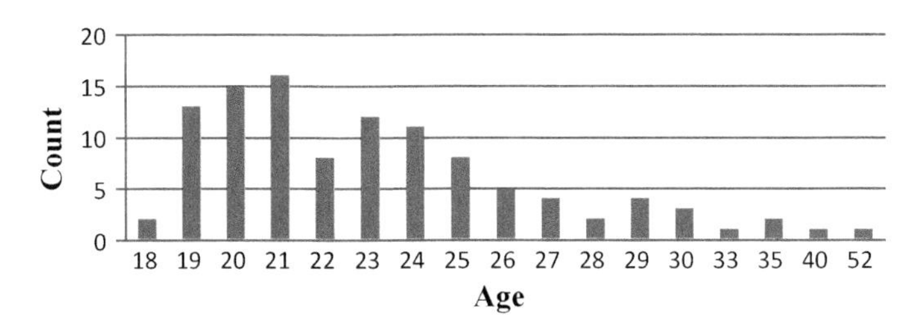

Fig. 3 Histogram of age variation among survey participants

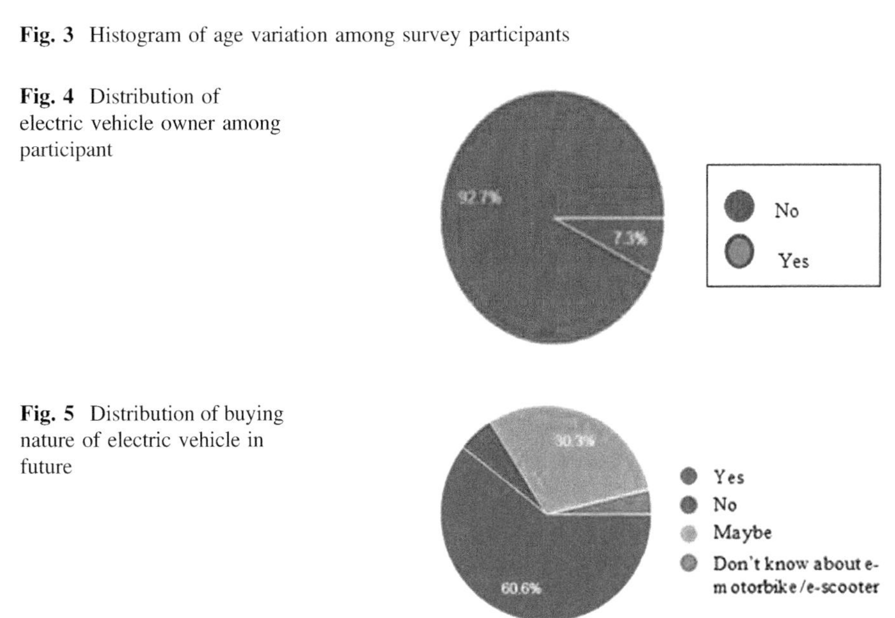

Fig. 4 Distribution of electric vehicle owner among participant

Fig. 5 Distribution of buying nature of electric vehicle in future

Percentage of participants is ready to buy e-scooter/e-motorbike in the near future: The above Fig. 5 distribution shows the nature of buying EV shortly. It can be observed from distribution 60.6% of participant ready to purchase EV soon and 30.3% of participants are ready to buy if we provide best feature to customer, that

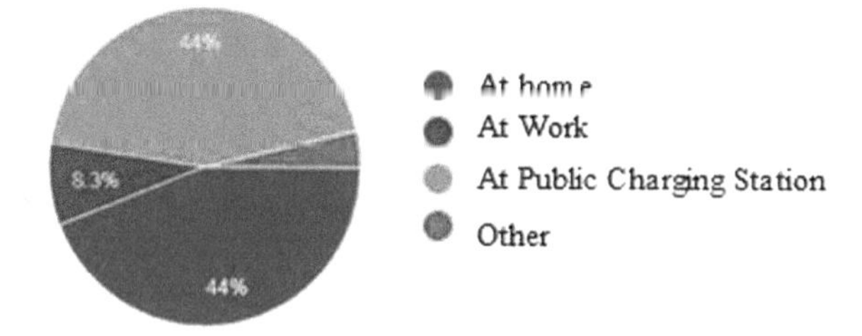

Fig. 6 Distribution of charging location choice

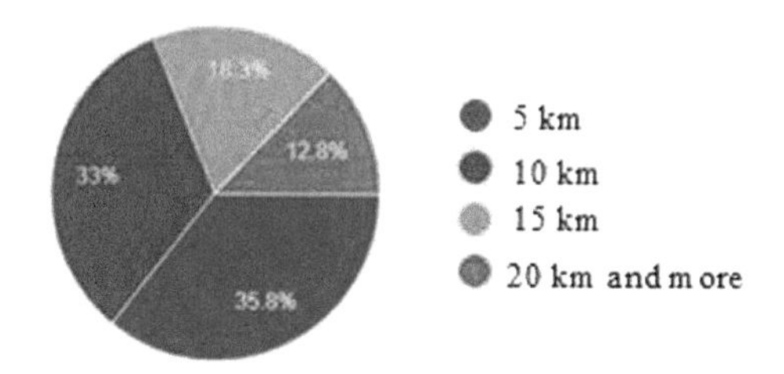

Fig. 7 Distribution of distance between two public charging locations

means the majority of participants are ready to buy an EV in the near future. 3.7% of the participant does not know about e-motorbike/e-scooter. 5.5% of participants are not ready to buy an EV in the near future.

Location of charging facility: Location of charging facility is one of the critical parameters in purchasing of an electric vehicle. In Fig. 6, it can be observed that 44% of participant wants charging facility at home. 44% of participant wants charging facility at a public charging station. So, it can be said that most of the participants either charging facility at home or a public charging station. 8.3% of participant wants charging facility at work as a percentage of this choice is very low; it can be ignored. So, the manufacturer must keep in mind that charging facility either at home or at a public charging location.

The distance between two public charging locations: Figure 7 shows the demand for the distance between two public charging stations of participants. The main hindrance that is occurring in purchasing of EV is the location of the charging station because the range of e-scooter/e-motorbike is quite low. So, it has to charge it regularly. It can be observed from the distribution that 68% of participants want distance between two public charging stations must be below 10 km. 86% of participants want distance between two public charging stations that must be below 15 km.

Maximum allowed waiting for 40 km range extension by the participant: In Fig. 8, distribution shows the maximum allowed waiting time by participants for 40 km range extension. 44% of participants wait up to 10 min for 40 km range extension. 82% of participants can expect maximum wait up to 20 min for 40 km range extension. So, manufacturers keep this in mind that battery must have fast charging quality.

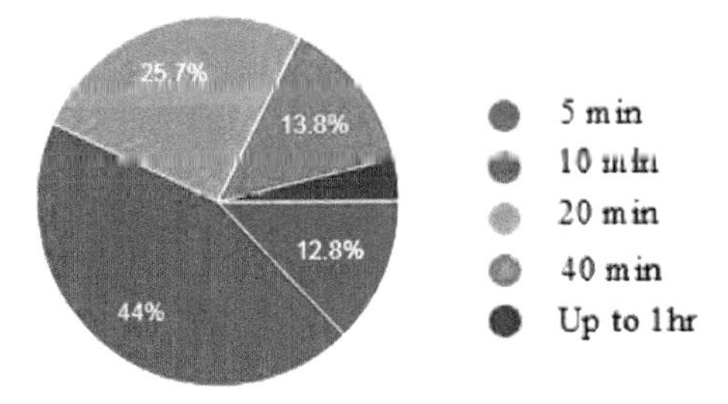

Fig. 8 Distribution of maximum allowed waiting time for 40 km range extension

This section of the paper shows the outcome of the survey and the current manufacturer's specification. After comparing real scenario and exiting product, it is beneficial to see where current manufacturers are lacking and in which segment they have to focus.

Comparison between actual availability to expectation: Once all the specifications are compiled and as well survey result is also compiled, then it became effortless to compare between what is the real situation and what people want. The analyzed data had been shown in Table 4 which lists the speed, range, and charging time. It can be observed from Table 4 that 56% of participants expect that speed of e-scooter/e-motorbike must be above 75 km/h and 34% of participants expected speed of e-scooter/e-motorbike must be above 50 km/h; in e-scooter segment, 70% of model that had been taken for analysis have speed below 50 km/h that means and 70% of available e-scooters do not satisfy customer demand. 18% of e-scooter has speed range between 51 and 75 km/h and only 5% of e-scooter have a hundred plus speed. In the e-motorbike segment, 85% of e-motorbikes have a speed of more than 100 km/h e-motorbike somewhat fulfills the demand of customer in speed criteria. 52% of participants expected the range of must to be above 70 km on per charge. 34% of participants want range between 50 and 70 km on per charge. In e-scooter, 55% have a range more than 50 km on per charge, and only 33% of e-scooters have a range above 70 km in one charge. It can be observed that only 33% of e-scooter fulfills the expectation of 52% of participants. In the e-motorbike segment, 29% of the model has ranged between 70 and 100 km and 61% of models have a range above 100 km. It can be observed that 90% of e-motorbikes are fulfilling the expectation of 90% of participants. Charging time of an electric vehicle is a critical parameter. 70% of participants want charging time from 2 to 4 h; only 12% of participants want charging time for more than 6 h. In the e-scooter segment, 28% of models have to charge time below 4 h that means only 28% of models fulfill the 70% of participants. 72% of models have charging time more than 4 h or, in other words, it can be said that 72% of models are not fulfilling the demand of the customer. In the e-motorbike segment, 44% of models have charging time below than 4 h. 27% of models have charging time for more than 8 h. It can be observed that only 44% of models satisfy the demand of 70% of participants. So, it can be observed from the above analysis that current manufacturers lacking in charging time most and e-scooter need improvement in speed, range, and charging time. E-motorbike fulfills the demand in speed and range, but it is also lacking in charging time.

Table 4 Statistical values of the standard specification

	E-motorbike	E-scooter	Expected
Speed (km/h)	0-50 51-75 75-100 100-151+ 3% 5% 7% 85%	0-50 51-75 75-100 100-150+ 5% 7% 18% 70%	25-50 10% 100-150 24% 50-75 34% 75-100 32%
Range (km)	30-50 50-70 71-100 100+ 3% 7% 29% 61%	30-50 50-70 71-100 100+ 12% 19% 14% 55%	30-50 14% 100+ 19% 50-70 34% 70-100 33%

(continued)

Table 4 (continued)

	E-motorbike	E-scooter	Expected
Charging time (h)			

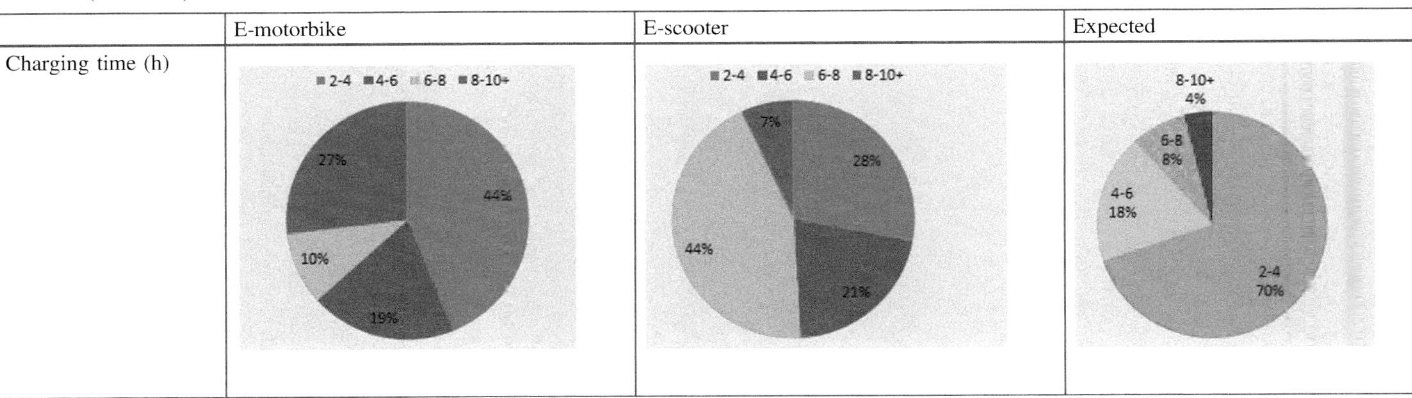

5 Conclusion

The automotive business in India is at a cross-street between proceeding on the standard course or making strong strides and jumping into the new domain, from individual portability to shared versatility and from conventional mobility to electric mobility. From survey and data interpretation of current manufacturer, it is very clear to observe that where a current manufacturer is lacking from the expectation of customers. Government of India recently launched a scheme known as FAME-2 providing an incentive for the buyer and manufacturer, but in a real scenario, very few people know about this. Current manufacturer is lacking in speed, range, and charging time. As India is a developing country and the population is also increasing with a significant rate, if a certain strong implementation does not come in the transportation system, then our future generation will face a number of problems.

References

1. Larminie J, Lowry J (2012) Electric vehicle technology explained. Wiley
2. Nag PK (2013) Engineering thermodynamics. Tata McGraw-Hill Education
3. Thomas CE (2009) Fuel cell and battery electric vehicles compared. Int J Hydrogen Energy 34(15):6005–6020
4. World Health Organization (2016) WHO global urban ambient air pollution database (update 2016). Diunduh, Geneva
5. Faster Adoption and Manufacturing of (Hybrid &) Electric Vehicles https://www.autocarindia.com/industry/fame-ii-scheme-for-ev-adoption-in-india-to-begin-October-1-409477. [Online]; Accessed 25 Feb 15, 2019
6. Dhar S, DTU UNEP, Cherla S A study of electric mobility for city of Hyderabad
7. Guttikunda SK, Goel R, Pant P (2014) Nature of air pollution, emission sources, and management in the Indian cities. Atmos Environ 95:501–510
8. Dhar S, Pathak M, Shukla PR (2017) Electric vehicles and India's low carbon passenger transport: a long-term co-benefits assessment. J Cleaner Prod 146:139–148
9. Bishop JD, Doucette RT, Robinson D, Mills B, McCulloch MD (2011) Investigating the technical, economic and environmental performance of electric vehicles in the real-world: a case study using electric scooters. J Power Sources 196(23):10094–10104
10. Hardt C, Bogenberger K (2019) Usage of e-Scooters in Urban environments. Transp Res Procedia 37:155–162
11. Zhang Z, Zhang X, Chen W, Rasim Y, Salman W, Pan H, Yuan Y, Wang C (2016) A high-efficiency energy regenerative shock absorber using supercapacitors for renewable energy applications in range extended electric vehicle. Appl Energy 178:177–188
12. Qiu C, Wang G (2016) New evaluation methodology of regenerative braking contribution to energy efficiency improvement of electric vehicles. Energy Convers Manag 119:389–398

Design for Creativity and Optimization

Experimental and Numerical Investigation of Dynamic Interconnected Anti-roll Suspension System in Automobile Application

S. Deepankumar, B. Saravanan, R. Gobinath, B. Balaji, K. Siva Suriya and V. Nitheesh

Abstract This research work extends on vehicles with dynamic interconnected anti-roll suspension (DIAS) systems. A hydraulic anti-roll system for a vehicle includes a hydraulic actuator, an anti-roll control module, and an anti-roll bypass valve. The first hydraulic actuator is adapted to be connected between the suspension and frame of the vehicle on the one side and the second hydraulic actuators adapted to be connected between the suspension and frame of the vehicle on its other side. The anti-roll control module stiffens the compress the first hydraulic actuator relative to the expansion take in the second hydraulic actuator and stiffens the compress the second hydraulic actuator relative to the expansion take in first hydraulic actuator. The anti-roll bypass valve is adapted to operate the stiffening of the anti-roll control module. Anti-roll bar as part of a vehicle suspension system is a high standard configuration widely used in vehicles to arrange the essential roll stiffness to enhance vehicle handling and safety during fast cornering. However, the defect of the anti-roll bar is apparent that they restraint the wheels travel on the uneven road surface and weaken the wheel/ground holding ability, particularly in articulation

S. Deepankumar (✉) · B. Saravanan · R. Gobinath · B. Balaji · K. Siva Suriya · V. Nitheesh
Department of Automobile Engineering, Bannari Amman Institute of Technology, Sathyamangalam 638401, India
e-mail: deepankumar@bitsathy.ac.in

B. Saravanan
e-mail: saravananb@bitsathy.ac.in

R. Gobinath
e-mail: gobinathr@bitsathy.ac.in

B. Balaji
e-mail: balaji@bitsathy.ac.in

K. Siva Suriya
e-mail: sivasuriya.au16@bitsathy.ac.in

V. Nitheesh
e-mail: nitheesh.au16@bitsathy.ac.in

BBVL. Deepak et al. (eds.), *Innovative Product Design and Intelligent Manufacturing Systems*, Lecture Notes in Mechanical Engineering,
https://doi.org/10.1007/978-981-15-2696-1_23

mode. Roll-plane hydraulically interconnected suspension (HIS) system, as a potential replacement of anti-roll bar, could effectively increase vehicle roll stiffness without compromising vehicle's flexibility in articulation mode.

Keywords Dynamic interconnected · Anti-roll · Suspension · Hydraulic · Anti-roll control module

1 Introduction

Interconnected suspensions have proven commercially successful—both in the past and at present—yet the attention they have received in the research community has not, in the author's opinion, been commensurate with their commercial success, for example, a number of widely cited review papers from the 1980s and 1990s focusing on road vehicle suspension made little or no mention of interconnected suspensions. A small number of relevant publications have appeared in the last decade, but the area remains largely underexplored most of the literature on interconnected suspensions either assumes that the interconnecting mechanism is ideal, or a nonlinear model is used, which requires ample computational effort to obtain an acceptable numerical solution. A methodology is therefore required for modeling these systems efficiently in the frequency domain while still accounting for non-ideal interconnections. In recent years, an Australian company, Kinetic Pvt. Ltd, has had a great deal of success in inventing, developing, and applying—to rally cars and passenger vehicles—a number of unique hydraulically interconnected suspension (HIS) systems. It is these kinetic suspensions that form the particular focus of this thesis, although the approaches used are intended to be relevant to virtually any suspension employing hydraulic interconnection (Fig. 1).

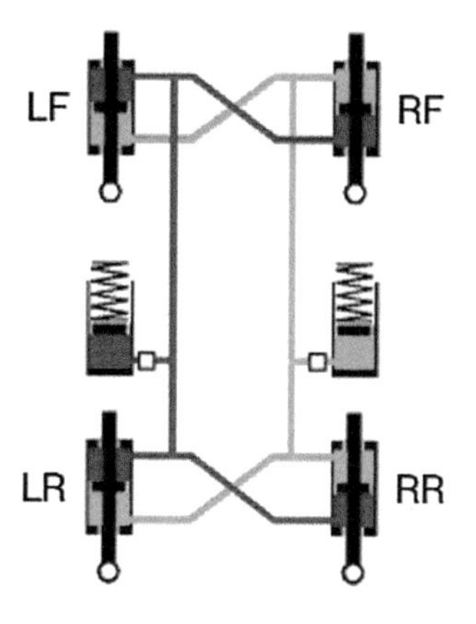

Fig. 1 Interconnected suspensions

1.1 Vehicle Dynamics

Vehicle dynamics is a field of engineering primarily based on classical mechanics. It is used to explore and understand the response of a vehicle in various in-motion situations. Vehicle dynamics plays an important role to determine the safety, handling the response, and ride quality of many different kinds of vehicles (Fig. 2).

1.1.1 Coordinate Frame

Before the analysis of the kinematics and dynamics of vehicle motion, an appropriate coordinate frame must be selected first for expressing the equations of motion. A vehicle coordinate frame B (C_{xyz}), of which the origin is attached to the mass center of the vehicle. The longitudinal axis that passes through C in the forward direction is considered as the x-axis. The y-axis is in the lateral direction from right to left in the viewpoint of the driver. The z-axis is in the vertical direction, perpendicular to the ground and opposite to the gravitational acceleration. Three angles, the roll angle ψ about the x-axis, the pitch angle θ about the y-axis, and the yaw angle ϕ about the z-axis, are employed to express the orientation of the vehicle coordinate frame. Another three motion variables are introduced and called roll rate, pitch rate, and yaw rate. These are the components of the angular velocity vector in the x, y, and z directions, respectively.

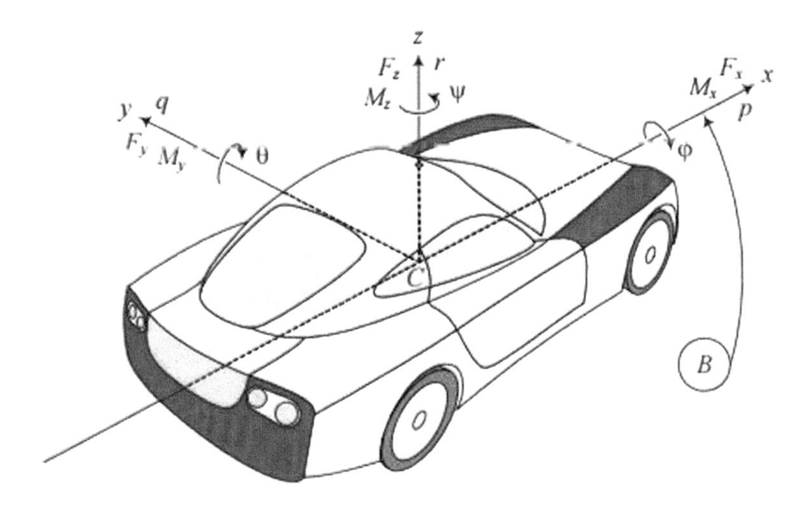

Fig. 2 Coordinate systems

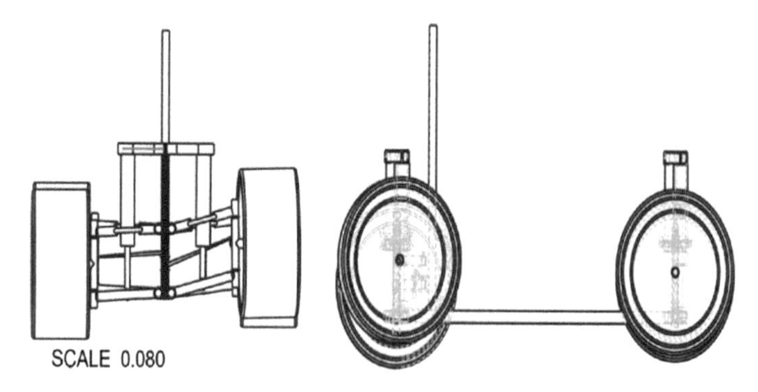

Fig. 3 Chassis design

2 Chassis Constructional Details

2.1 Chassis Design

The chassis consists of identical front and rear suspension geometry, i.e., double-wishbone parallel link suspension. The constructional arrangement can be explained as follows: A central load-carrying member is connected on both sides with two parallel wishbone links with a torsion bar apparatus consisting of cylindrical support at the central member and an anchoring joint at a further longitudinal distance into the frame. This way the torsion bar is twisted when the wishbone moves up and down, thereby providing spring stiffness for them to returning to mean position (Fig. 3).

The wishbones are further connected to kingpins with ball joint apparatus which is needed to facilitate rotation about two axes, i.e., steering axis and axis passing through wishbones. The lower wishbone is mounted with the anti-roll cylinder apparatus' lower end while the upper end of it is mounted to a pin in the frame. The kingpin is then mounted with the taper roller bearing above which are the tires are mounted. The third arm of the kingpin has provision for mounting of one end of the tie rod while the other end of it is connected to the steering column (Fig. 4).

3 Working of DIAS Systems

The cylinder piston assembly (actuators) is actually industrial purpose components and is bought from a used-product market at Coimbatore along with hydraulic lines, fittings, and flow control valves. Their mountings are for a different purpose, and hence, they have to be altered for mounting in a vehicle's wishbone. Here, the lower mounting is pin fixed to the lower wishbone and the upper one is to the chassis frame. The upper wishbone has to have enough space to allow the passage of cylinder piston assembly along with its lateral movements during movement.

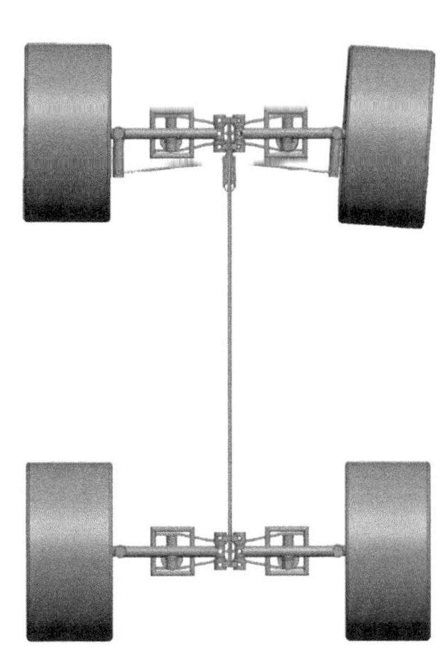

Fig. 4 Chassis design in Creo

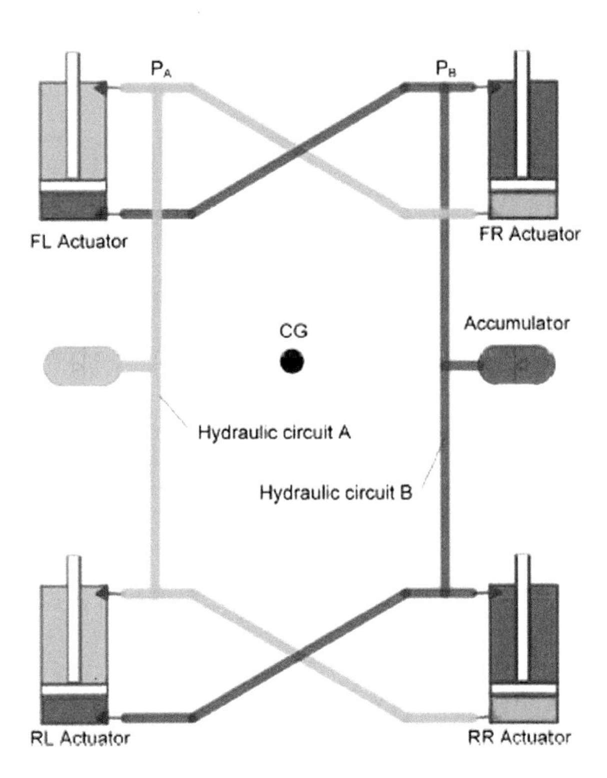

Fig. 5 Working of DIAS system

A hydraulic anti-roll system for a vehicle includes a two hydraulic actuator (one and two) and an anti-roll bypass valve. The X cylinder of the top chamber is connected to the Y cylinder of the bottom chamber, and the Y cylinder of the top chamber is connected to X cylinder of the bottom chamber. Before the flow control valves of both the hydraulic lines, two bypass lines connect runs in between the two hydraulic lines and thus connects the upper and lower chambers of the respective cylinders. The anti-roll module (which is here the setup consisting of the four flow control valves along with the ECU) is adapted to activate and deactivate the working of the system apart from ensuring ride comfort to some extent by gradually allowing relative cylinder displacements up to some extent. This setup is connected for the frontal end of the vehicle only, and the effects of setting up the system for the rear end and interconnecting the front and rear ends for anti-dive and anti-squat tendencies are not in the scope of the project. However, they are viable things to do and we can possibly take them up and incorporate in the project during the advanced stages of our project (Fig. 5).

4 Vehicle Specification

See Table 1.

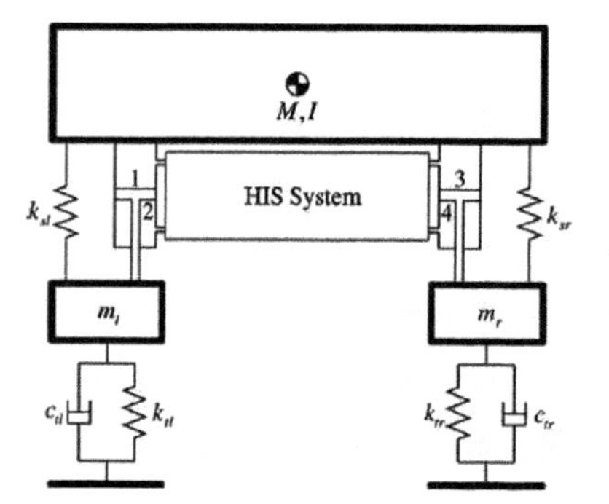

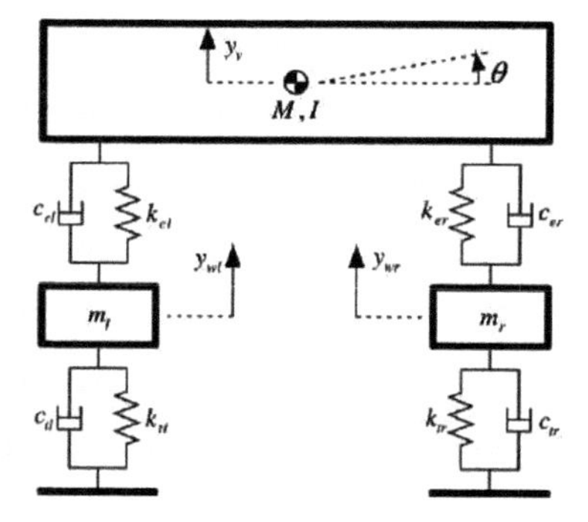

Table 1 Vehicle specification

S. No	Specifications	Details
1.	Overall length	1346 mm
2.	Overall width	736 mm
3.	Overall Height	838 mm
4.	Overall Weight	135 kg
5.	Wheelbase	930 mm
6.	Ground clearance	150 mm
7.	Suspension	Independent
8.	Tires	16*8–7 Tubed

5 Frequency Analysis of RHIS Under Warp Mode

The traction and control of the road vehicle are only through the contact patch of the tyres so that the contact between tyre and ground is vital for the vehicle's stability and safety especially under off-road rough terrain conditions (Figs. 6 and 7).

The comparison results of the tyre dynamic force under warp excitation, presented in Fig. 8, show that the vehicle has a lower tyre dynamic force similar to the vehicle in the whole frequency range, except for some difference presented near the first suspension natural frequency around 1.5 Hz owing to the phase change. This means that anti-roll bars not only increase the roll stiffness but also unfavorably increase the warp stiffness. Further, the tyre dynamic force varies to a much larger degree than the vehicle with the systems.

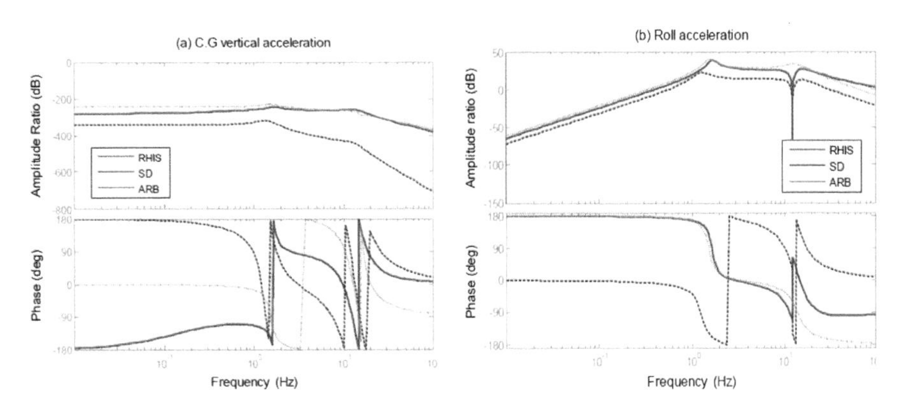

Fig. 6 Vehicle accelerations under warp excitation: **a** CG vertical and **b** roll

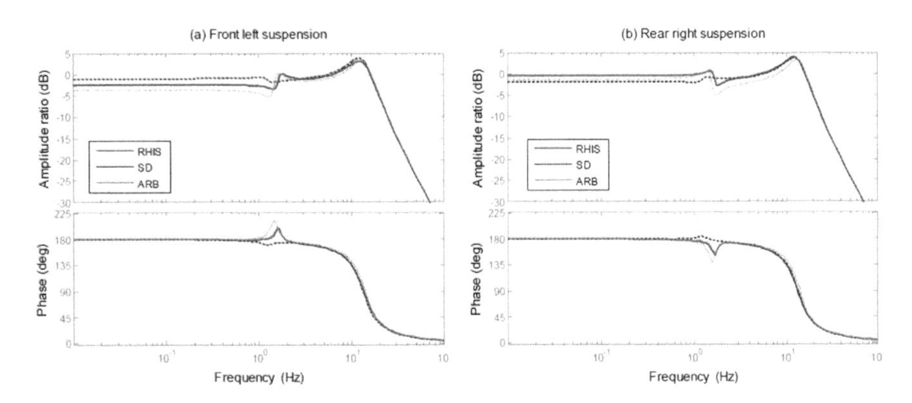

Fig. 7 Suspension deflection under warp excitation: **a** front left and **b** rear right

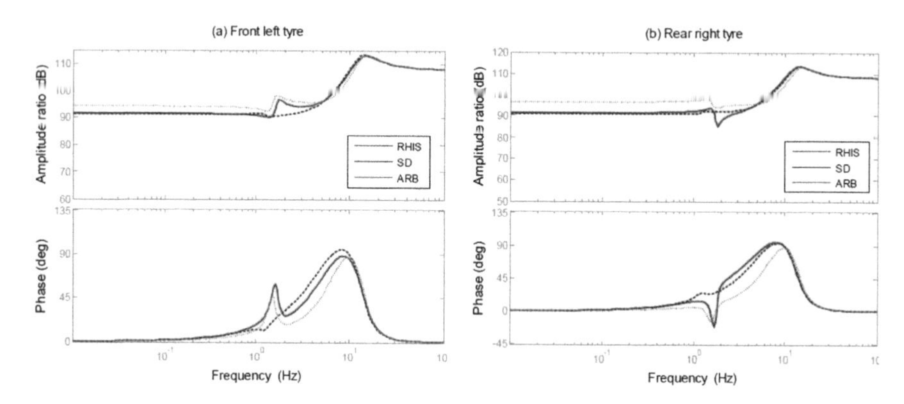

Fig. 8 Tyre dynamic forces under warp excitation: **a** front left and **b** rear right

6 Conclusion

This work employed a dynamically interconnected suspension (DIAS) and then studies the performance potential of such systems. The results of the study showed that this is possible if more amounts of ride comfort and anti-roll bar performance are compromised. The study showed that 90% load carried by mechanical springs and 10% load carried by the dynamically interconnected suspension (DIAS) are an optimized load distribution that will achieve possible strut sizes along with acceptable ride comfort and handling range. The reduction in anti-roll bar performance because of reduced distributed load taken by the dynamically interconnected suspension (DIAS) can be improved by incorporating the anti-roll bar. Fluid flow through interconnected pipe improves both static and dynamic properties of the suspension. In general, more ride comfort can be achieved from dynamically interconnected suspension (DIAS) as compared to the unconnected suspension system.

References

1. Eisele DD, Peng H (2008) Vehicle dynamics control with rollover prevention for articulated heavy truck
2. Shen X, Peng H (2003) Analysis of active suspension system with hydraulic actuators
3. Mohammadzadeh A, Haider S (2006) Analysis and design of vehicle suspension system using MAT lab and Simulink
4. Lu J, Messih D, Salib A, Harmison D (2007) An enhancement to an electronic stability control system to include a rollover control function
5. Smith W, Zhang N, Jeyakumaran J (2007) High frequency parameters sensitivity in hydraulically interconnected suspensions
6. Smith WA (2009) An investigation into the dynamic with hydraulic interconnected suspensions

7. Chu DF, Lu XY, Li GY Hendrick JK (2010) Rollover prevention for vehicles with elevated CG using active control
8. Ryan JS, Holbert TJ, Holden DJ, Peterson L (2011) Hydraulic anti-roll system
9. Xu G (2013) Experimental comparison of anti-roll bar with hydraulically interconnected suspension in articulate mode

Design and Modeling of Waste Container Package Considering Impact and Remote Handling Operation

Sunny Brar, Jaideep Gupta, Binu Kumar and K. M. Singh

Abstract In a radioactive waste management facility, solid waste is stored in stainless steel (SS) containers. Drop analysis of the container was performed to ensure the integrity of the container in case of the drop during the handling process. The simulation of drop test of the container from a height of 1.5 m with different orientations on a rigid base was carried out with the help of explicit dynamic method, and the results of stress and strain were analyzed based on which the dimensions and CAD modeling of the container were finalized. The design analysis and modeling of a radioactive waste SS container are done with the help of CAD tool. The container was designed in the square shape for the ease of handling it in every orientation with trunnions provided on each face for remote handling of the container which is operated by a remote handling crane.

Keywords CAD · CAA · Explicit dynamics · FEM · Remote handling

1 Introduction

The radioactive waste generated in the backend processing of nuclear fuel is transported and stored in SS container. The drop testing of the high-level radioactive waste transport container is a very important task for safety requirements and handling. During the handling operation of the waste container, there is a possibility of the sudden drop in the container. Hence, it is mandatory to ensure that in case of any accident the container does not fails to prevent the radioactive waste from exposure in the outer environment. With the advancement in computer

S. Brar (✉) · J. Gupta
Mechanical Engineering Department, NIT Kurukshetra, Kurukshetra, Haryana, India

B. Kumar · K. M. Singh
FRWMD, BARC, Mumbai, India
e-mail: binu@barc.gov.in

K. M. Singh
e-mail: kmsingh@barc.gov.in

BBVL. Deepak et al. (eds.), *Innovative Product Design and Intelligent Manufacturing Systems*, Lecture Notes in Mechanical Engineering,
https://doi.org/10.1007/978-981-15-2696-1_24

simulations, the finite element method (FEM) is applied for the analysis of the impact of the container with good accuracy. Bender et al. [1] have given a brief introduction about the development of computer simulations for the rigid body dynamics for the past 20 years. In addition, Weinstein et al. [2] proposed an approach for the dynamic modeling of the contact between the articulated rigid bodies, Furthermore, Smith et al. [3] presented simple generalized model for modeling the impact problems satisfying the five physical desiderata, i.e., beak away, symmetry preserved, energy bounded, momentum conserved, one-sided impulse and compared with few other models. Frano et al. [4] used Ansys program for the free drop test of IP2 package and validated with the experimental results seeing good conformity between both results further showing the advantages of the numerical simulations over the experimentation.

The high-level waste (HLW) generated from the radioactive waste processing plant cannot be directly disposed into the environment. This HLW is processed in the waste immobilization plant (WIP) and converted into glass material by the process of vitrification and are stored in the waste overpacks. This treated waste is then transported in these containers to the disposal sites where it is stored for a specific period. The present research is primarily focused on computer simulation of drop test of the radioactive waste drum package from a height of 1.5 m during handling in storage based on the explicit dynamics method. The waste drum made up of stainless steel is used to transport radioactive waste to the disposal location. During the transportation or handling of the waste container, there might be the possibility of impact/drop of the container with the floor or another container. So, it is essential to carry on the drop test of the container for different orientations. The obtained stress and deformation are analyzed so that the design and dimensions of the container are optimum according to the guidelines of the ASME code. Also, the container is provided with trunnions on each side for ease in remote handling of the containers which is operated by the over-hanged cranes.

2 Design Concept

Based on the design and analysis of radioactive waste, container modeling was carried out with the help of the CAD tool incorporating all safety requirements. The dimension and part details of the container are given in Table 1. The container is modeled in the form of a box having a square base with a flange at the top with a cover plate. A gasket is provided between the flange and the cover plate for maintaining leak tightening and the whole assembly is bolted.

Also, stiffeners are provided on the edges to provide support to the container. Mesh optimization was performed in order to produce an optimum model as the accuracy of FE model greatly depends on the quality of mesh generated [5–7]. Also, the weight of the container was optimized by selecting the optimum thickness of the walls of the container. The model is then analyzed for the drop test with the

Table 1 Dimensions of the container

S. no.	Part	Dimension (m)
1	Container	1.5 (L) × 1.5 (W) × 0.972 (H)
2	Container wall thickness	0.006
3	Flange	0.044 (outwards)
4	Flange	0.012 (thickness)
5	Cover plate	1.58 × 1.58
6	Cover plate	0.012 (thickness)
7	Stiffener	0.04 (dia) × 0.004 (thickness)

help of FE-based numerical simulation method. Based on the simulation analysis the design specification of the model was finalized.

3 Numerical Simulation

The numerical model was developed based on explicit dynamics method using finite element formulation. The model of the container as shown in Fig. 1 was discretized by taking the shell elements as the thickness of all the parts is very less and to reduce the computation time and the trunnions were modeled with solid elements. The total number of elements used was 63,339. The radioactive solid waste is fully packed inside the container and also, the mass of the waste is marginal in comparison to the mass of the container so the waste load can be assumed to be uniformly distributed on the container walls. In the simulation, the following assumptions were made: the solid radioactive waste is considered to be uniformly distributed in the container, mass of the radioactive waste is considered by increasing the density of the container to 10,000 kg/m^3, the material of the

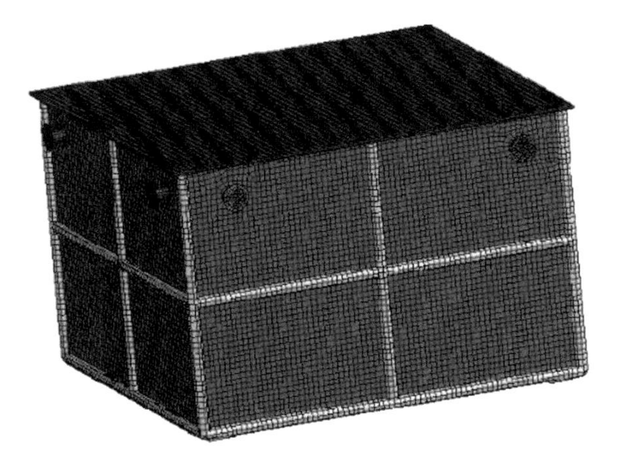

Fig. 1 FE mesh model of the container

Table 2 Properties of the container

S. no.	Properties	Value
1	Mass	1206.29 (Kg)
2	Elastic modulus (E)	2E+11 (Pa)
3	Density	10,000 (Kg/m^3)
4	Poisons ratio	0.27
5	Yield stress	3.1E+08 (Pa)
6	Tangent modulus	7.63E+10 (Pa)
7	Strain rate parameter (C)	40
8	Strain rate parameter (P)	5
9	Failure strain (%)	0.75

container has been considered as elastoplastic in order to include the strain hardening effect and the bolted joint of the top plate with flange was not considered in order to simplify the model.

The container was modeled using nonlinear material properties in which [8] Cowper–Symonds material model is used which are a simple elastoplastic model that include the strain hardening behavior, i.e., strain rate dependent material because of very high strain rates during the impact. The container is dropped on a rigid base. The container was dropped in various orientations as below,

A. Horizontal drop
B. Edge drop
C. Corner drop
D. Trunnion side drop
E. One container dropped over the other.

Plastic kinematic model properties of the container are given in Table 2. The simulation was carried out for a total of 0.03 s in order to simulate the impact on the base. The results of stress, strain and displacement data during the impact were obtained with the help of post-processing and the results are shown in the graphs in the result section. Based on the results the optimized CAD model of waste handling container was designed. The detail of the adopted method is discussed in detail below.

3.1 Explicit Dynamic Method

The FE code used for the simulating the model is based on the explicit dynamic method. In explicit methods, the values of $(n + 1)$ time step are dependent on the previous time steps, i.e., there are no dependencies on the current time step and hence require less memory and computation time. The most common scheme used in the explicit dynamics method is the central difference scheme [9].

4 CAD Modeling

The radioactive waste container as shown in Fig. 2 was modeled with help of extruding and shell feature with the dimensions given in Table 1 and holes on the sides were provided for the trunnions which were cut with the help of extruding cut feature. Then the flange and the cover plates with holes for M17 bolts were modeled with the given dimensions. The corner and edges of the container were smoothened and strengthened by applying the stiffeners. The container was modeled as a shell with a square base so that it can be handled easily in any orientation. Two trunnions on each side were provided for lifting the container through cranes on the site and during transportation.

5 Results

Figure 3 shows that the peak effective stress developed for the case C and D are greater than the yield strength of the material so there would be plastic deformation, and Fig. 4 shows the peak effective stress for the case E which is below the yield stress. As shown in Fig. 5, the plastic strain for the case C is less than the case D because of the stiffeners provided on the edges of the container, but for the case D, the plastic strain is continuously increasing to about 1% leading to the puncture of the trunnion and deformation of the flange. So, some stiffener is needed to be provided on the flange to protect it from failure. There is no plastic strain observed for case A, B and E as shown in Figs. 5 and 6.

From Fig. 7, it is observed that there is no bounce back for case B because of the toppling of the container after the impact so the translational K. E is transformed into rotational energy so there is lesser internal energy to be retained by the material

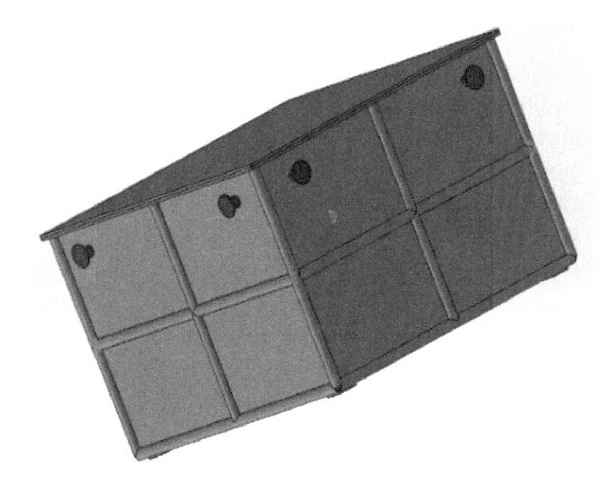

Fig. 2 CAD model of the radioactive waste container

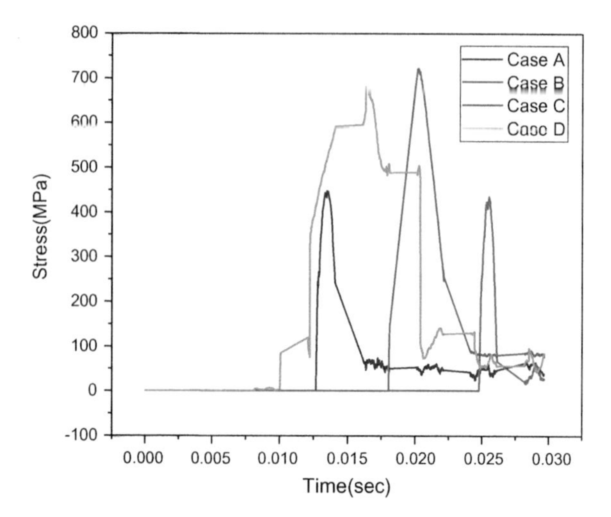

Fig. 3 VM stress for case A, B, C and D

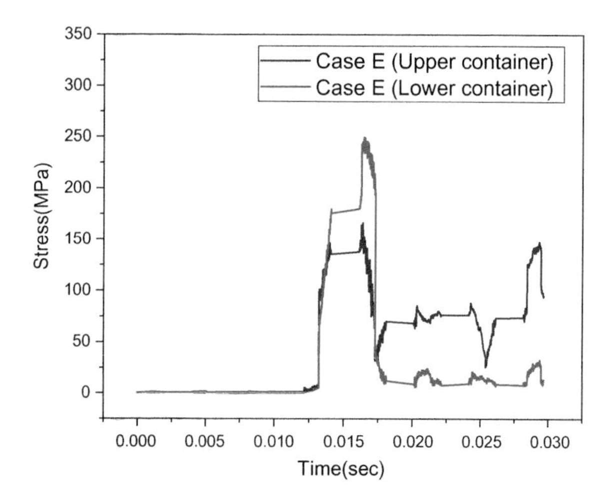

Fig. 4 VM stress for case E

of the container, and as a result, the less plastic strain will be observed. Similarly, for the case D, the displacement is in the form of steps and there is a negligible bounce back, so all the energy is dissipated in the form of deformation of the material causing the puncture of the trunnion and the deformation of the flange. For case C, there is bounce back but still, there is a plastic strain because the stress is highly localized at the corner, so less volume of material handles more energy leading to the plastic deformation. Also, from Fig. 7, for the two containers drop case the upper container bounces back and the lower container also jumps back little from the base in order to dissipate the energy so it might be the reason for no plastic strain observed in this case.

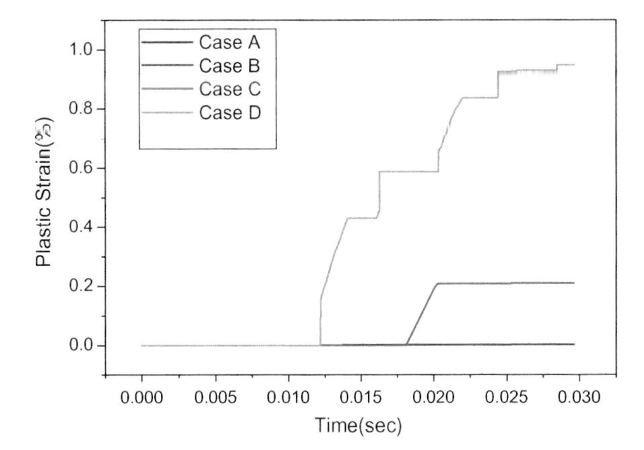

Fig. 5 Plastic strain for case A, B, C and D

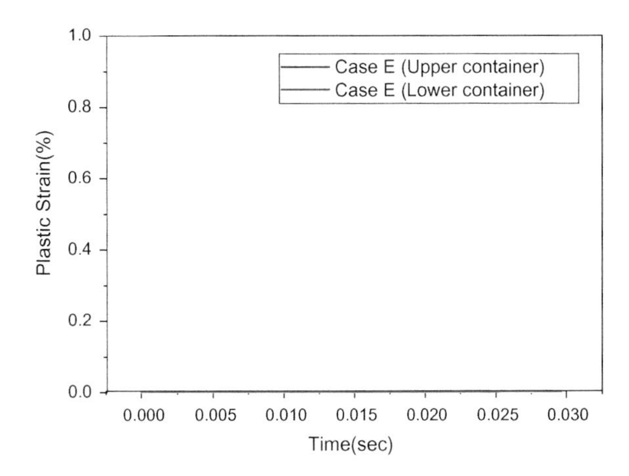

Fig. 6 Plastic strain for case E

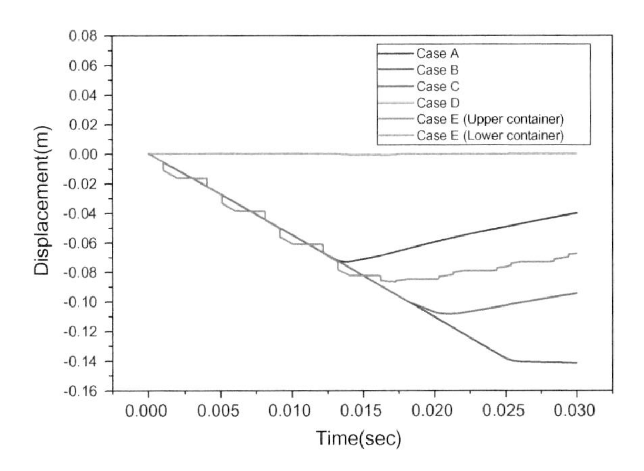

Fig. 7 Displacement of the container for case A, B, C, D and E

6 Conclusion

The CAD modeling of the radioactive container has been validated from the numerical simulation. The stresses and strains generated in the container are within the permissible limits for all the cases except case D. During the drop test on the trunnion side (case D), there is deformation of the flange, so a stiffener is provided on the flange. The container is dropped in different orientations in order to include all the possibilities of a drop of the container during the handling and loading operation. It may be concluded that the flange and the trunnion are the weakest parts which needed to be protected by proper changes in the design of the container. Hence, computer-aided analysis (CAA) is a very important step after modeling of the component and they both together contribute toward safe and optimum designing of the component. The square shape of the container is taken so that it seems the same from all sides so easy for handling operation. Trunnions are provided on each face for remote handling of the containers through cranes at the site and during transportation as these containers carry the radioactive waste so the whole process is automated in order to avoid human's exposure from the hazardous radioactive environment. Therefore, computer analysis act as validating a tool for the model designed with the help of computer-aided designing tool.

References

1. Bender J, Erleben K, Trinkle J (2014) Interactive simulation of rigid body dynamics in computer graphics. Comput Graph Forum 33(1):246–270
2. Weinstein R, Teran J, Fedkiw R (2006) Dynamic simulation of articulated rigid bodies with contact and collision. IEEE Trans Visual Comput Graph 12(3)
3. Smith B, Kaufman DM, Vouga E, Tamstorf R, Grinspun E (2012) Reflections on simultaneous impact. ACM Trans Graph (TOG), Article no 106
4. Franco RL, Pugliese G, Nasta M (2014) Structural performance of an IP2 package in free drop test conditions: numerical and experimental evaluation. Nucl Eng Des 280:634–643
5. Introduction to meshing. https://altairuniversity.com/wp-content/uploads/2014/02/meshing.pdf. Accessed 7 Apr 2019
6. Zavattieri PD, Buscaglia GC, Dari EA (1996) Finite element mesh optimization in three dimensions. Latin Am Appl Res 26:233–236
7. Geuzaine C, Remacle J-F (2009) A 3-D finite element mesh generator with built-in pre- and post-processing facilities. Int J Numer Methods Eng 79(11):1309–1331
8. Cowper G, Symonds P (1957) Strain hardening and strain-rate effects in the impact loading of cantilever beams. Technical report, Brown University Division of Applied Mathematics, Technical Report No. 28
9. Gokhale NS, Deshpande SS, Bedekar SV (2008) Practical finite element analysis, Finite to infinite publication, pp 321–349

Effect of Attack Angle on Lift and Drag of a Bio-Inspired Corrugated Aerofoil

Ashwini Biradar, Ashutosh Chandraker, Royal Madan, Shubhashis Sanyal and Shubhankar Bhowmick

Abstract Bio-inspired corrugated aerofoil has better flow physics compared to traditional aerofoil in addition to less wear and high strength. For identifying the aerodynamics characteristics, lift and drag, a study has been done by varying angle of attack (AOA). Both low and high AOAs are analyzed for identifying the optimal flapping condition at constant Reynold number. CFD analysis is performed using ANSYS Fluent. 2D analysis is performed using k-ε model, and sphere of influence is utilized to capture the flow field and to reduce the computational cost. Results report that low AOA performs better due to high lift-drag ratio and less flow separation.

Keywords Bio-inspired · Corrugated · Fluent · *k*-ε model · Aerofoil

1 Introduction

Bio-inspired or nature-inspired innovations are the ones where one can solve human problems by looking at the solutions available in nature. Nature design and mechanisms are unbeatable and reliable which can be realized by looking at the human mechanisms also. Many innovations have already been done till now in diversified areas like mimicking of kingfisher birds' beak to solve the aerodynamic problem of Japan bullet train, box fish to reduce the drag coefficient of passenger, where the reduction of drag coefficient not only reduces the fuel consumption but also reduces CO_2 emissions [1]. The other areas of bio-inspired are furniture design for comfort, textile, and architecture, to name a few, and with the current database available like bio-mimicry taxonomy identification of a creature available that solves a particular problem not only for biologist but for anyone. Furthermore, new methods have been developed like the law of system completeness [2] that helps in applying the concept with ease. Borrowing inspiration from insects and flies, design

A. Biradar · A. Chandraker · R. Madan · S. Sanyal · S. Bhowmick (✉)
Department of Mechanical Engineering, NIT Raipur, Raipur, Chattisgarh 492010, India
e-mail: sbhowmick.mech@nitrr.ac.in

BBVL. Deepak et al. (eds.), *Innovative Product Design and Intelligent Manufacturing Systems*, Lecture Notes in Mechanical Engineering,
https://doi.org/10.1007/978-981-15-2696-1_25

of small aerial vehicles, although quite complex in the subject, has been an interesting research domain. The fabrications of these pose the first major difficulty. The second one is the problem of extending a consistent power source for their operations. These vehicles perform at Reynolds number as low as 80,000 which in turn offer excessive drag to the conventional wings if used. Looking for a solution, one is readily found in nature. The observation of the flight of flies and insects offered insight into the corrugated nature of their wing crosssection. In the present case, dragon fly's wing cross section is analyzed for its better flow characteristics.

Different flight physics can be obtained by changing the thickness of corrugated, geometry of corrugated, and corner corrugated shapes. Two classifications of angle of attack (AOA) are available: when AOA < 10, it is considered as low angle of attack and for AOA > 10 it is known as high angle of attack. Two different corrugated aerofoil shapes were studied by New et al. [3] and the first one was based on typical dragonfly wing cross section investigated by Hu and Tamai [4], Murphy and Hu [5] named it as corrugated A, while the second one was based on a simplified dragonfly wing cross section studied by Levy and Seifert [6] known as corrugated B. Corrugated A has continuous peaks and valleys of different sizes, whereas corrugated B has 2 peaks with a valley in between and a smooth trailing hump at the end. Experimental investigation was carried out by New et al. [3] to study flow separation on different corrugated for both low and high angle of attacks at Re = 14,000, and found flow separation is high in corrugated A than corrugated B. Experimental analysis on a corrugated aerofoil for low Reynold number of Re = 34,000 to understand micro air vehicles (MAV) flight physics is performed taking mid cross section of dragonfly wings for analysis purposes. Both high and low angle of attacks were considered, and lift and drag coefficients were studied by Hu and Tamai [4]. Study for the range of Reynold number was carried out by Levy and Seifert [6] by varying as 2000 < Re < 8000, finding reveals as the Reynold number increases the drag coefficient decreases, and this trend is also valid for the present work. Experimental analysis Murphy and Hu [5] performed experimental analysis to understand drag and lift behavior of corrugated A type and found its performance better than smooth aerofoil for AOA > 8 at Re = 100,000. In [7], corrugated aerofoil shape given by Levy and Seifert [6] for its drag and lift characteristics at Re = 14,000 were analyzed. Two-dimensional CFD on an oscillating wing is carried out to study lift and drag coefficient under unsteady flow regime. Aerodynamic characteristic of corrugated aerofoil was compared with flat plate and the Eppler E61 profile for AOA of 0° and 5° with Re = 34,000 by Khurana and Chahl [8]. Optimum shapes of aerofoil were presented by using various optimization strategies and objective functions by Jain et al. [9].

Unsteady flow simulations on three different airfoils, namely NACA0010, corrugated A, and corrugated B airfoils, were studied by Ho and New [10] and their differences were compared with similar profiles, and experimentation was performed by Dwivedi and Bhargava [11]. Many studies were done by neglecting the wing flapping for simplification by considering one side of the wing to be fixed. In order to understand flapping behavior, the effect of both negative and positive angle of attacks is studied in the present work and their lift and drag ratios are compared.

The benefits of small flow separations in corrugated B than corrugated A because of larger recirculating regions in the corrugated B motivated to carry out the present study taking corrugated B shape for both high and low AOAs at constant Re = 14,000.

2 CFD Methodology

A 2D geometry of aerofoil of the chord length, $\boldsymbol{c}$ = 75 mm, as mentioned by Levy and Seifert [6] was created and studied using ANSYS Fluent® for analysis. The aerofoil was kept in the flow domain with inlet boundary at three times the chord length of the aerofoil (3$\boldsymbol{c}$) from the center of aerofoil while the outlet boundary was kept at 4$\boldsymbol{c}$ from the center of the aerofoil. For meshing, patch conformal mesh was used, and the entire domain was discretized using quadrilateral mesh. A quadrilateral mesh provides quality mesh with a lesser number of elements, and hence, it leads to low computational run-time and less numerical errors. A grid independency test is performed as mentioned in Table 1, and the result shows there is just 0.572% error in drag coefficient, thus indicating that the solution is sufficiently grid independent.

Formulation for turbulent kinematic energy is given in Eq. 1. [12]

$$\frac{\partial(\rho k)}{\partial t} + \frac{\partial(\rho k u_i)}{\partial x_i} = 2\mu_t E_{ij} E_{ij} - \rho\varepsilon + \frac{\partial}{\partial x_j}\left[\frac{\mu_t}{\sigma_k}\frac{\partial k}{\partial x_j}\right] \tag{1}$$

where

u_i represents velocity component in corresponding direction
E_{ij} represents component of rate of deformation
μ_t represents eddy viscosity.

A novel refinement approach was used by creating a sphere of influence around the corrugated as shown in Fig. 1, to capture the flow characteristics near the corrugated. Two spheres of influences (SOIs) of different radius were created with different refinements to optimize the computational accuracy and run-time.

Table 1 Grid independence test

S. No.	Mesh	No. of cells	No. of iterations	C_D	% error (CD)
1.	Coarse	1600	45	0.01481	–
2	Fine mesh with 1 SOI	10,630	89	0.013026	13.45
3	Fine mesh with 2 SOI	22,155	125	0.013101	0.572

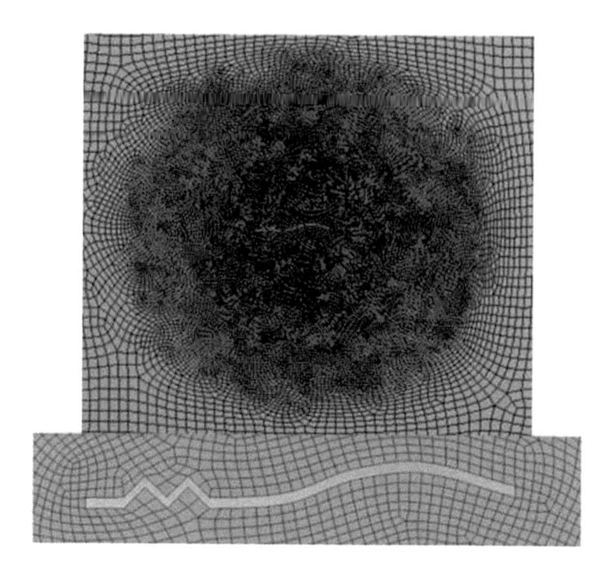

Fig. 1 Mesh with sphere of influence at the center and enlarged mesh view around the aerofoil of chord length ‘$\boldsymbol{c}$’

A coupled pressure-based solver was used to solve the flow around the aerofoil. The turbulence model, ***k*-ε**, model was used. This model being a turbulent model gives better results for low Reynolds number. Along with the model, a standard boundary function was provided to obtain the flow characteristics near the boundary as well. The fluid medium considered in the study is air at Reynolds number = 14,000. The boundary conditions used are uniform velocity at the inlet with a turbulence intensity of 1.1%, and zero-gauge pressure at the outlet, with no-slip wall condition on the aerofoil surface, top wall and bottom wall assumed.

3 Results and Discussion

In the present work, flow characteristics of bio-inspired corrugated aerofoil were studied for low Reynolds number and model used ***k*-ε** as it gives better results for low Reynolds number. In case of a positive angle of attack for 5°, 10°, and 15°, the lift coefficient is negative as shown in Fig. 2, and larger pressure is obtained at the upper corrugated surface which makes the aerofoil to bow down. In Fig. 3, contour plots of velocity distribution for different AOA are plotted while in Fig. 4, similar contour plots of pressure distribution accross the aerofoil is reported. The magnitude of pressure goes on increasing as the positive angle of attack increases. The lift coefficient is positive when the angle of attack is negative, reason being the pressure which acts at the bottom of the aerofoil forces the aerofoil to lift, and the magnitude of lift depends upon pressure acting at the downside which was maximum for −15° followed by −10° and minimum at −5° when negative AOA were compared; hence, the lift coefficient was high for −15°. Similarly, depending upon the direction of pressure acting on the aerofoil, the drag coefficient varies and its magnitude goes on

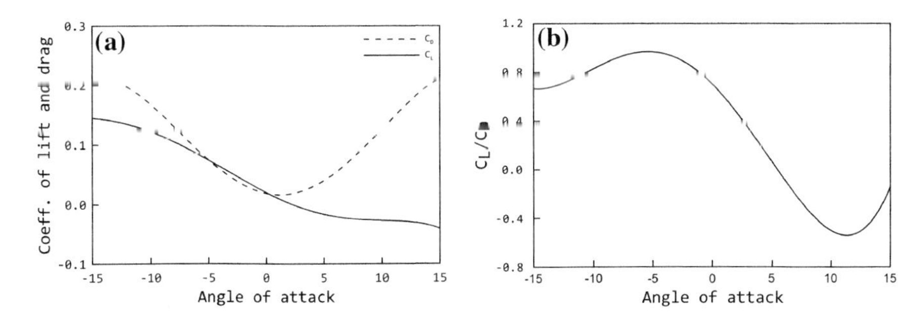

Fig. 2 **a** Coefficient of lift and drag versus AOA and **b** lift-to-drag ratio versus angle of attack

decreasing as shown in Fig. 4. The drag coefficient decreases when the angle of attack decreases from −15° to −5°, and it increases when the angle of attack increases from +5° to +15° as in this case the pressure acting at the upper surface of the wing forces the aerofoil downwards. The behavior of lift and drag coefficient for the angle of attack <0° were the same, and similar trend for angle >0° are obtained (Fig. 3; Table 2).

In order to select the optimum angle of attack, their ratios are compared and the effect of angle of attack of similar magnitude is compared. From Fig. 2, it can be seen that the C_L/C_D ratio is maximum for −5° and is minimum for 10°. The flapping angle 10° cannot be a choice as it has least C_L/C_D ratio compared to 5°. Comparing the ratios of 5° and 10° reveals that 5° has high C_L/C_D ratio than 10°, and −10° has low C_L/C_D ratio than −5°, but when the variation of C_L/C_D ratio is compared for −5° to 5° with C_L/C_D ratio at zero angle of attack, flapping angle of 5° is found to be an optimal angle. The 0° angle of attack separates the positive and negative angle of attack and can be considered as a critical point when the effect of AOA is analyzed. From Fig. 3, it can be seen that flow separation is high for 15°, i.e., for both 15° and −15°, compared to 10° and 5°. The flow is favorable for −5° compared to 5°. When compared with 15° AOA, where the flow is observed to be highly separated, at −15°, flow adheres to the aerofoil. In the case of AOA 15°, the flow is separated from the beginning and never adheres even after reaching the hump.

4 Conclusion

In the present work, it is found that low flapping angle of corrugated aerofoil performs better as a low angle of attack has an optimal lift–drag ratio. Flow separation is very high in case of AOA 15° compared to another angle of attack, and the flow never adheres back once separated from the tip of the aerofoil.

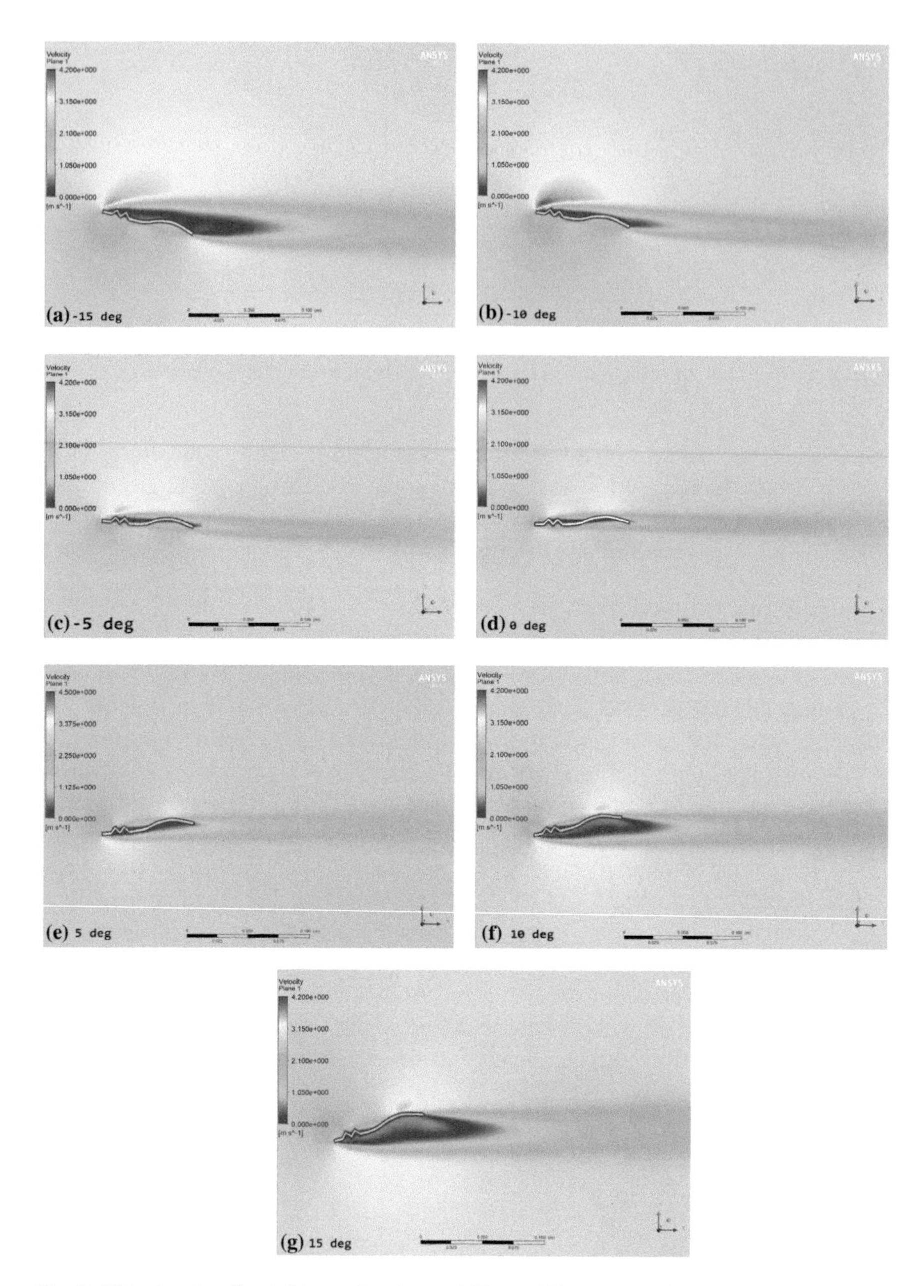

Fig. 3 Velocity plot (for AOA ranging from −15° to +15° at a step of 5°)

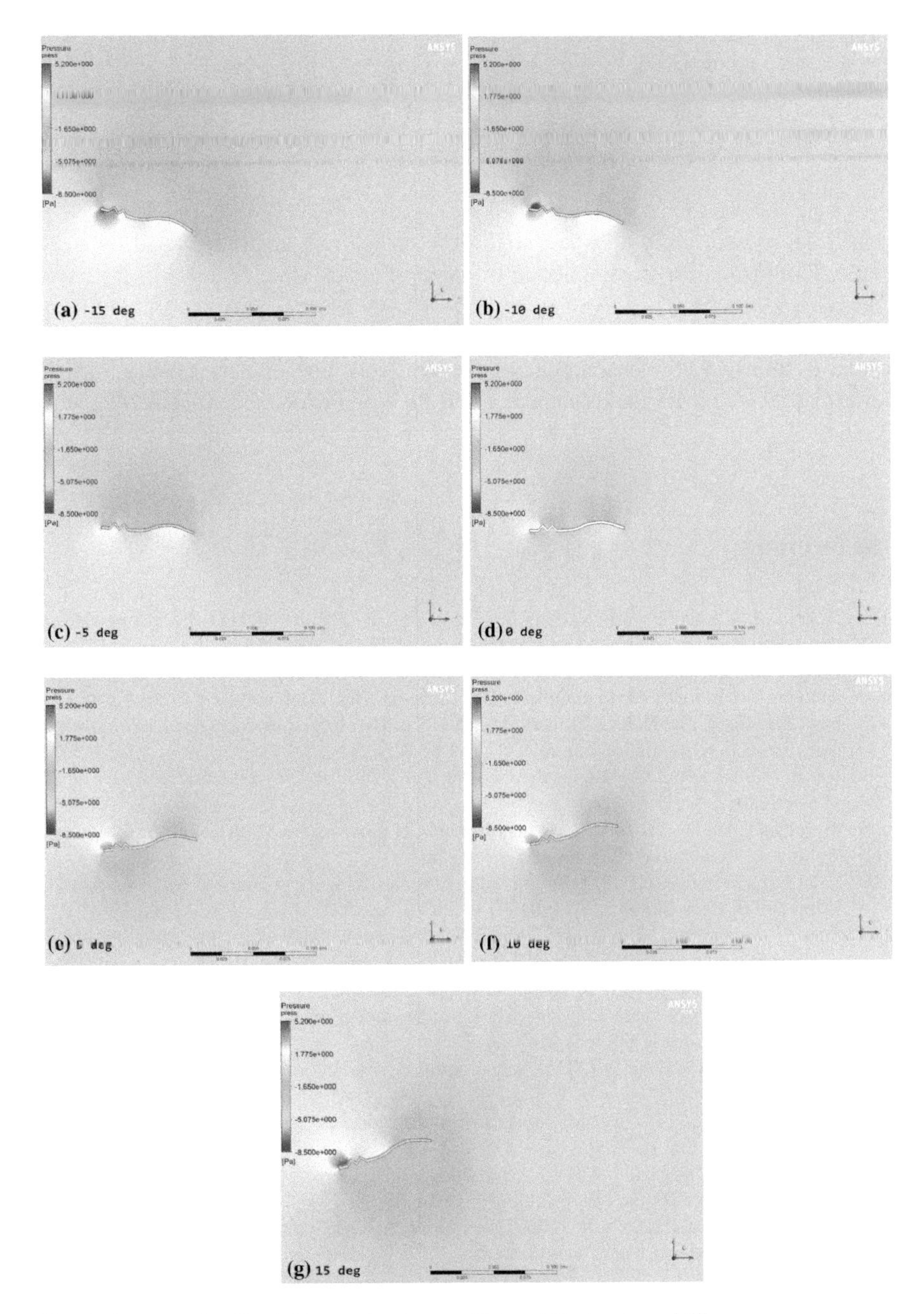

Fig. 4 Pressure plot (for AOA ranging from −15° to +15° at a step of 5°)

Table 2 Validation of present work with standard results

AOA	C_D (present)	C_D [10]	Experimental [11]
5	0.040709186	0.05653884	0.039085546
10	0.1249955	0.1086529	0.12192724
15	0.20928181	0.27826942	0.30875123

In the present case, 5° was found to be an optimal angle indicated by C_L/C_D ratio. Flow is almost streamlined in 0° angle of attack which is a static case, when flapping occurs both positive and negative AOAs are necessary for a flight; hence, by knowing the flow separations regions, the corrugated shapes can be modified in order to improve flow characteristics. Present work does not incorporate variation effect of AOA on Reynolds number. The flow behavior of current AOA considerations indicates favorable flow in the case of low AOA.

References

1. Kozlov A, Chowdhury H, Mustary I, Loganathan B, Alam F (2015) Bio-inspired design: aerodynamics of boxfish. Procedia Eng 105:323–328
2. Cohen YH, Reich Y, Greenberg S (2015) What can we learn from biological systems when applying the law of system completeness? Procedia Eng 131:104–114
3. New TH, Chan YX, Koh GC, Hoang MC, Shi S (2014) Effects of corrugated aerofoil surface features on flow-separation control. AIAA J 52:206–211
4. Hu H, Tamai M (2008) Bioinspired corrugated airfoil at low Reynolds numbers. J Airc 45:2068–2077
5. Murphy JT, Hu H (2010) An experimental study of a bio-inspired corrugated airfoil for micro air vehicle applications. Exp Fluids 49:531–546
6. Levy D-E, Seifert A (2009) Simplified dragonfly airfoil aerodynamics at Reynolds numbers below 8000. Phys Fluids 21:071901
7. Flint TJ, Jermy MC, New TH, Ho WH (2017) Computational study of a pitching bio-inspired corrugated airfoil. Int J Heat Fluid Flow 65:328–341
8. Khurana M, Chahl J (2013) Bioinspired corrugated airfoils for micro air vehicles. Presented at the SPIE smart structures and materials + nondestructive evaluation and health monitoring. San Diego, California, USA, April 8 2013
9. Jain S, Bhatt VD, Mittal S (2015) Shape optimization of corrugated airfoils. Comput Mech 56:917–930
10. Ho W, New T (2017) Unsteady numerical investigation of two different corrugated airfoils. Proc Inst Mech Eng [G] 231:2423–2437
11. Dwivedi YD, Bhargava V (2009) Aerodynamic characterization of bio inspired corrugated wings. MOJ Appl Bionics Biomech 3(1):1–10
12. Versteeg HK, Malalasekera W (2007) An introduction to Computational fluid dynamics: the finite volume method. Pearson Education

Design Optimization of Slag Pot Transfer Car

V. Naga Sudha, K. S. Raghuram, V. Savitri and A. Shanthi Swaroopini

Abstract A majority of the steel production industries rely on machinery or equipment for effective operation thereby avoiding or minimizing the accidents due to inevitable human errors. One such problem has been identified where the high temperature slag is damaging the power cables used to run the slag pot transfer car due to leakage or spillage. In order to overcome this, a new generation slag pot transfer car has been developed by incorporating modifications in the mast of the slag pot transfer car. A number of components such as canopy, mast, reeling drum, columns, and rollers are designed by assuming the dimensions based on existing industry requirements. The assembled components are then subjected to testing at full load conditions using a finite element analysis package where a suitable design or modification has been suggested taking factor of safety into consideration. This work is focused on indigenously designed and developed solutions for effective working of the industry.

Keywords Von Mises stress · Principal stress · FEA

1 Introduction

Visakhapatnam Steel Plant bestowed with modern technologies, and VSP has an installed capacity of 3 million tons per annum of liquid steel and 2.656 million tones of saleable steel. At VSP, there is an emphasis on total automation, seamless integration, and efficient upgradation, which result in wide range of long and structural products to meet stringent demands of discerning customers within India and abroad [1–4]. VSP products meet exacting international quality standards such as JIS, DIN, BIS, and BS. VSP has become the first integrated steel plant in the country to be certified to all three international standards for quality (ISO-9001), for environment management (ISO-14001) and for occupational health and safety

V. Naga Sudha (✉) · K. S. Raghuram · V. Savitri · A. Shanthi Swaroopini
Department of Mechanical Engineering, Vignan's Institute of Information Technology, Visakhapatnam 530040, India

BBVL. Deepak et al. (eds.), *Innovative Product Design and Intelligent Manufacturing Systems*, Lecture Notes in Mechanical Engineering,
https://doi.org/10.1007/978-981-15-2696-1_26

(OHSAS-18001) [5–7]. This covers quality systems of all operational, maintenance, and service units besides purchase systems, training and marketing functions spreading over four regional marketing offices, 24 branch offices, and stockyards located all over the country.

The research gap is that still optimization is necessary for the effective usage of transport of the slag in steel industries. This is to minimize the material handling cost and manhandling of production. Hence, this gap is to be filled by further researchers or mathematicians to fill this. Time of handling is also to be minimized so as to maximize the quantity of production.

2 Modeling

2.1 Need for Transfer Car

Steel is produced by LD process in Visakhapatnam Steel Plant, RINL. The molten metal collected after the LD process, and the impurities in the form of slag are collected in the slag pot. The slag pot is to be transported to slag yard. In order to transfer the slag pot, the slag pot transfer car plays a major role. The distance between the convertor sections to the slag yard is 84650 mm. By transferring the slag pot from convertor section to the slag yard, there is no provision to stop the process which is not recommended. This results in considerable savings in terms of time and money.

2.2 Design of Transfer Car Components

As there is a need of carrying the slag from convertor section to slag yard, the cable is being burnt during the transfer due to hot slag. The idea of changing the mast in the transfer car in SMS-2 was conceived by carrying out brainstorming session in plant design. The modification in transfer car was designed by taking into consideration the various constraints like available material cross section, manufacturing and testing facility, transfer car components' geometry and dimensions, and maximum height of lift of mast. The mast was designed to raise in level from the original length to modified length. In order to protect, new components were redesigned, namely canopy, reeling drum, columns, and rollers. The analysis of the model was carried out by simulation methods and found satisfactory. By innovative design of the transfer car modification, we could realize our dream of protecting the cables from being burnt come true with a ray of hope in commissioning our continuous casting machine SMS-2 at the earliest.

2.3 *Components of Transfer Car*

Various components of transfer car are canopy, reeling drum, columns, rollers, and mast. The transfer car along with its accessories forms the total assembly comprising the components like mast, canopy, columns, rollers, and reeling drum.

2.3.1 Design and Modeling of Canopy Structure

The canopy structure is one of the components used for the protection of mast under the convertor section. The canopy is placed under the convertor section. Canopy plays a major role in protection of the cable and the mast under the convertor section. This canopy is placed just below the convertor attached to the column. When the slag is ready to collect in the pot, the transfer car moves up to the canopy, such that the mast rests below it. This protects the cable and the mast from being damaged (Fig. 1).

2.3.2 Design and Modeling of Mast Component

Mast is one of the components of slag pot transfer car. As the transfer car is a self-propelled vehicle, the electric supply is supplied with the help of cables. The cable is supported by the mast. These cables are placed on the ground level, due to which the cable is being burnt. Modification is done to protect the cable from being burnt by raising the level of mast. This raise in level of height is at 4.5 m above the ground level (Fig. 2).

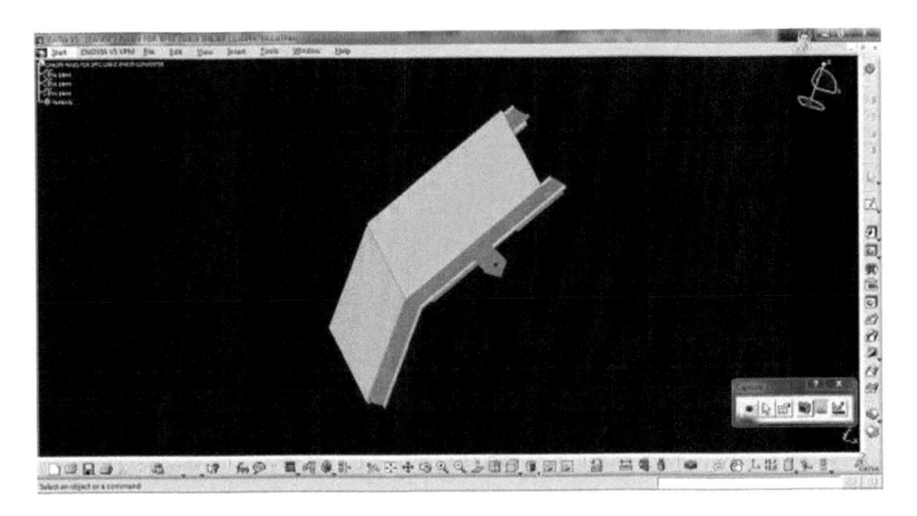

Fig. 1 Modeling of canopy structure

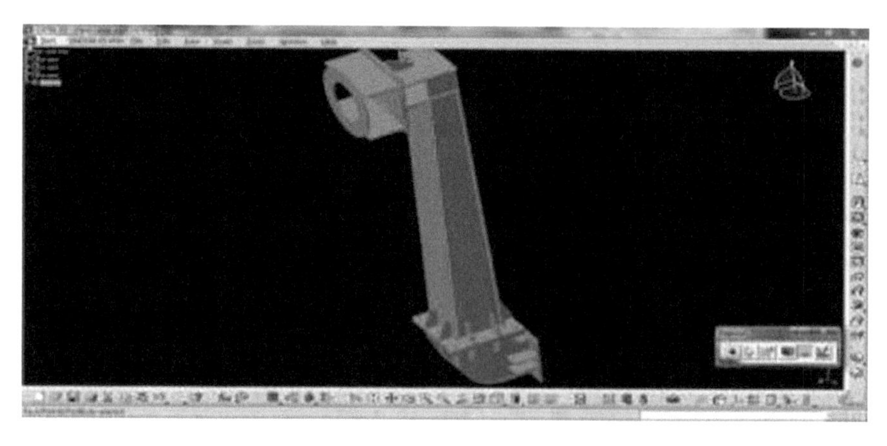

Fig. 2 Modeling of mast

2.3.3 Design and Modeling of Roller Structure

Roller supports the cable. When the cable winds or rewinds, the roller rotates such that the cable can move freely. Earlier the rollers are placed on the ground level, and this caused the cable to burn during the transfer of the slag. Now the rollers are placed on the columns (Fig. 3).

Fig. 3 Modeling of roller

2.3.4 Design and Modeling of Column

Column is the structure which gives support to the roller. The columns are placed aside to the rail at a distance of 1275 mm. The total number of columns required from convertor section to slag yard is 8. The distance between columns to column is 2000 mm (Fig. 4).

2.3.5 Design and Modeling of Reeling Drum

Reeling drum is the component used for carrying the cable in movement of the transfer car from convertor section to slag yard. Reeling drum rotates in two opposite directions for movement of the car. From convertor section, it rewinds the cable up to reeling drum, and after reeling drum, it winds the cable till it reaches the slag yard (Fig. 5).

3 Analysis

The following are the various steps involved in the analysis of each component used in transfer car accessories. The steps are as follows.

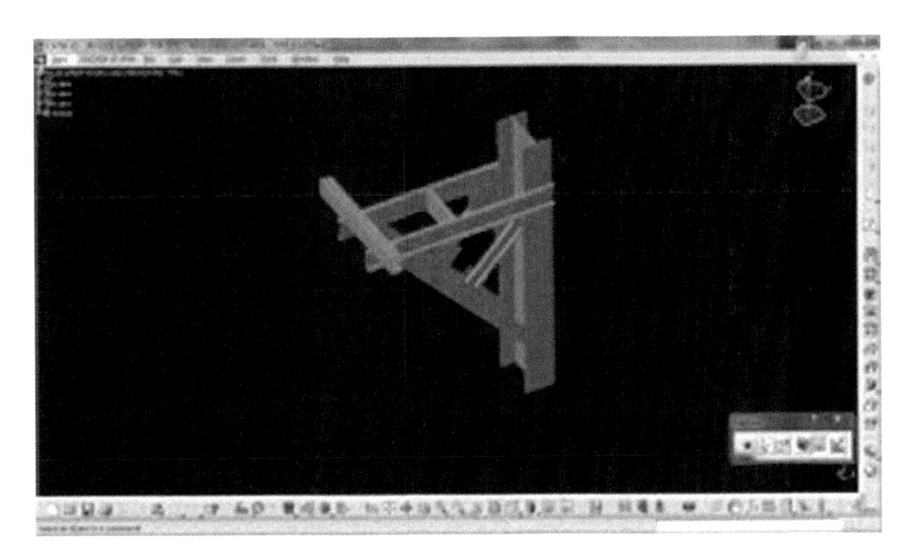

Fig. 4 Modeling of column

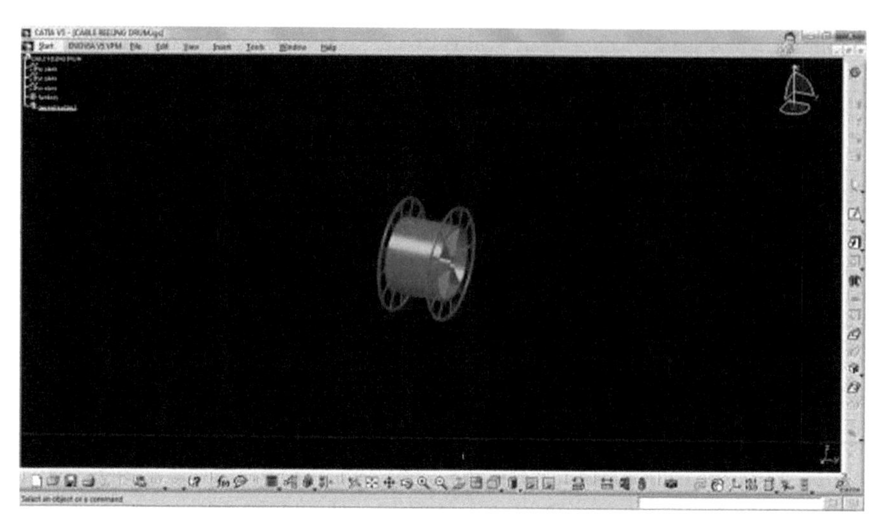

Fig. 5 Modeling of reeling drum

3.1 *Loads and Constraints on Canopy*

See Figs. 6 and 7.

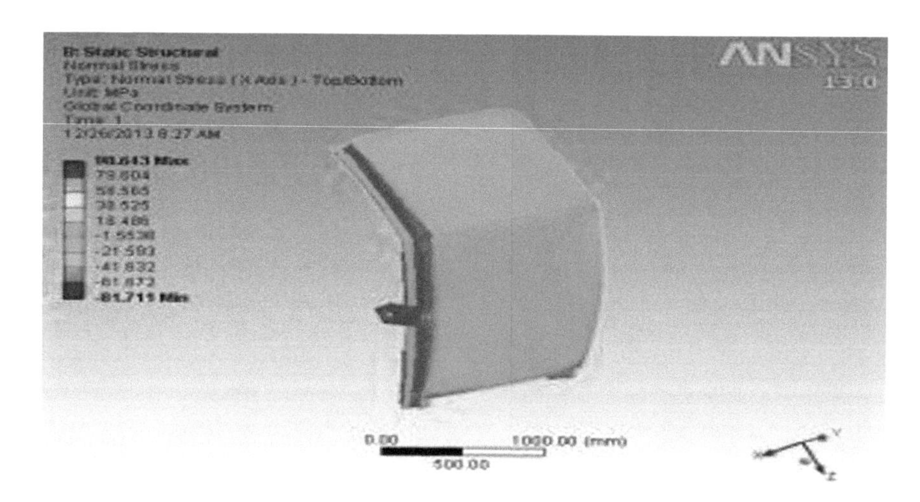

Fig. 6 Contour plots of stresses on canopy

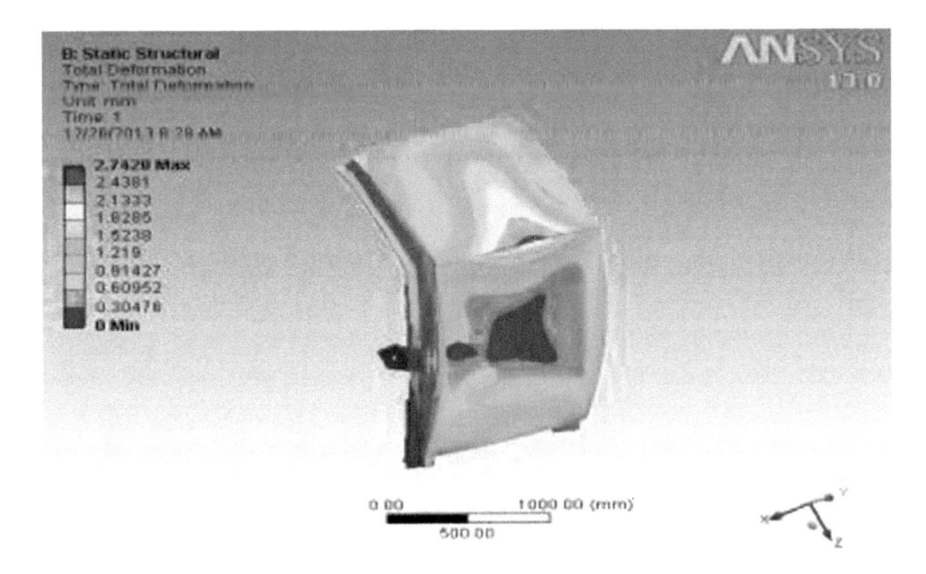

Fig. 7 Static analysis on canopy

3.2 *Load and Constraints on Mast*

See Figs. 8 and 9.

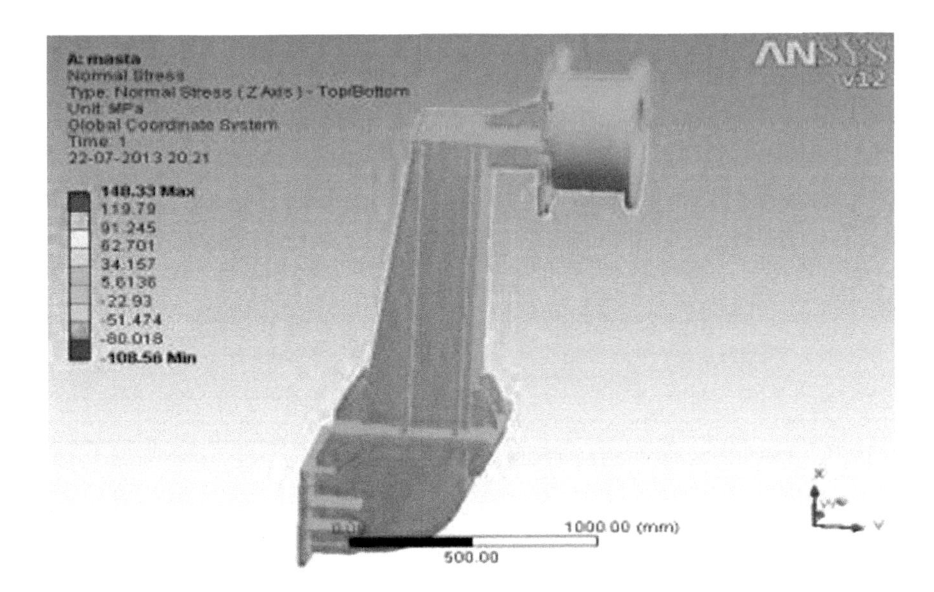

Fig. 8 Contour plots of stresses on mast

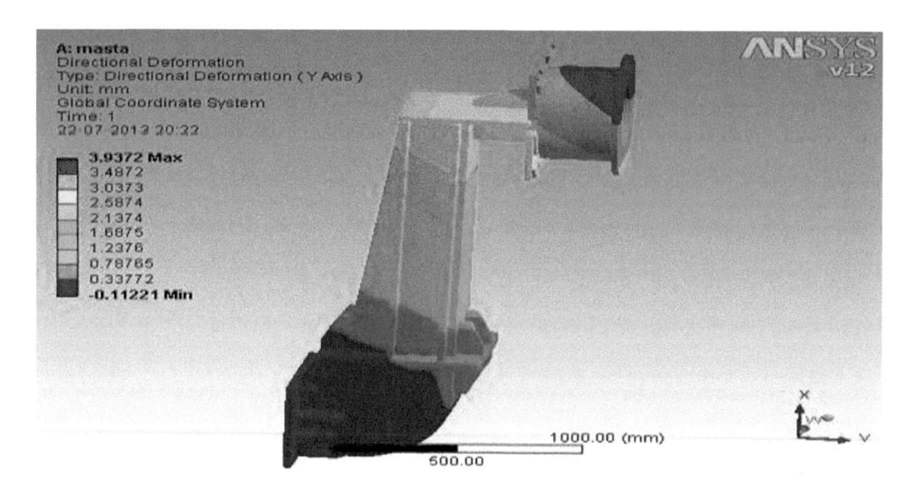

Fig. 9 Static analysis on mast

3.3 *Loads and Constraints on Roller*

See Figs. 10 and 11.

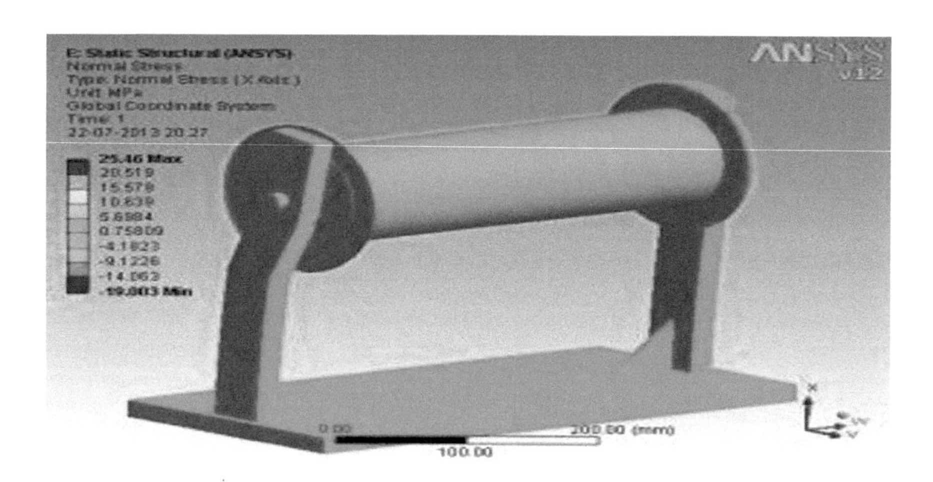

Fig. 10 Contour plots of stresses on roller

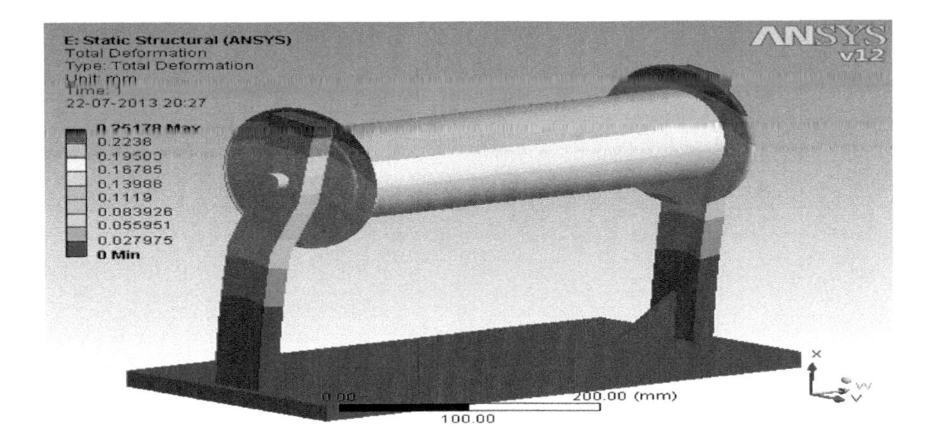

Fig. 11 Static analysis on roller

3.4 Loads and Constraints on Column

See Figs. 12 and 13.

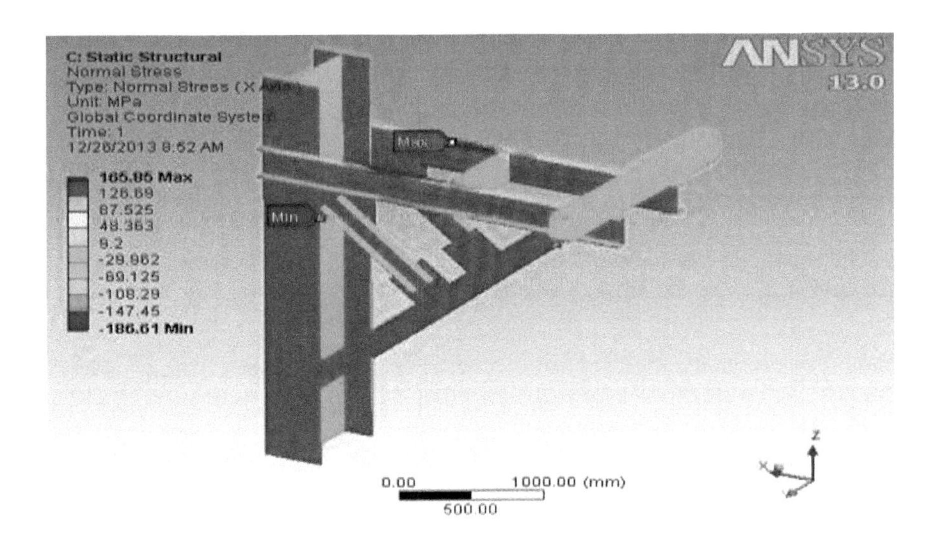

Fig. 12 Contour plots of stresses on column

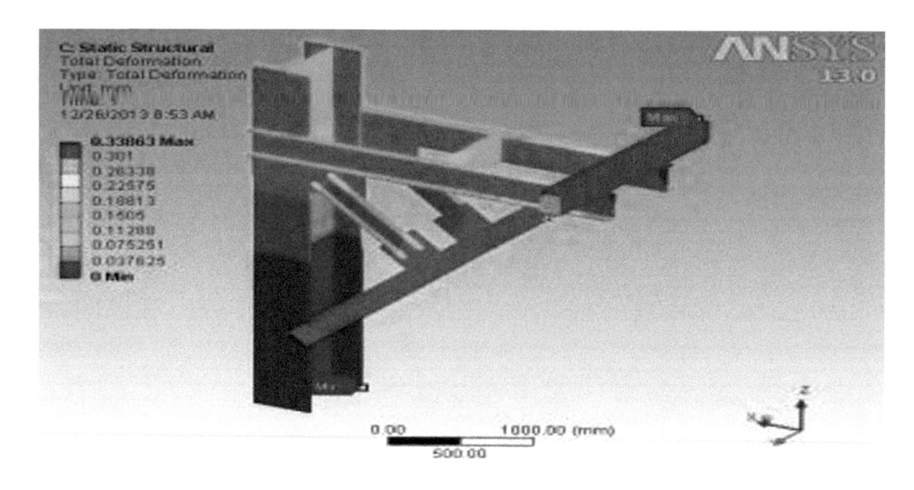

Fig. 13 Static analysis on column

3.5 Loads and Constraints on Reeling Drum

See Figs. 14 and 15.

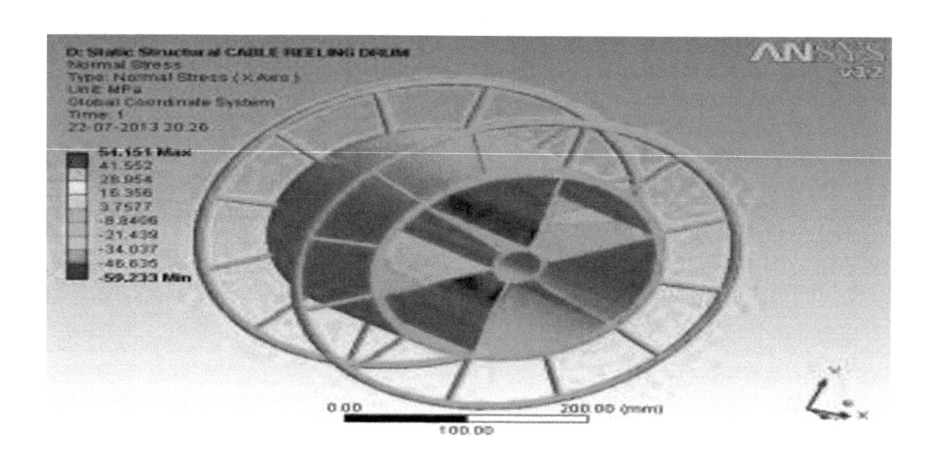

Fig. 14 Contour plots of stresses on reeling drum

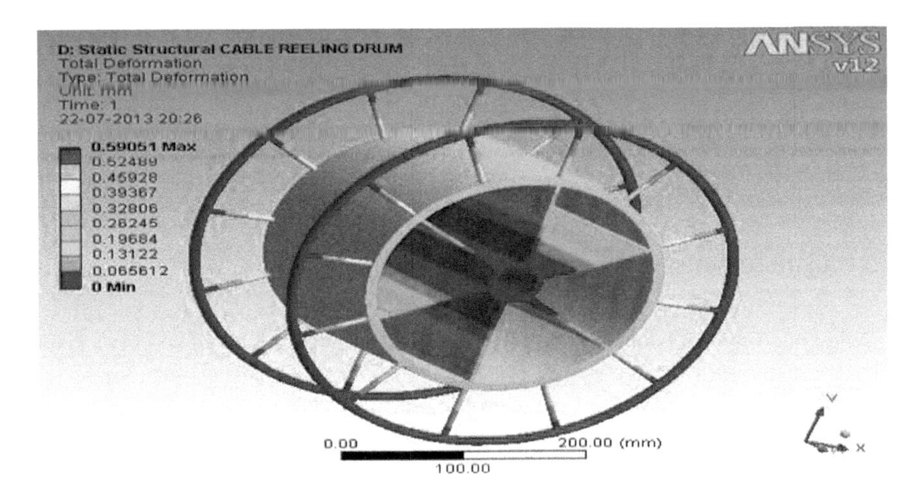

Fig. 15 Static analysis on reeling drum

4 Results and Discussions

Bending stress and deformation of different components like mast, canopy reeling drum, column, and rollers are computed using mathematical calculations, and the results are compared with maximum stress and deformation given by the ANSYS software. And the results obtained are shown in Tables 1, 2, 3, and 4.

Table 1 Comparison of mast

S. No.	Stress	Theoretical value MPa	ANSYS value MPa
1	Stress	140.83	148.33

Table 2 Simulation values of mast

S. No	Type of stress	ANSYS values (MPa)
1	Shear stress along *XY* plane	19.126
2	Shear stress along *YZ* plane	99.614

Table 3 Deformation values

Total deformation	3.9377 mm
Deformation (Along *X*)	0.44 mm
Deformation (Along *Y*)	0.93 mm
Deformation (Along *Z*)	0.55 mm

Table 4 ANSYS values of components

S. No.	Component	Total deformation (mm)	Von Mises stresses (MPa)
1	Canopy	2.748	162.08
2	Reeling drum	0.590	57.95
3	Roller	0.251	39.63

5 Conclusions

Different components in slag pot transfer car pathway are designed and get satisfied by the manager of plant design for future modification of slag pot transfer car. Numerical analysis is carried out with the aid of CATIA and ANSYS software to predict the stress induced on the mast as well as on the reeling drum, rollers, canopy, and columns. The theoretical calculation is compared with numerical results of mast using ANSYS. The values compared are nearer to the ANSYS results. 140.83 MPa with theoretical calculation and 148.38 MPa of ANSYS results for mast.

The following modifications are suggested based on the results of the present study. The mast is designed by changing the length, and also, the mast is protected from the hot slag with the help of canopy.

Due to the change in mast level, the height of the rollers is also changed by placing them on the columns. The damage of the cable is rectified and being protected from hot slag.

Acknowledgements I would like to express my special thanks to our guide Dr. O. Ram Mohan Rao, HR in RINL, Visakhapatnam, and also Dr. S. Rama Krishna, Associate Professor in Gayatri Vidya Parishad at Madurai. This research did not receive any specific grant from funding agencies in the public, commercial, or not-for-profit sectors.

References

Journals

1. Bawiskar P, Kamble S, Bhole K (2012) Design of tilting ladle transfer car for steel industries. World J Sci Technol 2(4):20–23
2. Yong SZ, Gazzino M, Ghoniem A (2012) Modeling the slag layer in solid fuel gasification and combustion—formulation and sensitivity analysis. World J Sci Technol 92:162–170
3. Raghuram KS et al (2016) Experimental analysis of stress, strain and deformation on different types of conveyor belt pulleys. An analysis conducted at Visakhapatnam Steel Plant. Aust J Basic Appl Sci 10(1):715–719

4. Yue Z, Qing S, Wang H (2009) Numerical modelling of a slag- metal behavior of smelting reduction process: ironmaking technology based on the hismelt. In: The ninth international conference on electronic measurement & Instrument, ICEMI', vol 90, pp 120–129
5. Muhmood L, Viswanathan NN, Seetharaman S (2011) Some investigations into the dynamic mass transfer at the slag–metal interface using sulfur: concept of interfacial velocity. Min Metals Mater Soc ASM Int 35:261–270
6. Channon WP, Urquhart RC, Howat DD (1998) The mode of current transfer between electrode and slag in the submBerged-arc furnace. J S Afr Inst Min Metall 91:405–409
7. Raghuram KS, Ajay Kumar D (2017) Stress analysis of traverse beam crane hook used in steel melting shops of steel plant by Ansys and Catia. Am J Mech Ind Eng 2(1):37–40

Design and Fabrication of DLP 3D Printer

Jaydev Gohil, Rajveersinh Gohil, Romil Gundaraniya, Mihir Prajapati and Savan Fefar

Abstract Digital light processing (DLP) 3D printing is an additive manufacturing (AM) process which is used to produce parts via photopolymerization process in which resin is cured by UV light. Vat photopolymerization is a form of AM. It has a liquid bath of a polymeric resin which is cured layer by layer through precise control with help of stepper motor and projection of a light source of DLP projector. Printing time, layer thickness, and lumens of the projector play an important role in the printing process. A series of specimens was designed, printed, and tested. Total printing time, layer thickness, and layer exposure time were analyzed. We used 365 nm wavelength photopolymer resin, BENQ MP515 projector having 2500 ANSI lumens. This paper shows the design and fabrication of DLP 3D printer with low cost and good accuracy.

Keywords 3D printing · Additive manufacturing · Digital light processing (DLP) · Vat photopolymerization

J. Gohil · R. Gohil · R. Gundaraniya · M. Prajapati · S. Fefar (✉)
Department of Mechanical Engineering, Indus University,
Ahmedabad, Gujarat 382115, India
e-mail: savanfefar.me@indusuni.ac.in

J. Gohil
e-mail: JaydevGohil.15.me@iite.indusuni.ac.in

R. Gohil
e-mail: RajveersinhGohil.15.me@iite.indusuni.ac.in

R. Gundaraniya
e-mail: RomilGundaraniya.15.me@iite.indusuni.ac.in

M. Prajapati
e-mail: MihirPrajapati.15.me@iite.indusuni.ac.in

BBVL. Deepak et al. (eds.), *Innovative Product Design and Intelligent Manufacturing Systems*, Lecture Notes in Mechanical Engineering,
https://doi.org/10.1007/978-981-15-2696-1_27

1 Introduction

There are various types of 3D printing. All types have their own sets of process and applications. They all have their various principles and advantages. Table 1 shows the different types of rapid prototyping. Additive manufacturing (AM) is the manufacturing process by which three-dimensional (3D) parts are produced using an additive approach [3]. Vat photopolymerization is a form of AM. Vat polymerization 3D printing uses a liquid photopolymer resin which is solidified under the light source [3]. There are two main technologies in vat polymerization: DLP and SLA. Basically, both use the resin but the major difference between them is the light source which cured the resin [1]. In order to understand DLP, first look at the process of its forerunner SLA. In DLP, the light source is a digital light projector screen which blinks the layer of part all at once. Therefore, all points of layer cured simultaneously, and the printing speed is increased and at the same time printing time is decreased [2]. On the other hand, SLA takes more time than DLP because it uses a point-to-point method to cure [6, 7]. Also, the accuracy of the part made by DLP is seen better than SLA. DLP 3D printer uses in dental, jewelry, art and other sectors which require high detailing and finish.

This paper is organized as follows: first basic introduction of vat polymerization, second its methodology, and next its machine building. Finally, we present testing, analysis, and conclusion.

1.1 Working Principle

3D printing is any of the various processes in which material is joined or solidified under computer control to create a 3D object [1]. DLP 3D printing uses light to solidify a liquid photopolymer (Fig. 1).

By changing the pattern of the light and incrementing the vertical position of the workpiece, the desired geometry is built up layer by layer. In this process, once the 3D model is sent to the printer, a vat of liquid polymer is exposed to light from a DLP projector under safelight conditions. The DLP projector displays the image of the 3D model onto the liquid polymer. The exposed liquid polymer hardens, the build plate moves down, and the liquid polymer is once more exposed to light. The process is repeated until the 3D model is complete, and the vat is drained of liquid, revealing the solidified model [5].

1.2 Methodology

For making any object in DLP,

Table 1 Comparative summary of rapid prototyping (3D printing) techniques [4]

Types of 3D printing technology	Fused deposition modeling	Stereolithography, digital light processing	Selective laser sintering	Material jetting, drop on demand	Direct metal laser sintering; electobeam melting
Materials	Thermoplastic filament	Photopolymer resin	Thermoplastic Powder	Photopolymer Resin	Metal Powder
Dimensional accuracy	±0.5 mm (lower limit ± 0.5 mm)	±0.5 mm (Lower limit ± 0.5 mm)	±0.3 mm (Lower limit ± 0.3 mm)	±0.1 mm	±0.1 mm
Applications	– Electrical housings; form and fit tastings; – jigs and fixtures; investment casting patterns	– Injection mold-like polymer prototypes; – jewelry (Investment Casting); dental applications; hearing aids	Functional parts; complex ducting (Hollow Designs); low run part production	Full color product prototypes; injection mold-like prototypes; low run injection molds	Functional metal parts (Aerospace and Automotive); medical; dental

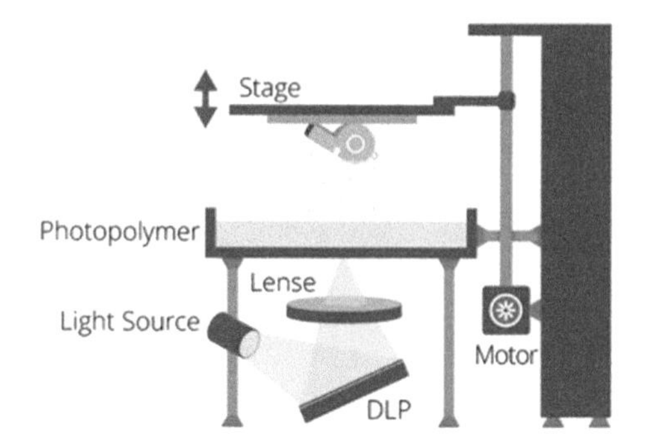

Fig. 1 Working of DLP 3D printer (*Source* www.think3d.in) [4]

(1) Create 3D design of an object in CAD software, (2) convert into .stl file, (3) creation workshop, (4) print, (5) 3D object, and (6) post-process.

2 Machine Building

2.1 Design of the Frame with Dimensions

The detailed design set-up is as shown in Fig. 2.

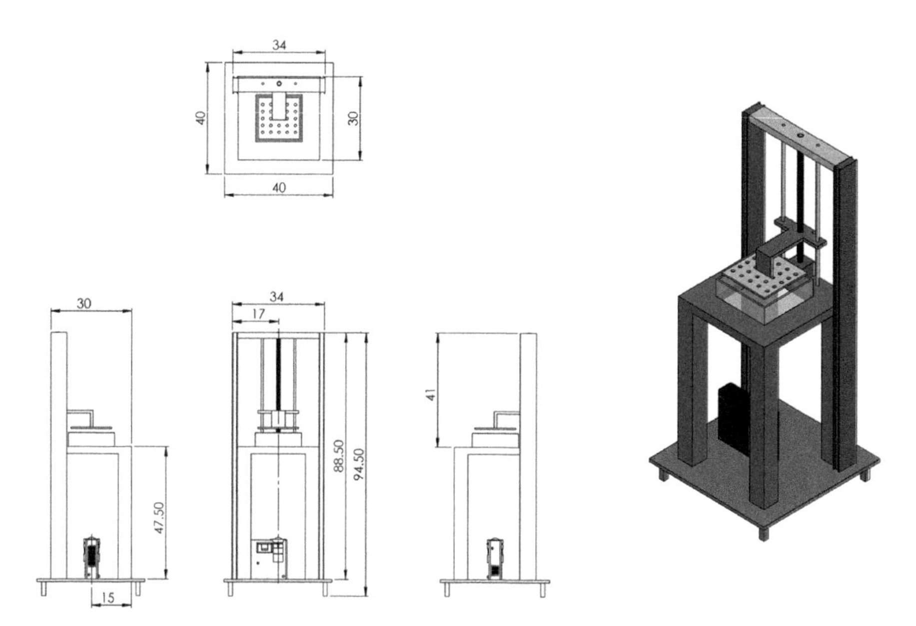

Fig. 2 Detailed design

2.2 Materials and Equipment

See Table 2.

Table 2 Material and equipment used in machine building

S. No.	Material & equipment	Specification	Qty. (Q)	Price (P)	TOTAL (Q*P) RS.
1	Projector	Model: BENQMP515 Resolution: 800*600 Lumens: 2500 Display type: DLP Throw ratio: 1.86:1–2.04:1	1	20,000	20,000
2	Resin	Company: Drop Shape Color: Orange Weight: 250 gm Wavelength: 365–405 Nm		4000	4000
3	Electronic components	(i) Nema17 Stepper Motor: 200 steps per mm Max. speed- 600–800 rpm (ii) Arduino Mega 2560 (iii) Motor drive circuit (iv) Power supply 12 V 20A	–	4323	4323
4	Aluminum frame	Aluminum from scrap	–	1000	100
5	Linear bearing	For 8 mm rod	2	250	500
6	SS Rod	8 mm	2	120	240
7	M8 Lead screw (300 mm)	Screw diameter: 8 mm Length: 300 mm Pitch: 1.25 mm Material: Stainless steel- lead screw, brass- nut	1	200	200
8	Flange nut		2	150	300
9	Flexible coupling		1	150	150
10	Vat and build plate	Vat crystal glass build plate aluminum	1	480	480
12	Bearing	8 mm	1	120	120
14	labor cost		–	450	450
GRAND TOTAL		30,863 INR			

2.3 Fabrication

First, we made our aluminum frame section from the scrap. The frame section looked good. We put stepper motor with an attached lead screw with the support of bearing on the top (Fig. 3).

But later on, we found from the above setup not up to the mark because it could not hold the weight of the build plate properly, and at the time of curing process, it could not have smooth movement. So, we changed that setup and used a linear channel for better up-down movement as in Fig. 4-i. We see the one channel which is held vertically to support, we put 3 mm glass at the bottom and stick vat on it, and we tested this setup but again we found the same problem which we found before. And also, we found that resin took more time to cure because of too much thickness of glass sheet and vat. So now, we put two channels facing each other and attached build plate and bearing on top (Fig. 4-ii). The vertical movement now seemed good. Then we attached plywood and cut it from the center as we stick our crystal glass vat with help of Araldite glue. Another remarkable point is that first, we used PCB board as our build plate. But the problem is that the layers would not stick to it properly. We also used Bakelite, SS plate, but the problem was not solved. Then we used an aluminum plate as our build plate; it is light in weight and also resin layer stick to it properly.

After the testing, the movement was good at earlier, but because of error in channel alignment and lack of support at the bottom later, we found some defaults. Therefore, we decided to change the setup. We used linear guild with two SS rods, and this time we gave support from the side all from down to up. Also, we put bearing at the top for proper lead screw movement. After testing, we found better results with smooth vertical movement. Here is the final setup which is given below in Fig. 4-iii. Also, at the time of the curing process, the build plate sticks to the vat after each layer, and sometimes layers of component stick to vat and affect the

Fig. 3 Frame section and initial design setup

Fig. 4 **i** Channel used for better movement. **ii** Fixed bearing at top of lead screw for stability of build plate. **iii** Used linear bearing and steel road for smooth movement and better stability

Fig. 5 Finished components made by our DLP 3D printer

accuracy of the finished product. So, we used non-stick material acrylic film on the vat. After using non-stick film, the layers would not stick to vat. Later on, we removed the color wheel from the projector for a better throw of UV light.

3 Results and Discussion

3.1 *Testing and Analysis*

After our 3D printer being complete in fabrication, we then do testing and analysis. We focused on printing time and accuracy and thickness of each layer of components that we made. We concluded the results of layer thickness and exposure time. We conclude that the cost of our DLP 3D printer is low compared to others in the market. Also, we got great accuracy with minimum printing time. Also, another aspect that we found is that if we removed UV filter from the projector, then our curing time will be decreased (Tables 3 and 4; Fig. 5).

3.2 *Environmental Parameters*

There are some environmental parameters that one should aware of before using the printer. The room temperature should be maintained at about 25–28 °C. And also, if printing process is done in the darkroom, then it would be easy for the resin to cure. Because if the resin exposes to the sunlight for a longer duration, then it may start to cure automatically at certain extent and this could affect the printing process.

4 Conclusion

From this paper, we conclude that after removing UV filter, the curing time is decreased by around 60%. Also, if we decreased cross-sectional area of projection or if your component is having a complex structure, then it is necessary to increase exposure time. Another point is that the thickness of vat should be minimum; otherwise, build plate will stick to the vat and also resin will take more time to cure. Therefore, with a minimum thickness of vat you do not need to use any non-stick

Table 3 Testing and analysis

No.	Layer thickness (mm)	Layer exposure (seconds)		First three-layer exposure (seconds)	
		Before	After	Before	After
1	0.05	50	20–25	100	55–60
2	0.1	80	35–40	130	70–100
3	0.15	100–110	50–60	160	85–120

Table 4 Comparison of cost estimation [1, 8–10]

Comparison	Micro protojet [8]	Formlabs from 2 [9]	B9Creation B9core550 [1]	Rapidshape S30 [10]	Our DLP printer
Built volume XY resolution	65 × 36 × 1 20 mm	145 × 145 × 175 mm	104 × 75 × 203 mm	50 × 31 × 80 mm	100 × 100 × 200 mm
Light source	LED projector	UV laser	LED projector	LED projector	DLP projector
Resolution (μ)	60	140	55	50	140
Resin per Kg.	150$	300$	150$	800$	114$
Cost	4995$	5500$	10,000$	20,000$	460$

film. If DLP projector has an adjustable resolution, it will affect your exposure times. If your distance between projector and vat is minimum, then curing time is increased. And also, from this research paper, we can conclude that DLP technology can replace the SLA technology with its great speed. And it can also replace FDM in many areas of work. Also, this research paper shows how to make DLP 3D printer at a low cost with good accuracy. Also, we conclude that DLP 3D printer has less maintenance with great reliability and consistency.

5 Future Scope

There is no doubt about the future of DLP 3D printer, because of its better capabilities than SLA in many industries. Also, in future, there might be a solution of having components with different color. This could be possible with the conveyor which has continued flow of various colored resin vat. For Example: firstly a build plate dip in red colored resin vat, after it being cured the red resin vat moves and blue colored resin vat comes and again the same process can repeat with other colors. There might be a problem to control the flow of vats. However, DLP technology has a great future scope.

References

1. Ibrahim A, Sa'ude N, Ibrahim M, Optimization of process parameter for digital light processing (DLP) 3d printing. Faculty of Mechanical Engineering, UniversitiTun Hussein Onn (UTHM), Malaysia, 86400 BatuPahat, Johor, Malaysia
2. Tyge E, Characterizing digital light processing (DLP) 3D printed primitives. Electrical Engineering, Technical University of Denmark, Lyngby, Denmark, Applied Mathematics and Computer Science, Technical University of Denmark, Lyngby, Denmark
3. Aznarte E, Ayranci C, Qureshi AJ, Digital light processing (DLP): anistropic tensile considerations. Department of Mechanical Engineering, University of Alberta, 10-203 DICE, 9211-116 Street NW, Edmonton, AB, Canada
4. Jasveer S, Jianbin X, Comparison of different types of 3D printing technologies. Department of Mechanical and Electrical Engineering, Nanjing University of Aeronautics and Astronautics
5. Liska R, Schuster M, Infuhr R, Turecek C, Fritscher C, Seidl B, Schmidt V, Photopolymers for rapid prototyping. J Coat Technol Res 4(4):505–510
6. Bangalore, Narendra DD (2014) "Studies on the process parameters of rapid prototyping technique" (Stereolithography) for the betterment of part quality. Int J Manuf Eng
7. Colombo P, Schmidt J, Franchin G, Zocca A, Gunster J (2017) Additive manufacturing techniques for fabricating complex ceramic components from preceramic polymers. Am Ceram Soc Bull 96(3):16–23
8. https://5.imimg.com/data5/HM/AM/MY-11147533/micro-protojet-dlp-3d-printer-cad-cam-jewellery.pdf
9. https://formlabs.com/3d-printers/form-2/
10. https://www.aniwaa.com/product/3d-printers/rapidshape-s30/

Design Evaluation of Cars Taillights in India Based on Novelty and Typicality

Adireddi Balaji and Dhananjay Singh Bisht

Abstract This paper presents a design research methodology to examine user perceptions about novelty and typicality in product design. This work approaches the concepts of novelty and typicality in product design through two different case studies. The first case study is a preliminary case study to explore and collect the descriptors related to novelty and typicality in car taillight designs in India using primary research. Using a survey, inputs from 72 design students were also collected regarding the most novel and most typical designs from among 100 taillight models. The second case study was conducted to assess the subjective perceptions about the five most novel and five most typical car models using descriptors of novelty and typicality found from the first case study. Nissan Leaf car taillights were found to be the most novel, and Chevrolet SRV car taillights were found to be most typical.

Keywords Design methodology · Design research · Novelty · Typicality · Car taillights

1 Introduction

The practice of industrial design typically involves not just a focus on the form of the product to be designed but research and analysis of the different factors that lead to aesthetic appeal as well. During such research, inputs may be acquired from consumer representatives, as well as subject matter experts. For customer inputs, user research could be conducted at different stages during product development. The results of such researches, given a specific product, could provide useful information on various actionable aspects of the product design that lead to qualitative excellence of product, enabling it to compete strongly in the market. The experts could provide valuable inputs coming from various perspectives of research

A. Balaji (✉) · D. S. Bisht
Department of Industrial Design, National Institute of Technology, Rourkela, Odisha 769008, India

BBVL. Deepak et al. (eds.), *Innovative Product Design and Intelligent Manufacturing Systems*, Lecture Notes in Mechanical Engineering,
https://doi.org/10.1007/978-981-15-2696-1_28

and practice such as those of product utility, service, style, and aesthetics. Both the users and experts could also provide valuable insights by evaluating new or existing product designs.

In the literature, a few product design properties have been found to have significant impact on product evaluations. Specific research has also been conducted in the past regarding aesthetics-related responses of users when considering the different product design properties. In this tradition, Loken and Ward [1] make a claim that prototypically or typicality is a major design property which defines the degree to which an object represents a category. Hekkert et al. [2] have discussed the joint effects of novelty and typicality on aesthetic preferences. They concluded that the design properties of typicality and novelty constitute two separate factors that are highly negatively correlated and tend to inhibit the effect of each other.

Several studies in the past have investigated the relationship between novelty and aesthetic preferences in product design. Hung and Chen [3] employed three fundamental dimensions of product semantics—trendiness, complexity, and emotion—and explored how changes in product semantics affected the judgments of product novelty and, in turn, the judgments of aesthetic preferences. Mukherjee and Hoyer [4] have studied the effects of "complexity" in the "novel" attribute during a product evaluation and found that novel attributes can be easily perceived as new technological innovations unknown to large number of customers. It was also found that the positive effects of novel attributes hold only in the case of low complexity products. Thurgood et al. [5] have explored the combined effects of "typicality" and "novelty" on "aesthetic pleasure" of product designs in terms of the influence of "safety and risk" perception. Some other specific information on product evaluation given the perspective of novelty and typicality has been provided in the following sub-sections.

1.1 Novelty

Novelty is often defined as the quality of being something new, original, or unusual. The term could be used to describe ideas, designs, methodologies, etc. In the context of industrial design, the term novelty could be used for describing an unusual or unique idea developed while developing a product or a breakthrough in terms of functionality of the product or aesthetic appeal of the product. The novelty of a product could be applied in the context of its appearance, utility, mode of interaction, etc. Berlyne [6] has classified novelty into two kinds: (i) absolute novelty—an object that has never been experienced before and (ii) relative novelty —an object that consists of a new combination of previously experienced elements. Similarly, Mugge and Schoormans [7] have classified novelty on the basis of apparent usability of the product as (i) functional novelty—defines the technological and functional development of a product often with a focus internal parts of product, and (ii) appearance novelty—brings the focus to the form of a product often described by the external parts of product. Novelty is also perceived in different manners by experts and non-experts [8]. Sluis-Thiescheffer et al. [9] describe the involvement of children in the design process in order to study novelty during the process of obtaining design solutions. Novel design solutions could come from involvement of diverse groups of participants (children, adults, old people, etc.) with varying parameters like gender, group size, power structures, etc. Radford and Bloch [10] found that products with high levels of visual product "newness" elicit more affective reactions than those with lower levels of newness.

1.2 Typicality

A semantically opposite term to novelty which is often described in the literature is typicality. Therefore, it could be claimed that the more novel a product is, the less typical it is for the customer and vice versa. Typicality provides an extent to which an object belongs to a category. According to Hung and Chen [3], typicality could be approached in three ways: (i) similarity to the ideal of the category; (ii) similarity to the central tendency of the category; and (iii) frequency of encounters with the object as a category member. Typicality, therefore, decreased with the increased familiarity with an object. This was demonstrated in Leder and Carbon [8] when studying aesthetic appreciation of car interiors. They found that participants changed from preferring a classical version to a more innovative version after repeated exposures to the different designs.

2 Methodology

The research work consists of two case studies. For the first study, a pilot survey was conducted using 72 design students who were asked to rate the most novel and typical stimuli presented to them. The stimuli provided were images of taillight models of a hundred motor cars (four-wheeled light motor vehicles) available in the Indian market. Car names were not provided to avoid brand biases and were numbered for identity. The survey was conducted in conformance to the protocols of the Institute Ethics Committee (IEC), NIT Rourkela (India). All the data collected were anonymized to ensure protection of participant identities.

The participants were explained the concept of novelty and typicality prior to the survey. Then, they were asked to choose the top ten most novel car taillights and ten most typical ones and to describe each selected taillight using a few adjectives. From the various adjectives collected, the top five most frequently used adjectives were identified for novelty as well as for typicality.

In the second case study, the results obtained from the first case study are used to conduct another survey in which participants were asked to rate each of the ten most novel and most typical car taillights on a descriptive scale. This scale was constructed using the top five adjectives of novelty and typicality. The descriptors were rated on a scale of 1–5 (1 = least; 5 = most) for the ten most novel and typical taillight stimuli.

2.1 *Flow of the Work*

Figure 1 provides a systematic overview of the methodology followed during this research work using a flowchart.

3 Results

On analysing the survey results, Figs. 2 and 3 show the top five most novel taillights and most typical taillights obtained from the user survey. Tables 1 and 2 list the five most frequently cited descriptor adjectives related to novelty and typicality, respectively, based on the analysis of user survey data. The top five descriptors for novelty were found to be attractive, sporty, modern, trendy, and futuristic. The top five descriptors for typicality were found to be boring, ordinary, compact, imitative, and common.

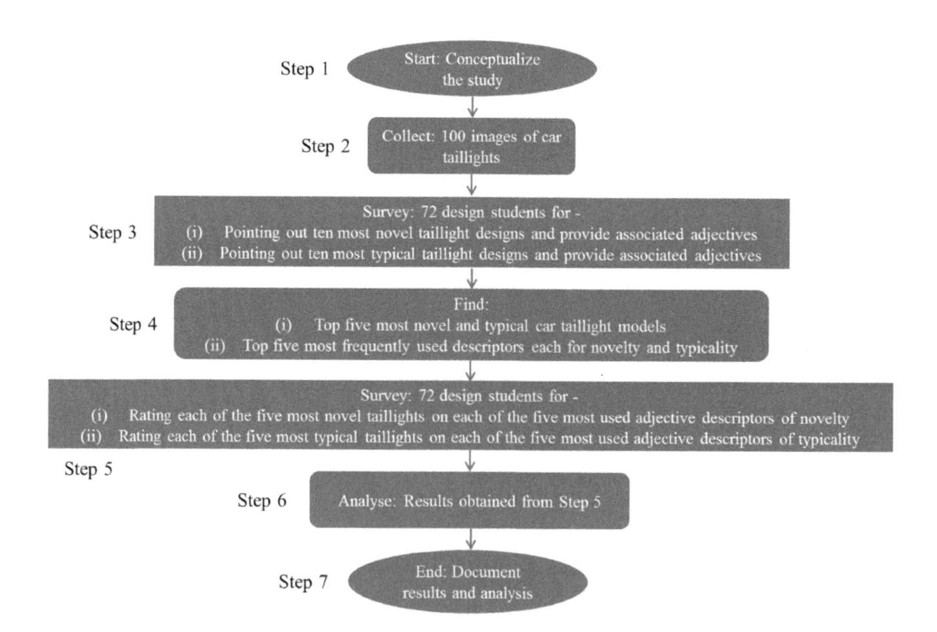

Fig. 1 Methodology followed during this research work

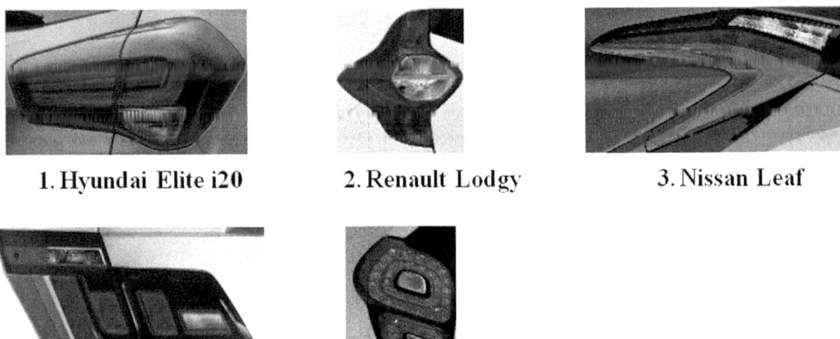

Fig. 2 Images of the five most novel car taillights

Fig. 3 Images of the five most typical car taillights

Table 1 Top five descriptors for novelty obtained from the survey

1. Attractive	2. Sporty	3. Modern	4. Trendy	5. Futuristic

Table 2 Top five descriptors for typicality obtained from the survey

1. Boring	2. Ordinary	3. Compact	4. Imitative	5. Common

3.1 Relationships Among the Car Taillights and Novelty Descriptors

Figure 4 provides the mean rating scores using five most cited novelty specific descriptors, for the five most novel car taillights indicated by the users. It can be interpreted that Nissan Leaf taillight is 41% more "attractive" than Chevrolet Beat taillight, 26% more "sporty" than Hyundai Elite i20 taillight, 25% more "modern" than Volkswagen Tiguan taillight, and 32% more "futuristic" than the Hyundai Elite i20 taillight. Also, Renault Lodgy taillight is 24% more "trendy" than Nissan leaf taillight. It is evident that Nissan Leaf taillight is the most novel design as four out of the five novelty descriptors have scored the highest for this design. Hyundai Elite i20 taillight has been found to be the least novel among the top five most novel car taillights. When discussing the rating scores in each of the top five novelty descriptors, it was found that in the case of descriptor "attractive", Nissan Leaf car taillight scored the highest and Chevrolet Beat taillight the lowest. In the case of descriptor "sporty", Nissan Leaf car taillight scored the highest, and Hyundai Elite i20 taillight scored the lowest. In the case of descriptor "modern", Nissan Leaf car taillight was found to be the most modern, and Volkswagen Tiguan taillight was the least modern. For the descriptor "trendy", Renault Lodgy car taillights was rated as the most trendy and Nissan Leaf car taillight was rated least trendy. For the descriptor "futuristic", Nissan Leaf car taillight was rated as the most futuristic, and Hyundai Elite i20 taillight was the least futuristic.

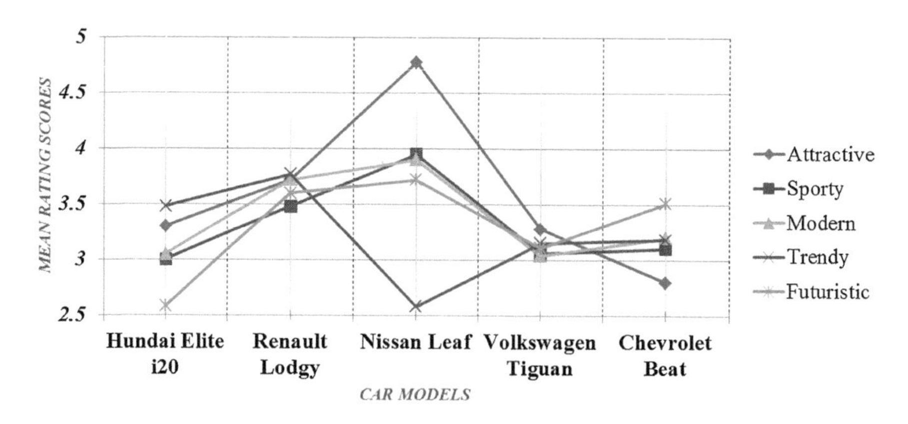

Fig. 4 Average ratings of most novel taillight designs using novelty descriptors

3.2 Relationships Among the Car Taillights and Typicality Descriptors

Figure 5 provides the mean rating scores using five most cited typicality specific descriptors, for the five most typical car taillights indicated by the users. It can be interpreted that Mahindra Verito taillight is 35% more "boring" than Nissan Micra taillight. Mahindra Verito taillight is 23% more "ordinary" than Chevrolet SRV taillight. Nissan Micra taillights is 21% more "compact" than Chevrolet SRV taillight. Maruti Suzuki Eco taillight is 22% more "imitative" than Chevrolet SRV taillight. Mahindra Bolero taillight is 34% more "common" than Nissan Micra taillight. It was found that Mahindra Verito taillight is the most typical design out of the five designs. When discussing the rating scores in each of the top five typicality descriptors, it was found that in the case of descriptor "boring", Mahindra Verito taillight was rated as most boring and Chevrolet SRV taillight, the least boring. In case of the descriptor "ordinary", Mahindra Verito taillight was rated as the most ordinary, and Chevrolet SRV taillight was the least ordinary. For the descriptor "compact", Nissan Micra car taillight was rated the most compact, and Chevrolet SRV taillights were rated the least compact. For the descriptor "imitative", Mahindra Bolero car taillight was rated as the most imitative and Chevrolet SRV car taillight, the least imitative. In the case of descriptor "common", Mahindra Bolero car taillight was rated as the most common, and Chevrolet SRV taillight was rated the least common.

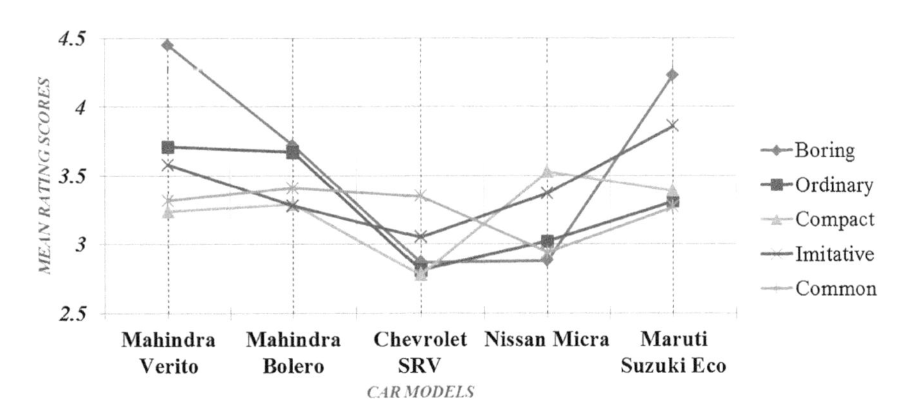

Fig. 5 Average ratings of most typical taillight designs using typicality descriptors

4 Discussion and Conclusions

4.1 Discussion

Based on the results summarized in Figs. 4 and 5, some interpretations can be made about the decisions made the users. Nissan Leaf taillight was rated more "attractive" possibly because it has a unique design in the form of horizontal V-shape which is not available in any other car taillight's design. This design can be claimed to be "sporty" as compared to others because of the abundance of sharp edges in its design and probably also because it occupies lesser space. This taillight shows a lot of variation in shape and curvatures as compared to the other existing taillights which perhaps makes it appear "modern" and "futuristic". Hyundai Elite i20 taillights are possibly one of the most easily visible designs due to the popularity of the model, and it being in the market for a long period of time. Probably, that is why these taillights have been rated as the least novel design among the five most novel designs. As Hyundai Elite i20 has been found to be least novel, it possibly could be considered as a relatively typical design among the most novel designs.

Mahindra Verito and Mahindra Bolero models have been in the market for a long period of time, and it may be possible that many have become too familiar with this design, which contributes to make it the most typical among all 100 designs. Based on the user preferences data from Fig. 5, it could be seen that the average ratings of these designs across the different descriptors of typicality are similar. Chevrolet SRV taillight seems to occupy much space, appears attractive, and seems to have distinct appearance and a kind of royal essence which makes it relatively uncommon or the least typical when compared to the other taillights. As Chevrolet SRV taillight has been found to be least typical, it can be considered as a relatively novel design among the most typical designs.

4.2 Conclusions

Based on the results, it can be concluded that taillights which have more sharp edges and with unique shapes have been rated as more novel, such as in the case of Nissan Leaf taillights. Taillights having more concave and convex shapes have been rated as less novel, just like most of the other taillights among the top five most novel designs. The taillights which have traditional shape like a variant of a square or a rectangle are likely to be seen as more typical, just like the case with Mahindra Verito. Taillights having different colour contrasts and contrasting shapes within a typical design could be perceived lesser typical, just like the case with Chevrolet SRV taillight. It was also observed that if a design is perceived as strongly novel, it is less likely to be considered be strongly typical and vice versa.

References

1. Loken B, Ward J (1990) Alternative approaches to understanding the determinants of typicality. J Consum Res 17(2):111–126
2. Hekkert P, Snelders D, and Van Wieringen PCW (2003) 'Most advanced, yet acceptable': typicality and novelty as joint predictors of aesthetic preference in industrial design. Br J Psychol 94(1):111–124
3. Hung W-K, Chen L-L (2012) Effects of novelty and its dimensions on aesthetic preference in product design. Int J Des 6(2):81–90
4. Mukherjee A, Hoyer WD (2001) The effect of novel attributes on product evaluation. J Consum Res 28(3):462–472
5. Thurgood C, Hekkert P, Blijlevens J (2014) The joint effect of typicality and novelty on aesthetic pleasure for product designs: influences of safety and risk. Congress of the International Association of Empirical Aesthetics
6. Berlyne DE (1970) Novelty, complexity, and hedonic value. Percept Psychophys 8(5): 279–286
7. Mugge R, Schoormans JPL (2012) Product design and apparent usability. The influence of novelty in product appearance. Appl Ergonomics 43(6):1081–1088
8. Leder H, Carbon C-C (2005) Dimensions in appreciation of car interior design. Appl Cogn Psychol Official J Soc Appl Res Mem Cogn 19(5):603–618
9. Sluis-Thiescheffer W, Bekker T, Eggen B, Vermeeren A, De Ridder H (2016) Measuring and comparing novelty for design solutions generated by young children through different design methods. Des Stud 43:48–73
10. Radford SK, Bloch PH (2011) Linking innovation to design: consumer responses to visual product newness. J Prod Innov Manag 28(s1):208–220

Design and Fabrication of Small-Scale Automatic Stamping Machine

R. Sundara Ramam and B. Harisankar

Abstract Stamping is one of the important processes that are to be performed in the packaging of industrial products. Stamping is the process used to print the text or symbol or trademark of companies' products on the paper or packaging boxes. As stamping is a manual process from its beginning till now, it takes more time and labor charges, we thought to automate the stamping process to reduce the time taken for the stamping process and to reduce the cost involved in labor. This leads to the invention of the automatic stamping machine. This works on the principle of rack and pinion mechanism and worm and worm wheel drive mechanism. The present work deals with the automatic stamping machine combined with the indexing table which automatically feeds the jobs for stamping. The main objective of this work is to help small-scale packaging industries which in turn reduce the time taken for stamping, reduce machinery cost, and increase productivity.

Keywords Rack and pinion · Worm and worm drive · Stamping · Trademark

1 Introduction

An ink stamp has high importance because it is not just an ink print. It may be an official signature of government officials which is needed for a work to be completed. It may be a trademark which gives so many details about the product and its company. Stamping is one of the oldest processes. The oldest tool used for stamping without ink is a seal, the first seal is known to be found during the ancient Mesopotamia (5000-3500 BC), and it is made up of stone or bone. Next, the seals are made with wax, and after that woodblock printing has been invented.

R. Sundara Ramam (✉) · B. Harisankar
Department of Mechanical Engineering, Vignan's Institute of Information Technology, Visakhapatnam, India
e-mail: sundar.ramam@gmail.com

B. Harisankar
e-mail: sankar340@gmail.com

BBVL. Deepak et al. (eds.), *Innovative Product Design and Intelligent Manufacturing Systems*, Lecture Notes in Mechanical Engineering,
https://doi.org/10.1007/978-981-15-2696-1_29

Woodblock printing is similar to our rubber stamps. The image is carved on wood and then be inked and applied on fabric or paper. As stamping is one of the old processes, even in this modern era stamping is still in manual practice by humans [1]. Humans feel bored when they do repetitive tasks, and it causes fatigue to the workers, which affects the performance of the operator and this causes work being completed late and with less accuracy.

In this modern world, the large-scale industries are using highly advanced technologies to reach high levels in the market [2]. But the small scale is not able to survive in this competitive market due to a number of problems like less labor, less investment, and due to lack of automation which can be done with less capital and only some of their operations being automated. Small-scale industries are accustomed to manual operations which increase their operation time, and products cannot be delivered on time [3]. Thus, these industries lose reputation and cannot run effectively. So to overcome the above-mentioned problems and to reduce the time taken for stamping with less cost and eliminate human work, automatic stamping machine has been invented [4].

Amol Kadam and his students have designed and fabricated an automatic stamping machine working on the mechanical structure of the Geneva mechanism and simple crank mechanism. He concluded that with the help of Geneva mechanism, stamping is done at the desired position and paper is fed on desired time. The design is purely mechanical and operates at low cost with low time consumption and maximum accuracy [5].

Mauton Gbededo and Olayinka Awopetu have designed an automatic stamping machine for small-scale industries which works on pneumatic systems which consist of air compressor, directional control valves, and air service unit have been fabricated and operated. Fluidsim software made by Festo Didactic using the cascade method was adopted in the design and simulation of pneumatic circuits. This machine was tested meeting the safety and regulatory requirement of food industries in accordance with the National Agency for Food, and Drug and Administration and Control in Nigeria. It can be operated by anyone with little energy required to drive the pedals [6].

A number of stamping machines have been designed with different mechanisms and power sources like the ones mentioned above. But every stamping designed is only up to the stamping action, so we are extending the idea by adding an indexing table to the automatic stamping machine whose objective is to automatically feed the work parts for stamping to be done. In our project, the automatic stamping machine has been designed by using rack and pinion mechanism for the stamping process and worm and worm drive for indexing operation. Automatic stamping machine has been designed and fabricated using the above mechanisms. The design is completely mechanically operated, and the cost of the machine is very low and less operating time with good accuracy.

2 Methodology

The sequence of operations is loading, stamping, indexing, and unloading.

2.1 Loading

Loading is the process of placing the workpieces in their respective positions on the working table to undergo the operation to be performed on them. The loading of workpieces is done manually in this machine. After placing the workpieces on the table, they are fixed by using holding devices to avoid vibrations during the operation.

2.2 Stamping

Stamping is the process of imprinting the images, symbols, texts, and logos on papers or packaged products by using rubber or acrylic stamps. In this machine, the stamp is attached to the pinion, and as the pinion moves over the rack the stamp attached to the pinion moves up and down. Thus, it automatically stamps each box with a certain time of interval.

2.3 Indexing

Indexing is the process of dividing the periphery of a cylindrical workpiece into an equal number of divisions with the help of an indexed crank and index plate. In this machine, the worm drive is used to index the table automatically. Generally, indexing is done by maintaining a worm drive ratio as 20:1, 40:1, etc. For 20 turns of the worm, worm gear completes one revolution [7].

2.4 Unloading

Unloading is the process of taking out the work parts from the table after the operation is performed. It is done manually after the operation is completed on all parts of the batch [8].

The mechanisms used are rack and pinion mechanism and worm and gear mechanism.

2.5 Rack and Pinion Mechanism

A rack can be best described as a gear of infinite radius. It works in conjunction with a small gear called opinion. The combination provides a means to convert the reciprocating motion into rotary motion and vice versa. The circular gear called pinion engages teeth on a linear gear bar called rack. Rotational motion applied to the pinion causes the rack to move relative to the pinion, thereby translating the rotational motion of the pinion into linear motion. Figure 1 shows the rack and pinion mechanism.

2.6 Worm and Worm Gear

A worm drive is a gear arrangement in which a worm is a cylindrical helical gear with one or more threads and resembles a screw thread, and worm wheel or worm gear is a cylindrical gear with flanks cut in such a way as to ensure contact with the flanks of the worm. A worm drive is used to accomplish large speed reduction in skew shafts, spiral gears with a small driver, and a larger follower are required [9]. Figure 2 shows the worm and gear mechanism.

Also, the load transmitted through these gears is limited. By using a worm gear having many teeth, considerable speed reduction can be affected, which is not possible through other gearing arrangements within a limited space. One of the major advantages of the worm drive is that they can transmit motion in 90°.

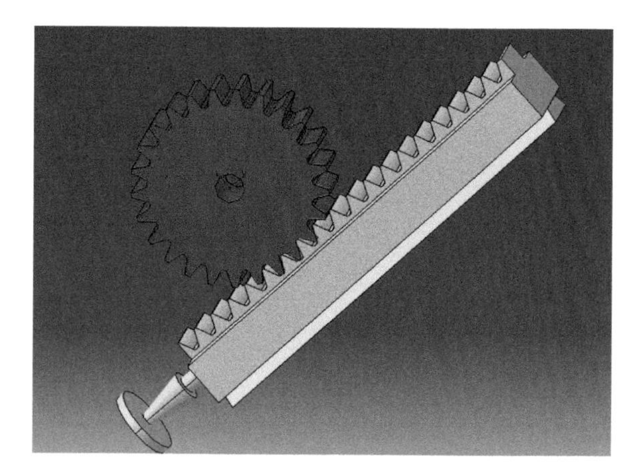

Fig. 1 Rack and pinion mechanism

Fig. 2 Worm and gear mechanism

3 Mechanism Specifications

The individual components' specifications of the stamping machine are given in Table 1.

Minimum pressure required to stamp is 4 PSI

1 PSI = 0.0703 kgf/cm^2 = 6896.43 N/m^2

4 PSI = 0.2812 kgf/cm^2 = 27,585.72 N/m^2

Power output from battery = $V*I$ = 12*7 = 84 W

Power utilized by motor = $T*w$ = 0.4905*(2*π*30)/60 = 1.54 W

Area of stamp = πr^2 = 3.14*$(0.025)^2$ = 4.908 cm^2 = 0.4908*10^{-3} m^2

Force exerted by motor = T/r = 0.4905/0.03 = 16.35 N

$$\text{Pressure exerted on stamp} = F/A = 16.35/\left(0.4908 * 10^{-3}\right)$$
$$= 33,312.958\,\text{N/m}^2 > 27,585.72\,\text{N/m}^2.$$

Therefore, the pressure that is exerted on the stamp is more than the required pressure to stamp.

Table 1 Dimensions of the components

Component	Dimension
Length of rack	0.22 m
Diameter of pinion	0.06 m
Speed of motor	30 RPM
Torque of motor	0.4905 N-m
Diameter of stamp	0.025 m
Battery voltage	12 V
Battery current	7 A
Diameter of table	0.28 m

4 Results and Discussion

The project design was modeled using the CATIA V5 Software on Windows 7 operating system by using different tools like a pad, helix, pocket, groove, and rib. The CATIA model diagram is shown in Fig. 3. Figure 4 shows the fabricated setup of the stamping machine.

The minimum pressure required to stamp is 4 Psi (0.276 bar). The size of the stamp used in this work is 0.0005 m^2. So, the minimum force required is 13.79 N to stamp on the box. The power supplied to the motor is 1.54 W, and the radius of the torque arm is 0.03 m. Therefore, the force exerted by the motor is 16.35 N which is greater than the minimum force required. After designing and fabricating the stamping machine using the above mechanisms, it can be used to stamp eight boxes per minute. The number of boxes stamped can be increased by increasing the speed of the motor and changing the Arduino setup. Parameters affecting the performance of the stamping machine are rack length, motor speed, gear teeth, and table diameter.

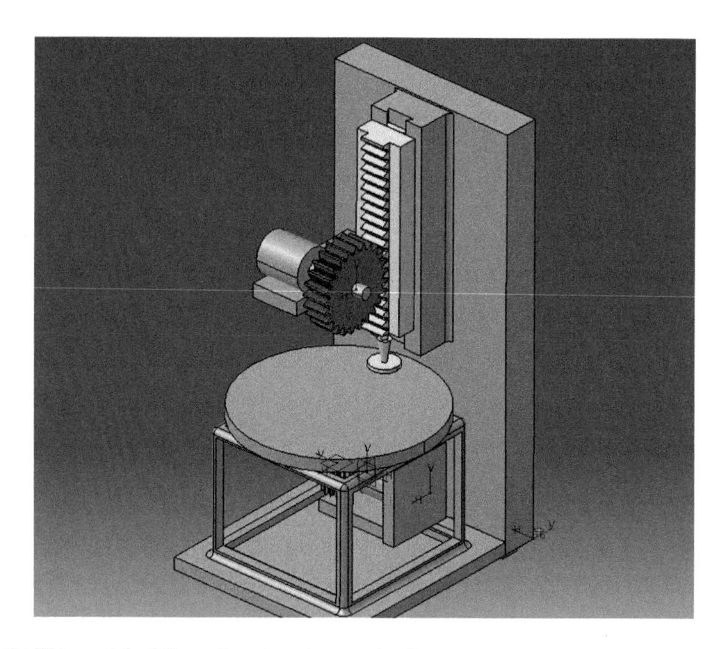

Fig. 3 CATIA model of the entire stamping mechanism

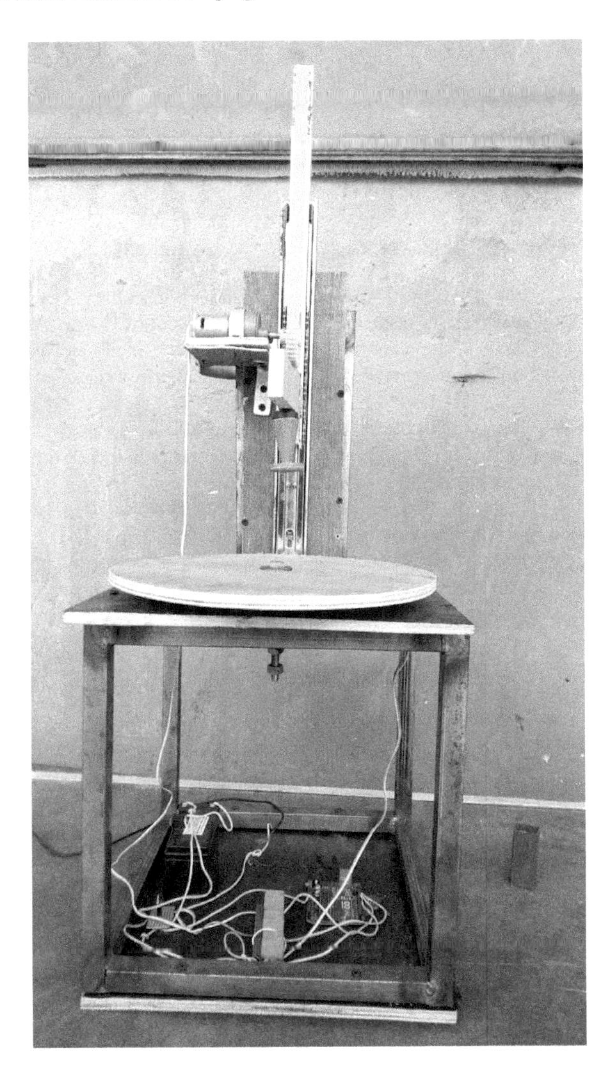

Fig. 4 Fabricated model of the stamping machine

5 Conclusions

The automatic stamping machine working on the principle of rack and pinion mechanism and worm and worm wheel mechanism has been designed and fabricated successfully. With the help of the worm and worm wheel, we are able to feed the work part automatically below the stamping unit at the desired time. Hence, we succeed to design and fabricate an automatic stamping machine structure which operates at low cost, low time consumption, and with maximum accuracy.

This machine can be used in food packaging industries, pharmaceutical industries, etc. Its performance can be increased by using advanced mechatronics structures. By replacing the stamping box at the end of the rack and pinion with punches, dies, forming tools, and supplying the required power to the machine can give rise to a number of applications.

References

1. Umarkar AP (2017) Design of stamping machine for use in industries. IJRMEE 4(7):49–52
2. de Larrea de la Peña FAL, Vértiz SVP, Berrios DJE, Huelsz JAP, Servín EM, Yousuf MA (2007) Automation at a stamping industry. In: Proceedings of the 7th WSEAS International Conference on Robotics, Control & Manufacturing Technology, Hangzhou, China, pp 257–262, April 15–17
3. Ravipothina (2015) Automatic pneumatic stamping machine. IJMETMR 2(7):85–91
4. Hussain A, Choudhary R, Khatri S, Nalawade CG (2017) Automatic pneumatic stamping and counting machine. IJSRD 4(12):538–541
5. Dhoble A, Gajare D, Vibhute S, Kadam A (2018) Design and fabrication of automatic stamping machine. IJSER 9(5):204–208
6. Gbededo M, Awopetu O (2017) Design and construction of automated stamping machine for small scale industries. IJERT 6(12):19–26
7. Welkar DS, Saindane LS, Nerker NS, Baviskar HR, Sonawane VP (2017) Automatic stamping and pad printing machine. IJARSE 6(2)
8. Salvi O, Pawar G, Mudshi S, Naik A, Gawade S (2017) Automatic bottle cap stamping machine for small scale bottle industries. IJESRT 6(4):492–497
9. Koppa P, Nagaraja N, Amith V, Mutalikdesai S, Vyasaraj T, Kumar GHP, Pandey RK, Vishwanath S (2016) Development and fabrication of electro pneumatic automatic stamping machine. IJIRSET 5(9):16188–16197

Smart Manufacturing, Precision Engineering, Metrology

Application of Fuzzy Controller for Multi-area Automatic Load Frequency Control System

Manmadha Kumar Boddepalli and Prema Kumar Navuri

Abstract This present work deals with the performance of the fuzzy logic controller (FLC) for a multi-area automatic load frequency control system. At the start, thermal, hydro, and gas units in each area are considered in a three areas' nine-unit power system and appropriate area participation factors are assumed for each unit. After that, fuzzy logic controllers having triangular member functions are incorporated as a secondary controller to attain a better performance of the system. The effectiveness of the FLC is observed while compared with a classical PID controller. In addition, the capability of considered FLC controller is also exhibited by considering random step load disturbance.

Keywords Automatic load frequency controller · Fuzzy logic controller · PID controller · Random step · Tie-line

1 Introduction

One of the most significant aspects of contemporary power system operation, control, and design is automatic load frequency controller. Reliability of the power system depends on the deviation in the system operating frequency and tie-line power within their predestine limits. Changes in system frequency and tie-line power lead to the overload of the transmission line and sometimes damage the equipment; load performance will be degraded. These changes can push the power system into an unstable state.

M. K. Boddepalli (✉)
Aditya Institute of Technology & Management, Tekkali, India
e-mail: manmadhakumarboddepalli@gmail.com

P. K. Navuri
Andhra University College of Engineering (A), Andhra University, Visakhapatnam, India

BBVL. Deepak et al. (eds.), *Innovative Product Design and Intelligent Manufacturing Systems*, Lecture Notes in Mechanical Engineering,
https://doi.org/10.1007/978-981-15-2696-1_30

The major role of the automatic load frequency controller is:

- Maintaining the system frequency within the predestine limits.
- Interchange of power between the two or more control areas within their specified limits.

The increasing size and structure of complex interconnected power system cause an increase in economic pressure and also affect the efficiency and reliability of the system. This discloses the significance of preservation of the power system frequency and tie-line power in their predestine values [1]. Conventional PI and PID controllers are recognized as fixed parametric controllers. Conversely, the power system is dynamic in nature. So these traditional controllers are incompetent to give their best retort in this concern. Lotfi A. Zadeh introduced fuzzy logic as a replacement. The fuzzy logic controller has been designed and anticipated in the vicinity of automatic load frequency control in two-area power system and its effectiveness also compared with conventional integral controller [2]. Robustness of fuzzy logic controller has been judged against the traditional PID controller in a single area and two areas of no reheat thermal power systems. For different loading conditions, the performance of change in system frequency and tie-line power is better by using fuzzy logic controller [3]. PID controller parameters have been optimized by firefly algorithm to study the two areas interconnected power system and compared its effectiveness with I, PI controllers [4].

Fuzzy controller operational performance has been addressed in [5], also used to calculate the area control error, and also addressed the difficulties in secondary frequency control. PI, PID controllers are used to comparing with fuzzy PID controller for single and two area systems to analyze both AGC and AVR problems. The superiority of fuzzy PID controller has been addressed [6]. 2-degrees-of-freedom of proportional–integral–derivative (2-DOFPID) controller [7] has been addressed to observe the performance of two-area thermal power system, and the behavior of the acknowledged system has been analyzed with differential evolution (DE) algorithm. PI, PID controller gains are optimized by firefly algorithm, and the performance of a two equal areas of non-reheat thermal system has been addressed. These results are compared with other optimization techniques and also observed [8] the performance of three unequal area thermal systems. The single area with an integral controller in each unit of a multi-source power system and multi-area multi-source power system with HVDC link has been addressed [9]. I, PI, and PID controller parameters are optimized by the DE technique. Firefly algorithm (FA) [10] has been used to optimize the hybrid fuzzy PID controller with derivative filter (PIDF) system; some of the system nonlinearities have been considered in the deregulated multi-area multi-source system.

The novelty of the present paper is the parameter values of fuzzy logic-based PID controller have been optimized by firefly algorithm. The proposed fuzzy PID controller shows its superiority when compared to the conventional PID controller for the same three areas' nine-unit power system. And it has been perceived that, in terms of peak overshoot, peak undershoot and settling time, the fuzzy logic-based

PID controller gives a better dynamic performance. The performance of the system has also observed for a random step load pattern.

2 Power System Under Study

A three areas' nine-unit power system has been addressed in the present analysis, where three similar areas have been acknowledged—each area consists of a thermal power plant with reheat turbine in the first unit, a hydropower plant with the governor in the second unit, and in the third unit, a gas turbine power plant has been considered as shown in Fig. 1.

3 Fuzzy Logic

For the application of indefinite and uncertain models, the fuzzy logic or set theory establishes the rules of nonlinear mapping to provide a platform to the system. The inputs of the fuzzy logic controller are fuzzified, and rules are used to infer the decision of the fuzzy controller. Finally, the outputs of the fuzzy control decision are produced by defuzzification. Major phases involved in fuzzy logic controller are: (1) fuzzification, (2) defining the rules, (3) inference engine, (4) defuzzification.

For this study, Mamdani fuzzy interface engine has been preferred. Triangular membership functions have been taken for the present study as shown in Fig. 2. NB (negative big), NS (negative small), Z (zero), PS (positive small), and PB (positive big) are the linguistic variables for both the inputs and the output of the controller. In the defuzzification center of gravity, the method has been used to find the outputs of the fuzzy logic controller. The two-dimensional rule base for error, error derivative, and fuzzy logic controller outputs is presented in Table 1.

4 Results and Discussions

Any controlled process in a system will anticipate stability and fast response, and in contrast to a real-time problem, both these requirements cannot at all be accomplished at the same time. It can be practicable only by selecting a good controller. Commonly, three control parameters are used in the firefly algorithm (FA) [11, 12]:

(i) α: the randomization parameter,
(ii) β: the attractiveness, and
(iii) γ: the absorption coefficient.

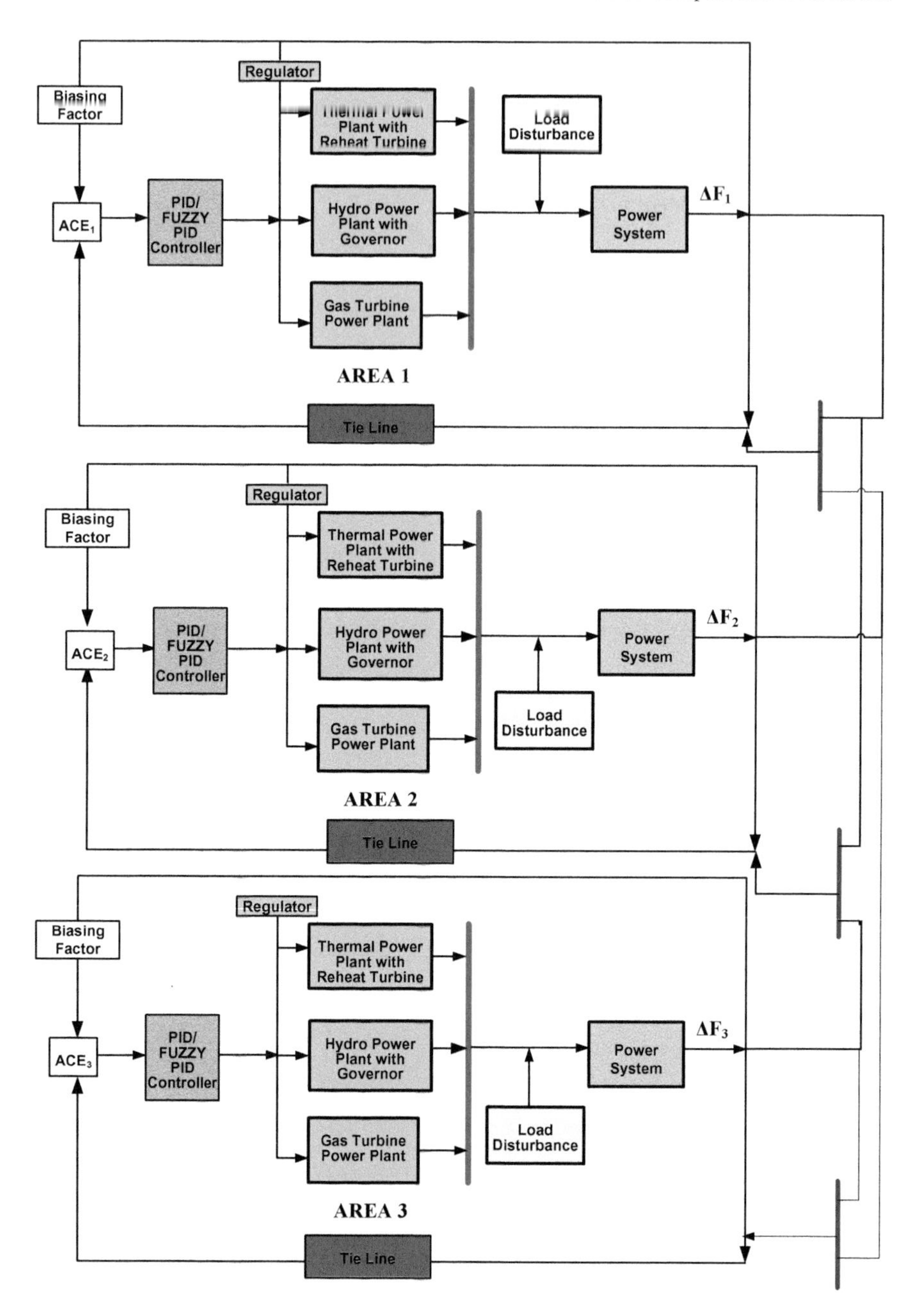

Fig. 1 Three areas' nine-unit power system with PID/fuzzy PID controller

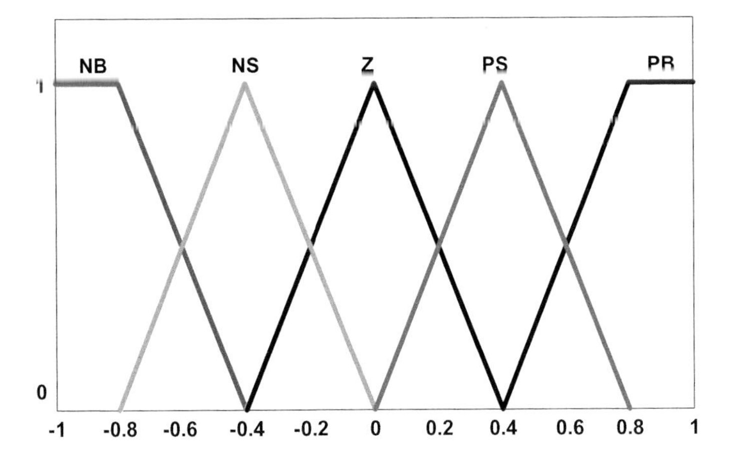

Fig. 2 Triangular membership function for input and output

Table 1 Fuzzy rule base

$\dot{e}$	e				
	NB	NS	Z	PS	PB
NB	NB	NB	NS	NS	Z
NS	NB	NS	NS	Z	PS
Z	NS	NS	Z	PS	PS
PS	NS	Z	PS	PS	PB
PB	Z	PS	PS	PB	PB

The fireflies' light intensity is given by:

$$I(r) = I_0 e^{-\gamma r} \tag{1}$$

Attractiveness function of firefly is given by:

$$\beta = \beta_0 e^{-\gamma r^2} \tag{2}$$

The distance between any two fireflies *i, j* can be expressed as:

$$r_{ij} = \left\| x_i - x_j \right\| = \sqrt{\sum_{k=1}^{d} \left(x_{i,k} - x_{j,k} \right)^2} \tag{3}$$

where '*d*' is the number of dimensions.

The controlled parameters are considered in the present work: number of fireflies considered = 4; maximum generation = 50; $\beta = 0.6$; $\alpha = 0.2$, and $\gamma = 0.4$. Three areas' nine-unit power system is tested for an increase of 20% step in the load in area-1 in MATLAB/Simulink environment.

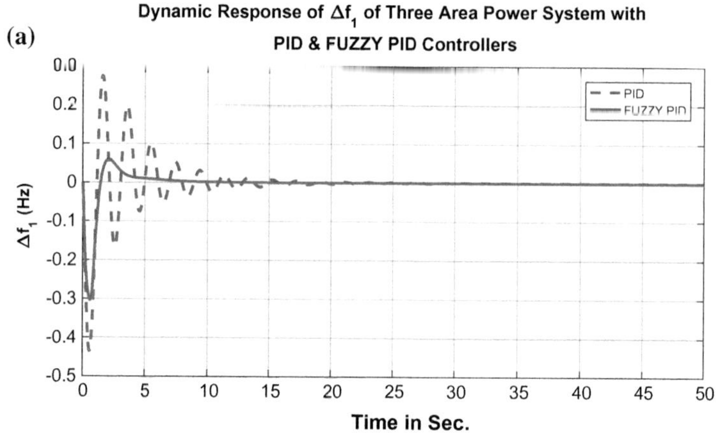

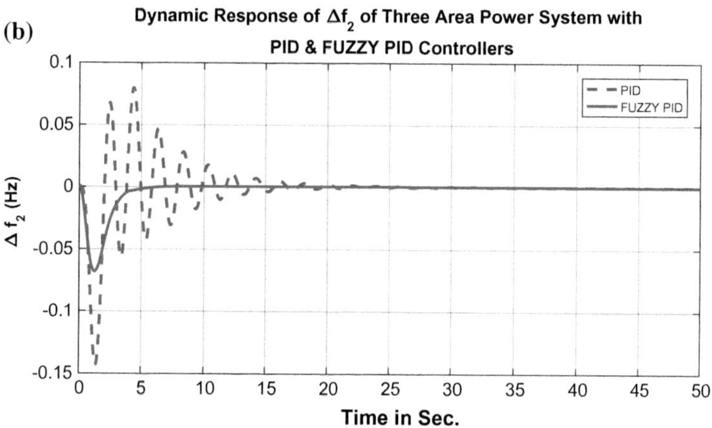

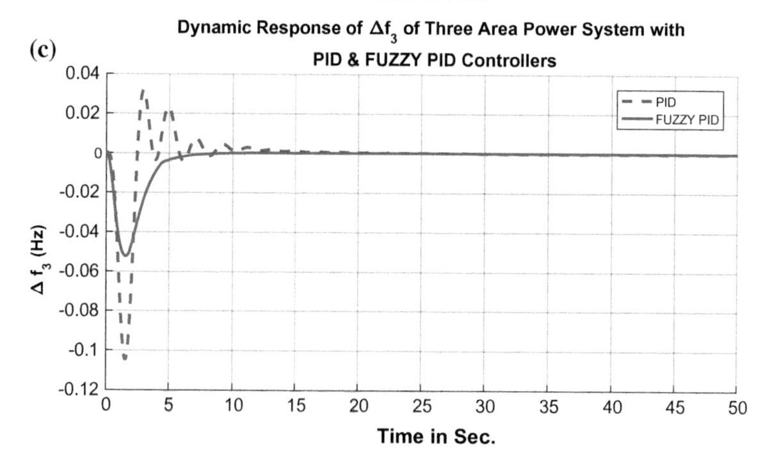

Fig. 3 **a** Dynamic response of Δf_1. **b** Dynamic response of Δf_2 **c** Dynamic response of Δf_3

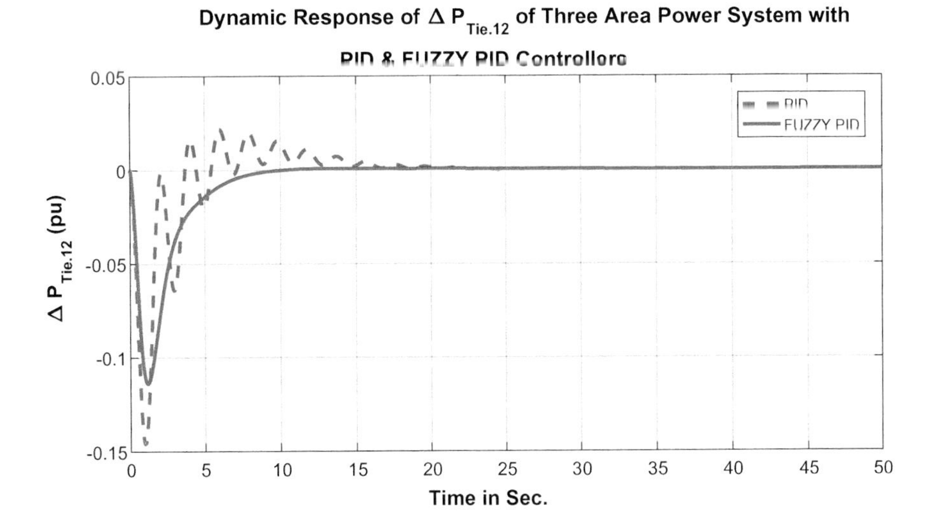

Fig. 4 Dynamic response of $\Delta P_{\mathrm{Tie.12}}$

With the change in step, the dynamic response of change in frequency of area-1 is shown in Fig. 3a, for area-2 in Fig. 3b and for area-3, it is shown in Fig. 3c.

Figure 4 exhibits the response of exchange of power ($\Delta P_{\mathrm{Tie.12}}$) between the tie lines of area-1 and area-2. Similarly, the system has also tested for area-2 with 20% step increase in the load.

The system also tested for a random step load pattern shown in Fig. 5, and the deviation in the frequency of area-1 has been shown in Fig. 6.

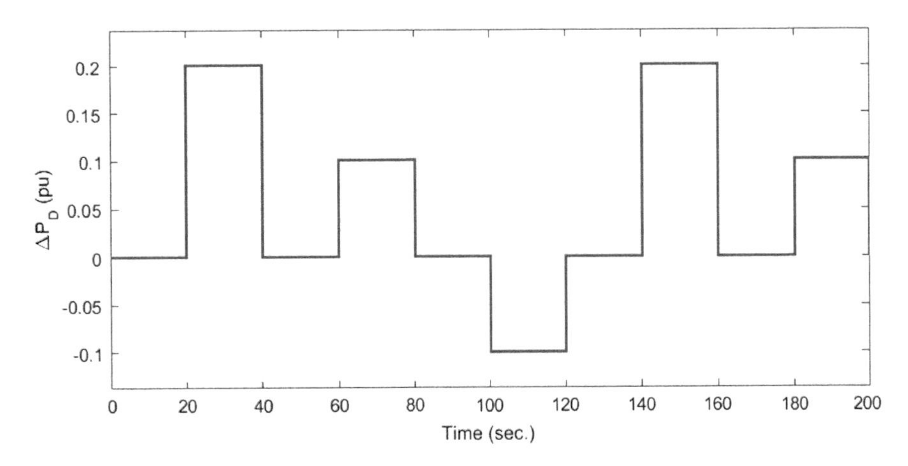

Fig. 5 Random step load pattern

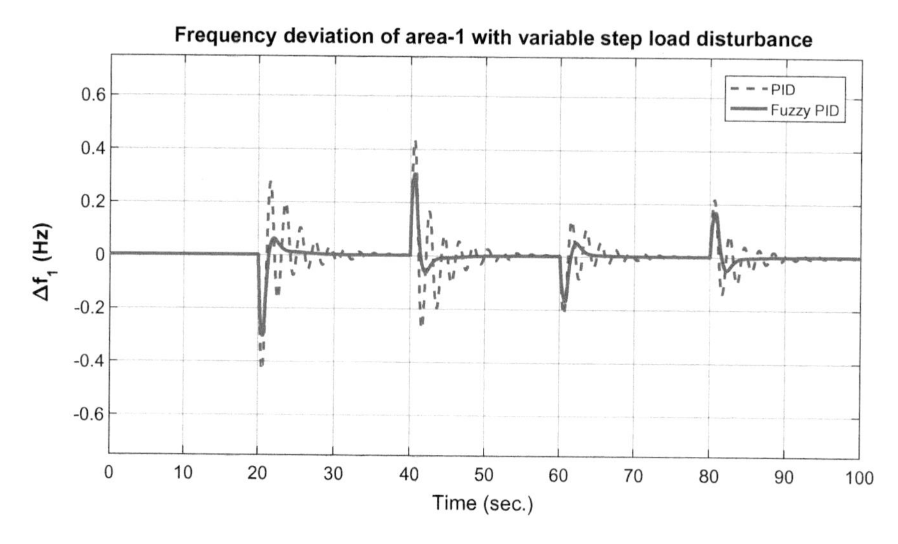

Fig. 6 Frequency deviation of area-1

Table 2 Performance index values

Parameters/controller		20% Disturbance in area-1			20% Disturbance in area-2		
		Peak undershoot	Peak overshoot	Settling time (Sec.)	Peak undershoot	Peak overshoot	Settling time (Sec.)
Δf_1	PID	−0.4356	0.2742	9.65	−0.2865	0.1345	12.4
	FUZZY PID	−0.3044	0.0590	3.365	−0.1397	0.0	3.32
Δf_2	PID	−0.1432	0.0675	8.595	−0.4525	0.1495	9.3
	FUZZY PID	−0.683	0.0	2.66	−0.3309	0.015	1.74
Δf_3	PID	−0.1048	0.0315	5.15	−0.2006	0.0481	5.45
	FUZZY PID	−0.05250	0.0	3.15	−0.1055	0.0	3.97
$\Delta P_{\mathrm{Tie.12}}$	PID	−0.1461	−0.0018	6.16	−0.0195	0.0608	5.04
	FUZZY PID	−0.1143	0.0	4.3	0.0	0.0438	2.14
$\Delta P_{\mathrm{Tie.23}}$	PID	−0.0077	0.0316	3.0	−0.1107	−0.0092	3.515
	FUZZY PID	0.0	0.0236	1.49	−0.0883	0.0	4.003
$\Delta P_{\mathrm{Tie.13}}$	PID	0.0013	0.216	1.335	0.0053	0.0413	1.76
	FUZZY PID	0.0	0.0171	0.0	0.0	0.0339	2.63

The performance index values like peak undershoot, peak overshoot, and settling times of area-1 and area-2 are provided in Table 2 for both PID and fuzzy PID controllers. Table 2 provides the performance index of fuzzy PID controller for the change of load in both the areas and exhibits its superiority over the conventional PID controller.

5 Conclusion

In this paper, a multi-area power system was considered which is having diverse sources of units such as thermal, hydro, and gas. The fuzzy PID controller is considered as a secondary controller, and its gains are optimized with the firefly algorithm. The supremacy of fuzzy PID was compared with the classical PID controller. The effectiveness of the proposed controller is also shown by having a random step load as a disturbance. From simulation results, it can be concluded that the fuzzy PID controller performed well and attained better performance index values. The system is stable even 20% of disturbance is applied in the load. In the future, present work can be extended by taking nonlinearities of the system into account. In the future, the present can be extended for deregulated power system and price estimation. The work can also be extended to consider all types of nonlinearities to have a real system.

References

1. Elgerd OI (2000) Electric energy systems theory—an introduction. Tata McGraw Hill, New Delhi
2. Indulkar CS, Raj B (1995) Application of fuzzy controller to automatic generation control. Electr Mach Power Syst 23(2):209–220. https://doi.org/10.1080/07313569508955618
3. Sambariya DK, Nath V (2016) Load frequency control using fuzzy logic based controller for multi-area power system. Br J Math Comput Sci 13(5):1–19, Article no. BJMCS.22899
4. Boddepalli MK, Navuri PK (2018) Design, and analysis of firefly algorithm based PID controller for automatic load frequency control problem. In: IEEE international conference on technologies for smart-city energy security and power (ICSESP-2018)
5. Chown GA, Hartman RC (1998) Design and experience with a fuzzy logic controller for automatic generation control (AGC). IEEE Trans Power Syst 13(3):965–970
6. Nath V, Sambariya DK (2015) Analysis of AGC and AVR for single area and double area power system using fuzzy logic control. Int J Adv Res Electr Electron Instrum Eng 4(7)
7. Sahu RK, Panda S, Rout UK (2013) DE optimized parallel 2-DOF PID controller for load frequency control of power system with governor dead-band nonlinearity. Int J Electr Power Energy Syst 49:19–33
8. Pradhan S, Sahu RK, Panda S (2014) Application of firefly algorithm for load frequency control of multi-area interconnected power system. Electr Power Compon Syst 42:1419–1430
9. Mohanty B, Panda S, Hota PK (2014) Controller parameters tuning of differential evolution algorithm and its application to load frequency control of multi-source power system. 54:77–85
10. Chandra Sekhar GT, Sahu RK, Baliarsingh AK, Panda S (2016) Load frequency control of power system under deregulated environment using optimal firefly algorithm. Electr Power Energy Syst 74:195–211
11. Yang XS (2010) Firefly algorithm, stochastic test functions, and design optimization. Int J Bio-inspired Comput 2:78–84
12. Yang XS, Hosseini SSS, Gandomi AH (2012) Firefly algorithm for solving non-convex economic dispatch problems with valve loading effect. Appl Soft Comput 12:1180–1186

Process Capability Improvement Using Internally Cooled Cutting Tool Insert in Cryogenic Machining of Super Duplex Stainless Steel 2507

D. Narayanan and T. Jagadeesha

Abstract The present work deals with the influence of cryogenic coolants LN_2 delivered through holes made on flank surface and rake surface of tungsten carbide cutting tool inserts in turning of super duplex stainless steel (SDSS) using in-house developed cryogenic setup. Experiments were conducted with the cryogenically treated tool, cryogenically treated tool with tempering and cryogenic coolant directly passed through a modified cutting tool insert. Results are compared with dry cutting conditions. The cutting conditions are low feed rate/high depth of cut, medium feed rate/medium depth of cut, and high feed rate/low depth of cut. The material removal rate and cutting speed is kept constant under all three cutting conditions. Microstructural study of the tool as received and cryogenically treated is examined using SEM. The population of harder tungsten carbide phase (gamma phase) is found to be more in the cryogenically treated tool. Due to tempering, the hardness of insert is improved by 8% which in turn increased tool life. By direct supply of LN_2 through modified cutting tool increased tool life by 23%, more than the cryogenically tempered tool. There are no appreciable changes in the temperature of the cutting tool under dry cutting and cryogenically treated inserts. However, there is a large difference observed in temperature of cutting tool when LN_2 is supplied through a modified insert directly, which in turn yielded high tool life.

Keywords Cryogenically treated inserts · Cryogenic machining · Tool wear · Cutting temperature

D. Narayanan · T. Jagadeesha (✉)
National Institute of Technology Calicut, Calicut, Kerala, India
e-mail: jagdishsg@nitc.ac.in

BBVL. Deepak et al. (eds.), *Innovative Product Design and Intelligent Manufacturing Systems*, Lecture Notes in Mechanical Engineering,
https://doi.org/10.1007/978-981-15-2696-1_31

1 Introduction

Machining operation for super duplex stainless steels requires higher cutting forces thus cause for tool wear at a rapid speed. The processing capability of machining is one of the ways to prolong the cutting tool life and it can be achieved with better tool design and cryogenics [1–3]. Favorable chip formation is achieved with the help of cryogenic cooling. [4–7]. While comparing, it was found that wet machining gave prior works have reported that cryogenic CO_2 has substantially reduced the cutting temperature and cutting forces when compared to LN_2 coolant (cryogenic), wet cutting and dry cutting, respectively [8]. This current study focuses on finding the most optimum parameter combination which gives higher tool life.

2 Experimentation

2.1 Material Selection

Super duplex stainless steel (SDSS) 2507 is used as the workpiece material (Ø 36 × 300 mm) for the experimentation. Properties of SDSS are given in Table 1.

2.2 Tool Selection and Modification

PVD-coated tungsten carbide inserts (CNMG 120408MT12) are used for the experimentation. Tools are purchased from Though Tech, Coimbatore. ISO DCLNL 2525 M12 tool holder is selected to hold tungsten carbide inserts. Holes for LN2 delivery are created in CNMG 120408MT12 tool by using Electric Discharge Machining (Electro cut, NIT Calicut). The rake surface of the tool was modified by making a hole having a diameter of 2 mm along with a depth of 2.4 mm. Modified cutting tool insert is shown in Fig. 1.

Table 1 Mechanical properties of duplex stainless steel 2507

Tensile strength (MPa)	Yield strength (MPa)	Hardness (HRC)
795	550	32

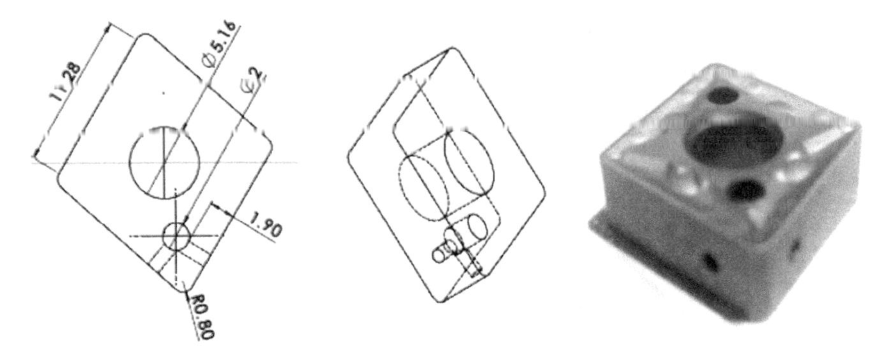

Fig. 1 Modified cutting tool insert

2.3 Experimental Setup

Schematic setup of the experimental setup is shown in Fig. 2 and in-house developed system is shown in Fig. 3a. LN_2 coolant from Dewar is pushed into the in-house designed transfer device using compressed air. Details of the transfer device are shown in Fig. 3b.

The cutting temperature was measured by using K-type thermocouple, and it was recorded in PC by using data acquisition system Agilent 34972A. The tool wear was measured using 3D Profilometer. Based on the initial trials and manufacturers recommendation, the cutting conditions are chosen. The details of the cutting parameters are given in Table 2. During machining, the modified cutting tool insert provided the coolant LN_2 from the rake and flank at a pressure of 2 bars in order to splatter at the tool-chip interface to achieve a maximum cooling effect. The two side

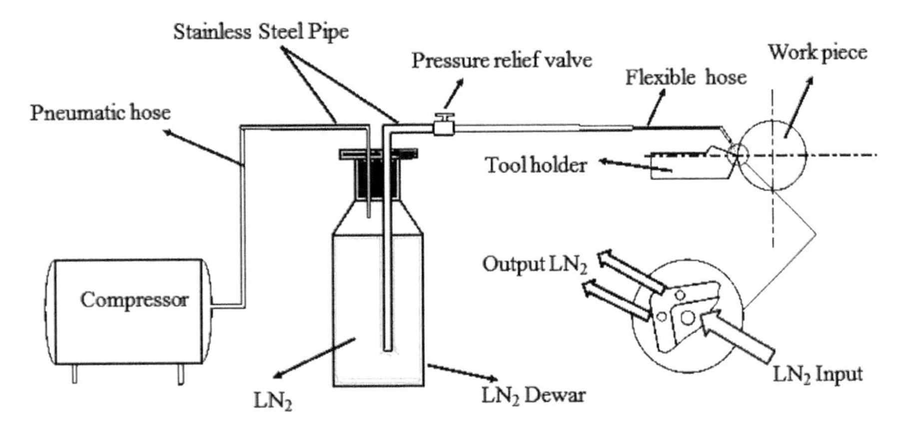

Fig. 2 Schematic representation of experimental setup

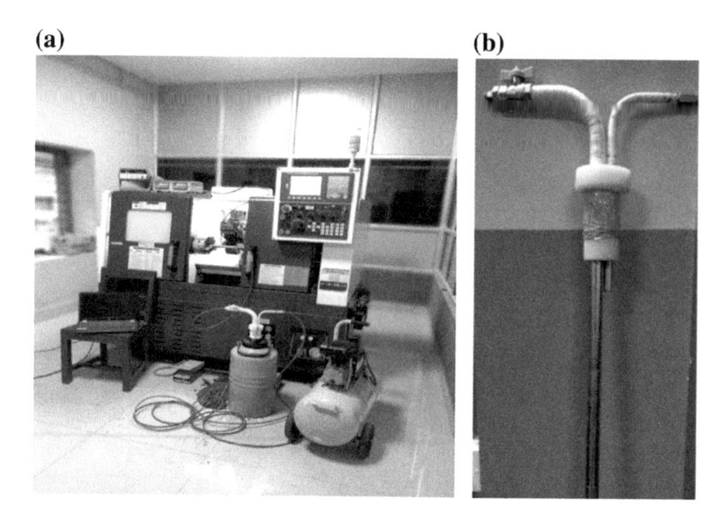

Fig. 3 **a** Experimental setup at NIT Calicut. **b** Transfer device

Table 2 Experimental conditions investigated

Cutting velocity (V_c) (m/min)	Feed (S_o) (rev/min)	Depth of cut (mm)	Material removal rate (cm^3/min)
113	0.35	1.2	47.460
113	0.26	1.6	47.008
113	0.21	2.0	47.460

holes guide a small quantity of LN_2 which would reach the minor and the major significant cutting edges to provide active cooling at the interface.

2.4 *Cryogenically Treated Tool Inserts*

Cryogenic treatment was given to the carbide tools, by soaking at −180 °C for a time duration of 24 h and then tempered in an electric muffle furnace at a temperature of 245 °C for 180 min followed by furnace cooling. Hardness is measured using micro-hardness tester (VMT-X7) with diamond indentor. Diameter and minor loads are 1 mm and 1 kg, respectively.

3 Results and Discussion

3.1 *Cutting Temperature*

Figure 4 shows the influence of feed rate and depth of cut on cutting tool temperature under dry conditions and using cryogenically treated inserts.

It is found at feed rate of 0.35 mm/rev and 1.2 mm depth of cut the temperature raised quickly than other cutting conditions and maximum temperature attained is 445 °C at a feed rate of 0.21 mm/rev and 2 mm, depth of cut temperature raise was slower than the other two cutting conditions. Therefore, it can be concluded that feed rate has more influence on cutting tool temperature compared to the depth of cut. Figure 5 shows the measurement of tool temperature during cryogenic machining at the maximum cutting condition. Tool temperature is—196 °C before machining. The tool temperature is −50 °C. There were no appreciable changes in temperature different under all three cutting conditions.

3.2 *Tool Wear*

Table 3 shows the tool flan wear under various cutting conditions.

Tool flank wear is found to be minimum when coolant is supplied directly through the modified tool insert. Tool flank wears using 3D Profilometer was captured under all cutting conditions. Typical tool wear at 0.35 mm/rev and

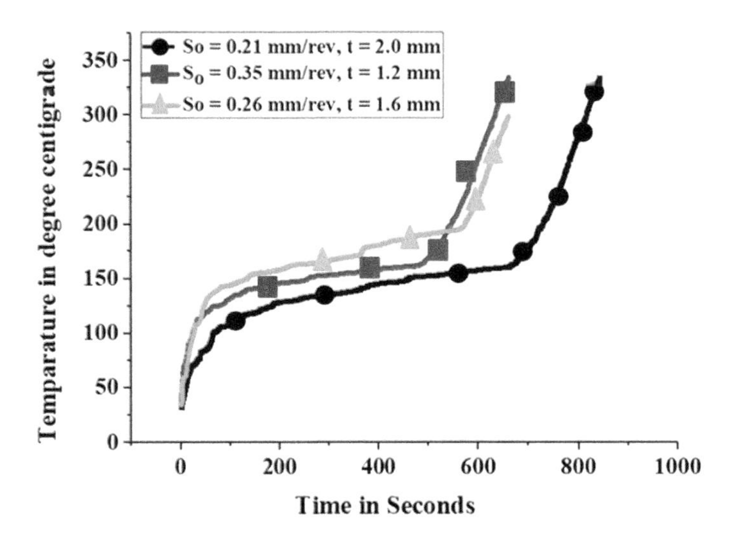

Fig. 4 Chip-tool interference temperature for each combination of dry machining and machining with cryogenically treated inserts

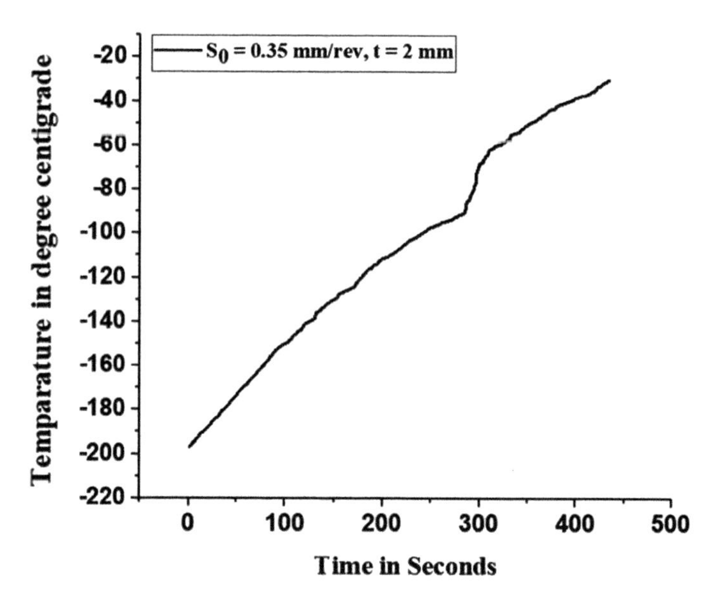

Fig. 5 Chip-tool interface temperature during cryogenic machining at the maximum cutting condition

Table 3 Tool flank wear for each cutting conditions

Cutting conditions	Tool flank wears (μm)			
	Dry	Cryogenically treated (without tempered)	Cryogenically treated (with tempered)	LN_2 coolant
S_o = 0.35 mm/rev, t = 1.2 mm	1084.758	394.966	322.741	247.37
S_o = 0.26 mm/rev, t = 1.6 mm	643.587	353.093	316.330	238.39
S_o = 0.21 mm/rev, t = 2.0 mm	288.353	207.975	199.70	164.43

1.2 mm depth of cut is shown in Fig. 6. It can be observed from Table 3, tool wear for the cryogenically treated tool is less compared to dry and non-treated tools. This is due to the hardness increase of the cutting tool insert due to cryogenic treatment. The hardness of as received sample, cryo-treated without tempering, cryo-treated with tempering is 1072 HV, 1166 HV, and 1094 HV, respectively. Figure 7 shows the SEM images of cutting tool inserts as received and cryo-treated conditions.

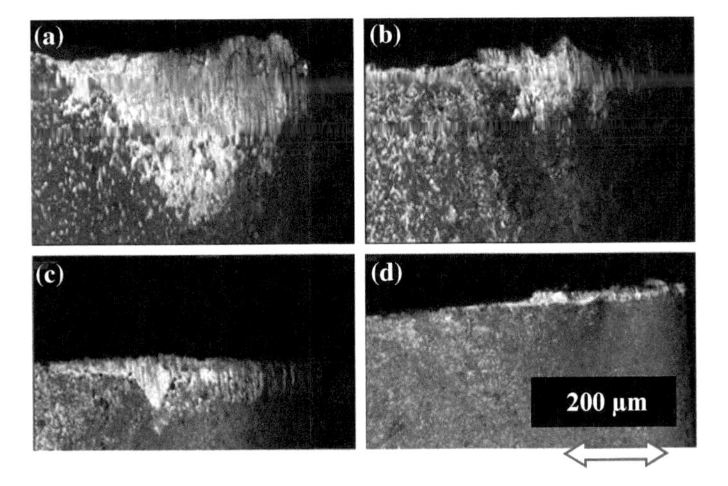

Fig. 6 SEM images of flank wear (worn-out cutting edges) after machining for 6 min at V_c = 113 m/min, S_o = 0.35 mm/rev and t = 1.2 mm machining condition with **a** dry machining, **b** cryogenically treated without tempered, **c** cryogenically treated with tempered, **d** cryogenic machining

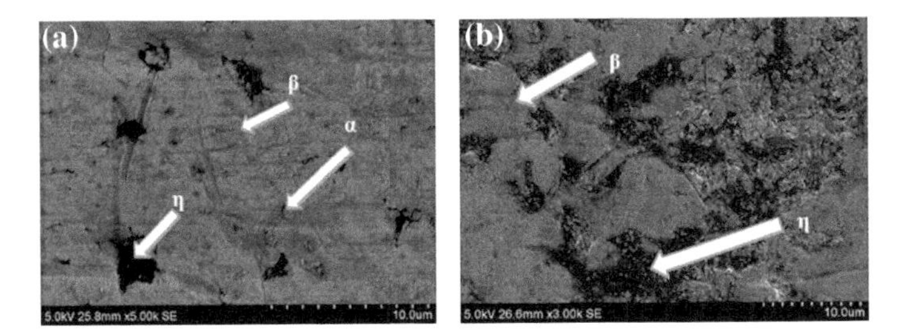

Fig. 7 SEM micrographs of cutting tool inserts **a** as received, **b** cryogenically treated

4 Conclusions

In-house experimental setup to machine super duplex stainless steel is developed. Machining was done with a cryogenically treated tool with/without tempering. Tool insert is modified to introduce LN2 to enhance the cooling rate at flank and rake surfaces. Feed rate has more influence on cutting tool temperature compared to the depth of cut under all cutting conditions. During the machining using LN2, there is no appreciable temperature difference under all cutting conditions. The modified tool gives less wear and prolongs tool life compared to dry and cryo-treated tools. Application of LN2 as a coolant reduces the amount of flank wear up to 77.19%, 37.36%, and 23.35% compared to dry machining, machining with cryogenically treated inserts without tempered and with tempered, respectively. There was an improvement in hardness of tool after cryo-treatment which was not tempered by

8.8%. This made the tool more brittle in nature, therefore, the cryo-treated tool was tempered to reduce brittleness plus to retain the achieved hardness. There was a drop of hardness value by 6.8% which reduced the brittle nature of the tool.

Acknowledgements This work has been funded by the Kerala State Council for Science, Technology and Environment, Sanction No. KSCSTE/1452/2018-TDAPdated 16 October 2018.

References

1. Deshpande RG, Venugopal KA (2015) Machining with cryogenically treated carbide cutting tool inserts. In: International conference on materials processing and characterization, pp 1814–1824
2. Reddy TVS, Ajaykumar BS, Reddy MV, Venkataram R (2007) Improvement of tool life of cryogenically treated P-30 tools. In: Proceedings of international conference on advanced materials and composites (ICAMC-2007) at National Institute for Interdisciplinary Science and Technology, CSIR. Trivandrum, India, pp 457–460
3. Yong AYL, Seah KHW, Rahman M (2007) Performance of cryogenically treated tungsten carbide tools in turning. Int J Adv Manuf Technol 46:2051–2056
4. Isik Y (2016) Using internally cooled cutting tools in the machining of difficult-to-cut materials based on Waspaloy. Adv Mech Eng 8(5):1–8
5. Paul S, Dhar NR, Chattopadhyay AB (2001) Beneficial effects of cryogenic cooling over dry and wet machining on tool wear and surface finish in turning AISI 1060 steel. J Mater Process Technol 116:44–48
6. Dhar NR, Paul S, Chattopadhyay AB Role of cryogenic cooling on cutting temperature in turning steel Trans. ASME J Mfg Sci Eng
7. Bermingham MJ, Kirsch J, Sun S, Palanisamy S, Dargusch MS (2011) New observations on tool life, cutting forces and chip morphology in cryogenic machining Ti-6Al-4V. Int J Mach Tools Manuf 51:500–511
8. Dilip Jerold B, Pradeep Kumar M (2013) The Influence of cryogenic coolants in machining of Ti-6Al-4V. ASME J Manuf Sci Eng

Experimental Investigation for Finding Defects on Epoxy-Coated Cantilever Beam Using Optical Method

M. Raghuraman and I. Ramu

Abstract This research paper is concentrated mainly on finding the defects in epoxy-coated cantilever beam surface using an optical method. The work primarily focuses on the application of the non-destructive method on a material which is non-reflective in nature and stainless steel material is chosen for the study. Identification of defects either a void or an inclusion in the epoxy resin coating applied on the cantilever beam can be identified using the grating lines on the specimen when it is kept under the illumination of He–Ne laser light and the application of lateral load on it.

Keywords Non-destructive · Testing · Intensity · Light · Epoxy resin · Coating · Image · Cantilever

1 Introduction

Considering an object in a three-dimensional space, the displacements and contour can be determined using an optical method. For a reflective surface, when it is illuminated with laser light, we can obtain the information by comparing the two states of the object, one before and after deformation as shown in Fig. 1. This method is analogous to the digital image correlation (DIC). The results obtained from the method can be compared with the finite element method.

For a highly reflective surface, for finite deformation, the total quantity of light reflected by area of study on the surface will remain the same before and after loading when there is no change in illumination. From the conservation principle of light, we can find defects in the specimen in addition to the slope and curvature associated with beam.

M. Raghuraman (✉) · I. Ramu
Vishnu Institute of Technology, Bhimavaram 534202, India
e-mail: raghuraman262@gmail.com

BBVL. Deepak et al. (eds.), *Innovative Product Design and Intelligent Manufacturing Systems*, Lecture Notes in Mechanical Engineering,
https://doi.org/10.1007/978-981-15-2696-1_32

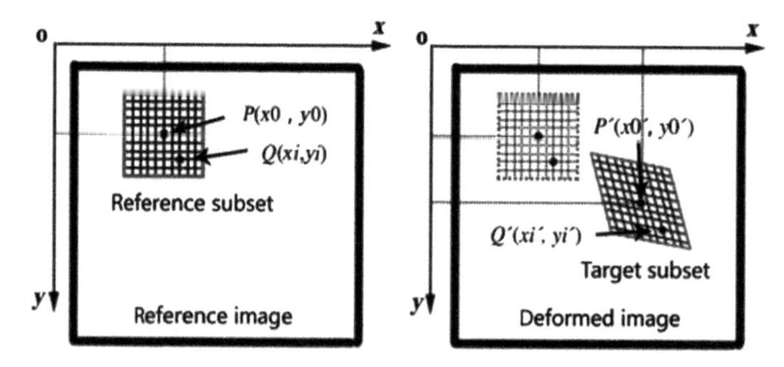

Fig. 1 Image showing the change in pixel positions

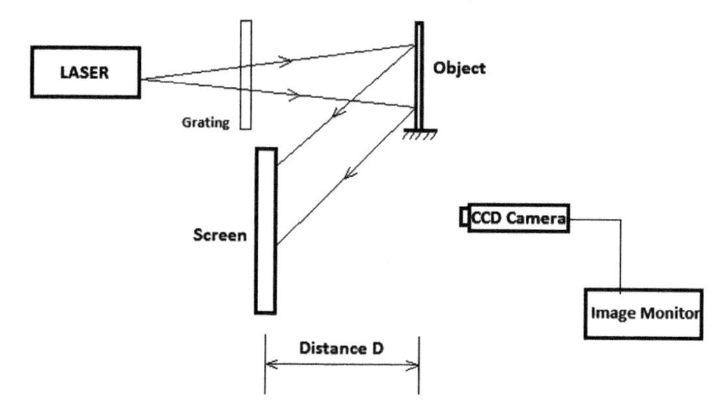

Fig. 2 Experiment set-up

The defects are usually a void or inclusion in the epoxy resin coating; these defects can be identified by employing a coarse grating between the laser light and the CCD camera which is similar to the process explained in [1]. These grating lines on the reflected cantilever beam images are used to analyse whether the defect is a void or an inclusion. The setup for the experiment shown in Fig. 2 is taken from Subramanian and Arunagiri [2].

1.1 Literature Review

Subramanian and Arunagiri [3] state that non-destructive testing can be applied to thin plates and diaphragms where we can find the defects in the plates when there is a lateral load applied to it. This lateral load causes bending and is related to the curvature. The slope and curvature are the functions of first- as well as the second-order integrations of the curvature function.

The proposed optical method which is presented in this paper is based on the work by Raghuraman et al. [4], where the work is focused on creating an epoxy-coated surface on stainless steel cantilever beam specimen so that the defects can be found out when the coated specimen is kept under illumination.

1.2 Relation Between the Curvature of Beam and Defects

The defects which are nothing but the material discontinuities can be found out using the curvature of a cantilever beam. When the cantilever beam is loaded laterally, it can be deformed away and towards the grating. This deformation produces a tension in the top fibres then spreading of the grating lines occurs; similarly, the deformation produces compression in the lower most fibre, and it makes the grating lines to squeeze, and these can be observed on the specimen [5] as shown in Fig. 3, and these are called hogging and sagging moments.

According to the theory of bending of beams and shells [6], the relationship between bending moment and curvature is given by

$$\sigma/y = M/I = E/R \tag{1}$$

where

σ is stress developed in the beam which has units of N/mm^2
y is neutral axis distance from the top or bottom layer and
moment of inertia, and Young's modulus is represented by I and E, respectively.

Theory of bending implies that the curvature in the beam $1/R$ is directly proportional to bending moment M as stated in [7]. Therefore, the curvature will be maximum at the point where the bending moment is maximum.

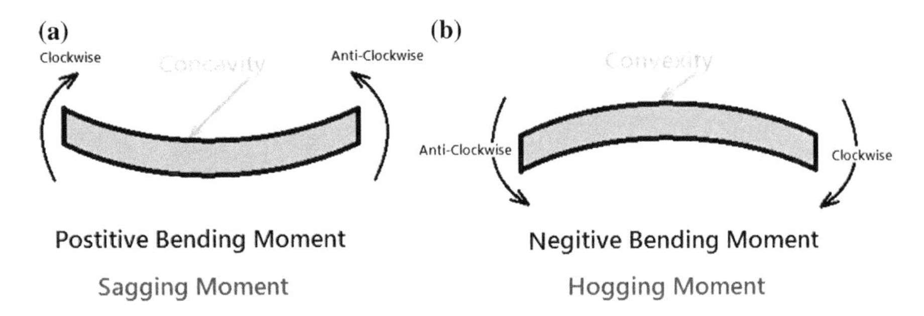

Fig. 3 Sagging and hogging moments

2 Experimental Validation

2.1 *Reflection Coating Process on Stainless Steel with Epoxy Resin*

Epoxy resin coating procedure on stainless steel cantilever beam.

- The epoxy resin is taken in a beaker and heated, and a thermometer is used where the resin temperature is measured.
- Araldite CY 230 is taken, and black pigment 20% of the weight is added and mixed thoroughly at 60 °C.
- Araldite—pigment mixture is cooled, continuously mixed for 5 min and 10% weight of hardener HY 951 is added to the mixture.
- Cantilever surface is cleaned with acetone—(CH3)2CO.
- Avoiding the air bubbles, gently pour the mixture on to the cantilever surface through a fibre glass mat filter.
- One edge of the good-quality surface of a transparent perspex sheet is kept on the resin hardener mixture, and the other edge is lowered squeezing the resin mixture to other edge.
- Deadweight pressure of 100 gm/cm^2 is applied to get a coating thickness of 40 μm.
- Curing is done for 24 h at 40 °C (or) can be done at room temperature for 72 h as stated in [8], and the perpex sheet is taken off.

2.2 *Optical Bench Set-Up*

The epoxy resin-coated specimen as shown in Fig. 4 is placed in a fixture as a cantilever beam in between the screen and laser source. The laser light is focused through the lens and made fall on the specimen which is at an angle to the screen. After passing through condensing lens, the reflection of the specimen is displayed in the screen. The charge-coupled device (CCD) camera is used to capture the image of the reflected specimen as shown in Fig. 5 which is taken from [9].

2.3 *Experiment Procedure*

The laser kept on the optical bench is switched on, and the whole set-up alignment is made to reduce errors, and the component positions are kept constant as per the specifications. A grid is placed in between the light source and the matte screen to get a grid pattern on the reflected beam image. The charge couple device camera is kept at a normal position to the laser light and screen, and the images are captured

Fig. 4 Epoxy resin-coated cantilever beam

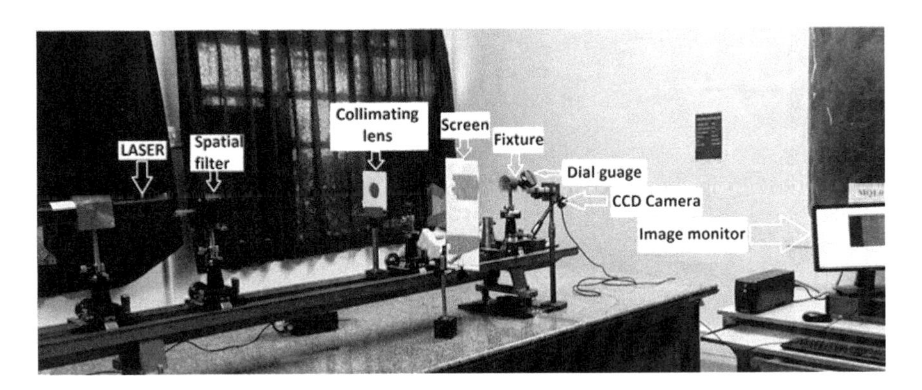

Fig. 5 Experimental set-up in the laboratory for image correlation technique

for loaded and unloaded conditions as stated in [10]. The specifications of the optical bench are shown in Table 1.

The epoxy-coated cantilever beam is placed in a fixture, and the beam is illuminated by He–Ne laser for the undeformed condition. Deformation of the cantilever beam is given by a dial gauge where the beam is given a deformed towards the grating in millimetres at the free end of the cantilever beam towards the light source so the length of the cantilever specimen will have a change which is smaller

Table 1 Specifications of optical bench

Description	Specifications
Length of bed	170 cm
Accuracy	0.01%
Guide ways	Lathe type
Laser type	He–Ne, 2 mW
Power supply	220 V AC, 50 Hz

than the unloaded specimen. Hence, there will be compression in the top layer and it causes squeezing of grid lines. Here, the tip displacement is given 1 mm at the free end and the images which are captured by CCD camera for deformed and undeformed states are taken and stored in the computer.

3 Results and Discussion

From Fig. 6, the images of the cantilever beam are captured using CCD camera placing a grid in between the laser light source and the beam specimen, and the images are taken for loaded and unloaded conditions. It is to be observed that the

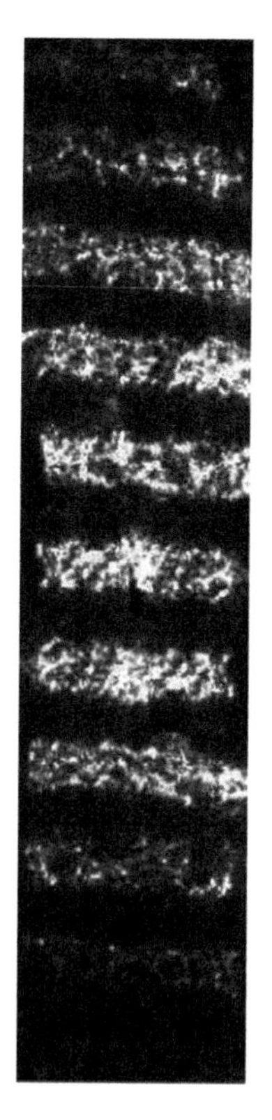
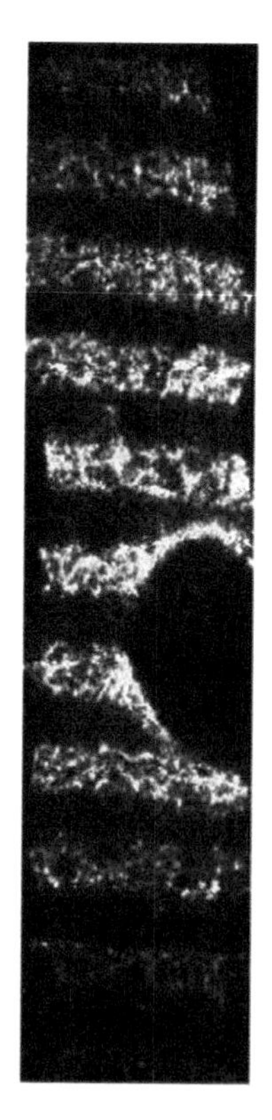

Fig. 6 Cantilever beam images taken from CCD camera before and after deformation

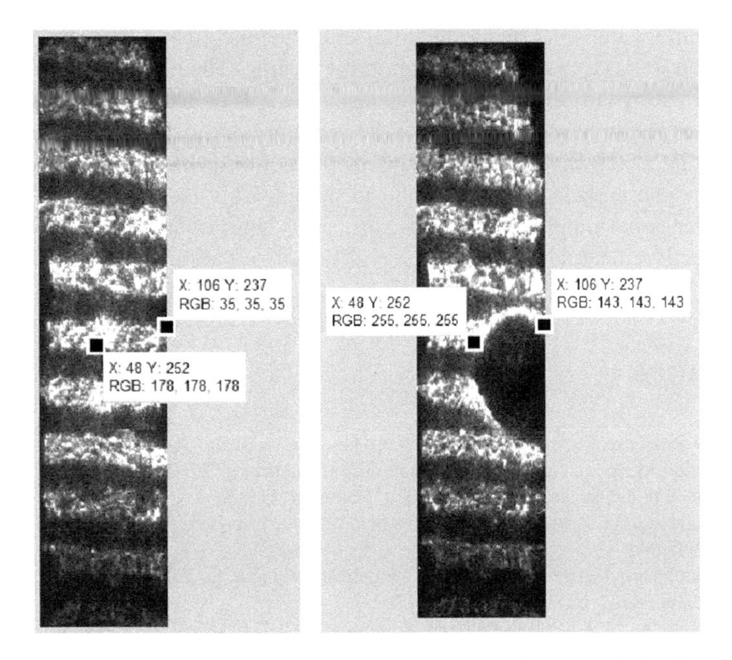

Fig. 7 Reflected images from the cantilever beam showing the change in pixel values

image which was captured for the unloaded specimen shows no sign of the void or air bubble in the specimen, whereas in case of the loaded specimen there is the sign of the grid lines spreading out from the specimen and there is a separation of the two grid lines by a large margin. This shows that the grid lines spreading out have the negative curvature.

The differences between the pixel values of deformed and undeformed greyscale images are shown in Fig. 7 where the value of a pixel on a grid line at a position 48×252 in undeformed cantilever beam image is 178 and for the same position in the deformed image it is 255. Similarly, at a position of 106×237, the values in undeformed and deformed images of the cantilever beam are 35 and 143.

The slope is given by the difference in the pixel values of undeformed and deformed images of the cantilever beam; by differentiating the slope values, the curvature is obtained. The values obtained in the curvature are negative; hence, the curvature will be negative.

4 Conclusion

The epoxy resin coating is applied on the cantilever beam, and the defects are found out to be a void by observing the change in grating lines with respect to the displacement applied at the tip of a cantilever beam. When the grating lines are spread out, the defect is said to be void, and when the grating lines are squeezed in the defects are said to be an inclusion in the coating. The nature of defects can also be identified by the deformation of the beam towards or away from the grating kept in between the light source, i.e. from the curvature of the beam. The main purpose of applying a reflective coating in this method is to make the specimen reflective then only the surface defects can be identified. This method can be applied to two-dimensional simply supported beams, and it helps in finding the defects on any non-reflective material without the use of destructive testing.

References

1. Huke P, Focke O, Falldorf C, Von Kopylow C, Bergmann RB (2010) Contactless defect detection using optical methods for non destructive testing. In: 2nd international symposium on NDT in aerospace, 22–24 Nov 2010. Hamburg, Germany
2. Subramanian G, Arunagiri R (1981) Reflecting grid method for curvature and twist. J Strain 17(3):87–88
3. Subramanian G, Arunagiri R (1981) An optical technique for detecting defects in thin plates and diaphragms. NDT Int 14(5):279–280
4. Raghuraman M, Chakravarthi,VM, Mummina V (2018) Optical grid method for contouring of 1-d non-reflective surfaces. In: IOP conference series: materials science and engineering, vol 377
5. Chiang FP, Bailangadi M (1975) A method for direct determination of small curvatures. J Appl Mech Trans ASME 42(1):29–31
6. Timoshenko S, Krieger WS (1959) Theory of plates and shells. 2nd edn. McGraw-hill, Newyork, pp 79–98
7. Popov Egor P (1976) Engineering mechanics of solids, 2nd edn. Prentice Hall, India, pp 60–67
8. Pradhan S, Pandey P, Mohanty S, Nayak SK (2016) Insight on the chemistry of epoxy and its curing for coating applications: A detailed investigation and future perspectives. J Polymer-Plast Technol Eng 55(8):862–877
9. Raghuraman M, Ramu I (2018) Image correlation technique for slope and curvature of a cantilever beam using light intensity. In: International conference on emerging trends in mechanical engineering, December 21 & 22, 2018. SRIT, Ananthapur, India
10. Subramanian G, Vijaya A (2017) Iterative intensity integration technique for contouring 2D reflective surfaces. J Optics Lasers Eng 93:92–99

Dynamic Analysis of Cracked FGM Cantilever Beam

Sarada Prasad Parida and Pankaj C. Jena

Abstract The development and presence of crack in engineering structures is a natural phenomenon. Now a day's, FGMS being a modern class of material. Due to its physical and mechanical properties varying through a particular geometrical regions, it is utilized in many of the structural applications. These structures like other metallic and composites are exposed to dynamic conditions. Hence, an FGM structure with initial cracks when subjected to dynamic environment, the performance of the structure is remarkably affected. So the study of the performance of the structures with cracks has been kept as a great area of interest. In this work, the effect of initial cracks at different positions with variable severity on dynamic property as a function of natural frequency is determined. For the purpose, a FGM cantilever beam of size 500 × 20 × 20 mm made of SUS304 and Si3N4 is simulated for free vibration in finite element method using.

Keywords FGM · Cantilever · Crack · Finite element

1 Introduction

A functionally graded material is initially developed as high class of thermally obligated material by the researchers to meet the need of variation of thermal environments up to a temperature difference of 1000° within a few millimeters of thickness by the researchers. Nowadays, on the ongoing process of advancement of material development, FG materials are designed according to the need. Starting from the application of parts of aerospace to electronic appliances, FG materials have taken a wide range of popularity and acceptance. The presence and development of crack are usual process; however, this leads to reduction of stiffness and consequently to change the dynamic characteristics of structure. The cracked

S. P. Parida · P. C. Jena (✉)
Department of Production Engineering, Veer Surendra Sai University of Technology, Burla, Odisha, India

BBVL. Deepak et al. (eds.), *Innovative Product Design and Intelligent Manufacturing Systems*, Lecture Notes in Mechanical Engineering,
https://doi.org/10.1007/978-981-15-2696-1_33

analysis in FGM structure have great importance and there is a need to evaluate FGM structure's serviceability and integrity.

Rizos et al. [1] proposed the scheme and investigated to identify the crack location and magnitude from vibration modes of a cantilever beam. Shen and Pierre [2] used Bernoulli–Euler beam theory to find natural modes of vibration with symmetric cracks. Liang et al. [3] proposed the scheme to detect the cracks in beam structures by calculating natural frequencies. Narkis [4] developed a scheme to identify the crack location of simply supported beam using algebraic model validated through FEA result. Krawczuk and Ostachowicz [5] analyzed cantilever beam, made from graphite-fiber-reinforced polyimide, with a transverse on-edge non-propagating open crack by attached spring method and finite element methods. Nandwana and Maiti [6] used rotational spring model to analyze a slender beam in the presence of an inclined edge, or internal normal and detected location of the crack based on the measurement of natural frequencies. Erdogan and Wu [7] investigated the surface crack problem for a plate with functionally graded properties. Hsu [8] used differential quadrature method to determine the natural frequencies and the mode shapes of the beam. Lin and Chang [9] developed an analytical method to study the dynamic response of a cracked cantilever beam exposed to a concentrated moving load modeled as two-span beam with Euler–Bernoulli beam theory. Loya et al. [10] investigated the bending natural frequencies of cracked beam. Kisa and Gurel [11] used a technique to find the natural frequencies of stepped circular-cracked beam. Aydin [12, 13] studied vibration characteristics of Timoshenko beams and Euler–Bernoulli beams with an arbitrary number of cracks. Yang and Chen [14] used an analytical method to calculate natural frequencies of cracked FGM beam based on Euler–Bernoulli theories and spring model of cracks. Authors also studied free and forced vibration of in homogeneous Euler–Bernoulli beams under an axial force and a transverse moving load. Ke et al. [15] studied effects of open edge cracks to vibration of FGM Timoshenko beam with different boundary conditions. Matbuly et al. [16] investigated the free vibration of an elastically support cracked FGM beam rests on Winkler–Pasternak foundation. Kitipornchai et al. [17] studied dynamic behavior of edge-cracked shear deformable functionally graded beams on an elastic foundation under a moving load. Wei et al. [18] established equations of motion with rotary inertia and shear deformation for cracked beam. Sherafatnia et al. [19] studied the mode shapes and natural frequencies of cracked beam using Euler–Bernoulli, Rayleigh, shear deformation, and Timoshenko theories. Khiem and Huyen [20] proposed a method to detect a single crack in functionally graded Timoshenko beam by measuring the natural frequencies. Sherafatnia et al. and Jena et al. [21, 22] conducted the modal study of cracked composite beams numerically and validated the result with the experimental findings.

A diversified variety of FG materials are elaborated along with different process techniques and applications in the literature. In this work, an effort is given to give a throughput on the concept of FG material along with the processing techniques and formulae of material properties.

2 Mathematical Model

For the study, a FGM cantilever beam with dimensions is considered as shown in Fig. 1. The mechanical properties of FGM beam are varied along the height of the beam according to power distribution formula as depicted by Eq. (1).

$$P(z) = [P_t - P_b]\left(\frac{2z+h}{2h}\right)^n + P_b \tag{1}$$

The modal characteristics of the composite beam can be incorporated Castigliano's theorem where the natural frequency and mode shape can be described as a function of strain energy release rate of a cracked FGM beam is given by

$$J = \sum_{i=1}^{n} \frac{1-v^2}{E} K_i^2 \tag{2}$$

where v, E, and K_i are Poisson's ratio, modulus of elasticity, and the stress intensity factors.

The flexibility influence co-efficient of cracked beam is given by

$$\frac{\partial U_i}{\partial F_j} = \frac{\partial^2}{\partial F_j \partial F_i} \int_0^a \left(\sum_{i=1}^{n} K\right)^2 d\xi = C_{ij} \tag{3}$$

3 Result and Discussion

The effect of an open crack on the dynamic behavior of FGM is studied. For the purpose, a FGM beam of size 500 × 20 × 20 mm made of SUS304 and Si_3N_4 is taken for the study in FEM software of ANSYS 14.0. The result of the cracked beam with an un-cracked beam is studied and compared. For these two relative terms, relative cracked position and relative crack severity (RCS) are used (Figs. 2, 3, 4 and 5).

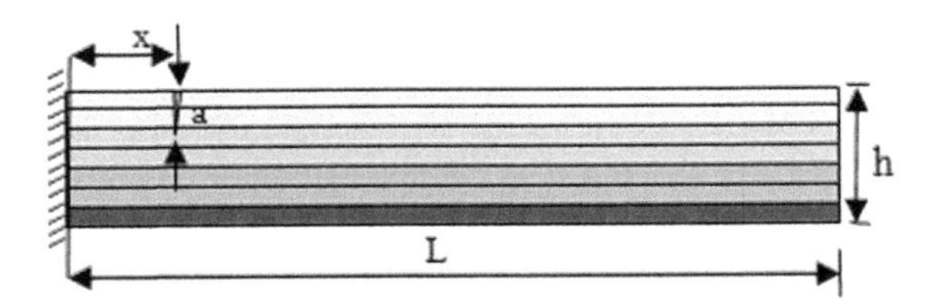

Fig. 1 A schematic view of FGM crack beam

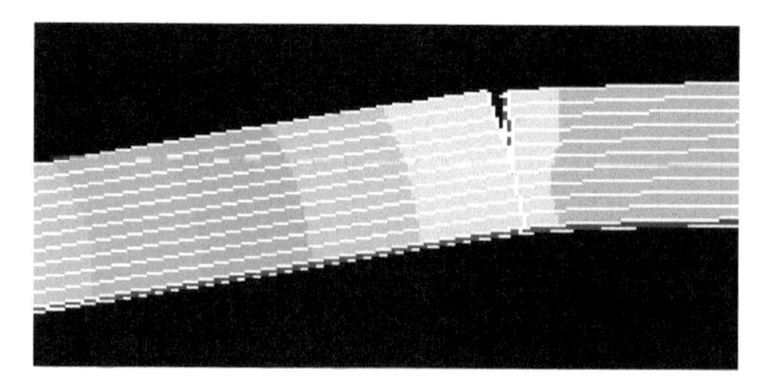

Fig. 2 A FEA model of crack beam

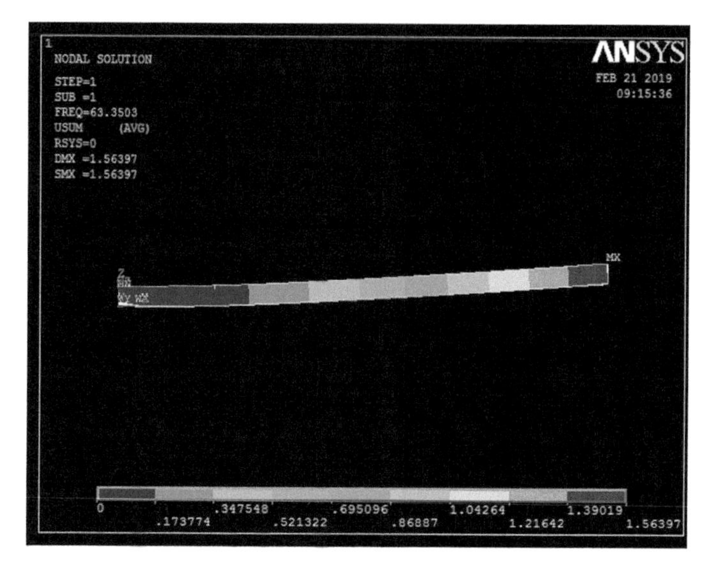

Fig. 3 First mode of vibration with RCP 0.2 and RCS 0.4

The relative crack position (RCP) and relative crack severity (RCS) are defined as,

$$\text{RCP} = \frac{\text{position of crack}}{\text{total length of beam}} = \frac{x}{L}$$

$$\text{RCS} = \frac{\text{height of crack}}{\text{total height of beam}} = \frac{a}{H}$$

In this work, the first four relative natural frequencies (RNF) that are the ratio of natural frequency of cracked beam to natural frequency of un-cracked beam is determined. The variation of RNF with RCP for fixed RCS and the relation of RNF

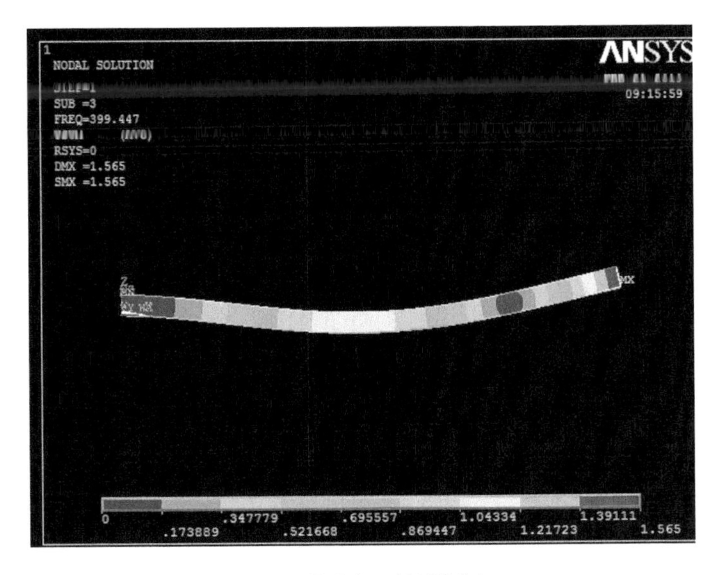

Fig. 4 Second mode of vibration with RCP 0.2 and RCS 0.4

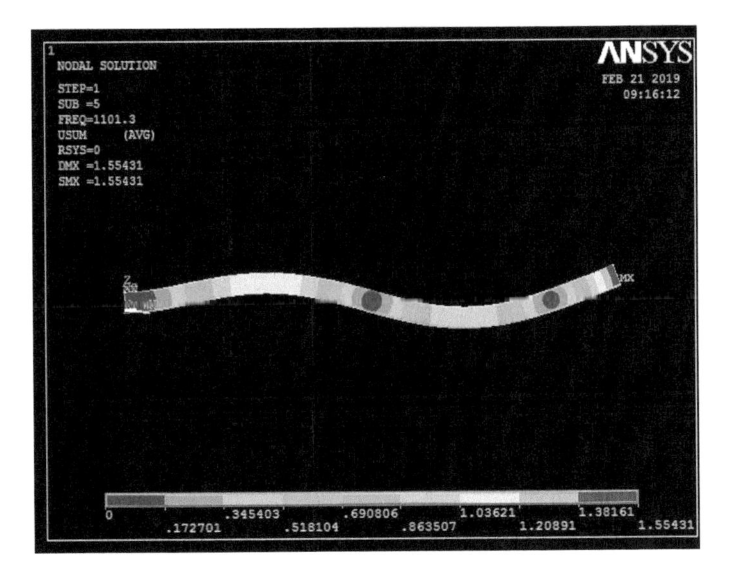

Fig. 5 Third mode of vibration with RCP 0.2 and RCS 0.4

with RCS for fixed RCP are examined and elaborated. For the cantilever beam, the RCPs of 0.05, 0.1, 0.2, 0.4, 0.6, and 0.8 is taken and for the RCSs of 0.1, 0.2, 0.3, 0.4, and 0.5 is considered for the current investigation.

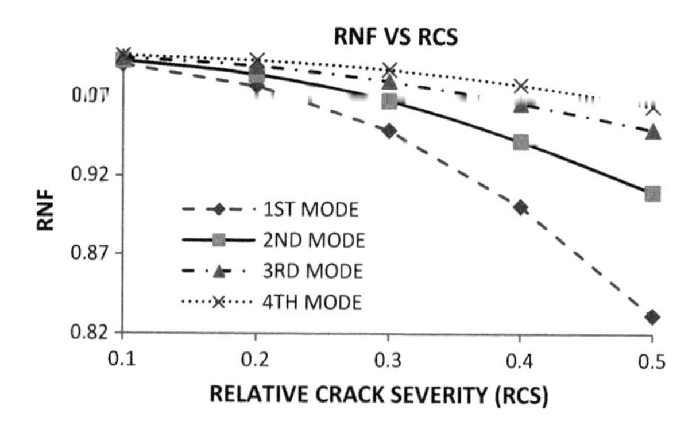

Fig. 6 Relation of RCS with RNF

It is quite evident that the increase in relative crack severity at any position decreases the stiffness of the beam and as a result there, is significant decrease in natural frequency as depicted in the table. Figure 6 shows the variation of RNF with RCS for a cantilever beam of RCP of 0.05. From the figure, it is clear that the first four relative natural frequencies decrease gradually and significantly. The same trends of variation of natural frequencies are obtained for other relative crack positions of cantilever beam as mentioned in Table 1.

The variations of RNF with the variation of relative crack position were also studied. Figure 7 shows the variation of RNF with RCP of a cantilever beam of RCS of 0.1. From the graph, it can be observed that there is a significant abrupt drop of RNF after the RCP of 0.4, i.e., 40% of total span of the cantilever beam. Then extending the crack position RCP after 0.8, there is an occurrence of slight increase in RNF. This is due to the decrease of total dead weight toward the free end beyond the crack. The same trends of variation of RNF are found as depicted in Table 1.

The variations of RNF with first four modes of five cracked samples with RCS of 0.1, 0.2, 0.3, 0.4, and 0.5 for RCP of 0.05, 0.1, 0.2, 0.4, 0.6, and 0.8 were studied. Figure 8 presents the variation of first four bending RNFs of five models cracked beams with RCP of 0.05. From the figure, it can be observed that the RNF increases with the bending modes and tends to uniform. However, another finding is that beam of sample −5 with RCS of 0.5 has higher deviation of RNF as compared to other beams hence implies increase in RCS leads to nonlinear variation of RNF.

Figure 9 shows the variation of RNF with bending modes of vibration of six samples of cracked beams with RCP of 0.05, 0.1, 0.2, 0.4, 0.6, and 0.8 for fixed RCS of 0.1. Samples −5 and 6 with RCP 0.6 and 0.8 have lowest RNF exhibiting lower stiffness of the beams.

Table 1 RNF of cracked composite beams

RCP	RNF	RCS				
		0.1	0.2	0.3	0.4	0.5
0.03	1	0.990296539	0.97674881	0.948579609	0.900844413	0.831435502
	2	0.992775552	0.984123263	0.967415395	0.941820817	0.909790299
	3	0.994686995	0.98952799	0.97990298	0.965811966	0.949179949
	4	0.996085011	0.993048897	0.987256312	0.977708533	0.963966123
0.1	1	0.986067845	0.974561555	0.950475231	0.908996911	0.847236767
	2	0.991837865	0.987831387	0.979882363	0.967053107	0.94974853
	3	0.995302995	0.994609995	0.993300993	0.991144991	0.988064988
	4	0.996923937	0.996883988	0.996724193	0.996444551	0.995885267
0.2	1	0.97953259	0.971313812	0.954465315	0.92424142	0.877182285
	2	0.991987043	0.99173131	0.991837865	0.991646066	0.9913264
	3	0.994301994	0.991606992	0.986293986	0.976514977	0.961268961
	4	0.993088846	0.986936721	0.974153084	0.949944072	0.910274848
0.4	1	0.977663481	0.978578151	0.967602105	0.954809974	0.927926614
	2	0.996291876	0.997932828	0.978838121	0.957974597	0.918826187
	3	0.991760992	0.992607993	0.982751983	0.972433972	0.94956495
	4	0.998082454	0.998721636	0.99085171	0.981184084	0.95985139
0.6	1	0.956374193	0.956652571	0.955565571	0.951244084	0.945954903
	2	0.958081153	0.960425369	0.951048504	0.915970505	0.877589293
	3	0.963193963	0.964656965	0.959035959	0.939323939	0.91953492
	4	0.964365612	0.964844998	0.962927453	0.955816555	0.947826782
0.8	1	0.970080995	0.970213556	0.970080995	0.969073532	0.969166324
	2	0.970718609	0.969546501	0.966584264	0.960148325	0.950600972
	3	0.972125972	0.966889967	0.955262955	0.932932933	0.897435897
	4	0.974832215	0.966962288	0.950303611	0.921540428	0.882710131

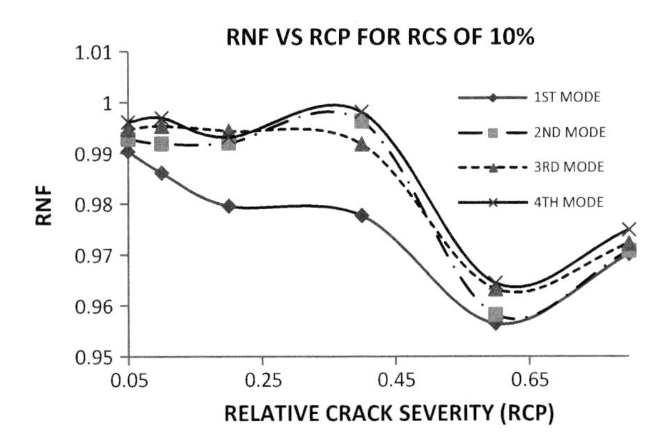

Fig. 7 Relation of RCP with RNF

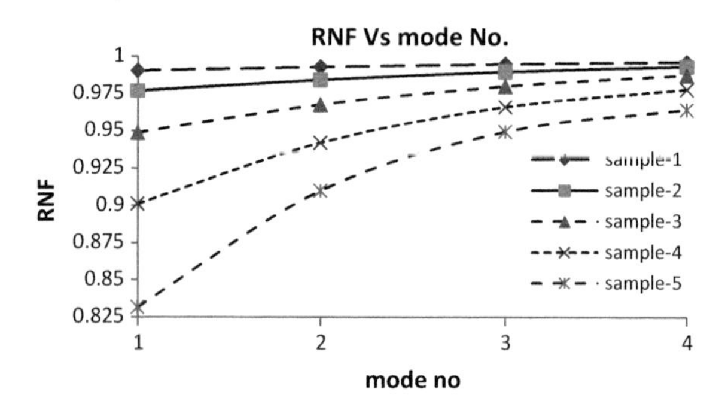

Fig. 8 Variation of RNF with mode number of cracked beams with RCP 0.05

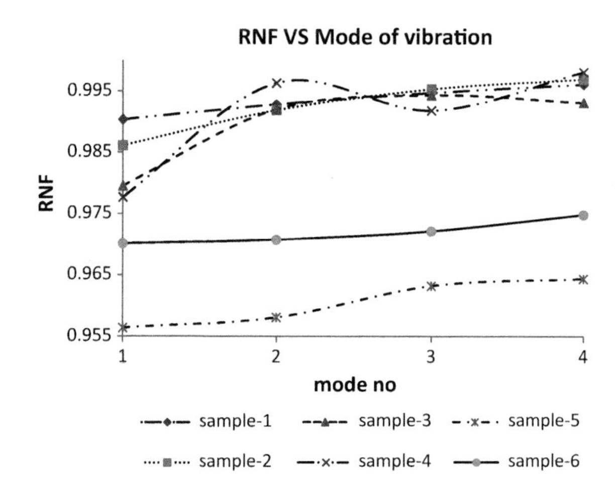

Fig. 9 Variation of RNF with mode number of cracked beams with RCS 0.5

4 Conclusion

The first four relative natural frequencies have remarkable changes than the other natural frequencies. The increase in relative crack severity at any position decreases the stiffness of the beam and as a result, there is significant decrease in natural frequency. There is abrupt drop of RNF after the RCP of 0.4, i.e., 40% of total span. After RCP of 0.8, there is an occurrence of slight increase in RNF. The increase in RCS leads to nonlinear variation of bending modes.

Acknowledgements The authors do hereby acknowledge the Department of Production Engineering, VSSUT, Burla, and TEQIP-III for providing the infrastructure facilities and financial support for carrying out this research work.

References

1. Rizos PF, Aspragathos N, Dimarogonas AD (1990) Identification of crack location and magnitude in a cantilever beam from the vibration modes. J Sound Vib 138:381–388
2. Shen MHH, Pierre C (1990) Natural modes of Bernoulli-Euler beams with symmetric cracks. J Sound Vib 138:115–134
3. Liang RY, Choy FK, Hu J (1991) Detection of cracks in beam structures using measurements of natural frequencies. J Franklin Inst 328(4):505–518
4. Narkis Y (1994) Identification of crack location in vibrating simply supported beams. J Sound Vib 172(4):549–558
5. Krawczuk M, Ostachowicz WM (1995) Modeling and vibration analysis of a cantilever composite beam with a transverse open crack. J Sound Vib 183:69–89
6. Nandwana BP, Maiti SK (1997) Modeling of vibration of beam in presence of inclined edge or internal crack for its possible detection based on frequency measurements. Eng Fract Mech 58:193–205
7. Erdogan F, Wu BH (1997) The surface crack problem for a plate with functionally graded properties. J Appl Mech 64:448–456
8. Hsu MH (2005) Vibration analysis of edge-cracked beam on elastic foundation with axial loading using the differential quadrature method. Comput Methods Appl Mech Eng 194:1–17
9. Lin HP, Chang SC (2006) Forced responses of cracked cantilever beams subjected to a concentrated moving load. Int J Mech Sci 48:1456–1463
10. Loya JA, Rubio L, Saez JF (2006) Natural frequencies for bending vibrations of Timoshenko cracked beams. J Sound Vib 290:640–653
11. Kisa M, Gurel MA (2007) Free vibration analysis of uniform and stepped cracked beams with circular cross sections. Int J Eng Sci 45:364–380
12. Aydin K (2007) Vibratory characteristics of axially loaded Timoshenko beams with arbitrary number of cracks. J Vib Acoust 129:341–354
13. Aydin K (2008) Vibratory characteristics of Euler-Bernoulli beams with an arbitrary number of cracks subjected to axial load. J Vib Control 14:485–510
14. Yang J, Chen Y (2008) Free vibration and buckling analyses of functionally graded beams with edge cracks. Compos Struct 83:48–60
15. Ke LL, Yang J, Kitipornchai S, Xiang Y (2009) Flexural vibration and elastic buckling of a cracked Timoshenko beam made of functionally graded materials. Mech Adv Mater Struct 16:488–502
16. Matbuly MS, Ragb O, Nassar M (2009) Natural frequencies of a functionally graded cracked beam using differential quadrature method. Appl Math Comput 215:2307–2316
17. Yan T, Kitipornchai S, Yang J, He XQ (2011) Dynamic behaviour of edge-cracked shear deformable functionally graded beams on an elastic foundation under a moving load. Compos Struct 93:2992–3001
18. Wei D, Liu YH, Xiang ZH (2012) An analytical method for free vibration analysis of functionally graded beams with edge cracks. J Sound Vib 331:1685–1700
19. Sherafatnia K, Farrahi GH, Faghidian SA (2014) Analytic approach to free vibration and bucking analysis of functionally graded beams with edge cracks using four engineering beam theories. Int J Eng 27(6):979–990
20. Khiem NT, Huyen NN (2017) A method for crack identification in functionally graded Timoshenko beam. J Nondestr Test Eval 32(3)
21. Jena PC, Parhi DR, Pohit G (2016) Dynamic study of composite cracked beam by changing the angle of bidirectional fibres. Iran J Sci Technol Trans A 40A1:27–37
22. Jena PC, Parhi DR, Pohit G (2014) Theoretical, numerical (FEM) and experimental analysis of composite cracked beams of different boundary conditions using vibration mode shape curvatures. Int J Eng Technol 6.2:509–518

Influences and Applications of Aluminum Addition on the Mechanical Properties of Pure Magnesium

Sagnik Sarma Choudhury, Neelabh Jyoti Saharia, Suvan Dev Choudhury and B. Surekha

Abstract Magnesium alloys are known for their high strength and lightweight that made them as a potential candidate to be used in different industrial components. Magnesium (Mg) and aluminum (Al) are two highly important lightweight metals that are used to improve the overall efficiency. However, the reactive nature of magnesium near its melting point makes it very difficult to produce alloys. This paper aims at the production of Mg–Al alloys of various concentrations using the die-casting methodology. The objective of the paper is to identify the optimum configuration of the magnesium–aluminum alloy to assure the highest productivity. For this, a different configuration of the alloys was produced at various concentrations of magnesium and aluminum. Further, various mechanical tests (such as, ultimate tensile test (UTS), hardness test, and impact test) are conducted to decide the optimum concentration of the alloy system that produced better mechanical properties.

Keywords Stir casting · Tensile strength · Yield strength · Hardness

1 Introduction

Recently, magnesium–aluminum alloys have found its huge applications in automotive industries due to its lightweight, good formability, and malleability with high electrical as well as thermal resistivity and high corrosion resistance [1]. Magnesium and its alloys are the most demanding implant materials in biomedical applications because of their self-dissolving capability inside the human body that results in the elimination of secondary surgeries to remove it [2]. Magnesium and aluminum are two highly important lightweight metals that are closed to each other according to various properties, which include strength, melting point, atomic

S. S. Choudhury (✉) · N. J. Saharia · S. D. Choudhury · B. Surekha
School of Mechanical Engineering, KIIT Deemed to be University, Patia, Bhubaneswar 751024, India

BBVL. Deepak et al. (eds.), *Innovative Product Design and Intelligent Manufacturing Systems*, Lecture Notes in Mechanical Engineering,
https://doi.org/10.1007/978-981-15-2696-1_34

weight, and elasticity. The density (ρ) of Mg and Al is equal to 1.7 g/cm^3 and 2.7 g/cm^3 [3], respectively. Compared with wrought magnesium alloy, the casting of magnesium alloys exhibits clear dynamic as well as economic advantages [4]. But, the casting of magnesium leads to numerous defects which needs special attention to obtain defect-free alloy. Ray et al. [5] found that AZ31B specimens showed very minute evidence of twining because of their fine-grained size, while ZEK100 specimens showed excessive twining that indicates that the metal is prone to shear localization. The alloy Mg–16Gd–2Ag–0.3Zr demonstrated the very high value of yield and tensile strength (YS: 328 MPa and UTS: 423 MPa) [6]. Moreover, Sajuri et al. [7] stated that the distribution and size of the casting defects can greatly influence the fatigue and tensile performance of the cast alloy. In high-pressure die casting, the AZ91D alloy had shown the defects like gas and shrinkage pore during tensile and fatigue test crack initiation [8]. Another research done by Dong et al. [9], the AlSiCuMgMn alloy manufactured using the said manufacturing process had shown a high yield strength (321 MPa), high ultimate tensile strength (425 MPa), and high ductility (11.3%) after solution treating. Further, Kumar et al. [10] found that the vacuum-assisted stir-casting process was found to be the most suitable process for producing the castings of magnesium and its composites. Mabuchi et al. [11] observed an increase in the tensile strength of Mg–Si alloy when the high weight percentage (>10%) of Si was added to the alloy. Additionally, tensile strength was seen to be increased when zinc and aluminum were added to the alloy. Later on, Aghayani et al. [12] performed experiments and found that the tensile strength was seen to be improved significantly when the ultrasonic treatment was performed on AZ91 magnesium alloy. Ding et al. [13] reported that equal channel angular extrusion (ECAE) has resulted in the formation of refined grains in AZ31 alloy which in turn improved the tensile strength of the alloy. Goh et al. [14] produced magnesium composite by reinforcing it with CNT after using the hot extrusion process. The composite produced using the above process had shown an improvement in both the tensile and yield strengths. Moreover, the ductility was also seen to be improved for 1.3 weight percentage of CNT. Chunjiang et al. [15] added cerium-rich rare-earth metal to ZK 60 and observed a considerable reduction in the grain size, that improved the ultimate and yield strength of the composite considerably. Further, Ninomiya et al. [16] reported an increase in the hardness of Mg–Al alloy when calcium was added to the composite at a temperature of 350 °C.

In the present study, an attempt is made to suggest the best configuration of industry-grade magnesium–aluminum alloy that will be produced at a cheaper rate in compared to processes like hot extrusion [17], extrusion [15], equal channel angular extrusion [13], and high-pressure die casting [8, 9]. Once the Mg–Al alloy at different configurations is fabricated by utilizing the die-casting methodology, the alloy is tested for the microstructure and mechanical properties.

2 Experimentation

2.1 *Materials*

In the present research, Mg and Al are used to produce the alloy system. Magnesium becomes one of the most needed materials as it has the property of high strength and lightweight. Aluminum serves to be the best-alloying elements with magnesium as the base metal. The various mechanical and chemical properties of magnesium and aluminum are given in Table 1 [18, 19], respectively.

2.2 *Experimental Procedure*

Rods of pure aluminum metal with a diameter of 10 mm were cut into pieces. It was then put into the crucible which was placed in the heating furnace. The furnace was first preheated to 700 °C. After 15 min, pure magnesium cubes were then put into the crucible, which was kept in the furnace. This delay was taken into consideration to accommodate the higher melting point of aluminum (that is 10 °C greater than the magnesium). This was followed by the addition of hexachloroethane (deoxidizer) to the melt so as to avoid the oxidization of the magnesium alloy.

Simultaneously, the cast iron mold with a rectangle-shaped cavity having a thickness of 5 mm was being preheated with oxyacetylene flame. It is important to note that magnetic stirrer as shown in Fig. 1 was used to accomplish the proper mixing of the metals. The molten metal was immediately poured into the mold to avoid any oxidation of the magnesium alloy (ref. to Fig. 2).

Four alloy configurations consisting of 10, 8, 6, and 4% aluminum in magnesium were fabricated after following the said procedure. To understand the microstructure of the fabricated alloys, the specimens are prepared and examined using an optical microscope. It can be observed that the formation of Mg–Al alloy leads to the development of $Al_{12}Mg_{17}$ solid solution and Magnesium in

Table 1 Properties of magnesium and aluminum

Sl. No.	Property	Magnesium	Aluminum
1	Density (g/cm^3)	1.738	2.7
2	Critical temperature (°C)	760	760
3	Melting point (°C)	650	660.37
4	Boiling point (°C)	1090	2327
5	Poisson's ratio	0.29	0.34
6	Young's modulus (GPa)	44–45.5	68.3
7	Tensile strength (MPa)	150	90

Fig. 1 Mixing of metals using a magnetic stirrer

Fig. 2 Pouring of alloy into the mold

hexagonal-closed packing (hcp) structure. Moreover, it can be also observed that the formation of $Al_{12}Mg_{17}$ solid solution decreases as the percentage of aluminum in magnesium decreases.

Further, the fabricated alloys were also tested for their tensile and impact strength. The universal testing machine TUF-C-1000 was employed to measure the tensile strength for which specimens with gauge length 25 mm, width 5.53 mm, and thickness 5 mm were taken. The impact strength was determined by using the Izod test. For this, the specimen size was prepared according to the ASTM E8M, standard which is 55 mm × 10 mm × 5 mm. This was then followed by testing the microhardness of the fabricated samples using the Zwick/Roell microhardness tester. It is to note that seven different indents were made with forces of 50, 100, 300, 500, 1000, and 2000 gf on each alloy configuration and corresponding hardness values were recorded. The dwell time for the indentation was kept at 10 s.

3 Results and Discussions

From the results, it has been observed that the alloy configuration with 94% Mg + 6% Al showed the highest value of tensile strength (176.49 MPa). Addition of aluminum in the magnesium metal increases the UTS and yield strength. This increase is due to the strong interfacial bonding between the aluminum and magnesium. It is to be noted that the 6% Al in magnesium is also the threshold percent where further addition of aluminum in magnesium will result in a decrease in the mechanical properties of the alloy. The high tensile strength could be the result of solid solution strengthened by $Al_{12}Mg_{17}$ solid solution as it provides the necessary geometric dislocation in the magnesium matrix. The detailed overview of the yield, tensile, and impact strengths are illustrated in Table 2. From the data, it is observed that the decrease in the impact strength is observed with the increase in the content of Al in the alloy system.

This is because, with the increase in Al content, the ductility of the alloy decreases and stress concentration areas increases. These results in the formation of small cracks which further lead to the propagation of those cracks. From the results of hardness test (ref. to Table 3), it has been observed that the highest average value was found to be equal to 77.42 HV for the 92% Mg + 8% Al alloy configuration (Fig. 3).

The optical microscopic images of the indentations obtained after the micro-hardness testing for various alloy configurations are shown in Fig. 4.

The highest value of the tensile strength was found to be higher than that of the Mg–6 wt%Si that was produced by MABUCHI et al. [11] whose value was 110 MPa. Further, Aghayani et al. [12] showed that the tensile strength of AZ91 alloy was found to be 165 MPa after treating it at 60% UST (Ultrasonic Treatment)

Table 2 Yield, tensile, and impact strength of the four alloys

Composition	Load at yield (kN)	Yield stress (N/mm^2)	Load at peak (kN)	Tensile strength (N/mm^2)	Impact strength (MPa)	Percentage elongation
90% Mg + 10% Al	2.94	113.61	3.75	144.94	1.8	2.36
92% Mg + 8% Al	3.06	112.11	4.38	155.04	1.8	2.74
94% Mg + 6% Al	3.34	110.83	5.32	176.49	1.7	7.26
96% Mg + 4% Al	1.89	80.58	2.76	117.67	2.9	5

Table 3 Microhardness (HV) for all the four alloys at different indentation forces

	50 gf	100 gf	200 gf	300 gf	500 gf	1000 gf	2000 gf	Average
90% Mg + 10% Al	78	68	82	66	69	68	63	70.57
92% Mg + 8% Al	99	93	78	75	60	73	64	77.42
94% Mg + 6% Al	59	57	54	49	59	50	58	55.14
96% Mg + 4% Al	42	39	56	45	45	41	42	44.28

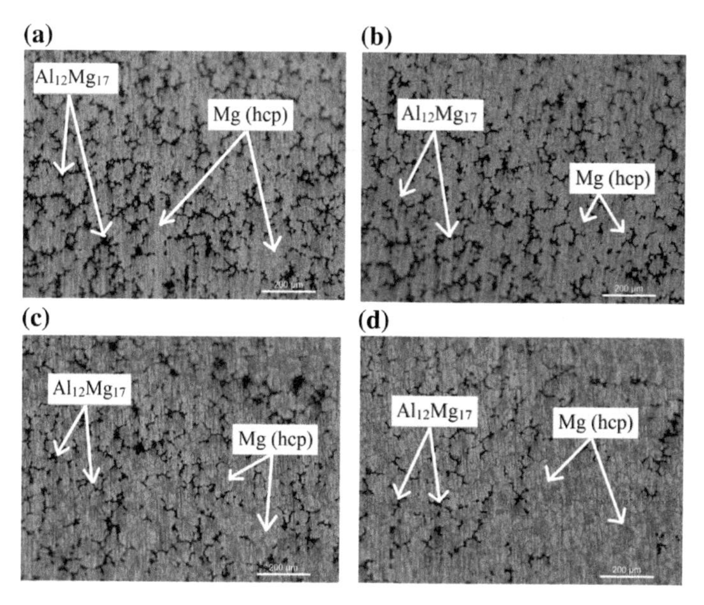

Fig. 3 Microstructures (200 μm) of **a** Mg90%–Al10%, **b** Mg92%–Al8%, **c** Mg94%–Al6%, **d** Mg96%–Al4%

power. This suggests that our material would be able to absorb more amount of energy under load.

Also, the highest value of hardness was found to be higher than that of Mg–3Al–5Ca alloy (70 HV) which was tested by NINOMIYA et al. [16]. It is also important to note that the hardness value of pure aluminum is found to be equal to 51.2 Dph [20]. Whereas the 94% Mg + 6% Al alloy that showed the highest value of tensile strength showed an average value of 55.14 HV. This suggests the very basic that a material cannot be hard and tough at the same time. Also, the hardness value was not more than 75 HV for Mg alloy with Nd and Zn content performed by Li et al. [21]. With the increase in aluminum content, there is an increase in the concentration of $Al_{12}Mg_{17}$ solid solution. The increase in the hardness value may be due to the variation of plastic deformation of the alloy with the presence of aluminum which is a hard metal (HV of pure Al = 76 ± 5) [22]. However, the hardness of an alloy also depends on the processing method.

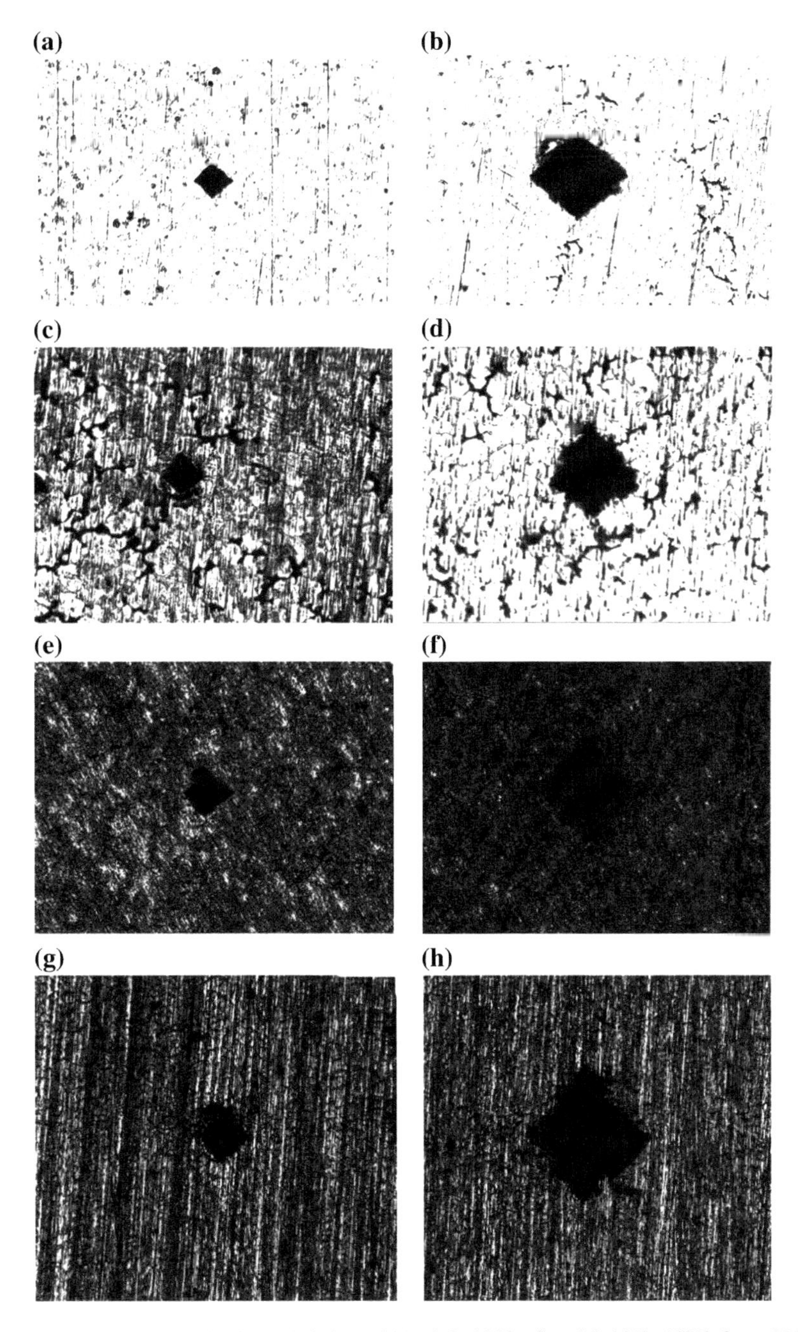

Fig. 4 Indentation at Mg90%–Al10% for **a** 200 gf, **b** 1000 gf, at Mg92%–Al8% for **c** 200 gf, **d** 1000 gf, at Mg94%–Al6% for **e** 200 gf, **f** 1000 gf, at Mg96%–Al4% for **g** 200 gf and **h** 1000 gf

4 Conclusion

In the present paper, the authors have attempted to produce magnesium–aluminum alloys at different configurations, namely 10% Al 90% Mg, 8% Al 92% Mg, 6% Al 94% Mg, and 4% Al 96% Mg. Manufacturing of magnesium alloys requires a highly inert atmospheric condition which is not very easy to achieve as the metal readily gets oxidized when it comes in contact with oxygen and turns into white ash.

After production, various mechanical tests (tensile test, impact test, and hardness test) were been carried out to sort out the best configuration out of the four fabricated configurations. It has been observed that the alloy with 6% Al 94% Mg has given the maximum tensile strength of 176.49 MPa. This value was comparable with other ASTM grade magnesium alloy. Keeping low production cost in mind, the alloy with 6% Al and 94% Mg could be used for industrial grade applications like the frame of airplane seats, the frame of bicycles, door panel and bonnets of car, etc. Moreover, it can be coated with hydroxyapatite powder (HAp) and use it as a biomedical implant as the HAP powder will make the alloy biocompatible [23].

References

1. Kulekci MK (2008) Magnesium and its alloys applications in automotive industry. Int J Adv Manuf Technol 39:851–865
2. Sillekens WH, Bormann D (2012) Biomedical applications of magnesium alloys, Woodhead Publishing Limited
3. Hirsch J, Al-Samman T (2013) Superior light metals by texture engineering: optimized aluminum and magnesium alloys for automotive applications. Acta Mater 61:818–843
4. Fu P, Peng L, Jiang H, Ding W, Zhai C (2014) Tensile properties of high strength cast Mg alloys at room temperature: a review. China Foundry 11:277–286
5. Ray AK, Wilkinson DS (2016) The effect of microstructure on damage and fracture in AZ31B and ZEK100 magnesium alloys. Mater Sci Eng, A 658:33–41
6. Pan F, Yang M, Chen X (2016) A review on casting magnesium alloys: modification of commercial alloys and development of new alloys. J Mater Sci Technol 32:1211–1221
7. Sajuri ZB, Miyashita Y, Hosokai Y, Mutoh Y (2006) Effects of Mn content and texture on fatigue properties of as-cast and extruded AZ61 magnesium alloys. Int J Mech Sci 48:198–209
8. Li X, Xiong SM, Guo Z (2016) Failure behavior of high pressure die casting AZ91D magnesium alloy. Mater Sci Eng A 672:216–225
9. Dong X, Yang H, Zhu X, Ji S (2019) High strength and ductility aluminium alloy processed by high pressure die casting. J Alloy Compd 773:86–96
10. Kumar A, Kumar S, Mukhopadhyay NK (2018) Introduction to magnesium alloy processing technology and development of low-cost stir casting process for magnesium alloy and its composites. J Mag Alloys 6:245–254
11. Mabuchi M, Kubota K, Higashi K (1996) Tensile strength, ductility and fracture of magnesium-silicon alloys. J Mater Sci 31:1529–1535
12. Khosro Aghayani M, Niroumand B (2011) Effects of ultrasonic treatment on microstructure and tensile strength of AZ91 magnesium alloy. J Alloys Compd 509:114–122

13. Ding SX, Lee WT, Chang CP, Chang LW, Kao PW (2008) Improvement of strength of magnesium alloy processed by equal channel angular extrusion. Scripta Mater 59:1006–1009
14. Goh CS, Wei J, Lee LC, Gupta M (2006) Simultaneous enhancement in strength and ductility by reinforcing magnesium with carbon nanotubes. Mater Sci Eng A 423:153–156
15. Ma C, Liu M, Guohua W, Ding W, Zhu Y (2003) Tensile properties of extruded ZK60-RE alloys. Mater Sci Eng, A 349:207–212
16. Ninomiya R, Ojiro T, Kubota K (1995) Improved heat resistance of Mg–Al alloy by the Ca addition. Acla Metal/Mater 43(2):669–674
17. Nakanishi M, Mabuchi M, Saito N, Nakamura M (1998) Tensile properties of the ZK60 magnsium alloy produced by hot extrusion of machined chip. J Mater Sci Lett 17:2003–2005
18. AZO materials, aluminium—specifications, properties, classifications and classes, 17 May 2005. https://www.azom.com/article.aspx?ArticleID=2863
19. AZO materials, magnesium, 29 July 2001. https://www.azom.com/article.aspx?ArticleID=618
20. Cahoon JR, Broughton WH, Kutzak AR (1971) The determination of yield strength from hardness measurements. Metal Trans 1980, 2
21. Li Z, Wang Q, Luo AA, Penghuai F, Peng L (2015) Fatigue strength dependence on the ultimate tensile strength and hardness in magnesium alloys. Int J Fatigue 80:468–476
22. Rashada M, Pana F, Tanga A, Asif M (2014) Effect of graphene nanoplatelets addition on mechanical properties of pure aluminum using a semi-powder method. Prog Nat Sci Mater Int 24:101–108
23. Shuyan Xu, Long Jidong, Sim Lina, Diong CH, Ostrikov K (Ken) (2005) RF plasma sputtering deposition of hydroxyapatite bioceramics: synthesis, performance, and biocompatibility. Plasma Process Polym 2:373–390

Abrasive Wear Behaviour of Sand Cast B_4C Particulate Reinforced AA5052 Metal Matrix Composite

Murlidhar Patel, Mukesh Kumar Singh and Sushanta Kumar Sahu

Abstract In the present work, particulate reinforced aluminium metal matrix composite has been developed by stir-casting processing technique in which AA5052 is reinforced with 5 wt% B_4C-p of 63 μm particles size. The density and abrasive wear of B_4C-p reinforced AA5052 MMC were investigated and compared with that of the AA5052 alloy. Abrasive wear test was carried out by pin-on-disc wear test method at different applied loads (5–15 N). The worn out surfaces are analysed by using optical microscopy. The results show that at all applied load, B_4C-p reinforced AA5052 MMC gives good tribological properties as well as lower density as compared to the unreinforced AA5052.

Keywords MMC · B_4C · AMMC · Wear · Density

1 Introduction

Metal matrix composite (MMC) is a type of engineering material developed by the macroscopic composition of reinforcement and metal as a matrix material. MMCs have outstanding mechanical, tribological and excellent corrosive resistance properties [1–3]. Aluminium metal matrix composite (AMMCs) is a lightweight material

---

M. Patel (✉)
Department of Mechanical Engineering, Institute of Technology, Guru Ghasidas Vishwavidyalaya (A Central University), Bilaspur, Chhattisgarh 495009, India
e-mail: murlidharpatel4@gmail.com

M. K. Singh
Department of Industrial and Production Engineering, Institute of Technology, Guru Ghasidas Vishwavidyalaya (A Central University), Bilaspur, Chhattisgarh 495009, India
e-mail: mukeshetggv@gmail.com

S. K. Sahu
Department of Mechanical Engineering, National Institute of Science and Technology, Berhampur, Odisha 761008, India
e-mail: sushanta.sahu@nist.edu

BBVL. Deepak et al. (eds.), *Innovative Product Design and Intelligent Manufacturing Systems*, Lecture Notes in Mechanical Engineering,
https://doi.org/10.1007/978-981-15-2696-1_35

and having very good strength to weight ratio. AMMCs are mostly used for light-weight applications like aerospace, automobiles, marine, etc. [4–6]. Rolling, forging and extrusion forming process can be performed in particulate reinforced AMMCs, and it give isotropic property when the particulates are distributed homogeneously into the matrix [5, 7]. Particulate reinforced AMMCs provide a good combination of mechanical, tribological and corrosion resistance properties at low cost which makes it more attractive than conventional materials [8].

The mechanical and tribological properties of B_4C-p reinforced AMMCs have been characterized by many researchers. The dry sliding friction and wear properties of B_4C-p reinforced AA5083 MMCs are analysed by Tang et al. [9]. AA5083 was reinforced with 5 and 10 wt% of B_4C-p by cryomilling and consolidation processes. Pin-on-disc dry sliding wear test of these AMMCs is done at loads ranging from 50 to 80 N and sliding speeds ranging from 0.6 to 1.25 m/s. They found that under the same test conditions, the wear rate of the AMMC with 10 wt% B_4C-p was approximately 40% lower than that of the AMMC with 5 wt% B_4C-p.

The effect of particles size on the abrasive wear of B_4C-p reinforced AMMCs are evaluated by Nieto et al. [10]. They used three different particle sizes of B_4C particulates, i.e. micrometric (μB_4C), submicron ($s\mu B_4C$) and nano-metric B_4C-p (nB_4C) to reinforce the AA5083. They observed AA5083 + nB_4C AMMC has superior wear resistance due to its high hardness and greater interfacial area, which hindered pull-out of nano-B_4C particles. The effect of nano and microparticle size of B_4C-p on the properties of stir and ultrasonic cavitation-assisted cast Al/B_4C-p MMCs are analysed by Harichandran and Selvakumar. They reported that the wear resistance of nano B_4C-p reinforced AMMC is greater than the micro B_4C-p reinforced AMMC. They noted that the ductility and strength of nanoparticulate MMCs are decreased beyond the 6 wt% due to the agglomeration of particles and porosity in the composite. The most interesting thing about B_4C-p reinforced AMMC is that it can be recyclable with better particulate distributions without losing the mechanical and tribological properties [11].

In this study, the effect of applied load on abrasive wear behaviour of the 5 wt% B_4C-p reinforced AA5052 MMC, the effect of B_4C-p on the density as well as porosity, and the pattern of wear on the worn-out surfaces after abrasive wear test have been investigated.

2 Experimental Studies

2.1 Materials and Method

The materials chosen for the present work is 5xxx series Al alloy in which Mg is the main alloying element. The 5xxx series is mainly used in marine applications due to its high corrosion resistance property. When the content of Mg element present in the Al is more than 3%, then the problem of stress corrosion cracking occurs at a temperature above 63 °C [12]. Hence, in this present work AA5052 has been

Table 1 Chemical composition of AA5052 by weight percentage

Elements	Mg	Cr	Fe	Mn	Si	Cu	Zn	Al
wt%	2.5	0.25	0.35	0.1	0.2	0.1	0.1	Remaining

chosen for the experiment due to its excellent corrosion resistance property with Mg content less than 3%. 5 wt% of B_4C particles are used as reinforcement material due to their high hardness, high modulus of elasticity. The particle size of the used B_4C-p is 63 μm. The bigger particle size of the reinforcement helps to gain a homogeneous mixture or uniform distribution of the particles in the matrix, small particle size reinforcements tend to agglomerate in the mixture. Small particulate reinforcements help to increase the strength of mixture greater than bigger particle size reinforcement [13]. 63 μm particles size is the intermediate size, so that is the reason for choosing 63 μm (230 mesh) size of B_4C-p for the present work. The chemical compositions of AA5052 are listed in Table 1.

The stir casting technique is used for the fabrication of AMMC for present work. The required quantity of AA5052 was taken in a graphite crucible. This clay-graphite crucible was set in a resistance heating furnace (Single phase, 7.5 kW, max. working temperature 850 °C) and the alloy is melted at 800 °C as shown in Fig. 1a.

Dross product formed and floating over the molten alloy was firstly removed, then the molten AA5052 was degassed by using solid hexachloroethane (C_2Cl_6) tablet because the molten AA5052 reacts with atmospheric oxides and degrades properties by oxidation [14, 15]. The calculated amount (5 wt%) of B_4C-p (preheated at 175 °C for 2 h) packed in Al foil was added into the molten AA5052 then with the help of graphite impeller stirred for 10 min at 600 rpm. Then the composite slurry in the crucible was taken from the furnace with the help of tong and poured into preheated and cleaned permanent cast iron mould of required shape as shown in Fig. 1b, but some casting defect occur in casted sample, i.e. gas porosity, cold shut, rough surface etc., as shown in Fig. 1c due to the small cross-sectional

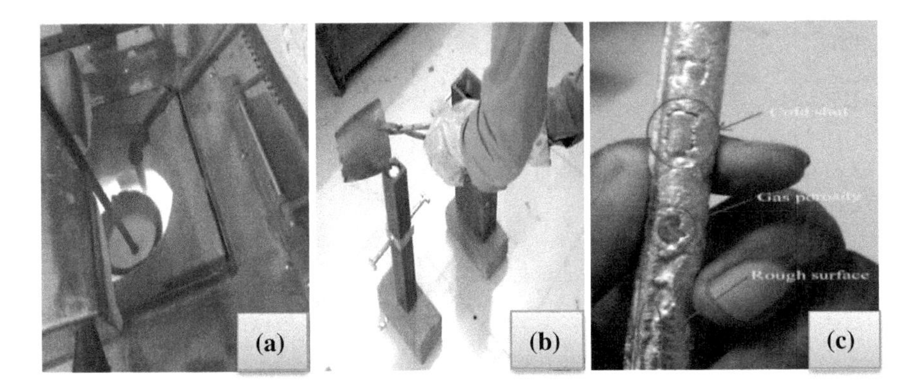

Fig. 1 **a** Melting of AA5052 in resistance furnace. **b** Pouring of composite slurry in permanent cast iron mould. **c** Casting defect in permanent mould cast sample

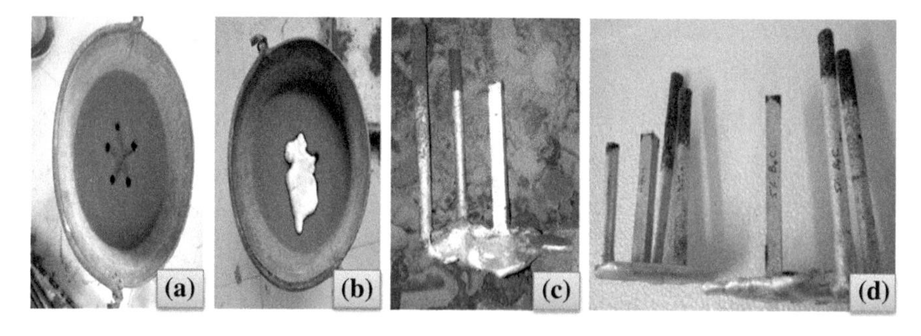

Fig. 2 **a** Cavities prepared for casting. **b** Poured composite slurry in cavities. **c** Developed AA5052 + 5 wt% B_4C-p MMC. **d** Developed as cast alloy and MMC

area and non-uniform preheating of permanent cast iron mould. To avoid these defects, sand mould was used. Prepared cavities of required shape for AA5052 + 5 wt% B_4C-p MMC is shown in Fig. 2a, poured composite slurry in the mould is shown in Fig. 2b and the developed B_4C-p reinforced AA5052 MMC is shown in Fig. 2c. The as-cast AA5052 and developed AA5052 + 5 wt% of B_4C-p MMC are shown in Fig. 2d.

2.2 *Density and Porosity Measurement*

The density of the as-cast AA5052 and the developed AA5052/B_4C-p MMC has been theoretically measured by rule of mixture formula, given by following Eq. (1) [2].

$$\frac{1}{\rho_c} = \frac{W_r}{\rho_r} + \frac{W_m}{\rho_m} \tag{1}$$

where: 'ρ_c', 'ρ_r' and 'ρ_m' are the density of the composite or the alloy, the reinforcement, and the matrix, respectively. 'W_r' and 'W_m' are the weight fraction of the reinforcement and the matrix, respectively.

The density of standard polished cylindrical specimens of 8 mm diameter and 30 mm height was experimentally determined by applying Archimedes' principle by using Eq. (2).

$$\rho_c = \frac{m_1}{m_2 - m_3} \times \rho_{H_2O} \tag{2}$$

where: 'ρ_c' is the density of the composites or alloy, 'm_1' is the mass of sample in the air 'm_2' is the mass of sample in the air with stand, 'm_3' is the mass of sample in water with stand and 'ρ_{H_2O}' is the density of distilled water (at 293 K) is 0.9982 g/cm^3.

The percentage of porosity presented in the as-cast AA5052 and the developed AA5052/B_4C p MMC was calculated by using the Eq. (3) [16].

$$\text{Porosity}(\%) = \left[1 - \frac{\text{Measured density}}{\text{Theoretical density}}\right] \times 100 \quad (3)$$

2.3 *Abrasive Wear Test*

At room temperature, abrasive wear test of the developed compositions was performed on a pin-on-disc wear tester supplied by Magnum Engineers, Bangalore. The test was conducted on the standard cylindrical samples of 8 mm diameter and 30 mm length of each composition. Before the experiment, the rotating disc (EN 31 steel disc) of 120 mm diameter was covered by 150 grade (100 μm) abrasive paper with the help of double-sided adhesive tape and the test samples were ground with 800-grit emery paper, then cleaned with water and acetone. Before the test, the weight of each test sample was measured by using an electronic weighing balance (accuracy ±0.001 g). The test sample was fixed in a sample holder and was abraded under the different applied loads of 5, 10 and 15 N for a constant sliding distance of 315 m. After the test of each sample, the test samples were removed from the sample holder, then cleaned with acetone to remove the debris in it and weighed to determine the weight loss during the test. The test samples after the abrasive wear test are shown in Fig. 3. During the abrasive wear test, the value of frictional force was noted for each 30 sec. and the average value of the frictional force (F) was calculated for each test sample.

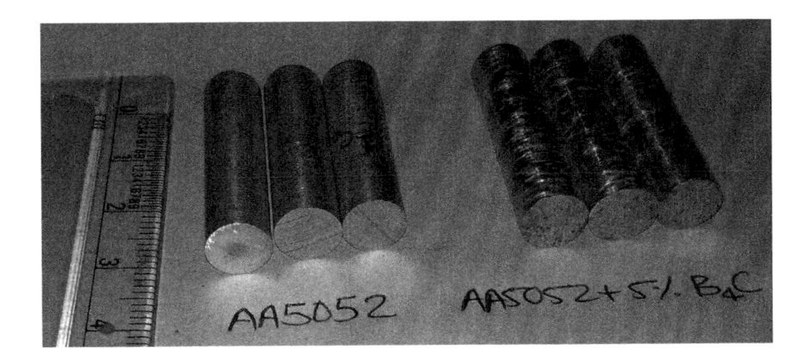

Fig. 3 Test samples after the abrasive wear test

The weight loss during the abrasive wear test, coefficient of friction (COF) at different applied loads (5–15 N) and abrasive wear rate of the developed compositions were calculated by using Eqs. (4), (5) and (6), respectively [17–19].

$$\Delta w = w_b - w_a \tag{4}$$

$$\mu = \frac{F}{P} \tag{5}$$

$$W_r = \frac{\Delta w}{S_d} \tag{6}$$

where: 'Δw' is weight loss of the test sample in g, 'w_b' and 'w_a' are the weights of the test sample before and after abrasion test in gram respectively, 'μ' is COF, 'F' is frictioanal force in Newton, 'P' is applied load in Newton, 'W_r' is wear rate in g/m, 'S_d' is the sliding distance in metre. After the abrasion test, the worn surface of the tested samples was observed by ZEISS AxioCam ERc 5 s optical microscope.

3 Results and Discussion

3.1 Density and Porosity

The measured density, theoretical density and the percentage porosity presented in as-cast AA5052 and AA5052/B_4C-p MMC are tabulated in Table 2.

It is noted that both measured and theoretical density decrease with the addition of B_4C-p. This is due to the density of B_4C-p (2520 kg/m^3) is lower than the density of AA5052 (2680 kg/m^3). A similar decrease in density with the addition of B_4C-p in Al alloy was found by Rao and Padmanabhan [20]. The difference between measured and theoretical density indicates that the presence of porosity in the casted compositions and it can affect the performance of cast compositions. From Table 2, it can be also concluded that the porosity presented in as-cast AA5052 and AA5052/B_4C-p is very small (<1%) due to the good casting process and also shows the suitability of the liquid-state processing technique.

Table 2 Density and porosity of the developed compositions

Cast composition	Measured density (kg/m^3)	Theoretical density (kg/m^3)	Porosity (%)
AA5052	2669.999	2680.00	0.373171
AA5052 + 5% B_4C-p	2651.828	2671.519	0.737071

3.2 *Tribological Test Results*

The abrasive wear tribological test have been done on P150 grade abrasive paper at three different applied loads (5, 10 and 15 N) and its effect on the weight loss, frictional force, coefficient of friction (COF) and wear rate are reported as follows.

3.2.1 Abrasive Weight Loss

The gain weight losses with respect to applied load (5–15 N) during the abrasive wear test are plotted on Fig. 4a and it is observed that at same applied loading condition, the weight loss of unreinforced AA5052 is higher than the AA5052/ B_4C-p MMC. It is also observed that weight loss of both compositions increases with increase in applied load. This is due to the minimum degree of penetration of abrasive particles on the surface of the test sample at lower applied load, that results in lower material removal. Degree of penetration increases with increase in applied load, so at higher applied load material removal from the test sample is high.

3.2.2 Abrasive Wear Rate

Effect of different applied loads (5–15 N) on the abrasive wear rate of casted compositions is presented in Fig. 4b. It is noted that the wear rate increases with increase in applied load. This is because the material loss increases with increase in applied loads. The wear rate of B_4C-p reinforced AA5052 MMC is lower than the unreinforced AA5052 due to the high hardness of reinforcement element and good bonding between the particulate reinforcement and matrix material.

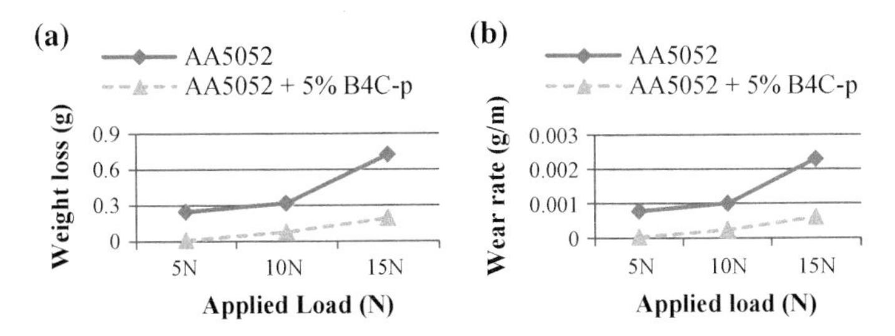

Fig. 4 **a** Effect of applied load on weight loss of developed compositions. **b** Effect of applied load on the abrasive wear rate of developed compositions

3.2.3 Frictional Force and Co-efficient of Friction

From Fig. 5a, it is observed that the average frictional force increases with increase in applied load. It is also noted that the frictional force decreases with the addition of B_4C-p into the AA5052 matrix. At the same applied load, the frictional force is higher for as-cast AA5052 as compared to AA5052/B_4C-p MMC. Figure 5b shows the variation of COF with respect to applied loads for the fabricated compositions. It is noted that COF decreases with increase in applied load. At the same applied load, the COF is higher for as-cast AA5052 as compared to AA5052/B_4C-p MMC.

3.2.4 Analysis of the Worn Surfaces

The worn out surfaces of the test samples during the abrasion test are observed by using an optical microscope. The micrographs of the worn surfaces are shown in Fig. 6. Figure 6a, b show the wear behaviour of the test samples of as-cast AA5052 and AA5052/B_4C-p MMC, respectively at applied load range of 5–15 N. From Fig. 6, it is observed that the abrasive groove marks are in the sliding direction, in which the depths increase with increase in applied load. It is also observed that delamination occurs at some points due to cracks between two abrasive grooves.

In AA5052/B_4C-p MMC some particles are pull-out from the matrix material due to high applied stress. It is also observed that the B_4C-p is restricting the groove formed in the direction of sliding and reduced the material loss. Optical micrographs of worn surfaces show that penetration of abrasive particles and delamination between two grooves are the prior reasons for wear in AA5052 samples. Penetration of abrasive particles and particles pull-out are the major reasons for wear in AA5052/B_4C-p MMC.

Brackish water aquaculture industry requires reliable, efficient, lightweight pumps that operate off-grid for water re-circulation and aeration. The water of the fish pond has high turbidity and often contains particulate matter. Wear resistance is a critical parameter for material selection ensuring high reliability of the pumps. So there is a need for replacement of high chrome cast iron and stainless steel slurry pump impellers by particulate reinforced AMMC. The developed 5%

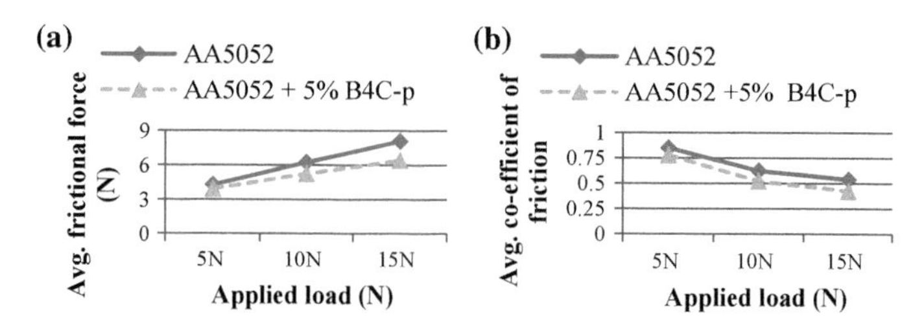

Fig. 5 **a** Effect of applied load on the frictional force. **b** Effect of applied load on the COF of developed compositions

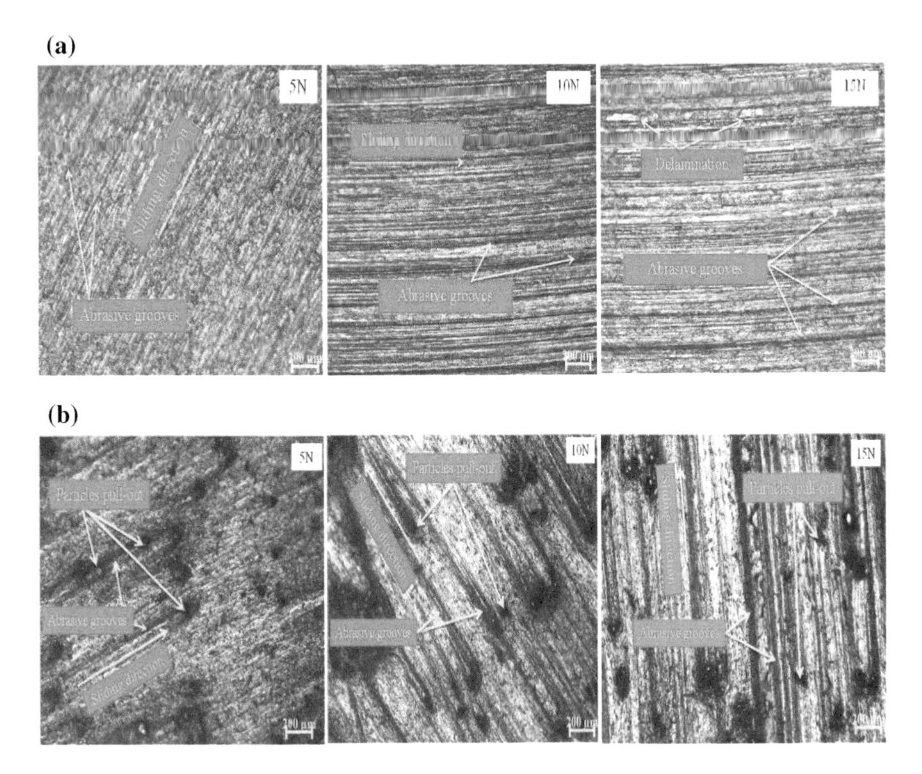

Fig. 6 Worn surfaces of the developed compositions at applied loads 5–15 N. **a** Worn surface of as-cast AA5052. **b** Worn surface of AA5052 + 5% B4C-p MMC

B_4C-p + AA5052 MMC can be used for the slurry pump impeller and casing because it has high abrasive wear resistance. The developed composite can also be used for many other industries, i.e. aerospace, automobile, sports due to their low density (lightweight) with good wear resistance property.

4 Conclusions

In this research, the AA5052 matrix has been reinforced with 5 wt% B_4C-p of 63 μm particles size by using stir casting liquid processing technique. The sand mould was used instead of permanent mould to avoid casting defects. The density of the AA5052 decreases up to 0.68% with the addition 5 wt% B_4C-p. Porosity is less than 1% in the developed compositions, which is 0.37% and 0.74% for the as-cast AA5052 and AA5052/B_4C-p MMC, respectively. Abrasive weight loss, abrasive wear rate and frictional of the developed compositions are increased with increasing the applied load. COF of the developed compositions is decreased with an increase in applied load. At all applied loading conditions, the abrasive weight loss, abrasive wear rate, frictional force and COF of the unreinforced AA5052 are higher than the AA5052/B_4C-p MMC. Optical micrographs of worn surfaces show that penetration of abrasive particles and delamination between two grooves are the primary reasons for wear in unreinforced AA5052. Penetration of abrasive particles and particle pull-out are the main reasons for wear in AA5052/B_4C-p MMC. Due to the good wear resistance property of AA5052/B_4C-p MMC, it can be used for slurry pump impeller and its casing.

References

1. Jones RM Mechanics of composite materials, 2nd edn. CRC Press
2. Kaw AK (2006) Mechanics of composite materials. CRC Press
3. Ozben T, Kilickap E, Orhan C, Çakir O (2008) Investigation of mechanical and machinability properties of SiC particle reinforced Al-MMC. J Mater Process Technol 198(1–3):220–225
4. Feng YC, Geng L, Zheng PQ, Zheng ZZ, Wang GS (2008) Fabrication and characteristic of Al-based hybrid composite reinforced with tungsten oxide particle and aluminum borate whisker by squeeze casting. Mater Des 29(10):2023–2026
5. Chen ZZ, Tokaji K (2004) Effects of particle size on fatigue crack initiation and small crack growth in SiC particulate-reinforced aluminum alloy composites. Mater Lett 58:2314–2321
6. Shorowordi KM, Laoui T, Haseeb ASMA, Celis JP, Froyen L (2003) Microstructure and interface characteristics of B4C, SiC and Al2O3 reinforced Al matrix composites: a comparative study. J Mater Process Technol 142:738–743
7. Karamis MB, Tasdemirci A, Nair F (2003) Failure and tribological behavior of the AA5083 and AA6063 composites reinforced by SiC particles under the ballistic impact. Compos Part A Appl Sci Manuf 34:217–226
8. Sujan D, Oo Z, Rahman ME, Maleque MA, Tan CK (2012) Physio-mechanical properties of aluminum metal matrix composites reinforced with Al_2O_3 and SiC physio-mechanical properties of aluminium metal matrix composites reinforced with Al_2O_3 and SiC. Int J Eng Appl Sci 6
9. Tang F et al (2008) Dry sliding friction and wear properties of B4C particulate-reinforced Al-5083 matrix composites. Wear 264:555–561
10. Nieto A, Yang H, Jiang L, Julie MS (2017) Reinforcement size effects on the abrasive wear of boron carbide reinforced aluminum composites. Wear 1–27
11. Guo Z, Li Q, Liu W, Shu G (2018) Evolution of microstructure and mechanical properties of Al-B 4 C composite after recycling. IOP Conf Ser Mater Sci Eng 409(1):1–8
12. Dix EH, Anderson WA, Shumaker MB (1959) Influence of service temperature on the resistance of wrought aluminum-magnesium alloys to corrosion. Corros Natl Assoc Corros Eng 15:19–26
13. Ozben T, Kilickap E, Orhan C (2008) Investigation of mechanical and machinability properties of SiC particle reinforced Al-MMC. J Mater Process Technol 198:220–225
14. Mahendra KV, Radhakrishna K (2010) Characterization of stir cast Al-Cu-(fly ash + SiC) hybrid metal matrix composites. J Compos Mater 44(8):989–1005
15. Annigeri UK, Veeresh Kumar GB (2017) Method of stir casting of aluminum metal matrix composites : a review. Mater Today Proc 4(2):1140–1146
16. Rao JB, Rao DV, Murthy IN, Bhargava N (2011) Mechanical properties and corrosion behavior of fly ash particles reinforced AA 2024 composites. J Compos Mater 42(12):1393–1404

17. Mishra P, Acharya SK (2010) Anisotropy abrasive wear behavior of bagasse fiber reinforced polymer composite. Int J Eng Sci Technol 2(11):104–112
18. Majhi S, Samantarai SP, Acharya SK (2012) Tribological behavior of modified rice husk filled epoxy composite. Int J Sci Eng Res 3(6):1–5
19. Kaushik NC, Sri Chaitanya C, Rao RN (2018) Abrasive grit size effect on wear depth of stir cast hybrid Al–Mg–Si composites at high-stress condition. Proc Inst Mech Eng Part J J Eng Tribol 232(6):672–684
20. Rao SR, Padmanabhan G (2012) Fabrication and mechanical properties of aluminium-boron carbide composites. Int J Mater Biomater Appl 2(3):15–18

Experimental Investigation and Empirical Modelling of FDM Process for Tensile Strength Improvement

Shrikrishna Pawar and Dhananjay Dolas

Abstract Fused deposition modeling (FDM) is a layer-by-layer manufacturing technology with the potential to create complex parts from thermoplastic materials. In order to avoid the failure of these parts under tensile loading, it is imperative to study the tensile test behavior of FDM parts. The quality of FDM parts mainly depends upon the accurate selection of process parameters. Thus, the identification of the FDM process parameters which significantly influence the quality of FDM processed parts is important. The present work focuses on experimental analysis to understand the effect of important parameters such as layer thickness, part build orientation, and infill density on the tensile strength of polycarbonate–acrylonitrile butadiene styrene (PC-ABS) material test specimen. This investigation provides insight into the dependency of tensile strength on process parameters and also develops a statistical predictive equation between input and output parameters. A full factorial experimental design was used to study the effects of process parameters on the tensile properties of the specimens.

Keywords Fused deposition modeling · Tensile strength · PC-ABS

1 Introduction

Additive manufacturing (AM) is the formalized term for what used to be called as rapid prototyping and what is popularly known as three-dimensional (3D) printing. The term rapid prototyping (RP) is used in many industries to explain a process for quickly producing a prototype in the form of a system or part before final release or commercialization. Users of RP technology understand that this term is inadequate as it is not considered recent advances and applications of the technology and hence recently adopted the American Society for Testing and Materials (ASTM) consensus standards now use the term additive manufacturing [1]. The basic principle of AM is that parts are made by adding material in layers; each layer is a thin cross

S. Pawar (✉) · D. Dolas
Jawaharlal Nehru Engineering College, Aurangabad, India

BBVL. Deepak et al. (eds.), *Innovative Product Design and Intelligent Manufacturing Systems*, Lecture Notes in Mechanical Engineering,
https://doi.org/10.1007/978-981-15-2696-1_36

section of the part derived from the original computer-aided design CAD model of the final part. Fused deposition modeling (FDM) is an additive manufacturing (AM) process in which a physical object is created directly from a CAD model using layer-by-layer deposition of a feedstock plastic filament material extruded through a nozzle [2–4]. This is one of the earliest types of AM processes, originally developed and marketed by Stratasys Company in the USA during the early 1990s. Basically, this technology is based on the principle of melting filament by adding heat and then extruding it through moving nozzle head for depositing it on the platform [5].

FDM built parts are finding application in many fields as end-user parts which are subjected to various types of mechanical loadings like tensile loading, compressive loading, or combined loading [6]. Many researchers [7–10] conducted studies to understand the relation between mechanical strength of FDM parts and process parameters affecting mechanical strength. Peng et al. [11] derived the relationship between output and input variables with second-order response surface methodology and fitness of which is validated by the artificial neural network. Optimization combining response surface methodology with fuzzy inference system and genetic algorithm concluded that optimal parameters are not confined to experimental values and can be found anywhere within feasible region. Aw [12] conducted a study to investigate the effect of infill density and printing pattern on the tensile, dynamic mechanical and thermoelectric properties of parts produced by the FDM process. The study is carried out on conductive acrylonitrile butadiene styrene/zinc oxide (CABS/ZnO) composite material. Experimental results were indicated significant improvement in tensile strength and Young's modulus, with an increase in infill density. Performance of line pattern, in terms of tensile strength, is of higher order as compared to the rectilinear pattern.

Onwubolu et al. [13] developed a functional relationship between five-process parameters including layer thickness, part orientation, raster angle, raster width, and air gap with response parameter tensile strength using group method for data handling for generalizing the prediction of the FDM process conditions. Differential evolution is also used to find out optimal process parameters. Jayanth [14] analyzed the influence of chemical treatment on surface roughness and tensile strength of FDM specimens of acrylonitrile butadiene styrene (ABS) material. Two chemicals, viz. acetone and 1, 2-dichloroethane, were used for the treatment. Surface finish after chemical treatment is improved radically, but tensile strength is reduced as specimen absorbs the chemicals during treatment which in turn makes it softer.

Raney [15] experimentally developed a functional relationship between infill percentage and build orientation and tensile strength of FDM specimens. The tensile strength of parts builds with the FDM process was compared to the nominal ABS filament strength values stated by the material supplier. Build orientation has a significant impact on the response variable as bonding strength varies with change in the build direction. Sood et al. [16] adopted weighted principal component analysis method for multi-response optimization with studying the effect of process variables, viz. layer thickness, part orientation, raster angle, raster width, and air gap on tensile, bending and impact strength. Furthermore, the optimal process condition is found out along with explaining the failure mechanism through SEM micrographs. Rajpurohit et al. [17] studied the relationship between tensile strength and process variables, viz. layer height, raster width, and raster angle. Furthermore, to understand the fracture mechanism, microscopic examination of fractured specimens is conducted. Panda et al. [18] attempted to study the effect of five processing parameters that are layer thickness, part build orientation, raster angle, raster width, and air gap on the tensile, flexural, and impact strength of test specimen.

2 Methodology

Factors are selected on the basis of a literature survey and also on the recommendation of industry experts. The ranges for parameters such as layer thickness and infill density are selected by referring printer manufacturer catalog and also by performing pilot experiments. Selected factors are defined as:

1. Layer thickness: Thickness of material layer deposited through the nozzle on build platform or distance traversed in the vertical direction between successive layers.
2. Infill density: Amount of material deposited for part manufacturing which depends upon selected infill pattern.
3. Build orientation: Part orientation in which part is manufactured in a build chamber.

In order to study the influence of selected process parameters on tensile strength, experiments are conducted with the full factorial design of experiments. Considering a full factorial design is foolproof design approach and contains all possible combinations of a set of factors to perform experimental runs. A factorial experiment can be analyzed using analysis of variance (ANOVA) or regression analysis. It is relatively easier to estimate the main effect of a factor. Other useful exploratory analysis tools for factorial experiments include main effects plots, interaction plots, and a normal probability plot of the estimated effects.

A number of experiments to be performed are decided with the help of full factorial method in Minitab 17 software. According to inputs in Minitab software for an optimum number of experiments, it provides 18 runs for various combinations of the different levels of these three factors as depicted in Table 1. For accurate determination of tensile strength (TS) of materials, a testing specimen is

Table 1 Input parameters with its levels

Parameters	Unit	Level 1	Level 2	Level 3
Layer thickness (LT)	mm	0.14	0.19	0.29
Infill density (ID)	Percentage	20	60	100
Build orientation (BO)	–	Horizontal	Vertical	–

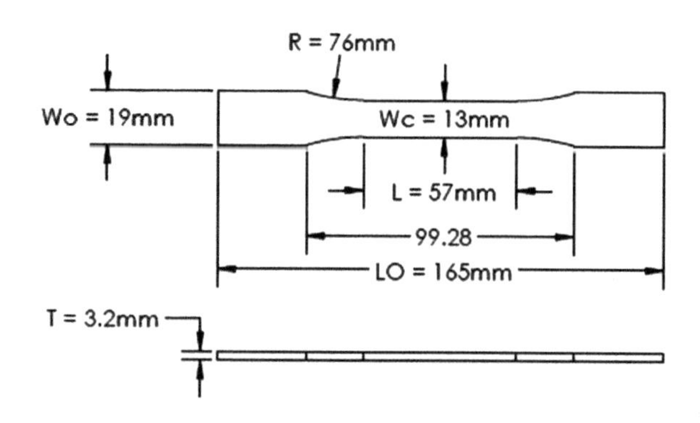

Fig. 1 Type-I specimen

Table 2 Experimental data

Run order	LT	ID	BO	TS
1	0.29	60	Horizontal	26.9880
2	0.14	20	Vertical	33.4184
3	0.29	20	Horizontal	23.3665
4	0.29	100	Horizontal	27.8798
5	0.19	60	Horizontal	21.9078
6	0.14	20	Horizontal	18.6234
7	0.29	60	Vertical	20.4433
8	0.29	100	Vertical	31.5711
9	0.14	100	Horizontal	32.9529
10	0.14	60	Horizontal	21.3699
11	0.29	20	Vertical	21.6815
12	0.19	60	Vertical	28.0326
13	0.19	20	Horizontal	21.8884
14	0.19	100	Vertical	34.6867
15	0.19	20	Vertical	29.0398
16	0.14	60	Vertical	29.9687
17	0.19	100	Horizontal	33.5342
18	0.14	100	Vertical	36.8358

required of the specified size and shape. The dimensions and the shape of the specimen for tensile test, flat and dumbbell-shaped, are as shown in Fig. 1.

3D model of the test specimen is modeled in computer-aided three-dimensional interactive application (CATIA) V5, and this model in Standard Tessellation Language (.STL) file is imported to Z-Suite software for slicing.

Selected factors are set according to the experimental run as depicted in Table 2. Specimens are manufactured for each run on Zortrax M200 FDM printer for tensile testing, and material used for manufacturing is PC-ABS in the form of filaments of 1.75 mm diameter. PC–ABS is one of the most widely used industrial thermoplastics, offering desirable properties of both materials, i.e., PC and ABS. This material has superior strength and a heat resistance of PC and excellent feature definition of ABS. Automotive, electronics, and telecommunications fields commonly found applications for PC–ABS blends. Its advantages include high mechanical durability and resistance to various factors, including high temperatures, UV rays, and chemicals.

All tests are conducted in accordance with ASTM D638-14 standard. During the tensile testing process, samples are tested on Autograph AS-IS 100KN, Make Shimadzu, universal testing machine of the constant rate of crosshead movement. The experimentally observed ultimate tensile strength is shown in Table 2.

3 Results and Discussion

For determining significant parameter, ANOVA is conducted and the *p*-value is checked, as exhibited in Table 3. If *p*-value of the parameter is less than 0.05, the significance of that corresponding term is established. For lack of fit, the *p*-value must be greater than 0.05. An insignificant lack of fit is desirable because it specifies that any term left out of the model is not significant and that the developed model fits well.

Percentage contribution of all input parameters is shown in Fig. 2. Analysis of variance (ANOVA) assumptions were also checked using diagnostic plots including normal probability plot and residuals versus fits as exhibited in Fig. 3 and Fig. 4, respectively. The residuals follow a normal distribution and the random

Table 3 ANOVA results

Source	Degree of freedom	The adjacent sum of square	The adjacent mean of the square	F-value	*p*-value
LT	2	42.34	21.17	1.60	0.242
ID	2	267.87	133.93	10.13	0.003
BO	1	76.74	76.74	5.81	0.033
Error	12	158.59	13.22		
Total	17	545.54			

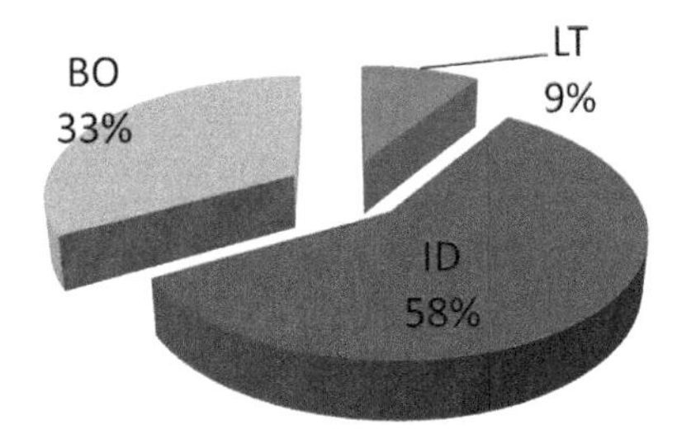

Fig. 2 Percentage contribution of input parameters

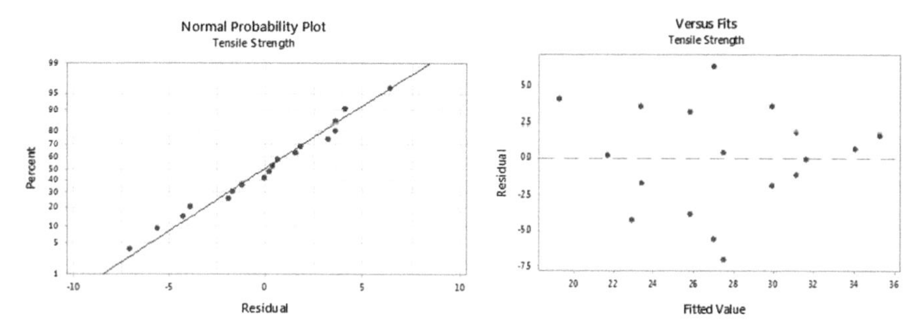

Fig. 3 Normal probability plot

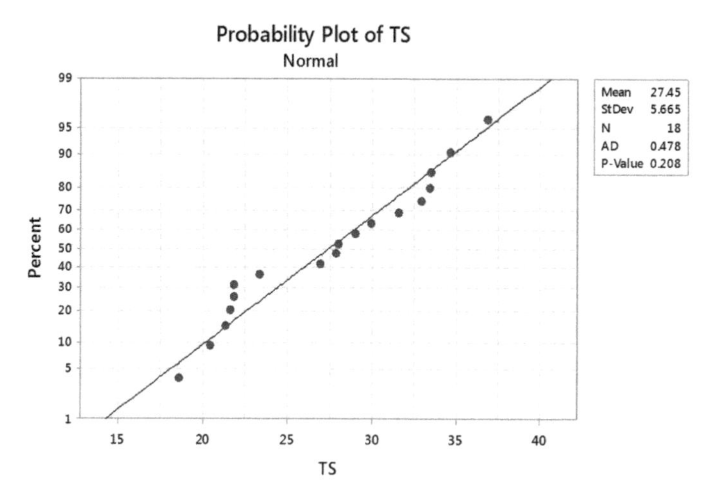

Fig. 4 Probability plot

scatter pattern of residuals in residuals versus fits about the zero specifies homogeneity of variance.

Also, the Anderson–Darling (AD) normality test is performed to validate the suitability of tensile strength model for end-user engineering applications. The p-value for the Anderson–Darling test is lesser than the selected significance level, i.e., 0.05 in this experimental work, so it is concluded that the data do not follow the normal distribution. In the selected model, the p-value is 0.208 indicating the normal distribution of the data.

Response surface method is used to investigate the relationship between tensile strength and input variables including layer thickness, infill density, and build orientation. Regression equation in uncoded units which indicates the best-fit model to experimentally observed values is as follows:

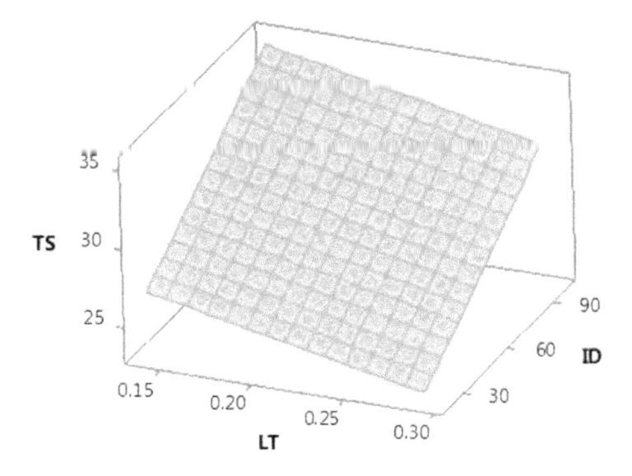

Fig. 5 Surface plot

- Horizontal build orientation

$$\text{Tensile Strength} = 24.23 - 24.3\,\text{Layer Thickness} + 0.1030\,\text{Infill Density} \tag{1}$$

- Vertical build orientation

$$\text{Tensile Strength} = 28.36 - 24.3\,\text{Layer Thickness} + 0.1030\,\text{Infill Density} \tag{2}$$

Surface plots are shown in Fig. 5, plotted for the understanding, the combined effect of two factors, viz. layer thickness and infill density, while keeping build orientation constant.

Tensile strength has an increasing trend with infill density, and it implies that with an increase in infill density, the value of tensile strength also increases. Highest of tensile strength is available when infill density is set at maximum level. Surface plots showing that layer thickness has a minimum effect on tensile strength.

4 Conclusion

In the present experimental work, an attempt has been made to investigate the effect of layer thickness, build orientation and infill density on the tensile strength in FDM built specimen. The results of statistically designed experimental runs establish the linear nature of FDM processed PC-ABS specimens. This study is one of the few published research works conducted for PC-ABS material. Statistical analysis indicated that the infill density has a dominant and statistically significant effect on tensile strength. Effects of all three factors and their interactions are explained using response surface plots which show the high extent of linearity indicating the linear relationship between process parameters, viz. layer thickness, infill density, and tensile strength, which is output response. This study was limited to two build orientations only; this paves the way for future work of investigating the influence of three build orientations along with microscopic analysis to understand fracture mechanism.

References

1. Gibson I (2015) Additive manufacturing technologies: 3D printing, rapid prototyping, and direct digital manufacturing, 2nd edn. Springer, New York
2. Kai C (2003) Rapid prototyping: principles and applications, 2nd edn. World Scientific Publishing Co. Pte. Ltd. Singapore
3. Ahn D, Kweon J, Kwon S, Song J, Lee S (2014) Representation of surface roughness in fused deposition modeling. J Mater Process Technol 209(2009):5593–5600
4. Hamza I (2018) Experimental optimization of fused deposition modeling process parameters: a Taguchi process approach for dimension and tolerance control. In: Proceedings of the international conference on industrial engineering and operations management, pp 1–11. IEOM Society International Paris
5. Vishwas M (2017) Experimental investigation using Taguchi method to optimize process parameters of fused deposition modeling for ABS and nylon materials, materials today: proceedings, pp 7106–7114. Elsevier Ltd
6. Motaparti K (2017) Experimental investigations of effects of build parameters on flexural properties in fused deposition modeling parts, virtual and physical prototyping, pp 1–15
7. Omar A (2015) Mohamed: optimization of fused deposition modeling process parameters: a review of current research and future prospects. Adv Manuf 3:42–53
8. Thrimurthulu K (2004) Optimum part deposition orientation in fused deposition modeling. Int J Mach Tools Manuf (44):585–594
9. Rodrıguez J (2003) Design of Fused-Deposition ABS Components for Stiffness and Strength. J Mech Des 125:545–551
10. Rezayat H (2014) Structure–mechanical property relationship in fused deposition modeling. Mater Sci Technol 1–9
11. Peng A (2014) Process parameter optimization for fused deposition modeling using response surface methodology combined with a fuzzy inference system. Int J Adv Manuf Technol 87–100
12. Aw Y (2018) Effect of printing parameter on tensile, dynamic mechanical and thermoelectric properties of FDM 3D printed CABS/ZnO composites. Materials 11(466):1–14
13. Onwubolu G Characterization and optimization of mechanical properties of ABS parts manufactured by the fused deposition modeling process. Int J Manuf Eng 1–13
14. Jayanth N et al (2018) Effect of chemical treatment on tensile strength and surface roughness of 3D-printed ABS using the FDM process. Virtual Phys Protot 13(3):155–163
15. Raney K (2016) Experimental characterization of the tensile strength of ABS parts manufactured by fused deposition modeling process. In: Proceedings of the international conference on advancements in aeromechanical materials for manufacturing, pp 7956–7961. Materials TodayDenver
16. Sood A (2011) Optimization of process parameter in fused deposition modeling using weighted principal component analysis. J Adv Manuf Syst 10(2):241–259
17. Rajpurohit S et al (2018) Impact of process parameters on tensile strength of fused deposition modeling printed crisscross polylactic acid. Int J Mater Metal Eng 12(2):52–57
18. Panda S (2009) Optimization of fuse deposition modeling (FDM) process parameters using bacterial foraging technique. Intell Inf Manag (1):89–97

Tensile Properties of Bamboo Fiber Filled with Rock Dust Filler Reinforced Hybrid Composites

T. Venkateswara Rao, V. Sankara Rao and K. V. Viswanadh

Abstract The field of material science turned out to be very well known and even minded with a colossal desire for composite materials that display the positive attributes of both the parts. Worldwide there has been a lot of interest to tailor the structure and composites of materials on sizes of micrometer scale. Hence, a systematic review on the preparation, properties, and applications of composites is extremely important. So here, we took bamboo fiber from natural fiber and carbon fiber from artificial fiber and prepared fiber-reinforced composites with the help of resin polyester. The fiber-reinforced composites with rock dust as filler material have also prepared. By using hand lay-up process, we prepared these composites. The thus formed polymer matrix is tested under various mechanical properties like a tensile test, bending test. Then the properties of fiber composites are compared for natural and artificial fibers, and also, the properties of composites are calculated in which filler materials are added. Being environmentally friendly, applications of composites offer new technology and business opportunities for several sectors of the aerospace, automotive, electronics, and biotechnology industries.

Keywords Bamboo fiber · Rock dust · Tensile properties · Polymer composite

1 Introduction

In the course of the most recent thirty years of composite materials, plastics and ceramics have been the overwhelming rising materials. The volume and number of uses of composite materials have grown reliably, infiltrating and overcoming new markets tirelessly. Present-day composite materials establish a huge extent of the

T. Venkateswara Rao (✉) · V. Sankara Rao · K. V. Viswanadh
Lakireddy Bali Reddy College of Engineering, Mylavaram, Andhra Pradesh, India
e-mail: tvrao722@lbrce.ac.in

K. V. Viswanadh
Andhra University, Visakhapatnam, India

BBVL. Deepak et al. (eds.), *Innovative Product Design and Intelligent Manufacturing Systems*, Lecture Notes in Mechanical Engineering,
https://doi.org/10.1007/978-981-15-2696-1_37

built materials advertises extending from ordinary items to modern specialty applications. While composites have officially demonstrated their value as weight-saving materials, the present test is to make them cost-effective.

Defoirdt et al. and Biswas et al. assessed the tensile properties of coir, bamboo, and jute fiber composites [1, 2]. Laua et al. [3], Sanjay [4] and Ramachandran et al. [5] analyzed different natural fibers for real-life engineering applications. Chandramohan et al. fabricated some natural fibers and evaluated the tensile and hardness properties [6]. Various synthetic fibers, their comparison is presented by Prasanth et al. [7]. Influence of fly ash fillers on mechanical and tribological properties of woven jute fiber-reinforced polymer hybrid composite is given by Manikandan et al. [8]. Venkateswara Rao et al. [9] investigate the mean tensile strength, tensile modulus, specific tensile strength, specific tensile modulus, mean flexural strength, flexural modulus, and impact strength of bamboo fiber filled with fly ash filler reinforced hybrid composites. The impacts of type of synthetic resins, percentage of natural polymers, and percentage of fly ash on tensile behavior of fly ash reinforced hybrid polymer matrix composite are studied by Kesarla et al. [10]. The effect of fiber loading on the performance of jute/bamboo fiber-reinforced epoxy-based hybrid composites is observed by Rama Krishna [11].

From the literature, it is seen that filler material rock dust has not utilized anyplace which is conservative and accessible in the market. The present paper focuses on this filler material which gives more strength than using fly ash. These materials are preferable while a selection of the vehicle dome and airplane structures due to their less weight as compared to metals.

2 Material Selection and Methodology

This chapter describes the details of the processing of the composites and the experimental procedures followed for their mechanical characterization.

2.1 *Material Selection*

From the literature, it was observed that the mechanical properties of the natural fiber-reinforced composites depend on many parameters, such as fiber strength, modulus, fiber length, and orientation; in addition to the fiber-matrix interfacial bond strength, strong fiber-matrix interface bond is critical for high mechanical properties of composites. In this paper, the tests were conducted based on three categories: (i) bamboo fiber reinforced with polyester resin, (ii) bamboo fiber filled with rock dust reinforced with polyester resin, and (iii) carbon fiber reinforced with polyester resin.

2.2 Fabrication Process

Hand lay-up technique is the simplest method of composite processing. The infrastructural requirement for this method is also minimal. The processing steps are quite simple. Thin plastic sheets are used at the top and bottom of the mold to get a good surface finish of the product. Then polyester resin in liquid form is mixed thoroughly in suitable proportion with a prescribed hardener along with the bamboo and carbon fibers and poured onto the surface of mold which is prepared on the mat as per ASTM standards. A rubber mold which has a set of five blocks with a dimension of 160 * 12.5 * 4 mm for is used for tensile composites fabrication. Here, rust dust has been selected as filler material. Within less time, place the weights over the mold for stiffness. After 24 h, remove the mold and take the specimens properly.

2.3 Compositions

2.3.1 Bamboo Fiber Composites

The composite samples of five different compositions of bamboo go from 12.5 to 24.5% with 3% variance that has been prepared for testing (Table 1).

2.3.2 Bamboo Fiber with Rock Dust Filler Material

In these composites in addition to bamboo fiber, 25% of rock dust filler material added to the polyester resin by the simple mechanical stirring and the mixture is poured into various blocks (Table 2).

Table 1 Composition for tensile pure bamboo specimen

S.No.	Specimen description
1	12.5% bamboo + 87.5% resin
2	15.5% bamboo + 84.5% resin
3	18.5% bamboo + 81.5% resin
4	21.5% bamboo + 78.5% resin
5	24.5% bamboo + 75.5% resin

Table 2 Composition for tensile bamboo and rock dust specimens

S.No.	Specimen description
1	12.5% bamboo 2 g + 62.5% resin + 25% rock dust
2	15.5% bamboo 2.5 g + 59.5% resin + 25% rock dust
3	18.5% bamboo 3 g + 56.5% resin + 25% rock dust
4	21.5% bamboo 3.5 g + 53.5% resin + 25% rock dust

Table 3 Composition for tensile pure carbon specimen

S.No.	Specimen description
1	12.5% carbon 2 g + 87.5% resin
2	15.5% carbon 2.5 g + 84.5% resin
3	18.5% carbon 3 g + 81.5% resin
4	21.5% carbon 3.5 g + 78.5% resin
5	24.5% carbon 4 g + 75.5% resin

2.3.3 Carbon Fiber Composites

Carbon fiber mixed with polyester resin by the simple mechanical stirring, and the mixture is poured into various blocks. The composite samples of five different compositions of bamboo go from 12.5 to 24.5% with 3% weight variance (Table 3).

3 Tensile Tests

The unidirectional composite specimens were made as per the ASTM D 638-89 to measure the tensile properties. The tensile tests are conducted for each specimen on a two-ton capacity tensometer. The load versus displacement plots were obtained for each specimen from the automatic computerized chart recorder inbuilt in the machine (Fig. 1).

3.1 Calculation of Tensile Properties

$$\text{Tensile stress}, \sigma_t = \frac{P}{A} \text{(for a rectangular cross section)} \quad (1)$$

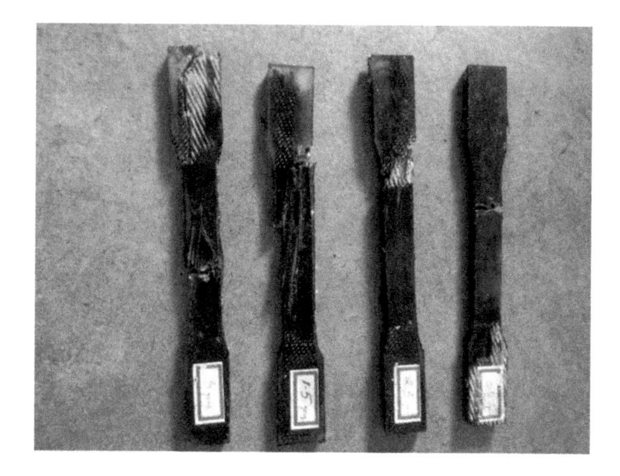

Fig. 1 Tensile specimens

$$\text{Tensile strain}, \varepsilon_t = \frac{\Delta L}{L} \tag{2}$$

$$\text{Tensile modulus}, E_t = \frac{\sigma t}{\varepsilon t} \tag{3}$$

4 Results and Discussion

From the above results, the stress versus strain graphs are drawn by taking the stress on y-axis and strain on the x-axis for bamboo specimens (Figs. 2 and 3 and Table 4).

The strain values are calculated at the same displacement for all specimens and at that corresponding points stress values are calculated. From the above graph, we can clearly understand that the 3.5 g of the bamboo composition shown good stress values. The 3.5 g bamboo specimen has shown highest values than other specimens.

From the suitable formulas, Young's modulus of specimens is calculated from the basis of load versus displacement graphs. The 3.5 g bamboo composition has shown highest Young's modulus when compared to the other specimen values (Table 5).

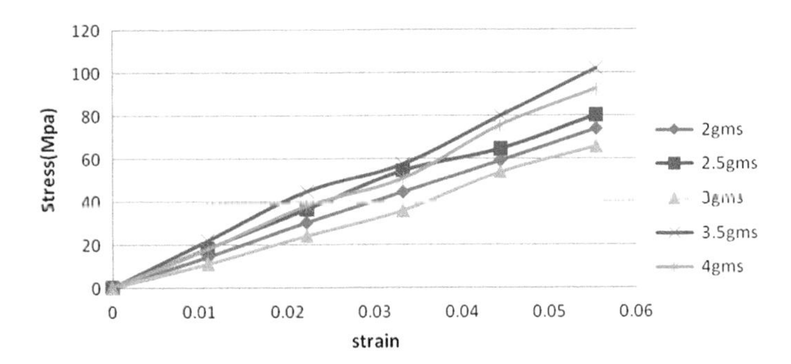

Fig. 2 Stress versus strain for pure tensile bamboo specimens

Fig. 3 Young's modulus versus weight of bamboo

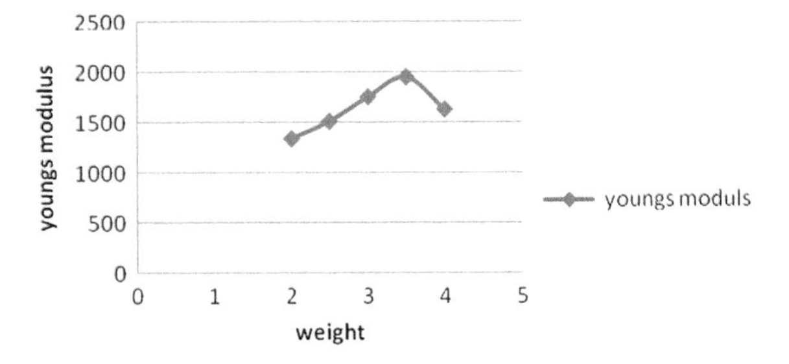

Table 4 Tensile properties for pure bamboo specimen

S.No.	Composition of bamboo (%)	Peak load (N)	Displacement (mm)	Tensile strength (N/mm^2)	Strain	Young's modulus (N/mm^2)
1	12.5 (2 g)	3697.2	2.77	73.944	0.0554	1334.729
2	15.5 (2.5 g)	4521.6	3.50	90.432	0.06	1507.2
3	18.5 (3 g)	5979.5	3.6	119.59	0.0682	1753.519
4	21.5 (3.5 g)	7433.7	3.81	148.674	0.0762	1951.102
5	25.5 (4 g)	6090.1	3.74	121.802	0.0748	1628.369

Table 5 Tensile properties for bamboo + rock dust specimen

S.No.	Composition of bamboo (%)	Peak load (N)	Displacement (mm)	Tensile strength (N/mm^2)	Strain	Young's modulus (N/mm^2)
1	12.5 (2 g)	3697.9	1.11	73.95	0.056	1320.5
2	15.5 (2.5 g)	4821.8	1.9	96.4	0.0625	1542.4
3	18.5 (3 g)	6079.4	2.36	121.88	0.0682	1747.21
4	21.5 (3.5 g)	7914.2	4.40	158.2	0.088	1846.97
5	25.5 (4 g)	6649.1	3.6	132.98	0.072	1797.72

From the above results, the stress versus strain graphs are drawn by taking the stress on y-axis and strain on the x-axis for bamboo + 4 g of rock dust specimens.

The strain values are calculated at the same displacement for all specimens and at that corresponding points stress values are calculated. From the above graph, we can clearly understand that the 3.5 g of bamboo + 4 g of rock dust composition has shown good stress values. The 3.5 g bamboo + 4 g rock dust specimen has shown highest values than other specimens.

From the suitable formulas, Young's modulus of specimens is calculated from the basis of load versus displacement graphs. The 3.5 g of bamboos + 4 g of rock dust composition has shown highest Young's modulus when compared to the other specimen values.

From the above results, the stress versus strain graphs are drawn by taking the stress on y-axis and strain on the x-axis for pure carbon specimens.

The strain values are calculated at the same displacement for all specimens and at that corresponding points stress values are calculated. From the above graph, we can clearly understand that the 2.5 g of carbon composition has shown good stress values. The 2.5 g carbon specimen has shown highest values than other specimens (Fig. 4).

From the suitable formulas, Young's modulus of specimens is calculated from the basis of load versus displacement graphs. The 2.5 g carbon composition is shown highest Young's modulus when compared to the other specimen values. It is observed that the tensile strength and tensile modulus values for the proposed filler material are high as contrasted with fly ash filler composites [9] (Figs. 5 and 6 and Table 6).

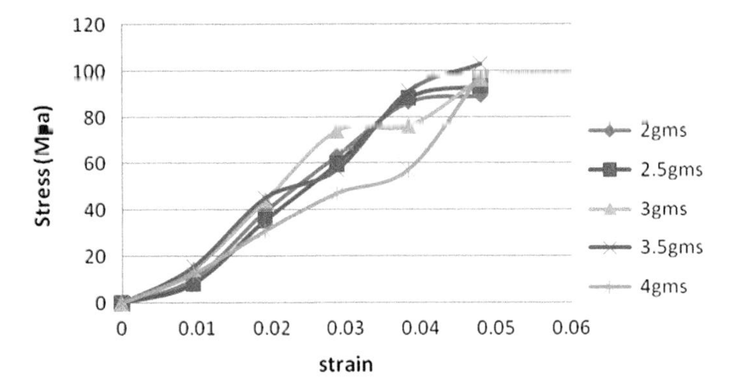

Fig. 4 Stress versus strain for bamboo + rock dust specimens

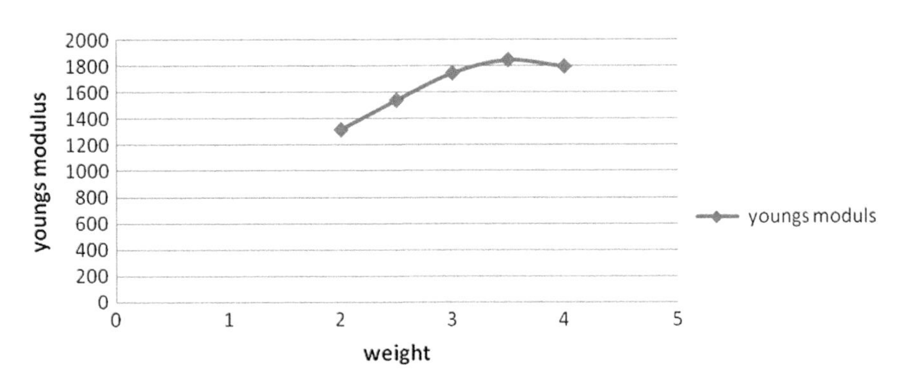

Fig. 5 Young's modulus versus weight of bamboo + 4 g rock dust of tensile specimens

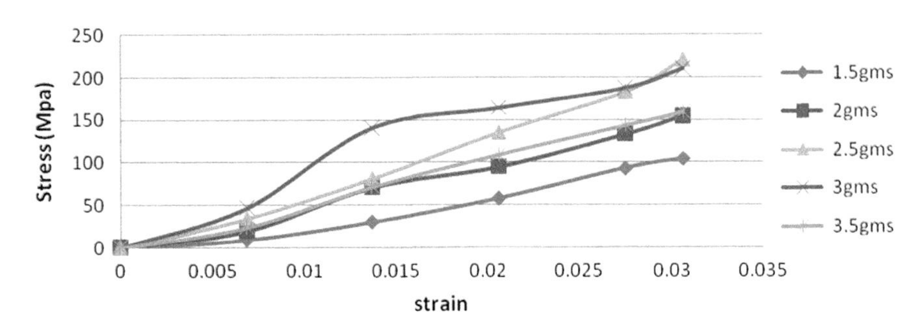

Fig. 6 Stress versus strain for pure tensile carbon specimens

Table 6 Tensile properties for pure carbon specimen

S.No.	Composition carbon (g)	Peak load (N)	Displacement (mm)	Tensile strength (N/mm^2)	Strain	Young's modulus (N/mm^2)
1	12.5 (2 g)	4658.3	3.2	93.48	0.0643	1453.81
2	15.5 (2.5 g)	9777.6	5.2	195.44	0.104	1879.2
3	18.5 (3 g)	13533	5.9	270.6	0.118	2288.13
4	21.5 (3.5 g)	11792	6.1	235.76	0.122	1932.45
5	25.5 (4 g)	7953.5	3.75	159.72	0.085	1879.05

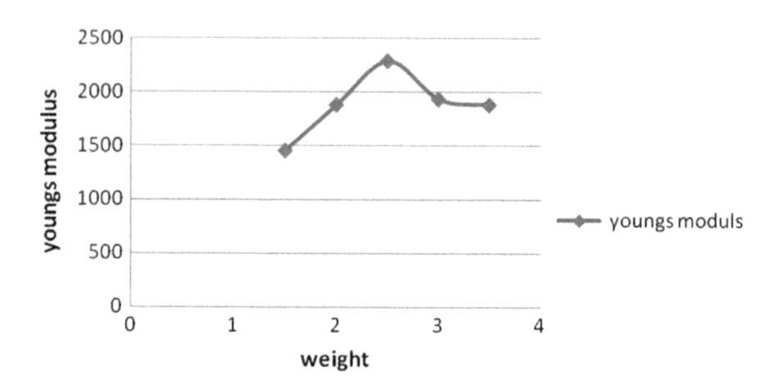

Fig. 7 Young's modulus versus weight of carbon

5 Conclusions

The tensile test values of various specimens are compared for bamboo, bamboo + rock dust, and carbon fiber specimens. The carbon fiber specimens showed the best tensile properties when compared to the remaining another set of specimens. The rock dust added bamboo specimens had shown more tensile properties than pure bamboo specimens but not up to properties of carbon fiber. Finally, the carbon fiber is shown the highest tensile properties than bamboo and bamboo + rock dust properties (Fig. 7).

The stone residue coarseness particles are not the uniform size which may impact the properties of the composites. Pulverization of rock dust to the nanoscale can be favored on to improve the forms future scope regarding the enhancement of this research work.

References

1. Defoirdt N (2010) Assessment of the tensile properties of coir, bamboo and jute fibre composites: part A, vol 4, pp 588–595
2. Biswas S, Shahinur S, Hasana M, Ahsana Q (2015) Physical, mechanical and thermal properties of jute and bamboo fiber reinforced unidirectional epoxy composites. Proc Eng 105:933–939

3. Laua K-T, Hunga P-Y, Zhu M-H, Huid D Properties of natural fibrecomposites for structural engineering application composites part B, vol 136, pp 222–233
4. Sanjay MR, Arpitha GR, Laxmana Naik L, Gopalakrishnan K, Yogesha B (2016) Applications Of natural fibers and its composites: an overview natural resources, vol 7, pp 108–114 (Published Online March 2016 In SciRes)
5. Ramachandran M, Bansal S, Fegade V, Raichurkar P (2015) Analysis of bamboo fibre composite with polyester and epoxy resin. Int J Text Eng Process 1(4). ISSN: 2395-3578
6. Chandramohan D, Marimuthu K Tensile and hardness tests on natural fiber reinforced polymer composite material (Ijaest) Int J Adv Eng Sci Technol 6(1):097–1
7. Prashanth S, Nithin S (2017) Fiber reinforced composites-a review journal of material sciences & engineering. J Mater Sci Eng 6:3
8. Manikandan V, Richard S, Chithambara Thanu M, Selwin Rajadurai (2015) J effect of fly ash as filler on mechanical & frictional properties of jute fiber reinforced composite. Int Res J Eng Technol (Target) 02(07)
9. Venkateswara Rao T et al (2014) Mechanical Properties of bamboo fibre filled with fly ash filler reinforced hybrid composites. Int J Eng Res Technol (IJERT) 3(9)
10. Kesarla H et al (2018) Study on Tensile behavior of fly ash reinforced hybrid polymer matrix composite. In: International conference on materials manufacturing and modelling, materials today proceedings, vol 5, Issue 5, Part 2, pp 11922–11932
11. Rama Krishna K et al Mechanical properties of bamboo and jute fibre filled with fly ash filler reinforced hybrid composites. Int J Sci Eng Technol Res

Effect of Eggshell Particulate Reinforcement on Tensile Behavior of Eggshell–Epoxy Composite

Manoj Panchal, G. Raghavendra, M. Omprakash, S. Ojha and B. Vasavi

Abstract This work examines the prospects of eggshells as a reinforcing agent in polymer composite. Initially, the scanning electron microscopy (SEM) was done on the eggshell particles for morphology analysis. The 4, 8 and 12 wt% of eggshell particulate epoxy composite samples were fabricated using hand lay-up technique. The tensile behavior of the eggshell reinforced epoxy composite was investigated according to ASTM standards. The results of the tensile behavior of eggshells reinforced epoxy composite show affirmative results when compared with bare epoxy. The 4 wt% eggshell particulate addition is found to be the optimum percentage for tensile strength and the tensile modulus.

Keywords Eggshells · Epoxy · Composite · Tensile strength · Tensile modulus

1 Introduction

A physically bonded material comprising two or more different constituents with distinct properties is called composite. Since the inception of the composite field, the synthetic fibers or mineral particles with a high modulus have been the point of attraction and have been used from decades; however, their high cost and environmental hazardous nature have led to growing research and development on biocomposites. A composite is called a biocomposite when one constituent is derived from renewable resources, such as avian products or cellulose fibers [1]. Chicken eggshell (ES) is an avian by a product that has daily huge stock worldwide. For example, in India, the annual production of eggs is about 63 billion units of eggs. The weight of an egg is approximately 62 g, on an average eggshell weighs 10% of the total weight of the egg, so the total eggshell is 390,600 tons per year. These eggshells are a total waste. Eggshell is usually composed of calcium carbonate (by weight) $\sim$94% [2]. As the calcium carbonate is major constituents of

M. Panchal · G. Raghavendra (✉) · M. Omprakash · S. Ojha · B. Vasavi
Department of Mechanical Engineering, NIT Warangal, Hanamkonda, Telangana, India
e-mail: raghavendra.gujjala@nitw.ac.in

BBVL. Deepak et al. (eds.), *Innovative Product Design and Intelligent Manufacturing Systems*, Lecture Notes in Mechanical Engineering,
https://doi.org/10.1007/978-981-15-2696-1_38

eggshells, the mechanical behaviors of calcium carbonate polymer composite and eggshells' polymer composites are reviewed here.

Toro et al. [1] examined the tensile behavior of the polypropylene/eggshell composite and did the comparative analysis of tensile properties with polypropylene composite reinforced with commercial talc and calcium carbonate filler. A positive result was obtained for tensile modules while the tensile strength was found to decrease with eggshells' particles reinforcement.

Chaithanyasai et al. [3] have examined the morphology and mechanical properties of eggshell particulate reinforced aluminum 6061 composite. Three different weight percentages of eggshells' particles were added in aluminum 6061alloys. The hardness was increased by 14% while density was reduced with the incorporation of eggshells' particles in the composite. Yashonant Sai et al. [4] have done a similar study for eggshell particulate reinforced aluminum 6351 alloy composite.

Ji et al. [5] evaluated the influence of eggshells' particles surface area in the epoxy resin. It was found that the eggshells' particulate reinforcement increases the toughness of the composite. It was claimed the positive change in toughness is due to the increase in surface area of eggshell particles. Yang et al. [6] investigated the mechanical behavior of polypropylene nano $CaCO_3$ composites. An appreciable increase in yield strength of $CaCO_3$ filled nanocomposites had been observed as compared to polypropylene. Maiti and Mahapatra [7] had also investigated the tensile and impact behavior of different weight percentage of polypropylene/$CaCO_3$ composite. It was found that the increase in $CaCO_3$ content results in increased tensile modulus while a decrease in tensile strength and breaking elongation was observed. It had been concluded that the increase in tensile modulus and elongation decrease was due to increased filler–polymer adhesion and better dispersion. Leong et al. [8] had examined the tensile and thermal behavior of the polypropylene/talc/$CaCO_3$ hybrid composite samples. It has been very well established through experimental results that the mechanical properties synergistically change at weight ratio 70:15:15 of polypropylene/talc/$CaCO_3$. Xie et al. [9] had prepared polyvinyl chloride (PVC)/calcium carbonate ($CaCO_3$) nanocomposites by in situ polymerization. The optimum mechanical properties were found to be achieved with the addition of 5 wt% of $CaCO_3$ nanoparticles. Wu et al. [10] had also prepared polyvinyl chloride/calcium carbonate composite with chlorinated polyethylene as an interfacial modifier. The mechanical, rheological and morphological properties have been investigated, and it has been observed from results that the tensile modulus and elongation at break increase with increase in $CaCO_3$ content. It was suggested that better dispersion of $CaCO_3$ could be the reason behind the increase in Young's modulus and elongation at break. Literature review suggests that very little work has been done on eggshells as a reinforcing agent. This work is an attempt to explore the possibilities of eggshell particles as reinforcing media. In the present work, the ability of eggshells as a reinforcing agent in tensile behavior is investigated.

2 Materials and Method

2.1 Raw Materials

Materials used for the preparation of eggshell filler reinforced epoxy polymer composite material are given below

- Epoxy resin LY 556.
- Hardener HY 951.
- Chicken eggshells.

2.2 Sample Preparation

Initially, the collected chicken eggshells' flakes were washed with water and ethanol for removal of contamination. Then, they were dried in sunlight for 4 days. Further, these dried eggshells were crushed into fine micro powder into the laboratory ball mill.

2.3 Composite Preparations

The rectangular composite plates of dimension $180 \times 280 \times 5\ \text{mm}^3$ were prepared using hand lay-up technique. Three combinations of eggshell particulate epoxy composite were prepared with 4, 8 and 12% weight percentage of eggshells' particulate addition. The prepared plates after adequate curing were cut in required sizes of dog bone-shaped tensile test specimen according to ASTM standards. The composite plates are shown in Fig. 1.

2.4 Scanning Electron Microscopy

The scanning electron microscopy was done on eggshell particles for morphological analysis. The SEM analysis was conducted on the TESCAN VEGA3 scanning electron microscope at NIT Warangal.

Fig. 1 Composite plates

2.5 *Tensile Test*

American Society for Testing and Materials (ASTM) D 3039-76 standards were used for the preparation of a flat dog bone which is shape specimen. Tests were conducted on the digital universal testing machine (UTM). In UTM, one end of the sample is fixed and load is applied through another movable end. The gradual tensile load is applied, and the elongation is measured. The machine automatically generates the stress–strain curve and load-elongation curve. The tensile behavior specimen can be studied from the obtained curves. The load rate of 2 mm/min was for testing. The tensile test specimen is shown in Fig. 2. The dimensions of samples are as follows: length—140 mm, width at center—10 mm and thickness—4 mm.

Fig. 2 Specimen for tensile test

3 Results and Discussion

3.1 *Scanning Electron Microscopy*

From Fig. 3, it can be observed that the particles are spherical and are well within micro range although the particles are agglomerated. The spherical shape provides better adhesion with resin.

3.2 *Tensile Test Analysis*

Figure 4a, b shows the tensile strength and tensile modulus of composite samples, respectively. It is evident from Fig. 4a when 4% of eggshell filler is added to epoxy, the tensile strength of composite increases by 48%. The reason behind this synergistic effect could be good bonding between filler and epoxy at their interface. When the load is applied to matrix, it transmits the load to the fillers and fillers disperse the load transmitted by matrix. For eggshell particulate reinforcement of 8 wt%, tensile strength decreases by 21% when compared with 4 wt% eggshells reinforced epoxy composite; however, the tensile strength is still higher than pure epoxy. When particulate reinforcement weight percentage increased to 12 wt%, the tensile strength further decreases almost 18% when contrasted with 4 wt% eggshells reinforcement. Still, the tensile strength is quantitatively the same as the bare epoxy. The formation of a cluster of particles during the milling, storage of

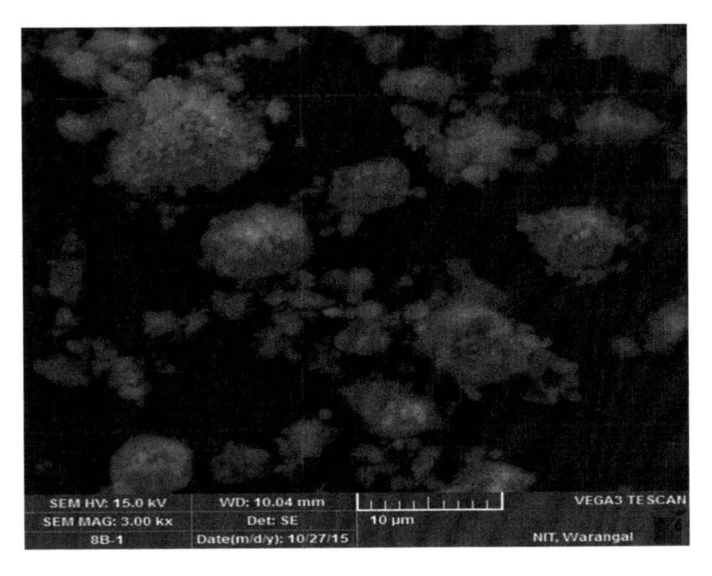

Fig. 3 SEM pictograph of the eggshell particle

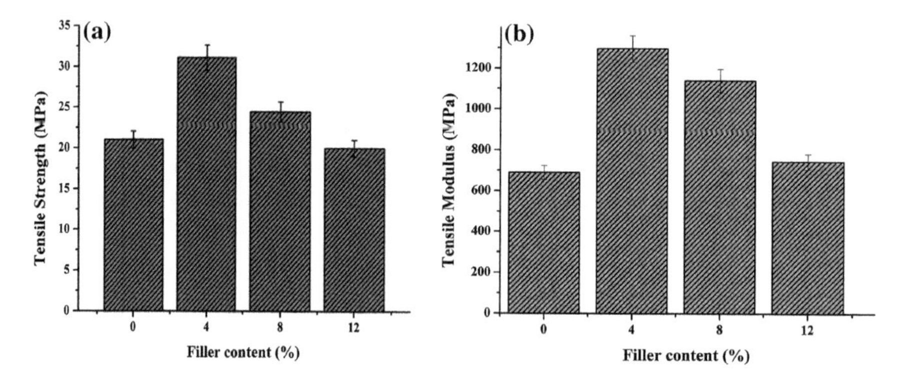

Fig. 4 **a** Variation of the tensile strength (MPa) with filler loading. **b** Variation of tensile modulus (MPa) with filler content

powder (same can be seen from SEM figure) and fabrication of composite, leading to the improper bonding between epoxy and eggshell particles could be the reason behind the reduction in tensile strength with the addition of eggshell particles more than 4 wt%.

From Fig. 4b, it is observed that with the addition of particulate reinforcement, the tensile modulus also increases as compared to pure epoxy like tensile strength. The tensile modulus increases when added up to 12 wt% eggshell particles, further increase in filler percentage results in a reduction in tensile modules when compared with bare epoxy. The highest tensile modulus is found with 4 wt% particulate reinforcement; however, the tensile modulus still higher when compared with bare epoxy. The tensile modulus for 12 wt% reinforcement is almost same as bare epoxy further increase in reinforcement will result in a detrimental effect on tensile modulus, which can be predicted from the trend observed in Fig. 4b. It can be observed from the figure that the tensile modulus increases by 87% when 4 wt% reinforcement is added when compared with bare epoxy. While in the case of 8 wt % and 12 wt% of eggshell reinforcement, the tensile modulus increases by 65% and 7%, respectively, as compared to bare epoxy addition. The reason behind the enhancement in tensile properties could be interfacial bonding between surface reactive groups such as amine/carbonyl group on eggshells' microparticles, and epoxy resin similar results are obtained by Ji et al. [5]. The eggshell microparticles have much large BET surface than macro particles results in better adhesion with epoxy resin [5]. Also, the results of tensile test analysis can be confirmed with machine graphs given in Figs. 5, 6 and 7. The machine graph also shows that the tensile behavior of composite is a combination of elastic and perfectly plastic material. The perfectly plastic behavior is due to the epoxy while the elastic behavior is due to the particles.

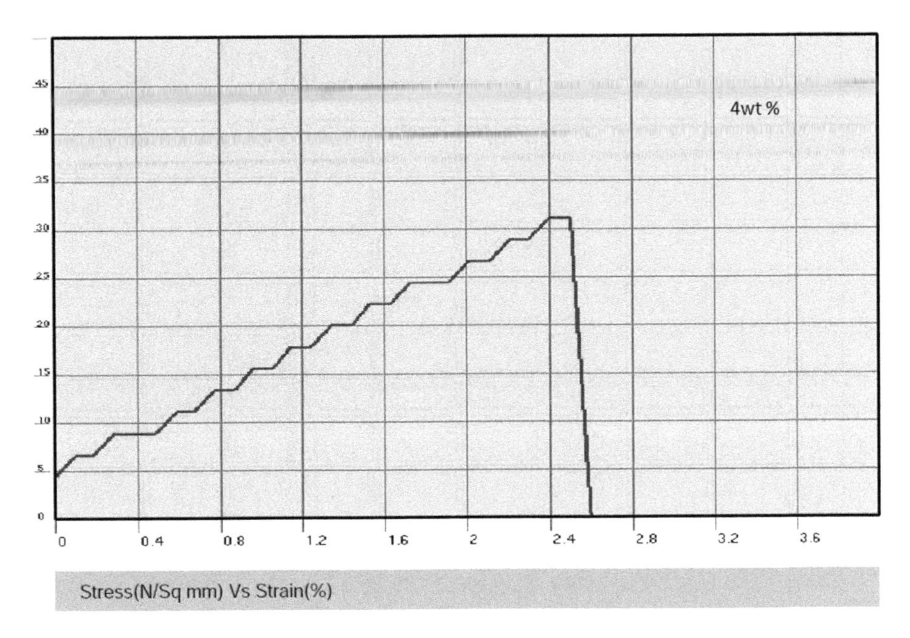

Fig. 5 Machine graph of tensile strength for 4 wt% eggshell–epoxy composite

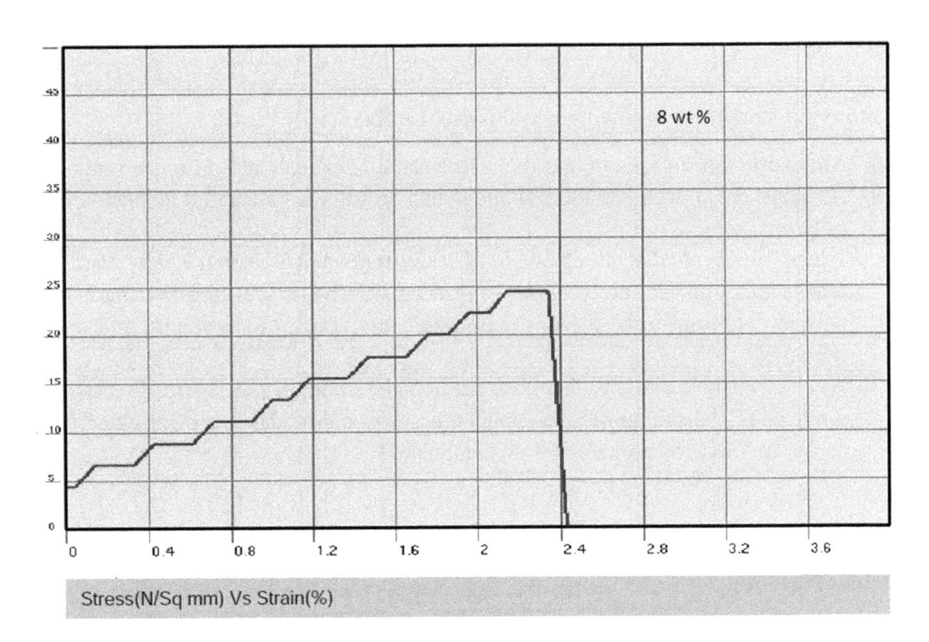

Fig. 6 Machine graph of tensile strength for 8 wt% eggshell–epoxy composite

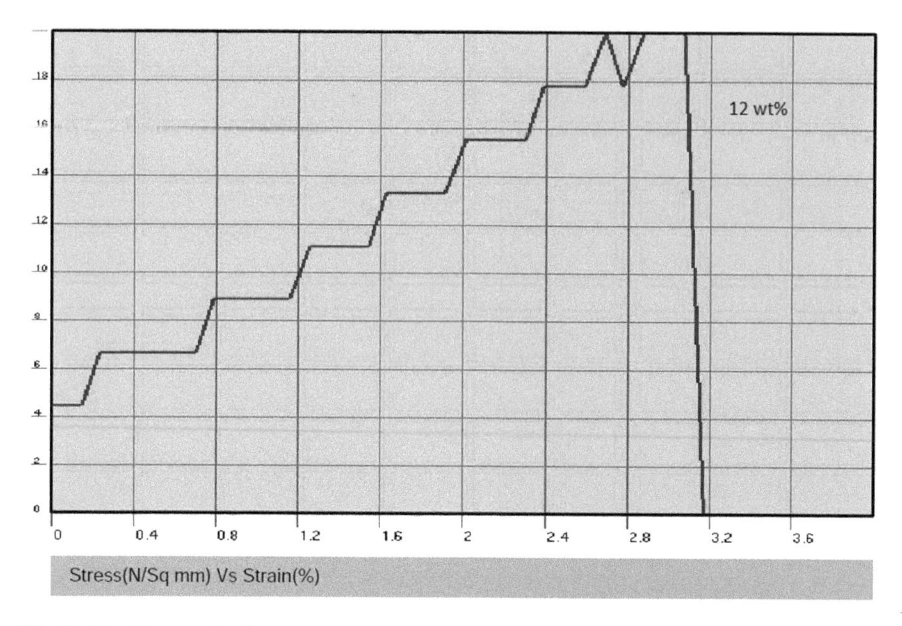

Fig. 7 Machine graph of tensile strength for 12 wt% eggshell–epoxy composite

4 Conclusion

From the study, the following conclusions can be drawn.

1. Microstructure figures reveal the size eggshell particles are in micro range and particles are a well-rounded shape which results in increased adhesion with polymer resin.
2. It was observed that as eggshell filler content increases, tensile strength and tensile modulus of epoxy composite increase. The 4 wt% eggshell particulate addition is found to be the optimum percentage for tensile strength and tensile modulus.
3. As the eggshell particles increase the tensile strength, the future direction for work could be preparation of nanocomposite and/or the use of another fabrication method for preparation of composite.

References

1. Toro P, Quijada R, Yazdani-Pedram M, Arias JL (2007) Eggshell, a new bio-filler for polypropylene composites. Mater Lett 61(22):4347–4350
2. Panchal M, Raghavendra G, Prakash MO, Ojha S, Bose PSC (2018) Moisture absorption behavior of treated and untreated eggshell particulate epoxy composites. Silicon 1–9

3. Chaithanyasai A, Vakchore PR, Umasankar V (2014) The microstructural and mechanical property study of effects of eggshell particles on the aluminum 6061. Proc Eng 97:961–967
4. Yashonant Sai RD, Mohamed FK, Vithun S Fabrication and characterization of aluminum alloy 6351-egg shell composite. Int J Latest Trends Eng Technol (2016). https://doi.org/10.21172/1.73.037
5. Ji G, Zhu H, Qi C, Zeng M (2009) Mechanism of interactions of eggshell microparticles with epoxy resins. Polym Eng Sci 49(7):1383–1388
6. Yang K, Yang Q, Li G, Sun Y, Feng D (2006) Morphology and mechanical properties of polypropylene/calcium carbonate nano composites. Mater Lett 60(6):805–809
7. Maiti SN, Mahapatro PK (1991) Mechanical properties of i-PP/CaCO3 composites. J Appl Polym Sci 42(12):3101–3110
8. Leong YW, Ishak ZM, Ariffin A (2004) Mechanical and thermal properties of talc and calcium carbonate filled polypropylene hybrid composites. J Appl Polym Sci 91(5):3327–3336
9. Xie XL, Liu QX, Li RKY, Zhou XP, Zhang QX, Yu ZZ, Mai YW (2004) Rheological and mechanical properties of PVC/CaCO3 nanocomposites prepared by in situ polymerization. Polymer 45(19):6665–6673
10. Wu D, Wang X, Song Y, Jin R (2004) Nanocomposites of poly (vinyl chloride) and nanometric calcium carbonate particles: Effects of chlorinated polyethylene on mechanical properties, morphology, and rheology. J Appl Polym Sci 92(4):2714–2723

Influence of Laser-Machined Micro-geometrical Features on the Surface Wettability of Stainless Steel 304

Chilaparapu Venkata Vamsi, Vimal Thomas and M. Govindaraju

Abstract Recently, hydrophobic and superhydrophobic natural surfaces have attained great attention due to their diverse properties. It was found that micro-geometrical features on these natural surfaces have a great influence on generating hydrophobicity. Butterfly and Cicada wings are commonly studied natural hydrophobic surfaces for bio-mimicking and their microstructures are simplified into micro-grooves and micro-pillars, respectively. In the present study, a nanosecond fiber laser is used to produce these micro-geometries on stainless steel AISI 304 surface and its effect on surface wettability is investigated. Experimental results show that fabricated micro-geometries have a wetting transition from hydrophilic to the hydrophobic state over time. In this study, a good hydrophobic surface with a large apparent contact angle of 120° is obtained with 150, 200 μm step size micro-groove geometry on day-29 and micro-pillar geometry with 200 μm step size attained the highest contact angle of 118° on day-4. The contact angle of the non-textured parent material surface is 75°.

Keywords Hydrophobicity · Stainless steel AISI 304 · Nanosecond laser · Micro-geometries · Contact angle

1 Introduction

Hydrophobicity refers to a property of surfaces by which they repel water. The phenomenon of hydrophobicity or the state of wetting is characterized by the contact angle (CA) of a water droplet on a solid surface. For a liquid droplet on an ideal flat surface, the angle between solid–liquid and liquid–air interfaces is known as Young's contact angle θ. When $\theta \leq 5°$ for a surface, it is super-hydrophilic

Chilaparapu Venkata Vamsi (✉) · M. Govindaraju
Department of Mechanical Engineering, Amrita School of Engineering, Amrita Vishwa Vidyapeetham, Coimbatore 64112, India

Vimal Thomas
Vikram Sarabhai Space Centre (VSSC), ISRO, Thiruvananthapuram 695022, India

BBVL. Deepak et al. (eds.), *Innovative Product Design and Intelligent Manufacturing Systems*, Lecture Notes in Mechanical Engineering,
https://doi.org/10.1007/978-981-15-2696-1_39

(fully wettable) and for $5° < \theta < 90°$, it is called a hydrophilic surface. For a non-wetting or hydrophobic surface, the condition is $90° < \theta < 150°$ and for a superhydrophobic surface (fully non-wettable), $\theta \geq 150°$.

Controlling wettability of a metallic surface has a wide range of applications like anti-icing [1] drag reduction [2], and corrosion resistance [3]. From previous studies, it is well known that hydrophobicity on a metallic substrate is governed by two major factors, i.e., surface roughness and surface energy [4]. Many researchers created hydrophobic surfaces by applying low surface energy coatings [5] using electrochemical deposition, chemical vapor deposition, and sol-gel method techniques. Coatings include many drawbacks like its durability, environmental hazards, low service life, and high production cost.

Various surface engineering techniques are used for the preparation of hydrophobic surfaces to control the wetting properties of metals including micro-milling [6], lithography [7], and wire-EDM [8]. Lithography is a slow and complicated multistep process with high production cost. Micro-EDM and micro-milling have the ability to produce complex microstructures but it has limitation to the scale of fabrication for industrial applications. Laser surface texturing is a non-contact, single-step, repeatable material processing technique with short processing time, precise control over shape and size of micro-features and has the ability to machine over a specific area of the substrate without affecting entire substrate surface with low risk of contamination.

The present investigation aims to achieve good hydrophobic surface on stainless steel 304 surface with simple micro-geometries inspired from natural hydrophobic surfaces. Many scholars used ultra-short pulse lasers such as femto [9] and picosecond lasers [10] to produce nano/micro-hierarchical structures on metallic surfaces for inducing hydrophobicity. Further, these lasers are highly expensive and require more processing time for fabrication. In this work, cost-effective nanosecond laser which is mostly used in industries is considered. It can produce micro-features with short processing time than the other fabrication methods which were discussed earlier. The selected material has a wide range of industrial applications, such as water pipelines, power generation, food processing, and marine industries, where wettability is a major concern. The effect of step size on different micro-geometries upon the wettability transition time was investigated. The results validate that relatively simple micro-geometries fabricated by nanosecond laser texturing can achieve good hydrophobicity on stainless steel surfaces without any assistance of chemical coatings.

2 Biological Mimicking of Natural Hydrophobic Surfaces

Many hydrophobic surfaces found in nature repel water for a biological advantage, for example, having a non-wettable wing surface butterfly and cicada can fly without drag. It was found that structures existing on micro and nanoscale on these natural surfaces have a prominent role in creating hydrophobicity. Butterfly and

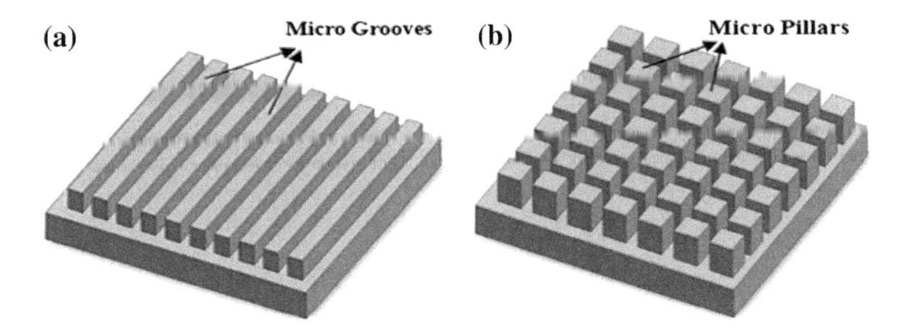

Fig. 1 Schematic diagram of bio-inspired designs: **a** micro-groove structure, **b** micro-pillar structure

cicada wings are the most commonly studied naturally occurring hydrophobic surfaces having distinct micro-geometrical features. Butterfly wings have shingle like microscale structures with ridges and grooves along its surface which directs anisotropic flow (directional dependent) [11] and it was simplified into micro-grooves shown in Fig. 1a and cicada wings are covered with orderly paired waxy pillars [12], which is simplified as micro-pillars in Fig. 1b for micropatterning on stainless steel AISI 304 surface using laser ablation technique.

3 Methodology

3.1 Sample Preparation

Stainless steel AISI 304 cylindrical samples with dimensions of 50 mm diameter × 20 mm thickness are used in this study. To remove oil and other contaminants from the specimen surface, samples were subjected to ultrasonic cleaning with acetone for 20 min and rinsed with de-ionized water at neutral pH to ensure a clean surface.

3.2 Experimental Setup and Fabrication

A 20 W, Q-switched, Nd: YAG, 1064 nm, nanosecond fiber laser machine is used to fabricate the designed micro-geometries on the samples. Following parameters are used for micro-pattering: power—4 W, pulse repetition rate—25 kHz, scanning speed—100 mm/s, and laser passes—500. The beam diameter of the laser is approximately 20 μm. The laser beam moves in x-direction and y-direction with the help of scan head. A suction unit is provided to evacuate the metallic fumes formed during machining.

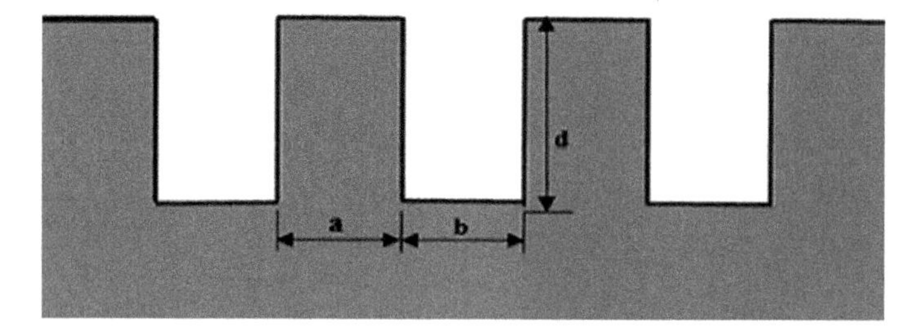

Fig. 2 Schematic of geometric parameters

The distance between two adjacent repeating features is the pitch of the pattern, and half of the pitch is termed as the step size. Width of the laser ablated area is denoted as *b*, and the unmachined island area is denoted by *a* and the depth of the ablated region, *d*, is fixed as 100 μm. In this study, $b/a = 1$. The value of *b* and *a* is varied from 50 to 500 μm for both the designs shown in Fig. 2. These patterns were engraved on the stainless steel samples over an area of 4 mm × 4 mm. An optimum 10 μm scan line spacing is given to ensure a completely machined surface. This aspect will be discussed in later sections.

3.3 Surface Wettability Measurement of Laser Textured Surface

Surface wettability of the textured surface is measured through contact angle measurement using image processing technique. The setup consists of a micro-pipette, charge-coupled device (CCD) camera and a computer with image processing software as shown in Fig. 3.

De-ionized water at neutral pH which does not affect the surface chemistry of patterned surface is used in the measurement. The sample is placed on a flat supporting bed in line with the camera. For samples with micro-grooves, camera and micro-grooves are positioned parallel for measurement. A droplet volume of 2 μL is dispensed onto the textured surface, and dwell time (approx. 10 s) is given for water droplet to settle on the textured surface before measurement. Image of the water droplet on a textured surface is captured by a CCD camera and analyzed using image processing software. After each measurement, water droplet is removed using lint-free paper. A set of 4 readings were taken to minimize errors and an average of these readings is taken as the contact angle (θ).

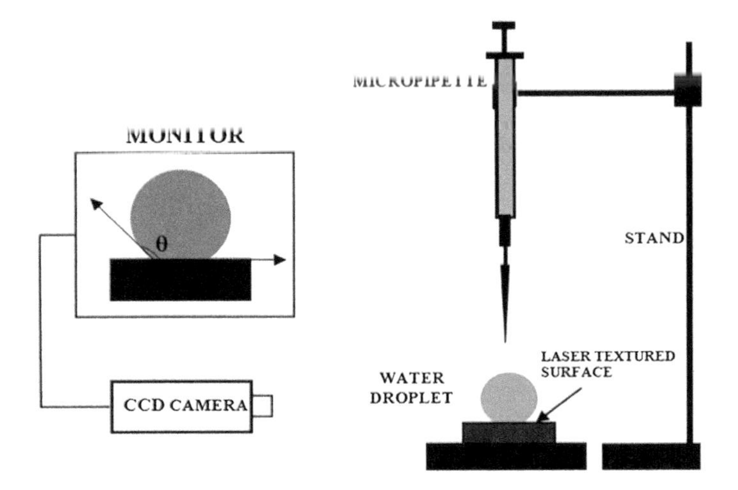

Fig. 3 Schematic of contact angle measurement system

4 Results and Discussion

4.1 Effect of Pulse Overlap and Scan Line Separation

Fabrication of micro-geometrical features is carried out through laser ablation technique (melting and vaporization) of the substrate surface. A laser beam with a spot diameter of ($\sim$20 μm) moves along the predefined scan line path as shown in Fig. 4, with pulse overlaps in lateral and transverse directions to create a textured surface with required micro-geometry on a substrate surface. It is recommended to have pulse overlap above 50% for obtaining a completely machined surface.

The pulse overlaps in the lateral direction shown in Fig. 5a is controlled by scanning speed, pulse frequency, and spot diameter. The mathematical equation of pulse overlaps in a lateral direction [13].

$$\text{Pulse overlap} = \frac{D - Di}{D} \times 100 \quad (1)$$

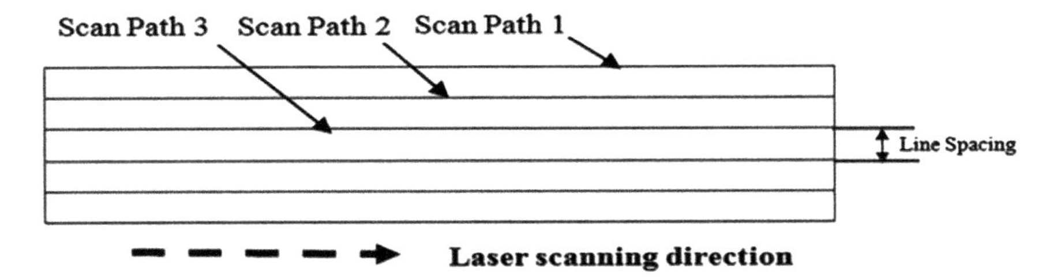

Fig. 4 Schematic of laser beam scan line path with line spacing

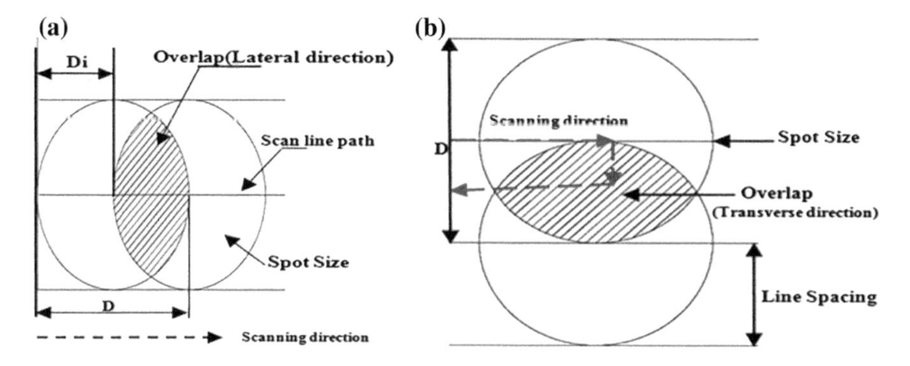

Fig. 5 Schematic representation of **a** lateral and **b** transverse overlaps with the laser scanning direction

where D = spot diameter (μm) $Di = \frac{v}{f}$, v = Scanning speed (mm/s), f = pulse frequency (Hz).

With the parameters, v—100 mm/s and f—25 kHz, lateral direction pulse overlaps of 80% is achieved.

In contrast, pulse overlap in the transverse direction can be controlled through scan line separation between two consecutive laser paths as shown in Fig. 5b. Experiments were conducted to determine pulse overlap in the transverse direction with a scan line spacing of 10, 20, 30, and 40 μm. From the microstructures Fig. 6, it is clear that for line spacing of 30 and 40 μm has some area was left unmachined. For 20 μm spacing, the width of these features decreased but complete machining is not achieved. A completely machined surface is observed with 10 μm scan line spacing. So, the Scan line separation should be less than or equal to the laser spot diameter for a greater pulse overlap. Hence line spacing of 10 μm is selected for this experiment.

4.2 *Effect of Number of Laser Passes on Groove Depth*

The micromorphology of textured samples was characterized by a surface profilometer with a resolution of 1.3 nm and tip size of 2 μm radius conisphere diamond (Form Talysurf PGI). The depth of the micro-feature is one of the key parameters in the micro-geometrical design. The depth is affected by laser power, scanning speed, and laser passes. Out of these, it was found that controlling the laser passes gave more control over depth. It was observed that there was a constant increase in depth along with laser passes as shown in Fig. 7a. Figure 7b shows the average depth obtained in a single laser pass. When the laser passes are less, the average depth obtained is higher and as the number of passes increases, the average depth tends to decrease. This could be due to a higher requirement of laser power to ablate the re-solidified layer.

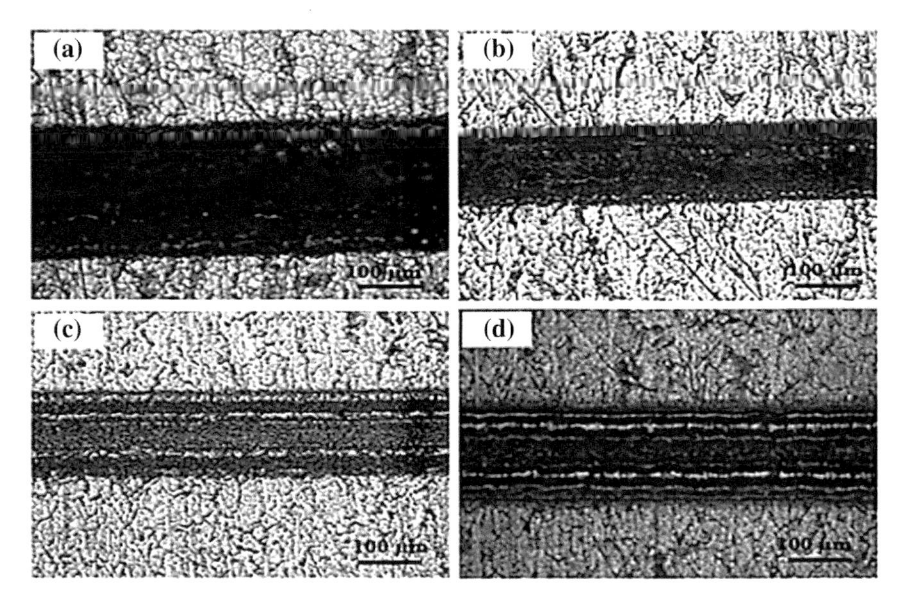

Fig. 6 Microscopic images of laser textured surface with scan line separations, **a** 10 μm, **b** 20 μm, **c** 30 μm, and **d** 40 μm

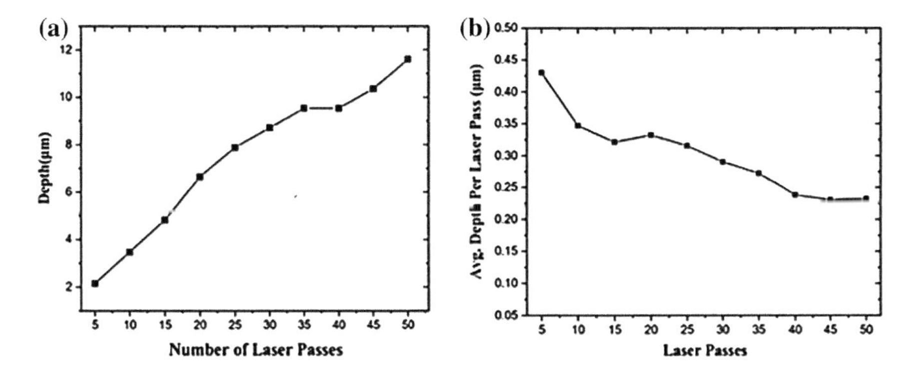

Fig. 7 Effect of laser passes on depth

4.3 *Effect of Microgroove Step Size on Contact Angle*

It is observed that wetting behavior of micro-grooves on stainless steel 304 surface has transformed after laser surface texturing with respect to time under ambient conditions. The graph shown in Fig. 8 demonstrates the CA transition of various textured microgroove surfaces. It is observed that shortly after fabrication, irrespective of step size micro-grooves exhibited hydrophilic property and their CAs

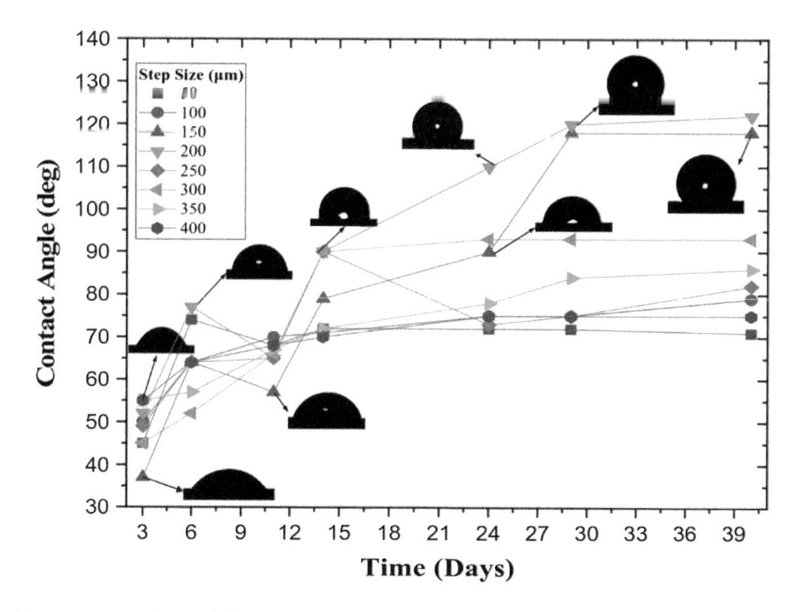

Fig. 8 Illustration of wettability transition of micro-grooves over time

are less than the base metal. However, wettability has altered from the hydrophilic to hydrophobic state with respect to time. The untextured base metal surface kept open to the ambient atmosphere and it exhibited the same hydrophilic state even after 40 days of observation. There is no sign of an increase in water CA on the textured micro-groove surfaces for the first three days after texturing and after that CAs started increasing progressively with time. The graph shows CAs tend to decrease after day-6, and thereafter, there was a significant change in CAs. On day-29, some textured micro-grooves attained hydrophobic state with the highest CA of 120° and 90° on 150, 200 μm and 250, 300 μm micro-groove step sizes, respectively, and has stabilized over time. But in contrast, micro-groove geometries with lower bound step size 50, 100 μm and upper bound step size 450, 500 μm remained hydrophilic even after 2 months. Results reveal that after attaining maximum CA there was no further increase in contact angles.

4.4 *Effect of Micro-pillar Geometry on Contact Angle*

The surface wettability of laser-textured micro-pillars on substrate surface has changed when exposed to an ambient atmosphere with time. Figure 9 plots the change of water CA on various micro-pillar step sizes with time as a function. Interestingly, on day-1, after fabrication micro-pillars with step sizes 150 and 200 μm has reached hydrophobic state. But other step sizes exhibited hydrophobicity with CAs even less than that of base metal. Moreover, the CAs of

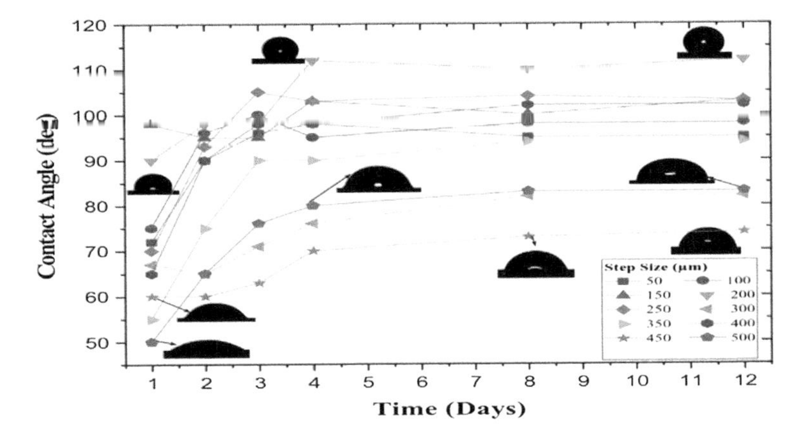

Fig. 9 Illustration of wettability transition of micro-pillars over time

micro-pillars have started increasing rapidly after day-1. Up to day-4, CAs of textured micro-pillar geometries increased and thereafter no further increment is observed. Micro-pillar geometry with step size 200 μm exhibited the highest water CA of 118° and step sizes 50, 100, 150, 200, 250, 350, and 400 μm has also reached hydrophobic state. But, in contrast, micro-pillars with upper bound step sizes 300, 450, and 500 μm are still hydrophilic after day-10. These CAs remain unchanged even after 2 months observation. It was observed that the hydrophobicity on micro-pillar geometry can be attained only with lower bound step sizes.

5 Conclusions

In this paper, nanosecond fiber laser is used to fabricate micro-geometries of grooves and pillars on the SS 304 surface and their influence on surface wettability is investigated. These surfaces exhibited enhanced hydrophobicity than the non-textured substrate surface. In spite of the hydrophilic state after fabrication, a wetting transition is observed in micro-geometries with time and the samples reached hydrophobic state under ambient conditions. Wettability on these patterned surfaces is investigated by measuring contact angles of a water droplet with volume 2 μL and results show that a hydrophobic surface with larger CA of 120° is obtained with 150, 200 μm micro-groove step size on day-29 and a CA of 118° is obtained on 200 μm micro-pillar geometry on day-4. A time-dependent phenomenon is observed in both the micro-geometries before reaching non-wetting state. In this work, we made practically simple geometries for laser surface texturing on stainless steel to attain hydrophobicity and further can be used in industries for large scale production. However, samples with larger step size remained hydrophilic due to the increase in the contact area between textured surface and water. Some studies say that achieving superhydrophobicity with a nanosecond laser may not be possible as these lasers limit the creation of hierarchical structure, unlike pico and femtosecond lasers. Furthermore, mathematical models have to be developed and micromorphology can be studied to understand the phenomenon in-depth.

References

1. Farhadi S, Farzaneh M, Kulinich SA (2011) Anti-icing performance of superhydrophobic surfaces. Appl Surf Sci 257:6264–6269. https://doi.org/10.1016/j.apsusc.2011.02.057
2. Truesdell R, Mammoli A, Vorobieff P, Van Swol F, Brinker CJ (2006) Drag reduction on a patterned superhydrophobic surface. Phys Rev Lett 97:1–4. https://doi.org/10.1103/PhysRevLett.97.044504
3. Rafieazad M, Jaffer JA, Cui C, Duan X, Nasiri A (2018) Nanosecond laser fabrication of hydrophobic stainless steel surfaces: the impact on microstructure and corrosion resistance. Materials (Basel) 11. https://doi.org/10.3390/ma11091577
4. Nishino T, Meguro M, Nakamae K, Matsushita M, Ueda Y (1999) La981727S.Pdf, pp 4321–4323. https://doi.org/10.1021/la981727s
5. Sankar S, Nair BN, Suzuki T, Anilkumar GM, Padmanabhan M, Hareesh UNS, Warrier KG (2016) Hydrophobic and metallophobic surfaces: Highly stable non-wetting inorganic surfaces based on lanthanum phosphate nanorods. Sci Rep 6:4–10. https://doi.org/10.1038/srep22732
6. Zhang X, Xu J, Lian Z, Yu Z, Yu H (2015) Influence of microstructure on surface wettability. In: International conference on advanced mechatronic systems (ICAMechS 2015), pp 217–222. https://doi.org/10.1109/ICAMechS.2015.7287112
7. Alrasheed S, Candeloro P, Allione M, Limongi T, Malara N, Francardi M, Miele E, Di Fabrizio E, Giugni A, Di Vito A, Raimondo R, Torre B, Schipani R, Mollace V (2015) Photolithography and micromolding techniques for the realization of 3D polycaprolactone scaffolds for tissue engineering applications. Microelectron Eng 141:135–139. https://doi.org/10.1016/j.mee.2015.02.030
8. Bae WG, Song KY, Rahmawan Y, Chu CN, Kim D, Chung DK, Suh KY (2012) One-step process for superhydrophobic metallic surfaces by wire electrical discharge machining. ACS Appl Mater Interf 4:3685–3691. https://doi.org/10.1021/am3007802
9. Long J, Pan L, Fan P, Gong D, Jiang D, Zhang H, Li L, Zhong M (2016) Cassie-state stability of metallic superhydrophobic surfaces with various micro/nanostructures produced by a femtosecond laser. Langmuir 32:1065–1072. https://doi.org/10.1021/acs.langmuir.5b04329
10. Wang A, Jiang L, Li X, Xie Q, Li B, Wang Z, Du K, Lu Y (2017) Low-adhesive superhydrophobic surface-enhanced Raman spectroscopy substrate fabricated by femtosecond laser ablation for ultratrace molecular detection. J Mater Chem B 5:777–784. https://doi.org/10.1039/c6tb02629j
11. Bixler GD, Bhushan B (2014) Rice- and butterfly-wing effect inspired self-cleaning and low drag micro/nanopatterned surfaces in water, oil, and air flow. Nanoscale 6:76–96. https://doi.org/10.1039/c3nr04755e
12. Wan Y, Lian Z, Liu Z, Yu H (2014) Cicada wing with adhesive superhydrophobicity and their biomimetic fabrication. In: 2014 IEEE international conference on mechatronics and automation, IEEE ICMA 2014, pp 780–785. https://doi.org/10.1109/ICMA.2014.6885796
13. Marczak J (2015) Micromachining and pattering in micro/nano-scale on macroscopic areas. Arch Metall Mater 60:2221–2234. https://doi.org/10.1515/amm-2015-0368

Generation of Slip Line Fields Incorporating BUE and Shear Zone to Model Machining Using MATLAB

Hridayjit Kalita and Kaushik Kumar

Abstract Slip line field theory for modeling of machining process provides an analytical method where unknown slip lines are first generated assuming an initial base slip line and then are later evaluated considering different cutting and boundary conditions. All the slip lines are inter-dependent on each other and the curvature of these slip lines is altered with a slight change in the condition of cutting. This enables an inclusive and robust model to develop which can take into account the curling of the chip, frictional condition at the tool rake face and elastic condition at the upper end of the shear zone. In the current paper, an attempt to construct the slip lines using MATLAB has been presented based on the model given by Fang and Dewhurst and incorporating an additional shear zone into it. This model can detect buildup edge (BUE) of a larger size which influences directly the cutting forces and the curvature of the slip lines. The slip lines generated are studied and the characteristic of these can be observed by altering the input parameter values such as hydrostatic pressure at a fixed location, the rake angle of the tool or the input slip line angular ranges.

Keywords Slip line field · Machining · Modeling · Shear zone · Buildup edge · Sticking region · Curvature · MATLAB

1 Introduction

Determination of machining forces has always been a matter of research subject due to its complex material behavior with multi-factor dependencies and therefore appropriate prediction of it is hard to obtain. A number of constitutive models based on experimental observations have been formulated which display a predictable nature of the material flow in machining that is inclusive to a selective number of materials or to all materials. The recent developments in the constitutive models in

H. Kalita · K. Kumar (✉)
Birla Institute of Technology, Mesra, Ranchi, India
e-mail: kkumar@bitmesra.ac.in

BBVL. Deepak et al. (eds.), *Innovative Product Design and Intelligent Manufacturing Systems*, Lecture Notes in Mechanical Engineering,
https://doi.org/10.1007/978-981-15-2696-1_40

machining including the tool-chip frictional interface models have been detailed in [1] where the data or relations for the modeling were obtained by conducting selective experiments.

These constitutive models take into consideration the strain values at each point in the complex plastic material regions and therefore can be numerically solved using Finite Element Method (FEM) models. The appropriateness of these FEM models depends on the size and number of the elements taken and the computational time required for a better effective modeling procedure. Due to the strain depending criteria in FEM models, the number of calculations to be obtained for each region in machining is large and time-consuming. Few examples of constitutive models incorporating the causes and effects of the BUE formation in machining can be taken from the literature [2–4] which tends to be the main concern of this paper since a number of machining factors such as fluctuation in the cutting forces, tool wear, temperature effect and surface finish of the work material are all dependent on the size of it. Constitutive models are based on a study on a fixed region in machining and do not provide an inclusive inter-dependent material flow model.

Slip lines are usually employed for defining plastic and plane strain conditions in rolling, strip drawing, slab extrusion, etc. [5, 6]. Since in orthogonal cutting of machining the plastic regions can be assumed to be in a plane strain condition and from the hill's feasibility approach of solving the slip lines [Hill's], slip lines can be very well suited even to machining. Slip line method provides an analytical method to analyze the material behavior under different boundary and cutting conditions. Here, strain values need not be found for every region in machining (as in the case of constitutive models) and stress values (or boundary conditions) at fixed predetermined locations can be used to modify the characteristics of the slip line fields.

Since upon application of cutting loads in machining, above the yield stress the material plastic strain becomes significant as compared to the elastic strain and a rigid plastic condition is attained. Due to this, the material is subjected to zero volume change within the rigid plastic region in machining, the behavior of which can be analyzed only in terms of the stress state in accordance with the Tresca and von mises criterion [7]. From Hencky's first theorem, it is well established that slip lines can be well defined in terms of hydrostatic and maximum shear flow stresses (independent of strain) which is another reason for the adoption of slip line field method in machining.

The first slip line model in machining was given by Dewhurst [8] which takes into account the chip curl effect by adopting a curvature in the primary shear line with a flat-faced tool. Lee and Shaffer [9] in their model incorporated a stagnation region at the tool-chip interface near the cutting edge but did not take into account the chip curl effect and assumed a straight chip. Moreover, the size detected in the Lee and Shaffer's model of the buildup edge (BUE) region (or the stagnation region) was small. In the current paper, the slip line model given by Fang and Dewhurst [10] which detects a BUE region much larger than the size detected by Lee and Shaffer has been utilized to model the slip line fields in machining for force determination and other outputs using MATLAB, though only the slip line characteristics had been discussed in this paper. The comparison between the Fang and Dewhurst model and the modified model has been discussed in detail in Sect. 2. The modeling procedure, results, and discussion have been described in the later sections.

2 Slip Line Field Model

Fang and Dewhurst's slip line field model [10] very well take into account a large area of the buildup of material on the tool rake surface which directly governs a number of machining outcomes as described in the previous section. It can take into account the chip curl effect and conforms to the velocity continuity and stress equilibrium conditions.

A modified model has been constructed with reference to the Fang and Dewhurst's model where a shear zone has been incorporated instead of a shear plane and the conditions were assumed for the maximum buildup edge size. The shear zone in the modified model can be denoted by a region bounded by two curved slip lines AB and A1B as shown in Fig. 1(right) which represents the velocity discontinuity across the shear zone. This idea of incorporation of a shear zone can be taken from the slip line model given by Fang [11] where a parallel curved shear zone was found to be dependent on the tool edge geometry (or roundness) and is affected by the frictional conditions and the location of the stagnation point. This model includes a large number of slip lines to define the complex material flow within the plastic region. In another analysis of the deformation region in machining using slip line field method by Jin and Altintas [12], incorporates two parallel sided shear zone which is curved upward (having the same curvature) due to material pressure during flow and the thickness of which depends on the rounded edge of the cutting tool as in the case of previously mentioned Fang's model. In the present model, a BUE region is incorporated instead of the tool rounded edge and a singularity condition has been taken at the front edge of the BUE at point C where a slip line fan region represented by BCD originates representing the material backflow due to surface adhesion on the rake face. The assumption of a singular region at point C eliminates the requirement of any field to be extended to the tool cutting edge which in this case has been taken as a pointed cutting edge with contrast to the rounded cutting edge model suggested in [12]. The remaining region at the cutting edge and rake surface of the tool has been assumed to be filled with work hardened or buildup edge (BUE) material in the present model represented by CHE. The length and height of the BUE can be represented by HE and CH in Fig. 1. The curvilinear quadrilateral region CDFE representing the material shear flow zone along the tool-chip interface (or the transition region between the chip and stagnant region material) is defined between the BUE and the region BCD.

The sliding and sticking lengths on the rake surface took in the model by Jin and Altintas [12] have been omitted with a constant value of the parameter ($\tau/k = 1$)

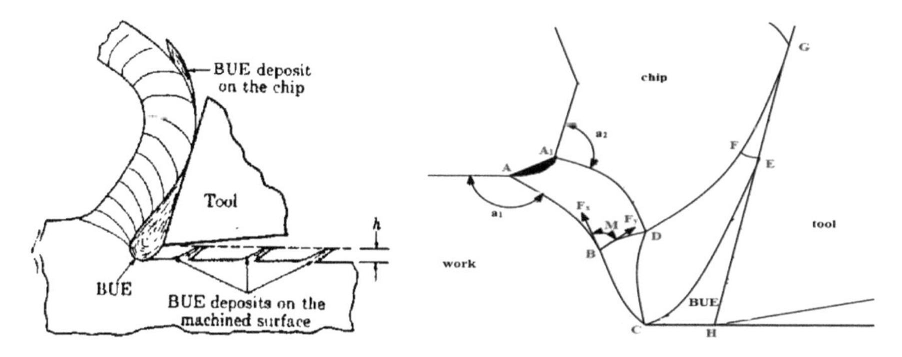

Fig. 1 BUE formation in machining (left) [15], modified slip line model (right)

taken in the present model throughout the length of the rake-chip interface. The region FGE is governed by the frictional condition in the sliding length of the chip on the rake surface (GE) and the slip line CE is assumed to meet at 'E' parallel to rake surface to avoid any singular region formation and to avoid requirement of an additional fan region as in the case of Fang's model [11]. The angle made by the slip line AB with the un-machined surface can be determined considering an elastic condition (or overstressing limit) at point 'A.'

Dewhurst-Collin's matrix technique [13] to solve for the slip lines using matrix operators has been employed and an iterative solution for evaluating the initial base slip line 'BD' is adopted. MATLAB functions were generated as will be described in detail in the next section to evaluate for the matrix operators, generation of Mikhlin and Cartesian co-ordinates. The input to the overall program includes the angular ranges (υ and θ as shown in Fig. 1), frictional condition ($\tau/k = 1$), where 'τ' is the shear stress along the tool rake interface with and 'k' is the maximum shear flow stress, hydrostatic pressure at A (P_{A}), rake angle of the tool the (γ) and the angular velocity (ω) of the chip. As given by Fang and Dewhurst [10], the slip line denoted by the angular range 'υ' has been found to directly influence the size of the BUE. With higher the value of the angular range 'υ' of the slip line 'FD,' larger is the size of the BUE with 'zero' value for no BUE formation. On the other hand, the slip line represented by the angular range 'θ' was found to have no effect on the size of the BUE.

3 Generation of the Slip Lines

The MATLAB program generated involves a number of functions created to define each aspect in the mathematical formulation of the slip lines. These functions are described in detail below.

3.1 Function for Defining the Angular Ranges of the Slip Line

In this function, the four angular ranges which are ψ, λ, υ and θ are defined with the maximum value of ψ given by the relation, $\psi = \gamma + \upsilon - \eta$, where γ is the rake angle, η is the angular range of the slip line 'BC' and υ is the angular range of the slip line GD. This condition is satisfied in order to ensure material flow into the slip line fan region BCD. The angular ranges can be related employing the expression for the overstressing angle (a_1) as shown in Fig. 1, found from the hodograph diagram given in [10] and is given by $a_1 = pi + \theta - \eta - \gamma - \upsilon + \psi$. The overstressing angles a_1 and a_2 can be obtained considering an elastic state of material stress at point 'A' and the condition is given in Eqs. 1 and 2:

$$\frac{\pi}{2} - 1 - 2a_1 \leq \frac{P_A}{k} \leq 2a_1 - \frac{3\pi}{2} + 1 \quad (1)$$

$$2\cos\left(a_2 - \frac{\pi}{4}\right) - 1 \leq \frac{P_{A1}}{k} \leq 1 + 2\left(a_2 - \frac{\pi}{4}\right) \quad (2)$$

The term $\frac{\pi}{2} - 1 - 2a_1 \leq \frac{P_A}{k}$ is of less significance and can be omitted as it is found to be satisfying in most of the cases [14].

3.2 Functions for the Generation of the Matrix Operators

The functions for each matrix operators given by [13] such as P^*, Q^*, R (reversion matrix operator) and S (shift operator) are generated to be used in the later functions for the construction of the slip lines of the modified model. The reversion operator is to be used when the intrinsic directions of the slip lines derived from the base slip line column vectors need to be reversed. Shift operator is employed when the origin of the slip lines needs to be translated through an angular distance 'ϕ' along the intrinsic direction of the slip line. The matrix forms of these operators can be given as below where 'ϕ' is the angle turned by the slip line during curving.

$$P_\varphi^* = \begin{vmatrix} \phi_0 & 0 & 0 \\ \phi_1 & \phi_0 & 0 \ldots \\ \phi_2 & \phi_1 & \varphi_0 \end{vmatrix}, \quad Q_\varphi^* = \begin{vmatrix} \phi_1 & \phi_2 & \phi_3 \\ \phi_2 & \phi_3 & \phi_4 \ldots \\ \phi_3 & \phi_4 & \phi_5 \end{vmatrix}, \quad R_\phi = \begin{vmatrix} \phi_0 & \phi_1 & \phi_3 \\ 0 & -\phi_0 & -\phi_1 \ldots \\ 0 & 0 & \phi_0 \end{vmatrix}$$

$$\text{and } S_\phi = \begin{vmatrix} \phi_0 & \phi_1 & \phi_2 \\ 0 & \phi_0 & \phi_1 \ldots \\ 0 & 0 & \phi_0 \end{vmatrix}$$

where $\phi_m = \frac{\phi^m}{m!}$.

3.3 Function for the Generation of the Operators for the Modified Slip Lines

The above-defined matrix operators are now utilized to construct the operators such as $P_{\eta\upsilon}, Q_{\eta\upsilon}, P_{\psi\eta}$ and $Q_{\psi\eta}$ for the generation of the modified slip line fields. The expressions for these are given by Eqs. 3–6:

$$Q_{\psi\eta} = R_{\psi}Q_{\eta} \tag{3}$$

$$P_{\psi\eta} = R_{\eta}P_{\psi} \tag{4}$$

$$P_{\eta\upsilon} = R_{\upsilon}P_{\eta} \tag{5}$$

$$Q_{\eta\upsilon} = R_{\eta}Q_{\upsilon} \tag{6}$$

Apart from the above-generated operators, a '*G*' operator to construct the slip line field over the sliding frictional region (EFG) has also been generated which is basically a straight boundary operator. The expression for the 'G_{λ}' is as given in Eq. 7:

$$G_{\lambda} = Q_{\upsilon\upsilon} + P_{\upsilon\upsilon}(I * \cos\lambda - J\sin\lambda)^{-1}(J\cos\lambda - I\sin\lambda) \tag{7}$$

where $\lambda = \frac{1}{2}\cos^{-1}(\tau/k)$ '*I*' is a unit matrix and $J = \begin{vmatrix} 0 & 0 & 0 & \\ 1 & 0 & 0 & \cdots \\ 0 & 1 & 0 & \end{vmatrix}$

$Q_{\upsilon\upsilon}$ and $P_{\upsilon\upsilon}$ are evaluated in the same way as already described in the previous function.

3.4 Function for the 1st Iteration of the Velocity Discontinuities Along the Slip Lines and the Slip Lines

For generating the velocity fields along the slip lines in the current model, firstly an initial slip line is chosen as the base slip line which in this case is the slip line 'BD' and the column matrix can be denoted as 'σ.' The velocity discontinuity along the slip line 'ABC' is taken as ρ_1 [10] (with contrary to the Jin and Altintas model) which is given a unit value and ρ_2 can be obtained in relation to ρ_1 from the hodograph diagram given in [10]. The velocity fields along the slip line GF, DF, EF and CD are represented as lowercase letters and are formulated as given by the Eqs. 8–10:

$$cd = P_{\psi\eta}\rho_1 + Q_{\psi\eta}\sigma \tag{8}$$

$$ef = P_{\upsilon} * cd + Q_{\upsilon}\rho_2 \tag{9}$$

$$df = P_{\eta}\rho_2 + Q_{\eta} * cd \tag{10}$$

Velocity field along the slip line GF is generated employing the G_{λ} operator.

$$gf = G_{\lambda} * ef \tag{11}$$

After the velocity discontinuity 'gf' is formulated, the radius vector 'GF' can be determined by the expression, $GF = \frac{gf}{\omega}$, where 'ω' is the angular velocity of the chip. The other slip line radius vectors are given in Eqs. 12–17:

$$EF = G_{\lambda} * GF \tag{12}$$

$$DF = \frac{df}{\omega} \tag{13}$$

$$CE = P_{\eta\upsilon} * FD + Q_{\eta\upsilon} * FE \tag{14}$$

$$CD = P_{\eta\upsilon} * FE + Q_{\eta\upsilon} * FD \tag{15}$$

$$BD = Q_{\psi\eta} * CD \tag{16}$$

$$BC = P_{\psi\eta} * CD \tag{17}$$

3.5 *Function for the Final Slip Lines and Velocity Discontinuities*

After the first iteration, the 'σ' is re-evaluated which can be expressed as $\sigma = \omega * BD$. The other slip line velocity fields and slip lines can be re-evaluated using Dewhurst and Collin's matrix technique as already shown above.

3.6 *Function for the Mikhlin Co-ordinates of the Slip Lines in the Modified Model*

Two functions have been generated for evaluating the coefficients of the Mikhlin co-ordinates and the Mikhlin co-ordinates. The equations for the Mikhlin co-ordinates can be given by Eqs. 18 and 19:

$$\bar{x}(\phi) = \sum_{n=0}^{\infty} t_n \left(\frac{\phi^n}{n!} \right) \quad (18)$$

$$\bar{y}(\phi) = \pm \sum_{n=0}^{\infty} t_n \left(\frac{\phi^{n+1}}{(n+1)!} \right) \quad (19)$$

The plus–minus sign in the y Mikhlin co-ordinates is for taking into account the anti-clockwise or clockwise curving of the slip lines. The coefficient (t_n) can be expressed in relation to the coefficient of the radius of curvature of the slip lines, $R(\phi) = \sum_{n=0}^{\infty} r_n \left(\frac{\phi^n}{n!} \right)$ as given by Eq. 20:

$$t_{n+1} - t_{n-1} = |r_n| \quad t_0 = 0, t_1 = |r_0| \quad (20)$$

3.7 *Function for the Cartesian Co-ordinates of the Slip Lines in the Modified Model*

Cartesian co-ordinates of the slip lines can be evaluated for the final plotting of the slip lines from its Mikhlin co-ordinates as given by Eqs. 21 and 22:

$$x = \bar{x}\cos\phi - \bar{y}\sin\phi \quad (21)$$

$$y = \bar{x}\sin\phi + \bar{y}\cos\phi \quad (22)$$

4 Results and Discussion

The above slip lines generated can be plotted in terms of Cartesian co-ordinates of the slip lines. In the present program, the input parameters that have to be initially fed into are already described in Sect. 2. As already mentioned in [10], the size of the BUE (or the length of the slip line CE) tends to increase in its size with an increase in the value of ‘υ’ is evident from Fig. 2 (left) as shown.

The curvature of the other slip lines (Fig. 2, right) is also observed to be varying with the change in the value of the ‘υ’ such as the decreasing value of the curvature of the slip line A1D with increasing ‘υ.’

With the increasing value of the hydrostatic pressure at ‘A,’ the curvature of the slip line A1D is found to be increasing as shown in Fig. 3.

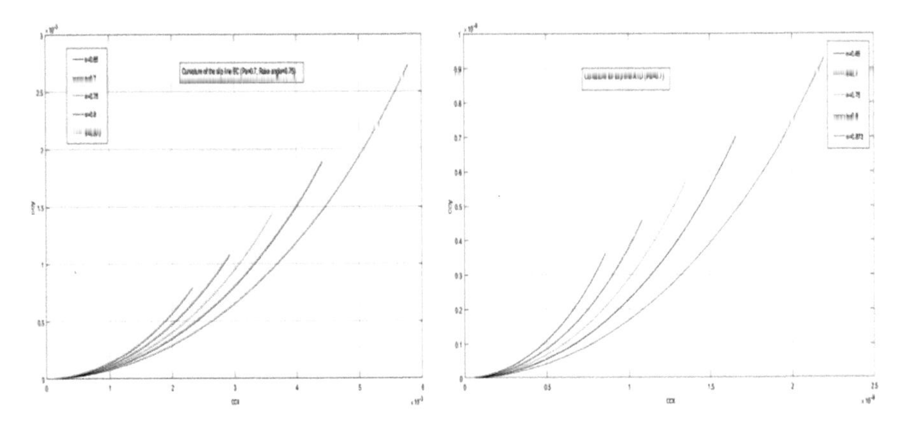

Fig. 2 Change in curvature of the slip line 'CE' with a change in values of 'υ' (left) and change in curvature of the slip line 'A1D' with a change in 'υ' (right)

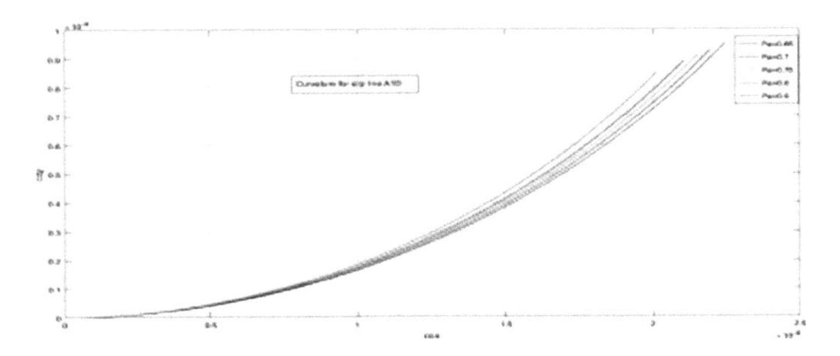

Fig. 3 Change in the curvature of the slip line

Now by the assumption taken of the velocity discontinuity of the slip line ABC to be 'ρ_1' (with contrary to the different curvature of the slip lines at the top upward curving and the bottom downward curving along the primary shear region in the rounded cutting edge slip line model [12]), the increase in curvature of the slip line A1D would also mean an increase in curvature of the slip line BC and eventually CD. This results in larger BUE height leading to larger cutting force.

The rake angle was found to have no effect on the slip lines except for the slip line BD, the curvature of which has a direct influence on the cutting forces. With an increase in rake angle, the curvature of the slip line BD is reduced and the length increases which can be observed from Fig. 4. There is a mention of a triangular region on the upper portion of the shear zone in the model given by Jin and Altintas [12], which is bounded by a stress-free surface on top and a quadrilateral region in the shear zone. The material in this region moves down due to downward flow at an angle of 45° to the free surface (as it is a frictionless surface). This region is omitted in the current model and replaced by a downward curvature of the slip line AA1, indicating a downward material pressure.

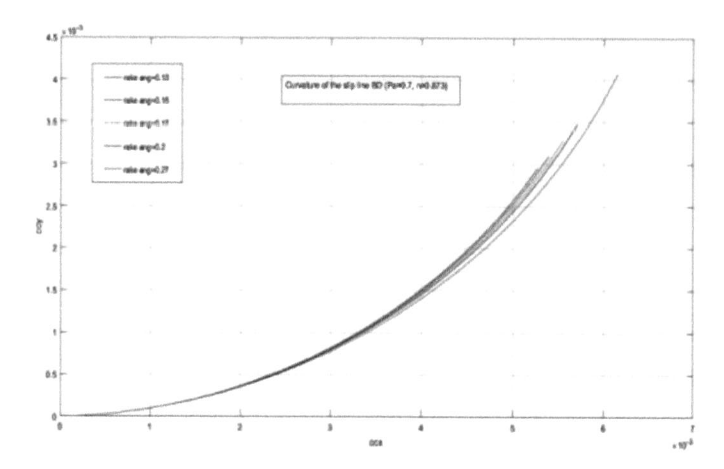

Fig. 4 Change in curvature of the slip 'A1D' with a change in hydrostatic pressure at A. Line 'BD' with a change in rake angle

5 Conclusion

The modeling of machining using slip line field theory provides an analytical method where unknown slip line fields are first assumed and later on evaluated for the given boundary conditions. This model can take into account all the complex material regions in machining and eliminates the need for different constitutive laws for defining complex flow through the regions of shear zone and tool-chip interface. In the current paper, a program is generated on MATLAB to generate the slip line fields for a modified model of the previously given Fang and Dewhurst model. A shear zone region is incorporated in the current model which is in accordance with the model given in [12] and a singular region is assumed at the front of the BUE. The characteristics (curvatures) of the slip lines with a change in boundary conditions in terms of the rake angle, hydrostatic pressure at A and angular range (υ) of the slip line FD have been discussed in this paper.

References

1. Melkote SN, Grzesik W, Outeiro J, Rech J, Schulze V, Attia H, Arrazola PJ, M'Saoubi R, Saldana C (2017) Advances in material and friction data for modeling of metal machining. CIRP Annals Manuf Technol 66:731–754
2. Uhlmann E, Henze S, Bromelhoff K (2015) Influence of the built-up edge on the stress state in the chip formation zone during orthogonal cutting of AISI1045. Proc CIRP 31:310–315
3. Atlati S, Haddag B, Nouari M, Moufki A (2015) Effect of the local friction and contact nature on the built-up edge formation process in machining ductile metals. Tribol Int 90:217–227

4. Gomez-Parra A, lvarez-Alco MA, Salguero J, Batista M, Marcos M (2013) Analysis of the evolution of the built-up edge and built-up layer formation mechanisms in the dry turning of aeronautical aluminium alloys. Wear 302:1209–1218
5. Quagliato L, Berti GA (2017) Temperature estimation, and slip-line force analytical models for the estimation of the radial forming force in the RARR process of flat rings. Int J Mech Sci 123:311–323
6. Quagliato L, Berti GA, Kim D, Kim N (2018) Slip line model for forces estimation in the radial-axial ring rolling process. Int J Mech Sci 138–139:17–33
7. Freddi F, Carfagni GR (2016) Phase-field slip-line theory of plasticity. J Mech Phys Solids 94:257–272
8. Dewhurst P (1978) On the non-uniqueness of the machining process. Proc R Soc London Ser A 360:587–610
9. Lee EH, Shaffer BW (1951) The theory of plasticity applied to a problem of machining. Trans ASME 18:405–413
10. Fang N, Dewhurst P (2005) Slip-line modeling of built-up edge formation in machining. Int J Mech Sci 47:1079–1098
11. Fang N (2003) Slip-line modeling of machining with a rounded-edge tool—part I: new model and theory. J Mech Phys Solids 51:715–742
12. Jin X, Altintas Y (2011) Slip-line field model of micro-cutting process with round tool edge effect. J Mater Process Technol 211:339–355
13. Dewhurst P, Collins IF (1973) A matrix technique constructing slip-line field solutions to a class of plane strain plasticity problems. Int J Numer Meth Eng 7:357–378
14. Fang N, Jawahir IS, Oxley PLB (2001) A universal slip-line model with non-unique solutions for machining with curled chip formation and a restricted contact tool. Int J Mech Sci 43:557–580
15. Some built-up edge (BUE) is normally encountered in machining (2010) Engineering, Shop Floor. http://web.mit.edu/2.670/www/Tutorials/Machining/physics/bue.gif

Review on Magnesium Alloy Processing

Pradipta Kumar Rout, Pankaj C. Jena, Girija Nandan Arka
and B. Surekha

Abstract Demand to design engineering components with lightweight materials is a big challenge for the twenty-first-century engineer. Many researchers of the world have been trying hard to innovate new lightweight and energy-efficient alloy materials to replace steel and other metallic alloys. So far magnesium alloys are found fit for the purpose. But its alloying process is a little bit tough as molten magnesium is more prone to oxidation. Therefore, to improve its corrosion resistance and mechanical properties, researchers have been trying to establish new alloy systems in order to fulfill requirements from different industries especially automobile, aerospace, and health sector. Magnesium melting techniques, different Mg–alloy systems, and effect of alloying elements on the alloy have been discussed in this paper.

Keywords Magnesium alloy · Flux · Phase diagram · Mechanical property

1 Introduction

Though ferrous alloys have superior mechanical properties, their strength and density ratio is not satisfactory for many engineering applications. Ferrous alloys are also more prone to oxidation and corrosion. Hence, researchers are showing

P. K. Rout (✉) · B. Surekha
Kalinga Institute of Industrial Technology, Deemed to Be University,
Bhubaneswar 751024, India
e-mail: pradiptakumar.rout@kiit.ac.in

B. Surekha
e-mail: surekhafme@kiit.ac.in

P. C. Jena
Veer Surendra Sai University of Technology, Burla, Odisha 768018, India

G. N. Arka
Centurion University, Bhubaneswar 752050, India
e-mail: girija.arka@cutm.ac.in

BBVL. Deepak et al. (eds.), *Innovative Product Design and Intelligent Manufacturing Systems*, Lecture Notes in Mechanical Engineering,
https://doi.org/10.1007/978-981-15-2696-1_41

more interest to explore new lightweight alloy and composite. There is a huge demand for high strength and lightweight materials from aerospace, automobile, defense, and health sectors. These sectors are focusing on low-cost machining, high durability with high strength, and low weight. Therefore, it becomes a gray area for researchers to fulfill the requirements of such industries.

Because of the above reasons, aluminum and its alloys are the first choices of engineers to design lightweight machine components. Aluminum alloys have been widely used and become more popular in the twenty-first century. Though aluminum is a very good lightweight material, still researchers have not stopped to find new materials. In this context, magnesium is lighter in weight than aluminum. It has a good strength–density ratio. Therefore, magnesium alloy is slowly replacing aluminum alloy. The density of magnesium is 1.739 g/cm^3 whereas aluminum has 2.7 g/cm^3. Thus, the weight of the magnesium is almost two-third that of aluminum. Magnesium is found abundantly in the earth's ocean. The magnesium crystal structure is closed packed hexagonal structure like zinc and titanium. So it is difficult to deform at room temperature because of less slip system at lower temperature. Its plastic deformation is more complicated than a cubic lattice structure of aluminum, steel, and copper. These alloys are used where great strength is not required but are used in thick section form or where high stiffness is required. Hence, housing for aircraft, rotating, or reciprocating machine parts are made of magnesium alloys. Many automotive companies like Toyota Motors, Ford Motor Company, Volkswagen, and General Motor are using magnesium alloys as spare parts of their vehicle [1]. Usage of magnesium alloys results in 22–70% weight reduction in an automobile vehicle [2]. Magnesium has excellent flowability and solidification characteristics over other materials. Also, magnesium has excellent damping capacity, and it is nonmagnetic and nontoxic.

2 Magnesium Alloys

Pure metals are rarely used for engineering applications. Alloying elements are added to pure metal to enhance its mechanical characteristics. Most compatible alloying elements for Mg are aluminum and zinc. Silicon, zirconium, manganese, and some rare-earth metals also can be used as an alloying element in magnesium alloys [3]. Significance of alloying elements in Mg alloys is given in Table 1 [4].

2.1 Mg–Al System

This is the most important binary alloy which is widely used in industries. Many researchers have already studied its solidus, liquidus, and solvus lines of Mg–Al system. The maximum solubility of Al in Mg ranges from about 2.1–12.6% by weight at room temperature and eutectic point (437 °C), respectively. Figure 1

Table 1 Significance of alloying elements in Mg alloy

Alloying element	It's significance
Aluminum	- Improves castability - Enhances corrosion resistance - Increases tensile strength and hardness
Zinc	- Refines grain structure - Increases tensile strength and hardness - Improves castability - Enhances corrosion resistance
Silicon	- Increases molten metal viscosity - Increases creep resistance - Forms Mg_2Si particles - But decreases corrosion resistance and castability
Manganese	- Improves corrosion resistance and reduces the iron effect - Enhances yield strength
Rare-earth metals	- Increases tensile strength and hardness

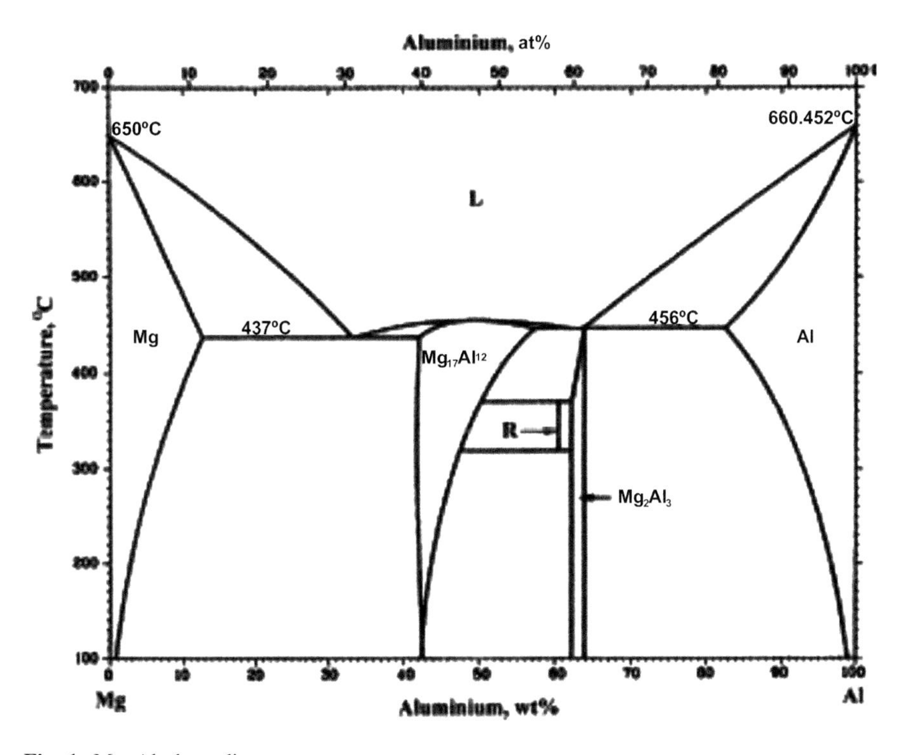

Fig. 1 Mg–Al phase diagram

shows the phase diagram of Mg–Al binary alloy. The phase diagram reveals that the β-phase is formed at a temperature of 437 °C with the Al content at about 33% by weight during cooling of casting [5]. Further, Eq. 1 clarifies eutectic reaction between primary magnesium (α-Mg) and β-phase ($Mg_{17}Al_{12}$).

$$\text{At } T = 437\,^{\circ}\text{C},\ \text{Mg(l)} + \text{Al(l)} = \text{Mg(solid solution)} + \text{Mg}_{17}\text{Al}_{12} \quad (1)$$

2.2 Designation Systems

As addressed discussed earlier, many metallic elements have a favorable effect on magnesium. Aluminum, beryllium, calcium, cerium, copper, lithium, manganese, nickel, neodymium, and some rare-earth metals can be used as alloying elements to make customized Mg alloys. Nomenclature of those alloys has been done by the different standard organization. The ASTM standard alloy designation system is widely used by the metal industry. Considering magnesium alloy AZ91E-T6, the first part of the designation, AZ, denotes that aluminum and zinc are the two principal alloying elements. The second part, 91, signifies the rounded-off percentages of aluminum and zinc, i.e., 9 and 1, respectively. The third part, E, indicates that this is the fifth alloy standardized with approximately 9% Al and 1% Zn as the principal alloying elements. Letter O and I have not used the ASTM designation system. The fourth part, T6, implies that the alloy is solution treated and artificially aged [3].

2.3 Melting Techniques

Molten magnesium does not react with iron; therefore, steel crucible is used to melt magnesium. But molten aluminum has a high affinity toward iron, so it reacts with steel crucible. To avoid reaction, boron nitride coating is provided to steel crucible [6]. Molten magnesium easily reacts with oxygen and burns with bright light. So, due care has to be taken during melting magnesium. Aluminum forms a continuous membrane on the top surface during its melting. So it does not react with atmospheric oxygen while melting. But molten magnesium forms a discontinuous and brittle membrane on its top surface and reacts with atmospheric oxygen. Protective gas or flux is used to avoid oxidation. Sulfur hexafluoride (SF_6) is an effective protective gas used for melting and casting of Mg alloy. Mixture gases like SF6 + air, SF_6 + CO_2, or SF_6 + air + CO_2 are used as protective gases in Mg melting and casting. The technique was developed in the year 1970s and became popular for its significant contribution. Though the precise mechanism of oxidation inhibitor property is still not so clear, it forms an MgF_2 layer and helps to seal the top surface [6]. Following reactions occurred during the interaction of SF_6 with molten magnesium.

$$2\text{Mg(l)} + \text{O}_2\text{(g)} \rightarrow 2\text{MgO(s)} \quad (2)$$

$$2\text{MgO(s)} + \text{SF}_6\text{(g)} \rightarrow 2\text{MgF}_2\text{(s)} + \text{SO}_2\text{F}_2\text{(g)} \quad (3)$$

MgF_2 tends to block the pores exist in loose formed MgO layer and makes it more protective. This nontoxic gas became more popular for Mg die casting and also has been used in ingot production. However, SF_6 is not environmentally friendly, and it has severe global warming potential, i.e., 24,000 times more than CO_2 gas. In addition to this, it has a long retention period of about 3200 years in the atmosphere [6].

3 Literature Review

Magnesium alloy is one of the promising lightweight materials widely used in present time for structural application owing to its excellent damping capacity, low density, and satisfactory strength–weight ratio. A lot of works are done by researchers to improve their mechanical properties. A lot of publications are already available on this concern. As the lightweight material has a top demand in current times, thus it is a gray area for the researcher to innovate newer magnesium alloy. Magnesium wrought alloys in the form of sheet and strips have limited use due to their inferior mechanical properties such as low creep and corrosion resistance and poor formability at room temperature. So, now it becomes essential to put more attention on new magnesium alloys.

Calcium improves formability and mechanical properties. Ullmann et al. added calcium in two magnesium wrought alloys ZAX210 and AZ31. They produced above alloy strips of 2 mm thickness by twin-rolled casting and hot rolling followed by intermediate and final heat treatment and found ZAX210 is a most promising alloy with higher ultimate tensile strength. ZAX210 exhibits excellent elongation of 30% at fracture [7]. Calcium is an element which is known for its contribution to strengthening Mg alloys in elevated temperature. Koltygin et al. describe Mg–Al–Ca–Zn system which can be melted with low chloride fluxes. They suggest an alloy of composition Mg–7%Al–4%Ca for the industrial use which possesses the small propensity to hot brittleness, good castability, and ultimate strength of 150 MPa with the percentage of elongation more than 3%. When annealing is done to the alloy at 500 °C during more than 20 h results in a change of Al_2Ca particles from lamellar to globular form, which rises ultimate strength by 50 MPa as compared to its original cast state [8]. Sn is highly soluble in Mg at high temperature. Solubility decreases with decreasing temperature which supports the alloy for precipitation hardening. Addition of Sn in Mg forms intermetallic phase Mg_2Sn. The melting point of Mg_2Sn phase is 770 °C; hence, the alloy has a positive effect in elevated temperature on its mechanical property.

Karakulak et al. investigated microstructure, hardness, and wear resistance by adding Si in Mg–5Sn and Mg–10Sn cast alloys. Silicon weight of 0, 0.5, 1.0, and 1.5 gm is added during the melting to the cast alloys. Metallographic study of all the alloys reveals that all the alloys show dendrite grain structure. α-Mg dendrites and intermetallic phase are found at the grain boundary. Addition of Si to the binary alloys increases intermetallic phase at the grain boundary. Further increase of Sn from 5 to 10% increases amount α-Mg dendrite grains. It is evident from the study that increasing the alloying element, both Sn and Si decrease grain size, which results in increasing of hardness according to the Hall-Petch equation. All Mg–Sn with 0.5% silicon addition shows maximum wear resistance [9]. The researchers have been searching for biodegradable and low elastic modulus materials for metallic implants in the human body especially for application for bone repair and cardiovascular stents. Nowadays Mg and its alloys are considered as potential implant material because of its suitable mechanical properties for hard tissue replacement. The metal has an excellent biological positive property and promotes metabolism reactions. Magnesium and magnesium alloys degrade very fast, which is still an issue to solve. Jiang et al. worked in this context. They prepared Mg–Zn–$Sn_{(x)}$ (x = 0, 1.0, 1.5 and 2.0 wt%) ingots in an induction furnace under argon neutral gas shielding environment. Further, the ingots were extruded into 16 mm diameter bars at 300 °C which shows both yield strength and UTS are increasing with increasing Sn content. Severe localized corrosion is found in Mg–4Zn alloy with large and deep holes whereas Mg–4Zn–1.0Sn and Mg–4Zn–1.5Sn experience a little corrosion hole. But corrosion affects conversely in Mg–4Zn–Sn2.0 alloy because of the increased number of Mg_2Sn phases formation. Further, Mg–4Zn–1.5Sn alloy presents a low mass loss rate of 0.45 mm/y without any cytotoxicity to osteoblasts [10].

Basically, Al is the main alloying elements in commercial Mg alloys. However, Mg–Al system exhibits poor mechanical properties at elevated temperature due to the low thermal stability of $Mg_{17}Al_{12}$ phase which precipitates at grain boundaries. Many authors have been addressed on this issue and tried to increase the strength of Mg alloys in different thermal conditions by adding new elements into this. By introducing new elements such as rare-earth elements (RE), Ca, Si, Sn, and Sb in Mg–Al matrix thermally stable intermetallic can be formed. In this verse, Majd et al. produced two alloys: Mg–4Al–2Sn (AT42) and Mg–4Al–2Sn–1RE (AT42 + 1RE). They investigated the microstructure of both AT42 and AT42 + 1RE alloys before age treatment and after age treatment. The microstructure of AT42 alloy shows α-Mg dendrite form and is rosette type. But the addition of RE elements changes the microstructure to dendritic structure with narrow and sharp arms, and after aging, the grain size of AT42 alloy is found 250 μm after 8 h of aging at 523 K (250 °C). It is indicated that there is no substantial grain growth after aging. RE containing alloy average grain size found 88 μm after aging 8 h at the same temperature. It loses its dendritic structure to equiaxed structure. Nevertheless, to say, due to the high thermal stability of RE containing intermetallic, they act as strong barriers to grain growth during aging which enhances the tensile properties [11].

Investigation Mg–RE alloys have been a great interest to researchers because of substantial improvement in mechanical strength. It shows high tensile strength and corrosion resistance, good ductility, and thermal stability [12]. High purity alloys, precession processing control (heat treatment and semisolid processes), innovative conversion and anodizing treatments, minor alloying condition, changing composition, and microstructure are the essential consideration to improve the corrosion resistance of Mg alloy [13]. Addition of ZrB2 particles with Mg94 Zn2.5 Y2.5 Mn1 alloy shows nearly equiaxed grains with uniform size and enhanced ductility. The alloy with 0.0075 wt% ZrB2 presents the best mechanical properties with ultimate tensile strength and plasticity of 225 MPa and 17.5%, respectively [14]. Zhou et al. investigated the microstructure and mechanical properties of Mg–1Nd–1Ce–Zr on the addition of Zn (0, 0.2, 0.5 and 1.0 wt%) followed by solution and age treated. In as-cast alloy, the disk shape and prismatic precipitates are the main strengthening phases.

When 0.5% Zn is added into the alloy, the basal precipitates become the main strengthening phases and the disk shape and prismatic precipitates vanish significantly. Only spare basal and point like precipitates exist on the addition of 1%Zn into the Mg–1Nd–1Ce–Zr alloy. Their report also revealed that Mg–1Nd–1Ce–0.5Zn–Zr exhibits highest strength with a yield strength of 136 MPa, the ultimate tensile strength of 237 MPa and elongation of 9%, which is higher than the minimum requirement of car wheels, i.e., 130 MPa–210 MPa–7%, respectively [14]. Janbozorgi et al. designed Mg–2Zn–1Gd–1Ca alloy for bio-medical use and investigated its microstructural and mechanical properties and effect of heat treatment. The report reveals that 500° and 150 °C are the optimum solutionizing and aging temperature, respectively, for the alloy. Dissolution of casting precipitates and production of lattice distortion in the solution-treated alloy lead to reduction of corrosion rate (to about 56%) and decrease in shear strength (by 6%) after the solution treatment at 500 °C. When the alloy is age treated at 150 °C, the corrosion rate is decreased by about 40% and the shear strength is increased by 21% because of the precipitation process and morphological changes [15].

4 Conclusion

There would be no further search for any lightweight material for structural application if polymer and its composite could maintain satisfactory strength and resist required temperature. During the last 100 years, magnesium offers a hope to researchers to replace heavyweight ferrous alloy by which energy-efficient machines can be fabricated. Researchers already achieved some remarkable success in its alloying process. Substantial improvement has been seen in its mechanical properties by adding different alloying elements into magnesium metal. Rare-earth elements, aluminum, calcium, and zinc-like elements can be considered as alloying elements for magnesium alloy design process. Some evidence shows that by heat treatment process, Mg alloy can improve its strength and behavior.

5 Future Scope

Magnesium is nontoxic in nature, environment-friendly, biodegradable, and 100% recyclable. Magnesium exhibits better strength to density ratio than other material. Again its excellent damping capacity and stiffness property make it special among all lightweight material. There are many issues still to solve like melting of magnesium with zero chance of oxidation, finding new compatible elements for magnesium alloy, designing new composite, functionally graded material, coating to prevent corrosion, and heat treatment to improve the mechanical property. Hence, the authors strongly recommend for further work in this area.

References

1. Kumar A, Kumar S, Mukhopadhyay NK (2018) Introduction to magnesium alloy processing technology and development of low-cost stir casting process for magnesium alloy and its composites. J Magnes Alloys 6.3, 245–254
2. Kulekci MK (2008) Magnesium and its alloys applications in the automotive industry. Int J Adv Manuf Technol 39(9-10):851–865
3. Avedesian MM, Baker H (eds) (1999) ASM specialty handbook: magnesium and magnesium alloys. ASM International
4. Kainer KU (ed) (2006) Magnesium alloys and technologies. Wiley
5. Emley EF (1966) Principles of magnesium technology
6. Luo AA (2013) Magnesium casting technology for structural applications. J Magnes Alloys 1.1, 2–22
7. Ullmann M et al (2019) Development of new alloy systems and innovative processing technologies for the production of magnesium flat products with excellent property profile. Proc Manuf 27:203–208
8. Koltygin A.V et al (2013) Development of a magnesium alloy with good casting characteristics on the basis of Mg–Al–Ca–Mn system, having Mg–Al2Ca structure
9. Karakulak E, Küçüker YB (2018) Effect of Si addition on microstructure and wear properties of Mg–Sn as-cast alloys. J Magnes Alloys 6.4, 384–389
10. Jiang W et al (2019) Effect of Sn addition on the mechanical properties and bio-corrosion behavior of cytocompatible Mg–4Zn based alloys. J Magnes Alloys
11. Majd AM et al (2018) Effect of RE elements on the microstructural and mechanical properties of as-cast and age hardening processed Mg–4Al–2Sn alloy. J Magnes Alloys 6.3, 309–317
12. Shahid R, Scudino S (2018) Microstructure and mechanical behavior of Al–Mg composites synthesized by reactive sintering. Metals 8(10):762
13. Esmaily M et al (2017) Fundamentals and advances in magnesium alloy corrosion. Prog Mater Sci 89:92–193
14. Zhou Y et al (2019) Precipitation modification in cast Mg–1Nd–1Ce–Zr alloy by Zn addition. J Magnes Alloys
15. Janbozorgi M, Taheri KK, Taheri AK (2019) Microstructural evolution, mechanical properties, and corrosion resistance of a heat-treated Mg alloy for the bio-medical application. J Magnes Alloys 7.1, 80–89

Thermogravimetric Analysis of Biochar from Arhar Fiber Powder Prepared at Different Pyrolysis Temperatures

M. Om Prakash, G. Raghavendra, Manoj Panchal and S. Ojha

Abstract Arhar is one of the most cultivated crops in Asian countries; arhar dal is the main yield from the arhar crop. After crop harvest, arhar stalks are mostly used as firewood which causes pollution and has a bad impact on the environment. The present research work is to convert these arhar stalk fibers into biochar. Arhar fiber powder is subjected to pyrolysis process to prepare biochar. Different parameters involve in the pyrolysis process, in the current study effect of temperature on biochar yield was investigated. SEM and TGA analyses are performed on arhar biochar to know the thermal stability of biochar. From the experiment results, it was found that temperature has a great impact on surface structure and thermal stability of biochar. The surface morphology of biochar materials was studied by SEM analysis, it was found that with the increment in pyrolysis temperature there was enhancement in porosity of biochar materials. From the TGA analysis, it was observed that biochar materials prepared at high temperature exhibited high thermal stability.

Keywords Arhar fiber powder · Biochar · Pyrolysis · Temperature · SEM analysis · TGA analysis

1 Introduction

The concept of biochar was evolved long back from ancient period. In olden days, biochar was primarily used as a source of fuel as it has less reactivity to oxygen when compared to direct materials raw source used as fuel [1]. Biochar has many features like large surface area, good porosity developed during pyrolysis, improves soil quality, nutrients and water holding capacity, reducing the fertilizer usage and

M. Om Prakash · G. Raghavendra (✉) · M. Panchal · S. Ojha
Mechanical Engineering Department, NIT, Warangal, Telangana, India
e-mail: raghavendra.gujjala@gmail.com

M. Om Prakash
e-mail: omprakash5424@gmail.com

BBVL. Deepak et al. (eds.), *Innovative Product Design and Intelligent Manufacturing Systems*, Lecture Notes in Mechanical Engineering,
https://doi.org/10.1007/978-981-15-2696-1_42

reduced greenhouse gases emission from soil [2–4]. Over the past decades, researches are extensively working on different biomass materials for developing effective biofuels which are cost-effective, environmentally friendly as fossil fuels endangered of shortage. Biomass generally used as a precursor for the production of biochar, biomass may be agriculture waste [4], forest waste and sewage waste. Biomass is converted to biochar by a process called pyrolysis which is thermal decomposition process in the presence of an inert atmosphere.

Pyrolysis is the process in which organic material is thermally decomposed in the absence of oxygen, i.e., inert atmosphere. Pyro means heat, lysis means breaking down; therefore, pyrolysis is nothing but breaking down large molecules into a number of small molecules. It involves different physical and chemical reactions. The end product obtained from pyrolysis is irreversible. Pyrolysis is the most familiar process adopted for recycling of waste plastic and tire materials. Biomass material contains three basic compositions cellulose, hemicellulose, and lignin. When biomass materials subjected to pyrolysis process, each constituent of biomass gets decomposed at different temperatures. Thermal energy supplied during the pyrolysis process will decompose the carbohydrate structure of biomass material leaving a carbonaceous residue, various gaseous. From the past investigation, it was observed biomass pyrolysis involves different individual stages that are moisture evolution, hemicellulose decomposition, and cellulose and lignin decomposition [5]. During pyrolysis of biomass materials, different types of gaseous will be released like CO_2, CO, H_2, C_2H_6, C_2H_4 [6]. Carbonaceous solid residue (biochar) resulted from pyrolysis has slower oxidation rate. Due to slower oxidation rate of biochar, it is mostly used as fuel. In the current research work, arhar fiber used as a precursor for the production of biochar. Arhar is one of the most cultivated crops in India. The main yield of arhar crop is arhar dal after harvesting the crop the arhar stalks are thrown away as waste material or used as firewood which has a bad environmental effect. These arhar stalks which are lignocellulose material have a high potential for production of biochar [5]. Arhar fibers earlier used as reinforcing material for the preparation of polymer composite [6]. But there is no research has done so far on arhar fiber used as a source for preparation of biochar.

Yang et al. [7] studied the characteristics of hemicellulose, cellulose and lignin pyrolysis. The study focused on what type of gaseous releasing by hemicellulose, cellulose and lignin by how much amount of energy consumption during pyrolysis. From the experimental investigation, it was revealed that decomposition of cellulose produced more CO than cellulose and lignin, hemicellulose generated higher CO_2 and lignin resulted in higher H_2 and CH_4. Lee et al. [8] compared biochar properties produced from different biomass residues by slow pyrolysis at 500 °C. In the study cocopeat, sugarcane, paddy straw, palm kernel shell, bagasse, and umbrella tree were subjected to pyrolysis at 500 °C, their physical and chemical properties are compared. From the study, it was reported that lignin and ash content have a considerable impact on carbon yield. The study also stated that biochar from wood stem and bagasse has good porosity and surface area which are favorable for soil application.

Angın [9] investigated the effect of pyrolysis temperature and heating rate on biochar produced from pyrolysis of safflower seed press cake. From the experimental study, it was observed that as the temperatures increase the yield of biochar decreased. In the study, temperatures are raised from 100 to 600 °C with heating rate 10°/min, 20°/min and 30°/min. As the heating rate increased, the volatile matter decreased resulting in decreasing in volume pores and surface area. It was concluded that temperature as a great influence on biochar yield and composition then heating rates. Fu et al. [10] studied how temperature influences the amount of gas produced and char structure of pyrolyzed agricultural residues. In this study, agricultural residues such as cotton straw, maize stalk, rice husk, and rice straw are pyrolyzed at a different temperatures from 600 to 1000 °C in steps of 100. From the study, it was observed as the pyrolyzed temperature increased gas yield increased while char yield was decreased. At higher temperatures, micropore size, mesopore size, and surface area increased till 900 °C and decreased with the further increase in temperature.

2 Materials and Methods

2.1 Raw Materials

Dried arhar stalks are collected from local farms; these stalks are dried in sunlight. Dried arhar stalks are chopped into small pieces and introduced into the laboratory ball mill. Chopped arhar stalks are milled in laboratory ball for an 8 h time which results in arhar fiber powder. Arhar fiber powder is used as a precursor for the production of biochar.

2.2 Biochar Production

Arhar fiber powder obtained through laboratory ball milling is subjected to pyrolysis for generating biochar. In the present work, carbolite Gero's tube furnace is used for the pyrolysis process. Arhar fiber powder filled in ceramic boats is introduced into tuber furnace and subjected to pyrolysis. An inert atmosphere is created by passing argon gas at a rate of 200 ml/min through the tube before the heater is switched on. Pyrolysis process was carried at different peak temperatures, i.e., 600, 700, and 800 °C. Temperatures are raised to peak temperature at a rate of 10 °C/min, after reaching peak temperature it is held for a 1 h time and then cooled to room temperature at the same rate as of heating. Biochar thus produced at different temperatures weighed and results are noted.

2.3 *SEM Analysis*

SEM analysis usually gives the surface structure of the materials. In the current study, SEM analysis was performed on raw arhar fiber powder and biochar obtained at different peak temperatures to know the porous structures. X-act oxford instruments machine was used for both SEM investigation.

2.4 *Thermogravimetric Analysis (TGA) Analysis*

Thermal stability of the biochar materials was studied by thermogravimetric analysis. The loss of weight of biochar materials, when subjected to heat, was analyzed. In the current study by using NETZSCH STA 2500 TGA machine at temperature rate of 27 °C/10 min up to 800 °C. Argon gas was used as protective and carrying gas.

3 Results and Discussions

Biochar obtained after pyrolysis at different temperatures weighed and the percentage of yield was calculated. Results were plotted by taking different pyrolysis temperature on x-axis and yield percentage on the y-axis. From Fig. 1, it can be observed that as the temperature increases yield percentage is decreasing similar type of results were found in Angın [9] investigation. This phenomenon of decreasing yields at elevated temperature due to high primary decomposition at high temperatures or secondary decomposition of biochar residue [10]. At low temperatures, the material is not fully decomposed so the yield of biochar is high [11]. Figure 1 yields percentage of biochar at different temperatures.

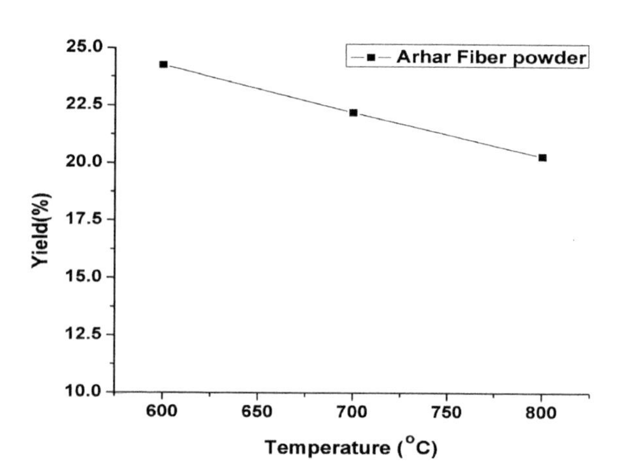

Fig. 1 Yield percentage of biochar at different temperatures

3.1 SEM Analysis

The effect of pyrolysis temperature on the surface topography of biochar yield was investigated through SEM analysis. Figure 2 shows the surface structure of raw arhar powder at 10 μm scale. From SEM, Fig. 2 can be clearly observed that there is no porous structure formation on the surface raw arhar fiber powder. When arhar powder subjected to pyrolysis process, the powder particles will start breaking and forms porosity. From SEM Fig. 3 biochar produced at 600 °C, it can be seen powder surface broken. It was observed from SEM Figs. 3 to 5 as the temperature increases micro-porosity and size and proportions of void increases. The porous formation is more at elevated temperatures which is evident from Figs. 4 and 5. As the pyrolysis temperature increases, volatile compounds were driven off and impurities were released from the materials result in porosity formation [12]. With the release of volatile compounds from material leads to substantial internal over pressure results formation of open structures [13]. Thus, as the temperature increases porosity increases.

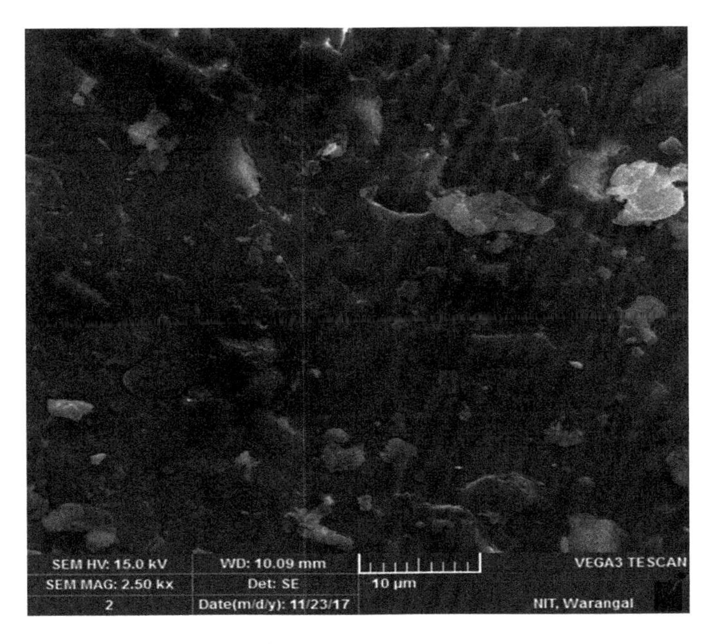

Fig. 2 SEM analysis arhar char powder

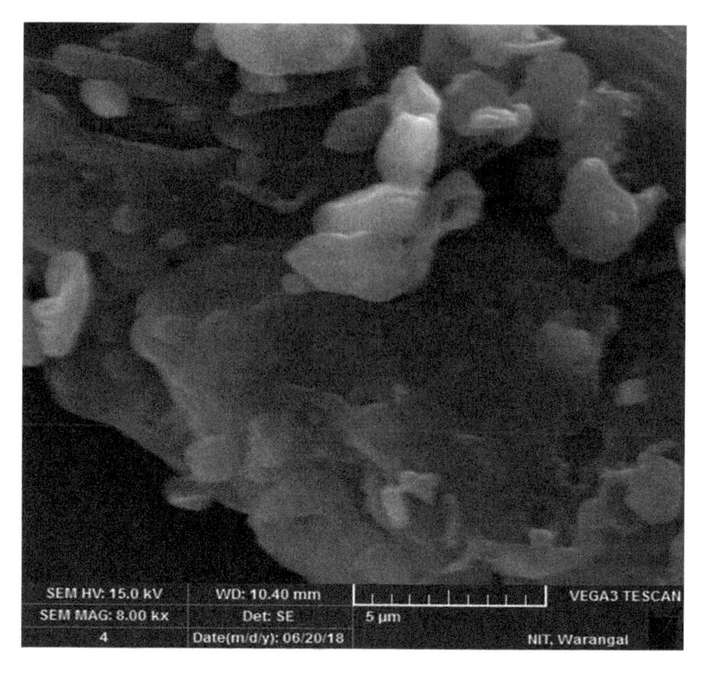

Fig. 3 SEM analysis of biochar at 600 °C

Fig. 4 SEM analysis of biochar at 700 °C

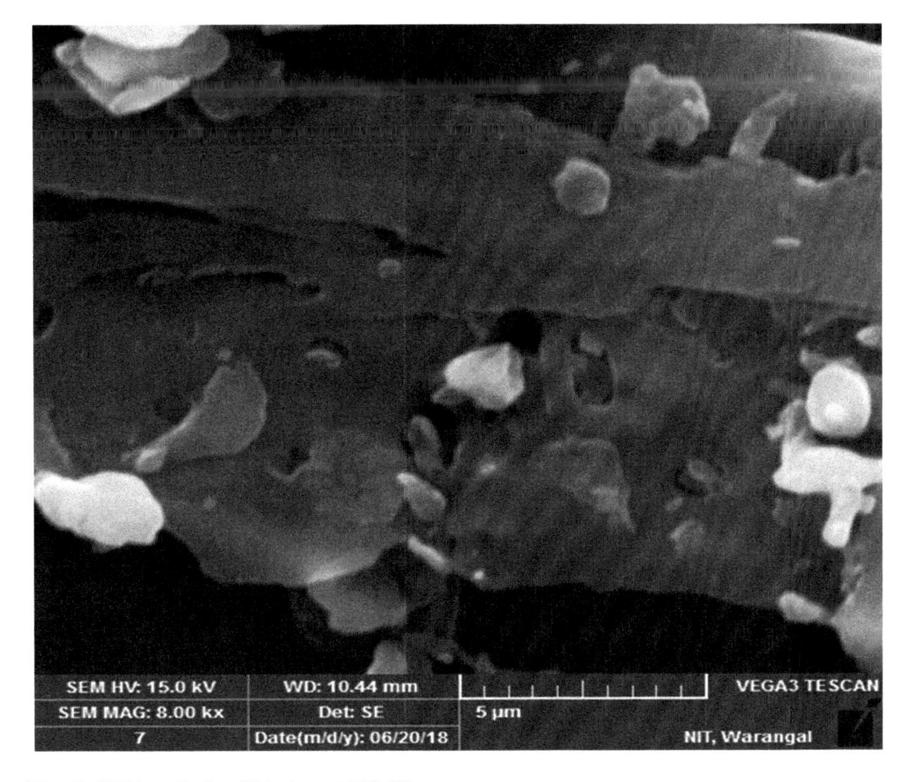

Fig. 5 SEM analysis of biochar at 800 °C

3.2 *Thermogravimetric Analysis (TGA) Analysis*

From Fig. 6 in all the biochar material the initial weight loss up to 150 °C, this is due to the loss of moisture present in the materials. The loss of mass after 150 °C in all biochar material is because of decomposition and degradation of impurities, volatile matter and also carbon into CO, CO_2, and CH_4 [14]. Though biochar is carbonaceous material, the carbon structure in biochar is disordered carbon structure [15]. The porous surface of biochar makes it reactive carbon material. From the figure, it is observed that biochar of 800 °C has lost less mass than biochar prepared at 600 and 700 °C temperatures. As the pyrolysis temperature increased, the thermal stability of the biochar materials also increased because of the presence of ordered carbon structures.

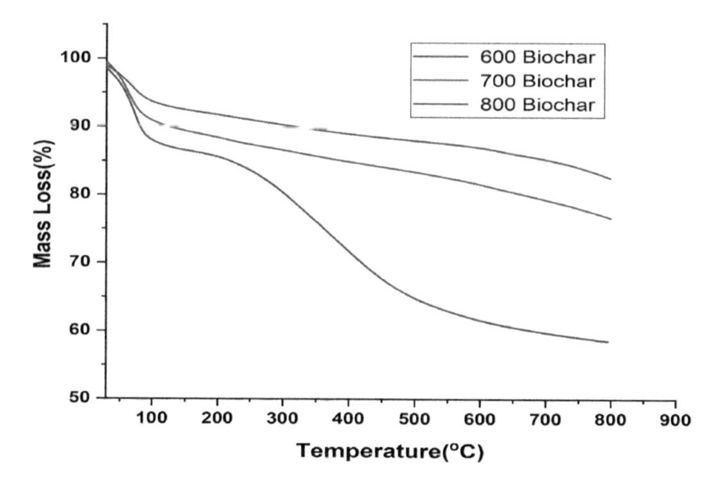

Fig. 6 Thermogravimetric analysis of biochar at different temperatures

4 Conclusions

As the temperature increases quality of arhar powder, biochar increases with decrease in quantity. As we observe porous micro-structure in biochar, arhar fiber further can be used as a resource for preparation of activated carbon. The amorphous structure of biochar makes it suitable for filtration and adsorbent applications. From the study, it can be concluded that biochar prepared at higher pyrolysis temperature have a greater thermal resistance.

References

1. Zimmerman AR (2010) Abiotic and microbial oxidation of laboratory-produced black carbon (biochar). Environ Sci Technol 44(4):1295–1301
2. Jeffery S, Verheijen FG, van der Velde M, Bastos AC (2011) A quantitative review of the effects of biochar application to soils on crop productivity using meta-analysis. Agric Ecosyst Environ 144(1):175–187
3. Lehmann J, Gaunt J, Rondon M (2006) Bio-char sequestration in terrestrial ecosystems—a review. Mitig Adapt Strat Glob J 11(2):395–419
4. McKendry P (2002) Energy production from biomass (part 1): overview of biomass. Bioresour Technol 83:37–46
5. Yang HP, Yan R, Chin T, Liang DT, Chen HP, Zheng CG (2004) TGA FTIR analysis of palm oil wastes pyrolysis. Energy Fuel 18:1814–21
6. Om Prakash M, Raghavendra G, Panchal M, Ojha S, Reddy BA (2018) Effects of environmental exposure on tribological properties of arhar particulate/epoxy composites. Polym Compos 39(9):3102–3109
7. Yang H, Yan R, Chen H, Lee DH, Zheng C (2007) Characteristics of hemicellulose, cellulose and lignin pyrolysis. Fuel 86(12):1781–1788

8. Lee Y, Park J, Ryu C, Gang KS, Yang W, Park YK, Hyun S (2013) Comparison of biochar properties from biomass residues produced by slow pyrolysis at 500 C. Bioresour Technol 140:196–200
9. Angın D (2013) Effect of pyrolysis temperature and heating rate on biochar obtained from pyrolysis of safflower seed press cake. Bioresour Technol 128:593–597
10. Fu P, Yi W, Bai X, Li Z, Hu S, Xiang J (2011) Effect of temperature on gas composition and char structural features of pyrolyzed agricultural residues. Bioresour Technol 102(17): 8211–8219
11. Katyal S, Thambimuthu K, Valix M (2013) Carbonisation of bagasse in a fixed bed reactor: influence of process variables on char yield and characteristics. Renew Energy 28(5):713–725
12. Ertaş M, Alma MH (2010) Pyrolysis of laurel (Laurus nobilis L.) extraction residues in a fixed-bed reactor: characterization of bio-oil and bio-char. J Anal Appl Pyrolysis 88(1):22–29
13. Onay O (2007) Influence of pyrolysis temperature and heating rate on the production of bio-oil and char from safflower seed by pyrolysis, using a well-swept fixed-bed reactor. Fuel Process Technol 88(5):523–531
14. Sun Y, Gao B, Yao Y, Fang J, Zhang M, Zhou Y, Yang L (2014) Effects of feedstock type, production method, and pyrolysis temperature on biochar and hydrochar properties. Chem Eng J 240:574–578
15. Yi Q, Qi F, Cheng G, Zhang Y, Xiao B, Hu Z, Xu S (2013) Thermogravimetric analysis of co-combustion of biomass and biochar. J Therm Anal Calorim 112(3):1475–1479

A Machine Learning Scheme for Tool Wear Monitoring and Replacement in IoT-Enabled Smart Manufacturing

Zeel Bharatkumar Patel and Sreekumar Muthuswamy

Abstract Tool wear monitoring is an important task in a smart manufacturing industry. Detecting worn-out tools and replacing them in time can increase the efficiency significantly. Various sensors are being used in machine tools to integrate them into a smart manufacturing setup. Continuously decreasing the cost of the sensors is encouraging the use of low-cost indirect methods for the task. Using multiple sensors increases the precision of estimating tool health over the single sensor-based approach. Appropriate mathematical models relating tool wear parameters and sensors data can be used here, but machine learning models become more suitable in a large variety of applications over normal mathematical models. This paper proposes a methodology for multi-sensor-based indirect tool wear monitoring system and presents a comparison of accuracy among various machine learning models. Standard references are used to generate dummy training and testing data. Python is used to create and test the models. In the end, it has been found that Naïve Bayes and support vector machine algorithms are yielding up to 97% accuracy. This is the initial work in the development of an IoT enabled and fully automated manufacturing setup.

Keywords Machine learning · Tool wear monitoring · Scikit-learn · AI and IoT · Smart manufacturing

Z. B. Patel · S. Muthuswamy (✉)
Centre for AI, IoT and Robotics, Department of Mechanical Engineering,
Indian Institute of Information Technology, Design and Manufacturing,
Vandalur-Kelambakkam Road, Kancheepuram, Chennai 600127, India
e-mail: msk@iiitdm.ac.in

Z. B. Patel
e-mail: smt17m010@iiitdm.ac.in

BBVL. Deepak et al. (eds.), *Innovative Product Design and Intelligent Manufacturing Systems*, Lecture Notes in Mechanical Engineering,
https://doi.org/10.1007/978-981-15-2696-1_43

1 Introduction

The present decade is the beginning of smart manufacturing-based automation, where there is a need for self-reliant automatic tool handling system. It includes measurement of tool wear, detection of fractures and intelligent tool replacement systems, among which tool wear monitoring is a slow and continuous process. Machining processes such as milling, drilling and turning can be monitored using such schemes. The milling process includes more complications among all because of multi-point cutting tools, and more axes are involved. Lathe and drilling processes can be monitored with more ease and fewer complexities comparatively [1]. Tool wear is mainly classified into flank wear and crater wear. From which, flank wear is addressed here because it is directly related to the main cutting edge of the tool. On the other hand, crater wear is generally affected by chip formation in the workpiece, which has been considered out of scope for the present work.

Based on the principle of monitoring, classification can be done as sensor-less and sensor-based monitoring systems. Taylor's tool life equation was used to predict the tool life, which belongs to the first category. Other techniques on sensor-less methods can be found at [2, 3]. Sensor-based approaches are further divided into direct or conventional and indirect methods [4]. Actual values of tool wear parameters are measured by direct methods, whereas, appropriate process parameters correlated with tool wear are measured by indirect methods. The methodology of direct measurement is costly and could be disturbed due to environmental condition changes. Indirect methods are less costly and easier to implement but require high computation, which is easily available now compared to previous years [5, 6]. Certain research is already carried out based on the application of laser and video-based online artificial vision systems for direct online tool condition monitoring [7, 8]. Again, high cost and varying nature due to illumination have prevented this approach to be implemented in real-time. The focus of this work is mainly on indirect methods. Some of them are represented in Table 1 [9].

Developing a promising scheme for tool wear monitoring is always been a difficult task because of a large variety of tool and workpiece materials, machining processes, tool types and much more complications [10]. A purely mathematical approach may perfectly work for a single set of configurations, but at the same time, it may become largely complicated when it has to deal with certain non-correlated parameters. The reason behind using conventional methods up to now is the lack of

Table 1 Indirect tool wear sensing methods

Process parameters	Measurement procedures	Transducers
Cutting forces	Force measurement	Strain gauge
Vibration of tool	Vibration measurement	Accelerometers
Power	Spindle power	Ampere meters
Surface roughness	Measurement of roughness	Mechanical stylus
Cutting temperature	Temperature	Thermo-couples

computational power, which is largely addressed in the past few years. Now, enough computational power and sensors, both are economical and could be implemented in each and every machine setup in a plant compared to the conventional methods. Still, machine learning is not yet been used in depth for monitoring specifically continuous tool wear. These algorithms are tested individually for similar purposes, where only binary estimation is done. This paper is an effort toward testing and comparing various machine learning algorithms for a setup where a large variety of operations are carried out. With enough data and correct decision parameters, tool wear can be estimated accurately at any point of time in an ongoing machining process. For large smart manufacturing plant, all the details can be monitored from anywhere using IoT attached with machine setups. In certain cases, decision making also can happen from the cloud rather than just monitoring.

2 Methodology

As described in Fig. 1, methodology can be divided into two subdivisions as training and online estimation. More about each step is explained in details in subsequent subsections.

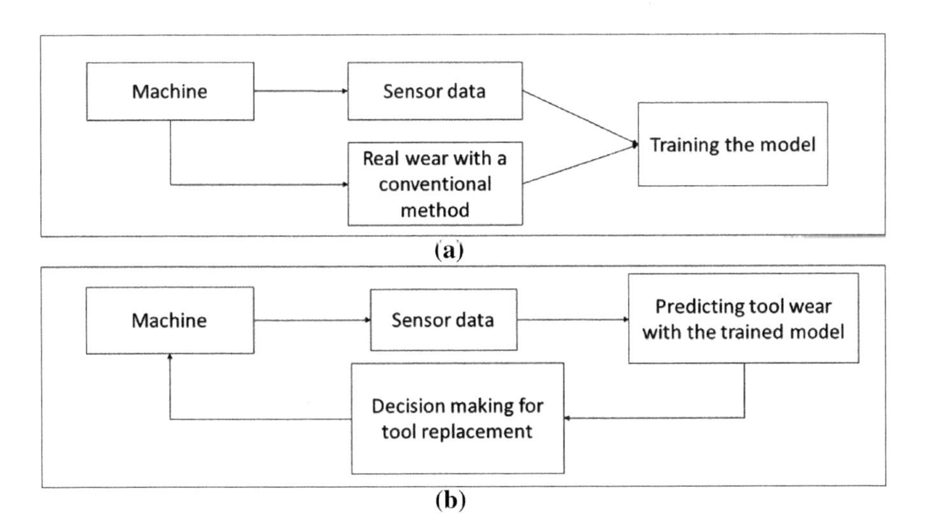

Fig. 1 Methodology **a** training **b** online estimation

2.1 Data Preparation

All sensors connected with the machine setup generate the data which can be collected from microcontrollers to sustainable storage with corresponding time-stamps. Real values of tool wear can be taken in synchronization, with the help of more accurate direct tool wear measuring methods, e.g., camera, measure of electrical resistance between tool edge and workpiece, etc. In this work, tool condition is estimated both in turning and milling operations. In the turning process, continuous tool wear was estimated using dummy data based on the reference. Mainly, three process parameters are taken into consideration for turning operation as inputs, which have good relationships with tool wear, i.e., longitudinal cutting force, transverse cutting force, spindle power.

Data for all of the above-mentioned parameters were generated such that they have similar characteristics with real data found at [11] with the help of Python random library. Python library uses uniform random distribution to provide well-distributed data over the span. Total of 105 training data points is generated with this process for all three input parameters. Cutting parameters considered for the data are cutting speed V_c = 140 m/min, feed rate S_0 = 0.22 mm/tooth and depth of cut t = 1.5 mm. The same is shown in Fig. 2.

2.2 Training the Model

The training process considered in this work is the same for all the models being used here. Scikit-learn library from Python is used to train the machine learning model [12]. Data were stored in csv files after generation. Python library csv was useful to handle the data. Optimum tuning parameters were set for each model, and they were individually saved to be used repeatedly. Characteristics of these models are such that any number of inputs can be mapped to any number of outputs. As stated earlier, a single sensor may not be sufficient to achieve good estimation accuracy. Multiple low-cost sensors attached to a machine are economical and also help in the improvement of accuracy. The input vector includes more than one parameter (sensor data), and output is scaler values of tool wear in μm. According to the scheme of a model, it fits the parameters to the target values and forms a mathematical relationship between them. Once training is completed, the model can estimate the corresponding tool wear values based on input sensor data in operation. For each combination of tool and workpiece, a separate model can be used after training. The accuracy of a model can be checked through cross-validation.

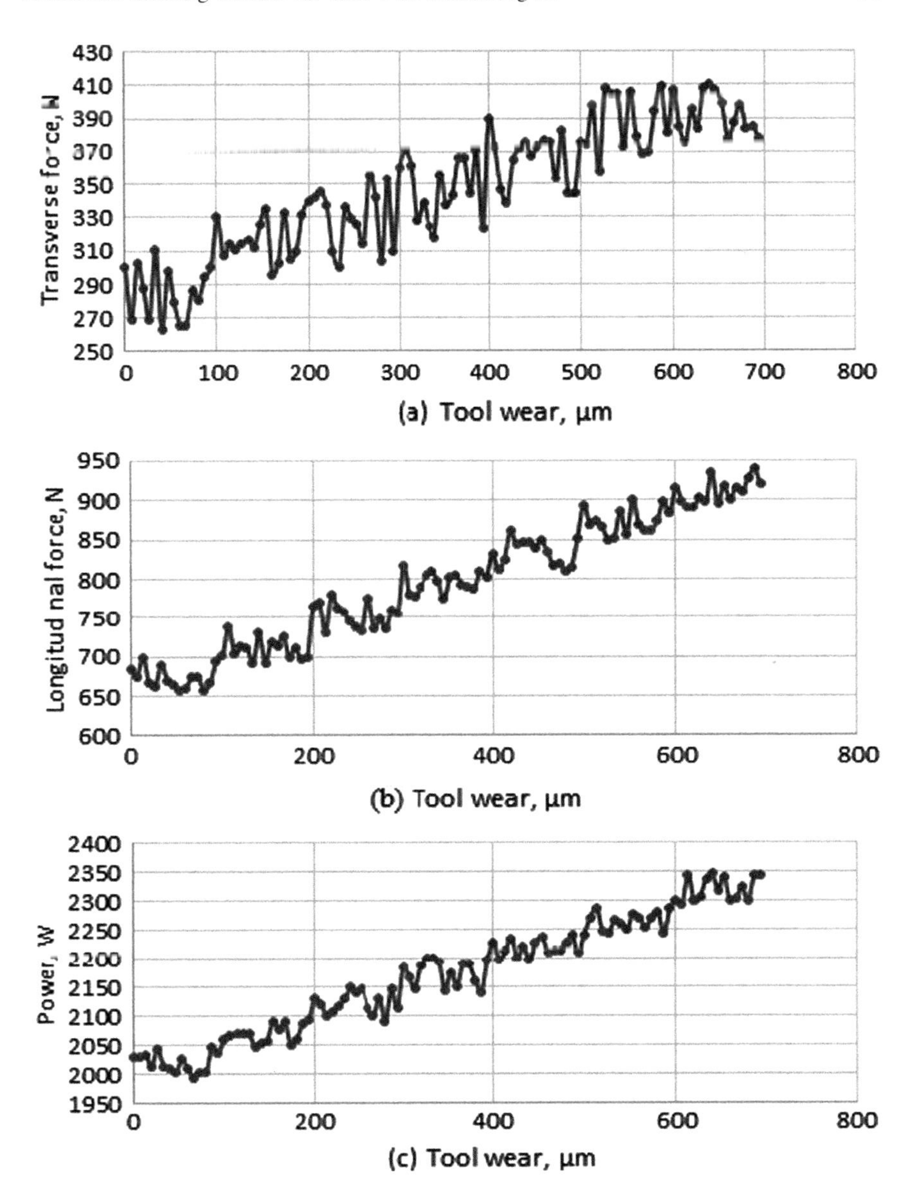

Fig. 2 Data generated based on the reference **a** transverse force versus tool wear **b** longitudinal force versus tool wear **c** power versus tool wear

2.3 Decision Making for Turning

After models are saved, they can be tested on test data. Test data were generated by adding noise in the data varying from 4 to 10% and taking samples from it.

Decision making is used here in the sense of what has to be done after tool wear is crossed the threshold limit. The threshold limit is the maximum allowable tool wear for a tool which can vary from one operation to another. The tool can be considered as worn after model detects that threshold value is crossed. The worn tool must be replaced to avoid low-quality products. Either a system can be configured such that it informs automatic tool changer via machine to machine communication or the task can be performed manually by the machine operator so that tool can be replaced or simply the information can be passed to the operator of respective machines.

2.4 Binary State Estimation of Milling Tool

Milling tools have multiple cutting edges, so estimating tool wear for it is a very difficult process. Binary state estimation can be done comparatively easy just about two states of the tool: worn and unworn. Data for this work were collected from [13]. S-shaped specimen was used to conduct the experiments. Total of 18 experiments was available, from which eight were done with an unworn tool and ten were done with a worn tool. There were 25,286 data rows containing 48 parameters in each of them. In which major parameters are

- Actual and reference x, y and z coordinates.
- Actual and reference x, y and z speeds and accelerations.
- Input–output power and current of each motor.

2.5 Machine Learning Models

This section gives a brief classification of machine learning algorithms used in this work. They belong to the category of supervised learning. Information and equations for each algorithm can be found in details at [12]. They are basically classified into classifiers and regressors. A classifier classifies predicted values into labels provided in the training step. It is best to use classifiers when target values are less in quantity, but also, they are safe toward unexpected predictions out of range. Regressors use specific mathematical models to fit the target values with input parameters. When data are having more complex relationships which can be formulated into mathematical relationships, regressors are suitable to use. Both regressors and classifiers are tested, but as per characteristics of data, regressors are more preferable to use. In this work, three classifiers and five regressors are used, i.e., decision tree, K-neighbor, random forest, support vector machine, Naïve Bayes, random forest, gradient boosting and multi-layer perception are used to serve the purpose.

3 Results

After training the models, they are saved for the estimation process. Each model was tested on randomly taken 20 data points with some added disturbance in a range from 4 to 10% in training data. Estimated values for one of the experiments are shown in Fig. 3. In sample data points, wear values are considered as real wear, and output of the model is considered as predicted wear. Real wear values are found to be close enough to the estimated values. The x-axis is having 20 different samples taken randomly and also predicted wear values. The y-axis is having tool to wear values in micron.

For classification, accuracy score is an average value calculated based on a number of results correctly predicted by the model over 10 trials. For regression, explained variance score is calculated which measures the proportion to which a mathematical model accounts for the variation. If y' is predicted output and y is target value, explained variance score can be given as following where Var is variance.

$$\text{explained variance}(y', y) = 1 - \frac{\text{Var}(y' - y)}{\text{Var}(y)} \quad (1)$$

It can be seen among classifiers that K-neighbor algorithm yields good average accuracy up to 93%, and after the application of the filter, it can go up to 95% in the best case. Among regressors, random forest and Naïve Bayes algorithms are yielding good average accuracy up to 95% and can yield the best accuracy of 97%. So, the best algorithm to use for the scheme is Naïve Bayes because of its adaptivity and tendency to fit the data correctly without overfitting. Support vector machine is a well-adopted algorithm and has good performance in this work. It can have promising performance if data are large enough, but it requires more training time. All the comparison values are represented in Tables 2 and 3. Similarly, milling operation data were trained on four classifiers, and results are represented in Table 4.

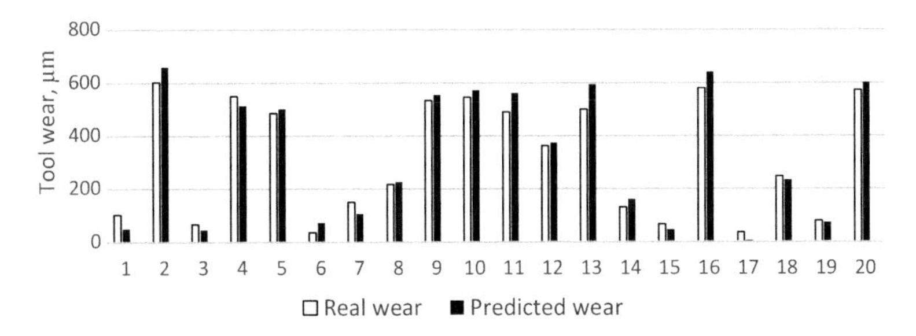

Fig. 3 Real versus predicted wear values

Table 2 Comparison among classifiers for turning

Classifier	Without filter		With filter	
	Average (%)	Best (%)	Average (%)	Best (%)
Decision tree	88	92	89	94
K-neighbor	92	94	93	95
Random forest	86	88	90	92

Table 3 Comparison among regressors for turning

Regressor	Without filter		With filter	
	Average (%)	Best (%)	Average (%)	Best (%)
MLP	80	92	89	94
Random forest	93	94	93	95
SVM	91	97	85	96
Naïve Bayes	94	98	95	97
Gradient boosting	91	93	91	95

Table 4 Comparison among classifiers for milling

Classifier	Average (%)	Best (%)
Decision tree	99.02	99.16
Gradient boost	97.47	97.52
Random forest	98.91	99.07
AdaBoost	95.63	96.04

4 Conclusion

In the process of monitoring tool wear, the important task is to replace tools on the correct time to promote good quality and optimized machining time. This can be done more easily with a machine learning approach. These methods are dynamic in nature and adaptive to the large variety of tools and machining processes. They also promise good accuracy to be implemented in manufacturing plants. Machine learning classifiers and regressors were tested here for the purpose of tool wear monitoring. Training and testing were carried out on the dummy data generated based on the reference. In the end, Naïve Bayes, support vector machine and gradient boosting were produced good accuracy up to 95–97%. For milling operation, classifiers are the best suitable choice because of binary classification. Decision tree classifier is giving the best accuracy up to 99.16% which is suitable to be implemented in real-life scenarios. This work can go further, many other direct and indirect parameters related with tool wear, e.g., temperature, hardness, materials, spindle speed, etc., can be included in data, and accuracy can be still improved with supporting a large variety of parameters.

Acknowledgements This research work is carried out with a financial grant of ICPS division of the Department of Science and Technology (DST), Government of India, Grant no: DST/ICPS/CPS-Individual/2018/769 (G), dated 18–12–2018.

References

1. Bernhard S (2002) On-line and indirect tool wear monitoring in turning with artificial neural networks: a review of more than a decade of research. J Mech Syst Signal Process 16:487–546
2. Sukhomay P, Stephan P, Burkhard H, Nico J, Surjya K (2011) Tool wear monitoring and selection of optimum cutting conditions with progressive tool wear effect and input uncertainties. J Intell Manuf 22(4):491–504
3. Muhammad R, Jaharah A, Mohd ZN, Haron CHC (2013) Online tool wear prediction system in the turning process using an adaptive neuro-fuzzy inference system. Appl Soft Comput 13(4):1960–1968
4. Mahardhika P, Eric D, Chow YL, Edwin L (2019) Metacognitive learning approach for online tool condition monitoring. J Intell Manuf 30(4):1717–1737
5. Dimla DE, Lister PM (2000) On-line metal cutting tool condition monitoring. I: force and vibration analyses. Int J Mach Tools Manuf 40:739–768
6. Tugrul O, Yigit K (2005) Predictive modeling of surface roughness and tool wear in hard turning using regression and neural networks. Int J Mach Tools Manuf 45(4–5):467–479
7. Castejon M, Alegre E, Barreiro J, Hernandez LK (2007) On-line tool wear monitoring using geometric descriptors from digital images. Int J Mach Tools Manuf 47(12–13):1847–1853
8. Jurkovic J, Korosec M, Kopac J (2005) New approach in tool wear measuring technique using CCD vision system. Int J Mach Tools Manuf 45(9):1023–1030
9. Dan L, Mathew J (1990) Tool wear and failure monitoring techniques for turning—a review. Int J Mach Tools Manuf 30:579–598
10. Andrew KS, Daming L, Dragan B (2006) A review on machinery diagnostics and prognostics implementing condition-based maintenance. Mech Syst Signal Process 20(7):1483–1510
11. Ghosha N, Ravib YB, Patrac A, Mukhopadhyayc S, Pauld S, Mohantyd AR, Chattopadhyayd AB (2007) Estimation of tool wear during CNC milling using neural network based sensor fusion. J Mech Syst Signal Process 21:466–479
12. Pedregosa F, Varoquaux G, Gramfort A, Michel V, Thirion B, Grisel O, Blondel M, Prettenhofer P, Weiss R, Dubourg V, Vanderplas J, Passos A, Cournapeau D, Brucher M, Perrot M, Duchesnay E (2011) Scikit-learn: machine learning in python. J Mach Learn Res 12:2825–2830
13. CNC Mill Tool Wear-Kaggle. https://www.kaggle.com/shasun/tool-wear-detection-in-cnc-mill. Last accessed 10 Mar 2019

Experimental Investigation of Vibration Response of Faulty Rotor Shaft Partially Submerged in Viscous Medium

Adik Yadao and Abhishek Kumar Kashyap

Abstract In the current investigation, it is taken an effort to determine the dynamic response of cracked rotor fixed at both ends partially immersed in the different viscous medium. The influence coefficient method is used to obtain dynamics response of faulty rotor shaft. The Navier-Stokes equation is applied to investigate the external fluid forces acted on the rotor shaft. The strain energy release rate at the crack section of the rotor has been used for determining the local stiffness and is dependent on the crack depth. The fluid viscosity, crack depth and the crack location are considered as main variable parameters. Also, the dynamic responses in two transverse directions (i.e., x-axis and y-axis) of the crack of the rotor shaft are numerically found. Further, experimental analysis has been done for authentication of the obtained numerical results.

Keywords Rotor · Multiple crack depths · Dynamic response · Viscous fluid

1 Introduction

The investigations have done as far as on the rotors with transverse crack are described in the literature review in detail, but no studies have been presented on the performance of the rotating faulty rotor in a viscous medium. Singh and Tiwari [1] have presented an experimental investigation to validate the algorithm (MCDLA) for the identification of multi-crack and localization in the simply supported rotor. Silania et al. [2] have described the vibration signature analysis with a breathing crack system. Yadykin et al. [3] have presented the numerical study of cantilever plate with additional mass submerged in the fluid. Liang et al.

A. Yadao (✉)
Department of Mechanical Engineering, G.H. Raisoni College of Engineering & Management, Pune, Maharashtra, India

A. K. Kashyap
Department of Mechanical Engineering, National Institute of Technology, Rourkela, Orissa, India

BBVL. Deepak et al. (eds.), *Innovative Product Design and Intelligent Manufacturing Systems*, Lecture Notes in Mechanical Engineering,
https://doi.org/10.1007/978-981-15-2696-1_44

[4] have used empirical added mass formulation for investigation of the frequencies and mode shapes of cantilever plates immersed in the fluid medium. Shahgholi et al. [5] have analyzed the dynamic behavior of a non-linear spinning simply supported shaft. Phan et al. [6] have analyzed the vibrations response of a rectangular cross-section cantilever beam which is immersed in a viscous medium under harmonic base excitation. Hossain et al. [7] have presented the experimental analysis to measure the vibration analysis of a cantilever beam immersed in fluid and air medium using polytech scanning vibrometer. In this study, the Navier-Stokes equation has been used to find the acting viscous forces on the cracked rotor. In addition, the influence coefficients method is utilized for calculating the vibration signature of the rotor shaft (in both transverse direction x-axis and y-axis). Finally, the numerical results are validated with experimental values.

2 Mathematical Background

2.1 Equation of Motion

The polar coordinates of the Navier-Stokes equation can be expressed as;

$$\frac{1}{r}\frac{\partial p}{\partial r} - n\left(\frac{1}{r^2}\frac{\partial^2 u}{\partial q^2} + \frac{1}{r}\frac{\partial u}{\partial r} - \frac{2}{r^2}\frac{\partial v}{\partial q} - \frac{u}{r^2} + \frac{1}{r^2}\frac{\partial^2 u}{\partial r^2}\right) + \frac{\partial u}{\partial t} = 0 \tag{1a}$$

$$\frac{1}{\rho r}\frac{\partial p}{\partial \theta} - v\left(\frac{\partial^2 v}{\partial r^2} + \frac{1}{r}\frac{\partial v}{\partial r} + \frac{2}{r^2}\frac{\partial u}{\partial \theta} - \frac{v}{r^2} + \frac{1}{r^2}\frac{\partial^2 v}{\partial \theta^2}\right) + \frac{\partial v}{\partial t} = 0 \tag{1b}$$

2.2 Analysis of Fluid Forces

The forces of the viscous fluid acting on the rotor in two different directions transverse direction (i.e., x-axis and y-axis direction) can be evaluated as:

$$F_x - \int_0^{2\Pi} (-\tau_{r\theta}\sin\theta + \tau_{rr}\cos\theta)R_1 \, d\theta = 0 \tag{2a}$$

$$F_y - \int_0^{2\Pi} (\tau_{r\theta}\cos\theta + \tau_{rr}\sin\theta)R_1 \, d\theta = 0 \tag{2b}$$

2.3 Analysis of Dynamic Response of the Rotor

Figure 1a and b illustrate the cross-section of crack element and coupling forces acting on the crack element of rotor fixed at both ends, respectively. Where the local stiffness matrix is given in Eq. 3.

$$k_S = \begin{bmatrix} k_{11} & k_{12} \\ k_{21} & k_{22} \end{bmatrix} = \begin{bmatrix} C_{11} & C_{12} \\ C_{21} & C_{22} \end{bmatrix}^{-1} \tag{3}$$

K_s can be calculated using the local stiffness matrix:

$$M_s \frac{d^2(x + \varepsilon \cos \omega t)}{dt^2} + EI \frac{d^4 x}{dz^4} = F_x \tag{4a}$$

$$M_s \frac{d^2(x + \varepsilon \sin \omega t)}{dt^2} + EI \frac{d^4 y}{dz^4} = F_y \tag{4b}$$

where

m_s = Rotor mass/unit length.
m = Fluid mass displaced by the rotor shaft/unit length.
EI = Bending stiffness of the shaft.

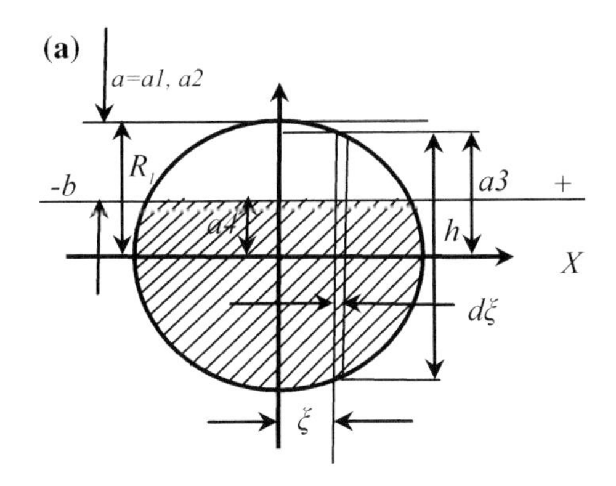

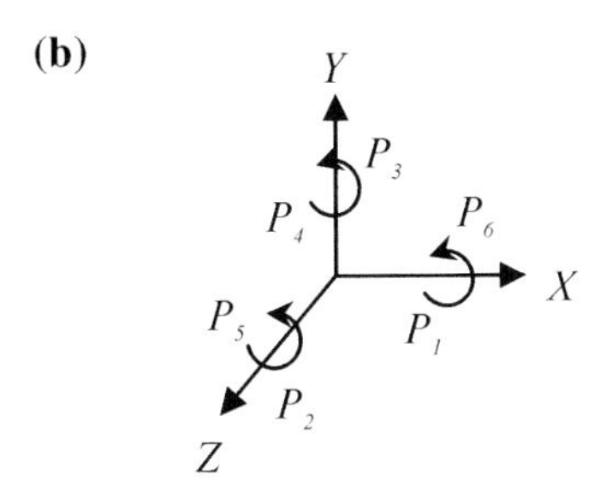

Fig. 1 **a** Cross-section of crack element **b** coupling forces on the crack element

$$(M_s + M\mathrm{Re}(H))\frac{\mathrm{d}^2x}{\mathrm{d}t^2} - M\omega\mathrm{Im}(H)\frac{\mathrm{d}x}{\mathrm{d}t} + K_s x = M_s\varepsilon\omega^2\cos\omega t \quad (5a)$$

$$(M_s + M\mathrm{Re}(H))\frac{\mathrm{d}^2y}{\mathrm{d}t^2} - M\omega\mathrm{Im}(H)\frac{\mathrm{d}y}{\mathrm{d}t} + K_s y = M_s\varepsilon\omega^2\sin\omega t \quad (5b)$$

$$\delta_1^* = \delta_{44}^* = \xi_{44}\left(\frac{L^*}{2}, \tau_1\right) + \xi_{55}\left(\frac{L^*}{2}, \tau_1\right) \quad (6)$$

$$\delta_2^* = \delta_{55}^* = \eta_{44}\left(\frac{L^*}{2}, \tau_1\right) + \eta_{55}\left(\frac{L^*}{2}, \tau_1\right) \quad (7)$$

where δ_1^* and δ_2^* is a dimensionless deflection of the multiple cracked rotor shaft (i.e., x-axis and y-axis direction).

3 Numerical Analysis and Discussion

In the current investigation, the sample being has got the following description, Mild steel rotor fixed at both ends with multiple cracked partially submerged in the viscous fluid medium.

The rotor under analysis has the following mechanical properties: Length $L = 1.2$ m, Young's modulus $E = 2.1 \times 1011$ N/mm^2, Material density $\rho = 7860$ kg/m^3, radius of rotor $R_1 = 0.01$ m, Relative crack depth $\beta_{1,2,3,4} = 0.125/0.150/0.225/0.250$, Relative crack location $\alpha_{1,2} = 0.29/0.416$, Gap ratio $q = 12$, viscosity of fluids $v = 2.9/0.541/0.0633$ and Stokes and mass density ratio $m^* = 0.143/0.136/0.127$. Figure 2a and b show the dynamic response in two different transverse directions (i.e., x-axis and y-axis direction) of the cracked rotor system fixed at both ends. It is observed that when crack depth is increased, there is decreasing in the frequency and amplitude of rotor shaft and also noted that as increase in the viscosity of the fluid, the amplitude of vibration of the cracked rotor is reduced. Moreover, it is observed that when the viscosity of the fluid is increased, the amplitude of vibration decreases. However, it is distinguished that the resonance frequencies for y-axis direction are lower than that of the x-axis direction of a crack in rotor shaft.

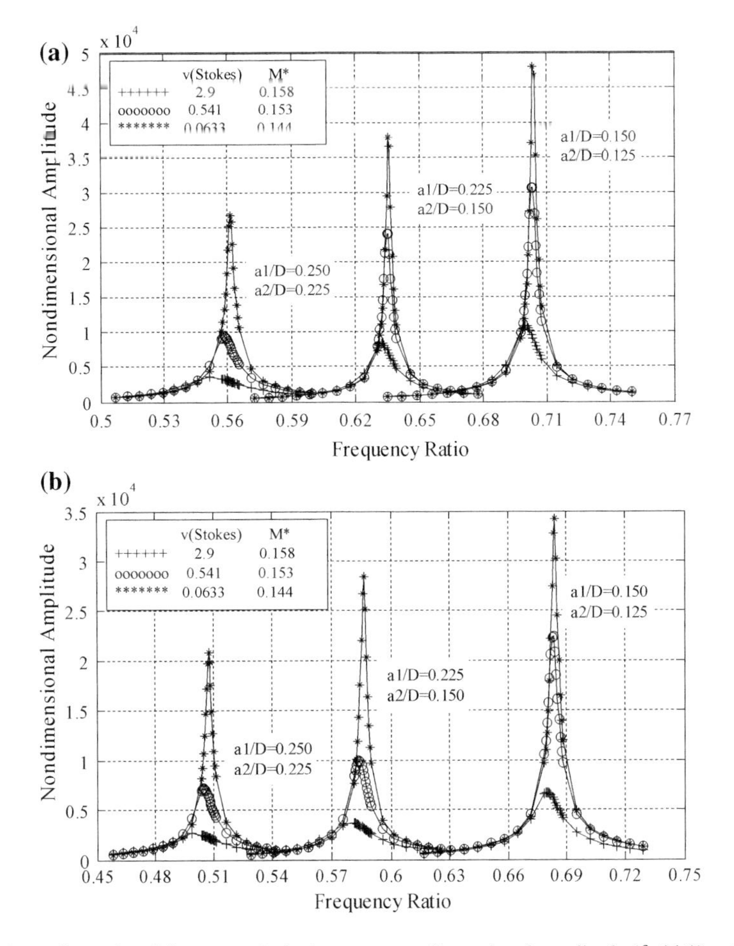

Fig. 2 Non-dimensional frequency (ω/ω_0) versus non-dimensional amplitude ($\delta n^*/\varepsilon^*$), mild steel rotor $R_1 = 0.01$ m, $L = 1.2$ m, $q = 12$, $L_1/L = 0.291$, $L_2/L = 0.416$, $R_D = 0.035$ m, $T_D = 0.018$ m, $M_D = 0.55$ kg **a** for x-axis direction **b** for y-axis direction

4 Experimental Analysis

An experiment held on the spinning rotor fixed at both ends with multiple transverse cracks submerged in the viscous fluid medium. For obtaining the dynamic response considering the altered viscosity of the fluid (i.e., lubricant oil 2T, palm oil, and seawater), multiple crack depths and crack locations on the rotor system. The schematic representation of the experimental test rig for the cracked rotor submerged in the viscous fluid has been illustrated in Fig. 3. The speed of the rotor shaft connected to the driving unit is controlled (i.e., motor) by the variac.

The three liquid-proof ultrasonic sensors are arranged around the rotor with a suitable space inside the liquid-filled container which is connected to the

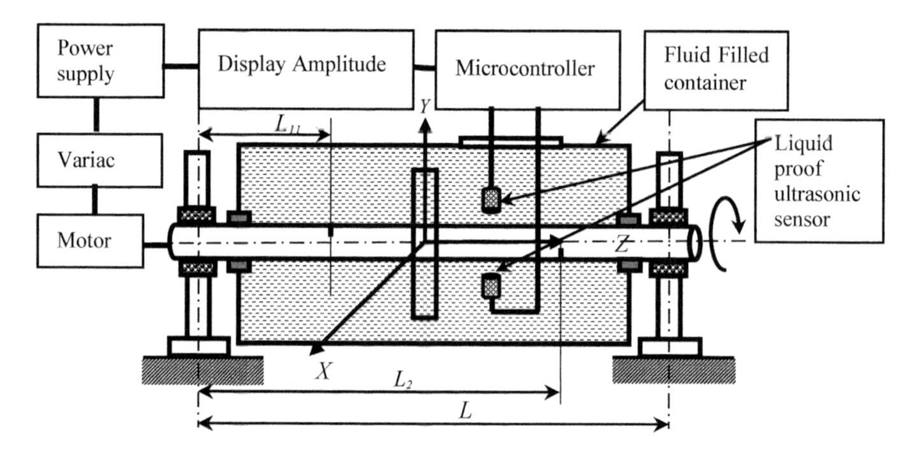

Fig. 3 Representation of experimental setup for the cracked rotor fixed at both ends submerged in viscous medium

Table 1 Comparison between theoretical and experimental results for crack locations $L_1/L = 0.291$, $L_2/L = 0.416$, crack depths $a_1/D = 0.150$, $a_2/D = 0.125$ and viscous medium ($v = 0.541$ stokes & $M^* = 0.136$)

S. No.	Theoretical analysis (non-dimensional amplitude ($\delta n^*/\varepsilon^*$))		Experimental analysis (non-dimensional amplitude ($\delta n^*/\varepsilon^*$))		Error in percentage (%)	
	X-axis direction	Y-axis direction	X-axis direction	Y-axis direction	X-axis direction	Y-axis direction
1	555.5878	573.1945	534.9032	555.2936	3.723	3.123
2	843.4156	882.8832	810.8681	861.005	3.859	2.478
3	1591.006	1723.734	1539.536	1661.127	3.235	3.632
4	7922.774	8358.822	7801.238	8196.242	1.534	1.945
5	12186.24	9791.495	11873.66	9437.826	2.565	3.612
6	21247.36	8976.235	20492.44	8663.323	3.553	3.486
7	17534.69	7234.973	16787.18	7036.590	4.263	2.742
8	10442.52	5761.224	10103.97	5619.209	3.242	2.465
9	2468.977	2136.468	2396.586	2067.353	2.932	3.235
10	1046.695	987.5314	1021.354	959.5645	2.421	2.832

microcontroller through the jumper wire and breadboard is used for determining the amplitude of rotor shaft. The ultrasonic sensors are used for identifying the amplitude of the spinning rotor shaft from the actual position (central axis) of the rotor shaft. Table 1 has been represented the comparison between the obtained results of theoretical and experimental analysis. The obtained results are in excellent agreement with each other.

5 Conclusions

In this work, we have presented numerical as well as experimental analysis of the dynamic response at both transverse directions (i.e., x-axis and y-axis) of the multiple cracked rotors fixed at both ends partially submerged in a viscous medium. There is a decrease in the amplitude and frequency of the rotor due to increase in viscosity and density of the viscous fluid. It is observed that as the crack depths are increased in crack locations, the frequency and amplitude of the rotor are also decreased. Also, it is found that the multiple cracked rotor in the x-axis direction exhibits greater stiffness as compared to the y-axis direction. In this work, the bearing effect is not taken into account as a simplifying assumption for the rotor system. In the future, the bearing effect will be taken into consideration for finding the dynamic behavior of the rotor.

References

1. Singh SK, Tiwari R (2014) Detection and localization of multiple cracks in a shaft system: an experimental investigation. Measurement 53:182–193
2. Silani M, Ziaei-Rad S, Talebi H (2013) Vibration analysis of rotating systems with open and breathing cracks. Appl Math Model 37:9907–9921
3. Yadykin Y, Tenetov V, Levin D (2003) The added mass of a flexible plate oscillating in a fluid. J Fluids Struct 17:115–123
4. Liang CC, Liao CC, Tai YS, Lai WH (2001) The free vibration analysis of submerged cantilever plates. Ocean Eng 28:1225–1245
5. Shahgholi M, Khadem SE, Bab S (2014) Free vibration analysis of a nonlinear slender rotating shaft with simply support conditions. Mech Mach Theory 82:128–140
6. Phan CN, Aureli M, Porfiri M (2013) Finite amplitude vibrations of cantilevers of rectangular cross sections in viscous fluids. J Fluids Struct 40:52–69
7. Hossain A, Humphrey L, Mian A (2012) Prediction of the dynamic response of a mini cantilever beam partially submerged in viscous media using finite element method. Finite Elem Anal Des 48:1339–1345

Influence of Double Elliptical Leaf Angle with Same Orientation and Direction to Evaluate Thermal Performance in Double Pipe Heat Exchanger

J. Bala Bhaskara Rao, B. Murali Krishna and K. Narendra

Abstract Double pipe heat exchanger is an apparatus utilized to transfer the energy between two liquids. To attain high performance of heat transfer rate in small-area passive methods are castoff. The heat transfer rate along the length of the heat exchanger is intended at different mass flow rates of water by experimentally and numerically. FLUENT analysis is conducted with different strips. In this FLUENT analysis, strip having two elliptical leafs at a distance of 50 mm along the length, the major and minor axes are in 2:1 with 1 mm thickness at altered angles between 0° and 180° at 10° intervals are arranged. These two elliptical leaves are having same orientation and same direction toward the length of the strip. Water is taken as a liquid at various turbulent regions in between 5000 and 20,000. From numerical results, increased rate of heat transfer rates is achieved with strip with elliptical leaves having 60° inclinations because of more turbulence and secondary flows are acquired.

Keywords Heat transfer · Pressure drop · Elliptical strip · Reynolds number · Leaf angle

1 Introduction

Heat exchangers are mostly utilized to transfer the energy between two different temperature fluids. These are mainly used in all process industries like chemical plants, power plants, etc. By using theoretical and practical analysis with simulations, the performance of the double pipe heat exchanger is studied and optimized the tube diameter [1, 2]. The heat transfer coefficient of the inner tube flow (circular cross section) is calculated using the standard correlations, and an attempt is made to enhance the rate of heat transfer in heat exchangers using Al_2O_3 nanofluid [3, 4].

J. Bala Bhaskara Rao (✉) · B. Murali Krishna
Department of Mechanical Engineering, SSCE, Kakinada, India

K. Narendra
Department of Mechanical Engineering, SISTAM, Srikakulam, India

BBVL. Deepak et al. (eds.), *Innovative Product Design and Intelligent Manufacturing Systems*, Lecture Notes in Mechanical Engineering,
https://doi.org/10.1007/978-981-15-2696-1_45

The horizontal circular tube heat exchanger, with air as the working fluid, is increased by means of rectangular inserts [5]. CFD analysis is conducted on heat exchanger with different strips, and it was observed that the heat transfer coefficient varied from 0.9 to 1.9 times that of the smooth tube with helical strips and corresponding friction factor increased by 1–1.7 times [5, 6]. The thermal performance is calculated by using circular and square strips inside heat exchangers, and heat exchanger with rectangle strip gives better results [7]. The analysis is carried out on double pipe heat exchanger with fin and without fin. The fins are shaped semicircular type arranged in an alternating way with spacing of 50 mm. The overall heat transfer coefficient using semicircular fin decreases more than 300% which shows that huge increases in overall heat transfer area [8]. In the present analysis, a double pipe heat exchanger with strips is used to evaluate thermal performance. The strips are designed with two elliptical leafs at a distance of 50 mm along the length, the major and minor axes are in 2:1 with 1 mm thickness at altered angles between 0° and 180° at 10° intervals are arranged. These two elliptical leaves are having the same orientation and same direction toward the length of the strip. The performance is calculated at different inclinations of elliptical leaves.

2 Experimental Setup

The double pipe heat exchanger with outer and inner pipes was taken as an experimental device (Fig. 1). The inner and outer tube diameters are 22 mm and 40 mm of 1 mm thickness and 750 mm of length and made up of stainless steel and copper, respectively. In this experiment, cold and hot fluids are circulated through the loop by using variable speed pumps. The analysis is conducted as shown in Fig. 2, Strips with two elliptical leafs at different inclinations with the same orientation and direction. Two streams of cold and hot fluids are circulated, and the mass flow rates and temperatures are measured by flow meters and thermocouples. The two fluids are moving in the opposite direction, and flow rates are controlled by valves. The thermophysical properties of fluids at bulk temperature and diameters of the heat exchanger are used to calculate different mass flow rates at various Reynolds number ranges of 5000–20,000.

The thermal performance of heat exchanger is calculated by the following equations

$$\text{Tube side Reynolds number } (\mathrm{Re}\, t) = \frac{\rho v D_e}{\mu} \tag{1}$$

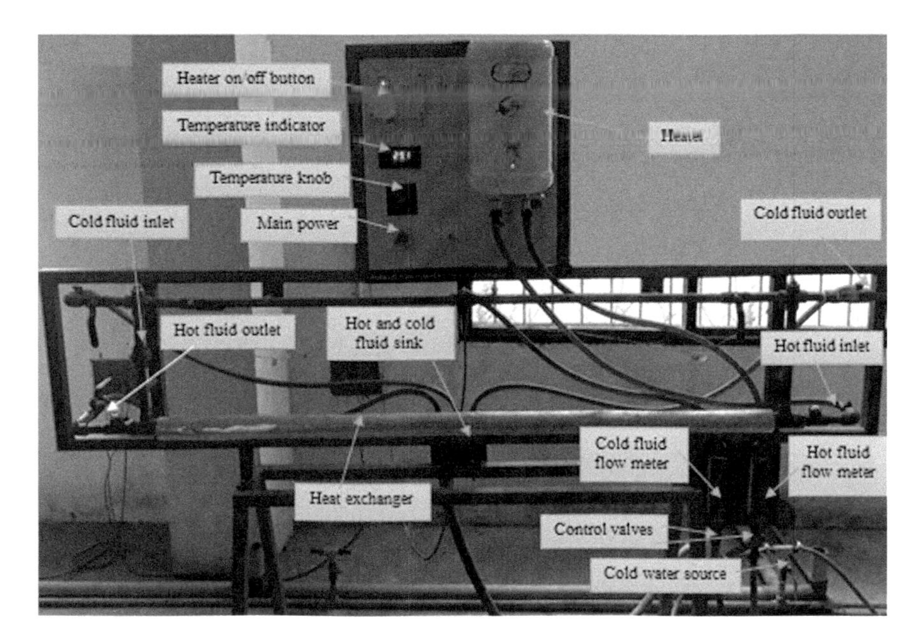

Fig. 1 Experimental setup of double pipe heat exchanger

where

ρ = Density of inner fluid in kg/m^3
V = Velocity of inner fluid in m/sec
D_e = Hydraulic diameter of pipe in meters

$$-\frac{4A_c}{P_h} = d$$

d = Inner diameter of inner pipe in meters
μ = Dynamic viscosity of inner fluid kg/m-s.

$$\text{Annual side pressure drop (Re } a) = \frac{\rho v D_e}{\mu} \tag{2}$$

where

ρ = Density of annual side fluid in kg/m^3
V = Velocity of annual side fluid in m/s
D_e = Hydraulic diameter of pipe in meters.

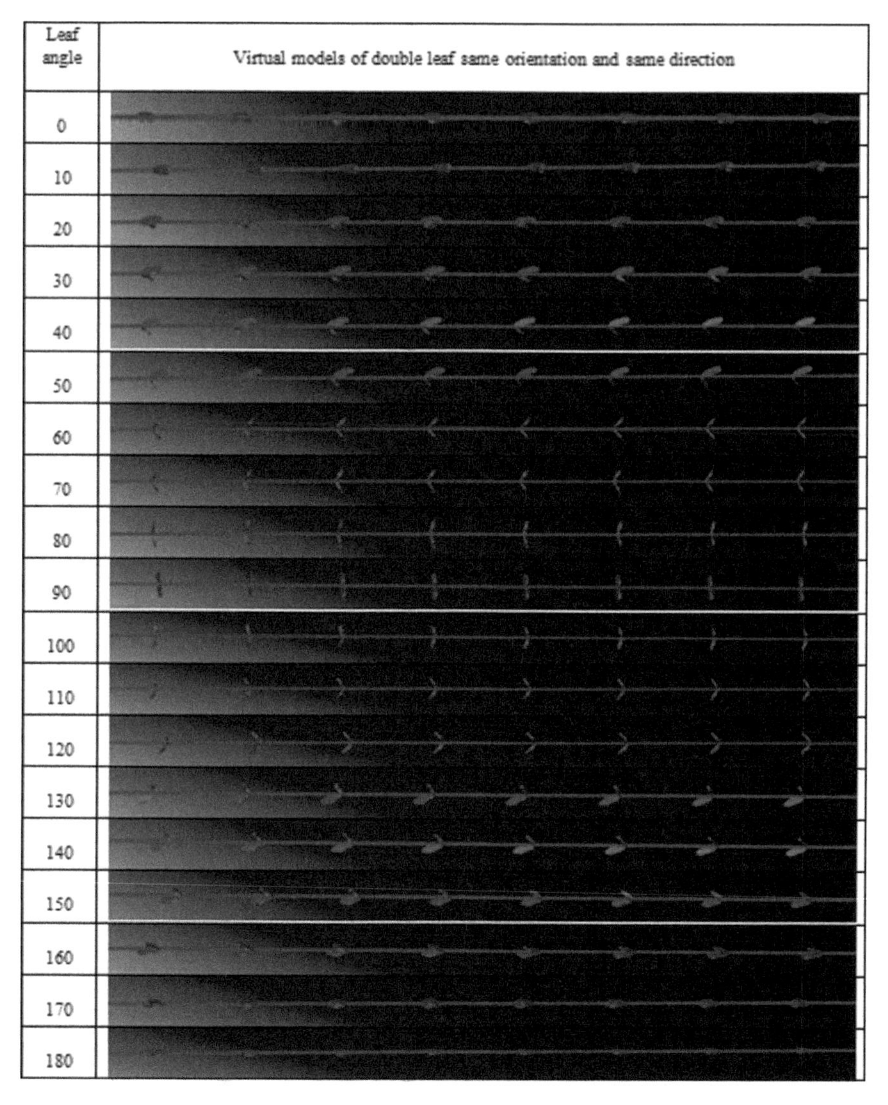

Fig. 2 Strip with two elliptical leaves at different angles

$$= \frac{4A_c}{P_h} = \frac{D_{i-}^2 d_0^2}{d_0}$$

$A_c = \frac{\pi}{4}\left(D_i^2 - d_0^2\right)$ & $P_h = \Pi d_0$
A_c = Minimum flow area in m^2
P_h = Wetted perimeter in meters
D_i = Inner diameter of outer pipe in meters

d_0 = Outer diameter of inner pipe in meters
μ = Dynamic viscosity of annual side fluid kg/m-s.

The heat transfer rate and pressure drops are premeditated from the initial records according to the flow rate, inlet temperature at shell and tube side, and outlet temperature at shell and tube side. The heat transfer rate is obtained by the following equation.

$$Q = AU\Delta T_m \tag{3}$$

where Q = average heat flux between the cold and the hot fluid in watts

$$= (Q_C + Q_h)/2$$

$$Q_C = m_c c_{pc}(T_{co} - T_{ci}) \tag{4}$$

$$Q_h = m_h c_{ph}(T_{hi} - T_{ho}) \tag{5}$$

At steady-state condition, the heat transfer rate is calculated. The enhancement of heat transfer can be obtained from the strip with three elliptical leaves which are kept in the inner tube. The strip edifice is such that it contains a circular rod and two elliptical shape leaves are fixed to the circular rod with a gap of 50 mm as shown in Fig. 2. The elliptical leaves quick ratio *a*/*b* is 2, same orientation (0°–180°), the same direction at different inclinations (0°–180°) with 10° regular intervals are attached to the strip to enhance heat transfer rate. At various mass flow rates and steady-state condition, the heat transfer rate is calculated.

3 Boundary Conditions

Boundary conditions are used according to the need of the model. The inlet and outlet conditions are specified as mass flow rate and pressure outlet. As this is a counterflow with two tubes, there are two inlets and two outlets. The heated water is circled in the inward pipe at a temperature of 348 K at various mass flow rates. At the inlet of the inward pipe nozzle, distinctive mass flow rates are given such as 0.032683, 0.065366, 0.098049, and 0.130731 kg/s. The cold fluid is gone through the annual side at 298 K with various mass flow rates such as 0.223883, 0.447766, 0.671649, and 0.895532 kg/s. The flow is considered to atmospheric pressure, and the outlets are defined as atmospheric pressure. The flow is assumed as an incompressible turbulent flow. No-slip condition is given for the inner surfaces of inward and external pipes. The flow is considered as normal to the boundary. At the inlet nozzles of both fluids, hydraulic diameter and turbulent intensity were stated. The walls are independently indicated with respective boundary conditions. Except the tube walls, each wall is set to zero heat flux condition.

4 Results and Discussions

To start the numerical investigation of temperature difference in fluid flow, the base circumstances must be analyzed. The primary state is that of fluid moving through a pipe with constant wall heat flux condition. In this investigation, turbulent incompressible fluid flow is measured in tube side with 19 strips with altered elliptical leaf angles to assess thermal analysis in the double pipe heat exchanger. The flow factors are altered by giving distinctive strips with elliptical leaves. Subsequently, superior heat transfer rate is created in the flow by inserting a supplementary surface area, and a mathematical model is restored for increasingly numerical analysis. The velocity, pressure, and temperature pattern are as discovered in Fig. 3 and exhibit the distribution of fluid near elliptical leaves.

The tendency of two elliptical leaves plays a basic role to appraise thermal performance of double pipe heat exchanger. The changed elliptical leaf angles from 0° to 180° at 10° intermissions are reflected to estimate heat transfer rate and pressure drop. Totally 19 models at four mass stream rates in the turbulent region are considered to explore the effectiveness of double pipe heat exchanger.

4.1 Heat Transfer Analysis

The heat transfer rate in the double pipe heat exchanger is considered by using modified strips with two elliptical leaves and at different tendencies. The employment of strip with two elliptical leaves in a pipe gives a passive process for advancement the heat transfer rate by generating swirl and auxiliary flows.

Fig. 3 Velocity, temperature, and pressure distribution along the length of heat exchanger

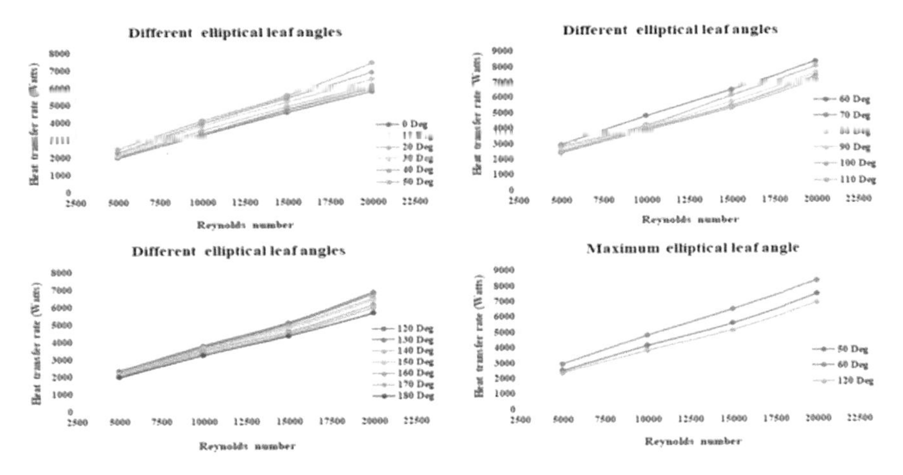

Fig. 4 Heat transfer variation at different elliptical leaf angles

These swirl streams are disquieting the boundary layer due to rapidly variances at the inner surface. Such sort of strips prompts turbulence and swirl flow which makes inside the boundary layer and provides improved effects of heat transfer rate.

The flow contiguous by the two elliptical leaves and inner walls is generally extraordinary. Temperature circulation for vector diagram in Fig. 4 assigns the heat transfer rate increment if elliptical leaf tendency changes from 0° to 60° and declines from 0° to 180°. The stream length increment if there should be an occurrence of 60° elliptical leaf angles and it prompts increment the heat transfer rate. At different two elliptical leaf angles (0°, 10°, 20°.... 180°), the temperature counters at the center of the heat exchanger are as revealed in Fig. 3. From all graphs (Fig. 5), it exhibits that the heat transfers rate increments with a heightening of Reynolds number. Temperature diffusion for temperature counters assigns the heat transfer rate development with increment in surface area and by presenting strips. Streamline distribution for elliptical leaf strip supplements of heat exchanger gives extra heat transfer rate because of creating of disorder over liquid by strip insert. The streamlines neighboring to the inside wall of the tubes keep running into a more prominent measure of the heat exchange because of strip inserts.

4.2 *Pressure Drop Analysis*

The pressure drop in inward pipe and annual side along its length is determined with strip having three elliptical leaves at various inclinations. Clearance space as stream zone is more for 60° declined gradually. These stream zones give less pressure drop as free space makes less turbulence. To maintain a strategic distance high-pressure drop in the heat exchanger, a uniform transfer of heat energy through

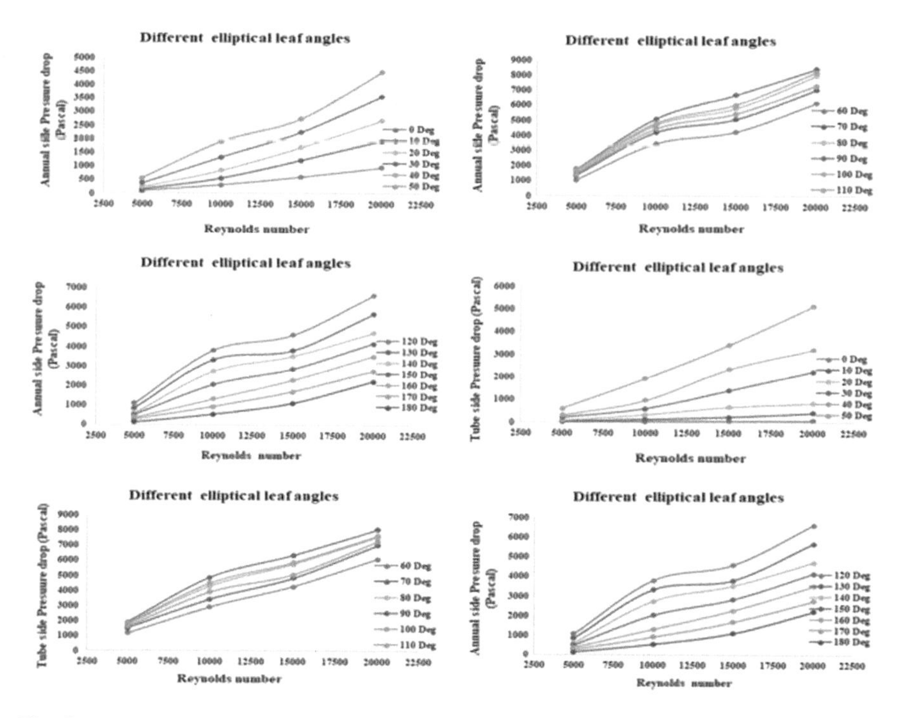

Fig. 5 Annual and tube side pressure drop variation at different elliptical leaf angles

the heat exchanger is required. With this view, the pressure drop with various measures is determined for the better procedure of the heat exchanger.

These deviations can be further additionally limited by expanding the number of components during meshing.

Thus, utilizing similar simulations for the tube with wire coil inserts is completed. The three elliptical leaf orientation and direction for space appropriation in the inner pipe impacts pressure drop by various zones.

By referring to Fig. 5, it can be seen that the pressure drop increases along the length with expanding the elliptical leaf angle up to 90° and in an increment of Reynolds number.

The vector shapes in the scientific model appeared in Fig. 5 for heat exchanger with strip show pressure drop variation in tube side with various elliptical leaf inclinations. From Fig. 5, it indicates the pressure drop increases with the increase of Reynolds number and it increases from 0° to 90° elliptical leaf angle and decreases from 90° to 180° elliptical leaf angle.

5 Conclusion

In this experiment, FLUENT analysis is directed on double pipe heat exchanger with adjusted inclined elliptical leaf angle strips to expand the thermal performance. By utilizing the finite volume approach to solve the governing partial differential equation, heat transfer and fluid flow are incompressible turbulent and the Reynolds numbers range from 15000 to 20000 were considered. The turbulent variant of the K-ϵ model was chosen to run the simulations. The strip having a circular rod and three elliptical shaped leaves is attached with a gap of 50 mm. the elliptical leaves major to minor axes ratio a/b is 2, same orientation (0°–120°–240°) and direction at altered inclinations (0°–180°) with 10° regular intervals are used to increases heat transfer rate. Considering the FLUENT results, the heat transfer rate increases from 0° elliptical leaf angles to 60° elliptical leaf angle and declines from 60° elliptical leaf angle to 180°. The heat transfer rate and pressure drop move with increase of Reynolds number. The novelty of this paper includes a passive method of strip insertion. This passive method has not been tried before for experimentation purpose. As we got better results at 60° angle, the future work may include the insertion of other shapes or at other different angles to obtain a better performance. The overall heat transfer rate increase when Reynolds number increases from 5000 to 20000, at 60 degree elliptical leaf angle it is 21% higher than the existing double pipe heat exchanger.

References

1. Koppula SB et al (2017) Study on various parameters in the design of double pipe heat exchanger on hot fluid side in inner pipe. Int J Adv Res Sci Eng (IJARSE) 06(12):1017–1028. ISSN: 2319-8354
2. Rakesh C et al (2017) Design and performance analysis of double pipe heat exchanger. Int J Innov Res Sci Eng Technol (IJIRSET) 6(7):12578–12584. ISSN (Online): 2319-8753 or ISSN (Print): 2347-6710
3. Mehrabian MA et al (2001) The overall heat transfer characteristics of a double pipe heat exchanger: comparison of experimental data with predictions of standard correlations. Trans Modell Simul 30:607–618. ISSN: 1743-355X
4. Durga Prasad PV et al (2015) Investigation of trapezoidal-cut twisted tape insert in a double pipe u-tube heat exchanger using Al_2O_3/water nanofluid. Proc. Mater. Sci. 10:50–63 (Elsevier)
5. Sunil J et al (2012) Experimental analysis of heat transfer enhancement in circular double tube heat exchanger using inserts. Int J Mech Eng Technol (IJMET) 3(3):306–314. ISSN (Print): 0976-6340, ISSN (Online): 0976-6359
6. Kale Shivam B et al (2017) Experimental analysis & simulation of double pipe heat exchanger. IJARIIE 3(2):2357–2367. ISSN (O): 2395-4396
7. Rao AV et al (2018) Numerical analysis of double pipe heat exchanger with and without strip. Int J Res Appl Sci Eng Technol (IJRASET) 6(VI):855–864. ISSN: 2321-9653
8. Ojha PK et al (2015) Design and experimental analysis of pipe in pipe heat exchanger. Int J Mod Eng Res (IJMER) 5(3):42. ISSN: 2249-6645

Artificial Intelligence in Manufacturing Process

Evaluation of Optimization Parametric Condition During Machining for Al-CSA Composite Using Response Surface Methodology

R. Sivasankara Raju, C. J. Rao, D. Sreeramulu and K. Prasad

Abstract This work aims to the optimization of cutting process parameters on the surface finish for Aluminum (Al)-coconut shell ash (CSA) composite reinforced with 15% volume prepared via stir casting route. The machining has been performed with H20TI tool. The experiments have been designed using response surface methodology (RSM), i.e., Box–Behnken design. The input process parameters are speed (*A*), feed (*B*) and depth of cut (*C*), whereas the output is surface finish (*Ra*). Better surface finish and reduction of tool wear can be achieved with deployment of lubricant. ANOVA has been performed and revealed feed is the most influencing parameter followed the depth of cut on surface finish.

Keywords ANOVA · AMCs · CSA · Machining · RSM

1 Introduction

The sector of manufacturing has been aimed at process concepts with the reduction of scrape rate with better surface finish via a reduction in manufacturing lead time (MLT) and production cost. Usually, the surface finish has been depending upon input process parameters such as feed, cutting speed and depth of cut [1, 2]. In the trend of the current scenario, AMCs are wildly used to automobile applications owing to a high specific strength and better modulus of elasticity [3–5]. Nowadays, most of the researches have been carried out based on the conventional particulate reinforcement like SiC, Al_2O_3, B_4C, BN, TiC, etc., but those were costly. Hence, focus must be given too easily available and inexpensive reinforcement particulates in the atmosphere [6–11]. However, AMCs are more difficult to machining as compared to base or its alloys. Hence, researchers are concerned with manufacturing concepts based on mathematical models, which recommended a reduction in

R. Sivasankara Raju
Associate Professor, Department of Mechanical Engineering, AITAM, Tekkali 532201, India

C. J. Rao · D. Sreeramulu · K. Prasad (✉)
Department of Mechanical Engineering, AITAM, Tekkali 532201, India

BBVL. Deepak et al. (eds.), *Innovative Product Design and Intelligent Manufacturing Systems*, Lecture Notes in Mechanical Engineering,
https://doi.org/10.1007/978-981-15-2696-1_46

scrap rate. Based on machining component with respect to strength and hardness of material, Insert varies for faster machining and better surface finish.

Kok and Ozdın [7] studied the tribological performance of Al-2024-Al_2O_3 composite. Bello et al. [12] studied on A1-CSA (2–10 wt%), fabricated with compo cast technique. The strength of composite has increased due to the distribution of fillers within the matrix and their recovery. Wear rate of Al6061/SiC (5–40 wt%) composite decreased with increasing SiC volume [13]. Siva Sankara Raju et al. [2, 4] studied the optimization of machinability properties of Al reinforced with coconut shell ash using the Taguchi Method. They revealed feed rate was the dominant parameter and then cutting speed followed by the depth of cut in the machining process of the Al-CSA composite. The objective of this study is to utilize coconut shell ash, which is a solid waste, easily available at low cost and possesses many environmental problems, as an effective filler material for the preparation of Al MMCs. The present study is to investigate the machinability of coconut shell ash particulates (CSA) reinforced in aluminum composite fabricated with 15% of Volume. ANOVA has been performed on surface finish and revealed feed rate was the dominant parameter in the machining process of the Al-CSA composite.

2 Testing Methods and Design of Experiments

Cast Al-CSA composites contain in situ produced CSA particulates have been developed by stir casting route varying amount of CSA volume in the molten melt, and the procedures sequences for preparation of reinforced particulates (i.e., CSA) and composite preparation are detailed in elsewhere [1, 3, 4, 14–16]. The experiments are designed using Box–Behnken with replicates of three. The total number of experiments is 15. Three selecting input parameters along levels illustrate in Table 1. The sample is shown in Fig. 1.

The turning operation was performed on the NC machine using H20TI (CNMG 120408) insert, shown in Fig. 2. The workpiece has a length of 200 mm with a diameter of 50 mm. The surfaces finish (*Ra*) has been measured by using SURTRONIC 25. The direction of the roughness measurement was perpendicular to the cutting vector. A randomized experimental run has been carried out to minimize the responses. Analysis of variance (ANOVA) has been used to investigate the individual, interaction and square effects of the process variables on the response. ANOVA has been performed using MINITAB 16 software.

Table 1 Machining process parameters with levels

Process variables	Low (−1)	Mid (0)	High (+1)
Speed (*A*)	175	200	225
Feed (*B*)	0.125	0.15	0.175
Depth of cut (*C*)	0.35	0.8	1.15

Fig. 1 Cast Al-CSA composites

Fig. 2 Machining process

3 Results and Discussions

The layout of experiments along response is shown in Table 2. The effects of variable parameters (i.e., *A*, *B* and *C*) with interaction on the surface finish (roughness) have elaborated using regression models as shown in Eq. (1).

$$\begin{aligned} Ra = 1.182 &- (A * 0.034) + (B * 0.185) + (C * 0.075) \\ &- (A^2 * 0.028) - (B^2 * 0.0242) - (C^2 * 0.0523) + (AB * 0716) \\ &- (AC * 0.146) - (BC * 0.079) \end{aligned} \quad (1)$$

The shown surface and contours plots have been indicating the influence of process variables and interactions on surface finish.

Contour and surface plots are represents interaction among process parameters on surface finish, which vide in multiple colors. Figure 3 shows interaction of

Table 2 Experimental layout with response (surface finish)

Std order	*A*	*B*	*C*	*Ra*
1	−1	−1	0	1.04944
2	1	−1	0	0.83908
3	−1	1	0	1.2763
4	1	1	0	1.3521
5	−1	0	−1	0.91505
6	1	0	−1	1.13809
7	−1	0	1	1.35443
8	1	0	1	0.99683
9	0	−1	−1	0.76597
10	0	1	−1	1.29563
11	0	−1	1	1.07475
12	0	1	1	1.28497
13	0	0	0	1.18176
14	0	0	0	1.18176
15	0	0	0	1.18176

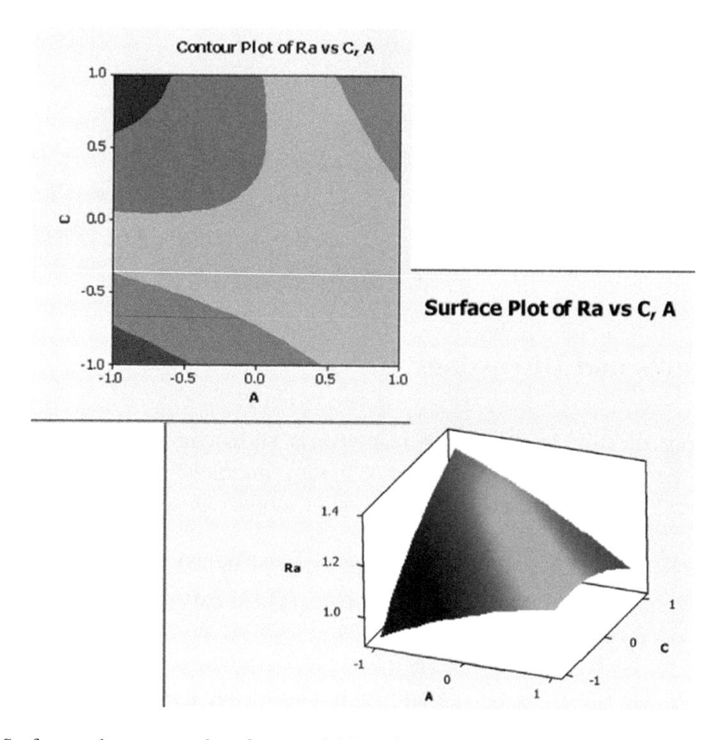

Fig. 3 Surface and contours plots for speed (*A*) and depth of cut (*C*) on *Ra*

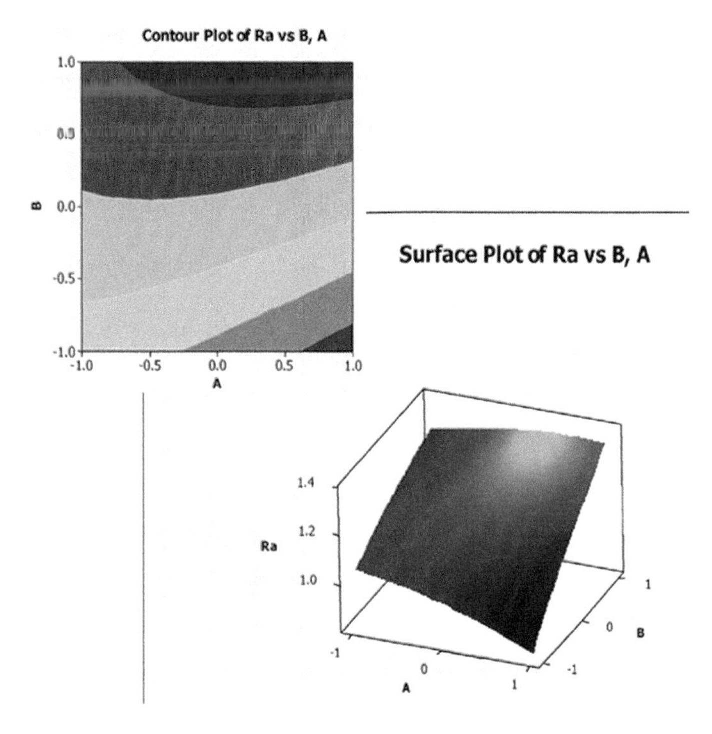

Fig. 4 Surface and contour plots for machining feed and speed

cutting speed, depth of cut and surface roughness increases with the increase of depth of cut at lower cutting speed. Figure 4 illustrates that surface plots show that as the cutting speed and feed increase, roughness increases, but if cutting speed increases, surface roughness decreased. In the interaction of feed and depth of cut (Fig. 5), roughness value increases as the values of both parameters increase.

Surface finish increases with the increase of feed and speed with its interactions. Similarly, other results are the same as given above.

Figure 6 indicates that cutting speed contributes less toward the surface finish. It can also be observed that the experiments performed minimum roughness is 0.6976, achieved at cutting speed of 175 m/min, feed of 0.125 mm/rev and depth of cut at 0.35 mm. The maximum value of the surface finish is 1.35443 μm. The interaction among feed rate and axial depth of cut has also a significant influence on roughness.

ANOVA (Table 3) revealed that feed is the maximum influence on surface finish, followed interaction of speed and depth of cut. Analysis of variance has been analyzed the adequacy of the model. Model is analyzed at a 95% level of confidence. The value of R^2_{Adj} is 91.52% for the current model which indicates that the model is adequate. As feed rate decreases from 0.15 mm/rev, the roughness starts decreasing, and when the depth of cut exceeds 0.8 mm, the surface finish decreased. Similarly, the actual model revealed that cutting speed has very less

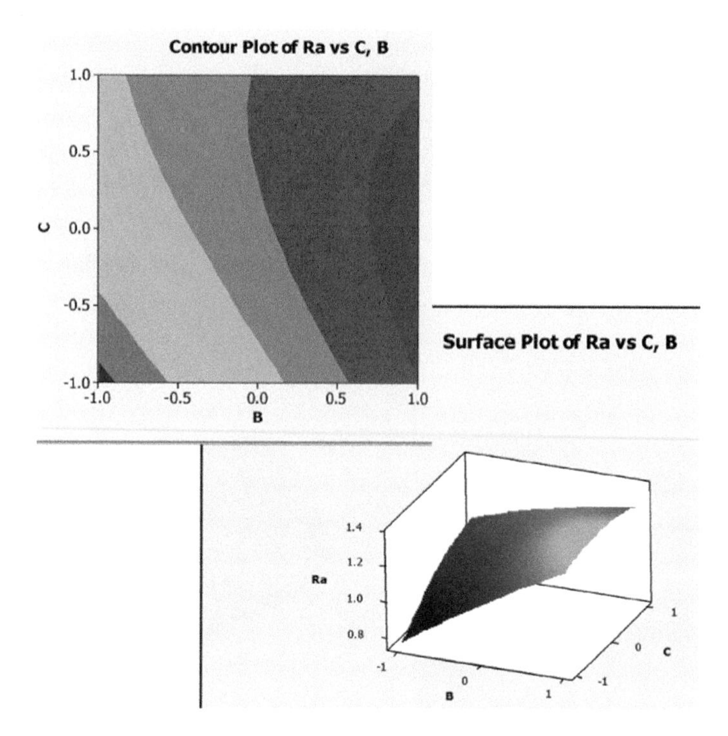

Fig. 5 Surface and contour plots for machining feed and depth of cut

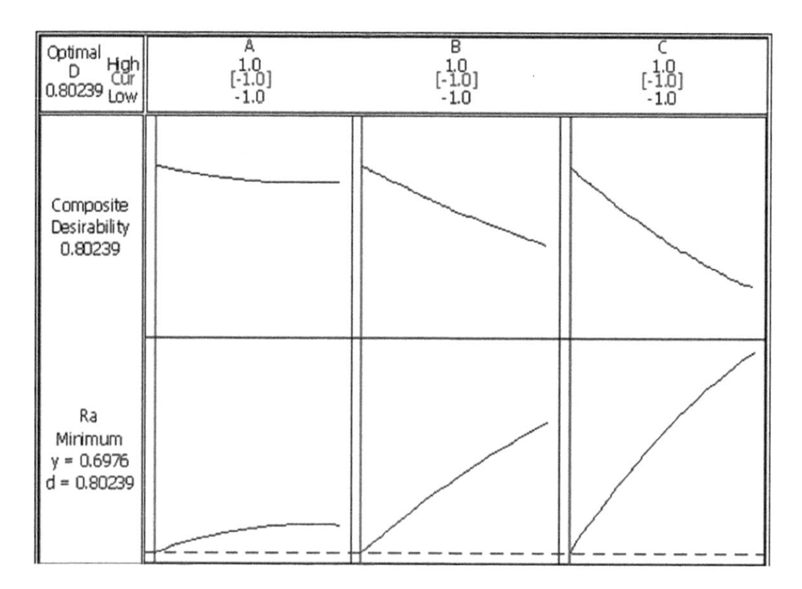

Fig. 6 The optimal condition of the surface finish

Table 3 ANOVA of surface finish

Source	DF	SS	Adj SS	Adj MS	F	Cont %
Regression	9	0.4510	0.4710	0.0523	7.81	96
Linear	3	0.3272	0.3272	0.1091	16.28	69
A	1	0.0091	0.0091	0.0091	1.35	2
B	1	0.2737	0.2737	0.2737	40.85	58
C	1	0.0444	0.0444	0.0444	6.63	9
Square	3	0.0136	0.0136	0.0045	0.67	3
A^2	1	0.0020	0.0030	0.0030	0.44	0
B^2	1	0.0015	0.0022	0.0022	0.32	0
C^2	1	0.0101	0.0101	0.0101	1.51	2
Interaction	3	0.1303	0.1303	0.0434	6.48	28
$A * B$	1	0.0205	0.0205	0.0205	3.06	4
$A * C$	1	0.0843	0.0843	0.0843	12.58	18
$B * C$	1	0.0255	0.0255	0.0255	3.81	5
Error	5	0.0506	0.5060	0.0500	7.46	11
Pure error	2	0.0335	0.0335	0.0067		7
Total	14	0.4710				

impact on the surface finish. The interaction of the cutting speed with a depth of cut evaluates that high values of both these parameters improve the surface quality, but from the main effect plot between *Ra* and D.O.C., Ra and cutting speed, it implies that D.O.C. has much impact on the *Ra* rather than cutting speed.

4 Conclusion

- This study has been aiming to find surface finish for Al-CSA composite. The feed rate has the highest influence on surface finish in the machining of Al-CSA composite followed by interaction of cutting speed and depth of cut.
- With an increase in depth of cut, surface finish increased then decreases.
- Cutting speed does not have much significance on the surface finish. However, interaction among cutting speed and depth of cut has substantial level on surface finish.
- Machining conditions, such as cutting speed (175 m/min), feed rate (0.125 mm/rev) and depth of cut (0.35 mm), can be used to achieve the minimum surface finish for Al-CSA composites are 0.6976.
- A regression equation was developed for correlates process parameters (speed, feed and depth of cut) and response (surface finish).

References

1. Sankara Raju RS, Panigrahi MK, Ganguly RI, Srinivasa Rao G (2018) Optimization of tribological behavior on Al-coconut shell ash composite at elevated temperature. In: IOP conference series: materials science and engineering, vol 314, p 012009
2. Siva Sankar Raju R, Srinivasa Rao G, Muralidhara Rao M (015) Optimization of Machinability properties on aluminium metal matrix composite prepared by in-situ ceramic mixture using coconut shell ash—Taguchi approach. Int J Concept Mech Civ Eng 3(2):17–21
3. Sankara Raju RS, Panigrahi MK, Ganguly RI, Srinivasa Rao G (2019) Tribological behavior of Al-1100-coconut shell ash (CSA) composite at elevated temperature. Tribol Int 129:55–66
4. Siva Sankara Raju R, Panigrahi MK, Ganguly RI, Srinivasa Rao G (2017) Investigation of tribological behavior of a novel hybrid composite prepared with Al-coconut shell ash mixed with graphite. Metall Mater Trans A 48(8):3892–3903
5. Raju SS, Thimothy P, Ch R (2018) Estimation of physical, mechanical and machinability properties of Al-MMCs reinforced with coconut shell ash particulates. Ceram Sci Eng 1(2):1–7
6. Baradeswaran A, Vettivel SC, Elaya Perumal A, Selvakumar N, Franklin Issac R (2014) Experimental investigation on mechanical behavior, modeling and optimization of wear parameters of B4C and graphite reinforced aluminum hybrid composites. Mater Des 63: 620–632
7. Kök M, Özdin K (2007) Wear resistance of aluminium alloy and its composites reinforced by Al_2O_3 particles. J Mater Process Technol 183(2–3):301–309
8. Raju SS, Senapathi AK, Rao GS (2017) Estimation of tribological performance of Al-MMC reinforced with a novel in-situ ternary mixture using grey relational analysis. Indian J Sci Technol 10(15):1–9
9. Altinkok N, Özsert I, Findik F (2013) Dry sliding wear behavior of Al_2O_3/SiC particle reinforced aluminium based MMCs fabricated by stir casting method. Acta Phys Pol A 124 (1):11–19
10. Rajesh S, Krishna AG, Murty PR, Duraiselvam M (2014) Statistical analysis of dry sliding wear behavior of graphite reinforced aluminum MMCs. Eng Technol 6:1110–1120
11. Sahin Y (2010) Abrasive wear behaviour of SiC/2014 aluminium composite. Tribiol Int 43:939–943
12. Bello SA, Raheem IA, Raji NK (2017) Study of tensile properties, fractography and morphology of aluminium (1xxx)/coconut shell micro particle composites. J King Saud Univ Eng Sci 29(3):269–277
13. Mishra AK, Srivastava RK (2016) Wear behaviour of Al-6061/SiC metal matrix composites. J Inst Eng Ser C 1–7
14. Raju SS, Rao GS, Siva BV (2018) Experimental studies of mechanical properties and tribological behaviour of aluminium composite reinforced with coconut shell ash particulates. Int J Mater Eng Innov 9(2):140–157
15. Rallabandi SR, Srinivasa Rao G (2019) Assessment of Tribological performance of Al-coconut shell ash particulate—MMCs using grey-fuzzy approach. J Inst Eng Ser C 100(1): 13–22
16. Raju RSS, Rao GS (2017) Assessment of tribological performance of coconut shell ash particle reinforced Al–Si–Fe composites using grey-fuzzy approach. Tribol Ind 39(3): 364–377

Optimization of Process Parameters in Resistance Spot Welding Using Artificial Immune Algorithm

Sudhakar Uppada, Subbarama Kousik Suraparaju, M. V. A. Raju Bahubalendruni and Sendhil Kumar Natarajan

Abstract Welding is one of the fundamental manufacturing processes and is used for manufacturing components or assemblies with great strength in minimal time. Resistance spot welding (RSW) is utilized often as an efficacious method of joining for different works, most commonly in automobile and other industrial processes. Recent researches in welding are trending toward the economical process with optimum productivity. It is laborious to formulate a mathematical model for the analysis of RSW parameters, because of obscureness during the process with many parameters especially with the property of less operating time. A novel optimization method based on artificial immune algorithm (AIA) is presented in this article to find the optimum set of welding parameters for an economical process which offers the highest load carrying capacity at low power consumption.

Keywords Resistance spot welding · Parametric optimization · Regression · AIA

1 Introduction

In the realm of manufacturing, welding is extensively utilized for joining the metals because of its less time of operation and flexibility to automation. Welding by spots is one among the different kinds of resistance welding that can be utilized as an alternative for screws, rivets, brazing, and soldering. This method of joining is significantly used for low carbon steel, copper, titanium, stainless steel, nickel, and aluminum peripherals [1]. In RSW, more than two metals were joined at once by sending current through them. This conducted current through electrodes produces heat and melts the metal at the spot when they pressed opposite to the metal that to be joined and thus, a spot weld is formed [2]. In recent researches, numerical and

S. Uppada
GMRIT, Rajam, Andhra Pradesh, India

S. K. Suraparaju (✉) ·
M. V. A.R. Bahubalendruni · S. K. Natarajan
National Institute of Technology – Puducherry, Karaikal (U.T. of Puducherry), India

BBVL. Deepak et al. (eds.), *Innovative Product Design and Intelligent Manufacturing Systems*, Lecture Notes in Mechanical Engineering,
https://doi.org/10.1007/978-981-15-2696-1_47

experimental methods have been applied to welding in order to create a relationship between process parameters and weld quality. Ates [3] proposed an ANN-based method for modeling the process parameters in gas metal arc welding (GMAW) and the recommended methodology predicted the mechanical properties of GMAW. The ultimate capacity of the weld was predicted based on investigational outcomes by using ANN [4]. An innovative and intelligent method for selecting process variables was developed to select parameters in GMAW [5]. Darwish and Al-Dekhial [6] investigated on a numerical model for welding aluminum sheets, but the outcome was not so gratifying, and later, it was advanced by Tseng [7], with the assistance of neural network to estimate the developed model. Later, genetic algorithm was used to attain optimum parameters. On the other hand, the effect of influential parameters in spot welding on the mechanical properties and features of the spot welds between 304 ASS sheets of the same thickness was reported. The correlation between the weld resistance and properties such as tensile strength and hardness was discussed [8]. The studies on the effect of process parameters on the titanium sheets strength that was joined by RSW reported that enhancing electrode force and the current time will improve the tensile and shear strength and also reported that the joint attained under the argon environment had comparatively improved strength [9]. A numerical study on the nugget formation in RSW concluded that the increase in welding time increases the sheet interface temperature rapidly and the development of nugget occurs. After the nugget development, the rate of temperature augmentation is declined [10]. Pouranvari [11] analyzed the failure of AISI304 spot welds under tensile—a shear test that conducted quasi-statically and concluded that failure occurs near the heat-affected zone. Panda et al. [12] investigated the influence of process parameters in welding and developed a relationship between predictors and responses. Later, the optimal process parameters for spot welding process were obtained by differential evolution algorithm for economic welding. There are several optimization procedures to accomplish the improved combination of parameters which lead to achieving the optimal and satisfactory output. This paper demonstrates the artificial immune algorithm (AIA) technique and its operation for welding parameters.

2 Experimentation

2.1 Materials and Machinery

The material utilized in this investigation was five BS 1050 aluminum sheets of 0.5–2.5 mm thickness at an increment of 0.5 mm thickness for each sheet, respectively. A Meruit spot welding machine of pedestal type (as shown in Fig. 1) including an AC power supply of single phase was used.

Fig. 1 Spot welding machine

2.2 Experimental Design

At the outset, the parameters like welding time, welding current, and force exerted by electrode were observed. The calculated resistances for 2.5- and 0.5-mm-thick aluminum sheet were about $3.691 \times 10^{-4}\ \Omega$ and $3.663 \times 10^{-4}\ \Omega$ correspondingly. Based on the relationship between the thickness and resistance of the sheet, the resistances of remaining sheets (2.0, 1.5, and 1.0 mm) were estimated as $3.685 \times 10^{-4}\ \Omega$, $3.678 \times 10^{-4}\ \Omega$, and $3.671 \times 10^{-4}\ \Omega$ correspondingly.

The experimental design with four influencing parameters at five levels was tabulated as Table 1. The power consumed during the operation shall be calculated by the following relation:

$$P = I^2RT \tag{1}$$

where

- I Welding current (A),
- R Welding resistance (Ω) and
- T Welding time (s),
- P Power (W).

Table 1 Process parameters with their levels

Factor	Symbol	Unit	Level				
			1	2	3	4	5
Welding current	C	A	13860.3	15236.315	16612.33	17988.345	19364.36
Electrode force	F	N	756.56	980.65	1204.746	1428.84	1652.93
Welding time	T	ms	75	140	205	270	335
Sheet thickness	ST	mm	0.5	1	1.5	2	2.5

Table 2 Process parameters and respective failure load

S. No.	Welding current (A)	Electrode force (N)	Welding time (ms)	Thickness (mm)	Failure load—*FL* (N)
1	15,236	1428.8	270	1.00	705.30
2	17,988	1428.8	140	1.00	831.00
3	15,236	980.7	270	2.00	706.30
4	15,236	980.7	140	1.00	737.70
5	17,988	980.7	270	1.00	973.00
6	17,988	1428.8	270	2.00	960.30
7	17,988	980.7	140	2.50	1197.00
8	15,236	1428.8	140	2.00	589.30
9	16,612	1204.7	205	1.50	911.85
10	17,988	980.7	140	1.00	936.00
11	15,236	1428.8	140	1.00	705.70
12	15,236	980.7	270	1.00	735.70
13	17,988	1428.8	270	1.00	918.70
14	15,236	980.7	140	2.00	758.30
15	17,988	1428.8	140	2.00	927.30
16	17,988	980.7	270	2.00	1053.30
17	15,236	1428.8	270	2.00	855.70
18	13,860	1204.7	205	1.50	681.00
19	19,364	1204.7	205	1.50	1157.30
20	16,612	756.6	205	1.50	1075.30
21	16,612	1652.9	205	1.50	762.00
22	16,612	1204.7	75	1.50	760.00
23	16,612	1204.7	335	1.50	973.30
24	16,612	1204.7	205	0.50	369.00
25	16,612	1204.7	205	2.50	772.70

From the above experimental design, 25 sets were taken for conducting experiments to estimate the average failure loads; for each combination of parameters, the outcomes were tabulated in Table 2.

It is observed that the highest failure load was 1197 N at the seventh set of combination as highlighted in Table 2 and the power consumed for the respective set to attain that failure load was 16720.06 W.

3 Artificial Immune Algorithm (AIA)

It is an optimization technique developed based on the functioning of human immunology. The human immune scheme consists of antibodies and the pathogen which are the key causes for illness; so as to heal these ailments, the antibodies are engendered. If the generated antibodies had the inability to cure or take away the pathogen, then the antibody with maximum fitness is cloned. The cloned antibody is mutated to attain the antibody with high affinity. These antibodies are thus selected for curing or eliminating the pathogen. The AIA technique was evolved as an imitation to the working of the human immune system and rejuvenated for several engineering applications [13–19]. This algorithm was utilized for transportation, scheduling, and flow shop problem till now [20–22]. MATLAB is used to create and execute code for optimization of process parameters. In this algorithm, the antibodies are denoted as binary codes which decrease the complication in mutation. This paper utilizes the inverse mutation technique.

3.1 Cloning

The principle of clonal selection illustrates the elementary characteristics of an immune response to antigen.

The key aspects in the selection of clones are

1. The improved cells are replicas of their parents (clone) subjected to mutation.
2. Differentiation and proliferation were due to mature cells' interaction with antigens.

3.2 Mutation

The formation of improved antibodies with changed affinity value is mutation. Here, inverse mutation is used.

For example, let *a* and *b* be the positions of antibody that are randomly generated.

$x = 2$, $y = 4$

Before mutation

1	0	0	1	0

After mutation

1	1	0	0	0

The criteria of AIA are as following:

1. Setting the size of the population
2. Finding the antibody fitness value
3. Selecting the antibody with the finest fitness
4. Conversion to the binary system from antibody
5. Cloning of particular antibody for some × no. of times
6. Mutating every clone
7. Attaining of mutated clone that having the finest fitness (F_b)
8. If the fitness of F_b is improved than the fitness of the initially selected body, then swap it with the latest
9. Repeat the steps from 2 to 9 for more iterations.

3.3 Regression Analysis

A multi-response regression equation was generated for each set of experiments. The generated equations are similar to the objective function used in AIA optimization. Since to getter better weld joint strength, the input parameters have to reduce their influence on the predictor. Therefore, the generated equations are to be minimized to attain better weld joint strength. The generated equations for various responses from the cubic model of the fitted line plot in regression analysis, viz. MINITAB 18 software, were

$$C = 26,860 - 48.02\,FL + 0.06074\,FL^2 - 0.000022\,FL^3, \tag{2}$$

subject to $13860.3 \leq C \leq 19364.36$

$$F = 123 + 4.754\,FL - 0.005960\,FL^2 + 0.000002\,FL^3, \tag{3}$$

subject to $756.56 \leq F \leq 1652.93$

$$T = 464.3 - 1.259\,FL + 0.001725\,FL^2 - 0.000001\,FL^3, \tag{4}$$

subject to $75 \leq T \leq 355$

$$ST = -1.207 + 0.00746\,FL - 0.000007\,FL^2 + 0.000000\,FL^3, \tag{5}$$

subject to $0.5 \leq ST \leq 2.0$.

4 Results and Discussions

The generated equations were solved by using a MATLAB code for artificial immune algorithm by following the steps stated earlier. Therefore, the optimized values shown in Table 3 of welding current, electrode force, welding time, and sheet thickness were 17,500 A, 750 N, 115 ms, and 2.5 mm, respectively, and the optimized values were investigated experimentally and calculated the failure load for the attained set of combination. The failure load obtained for the generated optimized values was satisfactory. The relationship is shown in Fig. 2.

Table 3 Representation of optimized results

Welding current (A)	Electrode force (N)	Welding time (ms)	Thickness (mm)	Failure load (N)	Power (W)
17,500	750	115	2.5	1196.41	12999.24

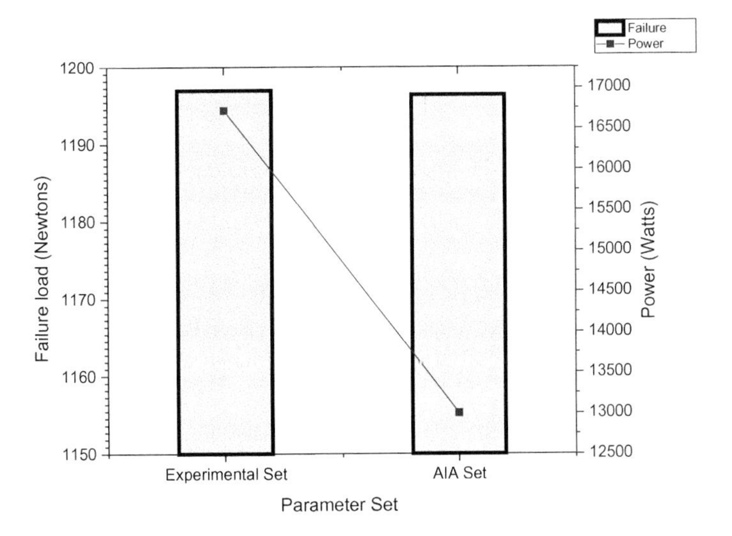

Fig. 2 Graphical representation of experimental setup in comparison with the optimized set

5 Conclusions

In the current study, an economic welding design method has been advanced to identify the quality of weld at several choices of process parameters at a low operating cost. The technique is based on the artificial immune algorithm motivated by the human immune system. In this optimization model, the optimum condition of welding was obtained by minimalizing the power consumption of the welding process which in turn reduces the operating cost. These parameters are utilized to attain the chosen welding quality at a minimum possible cost. The proposed method of optimization is based on single-objective optimization where it can only be able to optimize one parameter at one time. There is scope to extend this method for multi-objective optimization which all the parameters can be optimized in a single run. From the results, it is evident that the power consumed for at the optimum parameters for a failure load of 1196.41 N was only 12999.23 W but for the almost equal failure load of 1197 N at the 7th experimental run in Table 2 was 16720.06 W. Hence, the power consumption at an optimum set of parameters was low in which it enables the low-cost welding process so that the objective of optimization was achieved.

References

1. Jou M (2003) Real-time monitoring weld quality of resistance spot welding for the fabrication of sheet metal assemblies. J Mater Process Technol 132(1–3):102–113
2. Tsai CL, Papritan JC, Dickinson DW, Jammal O (1992) Modeling of resistance spot weld nugget growth. Weld J 71(2):47
3. Ates H (2007) Prediction of gas metal arc welding parameters based on artificial neural networks. Mater Des 28(7):2015–2023
4. Cevik A, Kutuk MA, Erklig A, Guzelbey IH (2008) Neural network modeling of arc spot welding. J Mater Process Technol 202(1–3):137–144
5. Kim IS, Son JS, Yarlagadda PKDV (2003) A study on the quality improvement of robotic GMA welding process. Robot Comput Integr Manuf 19(6):567–572
6. Darwish SM, Al-Dekhial SD (1999) Statistical models for spot welding of commercial aluminum sheets. Int J Mach Tools Manuf 39(10):1589–1610
7. Tseng HY (2006) Welding parameters optimization for an economic design using neural approximation and genetic algorithm. Int J Adv Manuf Technol 27(9–10):897–901
8. Bouyousfi B, Sahraoui T, Guessasma S, Chaouch KT (2007) Effect of process parameters on the physical characteristics of spot weld joints. Mater Des 28(2):414–419
9. Kahraman N (2007) The influence of welding parameters on the joint strength of resistance spot-welded titanium sheets. Mater Des 28(2):420–427
10. Hamedi M, Pashazadeh H (2008) Numerical study of nugget formation in resistance spot welding. Int J Mech 2(1):11–15
11. Pouranvari M (2011) Prediction of failure mode in AISI 304 resistance spot welds. Assoc Metall Eng Serbia Predict
12. Panda BN, Raju Bahubalendruni MVA, Biswal BB (2014) Optimization of resistance spot welding parameters using differential evolution algorithm and GRNN. In: 2014 IEEE 8th international conference on intelligent systems and control: green challenges and smart solutions, ISCO 2014—proceedings, pp 50–55

13. Lin WM, Gow HJ, Tsai MT (2011) An efficient hybrid Taguchi-immune algorithm for the unit commitment problem. Expert Syst Appl 38(11):13662–13669
14. Bakhouya M, Gaber J (2007) An immune inspired-based optimization algorithm: application to the traveling salesman problem. Adv Model Opt 9(1):105–116
15. Aickelin U, Dasgupta D (2013) Search methodologies : introductory tutorials in optimization and decision support techniques. Artificial immune systems. In: In: Burke EK, Kendall G (eds) Search methodologies—introductory tutorials in optimization and decision support technology, 2005, pp 1–29
16. Cisar P, Cisar SM, Markoski B (2014) Implementation of immunological algorithms in solving optimization problems. Acta Polytech Hungarica 11(4):225–239
17. Syahputra R, Soesanti I (2017) An artificial immune system algorithm approach for reconfiguring distribution network. AIP Conf Proc 1867
18. Wang M, Feng S, He C, Li Z, Xue Y (2017) An artificial immune system algorithm with social learning and its application in industrial PID controller design. Math Probl Eng 2017:1–13
19. Raju Bahubalendruni MVA, Deepak BBVL, Biswal BB (2016) An advanced immune-based strategy to obtain an optimal feasible assembly sequence. Assem Autom 36(2):127–137
20. Zhang Z, Liao M, Wang L (2012) Immune optimization approach for dynamic constrained multi-objective multimodal optimization problems. Am J Oper Res 02(02):193–202
21. Zandieh M, Fatemi Ghomi SMT, Moattar Husseini SM (2006) An immune algorithm approach to hybrid flow shops scheduling with sequence-dependent setup times. Appl Math Comput 180(1):111–127
22. Mohammad R, Akbarzadeh T, Davarzani Z, Khairdoost N (2012) Multiobjective artificial immune algorithm for flexible job shop scheduling problem. Int J Hybrid Inf Technol 5(3):75–88

An Evolutionary Algorithm-Based Damage Detection in Structural Elements

Sasmita Sahu, D. R. Parhi and B. B. Nayak

Abstract Damage in structural and rotating machine elements causes the local changes in the dynamical parameters of the system. To find the damage present in a system is becoming one of the important research topics in today's civil and mechanical engineering field. This topic of research can mainly be used in bridges, offshore platforms, plates, shells, beams, aerospace and composite structures and other large civil structures to detect structural damages by analyzing the dynamic features of the system. It has been observed due to any slight physical change, the stiffness of the system changes which changes the modal responses of the system. The aim of this research work is to derive a simple method for estimating the failure parameters (crack depth and crack location) in structures based on a data-driven subspace identification technique. These changes in the modal parameters can be used as the input variables to find out the damage severity. The responses (natural frequencies) were obtained using finite element analysis, and then, the differential evolution algorithm (a type of evolutionary algorithm) is used to detect and characterize these defects. This work proposes a robust computational application of the differential evolution algorithm that more accurately takes into account the natural evolution with initial point and produces a good converging result.

Keywords Damage · Natural frequencies · Differential evolution algorithm

Nomenclature

E Modulus of elasticity for the cantilever beam material
[K] Stiffness matrix
[M] Mass matrix
rcd Dimensionless crack depth

S. Sahu (✉) · B. B. Nayak
KIIT University, Bhubaneswar, India

D. R. Parhi
NIT Rourkela, Rourkela, India

BBVL. Deepak et al. (eds.), *Innovative Product Design and Intelligent Manufacturing Systems*, Lecture Notes in Mechanical Engineering,
https://doi.org/10.1007/978-981-15-2696-1_48

rcl	Dimensionless crack location
fnf	Dimensionless form of first natural frequency
snf	Dimensionless form of second natural frequency
tnf	Dimensionless form of third natural frequency

1 Introduction

An accurate, indirect and efficient method for damage identification and detection in structures remains one of the challenging problems in engineering. There a number of methods for the detection and localization of the damage in various structural elements like trusses, frames and beams. The traditional type of damage detection methods are not very convenient for online damage detection. For this type of problems, sometimes sophisticated optimization techniques are required. Evolutionary Algorithms can deal with above stated problems. This chapter focuses directly on the problem of crack detection by proposing a differential evolution algorithm. Here, mutation and crossover operations are sensible enough to find out the global maximum. Here is the problem, the input to the system is the relative values of the first three natural frequencies. The training of these vibration parameters gives the location of the crack.

Components of industrial machines are always in an overstressed condition as they are required to perform under a higher power and high speed of operation. So, it is now imperative to predict accurately the dynamics characteristics of the rotating machine members.

Traditional mathematical methods for damage identification and detection are becoming insufficient due to difficulty in modeling of highly nonlinear components. New techniques for modeling based on AI techniques have shown positive results for handling nonlinearities occurring in the problem. So, the inverse engineering method of damage detection and identification of any material of any section can be solved using computational intelligent techniques. The following section gives some of the literature based on the AI techniques for damage detection.

Different metaheuristic is described in this work, and one of them is the differential evolution algorithm. This evolutionary algorithm operates on real-valued optimization problems by evolving a population of possible solutions. It has been emerging as a powerful population-based stochastic search technique for various types of problems. This method does not require any encoding scheme, but its success largely depends on the appropriately choosing trial vector from the initial population of vectors. A lot of publications is present on the implementation and formulation of the differential evolution algorithm, but a few damage detection methodologies using differential evolution algorithm to date.

1.1 Finite Element Analysis Approach for Detection of Damage in the Cracked Beam Element

The Euler–Bernoulli beam model is assumed for the finite element formulation. The crack in this particular case is assumed to be an open crack, and the damping is not being considered in this theory. The geometry of the single crack in a cantilever beam has been shown in 'Fig. 1.' The following assumptions are made for finite element analysis of the cracked beam.

a. The crack is open, and it is not a breathing crack
b. The crack is uniform in the propagation
c. There is no shear deformation and rotary inertia effects.

The bending vibration (free) of a Euler–Bernoulli beam of a consistent rectangular cross area is given by the following differential equation as given in:

$$EI\frac{d^4y}{dx^4} - m\omega_i^2 y = 0 \tag{1}$$

where 'm' is the mass of the beam per unit length (kg/m), 'ω_i' is the natural frequency of the ith mode (rad/s), 'E' is the modulus of elasticity (N/m^2), and 'I' is the moment of inertia (m^4). By characterizing $\alpha^4 = \frac{m\omega_i^2}{EI}$, mathematical statement is adjusted as a fourth-arrange differential equation as follows:

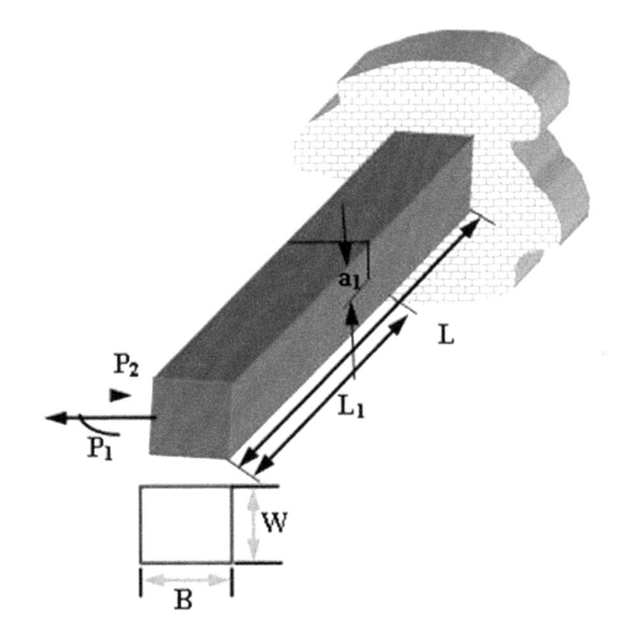

Fig. 1 Diagram for cracked cantilever beam with its dimensions

$$\frac{d^4y}{dx^4} - \alpha^4 y = 0 \tag{2}$$

The general solution to the equation is

$$y = A\cos\alpha_i x + B\sin\alpha_i x + C\cosh\alpha_i x + D\sinh\alpha_i x \tag{3}$$

where A, B, C, D are constants and 'λ_i' is a frequency parameter.

The governing differential equation for the system is given as

$$[M]\ddot{x} + [C]\dot{x} + [K]x = F\sin(\omega t) \tag{4}$$

But it is assumed that there is no damping and there is no external force applied on the system. So, the governing equation becomes

$$[M]\ddot{x} + [K]x = 0 \tag{5}$$

The equation of motion for natural frequencies for undamped free vibration is given in Eq. (5). To solve Eqs. (4 and 5), it is assumed that

$$\{x\} = \{\phi\}\sin\omega t \tag{6}$$

where

ϕ—The Eigenvector or mode shape, ω—Circular natural frequency, substituting the differential equation of assumed solution into the Eq. (5), the equation of motion will be changed to

$$\omega^2[M]\{\phi\}\sin\omega t + [K]\{\phi\}\sin\omega t = 0 \tag{7}$$

After simplification, it becomes

$$\left([K] - \omega^2[M]\right)\{\phi\} = 0 \tag{8}$$

The equation is called Eigen equation. The basic form of the Eigenvalue problem is

$$[A - \alpha I]x = 0 \tag{9}$$

A Square matrix,
α Eigenvalues,
I Matrix of Identity,
x Eigenvector.

In structural analysis, Eigen equation is written in terms of K, M and ω with $\omega^2 = \lambda$. There are two possible solutions for Eq. (8)

1. If $\left|([K] - \omega^2[M])\{\phi\}\right| \neq 0$ is a trivial solution where $\{\phi\} = 0$
2. If $([K] - \omega^2[M])\{\phi\} = 0$ is a non-trivial solution where $\{\phi\} \neq 0$

$$\left|([K] - \omega^2[M])\right| = 0 \tag{10}$$

$$|([K] - \alpha[M])| = 0 \tag{11}$$

The determinant is zero only at discrete Eigenvalues

$$\left|([K] - \omega^2[M])\right|\{\phi_i\} = 0, i = 1, 2, 3, \ldots \tag{12}$$

With the use of Hermitian shape functions, the two-noded beam element without a crack is achieved using the standard integration based on the variation in flexural rigidity. The elemental stiffness matrix of a cantilever beam without crack has been given below

$$[K^e] = \int [B(x)]^T EI[B(x)]\mathrm{d}x \tag{13}$$

$$[B(x)] = \left\{H_1''(x)H_2''(x)H_3''(x)H_4''(x)\right\} \tag{14}$$

Here, $H_1(x), H_2(x), H_3(x), H_4(x)$ represent the Hermitian shape functions or interpolation function.

2 Analysis of DEA for Damage Detection of Cracked Structural Elements

Differential evolution algorithm (DEA) compares the quality of trial vector with the target vector from which it receives parameters. Due to the above-mentioned characteristic of the algorithm, the crossover and selection process can be integrated more tightly as compared to other evolutionary algorithms. After the generation of the initial set of the population, the process of mutation, crossover and selection is repeated for several generations until the algorithm halts. In each generation, some improved individuals are included in the initial population which can explore the solution space in search for better solutions.

The above-mentioned three steps are repeated iteration after iteration until some specific termination criteria are satisfied.

The following steps describe the algorithm for differential evolution algorithm.

1. Define the variables used in the algorithm—population size N_p, crossover constant C_r, scaling factor F, maximum iterations G_{max} and decision variables D (i.e., 'rfnf,' 'rsnf,' 'rtnf,' 'rcd' and 'rcl')
2. The generation number is set at $G = 0$, population number $i = 1$ and the decision variable at $k = 1$
3. The parent vector is initialized uniformly in the random population space. The starting value of the 'k'th parameter in the 'i'th individual at the generation $G = 0$ is given as

$$x_{k,i}^{(0)} = x_k^{\min} + \text{rand}_k(0,1) \cdot \left(x_k^{\max} - x_k^{\min}\right),\ k = 1 \text{ to } D(5),\ i = 1 \text{ to } N_p \quad (15)$$

4. After initialization, find the fitness values of the candidate solutions in the initial population applying fitness evaluation function. The individual with the highest fitness value is selected as the target vector.
5. After the initialization of the population number, mutation of the population is started to produce a population of N_p mutant vectors.

$$X_i^{\prime(G)} = X_a^G + F\left(X_b^G - X_c^G\right), i = 1, \ldots, N_p \quad (16)$$

The indices a, b and c are chosen randomly. These indices are different from each other and from the base vector index 'i'. The scaling factor 'F' is a real and constant factor.

After the completion of the mutation operation, a binomial crossover is employed to generate trial vectors $\left(X_i''\right)$, by mixing the mutant vectors and target vector (X_i).

$$X_{k,i}^{\prime\prime(G)} = \begin{cases} \left(X_{k,i}^{\prime(G)}\right) & \text{if rand}_k(0,1) \leq C_r \text{or } k = k_{\text{rand}} \\ X_{k,i}^{(G)} & \text{Otherwise} \end{cases} \quad (17)$$

Crossover operation is applied to each pair of the target vector (X_i) and its corresponding mutant vector $\left(X_i'\right)$ to generate a trial vector $\left(X_i''\right)$: The probability of crossover $C_r \in (0, 1)$ is a user-defined value. The crossover operator replicates the 'j'th parameter of the mutant vector $\left(X_i'\right)$ to the corresponding parameter of the trial vector $\left(X_i''\right)$ if rand$_k$(0, 1) $\leq C_r$, otherwise, it is copied from the corresponding target vector $\left(X_i'\right)$.

6. Sometimes, the upper and lower value of the newly generated trial vectors exceeds the given value. So, they are randomly and uniformly initialized to the initial value given previously. The objective function values of all trial vectors are evaluated. Then, the selection operator (according to the fitness rank) determines the population by choosing between the trial vectors and their predecessors (target vectors).

a. If the trial vector $X_i^{\prime\prime(G)}$ has an equal or lower fitness function value (optimal) than that of its target vector $X_i^{(G)}$, it replaces the target vector in the next generation.

Otherwise, the target vector retains its place in the population for at least one more generation.

$$X_i^{G+1} = \begin{cases} X_i^{\prime\prime(G)} & \text{if} f(X_i^{\prime\prime(G)}) \leq f(X_i^{(G)}),\ i = 1, \ldots, N_\text{p} \\ X_i^{(G)} & \text{otherwise} \end{cases} \tag{18}$$

After the generation of a new population is established, the process of mutation, crossover and selection is repeated for several generations.

7. The algorithm stops after attaining the threshold values for the target vector.

3 Experimental Investigation

The outcomes from the proposed strategies have been contrasted with the outcomes of the finite element analysis. For further validation of the results, it is necessary to compare the results obtained from the proposed methods with experimental results. The procedures employed in the experimental inspection have been systematically organized in the segment. The experimental setup has been described pictorially in 'Fig. 2.' Various tests are performed on the test pieces using these setups, and the natural frequencies of the first, second and third mode of vibration are noted down.

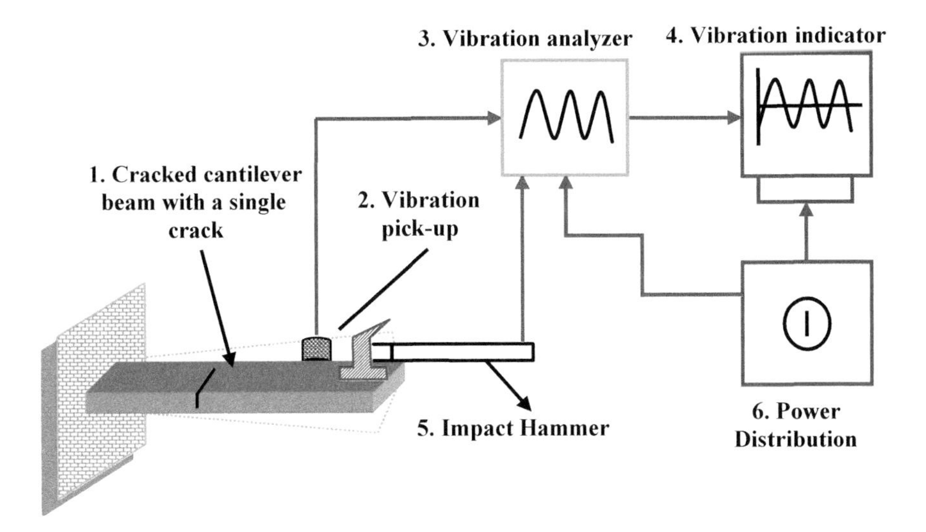

Fig. 2 Diagrammatic presentation of the experimental setup

For the experiments, aluminum alloy beam of 800 × 38 × 8 mm^3 has been used. Here, free vibration analysis of the beam element is performed. As the experiment is done at free vibration mode, so the beam element is excited using an impact hammer. Then, the vibration features from the whole length of the beam element are collected by the accelerometer or vibration pick-up. The vibration responses from the accelerometer which contains information about the modal frequencies are supplied to the vibration analyzer for further study. Then, the results from the analyzer are shown in vibration indicator. The transverse hairline cracks on the aluminum alloy beam are created using electrical discharge machining. The natural frequencies are obtained both for cracked and uncracked beam element.

3.1 Specifications of the Vibration Measuring Instruments

See Table 1.

Table 1 Description and specifications of the instruments used in the experimental setup

S.No.	Name of the instrument	Description
1	Beams	Cracked (Single crack) cantilever beams made from aluminum alloy with dimension 800 mm × 3.8 mm × 8 mm
2	Vibration pick-up	Type: 4513-001 Make: Bruel & kjaer Sensitivity: 10–500 mv/g Frequency range: 1 Hz–10 kHz Supply voltage: 24 V Operating temperature range: −50 to +100 °C
3	Vibration analyzer	Type: 3560L Product name: pocket front end Make: Bruel & kjaer Frequency: 7 Hz to 20 kHz range ADC bits: 16 Simultaneous channels: 2 inputs, 2 Tachometer input type: direct/CCLD
4	Vibration indicator	PULSE LabShop software version 12 Make: Bruel & kjaer
5	Impact hammer	Range: 222N Maximum force: 890 N Frequency range: 10 kHz Overall length: 122 mm
6	Power distribution	220 V power supply, 50 Hz

Table 2 Results for cantilever beam for FEA

S. No.	rfnf from exp. analysis	rsnf from exp. analysis	rtnf from exp. analysis	rcd from exp. analysis	rcl from exp. analysis	rcd using the DEA technique	rcl using the DEA technique	Percentage error rcd	Percentage error rcl	Total error
1	0.9973	0.9914	0.9995	0.3437	0.4687	0.3295	0.4493	4.12	4.13	4.12
2	0.9974	0.989	0.9999	0.375	0.5	0.3594	0.4794	4.14	4.12	4.13
3	0.9981	0.9982	0.9979	0.25	0.375	0.2397	0.3594	4.12	4.14	4.13
4	0.9988	0.9981	0.9989	0.2187	0.4062	0.2096	0.3894	4.15	4.14	4.14
5	0.9892	0.9996	0.9981	0.35	0.1875	0.3356	0.1797	4.1	4.12	4.11

Table 3 Results for cantilever beam for experimental analysis

S. No.	rfnf from FEA	rsnf from FEA	rtnf from FEA	rcd from FEA	rcl from FEA	rcd from the proposed technique	rcl from the proposed technique	Percentage error rcd	Percentage error rcl	Total error
1	0.9923	0.9912	0.9966	0.325	0.2187	0.3114	0.2096	4.16	4.18	4.17
2	0.9931	0.9926	0.9978	0.3	0.2062	0.2874	0.1976	4.17	4.16	4.16
3	0.9946	0.9942	0.9972	0.2875	0.2312	0.2755	0.2216	4.15	4.17	4.16
4	0.9959	0.9977	0.999	0.125	0.2187	0.1198	0.2096	4.16	4.15	4.15
5	0.9974	0.9977	0.9965	0.275	0.3625	0.2635	0.3474	4.17	4.16	4.16

4 Result Tables

The percentage error is found out using the following equations (Tables 2 and 3).

$$((FE \text{ result} - \text{result from the proposed soft computing method})/(FE \text{ result})) \times 100 \tag{19}$$

$$((\text{Result from Exp.} - \text{the result from the soft computing method})/(\text{Result from Exp.})) \times 100 \tag{20}$$

$$\text{Total error in } \% = (\% \text{ error in rcd} + \% \text{ error in rcl})/2 \tag{21}$$

5 Result and Discussion

A very small or minute change in the contour of the object changes its natural frequencies and modeshapes, which could be used as the index for damage detection in various objects. The research work describes the modeling of the cantilever beam with single crack (Fig. 1) is done using the change in the local stiffness and mass matrices which changes the natural frequencies and mode shapes of the object. For simplification of different analyses described in the work, Euler–Bernoulli type of beam element is used. First, the finite element analysis (FE) has been used to find out the first three natural frequencies for different crack depths and crack locations. Then, the dimensionless values of the first three natural frequencies are trained in the soft computing method proposed to find the hairline crack geometry (crack location, crack depth).

The proposed soft computing method is a type of inverse engineering problem. The first three natural frequencies which are the results from the finite element analysis are supplied to the proposed algorithm it provides the crack location. From the results, it could be observed that the algorithm is efficient enough to predict the damage severity. This work can also be used to predict the rest of the life of the structural element because the results directly give the crack depth and crack location.

6 Conclusions

This work describes an inverse method to predict the damage location (rcd, rcl) using a soft computing method. There are numerous soft computing methods existing, and many are being made. But it is not always possible to apply all these methods in every aspect of science and engineering and compare the efficiencies.

So, in this work, one of such methods has been applied in the field of damage detection in beams. The proposed soft computing method is a variety of evolutionary algorithm which used to design and develop the algorithm. Then, the natural frequencies are treated as inputs which are trained in the algorithm to predict the crack location which forms the inverse method. The results give a direct comparison of the outcomes with the finite element analysis (FEA) and experimental analysis. The results found to be converging toward the results of the FEA and experimental analysis. The errors are found to be within 4%. The errors are found out using the Eqs. 19–21.

References

1. Jaiswal NG, Pande DW (2015) Sensitizing The mode shapes of beam towards damage detection using curvature and wavelet transform. Int J Sci Technol Res 4(4):266–272
2. Tada H, Paris PC, Irwin GR (1973) The stress analysis of cracks handbook. Del. Research Corporation, Hellertown
3. Ahmed E, Mahmoud H, Marzouk H (2010) Damage detection in offshore structures using neural networks. Marine Struct 23(1):131–145
4. Ramanamurthy EVV, Chandrasekaran K, Nishant G (2011) Vibration analysis on a composite beam to identify damage and damage severity using finite element method. Int J Eng Sci Technol (IJEST) 3(7)
5. Chopade JP, Barjibhe RB (2013) Free vibration analysis of fixed free beam with theoretical and numerical approach method. Int J Innov Eng Technol 2(1):352–356
6. Peng ZK, Lang ZQ, Billings SA (2007) Crack detection using nonlinear output frequency response functions. J Sound Vib 301:777–788
7. Meshram NA, Pawar VS (2015) Analysis of crack detection of a cantilever beam using finite element analysis. Int J Eng Res Technol 4(4):713–718
8. Storn R, Price K (1997) Differential evolution—a simple and efficient heuristic for global optimization over continuous spaces. J Global Opt 11:341–359
9. Sreedhar Kumar AVS, Veeranna V, Durgaprasad B, Sarma BD (2013) A MATLAB GUI tool for optimization of FMS scheduling using conventional and evolutionary approaches. Int J Curr Eng Technol 3(5):1739–1744
10. Reed HM, Nichols JM, Earls CJ (2013) A modified differential evolution algorithm for damage identification in submerged shell structures. Mech Syst Signal Process 39:396–408
11. Morales JDV, Laier JE (2014) Assessing the performance of a differential evolution algorithm in structural damage detection by varying the objective function. Dynamics 81(188):106–115
12. Vincenzi L, Roeck GD, Savoia M (2013) Comparison between coupled local minimizers method and differential evolution algorithm in dynamic damage detection problems. Adv Eng Softw 65:90–100

Optimization of EDM Process Parameters on Aluminum Alloy 6082 by Using Multi-objective Genetic Algorithm

Sk. Md Riyaz, P. Srinivas and I. Ramu

Abstract Electric discharge machining is one of the most popular machines which are capable of machining geometrically complex and hard materials, that are precise and difficult to machine such as heat-treated tools, superalloys, heat-resistant steels, carbides, etc; these types of materials are being widely used in aerospace, automobiles, marine industries, etc. This paper aims to investigate the optimal set of process parameters of die sinker EDM on aluminum alloy 6082 with copper tube electrode by varying the input parameters such as pulse-on time (Ton), pulse-off time (Toff), current (I), and voltage (v). An L16 orthogonal array has been designed using the Taguchi method for input parameters to conduct experiments for getting the output variables such as metal removal rate (MRR) and tool wear rate (TWR). Based on the analysis, the experimental results have been carried out using ANOVA method. To find out the optimal set of values, multi-objective genetic algorithm (MOGA) is used to predict the experimental results. Upon comparing the experimental values and predicted values, an optimal value is obtained which is useful to the manufactures to get the high-performance rate of EDM to machine aluminum alloy 6082.

Keywords Electric discharge machining · Taguchi method · Multi objective genetic algorithm

1 Introduction

One of the non-traditional machining processes is electric discharge machining (EDM) which is also known as spark eroding machine. There are different types of EDM machines, and they are mainly die sinker EDM, wire EDM, EDM drill, etc. This paper mainly focuses on die sinker EDM which carries an electrode substance of 3–380 mm in size. As there is no physical contact between electrode and workpiece, a dielectric fluid acts as the medium substance which transfers the high

Sk. Md Riyaz (✉) · P. Srinivas · I. Ramu
Mechanical Department, Vishnu Institute of Technology, Bhimavaram 534202, India

BBVL. Deepak et al. (eds.), *Innovative Product Design and Intelligent Manufacturing Systems*, Lecture Notes in Mechanical Engineering,
https://doi.org/10.1007/978-981-15-2696-1_49

voltage of electric current from electrode to the workpiece and also floods away from the eroded material from the spark zone. Between electrode and the workpiece, a small gap is maintained which is known as a spark gap and it varies from 0.005 to 0.5 mm. The temperature produced at the spark zone is nearly 10,0000 C, but here the limitation is that both electrode and workpiece must have electrolytic substance. There are different types of optimization techniques, and this paper mainly focuses on the Taguchi method and multi-objective genetic algorithm technique.

Jamwal et al. [1] studied the recent developments in the electric discharge machining. Ramu et al. [2] for the optimization of CNC turning process parameters on stainless steel 316 Taguchi-based gray relational analysis were adopted. Vikram Reddy el al. [3] conducted an experiment on aluminum alloy 6082 by using electric discharge machining to find the optimal parameters of metal removal rate, tool wear rate, and surface roughness by using gray relational analysis. Chandramouli et al. [4] investigated on 17-4 PH stainless steel and electrode as copper tungsten by using EDM, and Taguchi L27 method is employed to get the maximum metal removal rate and minimum tool wear rate. Raghavendra et al. [5] conducted experiments on four steel materials as HE20, C45, EN47, and AISI P20 which are machined on EDM by using three electrodes as copper, brass, graphite to find the optimal values of MRR and TWR. Siddiqueea et al. [6] effectively used the optimization for deep drilling process of CNC lathe machine on AISI 321 austenite stainless steel bar by using the tool as solid carbide to the minimization of surface roughness. Laxman et al. [7] investigated the various process parameters of EDM on titanium superalloy, and to get optimal value, a mathematical model is developed. Backer et al. [8] optimized the surface roughness of EN31 material on EDM; response surface method (RSM) is employed for L31 orthogonal array to find the effect of various process parameters on a single response variable surface roughness. Prasanna et al. [9] observed the performance of EDM process parameters such as current and voltage and on response variables as metal removal rate and by using different types of materials such as Ti6A14V, HE15, 15CDV6, and M-250 on single copper electrode. Multilayer perceptron of neural network developed a model for process parameters to find the optimal value, and the maximum error rate has been reduced considerably when the neural network is optimized with the genetic algorithm. Umanath et al. [10] L25 orthogonal array is used to optimize the response variables such as MRR, TWR, RWR, and SR based on different input parameters such as current, voltage, pulse-on time, pulse-off time by using principal component analysis. The experiments are conducted on metal matrix composite AA7075–SiC where electrolyte as kerosene and electrode as a copper on EDM. From the literature survey, it is concluded that there is no experiment which is conducted on EDM by using an electrode as a copper tube on AA6082 material. This paper aims at increasing the performance of EDM on response variables by applying different types of input parameters. Taguchi L16 orthogonal array is used to find the experimental values, and multi-objective genetic algorithm (MOGA) is used to predict the response variables. To get the optimal value, the comparison is made between the experimental and predicted values.

2 Methodology and Experimentation

2.1 *Taguchi-Based DOE*

'Genichi Taguchi' developed Taguchi method or robust design method to improve the quality of the manufacturing goods and products, and nowadays, Taguchi method is widely used in various fields like engineering, biotechnology, marketing to improve their quality of the product by optimizing the process methods. Design of experiment (DOE) is used to convert the standard design into a robust design. In DOE, Taguchi is one of the best methods which use an orthogonal array to set a minimum number of experiments. There are different sets of combinations based on the level of design and number of factors, and the combinations of a different set of values are formed which are used as parameter tables and based on that table, responses are experimentally found. Analysis of variance (ANOVA) is an analysis tool used in statistics that splits the aggregate variability and found inside a data set into two parts, namely systematic factors which have a statistical influence on the given data set where random factors do not influence on the given data sets. To check the hypothesis in ANOVA, F-test is used and finds the parameter which is more affecting the output.

2.2 *Multi-objective Genetic Algorithm (MOGA)*

MOGA is one of the most widely used optimization techniques which is mainly used for 2 or more number of responses to get the optimal value, and a Pareto chat defines the performance of the MOGA based on the input variables. MOGA works on the basis of the population, where it defines a set of points in a design space by using the set of operations that are applied to the population. Here, the initial population is designed randomly by default, and then from the next generation of population, non-dominating rank and distance measure of individuals were used.

3 Experimental Setup

Here the experiment is conducted on the aluminum alloy 6082 which is one of the hardest materials, and it is difficult in machining on the conventional machines. The material determinations are Al 6082 amalgam of length 50 mm, the thickness of 6 mm, and width of 50 mm are taken to play out a test by utilizing anode as a copper container of measurements—100 mm of long and 2 mm of thickness—is taken. The experiment is performed on the die sinker EDM machine which is shown in Fig. 1. Aluminum alloy is nonferrous material which does not attract to

Fig. 1 Electric discharge machine

magnetic force. For that to hold the workpiece, an additional setup which is known as holders is used to hold the piece to the table as shown in Fig. 2.

A four-level machining parameter is designed to study the parameters such as pulse-on time, pulse-off time, current, and voltage on variables of metal removal rate and tool wear rate.

To calculate the metal removal rate and the tool wear rate, Eqs. (1) and (2) are used. The below formulas are taken from Laxman [7]. For metal removal rate,

$$\text{MRR} = \frac{(W_\text{i} - W_\text{f})}{t}\,\text{gms/min} \tag{1}$$

where W_i = initial weight of the workpiece in grams, W_f = final weight of the workpiece in grams, t = time for machining in minutes. For tool wear rate,

$$\text{TWR} = \frac{(E_\text{i} - E_\text{f})}{t}\,\text{gms/min}. \tag{2}$$

where E_i = initial weight of the electrode in grams, E_f = final weight of the electrode in grams, t = time for machining in minutes.

To perform the optimization, four different levels are design based on the four main factors which influence the metal removal rate and tool wear rate. Table 1 shows the machining factors and their levels.

Fig. 2 Machining with a copper tube on workpiece

Table 1 Factors and their levels

Factors	Units	Levels			
Current	Amps	15	20	25	30
Voltage	Volt	40	50	60	70
Ton	Microns	5	6	7	8
Toff	Microns	5	6	7	8

4 Results and Discussion

4.1 Taguchi Analysis

Based on the input parameters, L16 orthogonal array is a design based on the Taguchi method and experimentation is performed to get the performance variables. Where the metal removal rate (MRR) is calculated by using a mathematical Eq. (1), and the tool wear rate (TWR) is calculated by using Eq. (2). These values are then exported to worksheet and analyze the experimental values by using Taguchi analysis method to get the signal-to-noise ratio and delta values. As shown in Table 2, input parameters are listed on the left side of the table, and the output variables of metal removal rate and tool wear rate are listed on the right side along with the signal-to-noise ratios. By observing the results, the input parameters of the current 30, voltage 60, pulse-on time 5, and pulse-off time 6 are the values for a larger metal removal rate which shows better results.

From the Taguchi analysis, signal-to-noise ratio and mean results are obtained from metal removal rate. Response table is used to get the delta and rank values to identify the factors that have the largest effect on the performance variables. Table 3 shows the response tables for metal removal rate, and Fig. 3 shows that the graph for the signal-to-noise ratio larger is better.

Table 2 Analysis of Taguchi design

I	V	Ton	Toff	MRR	SNRA	TWR	SNRA
15	40	5	5	0.696	−3.152	0.005	45.794
15	50	6	6	0.755	−2.442	0.007	43.736
15	60	7	7	1.266	2.047	0.011	39.256
15	70	8	8	0.477	−6.435	0.003	50.938
20	40	6	7	0.785	−2.100	0.006	44.392
20	50	5	8	0.565	−4.960	0.004	47.082
20	60	8	5	0.971	−0.254	0.008	42.455
20	70	7	6	0.876	−1.154	0.008	41.838
25	40	7	8	0.657	−3.653	0.008	41.866
25	50	8	7	0.690	−3.227	0.005	46.057
25	60	5	6	1.470	3.345	0.011	38.847
25	70	6	5	0.625	−4.080	0.005	46.843
30	40	8	6	0.907	−0.844	0.007	42.590
30	50	7	5	0.505	−5.932	0.009	41.074
30	60	6	8	1.187	1.491	0.010	39.755
30	70	5	7	1.060	0.505	0.007	42.990

Table 3 Response table for MRR

Level	I	V	Ton	Toff
1	−2.4954	−2.4373	−1.0653	−3.3546
2	−2.1171	−4.1403	−1.7828	−0.2735
3	−1.9038	1.6575	−2.1729	−0.6939
4	−1.1949	−2.7911	−2.6902	−3.3893
Delta	1.3005	5.7977	1.6249	3.1158
Rank	4	1	3	2

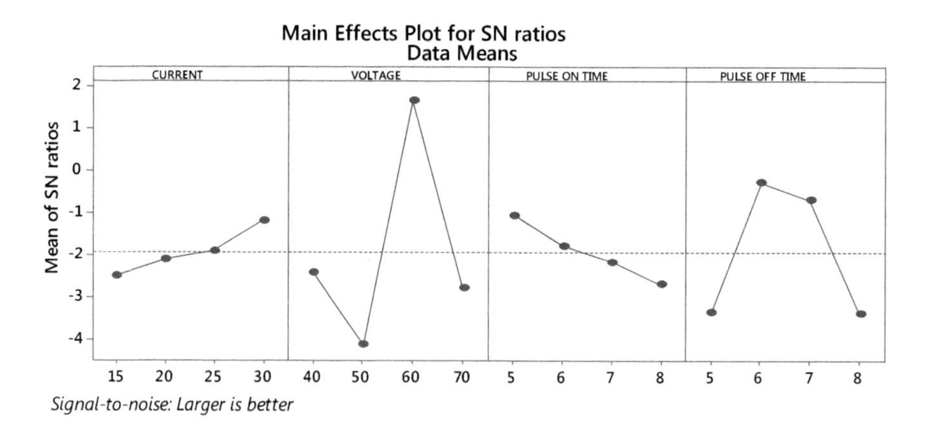

Fig. 3 Signal-to-noise ratio for MRR

Table 4 Response table for TWR

Level	I	V	Ton	Toff
1	44.93	43.66	43.68	44.04
2	43.94	44.49	43.68	41.75
3	43.4	40.08	41.01	43.17
4	41.6	45.65	45.51	44.91
Delta	3.33	5.57	4.5	3.16
Rank	3	1	2	4

The optimal values for tool wear rate are found at current 15, voltage 70, pulse-on time 8 and pulse-off 8 for signal to noise ratio: smaller is better. Signal-to-noise ratio and mean results are obtained for tool wear rate. Response tables and graphs are plotted to identify in which parameter affects the performance characteristics. Table 4 shows the response table for tool wear rate, and Fig. 4 shows that the signal-to-noise ratio for smaller is better.

Here, the effect of individual parameters cannot be decided by the Taguchi method. By using the ANOVA table percentage of contribution is calculated to define the effect of each parameter.

From the above ANOVA table and from the metal removal rate, it is clear that F-value is high for voltage and it contributes 66.8% of total percentage; it means that the voltage is the most affecting parameter followed by pulse-off time, pulse-on time, and current. Whereas for the tool wear rate, F-value is high for voltage and it contributes 52.2% of total percentage which means that it is the most affecting parameter followed by a pulse-on time, current, and pulse-off time.

4.2 *Multi-objective Genetic Algorithm*

The multi-objective genetic algorithm is one of the optimization techniques which are used to predict the experimental data. To do a multi-objective genetic algorithm, some basic input must need and those are regression equations for both metal removal rate and tool wear rate. Regression is generated by ANOVA from the Minitab. The regression of Eqs. (3) and (4) shows both MRR and TWR.

For metal removal rate (MRR), the regression equation is

$$y(1) = -(0.696 + 0.0082 * x(1) + 0.00589 * x(3) + 0.0015 * x(4)); \tag{3}$$

For tool wear rate (TWR), the regression equation is

$$\begin{aligned} y(2) = {} & 0.00571 + 0.000139 * x(1) + 0.000008 * x(2) \\ & - 0.000183 * x(3) - 0.000144 * x(4); \end{aligned} \tag{4}$$

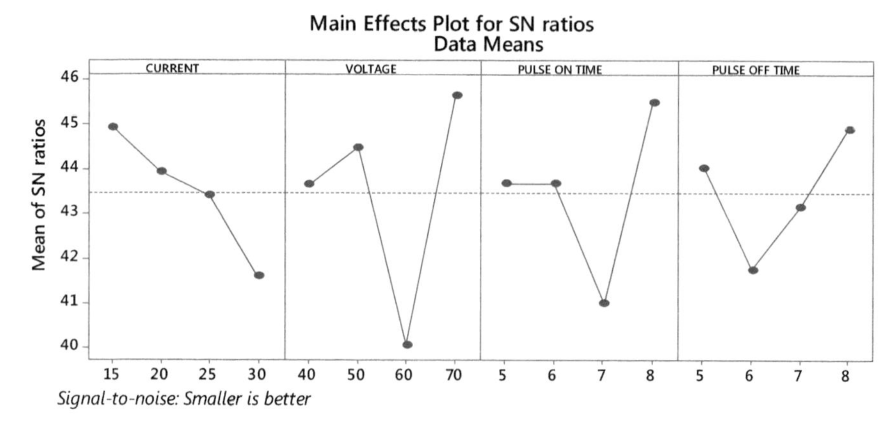

Fig. 4 Signal-to-noise ratio for TWR

By using Eqs. 3 and 4, a general program is written in MATLAB and optimization toolbox is used to do the multi-objective genetic algorithm. In the toolbox, a fitness function is provided to run the optimization and its constraints, and lower and upper bounds are mentioned. Upon changing some options to run the optimization, the solver is used to run the experiment and to view the results. As shown in Fig. 5, Pareto graph is plotted for two objectives and the solution for the multi-objective genetic algorithm is captured by the Pareto front. By running the experiment for n number of times, different values are obtained because the multi-objective genetic algorithm uses a random number of generations to get the results. So, by doing number of runs, we get the required predicted values, and this can be observed by comparing experimental values with predicted values as shown in Table 5.

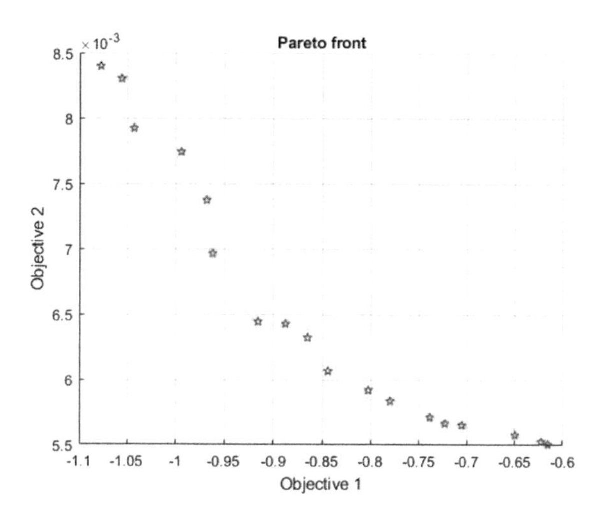

Fig. 5 Pareto graph for MOGA

Table 5 Validation table

S. No.	Initial machining parameters	Optimal machining parameters	
		Predicted value	Experimental value
Setting level	A3B3C1D2	A4B4C1D4	A4B4C1D4
MRR	1.47	1.078	1.56
Setting level	A1B4C4D4	A1B1C4D4	A1B1C4D4
TWR	0.003	0.006	0.0028

5 Conclusion

The present work is evaluated by the optimized process parameters by using Taguchi technique, and multi-objective genetic algorithm is to find the optimal set of parameters for higher metal removal rate and lower tool wear rate. As shown in Table 5, it is found that for metal removal rate and the optimal set of parameters are at current 30, voltage 70, pulse-on time 5, and pulse-off time 8. Similarly, for tool wear rate the optimal set of a parameter is found at current 15, voltage 40, pulse-on time 8, and pulse-off time 8. These values are obtained by signal-to-noise ratio graph of MRR and TWR. At the same time by doing ANOVA for MRR, it is found that voltage plays a major contribution of 66.8%, and for TWR, it is found that voltage plays a major contribution of 52.2% on performance parameters. A regression equation is also generated to predict the optimal value by MOGA. Based on the equation, runs are conducted and get different values for different generations. Finally, a Pareto chart shows the results of optimized values with the number of generations and these results are satisfying the experimental values. The experimental and predicted values are compared; there is a good agreement between them. The further work is carried out by considering two or more different aluminum alloy series, and the experiments are designed using the response surface method (RSM).

References

1. Jamwal A, Aggarwal A (2018) electro-discharge machining: recent developments and trends. Int Res J Eng Technol (IRJET) 5(02):433–448
2. Ramu I, Srinivas P (2018) Taguchi based grey relational analysis for optimization of machining parameters of CNC turning steel 316. In: International conference on mechanical, materials and renewable energy
3. Vikram Reddy V, Jawahar M (2016) Multi-objective optimization of parameters during EDM of aluminum alloy 6082 using grey relation analysis. Int J Latest Trends Eng Technol (IJLTET) 6(3):570–576
4. Chandramouli S, Eswaraiah K (2017) Optimization of EDM process parameters in Machining of 17-4 PH steel using Taguchi method. In: 5th international conference of materials processing and characterization (ICMPC 2016), pp 2040–2047

5. Raghavendra C, Venugopal Reddy V (2017) Optimization of various machining parameters of EDM by using genetic algorithm. Int J Adv Eng Res Dev 4(9):135–145
6. Siddiquee AN, Khan ZA (2014) Optimization of deep drilling process parameters of AISI 321 steel using the Taguchi method. In: 3rd international conference on materials processing and characterisation (ICMPC 2014), pp 1217–1225
7. Laxman J, Guru Raj K (2015) Mathematical modeling and analysis of EDM process parameters based on the Taguchi design of experiments. In: International conference on vibration problems (ICOVP-2015)
8. Backer S, Mathew C (2014) Optimization of MRR and TWR on EDM by Using Taguchi's method and ANOVA. Int J Innov Res Adv Eng (IJIRAE) 1(8):106–112
9. Prasannaa P, Sashank TVSSP (2017) Optimizing the process parameters of electrical discharge machining on AA7075—SiC alloys. In: International conference on advancements in aeromechanical materials for manufacturing (ICAAMM), pp 8517–8527
10. Umanath K, Devika D (2018) Optimization of electric discharge machining parameters on titanium alloy (ti-6al-4v) using Taguchi parametric design and genetic algorithm

Optimizations of Process Parameters for Friction Stir Welding of Aluminium Alloy Al 7050

Vineet Chak, V. M. S. Hussain and Mayank Verma

Abstract In the present study, an attempt is made to evaluate the effect of process parameters on mechanical properties of but welded similar plates of Al 7050 by friction stir welding (FSW) process. The process parameters play a significant role in achieving the desired characteristics and properties of welded joints. For the effective use of welded plates, welded joints must have adequate strength. The process parameters such as tool transverse speed and its rotation speed were varied for 03 levels and optimized by Taguchi technique. The optimum levels of process parameters were determined. It was concluded that process parameters play an important role in the mechanical properties and the micro-structure of the joints.

Keywords Friction stir welding · Al 7050 · Optimization · Taguchi array

1 Introduction

Aluminium alloys are gaining much attention as a potential material for aerospace and automobile sectors [1–3] due to the excellent properties offered by them as desired by these sectors. The unique properties are its high strength to weight ratio [4, 5] and also ballistic properties as required by the above-mentioned applications. But joining of Al 7050 by traditional welding process is difficult as these processes require melting of base metal and therefore properties are affected. Since the inception of friction stir welding (FSW) at the TWI, UK [6], the process is gaining much attention due to its characteristic property of not affecting base metal thermally [7–9]. The process does not involve melting of base metal [10–14] and hence can be used successfully for welding of materials that finds difficulty during joining by traditional process [8, 15, 16]. In the present investigation, an effort has been made to weld two similar Al 7050 plates by FSW and tensile, yield strength and

V. Chak (✉)
Department of Forge Technology, NIFFT, Hatia, Ranchi 834003, India

V. M. S. Hussain · M. Verma
Department of Manufacturing Engineering, NIFFT, Hatia, Ranchi 834003, India

BBVL. Deepak et al. (eds.), *Innovative Product Design and Intelligent Manufacturing Systems*, Lecture Notes in Mechanical Engineering,
https://doi.org/10.1007/978-981-15-2696-1_50

hardness of the weld joints is measured for different trials. Optical microscopy of weld zone was also carried out to study the effect of the FSW process at micro-structural level. Taguchi L9 orthogonal array [4, 7] has been utilized to carry out the DOE (design of experiments) and 09 weld samples were prepared as per DOE.

2 Experimental Procedure

Aluminium alloy Al 7050 having chemical composition as illustrated in Table 1 is used as the base metal for making the rectangular plates for carrying out the FSW. The chemical composition, as per test certificate provided by the supplier of the aluminium alloy Al 7050, is presented in Table 1. It is basically a Al–Zn–Mg alloy and is widely used in aerospace and automotive sectors.

Rectangular plates of dimensions 130 mm × 60 mm × 6 mm were cut using wire EDM followed by machining for cleaning the surfaces of the plates. A total of 18 plates of the above dimensions were taken. The above plates were kept in a combination of two and placed side by side in butt configuration over the FSW setup, for carrying out the welding. Total of nine welding passes was made by using different sets of parameters as designed by (L9) Taguchi array. HSS EN 31 (high carbon, high chromium tool steel) was selected as the FSW tool material. The dimensions and the actual tools used during experiments for making welding passes are presented in Fig. 1.

For experiments, rotational and transverse speed of the tool is taken as the variable parameters while the other parameters of the process were kept constant. The parameters which were kept constant are tool tilt angle as 3°, plunge depth as 5.85 mm, shoulder plunge depth as 0.05 mm and dwell time as 20 s. The variable

Table 1 Chemical composition of Al 7050

Elements	Zn	Mg	Cu	Fe	Si	Mn	Cr	Ti	Others	Al
% age	6.2	2.3	2.2–2.3	0.07	0.03	0.01	0.01	0.05	0.15	Rest

Parameter	Dimensions
Shoulder dia.	18 mm
Max. Dia. of pin	6 mm
Min. Dia. of pin	4 mm
Pin length	5.8 mm

Fig. 1 Tool and its dimensions used for welding

Table 2 Taguchi L9 array

Parametric design (L9 array)			Parameters	
Weld pass	Parameter 1	Parameter 2	Tool transverse speed (mm/min)	Tool rotation speed (rpm)
1	1	1	80	700
2	1	2	80	900
3	1	3	80	1100
4	2	1	100	700
5	2	2	100	900
6	2	3	100	1100
7	3	1	120	700
8	3	2	120	900
9	3	3	120	1100

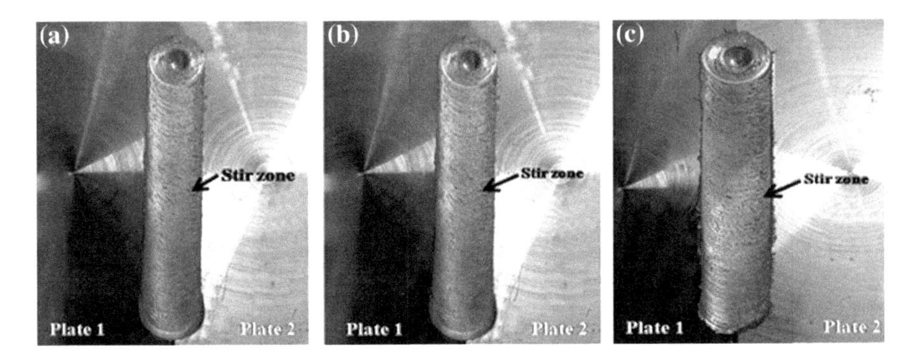

Fig. 2 Image of welding passes made on two similar Al 7050 plates

values of tool rotation and transverse speed for each weld pass were selected from Taguchi array as illustrated in Table 2. The variable values of tool rotation (700–1100 rpm) and transverse speed (80–120 mm/min) are set in the array and experiments were designed accordingly.

Total nos. of nine welding passes were made considering different sets of the parameter as per above L9 orthogonal array. Images of some welds are shown in Fig. 2. The welding samples obtained were then subjected to mechanical testing and metallurgical characterization.

3 Results

3.1 Optical Microscopy

The optical micrographs for different trials are shown in Fig. 3, considerable reduction in grain size and deformation is reported as result of severe plastic deformation in the stirred zone. This significant change in micro-structure in the welded zone has caused considerable changes in mechanical properties. The elimination of the grain structure of the base metal is confirmed by the obtained refined equiaxed structure.

3.2 Mechanical Testing

The mechanical testing such as ultimate tensile strength (UTS), yield strength (YS) and hardness (BHN) of the prepared welded specimens was carried out on respective testing equipment.

3.2.1 Tensile Testing

Tensile testing was done to determine ultimate tensile and yield strength of the friction stir welded specimens and base metal. The tensile specimens were prepared as per ASTM E8 with total length as of 100 mm and gauge length of 25 $\pm$ 0.1. Some of the initial and fractured tensile specimens are shown in Fig. 4.

The UTS and YS obtained with reference to every trial are reported in Table 3, and the corresponding graphs presenting the effect of process parameters on UTS and S/N ratio are shown in Figs. 5 and 6, respectively, the UTS value

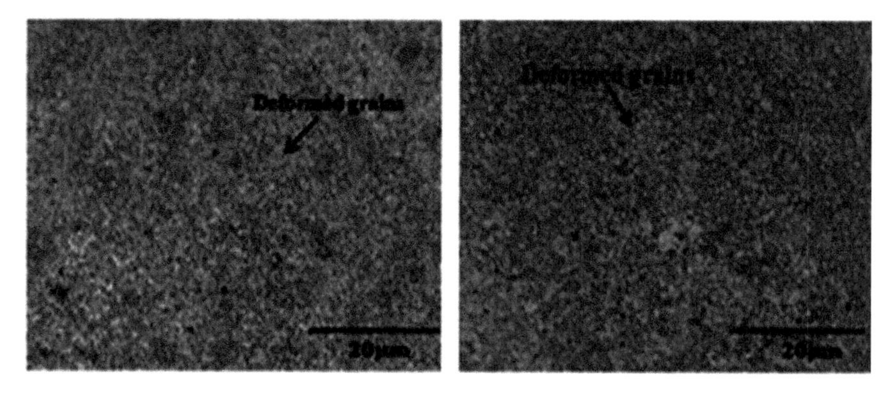

Fig. 3 Optical micrographs of the weld zone

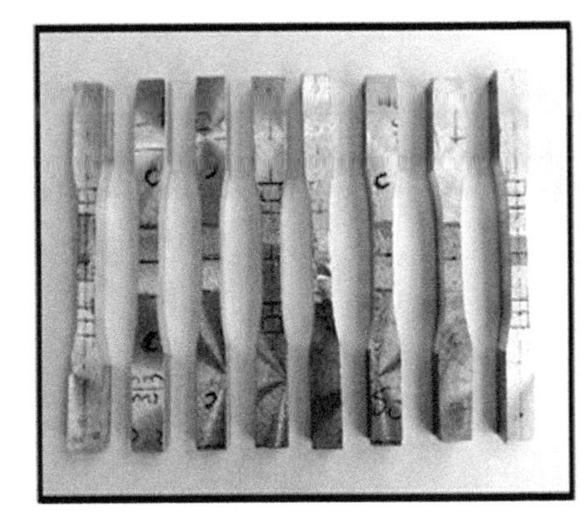

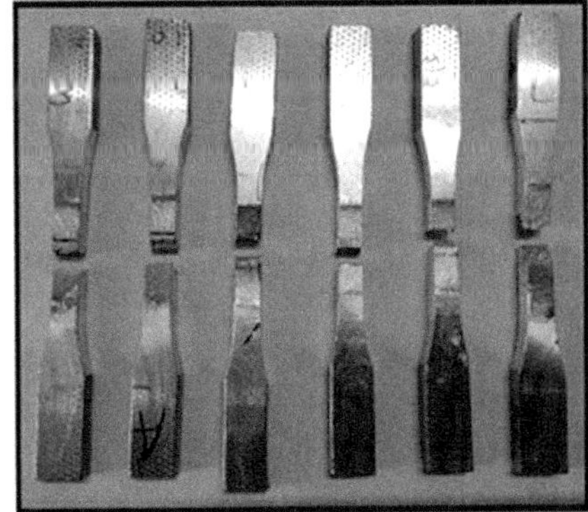

Fig. 4 Tensile specimens before and after tensile testing

Table 3 Yield and ultimate tensile strength for welded specimens

Trial No.	Yield strength (MPa)	UTS (MPa)
1	113.749	241.201
2	114.704	355.764
3	89.581	277.693
4	107.129	218.590
5	99.405	295.585
6	81.162	195.678
7	108.821	176.342
8	66.864	220.015
9	108.374	265.742

corresponding to trial no. 2 is reported to be maximum and the yield strength corresponding to trial no. 1 has achieved the maximum value.

As S/N (signal-to-noise ratio) response takes into account both the magnitude and the variation in response, the factor levels that correspond to the highest S/N ratio are termed as optimum. The effects of two welding parameters (tool transverse speed and tool rotation speed) on UTS are shown in Fig. 5. The variation of UTS with tool transverse speed is not uniform or linear; the UTS decreases with increase in tool transverse speed, tool transverse speed at 80 mm/min and tool rotation speed at 900 mm/min has contributed for a maximum value of ultimate tensile strength.

3.2.2 Yield Strength

The values of yield strength obtained for nine different weld specimens obtained are given in Table 3 and the respective plots for evaluating the effect of process parameters on yield strength and S/N ratio are illustrated in Figs. 7 and 8, the

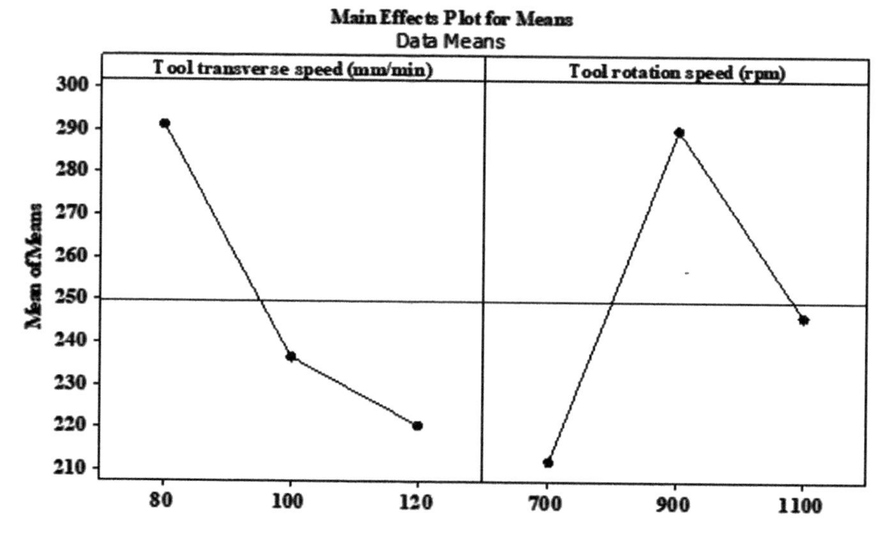

Fig. 5 Effect of process parameters on UTS

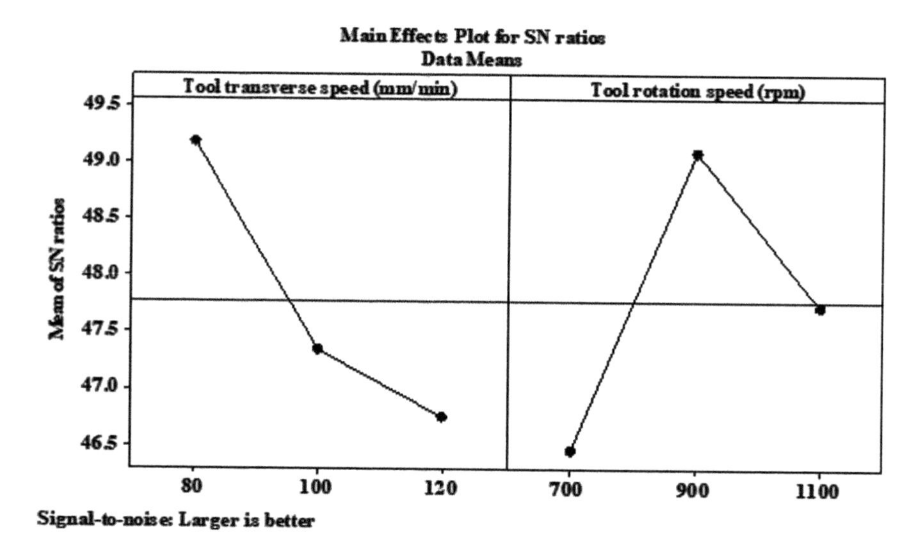

Fig. 6 Effect of process parameters on S/N

graphs obtained clearly demonstrate that tool transverse speed at 80 mm/min and rotation speed at 700 rpm have contributed to the highest values of yield strength (MPa). The values of S/N ratio are found highest for parameters levels corresponding to highest average response and therefore, these factor levels can be named as optimum.

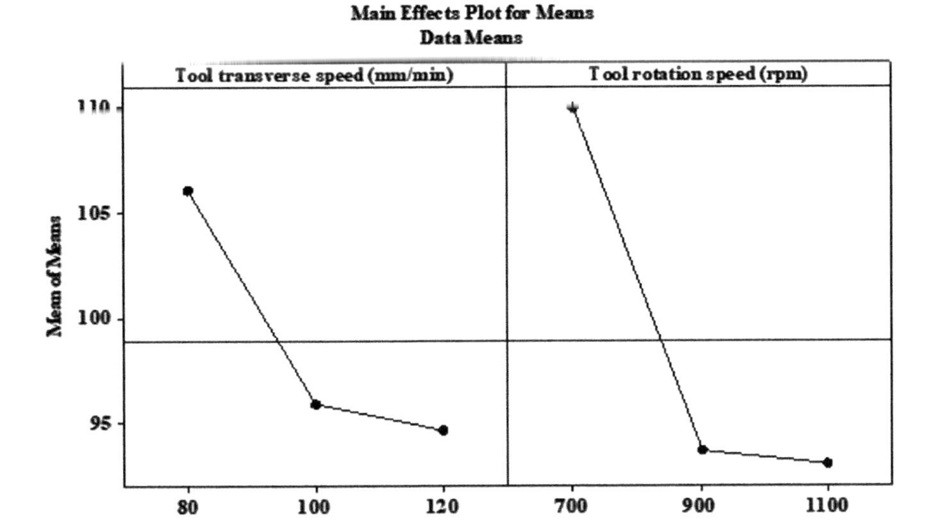

Fig. 7 Effect of process parameters on yield strength

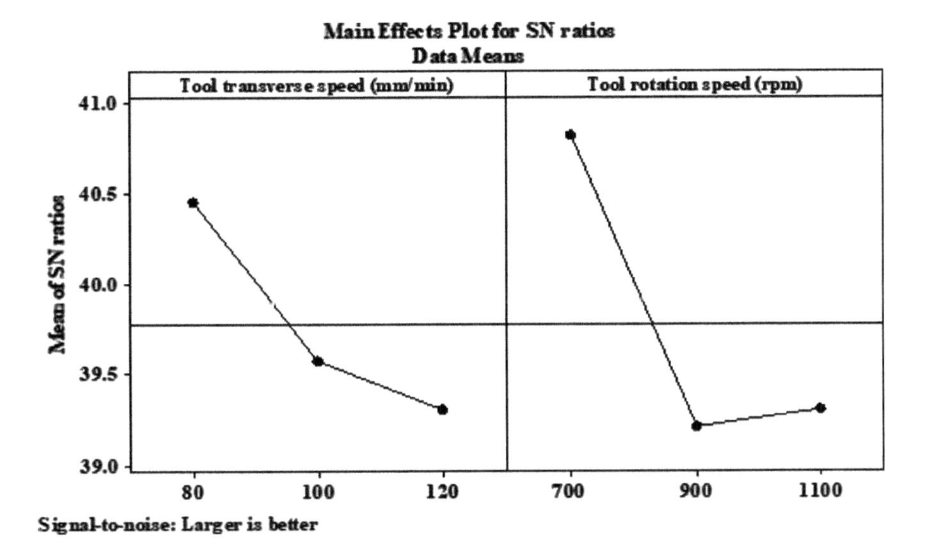

Fig. 8 Effect of process parameters on S/N

3.2.3 Hardness (BHN)

The hardness of weld samples was measured on the samples cut from the middle portion of the welded specimens. The harness was measured on three different locations and average response is reported in Table 4. The ball of 5 mm diameter

Table 4 Hardness of welded specimens

Trial	H_1	H_2	H_3	Avg.
1	80.4	103.7	113.6	99.2
2	83.8	113.6	95.1	97.5
3	87.3	83.8	113.6	94.9
4	74.2	95.1	113.6	94.3
5	87.3	124.8	80.4	97.5
6	74.2	74.2	95.1	81.7
7	87.3	103.7	113.6	101.5
8	95.1	83.7	103.7	94.2
9	103.7	74.2	95.1	91

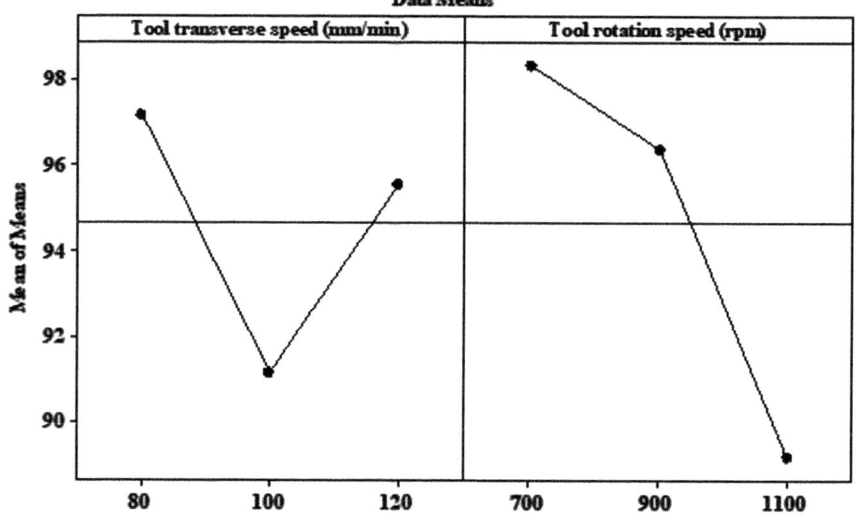

Fig. 9 Effect of process parameters on hardness

was taken and the indentation diameter was measured for three indents on each sample and hardness (BHN) is calculated according to the available standard formula.

The plots corresponding to different process parameters and values measured for hardness are shown in Figs. 9 and 10, the level 1 of tool transverse speed (mm/min) and tool rotation speed (rpm), i.e. 80 mm/min and 700 rpm, respectively, are found to be the optimum values as they have contributed for highest values of hardness recorded. The value of S/N ratio is found to be highest for those factor levels that correspond to the highest average response. Hence, these factor levels can be termed as optimum from the point of view of average response as well as S/N response.

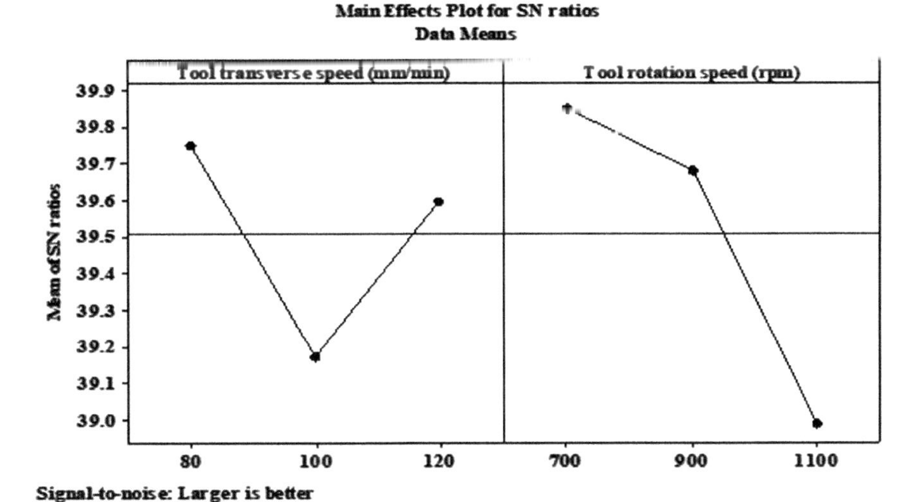

Fig. 10 Effect of process parameters on S/N

4 Conclusions

This paper has discussed the feasibility of welding Al 7050 plates by FSW technique. Taguchi method is utilized for determining the main effects, significant factors and optimum joining condition for the FSW process. Based on the results obtained it may be concluded that

- Friction stir welding process can be successfully used for welding of Al 7050 which otherwise is considered non-weldable by conventional welding techniques. Tensile strength, yield strength and hardness of welded plates were measured and are discussed.
- Optical micrographs have reported a considerable reduction in grain size and deformation of grains. The elimination of the grain structure of the base metal is confirmed by the obtained refined equiaxed structure.
- The optimum level of process parameters for the highest values of UTS, YS and hardness is reported.
- Friction stir welding can be utilized successfully for welding of materials that find difficulty with conventional fusion welding technique, due to the melting of metal associated with it. The FSW process does not affect the metal thermally and also checks the occurrence of defects due to fusion.

- The cost of FSW setup and complicacy in the selection of optimum process parameters levels for getting the desired welding properties limits the usage of this technique. FSW is still not being used for steel welding and finds limited applications in engineering applications.
- The emphasis may be given to use FSW for steel and other materials to extend its application areas.

References

1. Vignesh VR, Padmanaban R, Arivarasu M, Thirumalini S, Gokulachandran J, Ram MSSS (2016) Numerical modelling of thermal phenomenon in friction stir welding of aluminum plates. Mater Sci Eng 149:012208. https://doi.org/10.1088/1757-899X/149/1/012208
2. Yuan W, Mishra RS, Webb S, Chen YL, Carlson B, Herling DR, Grantg J (2011) Effect of tool design and process parameters on properties of Al alloy 6016 friction stir spot welds. J Mater Process Technol 211(6):972–977
3. Rodriguez RI, Jordon JB, Allison PG, Rushing T, Garcia L (2015) Microstructure and mechanical properties of dissimilar friction stir welding of 6061-to-7050 aluminum alloys. Mater Des 83(15):60–65
4. Effertz PS, Quintino L, Infante V (2017) The optimization of process parameters for friction spot welded 7050-T76 aluminium alloy using a Taguchi orthogonal array. Int J Adv Manufact Technol 91(9–12):3683–3695
5. Tebbe PA, Kridli GT (2004) Warm forming of aluminium alloys: an overview and future directions. Int J Mater Prod Technol 21(1–3):24–40
6. Dawes C, Thomas W (1995) TWI bulletin 6, November/December 1995, p 124
7. Bozkurt Y, Bilici KM (2013) Application of Taguchi approach to optimize of FSSW parameters on joint properties of dissimilar AA2024-T3 and AA5754-H22 aluminum alloys. Mater Des 51:513–521
8. Nandan R, DebRoy T, Bhadeshia HKDH (2008) Recent advances in friction-stir welding—process, weldment structure and properties. Prog Mater Sci, 53(6):980–1023
9. Sharma C, Upadhyay V, Dwivedi DK, Kumar P (2017) Mechanical properties of friction stir welded armor grade Al–Zn–Mg Alloy joints. Trans Nonferrous Metals Soc China 27(3):493–506
10. Sharma C, Dwivedi DK, Kumar P (2015) Influence of pre-weld temper conditions of base metal on microstructure and mechanical properties of friction stir weld joints of Al–Zn–Mg alloy AA7039. Mater Sci Eng A 620(3):107–119
11. Dawood HI, Mohammed KS, Rahmat A, Uday MB (2015) Effect of small tool pin profiles on microstructures and mechanical properties of 6061 aluminum alloy by friction stir welding. Trans Nonferrous Metals Soc China 25(9):2856–2865
12. Komarasamy M, Alagarsamy K, Ely L, Mishra RS (2018) Characterization of 3″ through-thickness friction stir welded 7050-T7451 Al alloy. Mater Sci Eng, A 716(14):55–62
13. Himmaraju P, Arkanti K, Reddy GCM, Tilak KBG (2016) Comparison of microstructure and mechanical properties of friction stir welding of Al 6082 aluminum alloy with different tool profiles. Mater Today Proc 3(10):4173–4181
14. Shah PH, Badheka VJ Friction stir welding of aluminium alloys: an overview of experimental findings—process, variables, development and applications. J Mater Des Appl 0(0):1–36. https://doi.org/10.1177/146442071668958

15. Threadgill PL, Leonard AJ, Shercliff HR, Withers PJ (2009) Friction stir welding of aluminium alloys. J Int Mater Rev 54:49–93| Published online: 18 Jul 2013
16. Gibson BT, Lammlein DH, Prater TJ, Longhurst WR, Cox CD, Ballun MC, Dharmaraj KJ, Cook GE, Strauss AM (2014) Friction stir welding: process, automation, and control. J Manuf Process 16(1):56–73

Analysis of Discharge Characteristics During EDM Process

Shailesh Dewangan, Sanjay Kumar Jha and S. Deepak Kumar

Abstract During the process of electric discharge machining (EDM), the spark is generated between tool and workpiece which is controlled by a servo-controller. This spark can be categorized into five different types and can be used to eliminate material and affect the quality of the surface. Different process parameters affect discharge characteristics in the EDM process, while machining, the various sets of current–voltage waveforms affect the value of EDM, which can be captured by signals. In this work, different EDM waveforms are conformed followed by the analysis of the experiment using wavelet transforms. The influence of different EDM parameters like discharge current (Ip), pulse-on time (T_{on}), and duty cycle (T_{au}) on the different discharge characteristics of EDM pulses has also been investigated.

Keywords Electrical discharge machining (EDM) · Discharge characteristics V–I waveforms · Wavelet transform

1 Introduction

In the EDM process, electrical sparks/discharges are used to eliminate metal from the surface of the workpiece. While machining, various discharge waveforms are generated, and this waveform affects the quality and productivity of a workpiece after the experiment. In this research, the different EDM waveforms are discussed followed by an overview of the wavelet transform. This conversion can be used to recognize the different waveforms of EDM.

Dauw et al. [1] developed an EDM discharge characteristics for EDM process analysis and online control and also analyzed pulse train detection with the help of EDM-PD data. Yu et al. [2] analyzed various waveforms (i.e., voltage–current

S. Dewangan (✉) · S. K. Jha · S. D. Kumar
Department of Production Engineering, Birla Institute of Technology Mesra, Ranchi, Jharkhand 835215, India
e-mail: shaileshdewangan123@gmail.com

BBVL. Deepak et al. (eds.), *Innovative Product Design and Intelligent Manufacturing Systems*, Lecture Notes in Mechanical Engineering,
https://doi.org/10.1007/978-981-15-2696-1_51

characteristics) of EDM using wavelet transform. The result indicates that after the conversion, the original data can easily distinguish the various machining parameters, thus, providing clear and useful clarification for the online control of EDM. Jiang et al. [3, 4] utilized wavelet techniques to analysis the EDM pulse characteristics. A reacquisition system used to process the system based on the digital signal was established for high-speed wavelet transmutes and related data calculations.

Recent research work is examining the effect of different machining variables in EDM on various discharge characteristics analyzed using wavelet transformation and using AISI P20 tool steel material [5]. Challenge was also made to correlate the EDM pulse characteristics with morphology the machine surface using scanning electron microscopic (SEM).

2 Discharge Characteristics in EDM

During the machining of EDM, the electrical discharge pulses are generated, and this pulse is characterized into five types such as an open, spark, transient arc, arc, and short pulses, based on the voltage–current (V–I) waveforms [6, 7]. Darning machining no discharge takes then the pulse is called open pulse, i.e., no passage of current. Spark implies when a decrease in voltage of two electrodes with appropriate discharge current is present. This corresponds to ideal machining in EDM. When debris is coming during machining between tool and workpiece, so sticking of materials to the electrodes then the result is resulting in a decrease in the gap, the discharge takes place before pick then Arc created. Generally, spark pulses are indicating that good quality of surface has been achieved with better surface finish because of good stability of sparking. Short pulses happen if the electrodes are polluted and are discharged owing to carbon deposited on the workpiece. In order to avoid surface damaging and poor machining, short pulses must be prevented [8]. Transient arc is characterized by extremely short ignition delay time.

3 Experimentation

In the current study, workpiece material was chosen AISI P20 tool steel with the copper electrode. A total of three parameters are selected such as pulse-on time (T_{on}), discharge current (Ip) and duty cycle (T_{au}). Whereas pulse-on time is four levels and discharge current and the duty cycle is two levels are selected. Taguchi mixed level design is used and eight experiments has been done in other work L_8 orthogonal array based on Taguchi design is selected. Machining parameters and their different ranges are shown in Table 1.

During machining, the spark is generated between tool and workpiece; this spark can be categories into five types such as a spark, open, transient arc, arc, and short pulses. These five types of discharge (pulse) characteristics are recognized by

Table 1 Machining variable and different range

Control variable						
Parameter	Symbol	Levels				Unit
		1	2	3	4	
Pulse-on time	T_{on}	50	300	650	900	μs
Discharge current	Ip	1	8			A
Duty cycle	T_{au}	70	90			%
Fixed parameter						
Voltage	V	45				V
Flashing pressure	Fp	0.3				Kgf/cm^2
Sensitivity	S	6				
Inter-electrode gap	I_{gap}	90				μm
Work time	T_w	0.6				s
Lift time	T_{up}	0				s

voltage–current (V–I) waveform. Each trial of the run was repetitive for five times and the signal was capture in the duration of 1 μs with the help of an oscilloscope (make: Tektronix 4000), that is attached to the EDM circuit. With the help of an oscilloscope to capture the exact voltage and current waveform during EDM, an oscilloscope is generally calibrated the voltage versus time plotted; this allows the measurement of peak-to-peak voltage of relative to time corresponding to several related signals. The current was remeasured by the voltage drop across the resistance (0.003 Ω) with a scale of 1A = 0.003 V, which was also attached to the EDM circuit. These two signals were shifted simultaneously to a double channel oscilloscope and the data was kept using the USB device. The waveforms of different experimental runs are shown in Fig. 1.

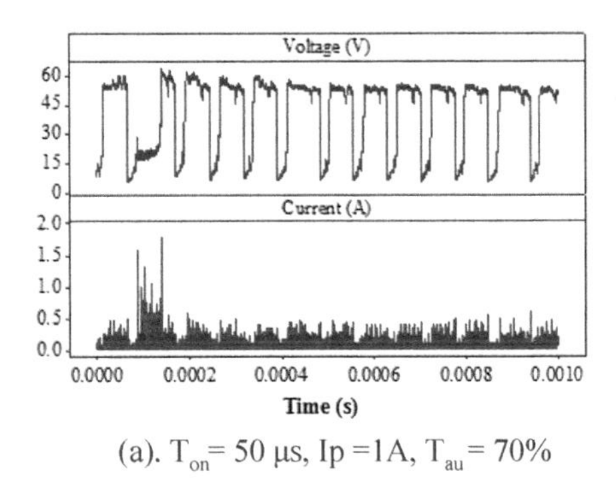

(a). T_{on}= 50 μs, Ip =1A, T_{au}= 70%

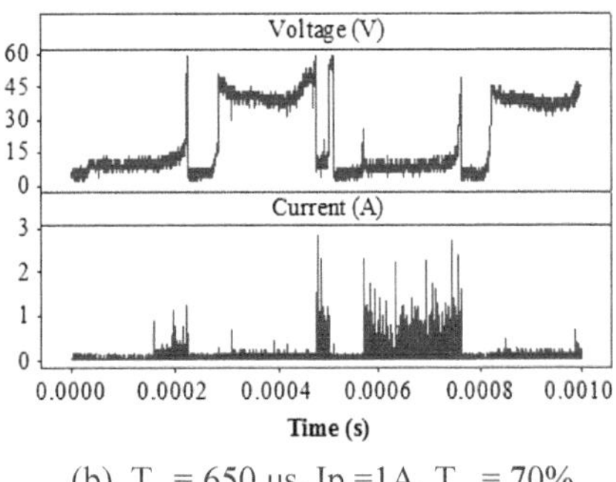

(b). T_{on}= 650 μs, Ip =1A, T_{au}= 70%

Fig. 1 V–I waveforms obtained during different experimental runs

4 Analysis of EDM Waveforms Using Wavelet Transform

Voltage–current waveform is one of the best techniques for given information about various discharge characteristics in the process of machining in EDM. Analysis of electrical signals is one of the best methods of wavelet transformation. This transformation is computing the signal data by using an exact signal and image.

Wavelet transform, on the other hand, is considerable similar for Fourier transform and that gives the information about time and frequency related data [2]. This information further used for filtering and recognition of the signal characteristics. For the present purpose, MATLAB toolbox is used to analyze 'wavelet transform' for boundary layer of Daubechies three-order wavelet (Db3) and decomposition level was kept at 3 (CA3). This type of work has also been implemented by previous researchers [4].

The frequency of occurrence (n) related data of wavelet transforms threshold value for various spark characteristics is provided in Table 2. It is given information related to the discharge/spark characteristics. This threshold value is detected respective spark characteristics for each set of runs that value is shown in Table 3. In this experiment noted that no short pulse was detected. The obtained signals were analyzed using the wavelet transform for both frequencies of occurrence and time of occurrence of various types of EDM pulses. The wavelet transform outcomes for every two different settings of runs are shown in Fig. 2.

ANOVA results (shown in Table 4) specify that parameters such as discharge current and duty cycle are not influencing for analysis of discharge characteristics in the EDM process. While machining no short pulse is recognized and ANOVA could not be considered. ANOVA results demonstration that only pulse-on time is a most affecting parameter for spark and arc pulses through EDM of AISI P20 tool steel. Low pulse-on time (50 μs) is given good spark characteristics the meaning that is surface quality was fine which is also described in resolidified layer machine surface.

This can be clarified that when machining with higher T_{on} quality of the surface is poor because it is problematic for dielectric to flush away from debris parts which in the sparking zone so a larger number of arc pulses activated. Since low value on T_{on} poorly affects MRR (i.e., productivity), it is suggested to improve the flushing system and for avoiding the formation of the arc during machining taken the large value of pulse-on time [9]. Therefore, it is determined that transient arc pulse, open

Table 2 Relationship of discharge characteristics and computed values (DB3)

Spark characteristics (μs)	Wavelet transform threshold
Open	≥ 150
Spark	80–150
Transient arc	70–80
Arc	20–70
Short	<20

Table 3 Spark characteristics detected for different set of runs

Run no.	T_{on}	Ip	T_{au}	Discharge characteristics				
	(μs)	(A)	(%)	Short	Arc	Transient	Spark	Open
1	50	1	70	0	0	1	37	26
2	50	8	90	0	0	1	39	27
3	300	1	70	0	0	0	4	10
4	300	8	90	0	0	0	16	6
5	650	1	70	0	3	6	13	13
6	650	8	90	0	3	7	12	12
7	950	1	70	0	9	8	6	30
8	950	8	90	0	10	4	3	15

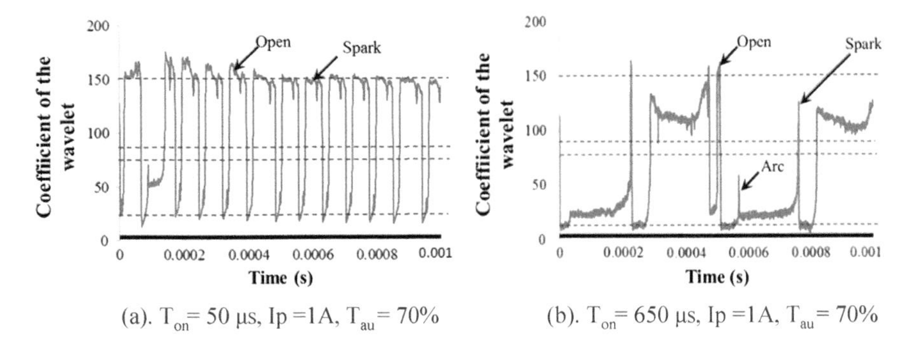

Fig. 2 Wavelet transforms results for different experimental runs

Table 4 Analysis of variance for pulse characteristics

Source of variation	DF	Spark		Open		Arc		Transient	
		P	% con.	P	% con.	P	% con.	P	% con.
T_{on}	3	0.029	94.39	0.162[a]	78.45	0.003[a]	99.59	0.125[a]	88.80
Ip	1	0.430[a]	0.89	0.330[a]	8.00	0.423[a]	0.10	0.609[a]	1.48
T_{au}	1	0.220[a]	2.88	0.474[a]	3.75	0.423[a]	0.10	0.609[a]	1.48
Residual error	2		1.85		9.80		0.21		8.24
Total	7								

[a]Insignificant factor, % con. = % Contribution

pulse, and short pulse are self-determining of the difference of Ip, T_{on}, and T_{au} while machining using EDM [10–12].

During the study of discharge characteristics in EDM, it is noted that Ignition delay time greatly influences the morphology of machine surface. Figure 3 exhibits

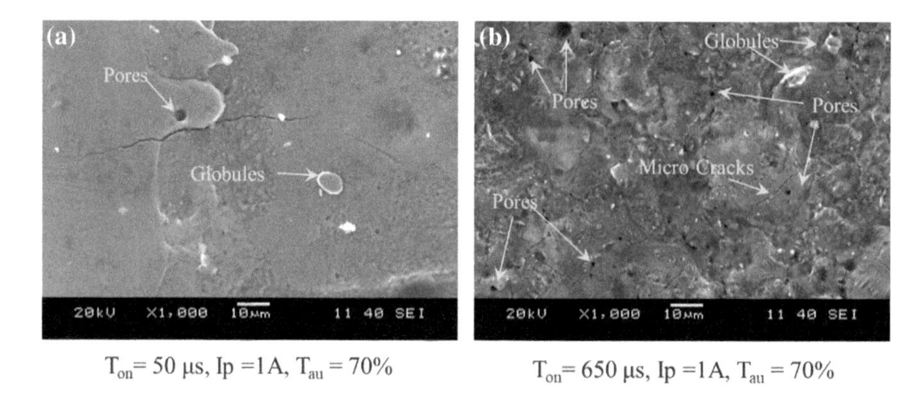

T_{on} = 50 μs, Ip = 1A, T_{au} = 70% T_{on} = 650 μs, Ip = 1A, T_{au} = 70%

Fig. 3 Morphology of machine surface for different parameters

EDM SEM image of machine surface obtained under different experimental conditions. Since experiment run number 1 clearly demonstrates restricted is the formation of primary spark pulses. Therefore, debris is presumed to be efficiently removed thus during the possibility of formation of the resolidified layer and other prominent surface defects. Figure 3a shows much smother machine surface morphology with less formation of micro surface cracks, globules, or pockmarks. With increase pulse on Time (T_{on}) beyond 50 μs, there will be reduction of a number of spark pulses, there is a gradual build-up which gave rise to the formation of recast layer microholes, pinhole, pockmarks is described in Fig. 3b.

5 Conclusions

In the current work, the novelty includes the influence of various EDM process parameters on discharge characteristics of AISI P20 tool steel. Different spark characteristics such as spark, open, arc, and transient arc during machining of AISI P20 tool steel is investigated. A short circuit pulse is not detected during the experiment. The discharge characteristics of spark pulse and arc pulse have a significantly influence on pulse-on time (T_{on}). While machining discharge current and duty cycle did not have any significant consequence on the pulse characteristics. While machining with the range of 50 μs of pulse-on time has obtained good spark characteristics. However, with the increase in pulse-on time has resulted in increase of arc pulses and decrease in surface quality. Open and transient arc pulses were found to be not affecting the discharge characteristics for variation of EDM parameters, under the selected operational range.

References

1. Dauw DF, Snoeys R, Dekeyser W (1983) Advanced pulse discriminating system for EDM process analysis and control. CIRP Ann Manuf Technol 32:541–549
2. Yu SF, Lee BY, Lin WS (2001) Waveform monitoring of electric discharge machining by wavelet transform. Int J Adv Manuf Technol 17:339–343
3. Jiang Y, Zhao W, Xi X, Gu L (2012) Adaptive control for small-hole EDM process with wavelet transform detecting method. J Mech Sci Technol 26:1885–1890
4. Jiang Y, Zhao W, Xi X, Gu L, Kang X (2012) Detecting discharge status of small-hole EDM based on wavelet transform. Int J Adv Manuf Technol 61:171–183
5. Huaa H, Yanga F, Yanga J, Caoa Y, Chunhui Lia C, Penga F (2018) Reanalysis of discharge voltage of RC-type generator in Micro-EDM. Conf Electro Phys Chem Mach Procedia CIRP 68:625–630
6. Ramesh S, Jenarthanan MP, Bhuvanesh Kanna AS (2018) Experimental investigation of powder-mixed electric discharge machining of Aisi P20 steel using different powders and tool materials. Multidiscip Model Mater 14:549–566
7. Obwald K, Lochmahr I, Schulze HP, Kroning O (2018) Automated analysis of pulse types in high speed wire EDM. Procedia CIRP 68:796–801
8. Tee KTP, Hosseinnezhad R, Brandt M, Mo J (2013) Pulse discrimination for electrical discharge machining with rotating electrode. Mach Sci Technol 17:292–311
9. Zhou M, Mu X, He L, Ye Q (2019) Improving EDM performance by adapting gap servo-voltage to machining state. J Manuf Process 37:101–113
10. Dewangan S, Gangopadhyay S, Biswas CK (2015) Multi-response optimization of surface integrity characteristics of EDM process using grey-fuzzy logic-based hybrid approach. Eng Sci Technol an Int J 18:361–368
11. Zahoor S, Mufti NA, Saleem MQ, Shehzad A (2018) An investigation into surface integrity of AISI P20 machined under the influence of spindle forced vibrations. Int J Adv Manuf Technol 96:3565–3574
12. Gamage JR, DeSilva AKM, Harrison CS, Harrison DK (2016) Process level environmental performance of electro discharge machining of aluminium (3003) and steel (AISI P20). J Clean Prod 137:291–299

Influence of FGM on the Parametric Instability of Skew Plates in Thermal Environment

I. Ramu, M. Raghuraman and M. Venu

Abstract The present work explains the influence of functionally graded material on parametric instability of skew plate in high thermal environments subjected to periodic loads. The graded material properties and temperature distribution are varied using power law. The basic kinematics of third-order shear deformation hypothesis is adopted to analyze the plate to be established in a finite element using quadrangular elements and these are updated into skew plate geometry by applying an appropriate transformation rule. A four-noded rectangular element is employed to create the work for finite element study of the instability regions of the skew plate in the parametric space by using Bolton's approach. Difference between the results of the current numerical solution for finite element approach and the results of literature are plotted. The effects of temperature difference, index of power law, and skew angle of FGM plate parametric instability discussed in detail.

Keywords Functionally graded materials · Skew plate · Thermo-mechanical analysis · Finite element method · Parametric instability regions

1 Introduction

The finest properties of high stiffness metals and virtuous thermal resistance of ceramics are both combines to form as a new novel material model. These material models advancement is very important, particularly in high thermal environment structural applications. These properties are combined with two or more material gives novel material called as functionally graded material (FGM). In this, the material properties are varying spatially with two or more constituent materials. These scales of material characteristics may reduce thermal and residual stresses. The usage of FG material has been increased in the areas of fines of swept wing missiles, wings, skew tails, and skew bridge panels. So, the extended research is

I. Ramu (✉) · M. Raghuraman · M. Venu
Department of Mechanical Engineering, Vishnu Institute of Technology, Bhimavaram 534202, India

BBVL. Deepak et al. (eds.), *Innovative Product Design and Intelligent Manufacturing Systems*, Lecture Notes in Mechanical Engineering,
https://doi.org/10.1007/978-981-15-2696-1_52

required to concede their dynamic durability properties with thermo-mechanical loading circumstances. It is demanded and need to recognize their dynamic stability characteristics of FGM skew plate with different loading conditions. The elastic structure on dynamic stability theory is developed by Bolotin [1].

Many works of literature illustrate that studies had been conducted explanation on natural frequency analysis of isotropic inclined plates. The problems of the skew plate are examined for numerous boundary conditions by Nair and Durvasula [2]. A double series of specific functions being used to the investigation of a beam with the incorporation of various boundary positions by applying the Ritz approach. The mixed finite element-differential quadrature approach purposes by Eftekhari and Jafari [3], to analyze the vibrations of skew plates with various boundary circumstances. The exact solutions for free vibration analysis of skew plate characteristics are determined through differential quadrature techniques were studied by Wang et al. [4]. Finite element approach for experimental and numerical analysis of vibration of the laminated composite plates are examined by Srinivasa et al. [5].

There are several researchers studied the thermo-mechanical characteristics of functionally graded material plates. Reddy and Chin [6] studied the dynamic thermoelastic response of functionally graded cylinders and plates. They have developed a thermo-mechanical coupling formulation and a finite element model. The studies about the higher-order shear deformation hypothesis for the response of static and analysis of vibration with FG material by Talha and Singh [7]. Kim [8] applied third-order shear deformation plate theory to formulate the theoretical model for vibrational characteristics of initially stressed functionally graded rectangular plates made up of metal and ceramic in the thermal environment. Talha and Singh [9] presented the thermo-mechanical-induced vibration characteristics of shear deformable FGM plates. Ramu and Mohanty [10, 11] studied the modal characteristics and stability analysis of plates with FG materials applying finite element approach. Influence of skewness on FGM plate was studied by Ramu et al. [12] using a numerical approximation method of finite element approach. Earlier investigations about the temperature-dependent FGM plates constrained to dynamic axial loads were mainly focused on the shear deformation approaches. So, the report presents the influence of functionally graded material properties on the instability of the skew plate with time-varying load has been systematically analyzed. Instability regions are generated by using frequencies obtained with boundaries based on Bolotin's system. The influence of the skew angle, aspect ratio, power law, and heat field about the skew plate stability in the dynamic condition is studied in detail.

2 Mathematical Formulation

2.1 Problem Statement

A conventional skew plate following in-plane biaxial cyclic load is shown in Fig. 1. The cyclic force is represented as

$$W(t) = W_s + W_t \cos(\psi t) \tag{1}$$

Here, ψ dynamic capacity of an element due to the excitation frequency, W_s is the dead, and W_t is the load component of dynamic amplitude.

2.2 Constitutive Law for Functionally Graded Material

The material creation is diverse from the base to the topmost surface. The uppermost facade ($z = t/2$) of the plate is ceramic-rich, whereas the base covering ($z = -t/2$) is metal-rich. Thermo-mechanical modeling and investigation of functionally graded materials require an effective property of FGM made of two integral materials. The active element attributes are represented

$$M(z) = R_c(z)C_v(z) + R_m(z)M_v(z) \tag{2}$$

where $M(z)$ denotes effective material property.

The amount divisions of the integral materials, metal $M_v(z)$, and ceramic $C_v(z)$ at every position z starting the center plane are described as observes:

$$C_v(z) + M_v(z) = 1 \tag{3}$$

The amount division of the ceramic particle material can be expressed as

$$C_v = \left(\frac{2 \times z + t}{2 \times t}\right)^n, \quad 0 \leq n \leq \infty \tag{4}$$

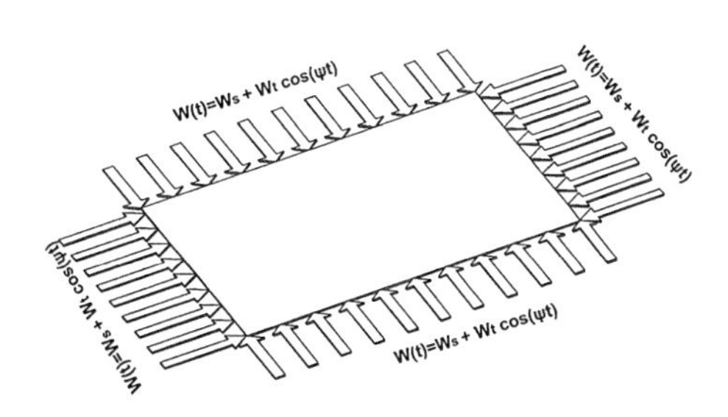

Fig. 1 Sketch of skew plate subjected to in-plane periodic loads

where n is the index, which designates the ceramic component of material modification simultaneously the thickness orientation of the skew plate.

2.3 Properties of the Material with Temperature Dependent

The material properties $M(T)$ like modulus of Young's E, the density of mass ρ, Poisson's ratio ν, and the thermal expansion of coefficient α, are recognized as temperature dependent. The temperature-dependent material characteristics are accomplished using the subsequent representation from Young (2005).

$$M(T) = P_0\left(P_{-1}T^{-1} + 1 + P_1T + P_2T^2 + P_3T^3\right) \tag{5}$$

Temperature measurement (T) in Kelvin and the unique properties are represented P_0, $P - 1$, P_2, and P_3 for the fundamental materials.

$T = T_0 + T(z)$, The temperature variation in the direction of thickness is represented as $T(z)$ and temperature at room is expressed as T_0.

The two material characteristics variation along the thickness of skew plates can be explained as follows

$$E(z,T) = E_m(T) + [E_c(T) - E_m(T)]\left(\frac{2z+t}{2t}\right)^n \tag{6}$$

2.4 Temperature Distribution

During this study, the temperature distribution is allowed in the thickness direction and it is assumed to be varied applying power law and kept as constant along XY directions of the plane of the skew plate.

$$T = T_0 + (T_t - T_0)\left(\frac{1}{2} + \frac{z}{t}\right)^n \tag{7}$$

where T_0 room temperature and T_t high temperature at the ceramic side.

2.5 Energy Equations

The absolute strain energy $\left(u_{st}^{(e)}\right)$ of the plate component conceding to the third-order shear deformation hypothesis displayed below

$$u_{st}^{(e)} = \frac{1}{2}\int\limits_{0}^{l}\int\limits_{0}^{w}\left[[N]^T\left\{\varepsilon^{(n)}\right\} + [M]^T\left\{\varepsilon^{(1)}\right\} + [P]^T\left\{\varepsilon^{(3)}\right\} + [Q^s]^T\left\{\gamma^{(n)}\right\} + [R^s]^T\left\{\gamma^{(2)}\right\}\right]\mathrm{d}s\,\mathrm{d}l \tag{8}$$

The elemental energy of kinetic for the plate is represented below

$$T^{(e)} = \frac{1}{2}\rho\int\limits_{A}\left(\dot{u}^2 + \dot{v}^2 + \dot{w}^2\right)\mathrm{d}A \tag{9}$$

where $\rho = \int_{-h/2}^{h/2}\rho(z)\mathrm{d}z$.

The plate is subjected to in-plane load, the work done can be denoted as:

$$WD^{(e)} = \frac{1}{2}\int\limits_{A}\left[W(t)\left(\frac{\partial w}{\partial x}\right)^2 + W(t)\left(\frac{\partial w}{\partial y}\right)^2\right]\mathrm{d}A \tag{10}$$

Here, $W(t)$ describes the applied cyclic load along the x- and y-axis, sequentially.

2.6 *Oblique Boundary Conversion*

The inclined plate boundaries may not be identical to global axes x and y. Consequently, it is necessary to establish the circumstance situations in words of the displacements w, θ_x, and θ_y. The local reference plane edge displacements w, θ_T, and θ_s, these are tangential and natural to the inclined edge, at such edges to specify the boundary conditions. The angled side conversion of displacement for a node nth is presented by

$$\begin{Bmatrix} u \\ v \\ w \\ \theta_x \\ \theta_y \end{Bmatrix} = \begin{bmatrix} \cos\phi & -\sin\phi & 0 & 0 & 0 \\ \sin\phi & \cos\phi & 0 & 0 & 0 \\ 0 & 0 & 1 & 0 & 0 \\ 0 & 0 & 0 & \cos\phi & -\sin\phi \\ 0 & 0 & 0 & \sin\phi & \cos\phi \end{bmatrix}\begin{Bmatrix} u_T \\ v_S \\ w \\ \theta_T \\ \theta_s \end{Bmatrix} \tag{11}$$

The association of conversion can be formulated as

$$u_n = Tr_{ij}\bar{u}_i \tag{12}$$

where u_n and $\bar{u}$ are the global displacement vectors under the global and local edge coordinate arrangements. This model is shown simply for five degrees of freedom per node.

2.7 Finite Element Analysis

For this analysis, one node at each corner of the rectangular element is employed and as presented in Fig. 6. Each edge of the component consists of four nodes 1, 2, 3, and 4 with each node has five degrees of freedom as the u, v are in-plane displacements, w is the transverse displacement and θ_x, θ_y represents the rotations about x- and y-axis.

$$\left\{Q^{(e)}\right\} = \left\{u_i, v_i, w_i, \theta_{xi}, \theta_{yi}\right\}_{i=1,2,3,4}$$

$$u = \sum_{i=1}^{4} N_i u_i, v = \sum_{i=1}^{4} N_i v_i, w = \sum_{i=1}^{4} N_i w_i, \theta_x = \sum_{i=1}^{4} N_i \theta_x^i, \theta_y = \sum_{i=1}^{4} N_i \theta_y^i, \tag{13}$$

The energy equations are employed to produce the mass, geometric, and stiffness matrices by using kinetic energy, work is done and potential energy minimization concept, simultaneously. The standard (T, S) coordinate system is formulated from the mapping of the normal coordinate system.

3 Equations of Motion for an Arbitrary

The arbitrary equation of the skew plate is formulated by adopting Hamilton's principle.

$$\delta \int_{t_1}^{t_2} \left(u^{(e)} - T^{(e)} + w^{(e)}\right) \mathrm{d}t = 0 \tag{14}$$

By distributing the skew plate into a number of components and congregating the component matrices, the kinetic energy and potential energy for a component can be formulated as

$$u^{(e)} = \frac{1}{2}\left\{Q^{(e)}\right\}^T \left[K_s^{(e)}\right]\left\{Q^{(e)}\right\} - \frac{1}{2}\left\{Q^{(e)}\right\}^T W(t)\left[K_g^{(e)}\right]\left\{Q^{(e)}\right\} \tag{15}$$

$$T^{(e)} = \frac{1}{2}\left\{Q^{\cdot(e)}\right\}^T \left[M^{(e)}\right]\left\{Q^{\cdot(e)}\right\} \tag{16}$$

Here, $K_s^{(e)} = K_{es}^{(e)} - K_t^{(e)}$ active component of matrix for stiffness, this is the difference of component matrix of stiffness $K_{es}^{(e)}$ and matrix of component thermal stiffness $K_t^{(e)}$, respectively.

Therefore replacing, periodic load $W(t) = \alpha P_{\mathrm{BK}} + \beta P_{\mathrm{BK}} \cos(\psi t)$ using α and β are called static and dynamic load factors, respectively.

$$[M]\{Q\} + \left([K_s] - \alpha P_{\mathrm{BK}} [K_g]_s\right)\{Q\} - \beta P_{\mathrm{BK}} \cos(\psi t) [K_g]_t \{Q\} = 0 \tag{17}$$

To avoid the instability for the importance of experimental analysis with principle boundaries are formulated as.

$$Q(t) = \sum_{k=1,3,..}^{\infty} \left[\{c_k\} \sin\frac{k\psi t}{2} + \{d_k\} \cos\frac{k\psi t}{2} \right] \tag{18}$$

The plus and minus symbol represents the two eigenvalues denote to generate the regions of instability. The formulated equations are used to produce the boundaries of instability.

$$\left|[K_s] - \left(\alpha \pm \frac{\beta}{2}\right) P_{\mathrm{BK}} \times [K_g] - \frac{\psi^2}{4}[M]\right| \{Q\} = 0 \tag{19}$$

4 Numerical Experimental Results and Discussions

4.1 Study of Correlation

The natural frequencies of a calculated from the present computational program were compared with these of Liew et al. (1993). The results were concluded to be in an extensive agreement. The results obtained displayed in Fig. 2.

In this example, FGM plate and temperature-dependent FGM plate are considered for the analysis, and the results obtained are shown in Fig. 3 and those are compared with Talha and Singh (2010). Figure 3 shows that increase power law

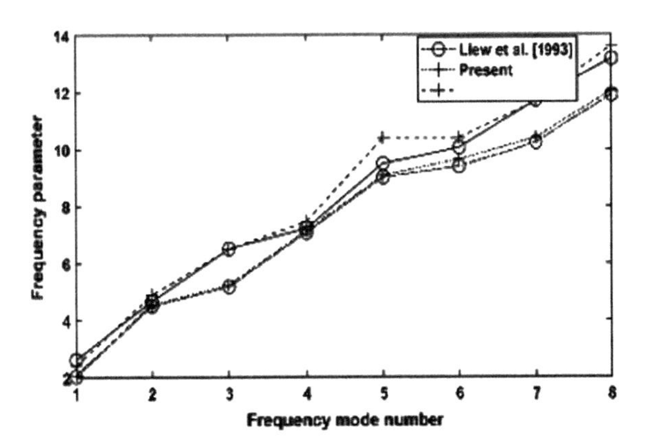

Fig. 2 Comparison with frequency parameters, λ of skew plates and $W/L = 1$, $h = 0.1$ m, poisons ratio 0.3

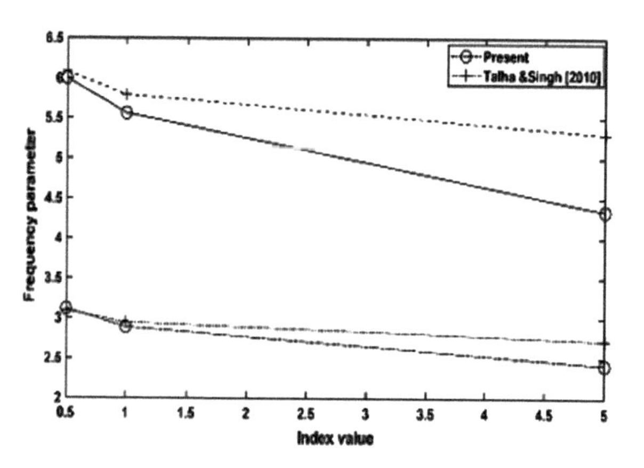

Fig. 3 Frequency parameter (ϖ) variation with the volume fraction index n for all sides clamped square (Al/Al_2O_3) FGM plates ($a/h = 10$)

index decreases the frequency parameter of FGM plate. There is immeasurable negotiation among existing results with those of Talha and Singh (2010). The discrepancies may be due to various types of shear deformation theory examined.

4.2 Parametric Instability

The skew plate subjected to dynamic loading with simply supported circumstances is analyzed by varying the index of power law about parametric instability as displayed in Fig. 4. The difference in temperature 1000 K is considered with thermal conditions arrangement based on power law is considered for this analysis.

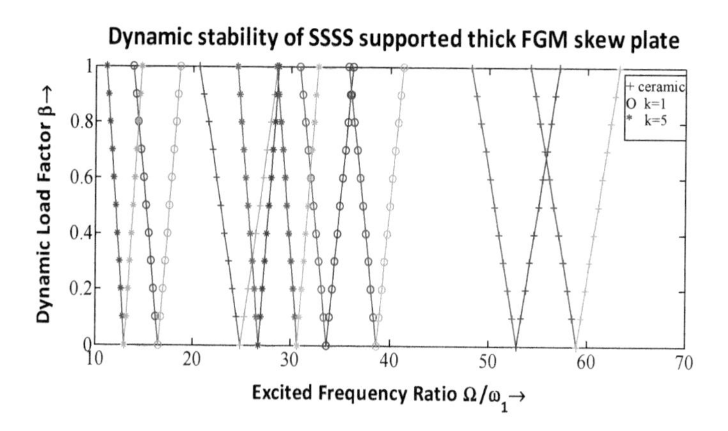

Fig. 4 Index value influence of parametric instability of skew plate ($L/W = 1$, $h/L = 0.15$, $\Phi = 15°$)

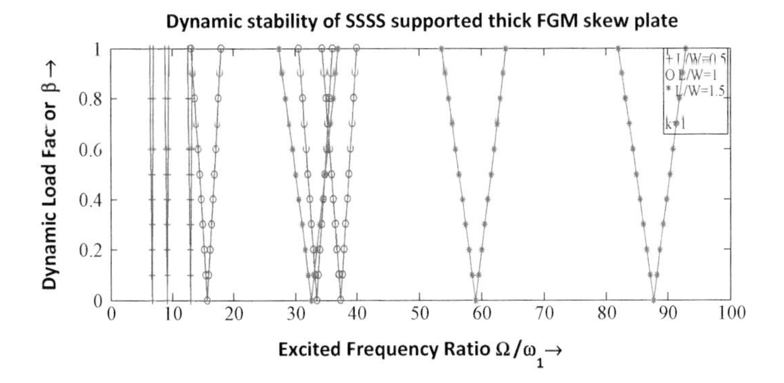

Fig. 5 Influence of aspect ratio on parametric instability of skew plate ($\frac{h}{L} = 0.15, \Phi = 15°$)

Lower the excitation frequency with an increase of index of power law, it causes increases the instability as demonstrated in the figure. The ratio of aspect is $L/W = 0.5$, 1, and 1.5 variations on the parametric instability of skew plates are shown in Fig. 5. The instability regions also reduce from wider to narrow area. Here, the thermal environment temperature change is 1000 K and power law index k is 1.

The variation of thermal environment profiles is presented in Fig. 6 about the parametric instability of skew plates. The ratio of thickness $h/L = 0.15$ also the index $k = 1$ is considered for these geometrical dimensions. The stability regions move of low excited frequency ratio further high moved frequency ratio in this diagram with the increase of skew angle, it enhances the instability of inclined plate.

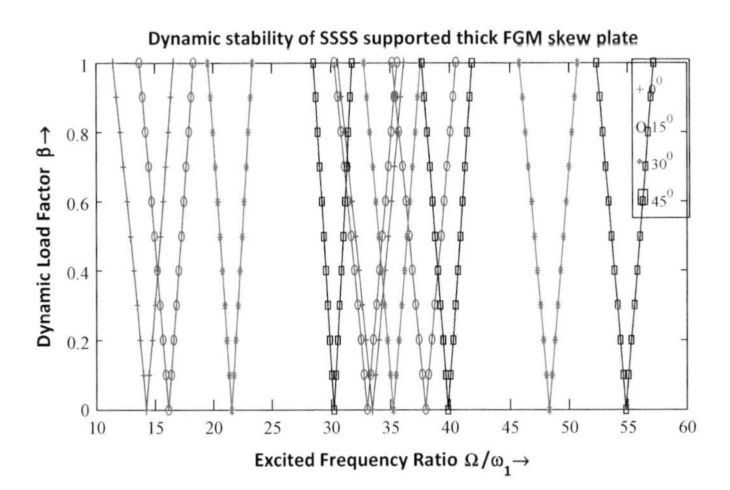

Fig. 6 Nonlinear temperature field of a skew plate under simply supported case ($L/W = 1$, $k = 1$, $h/L = 0.15$)

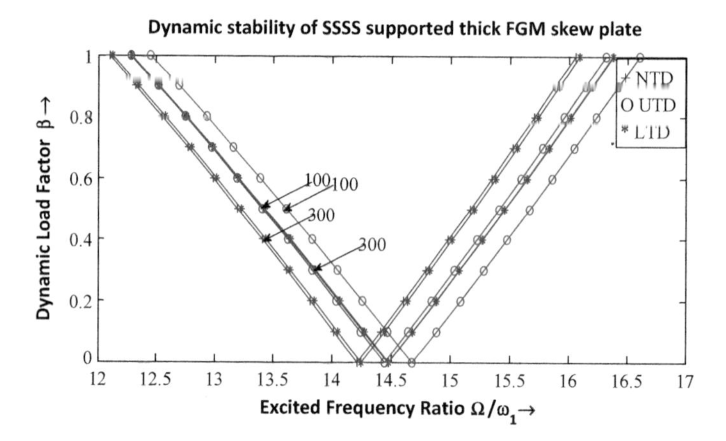

Fig. 7 Change of temperature with 100 and 300 for stability of skew plate ($L/W = 1, \Phi = 15°, k = 2$)

Figure 7 shows the parametric instability region for simply supported boundary conditions. It can be observed that with a growth in temperature the uncertainty zones change from high excited frequency ratio to low exited frequency ratio in the dynamic stability plot. In a high-temperature environment, the FGM skew plate exhibits instability for a small variation of dynamic loads. Figure 7 indicates the study the stability of FGM skew plate through temperature change at 1000 K and 3000 K with increase temperature difference reduces the excited frequency ratio also it reduces the dynamic stability. Here, the thermal environment index value 10 effects the dynamic stability region at a lower excited frequency than index 1 and 0 thermal environment fields.

5 Conclusions

The present work studied the influence of skewness and FG materials on the parametric instability plates under various thermal fields. In high thermal fields, the skewness also affects the parametric instability of plates with the variation of the bottom to top planes. With the rise in index of power law the excited frequency gets lowered, so the excited frequencies are shifted towards dynamic load axis. These lowered excited frequencies may cause the deterioration of the dynamic stability. The instability of the skew plate is enhanced overall with an increase of aspect ratio. The temperature-dependent FGM skew plate structure stiffness degrades with an increase in temperature difference, therefore, vibration frequencies reduce, lowers stiffness, and decreases the excitation frequencies. In a high-temperature environment, the skew structure may lose its features like natural frequencies and critical buckling loads enhance the parametric uncertainty. In the future, the operating conditions may include for further research related to the skew rotational foundation, and nonlinear analysis.

References

1. Bolotin VV (1962) The dynamic stability of elastic systems volume I, report, Aerospace Corporation, El Segundo, California
2. Nair PS, Durvasula S (1973) Vibration of Skew plates. J Sound Vib 26(1):1–19
3. Eftekhari SA, Jafari AA (2013) Modified mixed Ritz-DQ formulation for free vibration of thick rectangular and skew plates with general boundary conditions. Appl Math Model 37:12–13
4. Wang X, Wang Y, Yuan Z (2014) Accurate vibration analysis of skew plates by the new version of the differential quadrature method. Appl Math Model 38:926–937
5. Srinivasa CV, Suresh YJ, Prema Kumar WP (2014) Experimental and finite element studies on free vibration of skew plates. Int J Adv Struct Eng 6(48):1–11
6. Reddy N, Chin CD (1998) Thermo-mechanical analysis of functionally graded cylinders and plates. J Therm Stress 21:593–626
7. Talha M, Singh BN (2010) Static response and free vibration analysis of FGM plates using higher order shear deformation theory. Appl Math Model 34:3991–4011
8. Kim YW (2005) Temperature dependent vibration analysis of functionally graded rectangular plates. J Sound Vib 284:531–549
9. Talha M, Singh BN (2011) Thermo-mechanical induced vibration characteristics of shear deformable functionally graded ceramic–metal plates using the finite element method, Proceedings of the Institution of Mechanical Engineers, Part C: J Mech Eng Sci 225:50–65
10. Ramu I, Mohanty SC (2014) Modal analysis of functionally graded material plates using finite element method, 3rd international conference material processing and characterization, Hyderabad, India, Procedia Mater Sci 6:460–467
11. Ramu I (2015) On the dynamic stability of functionally graded material plates under parametric excitation. National Institute of Technology, Rourkela, India
12. Ramu I, Seshu KVGR, Venu M (2018) Natural frequency of skew FGM plates using finite element method, IOP Conf. Series: Mater Sci Eng 455:012024

Optimization of Input Control Variables in Electric Discharge Machining of Inconel-718

Rahul Davis, Abhishek Singh, Tanya Singh, Subham Chhetri, V. Vikali Sumi, Alomi P. Zhimomi and Stephen Dilip Mohapatra

Abstract With rising requests of ongoing building items, the controlling of surface texture alongside dimensional exactness turns out to be increasingly indispensable. It has been analyzed that the working of the machined components and properties, for example, appearance, resistivity against fatigue/wear/corrosion, grease, introductory resistance, capacity to hold weight, load conveying limit, and commotion decrease (if there should arise an occurrence of apparatuses), are largely extraordinarily impacted by surface texture. The anomalies superficially as variety in stature and spacing are termed as surface roughness usually. It is always very strenuous and costly to control this in manufacturing, no matter what process is employed. Thus, accuracy in dimension and surface roughness is one of the main factors required to consider machining variables of any machining operation. In this paper, a research is being conducted to obtain optimal settings of the various levels of the input control variables in the machining of Inconel-718 by electric discharge machining (EDM), for achieving minimum roughness of the machined surface (SR).

Keywords Design of experiment · Electric discharge machining · Inconel-718 · Taguchi method · ANOVA

1 Introduction

In electric discharge machining (EDM), cutting action is caused by an electric discharge between an electrode and the workpiece. This machining takes place in a dielectric fluid. The cutting tool does not have any direct contact with the workpiece, thus tool can be made of materials such as brass, copper, etc., which are soft

R. Davis (✉) · A. Singh
National Institute of Technology Patna, Patna 800005, India
e-mail: rahul.me18@nitp.ac.in

T. Singh · S. Chhetri · V. V. Sumi · A. P. Zhimomi · S. D. Mohapatra
Vaugh Institute of Agricultural Engineering & Technology, SHUATS, Prayagraj 211007, India

BBVL. Deepak et al. (eds.), *Innovative Product Design and Intelligent Manufacturing Systems*, Lecture Notes in Mechanical Engineering,
https://doi.org/10.1007/978-981-15-2696-1_53

and can be easily used. The tool electrode should have a high melting point and also should be a good conductor of electricity. The tool works in conjunction with the fluids like kerosene or mineral oil, which act as a coolant as well as the dielectric. The advantage of these coolants is that they wash every disintegrated material from the workpiece or instrument and keep up consistent protection from the current flow [1]. In this process, the machining process is done due to the electric spark generated between an electrode and the workpiece. It is caused by an electric discharge between the metal to be cut (+ve charge) and an electrode (−ve charge). A condenser that is in a parallel position to electrode and workpiece assumes responsibility for direct current through a resistor. As the condenser is charged, its potential ascents quickly to a specific value adequate to beat the dielectric fluid between the workpiece and electrode. The gap distance between an electrode and workpiece is servo-controlled and is about 0.025 mm. Sparking occurs where the gap is least, regardless of the tool area. This is because the current density is high at this point and is of enough power to disintegrate particles from the workpiece. Another variant of EDM, called Wire-cut EDM, can cut very complicated and delicate designs [2]. Since wire EDM has zero cutting forces, it has no residual stresses, and thus, there are only slight changes in the mechanical characteristics of the materials [3]. While sinker EDM uses an electrically charged electrode to implant shape on a metallic component. The electrode is used to burn the shape in the oil-immersed working metal [4]. Graphite can also be used as an electrode because of properties like machining and wearability (graphite). Through this process, even the complex shapes can be formed out of most rigid materials. However, this process is only limited to electrically conductive materials [5]. In recent research, machining performance with EDM has been compared with abrasive water jet machining (AWJM) and laser beam machining (LBM) for obtaining minimum roughness on the machined region of alloy 718. AWJM was concluded to be the best method followed by EDM, while LBM was strongly prohibited for the same [6]. Another comparative analysis was performed in the case of micro-EDM using EDM oil and powder (Al) mixed EDM oil in the machining of Inconel 706. The findings concluded that the addition of Al powder and concentration of additives can improve MRR significantly [7, 8]. In the present research work, EDM has been opted and used due to its ease of availability and operations could be performed efficiently at a relatively cheaper price.

1.1 Input and Output Control Variables of EDM

Generally, the following input control variables are used in an EDM:

a.	Discharge current (*I*p):	It is the current, also known as 'working current,' flowing through the machine
b.	*T* (on):	The time allowed for the current to flow per cycle (measured in micro-second)
c.	*T* (off):	The time between the sparks that enables the solidification of molten material and its flushing out of the tool–workpiece gap. If this time is too short, then the spark becomes unstable
d.	Voltage:	It is defined as the potential measured in volts
e.	Spark gap:	It is the arc gap between the tool electrode and the work material

Surface roughness, material removal rate, tool wear, tool life, etc. are the commonly chosen output response variables for EDM and other non-traditional machining methods. For the present research work, surface roughness has been picked as the output response variable. Each cutting or material removal tool leaves its own pattern on the surface which can be identified. This pattern is known as surface roughness, which can also be defined as the deviation in the direction of the real surface from the ideal form. Surface roughness is generally measured by root means square (RMS). RMS value is characterized as the root square of the arithmetic mean of the squared values of the ordinates of the surface estimated from a mean line.

2 Design of Experiment via Taguchi Method

The Taguchi experimental design pursues a three-step procedure:

Step 1: This is the first step and this step includes finding the total degree of freedom (DOF).
Step 2: Followed by this, the second step incorporates choosing a suitable orthogonal array utilizing the following regulations:

Rule 1: The number of runs in the orthogonal array should be equal to or less than the total degree of freedom of the experiment.
Rule 2: The chosen orthogonal array should suit the variable level amalgamations in the analysis.

Step 3: This step assigns the components to fitting segments again by utilizing the following rules:

Rule 1: Interactions should be assigned according to the interaction table and linear graph.

Rule 2: In the experiment, when the variable levels are not accommodated by the original orthogonal array due to its incapability, then some specific methods can be chosen to be used.

Between 'regular' experimental design and Taguchi's experimental design, numerous similarities can be found. It is assumed that higher-order interactions are nonexistent. Moreover, through the insight of the experimenters into the topic, they are approached to distinguish which interactions may be noteworthy before leading the test. The total degrees of freedom of the trial variables ought to be resolved in the Taguchi exploratory plan after these two steps. The degrees of freedom are the overall measure of information required so as to evaluate each of the effects to be examined [9–11]. One of the recent research articles targeted to achieve the best combination of input variables and a specific tool during machining of Inconel-718 with an integrated application of principal component analysis, fuzzy inference system (FIS), and Taguchi method, in order to have contented results in terms of the desired outcomes [12, 13].

3 Details of the Experiments

In the present work, the experiment trial runs were done altogether on an electric discharge machine (EDM) so as to examine the influence of various input control variables on the workpiece by ascertaining the profundity of the machined surface on the workpiece. It is critical to decide the surface unpleasantness of the material when the machining shows signs of improvement in the results [14]. Inconel-718 provides very good toughness at low temperatures, because of which it is used in jet engines, rocket engines, cryogenic applications, and gas turbine applications, and also has been selected as the work material for the present research. The components like nickel, chromium, iron, etc. are found sufficiently in Inconel-718, because of which it offers excellent resistivity against corrosion even at temperature constrains above 950 °C. Details of the properties of Inconel-718 specimen, chosen for the present work, are given in the following Table 1.

Table 1 Properties of Inconel-718 specimen material

Property	Value
Density	8.192 gm/cm^3
Melting range	1370–1430 °C
Ultimate tensile strength	1375 MPa
Ultimate tensile strength, at elevated temperature	1100 MPa
Thermal conductivity	0–100 °C
Hardness	40 (maximum)
Electrical resistivity	1210 μΩ-mm

Table 2 Input control variables for EDM

S.N.	Input control variables	Range of input control variables
1	Pulse on [T(on)]	0.25–3000
2	Pulse off [T(off)]	Duty cycle 1 to 32 linearly in the steps of 1
3	Discharge current (I_P)	0–50 A
4	Voltage (S_v)	1–120 vg

To machine the material through EDM, an electrode made up of copper was used to meet the requirements such as good electrical conduction, low cost, and fine surface finish [15, 16].

To understand the impact of input control variables such as I_P, T(on), T(off), S_V, in machining metals, where time had been set consistent for different hard-to-cut materials, investigations have been executed several times. Table 2 shows the input control variables of interest, chosen for the proposed EDM of Inconel-718.

4 Results and Discussion

Another research was carried out on Inconel-718 super alloy amid its machining via EDM using copper tool electrode and responses of the research were measured in terms of material removal rate (MRR), surface crack density (SCD), and white layer thickness (WLT) [17]. In the present work, the input control variables of EDM were varied at three different levels. Surface roughness (SR) was the measured response [18, 19]. Experimental trial runs were performed using the L18 orthogonal array. The depth of penetration (cut) was also observed. The following Table 3 presents 18 experiment trial runs on Inconel-718 specimens via electric discharge machine.

Figure 1 shows an EDM set-up on which EDM of Inconel-718 specimen was performed and Fig. 2 displays the workpiece after one of the machining trial runs using copper tool electrode.

In previous researches, in the machining of hard-to-cut materials via EDM, discharge current appears to be one of the most significant input variables with the greatest contribution among other variables. During machining of tool steel 55NiCrMoV7, it was found that the current influences surface roughness by more than 50% [20].

Table 4 shows the obtained response for signal to noise ratio using Minitab 18 software, which shows that discharge current (I_P) has influenced the roughness of machined surface as the most dominating variable, followed by voltage, T(off) and T(on). The same can be confirmed by observing the nature of the graph in Fig. 3. At the lower level of current, the SN ratio is higher, while at second level of T(off), first level of T(on), and third level of S_V, SN ratio is higher, and it is being higher, in turn, reduces the noise and thus helps in obtaining the optimal levels of the input control variables for obtaining minimum roughness. Table 4 shows the order of the

Table 3 Control log of EDM operations

S.N.	I_P	T(off)	T(on)	S_v	Depth of cut (mm)	Surface roughness (R_a)
1	4	7	50	30	0.285	1.16
2	4	7	75	40	0.260	1.13
3	4	7	100	50	0.200	0.83
4	4	8	50	30	0.295	1.28
5	4	8	75	40	0.385	1.01
6	4	8	100	50	0.285	0.99
7	4	9	50	40	0.380	1.17
8	4	9	75	50	0.305	1.18
9	4	9	100	30	0.240	1.25
10	10	7	50	50	0.355	1.33
11	10	7	75	30	0.725	1.23
12	10	7	100	40	0.445	1.35
13	10	8	50	40	0.710	1.11
14	10	8	75	50	0.485	1.26
15	10	8	100	30	1.125	1.30
16	10	9	50	50	0.780	1.15
17	10	9	75	30	1.195	1.24
18	10	9	100	40	1.105	1.42

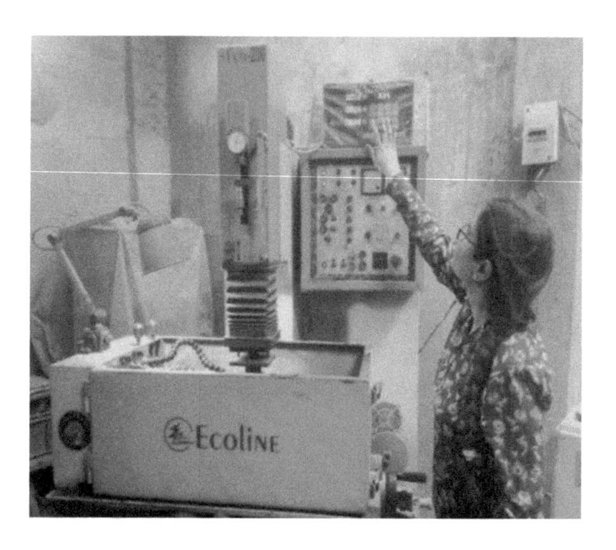

Fig. 1 EDM set-up

input variables affecting the response, while Table 5 shows the ANOVA for the same, and Tables 5 and 6 altogether depict that discharge current alone affected the response data significantly. ANOVA table has been basically used to present the significant effect or contribution of the input variable on the finally obtained response.

Fig. 2 Inconel-718 specimen after one of the trial runs on EDM

Table 4 Response table for signal to noise ratio

Level	I_P(amp)	T(off)	T(on)	S_V(volt)
1	−0.8461	−1.2690	−1.3606	−1.8862
2	−2.0225	−1.2227	−1.5658	−1.5118
3		−1.8113	−1.3766	−0.9051
Delta	1.1764	0.5886	0.2051	0.9811
Rank	1	3	4	2

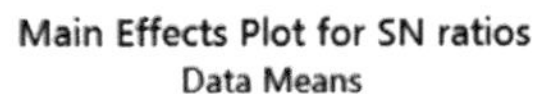

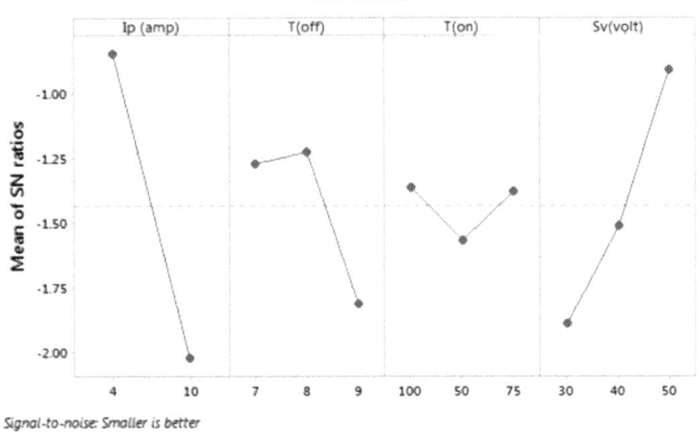

Fig. 3 Main effects plot for surface roughness

Table 5 Analysis of variance (ANOVA)

Source	DF	Adj SS	Adj MS	*F*-Value	*P*-Value
I_P(amp)	1	0.107339	0.107339	6.23	0.032
T(off)	2	0.020133	0.010067	0.58	0.576
T(off)	2	0.001900	0.000950	0.06	0.947
S_V(volt)	2	0.0044100	0.022050	1.28	0.320
Error	10	0.172378	0.017238		
Total	17	0.345850			

Table 6 Coefficients

Term	Coef	SE Coef	*T*-Value	*P*-Value	VIF
Constant	1.1883	0.0309	38.40	0.000	
*I*p(amp)					
4	−0.0772	0.0309	−2.50	0.032	1.00
T(off)					
7	−0.0167	0.0438	−0.38	0.711	1.33
8	−0.0300	0.0438	−0.69	0.509	1.33
T(on)					
100	0.0017	0.0438	0.04	0.970	1.33
50	0.0117	0.0438	0.27	0.795	1.33
S_V(volt)					
30	0.0550	0.0438	1.26	0.237	1.33
40	0.0100	0.0438	0.23	0.824	1.33

5 Conclusion

In this research, the ideal choice of machining variables for electric discharge machine was determined by the application of design of experiment technique using Taguchi method. For the particular experiment, L18 orthogonal array was selected. A progression of organized experiments was structured and arrayed; changes were made to the input control variables according to a framework structure. The analysis of the resultant impacts of these progressions on the output variable was executed. Therefore, minimum surface roughness could be procured with the optimal setting of the input control variables. Following conclusions can be summed up from the above work:

- Only discharge current I_P ($P < 0.05$), among all the input variables, was found to be significant for minimum surface roughness.
- Combination of the optimal levels of the input control variables for electric discharge machining of Inconel-718 for minimum surface roughness (SR) was, 4 A of I_P, 8 units *T*(off), 100 units of *T*(on), 50 V of S_v.

References

1. Mohan B, Muthuramalingam T (2015) A review on influence of electrical process parameters in EDM process. Arch Civ Mech Eng 15(1):87–95
2. Rahman M, Jahan M, Wong YS (2013) Micro-electrical discharge machining (Micro-EDM): micro-manufacturing: design and manufacturing of micro-products, vol 11. Springer-Verleg London, pp 333–371
3. Mahapatra SS, Amar P (2007) Optimization of wire electrical discharge machining (WEDM) process parameters using Taguchi method. Int J Adv Manuf Technol 34(9):911–925
4. Ho KH, Newman ST (2003) State of the art electrical discharge machining (EDM). Int J Mach Tools Manuf 43:1287–1300
5. Mohanty CP, Mahapatra SS, Singh MR (2017) An intelligent approach to optimize the EDM process parameters using utility concept and QPSO algorithm. Int J Eng Sci Technol 20(2): 552–562
6. Holmberg J, Berglund J, Wretland A, Beno T (2019) Evaluation of surface integrity after high energy machining with EDM, laser beam machining and abrasive water jet machining of alloy 718. Int J Adv Manuf Technol 100:1575–1591
7. Wang C, Qiang Z (2019) Comparison of Micro-EDM characteristics of inconel 706 between EDM oil and an Al powder-mixed dielectric. Adv Mater Sci Eng 1–11
8. Kumar S, Singh R, Singh TP, Sethi BL (2009) Surface modification by electrical discharge machining: a review. J Mater Process Technol 209(8):3675–3687
9. George PM, Raghunath BK, Manocha LM, Warrier AM (2004) EDM machining of carbon-carbon composite—a Taguchi approach. J Mater Process Technol 145(1):66–71
10. Her GM, Weng TF (2002) A study of the electrical discharge machining of semi-conductor $BaTiO_3$. J Mater Process Technol 122(1):1–5
11. Rao PS, Ramji K, Satyanarayana B (2016) Effect of Wire EDM conditions on generation of residual stresses in machining of aluminium T6 alloy. Alex Eng J 55:1077–1084
12. Rahul, Abhishek K, Datta S, Biswal BB, Mahapatra SS (2017) Machining performance optimisation during EDM of Inconel 718: a case experimental investigation. Int J Prod Qual Manag 21(4):460–489
13. Ramakrishnan R, Karunamoorthy L (2008) Modelling and multi-response optimization of inconel-718 on machining of CNC WEDM process. J Mater Process Technol 207:343–349
14. Bigot S, Valentincic J, Blatnik O, Junkar M (2006) Micro EDM parameters optimization, second international conference on multi material micro manufacture (4M).195–198
15. Anil K, Sachin M, Sharma C, Beri N (2012) Machining efficiency evaluation of cryogenically treated copper electrode in additive mixed EDM. Mater Manuf Processes 27:1051–1058
16. Balasubramanian P, Senthilvelan T (2014) Optimization of machining parameters in EDM process using cast and sintered copper electrodes. In: 3rd International conference on materials processing and characterization (ICMPC 2014), vol 6, pp 1292–1302
17. Sahu BK, Datta S, Mahapatra SS (2018) On electro-discharge machining of inconel 718 super alloys: an experimental investigation, ICMPC 2017. Mater Today Proc 5:4861–4869
18. Keskin Y, Halkac HS, Kizil M (2006) An Experimental study for determination of the effects of machining parameters on surface roughness in electrical discharge machining (EDM. Int J Adv Manuf Technol 28:1118–1121
19. Seref A (2011) Surface roughness prediction in machining castamide material using ANN. Acta Polytech Hung 8(2):21–32
20. Swiercz R, Oniszczuk-Swiercz D, Chmielewski T (2019) Multi-response optimization of electrical discharge machining using the desirability function. Micromachines 72:1–25

Navigational Control Analysis of Mobile Robot in Cluttered Unknown Environment Using Novel Neural-GSA Technique

Swadhin Sambit Das, Suranjan Mohanty, Adhir Kr. Behera, Dayal R. Parhi and Saroj Kr. Pradhan

Abstract A unique hybridized neural-GSA artificial intelligence strategy has been proposed in this current paper for the steerage of a wheeled mobile robot in an obstacle prone environment. In this work, a seven-layered back propagation neural network has been hybridized with GSA to synthesize a controller for the wheeled mobile robot. The inputs to the neural-GSA approach are front obstacle distance, left obstacle distance, right obstacle distance and target angle. The output from the neural network is intermediate steering angle. The inputs to the GSA system in neural-GSA technique are front obstacle distance, left obstacle distance, right obstacle distance and intermediate steering angle. The output from the GSA controller is final steering angle. During the research, several simulations are carried out. Using the proposed neural-GSA strategy as well as theoretical results, it has been found out that the robot can successfully navigate in an obstacle prone environment.

Keywords Neural network · Gravitational search algorithm (GSA) · Hybridized neural-GSA approach · Koala real robot · Simulation results

---

S. S. Das (✉) · S. Mohanty
Padmanava College of Engineering, Rourkela, Odisha, India

A. Kr. Behera
Rourkela Institute of Technology, Rourkela, Odisha, India

D. R. Parhi
National Institute of Technology, Rourkela, Odisha, India

S. Kr. Pradhan
College of Engineering and Technology, Bhubaneswar, Odisha, India
e-mail: skpradhan@cet.edu.in

BBVL. Deepak et al. (eds.), *Innovative Product Design and Intelligent Manufacturing Systems*, Lecture Notes in Mechanical Engineering,
https://doi.org/10.1007/978-981-15-2696-1_54

1 Introduction

Mobile robots are self-reliant agents which are capable of navigating intelligently anywhere using sensor-actuator manipulation techniques. This paper emphasizes on the path planning approach of a Koala mobile robot. Using a hybridized neural-GSA artificial intelligence strategy, a wheeled mobile robot can develop a pathway to arrive at the target, staying away from static as well as dynamic impediments.

In this modern world, mobile robots play an indispensable role as they can substitute humans in various applications such as industries and organizations, space, defense, transportation and hospitals and many other social sectors. Therefore, by implementing a mobile robot in place of human beings in a hazardous and unpredictable environment, the performance and efficiency with respect to cost optimization and time optimization are enhanced to a larger extent. Also, risk factor to human life is substantially reduced.

In robotics, robot path planning strategy is one amongst a few elementary difficulties. During the last few decades, various explorations on navigation of mobile robots using artificial intelligence methodologies have been carried out for single as well as multiple robots so that the mobile robots can reach the target by avoiding obstacles and without colliding with each other.

Pandey et al. [1] and Parhi et al. [2], in their research papers, have discussed about the analysis and review of guidance, navigation control and obstacle avoidance techniques for smooth steerage of autonomous mobile robots and underwater mobile robots in static and dynamic environments. In Paper [3], the kinematic control of mobile manipulators has been introduced where the manipulator and the mobile platform are integrated in such a way that the system is able to perform the constrained motion, while the platform is moving. In the articles [4, 5], researchers have discussed about artificial intelligence strategy like fuzzy systems and adaptive decision-making technique like fuzzy Bayesian reinforcement learning for robot soccer. Fuzzy logic and hybrid neural networks have been utilized in papers [6–8] for the real-time navigational path analysis of versatile robots. Authors have used heuristic rule-based hybrid neural network, fuzzy logic and evidence theory along with FNN and evolutionary fuzzy control as tools for locomotion of robots and multiple wheeled mobile robots in any unfamiliar obstacle prone scenario. The simulation outcomes have been tested with experimental results and are found to be in admissible settlement.

2 Reviews and Analysis of Various Artificial Intelligence Techniques

Phenomenal strategies have been developed for path generation, path planning and locomotion control of self-reliant mobile robots. These techniques include artificial immune system and ant-colony optimized recurrent neural networks which have been proposed in the articles [9–12]. In one of these papers, Suzuki et al. have implemented locomotion control using Neural Oscillator space for a seven-Linked Walking Robot Embedded with CPG. The outcomes proved the efficacy of these techniques. Papers [13–17] describe neural decoding, MLP and RBF-based neural networks, capacity margin index, innate immune-based path planning optimization methods for wheeled mobile robots in an actual densely populated environment. Papers [18, 19] focus on fuzzy wind-driven optimization algorithm and ANFIS rule-based control technique for navigational path planning and control of mobile robotic manipulators in clustered unknown or partially known environments.

Articles [20–28] propose artificially intelligent methods and hybridized intelligent approaches for the motion of mobile robots. A neural network control for a two-linked flexible robotic manipulator using assumed mode method has been successfully designed in one of these papers. In the other papers, radial basis function neural network and Sugeno fuzzy-based navigational control methodologies are used. The experimental outcomes and the simulated results were compared to show the efficiency of these techniques. Both the results very much match with each other. Some of these techniques have been suitably implemented in other practical applications as well. Authors of the papers [29–31] have discussed about foundations of ANN and continual lifelong learning with NN. New intelligent control techniques for the navigation planning of mobile robots have been recommended in the articles [32–39]. PSO, genetic algorithm and neuro-fuzzy control strategy have been successfully implemented for the trajectory analysis in three-RRR parallel robot and one-legged hopping robot.

Comprehensive studies on different artificial intelligence techniques like inverse kinematics application using MANFIS, adapted PSO, artificial immune system and recurrent neural network learned through group-based hybrid meta-heuristic algorithms are few intelligent techniques which have been introduced in the papers [40–44]. These techniques allow the mobile robot to navigate avoiding obstacles in a clustered environment. The outcomes, both experimental as well as simulation work, are verified and compared and were found to be satisfactory. In the papers [45–52], these intelligent techniques are applied successfully in other real-life engineering problems. Papers [53–57] discuss about immunized navigational controller, fuzzy-neuro-based and fuzzy logic-based controller which are used for path planning and motion control of wheeled mobile robots. In paper [58], a six-layered NN along with other AI techniques has been designed as path planner for a wheeled mobile robot.

Some of these papers solve some social problems using these artificial intelligence techniques. Articles [59–64] have implemented motion planning strategies using ANN and TLBO-based ANFIS for avoiding the obstacles and optimal path planning of wheeled mobile robots and quadrotor aerial robot. Articles [65–70] have presented various navigation strategies such as MANFIS and ANN for autonomous flying agent, hybrid parallel robot and humanoid robot. Papers [71–76] have used feedback control navigation and hybrid methodologies to develop control strategy for Takagi Sugeno fuzzy model. Simulated annealing technique has been hybridized so that the robot can successfully navigate avoiding hazards in a multiple obstacle-based scenario. Research papers [77–80] have used ANFIS and WNN approach with RBFNN control techniques for proper steering of mobile robots in cluttered surroundings.

3 Analysis of Seven-Layered Hybrid Neural-GSA Network Method for the Navigation of Mobile Robot

In this current paper, a seven-layered neural network has been synthesized for the navigational path analysis of a Koala mobile robot. The layers and their number of neurons are chosen provisionally to promote the learning of neural network. There are four neurons in the input layer out of which three measures the hurdles distances in front (FOD), to the left (LOD) as well as to the right (ROD) of the robot, and the last input neuron measures the target angle (TA). If no target is found in the territory, the fourth input becomes "zero." The first intermediate layer has eighteen neurons, the second intermediate layer has thirty-six neurons, the third intermediate layer has eighteen neurons, the fourth intermediate layer has seven neurons, and the fifth intermediate layer has three neurons. The output neuron is the intermediate steering angle. It also acts as one of the four inputs to the GSA controller. The three other inputs to the GSA controller proposed in this research work are FOD, ROD and LOD. The output from this GSA controller is the final steering angle. It governs the motion direction of the robot. The neural network is made to learn thousands of patterns to give an optimal solution for the robot in a naturally crowded surrounding. Figure 1 explains the neural-GSA network emphasizing the explanations of the neurons in addition to the input and output layers.

For neural network [81], the equations used are as follows:

$$W_{ji}(t+1) = W_{ji}(t) + \Delta W_{ji}(t+1) \tag{1}$$

$$\Delta W_{ji}(t+1) = K\Delta W_{ji}(t) + \lambda \delta_j^{[\text{layer}]} y_i^{\text{layer}-1} \tag{2}$$

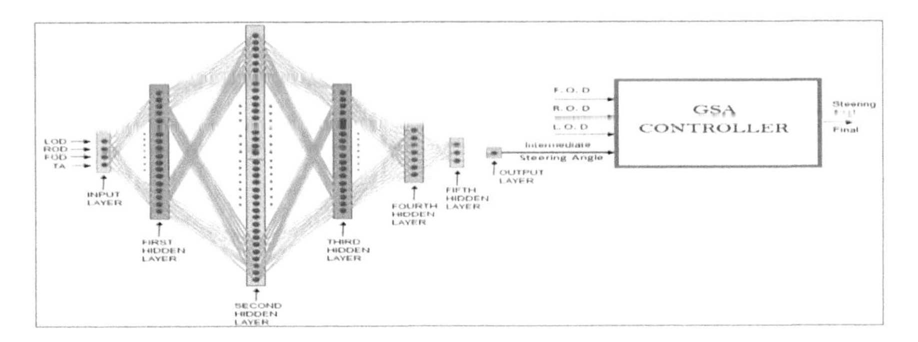

Fig. 1 Seven-layered hybrid neural-GSA controller for robot navigation

where K = Momentum coefficient (selected as 0.35), λ = Learning rate (selected as 0.25), t = Iteration number, each iteration consists of a learning pattern and correction of the weights $\delta^{[6]}$ = Error Gradient.

Also, for gravitational search algorithm [82], gravitational constant at iteration t for total iteration T is computed as follows:

$$G(t) = G_0 e^{-\alpha t/T} \tag{3}$$

4 Description of Robot Used in the Experiment

Koala robot is a powerful medium-size robot which can carry large appliances. It is designed for applications in real world. It has essential functionalities for practical applications like navigation and artificial intelligence. It has six wheels, multiple sensors which help to perform in both indoor and outdoor terrains. It has high-quality DC motors for better accuracy. It is adaptable with USB and Ethernet (Fig. 2).

Fig. 2 Different real views of Koala robot

5 Simulation and Experimental Results in Pictorial Form

The neuro-GSA network established here is tested in simulation and experimental mode. In simulation mode, a C program has been written, and the equivalent outputs are evaluated. In the experimental mode, the neuro-GSA network is implemented in Koala robot. Figure 3 illustrates six different robot positions, in simulation mode, when the robot moves from source to the target. Figure 3a shows the original position from where the robot starts. Figure 3b, e shows the transitional positions. Figure 3f shows the eventual position of the robot and analogous trajectory from source to target during simulation mode. The robot finds an optimum path which has been illustrated in Fig. 3f. The robot navigates from the source to the target evading eight obstacles.

Figure 4 illustrates six different robot positions, in experimental mode. Figure 4a shows the original position from where the robot starts. Figure 4b, e shows the transitional positions. Figure 4f shows the eventual position of the robot and analogous trajectory from source to target during experimental mode.

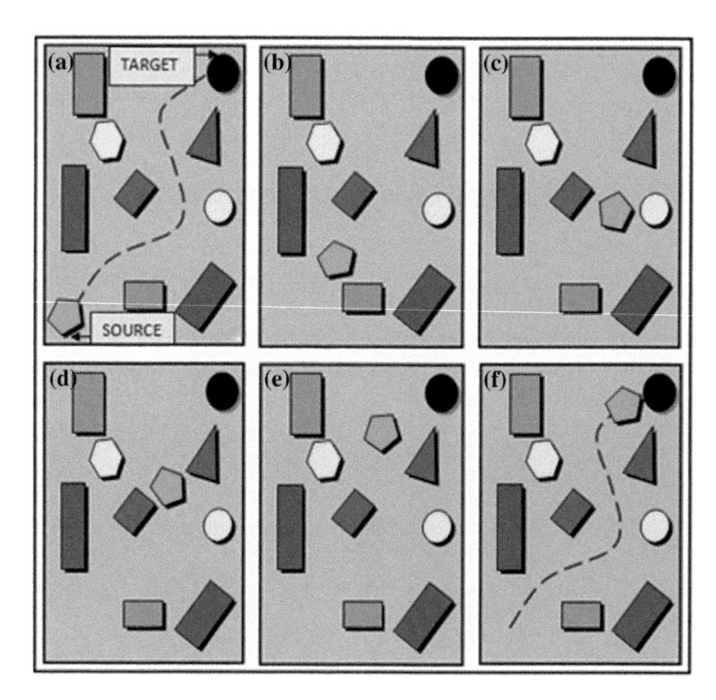

Fig. 3 Results from simulation mode

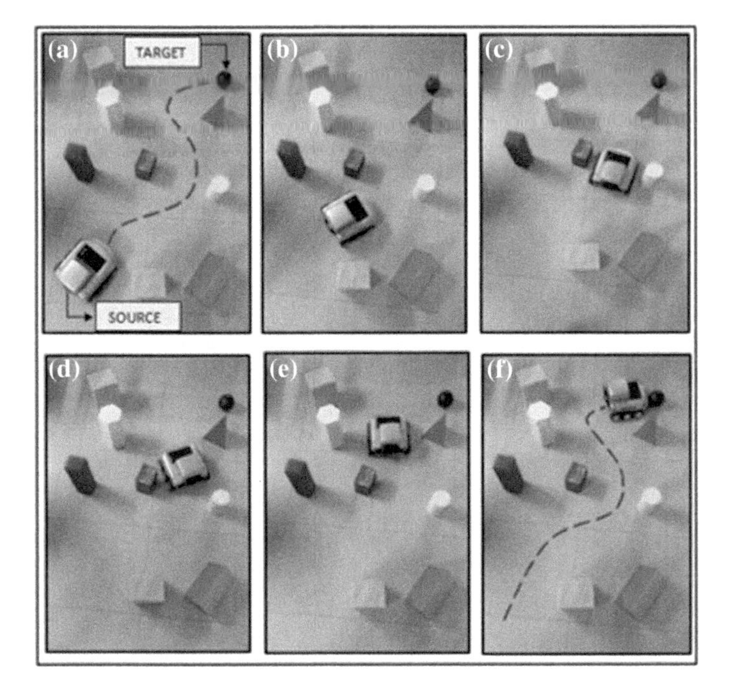

Fig. 4 Results from experimental mode

6 Simulation and Experimental Results in Tabular Form

Table 1 depicts the results for 23 exercises in respect of path length in simulation and experimental mode. Table 2 depicts the results for 23 exercises in respect of time taken in simulation and experimental mode. During comparison between simulation and experimental modes, the error is also found to be within 7% in both the cases. Also, appropriate actions have been taken to avoid the error and sleepage between wheels and floor during the experiment.

Table 1 Path travelled by robot modes

No. of exercise	Simulation path length in meter	Experiment path length in meter	% deviation	Average % deviation
1	8.53	9.08	6.06	
2	4.65	5.02	7.38	
3	9.42	10.06	6.37	
4	1.64	1.69	2.96	
5	5.76	6.01	4.16	

(continued)

Table 1 (continued)

No. of exercise	Simulation path length in meter	Experiment path length in meter	% deviation	Average % deviation
6	9.18	9.59	4.28	
7	2.06	2.2	6.37	
8	2.29	2.42	5.38	
9	6.71	7.17	6.42	
10	1.55	1.65	6.07	
11	6.22	6.6	5.76	
12	8.92	9.67	7.76	5.81
13	3.95	4 13	4.36	
14	2.49	2.7	7.78	
15	7.81	8.02	2.62	
16	3.91	4.11	4 87	
17	7.31	7.55	3.18	
18	3.68	3.92	6.13	
19	8.13	8.8	7.62	
20	5.68	6.15	7.65	
21	8.04	8.55	5.97	
22	7.99	8.48	5.78	
23	1.83	2	8.5	

Table 2 Time taken by robot modes

No. of exercise	Simulation time taken in millisecond	Experiment time taken in millisecond	% deviation	Average % deviation
1	106,625	117,029	8.9	
2	58,125	63,650	8.69	
3	117,750	126,586	6.99	
4	20,500	21,204	3.33	
5	72,000	76,580	5.99	
6	114,750	125,014	8.22	
7	25,750	27,406	6.05	
8	28,625	30,446	5.99	
9	83,875	92,140	8.98	
10	19,375	20,820	6.95	
11	77,750	84,603	8.11	
12	111,500	121,104	7.94	6.42
13	49,375	53,143	7.1	
14	31,125	32,527	4.32	
15	97,625	104,167	6.29	
16	48,875	52,520	6.95	

(continued)

Table 2 (continued)

No. of exercise	Simulation time taken in millisecond	Experiment time taken in millisecond	% deviation	Average % deviation
17	91,373	94,660	3.48	
18	46,000	47,248	2.65	
19	101,625	105,453	3.64	
20	71,000	77,622	8.54	
21	100,500	105,236	4.51	
22	99,875	106,104	5.88	
23	22,875	24,878	8.06	

7 Conclusion

In this present research, a novel investigation on navigational path analysis in an impediment prone environment has been carried out using neural-GSA technique for a wheeled mobile robot. The conclusions for the developed strategy are outlined on the basis of the investigation done above.

The inputs to the hybrid neural-GSA technique are FOD, LOD, ROD and TA. The interim output from the neural technique is the intermediate steering angle. The inputs to the GSA controller in the hybrid model are intermediate steering angle, FOD, LOD and ROD. The final output from the hybrid model is final steering angle. During the analysis, several results are drawn and are reported in tabular and pictorial forms for the parameters such as path length and time taken from the source to the target position during the navigation of mobile robots. The results are given for the simulation and experimental modes. A comparison is also made between the simulation and experimental results, and a deviation of below 7% is found between them. It is noted that the developed neuro-GSA network can be used successfully for route planning of mobile robot in a densely populated scenario. In future, other intelligent hybrid techniques will be analyzed to get a better path during navigation for the wheeled robot.

References

1. Pandey A, Pandey S, Parhi DR (2017) Mobile robot navigation and obstacle avoidance techniques: a Review. Int Rob Auto J 2(3):00022
2. Parhi DR, Kundu S (2012) Review on guidance, control and navigation of autonomous underwater mobile robot. Int J Artif Intell Comput Res (IJAICR) 4(1)
3. Deepak BBVL, Parhi DR, Prakash R (2016) Kinematic control of a mobile manipulator. In: Proceedings of the international conference on signal, networks, computing, and systems. Springer, India, pp 339–346

4. Rubio Y, Picos K, Orozco-Rosas U, Sepúlveda C, Ballinas E, Montiel O, Sepúlveda R (2018) Path following fuzzy system for a nonholonomic mobile robot based on frontal camera information. In: Fuzzy logic augmentation of neural and optimization algorithms: theoretical aspects and real applications. Springer, Cham, pp 223–240
5. Shi H, Lin Z, Zhang S, Li X, Hwang KS (2018) An adaptive decision-making method with fuzzy Bayesian reinforcement learning for robot soccer. Inf Sci 436:268–281
6. Chen G, Zhang W (2018) Comprehensive evaluation method for performance of unmanned robot applied to automotive test using fuzzy logic and evidence theory and FNN. Comput Ind 98:48–55
7. Chou CY, Juang CF (2018) Navigation of an autonomous wheeled robot in unknown environments based on evolutionary fuzzy control. Inventions 3(1):3
8. Parhi DR, Singh MK (2010) Heuristic-rule-based hybrid neural network for navigation of a mobile robot. Proc Inst Mech Eng, Part B J Eng Manuf 224(7):1103–1118
9. Deepak BBVL, Parhi AKJDR Path planning of an autonomous mobile robot using artificial immune system
10. Jena SP, Parhi DR, Jena PC Dynamic response of damaged cantilever beam subjected to traversing mass
11. Suzuki H, Lee JH, Okamoto S (2018) Locomotion control of 7-linked walking robot embedded with CPG using neural oscillator space. In: Proceedings of the international multi conference of engineers and computer scientists, vol 2
12. Juang CF, Yeh YT (2018) Multiobjective evolution of biped robot gaits using advanced continuous ant-colony optimized recurrent neural networks. IEEE Trans Cybern 48(6):1910–1922
13. Contreras-Vidal JL, Bortole M, Zhu F, Nathan K, Venkatakrishnan A, Francisco GE, Soto R, Pons JL (2018) Neural decoding of robot-assisted gait during rehabilitation after stroke. Am J Phys Med Rehabil 97(8):541–550
14. Manzoor S, Cho YG, Choi Y (2018) Neural oscillator based CPG for various rhythmic motions of modular snake robot with active joints. J Intell Robot Syst 1–14
15. Yadao AR, Singh RP, Parhi DR (2014) Influence of parameters of cracked rotor system on its vibration characteristics in viscous medium at finite region. In: Applied mechanics and materials, vol 592. Trans Tech Publications, pp 2061–2065
16. Singh A, Sahoo C, Parhi DR (2015) Design of a planar cable driven parallel robot using the concept of Capacity Margin Index. In: 2015 IEEE 9th international conference on intelligent systems and control (ISCO). IEEE, pp 1–7
17. Deepak BBVL, Parhi DR, Kundu S (2012) Innate immune based path planner of an autonomous mobile robot. Procedia Eng 38:2663–2671
18. Pandey A, Parhi DR (2017) Optimum path planning of mobile robot in unknown static and dynamic environments using Fuzzy-Wind Driven Optimization algorithm. Def Technol 13 (1):47–58
19. Mohanty PK, Parhi DR (2014) Navigation of autonomous mobile robot using adaptive network based fuzzy inference system. J Mech Sci Technol 28(7):2861–2868
20. Gao H, He W, Zhou C, Sun C (2018) Neural network control of a two-link flexible robotic manipulator using assumed mode method. IEEE Trans Ind Inform
21. Mirsky Y, Haddad Y, Rozenblit O, Azoulay R (2018) Predicting wireless coverage maps using radial basis networks. In: 2018 15th IEEE annual consumer communications and networking conference (CCNC). IEEE, pp 1–4
22. Qin Z, Chen X, Fu H, Hu S, Wang J (2018) Slope stability analysis based on the radial basis function neural network of the cerebral cortex. NeuroQuantology 16(5)
23. Parhi DR, Deepak BBVL (2011) Sugeno fuzzy based navigational controller of an intelligent mobile robot. Int J Appl Artif Intell Eng Syst 3(2):103–108
24. Parhi DR, Kundu S (2011) A hybrid fuzzy controller for navigation of real mobile robot. Int J Appl Artif Intell Eng Syst 3(1)
25. Sethi R, Parhi DRK, Senapati SK (2016) Fault detection of cracked beams like structure using artificial neural network (Ann) approach. Int J Res Eng Technol 5(13):1–4

26. Parhi DR, Behera AK (1998) The study of virtual mass and damping effect on a rotating shaft in viscous medium. J-Inst Eng India Part MC Mech Eng Div 109–113
27. Kumar PB, Parhi DR (2017) Vibrational characterization of a human femur bone and its significance in the designing of artificial implants. World J Eng 14(3):222–226
28. Sahu S, Parhi DR (2014) Automatic design of fuzzy rules using GA for fault detection in cracked structures. In: Applied mechanics and materials, vol 592. Trans Tech Publications, pp 2016–2020
29. Cangelosi A, Schlesinger M (2018) From babies to robots: the contribution of developmental robotics to developmental psychology. Child Dev Perspect
30. Parisi GI, Kemker R, Part JL, Kanan C, Wermter S (2018) Continual lifelong learning with neural networks: a review. arXiv preprint arXiv:1802.07569
31. Bielecki A (2019) Foundations of artificial neural networks. In: Models of neurons and perceptrons: selected problems and challenges. Springer, Cham, pp 15–28
32. Deepak BBVL, Parhi D (2012) PSO based path planner of an autonomous mobile robot. Open Comput Sci 2(2):152–168
33. Refoufi S, Benmahammed K (2018) Control of a manipulator robot by neuro-fuzzy subsets form approach control optimized by the genetic algorithms. ISA Trans 77:133–145
34. Sheng L, Li W (2018) Optimization design by genetic algorithm controller for trajectory control of a 3-RRR parallel robot. Algorithms 11(1):7
35. Azahar AH, Horng CS, Kassim AM, Abidin AFZ, Harun MH, Shah MBN, Annuar KAM, Manap M, Rizman ZI (2018) Optimizing central pattern generators (CPG) controller for one legged hopping robot by using genetic algorithm (GA). Int J Eng Technol 7(2.14):160–164
36. Kundu S, Parhi DR (2010) Fuzzy based reactive navigational strategy for mobile agent. In: 2010 international conference on industrial electronics, control and robotics (IECR). IEEE, pp 12–17
37. Parhi DR (2000) Navigation of multiple mobile robots in an unknown environment. Doctoral dissertation, University of Wales, Cardiff
38. Khan IA, Parhi DR (2015) Damage identification in composite beam by vibration measurement and fuzzy inference system. J Mech Des Vib 3(1):8–23
39. Das HC, Dash AK, Parhi DR, Thatoi DN (2010) Experimental validation of numerical and fuzzy analysis of a faulty structure. In: 2010 5th international conference on system of systems engineering (SoSE). IEEE, pp 1–6
40. Oda K, Takimoto M, Kambayashi Y (2018) Mobile agents for robot control based on PSO. In: ICAART (1), pp 309–317
41. Sancaktar I, Tuna B, Ulutas M (2018) Inverse kinematics application on medical robot using adapted PSO method. Eng Sci Technol Int J
42. Juang CF, Chang YC, Chung I (2018) Evolutionary hexapod robot gait control using a new recurrent neural network learned through group-based hybrid metaheuristic algorithm. In: Proceedings of the genetic and evolutionary computation conference companion. ACM, pp 111–112
43. Pandey A, Parhi DR (2014) MATLAB simulation for mobile robot navigation with hurdles in cluttered environment using minimum rule based fuzzy logic controller. Procedia Technol 14:28–34
44. Mohanty PK, Pandey KK, Parhi DR (2014) MANFIS approach for path planning and obstacle avoidance for mobile robot navigation. In: ICT and critical infrastructure: proceedings of the 48th annual convention of computer society of India, vol I. Springer, Cham, pp 361–370
45. Parhi DR, Sahu S (2017) Intelligent system for fault diagnosis of cracked beam. Int J Damage Mech 27(6):1056789517708019, 840–858
46. Thatoi DN, Nanda J, Das HC, Parhi DR (2012) Analysis of the dynamic response of a cracked beam structure. In: Applied mechanics and materials. vol 187. Trans Tech Publications, pp 58–62
47. Sutar MK, Parhi DRK (2010) Smart detection of damage in a cracked cantilever beam using artificial intelligence. Noise Vib Worldw 41(5):26–36

48. Parhi DR, Dash AK, Das HC (2011) Formulation of a genetic algorithm based methodology for multiple crack detection in a beam structure. Aust J Struct Eng 12(2):127–139
49. Parhi DR, Agarwalla DK (2012) Diagnosis of damaged Al cantilever beam subjected to free vibration by numerical and experimental method. IJAICR 4(2):71–77
50. Parhi DR, Agarwalla DK (2012) Determination of modified natural frequencies of fractured fixed-fixed beam by numerical and experimental method. IJAAIES 4(2):95–101
51. Behera RK, Pandey A, Parhi DR (2014) Numerical and experimental verification of a method for prognosis of inclined edge crack in cantilever beam based on synthesis of mode shapes. Procedia Technol 14:67–74
52. Jena SP, Parhi DR (2014) Dynamic deflection of a cantilever beam carrying moving mass. In: Applied mechanics and materials, vol 592. Trans Tech Publications, pp 1040–1044
53. Mahapatra S, Jha AK, Patle BK, Parhi DRK (2012) Fuzzy logic control of a WMR. In: 2012 international conference on computing, communication and applications (ICCCA). IEEE, pp 1–5
54. Jena PC, Pohit G, Parhi DR (2017) Fault measurement in composite structure by fuzzy-neuro hybrid technique from the natural frequency and fibre orientation. J Vib Eng Technol 5 (2):123–136
55. Thatoi DN, Das HC, Parhi DR (2012) Review of techniques for fault diagnosis in damaged structure and engineering system. Adv Mech Eng 4(327569):1–11
56. Parhi DR, Deepak BBVL, Mohana J, Ruppa R, Nayak M (2012) Immunised navigational controller for mobile robot navigation, 171–182
57. Kundu S, Parhi R, Deepak BBVL (2012) Fuzzy-neuro based navigational strategy for mobile robot. Int J Sci Eng Res 3(6):1–6
58. Das SS, Parhi DR, Mohanty S (2018) Insight of a six layered neural network along with other AI techniques for path planning strategy of a robot. In: Emerging trends in engineering, science and manufacturing (ETESM-2018), IGIT Sarang, India
59. Sahu S, Kumar PB, Parhi DR (2017) Design and development of 3-stage determination of damage location using Mamdani-adaptive genetic-Sugeno model. J Theor Appl Mech 55 (4):1325–1339
60. Kashyap SK, Parhi DR, Sinha A (2009) Artificial neural network-a tool for optimising mining parameters paper No. 377. In: Proceedings of the ISRM-sponsored international symposium on rock mechanics: rock characterisation, modelling and engineering design methods (SINOROCK2009)
61. Parhi DR, Yadao AR (2016) Analysis of dynamic behavior of multi-cracked cantilever rotor in viscous medium. Proc Inst Mech Eng, Part K: J Multi-Body Dyn 230(4):416–425
62. Baykal C, Bowen C, Alterovitz R (2018) Asymptotically optimal kinematic design of robots using motion planning. Auton Robot 1–13
63. Aouf A, Boussaid L, Sakly A (2018) TLBO-based adaptive neurofuzzy controller for mobile robot navigation in a strange environment. Comput Intell Neurosci
64. Ireland ML, Anderson D (2018) Optimisation of trajectories for wireless power transmission to a Quadrotor aerial robot. J Intell and Robot Syst 1–18
65. Mohanty PK, Parhi DR (2014) Path planning strategy for mobile robot navigation using MANFIS controller. In: Proceedings of the international conference on frontiers of intelligent computing: theory and applications (FICTA) 2013. Springer, Cham, pp 353–361
66. Parhi DR, Singh MK (2009) Navigational strategies of mobile robots: a review. Int J Autom Control 3(2–3):114–134
67. Mohanta JC, Parhi DR, Mohanty SR, Keshari A (2018) A control scheme for navigation and obstacle avoidance of autonomous flying agent. Arab J Sci Eng 43(3):1395–1407
68. Agarwalla DK, Parhi DR (2013) Effect of crack on modal parameters of a cantilever beam subjected to vibration. Procedia Eng 51:665–669
69. Sahoo B, Parhi DR, Priyadarshi BK (2018) Analysis of path planning of humanoid robots using neural network methods and study of possible use of other AI techniques, emerging trends in engineering, science and manufacturing, (ETESM-2018). IGIT, Sarang, India

70. Rastegarpanah A, Rakhodaei H, Saadat M, Rastegarpanah M, Marturi N, Borboni A, Loureiro RC (2018) Path-planning of a hybrid parallel robot using stiffness and workspace for foot rehabilitation. Adv Mech Eng 10(1):1687814017754159
71. Li Y, Wu J, Sato H (2018) Feedback control-based navigation of a flying insect-machine hybrid robot. Soft robotics
72. Sahu S, Kumar PB, Parhi DR (2017) Intelligent hybrid fuzzy logic system for damage detection of beam-like structural elements. J Theor Appl Mech 55(2):509–521
73. Pandey A, Parhi DR (2016) Autonomous mobile robot navigation in cluttered environment using hybrid Takagi-Sugeno fuzzy model and simulated annealing algorithm controller. World J Eng 13(5):431–440
74. Ranjan KB, Sahu S, Parhi Dayal R (2014) A new reactive hybrid membership function in fuzzy approach for identification of inclined edge crack in cantilever beam using vibration signatures. In: Applied mechanics and materials, vol 592. Trans Tech Publications, pp 1996–2000
75. Panigrahi PK, Ghosh S, Parhi DR (2014) Comparison of GSA, SA and PSO based intelligent controllers for path planning of mobile robot in unknown environment. J Electr Comput Electron Commun Eng 8(10):1523–1532
76. Sethi R, Senapati SK, Parhi DR (2014) Structural damage detection by fuzzy logic technique. In: Applied mechanics and materials. vol 592. Trans Tech Publications, pp 1175–1179
77. Mohanty PK, Parhi DR, Jha AK, Pandey A (2013) Path planning of an autonomous mobile robot using adaptive network based fuzzy controller. In: 2013 3rd IEEE international advance computing conference (IACC). IEEE, pp 651–656
78. Ghosh S, Kumar PP, Parhi DR (2016) Performance comparison of novel WNN approach with RBFNN in navigation of autonomous mobile robotic agent. Serbn J Electr Eng 13(2):239–263
79. Yadao AR, Parhi DR (2016) The influence of crack in cantilever rotor system with viscous medium. Int J Dyn Control 4(4):363–375
80. Yadao AR, Parhi DR (2015) Experimental and numerical analysis of cracked shaft in viscous medium at Finite region. In: Advances in structural engineering. Springer, New Delhi, pp 1601–1609
81. Haykin S (2009) Neural networks and learning machines, Third Edition, Prentice Hall
82. Rashedi E, Nezamabadi-Pour H, Saryazdi S (2009) GSA: a gravitational search algorithm. Inf Sci 179(13):2232–2248

Control Strategy of Mobile Robots Using Fuzzy-Gravitational Search Method and Review of Other Techniques

Suranjan Mohanty, Swadhin Sambit Das, Adhir Kumar Behera, Dayal R. Parhi and Saroj Kumar Pradhan

Abstract A systematic research methodology has been adapted using fuzzy-gravitational search algorithm for solving a complex navigational control problem of a mobile robot from source point to destination point while negotiating with obstacles. Inputs to the hybrid method are target angle, right obstacle distance, left obstacle distance, front obstacle distance. The interim output is interim steering angle. These interim inputs along with LOD, ROD, FOD are applied to gravitational search algorithm, and the outputs are final steering angle (SA). It is observed from the simulation and experimental results that the proposed technique is well suited for navigational control of robots in a densely populated environment. Keeping in view of the methodology used for control of robots, several artificial intelligence techniques are also discussed in the current paper.

Keywords Artificial · Control strategy · Gravitational search algorithm · Obstacles · Path length · Fuzzy logic

1 Introduction

Several researchers have discussed about various methodologies for navigational control of mobile robots. Some of them are discussed below. Papers [1–3] discuss on neuro-fuzzy technique, adaptive neuro-fuzzy inference system, fuzzy wind-driven optimization algorithm, artificial neural network for navigation and

S. Mohanty (✉) · S. S. Das
Padmanava College of Engineering, Rourkela, India

A. K. Behera
Rourkela Institute of Technology, Kalunga, India

D. R. Parhi
National Institute of Technology, Rourkela, India

S. K. Pradhan
College of Engineering and Technology, Bhubaneswar, India
e-mail: skpradhan@cet.edu.in

BBVL. Deepak et al. (eds.), *Innovative Product Design and Intelligent Manufacturing Systems*, Lecture Notes in Mechanical Engineering,
https://doi.org/10.1007/978-981-15-2696-1_55

control of mobile robots in cluttered environments. Papers [4–6] elaborate about matrix-binary codes based genetic algorithm, fuzzy-based reactive navigational strategy, cuckoo search algorithm, observation-based fuzzy sliding mode controller, robust genetic algorithm and fuzzy inference mechanism, type-2 fuzzy model, adaptive fuzzy and robust predictive controller, intelligent hybrid technique for planning path for mobile robots as well as other robots in an obstacle prone environment. Innate immune-based path planner, behavior-based navigation, hybrid fuzzy controller, intelligent fuzzy interference technique are some of the important methods used for navigation of mobile robots. These are discussed in various papers [7–9] by many researchers, but a systematic method is lacking for use of fuzzy-gravitational search method for navigation of mobile robot. For this, a methodology has been developed to apply fuzzy-gravitational search algorithm for navigation of mobile robot. Papers [10–12] narrate about artificial neural network, hybrid Takagi-Sugeno fuzzy model and simulated annealing algorithm controller, deep neural network, group-based hybrid metaheuristic algorithm, particle swarm optimization for navigation of mobile robots and other robots as well as in other engineering fields. Inverse kinematic model, forward kinematic model, artificial intelligence have been covered in papers [13–15]. Neural network model, radial basis function neural network, evolutionary radial basis function network, neural adaptive tracking control, innate immune-based path planner have been discussed in the papers [16–18]. The corresponding researchers have used these techniques for navigation of mobile robots, and also, they have verified the results in numeric forms. It has also been observed that fuzzy rules using GA, ant-colony optimization technique, MANFIS approach have been used in various engineering fields as discussed in papers [19–21]. Back propagation, artificial neural network, online fuzzy logic crack detection, knowledge-based and intelligent engineering system, Petri-potential fuzzy model are in limelight as discussed in papers [22–27]. Real-time navigation approach, neural network and a Pertinet model, Mamdani fuzzy control, circular back propagation and deep kohonen neural network have been narrated in papers [28–33]. Researchers have used these methods intelligently for movement of robots intelligently in various environments. Papers [34–36] discuss on fuzzy verification method, adaptive SOM neural network, six-layered neural network, neural network and finite element, fault detection by finite element for calculating steering angle for mobile robots and also in other engineering fields. Neural network method, finite element method, fuzzy controller, fuzzy-neuro-based navigational strategy, robust genetic algorithm and fuzzy interference mechanism, genetic algorithm-based motion estimation method have been used by many scientists and researchers in the papers [37–39]. Use of these techniques is beneficial for many researchers because applying these techniques they have found an obstacle-free path for robots. Genetic algorithm, improved genetic algorithm and fuzzy logic control, Mamdani-adaptive genetic-Sugeno model, adaptive shuffled frog-leaping algorithm, ANFIS approach, Sugeno fuzzy logic are being used by many researchers and scientists in the papers [40–42]. In these papers, they have derived many behaviors such as obstacle avoidance, target seeking using these AI techniques. Papers [43–45] discuss on type-2 FLC, PSO, BPNN-PSO, fuzzy logic-based PSO algorithm for finding obstacle-free path for mobile robots. PID-based PSO, Fuzzy controller, Petri-GA optimization have been cited in the papers [46–51]. Papers [52–54] depict on modal parameters and RBFNN technique, MLP and RBF-based neural network, improved multi-objective discrete bees algorithm, modified distributed bees algorithm for successful control of mobile robots from start position to goal position. In papers [55–57], a specific has been made on minimum rule-based ANFIS network controller, inverse weed optimization algorithm, adaptive neuro-fuzzy interference system, artificial immune system for solving navigational problems of wheeled mobile robots. Regression analysis and genetic algorithm, bacteria forging optimization, improved RBF have been narrated in the papers [58–60] for planning optimized route for a mobile robot and different types of robots. Papers [61–63] elaborate on probabilistic fuzzy controller, intelligent neuro controller, adaptive neuro-fuzzy interference system, fuzzy-neuro-hybrid technique, genetic controller for finding obstacle-free path for wheeled robots. Artificial intelligence techniques are very much useful for planning best routes for mobile robots. Papers [64–66] discuss on differential evolution, artificial immune system, firefly algorithm, ANN approach. These AI techniques are very much necessary for planning best routes for mobile robots. Multiple ANFIS approach, adaptive immune-based motion planner, singleton Takagi-Sugeno fuzzy inference methodology, WNN approach with RBFNN have been focused in papers [67–69] for finding solutions to different navigational problems of mobile robots. Papers [70–72] discuss on dynamically adaptive harmony search algorithm, rule-based neuro-fuzzy technique. FPA and BA metaheuristic controllers, fuzzy logic technique, intelligent hybrid controller, genetic algorithm and differential evolution algorithm, behavior-based algorithm have been discussed in papers [73–76] for behavior-based navigation. Cuckoo search algorithm, modified cuckoo search algorithm, artificial neural network have been narrated in the papers. These are some of the novel methods for finding obstacle-free path for wheeled robots and humanoid robots. Papers elaborate on finite element analysis, MANFIS controller, fuzzy-neuro controller. These methods are very helpful for designing obstacle-free path for humanoid robots. Many engineers and researchers have focused on calculating steering angle for mobile robots. Artificial potential field, chaotic artificial potential field method, parallel bacterial potential field, Sugeno fuzzy-based navigational controller are the most important methods to calculate steering angles for mobile robots which are given in papers. Papers deal with invasive weed optimization, artificial neural network, adaptive neuro-fuzzy controller. Papers [76–78] discuss on simulated annealing algorithm, mutual-coupled immune network planning algorithm, Cs-ANFIS approach, minimum rule-based fuzzy controller, adaptive neuro-fuzzy interference system, shuffled frog-leaping algorithm-based hybrid learning approach for control of wheeled robot in different environments. Fuzzy logic technique, swarm intelligence, advanced continuous ant-colony optimized recurrent neural network have been addressed in the papers. Papers depicted ant-colony optimization, neural network. Pheromone deposition is the criteria in ant-colony optimization to get the optimized path for mobile robots. Fuzzy logic controller, RBFNN, neuro-fuzzy navigation, binary particle swarm optimization and binary gravitational search algorithm are in limelight. Those techniques are addressed in papers. Papers deal with immunized navigation, fuzzy logic technology with hybrid membership functions, neuro-fuzzy approach, intelligent hybrid fuzzy logic system.

2 Analysis of Fuzzy-Gravitational Search Algorithm for Robot Navigation

2.1 Description of Inputs and Outputs

In the fuzzy controller, four inputs have been used. They are target angle, left obstacle distance, right obstacle distance and front obstacle distance. The interim output is interim steering angle. The interim output, LOD, ROD and FOD are fed to gravitational search algorithm controller. The final output is final steering angle (FSA).

'G' is the gravitational constant [79] used in the gravitational search algorithm and can be computed as follows:

$$G(t) = G_0 \mathrm{e}^{-\alpha t/T} \tag{1}$$

where G_0 and α are initialized at the start of the search.

Then, these values will be reduced as the search continues, t = iteration number at any time, T = total no. of iteration.

For calculating the updated velocity, the following formula is used

$$V_i(t+1) = \mathrm{rand}_i\, V_i(t) + a(t). \tag{2}$$

$V(t)$, $a(t)$ are the existing velocity and acceleration.

For calculating the updated position, the formula used is

$$X_i(t+1) = X_i(t) + V_i(t+1). \tag{3}$$

2.2 Corresponding Fuzzy-Gravitational Search Algorithm Figure in Final Form

See Figs. 1 and 2.

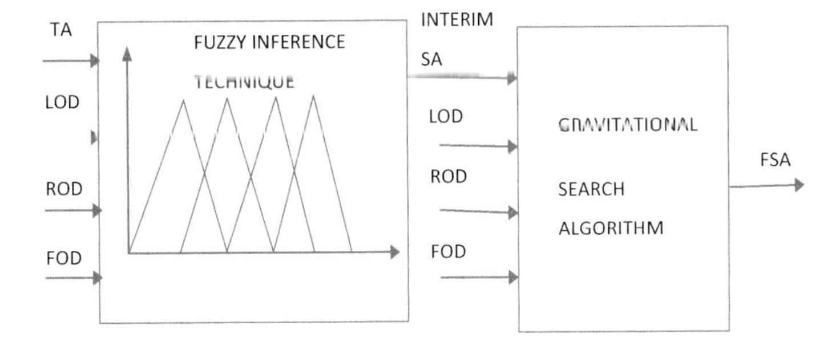

Fig. 1 View of hybrid controller

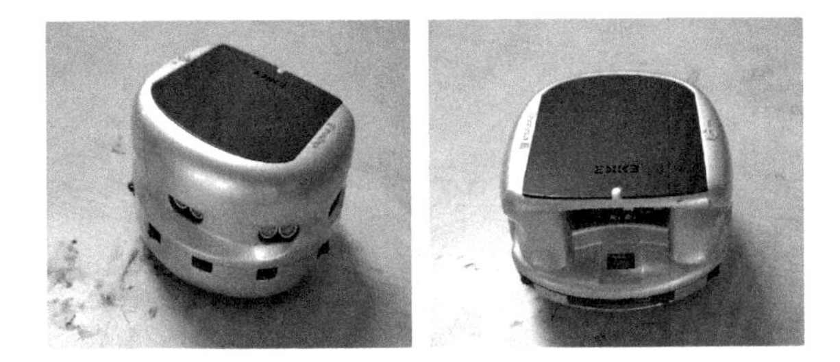

Fig. 2 Views of robot

3 Description of Robot Used for Experimental Purpose

3.1 Robot (Khepera-III)

The Khepera-III is made by K-team. The robot has the diameter of 12 cm XSCALE PXA-255 processor at 400 MHz frequency is used to run Linux operating system in the robot. The robot has several ultrasonic sensors and proximity sensors.

4 Simulation and Pictorial Results

See Figs. 3 and 4.

5 Experiment and Simulation of Path Length and Time Taken Result in Pictorial Form

See Tables 1 and 2.

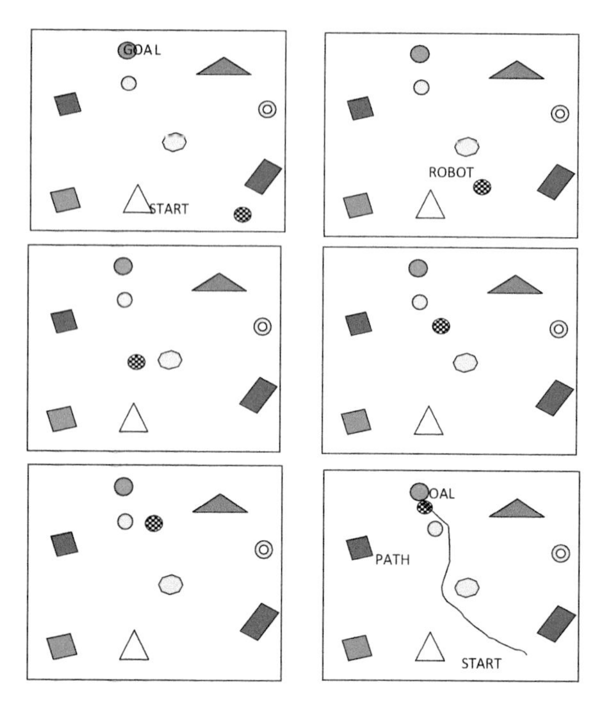

Fig. 3 Simulation results

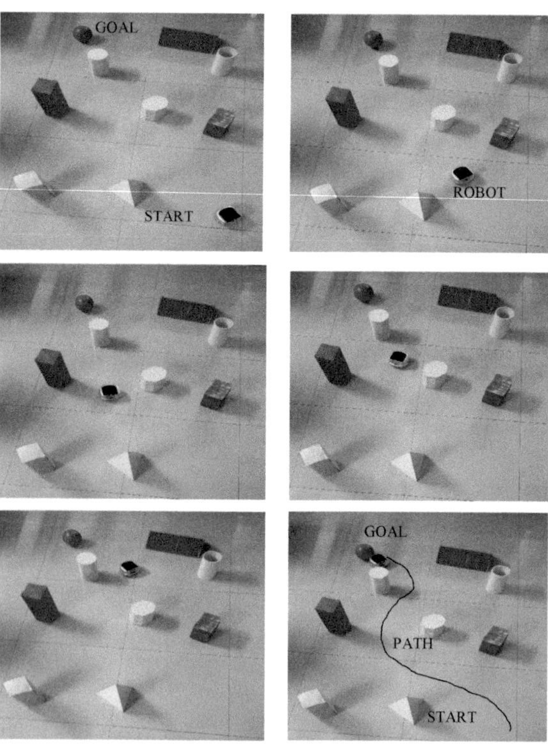

Fig. 4 Experimental results

Table 1 Path length from start to goal of robot

No. of exercise	Simulation path length in meter	Experiment path length in meter	% deviation	Average % deviation
1	3.24	3.42	5.27	4.84
2	3.49	3.77	7.43	
3	8.97	9.53	5.88	
4	5.01	5.26	4.76	
5	4.15	4.42	6.11	
6	7.97	8.3	3.98	
7	2.99	3.23	7.44	
8	6.11	6.63	7.85	
9	7.17	7.58	5.41	
10	3.13	3.24	3.4	
11	6.38	6.57	2.9	
12	2.06	2.23	7.63	
13	7.68	7.86	2.3	
14	4.78	5	4.4	
15	3.05	3.29	7.3	
16	6.18	6.44	4.04	
17	6.38	6.59	3.19	
18	9.27	9.48	2.22	
19	3.38	3.55	4.79	
20	5.1	5.25	2.86	
21	4.69	5.06	7.32	
22	4.14	4.27	3.05	
23	6.73	6.88	2.19	
24	8.22	8.59	4.31	

6 Discussion on Pictorial and Tabular Form

In this work, we have given six figures for simulation and six figures for experiment. This six set of figures show the various positions of robots while moving in a obstacles prone environment from start point to goal point. The corresponding experimental figures are also given in the figures, and they are showing the similar situations as those of the figures in simulation. Similarly, two tables are given as Tables 1 and 2. Table 1 represents the length of path in simulation and path length in actual experiment. Table 2 represents time taken simulation and time taken in actual experiment. The deviation has been calculated and is less than 6% in both cases. For both path length and time taken, twenty-four number of exercises have been given in both the tables.

Table 2 Time taken from start to goal by robot

No. of exercise	Simulation time taken in millisecond	Experiment time taken in millisecond	% deviation	Average % deviation
1	40,500	41,753	3.01	5.34
2	43,625	45,844	4.85	
3	112,125	117,557	4.63	
4	62,625	64,357	2.7	
5	51,875	54,003	3.95	
6	99,625	107,228	7.1	
7	37,375	40,839	8.49	
8	76,375	83,819	8.89	
9	89,625	93,467	4.12	
10	39,125	41,081	4.77	
11	79,750	86,133	7.42	
12	25,750	28,291	8.99	
13	96,000	100,915	4.88	
14	59,750	62,908	5.03	
15	38,125	39,244	2.86	
16	77,250	83,941	7.98	
17	79,750	83,082	4.02	
18	115,875	123,785	6.4	
19	42,250	44,390	4.83	
20	63,750	68,439	6.86	
21	58,625	60,884	3.72	
22	51,750	53,969	4.12	
23	84,125	86,817	3.11	
24	102,750	108,570	5.37	

7 Discussion on Cause of Deviation of Results in Simulation and Experiment

Simulation and experimental results differ by 6%. This is because of the reasons that some of the assumptions are taken during simulation. But during experiment, those assumptions are not taken. One of the measure factors of the deviation of simulation and experimental results is sleepage between robot wheel and floor. During movement, sleepage occurs between robot wheel and floor for which deviation between the simulation and experimental result is observed.

8 Conclusion

The current paper deals with fuzzy-gravitational search algorithm for movement of mobile robot in highly populated environment. Here, inputs are TA, LOD, ROD, FOD to the fuzzy logic system. The output is interim SA. This interim SA becomes input for gravitational search algorithm. The final output is final steering angle. A comparison has been given between simulation and experimental results. They are described in Figs. 3 and 4. A numerical comparison has also been given in twenty-four numbers in Tables 1 and 2. By comparison between tabular and pictorial form, the deviation is calculated to be within 6%. It has been found that using this method, the robot can move successfully in a densely populated environment. In future, other hybrid methods will be discussed for movement of mobile robots.

References

1. Kumar PB, Parhi DR (2017) Vibrational characterization of a human femur bone and its significance in the designing of artificial implants. World J Eng 14(3):222–226
2. Pandey A, Parhi DR (2017) Optimum path planning of mobile robot in unknown static and dynamic environments using Fuzzy-Wind Driven Optimization algorithm. Def Technol 13 (1):47–58
3. Kundu S, Parhi DR (2010) Fuzzy based reactive navigational strategy for mobile agent. In: 2010 international conference on industrial electronics, control and robotics (IECR). IEEE. pp 12–17
4. Jena SP, Parhi DR (2014) Dynamic deflection of a cantilever beam carrying moving mass. In: Applied mechanics and materials, vol 592. Trans Tech Publications, pp 1040–1044
5. Agarwalla DK, Parhi DR (2013) Effect of crack on modal parameters of a cantilever beam subjected to vibration. Procedia Eng 51:665–669
6. Parhi DR, Agarwalla DK (2012) Diagnosis of damaged Al cantilever beam subjected to free vibration by numerical and experimental method. IJAICR 4(2):71–77
7. Pandey A, Parhi DR (2016) Autonomous mobile robot navigation in cluttered environment using hybrid Takagi-Sugeno fuzzy model and simulated annealing algorithm controller. World J Eng 13(5):431–440
8. Juang CF, Chang YC, Chung I (2018) Evolutionary hexapod robot gait control using a new recurrent neural network learned through group-based hybrid metaheuristic algorithm. In: Proceedings of the genetic and evolutionary computation conference companion. ACM, pp 111–112
9. Deepak BBVL, Parhi DR, Prakash R (2016) Kinematic control of a mobile manipulator. In: Proceedings of the international conference on signal, networks, computing, and systems. Springer, India, pp 339–346
10. Parhi DR, Agarwalla DK (2012) Determination of modified natural frequencies of fractured fixed-fixed beam by numerical and experimental method. IJAAIES 4(2):95–101
11. Sutar MK, Parhi DRK (2010) Smart detection of damage in a cracked cantilever beam using artificial intelligence. Noise and Vib Worldw 41(5):26–36
12. Sahu S, Parhi DR (2014) Automatic design of fuzzy MF using genetic algorithm for fault detection in structural elements. In: 2014 Students Conference on Engineering and systems (SCES). IEEE, pp 1–5
13. Mirjalili S (2019) Evolutionary radial basis function networks. In: Evolutionary algorithms and neural networks. Springer, Cham, pp 105–139

14. Rahimi HN, Howard I, Cui L (2018) Neural adaptive tracking control for an uncertain robot manipulator with time-varying joint space constraints. Mech Syst Signal Process 112:44–60
15. Parhi DR, Yadao AR (2016) Analysis of dynamic behavior of multi-cracked cantilever rotor in viscous medium. Proc Inst Mech Eng, Part K: J Multi-Body Dyn 230(4):416–425
16. Sahu S, Parhi DR (2014) Automatic design of fuzzy rules using GA for fault detection in cracked structures. In: Applied mechanics and materials, vol 592. Trans Tech Publications, pp 2016–2020
17. Mohanty PK, Pandey KK, Parhi DR (2014) MANFIS approach for path planning and obstacle avoidance for mobile robot navigation. In: ICT and critical infrastructure: proceedings of the 48th annual convention of computer society of India-Vol I. Springer, Cham, pp 361–370
18. Zou W, Yao F, Zhang B, Guan Z (2018) Back propagation convex extreme learning machine. In: Proceedings of ELM-2016. Springer, Cham, pp 259–272
19. Konar A, Saha S (2018) EEG-gesture based artificial limb movement for rehabilitative applications. In: Gesture recognition. Springer, Cham, pp 243–268
20. Sailaja M, Prasad RDV (2018) Back propagation method of artificial neural networks for finding the position control of stanford manipulator and direct kinematic analysis of Elbow manipulator. Int J Emerg Res Manag Technol 7(1):46–56
21. Park MH, Jin BJ, Yun TJ, Son JS, Kim CG, Kim IS (2018) Control of the weld quality using welding parameters in a robotic welding process. J Achiev Mater Manuf Eng 87(1)
22. Hemanth DJ, Anitha J (2018) Brain signal based human emotion analysis by circular back propagation and Deep Kohonen Neural Networks. Comput Electr Eng 68:170–180
23. Versino C, Gambardella LM (2018) Learning fine motion in robotics: design and experiments. In: Recent advances in artificial neural networks. CRC Press, pp 149–176
24. Petrova O, Tabunshchyk G, Kapliienko T, Kapliienko O (2018) Fuzzy verification method for indoor-navigation systems. In: 2018 14th international conference on advanced trends in radioelecrtronics, telecommunications and computer engineering (TCSET). IEEE, pp 65–68
25. Li X, Zhu D (2018) An adaptive SOM neural network method to distributed formation control of a group of AUVs. IEEE Trans Ind Electron 65(10)
26. Das SS, Parhi DR, Mohanty S (2018) Insight of a six layered neural network along with other AI techniques for path planning strategy of a robot. In: Emerging trends in engineering, science and manufacturing (ETESM-2018), IGIT Sarang, India
27. Sahoo B, Parhi DR, Priyadarshi BK (2018) Analysis of path planning of humanoid robots using neural network methods and study of possible use of other AI techniques. In: Emerging trends in engineering, science and manufacturing, (ETESM - 2018), IGIT, Sarang, India
28. Sethi R, Senapati SK, Parhi DRK (2014) Analysis of crack in structures using finite element method. Analysis 2(2)
29. Behera RK, Pandey A, Parhi DR (2014) Numerical and experimental verification of a method for prognosis of inclined edge crack in cantilever beam based on synthesis of mode shapes. Procedia Technol 14:67–74
30. Chin CS, Lin WP (2018) Robust genetic algorithm and fuzzy inference mechanism embedded in a sliding-mode controller for an uncertain underwater robot. IEEE/ASME Trans Mechatron 23(2):655–666
31. Chae J, Jin Y, Sung Y, Cho K (2018) Genetic algorithm-based motion estimation method using orientations and EMGs for robot controls. Sensors 18(1):183
32. Sangdani MH, Tavakolpour-Saleh AR, Lotfavar A (2018) Genetic algorithm-based optimal computed torque control of a vision-based tracker robot: Simulation and experiment. Eng Appl Artif Intell 67:24–38
33. Tan X (2018) Design of autonomous navigation mobile robot based on improved genetic algorithm and fuzzy logic control. J Adv Oxid Technol 21(2)
34. Sahu S, Kumar PB, Parhi DR (2017) Design and development of 3-stage determination of damage location using Mamdani-adaptive genetic-Sugeno model. J Theor Appl Mech 55 (4):1325–1339

35. Thatoi DN, Nanda J, Das HC, Parhi DR (2012) Analysis of the dynamic response of a cracked beam structure. In: Applied mechanics and materials, vol 187. Trans Tech Publications, pp 58–62
36. Coronel-Escamilla A, Torres F, Gomez-Aguilar JF, Escobar-Jimenez RF, Guerrero-Ramírez GV (2018) On the trajectory tracking control for an SCARA robot manipulator in a fractional model driven by induction motors with PSO tuning. Multibody Sys Dyn 43(3):257–277
37. Gao G, Liu F, San H, Wu X, Wang W (2018) Hybrid optimal kinematic parameter identification for an industrial robot based on BPNN-PSO. Complexity
38. Maharuddin MF, Ghani NA, Jamin NF (2018) Two-wheeled LEGO EV3 robot stabilisation control using fuzzy logic based PSO algorithm. J Telecommun, Electron Comput Eng (JTEC) 10(2–5):149–153
39. Copot C, Muresan CI, Mac Thi T, Ionescu CM (2018) An application to robot manipulator joint control by using constrained PID based PSO. In: IEEE 12th international symposium on applied computational intelligence and informatics-SACI, pp 279–284
40. Behera RK, Parhi DR (2014) Validation of results obtained from different types of fuzzy controllers for diagnosis of inclined edge crack in cantilever beam by vibration parameters. J Mech Des Vib 2(3):63–68
41. Yadao AR, Parhi DR (2016) The influence of crack in cantilever rotor system with viscous medium. Int J Dyn Control 4(4):363–375
42. Liu J, Zhou Z, Pham DT, Xu W, Yan J, Liu A, Ji C, Liu Q (2018) An improved multi-objective disKineticcrete bees algorithm for robotic disassembly line balancing problem in remanufacturing. Int J Adv Manuf Technol 1–26
43. Liu J, Zhou Z, Pham DT, Xu W, Ji C, Liu Q (2018) Robotic disassembly sequence planning using enhanced discrete bees algorithm in remanufacturing. Int J Prod Res 56(9):3134–3151
44. Jena SP, Parhi DR, Mishra D (2015) Response of cracked cantilever beam subjected to traversing mass. In: ASME 2015 gas turbine India conference. American Society of Mechanical Engineers, pp V001T05A011–V001T05A011
45. Jena PC, Pohit G, Parhi DR (2017) Fault measurement in Composite Structure by fuzzy-neuro hybrid technique from the natural frequency and fibre orientation. J Vib Eng Technol 5 (2):123–136
46. Khan IA, Parhi DR (2015) Damage identification in composite beam by vibration measurement and fuzzy inference system. J Mech Des Vib 3(1):8–23
47. Sethi R, Parhi DRK, Senapati SK (2016) Fault detection of cracked beams like structure using artificial neural network (ANN) approach. Int J Res Eng Technol 5(13):1–4
48. Ghosh S, Kumar PP, Parhi DR (2016) Performance comparison of novel WNN approach with RBFNN in navigation of autonomous mobile robotic agent. Serbn J Electr Eng 13(2):239–263
49. Parhi DR, Behera AK (1997) Dynamic deflection of a cracked shaft subjected to moving mass. Can Soc Mech Eng, Trans 21(3):295–316
50. Sethi R, Senapati SK, Parhi DR (2014) Structural damage detection by fuzzy logic technique. In: Applied mechanics and materials. vol 592. Trans Tech Publications, pp 1175–1179
51. Mohanty PK, Parhi DR (2014) Path planning strategy for mobile robot navigation using MANFIS controller. In: Proceedings of the international conference on frontiers of intelligent computing: theory and applications (FICTA) 2013. Springer, Cham, pp 353–361
52. Singh A, Sahoo C, Parhi DR (2015) Design of a planar cable driven parallel robot using the concept of Capacity Margin Index. In: 2015 IEEE 9th international conference on intelligent systems and control (ISCO). IEEE, pp 1–7
53. Parhi DR, Deepak BBVL (2011) Sugeno fuzzy based navigational controller of an intelligent mobile robot. Int J Appl Artif Intell Eng Syst 3(2):103–108
54. Ranjan KB, Sahu S, Parhi Dayal R (2014) A new reactive hybrid membership function in fuzzy approach for identification of inclined edge crack in cantilever beam using vibration signatures. In: Applied mechanics and materials, vol 592. Trans Tech Publications, pp 1996–2000
55. Parhi DR, Behera AK (1998) The study of virtual mass and damping effect on a rotating shaft in viscous medium. J-Inst Eng India Part MC Mech Eng Div 109–113

56. Nanda J, Das S, Parhi DR (2014) Effect of slenderness ratio on crack parameters of simply supported Shaft. Procedia Mater Sci 6:1428–1435
57. Pandey A, Parhi DR (2014) MATLAB simulation for mobile robot navigation with hurdles in cluttered environment using minimum rule based fuzzy logic controller. Procedia Technol 14:28–34
58. Mahapatra S, Jha AK., Patle BK, Parhi DRK (2012) Fuzzy logic control of a WMR. In: 2012 International conference on computing, communication and applications (ICCCA). IEEE, pp 1–5
59. Parhi DR, Sahu S (2017) Clonal fuzzy intelligent system for fault diagnosis of cracked beam. Int J Damage Mech 27(6):840–858. 1056789517708019
60. Parhi DR, Dash AK, Das HC (2011) Formulation of a genetic algorithm based methodology for multiple crack detection in a beam structure. Aust J Struct Eng 12(2):127–139
61. Yadao AR, Singh RP, Parhi DR (2014) Influence of parameters of cracked rotor system on its vibration characteristics in viscous medium at finite region. In: Applied mechanics and materials, vol 592. Trans Tech Publications, pp 2061–2065
62. Mohanta JC, Parhi DR, Mohanty SR, Keshari A (2018) A control scheme for navigation and obstacle avoidance of autonomous flying agent. Arab J Sci Eng 43(3):1395–1407
63. Parhi DR, Deepak BBVL, Mohana J, Ruppa R, Nayak M (2012) Immunised navigational controller for mobile robot navigation 171–182
64. Thatoi DN, Das HC, Parhi DR (2012) Review of techniques for fault diagnosis in damaged structure and engineering system. Adv Mech Eng 4(327569):1–11
65. Sahu S, Kumar PB, Parhi DR (2017) Intelligent hybrid fuzzy logic system for damage detection of beam-like structural elements. J Theor Appl Mech 55(2):509–521
66. Mohanty PK, Parhi DR (2014) Navigation of autonomous mobile robot using adaptive network based fuzzy inference system. J Mech Sci Technol 28(7):2861–2868
67. Parhi DR, Kundu S (2011) A hybrid fuzzy controller for navigation of real mobile robot. Int J Appl Artif Intell Eng Syst 3(1)
68. Pandey A, Kumar S, Pandey KK, Parhi DR (2016) Mobile robot navigation in unknown static environments using ANFIS controller. Perspect Sci 8:421–423
69. Deepak BBVL, Parhi D (2012) PSO based path planner of an autonomous mobile robot. Open Comput Sci 2(2):152–168
70. Panigrahi PK, Ghosh S, Parhi DR (2014) Comparison of GSA, SA and PSO based intelligent controllers for path planning of mobile robot in unknown environment. World Acad Sci, Eng Technol, Int J Electr, Comput, Energ, Electron Commun Eng 8(10):1523–1532
71. Das HC, Dash AK, Parhi DR, Thatoi DN (2010) Experimental validation of numerical and fuzzy analysis of a faulty structure. In: 2010 5th international conference on system of systems engineering (SoSE). IEEE, pp 1–6
72. Parhi DR, Singh MK (2010) Heuristic-rule-based hybrid neural network for navigation of a mobile robot. Proc Inst Mech Eng, Part B: J Eng Manuf 224(7):1103–1118
73. Kashyap SK, Parhi DR, Sinha A (2009) Artificial neural network-a tool for optimising mining parameters paper no. 377. In: Proceedings of the ISRM-sponsored international symposium on rock mechanics: rock characterisation, modelling and engineering design methods (SINOROCK2009)
74. Parhi DR, Singh MK (2009) Navigational strategies of mobile robots: a review. Int J Autom Control 3(2–3):114–134
75. Yadao AR, Parhi DR (2015) Experimental and numerical analysis of cracked shaft in viscous medium at finite region. In: Advances in structural engineering. Springer, New Delhi, pp 1601–1609
76. Hurst J, Bull L (2006) A neural learning classifier system with self-adaptive constructivism for mobile robot control. Artif Life 12(3):353–380
77. Mohanty S, Parhi DR, Das SS (2018) Control strategy of a real mobile robot using singleton takagi sugeno fuzzy inference methodology within the frame work of artificial intelligence techniques. Emerging Trends in Engineering, Science and Manufacturing (ETESM-2018), IGIT Sarang, India

78. Choudhury S, Sahu S, Parhi DR (2011, July) Vibration analysis of cracked beam using genetic controller. In: 2011 International Conference on Process Automation, Control and Computing (pp. 1–1). IEEE
79. Rashedi E, Nezamabadi-pour H, Saryazdi S (2009) A gravitational search algorithm. Information Sciences 179(13):2232–2248

Application of PCA-TOPSIS Method for Selecting Optimal Welding Conditions in GMAW to Improve the Weld Quality

Amruta Rout, Golak Bihari Mahanta, BBVL. Deepak and Bibhuti Bhusan Biswal

Abstract Generally, the weld quality depends on the mechanical properties like ultimate strength and yield strength. Again the weld bead geometry mainly the depth of penetration affects these mechanical properties. These weld quality parameters weld bead geometry, weld bead strength are highly controlled or influenced by the welding process parameters like welding current, welding voltage, and wire feed rate, gas flow rate, and nozzle to workpiece distance in gas metal arc welding. As the process parameters affect the different welding performance parameters in different ways, single-objective optimization technique is not efficient enough to optimize all the parameters simultaneously. In this paper, a hybrid multi-objective approach, i.e., Principal Component Analysis (PCA) with the Technique for Order of Preference by Similarity to Ideal Solution (TOPSIS) approach has been used to obtain improved welding quality in optimal welding condition for gas metal arc welding (GMAW).

Keywords Weld quality · Welding process parameters · PCA · TOPSIS · GMAW

1 Introduction

Srivastava and Garg [1] utilized Box Behnken Design technique of response surface methodology to obtain mathematical model for optimizing weld bead height, bead width and depth of penetration individually for gas metal arc welding of mild steel material. Wire feed rate, gas flow rate, voltage, and travel speed have been considered as the input parameters for the process model. Ghosh et al. [2] proposed a gray-based Taguchi method for to study the effect of gas flow rate, current and nozzle to plate distance on weld quality in terms of percentage of elongation, yield strength and ultimate strength. Ghosh et al. [3] used Taguchi methodology and signal to noise ratio to analyze the effect of input parameters like gas flow rate,

A. Rout (✉) · G. B. Mahanta · BBVL. Deepak · B. B. Biswal
Department of Industrial Design, NIT Rourkela, Rourkela 769008, India

BBVL. Deepak et al. (eds.), *Innovative Product Design and Intelligent Manufacturing Systems*, Lecture Notes in Mechanical Engineering,
https://doi.org/10.1007/978-981-15-2696-1_56

nozzle to plate distance and welding current on percentage elongation and ultimate strength in MIG welding of ferritic stainless steel AISI409. Qin et al. [4] introduced an assistant arc with the main arc for reducing the welding defects like undercut and humping defect in high-speed tandem gas arc welding of stainless steel plate. Patil and Waghmare [5] used Taguchi's signal to noise ratio and ANOVA analysis for analyzing the effect of weld parameters like welding speed, current and voltage on ultimate tensile strength for MIG welding of AISI mild steel material. Sapakal and Telsang [6] presented effect of weld parameters like weld current, weld voltage and welding speed on performance parameter depth of penetration by Taguchi method involving orthogonal array, signal to noise ratio and ANOVA analysis. Benyounis and Olabi [7] gave brief review about the use of different design of experiment, numerical and statistical approaches for developing relationship between process parameters of welding process to the performance parameters of the process. Metzbower [8] used neural network technique to train input parameters chemical elements and weld cooling rate and developed model for output parameters of weld quality in terms of weld strength, reduction in area and elongation. Olabi et al. [9] investigated the effect of welding process parameters like laser power, focal point position and welding speed on mechanical properties of weld joint by response surface methodology in CO_2 laser welding. Hooda et al. [10] developed response surface model for predicting the tensile strength from input parameters consisting of wire speed, gas flow rate, welding current and voltage for MIG welding.

Literature highlights that immense effort attempted by pioneer researchers to optimize various process parameters during welding operation. Motivated by this, present work aims to add value to the previous research and proposes application of different multi-objective optimization technique as the researchers have optimized each performance parameter individually with respect to input process. The weld performance parameters need to be combined in an effective way as each process parameter affect different performance or output parameters differently. This paper presents a hybrid multi-objective optimization technique using PCA for combining several correlated weld quality parameters to small number of uncorrelated parameters and finally TOPSIS method to obtain combination of optimized level process parameters.

2 Experimental Details

In the present research work, AISI 1030 mild steel material has been used as workpiece material in gas metal work welding process. In welding, carbon wire (ER70S-6) of diameter 1.2 mm is utilized. The GMAW process is used for joining of steel plate with single V-groove and plate thickness of 10 mm. Argon gas has been used as shielding gas. Weld parameters like welding current, gas flow rate, wire feed rate, welding speed, distance between nozzle and workpiece have been considered in three different levels as shown in Table 1.

Table 1 Welding process parameters

Weld parameters	Unit	Level-1	Level-2	Level-3
Wire Feed Rate (WFR)	m/min	5	7	9
Welding Current (WC)	A	100	120	140
Gas Flow Rate (GFR)	lpm	12	16	20
Welding Speed (WS)	mm/min	150	180	210
Nozzle to plate distance (D)	mm	10	13	16

Generally, for the considered workpiece material, the welding process parameters are considered in a narrow range on trial and error basis. This range is further taken in 3 levels for obtaining optimal weld quality. The robotic arc welding set up and workpiece after welding have been shown by Fig. 1.

Taguchi L_{27} Orthogonal array has been chosen for design of experiment. The weld quality has been taken in terms of depth of penetration of weld (d), yield strength (YS) and ultimate strength (US) as shown in Table 2.

3 Proposed Methodology

3.1 *Principal Component Analysis*

Different researchers and engineers have pioneered different multi-response optimization techniques but because of anonymous correlations between the multi-performance characteristics (MPCs) many of the recommended methods enhance uncertainties. PCA is a commendable statistical methodology to resolve the correlation problem and to analyze the correlation among the MPCs [11, 12].

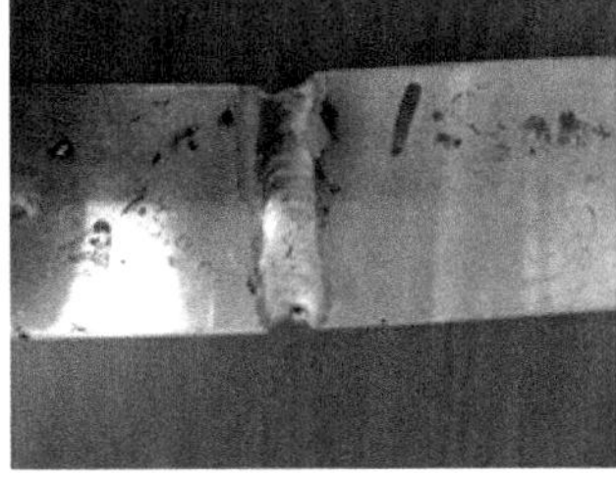

Fig. 1 GMAW by MOTOMAN MA1440A welding robot and workpiece after welding

Table 2 Welding process parameters

S. No	WFR (m/min)	WC (A)	GFR (lpm)	WS (mm/min)	D (mm)	d (mm)	YS (MPa)	US (MPa)
1	5	100	12	150	10	3.8	324.21	554.21
2	5	100	12	150	13	3.68	316.15	548.32
3	5	100	12	150	16	3.23	321.41	532.96
4	5	120	16	180	10	3.54	288.54	512.32
5	5	120	16	180	13	2.82	274.21	523.21
6	5	120	16	180	16	3.12	279.23	496.32
7	5	140	20	210	10	3.42	254.23	423.21
8	5	140	20	210	13	2.95	235.56	468.32
9	5	140	20	210	16	3.89	242.21	456.34
10	7	100	16	210	10	4.01	316.45	498.65
11	7	100	16	210	13	3.92	308.23	479.34
12	7	100	16	210	16	3.4	310.21	465.34
13	7	120	20	150	10	3.29	296.24	478.32
14	7	120	20	150	13	4.05	280.54	469.21
15	7	120	20	150	16	3.28	284.32	471.32
16	7	140	12	180	10	3.12	221.03	445.21
17	7	140	12	180	13	2.8	232.45	432.12
18	7	140	12	180	16	3.26	245.09	439.15
19	9	100	20	180	10	4.1	323.12	475.23
20	9	100	20	180	13	3.68	318.25	481.98
21	9	100	20	180	16	3.74	312.15	454.23
22	9	120	12	210	10	3.62	302.12	456.21
23	9	120	12	210	13	3.23	309.18	445.21
24	9	120	12	210	16	3.41	313.21	428.25
25	9	140	16	150	10	3.8	268.21	425.21
26	9	140	16	150	13	3.35	256.58	431.15
27	9	140	16	150	16	3.32	249.36	426.35

An adequate of information for deciding the optimal levels of control parameters is accommodated by the normalization of data. The original data are translated values ranging from values 0 to 1.

The steps of PCA are described as follows:

Step-1: Inspecting for correlation among each pair of quality characteristics:

Let $Q_i = \{X_0^*(i), X_1^*(i), \ldots, X_m^*(i)\}$, where $i = 1, 2, 3 \ldots n$

It is the normalized series of the ith quality characteristic. The correlation coefficient among two quality characteristics is evaluated by the following equation:

$$\rho_{ik} = \frac{\mathrm{Cov}(Q_j, Q_k)}{\sigma_{Qj} * \sigma_{Qk}} \tag{1}$$

here, $j = 1, 2 \ldots n,\ k = 1, 2, 3 \ldots n,\ j \neq k$ where, ρ_{jk} is correlation coefficient, σ_{Qj}, σ_{Qk} denotes standard deviation of the quality characteristics j and quality characteristics of k respectively.

Step-2: Computation of the principal component score:

Calculate the eigenvalue λ_k and the corresponding eigenvector $\beta_{kj} k = (1, 2, 3 \ldots n)$ from the correlation matrix and calculate the principal component scores of the comparative sequence and normalized reference sequence utilizing the equation shown below:

$$Y_i(K) = \sum_{j=1}^{n} X_i^*(j)\beta_{kj}, i = 0, 1, 2, 3 \ldots m, k = 1, 2, 3 \ldots n \tag{2}$$

Here, $Y_i(k)$ is the principal component score of the kth element in the ith series. $X_i^*(j)$ is the normalized value of the jth element in the ith sequence, and β_k is the jth element of the eigenvector β_k.

Step-3: Estimation of quality loss $\Delta_{0,i}(k)$:

Loss estimate $\Delta_{0,i}(\boldsymbol{k})$ is defined as the absolute value of the difference between ith experimental value for kth response and the desired (ideal) value. If responses are correlated, then on the contrary of using $[X_0(k)X_i(k)]$; $[Y_0(k)Y_i(k)]$ should be utilized for calculation of $\Delta_{0,i}(\boldsymbol{k})$.

3.2 Technique for Order Preference by Similarity to Ideal Solution (TOPSIS)

Hwang and Yoon in 1981 firstly came up with this advanced and efficient multi-response optimization technique. The solution that maximizes the benefit criteria and minimizes adverse criteria is known as positive ideal solution, whereas the solution that maximizes the adverse criteria and minimizes the benefit criteria is known as negative ideal solution [13, 14]. The steps involved for calculating the TOPSIS values are as follows:

Step 1: The alternatives are represented by row of this matrix and attributes are allocated to each column of the matrix. The decision-making matrix can be expressed as:

$$D = \begin{matrix} \\ A_1 \\ A_2 \\ \\ A_i \\ \\ A_m \end{matrix} \begin{bmatrix} x_{11} & x_{12} & \cdots & x_{ij} & \cdots & x_{1n} \\ x_{21} & x_{22} & \cdots & x_{2j} & \cdots & x_{2n} \\ \vdots & \vdots & \ddots & \vdots & \ddots & \vdots \\ x_{i1} & x_{i2} & \cdots & x_{ij} & \cdots & x_{in} \\ \vdots & \vdots & \ddots & \vdots & \ddots & \vdots \\ x_{m1} & x_{m2} & \cdots & x_{mj} & \cdots & x_{mn} \end{bmatrix} \quad (3)$$

Here, $A_i(i = 1, 2 \ldots m)$ represents the possible alternatives; $X_j(j = 1, 2 \ldots n)$ represents the attributes relating to alternative performance, $j = 1, 2 \ldots n$ and x_{ij} is the performance of A_i with respect to attribute X_j.

Step 2: Normalization of decision matrix is performed in this step. This can be obtained by formula as follows:

$$r_{ij} = \frac{x_{ij}}{\sqrt{\sum_{i=1}^{m} x_{ij}^2}} \quad (4)$$

Here, r_{ij} represents the normalized performance of A_i with respect to attribute X_j.

Step 3: Development of weighted normalized decision matrix, $V = \left[v_{ij}\right]$ can be found as:

$$V = w_j r_{ij} \quad (5)$$

$$\sum w_j = 1$$

Step 4: The ideal and negative ideal solution can be represented as:

(a) The positive ideal solution:

$$\begin{aligned} A^+ &= \left\{ \left(\max v_{ij} | j \in J \right), \left(\min v_{ij} | j \in J' \right) \right\} \\ &= \left\{ v_1^+, v_2^+, \ldots v_j^+, \ldots, v_n^+ \right\} \end{aligned} \quad (6)$$

(b) The negative ideal solution:

$$\begin{aligned} A^- &= \left\{ \left(\min v_{ij} | j \in J \right), \left(\max v_{ij} | j \in J' \right) \right\} \\ &= \left\{ v_1^-, v_2^-, \ldots v_j^-, \ldots, v_n^- \right\} \end{aligned} \quad (7)$$

Here, $J = \{j = 1, 2, \ldots n | j\}$: Associated with the beneficial attributes, $J' = \{j = 1, 2 \ldots n | j\}$: Associated with the non-beneficial attributes.

Step 5: The separation of each alternative from the ideal solution is obtained by *n*-dimensional Euclidean distance from the following equations:

$$S_i^+ = \sqrt{\sum_{j=1}^{n}\left(v_{ij} - v_j^+\right)^2}, i = 1, 2, \ldots m \quad (8)$$

$$S_i^- = \sqrt{\sum_{j=1}^{n}\left(v_{ij} - v_j^-\right)^2}, i = 1, 2, \ldots m \quad (9)$$

Step 6: Calculation the relative closeness to the ideal solution and is represented as:

$$C_i^+ = \frac{S_i^-}{S_i^+ + S_i^-} \quad (10)$$

Table 3 Quality loss estimates of the major principal components

S. No	MPC_1	MPC_2	MPC_3
1	0.118462	0.169385	0.112846
2	0.242196	0.205471	0.119138
3	0.446875	0.38053	0.371206
4	0.624665	0.0939	0.107255
5	0.955623	0.555007	0.212285
6	0.915392	0.246722	0.247027
7	1.260375	0.305557	0.243171
8	1.381783	0.191438	0.099969
9	1.019315	0.400584	0.132062
10	0.314258	0.238641	0.196636
11	0.482201	0.289176	0.242618
12	0.73015	0.067979	0.490235
14	0.811366	0.059611	0.378713
15	0.653628	0.418916	0.03796
16	0.921787	0.027303	0.323518
17	1.504692	0.026478	0.025261
18	1.608945	0.0878	0.267785
19	1.317506	0.133717	0.172253
20	0.331275	0.41012	0.302644
21	0.499915	0.138605	0.387385
22	0.629812	0.317355	0.428842
23	0.734213	0.240684	0.390986
24	0.885803	0.076625	0.618536
25	0.858395	0.265746	0.649825
26	1.012086	0.507832	0.204127
27	1.239944	0.2245	0.253928

Step 7: Ranking of the preference order. The best choice can be obtained by alternative with largest relative coefficient.

4 Results and Discussion

The result of Principal Component Analysis has been obtained as the major principal components and the final quality loss estimate values of these are computed and are shown in Table 3.

Then TOPSIS method has been applied, and the final values of closeness coefficient and the final optimal setting have been selected by the ranking of their values are shown in Table 4. The optimal setting for the weld process parameters is given as 5 m/min wire feed rate, 120 A welding current, 16 lpm gas flow rate and 180 mm/min of welding speed.

Table 4 Closeness coefficient and the corresponding ranking

S. No	S_i^+	S_i^-	C_i^+	Rank
1	0.114325529	0.176151	0.606421	10
2	0.051043765	0.166161	0.764997	2
3	0.089366041	0.106712	0.544232	15
4	**0.041839977**	**0.169919**	**0.802417**	**1**
5	0.142042472	0.101046	0.415675	26
6	0.088192456	0.120185	0.576765	13
7	0.111745974	0.105221	0.484964	23
8	0.096385461	0.142995	0.597354	11
9	0.109664239	0.120106	0.522722	20
10	0.062556586	0.14908	0.704415	3
11	0.080265346	0.130344	0.618889	6
12	0.105426972	0.133377	0.558522	14
13	0.087621213	0.140186	0.615371	8
14	0.099157113	0.146132	0.595753	12
15	0.082827541	0.148675	0.642218	5
16	0.095375326	0.179167	0.652603	4
17	0.115061304	0.135085	0.540024	16
18	0.091446379	0.141137	0.606823	9
19	0.107703129	0.118483	0.52383	19
20	0.083606355	0.135292	0.618059	7
21	0.113395709	0.098657	0.465248	3
22	0.10020033	0.109185	0.521456	21
23	0.134100294	0.122892	0.478194	24
24	0.149758161	0.085253	0.36276	27
25	0.133757169	0.101522	0.431497	25
26	0.101720306	0.115572	0.531873	17
27	0.104796275	0.116586	0.526628	18

5 Conclusion

The preceding study employs a hybrid optimization technique combining PCA with TOPSIS technique for optimization of performance parameters, thereby determining an optimal machining condition to produce desired weld performance parameters in robotic arc welding, as a case study. The integrated approach highlighted in this presentation can be applied for continuous quality improvement and off-line quality control in any production processes which involve multiple response features.

References

1. Srivastava S, Garg RK (2017) Process parameter optimization of gas metal arc welding on IS: 2062 mild steel using response surface methodology. J Manuf Process 25:296–305
2. Ghosh N, Pal PK, Nandi G (2016) Parametric optimization of MIG welding on 316L austenitic stainless steel by grey-based taguchi method. Procedia Technol 25:1038–1048
3. Ghosh N, Rudrapati R, Pal PK, Nandi G (2017) Parametric optimization of gas metal Arc welding process by using Taguchi method on ferritic stainless steel AISI409. Mater Today Proc 4(2):2213–2221
4. Qin G, Meng X, Fu B (2015) High speed tandem gas tungsten arc welding process of thin stainless steel plate. J Mater Process Technol 220:58–64
5. Patil SR, Waghmare CA (2013) Optimization of MIG welding parameters for improving strength of welded joints. Int J Adv Eng Res Stud 2(4):14–16
6. Sapakal SV, Telsang MT (2012) Parametric optimization of MIG welding using Taguchi design method. Int J Adv Eng Res Stud 1(4):28–30
7. Benyounis KY, Olabi AG (2008) Optimization of different welding processes using statistical and numerical approaches–a reference guide. Adv Eng Softw 39(6):483–496
8. Metzbower EA, DeLoach JJ, Lalam SH, Bhadeshia HKDH (2001) Neural network analysis of strength and ductility of welding alloys for high strength low alloy shipbuilding steels. Sci Technol Weld Joining 6(2):116–124
9. Olabi AG, Alsinani FO, Alabdulkarim AA, Ruggiero A, Tricarico L, Benyounis KY (2013) Optimizing the CO_2 laser welding process for dissimilar materials. Opt Lasers Eng 51(7):832–839
10. Hooda A, Dhingra A, Sharma S (2012) Optimization of MIG welding process parameters to predict maximum yield strength in AISI 1040. Int J Mech Eng Robot Res 1(3)
11. Tong LI, Wang CH, Chen HC (2005) Optimization of multiple responses using principal component analysis and technique for order preference by similarity to ideal solution. Int J Adv Manuf Technol 27:407–414
12. Chakravorty R, Gauri SK, Chakraborty S (2012) Optimization of correlated responses of EDM process. Mater Manuf Process 27:337–347
13. Athawale VM, Chakraborty S (2010) A TOPSIS method based approach to machine tool selection. In: Proceedings of the international conference on industrial engineering and operations management, Dhaka, Bangladesh
14. Lan TS (2009) Taguchi optimization of multi-objective CNC machining using TOPSIS. Inf Technol J 8(6):917–922

Experimental Analysis on Composite Material Using Multiple Electrodes by EDM Process

Subhashree Naik, Debabrata Dhupal and Bijoy Kumar Nanda

Abstract High-performance lightweight aluminium metal matrix composite (MMC) is used in the aerospace and automotive industries in recent years. This work suggests an experimental analysis of the various machining parameters for the electrical discharge machining (EDM) on aluminium metal matrix composite (AL-22%-SiC). A Box–Behnken design (BBD) of response surface methodology (RSM) has been used to calculate the responses such as the tool wear rate (TWR) and surface roughness (Ra) using copper and brass electrodes with input machining process parameters such as low voltage current (LVC), high voltage current (HVC), pulse-on time (T_{on}), pulse-off time (T_{off}) and flushing pressure (FP). Quadratic regression models are established for the individual responses, and response surface methodology has been applied to the invention of the optimum process parameter settings.

Keywords AlSiC MMC · EDM · BBD

1 Introduction

Aluminium is more comprehensively used because of its lightweight, particularly in the aerospace industry. EDM has been considered to be one of the greatest operational methods that can specifically machine metals and contours which would be problematic to produce with normal machines. The reinforced silicon carbide (SiC) particle is used to improve the strength of MMC as it has high strength and melting point. Normally, aluminium-based composites have an important interest in industries for their higher strength to weight ratio.

Many researchers in the past studied the performance of EDM, developed empirical models and different optimization techniques of machining parameters,

S. Naik (✉) · D. Dhupal
Department of Production Engineering, VSS University of Technology, Burla, Odisha, India

B. K. Nanda
Department of Mechanical Engineering, NIT, Rourkela, Odisha, India

BBVL. Deepak et al. (eds.), *Innovative Product Design and Intelligent Manufacturing Systems*, Lecture Notes in Mechanical Engineering,
https://doi.org/10.1007/978-981-15-2696-1_57

performed numerical and experimental analysis and made the necessary modifications. The names and works of some scientists, researchers and their contributions were briefly discussed. The machining appearances of Ti–6Al–4V through Cu–SiC composites to analyse the microstructure, surface topography and micro-hardness using EDM as an alternative process of machining titanium alloy have been studied [1]. Aditional quantities of 'Cu' has been introduced to MMC (ZrB2-Cu) for enhancing the moderate values of wear resistance, the thermal and electrical conductivity of tool materials of EDM process [2]. The surface characteristics of EDM were studied by conducting an experimental study on the composites with input machining parameters and the average crater diameter formed on the machined surface. The presence of bubbles and the development of recast layers of the composites at several machining situations were also detected [3]. After that, another researcher studied using the same machining parameters to obtain the power consumed and surface roughness during the EDM process and optimized them [4]. The complete enquiry of machining characteristics, surface integrity and material removal mechanisms of the innovative ceramic composite have been performed on Al_2O_3–SiCw–TiC using EDM process [5]. Another researcher using some innovative technique known as vacuum-assisted debris removal system for deep-hole EDM drilling by using the developed set-up, proposed a novel computationally fluid dynamics model [6]. The best EDM process parameters using Al-7075-based SiC_p (SiC particle) reinforced MMC with the copper tool, and then the tool wear rate and surface roughness have been studied [7]. A systematic and comprehensive investigation have been done using five dielectric fluids such as gaseous dielectrics, air and oxygen, and liquid dielectrics, de-ionized water, kerosene and water-in-oil emulsion. The recast material in the craters has been studied by metallographic method and relate the volume removed material and removal efficiency in different dielectrics [8]. The EDM machinability characteristics of Al/SiC composite (prepared through powder metallurgy route) have been carried out [9]. The outcome responses has been studied and developed the complete mathematical model for associating the collaboration and greater order impacts on various EDM parameters through RSM by utilizing relevant tentative data [10]. The researcher has calculated the effect of machining parameters on metal removal rate (MRR), tool wear rate (TWR), radial overcut (ROC) on EDM of Al–4Cu–6Si alloy–10%–SiCP composites [11]. The material removal rate and electrode wear on the die-sinking EDM of reaction-bonded silicon carbide (SiSiC) were found out by the effect of the five scheme features [12]. The process parameters like metal removal rate (MRR), tool wear rate (TWR), taper (T), radial overcut (ROC), and surface roughness (SR) on electric discharge machining (EDM) of Al–10%–SiC_P were obtained using Taguchi-based multi-response optimization technique [13].

This research work has been studied the variation of TWR and Ra using copper and brass electrodes also the effect of machining parameters on the machined surface. Although aluminium is a soft material, machining of aluminium and its composite are not an easy task. Aluminium matrix composites are very problematic to a machine with conventional machining procedure, so machining with EDM would be a better alternative. In this paper, an approach is made to establish suitable machining parameters for machining aluminium metal matrix composite with copper and brass electrodes by BBD of RSM. These works mainly focus on minimizing the TWR and Ra using different electrodes.

2 Experimental Procedure

The Electrical Discharge Machine (model: ECOWIN PS 50ZNC) with servo head (constant gap) and positive polarity for electrode were used to conduct the experiments. The commercial grade paraffin oil has been used as dielectric fluid as shown in Fig. 1.

Two types of EDM tool were taken such as copper and brass as electrode material having a cylindrical shape of 9 mm diameter as shown in Fig. 1a, b. The Al-22% SiC metal matrix was selected as the base material for the experimental work. After machining, the workpiece using two different type electrodes is shown in Fig. 1c, d.

2.1 Experimental Plan

The experiment is shown as per BBD of RSM to analyse the difficulties with numerous variables as it achieves the non-sequential experimentations having few design points by considering LVC, HVC, T_{on}, T_{off} and FP each having three levels. The contribution of machining process factors with their codes and standards at altered levels are shown in Table 1.

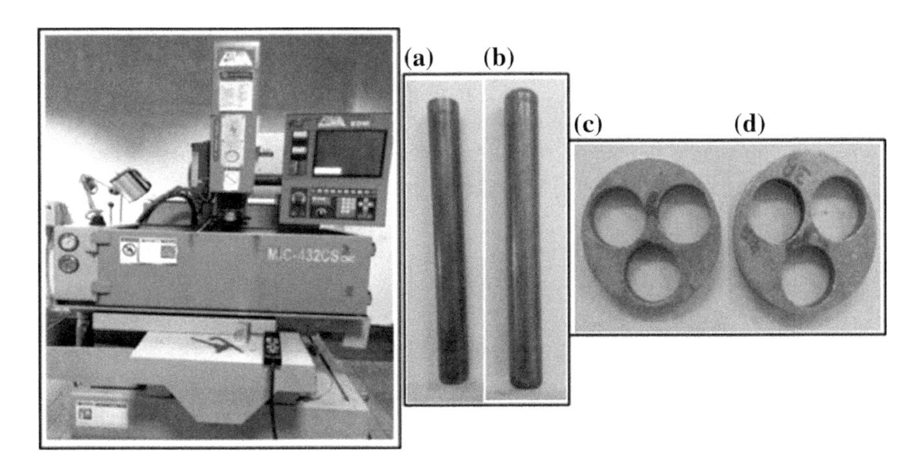

Fig. 1 Experimental set-up with EDM tool. **a** Copper, **b** brass and machined workpiece, using **c** copper tool, **d** brass tool

Table 1 Parameters with coded values at different levels

S. No.	Parameters	Low level (−1)	Mid level (0)	High level (+1)
1	LVC (amp)	5	10	15
2	HVC (volt)	1	1.5	2.0
3	T_{on}(μs)	100	200	300
4	T_{off}(μs)	10	20	30
5	FP (kg/cm^2)	0.2	0.4	0.6

3 Results and Discussions

The forty-six numbers of trial experiments have been carried out for the investigation to measure the responses as shown in Table 2.

3.1 *Adequacy Test of Responses*

It is a method towards the pattern of experimentation for construction and optimizing the practical model. The Minitab software is used for the following experimental design and for performing the analysis of variance. The RSM has been an assembly of mathematical and statistical procedures that are beneficial for the modelling and analysis of problems in which numerous variables influence the response of attention, and the objective is to optimize the responses. Regression analysis helps in recovering a response for independent input parameters. Tool wear rate (TWR) and surface roughness (Ra) of aluminium MMC are a function of the effort process parameters such as low voltage current (*A*), high voltage current (*B*), pulse-on time (*C*), pulse-off time (*D*) and flushing pressure (*E*). In RSM, the above-independent input parameters can be shown by:

$$X = f(A, B, C, D, E) \pm \in \quad (1)$$

where *X* represents the response surface and ∈ represents the error seen in the responses. The general form of the second-order polynomial equation RSM model is given below:

$$X = \beta_0 + \sum \beta_i y_i + \sum \beta_{ii} y_i^2 + \sum \beta_{ij} y_i y_j \pm \in \quad (2)$$

Table 2 Experimental results of TWR and Ra for copper and brass tools

Run	*A*	*B*	*C*	*D*	*E*	Copper		Brass	
						TWR	Ra	TWR	Ra
1	−1	−1	0	0	0	0.000001	0.044	0.0002588	0.017
2	1	−1	0	0	0	0.000008	0.023	0.0014795	0.075
3	−1	1	0	0	0	0.000001	0.039	0.0003513	0.021
4	1	1	0	0	0	0.00001	0.015	0.0013432	0.078
5	0	0	−1	−1	0	0.000013	0.016	0.0008591	0.038
6	0	0	1	−1	0	0.000001	0.073	0.0006995	0.052
7	0	0	−1	1	0	0.000011	0.065	0.0007007	0.043
8	0	0	1	1	0	0.000007	0.041	0.0005933	0.038
9	0	−1	0	0	−1	0.000004	0.004	0.0007376	0.045
10	0	1	0	0	−1	0.000007	0.003	0.0007714	0.04
11	0	−1	0	0	1	0.000007	0.002	0.0006922	0.039
12	0	1	0	0	1	0.000006	0.007	0.0007647	0.047
13	−1	0	−1	0	0	0.000004	0.052	0.0003742	0.02
14	1	0	−1	0	0	0.000025	0.095	0.001285	0.07
15	−1	0	1	0	0	0.000002	0.178	0.0001113	0.018
16	1	0	1	0	0	0.000003	0.021	0.0010023	0.075
17	0	0	0	−1	−1	0.000002	0.008	0.0006743	0.042
18	0	0	0	1	−1	0.000008	0.016	0.0003256	0.046
19	0	0	0	−1	1	0.000005	0.006	0.0007832	0.045
20	0	0	0	1	1	0.000002	0.012	0.0006395	0.035
21	0	−1	−1	0	0	0.000017	0.039	0.0008645	0.041
22	0	1	−1	0	0	0.000015	0.021	0.00088	0.04
23	0	−1	1	0	0	0.000004	0.054	0.0007056	0.036
24	0	1	1	1	1	0.000005	0.038	0.0006739	0.041
25	−1	0	0	−1	0	0.000001	0.058	0.0002364	0.021
26	1	0	0	−1	0	0.000003	0.038	0.0012573	0.076
27	−1	0	0	1	0	0.000001	0.073	0.0002474	0.017
28	1	0	0	1	0	0.000008	0.048	0.0013156	0.075
29	0	0	−1	0	−1	0.000012	0.012	0.0009197	0.038
30	0	0	1	0	−1	0.000005	0.071	0.0007441	0.039
31	0	0	−1	0	1	0.000013	0.035	0.0008609	0.041
32	0	0	1	0	1	0.000003	0.023	0.0006079	0.037
33	−1	0	0	0	−1	0.000001	0.052	0.0001219	0.02
34	1	0	0	0	−1	0.000011	0.012	0.0015142	0.07
35	−1	0	0	0	1	0.000001	0.031	0.0001231	0.022
36	1	0	0	0	1	0.000009	0.035	0.0014966	0.074
37	0	−1	0	−1	0	0.000004	0.011	0.0009716	0.034
38	0	1	0	−1	0	0.000005	0.005	0.0008972	0.042
39	0	−1	0	1	0	0.000005	0.023	0.0007859	0.045

(continued)

Table 2 (continued)

Run	*A*	*B*	*C*	*D*	*E*	Copper		Brass	
						TWR	Ra	TWR	Ra
40	0	1	0	1	0	0.000006	0.013	0.0008425	0.043
41	0	0	0	0	0	0.000004	0.128	0.0007429	0.041
42	0	0	0	0	0	0.000004	0.107	0.0004029	0.04
43	0	0	0	0	0	0.000005	0.145	0.0004058	0.042
44	0	0	0	0	0	0.000004	0.102	0.0004651	0.039
45	0	0	0	0	0	0.000005	0.134	0.0004838	0.038
46	0	0	0	0	0	0.000005	0.087	0.0007701	0.043

The above polynomial regression equation can be stated as:

$$\begin{aligned} X = {} & \beta_0 + \beta_1(A) + \beta_2(B) + \beta_3(C) + \beta_4(D) + \beta_5(E) + \beta_{12}(A * B) \\ & + \beta_{13}(A * C) + \beta_{14}(A * D) + \beta_{15}(A * E) + \beta_{23}(B * C) + \beta_{24}(B * D) \\ & + \beta_{25}(B * E) + \beta_{34}(C * D) + \beta_{35}(C * E) + \beta_{45}(D * E) + \beta_{11}(A)^2 \\ & + \beta_{22}(B)^2 + \beta_{33}(C)^2 + \beta_{44}(D)^2 + \beta_{55}(E)^2 \end{aligned} \tag{3}$$

where β_0 represents the average of the responses; LVC, HVC, T_{on}, T_{off} and FP are the independent variables; β_1, β_2, β_3, β_4 and β_5 are the linear coefficients; β_{12}, β_{13}, β_{14}, β_{15}, β_{23}, β_{24}, β_{25}, β_{34}, β_{35} and β_{45} are the interaction coefficients; β_{11}, β_{22}, β_{33}, β_{44} and β_{55} are the quadratic coefficients. The total values of the coefficient were designed by using Minitab software.

$$\begin{aligned} & \text{TWR}_{\text{copper}} = 1.24780 \times 10^{-5} + 2.3580 \times 10^{-6} A - 1.84892 \times 10^{-5} \\ & B - 1.59390 \times 10^{-7} C + 1.64525 \times 10^{-7} D + 4.02338 \times 10^{5} \\ & E + 2.65873 \times 10^{-7} AB - 1.03725 \times 10^{-8} AC + 2.6901 \times 10^{-8} \\ & AD - 5.92826 \times 10^{-7} AE + 1.90879 \times 10^{-8} BC + 1.42535 \times 10^{-8} \\ & BD - 8.14950 \times 10^{-6} BE + 1.99413 \times 10^{-9} CD - 4.83317 \times 10^{-8} \\ & CE - 1.06668 \times 10^{-6} DE - 8.41166 \times 10^{-9} A^2 + 5.15491 \times 10^{-6} \\ & B^2 + 4.11165 \times 10^{-10} C^2 - 8.33774 \times 10^{-9} D^2 + 1.01036 \times 10^{-5} E^2 \end{aligned} \tag{4}$$

$$
\begin{aligned}
\mathrm{Ra}_{\mathrm{copper}} = & -1.21329 + 0.026583A + 0.66050B + 2.45812 \times 10^{-3}C + 0.021892 \\
& D + 1.18469 \times 10^{-3}E - 3.00000 \times 10^{-4}AB - 1.00000 \times 10^{-4}AC - 2.50000 \times 10^{-5} \\
& AD + 0.11000AE + 1.00000 \times 10^{-5}BC - 5.00000 \times 10^{-5}BD + 0.015000 \\
& BE - 2.05000 \times 10^{-5}CD - 9.12500 \times 10^{-4}CE - 2.50000 \times 10^{-4} \\
& DE - 6.51667 \times 10^{-4}A^2 - 0.22383B^2 - 1.48750 \times 10^{-6} \\
& C^2 - 4.22917 \times 10^{-4}D^2 - 1.42188E^2
\end{aligned}
\tag{5}
$$

$$
\begin{aligned}
\mathrm{TWR}_{\mathrm{brass}} = & \; 2.81844 \times 10^{-3} + 3.63612 \times 10^{-5}A - 2.01139 \times 10^{-3} \\
& B - 3.70386 \times 10^{-6}C - 6.18657 \times 10^{-5}D - 1.83057 \times 10^{-3} \\
& E - 2.28832 \times 10^{-5}AB - 9.90200 \times 10^{-9}AC + 2.36585 \times 10^{-7} \\
& AD - 4.70425 \times 10^{-6}AE - 2.35400 \times 10^{-7}BC + 4.54860 \times 10^{-6} \\
& BD + 9.68425 \times 10^{-5}BE + 1.30747 \times 10^{-8}CD - 9.67225 \times 10^{-7} \\
& CE + 2.56305 \times 10^{-5}DE + 5.39795 \times 10^{-9}A^2 + 7.07053 \times 10^{-4} \\
& B^2 + 8.19376 \times 10^{-9}C^2 + 7.75190 \times 10^{-7}D^2 + 1.82872 \times 10^{-3}E^2
\end{aligned}
\tag{6}
$$

$$
\begin{aligned}
\mathrm{Ra}_{\mathrm{brass}} = & \; 4.16667 \times 10^{-4} - 3.87500 \times 10^{-4}A - 0.012500B + 7.64583 \times 10^{-5} \\
& C + 1.70000 \times 10^{-3}D - 0.011250E - 1.00000 \times 10^{-4}AB + 3.50000 \times 10^{-6} \\
& AC + 1.50000 \times 10^{-5}AD + 5.00000 \times 10^{-4}AE + 3.00000 \times 10^{-5} \\
& BC - 5.00000 \times 10^{-4}BD + 0.032500BE - 4.75000 \times 10^{-6} \\
& CD - 6.25000 \times 10^{-5}CE - 1.75000 \times 10^{-3}DE + 2.40000 \times 10^{-4} \\
& A^2 + 2.33333 \times 10^{-3}B^2 - 8.33333 \times 10^{-8} \\
& C^2 + 1.25000 \times 10^{-5}D^2 + 6.25000 \times 10^{-3}E^2
\end{aligned}
\tag{7}
$$

3.1.1 Model Adequacy Checking

There are different criterions for the inspection of the suitability of the model. The first criterion is the p-value. The p-value of TWR and Ra for both brass and copper is 0.0001 (<0.05) which specifies that the relationships in this model are significant. One more condition that is frequently used to demonstrate the accurateness of a fitted model is the determination of R^2 and adj. R^2. The determination of R^2 and adj. R^2 specifies the goodness-of-fit suitable for the model. In both cases, the intended values of R^2 and adjusted determination R^2 are more than 80 and 70%, correspondingly, which designates a high consequence of the model. The results of the ANOVA have been shown in Table 3.

Table 3 ANOVA table for copper and brass tool

Responses	*p*-value	Lack of fit		*R*-squared		Adj *R*-squared	
		Copper	Brass	Copper	Brass	Copper	Brass
TWR	<0.0001	0.0319	0.7812	0.9571	0.9170	0.9228	0.8505
Ra	<0.0001	0.9573	0.1381	0.9469	0.9839	0.9044	0.9711

3.2 Investigation of Response Surface Plot

The different surface plots of TWR and Ra for both tools are analysed with the input variables which have been shown in Fig. 2a–d. The graph Fig. 2a shows that, the TWR increases with increase in LVC and T_{on} shown in Fig. 2a but decreases at the mid-values of FP. Fig. 2b shows that Ra increases with increase in HVC, it is extreme at the mid-values of FP and then reduces. So, the lower value of LVC and HVC can be suggested for low TWR with the good surface finish. On behalf of the brass electrode in Fig. 2c, TWR rises with HVC and T_{on} and becomes extreme at the mid-values of T_{off} time. It is detected from Fig. 2d that Ra rises with T_{on} and reaches the lowest at the mid-values of LVC, and after that, it decreases. For achieving low TWR with smaller roughness, smaller values of LVC and T_{on} can be suggested.

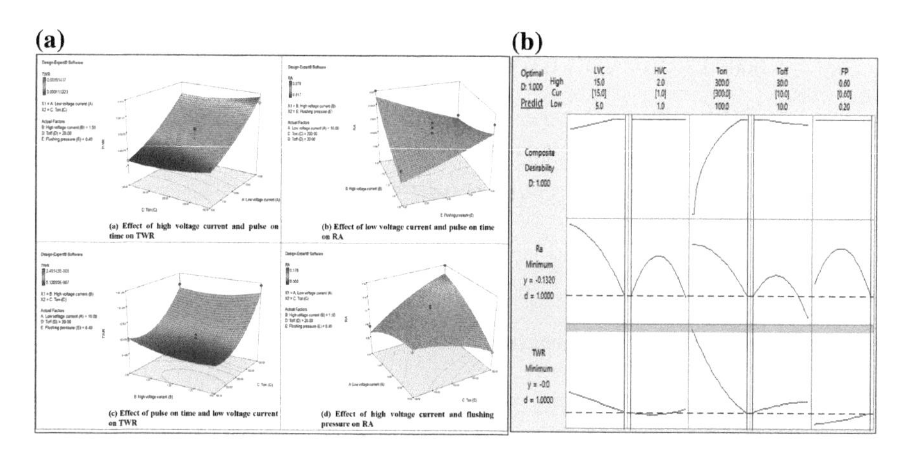

Fig. 2 **a** Response surface plot, **b** multi-objective optimization results for minimum tool wear rate and surface roughness

4 Optimization Using Response Surface Method

The multi-objective optimization is performed to optimize the two responses, i.e, TWR and Ra, on Al/SiC MMC composite for which Minitab software is utilized. TWR and Ra are optimized together, and the outcomes are shown in Fig. 2b. The recent best parameter situations are LVC of 15 amp. HVC of 1 V, T_{on} of 300 μs, T_{off} of 10 μs and FP of 0.6 kg/cm^2 were reaching the predicted minimum TWR. The value of composite desirability (D) has been taken as 1. The investigation has been approved at the above optimum parametric setting for minimum TWR and Ra. It is experimentally detected the finest values of TWR and Ra on Al/SiC MMC composite which are 0.000001 mg/s and 0.002 μm, respectively.

5 Conclusion

This investigational work evaluates the feasibility of machining Al-22%-SiC composites using EDM machine with copper and brass electrodes. The following conclusions can be drawn from the results.

(a) For higher current and pulse-on time situations, TWR has been found to be maximum.
(b) The effect of EDM machining with copper electrode has been produced lower tool wear rate than the brass electrode.
(c) The TWR and Ra also significantly affected by the electrode material. For copper electrode: TWR and Ra are less, and for brass electrode, TWR and Ra are more.
(d) The TWR and Ra were less for the intermediate values of flushing pressure.

Future work can be extended to the optimization of the control factors that affecting tool wear rate and surface roughness by considering other factors such as duty cycle, tool materials, dielectric fluids and spark gap for a comprehensive understanding of EDM process.

References

1. Li L, Feng L, Bai X, Li ZY (2016) Surface characteristics of Ti–6Al–4V alloy by EDM with Cu–SiC composite electrode. Appl. Surf. Sci. 388:546–550
2. Khanra AK, Sarkar BR, Bhattacharya B, Pathak LC, Godkhindi MM (2007) Performance of ZrB_2–Cu composite as an EDM electrode. J Mater Process Technol 183:122–126
3. Kumar SS, Uthayakumar M, Kumaran ST, Varol T (2018) Investigating the surface integrity of aluminium based composites machined by EDM. Def Technol 1–6
4. Mahanta S, Chandrasekaran M, Samanta S (2018) GA based optimization for the production of quality jobs with minimum power consumption in EDM of hybrid MMCs. Mater Today Proc 5:7788–7796

5. Patel KM, Pandey PM, Venkateswara Rao P (2009) Surface integrity and material removal mechanisms associated with the EDM of Al_2O_3 ceramic composite. Int J Refract Met Hard Mater 27:892–899
6. Tanjilul M, Ahmed A, Senthil Kumar A, Rahman M (2018) A study on EDM debris particle size and flushing mechanism for efficient debris removal in EDM-drilling of Inconel 718. J Mater Process Technol 255:263–274
7. Bodukuri AK, Chandramouli S, Eswaraiah K, Laxman J (2018) Experimental Investigation and optimization of EDM process parameters on aluminum metal matrix composite. Mater Today Proc 5:24731–24740
8. Zhang Y, Liu Y, Shen Y, Ji R, Li Z, Zheng C (2014) Investigation on the influence of the dielectrics on the material removal characteristics of EDM. J Mater Process Tech 214:1052–1061
9. Puhan D, Mahapatra SS, Sahu J, Das L (2013) A hybrid approach for multi-response optimization of non-conventional machining on AlSiC p MMC. Measurement 46:3581–3592
10. Habib SS (2009) Study of the parameters in electrical discharge machining through response surface methodology approach. Appl Math Model 33:4397–4407
11. Dhara S, Purohit R, Saini N, Sharma A, Hemath Kumar G (2007) Mathematical modeling of electric discharge machining of cast Al–4Cu–6Si alloy–10 wt% SiCP composites. J Mater Process Technol 194:24–29
12. Luis CJ, Puertas I, Villa G (2005) Material removal rate and electrode wear study on the EDM of silicon carbide. J Mater Process Technol 165:889–896
13. Singh PN, Raghukandan K, Pai BC (2004) Optimization by grey relational analysis of EDM parameters on machining Al–10%SiCP composites. J Mater Process Technol 156:1658–1661

Revelence of Multiple Breathing Cracks on Fixed Shaft Using ANFIS and ANN

J. Nanda, L. D. Das, S. Choudhury and D. R. Parhi

Abstract The article is focused on the revelation of using AI technique to detect cracks in shaft using ANFIS, ANN based on error percentage comparing with experimental work. The effectiveness of crosswise loaded fixed-fixed shaft with multiple cracks is contemplated using theoretical and experimental analysis in the article. The presence of fractures with its positions and dimensions on vibration domain is identified considering the consequence of curtailment in stiffness. The fundamental frequencies including their mode patterns in varying positions and intensities are estimated. These outcomes received from the analytical model have applied by ANFIS and ANN utilizing the modal frame. The parameters so as first three fundamental frequencies including their mode shapes for several positions and intensities of the shaft are rendered to ANFIS and ANN network separately. The test model including the sustainable error checks the authenticity of the AI techniques (ANFIS and ANN design). It is concluded that the rate of average error in ANFIS and ANN based on the experimental investigation is 2.17% and 3.5%, respectively. The existing approach is simple for preparing a condition-monitoring model of the shaft applying in a structure.

Keywords Crack detection · Vibration responses · Multi cracks · Shaft · ANFIS and ANN

J. Nanda (✉) · S. Choudhury
ITER, S'O'A University, Bhubaneswar, Odisha 751030, India
e-mail: jajneswarnanda@soa.ac.in

S. Choudhury
e-mail: sasankachoudhury@soa.ac.in

L. D. Das
Department of Mechanical Engineering, VSSUT Burla, Burla, Odisha, India
e-mail: das.layatitdev_mech@vssut.ac.in

D. R. Parhi
Department of Mechanical Engineering, N.I.T. Rourkela, Rourkela, Odisha, India

BBVL. Deepak et al. (eds.), *Innovative Product Design and Intelligent Manufacturing Systems*, Lecture Notes in Mechanical Engineering,
https://doi.org/10.1007/978-981-15-2696-1_58

1 Introduction

An investigation based on structural health monitoring provides the online tool for crack identification in recent engineering problem. Therefore, evidence of identifying crack in structural analysis is a decisive prevalence. Different advanced approaches are being contrived so far as to quantify the crack and estimate its response utilizing modal parameters. Most of these groundworks are on frequencies of vibration along with their inquiry based on shifting of modal parameters. Yu et al. [1] have refined a damage perceptional method by experiment for the laminated composite shell. The vibration parameter is conveniently authenticated using ANN. Pawar et al. [2] have improved a fuzzy genetic pattern to analyze the intensity of crack taking the help of the composite crack mode. An effective vibration controller has suggested by Hossain et al. [3] to differentiate between genetic algorithm and adaptive neuro-fuzzy inference system for detecting the defect in a system. A mathematical model devised by Saridakis et al. [4] for determining crack parameters such as position and intensity with the non-dimensional angle for two transverse cracks from the fixed end of the shaft along the longitudinal direction. Taghi and Baghmisheh [5] introduced a damaged crack beam design for judgment of crack parameters using genetic algorithms using Eigen frequencies with fracture parameters of the model. Rafiee et al. [6] have formulated a model for detecting the defect in gearbox adopting genetic algorithm and applying artificial neural systems. A new sliding window based on fuzzy logic system (FLS) represented by Chandrasekhar et al. [7] for the delivering of crack parameters. Panigrahi et al. [8] developed a real function concerning evidence of identifying the fatigue crack near a beam of identical strength by introducing both GA and residual force approach. Haryanto et al. [9] improved a computational approach for fault diagnosis of the damaged structure using the neural network. Lekhy and Novak [10] have delineated a methodology based on damage detection by ANN in addition with stochastic analysis. Rubio et al. [11] have implemented the close order polynomial to formulate the compliance functions of the fractured shaft to measure the modal reply both in bending and in torsion.

This paper enacts two inverse approaches for applying the consequence of multi-transverse cracks present in a fixed-fixed shaft with unvarying axial and twisting load. The essential parameters so as intensity (bi/*D*), position (Li/*L*) including changes in fundamental natural frequencies (Ω_1, Ω_2, Ω_3) are observed with their changes in average modes (η_1, η_2, η_3).

2 Logical Analyses of Crack Shaft

The basic elements frequently used in a given structure are the shaft, beam or plate. The position of damage with its severities depends on in the end condition of the structural element along with its properties involve with it. The response of damage is examined by linear fracture mechanics (LEFM). Stress concentration is the essential parameter that compromises with the progress of fracture. Strain energy release rate is used for computing Eigen frequencies including its mode patterns for various crack intensities of steel shaft (fixed-fixed) as presented below.

2.1 To Estimate Dimensionless Compliance from Crack Shaft

Figure 1 signifies a multi-fractured shaft (fixed at both ends) of diameter D. Two surface cracks of intensity b_1 and b_2 with positions L_1 and L_2 are presented from one of the fixed ends (Fig. 2). It is constrained with axial load F_1 and twisting load F_2 as shown in Fig. 1. A local flexibility has interposed due to the presence of the cracks with the order of 2 × 2 matrixes.

The strain energy released (J_e) is obtained from stress concentration ($C_{i,j}$) at each fractured section using fracture mechanics.

$$\begin{aligned} J_e &= \frac{1}{Y'}(C_{11} + C_{12})^2 \\ \frac{1}{Y'} &= \frac{1-\nu^2}{Y} \quad \text{(Condition for plane strain)} \\ \frac{1}{Y'} &= \frac{1}{Y} \quad \text{(Condition for plane stress)} \end{aligned} \tag{1}$$

γ and Y denote the Young's Modulus and Poisson ratio of the shaft.

Stress concentration at L_1 (Fig. 2) is C_{11} and C_{12} which is obtained from fracture mechanics as given below.

$$C_{11} = \frac{F_1}{\pi R^2}\sqrt{\pi h}\left(P_1\left\{\frac{h}{w}\right\}\right), C_{12} = \frac{2F_2}{\pi R^4} \times w\sqrt{\pi h}\left(P_2\left\{\frac{h}{w}\right\}\right) \tag{2}$$

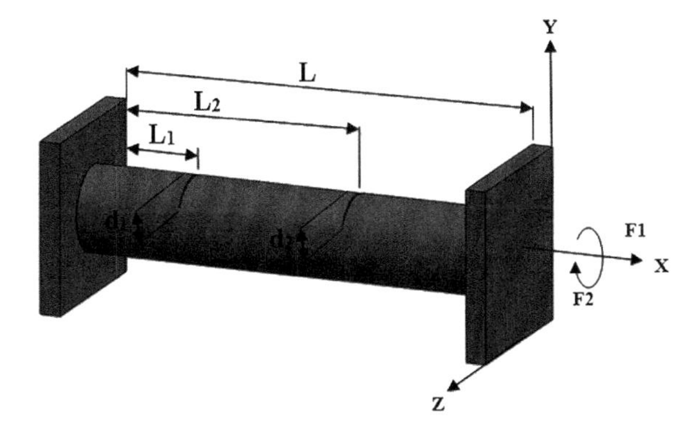

Fig. 1 Shaft with multiplied smacks

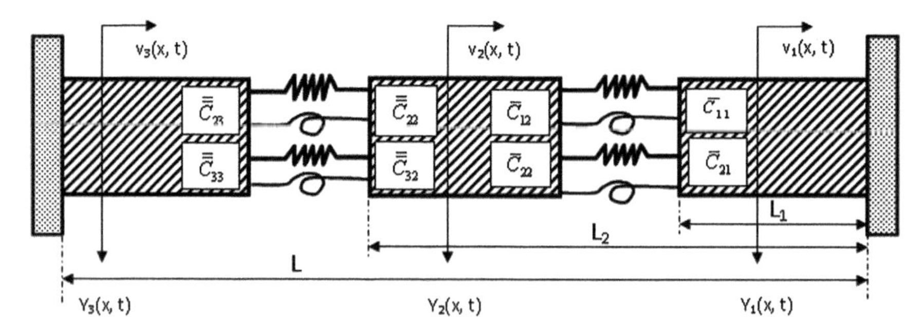

Fig. 2 Crack model

The new determined functions $p_i\{\frac{h}{w}\}$ are representing as to calculate stress concentration ($C_{i,j}$) at two locations are taken from crack intensity (h_i/w).

The rate of strain energy released (J_e) is presented by the new determined functions and as given below.

$$J_e = \frac{1}{E'}\left(\frac{F_1}{\pi R^2}\sqrt{\pi h}\left(P_1\left\{\frac{h}{w}\right\}\right) + \frac{2F_2}{\pi R^4} \times w\sqrt{\pi h}\left(P_2\left\{\frac{h}{w}\right\}\right)\right)^2 \tag{3}$$

Local stiffness matrix $C_{i,j}$ is obtained as the inverse of flexibility influence coefficient $S_{i,j}$ presented by the Eq. 4 (for i, j = 1, 2, 3), respectively.

$$C = [C]_{i,j} = [S]_{i,j}^{-1} \tag{4}$$

2.2 *Mathematical Formulation for Displacement and Mode Shape of the Shaft with Two Cracks*

Steel cracked shaft as described in Fig. 2. It is implemented for obtaining amplitudes/natural frequency. Both for longitudinal as well as bending vibration intended for the fracture part 1, part 2 and part 3 are v_i (x, t) and y_i (x, t), (i = 1, 2, 3). The standard functions of the subsequent crack parts for the specified model are presented in the following Equation.

$$\bar{v}_1(\bar{x}) = A_1\cos(\bar{C}_v\bar{x}) + A_2\sin(\bar{C}_v\bar{x}) \\ \ldots\ldots\ldots\ldots\ldots\ldots\ldots\ldots\ldots\ldots \tag{5}$$

$$\bar{y}_1(\bar{x}) = A_7\cos(\bar{C}_y\bar{x}) + A_8\sin(\bar{C}_y\bar{x}) + A_9\cos(\bar{C}_y\bar{x}) + A_{10}\sin(\bar{C}_y\bar{x}) \\ \ldots\ldots\ldots\ldots\ldots\ldots\ldots\ldots\ldots\ldots\ldots\ldots\ldots\ldots\ldots\ldots \tag{6}$$

$$\bar{x} = \frac{x}{l}, \bar{v} = \frac{v}{l}, \bar{y} = \frac{y}{l}, \ \alpha_1 = \frac{L_1}{l}, \ \alpha_2 = \frac{L_2}{l}$$

The Eighteen constants $A_1, A_2 \ldots A_{18}$ of the cantilever shaft at four positions were determined by using following four boundary conditions.

At fixed end (1) $\bar{v}_1(0) = 0; \quad \bar{y}_1(0) = 0; \quad \bar{y}_1'(0) = 0$

At fixed end (2) $\bar{v}_3'(1) = 0, \bar{y}_3''(1) = 0, \bar{y}_3'''(1) = 0$

At earliest crack position

$$\bar{v}_1(\alpha_1) = \bar{v}_2(\alpha_1), \bar{y}_1(\alpha_1) = \bar{y}_2(\alpha_1), \bar{y}_1''(\alpha_1) = \bar{y}_2''(\alpha_1), \bar{y}_1'''(\alpha_1) = \bar{y}_2'''\alpha_1)$$

At subsequent crack position

$$\bar{v}_2(\alpha_2) = \bar{v}_3(\alpha_2); \bar{y}_2(\alpha_2) = \bar{y}_3(\alpha_2), \bar{y}_2''(\alpha_2) = \bar{y}_3''(\alpha_2), \bar{y}_2'''(\alpha_2) = \bar{y}_3'''(\alpha_2)$$

The axial displacement is taking place at crack places due to discontinuity on either side of it using force and moment balance equations.

Taking all boundary conditions together with all normal functions from Eqs. (5) to (6) as quoted before, capitulate to form normal equation of matrix in the order of 18×18 for the given cracked shaft.

That is

$$|R| = 0 \tag{7}$$

The expression for the assigned matrix R is a combination of the fundamental frequency, crack position (α_i, α_j) including the bounded stiffness. Again, confined stiffness is the combination of crack intensity $(\beta_i \,\&\, \beta_j))$.

The physical properties of the un-cracked and cracked shaft are tabulated in Table 1 by considering the theoretical data for different crack section with physical properties of the same shaft. Effect of crack position and its intensity on Eigenvalues have been proposed as presented in Figs. 3a–c and 4. Some of the examples theoretical results are given in Table 2.

Figure 3a–c represents the amplitude of vibration with the position from one of the fixed ends of the shaft for crack positions ($\alpha_1 = 0.2$, $\alpha_2 = 0.8$) with the following intensities ($\beta_1 = 0.05$, $\beta_2 = 0.1$).

Table 1 Geometry of specimen

Shaft young's modulus (Y)	Shaft density (ρ)	Poisson's ratio (μ)	Experimental length (L)	Diameter (D)
2.1E + 06 MPa	7830 kg/m^3	0.3	1000 mm	10 mm

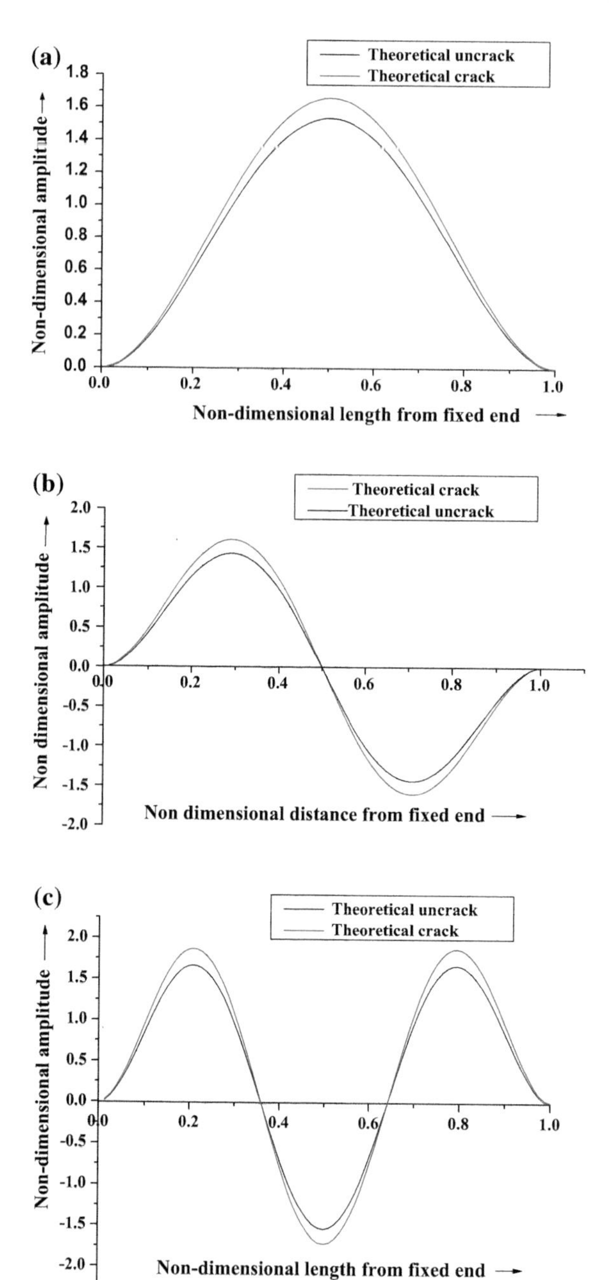

Fig. 3 **a** Amplitude versus position of crack from the fixed end (first mode of vibration), $\alpha_1 = 0.2$, $\alpha_2 = 0.8$, $\beta_1 = 0.05$, $\beta_2 = 0.1$, **b** amplitude versus position of crack from the fixed end (second mode of vibration) $\alpha_1 = 0.2$, $\alpha_2 = 0.8$, $\beta_1 = 0.05$, $\beta_2 = 0.1$, **c** amplitude versus position of crack from the fixed end (third mode of vibration) $\alpha_1 = 0.2$, $\alpha_2 = 0.8$, $\beta_1 = 0.05$, $\beta_2 = 0.1$

Table 2 Fundamental frequency for fixed-fixed shaft at various crack position and intensity from theoretical results

S. No.	Standard crack position with intensity				% change in fundamental frequency			% change in maximum mode sample		
	$\alpha_{n,1}$	$\alpha_{n,2}$	$\beta_{n,1}$	$\beta_{n,2}$	$\Omega_{n,1}$	$\Omega_{n,2}$	$\Omega_{n,3}$	$\eta_{n,1}$	$\eta_{n,2}$	$\eta_{n,3}$
1	0.1	0.8	0.1	0.1	0.0082	0.0114	0.0009	0.0900	0.1590	0.3338
2	0.1	0.8	0.1	0.3	0.0020	0.1366	0.3928	0.1073	0.2869	1.3613
3	0.1	0.8	0.1	0.5	0.0255	0.6039	1.5832	0.1573	0.6849	4.6291
4	0.1	0.8	0.3	0.1	0.2612	0.0353	0.0005	0.0835	0.2293	0.3065
5	0.1	0.8	0.3	0.3	0.2755	0.1848	0.3936	0.0575	0.1967	0.3323
6	0.1	0.8	0.3	0.5	0.3027	0.6520	1.5787	0.1460	0.7510	4.5974
7	0.1	0.8	0.5	0.1	1.0837	0.1821	0.0045	0.0567	0.4358	0.3227
8	0.1	0.8	0.5	0.3	1.1101	0.3353	0.4006	0.0704	0.5624	1.2785
9	0.1	0.8	0.5	0.5	1.1349	0.7940	1.5937	0.0583	0.2171	0.2336
10	0.35	0.65	0.1	0.1	0.0074	0.0211	0.0243	0.0854	0.1475	0.2648
11	0.35	0.65	0.1	0.3	0.1563	0.4014	0.0243	0.0370	0.2101	0.3212
12	0.35	0.65	0.1	0.5	0.6624	1.5153	0.0179	0.1010	0.4043	0.4913
13	0.35	0.65	0.3	0.1	0.1572	0.3991	0.0237	0.0385	0.2124	0.3046
14	0.35	0.65	0.3	0.3	0.3212	0.7827	0.0228	0.0121	0.1718	0.2299
15	0.35	0.65	0.3	0.5	0.8235	1.9328	0.0153	0.1535	0.3824	0.4006
16	0.35	0.65	0.5	0.1	0.6698	1.5280	0.0173	0.1031	0.4058	0.4914
17	0.35	0.65	0.5	0.3	0.8175	1.9134	0.0164	0.1529	0.3776	0.3983
18	0.35	0.65	0.5	0.5	1.2989	3.1424	0.0109	0.3025	0.3444	0.1205
19	0.45	0.55	0.1	0.1	0.0159	0.0157	0.0221	0.1089	0.1442	0.2548
20	0.45	0.55	0.1	0.3	0.3362	0.0480	0.3700	0.2547	0.3757	0.7527
21	0.45	0.55	0.1	0.5	1.3619	0.2538	1.4438	0.7389	1.0793	2.1958
22	0.45	0.55	0.3	0.1	0.3443	0.0505	0.3766	0.2557	0.3760	0.7534
23	0.45	0.55	0.3	0.3	0.6669	0.1198	0.7162	0.3893	0.0912	0.1214
24	0.45	0.55	0.3	0.5	1.6687	0.3324	1.7548	0.8447	0.8023	1.6158
25	0.45	0.55	0.5	0.1	1.3454	0.2524	1.4282	0.7301	1.0634	2.1712
26	0.45	0.55	0.5	0.3	1.6647	0.3311	1.7499	0.8381	0.7910	1.5987
27	0.45	0.55	0.5	0.5	2.5901	0.5650	2.6647	1.1835	0.0931	0.2369

$\Omega_{n,i}$, $\eta_{n,i}$, $\alpha_{n,i}$, $\beta_{n,i}$ % change in theoretical frequency, mode shape, crack position and crack intensity

3 Application of Artificial Neural Network (ANN) for Crack Detection

The present ANN model has been selected for determining the result of two crack intensities and two crack positions of a fixed-fixed shaft utilizing six input parameters (x_i). In this present work, the 6-20-4 feed-forward NN model with tangent sigmoid activation function is selected. In structure of the model, the input layer consists of six nodes, the single hidden layer consists of 20 nodes, and the

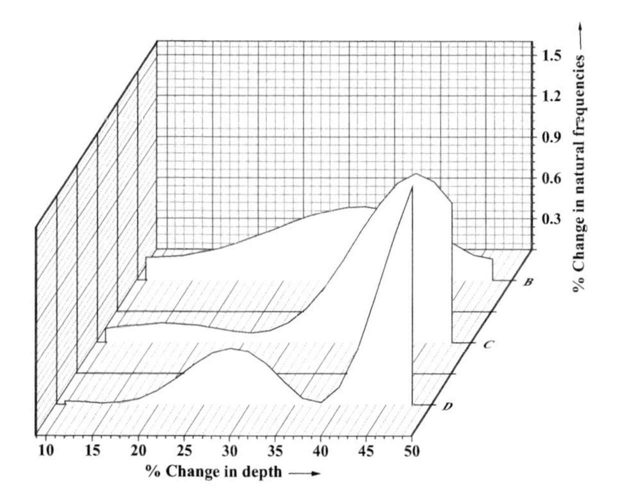

Fig. 4 Frequencies (*B*, *C*, *D*) versus crack intensity for $\alpha_1 = 0.2$, $\alpha_2 = 0.8$, $\beta_1 = 0.2$, $\beta_2 = 0.1, 0.12, 0.14 \ldots 0.48$ *B*, *C* and *D* are first, second as well as third fundamental frequency

output layer consists of four nodes with linear activation function. The neural network structure used in this work is presented in Fig. 5.

In this research study, MATLAB R2013 (Math Works, USA) software with NN toolbox is utilized for checking, testing and training the neural network. Here, 58 data are practiced as training, and rest 31 data are used as testing. The tangent sigmoid transfer function is used between the input layer and hidden layer. Similarly, the linear activation function is used between the hidden layer and output layer separately. Then, the learning rate is set to 1.9. The key codes are listed as below:

```
net = newff([−1 1; −1 1; −1 1; −1 1; −1 1; −1 1] [20, 4], {'tansig', 'tansig', 'traingdm')};
net.trainparam.show = 1000;
net.trainparam.h = 1.9;
net.train.param.epochs = 3000;
net.trainparam.goal = 0.001;
net = init(net); net = train (net, pn, tn);
```

The leaning FFBP network is trend up with theoretical data for training, testing. Few examples have been presented in Table 3.

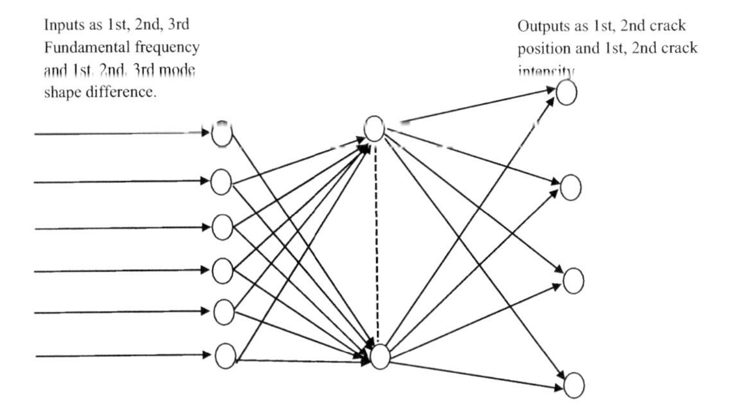

Fig. 5 Schematic representation of ANN used in this work

4 Adaptive Neuro-Fuzzy Inference System (ANFIS)

The structure of ANFIS can be used to classify nonlinear elements the control system, so also expose nonlinear functions and foretell a disorganized timeline. The concepts of fuzzy set theory, fuzzy if-then rule and fuzzy reasoning are combined governed by fuzzy inference system (FIS). The primary arrangement of FIS consists precisely of three conceptual components. They are, rule base, which contains, respectively, a judicious selection of fuzzy rules: comprehensive databases, which carefully define the membership function used in fuzzy rules: a reasoning mechanism, which properly performs the inference procedure upon the specific rules and given fact to typically derive a reasonable output or logical conclusion.

4.1 ANFIS Architecture used for Obtaining the Value of Crack Intensities and Crack Positions

ANFIS is a computing framework, which uses neural network-like structure. It is composed of five panels, which execute fuzzy inference system as guides in Fig. 6. ANFIS model used in this operation with six inputs (1st, 2nd, 3rd fundamental frequency) accompanying with their amplitude (1st, 2nd, 3rd amplitude) as I_1, I_2, I_3, I_4, I_5 and I_6. The ANFIS structure is considered as a first-order Sugeno model of containing (729 = 3 ^ 6) rules. The first three fundamental frequency along with their amplitudes received from the analytical method are handled as preparation data to carefully instruct the immune ANFIS system including the Gaussian membership function along with a composite planted algorithm. The key codes are listed as follows in Table 4.

Table 3 Training and testing for input data of designed (FFBN) ANN for fixed-fixed shaft

S. No.	Input data										Output data			
	Test points				% change in fundamental frequency			% Change in maximum mode shape			FFBP (ANN)			
	$\alpha_{n,1}$	$\beta_{n,1}$	$\alpha_{n,2}$	$\beta_{n,2}$	$\Omega_{n,1}$	$\Omega_{n,2}$	$\Omega_{n,3}$	$\eta_{n,1}$	$\eta_{n,2}$	$\eta_{n,3}$	$\alpha_{a,1}$	$\beta_{a,1}$	$\alpha_{a,2}$	$\beta_{a,2}$
1	0.10	0.80	0.20	0.20	0.0854	0.0536	0.1293	0.0355	0.1312	0.4341	0.0964	0.7743	0.1928	0.1933
2	0.15	0.75	0.20	0.20	0.0112	0.1030	0.1856	0.0361	0.1869	0.2107	0.1449	0.725	0.1935	0.1947
3	0.20	0.70	0.20	0.20	0.0020	0.1953	0.1848	0.0199	0.2137	0.5591	0.1943	0.6776	0.1939	0.1943
4	0.25	0.65	0.20	0.20	0.0390	0.2493	0.1181	0.0077	0.2355	0.0505	0.2424	0.6293	0.1944	0.1928
5	0.30	0.60	0.20	0.20	0.0954	0.2210	0.0699	0.0199	0.3944	0.3853	0.29	0.582	0.1952	0.1929
6	0.35	0.55	0.20	0.20	0.1621	0.1468	0.1124	0.0703	0.1302	0.2643	0.339	0.5325	0.1951	0.1933
7	0.40	0.50	0.20	0.20	0.2049	0.0613	0.1972	0.1631	0.1488	0.2020	0.3873	0.4838	0.194	0.193
8	0.10	0.80	0.30	0.30	0.2779	0.1865	0.3934	0.0395	0.2624	1.1250	0.0967	0.7745	0.2923	0.2908
9	0.15	0.75	0.30	0.30	0.0803	0.3375	0.5618	0.0403	0.4341	0.4993	0.1458	0.7256	0.2918	0.2913
10	0.20	0.70	0.30	0.30	0.0568	0.5866	0.5599	0.0039	0.4942	1.4889	0.1941	0.6769	0.2917	0.2948
11	0.25	0.65	0.30	0.30	0.1577	0.7317	0.3698	0.0377	0.5546	0.0527	0.2434	0.6287	0.2916	0.2898
12	0.30	0.60	0.30	0.30	0.3241	0.6630	0.2444	0.0043	0.9923	0.9794	0.2896	0.5791	0.2914	0.2911
13	0.35	0.55	0.30	0.30	0.4966	0.4506	0.3579	0.1352	0.2669	0.6497	0.337	0.5322	0.2912	0.2902
14	0.40	0.50	0.30	0.30	0.6294	0.2172	0.6012	0.4031	0.3112	0.4597	0.3853	0.482	0.2927	0.293
15	0.10	0.80	0.40	0.40	0.6145	0.4242	0.8619	0.0497	0.4921	2.4407	0.0946	0.7734	0.3877	0.3911
16	0.15	0.75	0.40	0.40	0.2041	0.7551	1.2093	0.0479	0.8924	1.0036	0.1458	0.7253	0.386	0.392
17	0.20	0.70	0.40	0.40	0.1509	1.2522	1.1965	0.0425	0.9859	3.1410	0.1921	0.6757	0.3865	0.3881
18	0.25	0.65	0.40	0.40	0.3636	1.5716	0.8034	0.1134	1.1267	0.0566	0.2454	0.6293	0.3864	0.3913
19	0.30	0.60	0.40	0.40	0.7153	1.4632	0.5500	0.0440	2.1094	2.0318	0.2896	0.5801	0.3886	0.3911
20	0.35	0.55	0.40	0.40	1.0673	0.9863	0.7808	0.2475	0.5209	1.3315	0.337	0.532	0.3874	0.3911
21	0.40	0.50	0.40	0.40	1.3399	0.4849	1.2646	0.7951	0.5771	0.8586	0.3864	0.4837	0.3858	0.3923

(continued)

Table 3 (continued)

S. No.	Input data										Output data			
	Test points				% change in fundamental frequency			% Change in maximum mode shape			FFBP (ANN)			
	$\alpha_{n,1}$	$\beta_{n,1}$	$\alpha_{n,2}$	$\beta_{n,2}$	$\Omega_{n,1}$	$\Omega_{n,2}$	$\Omega_{n,3}$	$\eta_{n,1}$	$\eta_{n,2}$	$\eta_{n,3}$	$\alpha_{a,1}$	$\beta_{a,1}$	$\alpha_{a,2}$	$\beta_{a,2}$
22	0.10	0.80	0.50	0.50	1.1409	0.7969	1.5956	0.0608	0.8640	4.5629	0.0963	0.7745	0.4849	0.4863
23	0.15	0.75	0.50	0.50	0.4034	1.3953	2.1750	0.0557	1.6426	1.8068	0.1478	0.7265	0.4846	0.4851
24	0.20	0.70	0.50	0.50	0.3126	2.3362	2.1865	0.1115	1.8702	5.7616	0.1961	0.6776	0.4838	0.4821
25	0.25	0.65	0.50	0.50	0.6849	2.8761	1.4586	0.2250	2.0451	0.0537	0.2474	0.6285	0.4834	0.485
26	0.30	0.60	0.50	0.50	1.3058	2.6842	1.0071	0.0927	3.8455	3.5967	0.2856	0.5784	0.4828	0.4860
27	0.35	0.55	0.50	0.50	1.9437	1.8663	1.4419	0.4210	0.9795	2.4367	0.3382	0.5321	0.4848	0.4867

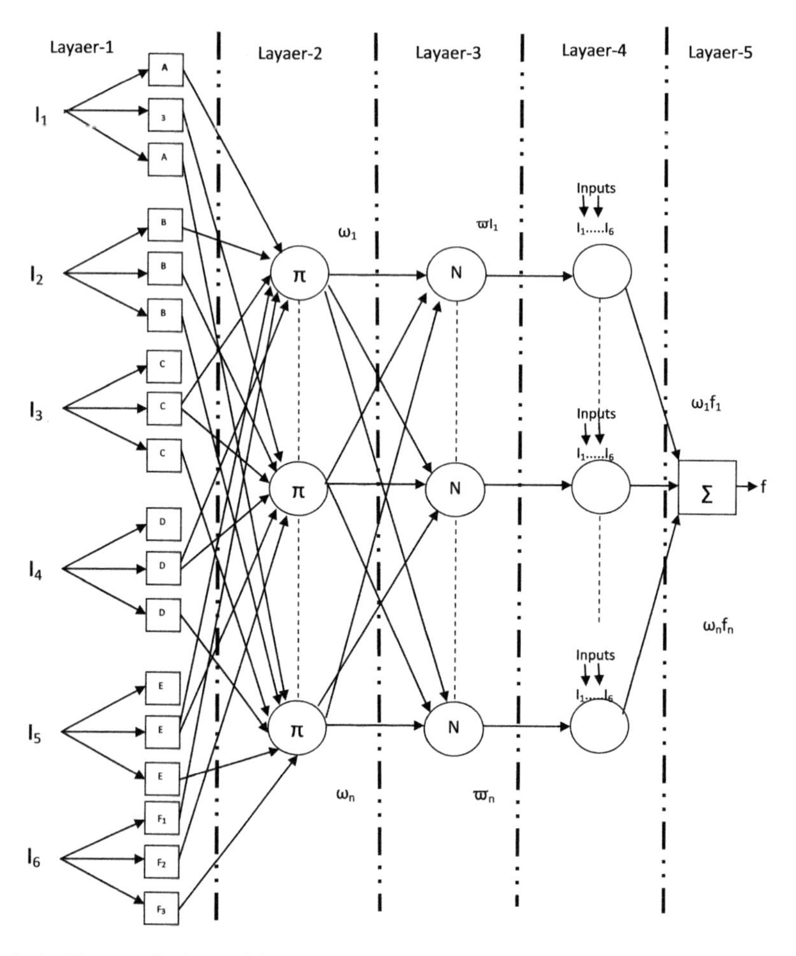

Fig. 6 Architecture for inputs (six) with membership functions (three) of the present model

4.2 *Comparison of Performance Between ANN and ANFIS*

The optimal performance of ANN and ANFIS design is properly compared using mean square error (MSE).

$$\mathrm{MSE} = \frac{\sum_{i=1}^{N} (P_{\mathrm{i}} - A_{\mathrm{i}})^2}{N} \tag{8}$$

The corresponding study for training data of ANFIS and ANN using MSE analysis is carried out. These are described in Fig. 7a–d. Identically, the comparative analysis can be properly made for testing data of ANN and ANFIS network.

Table 4 Key codes for ANFIS

Six inputs	First, second, third fundamental frequency and first, second, third amplitude
Four outputs	First crack intensity, second crack intensity, first crack position, second crack position
For each input parameter, three membership function used	Sugeno-type model
Total numbers of nodes	1503
Total numbers of linear parameters	729
Total numbers of nonlinear parameters	36
Total numbers of parameters	765
Total numbers of training data pairs	58
Total numbers of checking data pairs	89
Total numbers of fuzzy rules used	729

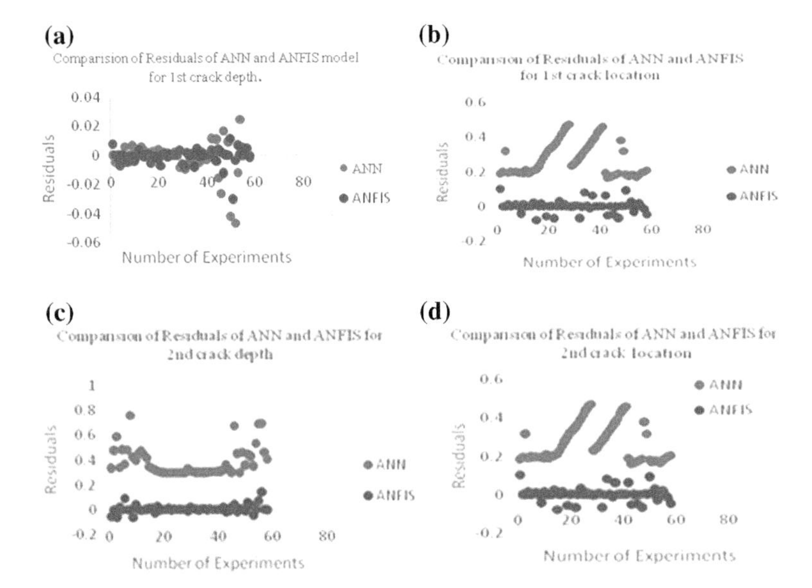

Fig. 7 **a** Comparison for training data of ANN and ANFIS for first crack intensity, **b** comparison for training data of ANFIS and ANN for first crack position, **c** comparison for training data of ANN and ANNFIS for second crack intensity, d comparison for training data of ANN and ANNFIS for second crack position

It may observed from the Figures that the MSE of training data collected from ANFIS model is acceptably low in a range between 0 and 0.02 as compared to the data obtained from ANN model which have in a range of 0.1–0.5. Therefore, the ANFIS model considered here in this work is more flexible as well as more acceptable than the ANN model considered.

The implemented ANFIS model for standard practice and experimenting input data for crack positions and proportional intensities are tabulated in Table 5.

5 Experimental Analysis

Experiment on the various specimens is carried to know the fundamental frequencies including modal profiles for multiple crack intensities on steel shaft of length 1000 mm and diameter 150 mm. The shaft is clamped at both ends by a clamping device. One of the magnified views of crack position in fixed-fixed is presented Fig. 8a. The laboratory setup is displayed in Fig. 8b. A fitting signal device known as the function generator excited the specimen. The shaft fluctuated at three principal modes of oscillation. Modal frequency of the shaft in the experimental setup is suitably presented.

6 Discussions

In the current investigation, an approach is made to improve accuracy suitably choosing a better AI technique in fault diagnosis based on percentage error in the results from ANN and ANFIS method. The changes in response are affected by the amplitude, fundamental frequencies which have been represented in Fig. 3a–c. It reveals that amplitude increases due to the appearance of crack. A relation is presented in Fig. 4 between intensity of crack and fundamental frequencies. Schematic representation of ANN used in this work is presented in Fig. 5. Architecture for inputs with membership functions of ANFIS system is presented in Fig. 6. Comparison for training data of ANN and ANFIS for different crack intensity and positions presented in Fig. 7a–d. Actual experimental setup for fixed shaft with measurement of modal parameters along with crack position of fixed-fixed shaft is presented in Fig. 8.

6.1 Comparison of Results Between ANFIS, ANN and Experimental Analysis

The genuineness of ANFIS/ANN model is examined taking experimental crack positions and intensity ($\alpha_{e,i}$, $\beta_{e,i}$) for the given shaft with same network planning. The outputs of networks are compared with experimental results for same crack intensity and position for the fixed-fixed shaft that have been shown in Table 6.The average error for both ANFIS/ANN can be determined using the Eq. 9 as given

Table 5 Training and testing input data for ANFIS network for fixed-fixed shaft

S. No.	Input data										Output data			
	Test points				% change in fundamental frequency			% change in maximum mode shape			ANFIS			
	$\alpha_{n,1}$	$\beta_{n,1}$	$\alpha_{n,2}$	$\beta_{n,2}$	$\%\Omega_{n,1}$	$\%\Omega_{n,2}$	$\%\Omega_{n,3}$	$\%\eta_{n,1}$	$\%\eta_{n,2}$	$\%\eta_{n,3}$	$\alpha_{f,1}$	$\beta_{f,1}$	$\alpha_{f,2}$	$\beta_{f,2}$
1	0.10	0.80	0.20	0.20	0.0854	0.0536	0.1293	0.0355	0.1312	0.4341	0.0971	0.7799	0.1942	0.1958
2	0.15	0.75	0.20	0.20	0.0112	0.1030	0.1856	0.0361	0.1869	0.2107	0.1459	0.7302	0.1949	0.1962
3	0.20	0.70	0.20	0.20	0.0020	0.1953	0.1848	0.0199	0.2137	0.5591	0.1958	0.6825	0.1953	0.1978
4	0.25	0.65	0.20	0.20	0.0390	0.2493	0.1181	0.0077	0.2355	0.0505	0.2442	0.6339	0.1958	0.1953
5	0.30	0.60	0.20	0.20	0.0954	0.2210	0.0699	0.0199	0.3944	0.3853	0.2921	0.5862	0.1966	0.1944
6	0.35	0.55	0.20	0.20	0.1621	0.1468	0.1124	0.0703	0.1302	0.2643	0.3414	0.5364	0.1965	0.1948
7	0.40	0.50	0.20	0.20	0.2049	0.0613	0.1972	0.1631	0.1488	0.2020	0.3901	0.4873	0.1954	0.1937
8	0.10	0.80	0.30	0.30	0.2779	0.1865	0.3934	0.0395	0.2624	1.1250	0.0974	0.7801	0.2944	0.2921
9	0.15	0.75	0.30	0.30	0.0803	0.3375	0.5618	0.0403	0.4341	0.4993	0.1469	0.7308	0.2939	0.2939
10	0.20	0.70	0.30	0.30	0.0568	0.5866	0.5599	0.0039	0.4942	1.4889	0.1955	0.6818	0.2938	0.2956
11	0.25	0.65	0.30	0.30	0.1577	0.7317	0.3698	0.0377	0.5546	0.0527	0.2452	0.6333	0.2937	0.2907
12	0.30	0.60	0.30	0.30	0.3241	0.6630	0.2444	0.0043	0.9923	0.9794	0.2917	0.5833	0.2935	0.2921
13	0.35	0.55	0.30	0.30	0.4966	0.4506	0.3579	0.1352	0.2669	0.6497	0.3394	0.5361	0.2933	0.2918
14	0.40	0.50	0.30	0.30	0.6294	0.2172	0.6012	0.4031	0.3112	0.4597	0.3881	0.4855	0.2948	0.2940
15	0.10	0.80	0.40	0.40	0.6145	0.4242	0.8619	0.0497	0.4921	2.4407	0.0953	0.7790	0.3905	0.3922
16	0.15	0.75	0.40	0.40	0.2041	0.7551	1.2093	0.0479	0.8924	1.0036	0.1469	0.7305	0.3888	0.3931
17	0.20	0.70	0.40	0.40	0.1509	1.2522	1.1965	0.0425	0.9859	3.1410	0.1935	0.6806	0.3893	0.3910
18	0.25	0.65	0.40	0.40	0.3636	1.5716	0.8034	0.1134	1.1267	0.0566	0.2472	0.6339	0.3892	0.3932
19	0.30	0.60	0.40	0.40	0.7153	1.4632	0.5500	0.0440	2.1094	2.0318	0.2917	0.5843	0.3914	0.3933
20	0.35	0.55	0.40	0.40	1.0673	0.9863	0.7808	0.2475	0.5209	1.3315	0.3394	0.5359	0.3902	0.3941
21	0.40	0.50	0.40	0.40	1.3399	0.4849	1.2646	0.7951	0.5771	0.8586	0.3892	0.4872	0.3886	0.3941

(continued)

Table 5 (continued)

S. No.	Input data										Output data			
	Test points				% change in fundamental frequency			% change in maximum mode shape			ANFIS			
	$\alpha_{n,1}$	$\beta_{n,1}$	$\alpha_{n,2}$	$\beta_{n,2}$	$\%\Omega_{n,1}$	$\%\Omega_{n,2}$	$\%\Omega_{n,3}$	$\%\eta_{n,1}$	$\%\eta_{n,2}$	$\%\eta_{n,3}$	$\alpha_{f,1}$	$\beta_{f,1}$	$\alpha_{f,2}$	$\beta_{f,2}$
22	0.10	0.80	0.50	0.50	1.1409	0.7969	1.5956	0.0608	0.8640	4.5629	0.0970	0.7801	0.4884	0.4894
23	0.15	0.75	0.50	0.50	0.4034	1.3953	2.1750	0.0557	1.6426	1.8068	0.1489	0.7317	0.4881	0.4869
24	0.20	0.70	0.50	0.50	0.3126	2.3362	2.1865	0.1115	1.8702	5.7616	0.1976	0.6825	0.4873	0.4842
25	0.25	0.65	0.50	0.50	0.6849	2.8761	1.4586	0.2250	2.0451	0.0537	0.2492	0.6331	0.4869	0.4866
26	0.30	0.60	0.50	0.50	1.3058	2.6842	1.0071	0.0927	3.8455	3.5967	0.2877	0.5826	0.4863	0.4872
27	0.35	0.55	0.50	0.50	1.9437	1.8663	1.4419	0.4210	0.9795	2.4367	0.3406	0.5360	0.4883	0.4887

$\Omega_{n,i}$, $\eta_{n,i}$, $\alpha_{n,i}$, $\beta_{n,i}$ Theoretical frequency, mode shape, position and intensity; $\alpha_{f,i}$, $\beta_{f,i}$: ANFIS model position and intensity for fixed-fixed shaft

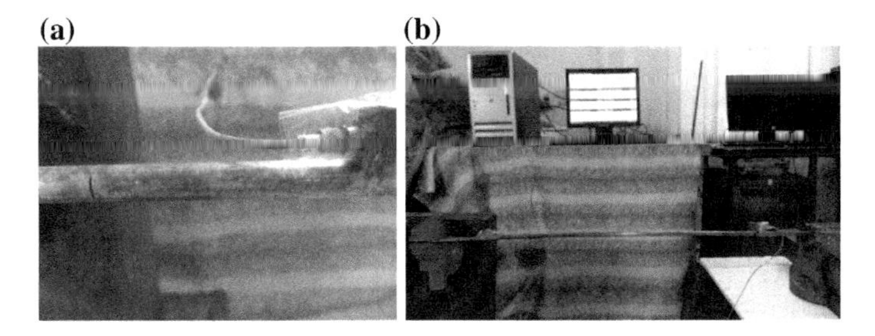

Fig. 8 **a** Experimental setup of crack position in fixed-fixed, **b** actual experimental setup for fixed-fixed shaft

below. However, in common ANFIS, results are noted to be more accurate in comparison with ANN.

$$\text{Percentage of deviation} = \left| \frac{\begin{array}{c}\text{Experimental crack location/depth fixed} - \text{fixed shaft} - \\ \text{ANFIS/ANN model based crack location/depth fixed} - \text{fixed shaft}\end{array}}{\text{Experimental crack location/depth cantilever shaft}} \right| \times 100 \quad (9)$$

Figure 9a and b presents the proficiency of calculating the error between two different network architectures (ANFIS, ANN) with respect to experimental results for crack positions and intensities.

7 Conclusions

The results obtained from the analysis using modified procedures for the prophecy of the crack characteristics for the fixed-fixed shaft are presented in the current paper. The vibration responses obtained from the theoretical investigation are supplied to the ANN and ANFIS model to prepare a robotics inverse technique for crack detection in the shaft. A comparison of results in between ANN and ANFIS is examined using MSE as presented in Fig. 9. ANFIS network produces the reliable performance applying root mean square error (RMSE) value that indicates that ANFIS model is more reasonable to predict crack than ANN. Crack positions and intensities accomplished from output of developed ANFIS model are good in accordance with experimental results (2.407% error in position and 1.95% error in intensity). In the same way, the outcomes of ANN model are presented 3.2% error in position and 3.25% error in intensity. It remarked that the newly designed model

Table 6 Comparison of % errors in ANFIS versus test values for fixed-fixed shaft

Input				Output				% variation in inaccuracy				Output				% variation in inaccuracy			
Experimental results				Trial values (ANFIS) outcome				Inaccuracy in Trial values (ANFIS) outcome				Trial values (ANN) outcome				Inaccuracy in trial values (ANN)			
$\alpha_{e,1}$	$\beta_{e,1}$	$\alpha_{e,2}$	$\beta_{e,2}$	$\alpha_{f,1}$	$\beta_{f,1}$	$\alpha_{f,2}$	$\beta_{f,2}$	Error in $\alpha_{f,1}$	Error in $\beta_{f,1}$	Error in $\alpha_{f,2}$	Error in $\beta_{f,2}$	$\alpha_{f,1}$	$\beta_{f,1}$	$\alpha_{f,2}$	$\beta_{f,2}$	Error in $\alpha_{f,1}$	Error in $\beta_{f,1}$	Error in $\alpha_{f,2}$	Error in $\beta_{f,2}$
0.1	0.1	0.8	0.1	0.0976	0.0975	0.7759	0.0979	2.40	2.50	3.01	2.10	0.0968	0.0969	0.7746	0.0968	3.18	3.11	3.17	3.16
0.1	0.1	0.8	0.3	0.0973	0.1004	0.7764	0.2932	2.70	−0.40	2.95	2.27	0.0968	0.0969	0.7747	0.2906	3.19	3.12	3.16	3.14
0.1	0.1	0.8	0.5	0.0983	0.0975	0.7761	0.4914	1.70	2.50	2.99	1.72	0.0969	0.0968	0.7745	0.4841	3.12	3.20	3.18	3.17
0.1	0.3	0.8	0.1	0.0968	0.2923	0.776	0.0978	3.20	2.57	3.00	2.20	0.0968	0.2907	0.7751	0.0969	3.17	3.11	3.12	3.15
0.1	0.3	0.8	0.3	0.0987	0.2922	0.7757	0.2922	1.30	2.60	3.04	2.60	0.0968	0.2893	0.7723	0.2902	3.16	3.56	3.46	3.26
0.1	0.3	0.8	0.5	0.0985	0.3025	0.7756	0.494	1.50	−0.83	3.05	1.20	0.0968	0.2892	0.7749	0.4833	3.19	3.59	3.13	3.34
0.1	0.5	0.8	0.1	0.0976	0.4888	0.7762	0.0979	2.40	2.24	2.98	2.10	0.0968	0.4834	0.7735	0.0967	3.21	3.32	3.31	3.32
0.1	0.5	0.8	0.3	0.0978	0.4874	0.7758	0.2918	2.20	2.52	3.03	2.73	0.0968	0.4813	0.7716	0.2904	3.22	3.74	3.55	3.21
0.1	0.5	0.8	0.5	0.0973	0.4891	0.7755	0.4893	2.70	2.18	3.06	2.14	0.0968	0.4829	0.7734	0.4837	3.23	3.42	3.33	3.27
0.35	0.1	0.65	0.1	0.3406	0.0981	0.645	0.0976	2.69	1.90	0.77	2.40	0.3387	0.0966	0.6287	0.0967	3.24	3.40	3.28	3.26
0.35	0.1	0.65	0.3	0.3407	0.0994	0.6372	0.2925	2.66	0.60	1.97	2.50	0.3392	0.0967	0.6297	0.2905	3.10	3.31	3.13	3.17
0.35	0.1	0.65	0.5	0.3411	0.0985	0.6376	0.4942	2.54	1.50	1.91	1.16	0.3394	0.0967	0.6302	0.4846	3.02	3.32	3.04	3.08
0.35	0.3	0.65	0.1	0.3415	0.2928	0.6392	0.0979	2.43	2.40	1.66	2.10	0.3393	0.2891	0.6304	0.0970	3.05	3.65	3.02	3.03
0.35	0.3	0.65	0.3	0.3418	0.2926	0.6408	0.2928	2.34	2.47	1.42	2.40	0.3392	0.2904	0.6300	0.2905	3.09	3.19	3.08	3.18
0.35	0.3	0.65	0.5	0.341	0.3014	0.6342	0.494	2.57	−0.47	2.43	1.20	0.3390	0.2909	0.6293	0.4839	3.14	3.04	3.19	3.21
0.35	0.5	0.65	0.1	0.3412	0.4899	0.635	0.0977	2.51	2.02	2.31	2.30	0.3389	0.4840	0.6274	0.0969	3.17	3.20	3.47	3.07
0.35	0.5	0.65	0.3	0.3422	0.4932	0.6353	0.2927	2.23	1.36	2.26	2.43	0.3390	0.4845	0.6298	0.2909	3.13	3.10	3.11	3.03
0.35	0.5	0.65	0.5	0.3413	0.4894	0.6301	0.494	2.49	2.12	3.06	1.20	0.3383	0.4831	0.6285	0.4834	3.33	3.37	3.31	3.32
0.45	0.1	0.55	0.1	0.4387	0.0987	0.5523	0.0978	2.51	1.30	−0.42	2.20	0.4361	0.0967	0.5326	0.0968	3.10	3.27	3.16	3.16
0.45	0.1	0.55	0.3	0.4389	0.099	0.5331	0.2916	2.47	1.00	3.07	2.80	0.4361	0.0969	0.5334	0.2909	3.08	3.09	3.02	3.03
0.45	0.1	0.55	0.5	0.4395	0.0998	0.533	0.4938	2.33	0.20	3.09	1.24	0.4362	0.0970	0.5298	0.4837	3.07	3.01	3.67	3.27
0.45	0.3	0.55	0.1	0.4426	0.2924	0.5331	0.0981	1.64	2.53	3.07	1.90	0.4340	0.2903	0.5307	0.0967	3.55	3.25	3.50	3.32

(continued)

Table 6 (continued)

Input				Output				% variation in inaccuracy				Output				% variation in inaccuracy			
Experimental results				Trial values (ANFIS) outcome				Inaccuracy in Trial values (ANFIS) outcome				Trial values (ANN) outcome				Inaccuracy in trial values (ANN)			
$\alpha_{e,1}$	$\beta_{e,1}$	$\alpha_{e,2}$	$\beta_{e,2}$	$\alpha_{f,1}$	$\beta_{f,1}$	$\alpha_{f,2}$	$\beta_{f,2}$	Error in $\alpha_{f,1}$	Error in $\beta_{f,1}$	Error in $\alpha_{f,2}$	Error in $\beta_{f,2}$	$\alpha_{f,1}$	$\beta_{f,1}$	$\alpha_{f,2}$	$\beta_{f,2}$	Error in $\alpha_{f,1}$	Error in $\beta_{f,1}$	Error in $\alpha_{f,2}$	Error in $\beta_{f,2}$
0.45	0.3	0.55	0.3	0.4401	0.2917	0.5343	0.2917	2.20	2.77	2.85	2.77	0.4358	0.2900	0.5316	0.2897	3.15	3.32	3.35	3.43
0.45	0.3	0.55	0.5	0.4413	0.2918	0.5343	0.4914	1.93	2.73	2.85	1.72	0.4356	0.2907	0.5323	0.4833	3.21	3.09	3.22	3.33
0.45	0.5	0.55	0.1	0.4398	0.4872	0.5326	0.0976	2.27	2.56	3.16	2.40	0.4354	0.4839	0.5321	0.0967	3.24	3.21	3.26	3.27
0.45	0.5	0.55	0.3	0.4423	0.4859	0.5337	0.2918	1.71	2.82	2.96	2.73	0.4358	0.4811	0.5329	0.2906	3.15	3.77	3.11	3.13
0.45	0.5	0.55	0.5	0.4421	0.486	0.533	0.489	1.76	2.80	3.09	2.20	0.4359	0.4828	0.5327	0.4842	3.14	3.44	3.15	3.16

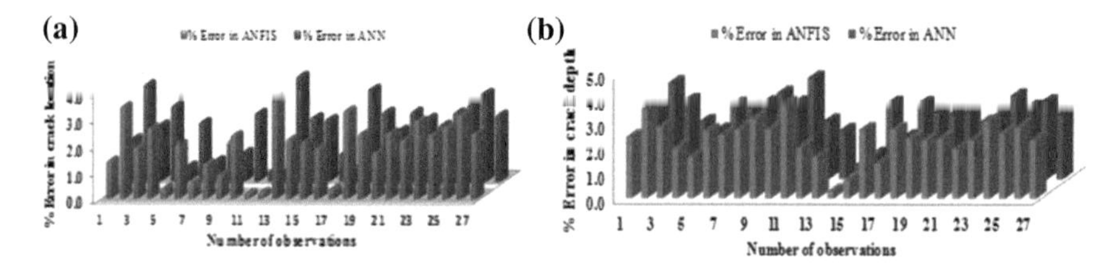

Fig. 9 **a** % error in crack position between ANFIS and ANN network for fixed-fixed shaft respect to experimental results, **b** % error in crack depth between ANFIS and ANN network for fixed-fixed shaft respect to experimental results

(ANFIS) predicts the position and restrictive of cracks (intensity) with more accuracy than any other AI techniques. This paper can be suitably promoting for online condition monitoring of the dynamically vibrating structures.

References

1. Yu L, Cheng L, Yam LH, Yan YJ, Jiang JS (2009) Experimental validation of vibration-based damage detection for static laminated composite shells partially filled with fluid. Compos Struct 79(2):288–299
2. Pawar PM, Reddy KV, Ganguli R (2007) Damage detection in beams using spatial fourier analysis and neural networks. J Intell Mater Syst Struct 18:347–360
3. Hossain MA, Madkour AAM, Dahal KP, Yu H (2008) Comparative performance of intelligent algorithms for system identification and control. J Intell Syst 17:313–329
4. Saridakis KM, Chasalevris AC, Papadopoulos CA, Dentsoras AJ (2008) Applying neural networks, genetic algorithms and fuzzy logic for the identification of cracks in shafts by using coupled response measurements. Comput Struct 86(11–12):1318–1338
5. Taghi M, Baghmisheh V (2008) Crack detection in beam-like structures using genetic algorithms. Appl Soft Comput 8(2):1150–1160
6. Rafiee J, Tse PW, Harifi A, Sadeghi MH (2009) A novel technique for selecting mother wavelet function using an intelli gent fault diagnosis system. Expert Syst Appl 36(3):4862–4875
7. Chandrashekhar M, Ganguli R (2009) Uncertainty handling in structural damage detection using fuzzy logic and Probabilistic simulation. Mech Syst Signal Process 23(2):384–404
8. Panigrahi SK, Chakraverty S, Mishra BK (2009) Vibration based damage detection in a uniform strength beam using genetic algorithm. Meccanica 44:697–710
9. Haryanto I, Setiawan JD, Budiyono A (2009) Structural damage detection using randomized trained neural networks. Stud Comput Intell 192:245–255
10. Lekhy D, Novak D (2009) Neural network based damage detection of dynamically loaded structures. Commun Comput Inf Sci 43:17–27
11. Rubio L, Muñoz Abella B, Loaiza G (2011) Static behavior of a shaft with an elliptical crack. Mech Syst Signal Process 25:1674–1686

Optimization of Process Parameters in Electro Discharge Machine Using Standard Deviation, MULTIMOORA and MOOSRA Methods

J. Anitha and Raja Das

Abstract Nowadays in the extremely competitive market, an industrial product should be manufactured in minimum duration with high precision and quality. To attain these, the process parameters are regulated to get the desired output based on the requirement. The process parameters play an important role in defining the surface quality and material removal rate. Electric discharge machine (EDM) is one of the most striking alternative in the industry based on its attributes. Four input parameters, namely voltage, pulse-on time, current and duty cycle, are considered. The effect of these parameters on material removal rate (MRR) and surface roughness (Ra) is studied based on the experimental results. Multi-objective optimization techniques help a decision-maker to select a best alternative from a set of alternatives which have conflicting objectives. In this paper, standard deviation is used to find the relative importance of the attributes. The weights obtained using standard deviation are MRR is 0.53 and Ra is 0.47. Further new multi-objective optimization techniques, namely full multiplicative form of MOORA (MULTIMOORA) and multi-objective optimization based on simple-ratio analysis (MOOSRA), are applied to obtain the optimum process parameters. The process parameters corresponding to run number 6 are voltage 50 units, current 15 units, duty cycle 50 and pulse-on time 100 units, are the best parametric combination to achieve maximum MRR and minimum surface roughness.

Keywords EDM · Material removal rate · Surface roughness · Standard deviation · Multi-objective optimization—MULTIMOORA · MOOSRA

J. Anitha (✉)
School of Advanced Science, VIT University, Vellore, Tamilnadu, India

R. Das
VIT University, Vellore, Tamilnadu, India

BBVL. Deepak et al. (eds.), *Innovative Product Design and Intelligent Manufacturing Systems*, Lecture Notes in Mechanical Engineering,
https://doi.org/10.1007/978-981-15-2696-1_59

1 Introduction

Electric discharge machining is the most non-conventional machining process which is used widely after grinding, milling and turning. Materials which are difficult-to-machine such as nickel-based alloys, titanium alloys, and hardened steels can be machined using EDM. This process of EDM is complex, which is affected by many parameters. It is difficult to select the optimum parametric combination that gives higher material removal rate (MRR) with a decent surface finish. MRR and surface finish are quite contradictory in nature. For better surface quality, the surface roughness should be minimized and for high productivity MRR has to be maximized.

Many problems in engineering involve simultaneous optimization which has conflicting objectives. Such problems are known as multi-objective optimization problems. One of the approaches to solve these multi-objective problems Jaimes et al. [1] is multi-criteria decision-making (MCDM). MOORA method was first developed by Brauers and Zavakdas [2] and was applied to privatization in transition economy. Chakraborty [3] applied MOORA method to find the solutions of problems in manufacturing industry. Brauers et al. used MOORA method to evaluate road design alternatives [4] and contractor rankings [5]. MOORA along with standard deviation method was applied by Joseph and William [6] for optimizing gas metal welding process parameters. Muniappan et al. [7] used standard deviation and MOORA methods for optimization of WEDM process parameters. Brauers [8] used MULTIMOORA method for fighter planes. Brauers [9] used optimization methods for a stakeholder society. Karande and Chakraborty [10] used fuzzy-MOORA for ERP system selection. Kracka et al. [11] applied MULTIMOORA for ranking heating losses in a building. An application of MOORA and MOOSRA methods were used for developing an efficient decision support system for non-traditional machine selection by Sarkar et al. [12]. Esra and Aysegul [13] applied MULTIMOORA and MOOSRA for laptop selection. Kumar and Ray [14] applied MOOSRA method for selection of material under conflicting situation. Based on the literature, a lot of work has been done using different MCDM methods, but the relative importance of performance measures is not considered. The novelty of this paper is that the importance of performance measures like MRR and Ra is found from experimental data using standard deviation.

In this study, standard deviation along with MOORA, MULTIMOORA, and MOOSRA methods are used to optimize the input parameters. Standard deviation is used to find the relative importance of performance measures. Later different MCDM methods which are quite simple and easy to calculate are applied to select the best alternative.

2 Experimental Environment and Methods

2.1 Experimental Set Up

The experiment is conducted on Electronica Elektra plus PS 50ZNC Processing Machine. A rectangular-shaped AISI D2 tool steel is used as the workpiece material which has a thickness of 4 mm (with negative polarity), and density 7.7 g/cc. A 30 mm diameter electrolytic copper with a positive polarity is used as electrode material. A viable grade EDM oil has a specific gravity of 0.76 and freezing point of 94 °C. Side flushing technique with 0.3 kg f/cm^2 pressure is maintained. The four input parameters and the experimental values are shown in Table 1.

2.2 Optimization Problem

If MRR = f_1(Ip, Ton, τ, V) and Ra = f_2(Ip, Ton, τ, V), then the multi-objective optimization problem is

Maximize f_1(Ip, Ton, τ, V) and
Minimize f_2(Ip, Ton, τ, V)

Table 1 Experimental values

Run	Ip (A)	Ton (μs)	τ	V	EXP_MRR	EXP_Ra
1	10	75	66.5	45	10.35	5.55
2	10	75	66.5	45	9.04	5.98
3	15	50	83	50	29.163	8.43
4	15	50	50	50	33.103	7.43
5	5	100	50	50	4.349	5.59
6	15	100	50	50	51.09	8.1
7	10	75	66.5	45	8.95	6.12
8	15	100	50	40	51.004	10.93
9	5	100	83	50	6.972	5.7
10	15	100	83	40	33.023	12.49
11	5	50	83	50	14.12	5.19
12	15	100	83	50	33.11	9.01
13	15	50	50	40	29.737	10.49
14	10	75	66.5	45	8.42	6.54
15	10	75	83	45	9.356	7.13
16	10	75	66.5	50	11.007	6.35
17	15	75	66.5	45	33.084	9.68
18	5	75	66.5	45	5.361	6.07

Subject to $5 \le \text{Ip} \le 15$

$50 \le \text{Ton} \le 100$
$50 \le \tau \le 83$
$40 \le V \le 50$

And Ip, Ton, τ, V € Z

2.3 Weight Calculation Using Standard Deviation Concept

Standard deviation is related to calculation of unbiassed assignment of weights.

Step 1 Calculate

$$X_{ij}^{1} = \frac{X_{ij} - \min_{1<j<n} X_{ij}}{\max_{1<j<n} X_{ij} - \min_{1<j<n} X_{ij}} \quad (1)$$

where max X_{ij} is the maximum, and min X_{ij} is the minimum value for the measure (j)

Step 2 Calculate standard deviation using Eq. (2)

$$\text{SDV}_j = \sqrt{\frac{\sum_{i=1}^{m}\left(X_{ij} - \overline{X_j^i}\right)^2}{m}} \quad (2)$$

where X_{ij} is the average of the values for the jth measure, where j = 1, 2, 3 … n.

Step 3 Calculate weights using Eq. (3)

$$w_j = \frac{SDV_J}{\sum_{j=1}^{n} SDV_j} \quad (3)$$

where w_j are the weights for j = 1, 2, 3…n.

2.4 MOORA Method

MOORA is one of the MCDM methods which are used to select the best option from number of selections. This method was first presented by Brauers [9] to solve intricate decision-making problems in the manufacturing industry. The MOORA method [15–17] begins with a decision matrix that represents various substitutes and objectives.

Step 1 Describing the problem and establishing the objectives is the first step. In the present work, MRR must be maximized and Ra must be minimized.

Step 2 Next step is to create a decision matrix based on the experimental results of the number of output parameters. Represent the decision matrix using X.

$$X = \begin{bmatrix} X_{11} & X_{12} & \cdots & X_{1n} \\ X_{12} & X_{12} & \cdots & X_{2n} \\ \vdots & \vdots & & \vdots \\ X_{m1} & X_{m2} & & X_{mn} \end{bmatrix} \quad (4)$$

Step 3 The performance of the ith alternate on jth feature is normalized using Eq. (5)

$$x_{ij}^{*} = \frac{x_{ij}}{\sqrt{\sum_{i=1}^{m} x_{ij}^{2}}} \quad (5)$$

where $j = 1, 2, 3\ldots\ n$.

Step 4 In the multi-objective optimization process, these normalized values are added if it is a case of maximization and subtracted if it is a case of minimization. Hence, the optimized values are calculated using the Eq. (6)

$$y_i = \sum_{j=1}^{g} x_{ij}^{*} - \sum_{j=g+1}^{n} x_{ij}^{*} \quad (6)$$

where g are the number of features which have to be maximized, and the number of features which has to be minimized are $(n - g)$.

Step 5 It is quite often detected that some features are more significant than the other features. To give more significance to such features, it is multiplied with corresponding weight. The optimized problem is calculated using Eq. (7)

$$y_i = \sum_{j=1}^{g} w_j x_{ij}^{*} - \sum_{j=g+1}^{n} w_j x_{ij}^{*} \quad (7)$$

for $j = 1, 2, 3\ \ldots\ n$, where w_j is the weight of the jth attribute, which is determined using standard deviation.

Step 6 The y_i values can be positive or negative depending on the maximizing attributes and minimizing attributes in the decision matrix. Rank the values from highest to the lowest. The best alternative has the highest value of y_i, and the worst value has the lowest y.

2.5 The Reference Point Approach of MOORA Method

The normalized values of *i*th alternative on *j*th criterion are used in reference point approach which is calculated using Eq. (2). Min–max method is the most suitable method according to Brauers and Zavadskas [18, 19].

$$r_j = \begin{cases} \dfrac{\max x^*_{ij} \quad \text{if criteria is to be maximized}}{\min x^*_{ij} \quad \text{if the criteria is to be minimized}} \end{cases} \tag{8}$$

Reference point formula is

$$\min_i \left\{ \max_j \left| r_j - x^*_{ij} \right| \right\} \tag{9}$$

If importance is given to a criterion the formula is

$$\min_i \left\{ \max_j \left| w_j r_j - w_j x^*_{ij} \right| \right\}. \tag{10}$$

2.6 The Full Multiplicative Form

Miller and Starr (1969) developed this method.

The formula is given as

$$U_i = \frac{A_i}{B_i} \tag{11}$$

where $A_i = \pi^g_{j=1} x_{ij}$ and $B_i = \pi^n_{j=g+1} x_{ij}$

Where g and $(n - g)$ are the criteria which have to be maximized and minimized according to Brauers [19].

2.7 MOOSRA Method

This method is similar to MOORA method. It is a simple ratio of the sum of normalized values of maximization to the sum of normalized values of minimization. The formula given by Kumar and Ray [14] is

$$y_i^* = \frac{\sum_{j=1}^{g} x_{ij}^*}{\sum_{j=g+1}^{n} x_{ij}^*} \tag{12}$$

If importance is given to the criteria, then the formula is

$$y_i^* = \frac{\sum_{j=1}^{g} w_j x_{ij}^*}{\sum_{j=g+1}^{n} w_j x_{ij}^*} \tag{13}$$

Ranking of alternatives is done and the one with the highest value is the best.

3 Results and Discussion

3.1 Allocation of Weights

The weights are calculated for MRR and Ra using formula (1), (2) and (3). The weight for MRR is 0.53 and for surface roughness is 0.47. Table 2 shows the standardized values of MRR and Ra.

Table 2 Standardized MRR and Ra

Run	X_{ij}' (MRR)	X_{ij}'' (Ra)	Standardized MRR	Standardized Ra
1	0.128	0.049	0.3065	0.3314
2	0.100	0.108	0.3383	0.2671
3	0.531	0.444	0.0228	0.0328
4	0.615	0.307	0.0045	0.1012
5	0.000	0.055	0.4651	0.3251
6	1.000	0.399	0.1011	0.0512
7	0.098	0.127	0.3405	0.2476
8	0.998	0.786	0.1000	0.0260
9	0.056	0.070	0.3917	0.3082
10	0.613	1.000	0.0047	0.1406
11	0.209	0.000	0.2237	0.3906
12	0.615	0.523	0.0044	0.0103
13	0.543	0.726	0.0193	0.0102
14	0.087	0.185	0.3539	0.1937
15	0.107	0.266	0.3305	0.1291
16	0.142	0.159	0.2911	0.2172
17	0.615	0.615	0.0045	0.0001
18	0.022	0.121	0.4361	0.2545
Standard deviation			0.6273	0.5654

3.2 Result Analysis

The decision matrix is normalized using Eq. (5) in MOORA method, and the normalized values are weighted using Eq. (7). The overall performance is ranked which is seen in Table 3. According to MOORA method, run 6 is the best parametric combination.

The weighted normalized values are used as the initial step in reference point approach. Table 4 shows the deviations from the reference points which are calculated using Eqs. (8) and (10). Run 6 is the best combination according to the reference point of MOORA.

Table 5 shows the ranking of alternatives using MULTIMOORA which is calculated using Eq. (11). Run 6 is the best combination according to MULTIMOORA.

According to MOOSRA method, run 6 is the best parametric combination which is calculated based on Eq. (13). MOOSRA method is shown in Table 3.

Figure 1 shows the ranking of alternatives using all the methods. Hence, run 6 is the best combination using all the methods.

Table 3 Normalized decision matrix, MOORA ranking and MOOSRA ranking

	Normalized		Weighted		Σmax − Σmin	MOORA rank	Ratio	MOOSRA rank
Run	MRR	Ra	MRR	Ra				
1	0.0938	0.1665	0.0497	0.0783	0.0285	10	0.6353	10
2	0.0820	0.1794	0.0434	0.0843	0.0409	12	0.5150	12
3	0.2644	0.2530	0.1401	0.1189	0.0212	6	1.1786	5
4	0.3001	0.2230	0.1591	0.1048	0.0543	3	1.5178	3
5	0.0394	0.1677	0.0209	0.0788	0.0579	17	0.2650	18
6	0.4632	0.2431	0.2455	0.1142	0.1312	1	2.1488	1
7	0.0811	0.1836	0.0430	0.0863	0.0433	13	0.4982	13
8	0.4624	0.3280	0.2451	0.1542	0.0909	2	1.5897	2
9	0.0632	0.1710	0.0335	0.0804	0.0469	14	0.4167	16
10	0.2994	0.3748	0.1587	0.1762	0.0175	9	0.9007	9
11	0.1280	0.1557	0.0678	0.0732	0.0054	8	0.9269	8
12	0.3002	0.2704	0.1591	0.1271	0.0320	4	1.2519	4
13	0.2696	0.3148	0.1429	0.1479	0.0051	7	0.9658	7
14	0.0763	0.1962	0.0405	0.0922	0.0518	15	0.4386	15
15	0.0848	0.2140	0.0450	0.1006	0.0556	16	0.4470	14
16	0.0998	0.1905	0.0529	0.0896	0.0367	11	0.5905	11
17	0.2999	0.2905	0.1590	0.1365	0.0224	5	1.1644	6
18	0.0486	0.1821	0.0258	0.0856	0.0599	18	0.3009	17

Table 4 Deviations from the reference points

Run	Normal MRR	Normal Ra	W_jN_i	W_jN_j	$\left\|w_jr_j - w_jx''_{ij}\right\|$	$\left\|w_jr_j - w_jx'_{ij}\right\|$	Max value	Rank
1	0.0938	0.1663	0.0497	0.0783	0.1958	0.0051	0.1958	11
2	0.0820	0.1794	0.0434	0.0843	0.2021	0.0112	0.2021	13
3	0.2644	0.2530	0.1401	0.1189	0.1054	0.0457	0.1054	8
4	0.3001	0.2230	0.1591	0.1048	0.0864	0.0316	0.0864	4
5	0.0394	0.1677	0.0209	0.0788	0.2246	0.0057	0.2246	18
6	0.4632	0.2431	0.2455	0.1142	0.0000	0.0411	0.0411	1
7	0.0811	0.1836	0.0430	0.0863	0.2025	0.0131	0.2025	14
8	0.4624	0.3280	0.2451	0.1542	0.0004	0.0810	0.0810	2
9	0.0632	0.1710	0.0335	0.0804	0.2120	0.0072	0.2120	16
10	0.2994	0.3748	0.1587	0.1762	0.0868	0.1030	0.1030	7
11	0.1280	0.1557	0.0678	0.0732	0.1777	0.0000	0.1777	9
12	0.3002	0.2704	0.1591	0.1271	0.0864	0.0539	0.0864	4
13	0.2696	0.3148	0.1429	0.1479	0.1026	0.0748	0.1026	6
14	0.0763	0.1962	0.0405	0.0922	0.2050	0.0191	0.2050	15
15	0.0848	0.2140	0.0450	0.1006	0.2005	0.0274	0.2005	12
16	0.0998	0.1905	0.0529	0.0896	0.1926	0.0164	0.1926	10
17	0.2999	0.2905	0.1590	0.1365	0.0865	0.0633	0.0865	5
18	0.0486	0.1821	0.0258	0.0856	0.2197	0.0124	0.2197	17

Table 5 Ranking of alternatives using MULTIMOORA

Run	MRR	Ra	Ui = MRR/Ra	Ranking
1	10.35	5.55	1.8649	10
2	9.04	5.98	1.5117	12
3	29.163	8.43	3.4594	5
4	33.103	7.43	4.4553	3
5	4.349	5.59	0.7780	18
6	51.09	8.1	6.3074	1
7	8.95	6.12	1.4624	13
8	51.004	10.93	4.6664	2
9	6.972	5.7	1.2232	16
10	33.023	12.49	2.6440	9
11	14.12	5.19	2.7206	8
12	33.11	9.01	3.6748	4
13	29.737	10.49	2.8348	7
14	8.42	6.54	1.2875	15
15	9.356	7.13	1.3122	14
16	11.007	6.35	1.7334	11
17	33.084	9.68	3.4178	6
18	5.361	6.07	0.8832	17

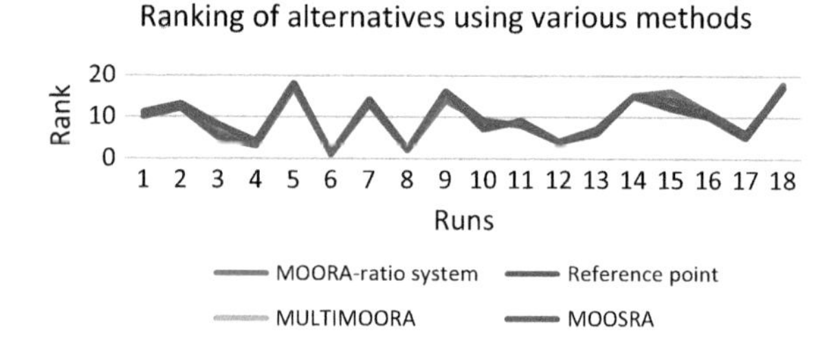

Fig. 1 Ranking of alternatives using various methods

4 Conclusion

In this paper, multi-objective optimization problem is solved using MOORA, MULTIMOORA and MOOSRA methods. These methods are suitable to find the best alternative from a set of alternatives. In future studies, the EDM problem can be solved using other MCDM methods and their performance can be compared using the existing methods. Further MULTIMOORA and MOOSRA methods can be applied to any multi-objective optimization problem.

These methods can be used for both qualitative and quantitative attributes simultaneously, but the limitation is, that they are not so effective when the decision matrix contains large amount of qualitative attributes.

The results show that the best parametric combination is Run 6. Hence, the optimized values are current 15 units, pulse-on time 100 units, duty cycle 50 and voltage 50 units that give maximum MRR as 51.09 units and minimum surface roughness of 8.1 units.

References

1. Jaimes AL, Martinez SZ, Coello CAC (2011) An introduction to multi-objective optimization techniques. In: Gaspar-Cunha A, Covas JA (eds) Optimization in polymer processing. Nova Science, New York, pp 29–57
2. Brauers WKM, Zavadskas EK (2006) The MOORA method and its applications to privatization in some transition economy. Control Cybern 35:445–469
3. Chakraborty S (2011) Applications of the MOORA method for decision making in manufacturing environment. Int J Adv Manuf Technol 54(9–12):1155–1166
4. Brauers WKM, Zavadskas EK, Pedschus F, Turkis Z (2008) Multi-objective decision making for road design. Transport 23:183–193
5. Brauers WKM, Zavadskas EK, Turkis Z, Viutiene T (2008) Multi-objective contractors ranking by applying the MOORA method. J Bus Econ Manag 9:245–255
6. Achebo J, Odinikuku WE (2015) Optimization of Gas metal arc welding process parameters using standard deviation and MOORA. J Miner Mater Charact Eng 2:298–308
7. Muniappan A, Raj JA, Jayakumar V, Prakash RS, Sathyaraj R (2018) Optimization of WEDM process parameters using standard deviation and MOORA method. In: 2nd International conference on advances in Mechanical Engineering (ICAME)
8. Brauers WKM (2002) The multiplicative representation for multiple objectives optimization with an application for arms procurement. Nav Res Log 49:327–340

9. Brauers WKM (2004) Optimization methods for a stakeholder society. A revolution in economic thinking by multiobjective optimization: non-convex optimization and its applications. Kluwer, Boston
10. Karande P (2012) Chakraborty: a fuzzy-MOORA approach for ERP system selection. Decis Sci Lett 1(1):11–22
11. Karcka M, Brauers WKM, Zavadskas EK (2010) Ranking heating losses in a building by applying the MULTIMOORA. Eng Econ 21(4):352–359
12. Sarkar A, Panja SC, Das D, Sarkar B (2015) Developing an efficient decision support system for non-traditional machine selection: an application of MOORA and MOOSRA. Prod Manuf Res 3(1):324–342
13. Adah EA, Isik AT (2016) The multi-objective decision-making methods based on MULTIMOORA and MOOSRA for the laptop selection problem. J Ind Eng Int 13(2):229–237
14. Kumar R, Ray A (2015) Selection of material under conflicting situation using simple ratio optimization technique. In: Das et al (eds) Proceedings of fourth international conference on soft computing for problem solving, advances in intelligent systems and computing, vol 335, pp 513–519
15. Brauers WKM, Zavadskas EK (2009) Multi-objective optimization with discrete alternatives on the basis of ratio analysis. Intellect Econ 2(6):30–41
16. Kalibatas D, Turskis Z (2008) Multicriteria evaluation of inner climate by using MOORA method. Inf Technol Control 37:79–83
17. Lootsma FA (1999) Multicriteria decision analysis via ratio and difference judgement. Springer, London
18. Brauers WKM, Zavadskas EK (2009) Robustness of the multiobjective MOORA method with a test for the facilities sector. Technol Econ Dev Econ 15(2):352–375
19. Brauers WKM, Zavadskas EK (2012) Robustness of MULTIMOORA: a method for multi-objective optimization. Informatica 23(1):1–25

Analysis of Smart Front-End Structure in Frontal Crash Mitigation

Soumitra Singh Kachhwaha, Mukesh Kumar Singh, Dhaneshwar Prasad Sahu and Nohar Kumar Sahu

Abstract Bumper system of a vehicle absorbs the kinetic energy of full-frontal and offset collision. The bumper of a vehicle is the most prominent component considering the safety of the occupant. Design of new bumper system should be done in a way in which the deceleration of the occupant and intrusion in the passenger compartment must reduce. A new way to improve safety in the frontal collision is the implementation of the hydraulic damper in the bumper system of the vehicle. Magnetorheological (MR) fluid damper is suitable for absorbing high-speed impact energy. Implementation of MR damper in the bumper system will reduce the occupant injury. A dynamic model of a vehicle has been developed. This model has four degrees of freedom and damping provided by an airbag is also taken into account, thus giving a more accurate prediction of response during the collision. A new configuration of MR damper has also been proposed.

Keywords Adaptive front bumper system · Magnetorheological damper · Smart front structure

1 Introduction

The frontal collision of a vehicle is very dangerous for its occupants. In case of a frontal collision, the bumper system collides with another object [1]. So, a bumper design is quite interesting as well as a useful topic for a design engineer. A bumper system of a vehicle consists of three main parts: fascia, energy absorber, and a bumper beam [2]. Facia reduces the drag force on the vehicle as well as serves the purpose of improving the aesthetic appeal of the vehicle. The energy absorber is used for mitigating the impact force during the collision. The bumper beam is the main component of a bumper system which absorbs most of the kinetic energy in both low- and high-speed collision [3–5]. There have been quite a few works

S. S. Kachhwaha (✉) · M. K. Singh · D. P. Sahu · N. K. Sahu
Department of Mechanical Engineering, Institute of Technology, Guru Ghasidas Vishwavidyalaya, Bilaspur, Chhattisgarh, India

BBVL. Deepak et al. (eds.), *Innovative Product Design and Intelligent Manufacturing Systems*, Lecture Notes in Mechanical Engineering,
https://doi.org/10.1007/978-981-15-2696-1_60

related to the development of new lighter and strong material for improving the property of a bumper beam by scientists, researchers, and design engineers. Most of the research in design and analysis of a bumper system has been done on the use of composite materials, although there have been some new controllable structures using hydraulic damper. Evans [6] gave a perspective to improve performance of the bumper including energy-absorbing ability, protection against damage, and response to change in temperature. He compared different bumper beam materials at a temperature range of −30 to 60 °C by using a 16 km per hour swinging barrier impact. King et al. [7] examined the low-velocity impact of automobile bumper system systems. They used data from 136 vehicle-to-vehicle and 58 vehicle-to-barrier aligned impacts. Bumper system of a vehicle includes a front and rear component of a vehicle which dissipates the kinetic energy by deflection in case of low-speed collision and by deformation during high-speed collision [8, 9].

Magnetorheological (MR) fluid damper is one of the best candidates for implementation in the bumper system of an automobile. The magnetorheological fluid is a type of smart fluid in which the application of a magnetic field can change its apparent viscosity [10–15]. MR fluids are dispersed in a carrier medium made of micron-sized ferrous [9]. The carrier medium of MR fluid is mostly hydrocarbon oil or silicon oil or water and carbonyl iron may be the ferrous particle [16–18]. The analysis of MR fluid is done mostly with Bingham plastic model, but the Herschel–Bulkley model is also a promising model [11]. Wang and Liao [19] have studied the characteristics of magnetorheological fluid with the help of parametric dynamic modeling to improve the performance of these dampers. Lee and Choi [20] came up with a cylindrical magnetorheological damper and evaluated it by taking the velocity of the piston as a criterion.

Lee et al. [21] used a Herschel–Bulkley model for analyzing impact dampers based on MR fluid and electrorheological fluid. Song et al. [22] used Bingham model to analyze impact damper based on electrorheological fluid. Choi and Wereley [23] have investigated the reduction of vibration and impact load using shock-absorbing system based on magnetorheological fluid. Wang and Xia [24] used MR damper to control the ground resonance of a helicopter. Bajkowski et al. [25] have analyzed MR fluid damper including the effect of temperature and friction by using a lumped mass model analysis. Zhu et al. [26] tested dropping of landing gear by using damper based on magnetorheological fluid. Tu et al. [27] used the finite element method to analyze the smart damper as a shock-absorbing unit in an automobile. Kazemi et al. [28] came up with a methodology to predict the behavior of magnetorheological fluid by using a hydraulic model. MR fluid damper has many applications due to its special nature and is one of the most feasible materials for absorbing high-speed impact in a collision. Design and analysis of the new bumper system are very important for the safety of the occupants. However, there is a lack of research in this area. Hence, it is a very fascinating area of research.

2 Dynamic Model Development

Mathematical model of the vehicle-to-barrier collision has been developed after reviewing approximately 100 research papers on the frontal crash analysis. The mathematical model developed for analysis is having four degrees of freedom to have a precise idea about the response of the system. The developed mathematical model has the inclusion of bumper system, engine and engine room, the body of the vehicle, and occupants. It has also accounted for passive restraint provided by the seatbelt as well as an airbag in the vehicle. The restraint provided by the airbag has been assumed to be linear damping. Magnetorheological damper used reduces the impact forces of collision, and the net force acting on the vehicle body is considered in the development of the dynamic model. The mathematical model of the vehicle in the event of collision is shown in Fig. 1.

The figure shows four lumped mass dynamic model of the vehicle system.

Where m_a = mass of the bumper system, m_b = mass of engine and engine room, m_c = mass of vehicle body, m_d = mass of occupants, F_{bd} = force transmitted to vehicle body after shock absorption by the damper, and c_3 = damping provided by the airbag. Applying Newton's second law in the above model gives equations of motion:

$$m_a\ddot{x}_a + 2k_a(x_a - x_b) = 2F_{bd} \tag{1}$$

$$m_b\ddot{x}_b - 2k_a x_a + x_b(2k_a + 2k_b) - 2k_b x_c = 0 \tag{2}$$

$$m_c\ddot{x}_c + c_3\dot{x}_c - c_3\dot{x}_d - 2k_b x_b + x_c(2k_b + k_c) - k_c x_d = 0 \tag{3}$$

$$m_d\ddot{x}_d - c_3\dot{x}_c + c_3\dot{x}_d - k_c x_c + k_c x_d = 0 \tag{4}$$

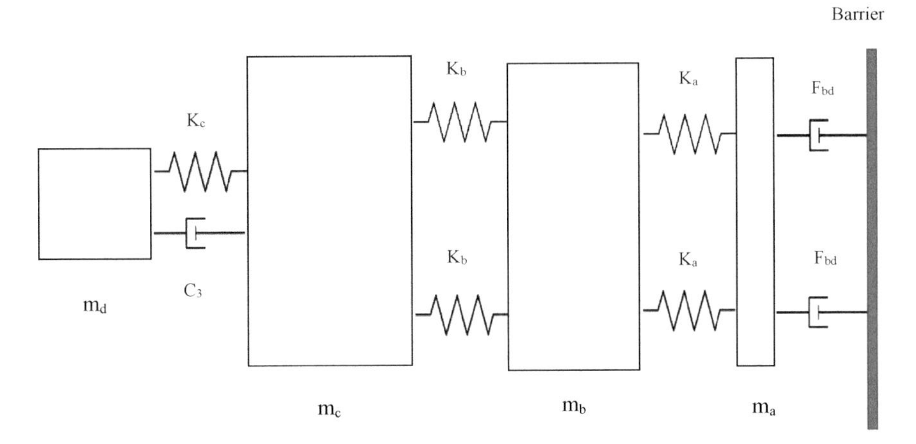

Fig. 1 Mathematical model of the vehicle

Above equations in matrix form can be written as

$$[m]\ddot{\vec{x}} + [c]\dot{\vec{x}} + [k]\vec{x} = \vec{F}$$

where

$$[m] = \begin{bmatrix} m_a & 0 & 0 & 0 \\ 0 & m_b & 0 & 0 \\ 0 & 0 & m_c & 0 \\ 0 & 0 & 0 & m_d \end{bmatrix}$$

$$[c] = \begin{bmatrix} 0 & 0 & 0 & 0 \\ 0 & 0 & 0 & 0 \\ 0 & 0 & c_3 & -c_3 \\ 0 & 0 & -c_3 & c_3 \end{bmatrix}$$

$$[k] = \begin{bmatrix} 2k_a & -2k_a & 0 & 0 \\ -2k_a & (2k_a + 2k_a) & -2k_b & 0 \\ 0 & -2k_b & (2k_b + k_c) & -k_c \\ 0 & 0 & -k_c & k_c \end{bmatrix}$$

$$\vec{F} = \begin{bmatrix} 2F_{bd} \\ 0 \\ 0 \\ 0 \end{bmatrix}$$

Taking equation $[m]\ddot{\vec{x}} + [c]\dot{\vec{x}} + [k]\vec{x} = \vec{F}$ and considering a special system,

$$[c] = a[m] + b[k] \tag{5}$$

The above equation gives,

$$m\ddot{\vec{x}} + [a[m] + b[k]]\dot{\vec{x}} + [k]\vec{x} = \vec{F}$$

putting $\vec{x}(t) = [X]\vec{q}(t)$ in above the equation

$$[m][X]\ddot{\vec{q}}(t) + [a[m] + b[k]][X]\dot{\vec{q}}(t) + [k][X]\vec{q}(t) = \vec{F}(t) \tag{6}$$

Premultiplying by $[X]^T$ leads to,

$$[X]^T[m][X]\ddot{\vec{q}}(t) + [X]^T[a[m] + b[k]][X]\dot{\vec{q}}(t) + [X]^T[k][X]\vec{q}(t) = [X]^T\vec{F}(t)$$

If eigenvectors are normalized

$$[I]\ddot{\vec{q}}(t) + \left[a[I] + b\left[\omega^2\right]\right]\dot{\vec{q}}(t) + \left[\omega^2\right]\vec{q}(t) = \vec{Q}(t) \tag{7}$$

In non-matrix form, it can be written as

$$\ddot{q}_i(t) + \left(a + b\omega_1^2\right)\dot{q}_i(t) + \omega_1^2 q_i(t) = Q_i(t) \tag{8}$$

where ω_i is the natural frequency of the damped system and $i = 1, 2, 3, \ldots, n$.
Now assuming,

$$a + \omega_i b = 2\zeta_i\omega_i \tag{9}$$

where ζ_i = modal damping ratio (For ith normal mode)

$$\ddot{q}_i(t) + 2\zeta_i\omega_i\dot{q}_\mathrm{i}(\mathrm{t}) + \omega_\mathrm{i}^2 q_i(t) = Q_i(t) \tag{10}$$

The above equation is similar to viscously damped single degree of freedom system and can be solved accordingly.

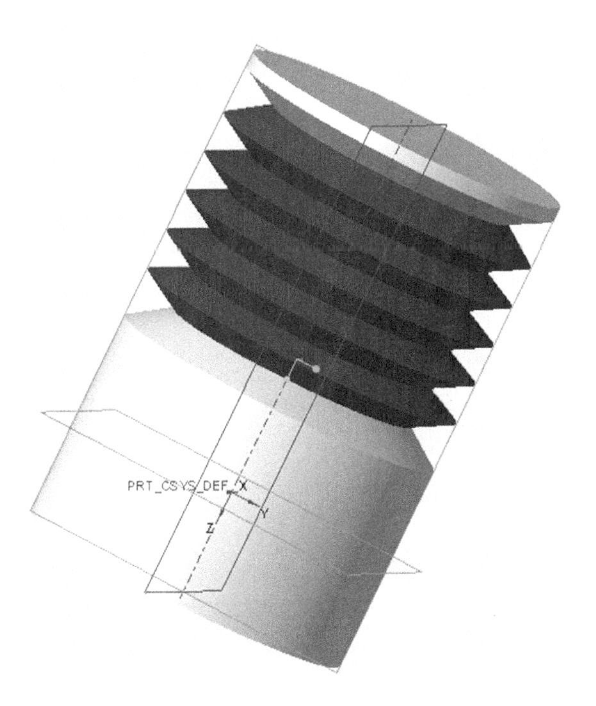

Fig. 2 Configuration of MR damper

3 Modeling of Magnetorheological Damper

Figure 2 shows the proposed magnetorheological damper configuration modeled using CREO™. This model can be installed at the frontal bumper system of a vehicle. The magnetorheological damper has MR fluids as damping or viscous material. Magnetorheological dampers do not require any movable mechanical parts for changing force characteristic of the damper. For increasing the damping force, they use unique characteristics of an MR fluid, which is controllability of apparent viscosity by varying the applied external magnetic field.

In this MR damper, bellows in place of piston and spring is used. Also, the electromagnets can be fixed at the inner part of the damper which will help with the wiring requirements. This model can be readily analyzed using Bou-Wen model (Fig. 3).

4 Conclusion

Safety has always been a matter of concern in the automobile industry. New systems have been developed for improving the crashworthiness of the vehicle. Some very innovative designs are discussed which will help researchers and designers to have a good understanding of design for further improvement. These novel designs

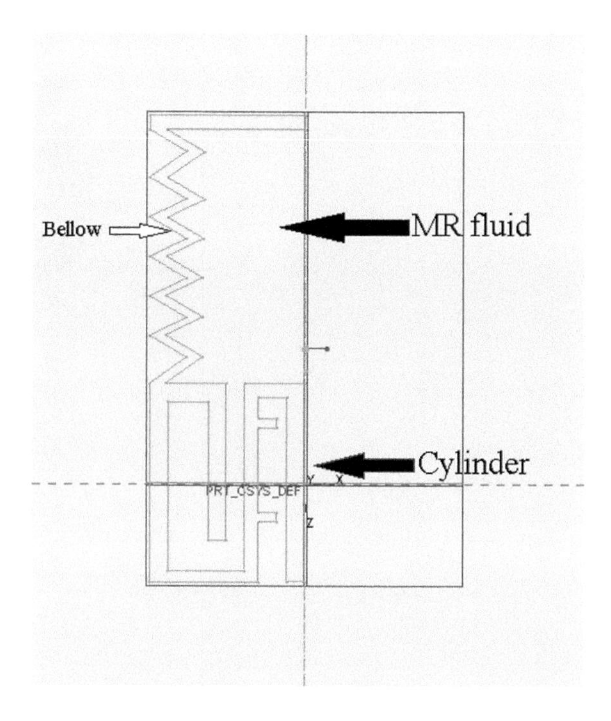

Fig. 3 Configuration of MR damper

will result in increased safety of occupants in the event of a collision. In this paper, dynamic modeling of a vehicle in a collision is carried out. A four-degree-of-freedom mathematical model is developed. This model also features a damping element for airbags. A new configuration of the magnetorheological damper has also been developed. This MR damper can be used in the bumper system of a vehicle to improve its crashworthiness. More degrees of freedom can be provided in the mathematical model for more accuracy.

References

1. Thornton PH, Jeryan RA (1988) Crash energy management in composite automotive structures. Int J Impact Eng 7(2):167–180
2. Sapuan SM, Maleque MA, Hameedullah M, Suddin MN, Ismail N (2005) A note on the conceptual design of polymeric composite automotive bumper system. J Mater Process Technol 159(2):145–151
3. Davoodi MM, Sapuan SM, Yunus R (2008) Materials & design conceptual design of a polymer composite automotive bumper energy absorber 29:1447–1452
4. Zeng F, Xie H, Liu Q, Li F, Tan W (2016) Design and optimization of a new composite bumper beam in high-speed frontal crashes. Struct Multidiscip Optim 53(1):115–122
5. Zakaria M, Mohammad B (2010) Simulation of an active front bumper system for frontal impact protection simulation of an active front bumper system for frontal impact protection (June)
6. Evans D (1998) SAE TECHNICAL consistency of thermoplastic bumper beam impact performance. Engineering 724
7. King DJ, Siegmund GP, Bailey MN (1993) Automobile bumper behavior in low-speed impacts 41 2
8. Davoodi MM, Sapuan SM, Aidy A, Abu Osman NA, Oshkour AA, Wan Abas WAB (2012) Development process of the new bumper beam for passenger car: a review. Mater Des 40:304–313
9. Imthiyaz Ahamed T, Sundarrajan R, Prasaath GT, Raviraj V (2014) Implementation of Magneto-rheological dampers in bumpers of Automobiles for reducing impacts during accidents. Procedia Eng 97:1220–1226
10. Guan XC, Guo PF, Ou JP (2011) Modeling and analyzing of hysteresis behavior of magneto rheological dampers. Procedia Eng 14:2756–2764
11. Wang X, Gordaninejad F (1999) Flow analysis of field-controllable, electro- and magneto-rheological fluids using the Herschel-Bulkley model. J Intell Mater Syst Struct 10 (8):601–608
12. Kumar HNA, Shilpashree DJ, Adarsh MS, Amith D, Kulkarni S (2016) Development of smart squeeze film dampers for small rotors. Procedia Eng 144:790–800
13. Orečný M, Segľa Š, Huňady R, Ferková Ž (2014) Application of a magneto-rheological damper and a dynamic absorber for a suspension of a working machine seat. Procedia Eng 96:338–344
14. Ahmed GMS, Reddy PR, Seetharamaiah N (2014) FEA based modeling of Magneto rheological damper to control vibrations during machining. Procedia Mater Sci 6 (Icmpc):1271–1284
15. Milecki A, Sedziak D, Ortmann J, Hauke M (2005) Controllability of MR shock absorber for vehicles. Int J Veh Des 38(2/3):222–233
16. Bica I, Liu YD, Choi HJ (2013) Physical characteristics of magnetorheological suspensions and their applications. J Ind Eng Chem 19(2):394–406

17. Gordeev BA, Dar'enkov AB, Okhulkov SN, Plekhov AS (2016) Magnetorheological fluids examination for antivibration mounts at impact loads. Procedia Technol 24:406–411
18. Graczykowski C, Pawłowski P (2017) Exact physical model of magnetorheological damper. Appl Math Model 47:400–424
19. Wang DH, Liao WH (2011) Magnetorheological fluid dampers: a review of parametric modeling. Smart Mater Struct 20(2)
20. Lee H-S, Choi S-B (2000) Control and response characteristics of a magneto-rheological fluid damper for passenger vehicles. J Intell Mater Syst Struct 11(1):80–87
21. Lee DY, Choi YT, Wereley NM (2002) Performance analysis of ER/MR impact damper systems using the Herschel-Bulkley model. J Intell Mater Syst Struct 13(7–8):525–531
22. Song HJ, Choi SB, Kim JH, Kim KS (2002) Performance evaluation of ER shock damper subjected to impulse excitation. J Intell Mater Syst Struct 13(10):625–628
23. Choi Y-T, Wereley NM (2005) Biodynamic response mitigation to shock loads using magnetorheological helicopter crew seat suspensions. J Aircr 42(5):1288–1295
24. Wang W, Xia P (2007) Adaptive control of helicopter ground resonance with magnetorheological damper. Chin J Aeronaut 20(6):501–510
25. Bajkowski J, Nachman J, Shillor M, Sofonea M (2008) A model for a magnetorheological damper. Math Comput Model 48(1–2):56–68
26. Zhu S, Wang P, Tian J (2011) Experimental research on aircraft landing gear drop test based on MRF damper. Procedia Eng 15:4712–4717
27. Tu F, Yang Q, He C, Wang L (2012) Experimental study and design on automobile suspension made of magneto-rheological damper. Energy Procedia 16:417–425
28. Kasemi B, Muthalif AGA, Rashid MM, Fathima S (2012) Fuzzy-PID controller for semi-active vibration control using magnetorheological fluid damper. Procedia Eng 41 (Iris):1221–1227

Optimization and Simulation

Enhancement of Line-Based Voltage Stability of Energy System with Thyristor Controlled Series Capacitor Using Cuckoo Search Algorithm

B. Venkateswara Rao, B. Sateesh, R. Uma Maheswari, G. V. Nagesh Kumar and P. V. S. Sobhan

Abstract Preserving stable conditions on encountering with small disturbances under normal or slightly overloaded conditions is termed as voltage stability. Maintaining voltage stability is one of the leading factors for energy system networks. In this paper, new line established voltage stability index entitled fast voltage stability index (FVSI) is proposed for optimal placement of Thyristor Controlled Series Capacitor (TCSC). Optimal tuning of TCSC is obtained using cuckoo search algorithm (CSA) to increase the voltage stability of the energy system established on minimization of total voltage deviation of the system. The CSA is coded in MATLAB and the performance is tested on IEEE 30 bus system with voltage deviation minimization as an objective function. TCSC is a series-connected device in the flexible alternating current transmission system (FACTS) family. It was capable of controlling the power flow through the line and also controls the line-based voltage stability. In this paper, TCSC is merged in CSA-based Power Flow to optimize the total voltage deviation. Results attained by CSA are related to that attained by genetic algorithm (GA) in both without and with TCSC conditions. These results show that CSA produces better results compared to GA for solving optimal tuning of TCSC.

Keywords FACTS device · Cuckoo search algorithm · Optimal tuning · TCSC

B. Venkateswara Rao (✉)
V R Siddhartha Engineering College, Vijayawada, India

B. Sateesh · R. Uma Maheswari
Vignan's Institute of Information Technology, Visakhapatnam, India

G. V. Nagesh Kumar
JNTUA College of Engineering, Pulivendula, India

P. V. S. Sobhan
Vignan University, Vadlamudi, Guntur, India

BBVL. Deepak et al. (eds.), *Innovative Product Design and Intelligent Manufacturing Systems*, Lecture Notes in Mechanical Engineering,
https://doi.org/10.1007/978-981-15-2696-1_61

1 Introduction

Voltage collapse and instability have been measured as main hazards to the current energy system networks caused by their heavily loaded operation. Due to increasing usages of inductive loads, losses in the transmission system enhanced and voltage profile values deviated from the prescribed value which also causes to increase the cost of the real power generation [1]. So for avoiding these problems, proper reactive energy compensation should be done in transmission systems. Proper reimbursement of reactive energy in system recovers the stability of the ac system which is achieved by suitable utilization of lines with installing FACTS devices. Among various series FACTS devices, TCSC is known for its power transferable abilities and stability of the line [2].

This problematic issue has been revealed in several ways in different journals. To locate the size and locality of FACTS devices, M. Saravanan et al. took into account the load ability of the system and formulated in the particle swarm optimization (PSO) algorithm [3]. Similarly, the cuckoo search algorithm was applied for optimal location of Static VAR Compensator (SVC) to improve the performance of the energy system by KhaiPhuc Nguyen et al. [4]. Researchers also enhanced adaptive differential evolution algorithm for an optimal solution [5, 6]. Quadratic programming and outdated optimization approaches like an interior point, L.P., nonlinear programming explain the difficulty in active power reallocation [7]. The disadvantage of the stated techniques is the struggle in acquiring the global minimum owing to several local smallest amounts which happen in them. Heuristic optimization outfits have been inspected as evolutionary and GA, PSO, ant colony optimization, firefly algorithm; GSA and bat search algorithm [8] are used to explain the trouble. An analysis using cuckoo search is scrutinized and pragmatic to IEEE 30 bus system for optimization of voltage deviation and sizing of TCSC specifications.

This paper uses FVSI for insertion of the TCSC at a suitable position. Once the place for mounting TCSC is resolute, its optimal fine-tuning is attained using cuckoo search algorithm. It is instigated on a single objective function in order to acquire the Optimal Power Flow. The objective function comprises total voltage magnitude deviations. Results are figured for cuckoo search algorithm constructed OPF without and with TCSC using MATLAB. Effects achieved using the CSA are then compared with the genetic algorithm (GA).

2 Problem Origination

As the foremost objective of the effort is to decide the optimal site for placement of TCSC in energy system network to minimize the losses and control power flow through the lines. The following performance index is selected.

2.1 Fast Voltage Stability Index

The goal here is to find the best location for placement of TCSC. Fast voltage stability index is a line-based approach which gives idea behind the placement of TCSC in the line between the buses. TCSC is placed among the weakest line in the transmission system to enhance the power flow through lines. FVSI index is calculated by using Eq. 1.

$$\mathrm{FVSI}_{ij} = \frac{4Z^2 Q_j}{V_i^2 X} \tag{1}$$

Z—transmission line impedance, Q_J—receiving end reactive energy, X_{ij}—transmission line reactance, V_j—voltage at sending end.

The value of the index determines the weakest line in the system. This technique is applicable to undefined lines in the energy system. The value of FVSI fluctuates between 0 and 1. If the index is close to 0 at a line, then the line is considered as stable. If the index is near to unity, then the line is near to collapse in the system. The highest value of FVSI indicates the weakest line in the system. And that is the optimal location for TCSC placement in the transmission system.

2.2 Objective Function

To have a stable system, the voltage deviation at all buses must be less. The voltage deviation (VD) [9] is given in (2).

$$F_{\mathrm{VD}} = \min(\mathrm{VD}) = \min\left(\sum_{k=1}^{N} \left(V_k - V_k^{\mathrm{ret}}\right)^2\right) \tag{2}$$

where F_{VD} is the function of voltage deviation. It is the difference between the actual voltage and a reference voltage, which should be minimum for better operation of the energy system. It is calculated for all buses.
V_k—voltage of kth bus, V_k^{ref}—mention voltage of kth bus.

Subject to following limitations

$$\sum_{i=1}^{N} P_{\mathrm{G}i} = \sum_{i=1}^{N} P_{\mathrm{D}i} + P_{\mathrm{L}} \tag{3}$$

i is $1, 2, 3, \ldots, N$ and N = number of buses, $P_{\mathrm{G}i}$ = real power generation, P_{L} = load, $P_{\mathrm{D}i}$ = real power demand.

$$V_{Gi}^{\min} \leq V_{Gi} \leq V_{Gi}^{\max} \tag{4}$$

$$P_{Gi}^{\min} \leq P_{Gi} \leq P_{Gi}^{\max} \tag{5}$$

TCSC Limits

$$X_{TCSC}^{\min} \leq X_{TCSC} \leq X_{TCSC}^{\max} \tag{6}$$

3 Modeling of TCSC

TCSC structure recommended by Vithaythil and team is a way of "speedy correction of network impedance". Separately from governing line power allocation ability, TCSC also boosts a system's stability [9, 10]. A model of TCSC is denoted in Fig. 1. It involves a series of recompensing capacitor shunted by TCR. Thyristor attached in TCSC unit allows in managing the flattering reactance beside specified system deviations. TCSC controls power flow [11–14]. Here reactance of the transmission line is adjusted by TCSC. As seen in Fig. 2, the impedance equations are written as in Eqs. 7 and 8

$$Z_{\text{line}} = R_{\text{line}} + X_{\text{line}} \tag{7}$$

$$X_{ij} = X_{\text{line}} + X_{\text{TCSC}} \tag{8}$$

XTCSC is reactance of TCSC, for escape over compensation range of the TCSC is varied from −0.7 X_{line} to 0.6 X_{line} [15, 16].

The real and imaginary power equations at bus k are:

$$P_i = V_i V_j B_{ij} \sin(\theta_i - \theta_j) \tag{9}$$

$$Q_i = -V_i^2 B_{ii} - V_i V_j B_{ij} \cos(\theta_i - \theta_j) \tag{10}$$

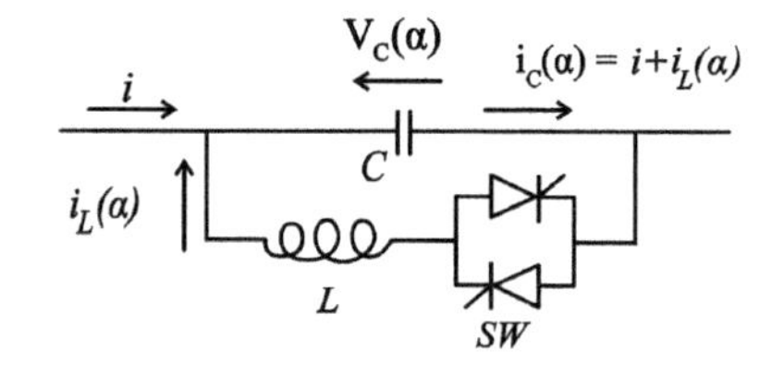

Fig. 1 TCSC classical model

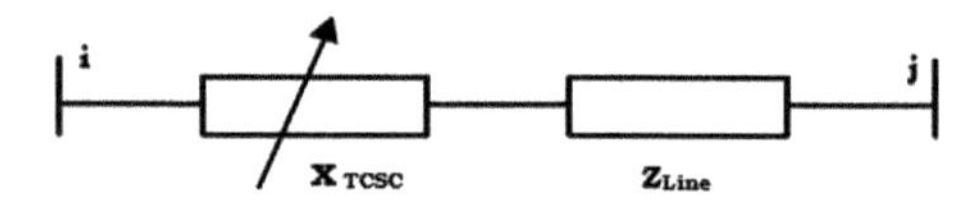

Fig. 2 Schematic diagram of TCSC

4 Cuckoo Search Algorithm

CSA is established by Yang and Deb. It is a population-based optimization algorithmic rule. It works on litter interdependency of selected cuckoo species in the atmosphere. It fantasizes few feminine Cuckoo birds to put their egg into the neighbor's nest. This method ponders on the likelihood of the host bird which moves to leave the Cuckoo egg. Contemporary research on CSA proves higher results as matched to alternative strategies. The virtual code of the CSA is offered in [17–20].

Equation (11) generates a few sets of solutions.

$$x_i^{t+1} = x_i^t + \alpha * \mathrm{Levy}(y) \tag{11}$$

Equation 7 is the stochastic expression of an arbitrary walk. Its subsequent steps are influenced by current position and evolution probability. α is the step size, the product funds entry sensible multiplications. Levy flight offers arbitrary walk and arbitrary step span is pinched from Levy sharing:

$$\mathrm{Levy} \neq t^{-\lambda}, (1 < \lambda < 3) \tag{12}$$

It is recurring to a maximum number of periods is covered. The primary set of nests fluctuates from 15 to 40 but n is 20 and Pa is 0.25 are appropriate values for maximum optimization complications.

5 Results and Analysis

In imperative to establish the show of the CSA in OPF with TCSC, IEEE 30 bus is considered. An OPF program using CSA to minimize full voltage deviation is written using MATLAB without TCSC and additionally stretched with it. A MATLAB program is veiled for system and the reports are presented and examined. Results obtained with CSA were compared with GA. The test system inputs are specified in Table 1. Table 2 shows the input values of GA. Table 3 presents the generator characteristics of the 30 bus system.

Here, bus no 1 is slack bus and bus numbers 2, 5, 8, 11 and 13 are reserved as generator buses, continuing are the load buses. This system has 41 interrelated lines. A MATLAB software is utilized and the results obtained are evaluated. FVSI values for the IEEE 30 bus system short of TCSC are publicized in Table 4.

Table 1 CSA specifications

Factors	Capacity
Nest count	20
Iteration count	100
Finding the rate of foreign eggs	0.25

Table 2 G.A. specifications

Factors	Capacity
Inhabitants size	20
The highest count of generations	100
Crossover segment	0.8
Migration segment	0.2
Migration pause	20

Table 3 Generator individualities in IEEE 30 bus system

Generator bus no	a	b	c	P_G^{min}	P_G^{max}
1	0.00375	2	0	50	300
2	0.0175	1.75	0	20	80
5	0.0625	1	0	15	50
8	0.00834	3.25	0	10	35
11	0.025	3	0	10	30
13	0.025	3	0	12	40

From Table 4, it can be understood that the total FVSI value is maximum for line number 13. So in this study, TCSC is placed at line number 13 to expand the line-based voltage stability. Table 5 designates the size of TCSC apparatus and signifies total FVSI value and total voltage deviation of the system for GA method without and with TCSC and CSA optimization without and with TCSC.

Table 5 shows CSA-based optimization contributes the size of the TCSC is 0.2123 p.u. and placing this TCSC in 13th line total FVSI value for all lines is

Table 4 Total FVSI value for 30 bus system without TCSC

Severity rank	Line number	FVSI value	Severity rank	Line number	FVSI value
1	13	0.334	6	14	0.1316
2	5	0.1875	7	36	0.1231
3	6	0.1868	8	3	0.1168
4	15	0.1436	9	12	0.1056
5	2	0.1364	10	9	0.0612

Table 5 Evaluation of total FVSI value and total voltage deviation for 30 bus system located TCSC at line number 13

	Power flow solution	Total FVSI value of total lines in p.u.	Total voltage deviation for buses in p.u.	Size of the TCSC in p.u.
GAOPF [12]	Without TCSC	2.9823	1.4532	–
	With TCSC	2.086	0.6254	0.2864
CSAOPF	Without TCSC	2.3804	1.3081	–
	With TCSC	1.8275	0.4837	0.2123

compact to 1.8275 p.u. from 2.3804 p.u. in without TCSC state. It specifies that line-based stability has been upgraded. The size of the TCSC in CSA-based optimization is 0.2123 p.u. which is less when associated with GA-based optimization. From this table, it has been experiential that CSA is finer to GA because of its global optimization. Figures 3 and 4 are the convergence characteristics of the voltage deviation using CSA without and with TCSC.

From Fig. 3, it has been detected that the objective function value that is total voltage deviation is optimized with 1.3081 p.u., and it takes nearly 80 iterations to

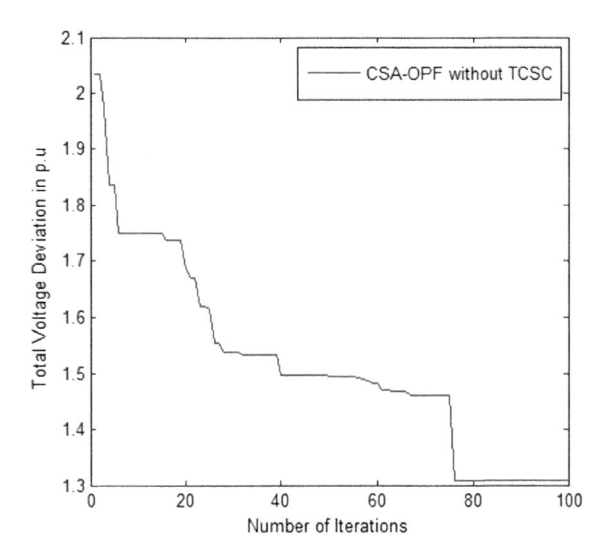

Fig. 3 Convergence individualities of voltage abnormality with CSAOPF without TCSC

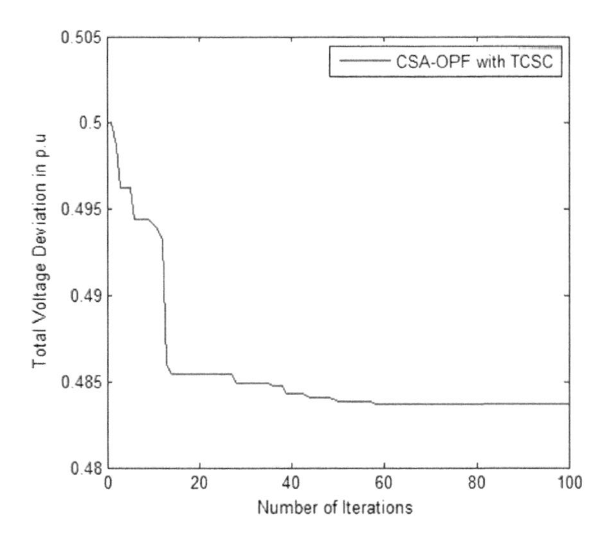

Fig. 4 Convergence individualities of voltage abnormality with CSAOPF with TCSC

converge. From Fig. 4, it is detected that after combining TCSC in cuckoo search algorithm condense the number of iteration to 60 to converge and the objective function value has been optimized to 0.4837 p.u.

Figure 5 demonstrates without and with TCSC, the voltage summary of 30 bus with CSA-based optimization. It specifies that by integrating the TCSC in line 13 associated with bus no 9 and bus no 11 in cuckoo search algorithm advances the voltage contour of the system.

From Fig. 5, it has been observed that bus 30 has the low voltage which is 0.9171 p.u., after incorporating the TCSC in cuckoo search algorithm it is witnessed that voltage at bus 30 increased to 0.9562 p.u. which is happened because of series compensation is done by the TCSC. Figure 6 denotes the FVSI values for all line in 30 bus system. This shows including TCSC in the system, FVSI values have been condensed which indicates that the line-based voltage stability has been upgraded. Figure 6 shows that FVSI value at line 13 is maximum which is 0.334 p.u. after incorporating TCSC, FVSI value at line 13 has been compact to 0.2774 p.u. This designates that after incorporating TCSC, line-based voltage stability has been upgraded.

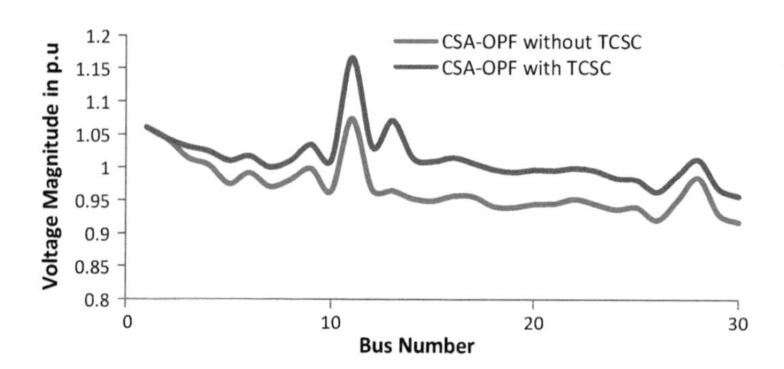

Fig. 5 Comparison of voltage without and with TCSC

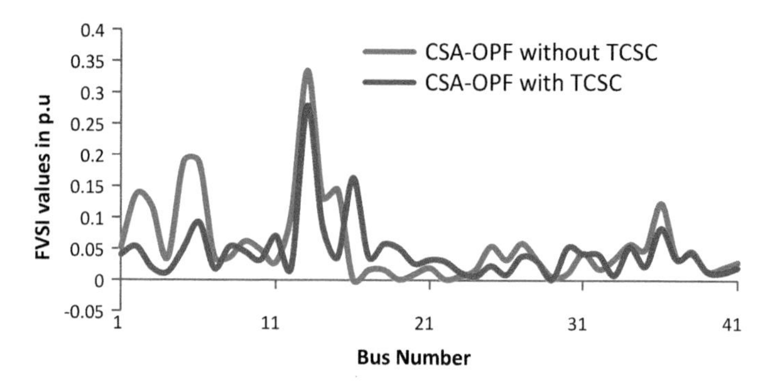

Fig. 6 Comparison of the FVSI values without and with TCSC

6 Conclusion

The research proposes a fast voltage stability index for placement of TCSC and application of CSA to discover the best tuning of TCSC based on minimization of the total voltage deviation. Results achieved are differentiated for CSA and GA. CSA is totally overriding and succeeded for formative optimal tuning of the TCSC device. Affording to case studies, the cuckoo search continuously delivers improved result with advanced performance. Results attained for IEEE 30 bus system by an employed method without and with TCSC are associated and interpretations disclose that total voltage deviation and FVSI values are enhanced with TCSC. Results acquired are helpful and illustrate that TCSC is the greatest active devices that can meaningfully progress the steadiness of the energy system.

References

1. Hingorani NG, Gyugyi L (2000) Understanding FACTS: concepts and technology of flexible AC transmission system. IEEE Press
2. Acha E, Fuerte-Esquivel CR, Ambriz-Perez H, Angeles-Camacho C (2004) FACTS modelling and simulation in power networks. Wiley
3. Saravanan M, Slochanal SMR, Venkatesh P, Abraham JPS (2007) Application of particle swarm optimization technique for optimal location of FACTS devices considering the cost of installation and system loadability. Electr Power Syst Res 77:276–283
4. Nguyen KP, Fujita G, Dieu VN (2016) Cuckoo search algorithm for optimal placement and sizing of static VAR compensator in large scale power systems. JAISCR, 6:59–68
5. Vadivelu KR, Reddy KBN, Marutheeswar GV (2015) Optimal reactive power planning using a new improved differential evaluation incorporating FACTS. J Electr Eng (JEE) 15:246–254
6. Elhameed MA, Elkholy MM (2016) Optimal power flow using cuckoo search considering voltage stability. WSEAS Trans Power Syst 11:18–26
7. Abdel-Moamen MA, Padhy NP (2003) Optimal power flow incorporating FACTS devices-bibliography and survey. In: Proceedings of 2003 IEEE PES transmission and distribution conference and exposition
8. Gerbex S, Cherkaoui R, Germond AJ (2001) Optimal location of multi-type FACTS devices in a power system by means of genetic algorithms. IEEE Trans Power Syst 16:537–544
9. Rao BV, Nagesh Kumar GV (2015) A comparative study of BAT and firefly algorithms for optimal placement and sizing of static VAR compensator for enhancement of voltage stability. Int J Energy Optim Eng (IJEOE) 4(1):68–84
10. Rao BV, Kumar GN (2014) Optimal location of thyristor controlled series capacitor for reduction of transmission line losses using BAT search algorithm. WSEAS Trans Power Syst 9:459–470
11. Singh JG, Singh SN, Srivastava SC (2007) Enhancement of power system security through optimal placement of TCSC and UPFC. In: Power engineering society general meeting, Tampaz, pp 1–6
12. Rao BV, Nagesh Kumar GV (2014) Optimal location of thyristor controlled series capacitor to enhance power transfer capability using firefly algorithm. Electr Power Compon Syst 42(14):1541–1552
13. Yang XS (2010) Firefly algorithm, Levy flights, and global optimization. In: Research and development in intelligent systems XXVI, Springer, London, UK, pp 209–218

14. Rao BV, Nagesh Kumar GV (2014) Sensitivity analysis based optimal location and tuning of static VAR compensator using firefly algorithm to enhance power system security. Indian J Sci Technol 7(8):1201–1210
15. Deepak Chowdary D, Nagesh Kumar GV (2008) DVR with sliding mode control to alleviate voltage sags on distribution system for three-phase short circuit fault. In: Proceedings of third IEEE international conference on industrial and information systems, IIT, Kharagpur, India, 08–10 Dec
16. Sravan Kumar B, Suryakalavathi M, Nagesh Kumar GV (2016) Optimal location of thyristor controlled series capacitor to improve power system performance using line based composite index. In: Proceedings of 6th IEEE international conference on power systems (ICPS 2016), Indian Institute of Technology Delhi, 4–6 March 2016
17. Deepak Chowdary D, Nagesh Kumar GV (2010) Mitigation of voltage sags in a distribution system due to three phase to ground fault using DVR. Indian J Eng Mater Sci 17(2):113–122
18. Yang X-S (2012) Flower pollination algorithm for global optimization. In: Unconventional computation and natural computation 2012, Lecture Notes in Computer Science, vol 7445, pp 240–249
19. Yang XS, Deb S (2010) Engineering optimization by cuckoo search. Int J Math Model Numer Optim 1(4):330–343
20. Nguyen KP, Fujita G, Dieu VN (2016) Cuckoo search algorithm for optimal placement and sizing of static var compensator in large-scale power systems. JAISCR 6(2):59–68

Follicle Detection in Digital Ultrasound Images Using BEMD and Adaptive Clustering Algorithms

M. Jayanthi Rao and R. Kiran Kumar

Abstract Ultrasound imaging is one of the techniques used to study inside the human body with images generated using high-frequency sounds waves. The applications of ultrasound images include an examination of human body parts such as kidney, liver, heart, and ovaries. This paper mainly concentrates on ultrasound images of ovaries. Monitoring of follicle is important in human reproduction. This paper presents a method for follicle detection in ultrasound images using adaptive data clustering algorithms. The main requirements for any clustering algorithm are the number of clusters *K*. Estimating the value of *K* is a difficult task for given data. This paper presents an adaptive data clustering algorithm which generates accurate segmentation results with simple operation and avoids the interactive input *K* (number of clusters) value for segmentation of ultrasound image. The qualitative and quantitative results show that adaptive data clustering algorithms are more efficient than normal data clustering algorithms in segmenting the ultrasound image. After segmentation, using the region properties of the image, the follicles in the ovary image are identified. The proposed algorithm is tested on sample ultrasound images of ovaries for identification of follicles and with the region properties, the ovaries are classified into three categories, normal ovary, cystic ovary, and poly-cystic ovary with its properties. The experiment results are compared qualitatively with inferences drawn by medical expert manually and this data can be used to classify the ovary images.

Keywords Ovarian classification · Image processing · Histogram equalization · Bi-dimensional empirical mode decomposition

---

M. Jayanthi Rao (✉) · R. Kiran Kumar
Department of Computer Science, Krishna University, Machilipatnam, India

BBVL. Deepak et al. (eds.), *Innovative Product Design and Intelligent Manufacturing Systems*, Lecture Notes in Mechanical Engineering,
https://doi.org/10.1007/978-981-15-2696-1_62

1 Introduction

Polycystic ovarian syndrome (PCOS) is one of the common disorders seen in females of reproductive age. The disorder is characterized by the formation of many follicular cysts in the ovary. The main cause of this disorder in females is due to menstrual problems, hirsutism, endocrine abnormalities, acne, obesity, etc. [1]. The detection of the ovarian follicle is done using ultrasound images of ovaries. Object recognition in an ultrasound image is a challenging task which includes the detection of follicles in the ovary, growth of the foetus, monitoring of proper development of the foetus, and presence of tumour [2]. Nowadays the diagnosis performed by doctors is to manually count the number of follicular cysts in the ovary, which is used to judge whether PCOS exists or not. This manual counting may lead to problems of variability, reproducibility, and low efficiency. Automating this mechanism will resolve these problems. In literature, less work is done in automating PSOC diagnosis [3, 4]. In this paper, an algorithm for identification of follicles in ovarian images is presented using adaptive data clustering algorithms. The experiment results are compared with inferences drawn by medical expert manually and identify the ovary images more accurately than manual segmentation. The ovary images used in this paper are taken from www.radiologyassist.nl and www.e-ultrasonography.org.

In the literature, the authors have employed various techniques for ultrasound image processing, as shown below:

1. Preprocessing—Gaussian low pass filter—homogeneous region growing mean filter (HRGMF)—contourlet transform
2. Segmentation—optimal thresholding—edge-based method—watershed transform—scanline thresholding—active contour method
3. Feature extraction—geometric features—texture features
4. Classification—3σ interval-based classifier.—K-NN classifier—linear discriminant classifier—fuzzy classifier—SVM classifier (Figs. 1 and 2).

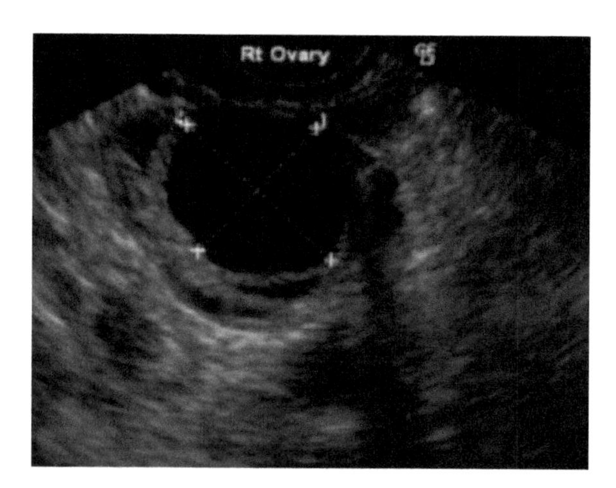

Fig. 1 Normal ovary

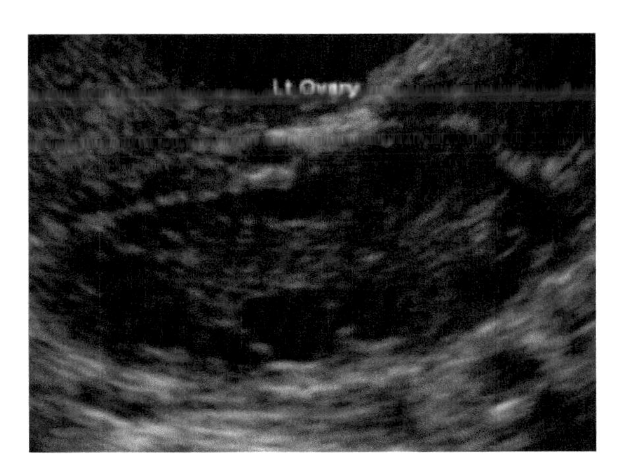

Fig. 2 PCOS affected ovary

2 Noise Removal Using Bi-dimensional Empirical Mode Decomposition

If the ultrasound image contains noise, the quality of the edges extracted from the image will be poor. This edge information is the primary source for the extraction of regions from an ultrasound image. The noise in the ultrasound image is removed using a nonlinear filter developed with bi-dimensional empirical decomposition and wavelets. The shifting process used to obtain IMFs on a 2-D signal (image) using BEMD [5–7]. The BEMD-DWT filtering process is as follows

(a) Apply 2-D EMD to generate IMFS.
(b) The first few intrinsic mode functions are high-frequency components. These IMFs are denoised with DWT. These denoised IMFs are represented using DNIMFs.
(c) The denoised image RI is reconstructed by

$$\mathrm{RI} = \mathrm{DNIMF1} + \sum_{i=2}^{k} \mathrm{IMF}_i \qquad (1)$$

The flow diagram of BEMD-DWT filtering is shown in Fig. 3.

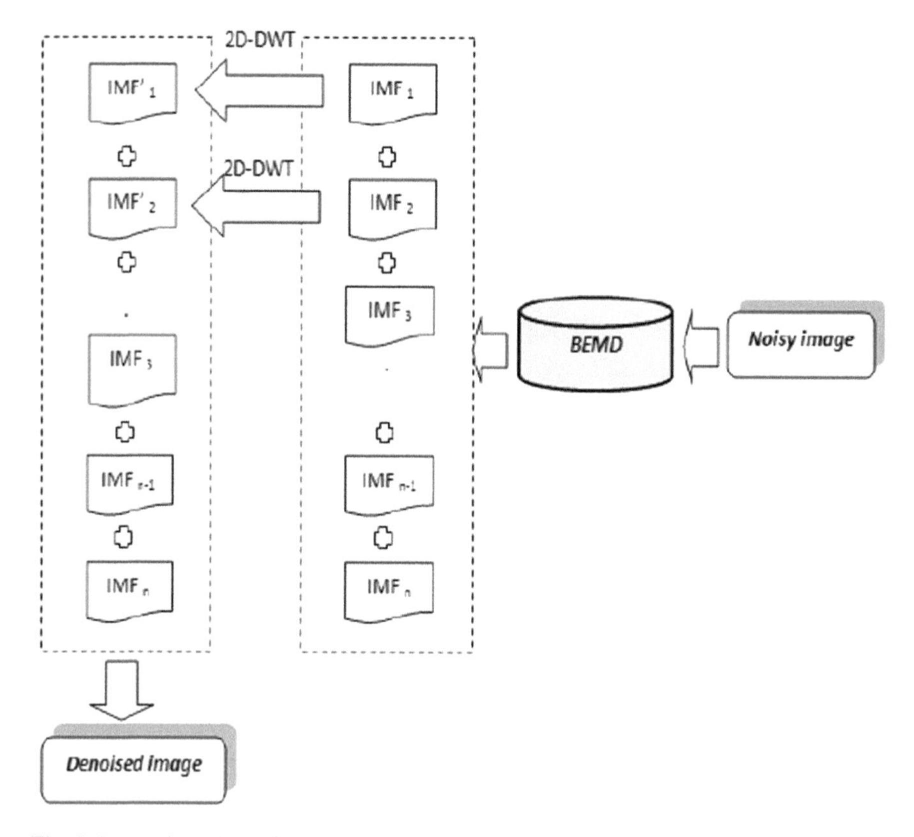

Fig. 3 Image denoising using BEMD + WAVELET

3 Clustering Algorithms for Segmentation of Ultrasound Image

Fuzzy *C*-Means Algorithm [8] is shown in Fig. 4. The fuzzifier m determines the level of cluster fuzziness. A large m results in smaller membership values, and hence fuzzier clusters. In the limit $m = 1$, the memberships, converge to 0 or 1, which implies a crisp partitioning. In the absence of experimentation or domain knowledge m is commonly set to 2.

4 Adaptive Data Clustering Algorithms

In adaptive data clustering algorithms, the required K-value is estimated using the maximum connected domain algorithm [9, 10]. Each time, the K-value is compared to the connected domain. If not equal, increment the K-value until it matches.

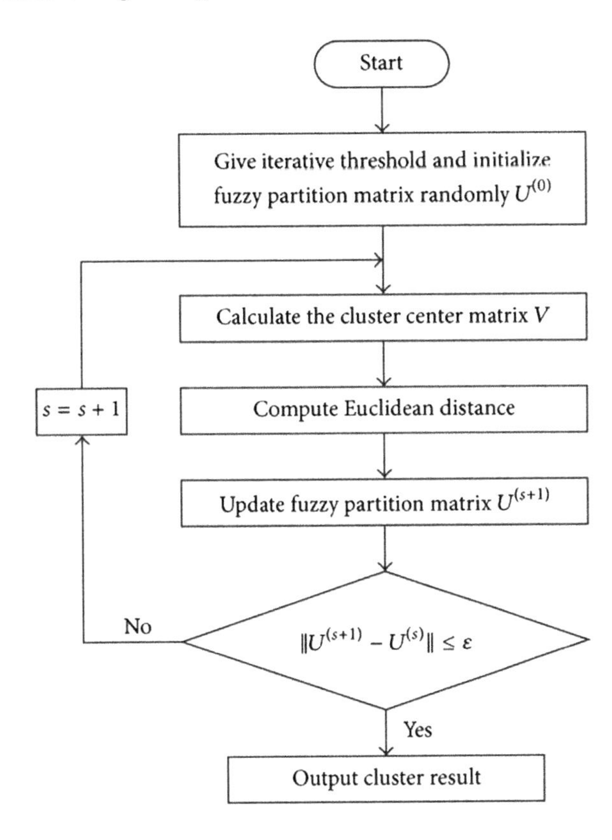

Fig. 4 FCM algorithm

The adaptive k-means clustering algorithm is shown in Fig. 5. The same algorithm can be extended to K-medoids and FCM.

5 Follicle Identification in Ovary Ultrasound Image

The main steps of the identification of follicles in the ovary image are:

1. The ultrasound image is processed by BEMD-based contrast enhancement algorithm.
2. Adaptive clustering algorithm to segment the contrast-enhanced ovary image.
3. Convert the segmented image into binary image and edges are extracted using BEMD by taking the first IMF and applying the thresholds on first IMF, the edge information is extracted.
4. Use infill function to fill the holes. These holes denote follicles in ovary image.
5. Using the region properties, we can extract the features of follicles such as major axis length, minor axis length, area, and centroids. With these features, ovaries are classified into the normal ovary, cystic ovary, and polycystic ovary.

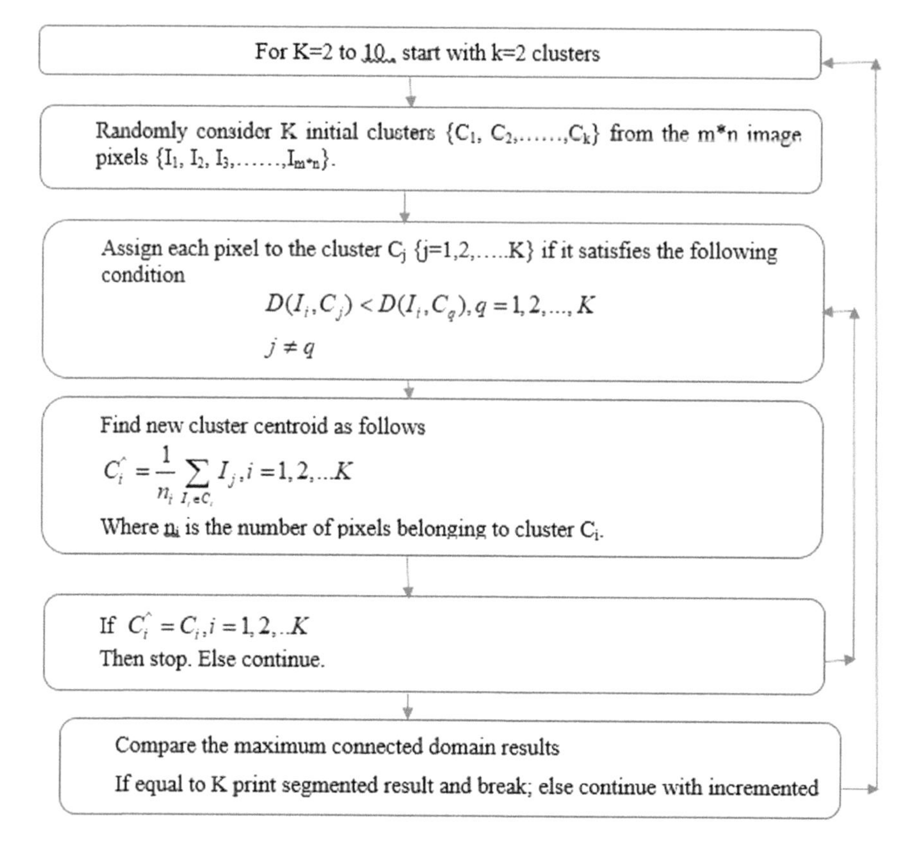

Fig. 5 Adaptive data clustering algorithms

6 Experimental Results

In this section, the proposed method is used to detect the follicles in ovary ultrasound images. Ovary images are obtained from the publicly available websites www.radiologyinfo.com [11, 12], www.ovaryresearch.com [13]. The image 1 and image 2 is taken from dataset D1 with size 256 * 256. The qualitative analysis of the proposed method is shown in Fig. 6 compared with manual segmentation. Table 1 shows the quantitative evaluations of adaptive clustering algorithms on ovary image 1 using MSE [8].

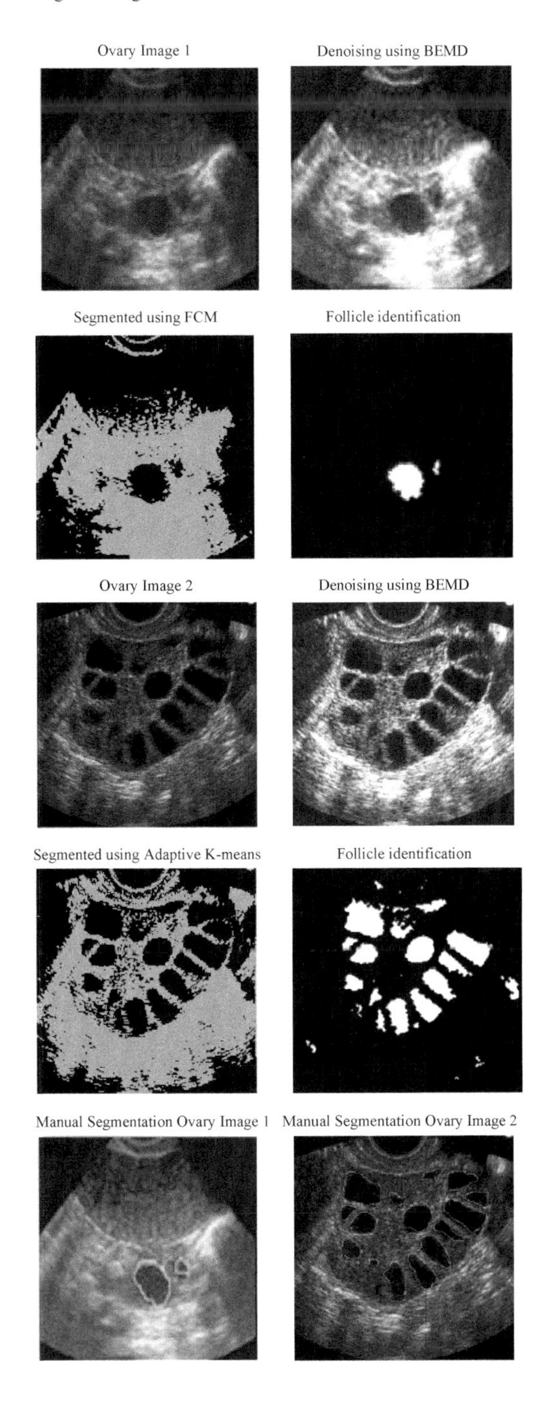

Fig. 6 Follicle detection in the ovary ultrasound image

Table 1 MSE values

Method/ovary image 1	Normal clustering	Adaptive clustering
K-means	95.8	94.5
K-medoids	94.2	93.2
Fuzzy *C*-means	90.2	89.6

7 Conclusion

The information about the status of the female reproductive system is important for problems related to fertility and family planning. The ultrasound imaging is an effective tool in infertility treatment. Monitoring the follicles in ultrasound images in terms of number, size, shape, and position is important in the human reproductive system. In this paper, a methodology of identification of follicles in ultrasound images is presented. The proposed method uses bi-dimensional empirical mode decomposition procedure to generate IMFs from the ultrasound image. Using wavelets for the first few IMFS and then combining IMFs will generate a denoised image. After enhancement, the image is segmented using adaptive *k*-means algorithm. By using region properties, the follicle regions in the ultrasound image are identified. This proposed method is compared qualitatively with manual segmentation. In the future, this follicle information is used to classify the ovary images into three classes, normal ovary, cystic ovary, and polycystic ovary using the region properties of identified follicles.

References

1. Deng Y, Wang Y, Chen P (2008) Automated detection of polycystic ovary syndrome from ultrasound images. In: 30th annual international IEEE EMBS conference, IEEE 2008, pp 4772–4776
2. Kiruthika V, Ramya MM (2014) Automated segmentation of ovarian follicle using *K*-means clustering. In: 2014 fifth international conference on signal and image processing, pp 137–142
3. Mehrotra P, Chakraborty C, Ghoshdastidar B (2011) Automated ovarian follicle recognition for polycystic ovary syndrome. In: Proceedings of international conference on image information processing, ICIIP, 2011
4. Deshpande SS, Wakankar A (2014) Automated detection of polycystic ovarian syndrome using follicle recognition. In: Proceedings of international conference on advanced communication control and computing technologies, IEEE 2014, pp 1341–1347
5. Saichandana B, Harikiran J, Srinivas K, KiranKumar R (2016) Application of BEMD and hierarchical image fusion in hyperspectral image classification. Int J Comput Sci Inf Secur 14 (5):437–445
6. Harikiran J et al (2015) Multiple feature fuzzy *C*-means clustering algorithm for segmentation of microarray image. IAES Int J Electr Comput Eng 5(5):1045–1053
7. Harikiran J et al (2012) Fuzzy *C*-means with Bi-dimensional empirical mode decomposition for segmentation of microarray image. Int J Comput Sci Issues 9(5):273–279

8. Harikiran J, Lakshmi PV, Kiran Kumar R (2014) Fast clustering algorithms for segmentation of microarray images. Int J Sci Eng Res 5(10):569–574
9. Zuo W (2006) Research on connected region extraction algorithms [J]. Comp. Appl. Softw. 23(1):97–98
10. Harikiran J, Lakshmi PV, Kiran Kumar R (2015) Multiple feature fuzzy C-means clustering algorithm for segmentation of microarray image. IAES Int J Electr Comput Eng 5(5): 1045–1053
11. Hiremath PS, Tegnoor JR Follicle detection and ovarian classification in digital ultrasound images of ovaries. INTECH. http://dx.doi.org/10.5772/56518
12. Rihina S et al (2013) Automated algorithm for ovarian cysts detection in ultrasonogram. In: Proceedings of international conference on advances in biomedical engineering IEEE, 2013
13. Saichandana B, Srinivas K, Kiran Kumar R (2014) Clustering algorithm combined with Hill climbing for classification of remote sensing image. Int J Electr Comput Eng 4(6):923–930

Optimal Allocation of Solar DGs in Distribution Network

Purnachandra Rao Thota, Srikanth Khandavalli, Lakshmi Narayana Vital Muktevi and Hari Vamsi Valluri

Abstract Solar photovoltaic (PV) systems are increasing in the power system day by day. These micropower sources are located in the distribution system at low voltage levels. These are also called type 1 distributed generations (DGs) as they generally supply real power to the system. Optimal location and sizing of the DGs are important as it will influence the power loss and voltage profile in the distribution network. The optimal location is found using loss sensitivity factor and sensitivity indices. The optimal size is found using successive sizing method and sensitivity factor method. Therefore, the present work develops an algorithm to solve the optimal allocation of DGs for IEEE 33 radial distribution test system. The power losses and voltage profile of the distribution network are found using forward and backward sweep methods. The power loss and voltage profile of the test system are calculated and presented without and with DGs.

Keywords Distributed generators (DG) · Photovoltaic (PV) cells · Successive sizing method (SSM) · Sensitive factor method (SFM) · Microgrid

1 Introduction

The demand for electricity is increasing tremendously in India [1]. With the rising demand, a competitive environment has been created in the sale of power. The conventional vertically integrated utility has been changed and is unbundled to one or more generations, transmission and distribution companies [2]. Apart from this, the distribution companies utilize the power supplied by the small generators. Introduction of microgrids into the system is necessary to meet the increasing demand. This distributed generators (DG) comprises of both the renewable and conventional energy sources. With the integration of various types of sources into the distribution systems, the primary distribution substation will not be the only

P. R. Thota · S. Khandavalli (✉) · L. N. V. Muktevi · H. V. Valluri
EEE Department, VR Siddhartha Engineering College, Kanuru,
Vijayawada, AP 520007, India

BBVL. Deepak et al. (eds.), *Innovative Product Design and Intelligent Manufacturing Systems*, Lecture Notes in Mechanical Engineering,
https://doi.org/10.1007/978-981-15-2696-1_63

source. Hence, the optimal choice of size and location of DGs should be considered for technical and economic benefits like power loss reduction and voltage profile improvement. Rao et al. [1] described a new approach for the allocation of DGs in the two modes of operation. The optimal location is determined by loss sensitivity factors and a new technique successive sizing method has been used for the optimal choice of DGs. The optimal placement and sizing of the DGs will depend on the type of DG also [7]. Hung et al. [2] have derived analytical expressions for four types of DG. The results obtained by the analytical approach are validated with the exhaustive load flow method.

The power factors of DG injecting active and reactive power play a vital role in reducing the losses has presented in [3] with improved analytical expressions. LSF method for injecting real power type of DG is also presented. Rambabu et al. [5] proposed the optimal placement of DG considering power stability index to increase loading capacity by injecting real power. The optimal placement of DG with sensitivity factors considering voltage stability and loss indices is described in [5]. In the present work, an attempt is made to identify the optimal location of DGs which can improve the power stability and line stability considering sensitivity factors. The optimal sizing is done using successive sizing method and sensitivity factor method. Mehta et al. [6] analyzed the selection of the best type of DGs and their penetration for voltage profile improvement and stability.

2 Distributed Generation and Microgrid

Distribution generators are defined as the small-sized generators operated in parallel or in an isolated connection with electric distribution networks [1]. Power electronic-based converters are used for the interface of fuel cells and solar cells into the grid. The converters provide real power to local loads and deliver reactive power to stabilize load voltages. DG is used for industrial, commercial and residential loads to supply power. Distribution generation can inject active power or reactive power based on the availability of resources.

Microgrids are small-scale, low voltage, networks to provide electrical loads for residential, commercial, agriculture and industrial loads [3]. Microgrid is basically an active distribution network as it has small generating sources at the load centers [4]. The DGs are categorized into different types based on the type of power injection capability [1].

Type 1 DG: Photovoltaic (PV) cells, microturbines that are tied to the grid via converters and inverters are called type 1 DGs. They inject real power only to the grid. Most of the DGs are based on PV technology. Solar PV cell converts solar power to electrical power in DC. This DC will be converted to AC at the system's frequency and voltage level using PV inverter. This inverter should also synchronize the solar PV system to the distribution system. The PV system can inject real power to the system. These types of DGs are called type-1 DGs. The solar PV

inverter interconnects the DC system to the AC grid to maintain the power as constant.
Type 2 DG: The DG delivering reactive power comes under this category. The synchronous compensators that inject reactive power such as gas turbines are examples of this type.
Type 3 DG: It injects both active and reactive power. Synchronous generators which inject real and imaginary power come under this category.
Type 4 DG: It is capable of injecting real power and consuming reactive power. Induction generators coupled with windmills inject active power and consumes reactive power are categorized under this type of DG.

As there is a rapid increase in solar power plants which are used as DGs in the distribution network. The present work considers a distribution system with solar PV DGs and tries to solve the problem of optimal allocation of type 1 DGs.

3 Problem Formulation

The objective function to minimize the total real power loss given by the exact loss formula while meeting the voltage constraints [1, 7]

$$P_L = \sum_{i=1}^{n}\sum_{j=1}^{n} A_{ij}(P_iP_j + Q_iQ_j) + B_{ij}(Q_iP_j - Q_jP_i)$$

where

$A_{ij} = \frac{R_{ij}}{V_iV_j}\cos(\delta_i - \delta_j)$ and $B_{ij} = \frac{R_{ij}}{V_iV_j}\sin(\delta_i - \delta_j)$.
P_i and Q_i are the active and reactive powers at ith bus, respectively.
P_j and Q_j are the active and reactive powers at jth bus, respectively. $V_i\angle\delta_i$ is the complex voltage at the ith bus.

$R_{ij} + jX_{ij} = Z_{ij}$ is the ijth element of $[Z_{bus}]$ the impedance matrix.
'n' is the number of buses.
Constraints: $V_{min} < V < V_{max}$.

Type 1 DG

The optimal size of DG at each bus 'i' for the minimum loss is [2].

$$P_{DGi} = P_{Di} - \frac{1}{A_{ii}} \sum_{\substack{j=1 \\ j \neq i}}^{n} (A_{ij}P_j - B_{ij}Q_j)$$

3.1 Sensitivity Factors

The candidate nodes for the placement of DGs are determined by using the sensitivity factors.

Loss sensitivity factors [1]:

Loss sensitivity factor gives the change in power loss with change in power injections at a bus.

Loss sensitivity factor method has been extensively used to solve the capacitor allocation problem

$$\mathrm{LSF1} = \frac{\partial P_{\mathrm{lineloss}}(j)}{\partial P_{\mathrm{eff}}} = \frac{2 * P_{\mathrm{eff}}(j) * R_{ij}}{V_j^2}$$

$$\mathrm{LSF2} = \frac{\partial P_{\mathrm{lineloss}}(j)}{\partial Q_{\mathrm{eff}}} = \frac{2 * Q_{\mathrm{eff}}(j) * R_{ij}}{V_j^2}$$

where

P_{eff}(j) = Total effective active power supplied beyond the node 'j'.

$Q_{\mathrm{eff}}(j)$ = Total effective reactive power supplied beyond node 'j'.

Power stability index (PSI) [8]:

It is derived from finding the most optimum site for DGs by calculating the most critical bus in the system that can affect system voltage stability when the load increases beyond a certain limit.

$$\mathrm{PSI}_{ij} = \frac{4P_jR_{ij}}{[|V_i|\cos(\theta - \delta)]^2}$$

where

$$\delta = \delta_i - \delta_j$$

Fast voltage stability index (FVSI) [8]:

A fast voltage stability index is developed for the placement of DG to determine the most critical bus in the system. The index value lies in between zero to unity for the stable condition.

$$\mathrm{FVSI}_{ij} = \frac{4Z_{ij}^2Q_j}{V_i^2X_{ij}}$$

Line stability index (LSI) [8]:

Line stability index is used as one of the stability indices. The line stability index for a line should be always less than unity.

$$\mathrm{LSI}_{ij} = \frac{4Q_j X_{ij}}{[|V_i| \sin(\theta - \delta)]^2}$$

3.2 Methodology

The optimal sizing and location of the DGs in the distribution system are obtained by successive sizing method (SSM) and sensitive factor method (SFM) [1].

Multiple DGs are placed considering the sensitivity factors. The following steps give the procedure for obtaining the optimal location of DG [9]

1. Sensitivity factors are calculated for all buses for the base case without installing DG.
2. Give weight of 20% for each sensitivity factor.
3. The buses are arranged in descending order of the obtained values of sensitivity factors to form a priority list.
4. The highest-ranked buses in the list are preferred for DG placements.

Successive Sizing Method (SSM)

The sizing of DG is necessary in order to have reduced losses. Inappropriate location and the size of DG lead to an increase in losses. The algorithm given below explains the procedure for optimal sitting and sizing of DG using successive sizing method.

1. Load flow analysis is performed for the base case without any DG units and losses are determined.
2. Select the type of DG to be sited into the system.
3. The top-ranked bus from the priority list is selected, and then the size of the DG is determined.
4. The computed size of DG is placed at the optimal location.
5. Again load flow analysis is done to find the losses and size of the DG.
6. The above steps are repeated until the required number of DG units are placed at the desired location.
7. The sizes of DGs are updated iteratively while considering the voltage constraints and minimizing distribution losses.

Sensitivity Factor Method (SFM)

The algorithm for optimal placement and sizing of DG by the sensitivity factor method is described below. This method gives the optimum size of DG by changing the obtained DG value of small steps from the analytical expressions.

1. Run load flow for the base case using backward and forward sweep method without the inclusion of DG units and find losses.
2. Decide the type of DG and number of DG units to be placed in the system.
3. Select the topmost priority bus for the placement of DG.
4. Calculate the optimal size of DG.
5. Now change this size of the DG in small steps and run the load flow to calculate the loss. Select and store the optimal size of the DG that gives the minimum loss.
6. Repeat the above steps for the presumed number of DG units.
7. Stop if the new iteration loss is greater than the previous iteration loss and store the previously obtained value.

4 Results and Discussions

The standard IEEE 33 bus radial distribution system [RDS] [1, 10] is used as the test system in this paper. The load flow of RDS without any DG units has been done initially and the losses are obtained as 202.7 kW.

Optimal Placement and Sizing of Multiple DG

The optimal location determined by using sensitivity factors and optimal sizing determined by using successive sizing method and sensitivity factor method. Ranking of buses based on the sensitivity index is done, and is given in Table 1 below.

From Table 1 shown above, it is observed that the DGs are to be placed at those buses which are ranked top.

Table 1 Priority list for placement of multiple DG

Rank	Bus	Index	Rank	Bus	Index	Rank	Bus	Index
1	30	0.0116	12	4	0.0047	23	2	0.0016
2	24	0.01	13	5	0.0042	24	18	0.0016
3	6	0.0088	14	7	0.004	25	12	0.0015
4	25	0.0086	15	9	0.0039	26	16	0.0014
5	3	0.0076	16	20	0.0039	27	26	0.0013
6	28	0.0062	17	10	0.0038	28	15	0.0011
7	8	0.0061	18	14	0.003	29	21	0.001
8	29	0.0061	19	17	0.0027	30	11	0.0008
9	31	0.0051	20	23	0.0024	31	33	0.0009
10	32	0.0051	21	22	0.0017	32	19	0.0005
11	13	0.0049	22	27	0.0017	33	1	0

Successive Sizing Method

The results obtained for the placement of three DG cases by successive sizing method of type 1 DG has been tabulated in Table 2.

Figure 1 shows the voltage profile of the IEEE 33 bus radial distribution system with type 1 DG. It shows that with the placement of DG the voltage profile has been improved.

With the placement of three DGs of type 1 DG at three different locations the losses are reduced to 89.4 kW with a loss reduction of 55.89% can be inferred from Fig. 2.

Sensitivity Factor Method

Based on the simulation results obtained Table 3 represents the optimal sizes for the minimum loss by sensitivity factor method for type 1 DG placement.

Figure 3 shows the voltage profile for the type 1 DG placement by the sensitivity factor method. Voltage has been improved at all the buses with three DG placement than that of other cases.

The losses are 89.4 kW with three DGs placement and the loss reduction is 59.89% which can be seen in Fig. 4.

Comparison of Successive Sizing Method and Sensitivity Factor Method

Table 4 shows the comparison between the optimal placements of DGs. The losses are more reduced with the optimal placement of three DGs.

The optimal sizing of DGs at the optimal locations is determined by using the successive sizing method and sensitivity factor method for microgrids. The type 1 DG of injecting real power is installed. After injecting it the losses are reduced and the voltage profile is improved as seen in Fig. 5.

Table 2 Type 1 DG placement by successive sizing method

Case	Installed DG size (MW)			Total size (MW)	Power loss (KW)	Loss reduction (%)
	Bus 30	Bus 24	Bus 6			
No DG	–	–	–	–	202.7	
1 DG	1.468	–	–	1.468	117.8	41.89
2 DG	1.3399	1.2268	–	2.5667	98.8	51.25
3 DG	0.5298	0.7488	1.1203	2.3989	89.4	55.89

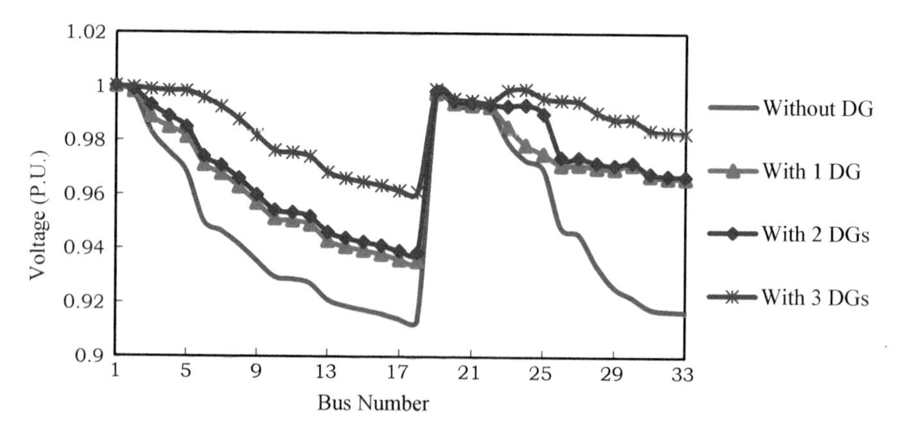

Fig. 1 Voltage profile of type 1 DG placement by SSM

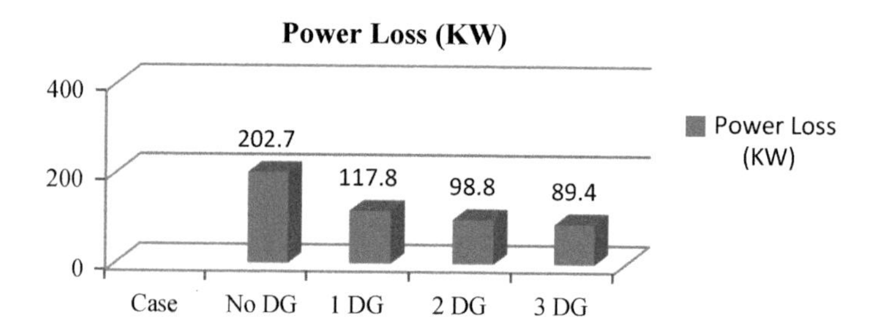

Fig. 2 Power loss of type 1 DG placement by SSM

Table 3 Type 1 DG placement by sensitivity factor method

Case	Installed DG size (MW)			Total size (MW)	Power loss (KW)	Loss reduction (%)
	Bus 30	Bus 24	Bus 6			
No DG	–	–	–	–	202.7	–
1 DG	1.468	–	–	1.468	117.8	41.89
2 DG	1.288	1.4461	–	2.7541	99.5	50.91
3 DG	0.568	0.7461	1.5835	2.8976	81.3	59.89

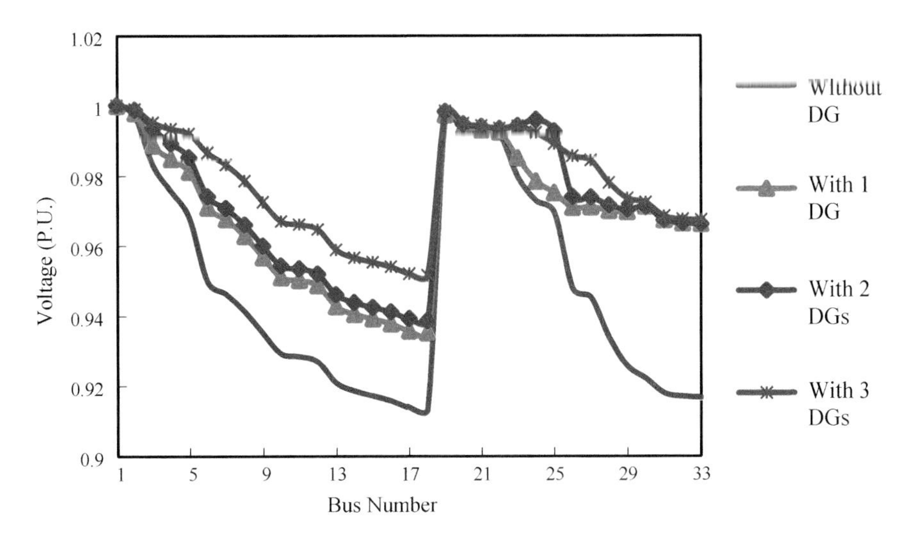

Fig. 3 Voltage profile of type 1 DG placement by SFM

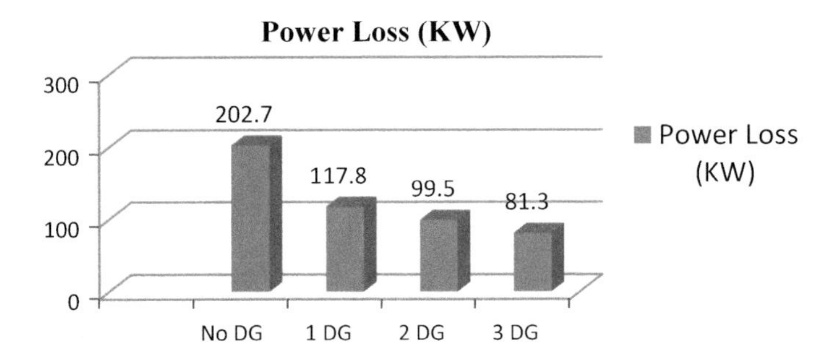

Fig. 4 Power loss of type 1 DG placement by SFM

Table 4 Comparison of SSM and SFM for type 1 DG placement

Case	Approach	Installed DG size (MW)			Total size (MW)	Power loss (kW)	Loss reduction (%)
		Bus 30	Bus 24	Bus 6			
No DG	–	–	–	–	–	202.7	–
1 DG	SFM	1.468	–	–	1.468	117.8	41.89
2 DG	SSM	1.3399	1.2268	–	2.5667	98.8	51.25
	SFM	1.288	1.4461	–	2.7541	99.5	50.91
3 DG	SSM	0.5298	0.7488	1.1203	2.3989	89.4	55.89
	SFM	0.568	0.7461	1.5835	2.8976	81.3	59.89

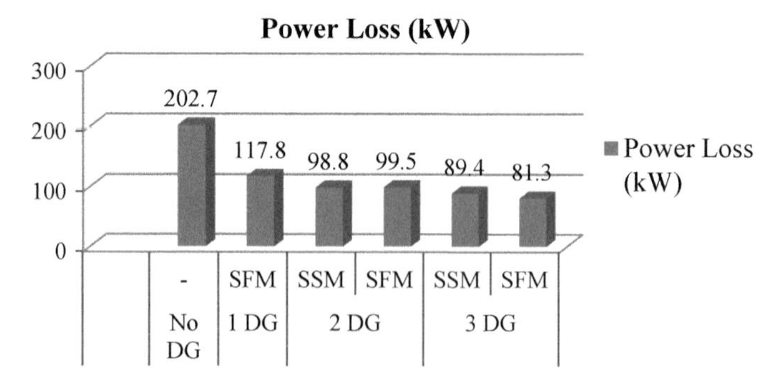

Fig. 5 Power loss of type 1 DG placement both SSM and SFM

5 Conclusion

Optimal allocation of DGs in the distribution system is important to improve the voltage profile and reduce the power loss of the distribution network. The present work solves the optimal allocation of type 1 DGs in IEEE 33 RDS with multiple DGs using SSM and SFM. The optimal results obtained by these methods are presented in crisp tabular forms. The tabular forms clearly show that the power loss is drastically reduced and the voltage profile is significantly improved. The power loss is reduced by 59% for the test system if the DGs are allocated using SFM method. The problem of multiple types 1 DG allocation for IEEE 33 RDS is solved using SSM and SFM methods. The same methodology can also be applied to another type of DGs for future works.

References

1. Thota PR, Venkata Kirthiga M (2012) Optimal siting & sizing of distributed generators in micro-grids. In: Proceedings of annual IEEE India conference in 2012, pp 730–735
2. Hung DQ, Mithulananthan N, Bansal RC (2010) Analytical expressions for DG allocation in primary distribution networks. IEEE Trans Energy Convers 25(3):814–820
3. Hung DQ, Mithulananthan N (2013) Multiple distributed generator placement in primary distribution networks for loss reduction. IEEE Trans Ind Electron 60(4):1700–1708
4. Gupta P, Pandit M, Kothari DP (2014) A review on optimal sizing and siting of distribution generation system. In: Proceedings of 6th IEEE power India international conference, 2014
5. Rambabu T, Venkata Prasad P (2014) Optimal placement and sizing of DG based on power stability index in radial distribution system. In: Proceedings of IEEE international conference on smart electric grid, 2014
6. Mehta P, Bhatt P, Pandya V (2018) Optimal selection of distributed generating units and its placement for voltage stability enhancement and energy loss minimization. Ain Shams Eng J 9:187–201
7. Kansal S, Kumar V, Tyagi B (2016) A hybrid approach for optimal placement of multiple DGs of multiple types in distribution networks. Electr Power Energy Syst 75:226–23

8. Parizad A, Khazali A, Kalantar M (2010) Optimal placement of distributed generation with sensitivity factors considering voltage stability and losses indices. In: Proceedings of 18th Iranian conference on electrical engineering, 2010
9. Prakash DB, Lakshminarayana C (2016) Multiple DG placements in distribution system for power loss reduction using PSO algorithm. In: The technology of global colloquium in recent advancement and effectual researches in engineering, science and technology, vol 25, pp 785–792
10. Murthy VVSN, Kumar A (2013) Comparison of optimal DG allocation methods in radial distribution systems based on sensitivity approaches. Electr Power Energy Syst 53:450–467

Hybridization of Particle Swarm Optimization with Firefly Algorithm for Multi-objective Optimal Reactive Power Dispatch

Manasvi Kunapareddy and Bathina Venkateswara Rao

Abstract Reactive power management is very crucial for stable operation of the system. The ultimate aim of reactive power dispatch (RPD) is to set the control variables to its optimal values to minimize the objective function of real power losses in lines and voltage deviation satisfying all the equality and inequality constraints. The multi-objective function is also proposed to solve both the objective functions simultaneously. This paper presents hybridization of two optimization techniques, one is particle swarm optimization (PSO) and the other is firefly algorithm (FA) represented as hybridization of particle swarm optimization with firefly algorithm (HPSOFA), which is used to yield a better result. This hybridization is carried out in MATLAB for IEEE 14 and 30 bus systems.

Keywords Firefly algorithm · HPSOFA · Multi-objective · PSO · Real power losses · Voltage deviation

1 Introduction

Optimal reactive power dispatch has been drawing the attention of many researchers as it ensures the economic power system operation and voltage security. This reactive power optimization mainly works on the downfall of real power and also improving the voltage profile. The control variables for RPD are reactive power sources, transformer tap settings, and capacitor bank output. These variables are tuned by satisfying their equality and inequality constraints [1–4].

Alam and De [5] applied hybrid loop genetic-based algorithm for the dispatch of the reactive power, achieving fast convergence and accurate result under less computational time which was tested on IEEE 14 bus system. K. Lenin, B. Ravindhranath Reddy, M. Suryakalavathi performed optimal reactive power dispatch with the main objective of minimizing real power transmission line losses by hybridizing backtracking search algorithm with differential evolution algorithm in

M. Kunapareddy · B. V. Rao (✉)
VR Siddhartha Engineering College, Vijayawada, India

BBVL. Deepak et al. (eds.), *Innovative Product Design and Intelligent Manufacturing Systems*, Lecture Notes in Mechanical Engineering,
https://doi.org/10.1007/978-981-15-2696-1_64

[6]. This hybridization improves the robustness of the system and was tested on IEEE 30, 57 bus systems.

Basu [7] presented quasi-oppositional differential evolution for optimal reactive power dispatch to improve the quality and effectiveness of a solution and tested for IEEE 30, 57, 118 bus test systems. Yuancheng Li, Yiliangwang, and Bin Li stated a hybrid approach for solving the problem of reactive power dispatch by considering differential evolution and ant colony optimization [8].

All the above-stated methods, however, lack the efficiency when finding the solution in global search space. In this paper, authors propose the hybridization of particle swarm optimization with firefly algorithm. Firefly algorithm is advantageous due to its ability of automatic subdivision and dealing with multimodality. Firefly combined with PSO enhances the performance providing optimal result and is carried out in MATLAB environment for IEEE 14 and 30 bus systems.

2 Problem Formulation

The primary objective of RPD [9, 10] is to attenuate the real power transmission line losses in the system and is given by Eq. (1)

$$F_{\mathrm{L}} = \mathrm{Min}(P_{\mathrm{L}}) = \sum_{k=1}^{N_l} g_K \left[V_a^2 + V_b^2 - 2V_a V_b \cos(\delta_a - \delta_b)\right] \quad (1)$$

where N_l represents the number of transmission lines, g_k is the conductance at the branch k, voltages at bus a, b are represented by V_a, V_b. The angle difference between the buses a, b is expressed as $\delta_a - \delta_b$.

The other objective which is also very crucial for the effective performance of the system is the minimization of total voltage deviation which is formulated as presented in Eq. (2)

$$F_{\mathrm{VD}} = \mathrm{Min}\{V_{\mathrm{deviation}}(a,b)\} = \sum_{n=1}^{N_b} \left|V_n - V_{\mathrm{spec}}\right| \quad (2)$$

The constraints that must be satisfied to meet the objectives are classified as equality and inequality constraints. For multi-objective type optimization, both the objectives of losses and voltage deviation are optimized simultaneously by considering equal weight age factor for both the objectives. This multi-objective problem formulation is stated as

$$F = 0.5 * F_{\mathrm{L}} + 0.5 * F_{\mathrm{VD}} \quad (3)$$

2.1 Equality Constraints

These are basic load flow equations stating the power generated must be equal to the power demand and losses.

$$P_{Gi} - P_{Di} - V_i \sum_{i=1}^{N_b} V_j \begin{pmatrix} G_{ij} & \cos\theta_{ij} \\ +B_{ij} & \sin\theta_{ij} \end{pmatrix} = 0 \quad (4)$$

$$Q_{Gi} - Q_{Di} - V_i \sum_{i=1}^{N_b} V_j \begin{pmatrix} G_{ij} & \sin\theta_{ij} \\ B_{ij} & cos\theta_{ij} \end{pmatrix} = 0 \quad (5)$$

2.2 Inequality Constraints

$$V_{Gi}^{\min} \leq V_{Gi} \leq V_{Gi}^{\max},\ i \in N_g \quad (6)$$

$$V_{Li}^{\min} \leq V_{Li} \leq V_{Li}^{\max},\ i \in N_l \quad (7)$$

$$Q_{Gi}^{\min} \leq Q_{Gi} \leq Q_{Gi}^{\max},\ i \in N_t \quad (8)$$

$$T_i^{\min} \leq T_m \leq T_i^{\max},\ i \in N_c \quad (9)$$

$$Q_{ci}^{\min} \leq Q_{ci} \leq Q_{ci}^{\max},\ i \in N_c \quad (10)$$

3 Methodology

3.1 Particle Swarm Optimization

Kennedy and Eberhart introduced particle swarm optimization [1] in the year 1995. They got inspired by the social behavior of birds and schooling of fish. In PSO, the birds or fishes are expressed as particles P, which moves in certain search domain having a certain velocity V. The position of the particle gets updated by considering Eqs. (11) and (12). PSO is extremely used for its simplicity and a fact that only a few parameters are to be adjusted.

The velocity between the two particles i, j are represented in Eq. (11) as

$$V_{ij}^{k+1} = w \times V_{ij}^k + c_1 \times r_1 \times \left(Pbest_{ij}^k - Z_{ij}^k\right) + c_2 \times r_2 \times \left(Gbest_j^k - Z_{ij}^k\right) \quad (11)$$

The new position of the particles is obtained as shown in Eq. (12)

$$Z_{ij}(\lambda + 1) = Z_{ij}(\lambda) + V_{ij}(\lambda + 1) \tag{12}$$

3.2 *Firefly Algorithm*

X. S. Yang, a famous mathematician was the designer of this firefly algorithm. FA was introduced in 2008 mimicking the flashing nature of fireflies. The main concept of fireflies that grabbed the attention is they use their own light to communicate and attract their mate [11]. The three important features of FA are: (1) they are attracted toward each other irrespective of their gender (2) the attractiveness between them is proportional to the distance, which says that as the distance increase attraction decreases and therefore the light exhibited by the firefly is limited to the few hundreds of meters; and (3) fitness function is calculated to determine the brightness of a firefly. The distance between a firefly and another firefly is calculated as

$$r_{ab} = \sqrt{\sum_{k=1}^{d} \left(x_{a,k} - x_{b,k}\right)^2} \tag{13}$$

When the light intensity of firefly a is less compared to firefly b, then a is moved toward b updating the particle position as

$$x_a = x_a + [\beta(r)](x_b - x_a) + \alpha(\text{rand}) \tag{14}$$

3.3 *Hybridization of PSO and FA (HPSOFA)*

The results obtained through firefly cannot yield better solutions due to its oscillatory nature whereas PSO is fitness function dependent with finely tuned control parameters. The demerits in the firefly algorithm which causes unbalance in exploration and exploitation can be overcome with the hybridization of FA with another heuristic method, likely PSO. In this HPSOFA, local search capability of PSO is combined with global thinking of FA achieving balance between local and global search ability. The velocity equation in PSO is modified by replacing acceleration constants with FA parameters.

$$V_i^{k+1} = wV_i^k + \beta_0 e^{-\gamma r_{ij}^2}(P\text{best} - Z_i) + \alpha(\text{rand} - 1/2)(G\text{best} - Z_i) \tag{15}$$

3.3.1 Implementation steps of HPSOFA

1. Initialize the parameters such as size, alpha, beta, and gamma.
2. Randomly initialize the velocity and position of the particles within the given limits.
3. Calculate the fitness, distance between the fireflies and determine the *P*best, *G*best values.
4. While (maximum number of iterations).
5. Calculate the new velocity of the particles using the Eq. (15).
6. Update the position of the particles considering Eq. (12) using the new velocities.
7. Calculate the fitness function with the new values.
8. End while.

4 Simulation Results

The tuning parameters along with its values are shown in Table 1 and limits of control variables that must be satisfied while performing the optimal reactive power dispatch for all the objectives are shown in Table 2.

4.1 IEEE 14 Bus

The bus, line and generator data of this system is adapted from [12]. This system consists of 5 generators located at 1, 2, 3, 6, 8 buses, 3 tap setting transformers at 4-7, 4-9, 5-6 buses, and 2 shunts placed VAR compensating devices at 9, 14 bus. The total real and reactive power demand of the system is given as 259 MW and 73.50 MVAr, respectively.

Table 1 Tuning parameters of PSO and FA

PSO	FA	HPSOFA
$c1 = 2$	alpha = 0.2	alpha = 1.5
$c2 = 2$	beta 0 = 0.1	beta 0 = 0.2
$w = 0 \leq w \geq 1$	gamma = 1	gamma = $0–\infty$

Table 2 Limits of control parameters

Control parameters	Min.–Max. in p.u.
Generator voltages	0.9–1.1
Tap settings of transformers	0.95–1.05
Shunt capacitors	0–0.20

4.1.1 Diminution of Real Power Loss

The primary objective of RPD is mainly diminution of real power loss. This objective is imposed on the IEEE systems using the techniques PSO, FA, and HPSOFF as shown in Table 3. There are 10 control variables that are needed to be controlled to obtain the optimum value such that the value is within the minimum and maximum limits. The voltages of generator and tap setting value of the transformer are represented in per unit, whereas the reactive power values of compensation devices are depicted in MVAr. The comparison of losses with other optimization methods is shown in Table 7.

4.1.2 Improvement of Voltage Profile

Improving the voltage at the buses is considered as another objective. The motive here is to keep the voltages at a nearly nominal value of 1 p.u., thereby decreasing the total voltage deviation in the system. The voltage deviation obtained after the application of optimization techniques is shown in Table 4.

4.1.3 Multi-objective Function

In this case, both the objectives of minimization of real power loss and voltage deviation are controlled. Though the output may not be as optimal as a single objective but simultaneous control of both the objectives is its main advantage. The optimal values of control variables along with its convergence curves are shown in Table 5 and Fig. 1, respectively.

Table 3 Real power losses for the IEEE 14 bus system

Objective	PSO	FA	HPSOFA
RPL in MW	**12.6107**	**12.5837**	**12.2484**
TVD in p.u.	0.722	0.7234	1.2086

RPL real power loss, *TVD* total voltage deviation, *CV* control variable
The bold indicates in table shows that optimization of particular objective

Table 4 Total voltage deviation of IEEE 14 bus system

Objective	PSO	FA	HPSOFA
TVD in p.u.	**0.0634**	**0.0600**	**0.0556**
RPL in MW	15.646	15.574	15.340

The bold indicates in table shows that optimization of particular objective

Table 5 Multi-objective function for the IEEE 14 bus system

CV	PSO	FA	HPSOFA	CV	PSO	FA	HPSOFA
Vg1	1.0334	1.0193	1.0155	T5-6	0.9761	1.0267	1.0323
Vg2	1.0133	1.0101	1.0159	Q9	18.7333	18.5362	18.5983
Vg3	1.0019	1.0044	0.9998	Q14	12.0971	10.6494	14.0286
Vg6	1.0137	1.0157	1.0124	RPL in MW	14.30	14.69	14.84
Vg8	1.0018	1.0000	0.9996	TVD in p.u.	0.085	0.0680	0.0662
T4-7	0.9913	1.0001	0.9911	**Optimal value in p.u.**	**0.1147**	**0.1075**	**0.1073**
T4-9	0.9987	0.9771	0.9839				

Optimal value for HPSOFA is stated as 0.5 * 0.1484 + 0.5 * 0.0662 = 0.1073
The bold indicates in table shows that optimization of particular objective

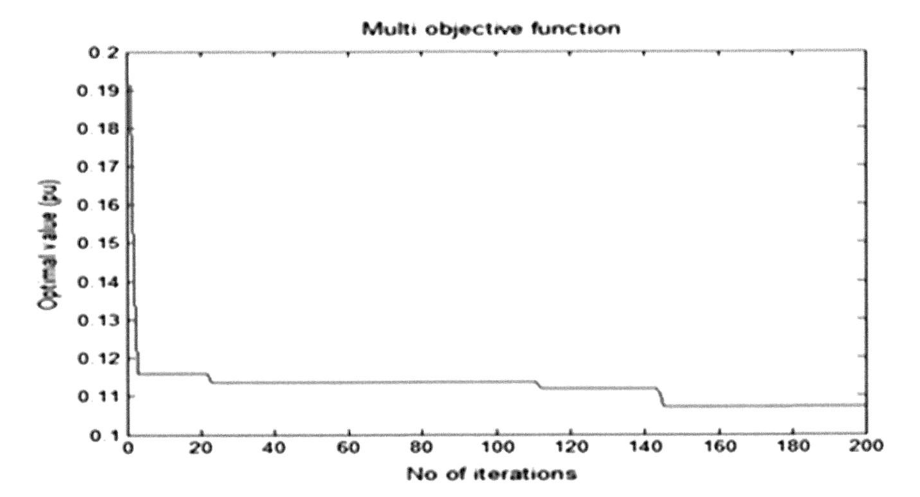

Fig. 1 Convergence characteristics of HPSOFA for the multi-objective function of IEEE 14 bus system

4.2 IEEE 30 Bus

The bus, line and generator data for this system is extracted from [12]. It comprises 6 generators located at 1, 2, 5, 8, 11, 13 buses, 4 tap setting transformers at 4-12, 6-9, 6-10, 27-28 and 3 reactive power compensation devices at 3, 10, 24 buses. The real and reactive power demand was 283.4 MW and 126.2 MVAr, respectively.

4.2.1 Declination of Real Power Loss

The 13 control variables in IEEE 30 bus system are tuned to attain the optimal value for the objective of minimization of real power loss. The control variables along with the applied optimization methods are shown in Table 6. The comparison of the

Table 6 Real power losses for the IEEE 30 bus system

Objective	PSO	FA	HPSOFA
RPL in MW	**16.0494**	**16.0196**	**15.9234**
TVD in p.u.	1.1023	1.0572	1.0426

The bold indicates in table shows that optimization of particular objective

Table 7 Comparison of real power loss with different optimization techniques

Technique	Power loss (MW)	
	IEEE 14 bus	IEEE 30 bus
DE [8]	–	16.3511
ABC [8]	–	16.9291
DE-ABC [8]	–	16.2165
SARGA [10]	13.21643	16.0912
PSO	12.9691	16.0494
FA	12.5837	16.0196
HPSOFA	12.2484	15.9234

real power losses for IEEE 14 and 30 bus systems with different optimization techniques is shown in Table 7.

4.2.2 Improvement of Voltage Profile

The voltage values at the 30 buses need to be improved to the nearer value of 1 p.u. The values of the voltages at the buses with optimization methods are displayed in Table 8.

4.2.3 Multi-objective Function

This function is advantageous since it optimizes both the objective functions. The values of control variables that are obtained within the limits are shown in Table 9 along with its convergence characteristics shown in Fig. 2. Optimal value for HPSOFA is stated as $0.5 * 0.1889 + 0.5 * 0.2269 = 0.2079$.

Table 8 Total voltage deviation of IEEE 30 bus system

Objective	PSO	FA	HPSOFA
TVD in p.u.	**0.2251**	**0.2061**	**0.2049**
PRL in MW	21.1669	20.0217	20.9124

The bold indicates in table shows that optimization of particular objective

Table 9 Multi-objective function for the IEEE 30 bus system

CV	PSO	FA	HPSOFA	S.NO.	PSO	FA	HPSOFA
Vg1	1.0579	1.0171	1.0293	T6-10	0.9675	0.9963	0.9590
Vg2	1.0296	1.0146	1.0136	T27-28	0.9847	0.9639	0.9596
Vg5	0.9788	1.0042	1.0042	Q3	5.6850	8.8785	17.9066
Vg8	1.0462	1.0126	1.0095	Q10	1.5426	19.1389	18.2668
Vg11	0.9954	0.9975	1.0002	Q24	12.2609	20	20
Vg13	1.0360	1.0355	1.0162	RPL in MW	18.83	19.50	18.89
T4-12	0.9672	1.0041	1.0103	TVD in p.u.	0.4436	0.23	0.2269
T6-9	0.9839	0.9933	0.9500	Optimal value in p.u.	**0.31245**	**0.2125**	**0.2079**

The bold indicates in table shows that optimization of particular objective

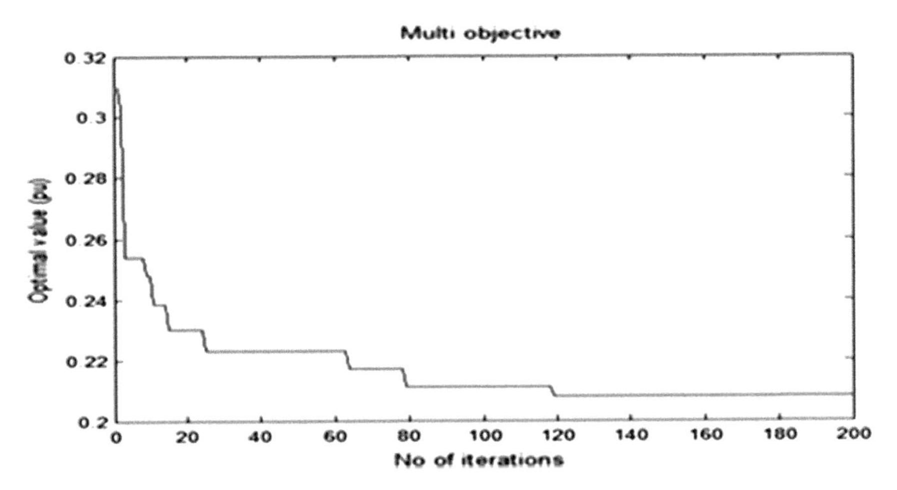

Fig. 2 Convergence characteristics of HPSOFA for multi-objective of IEEE 30 bus

5 Conclusion

The reactive power dispatch with HPSOFA performs two objectives of minimizing real power loss and voltage deviation along with multi-objective function. It is evident from the results that values of control variables with HPSOFA yield better output compared to PSO and FA. This method is tested on IEEE 14 and 30 bus systems. It is also observed that convergence speed also increases along with accuracy. Since this work is limited to the basic IEEE system it can be further carried out using FACTS device which results in improving the performance by reducing the losses further. In addition, solar and wind generation can also be considered for effective utilization of renewable energy combined with thermal energy.

References

1. Pandya S, Roy R (2015) Particle swarm optimization based optimal reactive power dispatch. In: International conference on electrical, computer & communication technologies 2015, IEEE, Coimbatore, pp 1–5
2. Das T, Roy R (2018) Optimal reactive power dispatch using JAYA algorithm. In: 2018 emerging trends in electronic devices and computational techniques, IEEE, Kolkata, pp 1–6
3. Zhang Y, Chen C, Lee C (2018) Solution of the optimal reactive power dispatch for power systems by using a novel charged system search algorithm. In: 2018 7th international symposium on next generation electronics, IEEE, Taipei, pp 1–4
4. Sulaiman MH, Mustaffa Z (2017) Cuckoo search algorithm as an optimizer for optimal reactive power dispatch problems. In: 2017, 3rd ICCAR, IEEE, Nagoya, pp 735–739
5. Alam MS, De M (2016) Optimal reactive power dispatch using a hybrid loop genetic algorithm. 2016 NPSC, IEEE, Bhubaneswar, pp 1–6
6. Lenin K, Reddy BR Kalavathi MS (2016) Hybridization of backtracking search optimization algorithm with differential evolution algorithm for solving the reactive power problem. Int J Adv Intell Paradig 8(3):355–364
7. Basu M (2016) Quasi oppositional differential evolution for optimal reactive power dispatch. Int J Electr Power Energy Syst 78:29–40
8. Wang Y, Li Y, Li B (2013) A hybrid artificial bee colony assisted differential evolution algorithm for optimal power flow. Electr Power Syst Res 52:25–33
9. Mehdinejad M, Mohammadi-Ivatloo B, Dadashzadeh-Bonab R, Zare K (2016) The solution of ORPD of power systems using hybrid particle swarm optimization and imperialist competitive algorithms. Int J Electr Power Energy Syst 83:104–116
10. Subbaraj P, Narayana PR (2009) Optimal reactive power dispatch using a self-adaptive real-coded genetic algorithm. Electr Power Syst Res 79:374–381
11. Rao BV, Nagesh Kumar GV (2015) A comparative study of BAT and firefly algorithms for optimal placement and sizing of static VAR compensator for enhancement of voltage stability. Int J Energy Optimization Eng, 4:68–84
12. Power system test case archive. https://labs.ece.uw.edu/pstca/

Optimal Scheduling of Hydrothermal Plant Using Particle Swarm Optimization

P. Sowmith, R. Madhusudhanrao and N. Gouthamkumar

Abstract This paper presents a particle swarm optimization (PSO) technique for solving hydrothermal problem by considering demand constraint, thermal generator constraint, reservoir capacity constraint, hydrogenerator constraint and water discharge constraint. The possibility examination of the proposed method is exhibited on one hydro-plant and a steam plant by utilizing MATLAB. The comparison of the simulation results shows that PSO has a better performance than the other programs like gravitational search algorithm, genetic algorithm, classical evolutionary programming, simulated annealing approach and fast evolutionary programming.

Keywords Hydrothermal coordination · Hydrothermal scheduling · Particle swarm optimization · Power generation

1 Introduction

The optimal scheduling of a power plant generation can be determined as total generation obtained from each generating plant when summed up resulting in minimal cost by satisfying the minimum and maximum limitations. The main objective of hydrothermal scheduling is based on the release of the amount of water from each reservoir to generate sufficient power such that the total fuel cost of thermal power plants is reduced. The complexity of hydrothermal scheduling is mainly affected by two issues: the vulnerability of inflows and the hydraulic coupling between hydro-plants.

Wang proposed various PSO methods which are effectively implemented for resolving the short-term hydrothermal plant scheduling complexity [1]. The efficiency of anticipated techniques is demonstrated by comparison with that of a prominent example used by other evolutionary algorithms. A comparison is also made between different PSO techniques. Banerjee investigated [2] a here and now

P. Sowmith (✉) · R. Madhusudhanrao · N. Gouthamkumar
Department of Electrical and Electronics Engineering, V R Siddhartha Engineering College, Vijayawada, A.P. 520007, India

BBVL. Deepak et al. (eds.), *Innovative Product Design and Intelligent Manufacturing Systems*, Lecture Notes in Mechanical Engineering,
https://doi.org/10.1007/978-981-15-2696-1_65

hydro-wind-thermal scheduling in light of molecule swarm enhancement method in which a perfect hourly program of energy time in a hydro-thermal-wind power system applying PSO procedure is presented with corresponding simulation results. He focused only on the single objective of the issue.

Yuan focused on multi-objective best possible scheduling in the wind-hydro-thermal system [3]. He has chosen three cases to exhibit different requirements like economic, efficient and efficient environmental power production. Mandal presented his work on particle swarm optimization where valve-point loading effect is also taken as one of the nonlinearities of objective function [4]. The PSO technique is compared to two other techniques called simulated annealing and evolutionary programming where the proposed technique gave close to the finest solutions. GSA is not taken into consideration in his work. A. J. Wood and B. F. Wollenberg "Power Generation Operation and Control" have made a detailed study of hydrothermal scheduling with various methods, discussing the pros and cons of each one over other with various standard examples [6].

The hydrothermal coordination (HTC) issue is generally tackled by the deterioration of the unusual difficulty into short-, medium- and long-term problems every one considering the suitable viewpoints for its chance stride and skyline of study. It is additionally fundamental to think about two essential parts of the hydrosystem:

1. The accessible water amount (water inflows) is stochastic in nature.
2. The choice for the energy distributed to hydro-units is deterministic.

Classification of Hydrothermal Scheduling Problem:

1. Long-range issue
2. Short-range issue.

2 Problem Formulation

2.1 *Objective Function*

The main intent of HTS is to decide the succession of hydro discharges which will limit the estimated operating cost of thermal plants along the scheduling horizon. The nonlinear compelled HTS planning enhancement issue is subjected to an assortment of requirements like the changing demand of system, the time combination impact of hydro-plant, the time fluctuating hourly supply inflows, operating limits of hydrothermal plant system misfortunes, the storage limits of reservoir, water discharge limits, continuity of hydraulic constraints and early and last supply storage limits. The objective function considered here is applied for a minimization problem as our motive is to minimize the cost function. The target work is communicated numerically as

$$\text{Minimize} \quad f_{\text{total}} = \sum_{t=1}^{Z} \sum_{i=1}^{N_{\text{th}}} f_i(\text{PT}_{i,t}) \tag{1}$$

where N_{th} is the number of steam plants and $\text{PT}_{i,t}$ is the generation of ith steam plant in MW at tth interval.

2.2 Constraints

(a) The constraint of demand:
The total power demand constraint is equal to the summation of power generated from thermal and hydro-plants.

$$\sum_{i=1}^{N_{\text{th}}} \text{PT}_{i,t} + \sum_{i=1}^{N_h} \text{PH}_{i,t} = \text{PD}_t \quad \text{for } t = 1, 2, \ldots, Z \tag{2}$$

(b) The constraint of thermal plant:
The generation of the power from the ith thermal plant lies in between its upper and lower limits.

$$\text{PT}_{i\min} \leq \text{PT}_{i,t} \leq \text{PT}_{i\max} \tag{3}$$

(c) The constraint of hydro-plant:
The power generation of ith hydro-plant lies in between its upper and lower limits.

$$\text{PH}_{i\min} \leq \text{PH}_{i,t} \leq P \tag{4}$$

(d) The constraint of reservoir capacity:
The quantity of ith reservoir storage during its operation lies in between the minimum and maximum capacity limits.

$$V_{i\min} \leq V_{i,t} \leq V_{i\max} \tag{5}$$

(e) The constraint of water discharge:
The discharge of water through turbines is limited by minimum and maximum operating bounds.

$$q_{i\min} \leq q_{it} \leq q_{i\max} \tag{6}$$

(f) Hydraulic continuity constraint:

$$V_{i(t+1)} = V_{it} + \sum_{u=1}^{R_u} \left[q_{u(t-\tau)}\right] - q_{i(t+1)} + r_{i(t+1)} \tag{7}$$

where τ is the water delay time at tth interval between ith reservoir and its upstream u at tth interval and R_u is the set of upstream units directly above the ith hydro-plant.

The generation of hydropower PH_{it} is considered a function of discharge rate and storage volume.

$$\mathrm{PH}_{it} = c_{1i}V_{it}^2 + c_{2i}q_{it}^2 + c_{3i}(V_{it}q_{it}) + c_{4i}V_{it} + c_{5i}q_{it} + c_{6i} \tag{8}$$

where $c_{1i}, c_{2i}, c_{3i}, c_{4i}, c_{5i}$ and c_{6i} are coefficients of given hydropower plant.

The discharge rate of hydrothermal plant is

$$q = a + bP_{g1} + \mathrm{CP}_{g1}^2 \tag{9}$$

The total discharge is taken as

$$q^{\mathrm{tot}} = \sum_{i=1}^{\mathrm{NT}} np(qc) \tag{10}$$

3 Methodology

PSO was developed by Eberhart and Kennedy in 1995, which is a well familiar method known to resolve comprehensive nonlinear problems effectively. PSO is one of the classifications of the evolutionary computations used to solve the problems of optimization. It is a population-based search procedure provoked by natural behavior such as bird flocking and fish schooling. The inspection leads to the assumption that every information is communal inside the flocking carried out by birds in search of the food. This hypothesis is a basic concept of PSO which is developed through simulation of a flock of birds in two-dimensional space. For every iteration, the velocities of the individual particles are stochastically adjusted according to the chronological best position for the particle itself and the region particle best position. Both the particle best and the region best are also called global best, and it is consequent according to the user-defined fitness function. A solution which is optimal or near optimal is obtained to the natural moment of the particles as shown in below Fig. 1.

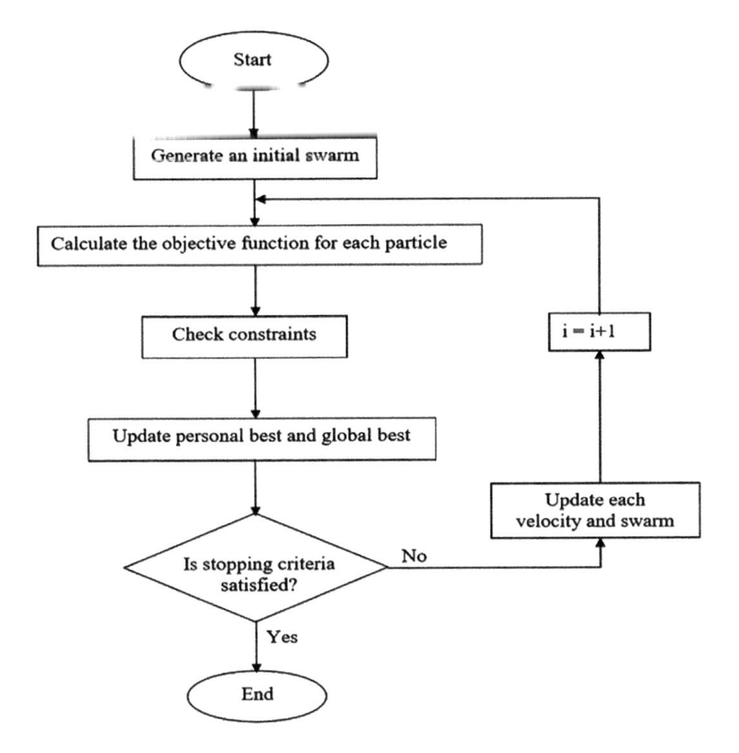

Fig. 1 Computational framework of PSO

PSO can generate good-quality solutions with a reduced amount of computational time and more constant convergence characteristics where most analytical methods fail to converge. The velocities and positions are keeping up to date of all the particles using the below equations, respectively,

$$V_{ij}^{K+1} = w * V_{ij}^{k} + c_1 * r_1 * \left(Pbest_{ij}^{k} - X_{ij}^{k}\right) + c2 * r2 * \left(Gbest_{j}^{k} - X_{ij}^{k}\right) \quad (11)$$

$$X_{ij}^{k+1} = X_{ij}^{k} + V_{ij}^{k+1} \quad (12)$$

4 Simulation Results

Test case: The considered hydro-plant and a steam plant load characteristics are given here.

Steam plant:

The total fuel cost function of equivalent steam system is $H = 575 + 9.2P_s + 0.00184P_s^2$ (MBtu/h). The power generation of the steam plant must lie between maximum and minimum limits of 1500 and 150 MW.

Table 1 Load pattern

Operation time	Load demand		
	Day 1 (MW)	Day 2 (MW)	Day 3 (MW)
24.00–12.00	1200	1100	950
12.00–24.00	1500	1800	1300

Hydro-plant:

The water discharge functions of the given hydropower plant are $q = 300 + 4.97 P_{\mathrm{H}}$ acre-ft/h; the upper and lower limits of power generation in the hydro-plant are 0 and 1000, respectively.

$q = 5300 + 12(P_{\mathrm{H}} - 1000) + 0.05(P_{\mathrm{H}} - 1000)^2$ acre-ft/h,
$1000 < P_{\mathrm{H}} < 1100$ MW.

Hydro-reservoir:

The initial volume of reservoir is 100,000 acre-ft at the end of schedule it must have 60,000 acre-ft. The upper and lower limits of reservoir volume are 60,000 acre-ft and 120,000 acre-ft; there is a steady inflow into the reservoir of 2000 acre-ft/h in excess of the entire three-day period as shown in Table 1.

Result:

Fuel cost = 709,530 \$/h with the computational time 3.814283 s though the obtained P_{H} and P_{S} are from random variables, the constraints and load demand are met with the generation in minimum number of iterations using PSO as shown in Table 2, and Fig. 2 shows the convergence characteristics of cost minimization of hydrothermal scheduling.

Table 3 gives the comparison of fuel cost of hydrothermal scheduling system with the former works and proposed PSO method. The obtained result of the proposed technique is compared to other heuristic techniques [6–9].

Table 2 Result for test case

Interval	P_{H}	P_{S}	q	v	P_{G}	Time (s)	Cost (\$/h)
P_1	335.35	864.65	1996.7	100,040	1200	3.814283	709,530
P_2	632.94	867.06	3475.7	82,331	1500		
P_3	229.94	870.06	1472.8	88,658	1100		
P_4	935.67	864.33	4980.3	52,895	1800		
P_5	106.61	843.39	859.84	66,577	950		
P_6	446.29	853.71	2548	60,000	1300		

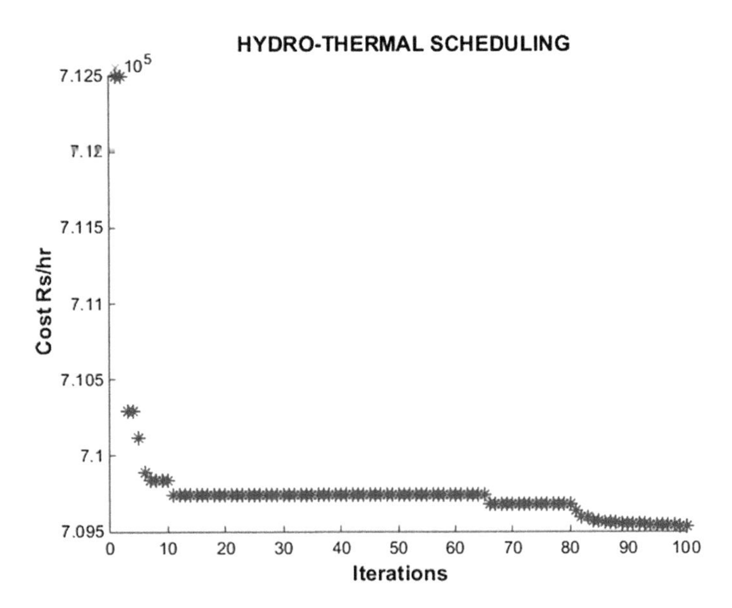

Fig. 2 Cost minimization of HTS using PSO

Table 3 Comparison of the proposed technique with other techniques

S. No.	Year	Method	Cost ($)	Simulation time (s)
1	1984	GS [9]	709,877.38	–
2	1994	SA [7]	709,874	907
3	2003	IFEP [8]	709,862.05	59.7
4	2006	GAF [6]	709,863.70	–
5	2006	CEP [6]	709,862.65	–
6	2006	FEP [6]	709,864.54	–
7	2006	PSO [6]	709,862.048	–
8	2019	Proposed PSO	709,530	3.814283

5 Conclusion

This paper presents the proposed method to minimize the fuel cost of thermal plants; the hydro-plant is integrated with the thermal plant system which is nothing but hydrothermal scheduling, and this scheduling can be done in various ways using conventional methods and heuristic methods. But considering the drawbacks of conventional methods, the heuristic method called PSO is considered sequentially to get better convergence rate and to meet up the load demand at least amount of total fuel cost.

References

1. Wang Y, Zhou J, Zhou C (2012) An improved self-adaptive PSO technique for short-term hydrothermal scheduling. Electr Power Energy Syst 39:2288–2295
2. Banerjee S (2016) Short term hydro-thermal-wind scheduling based on particle swarm optimization technique. Electr Power Energy Syst 81:275–288
3. Yu B, Yuan X (2007) Short-term hydro-thermal scheduling using particle swarm optimization method. Energy Convers Manag 48:1902–1908
4. Mandal KK, Basu M (2007) Particle swarm optimization technique based short-term hydrothermal scheduling. Energy Convers Manag 8:1392–1399
5. Ying Lee T (2007) Short term hydroelectric power system scheduling with wind turbine generators using the multi-pass iteration particle swarm optimization approach. Energy Convers Manag 49:751–760
6. Sinha N, Lai L-L (2006) Meta-heuristic search algorithms for short-term hydrothermal scheduling. In: International conference on machine learning and cybernetics, Dalian, 2006
7. Wong KP, Wong YW (1994) Short-term hydrothermal scheduling, part-I: simulated annealing approach. IEE Proc, Part C 141(5):497–501
8. Chakrabarti R, Sinha N, Chattopadhyay PK (2003) Fast evolutionary programming techniques for short-term hydrothermal scheduling. IEEE Trans PWRS 18(1):214–219
9. Wood AJ, Wollenberg BF (2012) Power generation, operation and control, 2nd edn. Wiley, New Delhi

A New Algorithm for Reduction of High Order Commensurate Non-integer Interval Systems

Kalyana Kiran Kumar, Kurman Sangeeta and Chongala Prasad

Abstract This note presents a novel methodology for reduction of high order linear time-invariant commensurate non-integer interval systems. It is shown first that the fractional-order interval system is reconstructed to integer interval system and further a hybrid technique is applied as a model reduction scheme. In this scheme, the reduced denominator is acquired by applying a modified least square method and the numerator is achieved by time moment matching. This formulated reduced interval integer model is reconverted to a reduced fractional interval model. As a final point, the results of a numerical illustration are verified to show the relevance and superiority of the proposed technique.

Keywords Non-integer interval systems · Kharitonov's theorem · Model reduction · Commensurate fractional

1 Introduction

A structure whose parameters are two-dimensional having lower and upper bounds is defined as an uncertain system. For example, interval parameters $\left[a_i^-, a_i^+\right]$ and $\left[b_i^-, b_i^+\right]$ mean, the parameters a_i and b_i can take independently any values in respective intervals $\left[a_i^-, a_i^+\right]$ and $\left[b_i^-, b_i^+\right]$. The width of the interval coefficient is equal to the difference between the upper and the lower boundary parameters. Some coefficients of an interval system may have zero interval width, i.e., equal values for both lower and upper bounds. These systems are successfully handled with interval arithmetic. The interval system having all the coefficients with zero width is called a fixed system. Moore's first book printed on interval analysis was the product of his Ph.D. thesis. He primarily focused on solutions for ordinary differential equations problems.

K. Kiran Kumar (✉) · K. Sangeeta · C. Prasad
Aditya Institute of Technology and Management, Tekkali, India

BBVL. Deepak et al. (eds.), *Innovative Product Design and Intelligent Manufacturing Systems*, Lecture Notes in Mechanical Engineering,
https://doi.org/10.1007/978-981-15-2696-1_66

Recently, fractional-order control systems are fetching progressively prominent in the control community. Fractional systems are governed by fractional-order derivatives and integrals which are basically infinite order linear operators. These fractional-order systems extend the notion of our integer order concepts in control and improvise on the existing results. So in essence, both fractional calculus and control systems have been applied to improve the performance of existing control theories [1]. The processes that we want to estimate the real-world objects are in general of fractional order [2–4]. These fractional-order systems on applying Laplace transform with zero initial conditions obtain a transfer function of unlimited order and therefore require infinite memory [5]. In order to reduce the infinite memory occupied by the fractional systems, it is possible to apply the concept of approximations [6]. Some of the frequency-based domain reduction methods have been suggested [7–9]. These approaches are used to generate a reduced fractional model whose characteristics are akin to the original fractional system. Since the obtained model is compact compared to the complex original fractional system. These approaches are called as model reduction techniques.

MOR is a well-recognized and long-standing concept applied for systems of integer order but an insufficient work being implemented for systems fractional order. [10]. this instigated us to select this particular zone and put an effort to introduce a unique reduction methodology for the commensurate non-integer interval systems [7, 11, 12]. The present study has been put on display with three examples that the determined reduced fractional interval model by the recommended method approximates well for stable class of high order nonminimum, non-integer fractional-order integer systems. To the author's knowledge, the proposed reduction technique applied to commensurate fractional-order interval system has not been proclaimed so far. The concept and the process endorsed in the proposed scheme would be a foundation stone for the evolution of new MOR schemes for fractional interval systems in the future. Two different theories, time moment matching and least square optimization, are blended together in this method [13, 14] to get the reduced model. The paper is organized as follows: To make a proper background, the proposed method is explained in Sect. 2. The capability of the proposed approach is shown in Sect. 3 through three test systems and the paper are concluded in Sect. 4.

2 Proposed Method

Step 1: Let the high order fractional-order interval system (FOIS) represented as a transfer function as cited in [15]

$$G(s)\frac{\sum_{i=0}^{m}\left[n_i^-, n_i^+\right]s^{\beta_i}}{\sum_{i=0}^{n}\left[d_i^-, d_i^+\right]s^{\alpha_i}} \tag{1}$$

where $\beta_m > \beta_{m-1} > \cdots > \beta_1 > 0$ and $\alpha_n > \alpha_{n-1} > \cdots > \alpha_1 \geq 0$.

If all the powers of the derivatives of the above equation are integer multiples of Υ, that is, α_i, $\beta_i = i\,\Upsilon$. Such that $\Upsilon \in R+$ and $i = 0, 1, 2, 3\ldots$ then it is considered as commensurate fractional-order interval system. Therefore, Eq. (1) can be formulated as

$$G(s) = \frac{\sum_{i=0}^{m}\left[n_i^-, n_i^+\right]s^{\Upsilon i}}{\sum_{i=0}^{n}\left[d_i^-, d_i^+\right]s^{\Upsilon i}} = \frac{\sum_{i=0}^{m}\left[n_i^-, n_i^+\right]\left(s^{\Upsilon}\right)^i}{\sum_{i=0}^{n}\left[d_i^-, d_i^+\right]\left(s^{\Upsilon}\right)^i} \tag{2}$$

By substituting $s^{\Upsilon} = \lambda$, in the above equation becomes integer order interval transfer function as,

$$G(\lambda) = \frac{\sum_{i=0}^{m}\left[n_i^-, n_i^+\right]\lambda^i}{\sum_{i=0}^{n}\left[d_i^-, d_i^+\right]\lambda^i} \tag{3}$$

where n = order of the system > m, i = 0, 1, 2, 3, … n and $\left[n_i^-, n_i^+\right]$, $\left[d_i^-, d_i^+\right]$ are the specified lower and upper boundaries of the ith perturbation.

Step 2: Formulation of four Kharitonov transfer functions from the above equation as

$$K_1(\lambda) = \frac{n_0^- + n_1^-\lambda + n_2^+\lambda^2 + n_3^+\lambda^3 + \cdots}{d_0^- + d_1^-\lambda + d_2^+\lambda^2 + d_3^+\lambda^3 + \cdots} \tag{4}$$

$$K_2(\lambda) = \frac{n_0^+ + n_1^+\lambda + n_2^-\lambda^2 + n_3^-\lambda^3 + \cdots}{d_0^+ + d_1^+\lambda + d_2^-\lambda^2 + d_3^-\lambda^3 + \cdots} \tag{5}$$

$$K_3(\lambda) = \frac{n_0^+ + n_1^-\lambda + n_2^-\lambda^2 + n_3^+\lambda^3 + \cdots}{d_0^+ + d_1^-\lambda + d_2^-\lambda^2 + d_3^+\lambda^3 + \cdots} \tag{6}$$

$$K_4(\lambda) = \frac{n_0^- + n_1^+\lambda + n_2^+\lambda^2 + n_3^-\lambda^3 + \cdots}{d_0^- + d_1^+\lambda + d_2^+\lambda^2 + d_3^-\lambda^3 + \cdots} \tag{7}$$

The above four nth order Kharitonov transfer functions can be generalized as

$$K_w(\lambda) = \frac{\sum_{i=0}^{n-1} x_{wi}\lambda^i}{\sum_{i=0}^{n} y_{wi}\lambda^i} = \frac{x_{w0} + x_{w1}\lambda + x_{w2}\lambda^2 + \cdots + x_{w(n-1)}\lambda^{n-1}}{y_{w0} + y_{w1}\lambda + y_{w2}\lambda^2 + \cdots + y_{wn}\lambda^n} \tag{8}$$

such that equate the coefficients of Eqs. (4)–(7) individually with Eq. (8) at $w = 1$, 2, 3, 4, respectively.

Step 3: Determination of the denominator coefficients of the reduced order interval model.

If $K_w(\lambda)$ is extended about $\lambda = 0$, then the time moment proportions, v_{wi}, are assumed to be

$$K_w(\lambda) = \sum_{i=0}^{\infty} v_{wi}\lambda^i = v_{w0}\lambda^0 + v_{w1}\lambda^1 + v_{w2}\lambda^2 + \cdots + v_{w\infty}\lambda^{\infty} \quad (9)$$

Let the corresponding rth order generalized reduced models of $K_w(\lambda)$ represented as

$$K_{wr}(\lambda) = \frac{p_{wr}(\lambda)}{q_{wr}(\lambda)} = \frac{\sum_{i=0}^{r-1} p_{wi}\lambda^i}{\sum_{i=0}^{r} q_{wi}\lambda^i} = \frac{p_{w0} + p_{w1}\lambda + \cdots + p_{w(r-1)}\lambda^{r-1}}{q_{w0} + q_{w1}\lambda + \cdots + q_{w(r-1)}\lambda^{r-1} + \lambda^r} \quad (10)$$

where $w = 1, 2, 3, 4$.

The minimum time moments required for a reduced model to retain is $2r$ time moments, 'r' indicates the order of the reduced model and the coefficient q_{wi}, p_{wi} in Eq. (10) are derived from following a set of equations as:

$$\left.\begin{array}{l}
p_{w0} = q_{w0}v_{w0} \\
p_{w1} = q_{w1}v_{w0} + q_{w0}v_{w1} \\
\vdots \quad \vdots \quad \vdots \quad \vdots \quad \vdots \\
p_{w(r-1)} = q_{w(r-1)}v_{w0} + \ldots + q_{w0}v_{w(r-1)} \\
-v_{w0} = q_{w(r-1)}v_{w1} + \ldots + q_{w1}v_{w(r-1)} + q_{w0}v_{w(r)} \\
-v_{w1} = q_{w(r-1)}v_{w2} + \ldots + q_{w1}v_{wr} + q_{w0}v_{w(r+1)} \\
\quad \vdots \quad \vdots \quad \vdots \qquad \vdots \quad \vdots \quad \vdots \qquad \vdots \quad \vdots \\
-v_{w(t-r-1)} = q_{w(r-1)}v_{w(t-r)} + \ldots + q_{w0}v_{w(t-1)}
\end{array}\right\} \quad (11)$$

Equating Eqs. (9) and (10) the linear set of equations to preserve 't' time moments of the original system, the solution of represented as

$$\underbrace{\begin{bmatrix} v_{w(r)} & v_{w(r-1)} & \cdots & v_{w1} \\ v_{w(r+1)} & v_{w(r)} & \cdots & v_{w2} \\ \vdots & \vdots & \cdots & \vdots \\ \vdots & \vdots & \cdots & \vdots \\ \vdots & \vdots & \cdots & \vdots \\ v_{w(t-1)} & v_{w(t-2)} & \cdots & v_{w(t-r)} \end{bmatrix}}_{H_w} * \underbrace{\begin{bmatrix} q_{w0} \\ q_{w1} \\ \vdots \\ \vdots \\ \vdots \\ \vdots \\ q_{w(r-1)} \end{bmatrix}}_{Q_w} = \underbrace{\begin{bmatrix} -v_{w0} \\ v_{w1} \\ \vdots \\ \vdots \\ \vdots \\ \vdots \\ -v_{w(t-r-1)} \end{bmatrix}}_{V_w} \quad (12)$$

(Or) $H_w * Q_w = V_w$ is the matrix-vector form, where $w = 1, 2, 3, 4$. The 'Q_w' vector can be solved in the least squares sense using the generalized inverse method. This gives the denominator vector estimate 'Q_w' as:

$$Q_w = \left(H_w^T H_w\right)^{-1} H_w^T V_w \quad (13)$$

If the coefficients 'q_{wi}' of 'Q_w' vector obtained by the solution of Eq. (12) does not constitute a stable denominator, then by adding an additional equation to this set to get stable denominator, then the solution of the linear set is:

$$\begin{bmatrix} v_{w(t)} & v_{w(t-1)} & \cdots & v_{w(t-r+1)} \end{bmatrix} \text{ and } \begin{bmatrix} -v_{w(t-r)} \end{bmatrix} \quad (14)$$

This process is continued until the Q_w vector gets stable and accurate denominator coefficients.

Finally, the reduced denominator obtained as

$$q_{wr}(\lambda) = q_{w0} + q_{w1}\lambda + \cdots + q_{w(r-1)}\lambda^{r-1} + \lambda^r \quad (15)$$

Step 4: Determination of numerator coefficients of reduced model substitute the reduced denominator coefficients in the Eq. (10) and replace it about $\lambda = 0$. Then, the time moments proportions, f_{wi} of the reduced model, is given as

$$K_{wr}(\lambda) = \sum_{i=0}^{\infty} f_{wi}\lambda^i \quad (16)$$

By equating the time moment proportionals of original system Eq. (9) with reduced model Eq. (16), the reduced numerator coefficients are found thus the reduced numerator is obtained as

$$p_{wr}(\lambda) = p_{w0} + p_{w1}\lambda + \cdots + p_{w(r-1)}\lambda^{r-1}$$

Thus, the corresponding four rth order reduced models obtained from the above as:

$$K_{1r}(\lambda) = \frac{p_{10} + p_{11}\lambda + \cdots + p_{1(r-1)}\lambda^{r-1}}{q_{10} + q_{11}\lambda + \cdots + q_{1(r-1)}\lambda^{r-1} + \lambda^{r}}$$

$$K_{2r}(\lambda) = \frac{p_{20} + p_{21}\lambda + \cdots + p_{2(r-1)}\lambda^{r-1}}{q_{20} + q_{21}\lambda + \cdots + q_{2(r-1)}\lambda^{r-1} + \lambda^{r}}$$

$$K_{3r}(\lambda) = \frac{p_{30} + p_{31}\lambda + \cdots + p_{3(r-1)}\lambda^{r-1}}{q_{30} + q_{31}\lambda + \cdots + q_{3(r-1)}\lambda^{r-1} + \lambda^{r}}$$

$$K_{4r}(\lambda) = \frac{p_{40} + p_{41}\lambda + \cdots + p_{4(r-1)}\lambda^{r-1}}{q_{40} + q_{41}\lambda + \cdots + q_{4(r-1)}\lambda^{r-1} + \lambda^{r}}$$

Step 5: From the above four transfer functions the minimum and maximum values of the coefficients are considered to formulate the reduced interval model represented as

$$G_r(\lambda) = \frac{[p_{w0\min}, p_{w0\max}] + [p_{w1\min}, p_{w1\max}]\lambda + \cdots + [p_{wr-1\min}, p_{wr-1\max}]\lambda^{r-1}}{[q_{w0\min}, q_{w0\max}] + [q_{w1\min}, q_{w1\max}]\lambda + \cdots + [q_{wr\min}, q_{wr\max}]\lambda^{r}}$$
$$= \frac{\sum_{i=0}^{r-1}[p_i^-, p_i^+]\lambda^i}{\sum_{i=0}^{r}[q_i^-, q_i^+]\lambda^i}$$

Step 6: Re-convert the above integer order interval transfer function into its fractional commensurate interval form of the transfer function by substitute the $\lambda = s^{\Upsilon}$

$$G_r(s) = \frac{[p_0^-, p_0^+] + [p_1^-, p_1^+]s^{\Upsilon} + \cdots + [p_{r-1}^-, p_{r-1}^+]s^{\Upsilon r-1}}{[q_0^-, q_0^+] + [q_1^-, q_1^+]s^{\Upsilon} + \cdots + [q_{r-1}^-, q_{r-1}^+]s^{\Upsilon r-1} + [q_r^-, q_r^+]s^{\Upsilon r}} \quad (17)$$

The integral performance indices of the original system related to its reduced model are expressed to calculate and measure the goodness of the reduced order models, by means of the relative integral square error criterion, which are given as

$\text{ISE}_{\text{impulse}}$, $\text{RISE}_{\text{impulse}}$, ISE_{step}, $\text{RISE}_{\text{step}}$, IAE_{step}, $\text{ITAE}_{\text{step}}$ and are defined as follows

$$\text{ISE}_{\text{impulse}} = \int_0^{t\text{sim}} [g(t) - g_r(t)]^2 \mathrm{d}t \tag{18}$$

$$\text{RISE}_{\text{impulse}} = \int_0^{t\text{sim}} [g(t) - g_r(t)]^2 \mathrm{d}t \Big/ \int_0^{t\text{sim}} g^2(t) \mathrm{d}t \tag{19}$$

$$\text{ISE}_{\text{step}} = \int_0^{t\text{sim}} [y(t) - y_r(t)]^2 \mathrm{d}t \tag{20}$$

$$\text{RISE}_{\text{step}} = \int_0^{t\text{sim}} [y(t) - y_r(t)]^2 \mathrm{d}t \Big/ \int_0^{t\text{sim}} [y(t) - y(\infty)]^2 \mathrm{d}t \tag{21}$$

$$\text{IAE}_{\text{step}} = \int_0^{t\text{sim}} |y(t) - y_r(t)| \mathrm{d}t \tag{22}$$

$$\text{ITAE}_{\text{step}} = \int_0^{t\text{sim}} t|y(t) - y_r(t)| \mathrm{d}t \tag{23}$$

where $g(t)$ and $y(t)$ are the impulse and step responses of the original system, respectively, and $g_r(t)$, $y_r(t)$ are that of their approximants. $t$sim indicates the time required for the responses to reach the final steady state value.

3 Examples and Graphs

In this section, three test systems are deliberated and then these systems are reduced by the proposed method to show the superiority and effectiveness of this proposed new algorithm and the results are successfully verified using MATLAB. A step-by-step procedure is given for test systems.

Test system-1

Let us consider a commensurate fractional-order interval system as:

$$G_1(s) = \frac{[1,1]s^{2.4} + [7,8]s^{1.6} + [24,25]s^{0.8} + [24,25]}{[1,1]s^{3.2} + [10,11]s^{2.4} + [35,36]s^{1.6} + [50,51]s^{0.8} + [24,25]} \tag{24}$$

By substituting $\lambda = s^{0.8,}$ a fourth-order integer interval transfer function framed out in terms of λ as:

$$G_1(\lambda) = \frac{[1,1]\lambda^3 + [7,8]\lambda^2 + [24,25]\lambda^1 + [24,25]}{[1,1]\lambda^4 + [10,11]\lambda^3 + [35,36]\lambda^2 + [50,51]\lambda^1 + [24,25]} \tag{25}$$

Using Eqs. (4)–(17) the proposed reduced fractional interval model obtained as

$$G_r(s) = \frac{[0.34444, 0.87745]s^{0.8} + [1.294117, 5.277799]}{[1,1]s^{1.6} + [2.27941, 5.8333525]s^{0.8} + [1.294117, 5.277799]} \tag{26}$$

The trajectories of the original fractional interval system and the proposed reduced fractional interval model are compared in Fig. 1 and 2 for four-time moments shown in Table 1. The integral performance indices showed in Table 2 which are given as $ISE_{impulse}$, $RISE_{impulse}$, ISE_{step}, $RISE_{step}$, IAE_{step}, $ITAE_{step}$ of the original system related to its reduced model are expressed to strengthen the superiority of the proposed method resulting in fewer error values. It can be observed from Figs. 1 and 2 that the proposed method generates stable reduced fractional

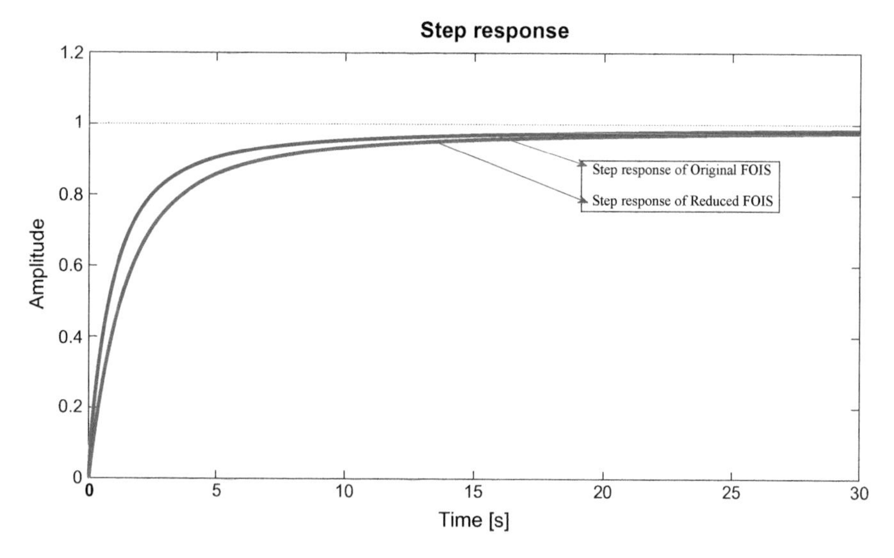

Fig. 1 The lower boundary step response of the reduced fractional order model and original system

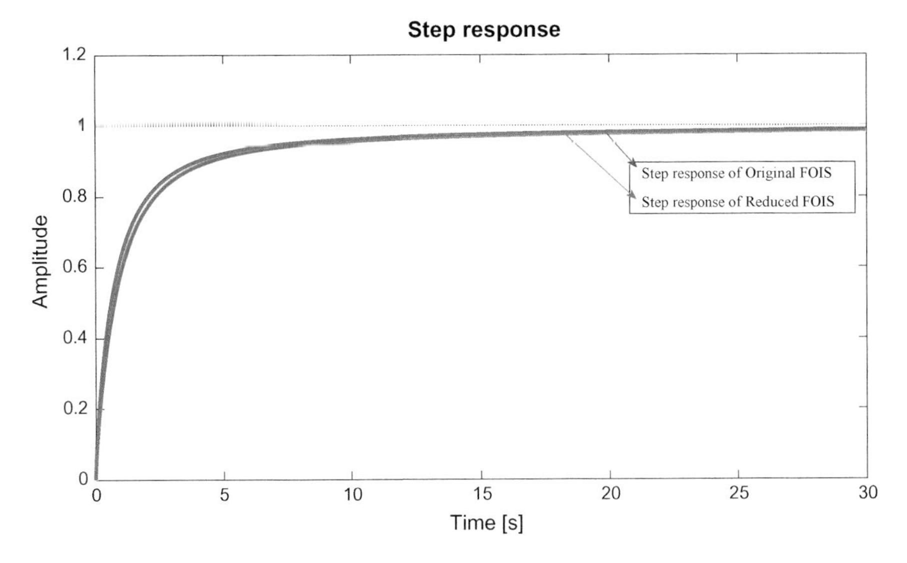

Fig. 2 The upper boundary step response of the reduced fractional order model and original system

Table 1 Time moments of four Kharitonov equations

$K_1(\lambda)$	$K_2(\lambda)$	$K_3(\lambda)$	$K_4(\lambda)$
$v_1 = 1$	$v_1 = 1$	$v_1 = 1$	$v_1 = 1$
$v_2 = -1.0833$	$v_2 = -1.04$	$v_2 = -1.04$	$v_2 = -1.0833$
$v_3 = 1.09028$	$v_3 = 1.0016$	$v_3 = 0.96$	$v_3 = 1.135417$
$v_4 = -1.06308$	$v_4 = -0.947264$	$v_4 = -0.86$	$v_4 = -1.162761$

Table 2 Performance indices of commensurate fractional-order interval systems

Examples		Performance indices					
		ISE for impulse response	RISE for impulse response	ISE for step response	RISE for step response	IAE for step response	ITAE for step response
Example 1	LB	0.052081	0.117601	0.44806	0.078574	0.411871	0.770593
	UB	0.009488	0.020671	0.004004	0.007276	0.100991	0.153221

interval order models with a good approximation for a stable original high order interval system. The transient and the steady state step response of the reduced model obtained by matching four-time moments closely follow the original high order commensurate fractional interval system. The proposed method provides the flexibility to the user in selecting the number and type of extra constraints to be taken into account for better approximations.

4 Conclusion

It is significant to study the behavior of non-integer interval system as they have a huge demand in applying for various control applications. Investigation still needs to be done vast as a part of research in this field. This approach tried to give solutions to some queries related to non-integer interval systems. This note recommended by the author states about the reduction of commensurate non-integer interval system. A blended model order reduction mechanism is implemented to extract a diminishing fractional non-integer model which constitutes a least squares and time moment matching algorithms. As the method associated with matrix-vector algebra, its estimation is clear, efficient and also develops a precise stable reduce interval model. Simulation results validate the dominance of this scheme where the original system may be retrieved with the approximate model thereby shorten the design procedure. Extending the results of this paper to be applicable for a designing of controllers to fractional-order interval systems can be noticed as an interesting research topic as future scope of work. The beauty of this technique is that it provides a great advantage to the engineering practitioners working with large and complex real-time systems.

References

1. Monje CA, Chen Y, Vinagre BM, Xue D, Feliu-Batlle V (2010) Fractional-order systems and controls fundamental and applications advances in industrial control. Springer
2. Maiti D, Konar A (2008) Approximation of a Fractional order system by an integer model using particle swarm optimization technique. In: IEEE sponsored conference on computational intelligence, control and computer vision in robotics and automation, pp 149–152
3. Saxena S, Hote YV (2013) Load frequency control in power systems via internal model control scheme and the model-order reduction. IEEE Trans Power Syst 28(3):2749–2757
4. Oprzedkiewicz K, Mitkowski W, Gawin E (2015) Application of fractional order transfer functions to the modeling of high-Order systems. In: 20th International conference on methods and models in automation and robotics MMAR, 24–27
5. Valerio D, Sa da Costa J (2011) Introduction to single-input, single output fractional control. IET Control Theory Appl 5(8):1033–1057
6. Chen YQ, Petráš I, Xue D (2009) Fractional order control-a tutorial. In: American control conference, pp 1397–1411
7. Tavakoli-Kakhki M, Haeri M (2009) Model reduction in commensurate fractional-order linear systems. ProF. IMechE Part I: J Syst Control Eng 223
8. Pollok A, Zimmer D, Casella F (2015) Fractional-Order Modelling in Modelica. In: Proceedings of the 11th international Modelica conference, Versailles, France, 21–23. ISBN: 978-91-7685-955-1
9. Schilders WH, Van der Vorst HA, Rommes J (2008) Model order reduction theory, research aspects and applications. Springer
10. Sahaj, Saxena., Yogesh., V., Arya, Pushkar Prakash. "Reduced-Order Modeling of Commensurate Fractional-Order Systems", 14th International Conference on Control, Automation, Robotics & Vision, Phuket, Thailand (2016)

11. Lanusse P, Benlaoukli H, Nelson-Gruel D, Oustaloup A (2008) Fractional-order control and interval analysis of SISO systems with time-delayed state. IET Control Theory Appl 2(1)
12. Sunaga T (2009) Theory of interval algebra and its application to numerical analysis. Jpn J Indust Appl Math 26:125–143
13. Kiran Kumar K, Sastry GVKR (2012) A new method for order reduction of high order interval systems using least squares method. Int J Eng Res Appl 2(2):156–160
14. Kiran Kumar K, Sastry GVKR (2012) An approach for interval discrete-time systems using least squares method. Int J Eng Res Appl 2(5):2096–2099
15. Sondhi S, Hote YV (2015) Relative stability test for fractional order interval systems using Kharitonov's theorem. J Control, Autom Electr Syst 27(1):1–9

Statistical Modeling and Optimization of Al-MMCs Reinforced with Coconut Shell Ash Particulates

K. Vikash Kumar and R. Sivasankara Raju

Abstract This study has been described on the optimization of tribological characteristics of Al-coconut shell ash (CSA)-reinforced composite prepared with stir-casting route. In this study, three operating variables (i.e., load, sliding speed, and % of CSA) and three responses (i.e., wear (μm), wear rate (mm^3/m), and coefficient of friction) are considered. The experiments are designed using L_{27} orthogonal array (full factorial design). The influence of each parameter on the response is established using response tables and response graphs. Based on experimental data, the model equations for each response were developed with multiple linear regressions. The models give the factor effects of individual parameters. Interaction effects provide additional information to understand the detailed behavior of parameters. It revealed that load is the most influencing effect on wear behavioral responses. An optimum parameter combination was obtained by using a genetic algorithm.

Keywords MMC · Design of experiments · Genetic algorithm · Regression

1 Introduction

Aluminum matrix composites (AMCs) are gained extensive concentration owing to better specific strength, and exceptional tribological and machinability properties in excess of the unitary alloys [3, 10, 11, 14]. However, AMCs have better-quality properties, but it is unable to expose due to intricacy in fabrication and characterization. In general, stir-casting technique is the cheapest and widely used in mass production for automobile industries. Particulate reinforced composites have exhibited economical worth in fabrication and design of components [3, 15, 17].

K. Vikash Kumar (✉)
Department of Mechanical Engineering, GIET University, Gunupur, Odisha 765022, India

R. Sivasankara Raju
Department of Mechanical Engineering, Aditya Institute of Technology and Management, Tekkali, Andhra Pradesh 532201, India

BBVL. Deepak et al. (eds.), *Innovative Product Design and Intelligent Manufacturing Systems*, Lecture Notes in Mechanical Engineering,
https://doi.org/10.1007/978-981-15-2696-1_67

Wear is a linear dimensional loss in the material occurred due to abrasion. Extensive studies on the tribological performance of AMC are considering with the effect of variables such as particulate size, shape, and volume of reinforcements, and load/pressure, velocity, sliding distance, and other extensive operating conditions. Similarly, the tribological behavior of composite has also been enhanced due to the distribution of reinforcement in the matrix which influences the effect of interfacial bonding between matrix and reinforcement [1–3, 8, 9]. Sahin [11] optimized the wear behavior of Al-SiC-MMCs and reported that the effect of abrasive particles' size and volume played a key role in the abrasion mechanism. Kok and Ozdin [3] studied the tribological performance of Al-2024-Al_2O_3 composite. The present work is aimed to determine the optimal tribological performance of Al-1100-CSA composites reinforced with 5, 10, and 15 volume %. Moreover, the response table and response graphs are used to investigate the significant parameters and the effect on the tribological performance of CSA composite.

2 Experimental Procedure

The commonly available coconut is one of the most important nature filler. Coconut shell of filler contains high lignin and prevents as weather resistant. Therefore, coconut shell is considered as reinforcement.

2.1 *Preparation of Coconut Shell Ash (CSA) Reinforcement Particles*

The coconut shell is collected from nearby areas, where a huge amount of the waste generated and a big problem for disposal. Shells are crushed in a jaw crusher followed hammer mill to make courser size of coconut dust powder. The powder is packed in a graphite crucible and fired in an electric resistance furnace at 1450 °C. The process of CSA has been explained elsewhere [4, 5, 12, 14]. The burnt CSA is sieved with a size of $\leq$ 240 BSS mesh (63 μm) and used as reinforcement particles. CSA is examined by X-ray diffraction (XRD), and chemical elements are determined and listed in Table 1.

Table 1 Chemical composition of CSA

Elements	Al_2O_3	CaO	Fe_2O_3	MgO	K_2O	MnO	Na_2O	SiO_2	ZnO	Fixed C
% of Wt.	18.68	0.67	3.42	18.2	0.59	0.25	0.47	49.2	0.32	4.26

2.2 Preparation of Composite

Preparation of AMC with virgin alloy (Table 2) is preheated at 450 °C. The preheated alloy is transferred to the bottom pouring furnace and raised to 690 °C. The preheated CSA charged into the melted alloy. After charging, the molten metal is mixed with the help of a motor-driven stirrer in the presence of argon gas at a speed of 600 rpm for 9 min for homogeneous distribution of reinforced particles. The molten metal is superheated at 720 °C and poured into a preheated (at 300 °C) cast iron mold of the size of 100 × 20 × 40 mm^3.

2.3 Optical Microscopic Studies

Microstructural of cast composites are examined with standard technique (ASTM E-3) and etched with Keller's reagent. Etched samples are examined by Infinity Lite Model no: XJL-17.

2.4 Testing Methods

Cast composites of hardness values are measured (model: DHV 1000) with an applied load of 100 N. Tensile properties of composites (Al-CSA) have been tested with Hounsfield tensometer (model: ETM-ER3/772/12). Al-CSA composites of abrasion resistance are tested using a pin-on-disk tester (TR-201 LE-PHM 400). The samples are cleaned with acetone. The test samples having 8 mm diameter and 35 mm length, which slide against steel counter-face (material EN-31) of track diameter 50 mm. Wear rate is the ratio of volume loss per unit sliding distance. Volume loss is calculated from weight loss to the density of sample [4, 5, 14]. Coefficient of friction is measured as frictional forces per unit normal load [6, 7, 12].

2.5 Design of Experiments

In this work, the experiments are carried out based on a full factorial design which consists of 27 experimental runs for three factors at three levels. The three operating variables such as load (*L*), % of coconut shell ash (*R*), and sliding distance (*D*) at

Table 2 Chemical composition of Al-1100

Elements	Cu	Zn	Mn	Si	Fe	Al
% of Wt.	0.05–0.2	0.1	0.05	0.7–0.85	0.25–0.75	Balance

Table 3 Parameters and their levels

Factor (units)	Symbol	−1	0	+1
Load (N)	*L*	10	30	50
% of CSAp (% of vol.)	*R*	5	10	15
Sliding distance (m)	*D*	1000	2000	3000

Table 4 Experimental layout with results

Run	*L*	*R*	*D*	*W*	Wr	C.F	Run	*L*	*R*	*D*	*W*	Wr	C.F
1	−1	−1	−1	191	5.556	0.265	15	0	0	1	460	2.436	0.226
2	−1	−1	0	193	2.815	0.201	16	0	1	−1	268	5.808	0.355
3	−1	−1	1	385	2.173	0.188	17	0	1	0	312	3.115	0.292
4	−1	0	−1	185	5	0.224	18	0	1	1	426	2.321	0.28
5	−1	0	0	180	2.423	0.185	19	1	−1	−1	480	7.111	0.663
6	−1	0	1	210	2.159	0.197	20	1	−1	0	521	4.167	0.501
7	−1	1	−1	165	3.462	0.249	21	1	−1	1	685	4.012	0.391
8	−1	1	0	175	2.327	0.234	22	1	0	−1	415	6.923	0.582
9	−1	1	1	268	1.936	0.271	23	1	0	0	505	3.962	0.445
10	0	−1	−1	380	6.333	0.412	24	1	0	1	525	2.846	0.359
11	0	−1	0	391	3.222	0.299	25	1	1	−1	398	6.615	0.565
12	0	−1	1	478	2.704	0.238	26	1	1	0	468	3.654	0.453
13	0	0	−1	310	6.5	0.351	27	1	1	1	478	2.641	0.392
14	0	0	0	285	3.115	0.263							

three different levels as given in Table 3. Experiments were carried out in an arbitrary way in order to evade discrimination. Table 4 has given the tribological performances such as wear (*W*), wear rate (Wr) ((mm^3/m)*10^{-3})), and coefficient of friction (C.F).

3 Results and Discussion

3.1 Characterization of CSAp

Figure 1 XRD spectrum of reinforced material shows various compounds which exist in coconut shell ash. CSA is basically alumina silicate. Table 1 presents components such as SiO_2, Al_2O_3, and Fe_2O_3 which are major constituents in CSA, whereas MgO, CaO, and K_2O are minor. The XRD peaks indicate the presence of carbon, apparently in the form of graphite. In-situ mixing of components such as SiO_2, Al_2O_3, and graphite (Gr) are able to improve thermo-mechanical properties by thermal treatment [7, 14].

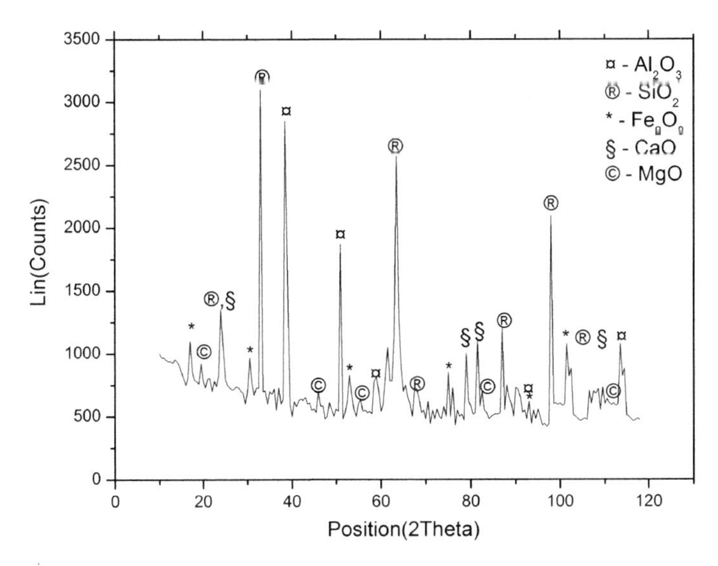

Fig. 1 XRD of coconut shell ash particulate

3.2 *Mechanical Properties of Al-CSA-MMC*

Hardness and elongation are proportional sense, with a further in aid as noted in Fig. 2a, which is expected to strain to harden of the composite. The tensile strength and elongation of CSA composites are vided in Fig. 2b, and this caused to grain refinement.

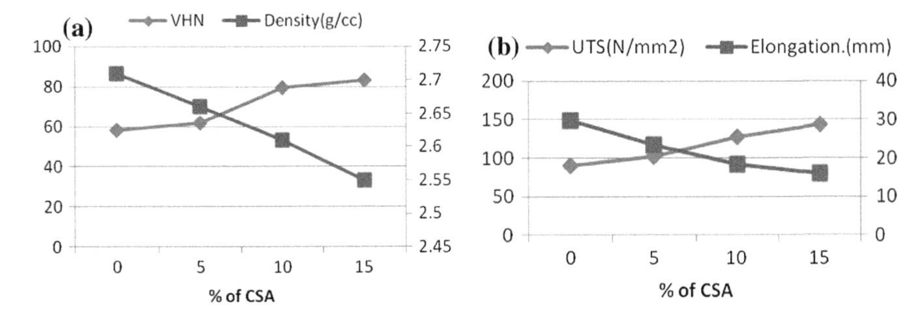

Fig. 2 Mechanical properties of Al-CSA-MMCs

3.3 Microstructural Studies

The microstructures of the CSA-MMCs are shown in Fig. 3. The share of grains in the microstructure is felt to be uniform. Likewise, noted that with a rise in the pct. of CSA, the grain diameter declines, which enhances the strength and drops the ductility [13].

3.4 Analysis of the Process Parameters and Interactions

Design of experiments is aiming to obtain the relation between operating variables and tribological performances. From the full factorial experimental design, mean of level is computed and given in Table 4. The influence of each variable parameter (i.e., *L*, *R*, and *D*) on the tribological performances (i.e., Wr and C.F) is carried out using level mean analysis. From the ANOVA (Vide in Table 5), load (74.83%) is the highest influencing parameter followed by sliding distance (14.11%) and %of CSAp (5.58%). The contributions of overall factors are performed by 97.3% of the

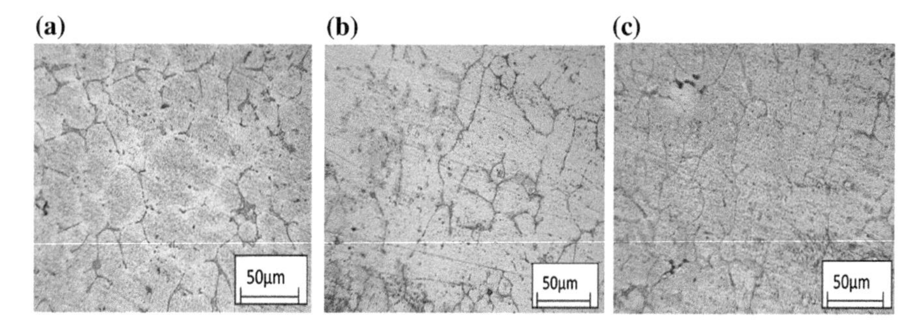

Fig. 3 Microstructures of. **a** Al-5%CSA, **b** Al-10%CSA, and **c** Al-15%CSA-MMC

Table 5 ANOVA of wear

Source	DF	SS	Adj SS	MS	*F*	*P*	% Cont.
L	2	254.008	254.008	127.004	111.	0	74.80%
R	2	18.958	18.958	9.479	8.33	0.011	5.58%
D	2	47.915	47.915	23.958	21.05	0.10%	14.11%
*L**R*	4	0.769	0.769	0.192	0.17	0.948	0.23%
*L**D*	4	5.891	5.891	1.473	1.29	0.349	1.73%
*R**D*	4	2.819	2.819	0.705	0.62	0.661	0.83%
Error	8	9.104	9.104	1.138			
Total	26	339.464					

total variance in sliding wear. It is clear that there is no strong interaction among parameters.

ANOVA Table 6 shows that sliding distance is the highest influencing parameter on the wear rate of Al CSA composites with 75.33% of the contribution. The load is the second influencing factor with 45.25 Fisher value. The overall contribution is 98.4% on the wear rate of Al-CSA composites. Similarly, ANOVA (Table 7) revealed that load (76.25%) is the most influencing parameter followed by sliding distance (15.11%). The combined effect of load and sliding distance is one of the influencing factors with 21.71 Fisher value. The contributions of overall factors are performed by 97.9% of the total variance in the coefficient of friction.

A level mean is the average value of response at a specific level. The interactions of input parameters are also having a significant effect on responses. The influence of process parameters (i.e., *L*, *R*, and *D*) on the tribological performance (wear (W), wear rate (Wr), and coefficient of friction (C.F)) is observed, which is shown in Fig. 4.

From response graphs, it is identified that *L* is the dominating factor on the responses such as wear and wear rate, whereas *D* is dominating parameter on C.F. The tribological performances (i.e., wear, Wr, and C.F) increase with increasing load. While the load increases, the composite losses its ability to carry the load due to fracture of reinforced particles and formation of debris, which causes the loss of material vis-a-vis plowing, delamination, and plastic deformation [14, 16]. Wear is increasing with the increase of sliding distance and decrease of the addition of reinforcement (CSA) due to hard ceramic compounds present in CSA which leads to greater bonding of the matrix. The addition of CSA particles has a minor effect on C.F, and better value is observed at 10% of CSA, whereas more the value of the sliding distance will lessen the value of C.F. The optimal conditions for wear, Wr, and C.F are $L_1R_3D_1$, $L_1R_3D_3$, and $L_1R_2D_3$, respectively.

Table 6 ANOVA of wear rate

Source	DF	SS	Adj SS	MS	*F*	*P*	% Cont.
L	2	61.309	61.309	30.655	45.25	0	18.05%
R	2	11.745	11.745	5.873	8.67	0.01	3.46%
D	2	255.793	255.793	127.896	188.8	0	75.33%
*L*R*	4	1.979	1.979	0.495	0.73	0.596	0.58%
*L*D*	4	1.27	1.27	0.317	0.47	0.758	0.37%
*R*D*	4	2.058	2.058	0.514	0.76	0.58	0.61%
Error	8	5.419	5.419	0.677			
Total	26	339.573					

Table 7 ANOVA of coefficient of friction

Source	DF	SS	Adj SS	MS	F	P	% Cont.
L	2	200.041	200.041	100.021	1120.53	0	76.25%
R	2	5.357	5.357	2.679	30.01	0	2.04%
D	2	39.638	39.638	19.819	222.03	0	15.11%
$L*R$	4	3.703	3.703	0.926	10.37	0.003	1.41%
$L*D$	4	7.751	7.751	1.938	21.71	0	2.95%
$R*D$	4	5.15	5.15	1.288	14.42	0.001	1.96%
Error	8	0.714	0.714	0.089			
Total	26	262.355					

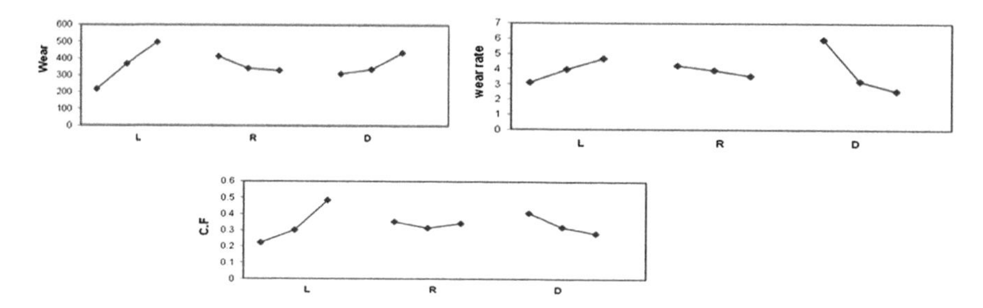

Fig. 4 Response graphs

3.5 *Modeling*

Second-order models for wear, Wr, and C.F are developed using full factorial design experimental data. The input data to SPSS software is provided in a coded form of factors, i.e., −1 to +1. The models developed were given by:

$$\begin{aligned}\text{Wear} = {} & 317.704 + 140.167L \\ & - 14.44\,R + 62.389\,D + 28.44\,R^2 + 35.94\,D^2\left(R^2 = 0.942\right)\end{aligned} \tag{1}$$

$$\begin{aligned}\text{Wear rate} = {} & 3.2 + 0.782L - 0.345R - 1.671D + 1.052D^2 \\ & - 0.283LD(R^2 = 0.965)\end{aligned} \tag{2}$$

$$\begin{aligned}\text{CF} = {} & 0.263 + 0.13L - 0.004R - 0.062D + 0.052L^2 + 0.033R^2 + 0.026D^2 \\ & - 0.020LR - 0.049LD + 0.025RD\left(R^2 = 1.00\right)\end{aligned} \tag{3}$$

The R^2 value of 0.942 specifies that 94.2% of the variation in the wear has explained by the model. Similarly, the R^2 value of Wr and C.F is 0.942 and 1.00, respectively. From the regression equation, the interaction between C.F. with LR, LD, and RD is identified. Further, the significant interaction between Wr with LD also observed. From the above equations, it can be observed that the L, R, and

D and their square terms and interaction terms also coming into the model. The square terms of parameters indicate the quadratic nature of the response. The three individual responses are not having the same optimum input values. To satisfy all the responses, it requires multi response optimization and has been performed by using a genetic algorithm. The optimum values of *L*, *R*, and *D* are 10, 13.61 and 1795, respectively, and corresponding values of response wear, Wr, and C.F are 165.28, 3.32, and 0.42.

4 Conclusion

1. Aluminum 1100 alloy CSA composites (i.e., Al-5%CSA, Al-10%CSA, and Al-15% CSA) are prepared indigenously with a laboratory scale by the stir-casting process.
2. Full factorial experiments for three process parameters at three levels were conducted to examine the tribological performance (i.e., wear, Wr, and C.F) on CSA composites.
3. The wear resistance of composite increases with increasing CSA volume and decreases with increasing load.
4. Load (*L*) is the dominating factor on the responses such as wear and wear rate (Wr), whereas sliding distance (*D*) is the dominating parameter on the coefficient of friction (C.F).
5. Optimum parameters were predicted through genetic algorithm technique (load = 10 N, percentage of coconut shell ash = 13.61, and sliding speed = 1795 m) to achieve the best responses.

References

1. Hatch JE (1984) Aluminum properties and physical metallurgy. In: ASM (ed) ASM International., OH, USA, pp 1–24. https://doi.org/10.1361/appm1984p001
2. Kato K (2000) Wear in relation to friction—a review. Wear 241:151–157. https://doi.org/10.1016/s0043-1648(00)00382-3
3. Kök M, Özdin K (2007) Wear resistance of aluminum alloy and its composites reinforced by Al_2O_3 particles. J Mater Process Technol 183(2–3):301–309. https://doi.org/10.1016/j.jmatprotec.2006.10.021
4. Rallabandi SR, Gunji SR (2019) Assessment of tribological performance of Al-coconut shell ash particulate—MMCs using grey-fuzzy approach. J Inst Eng Ser C 100, 1:13–22. https://doi.org/10.1007/s40032-017-0388-4
5. Raju RSS, Rao GS (2017) Assessment of tribological performance of coconut shell ash particle reinforced Al-Si-Fe composites using grey-fuzzy approach. Tribol Ind 39(3):364–377. https://doi.org/10.24874/ti.2017.39.03.12
6. Raju SS et al (2017) Estimation of tribological performance of Al-MMC reinforced with a novel in-situ ternary mixture using grey relational analysis. Indian J Sci Technol 10(15):1–9. https://doi.org/10.17485/ijst/2017/v10i15/113825

7. Raju SS, Rao GS (2017) Assessments of desirability wear behaviour on Al-coconut shell ash—metal matrix composite using grey—fuzzy reasoning grade. Indian J Sci Technol 10(15):1–11. https://doi.org/10.17485/ijst/2017/v10i15/113826
8. Rodopoulos CA, Wessel JK (2004) Metal matrix composites. In: Wessel JK (ed) Handbook of advanced materials: enabling new designs. Wiley Online Library, pp 99–117. https://doi.org/10.1002/0471465186.ch6
9. Rohatgi PK (1993) Metal-matrix composites. Def Sci J 43(4):323–349. https://doi.org/10.14429/dsj.43.4336
10. Rohatgi PK, Schultz B (2007) Lightweight metal matrix nanocomposites—stretching the boundaries of metals. Mater Matters 2(4):1–6
11. Sahin Y (2010) Abrasive wear behavior of SiC/2014 aluminum composite. Tribol Int 43:939–943. https://doi.org/10.1016/j.triboint.2009.12.056
12. Sankara Raju RS et al (2018) Optimization of tribological behavior on Al-coconut shell ash composite at elevated temperature. IOP Conf Ser Mater Sci Eng 314:012009. https://doi.org/10.1088/1757-899X/314/1/012009
13. Raju SS et al (2019) Wear behavioral assessment of Al-CSAp-MMCs using grey-fuzzy approach Measurement. 140: 254–268. https://doi.org/10.1016/j.measurement.2019.04.004
14. Raju SS et al (2017) Investigation of tribological behavior of a novel hybrid composite prepared with al-coconut shell ash mixed with graphite. Metall Mater Trans A 48(8):3892–3903. https://doi.org/10.1007/s11661-017-4139-1
15. Venkata Siva SB et al (2013) Machinability of aluminum metal matrix composite reinforced with in-situ ceramic composite developed from mines waste colliery shale. Mater Manuf Process 28(10):1082–1089. https://doi.org/10.1080/10426914.2013.811734
16. Venkata Siva SB et al (2013) Preparation of aluminum metal matrix composite with novel in situ ceramic composite particulates, developed from waste colliery shale material. Metall Mater Trans B 44(4):800–808. https://doi.org/10.1007/s11663-013-9832-x
17. Venkata Siva SB et al (2015) Quantitative studies on wear behavior of Al-(Al_2O_3-SiC-C) composite prepared with in situ ceramic composite developed from colliery waste. J Eng Tribol 229(7):823–834. https://doi.org/10.1177/1350650115570696

Automatic Generation Control of Multi-area System Incorporating Renewable Unit and Energy Storage by Bat Algorithm

Subhranshu Sekhar Pati, Aurobindo Behera and Tapas Kumar Panigrahi

Abstract The proposed work demonstrates the automatic generation control (AGC) of two area systems. Both areas contain conventional plants such as thermal–hydro–diesel, as well as renewable solar PV and geothermal units with proper physical constraints. Energy storage device named ultra-capacitor (UC) has also been incorporated in the system to support the stability and damp out the oscillations. Proportional–integral–derivative with filter (PIDN) optimized through bat algorithm is used, and the response is compared with other integer-order controllers. The system is tested under three distinct cases [i.e., SLP of 1% in area-1 (Case I), 1% area-2 (Case II), and 1% in area-1, 2 (Case III)]. The analysis points out the superior performance of PIDF to that of other controllers. The impact of geothermal plant and energy storage on system performance is investigated, and it can be inferred that the inclusion of such devices would enhance the performance of the model to a great extent.

Keywords Automatic generation control (AGC) · Bat algorithm · Integral square error (ISE) · Renewable sources

1 Introduction

With the rise in industry and population, the demand for electrical energy also goes on increasing. So for smooth operation, the frequency persists throughout time. Here, AGC is used to maintain the frequency of the system, as well as tie-line power deviations within the defined limit. Adverse effects may be caused due to an

S. S. Pati
International Institute of Information Technology, Bhubaneswar, Odisha 751003, India

A. Behera (✉)
Cambridge Institute of Technology, Ranchi, Jharkhand 835103, India
e-mail: C115002@iiit-bh.ac.in

T. K. Panigrahi
Parala Maharaja Engineering College, Berhampur, Odisha 761003, India

BBVL. Deepak et al. (eds.), *Innovative Product Design and Intelligent Manufacturing Systems*, Lecture Notes in Mechanical Engineering,
https://doi.org/10.1007/978-981-15-2696-1_68

imbalance between the generation and the load demand. Hence, AGC is one of the vital mechanisms in the aspect of generation–demand balance. Elgerd et al. [1] established the concept of applying AGC to the traditional system having thermal unit only. Nanda et al. [2] did exploration in a traditional scenario with hydro and thermal power stations, also integrating generation rate constraint (GRC) and reheat turbine as restraints. Gozde et al. [3] used a similar system as in [2] but considered another nonlinearity called governor dead-band (GDB). Bhatt et al. [4] considered the multi-area AGC having sources such as hydro–thermal–diesel. Saikia et al. [5] worked on an interconnected AGC applied to two, three, and five area systems. The analyses were performed on a lone reheat turbine and GRC. Later, a multi-source system incorporates dissimilar sources which present a more realistic scenario of the power sector such as nuclear, gas, and combined cycle gas turbine. Therefore, the exploration of the remodeled system is limited to traditional sources. Further, the addition of non-conventional sources might direct current research toward different scenarios. Further inclusion of wind turbine and the photovoltaic cell can be found in numerous research articles [5, 6]. Geothermal energy (GE) is another probable source of renewable energy for the generation of electrical energy and space heating. GE is a form of heat energy automatically stored inside the earth, therefore harnessed. GE being a discrete electricity generating unit has not yet been studied along with AGC. Therefore, the GE plant along with other non-conventional sources will be effective in the AGC study.

Literature surveys reveal that peak overshoot/undershoot and settling time of the system can be minimized by the use of FACTS and energy storage devices. However, a particular group of FACTs devices like synchronous static series compensators, series capacitor controlled by thyristors, interline power flow controllers are the most efficient for maintaining the system stability by reducing the inter-area oscillations [7]. On a similar note, use of energy storage systems such as ultra-capacitor (UC) and battery [8] can also support during the small load fluctuation by adding or subtracting the real power at a faster rate; it further attenuates the oscillation in power system and enhances the stability of the overall system. With the use of diverse constraints like GDB, GRC and boiler dynamics (BD) in the system for deriving realistic response make the system highly complex. Hence, it is recommended to use a controller like integral (I), proportional–integral (PI), and proportional–integral–derivative (PID) in the system to maintain the frequency deviation and change in tie-line power under the steady-state limits [9].

In recent papers, it is observed that more efficient and advanced controller like proportional–integral–derivative with filter (PIDN) and a controller having 2 degrees of freedom (2DOF) are bought together [10]. Apart from this controller, the cascade controller is also well effective to curb the system disturbance when nonlinearity present in the system. For the active contribution of the secondary controller, the gain parameter should be trained through an optimization algorithm. For this tenacity, researchers applied the classical technique. In this technique, only one parameter can be optimized while another parameter to be fixed with subjected to the performance characteristics. With an increase in a number of parameters, the classical algorithm cannot be applied to tune the system parameter also the time consumption in the process rises and becomes incompetent. To evade the complications prompted by the classical method, the recent researches are applied with diverse metaheuristic techniques such as genetic, differential evolution, flower pollination, bacteria foraging, cuckoo search, and bat algorithm [7–12].

In the present work, a combination of conventional and non-conventional sources is tested with a PIDN controller (PID having N-filter). The controller setting is tuned using the bat algorithm. The ISE objective function is applied to guide the optimization process toward the optimal solution. The performance of the same is tested against other controllers such as PID and 2DOFPID.

2 System Specification

Renewable energy-based two-area system is considered for the proposed work as shown in Fig. 1. The thermal, hydro, and solar photovoltaic plant is common for both areas. Apart from these plants, a new type of power plant in the energy sector

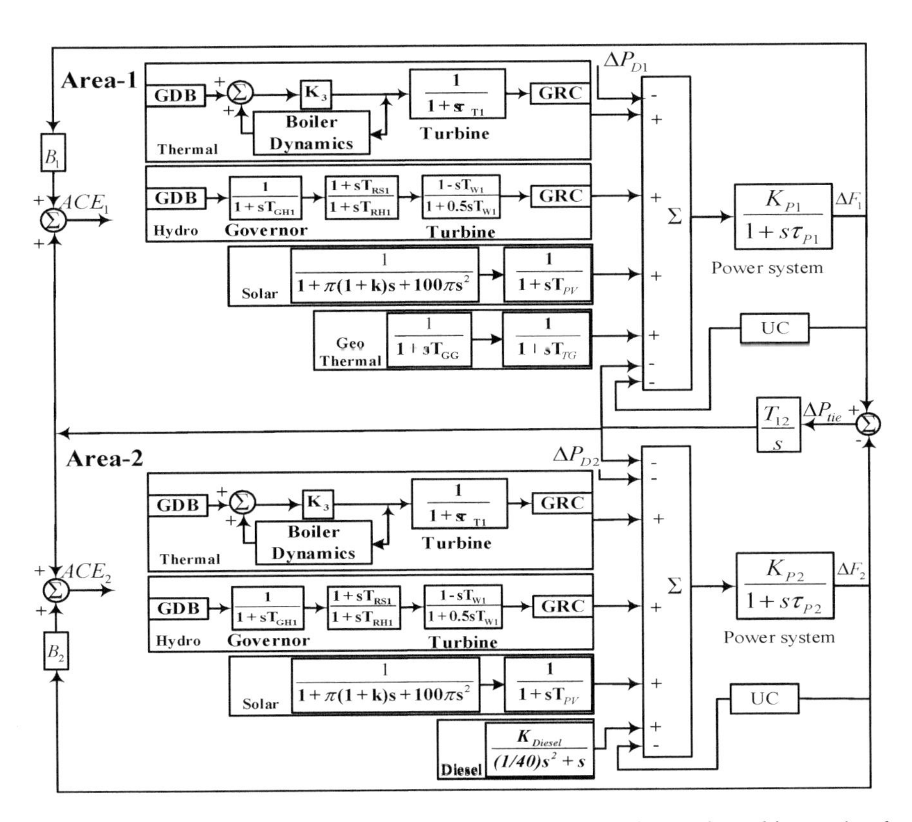

Fig. 1 System configuration diagram considering thermal–hydro–solar–geothermal in area-1 and thermal–hydro–solar–diesel in area-2

named as GE and diesel power plant is considered for study in area-1 and -2, respectively. However, appropriate nonlinearities like boiler dynamics (BD), generation rate constant (GRC), and governor dead band (GDB) are also included in the system model [10, 11]. Maximum power point t (MPPT) coordinated solar power plant is also put their footprint on both the areas. Diesel operated generating unit is also adopted in the area-2 to make the system more realistic. Moreover, renewable-based geothermal power plant (GTPP) generating unit is designed and used in the system. The generalized configuration of GTPP is quite similar non-reheat thermal power plant as there is no need for re-heater in case of GTPP. The modeling of GTPP is well presented in [8]. The simplified transfer function used in the work is depicted in the block diagram of the overall system as shown in Fig. 1, in which the value of the time constant is solely followed by the characteristics of main inlet volumes and steam chest.

$$\mathrm{TF_{UC}} = \frac{K_{\mathrm{UC}}}{1 + sT_{\mathrm{UC}}} \tag{1}$$

For maximizing of all generating unit, an energy storage device called UC included in both areas. UC is a double layer capacitor where a charge is stored in its layer made up of microporous material such as activated carbon and becoming one of the emerging areas of energy storage. Apart from storing, it also has a large power density means more energy can be accommodated in a small place and higher efficiency compared to a normal battery. UC is more efficient when it operated combined with battery. The expression of UC is depicted in Eq. (1).

3 Controller and Objective Function

For better control of the system, two robust controllers named PIDN are employed. PIDN controller is nothing but an extension of a PID controller with an additional filter (N). The filter is effective to minimize the unwanted noise and high-order frequency produced during the system operation. Proportional gain (K_{p}), integral gain (K_{i}), and derivative gain (K_{d}) are the major controller parameters. The block diagram of the PIDN controller is presented in Fig. 2, and in Eq. (2), the transfer function is illustrated.

$$\mathrm{TF_{PIDN}} = \left[K_{\mathrm{p}} + K_{\mathrm{i}}\left(\frac{1}{s}\right) + K_{\mathrm{d}}\left(\frac{Ns}{s+N}\right)\right] \tag{2}$$

Area control error (ACE) of all areas is the initiating signal, and the controller output signal is fed to the respective area. To apply the modern heuristic optimization technique, it is required to formulate a suitable objective function.

Based on different performance criteria, an objective function is designed keeping in mind the system specification and constraints [11]. However, integral

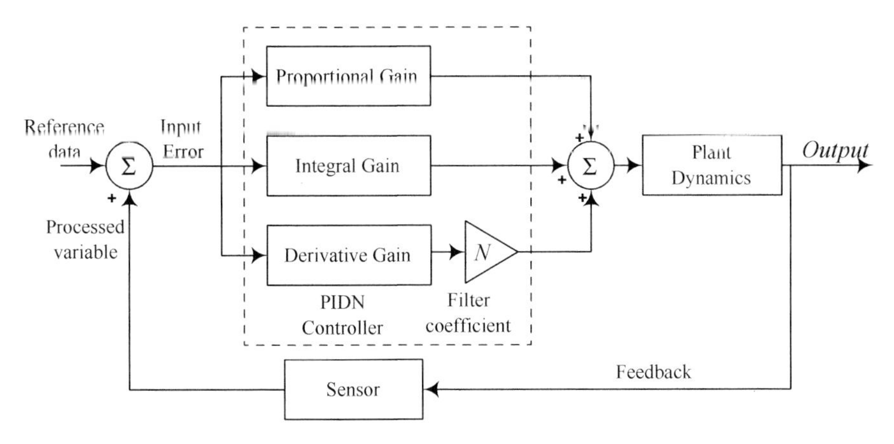

Fig. 2 Controller design diagram describing the flow of error signal and generation of the control signal

square error (ISE) provides a better result compared to integral time absolute error (ITAE) and integral absolute error (IAE). That is why ISE is applied as performance criteria in the design process of the bat algorithm.

$$\mathrm{TF}_{\mathrm{ISE}} = \int_{0}^{t_{\mathrm{sim}}} (\Delta F_1 + \Delta F_2 + \Delta P_{\mathrm{tie}})^2 \cdot \mathrm{d}x \quad (3)$$

4 Optimization Technique

An efficient nature-based meta-heuristic optimization technique called bat algorithm proposed by Yang [13] is introduced for the tuning of the controller parameter. The algorithm was designed by taking the echolocation behavior of bats and proved its effectiveness in the diverse multi-disciplinary research field. Bats are the only mammals in the whole aviary kingdom that have the capability of echolocation [13]. Generally, they produce the sound signal termed as sonar to search their food and their commuting way in the nighttime. The signals emitted by bats are normally in high loudness but very short frequency which can easily bounce back after hitting nearby objects. These rebounded modulated waves are detected by bats so that they can easily identify the foods or pray. However, this behavioral aspect of the bat can be used to formulate an algorithm that can be applied in the appropriate objective function for a better approximation of the controller parameter.

For a better understanding of the echolocation characteristics, simplified rules can be considered in the process-making steps of optimization algorithm [14].

1. For sensing the distance, all bats apply the echolocation characteristics, and predominantly, we assume that all bats know their food/prey and surrounding obstructers during the commutation.
2. When bats looked by prey, they must fly arbitrarily with fixed velocity u_i at position P as well as frequency f_{low} (means varying wavelength) and loudness l_{o}.
3. It is obvious that the loudness can be varied randomly but we vary (positively) from high value l_{o} to low constant value l_{low}.

The range of "f", f_{low} to f_{high} indicates the range of wavelength (from w_{low} to w_{high}) because $w \times f$ is a constant value. That is why it is advisable to change one variable (frequency) keeping another variable (wavelength) to a fixed number. For simple approximation in implementation, low value of frequency is discarded and only high frequency is taken into account as it corresponds to shorter wavelength and travels only a few meters. In fact, sound waves emitted by bat also travel 1–2 m. The mathematical expressions involved in the optimization algorithm are depicted in Eqs. (4)–(7). For finding the updated solution (i.e., position and velocity), Eqs. (4) and (5) are used.

$$f_i = f_{\text{low}} + \left(f_{\text{high}} - f_{\text{low}}\right) \times c \quad (4)$$

$$u_i^0 = u_i + (e_i - e_0) \times f_i \quad (5)$$

in which c is a constant random number in the range from 0 to 1. e_i is the latest best solution available. Then by using Eqs. (4) and (5), new better solution can be formulated as given below:

$$e_i^t = e_i + u_i \quad (6)$$

Once the current best solution is isolated from the existing solution, then again a new solution for each individual (bat) can be composed by adopting Eq. (7).

$$e_{\text{new}} = e_{\text{old}} + l^t \times c \quad (7)$$

where c is a constant value and it refers to average loudness. In Fig. 3, the step-by-step application of the optimization technique to the discussed problem is described. Here, the controller parameters are obtained through the optimization technique and using "sim('model', config.set)", the interface is executed to address the actual problem of AGC.

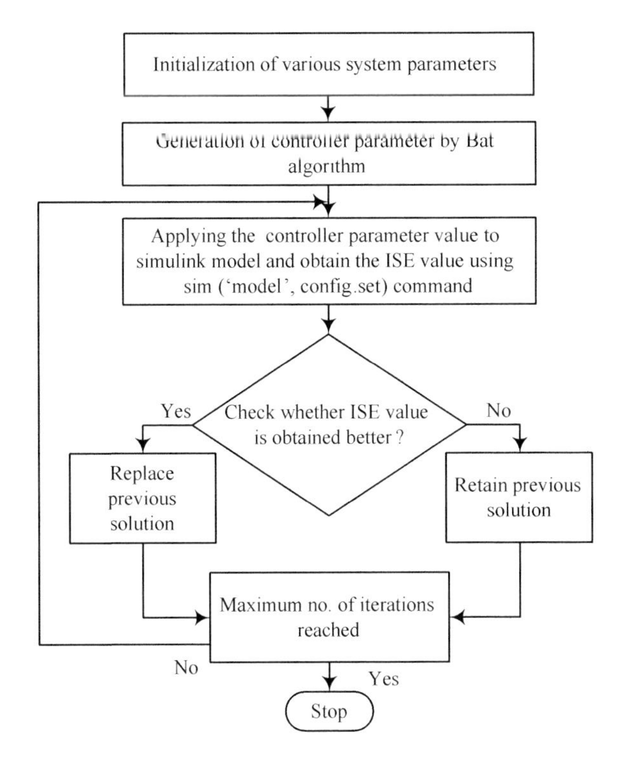

Fig. 3 Flowchart describing the implementation of the optimization technique to the discussed problem

5 Result and Analysis

Two area systems were considered for thermal, hydro, solar, diesel, and GTPP along with UC is considered for the investigation. Two PIDN controllers tuned through BA are employed in the proposed work. In the tuned controller parameter for PID, 2-Degree-of-Freedom PID (2DOFPID) and PIDN are displayed in Table 1. The range within which the parameters of the controller are to be varied is $0 < K_p$, K_i, $K_d < 2$; $0 < N < 500$.

1.9051, 1.5221, 0.8065, 429 are the PIDN controller parameter and are applied to the system under three different cases named as Case I (load perturbation of 1% in area-1 only), Case II (1% in area-2 only), Case III (1% in both area-1 and 2). In Fig. 4, comparative analyses of different controllers show the effectiveness of the PIDN controller and the graphical representation of 3 cases is displayed in Fig. 5. This is further emphasized by the mathematical representation regarding the system performance in terms of maximum over/undershoots and time of settling presented in Table 1.

Table 1 System response parameters ΔF_1, ΔF_2, and ΔP_{tie} for different values of ΔP_D in a step loading condition

Cases \Controller	ΔF_1		ΔF_2		ΔP_{tie}	
	Under shoot $\times 10^{-3}$	Settling time	Under shoot $\times 10^{-3}$	Settling time	Under shoot $\times 10^{-3}$	Settling time
Case I: PIDN	11.09	5.651	4.144	9.628	1.945	10.71
Case II: PIDN	9.982	6.637	3.730	9.518	1.750	9.015
Case III: PID	15.57	7.958	7.505	10.24	2.766	14.77
Case III: 2DOFPID	10.87	8.253	5.232	10.51	1.348	11.56
Case III: PIDN	10.26	6.911	4.358	9.379	1.695	10.63

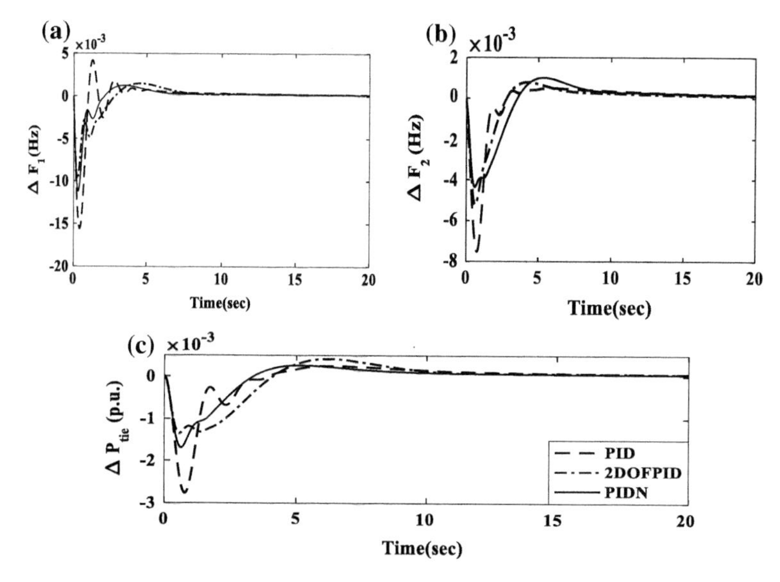

Fig. 4 Step load variation 1% in both areas for testing the system **a** frequency in area-1 **b** frequency in area-2 **c** power exchange within areas

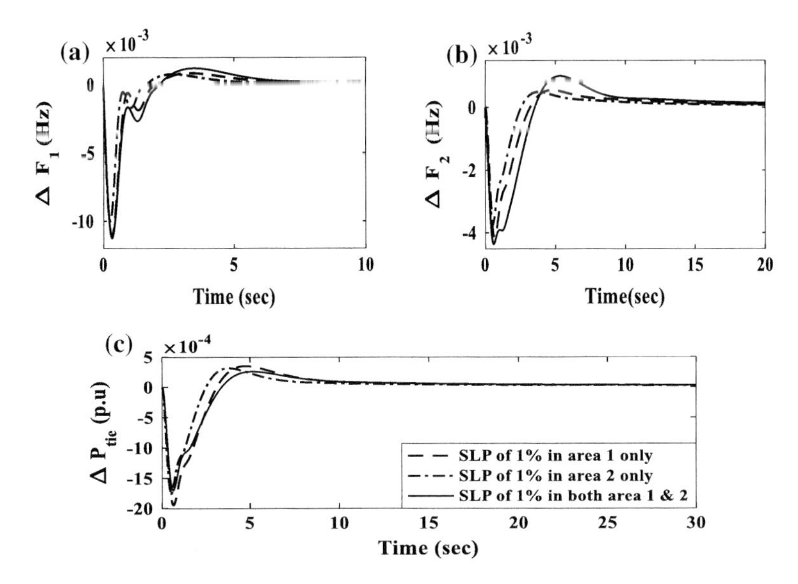

Fig. 5 Testing the system under various cases of load variation

6 Conclusion

The extensive analysis incorporating renewable generating units like solar plant and GTPP with energy storage device as UC has been performed in the paper. The PIDN controller is successfully tuned through bat algorithm. The key points of the analysis summarized as:

1. The superior performance of the proposed controller over other controllers is shown in Table 1.
2. The energy storage excellent supports the system during small load fluctuation and maximizes the efficiency of the generating units.

References

1. Elgerd OI (1983) Electric energy systems theory: an introduction, 2nd edn. Tata McGraw-Hill, New Delhi
2. Nanda J, Mishra S, Saikia LC (2009) Maiden application of bacterial foraging based optimization technique in multiarea automatic generation control. IEEE Trans Power Syst 24 (2):602–609
3. Gozde H, Taplamacioglu MC (2011) Automatic generation control application with craziness based particle swarm optimization in a thermal power system. Int J Electr Power Energy Syst 33(1):8–16

4. Bhatt P, Roy R, Ghoshal SP (2010) GA/particle swarm intelligence based optimization of two specific varieties of controller devices applied to two area multiunits automatic generation control. Int J Electr Power Energy Syst 32(4):299–310
5. Saikia LC, Nanda J, Mishra S (2011) Performance comparison of several classical controllers in AGC for the multiarea interconnected thermal system. Int J Electr Power Energy Syst 33 (3):394–401
6. Hossain MS, Madlool NA, Rahim NA, Selvaraj J, Pandey AK, Khan AF (2016) Role of smart grid in renewable energy: an overview. Renew Sustain Energy Rev 60:1168–1184
7. Morsali J, Zare K, High MT (2016) Performance comparison of TCSC with TCPS and SSSC controllers in AGC of the realistic interconnected multisource power system. Ain Shams Eng J 7(1):143–158
8. Tasnin W, Saikia LC (2018) Performance comparison of several energy storage devices in deregulated AGC of a multi-area system incorporating geothermal power plant. IET Renew Power Gen 12(7):761–772
9. Javad M, Kazem Z, Mehrdad TH (2017) Applying fractional order PID to design TCSC based damping controller in coordination with automatic generation control of interconnected multisource power system. Eng Sci Technol Int J 20(1):1–17
10. Sahu RK, Panda S, Rout UK (2013) DE optimized parallel 2DOF PID controller for load frequency control of power system with governor dead band nonlinearity. Int J Electr Power Energy Syst 49:19–33
11. Sekhar GC, Sahu RK, Baliarsingh AK, Panda S (2016) Load frequency control of power system under a deregulated environment using optimal firefly algorithm. Int J Electr Power Energy Syst 74:195–211
12. Barisal A, Panigrahi T, Mishra S (2017) A hybrid PSO-LEVY flight algorithm based fuzzy PID controller for automatic generation control of multi area power systems: fuzzy based hybrid PSO for automatic generation control. Int J Power Energy Convers 6:42–63
13. Yang XS (2010) A new metaheuristic bat-inspired algorithm. In: Nature inspired cooperative strategies for optimization (NICSO 2010). Springer, Berlin, Heidelberg, pp 65–74
14. Biswal S, Barisal AK, Behera A, Prakash T (2013) Optimal power dispatch using BAT algorithm. In: 2013 IEEE international conference on energy-efficient technologies for sustainability (ICEETS), pp 1018–1023

Disassembly Sequence Planning Methodology for EOL Products Through a Computational Approach

Anil Kumar Gulivindala, Vykunta Rao Matta and M. V. A. Raju Bahubalendruni

Abstract Minimization of adverse environmental effect by generated e-waste day to day became challenging in different sectors of both developed and underdeveloping countries. Promoting 3'Rs policy such as reuse, resale, and remanufacture from the EOL products found as an only possible solution for the encountered challenge. An efficient disassembly sequence plan is needed to perform necessary operations and sorting out the relevant parts from EOL products. In order to achieve this, different existing methods have been studied and observed that subassembly identification is most essential in disassembly sequence planning to formulate an efficient solution. But the involvement of more computational effort in SI-based DSP got less research interest. Part concatenation method in ASG proved for generation in ample amount of subassemblies besides ASP. In this paper, a novel attempt has been made by implementing PCM to perform DSP. The results indicated that the method has tremendous workability not only in DSP but also extendable to PDSP, SDSP, and CDSP. The working of PCM on various classifications in DSP is explained with a case study and described well with suitable illustrations.

Keywords Disassembly · Disassembly sequence planning · Part concatenation method

1 Introduction

Electronic products usage in human life increasing year on year and extending from global developed countries to underdeveloped countries also. With the development of new technologies and aesthetic appreciation, the survival of product shortened rapidly. Most of the products are being discarded directly into landfill which raises

A. K. Gulivindala (✉) · V. R. Matta
NIT Puducherry, Karaikal 609609, India

M. V. A. Raju Bahubalendruni
GMRIT, Rajam, Srikakulam 532127, India

BBVL. Deepak et al. (eds.), *Innovative Product Design and Intelligent Manufacturing Systems*, Lecture Notes in Mechanical Engineering,
https://doi.org/10.1007/978-981-15-2696-1_69

e-waste resources. These conditions led adverse to both the environment and human health also [1]. Promoting 3'Rs policy from e-waste such as reuse, remanufacture, recycle found an effective solution for minimization of e-waste and costs involved in the production. However, in most of the underdeveloped countries carrying out these processes by un-organized sectors only. Most of them using destructive operations to carry out these processes in less time caused an inability indirect usage of useful parts extracted from EOL. Researchers opinioned that lack of proper knowledge about the product and its structure causes severe health effects on the workers also and opinioned that an efficient disassembly is needed to carry out these operations in less time, low cost, and low environmental effect [2].

Based on the objectives of operations involved, disassembly classified into (1) complete disassembly, (2) partial disassembly, and (3) selective disassembly. Because multi-objective achievement through a single disassembly sequence plan is not possible [3]. Disassembly sequence planning is a common practice in research started long back and it is carrying on an assumption called reverse of assembly sequence planning [4]. However, that assumption is suitable for complete disassembly at some limitations. Dini proved that the reverse of assembly sequence planning would not give the valid disassembly sequence for all products. Because the virginity of parts lost at EOL stage and various destructive operations need to carry out different connections (rivets, adhesive bonds, etc.) which alter the parts shape drastically [5]. Smith opinioned that complete disassembly is expensive always and impractical because only some parts in the EOL product have higher environmental impacts or higher recycling values. Disassembly is essential at different stages at product life cycle such as repair and maintenance, part extraction from EOL, toxic material (parts) sort out for safe disposal [6].

Most of the existing methodologies followed the same assumption and generated a single optimal solution. It can be known that fulfillment of these multi-objectives with a single solution is impossible and sometimes it is impractical also. Subassembly identification is most essential in the methodology to formulate a practical and optimal solution in selective and complete DSP [7]. Smith also agreed that subassembly identification is a key step in partial DSP besides selective DSP and complete DSP [8]. Researchers implicated different predicates such as contact, collision, consistency, and connection feasibility to sort out the infeasible solutions generated during the planning stage. DSP also considered as NP-hard problem due to apparent multiple complexities with an increase in part count and predicates which led to non-consideration of some predicates [9, 10]. Despite generated solution proved with optimality for a particular product, the methodologies failed to meet the various requirements and validity for product variability [11].

Besides research difficulties, product representation stood as a key step in sequence planning which worked out extensively in the past research. AND/OR graphs are mostly preferred which consists of nodes and hyper arcs representing for parts and relations between them in the assembled product [12–14]. Theoretically, these graphs are good at representing product information for all disassembly sequences but the complexity rises with an increase in the part count which needs more space and time to storage [15]. Matrix representation is also one kind of representation which mostly opting for recent research due to its ease of storage and applicability of methodology to DSP [16, 17]. However, the need for different kind of matrices to represent various properties rises with the need for quality which adversely affects the time consumption.

The following research gaps are identified in the cited literature:

- Ample amount of solution space is not available to apply multiple optimality criteria.
- More time and space consumption to store product information and solution generation.
- The validity of the solution is not guaranteed due to non-consideration of necessary assembly predicates.
- Optimality of solution is confined to some extent only.
- Need of generalized method which could solve all types of classified DSPs, i.e., CDSP, SDSP, and PDSP.

In order to meet the stated objectives for various types of DSP on overcoming the addressed research gaps, application of an efficient method named as part concatenation method is proposing in this research. The paper is organized as Sect. 1 for an introduction about EOL management problem and a potential solution called disassembly on citing various methodologies from the literature. Section 2 used to explain the working of part concatenation method in generating solutions. Section 3 used to establish the working of the method on a preferred case study. Section 4 for selection criteria involved in identifying the optimal solution. Section 5 is for the conclusion of the article discussing the future scope of the research interest.

2 Methodology

A disassembly sequence plan (DSP) for a product provides an ordered set of parts and disassembly directions based upon functional, physical, environmental, and economic constraints to satisfy various objectives [18]. DSP is classified into three types such as complete disassembly sequence plan (CDSP), partial disassembly sequence plan (PDSP), and selective disassembly sequence plan (PDSP).

2.1 PCM and Its Working

PCM proved its versatility in generating an ample amount of optimal and practically feasible solution in ASP on satisfying various assembly constraints. The generosity of solution lies under the testing with necessary assembly predicates namely contact, collision-freeness, consistency, and connector feasibility before formulating the sequence. With the unique feature subassembly detection, while the formulation of a solution made PCM as unique to other ASP methods. A novel attempt has been made to test the applicability of PCM in DSP.

As mentioned afore, it is not a feasible assumption to consider always the DSP is reverse of the ASP. But some basic principles can be adopted such as extraction and analysis of product information in the matrix format. Later, the feasibility criteria can be applied to the generated set of solutions to examine their validity. The steps involved in the solution generation are shown in Fig. 1.

Obtain necessary product information in the form of matrices such as liaison, geometric feasibility, stability, and mechanical feasibility to test the presence of

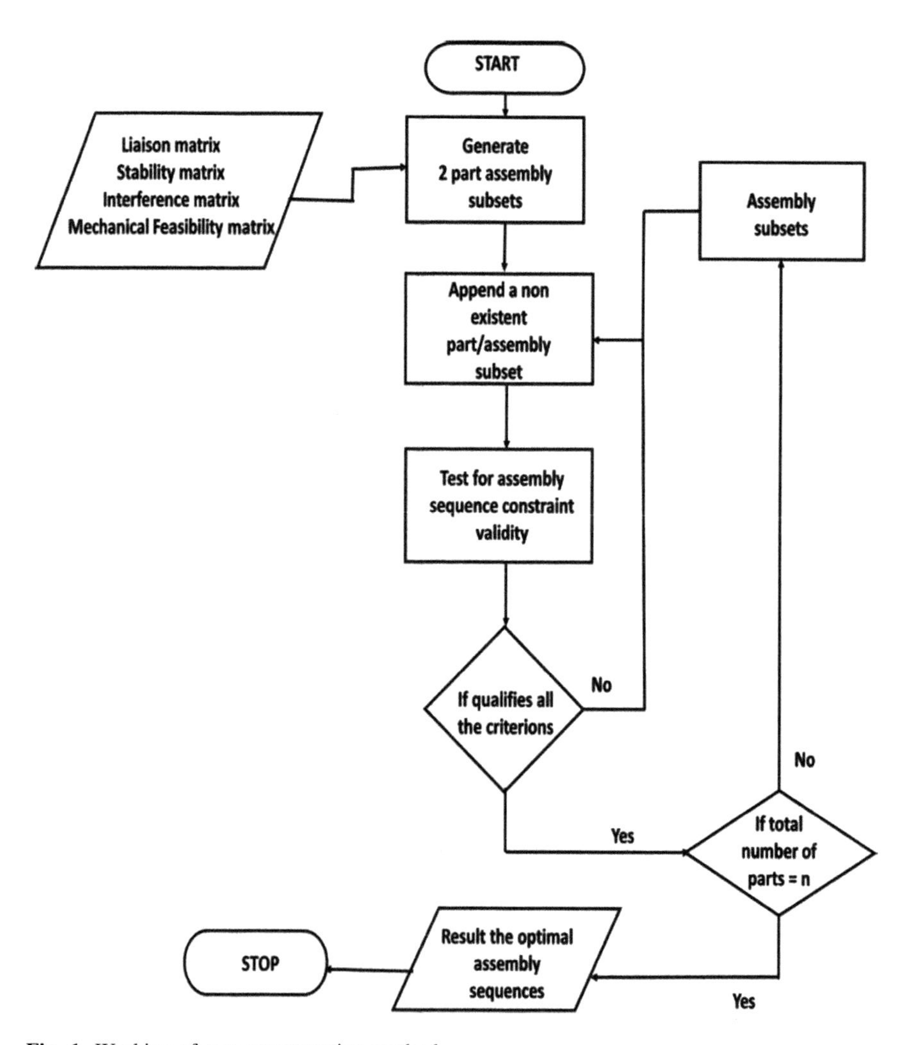

Fig. 1 Working of part concatenation method

contact, collision-free between operations, consistency between the parts and feasibility of placing connectors before applying the methodology [10, 17, 19–22].

Most suitable DSP is selected among the different classified types of disassembly such as CDSP, SDSP, and PDSP depending on the goals behind the disassembly and capability of product materials under recycle, remanufacture, and reuse indexes. At the present study, indexes are not considering only the type of 'R' in 3'R taken manually.

Case-1: CDSP
If the product having higher recyclable material compared to both reuse and remanufacture at EOL, then CDSP should be chosen. Select the one best sequence in which the overall disassembly should be completed in a minimal number of levels. The assumption of the reverse of ASP lead to DSP can be followed and corresponding operations could be carried out.

Case-2: SDSP
If the product having more reusable parts other than recyclability and remanufacturability, then SDSP could be the better choice. These types of instances arise while the condition of the product is good but discarded due to change in trend, minor damages, etc. Subassembly detection is very much needed at this stage to identify the safest path in a minimum number of levels to discord the part in the final product.

Case-3: PDSP
If the product having remanufacturability compared to recyclability and reusability, then optimal stopping point of disassembly should be known. Many cost parameters should be analyzed in this case to discover the optimal stopping point and these cases not discussed briefly here. Subassembly detection besides DSP is needed at this stage and destructive operations can be carried out up to a certain extent.

3 Case Study

Single-hole punch assembly is selected as a case study to demonstrate the working of the PCM in DSP generation. Kara used liaison, geometric feasibility, and precedence criteria to perform DSP generation for the single-hole punch assembly [16]. Figure 2 depicts the precedence graph of the single-hole punch assembly used in the Kara approach. Only single solution was generated by the followed approach and it confined only for the selective disassembly.

In this graph, a node represents for the part and the line represent for precedence relation between the assembly parts. Here, node-1 represents first part such as pivot in the disassembly operation and similarly for other existing parts also. Part name and respective code of the selected case study are provided in Table 1.

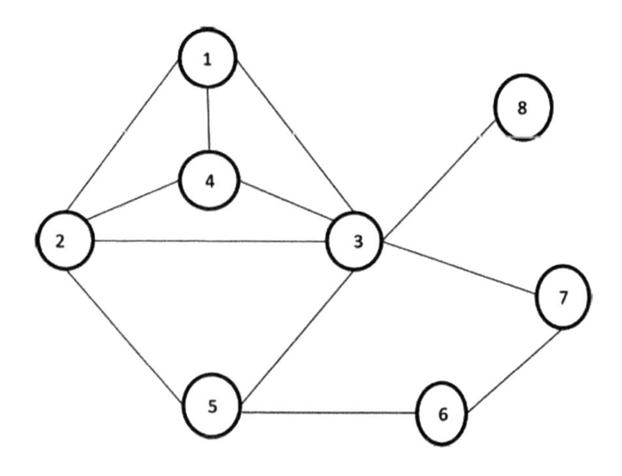

Fig. 2 Precedence graph of single-hole punch assembly

Table 1 Part listing of single-hole punch assembly at its EOL stage

Part name	Code	Material type	Reuse	Recycle	Remanufacture
Pivot	1	Scrap steel	–	✓	–
Handle	2	Scrap steel	–	✓	–
Body	3	Scrap steel	–	✓	–
Spring	4	Spring steel	✓	–	✓
Punch driver	5	Scrap steel	–	✓	–
Punch	6	Tool steel	✓	–	✓
Base	7	Scrap steel	–	✓	–
Cover	8	Plastic	–	✓	–

The 3'Rs applicability also enlisted in Table 1 for the selected case study shown in Fig. 2. After analyzing 3'Rs, the complete disassembly is recommended because disassembly with destructive operations also acceptable for this product because only one part (punch) have the potential for reuse/remanufacture. Among the assembly predicates, mechanical feasibility is not needed to consider this product due to less part count and pivot, the punch driver has a significant role in the functioning of the product more than a connector. After the extraction of necessary data possible a number of solutions are generated with respect to part relations.

This kind of product configurations mostly opts for the linear disassembly sequence plans. However, part concatenation method also generated linear disassembly plans besides a possible number of subassemblies represented in Table 2. Two types of braces used to differentiate subassemblies and individual parts. '(' for permanently stable subassembly/part and '[' partially stable subassembly/part.

Table 2 Generated DSPs by PCM

Solution	Directional changes
[1 [5 [4 (3 7 8 6) 2]]]]	3
[1-[5-[2-(8-3-4-7)-6]]]]	4
(8-[1-[5-[4-(3-7-6)-2)]]]]	5
(8-(1-[5-[2-[6-(4-3-7)]]))	4

4 Selection Criteria

Optimal sequence selected by which possess minimum directional changes, i.e., [1-[5-[4-(3-7-8-6)-2]]]]. Because pivot and punch driver needed only a single tool to carry out the operation. Only spring and punch having reuse and remanufacturing capability out of eight parts in the assembly. Destructive operations could be employed to finish the total in less span of time then complete DSP with more number of linear operations opted as the best solution among four different solutions.

5 Conclusion

The potentiality of DSP to solve the current stated problem such as EOL minimization have been explained with a case study and proposed a method to carry out the optimal number of operations. Different types of disassembly classified such as complete disassembly, selective disassembly, and partial disassembly and explained the selection of disassembly according to requirement. Importance of subassembly detection in solution generation is highlighted and the possibility of parallel disassembly operations also identified.

In this paper, a novel attempt has been made to test the applicability of ASP named as PCM in DSP. A real-time product called single dot punch assembly has chosen as a case study to apply the PCM. However, a selected case study of EOL product was proved the applicability of methodology in complete DSP. The possibility to extend other DSP's also mentioned with suitable illustrations. Ample amount of solutions availability offered to apply selection criteria on testing the most optimal solution based on a number of levels to finish the operations. The author intends to extend the optimality criteria on adding directional changes, tool changes, tool manipulations, and the cost involved in each operation filters. Besides this optimality criteria author also intends to develop a framework to analyze the nature of operations such as destructive and non-destructive to meet the safety of workers in ergonomics point of view.

References

1. Kiddee P, Naidu R, Wong MH (2013) Electronic waste management approaches: an overview. Waste Manage 33(5):1237–1250
2. Harivardhini S, Krishna KM, Chakrabarti A (2017) An integrated framework for supporting decision making during early design stages on end-of-life disassembly. J Clean Prod 168:558–574
3. Smith SS, Chen WH (2011) Rule-based recursive selective disassembly sequence planning for green design. Adv Eng Inform 25(1):77–87
4. Ilgin MA, Gupta SM (2010) Environmentally conscious manufacturing and product recovery (ECMPRO): a review of the state of the art. J Environ Manage 91(3):563–591
5. Santochi M, Dini G, Failli F (2002) Disassembly for recycling, maintenance and remanufacturing: state of the art and perspectives. In: AMST'02 advanced manufacturing systems and technology. Springer, Vienna, pp 73–89
6. Harivardhini S, Chakrabarti A (2016) A new model for estimating end-of-life disassembly effort during the early stages of product design. Clean Technol Environ Policy 18(5):1585–1598
7. Dini G, Santochi M (1992) Automated sequencing and subassembly detection in assembly planning. CIRP Ann 41(1):1–4
8. Smith S, Smith G, Chen WH (2012) Disassembly sequence structure graphs: an optimal approach for multiple-target selective disassembly sequence planning. Adv Eng Inform 26 (2):306–316
9. Sinanoğlu C, Rıza Börklü H (2005) An assembly sequence-planning system for mechanical parts using a neural network. Assembly Autom 25(1):38–52
10. Bahubalendruni MR, Kumar GA (2018) Practically feasible optimal assembly sequence planning with tool accessibility. In: IOP conference series: materials science and engineering, vol 390, no 1. IOP Publishing, Kancheepuram, pp 12–26
11. Bahubalendruni MR, Biswal BB, Kumar M, Deepak BBVL (2016) A note on mechanical feasibility predicate for robotic assembly sequence generation. In: CAD/CAM, robotics and factories of the future. Springer, New Delhi, pp 397–404
12. O'shea B, Kaebernick H, Grewal SS (2000) Using a cluster graph representation of products for application in the disassembly planning process. Concurr Eng 8(3):158–170
13. De Mello LH, Sanderson AC (1991) A correct and complete algorithm for the generation of mechanical assembly sequences. IEEE Trans Robot Autom 7(2):228–240
14. Baldwin DF, Abell TE, Lui MC, De Fazio TL, Whitney DE (1991) An integrated computer aid for generating and evaluating assembly sequences for mechanical products. IEEE Trans Robot Autom 7(1):78–94
15. Abdullah MA, Ab Rashid MFF, Ghazalli Z (2018) Optimization of assembly sequence planning using soft computing approaches: a review. Arch Comput Methods Eng 0(0):1–14
16. Kara S, Pornprasitpol P, Kaebernick H (2005) A selective disassembly methodology for end-of-life products. Assembly Autom 25(2):124–134
17. Bahubalendruni MR, Gulivindala A, Kumar M, Biswal BB, Annepu LN (2019) A hybrid conjugated method for assembly sequence generation and explode view generation. Assembly Autom 39(1):211–225
18. Wang X, Qin Y, Chen M, Wang CT (2005) End-of-life vehicle recycling based on disassembly. J Central South Univ Technol 12(2):153–156
19. Bahubalendruni MR, Biswal BB (2017) A novel concatenation method for generating optimal robotic assembly sequences. Proc Inst Mech Eng Part C J Mech Eng Sci 231(10):1966–1977
20. Bahubalendruni MR, Biswal BB (2015) An intelligent method to test feasibility predicate for robotic assembly sequence generation. In: Intelligent computing, communication, and devices. Springer, New Delhi, pp 277–283

21. Bahubalendruni MR, Biswal BB (2016) Liaison concatenation—a method to obtain feasible assembly sequences from the 3D-CAD product. Sadhana 41(1):67–74
22. Bahubalendruni MR (2018) An efficient method for exploded view generation through assembly coherence data and precedence relations. World J Eng 15(2):248–253

Optimization of Patch Size Using Response Surface in Asymmetric Patch Repair

Amol Rasane, Prashant Kumar and Mohan Khond

Abstract Cracks developed in thin metallic structures propagate with applied loads and result in catastrophic failure. A potential method of crack repair is by adhesively bonding FRP composite patch on the cracked region. The single-sided asymmetric repairs are commonly employed which produces unbalanced forces causing patch failure by separation at the bonded interface; largely influenced by the length and width of the patch. An optimum size of the patch is necessary due to limited bonding space and increased weight of the structure. The study of single-sided repair of a centre crack in a thin aluminium alloy sheet using carbon fibre reinforced polymer composite patch was carried out. The interface separation was numerically simulated by employing the cohesive zone material model in ANSYS 15.0. The failure stress and the mode of failure were obtained in the numerical simulations. The optimum patch size was obtained through the response surface methodology.

Keywords Cohesive zone model · Optimum size · Response surface

1 Introduction

Thin sheets of aluminium alloys are used in many structural applications such as in aircraft, automobile, construction and many other industries. During service, the structure may damage and develop cracks which grow with the applied loads. Composite patching in which a patch, made of polymer composite, is adhesively bonded onto the cracked region and has a potential for effective crack repair. The repair of metallic structures using composite materials was first introduced in Australia in the early 1970s and later in the USA in the early 1980s for the repair of military and civil aircraft [1]. Further research was carried out by Ratwani [2], Okafor et al. [3], Davis et al. [4], Grabovac and Whittaker [5] and Wang et al. [6],

A. Rasane (✉) · P. Kumar · M. Khond
College of Engineering, Pune, India

BBVL. Deepak et al. (eds.), *Innovative Product Design and Intelligent Manufacturing Systems*, Lecture Notes in Mechanical Engineering,
https://doi.org/10.1007/978-981-15-2696-1_70

confirming the satisfactory performance of the polymer composite patches in increasing the strength of the thin cracked structural sheets. This technique offers several advantages, such as high strength-to-weight ratio, better structural integrity, easy formability and rapid installation. In many practical applications, only one side of the cracked structure is available for bonding and the use of single-sided patch becomes necessary. The asymmetric patch has a tendency to separate out from the substrate at the bonded interface because of the stresses developed due to the applied loads. Kwon and Lee [7] proposed an analytical model for the single-sided composite patch repair. Belhouari et al. [8], Pandey and Kumar [9], Gu et al. [10], Kasavajhala and Gu [11], da Silva and Campilho [12], Errouane et al. [13], Ribeiro et al. [14] and Ramkrishna et al. [15] used the finite element analysis (FEA) for studying the mechanical response of the repair patches and found it to be an effective technique for simulating the interface separation mechanism. Alfano et al. [16, 17], Xie et al. [18], Fekih et al. [19], Valoroso and de Barros [20] successfully used the cohesive zone material model (CZM) for studying the effect of different factors on the interface separation. Kumar and Hakeem [21], Yala and Megueni [22], Ouinas et al. [23] and Ramji et al. [24] worked on patch optimization in terms of the important parameters such as the patch shape and type of adhesive on the basis of stress intensity factor through numerical simulations. Several other researchers have also contributed to the knowledge base of the composite patching technique [25]. A comprehensive study of the interface failure is necessary as the bonded repair has the potential in several applications involving the repair of primary structures. The patch separation is largely influenced by the patch length and width. The space available for bonding and the increased weight of the structure due to the patch are also important considerations. Therefore, the use of a patch of optimum size is necessary for an economical, efficient and feasible crack repair.

In this work, the study of the single-sided crack repair in a thin aluminium alloy sheet using a patch of single-ply FRP composite was carried out. The uniaxial tension tests were conducted to determine the failure stress and the mode of failure for different patch sizes. The experiments were numerically simulated using the cohesive zone material (CZM) model in the finite element method in ANSYS 15.0 [26–29]. The numerical results were compared with those obtained in the experiments and the difference was found less than 7%. Thus, the credibility of the numerical simulation was established. The length and width of the patch were varied and twenty-five numerical simulations were performed. The optimum size of the patch was obtained through the response surface methodology by using the results of the numerical analyses. In the earlier works of other researchers, the reduction in crack tip stress intensity factor, J-integral and the variation in stresses have been used for optimization of patch configuration. In the current work, the optimization is carried out with the consideration of minimum patch area and maximum failure stress which is a realistic approach, resulting in an economical crack repair.

2 Specimen

Figure 1a shows the schematic of the patched specimen of size 400 mm × 60 mm × 1 mm with X_1, X_2 and X_3 as the coordinate axes along the length, width and thickness direction, respectively. Aluminium alloy 6061-T6 was used as the skin material. A through-the-thickness centre crack of length $2a$ = 25 mm was made in the specimen with O as its centre point and OT as half the crack length. The patch was made of carbon fibre reinforced polymer composite (CFRP) with a reinforcement of unidirectional (UD) carbon fabric of 160 gsm. A thin separating ply made of glass fibre reinforced polymer composite (GFRP) having the unidirectional glass fabric of 76 gsm as reinforcement and was inserted in between the CFRP ply and the aluminium skin. Epoxy resin was used as the matrix material for the composite. The patch was bonded only on one side of the aluminium alloy skin.

A multi-linear isotropic hardening plasticity model was used for aluminium while the patch was modelled as the linear orthotropic material. The elastic constants of the unidirectional patches were obtained using the rule of mixture and Halpin-Tsai relations [30].

3 Numerical Simulation

Exploiting the planes of symmetry, just a quarter portion ORNM of the centre-cracked specimen was modelled as shown in Fig. 1b. In the quarter model, OR is half the specimen length while OM half the specimen width. The crack plane OT was modelled as a free surface. Two layers of FRP patch were modelled on the cracked region of the specimen; GFRP as the first layer and CFRP as the subsequent layer. In Fig. 1b, OABC is the quarter portion of the GFRP ply, 0.12 mm

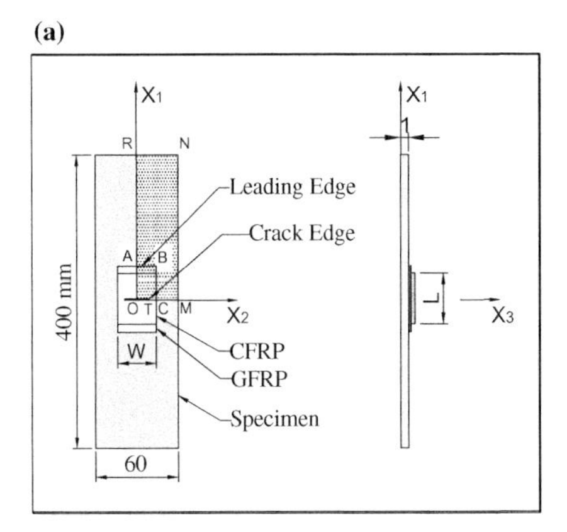

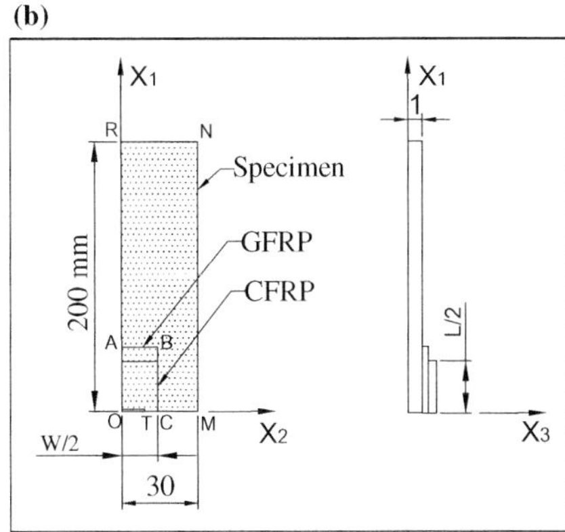

Fig. 1 **a** Patched specimen, **b** quarter model of the patched specimen

thick; with OA and OC as half its length and width, respectively. The GFRP ply was 20 mm longer than the CFRP ply. Thus, the CFRP ply, which was of 0.22 mm thickness, was 10 mm shorter than the GFRP ply in the quarter model. The width of both GFRP and CFRP plies was kept equal. The edge OT was termed as crack edge while edge AB as leading edge of the patch. The aluminium alloy skin, the GFRP ply and the CFRP ply were meshed with SOLID186 elements. An appropriate mesh sizes in the length, width and thickness direction were determined through separate convergence studies. Fine mesh was used near the areas prone to failure, i.e. the crack edge and leading edges of the patch. The model was symmetric about the X_1 (along the length) axis and X_2 (along the width) axis. The symmetry boundary conditions were used on the planes OR and OM. A constant displacement was applied at the free end of the quarter model (RN) and it was allowed to move in longitudinal (loading) direction with a gradually increasing displacement. The finite element model was solved using sparse direct solver. The separation area, induced load and the interfacial stresses were monitored at various stages of the loading.

4 Cohesive Zone Material (CZM) Model

The static strength of the patch on a crack is dependent mainly on the interface between the skin material and the patch. The separation at the interface between the skin and patch is the reason for failure of the patch, ultimately leading to failure of the skin. The numerical analysis of this phenomenon is carried out by the technique called the cohesive zone material (CZM) model. The cohesive zone is a small zone in front of crack tip in which gradual failure takes place as shown in Fig. 2. In fact, the interface of the two materials is degraded with the variation of material properties at the interface. The mechanical response of the cohesive interface is modelled with the traction-separation law that relates the traction field (T) with separation ($\boldsymbol{\delta}$). As the material deforms within cohesive zone, the traction first increases until a maximum (T^{max}) is reached and then subsequently reduces to zero which results in complete separation ($\boldsymbol{\delta}^{\mathbf{c}}$). The bilinear traction-separation law in normal and tangential direction was used in the numerical simulation of this work [27].

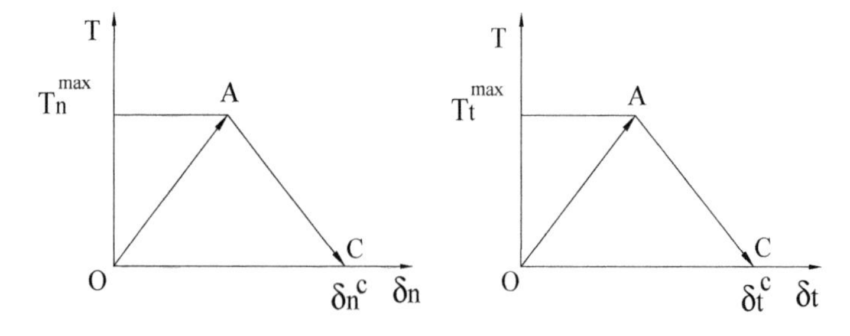

Fig. 2 Traction-separation curves in normal and tangential direction

At the interface, a high shear stress σ_{13} is developed, especially close to the leading and crack edges of the patch which transfers the induced load in the skin to the patch. A high normal stress σ_{33}, which is known as peel stress, is also developed at the interface because of the bending effect of the asymmetric patch [31]. The energy that causes the mode-I separation of the patch is given by the area under curve OAC (Fig. 2) and is determined through Eq. (1) [27]. It was taken to be equal to the critical energy release rate, also termed as critical fracture energy, G_{Ic} [32, 33]. Thus,

$$G_{\mathrm{Ic}} = \frac{1}{2}\left(T_n^{\max}\right) \times \left(\delta_n^{\mathrm{c}}\right). \tag{1}$$

Similarly, the critical fracture energy in mode-II, G_{IIc} [32, 33] is the area under curve OAC in second sketch of Fig. 2, given by Eq. (2) [27]. Thus,

$$G_{\mathrm{IIc}} = \frac{1}{2}\left(T_t^{\max}\right) \times \left(\delta_t^{\mathrm{c}}\right). \tag{2}$$

The normal and tangential interfacial stresses together causes the separation at the skin-GFRP interface. Thus, a mixed-mode criterion (MMC) of failure consisting of mode-I and mode-II separation was employed [27] and it was specified by the energy criterion based on power law as per the inequality (3) [27]:

$$\mathrm{MMC} = \left(\frac{G_{\mathrm{I}}}{G_{\mathrm{Ic}}}\right)^2 + \left(\frac{G_{\mathrm{II}}}{G_{\mathrm{IIc}}}\right)^2 \geq 1. \tag{3}$$

5 Results and Discussion

5.1 Results of 25 Numerical Simulations

The stress induced in the skin is reduced due to the patch. The crack tip in the patched specimen does not grow until the patch is separated from the skin. The numerical simulations were performed for different configurations of single-ply patch by varying its length (L) and width (W). The L_{25} Taguchi Orthogonal Array consisting of two factors, patch length (L) and patch width (W), was employed for deciding the patch configurations [34, 35]. Five levels of patch length, L (30, 40, 50, 60 and 70 mm) and five levels of patch width, W (30, 36, 42, 48 and 54 mm) were considered. The numerical simulations of the interface separation for these 25 patch configurations were carried out. The separation areas were determined for all the cases and the trend of patch separation was studied through the failure stress ($\sigma_{\mathrm{f}}^{\infty}$). The results of all the twenty-five numerical experiments are summarized in Table 1.

Table 1 Failure stress for 25 patch configurations

S. No.	L (mm)	W (mm)	σ_f^∞ (MPa)	Failure initiated at
1	30	30	268.9	Crack edge
2	40	30	285.2	Crack edge
3	50	30	294.3	Crack edge
4	60	30	301.4	Crack edge
5	70	30	303.7	Crack edge
6	30	36	279.6	Crack edge
7	40	36	304.9	Crack edge
8	50	36	327.0	Crack edge
9	60	36	338.5	Crack edge and leading edge
10	70	36	338.4	Crack edge and leading edge
11	30	42	286.8	Crack edge
12	40	42	320.5	Crack edge
13	50	42	338.6	Crack edge and leading edges
14	60	42	338.6	Crack edge and leading edge
15	70	42	338.7	Crack edge and leading edge
16	30	48	292.1	Crack edge
17	40	48	332.8	Crack edge and leading edge
18	50	48	338.7	Crack edge and leading edge
19	60	48	338.9	Crack edge and leading edge
20	70	48	338.9	Crack edge and leading edge
21	30	54	296.1	Crack edge
22	40	54	337.9	Crack edge and leading edge
23	50	54	338.9	Crack edge and leading edge
24	60	54	338.9	Crack edge and leading edge
25	70	54	339.1	Crack edge and leading edge

5.2 *Response Surface*

The response surface methodology (RSM) was employed to arrive at an optimum patch configuration [36]. A response surface explores the relationship between the response and the associated control variables [37]. Using Table 1 and design exploration module of ANSYS, a response surface was generated with patch length (L) and patch width (W) as the control variables and failure stress (σ_f^∞) as the response. Figure 3 shows the response surface.

Though the response surface increases with the patch length and width, it becomes nearly flat for longer and wider patches. The marginal return for failure stress diminishes with the increase in the patch area. There could be many different points on the response surface having high values of the failure stress and the optimum point could not be determined. The optimum size of the patch was then determined using the response curve.

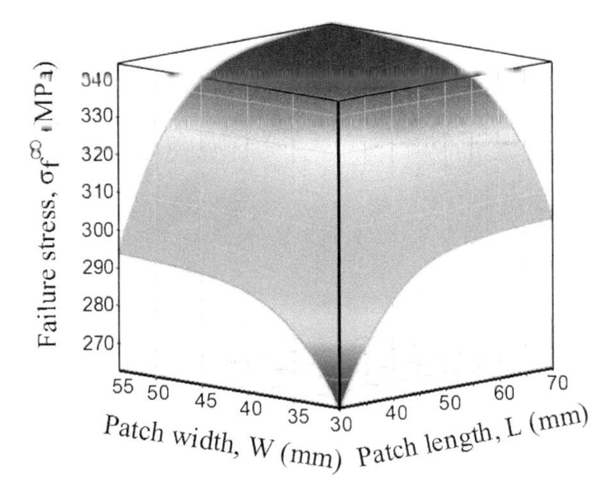

Fig. 3 Response surface for single-ply patch

5.3 *Response Curve*

A relation between the variation in the failure stress (σ_f^∞) with the patch area (A) was developed using the results of 25 numerical simulations. The relation is shown in Fig. 4.

The rate of increase in the failure stress with the increase in patch area is significant up to point P_1. Beyond this point, there is only a marginal increase in the failure stress even if the patch area is increased. Point P_1 corresponds to failure stress of 338.6 MPa. To find the optimum length and width of the patch, the response surface is used which provides various combinations of length and width for $\sigma_f^\infty = 338.6$ Mpa. Using these values, the variation of patch area with the patch width was obtained and is shown in Fig. 5. The resultant relation is termed as the response curve.

It is observed that as the patch width is increased, the patch area reduces till point P_2. With the further increase in the patch width, the patch area is found to increase.

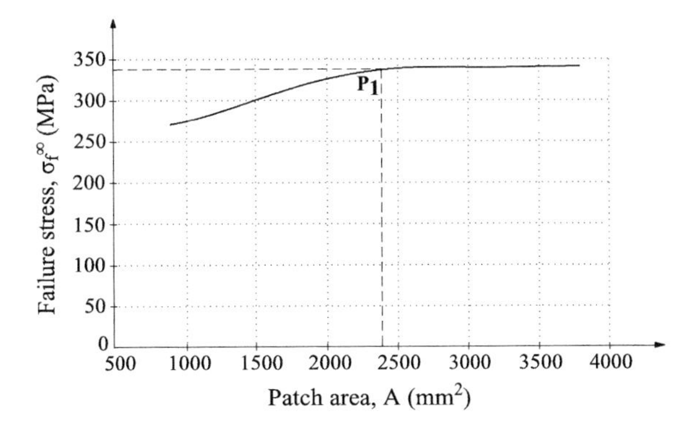

Fig. 4 Failure stress versus patch area for single-ply patch

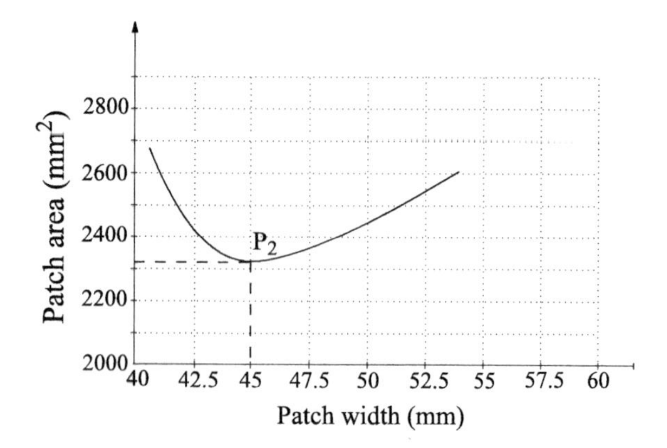

Fig. 5 Response curve for single-ply patch at $\sigma_f^\infty = 338.6\,\text{MPa}$

Table 2 Optimum size of single-ply patch

L (mm)	W (mm)	A (mm^2)	Aspect ratio, L/W
51.6	45.0	2320	1.15

The area of the patch is minimum at point P_2 and is considered as the optimum patch area, $A = 2320$ mm^2. The optimum width for this patch area is $W = 45.0$ mm, and then the patch length was determined as $L = 51.6$ mm. Thus, the optimum size of the single-ply patch was obtained and is given in Table 2.

For the patches with short length and width, the patch fails at a stress less than 320 MPa. Though the failure stress is high ($\sigma_f^\infty > 336\,\text{MPa}$) at an increased length and width, the bonding area of the patch also increases considerably. With the optimum patch size, the patch fails at a high stress and the area of the patch is also minimum, thus, providing a trade-off between the patch area and failure stress.

6 Conclusion

The centre crack in an aluminium alloy skin was repaired using a patch of FRP composite under uniaxial tensile load. The crack tip was not found to grow until the patch was separated from the skin. The patch separates at the skin-patch interface through the mixed mode consisting of mode-I and mode-II. The separation was found to initiate at the crack edge at low applied stress for all the patch configurations under consideration. In some cases, the separation was also initiated at high applied stress at the leading edge of the patch. A response surface was generated using the patch failure stress and the patch dimensions. A response curve was drawn from the response surface. The optimum patch size corresponding to the minimum patch area was obtained from the response curve as: 51.6 mm (L) × 45.0 mm (W). The current work was carried out under uniaxial quasistatic tensile load. In future, the methodology can be extended to determine the optimum patch size under fatigue loading conditions.

References

1. Baker AA, Jones R (1988) Bonded repair of aircraft structures. Kluwer Academic Publishers, UK
2. Ratwani MM (2000) Repair options for airframes. Aging aircraft fleets: structural and other subsystem aspects, RTO AVT lecture series EN-015, Bulgaria
3. Okafor AC, Singh N, Enemuoh UE, Rao SV (2005) Design, analysis and performance of adhesively bonded composite patch repair of cracked aluminium aircraft panels. Compos Struct 71:258–270
4. Davis MJ, Janardhana M, Wherrett A (2006) Adhesive bonded repair technology: supporting aging aircraft. Ageing Aircraft Users Forum, Brisbane
5. Grabovac I, Whittaker D (2009) Application of bonded composites in the repair of ships structures—a 15 year service experience. Compos Part A 40(9):1381–1398
6. Wang Z, Zhang Y, Xue J, Yang W (2013) Effect of accelerated ageing on the mechanical properties of composite repaired aluminium plates. In: International conference on quality, reliability, risk, maintenance, and safety engineering (QR2MSE), China, pp 863–866
7. Kwon YW, Lee WY (2013) Analytical model for single-side patch design of composite repair. Naval Post Graduate School, California, NPS-MAE-13-001
8. Belhouari M, Bouiadjra BB, Megueni A, Kaddouri K (2004) Comparison of double and single bonded repairs to symmetric composite structures: a numerical analysis. Compos Struct 65(1):47–53
9. Pandey PC, Kumar S (2010) Adhesively-bonded patch repair with composites. Defence Sci J 60(3):320–329
10. Gu L, Kasavajhalay A, Zhaoz S (2011) Finite element analysis of cracks in aging aircraft structures with bonded composite-patch repairs. Compos Part B 42:505–510
11. Kasavajhala ARM, Gu L (2011) Fracture analysis of Kevlar-49/epoxy and e-glass/epoxy doublers for reinforcement of cracked aluminium plates. Compos Struct 93(8):2090–2095
12. da Silva LFM, Campilho RDSG (2012) Advances in numerical modelling of adhesive joints. In: Advances in numerical modelling of adhesive joints. Springer briefs in applied sciences and technology. Springer, Berlin, Heidelberg
13. Errouane H, Sereir Z, Chateauneuf A (2014) Numerical model for optimal design of composite patch repair of cracked aluminium plates under tension. Int J Adhes Adhes 49:64–72
14. Ribeiro TEA, Campilho RDSG, da Silva LFM, Goglio L (2016) Damage analysis of composite–aluminium adhesively bonded single-lap joints. Compos Struct 136:25–33
15. Ramakrishna C, Balu J, Rajashekar S, Sivateja N (2017) Finite element analysis of the composite patch repairs of the plates. Int J Eng Res Appl 7(2):10–18
16. Alfano G (2006) On the influence of the shape of the interface law on the application of cohesive-zone models. Compos Sci Technol 66(6):723–730
17. Alfano M, Furgiuele F, Leonardi A, Maletta C, Paulino GH (2007) Cohesive zone modelling of mode I fracture in adhesive bonded joints. Key Eng Mater 348–349:13–16
18. Xie D, Garg M, Huang D, Abdi F (2008) Cohesive zone model for surface cracks using finite element analysis. In: 49th AIAA/ASME/ASCE/AHS/ASC structures, structural dynamics and materials conference
19. Fekih SM, Albedah A, Benyahia F, Belhouari M, Bachir Bouiadjra B, Miloudi A (2012) Optimisation of the sizes of bonded composite repair in aircraft structures. Mater Des 41:171–176
20. Valoroso N, de Barros S (2013) Adhesive joint computations using cohesive zones. Appl Adhes Sci 1(1):1–9
21. Kumar MA, Hakeem SA (2000) Optimum design of symmetric composite patch repair to centre cracked metallic sheet. Compos Struct 49(3):285–292
22. Yala AA, Megueni A (2009) Optimisation of composite patches repairs with the design of experiments method. Mater Des 30(1):200–205

23. Ouinas D, Achour B, Bouiadjra BB, Taghezout N (2013) The optimization thickness of single/double composite patch on the stress intensity factor reduction. J Reinf Plast Compos 32(9):654–663
24. Ramji M, Srilakshmi R, Bhanu Prakash M (2013) Towards optimization of patch shape on the performance of bonded composite repair using FEM. Compos B Eng 45(1):710–720
25. Duong CN, Wang CH (2007) Composite repair, theory and design. Elsevier Publications, The Netherlands
26. Logan DL (2007) A first course in the finite element method. Thomson Publishers, USA
27. ANSYS, Inc. (2015) ANSYS theory reference. ANSYS, USA
28. ANSYS, Inc. (2015) ANSYS design explorer user guide. ANSYS, USA
29. ANSYS, Inc. (2015) ANSYS workbench user's guide. ANSYS, USA
30. Agarwal BD, Broutman LJ (1990) Analysis and performance of fiber composites. Wiley, USA
31. Gilat A, Goldberg RK, Roberts GD (2007) Strain rate sensitivity of epoxy resin in tensile and shear loading. J Aerosp Eng 20(2):75–89
32. Joshi S (2015) Deterioration of toughness and shear strength of the bonded interface between aluminium skin and composite patch. Master's thesis, College of Engineering Pune, India
33. Kumar P (2009) Elements of fracture mechanics. Tata McGraw Hill, India
34. Montgomery D (2001) Design and analysis of experiments. Wiley, USA
35. Phadke MS (2009) Quality engineering using robust design. Pearson Publications, USA
36. Myers RH, Khuri AI, Carter WH (1989) Response surface methodology: 1966-1988. Technometrics 31(2):137–157
37. Myers R, Montgomery D, Anderson-Cook CM (2009) Response surface methodology. Wiley, USA

Swarm-Inspired Task Scheduling Strategy in Cloud Computing

Ramakrishna Goddu and Kiran Kumar Reddi

Abstract Cloud computing is the most emerging technology which provides sharing of computing resources and data storage through virtualization concept. However, managing plenty of virtualized resources made scheduling a difficult task in cloud computing. Task scheduling must be done in such a way that it must satisfy customer requirements and maintain the quality of service (QoS). In this paper, we proposed a method for resource allocation based on particle swarm optimization (PSO) algorithm and with two objectives which produce optimal task scheduling. The first objective is related to virtual machine processing, and the second objective is related to the time elapsed to complete the given task. Based on the throughput of these objectives, the virtual machines are allotted to the resources.

Keywords Cloud computing · Task scheduling · Particle swarm optimization · Virtual machines

1 Introduction

Cloud computing provides sharing of computing resources and data storage and allows its users to access information to utilize its services over the Internet and central remote servers on demand [1]. The resources are shared between cloud clients through virtualization technique. Virtualization divides a single computational resource into multiple independent execution environments, also known as virtual machines [2]. Once the partition is done, each logical server behaves like a physical server that runs the operating system automatically, and this is the main flexibility in the virtualization concept. In cloud computing, the resource allocation to the virtual machine is the main task of the management system, because the virtual machines' efficiency reflects the performance and cost of the system. An inefficient virtual machine allocated a resource will decrease performance and

R. Goddu (✉) · K. K. Reddi
Department of Computer Science, Krishna University, Machilipatanam, Andhra Pradesh, India

BBVL. Deepak et al. (eds.), *Innovative Product Design and Intelligent Manufacturing Systems*, Lecture Notes in Mechanical Engineering,
https://doi.org/10.1007/978-981-15-2696-1_71

increases the cost of the system [3]. The primary objective is to obtain optimal output and to solve large-scale computation problems by using resource infrastructure of resource allocation [4, 5]. The cloud provider can guarantee the quality of service (QoS) to its users with maximum resource utilization and minimum power consumption. QoS denotes the degree of satisfaction of a customer for a service depending on the level of performance, reliability, and availability [6]. To implement QoS in cloud computing, a task scheduling algorithm is used [7]. Assigning jobs to particular resources at a particular time is known as scheduling, and scheduling of tasks plays a major role in cloud computing. Improper scheduling may lead to a reduction in system performance. The main objective of task scheduling is to increase performance and to decrease the task completion time by developing an efficient scheduling algorithm. Based on the priority, the task is allocated to the virtual machine and then it is mapped to a suitable physical machine. The virtual machine is selected according to the task necessities like CPU power and cost; after that, the task scheduler allocates the task to the selected virtual machine [3]. Particle swarm optimization is a social-inspired evolutionary algorithm, which has been implemented for solving various engineering problems [8–10]. Past studies considered particle swarm optimization (PSO)-based task scheduling is more efficient for dynamic task scheduling. Juana et al. [5] proposed a PSO-based task scheduling algorithm to overcome the problems of cloud computing. They developed a cost vector model which measures scheduling scheme cost, and solution was developed based on input task and QoS parameters. This method is proven to be effective, but it is with more complexity. Krishnasamy [11] proposed a hybrid particle swarm optimization task scheduling algorithm which decreases the average operation time, increases the usage of resources, and provides the resources according to the user tasks. Alkayal et al. [3] developed a PSO-based multi-object task scheduling by introducing a new approach to ranking. Here, the tasks are allocated to virtual machines according to the rank, which decreases the waiting time and increases system performance. To find a solution to the problems of cloud computing in task scheduling, Dordaie and Navimipour [12] proposed a hybrid particle swarm optimization and hill climbing algorithm which was properly scheduled, but it takes more time for task completion.

In this paper, we proposed a PSO-based task scheduling algorithm by considering two objectives. The first objective is related to virtual machine processing, and the second objective is related to the time elapsed to complete the given task. Based on the throughput of these objectives, the virtual machines are allotted to the resources. The paper is illustrated in the following sections. Section 2 describes the basic particle swarm optimization model, Sect. 3 describes the fitness function evaluation, Sect. 4 describes results and evaluation, and Sect. 5 describes conclusion.

2 Particle Swarm Optimization

Particle swarm optimization (PSO) is an optimization method which is inspired by the functionalities of bird schooling or fish schooling. PSO is more advantageous and easier in its implementation as compared to a genetic algorithm. In PSO, a swarm of birds is considered as population and each bird is treated as an individual. Each individual has its own position, velocity, and fitness values while moving towards the destination position [13]. The fitness value of the particle/individual is to be evaluated in order to update its position and velocity. Thereby, particles of the population move in the search space after knowing the global best position ever achieved earlier. Let us consider swarm of birds is flocking for the food and there are '*n*' of birds also known as individuals in the swarm as represented with $\{\text{Set}_{\text{Swarm}}\} = \{\text{ind}_1, \text{ind}_2, \ldots, \text{ind}_n\}$.

Each individual has its own position and velocity as represented with $\{\text{Set}_{\text{Position}}\} = \{X_1, X_2, \ldots, X_n\}$ and $\{\text{Set}_{\text{velocity}}\} = \{V_1, V_2, \ldots, V_n\}$

During the fly, each individual imagines moving to its next best position. It means, for a swarm with '*n*' number set will have the same number on position best values. Among all the position values, the swarm decides the global best position and every individual tries to approach the global best position as represented in Fig. 1.

If X_i and V_i represent the position and velocity of a particle in *i*th iteration, the particle will update its position and velocity [12, 13], i.e. X_{i+1} and V_{i+1} after finding the G_{best} as represented in Eqs. (1) and (2).

$$V_{i+1} = V_i + C_1 * R_1 * (P_{\text{best}} - X_i) + C_2 * R_2 * (G_{\text{best}} - X_i) \tag{1}$$

$$X_{i+1} = X_i + V_{i+1} \tag{2}$$

where

P_{best}	Position best of the current particle,
G_{best}	Global best of the swarm which is not achieved earlier,
C_1 and C_2	Cognitive parameters and R_1 and R_2 = random parameters,

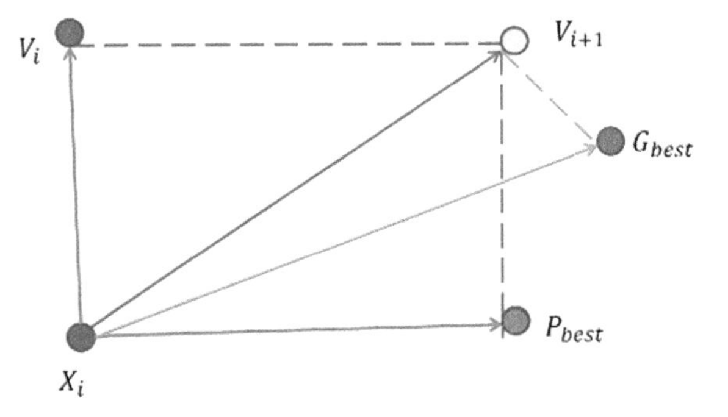

Fig. 1 Particle movement representation in PSO algorithm

X_i and V_i Position and velocity of individual particle in ith iteration,
X_{i+1} and V_{i+1} Updated position and velocity of individual particle in $(i + 1)$th iteration.

3 Development of Fitness Function

The present study aims towards the selection of virtual machines for the assigned tasks based on the total execution time. The total execution time depends on two factors:

(i) Virtual machine processing cost,
(ii) Time to complete the tasks.

Therefore, the virtual machine is to be selected which takes less virtual machine processing cost and less time to complete the tasks. The algorithm for fitness function generation is as follows:

```
Step-1: Fitness generation,
Step-2: Objective-1: cost for the processing of virtual machine as per Eq. (3),
Step-3: Objective-2: time for completing the task by the virtual machine as per Eq. (4),
Step-4: Combine objective-1 and objective-2 to form final fitness function as per Eq. (5),
Step-5: Define user constraints /*weights identification*/,
Case-1: Equal preference for objective-1 and objective-2,
Case-2: More preference for objective-1 and less preference for objective-2,
Case-3: More preference for objective-1 and less preference for objective-2,
Step-6: Choose any case from Step-5,
Step-7: End with a final fitness function.
```

The total execution time depends on virtual machine processing cost which is directly proportional to this factor, i.e.

$$C_{\text{exc}} \propto C_{\text{amp}} \tag{3}$$

where C_{exc} is the total execution time.

Also, C_{exc} varies with task completion time which is indirectly proportional to this factor, i.e.

$$C_{exc} \alpha \frac{1}{t_{complete}} \quad (4)$$

where $t_{complete}$ is the time required to complete the specific assigned task.

From Eqs. 3 and 4, the fine fitness function will be represented as shown in Eq. 5.

$$C_{exc} = w_1 * C_{vmp} + w_2 * \frac{1}{t_{complete}} \quad (5)$$

Such that $w_1 + w_2 = 1$,

where w_1 and w_2 are user-defined weights which give priorities to factors of total execution time.

Case-1: If equal performance is given for the both C_{vmp} and $t_{complete}$, then $w_1 = w_2 = 0.5$.

Case-2: If C_{vmp} is primary and $t_{complete}$ is secondary objective, then $w_1 = 0.6$ and $w_2 = 0.4$.

Case-3: If $t_{complete}$ is primary and C_{vmp} is secondary objective, then $w_1 = 0.4$ and $w_2 = 0.6$.

Here, C_{vmp} and $t_{complete}$ are calculated as follows:

i.e. C_{vmp} = (virtual machine processing time) * (virtual machine processing price).

$$C_{vmp} = VM_{pt} * VM_{pp} \quad (6)$$

And $t_{complete}$ = task length/virtual machine processing speed.

i.e.

$$t_{complete} = TL/VM_{ps} \quad (7)$$

And $VM_{pt}(i,j) = \frac{TL(i)}{VM_{ps}(j)} * 10$

Here, $VM_{pt}(i,j)$ gives the time to complete the ith task by jth virtual machine and this parameter value in seconds.

4 Proposed Methodology

The current methodology uses the particle swarm optimization to allocate the suitable virtual machine to the assigned tasks. In this, each particle is considered as the task assigned to a specific virtual machine. Therefore, there will be '$m \times n$'

number of particles corresponding to '*m*' number of virtual machines and '*n*' number of tasks. As the number of tasks/virtual machines increases, the probability of assigning tasks to virtual machines is raised and it is difficult to assign tasks to specific machines. In order to avoid this difficulty, the proposed PSO-based strategy is implemented to minimize the problem complexity and converges to a global solution. This approach initializes all the tasks and virtual machines to generate random particles, as each task to the corresponding virtual machine. These tasks are initialized with their current position, velocity. Later, all these particles are sent for fitness evaluation as represented in Eq. (5). The fitness values of all particles are stored as their current positional best values. Among the number of positional best values, one global best value will be selected.

From the obtained position and global best values, each particle's position and velocity are updated using Eqs. (1) and (2). This process will continue until the maximum number of iteration achieved. The flow chart of the proposed methodology is represented in Fig. 2.

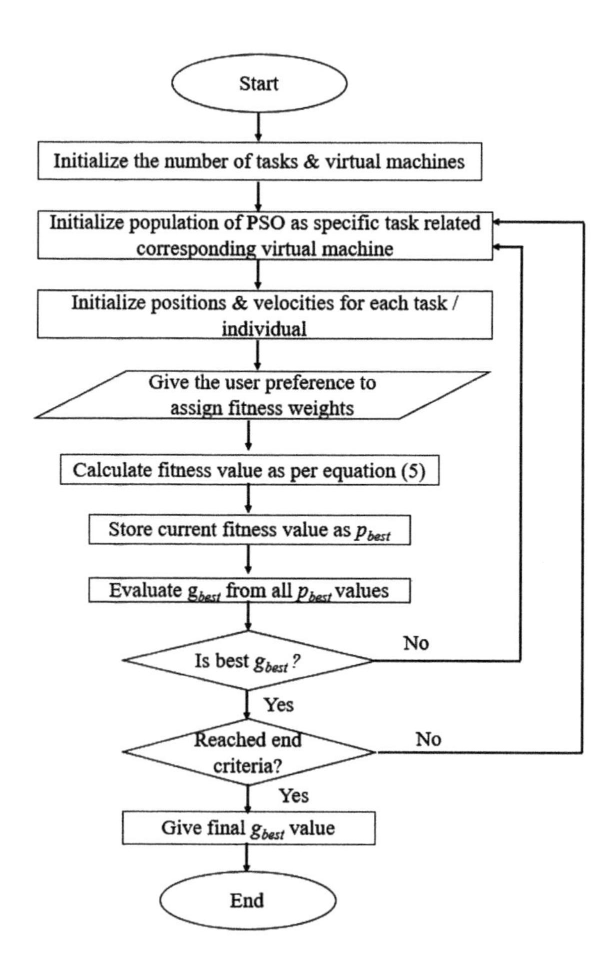

Fig. 2 Flow chart of proposed PSO-based task scheduling

5 Results and Discussion

The proposed method is implemented to four virtual machines for five tasks which are considered in [3]. The details of the considered data are represented in Tables 1 and 2. In this study, four virtual machines are considered and each of its processing speed and price is illustrated in Table 1.

To perform tasks by virtual machines, five numbers of tasks have been considered and each task length is represented in Table 2.

The algorithm has been implemented for the three cases, and the simulation results are presented accordingly. For all the three cases, the following PSO parameters have been considered during the implementation.

Population: 20 (no. of tasks * no. of virtual machines),
Cognitive parameters: $C_1 = C_2 = 2$ (for mathematical simplicity),
Random parameters: $R_1 = R_2 = 1$ (for mathematical simplicity),
Maximum no. of iterations: 20.

Case-1: Equal preference given for the both C_{vmp} and t_{complete}, i.e. $w_1 = w_2 = 0.5$.

It is observed from Fig. 3 the virtual machine-4 performs well as compared to other VMs' performance. Later on, the performance of VM-3, VM-2, and VM-1 is in descending order. Therefore, according to task priority, the virtual machines are to be selected for this case in order of 'VM-4 → VM-3 → VM-2 → VM-1'.

Case-2: C_{vmp} is primary, and t_{complete} is secondary objective, i.e. $w_1 = 0.6$ and $w_2 = 0.4$.

It is observed from Fig. 4 the virtual machine-1 performs well as compared to other VMs' performance. Later on, the performance of VM-3, VM-4, and VM-2 is in descending order. Therefore, according to task priority, the virtual machines are to be selected for this case in order of 'VM-1 → VM-3 → VM-4 → VM-2'.

Case-3: If t_{complete} is primary and C_{vmp} is secondary objective, then $w_1 = 0.4$ and $w_2 = 0.6$.

It is observed from Fig. 5 the virtual machine-3 performs well as compared to other VMs for the task lengths 1000. While increasing the task length further, virtual machine-1 is performing better as compared to other VMs, whereas VM-2 and VM-4 are almost performing equally for all the tasks. Therefore, according to task

Table 1 Virtual machine details for validation

Virtual machine	VM_1	VM_2	VM_3	VM_4
Processing speed (VM_{ps}) (MIPS)	250	500	750	1000
Processing price (VM_{pp}) ($)	0.02	0.047	0.148	0.200

Table 2 Task details for validation

Task name	Task-1	Task-2	Task-3	Task-4	Task-5
Task length	500	1000	1500	2000	2500

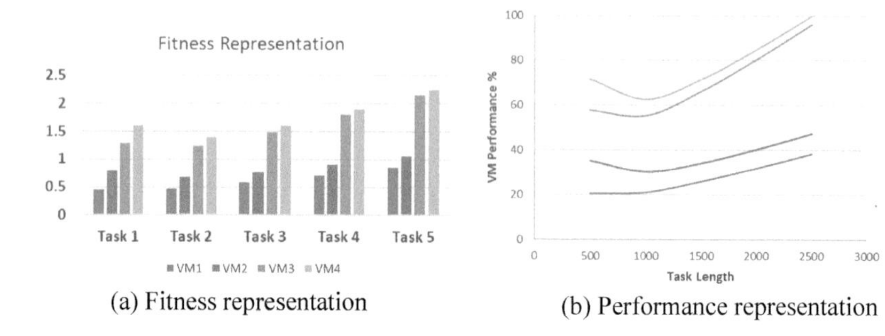

(a) Fitness representation (b) Performance representation

Fig. 3 Proposed task scheduling for case-1

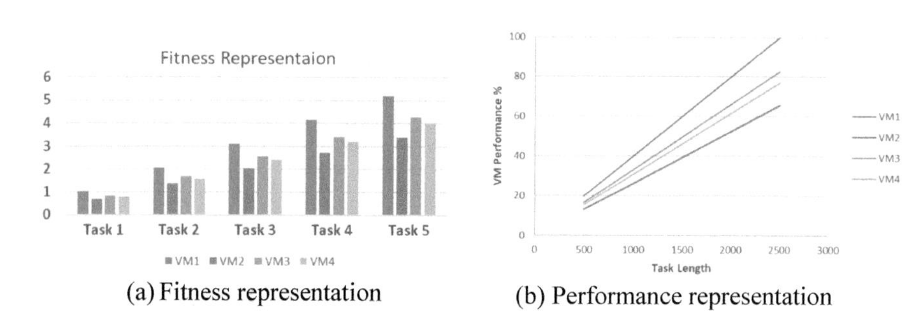

(a) Fitness representation (b) Performance representation

Fig. 4 Proposed task scheduling for case-2

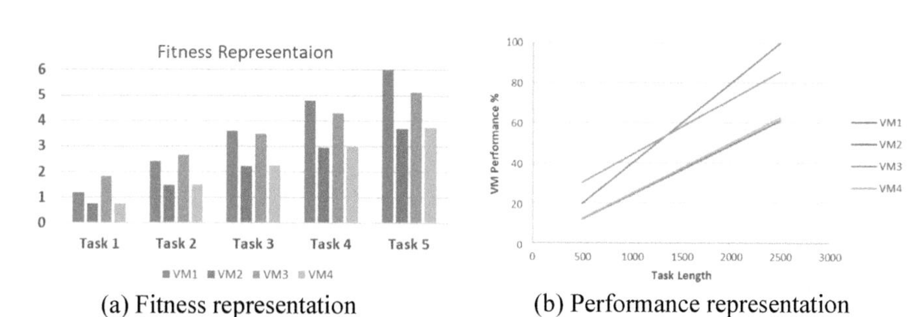

(a) Fitness representation (b) Performance representation

Fig. 5 Proposed task scheduling for case-3

priority, the virtual machines are to be selected for this case in order of 'VM-1 → VM-3 → VM-4 (or) VM-2' (for task length < 1500) and 'VM-3 → VM-1 → VM-4 (or) VM-2' (for task length > 1500).

6 Conclusion

In this research work, a novel strategy has been addressed for resource allocation based on particle swarm optimization (PSO) algorithm. The resource allocation is done with two objectives which produce optimal task scheduling. The first objective is related to virtual machine processing, and the second objective is related to the time elapsed to complete the given task. Based on the throughput of these objectives, three case studies have been identified: (i) equal priority to virtual machine processing and task execution period, (ii) more priority to virtual machine processing and less priority to task execution period, and (iii) less priority to virtual machine processing and more priority to task execution period. For each case, results are presented for the allocation of the virtual machine to the available resources. To obtain more efficient results in less processing time, the proposed algorithm may be incorporated with adaptiveness.

References

1. Tilak S, Patil D (2012) A survey of various scheduling algorithms in cloud environment. Int J Eng Invent 1(2):36–39
2. Buyya R, Vecchiola C, Selvi ST (2013) Mastering cloud computing: foundations and applications programming. Newnes
3. Alkayal ES, Jennings NR, Abulkhair MF (2016) Efficient task scheduling multi-objective particle swarm optimization in cloud computing. In: 2016 IEEE 41st conference on local computer networks workshops (LCN workshops). IEEE, pp 17–24
4. Vinothina V, Sridaran R, Ganapathi P (2012) A survey on resource allocation strategies in cloud computing. Int J Adv Comput Sci Appl 3(6):97–104
5. Juan W, Fei L, Aidong C (2012) An improved PSO based task scheduling algorithm for cloud storage system. Adv Inf Sci Serv Sci 4(18):465–471
6. Ardagna D, Casale G, Ciavotta M, Pérez JF, Wang W (2014) Quality-of-service in cloud computing: modeling techniques and their applications. J Internet Serv Appl 5(1):11
7. Wu X, Deng M, Zhang R, Zeng B, Zhou S (2013) A task scheduling algorithm based on QoS-driven in cloud computing. Proc Comput Sci 17:1162–1169
8. Kassarwani N, Ohri J, Singh A (2019) Performance analysis of dynamic voltage restorer using improved PSO technique. Int J Electron 106(2):212–236
9. Arumugam P, Panchapakesan M, Balraj S, Subramanian RC (2019) Reverse search strategy based optimization technique to economic dispatch problems with multiple fuels. J Electr Eng Technol 1–7
10. Ghosh P, Karmakar A, Sharma J, Phadikar S (2019) CS-PSO based intrusion detection system in cloud environment. In: Emerging technologies in data mining and information security. Springer, Singapore, pp. 261–269
11. Krishnasamy K (2013) Task scheduling algorithm based on hybrid particle swarm optimization in cloud computing environment. J Theoret Appl Inf Technol 55(1)
12. Dordaie N, Navimipour NJ (2017) A hybrid particle swarm optimization and hill climbing algorithm for task scheduling in the cloud environments. ICT Express
13. Deepak BBVL, Parhi DR, Raju BMVA (2014) Advance particle swarm optimization-based navigational controller for mobile robot. Arab J Sci Eng 39(8):6477–6487

Nonlinear Behaviour of Fixed-Fixed Beam with a Moving Mass

Anwesa Mohanty, Rabindra Kumar Behera and S. K. Pradhan

Abstract This study addresses the coupled nonlinear behaviour of a fixed-fixed beam under the travelling mass. Because of the beam and mass interaction phenomenon, coupling terms are more likely to arise which results in kinematic nonlinearities in the system. The major focus of this paper is to develop a theoretical model by introducing nonlinearities in the system. Later analysis of modal amplitude, mass position and tip deflection are done. For the beam modelling, Euler–Bernoulli beam assumptions are taken for consideration. Initially, a coupled mathematical model of the mentioned system is derived by using Hamilton's principle. Afterwards, the Galerkin discretization technique followed by the perturbation method is implemented in the mathematical system to analyse the dynamic characteristics of the desired system. Then, MATLAB ODE solver is used to plot various graphs for variation of amplitude and deflection with respect to time in case of both beam and mass. Under the internal resonance condition, the time-response curves are plotted to analyse the beating phenomenon for the beam and mass.

Keywords Fixed-fixed beam · Nonlinear analysis · MATLAB

1 Introduction

The research concerning the traversing load and mass problem is voluminous. The dynamic analysis of moving object acquires an important place in the field of research due to its practical implementation in today's world. Applications of moving mass problems are generally manifested in the area of transportation. Rails, bridges, runways and overhead cranes are some examples of structural elements to

A. Mohanty (✉) · R. K. Behera
Department of Mechanical Engineering, National Institute of Technology, Rourkela, Rourkela, Odisha 769008, India

S. K. Pradhan
College of Engineering & Technology, Bhubaneswar, Odisha 751029, India

BBVL. Deepak et al. (eds.), *Innovative Product Design and Intelligent Manufacturing Systems*, Lecture Notes in Mechanical Engineering,
https://doi.org/10.1007/978-981-15-2696-1_72

carry the moving mass. Since the lateral stage of the last decades, the challenge to design these systems has got the attention of many researchers. A small contribution towards the solution of the problem was made by different investigators. A broad discussion constituting various types of problem with moving mass is presented in the book by Fryba [1]. Yang presented a dynamic study of vehicle and bridge interaction to analyse the railway bridge behaviour due to the passing of high-speed train [2]. The dynamic analysis of simply supported beam (SSB) containing moving load using the finite element method was done by Olsson [3]. Further, Foda and Abduljabbar [4] used the Green function approach to determine the behaviour of SSB subjected to moving mass. Ye and Chen [5] investigated the effect of moving a load on the dynamic behaviour of SSB by changing various parameters. Abdelghyany et al. [6] analysed the dynamic characteristics of SSB under the action of moving load having a linear viscoelastic foundation.

According to various recent studies, the structures are mostly affected by abrupt changes in masses. Hence, the inertia effect of mass is unavoidable during the study of a dynamic behaviour problem. Influence of speed of moving mass on the structures with different boundary conditions was analysed by Dehestani et al. [7]. Şimşek [8] studied the vibration response of a functionally graded beam with travelling mass implementing different theories of the beam. A theoretical model of single-span beam with suspended moving mass was designed, and the dynamic analysis of that system was presented by He [9]. In recent year, Zupan and Zupan [10] presented a numerical analysis of three-dimensional beam under moving mass considering geometrical nonlinearity. Double-beam system considering viscously damped was analysed under harmonic loads by Wu and Gao [11]. Lai et al. [12] studied the dynamic behaviour of multiple SSB excited by the moving load. From the previous studies, it is realized that there is further scope to analyse the dynamic behaviour of moving beam–mass system. From the previous studies, it is realized that there is further scope to analyse the dynamic behaviour of moving beam–mass system. Hashemi and Khaniki [13] studied the dynamic behaviour of nanobeam under the influence of moving nanoparticle. Haitao et al. [14] introduced the viscoelastic foundation to the Euler–Bernoulli beam to study its dynamic characteristics. Heshmati and Yas [15] studied the dynamic behaviour of functionally graded MWCNT subjected to multi-moving load. The present paper is based on the dynamical characteristics of the beam with a fixed-fixed boundary condition having a spring–mass system.

2 Problem Formulation

Euler–Bernoulli beam theory is considered in the present analysis. It is presumed that there is always contact between moving mass and beam. Figure 1 shows the uniform beam–mass system with the fixed-fixed condition.

The length of the beam is L, the cross-sectional area of a mass density is ρ, and flexural rigidity is EI. The mass of the moving body is M which moves with a

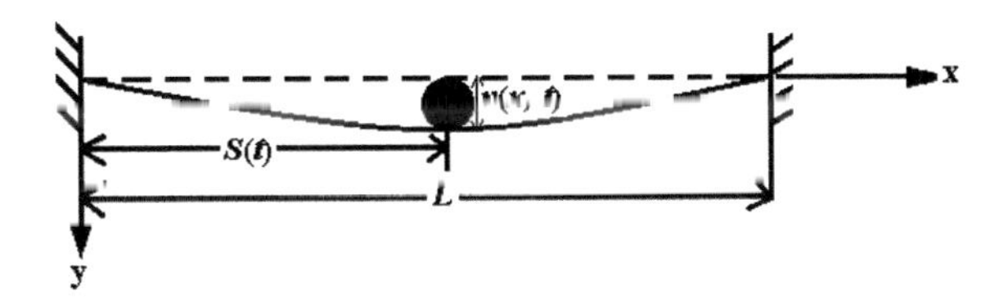

Fig. 1 Schematic diagram of the fixed-fixed beam with moving mass

uniform velocity (v). Mass is connected with a linear spring of stiffness 'k' with one fixed end. Here all these dimensionalized parameters are normalized by dividing required parameters for the proper calculation. Therefore, $s = S/L$, $A = a/L^2$, $m = M/\rho AL$, and $\omega^2 = k\rho Al^4/mEI$.

3 Theoretical Analysis and Results

3.1 Displacement Field

The Euler–Bernoulli beam model is assumed for the theoretical study. Here, $v(x, t)$ measures the vertical displacement of the reference axis. U, V and W denote the displacements in the undeformed x-, y- and z-direction, respectively. The following displacement field is assumed to be:

$$\begin{aligned} U &= -y\sin(\theta), \sin(\theta) = \left(\frac{\partial}{\partial x}v(x,t)\right) \\ V &= v(x,t), \cos(\theta) = 1 \\ W &= 0 \end{aligned} \tag{1}$$

3.2 Equation of Motion

The nonlinearity of the problem arises due to the interaction between the beam and spring–mass. Kinetic and potential energy for beam and mass system are given as

$$\begin{aligned} T_b &= \frac{1}{2}\int_0^l (v_t)^2 \mathrm{d}x,\ V_b = \frac{1}{2}\int_0^l (v_{xx})^2 \mathrm{d}x, \\ T_m &= \frac{1}{2}m\left[(s_t)^2 + (v_t)^2 + 2s_t\sin(\theta)v_t\right]_{X=s}, \\ \text{Potential energy for spring, } V_s &= \frac{1}{2}m\omega^2(s - s_e)^2 \end{aligned} \tag{2}$$

'ω' is the frequency of the spring–mass subsystem.

The equations of motion of the system can be derived by using Hamilton's principle as

$$s_{tt} + \omega^2(s - s_e) + \left[\phi_i(\phi_j)_x\right]_{x=s}\{(\alpha_i)_{tt}\alpha_j\} = 0 \tag{3}$$

$$\begin{aligned} m&\left\{\begin{array}{l} \left[\phi_i(\phi_j)_{xx}\right]_{x=s}(\alpha_j)_{tt} + s_{tt}\left[\phi_i(\phi_j)_{xx}\right]_{x=s}\{\alpha_j\} + \\ 2\dot{s}\left[\phi_i(\phi_j)_x\right]_{x=s}(\alpha_j)_t + (s_t)^2\left[\phi_i(\phi_j)_{xx}\right]_{x=s}\{\alpha_j\} \end{array}\right\} \\ &+ \left[\int_0^l \phi_i\phi_j\,\mathrm{d}x\right](\alpha_j)_{tt} + \left[\int_0^l (\phi_i)_{xx}(\phi_j)_{xx}\mathrm{d}x\right]\{\alpha_j\} = 0 \end{aligned} \tag{4}$$

Here subscripts 'x' and 't' denote the derivative with respect to distance and time. The boundary condition for the fixed-fixed beam is used to get the eigenfunction which is as follows:

$$\begin{aligned} \phi_i = &(\sinh(k_i x) - \sin(k_i x)) + \\ &\frac{\sinh(k_i l) - \sin(k_i l)}{\cos(k_i l) - \cosh(k_i l)}(\cosh(k_i x) - \cos(k_i x)) \end{aligned} \tag{5}$$

For the analysis purpose, first mode of the fixed-fixed beam is taken into consideration. Considering the first mode (ϕ_1) of the desired beam as a basis function, Eqs. (3) and (4) reduce to:

$$s_{tt} + \omega^2 s + c_1(\alpha_1)_{tt}\alpha_1 = 0 \tag{6}$$

$$(\alpha_1)_{tt} + \omega_1^2\alpha_1 + mc_2 s_{tt}\alpha_1 + 2mc_2 s_t(\alpha_1)_t + 2mc_2 s(\alpha_1)_{tt} = 0 \tag{7}$$

where

$$\begin{aligned} c_1 = \phi_1(\phi)_x\big|_{X=s_e},\; c_2 &= \frac{\phi_1(\phi)_x\big|_{X=s_e}}{\int_0^l(\phi_1)^2\mathrm{d}x + m(\phi_1)^2\big|_{x=s_e}}, \\ (\omega_1)^2 &= \frac{\int_0^l(\phi_1)_{xx}^2\mathrm{d}x}{\int_0^l(\phi_1)^2\mathrm{d}x + m(\phi_1)^2\big|_{x=s_e}} \end{aligned} \tag{8}$$

Here ω is the frequency of moving mass, and ω_1 is the first frequency of the beam. Relationship between them is:

$$\omega = 2\omega_1 + \varepsilon\sigma \tag{9}$$

where 'σ' is a small detuning parameter.

While σ is zero, we get perfect internal resonance 1 : 2. To obtain the analytical solution for the derived mathematical model, the method of multiple scales (MMS) is used having a timescale $T_n = \varepsilon^n t$, where 'ε' is scaling parameter and $n = 0, 1, 2, \ldots$; $T_0, T_1, T_2, \ldots$ are small timescales. Here two timescales are considered, i.e. T_0 and T_1, as the nonlinearities have a very small effect. The solution is assumed of the form

$$\begin{aligned} s(t;\varepsilon) &= s_0(T_0,T_1) + \varepsilon s_1(T_0,T_1) + \ldots \\ \alpha_1(t;\varepsilon) &= \alpha_0(T_0,T_1) + \varepsilon\alpha_1(T_0,T_1) + \ldots \end{aligned} \tag{10}$$

After simplifying Eqs. (6) and (7) and implementing this MMS technique [16], the solution for the required equation can be written as:

$$\begin{aligned} \frac{\partial p_1}{\partial T_1} &= \frac{1}{4}\frac{c_1 p_2^2 \omega_1^2}{\omega}\sin(2\varphi_2 - \varphi_1 - \sigma T_1), \\ p_1\frac{\partial \varphi_1}{\partial T_1} &= -\frac{1}{4}\frac{c_1 p_2^2 \omega_1^2}{\omega}\cos(2\varphi_2 - \varphi_1 - \sigma T_1), \\ \frac{\partial p_2}{\partial T_1} &= -\frac{1}{4}\frac{mc_2\left(\omega^2 - 2\omega\omega_1 + 2\omega_1^2\right)}{\omega_1} p_1 p_2 \sin(2\varphi_2 - \varphi_1 - \sigma T_1), \\ p_2\frac{\partial \varphi_2}{\partial T_1} &= -\frac{1}{4}\frac{mc_2\left(\omega^2 - 2\omega\omega_1 + 2\omega_1^2\right)}{\omega_1} p_1 p_2 \cos(2\varphi_2 - \varphi_1 - \sigma T_1) \end{aligned} \tag{11}$$

where p_1, φ_1 are the modal amplitudes and phase angle for moving mass,p_2, φ_2 are the amplitudes and phases for the beam deflection.

4 Results

Analysis of beam under fixed-fixed condition is done by using the perturbation method. Only first mode for the mentioned boundary condition is taken for the analysis purpose. Natural frequencies are calculated by using Eq. (9). The initial values for mass displacement (s_0) and tip deflection (v_{t_0}) are taken as 0.00001 and 0.1, respectively. p_{2_0} is taken as $0.5v_{t_0}$. To maximize accuracy and efficiency, the final ordinary differential Eq. (11) is solved using the Jacobi iteration method. Stiff ODE solver is used to plotting the modal amplitude and the variation of mass and beam deflection. The time response in figures shows the beating phenomenon for the mass and the beam by changing different parameter under internal resonance case. As mentioned in Nonlinear Oscillation by Nayfeh [16], the detuning parameter is calculated. Various graphs are plotted for different detuning parameter σ and with $\varepsilon = 1$.

Figure 2 shows the response curve for moving mass at equilibrium position and beam response due to the effect of moving mass for the required system. Here the equilibrium position (s_e) of moving mass is 0.9, and nondimensionalized mass ratio

$m = 0.1$ is taken. Figure 2a, b presents the fluctuation of modal amplitude (p_1) in case of mass and beam (p_2), respectively. The value σ is used to compare various solutions. In this case, it is considered as -0.0085. Figure 2c, d shows the numerical solutions obtained from Eqs. (6) and (7) by using stiff ode23s solver. In Fig. 2, maximum deflection of mass is 0.16 and minimum is 0, whereas in case of beam maximum deflection is 0.1 and minimum deflection is 0.04.

Figure 3 shows a similar response curve by taking the mass ratio as 0.1 and $\sigma = 0$. It is found that for the same initial value by changing the parameters like mass ratio and σ, the value of amplitude changes. It is found that after time period 250, the amplitude decreases gradually. In this case, maximum deflection obtained is 0.125, minimum deflection is 0 for the mass, and in case of beam, maximum deflection is 0.1 and the minimum is 0.06. It can be inferred that for a small change in the detuning parameter, there is no significant change in maximum deflection of the beam. It divulges from the comparison of Figs. 2 and 3 that time period decreases by reducing the nondimensionalized mass ratio and detuning parameter σ.

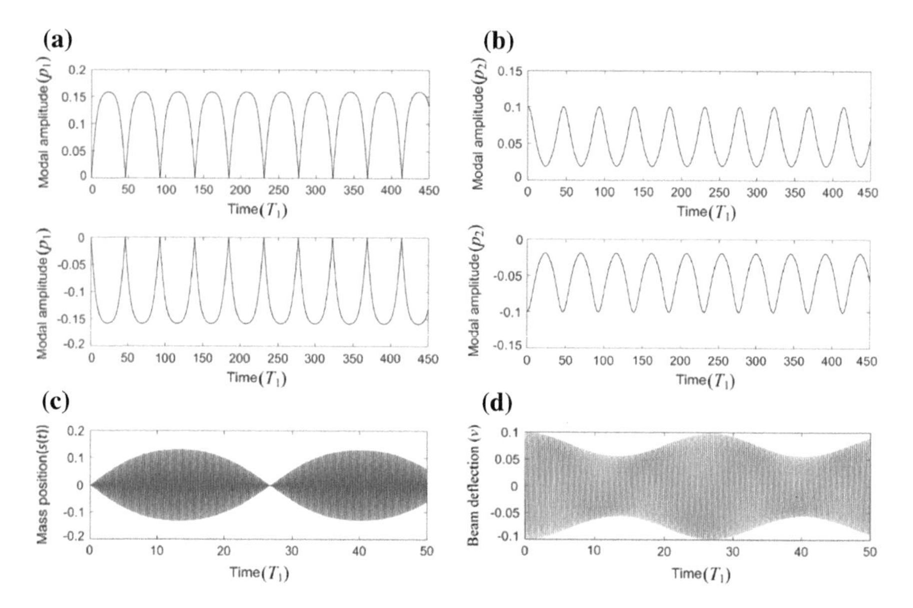

Fig. 2 $m = 0.1$, $s_e = 0.9$, $s_{1_0} = 0.00001$, $v_{t_0} = 0.1$, **a**, **b** perturbation solution $\sigma = -0.0085$ for mass and beam, respectively, **c** mass position and **d** beam deflection

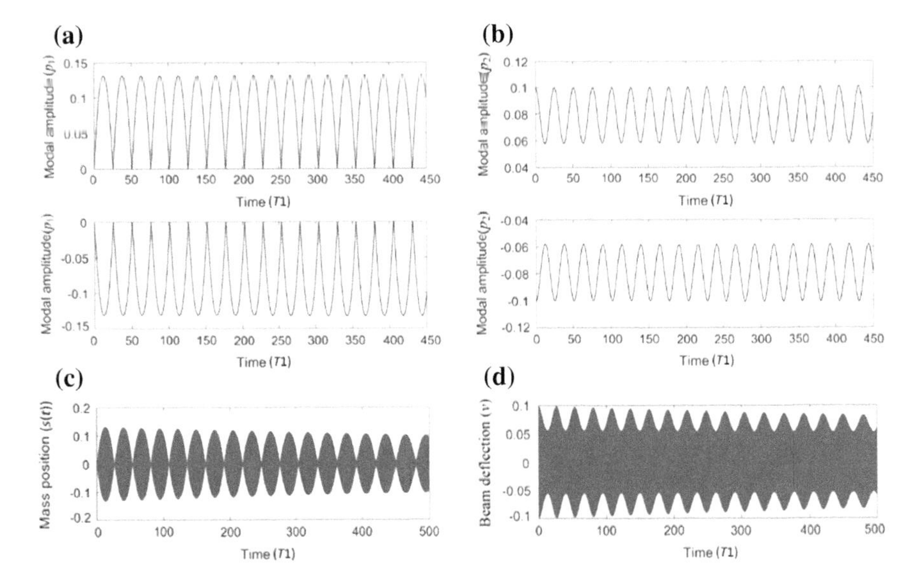

Fig. 3 $m = 0.1$, $s_e = 0.9$, $s_{1_0} = 0.00001$, $v_{t_0} = 0.1$, $\sigma = 0$. **a** Perturbation solution for mass, **b** perturbation solution for beam, **c** mass position and **d** beam deflection

5 Conclusion

1. Perturbation method can be successfully used for the dynamic analysis of the moving mass system.
2. The numerical results reveal that for a small value of σ, the system works near the resonance condition.
3. The minimum amplitude variation of the beam deflection never goes to zero. However, in the case of moving mass, minimum variation of amplitude tends to zero.
4. Both mass position and beam deflection appear dark due to small time steps and very high-frequency variation.

References

1. Fryba L (1999) Vibration of solids and structures under moving loads. Thomas Telford Publishing
2. Yang Y-B, Yau J-D (1997) Vehicle-bridge interaction element for dynamic analysis. J Struct Eng 123(11):1512–1518
3. Olsson M (1991) On the fundamental moving load problem. J Sound Vib 145(2):299–307
4. Foda MA, Abduljabbar Z (1998) A dynamic green function formulation for the response of a beam structure to a moving mass. J Sound Vib 210(3):295–306

5. Ye Z, Chen H (2009) Vibration analysis of a simply supported beam under moving mass based on moving finite element method. Front Mech Eng China 4(4):397–400
6. Abdelghany SM, Ewis KM, Mahmoud AA, Nassar MM (2015) Dynamic response of non-uniform beam subjected to moving load and resting on non-linear viscoelastic foundation. Beni-Suef Univ J Basic Appl Sci 4(3):192–199
7. Dehestani M, Mofid M, Vafai A (2009) Investigation of critical influential speed for moving mass problems on beams. Appl Math Model 33(10):3885–3895
8. Şimşek M (2010) Vibration analysis of a functionally graded beam under a moving mass by using different beam theories. Compos Struct 92(4):904–917
9. He W (2018) Vertical dynamics of a single-span beam subjected to moving mass-suspended payload system with variable speeds. J Sound Vib 418:36–54
10. Zupan E, Zupan D (2018) Dynamic analysis of geometrically non-linear three-dimensional beams under moving mass. J Sound Vib 413:354–367
11. Wu Y, Gao Y (2015) Analytical solutions for simply supported viscously damped double-beam system under moving harmonic loads. J Eng Mech
12. Lai Z, Jiang L, Zhou W (2018) An analytical study on dynamic response of multiple simply supported beam system subjected to moving loads. Shock Vib 2018:1–14
13. Hashemi SH, Khaniki HB (2018) Dynamic response of multiple nanobeam system under a moving nanoparticle. Alexandria Eng J 57(1):343–356
14. Yu H, Yang Y, Yuan Y (2018) Analytical solution for a finite Euler-Bernoulli beam with single discontinuity in section under arbitrary dynamic loads. Appl Math Model 60:571–580
15. Heshmati M, Yas MH (2013) Dynamic analysis of functionally graded multi-walled carbon nanotube-polystyrene nanocomposite beams subjected to multi-moving loads. Mater Des 49:894–904
16. Nayfeh AH, Mook DT (2008) Nonlinear oscillations. Wiley

Finite Element Dynamic Study of Inclined Beam Subjected to Moving Point Load

Suraj Parida, Sudhansu Meher and R. K. Behera

Abstract This paper showcases the response analysis of the inclined beam under consideration of concentrated load moving at a constant velocity. Finite element method is utilized for the formulation of the problem. A convergence study is made for natural frequencies of pinned-pinned (P-P) beam. The case is extended for the horizontal and inclined beam. Newmark integration method is implemented for dynamic vibration of the structure. The numerical results are obtained using MATLAB code. Influence of damping on dynamic magnification factor (DMF) due to moving load is investigated. Effect of velocity parameter on load is studied as well. Results are extended in graphical form with respect to vertical dynamic displacement of the beam for a different angle of inclination.

Keywords Finite element method · Newmark's integration method · Inclined beam · Moving the load

1 Introduction

The dynamic response for forced vibrations of beam type structure under the act of concentrated loads is widely used in aerospace and mechanical applications. Many studies have been done in the past few years to observe and predict the behavior of isotropic structure as well as FGM made system. Using perturbation and finite difference methods, motion and impact of mass and velocity of the pinned-pinned beam have been studied subjected to moving load [1–3]. To study infinite beam with multiple spans, modal analysis was done [4], whereas, for finite beam, Green's function approach was preferred [5]. In various cases, to understand simply supported [6–8] and cantilever beam [9], finite element method was preferably used for Euler–Bernoulli and Timoshenko beam [10, 11] and is compared with different

S. Parida (✉) · S. Meher · R. K. Behera
Department of Mechanical Engineering, National Institute of Technology, Rourkela, Rourkela 769008, India

BBVL. Deepak et al. (eds.), *Innovative Product Design and Intelligent Manufacturing Systems*, Lecture Notes in Mechanical Engineering,
https://doi.org/10.1007/978-981-15-2696-1_73

methods such as Galerkin's [12] and finite difference technique [13, 14] for verification. The further extension has been done by introducing multi-span beams excited by moving loads having constant velocity for different beam theories [15, 16]. Some researchers have studied the dynamic behavior of the beam subjected to soil foundation [17–19]. It is realized from the review of literature that most of the dynamic analyses are limited to horizontal beams. Hence, there is a scope to demonstrate the effect of dynamic magnificent factor (DMF) with respect to damping effect on inclined beam. Initially, the convergence for natural frequencies of the pinned-pinned beam is done. For the validations of present work, obtained results are compared with ANSYS results. Time histories for dynamic vertical displacements of the midpoint of P-P beam due to traversing load are obtained for different values of loads and angle of inclinations.

2 Modeling and Formulation

An inclined beam of length L with the angle of inclination θ, subjected to vertical load F with uniform velocity $v = L/\tau$, where τ represents the time taken by the vertical load to cover the beam. As it is an inclined beam, axial deformation is also considered along with vertical displacement and rotation for the stiffness matrix of beam element takes the form K as:

$$K = \begin{bmatrix} \frac{A_eE}{L_e} & 0 & 0 & -\frac{A_eE}{L_e} & 0 & 0 \\ 0 & \frac{12EI_e}{L_e^3} & \frac{6EI_e}{L_e^2} & 0 & -\frac{12EI_e}{L_e^3} & \frac{6EI_e}{L_e^2} \\ 0 & \frac{6EI_e}{L_e^2} & \frac{4EI_e}{L_e} & 0 & -\frac{6EI_e}{L_e^2} & \frac{2EI_e}{L_e} \\ -\frac{A_eE}{L_e} & 0 & 0 & \frac{A_eE}{L_e} & 0 & 0 \\ 0 & -\frac{12EI_e}{L_e^3} & -\frac{6EI_e}{L_e^2} & 0 & \frac{12EI_e}{L_e^3} & -\frac{6EI_e}{L_e^2} \\ 0 & \frac{6EI_e}{L_e^2} & \frac{2EI_e}{L_e} & 0 & -\frac{6EI_e}{L_e^2} & \frac{4EI_e}{L_e} \end{bmatrix} \tag{1}$$

where

A_e Cross-section area of beam element,
E Elastic stiffness of element,
I_e Beam element's moment of inertia about Z-axis,
L_e Element length of the beam.

Similarly, the consistent mass matrix for 3-DOF beam element is obtained as:

$$M = \begin{bmatrix} 2a & 0 & 0 & a & 0 & 0 \\ 0 & 156b & 22L_e^2b & 0 & 54b & -13L_eb \\ 0 & 22L_e^2b & 4L_e^2b & 0 & 13L_eb & -3L_e^2b \\ a & 0 & 0 & 2a & 0 & 0 \\ 0 & 54b & -13L_eb & 0 & 156a & -22L_e^2b \\ 0 & 13L_eb & -3L_e^2b & 0 & -22L_e^2b & 4L_e^2b \end{bmatrix} \quad (2)$$

where $a = \frac{\rho A_e L_e}{6}$ and $b = \frac{\rho A_e L_e}{420}$.

Using Eqs. (1) and (2), overall matrices of the beam [M] and [K] are formed by assembling its element matrices, e.g., M and K matrices of "n" number of elements, showing connectivity of it. Equation of motion for the dynamic system can be written as:

$$[M]\{\ddot{U}\} + [K]\{\dot{U}\} = \{F(t)\} \quad (3)$$

where $\ddot{U}$, $\dot{U}$, $F(t)$ are acceleration, velocity and load vectors of the beam, respectively. Natural frequencies of the beam are obtained under free vibration condition. Due to inclination, vertical load F acting on the ith node of the beam element is considered in the x- and y-direction as $F\sin\theta$ and $F\cos\theta$, respectively, as shown in Fig. 1.

Time histories of point load F for the ith node is shown in Fig. 2. The load is moving at constant velocity v, completing beam span in time $\tau = T/\alpha$ where T is time duration of first natural frequency and α is velocity parameter. When load reached the ith node, time $\tau_1 = x_i/v$.

For analysis, Newmark integration method is used, considering integration parameters of constant average acceleration, $\beta = 1/4$ and $\gamma = 1/2$. Velocity parameter is considered, $\alpha = T/\tau$.

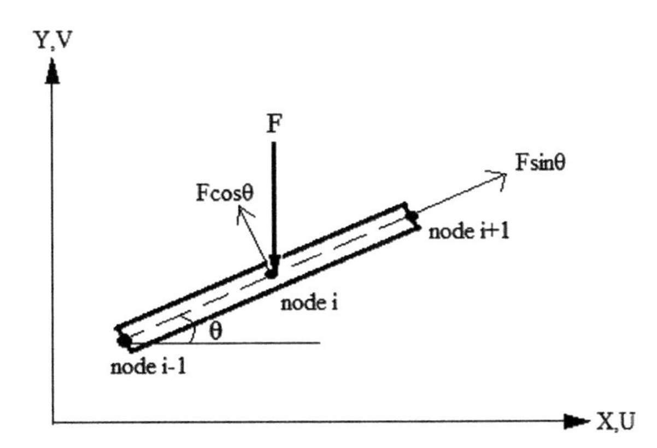

Fig. 1 Moving load analysis at the ith node

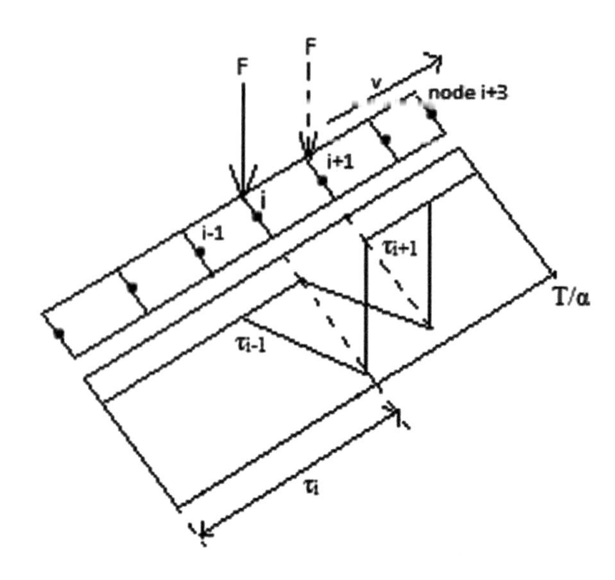

Fig. 2 Time histories of point load for the *i*th node

3 Numerical Results and Discussions

For numerical results, a rectangular beam of length 1 m is considered with width 0.05 m and height 0.01 m, mass density 7860 kg/m^3 and modulus of elasticity 206×10^9 kg/m^3. In order to validate the response of inclined beam subjected to moving point load, the angle of inclination is set as zero and compared with results obtained in ANSYS as shown in Fig. 3.

Figure 4 shows a graphical representation of frequencies versus number of elements in mess for P-P condition of the beam under free vibration. Convergence study of the first six frequencies is done. The maximum number of elements is taken as 120 in order to get convergent and accurate values of first six frequencies.

Figure 5 shows the effect of damping on dynamic behavior when the load is applied on the midpoint of the beam, considering various values of velocity parameters. The values of DMF decrease considerably due to the effect of damping, for a given value of "α".

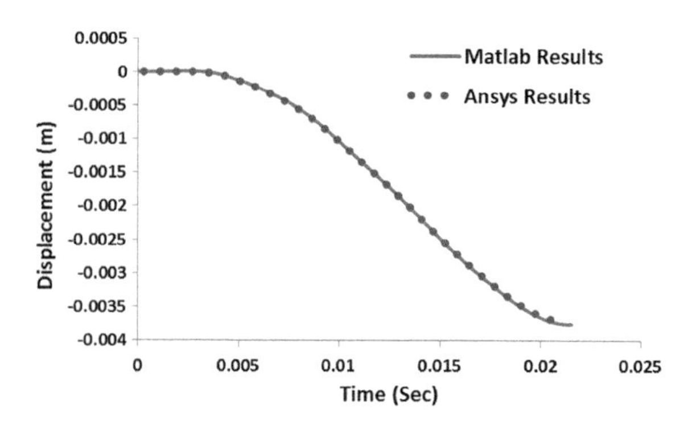

Fig. 3 Displacement versus time for the beam mass system

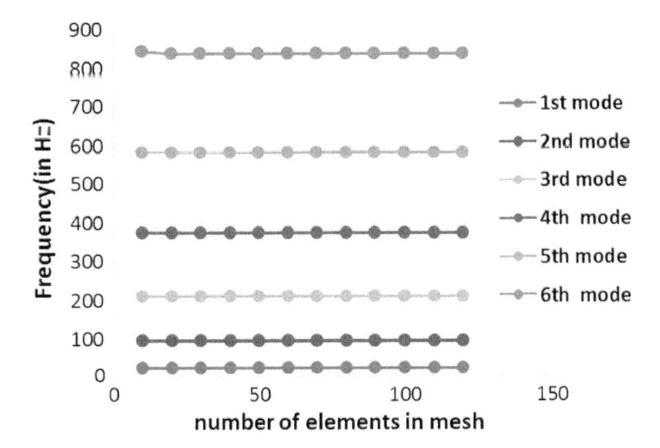

Fig. 4 Convergence of frequencies with no of elements in mesh for the P-P beam

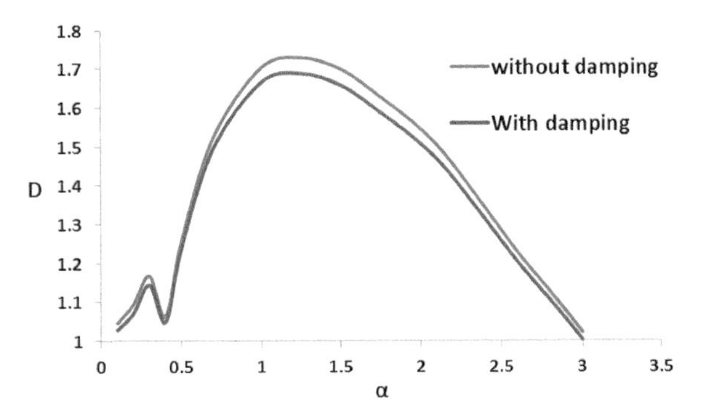

Fig. 5 Effect of damping on DMF due to load on the midpoint of the beam

Figure 6 depicts the variation of vertical displacements at midpoint w.r.t time for $\alpha = 1.6$ and $\theta = 15°$. One can visualize the value of the load is directly proportional to vertical displacement caused by the moving load. It is further observed that for particular velocity, by increasing load, midpoint deflection decreases. Figure 7 shows the effect of variation of the angle of inclination on a deflection at uniform velocity when $\alpha = 0.5$.

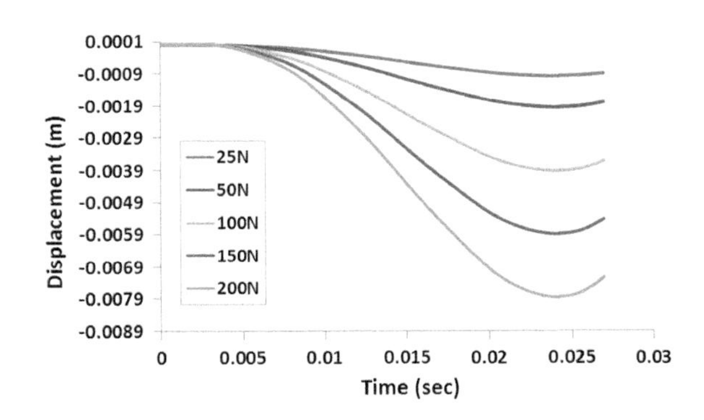

Fig. 6 Time histories for dynamic vertical displacements of the midpoint of P-P beam due to varying load, when $\alpha = 1.6$ and $\theta = 15°$

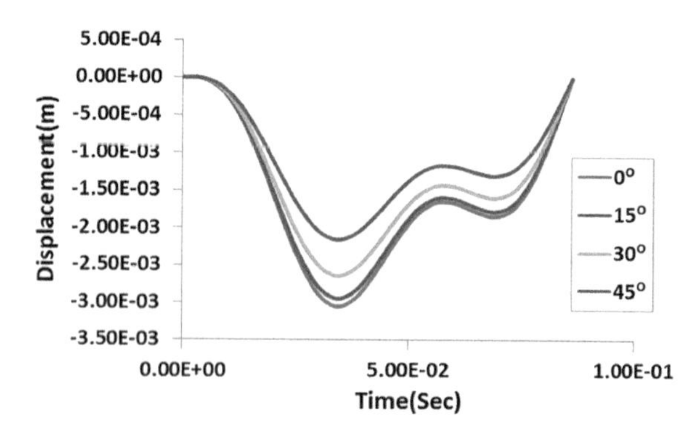

Fig. 7 Time histories for dynamic vertical displacements of the midpoint of P-P beam due to the varying angle of inclination, when $\alpha = 0.5$

4 Conclusion

This study comprises the dynamic study of an inclined beam subjected to uniformly moving point load. For this purpose, a pinned-pinned beam is taken. To validate the results, convergence studies are done. The MATLAB results are compared with values obtained in ANSYS for zero angles of inclination. Time histories plots are successfully presented for dynamic vertical displacement to study the effect of variation of load and a different angle of inclination of the beam. Effect of damping on DMF is studied when the load is acting on the midpoint of the beam. Application of inclined beam is found in car ramp where elevation is provided for effective service and maintenance of undercarriage of vehicles.

References

1. Xu X, Xu W, Genin J (1997) A non-linear moving mass problem. J Sound Vib 204(3):495–504
2. Michaltsos G, Sophianopoulos D, Kounadis AN (1996) The effect of a moving mass and other parameters on the dynamic response of a simply supported beam. J Sound Vib 191 (3):357–362
3. Frýba L (1973) Vibration of solids and structures under moving loads. Springer, Netherlands
4. Michaltsos GT (2002) Dynamic behaviour of a single-span beam subjected to loads moving with variable speeds. J Sound Vib 258(2):359–372
5. Foda MA, Abduljabbar Z (1998) A dynamic green function formulation for the response of a beam structure to a moving mass. J Sound Vib 210(3):295–306
6. Ju S-H, Lin H-T, Hsueh C-C, Wang S-L (2006) A simple finite element model for vibration analyses induced by moving vehicles. Int J Numer Meth Eng 68(12):1232–1256
7. Kidarsa A, Scott MH, Higgins CC (2008) Analysis of moving loads using force-based finite elements. Finite Elem Anal Des 44(4):214–224
8. Lin Y-H, Trethewey MW (1990) Finite element analysis of elastic beams subjected to moving dynamic loads. J Sound Vib 136(2):323–342
9. Siddiqui SAQ, Golnaraghi MF, Heppler GR (2000) Dynamics of a flexible beam carrying a moving mass using perturbation, numerical and time-frequency analysis techniques. J Sound Vib 229(5):1023–1055

10. Lou P, Dai G-L, Zeng Q-Y (2006) Finite-element analysis for a Timoshenko beam subjected to a moving mass. Proc Inst Mech Eng Part C J Mech Eng Sci 220(5):669–678
11. Wang R-T, Chou T-H (1998) Non-linear vibration of Timoshenko beam due to a moving force and the weight of beam. J Sound Vib 218(1):117–131
12. Lee HP (1998) Dynamic response of a Timoshenko beam on a winkler foundation subjected to a moving mass. Appl Acoust 55(3):203–215
13. Esmailzadeh E, Ghorashi M (1997) Vibration analysis of a Timoshenko beam subjected to a travelling mass. J Sound Vib 199:615–628
14. Esmailzadeh E, Ghorashi M (1995) Vibration analysis of beams traversed by uniform partially distributed moving masses. J Sound Vib 184(1):9–17
15. Ichikawa M, Miyakawa Y, Matsuda A (2000) Vibration analysis of the continuous beam subjected to a moving mass. J Sound Vib 230(3):493–506
16. Wang R-T (1997) Vibration of multi-span Timoshenko beams to a moving force. J Sound Vib 207(5):731–742
17. Thambiratnam D, Zhuge Y (1996) Dynamic analysis of beams on an elastic foundation subjected to moving loads. J Sound Vib 198(2):149–169
18. Hien TD, Hung ND, Kien NT, Noh HC (2019) The variability of dynamic responses of beams resting on elastic foundation subjected to vehicle with random system parameters. Appl Math Model 67:676–687
19. Beskou ND, Muho EV (2018) Dynamic response of a finite beam resting on a Winkler foundation to a load moving on its surface with variable speed. Soil Dyn Earthq Eng 109:222–226

Determination of Flow Characteristics in Fire-Tube Boiler by Numerical Simulation

A. Hari Kishan, Muppidi Chaitanya and P. Uma Maheswara Rao

Abstract This research work provides the thermal analysis of fire-tube boilers used in thermal power plants. For simulation purpose, a small-scale prototype of the original fire-tube boiler is designed using SolidWorks. The numerical simulation of the designed model is carried out in ANSYS Fluent. The initial section of the project represents the pressure and temperature variations along the length of the boiler for different water velocities (25, 30, 35 and 40 m/s). The later section deals by changing the boiler casing material between steel, brass and stainless steel to study the pressure and temperature variations at a constant water velocity of 30 m/s. Based on the results, the best boiler shell material among these three has been identified. The results have been provided in the form of pressure and temperature contours as obtained from the CFD analysis.

Keywords Fire-tube boiler · Boiler shell

1 Introduction

Boilers are pressure vessels which are used for heating water or generating steam for the heating facility in industries and to generate electricity through steam turbines. A boiler is an enclosed vessel that provides a means for combustion and transfers heat to water until it becomes hot water or steam. The hot water or steam under pressure is then usable for transferring the heat to a process. Water is a useful and cheap medium for transferring heat to a process. When water is boiled into steam, its volume increases about 1,600 times, producing a force that is almost as explosive as gunpowder. This causes the boiler to be extremely dangerous equipment and should be treated carefully. Liquid is heated upto the gaseous state, and this process is called evaporation (Fig. 1).

A. Hari Kishan (✉) · M. Chaitanya · P. Uma Maheswara Rao
Vignan's Institute of Information Technology, Duvvada, Vizag 530049, India

BBVL. Deepak et al. (eds.), *Innovative Product Design and Intelligent Manufacturing Systems*, Lecture Notes in Mechanical Engineering,
https://doi.org/10.1007/978-981-15-2696-1_74

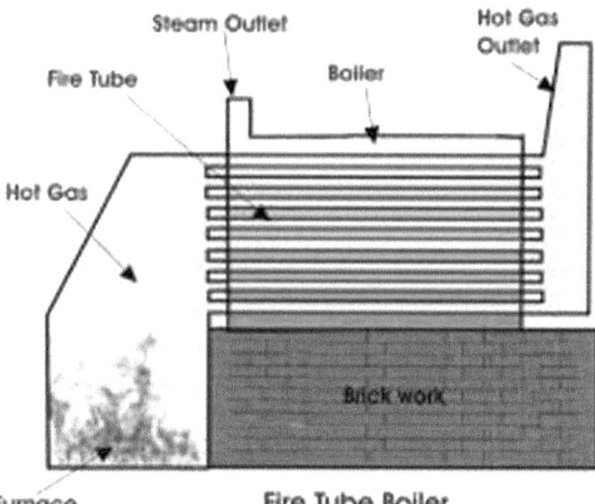

Fig. 1 Boiler system [1]

2 Literature Survey

Kodak et al. [2] designed and developed a non-IBR vertical fire-tube boiler. This boiler works on the thermodynamic principle. The main advantage of a vertical boiler is that it is economically comparatively cheap and simple in design. The main objective is to increase the efficiency of the boiler by increasing temperature. The flue gas passes inside boiler tubes, and heat is transferred to water on the shell side [3]. A dynamic model has been developed for the analysis of boiler performance, and MATLAB has been applied for integrating it. The mathematical model developed is based on the first principles of mass, energy and momentum conservations [4]. In the model, the two parts of the boiler (fire/gas and water/steam sides), the economizer, the superheater and the heat recovery are considered [5].

Senthur et al.'s work is on pressures and increased temperatures to which boiler components are exposed that initiate deformations and stresses and can lead to a construction breakdown [6]. So, strength calculations for most reliable boiler elements are standardized and subject to supervising inspection control. Norms such as EN 12952-3 and EN 12953-3 state the allowed stresses for a given temperature and bring explicit formulas for strength calculation through determining wall thickness for pressurized elements [7].

3 Methodology

In the present paper, modeling of the fire-tube boiler is done for analysis. The diameter of the considered boiler (inner/outer) is 300/350 mm, and diameter of fire tube (inner/outer) is 35/50 mm. Modeling of the fire-tube boiler is done by the software CATIA V5R16. In the process of modeling the boiler geometry, it is

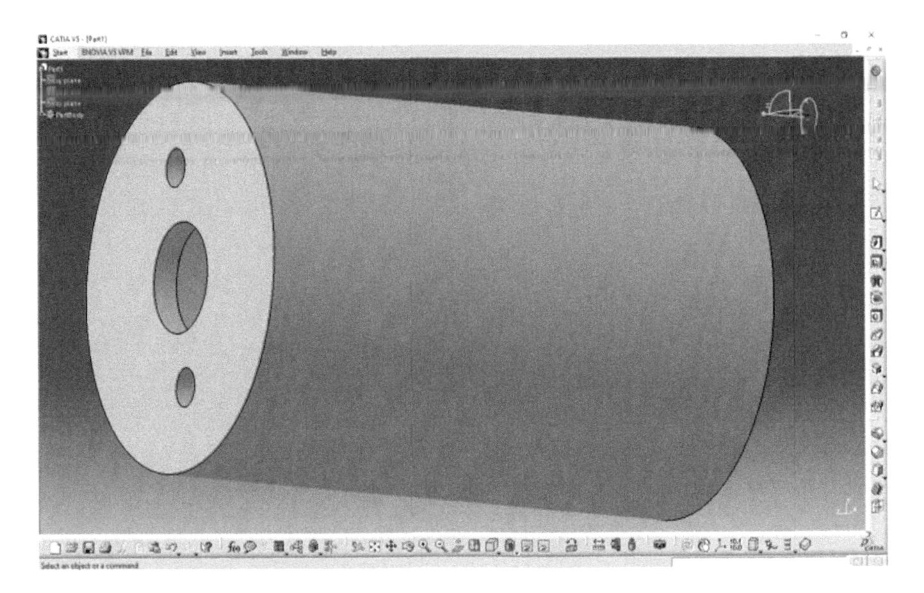

Fig. 2 Fire-tube boiler design

necessary to define the type of material that we are using for analysis. The surface model is created by enclosing the entire surfaces which are generated as shown in Fig. 2. Dimensions of the boiler are given in Table 1.

After convergence of meshing, 95,486 tetrahedral elements are formed. Figure 3 shows the actual meshed model of a fire-tube boiler. The solution is stopped when the changes in solution variables from one iteration to the next iteration are negligibly shown in Fig. 4. The process is continued for the varying of different flow speeds of 25 m/s, 35 m/s, 40 m/s and 45 m/s, respectively, for all these iterations' pressure and temperature variations along the length of the boiler and for different water velocities.

Table 1 Dimensions of boiler

S. no.	Feature	Dimension (mm)
1	Length of boiler	550
2	Diameter of boiler (inner/outer)	300/350
3	The diameter of the fire tube (inner/outer)	35/50

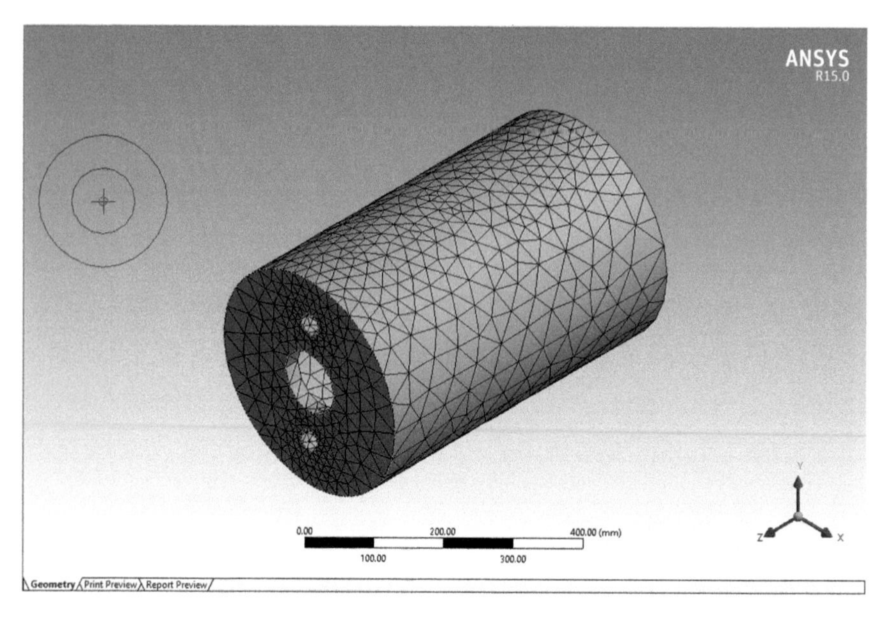

Fig. 3 Meshed model of fire-tube boiler

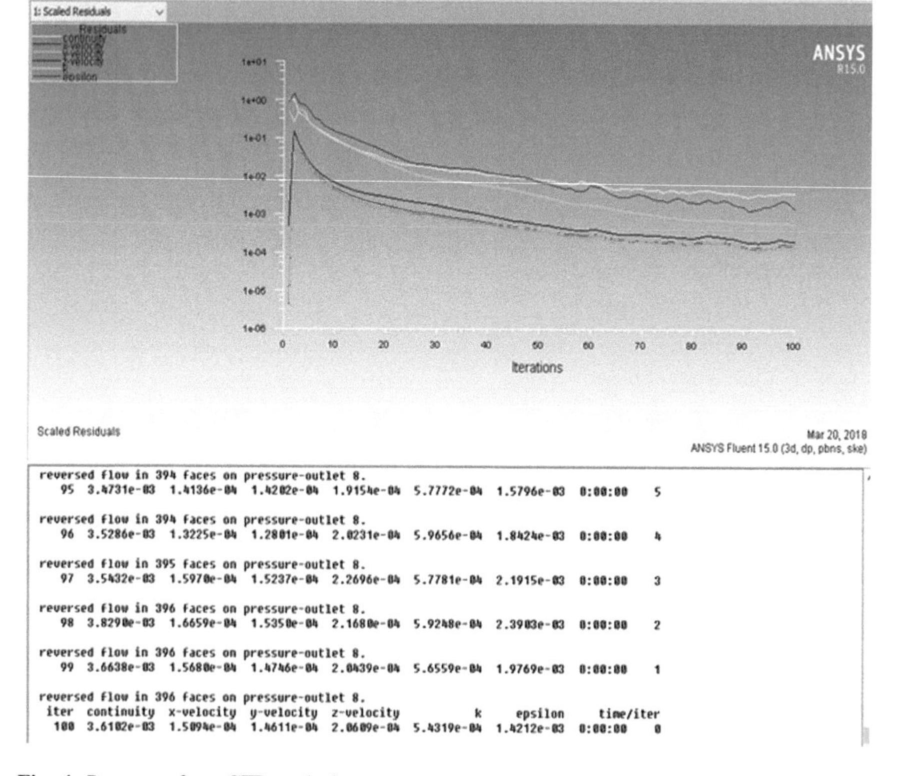

Fig. 4 Response from CFD analysis

4 Results

In order to study the variations in pressure and temperature during varying of the speed of water flow (25, 30, 35 and 40 m/s) in fire-tube boiler from the basis of the literature survey, the flow of water and hot exhaust gases through the fire-tube boiler has been simulated using ANSYS Fluent. In the first case, the inlet velocity of the water has been varied and the pressure distribution across the boiler shell has been studied. In the second case, the boiler casing material has been varied to identify the loss of heat through the boiler.

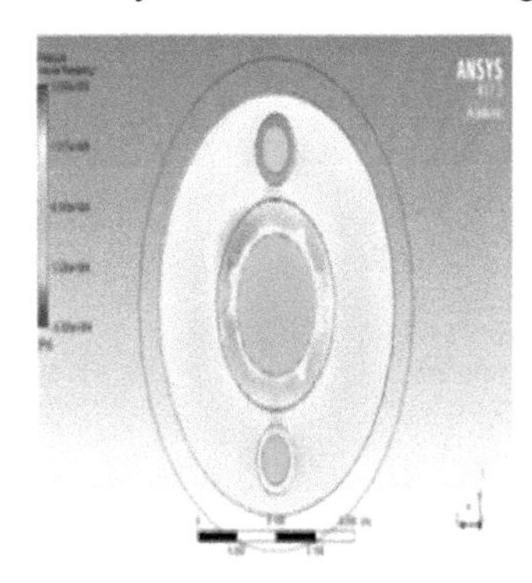

Velocity 25 m/s

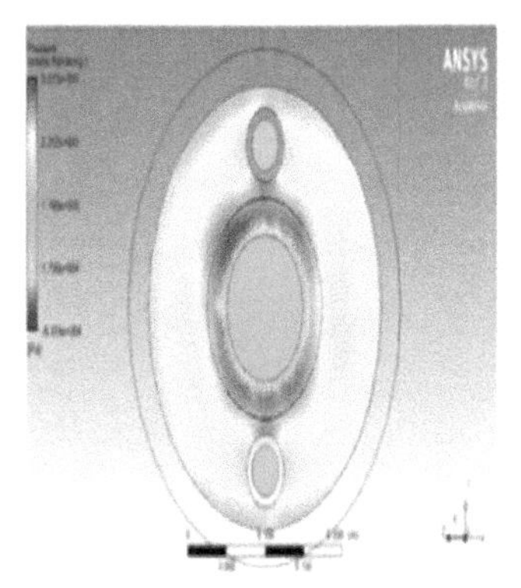

Velocity 30m/s

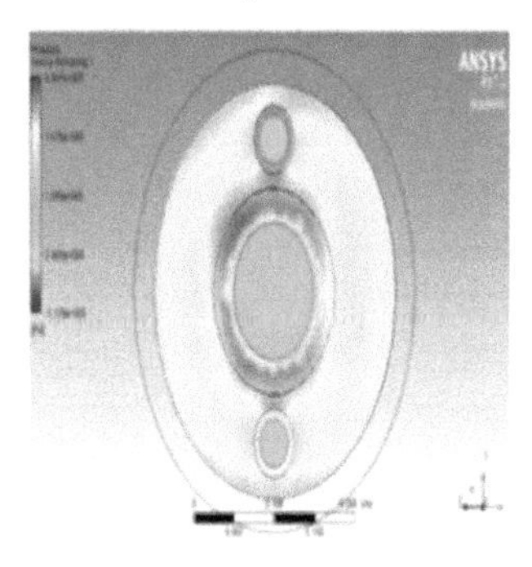

Velocity 35 m/s

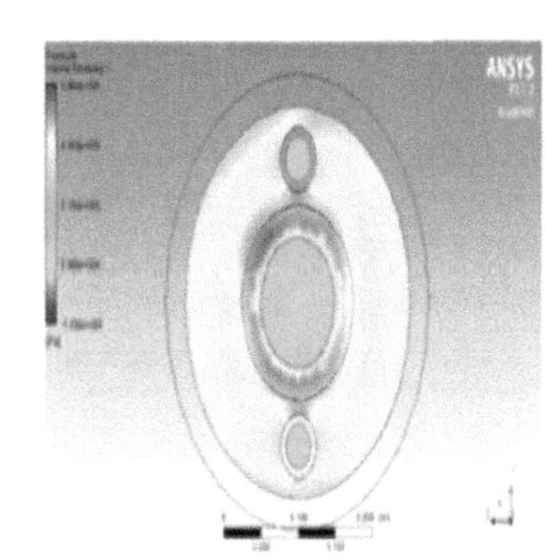

Velocity 40 m/s

The inlet velocity of the water into the boiler is varied between 25, 30, 35 and 40 m/s. The pressure induced inside the boiler is determined by simulation and presented in the form of contours which represent the variation of minimum and maximum pressure with velocity in the form of a graph. These results are plotted in Fig. 5. Pressure variations are predicted and plotted over an *XY* graph taking *X*-axis as the speed of water and *Y*-axis as pressure at the wall of fire-tube boiler. As by changing the casing material of fire-tube boiler, the following results were obtained.

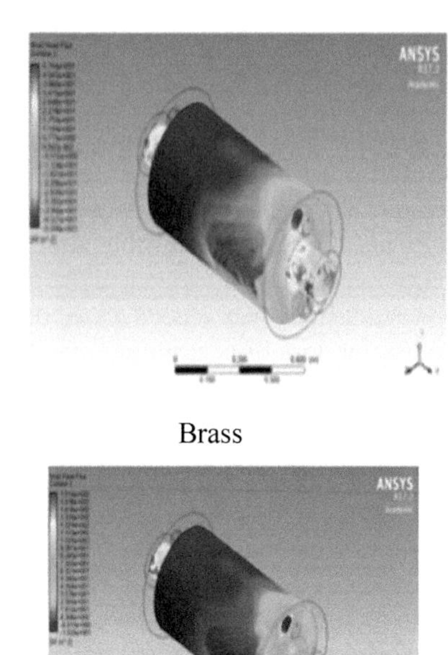

Brass

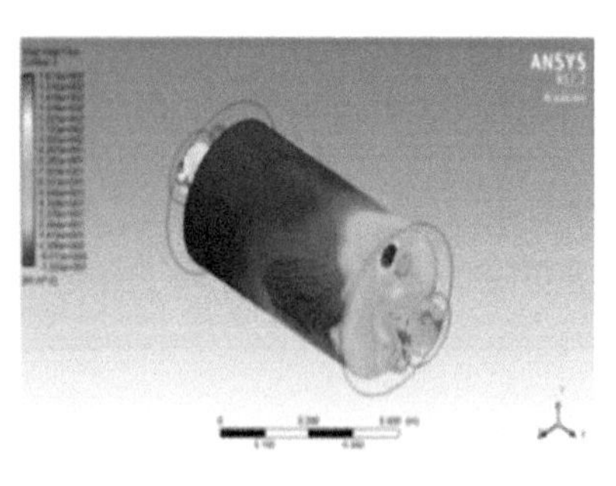

Stainless steel

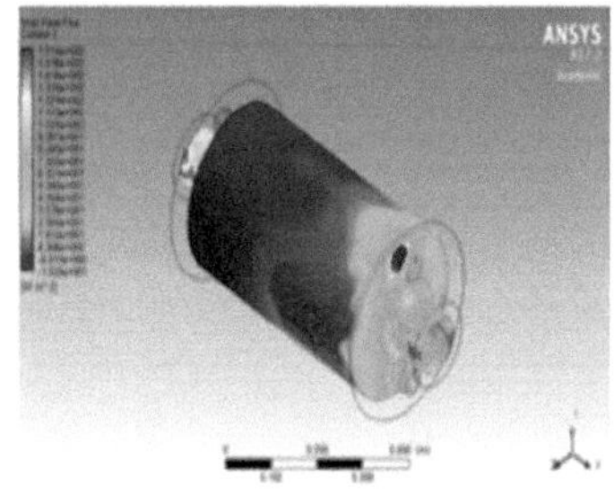

Steel

While selecting the boiler casing material, the key factor is the loss of heat through the boiler casing. Greater the loss of heat, lower is the efficiency of the boiler. Among the three materials tested, brass is having a very high negative heat flux of 50.94 W/m^2. Steel and stainless steel are nearly having the same negative heat flux value. Stainless steel is having the lowest negative heat flux of 15.32 W/m^2. These results are given in Table 2.

Fig. 5 Pressure versus water flow in fire-tube boiler

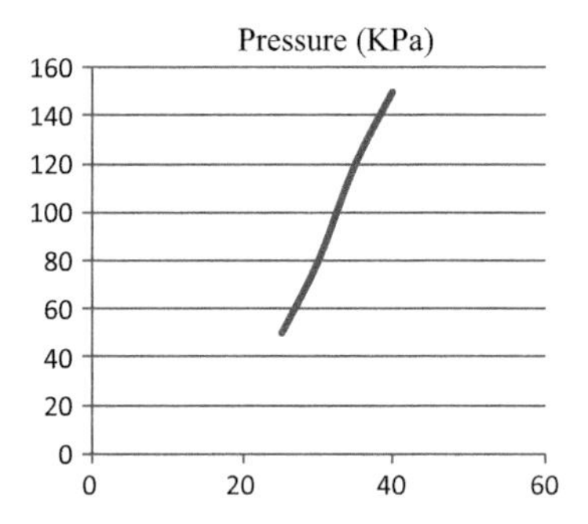

Table 2 Heat flux variation with material variation

Boiler casing material	Negative heat flux (W/m^2)
Steel	15.51
Brass	50.94
Stainless steel	15.32

5 Conclusion

The pressure–temperature variation is observed while varying water flow speed over a range of values, and casing material is changed to know the best material. In order to run the boiler at greater efficiency, the following points need to be considered.

- It is recommended to keep the velocity of the water as low as possible to reduce the induced pressure and increase the heat transfer.
- Among the three boiler casing materials, stainless steel has minimum heat loss from the boiler casing.
- The preferability order among the three materials is—stainless steel > steel > brass.
- From the results, it can be observed that the velocity of water increases inside the boiler and the pressure encountered during the flow also increases. Greater pressure implies greater stresses on the walls of the boiler. Hence, they are more susceptible to failure.
- Under high-pressure condition, as the velocity of the water increases, lesser heat will be transferred from the hot gases to the water. On varying the boiler casing material, significant changes can be seen in the amount of heat that is being lost through the boiler shell. This represents the heat flux contours, as obtained by varying the boiler casing material.
- While selecting the boiler casing material, the key factor is the loss of heat through the boiler casing. Greater the loss of heat, lower is the efficiency of the boiler. Among the three materials tested, brass is having a very high negative heat flux of 50.94 W/m^2. Steel and stainless steel are nearly having the same negative heat flux value. Stainless steel is having the lowest negative heat flux of 15.32 W/m^2.

References

1. https://www.google.com/search?tbm=isch&q=FIRETUBE+BOILER+&chips=q:fire+tube+boiler,g_1:construction:AHB07T5wpv4%3D&usg=AI4_kSbIE4ZT7hu8LPSAm6FWV2WJQW2Jw&sa=X&ved=0ahUKE
2. Kodak SS, Kshirsagar AM, Phadatare SM, Sawant Kedar AV, Ghatge DA, Sawant-Kedar JY, Wayse RD (2016) Design, development and analysis of non-IBR vertical fire tube boiler for improving the efficiency. Int J Eng Trends Technol 36:144–150

3. Ortiz FJG (2011) Modeling of fire-tube boilers. Appl Therm Eng 31(16):34–63
4. Liu SW, Wang WZ, Liu CJ (2017) Failure analysis of the boiler water-wall tube. Case Stud Eng Fail Anal 9:35–38
5. Akkinepally B, Shim J, Yo K (2017) Numerical and experimental study on biased tube temperature problem in tangential firing boiler. Appl Therm Eng 126:92–99
6. Senthur N, Pradeep Kumar AR, Annamalai K, Prem Kartik Kumar SR (2015) Effect of emulsified fuel on performance and emission characteristics in DI diesel engine. J Chem Pharm Sci 7:215–218
7. Reddy SVK, Premkartikkumar SR, Gopidesi RK, Kautkar NU (2017) A Review on nano coatings for IC engine applications. Int J Mech Eng Technol 8(9):70–76

Application of VIKOR for the Selection of Material for the Green and Sustainable Construction

Sonu Rajak, Prabhakar Vivek and Sanjay Kumar Jha

Abstract The objective of this article is to depict the appropriate techniques to understand the basic problems faced by the engineers to select the proper material and proportion it to enable a structure or machine to perform its function efficiently. Selection of material is a challenging task in the construction industry due to many alike materials available in the market. In order to overcome this challenge, multi-criteria decision analysis (MCDA) method is used, which yields a rank wise output which eases down the material selection procedure. Vlsekriterijuska optimizacijia I komoromisno resenje (VIKOR) is a robust decision-making method. VIKOR method focuses on ranking with finding the compromise solution and selects the best alternatives in the existence of conflicting criteria. This procedure can aid civil engineers in the selection of suitable sustainable materials for sustainable construction. The best solution was determined by the smallest VIKOR index. A comprehensive analysis of this method is illustrated in this paper, for the selection of best-suited construction material.

Keywords Sustainable construction · Decision-making · Selection of material · VIKOR

1 Introduction

The modern economies are endeavoring to reconcile ambitious equilibrium between eco-friendly material and infrastructure constructions over the matter of life cycle. Today, one of the basic problems facing the engineer is to select the proper material and proportion it to enable a structure or machine to perform its function efficiently. For this purpose, it is essential to determine the strength, stiffness, and other properties of materials. "Sustainable development is to meet the needs of the present generation without hampering the needs of the future generation" [5]. The wise

S. Rajak (✉) · P. Vivek · S. K. Jha
Department of Production Engineering, Birla Institute of Technology Mesra, Ranchi 835215, India

BBVL. Deepak et al. (eds.), *Innovative Product Design and Intelligent Manufacturing Systems*, Lecture Notes in Mechanical Engineering,
https://doi.org/10.1007/978-981-15-2696-1_75

choice is to adopt "the green building" practice. Green building is creating structures and construction and handling the processes that are resource-efficient and environmentally sustainable and throughout the life cycle. Sustainable development has been studied by authors in all the fields such as social sustainability for automobile sector by Rajak and Vinodh [5], transport sustainability by Rajak et al. [6], and ecological sustainability for the construction industry by Bamgbade et al. [10].

Material selection for construction plays a crucial role which directly influences the sustainability of the building. On the contrary, the selection of inappropriate materials may lead to the destruction of the environment and adversely influence economic, environment, and social aspect of building. There are a vast number of raw materials in use today, for example, plastic, there are 58 families of plastic with over one thousand grade of plastics for specific applications and it becomes strenuous to decide which material would be more productive to use and with a lesser impact on the environment. Thus, thereby, it becomes necessary to have a proper material selection in order to promote conservation of dwindling non-renewable resource. The condition of being protected from harm or other non-desirable outcomes is the state of "Safety" and when comes to the risk of life then it becomes crucial to consider the safety aspect. The material selected should endure all types of disruptions and bestow safety to the users till its lifetime. Assessment of drivers and growth of sustainable construction has been studied by Oke et al. [7]. Drivers indicators framework has been developed by Opon and Henry [9] for the concrete sustainable material. Greenhouse gas-based approach has been adopted by Hossain et al. [8] for sourcing of sustainable material for the construction.

Sustainable construction becomes important as it simply not only provides employment opportunities but also fosters community involvement. Moreover, it has environmental benefits too as hiring locals in greater percentage would certainly promote the greater ability to walk and take public transit to work than investing in vehicles for transportation and thus results in reduction in air pollution including greenhouse gases.

2 Problem Formulation

This analysis work is predicted on the selection of best-suited construction material for civil or construction work. The significant part of the construction values consists of construction material price. An effective process of construction material management may improve work efficiency. Thus, when selecting construction materials, it is very important that well considerable decision should be made.

The conceptual model is developed for the material selection for green and sustainable construction and is shown in Table 1.

Table 1 Conceptual model for the material selection for the green and sustainable construction

S No	Dimensions	Criteria	References
1	Economics	Initial cost (EC1)	[4]
		Maintenance cost (EC2)	
		Disposal cost (EC3)	
		Revenue (EC4)	
		Tax contribution (EC5)	
2	Environment	Recycling and reuse (EN1)	[3]
		Transportation (EN2)	
		Waste management (EN3)	
		CO_2 emission (EN4)	[1]
		Raw material extraction (EN5)	
3	Society	Sustainable material (S1)	
		Safety against failure (S2)	
		Working with local communities and road users (S3)	
		Operational life (S4)	
		Aesthetics (S5)	
4	Technical	Strength (T1)	
		Serviceability (T2)	[2]
		Thermal insulation (T3)	
		Weathering resistance (T4)	
		Earthquake and lightning resistance (T5)	

3 Method

This study used the VIKOR method for selecting the best material for sustainable construction. VIKOR is an MCDM method which was originally developed by Opricovic [11] to solve the decision-making problem with conflicting and non-commensurable criteria, assuming that:

VIKOR method is a robust tool for the decision-making and it is having some advantage over the other decision-making methods such as the VIKOR method gives the compromising solution which cannot be obtained by other decision-making methods. VIKOR method is given in the subsequent steps:

Step 1: Determine the normalized decision matrix.

The normalized decision matrix can be calculated as:

$$N = \left[x_{ij}\right]_{r \times p} \tag{1}$$

where $D_{ij} = \frac{x_{ij}}{\sqrt{\sum_{i=1}^{k} x_{ij}^2}}$, $i = 1, 2, \ldots, r; j = 1, 2, \ldots, p$; and x_{ij} is the execution of alternative with respect to the jth factor.

Step 2: Find out the ideal and negative-ideal solution.

The ideal solution I^* and the negative-ideal solution I^- are determined by the Eq. 2, as follows:

$$\begin{aligned} I^* &= \left\{ \left(\max D_{ij} | j \in J \right) \text{ or } (\min D_{ij} | j \in J'), | i = 1, 2, \ldots, k \right\} \\ &= \left\{ D_1^*, D_2^*, \ldots, D_j^*, \ldots, D_n^* \right\} \end{aligned} \tag{2}$$

$$\begin{aligned} I^- &= \left\{ \left(\min D_{ij} | j \in J \right) \text{ or } (\min D_{ij} | j \in J'), | i = 1, 2, \ldots, k \right\} \\ &= \left\{ D_1^-, D_2^-, \ldots, D_j^-, \ldots, D_n^- \right\} \end{aligned}$$

where $J = \left\{ j = 1, 2, \ldots, n | D_{ij}, a \, \text{higher value is desired} \right\}$

$$J' = \left\{ j = 1, 2, \ldots, n | D_{ij}, \text{ a lower value is desired} \right\}$$

Step 3: Determine the utility measure and the regret measure.

The utility measure (U_i) and the regret measure (R_i) can be determined by Eqs. 3 and 4, respectively, as follows:

$$U_i = \sum_{j=1}^{n} w_j \left(D_j^* - D_{ij} \right) / \left(D_j^* - D_j^- \right) \tag{3}$$

$$R_i = \text{Max}_j \left[w_j \left(D_j^* - D_{ij} \right) / \left(D_j^* - D_j^- \right) \right] \tag{4}$$

where w_j is the weight of the jth factor

Step 4: Determine the VIKOR index

The VIKOR index can be determined by the Eq. 5:

$$V_i = \mu \left[\frac{U_i - U^-}{U^* - U^-} \right] - (1 - \mu) \left[\frac{R_i - R^-}{R^* - R^-} \right] \tag{5}$$

where, $i = 1, 2, \ldots, k$; $U^- = \min_i U_i$; $U^* = \max_i U_i$; $R^- = \min_i R_i$; $R^* = \max_i R_i$ and μ is the weight of the maximum group utility. Value of μ is varying from 0 to 1; but generally, the value of μ is taken as 0.5.

Step 5: Rank the order of preference.

Arrange the alternative in ascending order according to the value of V_i, U_i, and R_i. Thus, three ranking lists can be determined. Thus, the best solution can be determined by the smallest VIKOR (V_i) value.

Step 6: Proposing compromise solution.

For alternative D_k, which is the best ranked by the VIKOR index V_i, if the following two cases are satisfied:

(i) C1: "Acceptable advantage" $V(D_1) - V(D_k) \geq (1/(r-1))$, where D_1 is the second-best alternative in the ranking list by V_i and r, total number of alternatives.

(ii) C2: "Acceptable stability in decision making." Alternative D_k also must be the best rank by U_i or/and R_i. If one of the conditions is not satisfied, then a set of compromise solutions is proposed, which consists of:

- Alternative D_k and D_1 if only condition C2 is not satisfied
- Alternative D_k, D_1 … D_p if condition C1 is not satisfied; and D_p is calculated using the relation $V(D_p) - V(D_1) \approx (1/(r-1))$.

4 Application of the Proposed Framework

To validate the proposed framework, we send the proposal to the top ten multinational construction companies situated in south India. We got a positive response from the eight organizations. Since this study is based on opinions of decision-makers, we added three more players apart from the construction company such as the client, material suppliers, and the consultant. As per the discussion with an expert, the material selected for sustainable construction is brick. Brick is the main material of any construction such as residential, bridges, corporates, etc., and method adopted for this study has been shown in Fig. 1.

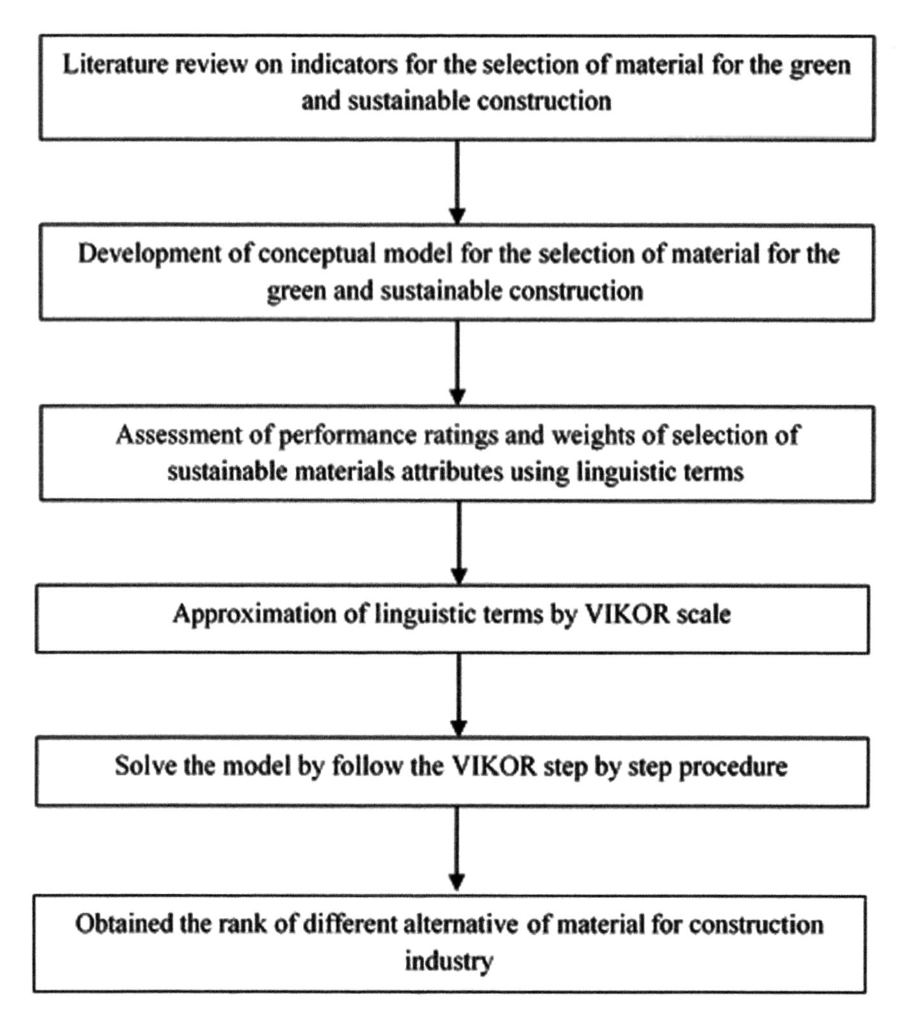

Fig. 1 Method adopted for the study

4.1 Evaluating the Best Sustainable Material Using VIKOR

After the discussion with decision-makers and experts, the decision matrix obtained for five different types of bricks is shown in Table 2. Decision matrix is normalized with Eq. 1 and after that, we followed all the steps of VIKOR and found the value of V_i, U_i and R_i which is shown in Table 3.

Table 2 Decision matrix

Weight	0.08	0.05	0.02	0.07	0.02	0.04	0.05	0.02	0.02	0.03	0.06	0.09	0.03	0.05	0.05	0.11	0.07	0.05	0.05	0.05
Category	EC1	EC2	EC3	EC4	EC5	EN1	EN2	EN3	EN4	EN5	S1	S2	S3	S4	S5	T1	T2	T3	T4	T5
A	2	3	2	4	2	2	3	1	2	3	5	3	2	4	3	3	2	2	2	3
B	4	5	3	5	3	2	3	4	1	2	4	2	1	2	2	4	4	3	3	2
C	3	2	4	3	5	4	2	3	4	1	2	2	2	5	4	2	3	1	2	1
D	2	3	1	2	2	1	4	3	3	4	3	3	3	2	5	3	3	2	3	1
E	1	4	3	4	2	3	3	2	1	1	2	2	1	3	4	2	2	1	1	2

Table 3 Value of V_i, U_i and R_i

Category	V_i	U_i	R_i	V_i rank	U_i rank	R_i rank
A	0.416482	0.068023	0.017138	1	3	1
B	0.506133	0.085107	0.428024	3	4	3
C	0.633502	0.108395	1	5	1	5
D	0.439399	0.06659	0.052797	2	2	2
E	0.63066	0.108395	0.993452	4	1	4

5 Conclusions

Selection of material is a difficult task in the construction industry. Material selected for sustainable construction should be sustainable itself. Appropriate sustainable indicators are selected from the literature survey as well as discussions with experts and decision-makers. Recycling and reuse indicators are found to be the most influential sustainable indicator. The most appropriate material found is wool brick among other alternatives. The conduct of the study provided guidelines for practitioners to systematically quantify and analyze the result for the material selection of the construction organization. The approach could be applied to any type of organizations.

6 Limitation and Future Scope

This research was based on human judgment and respondents from the group of people. The perspective of material election for sustainable construction may change from one group to another group. To obtain the robust result, statistical validation and sensitivity analysis may be performed in the future.

References

1. Saadah Y, AbuHijleh B (2010) Decreasing CO_2 emissions and embodied energy during the construction phase using sustainable building materials. Int J Sustain Build Technol Urban Dev 1(2):115–120
2. Balendran RV, Rana TM, Maqsood T, Tang WC (2002) Application of FRP bars as reinforcement in civil engineering structures. Struct Surv 20(2):62–72
3. Zhang L (2013) Production of bricks from waste materials–a review. Constr Build Mater 47:643–655
4. Govindan K, Shankar KM, Kannan D (2016) Sustainable material selection for construction industry–a hybrid multi criteria decision making approach. Renew Sustain Energy Rev 55:1274–1288

5. Rajak S, Vinodh S (2015) Application of fuzzy logic for social sustainability performance evaluation: a case study of an Indian automotive component manufacturing organization. J Clean Prod 108:1184–1192
6. Rajak S, Parthiban P, Dhanalakshmi R (2016) Sustainable transportation systems performance evaluation using fuzzy logic. Ecol Indic 71:503–513
7. Oke A, Aghimien D, Aigbavboa C, Musenga C (2019) Drivers of sustainable construction practices in the Zambian Construction Industry. Energy Procedia 158:3246–3252
8. Hossain MU, Sohail A, Ng ST (2019) Developing a GHG-based methodological approach to support the sourcing of sustainable construction materials and products. Resour Conserv Recycl 145:160–169
9. Opon J, Henry M (2019) An indicator framework for quantifying the sustainability of concrete materials from the perspectives of global sustainable development. J Clean Prod 218:718–737
10. Bamgbade JA, Kamaruddeen AM, Nawi MNM, Adeleke AQ, Salimon MG, Ajibike WA (2019) Analysis of some factors driving ecological sustainability in construction firms. J Clean Prod 208:1537–1545
11. Opricovic S (1998) Multicriteria optimization of civil engineering systems. Fac Civ Eng, Belgrade 2(1):5–21

Application of an MCDM Tool for Selection of 3D Bioprinting Processes

Sumanta Mukherjee and Jay Prakash Kumar

Abstract Bioprinting deals with layer-by-layer generation of organs or parts of organs by additive manufacturing, and the materials used for the purpose contain living cells. Different bioprinting processes offer different capabilities in terms of cell viability, the structural integrity of the printed parts, physical properties of the materials that can be handled, etc. So, for a given type of bioprinting material, selection of the most suitable bioprinting process to optimize the biological as well as the mechanical properties requires trade-offs. In this paper, a fuzzy-TOPSIS MCDM process has been applied for this purpose. Characteristics of various bioprinting processes have been summarized and sorted before fuzzification, and for any set of required characteristics, the MCDM tool can identify the ideal solution, i.e., the solution having the least Euclidean distance from the positive ideal solution and the largest Euclidean distance from the negative ideal solution using TOPSIS method.

Keywords Bioprinting · Fuzzy-TOPSIS · MCDM

1 Introduction

1.1 3D Bioprinting

Additive manufacturing (AM) processes offer unique flexibilities in terms of the design of the objects as well as the types of materials that can be used for manufacturing those objects. Capitalizing on such capabilities, researchers are attempting to model and manufacture the most complex machines ever known—human organs. This requires layer-by-layer deposition of natural or synthetic biomaterials along with living cells to build up the intended part of the organ. Such 3D printing of biological objects is termed 3D bioprinting. Since bioprinting involves dealing

---

S. Mukherjee (✉) · J. P. Kumar
Production Engineering Department, BIT Sindri, Dhanbad, Jharkhand, India

BBVL. Deepak et al. (eds.), *Innovative Product Design and Intelligent Manufacturing Systems*, Lecture Notes in Mechanical Engineering,
https://doi.org/10.1007/978-981-15-2696-1_76

with living cells, AM processes relying upon high-temperature melting or sintering of materials are clearly unsuitable for this purpose. Out of the seven categories of AM as designated by ASTM, three types of processes, namely material deposition [1], material jetting [2, 3] and vat polymerization [4], have already been successfully adopted for bioprinting.

Bioprinting using Material Deposition. This technique is based on the continuous deposition of bio-ink or hydrogel seeded with cells. Pneumatic or hydraulic piston systems are commonly used to extrude the material through nozzles for controlled deposition. Researchers have developed aortic valves [5], vascular trees with branching [6] and pharmo-kinetic and tumor models [7] using extrusion-based bioprinting. Such processes offer many crucial advantages like the ability to handle multiple types of materials and cells in combinations, sufficiently high cell viability and structural integrity of the generated structures. However, these processes do not fare well in terms of printing resolution and printing speed [8].

Bioprinting using Material Jetting. In jetting-type bioprinting, picolitre-sized bio-ink droplets are jetted through an array of micro-nozzles using a wide variety of techniques including piezoelectric actuation, pneumatic actuation, thermal method and laser-induced droplet formation. Keriquel et al. have proposed bioprinted bone-like structures which replete with hydroxyapatite nanoparticles and human mesenchymal stem cells through the jetting process [9]. The process has also been demonstrated to be able to develop cardiac tissue with beating cellular response [10] and neural tissue [11]. Jetting-based bioprinting processes are very fast, have high cellular viability and have extremely high resolution, but the structural integrity of the developed structures is usually inferior to that of objects developed through extrusion-based or vat polymerization processes.

Bioprinting using Vat Polymerization. In processes like stereolithography (SLA) and direct light processing (DLP), the liquid polymer is cross-linked layer-by-layer by localized photo-polymerization to create solid constructs. Loading the precursor liquid polymer with cells results in encapsulation of cells post-curing. Different types of cells including mice fibroblasts, human adipose-derived stem cells (hADSCs) and human umbilical vein endothelial cells (hUVECs) have been used to develop a stable 3D structures laden with cells [12]. These processes also offer a good dimensional resolution, speed and cellular viability with sufficient structural integrity. The preceding discussion elucidates that various bioprinting processes have their own advantages and limitations, and the selection of the bioprinting process for any given application requires a thorough comparison of the process characteristics of all the available options. Thus, well-informed trade-offs need to be done to identify the best suitable bioprinting process based on the required physical or mechanical properties.

1.2 Multi-Criteria Decision Making

Multi-criteria decision-making (MCDM) tools are computer programs that aid in such decision making under conflicting requirements. The literature review evidenced use of MCDM tools to select general AM systems for manufacturing of 3D objects [13–16], since bioprinting processes involve a number of parameters like bio-ink viscosity and cell viability, which are not considered for other AM processes, the MCDM tool for selection of bioprinting process needs to incorporate such specific requirements. No existing literature was found that has dealt with the development of MCDM tool for bioprinting processes, and hence, the presented work is an attempt toward addressing that gap. The applied tool relies on a fuzzy-TOPSIS framework, as some of the involved parameters were qualitative, and no quantifiable data was available in the literature. Adoption of a simple process like technique for order of preference by similarity to ideal solution (TOPSIS) helped to accommodate a sufficient number of parameters with no significant complexity, making the tool suitable for use by people without much exposure of the different bioprinting process mechanisms.

2 Proposed Model

The goal of TOPSIS is to assign rankings for a selection of possible alternatives by finding out the Euclidean distance of the alternatives from the positive ideal solution (PIS) and the negative ideal solution (NIS). The PIS is identified as the alternative that maximizes the benefit criteria while minimizing the cost criteria, and the alternative with minimum benefit criteria and maximum cost criteria is the NIS. Considering bioprinting applications, cell viability, mechanical integrity of the printed objects, printing speed, maximum printing volume, etc., can be identified as benefit criteria, whereas the gelation/cross-linking time, process cost, etc., can be recognized as the cost features. Table 1 represents the typical values of different bioprinting process characteristics, as found from literature.

As can be seen from the table, some of the process characteristics are expressed linguistically, making it difficult to combine with other characteristics. Therefore, all of those characteristics were fuzzified using a scale of 1–9 according to the succeeding table (Table 2).

The triplet $\tilde{a} = (a, b, c)$ denotes the triangular fuzzy number $\tilde{a}$, and the membership function is

$$\mu_{\tilde{a}} = \begin{cases} \frac{x-a}{b-a} & \text{if } a \leq x \leq b \\ \frac{c-x}{c-b} & \text{if } b \leq x \leq c \\ 0 & \text{Otherwise} \end{cases} \quad (1)$$

Table 1 Different bioprinting processes and their characteristics

Process characteristics		Bioprinting process			
		Material jetting		Vat polymerization	Material deposition
		Ink-jet based	Laser based		
Process capabilities	Structural integrity	Low	Low	High	High
	Resolution	High to medium (20–100 μm)	Medium ($\sim$100 μm)	High to medium (50–100 μm)	Low (200–500 μm)
	Printing speed	Fast	Medium	Medium	Slow
	Cellular viability	High (>85%)	Very high (>95%)	Medium to high (70–85%)	Low to medium (40–80%)
Input materials' properties	Viscosity (mPa/s)	Low ($\sim 10^{0-1}$)	Low to very high ($\sim 10^{1-6}$)	Low to medium ($\sim 10^{2-3}$)	Very low to low ($\sim 10^{0-2}$)
	Cell density	Low ($\sim 10^5$/ml)	Medium ($\sim 10^6$/ml)	Medium ($\sim 10^6$/ml)	High ($\sim 10^7$/ml)
	Cross-linking time	High	High	Low	Medium
Materials/process cost		Low	High	Medium	Medium

Table 2 Linguistic levels and their corresponding fuzzy numbers

Level of process characteristics	Abbreviation	Fuzzy number
Very low	VL	1, 1, 3
Low	L	1, 3, 5
Medium	M	3, 5, 7
High	H	5, 7, 9
Very high	VH	7, 9, 9

where a, b and c are real numbers with $a < b < c$ and $\mu_{\tilde{a}}(x) = 1$ at b. For example, if a researcher would like to use a low-viscosity liquid polymer with moderate cross-linking time and medium cell loading density to obtain mechanically rigid structures with medium-sized features, the cell viability as high as possible and lowest process cost, and the weights of the different characteristics can be tabulated as follows (Table 3).

It can be identified that out of the mentioned characteristics, viscosity, cross-linking time and process cost are the three cost parameters, whereas the rest are beneficial characteristics, as identifying the bioprinter that will be able to work with lowest viscous materials requiring lowest curing time will be fit for processing the liquid polymer mentioned here. So, the normalized fuzzy decision matrix obtained using linear normalization is presented in Table 4. Considering the fuzzy weights of the process characteristics, the following weighted fuzzy-normalized decision matrix can be obtained (Table 5).

Table 3 Fuzzy weights of the process parameters

Process characteristics		Weight	Fuzzified weight
Process capabilities	Structural integrity	High	5, 7, 9
	Resolution	Medium	3, 5, 7
	Printing speed	Very low	1, 1, 3
	Cellular viability	Very high	7, 9, 9
Input materials properties	Viscosity (mPa/s)	High	5, 7, 9
	Cell density	Medium	3, 5, 7
	Cross-linking time	Medium	3, 5, 7
Materials/process cost		Very high	7, 9, 9

A^+ and A^- in Table 5 are the selected best and worst performances, respectively, for each of the alternatives. So, they represent the fuzzy positive ideal solution (FPIS) and the fuzzy negative ideal solution (FNIS), respectively, in our election. The distance of each of the alternatives in terms of individual process characteristics from the FPIS and the FNIS can be calculated from the concept of distance between fuzzy numbers. Accordingly, the distance between two fuzzy numbers $\tilde{a} = (a, b, c)$ and $\tilde{b} = (a', b', c')$ is given by

$$d(\tilde{a}, \tilde{b}) = \sqrt{\frac{1}{3}\left[(a - a')^2 + (b - b')^2 + (c - c')^2\right]} \tag{2}$$

Thus, the distance of each of the individual performance metric from the FPIS and the FNIS is calculated and tabulated in Tables 6 and 7, respectively (Table 8).

3 Conclusions

The work presents a fuzzy-TOPSIS approach to assist researchers in the selection of the most suitable bioprinting process with a given liquid polymer. Once the properties of the input polymer and the required properties of the printed structure are identified, the fuzzy-TOPSIS tool can calculate the best and the worst possible properties. Therefore, the bioprinting process having the highest distance from the worst properties and least distance from the best properties is selected following the TOPSIS method. As a case study, a hypothetical liquid polymer with low viscosity, moderate cross-linking time and medium cell loading density was selected to generate structures with high integrity and the medium resolution with the aim to maximize cell viability and minimize process cost. From the adopted approach, vat polymerization was found to be most suitable for processing the polymer.

Table 4 Normalized fuzzy decision matrix

Specification	Structural integrity	Resolution	Printing speed	Cellular viability	Viscosity (mPa/s)	Cell density	Cross-linking time	Materials/ process cost
Ink-jet based	0.111, 0.333, 0.556	0.333, 0.667, 1.000	0.556, 0.778, 1.000	0.556, 0.778, 1.000	0.200, 0.333, 1.000	0.111, 0.333, 0.556	0.111, 0.143, 0.200	0.200, 0.333, 1.000
Laser based	0.111, 0.333, 0.556	0.333, 0.556, 0.778	0.333, 0.556, 0.778	0.778, 1.000, 1.000	0.111, 0.167, 1.000	0.333, 0.556, 0.778	0.111, 0.143, 0.200	0.111, 0.143, 0.200
Vat polymerization	0.556, 0.778, 1.000	0.333, 0.667, 1.000	0.333, 0.556, 0.778	0.333, 0.667, 1.000	0.143, 0.250, 1.000	0.333, 0.556, 0.778	0.200, 0.333, 1.000	0.143, 0.200, 0.333
Material deposition	0.556, 0.778, 1.000	0.111, 0.333, 0.556	0.111, 0.333, 0.556	0.111, 0.444, 0.778	0.200, 0.500, 1.000	0.556, 0.778, 1.000	0.143, 0.200, 0.333	0.143, 0.200, 0.333

Table 5 Weighted fuzzy-normalized decision matrix

Specification	Structural integrity	Resolution	Printing speed	Cellular viability	Viscosity (mPa/s)	Cell density	Cross-linking time	Materials/process cost
Ink-jet based	0.556, 2.333, 5.000	1.000, 3.333, 7.000	0.556, 0.778, 3.000	3.889, 7.000, 9.000	1.000, 2.333, 9.000	0.333, 1.667, 3.889	0.333, 0.714, 1.400	1.400, 3.000, 9.000
Laser based	0.556, 2.333, 5.000	1.000, 2.778, 5.444	0.333, 0.556, 2.333	5.444, 9.000, 9.000	0.556, 1.167, 9.000	1.000, 2.778, 5.444	0.333, 0.714, 1.400	0.778, 1.285, 1.800
Vat polymerization	2.778, 5.444, 9.000	1.000, 3.333, 7.000	0.333, 0.556, 2.333	2.333, 6.000, 9.000	0.714, 1.750, 9.000	1.000, 2.778, 5.444	0.600, 1.667, 7.000	1.000, 1.800, 3.000
Material deposition	2.778, 5.444, 9.000	0.333, 1.667, 3.889	0.111, 0.333, 1.667	0.778, 4.000, 7.000	1.000, 3.500, 9.000	1.667, 3.889, 7.000	0.429, 1.000, 2.333	1.000, 1.800, 3.000
A^+	2.778, 5.444, 9.000	1.000, 3.333, 7.000	0.556, 0.778, 3.000	5.444, 9.000, 9.000	1.000, 2.333, 9.000	1.667, 3.889, 7.000	0.600, 1.667, 7.000	1.400, 3.000, 9.000
A^-	0.556, 2.333, 5.000	0.333, 1.667, 3.889	0.111, 0.333, 1.667	0.778, 4.000, 7.000	0.556, 1.167, 9.000	0.333, 1.667, 3.889	0.333, 0.714, 1.400	0.778, 1.285, 1.800

Table 6 Distances from FPIS

Specification	Structural integrity	Resolution	Printing speed	Cellular viability	Viscosity (mPa/s)	Cell density	Cross-linking time	Materials/ process cost
Ink-jet based	3.195	0.000	0.000	1.463	0.000	2.338	3.283	0.000
Laser based	3.195	0.954	0.426	0.000	0.721	1.169	3.283	4.288
Vat polymerization	0.000	0.000	0.426	2.495	0.375	1.169	0.000	3.540
Material deposition	0.000	2.074	0.851	4.114	0.674	0.000	2.723	3.540

Table 7 Distances from FNIS

Specification	Structural integrity	Resolution	Printing speed	Cellular viability	Viscosity (mPa/s)	Cell density	Cross-linking time	Materials/ process cost
Ink-jet based	0.000	2.074	0.351	2.749	0.721	0.000	0.000	4.288
Laser based	0.000	1.169	0.426	4.114	0.000	1.169	0.000	0.000
Vat polymerization	3.195	2.074	0.426	1.864	0.349	1.169	3.283	0.765
Material deposition	3.195	0.000	0.000	0.000	1.371	2.338	0.566	0.765

Table 8 Calculated distances, closeness coefficient and ranking of the processes

Bioprinting process	Ink-jet based	Laser based	Vat polymerization	Material deposition
Total distance from FPIS (d^+)	10.278	14.035	8.005	13.976
Total distance from FNIS (d^-)	10.683	6.877	13.123	8.234
Closeness coefficient $d^-/(d^+ + d^-)$	0.510	0.329	0.621	0.371
Ranking	2	4	1	3

References

1. Pati F, Jang J, Lee JW, Cho DW (2015) Extrusion bioprinting. In: Essentials 3D biofabrication translation, pp 123–152. https://doi.org/10.1016/b978-0-12-800972-7.00007-4
2. Iwanaga S, Arai K, Nakamura M (2015) Inkjet bioprinting. In: Essentials 3D biofabrication Translation, pp 61–79. https://doi.org/10.1016/b978-0-12-800972-7.00004-9
3. Catros S, Keriquel V, Fricain JC, Guillemot F (2015) In vivo and in situ biofabrication by laser-assisted bioprinting. In: Essentials 3D biofabrication translation, pp 81–87. https://doi.org/10.1016/b978-0-12-800972-7.00005-0
4. Raman R, Bashir R (2015) Stereolithographic 3D bioprinting for biomedical applications. In: Essentials 3D biofabrication translation, pp 89–121. https://doi.org/10.1016/b978-0-12-800972-7.00006-2
5. Duan B, Hockaday LA, Kang KH, Butcher JT (2013) 3D bioprinting of heterogeneous aortic valve conduits with alginate/gelatin hydrogels. J Biomed Mater Res Part A 101A:1255–1264. https://doi.org/10.1002/jbm.a.34420
6. Li J, Chen M, Fan X, Zhou H (2016) Recent advances in bioprinting techniques: approaches, applications and future prospects. J Transl Med 14:271. https://doi.org/10.1186/s12967-016-1028-0
7. Norotte C, Marga FS, Niklason LE, Forgacs G (2009) Scaffold-free vascular tissue engineering using bioprinting. Biomaterials 30:5910–5917. https://doi.org/10.1016/j.biomaterials.2009.06.034
8. Poomathi N, Singh S, Prakash C, Patil RV, Perumal PT, Barathi VA, Balasubramanian KK, Ramakrishna S, Maheshwari NU (2018) Bioprinting in ophthalmology: current advances and future pathways. Rapid Prototyp J RPJ-06-2018-0144. https://doi.org/10.1108/rpj-06-2018-0144
9. Keriquel V, Oliveira H, Rémy M, Ziane S, Delmond S, Rousseau B, Rey S, Catros S, Amédée J, Guillemot F, Fricain J-C (2017) In situ printing of mesenchymal stromal cells, by laser-assisted bioprinting, for in vivo bone regeneration applications. Sci Rep 7:1778. https://doi.org/10.1038/s41598-017-01914-x
10. Wang Z, Lee SJ, Cheng H-J, Yoo JJ, Atala A (2018) 3D bioprinted functional and contractile cardiac tissue constructs. Acta Biomater 70:48–56. https://doi.org/10.1016/j.actbio.2018.02.007
11. Seol Y-J, Kang H-W, Lee SJ, Atala A, Yoo JJ (2014) Bioprinting technology and its applications. Eur J Cardio-Thoracic Surg 46:342–348. https://doi.org/10.1093/ejcts/ezu148
12. Derakhshanfar S, Mbeleck R, Xu K, Zhang X, Zhong W, Xing M (2018) 3D bioprinting for biomedical devices and tissue engineering: a review of recent trends and advances. Bioact Mater 3:144–156. https://doi.org/10.1016/j.bioactmat.2017.11.008
13. Prabhu SR, Ilangkumaran M (2019) Selection of 3D printer based on FAHP integrated with GRA-TOPSIS. Int J Mater Prod Technol 58:155. https://doi.org/10.1504/IJMPT.2019.097667

14. Liao S, Wu M-J, Huang C-Y, Kao Y-S, Lee T-H (2014) Evaluating and enhancing three-dimensional printing service providers for rapid prototyping using the DEMATEL Based Network Process and VIKOR. Math Probl Eng 2014:1–16. https://doi.org/10.1155/2014/349348
15. Al Ahmari A, Ashfaq M, Mian SH, Ameen W (2019) Evaluation of additive manufacturing technologies for dimensional and geometric accuracy. Int J Mater Prod Technol 58:129. https://doi.org/10.1504/ijmpt.2019.097665
16. Anand MB, Vinodh S (2018) Application of fuzzy AHP—TOPSIS for ranking additive manufacturing processes for microfabrication. Rapid Prototyp J 24:424–435. https://doi.org/10.1108/rpj-10-2016-0160

Optimization of Texture Geometry for Enhanced Tribological Performance in Piston Ring-Cylinder Liner Contact Under Pure Hydrodynamic and Mixed Lubrication

Peddakondigalla Venkateswara Babu, Ismail Syed and Beera Satish Ben

Abstract Surface texture can be positive (protruded out of the surface) or negative (recessed into the surface). In the present work, a theoretical model is presented to study the impact of positive texturing on tribological performance by solving a modified Reynolds equation and asperity contact model under steady-state conditions. Positive surface textures are introduced on a flat piston ring surface under simulated conditions of piston ring-cylinder liner contact. The influence of various parameters such as texture shape and area density is investigated under pure hydrodynamic as well as mixed lubrication conditions. The results indicate that elliptical-shaped textures are found to be superior in improving the tribological performance parameters such as load support and friction coefficient.

Keywords Friction reduction · Load support · Piston ring-cylinder liner contact · Surface texturing

1 Introduction

In an internal combustion engine (IC), fuel consumption rate and emissions are crucial factors that affect engine efficiency. In recent years, many researchers and internal combustion design engineers have focused on reducing friction to decrease fuel consumption rate and emissions. In the case of automobiles, friction losses in engine constitute 11.5% of the total fuel energy, of which the piston ring-cylinder liner (PR-CL) contact accounts for 38–65% [1]. The emissions might be decreased by 1 g/km, when the frictional losses of the PR-CL contact decrease by 5% in an IC engine [2]. Surface texturing is a potential method to improve the tribological

P. Venkateswara Babu · I. Syed (✉) · B. Satish Ben
Department of Mechanical Engineering, National Institute of Technology Warangal, Telangana 506004, India
e-mail: venky5041@gmail.com

BBVL. Deepak et al. (eds.), *Innovative Product Design and Intelligent Manufacturing Systems*, Lecture Notes in Mechanical Engineering,
https://doi.org/10.1007/978-981-15-2696-1_77

performance of sliding surfaces [2, 3]. Based on the previous studies, it is found that surface texturing can improve the tribological behaviors of sliding surfaces [2, 3].

Surface texturing has been successfully applied to many engineering applications including mechanical seals, journal bearings and piston rings [4–6]. In recent studies, Gadeschi and his colleague performed numerical simulations to study the effect of laser-surface-textured dimple parameters such as depth, area density and distribution pattern of PR-CL contact [7]. In their work, a hydrodynamic lubrication model was considered. In most of the previous works, the negative textures (such as dimples and grooves) have been introduced on the piston ring or liner surfaces to ensure their beneficial effects on tribological behaviors under different operating scenarios. Therefore, it is worthwhile to study the influence of positive texturing on the lubrication and tribological behaviors of PR-CL contact under different lubrication regimes. The present work addresses the impact of positive texturing on the tribological characteristics such as load support and friction coefficient of PR-CL contact under pure hydrodynamic and mixed lubrication conditions.

2 Theory

In a real engine scenario, the piston ring is convex in shape with a suitable ring radius. In the present work, a small fraction of piston ring is considered by assuming the fluid film thickness is much smaller than the ring curvature [8]. Therefore, the effect of the ring's curvature is neglected. Hence, in the present work, a flat PR-CL contact is considered for the analysis which is presented in Fig. 1. As illustrated in Fig. 1, the textures are distributed uniformly on the ring surface. The textured surface (i.e., ring surface) is stationary while the moving surface (i.e., liner surface) slides along x-axis direction with a sliding velocity U. In general, PR-CL contact operates under different lubrication regimes. The oil entrained into the clearance between ring and liner generates hydrodynamic pressure due to the wedge effect; as a result, the lubricant carries the total load itself. Under situations such as the moment of reversal approaches, the oil starvation occurs due to which inadequate hydrodynamic pressures would be developed. Therefore, some portion of the load is carried by the asperity interactions of sliding surfaces. The generated hydrodynamic load (W_h) and asperity load (W_a) is shown in Fig. 1.

2.1 Reynolds Equation

In the present study, a two-dimensional Reynolds equation is employed to determine the fluid film pressure. The effect of surface irregularities on the fluid flow is taken into account by adopting the Patir and Cheng's flow factor model [9, 10] as:

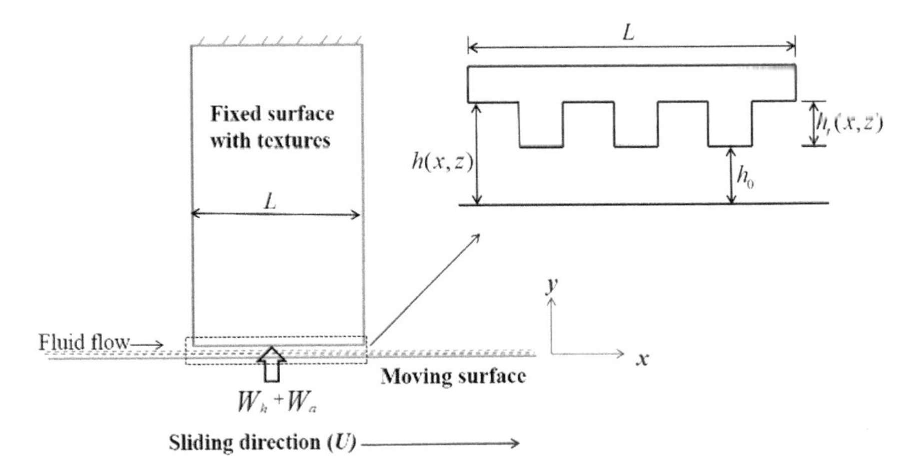

Fig. 1 Schematic view of textured piston ring-cylinder liner contact

$$\frac{\partial}{\partial x}\left(\phi_x \frac{h^3}{\mu}\frac{\partial p}{\partial x}\right) + \frac{\partial}{\partial z}\left(\phi_z \frac{h^3}{\mu}\frac{\partial p}{\partial z}\right) = 6U\phi_c\frac{\partial h}{\partial x} + 6U\sigma\frac{\partial \phi_s}{\partial x} \tag{1}$$

where p is the fluid film pressure, h is the mean film thickness, μ is the dynamic viscosity of lubricant, σ is the composite roughness of contact surfaces, U is the sliding velocity of the moving surface (i.e., cylinder liner), $\o_x$ and $\o_z$ are pressure flow factors in x- and z- direction, respectively, and $\o_s$ is the shear flow factor in the direction of motion. It is important to mention that all these flow factors depend on oil film ratio h/σ. These factors can be evaluated based on Patir and Cheng's flow model [9, 10]. The contact factor $\o_c$ is a function of the statistical distribution of roughness heights, and it can be determined according to studies of Meng et al. [11].

2.2 Film Thickness Equation

The computational domain along with the geometry of the textures considered is shown in Fig. 2. The symbol h_0 represents the minimum oil film thickness. From Fig. 2, the equation for oil film thickness is given as:

$$h(x,z) = \begin{cases} h_0 & (x',z') \in \Omega \\ h_0 + h_t(x,z) & (x',z') \notin \Omega \end{cases} \tag{2}$$

It is noted that the terms x', z' represent local coordinates of the texture as shown in Fig. 2 and the symbol Ω represents the presence of texture inside the unit cell. The area density of texture S_p is the ratio of the area occupied by texture (A_t) and the

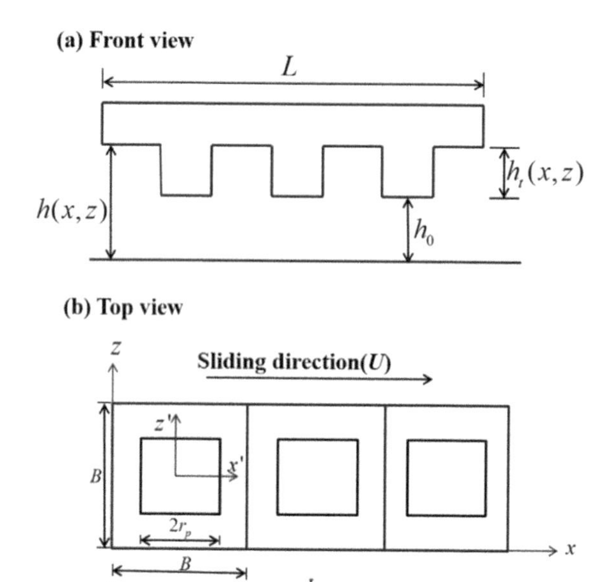

Fig. 2 Computational domain considered in mathematical modeling

area of the unit cell (A_u). As various texture shapes are considered in the present study, half of the base length (r_p) can be different for different configurations. The maximum possible area density (S_p) for various texture shapes such as square, triangular, circular and elliptical is 0.95, 0.3, 0.75 and 0.35, respectively. Accordingly, in order to study the effect of texture shape, the maximum possible area density for all shapes is 0.3. Therefore, in the present analysis, the area density is varied in between 0.05 and 0.3, with an increment of 0.05 for all the texture shapes.

2.3 Boundary Conditions

The boundary conditions considered in the present analysis are given as:

In x-direction,

$$p(x = 0, z) = p(x = L, z) = 0 \quad (3)$$

In z-direction,

$$p(x, z = 0) = p(x, z = B); \frac{\partial p}{\partial z}(x, z = 0) = \frac{\partial p}{\partial z}(x, z = B) \quad (4)$$

Reynolds cavitation condition,

$$\text{when } p < 0; p = 0 \text{ and } \frac{\partial p}{\partial x} = 0 \tag{5}$$

2.4 Asperity Contact Model for Rough Surfaces

In mixed lubrication condition, the total load is carried by both lubricant and asperities. These pressures can collectively contribute to the total load support of the system. In the present work, these contact pressures are calculated by adopting a well-known Greenwood–Tripp asperity contact model [12].

2.5 Load Support and Friction Equations

The total load support, W of the contact system, is the sum of the load carried by the fluid film and the load carried by the asperities due to mechanical interactions; thus,

$$W = \int_0^L \int_0^B p\mathrm{d}x\mathrm{d}z + \int_0^L \int_0^B p_{\mathrm{asp}}\mathrm{d}x\mathrm{d}z \tag{6}$$

Friction coefficient (*f*) of the sliding pair can be calculated using the equation:

$$f = \frac{F}{W} \tag{7}$$

where *F* is the total friction force which is the combination of dissipative forces due to hydrodynamic and asperity interaction effects. The equation of the total friction force is:

$$F = \int_0^L \int_0^B \left[\frac{\mu U}{h}(\phi_f + \phi_{fs}) + \phi_{fp}\frac{h}{2}\frac{\partial p}{\partial x}\right] \mathrm{d}x\mathrm{d}z + \int_0^L \int_0^B (C_f p_{\mathrm{asp}})\mathrm{d}x\mathrm{d}z \tag{8}$$

where $ø_{fs}$ and $ø_{fp}$ are the shear stress factors along the sliding direction and $ø_f$ is the friction factor. These factors are determined based on Patir and Cheng's average flow model [9, 10]. C_f is the boundary friction coefficient which is adopted as 0.12 in the present study.

Table 1 Parameters considered in simulation

Parameter	Value
Elastic modulus of liner surface, E_1	215 GPa [14]
Elastic modulus of ring surface, E_2	200 GPa [14]
Poisson's ratio of liner surface, ν_1	0.3
Poisson's ratio of ring surface, ν_2	0.3
Dynamic viscosity of fluid, μ	0.121 Pa s
Sliding velocity, U	4 m/s
Size of the unit cell, B	1 mm
The axial width of the ring surface, L	3 mm
The area density of texture, S_p	0.05–0.3
Height of texture, h_t	5–20 μm

3 Results and Discussion

Reynolds Eq. (1) is discretized by using finite difference method, and it leads to a set of linear algebraic equations for nodal pressures. These algebraic equations are solved by using point-successive under-relaxation method with Gauss–Siedel iterative scheme by satisfying boundary conditions. The material and texture parameters used in the numerical simulations are presented in Table 1. The present analysis considered the number of textures in the axial direction as 3. It is worthwhile to mention that when altering the texture area density, the size of the texture $2r_p$ changes while the unit cell size B remains constant. The average roughness (R_a) of the ring (σ_1) and liner (σ_2) surfaces is considered as 0.417 μm and 0.351 μm, respectively, which were adopted from the study of Venkateswara Babu et al. [13]. Therefore, the composite roughness is obtained as 0.546 μm.

In the present work, the numerical simulations are performed by assuming a constant minimum film thickness value under two cases: (1) $h_0 = 3$ μm, which indicates that interacting surfaces are under pure hydrodynamic lubrication ($h_0/\sigma \geq 4$); (2) $h_0 = 1$ μm, which means interacting surfaces are under mixed lubrication ($h_0/\sigma < 4$).

3.1 Validation of the Present Model

In order to validate the present model, a comparison of numerical results with Gu et al. [14] has been performed. The same parameters have been considered as given in [14] for validation. Gu et al. [14], has performed simulations with different crown heights by considering a one-dimensional average flow model with fully flooded inlet conditions. Based on their study, the hydrodynamic load support percentage of PR-CL contact is more in the case of higher crown heights when compared with lower crown height values. As can be seen from Fig. 3, the present model is in good agreement with the model presented by Gu et al. [14].

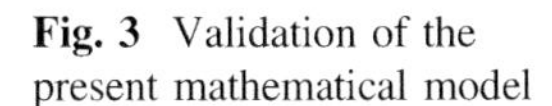

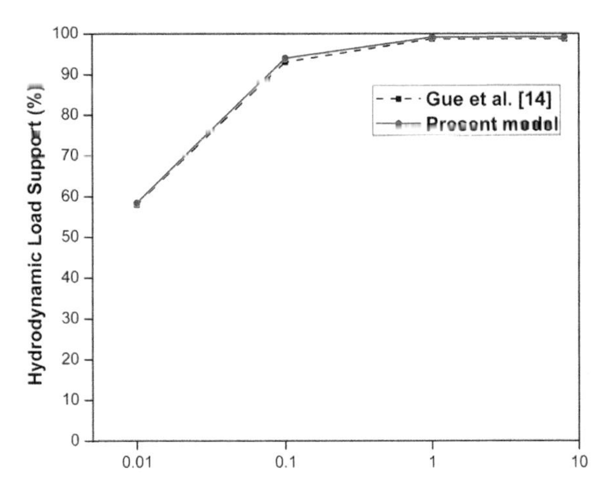

Fig. 3 Validation of the present mathematical model

3.2 *Textured Ring Surface Under Pure Hydrodynamic Lubrication* ($h_0/\sigma \geq 4$)

The effect of texture area density for different texture shapes on the load support at a constant texture height of 10 μm is presented in Fig. 4a. As can be seen from Fig. 4a, the load support increases with the increase of area density for all texture shapes. It is found that at $S_p = 0.3$, texture's shape effect is more dominant in generating the higher hydrodynamic pressure which resulted in maximum load support. Furthermore, it seems that among all the texture shapes, elliptical textures generate maximum load support.

The effect of texture area density on the friction coefficient is shown in Fig. 4b. It is observed that initially the friction coefficient reduces and then increases as the area density varies from 0.05 to 0.3. For all the texture shapes, a similar trend of friction coefficient is observed. Moreover, there exists an optimum area density that

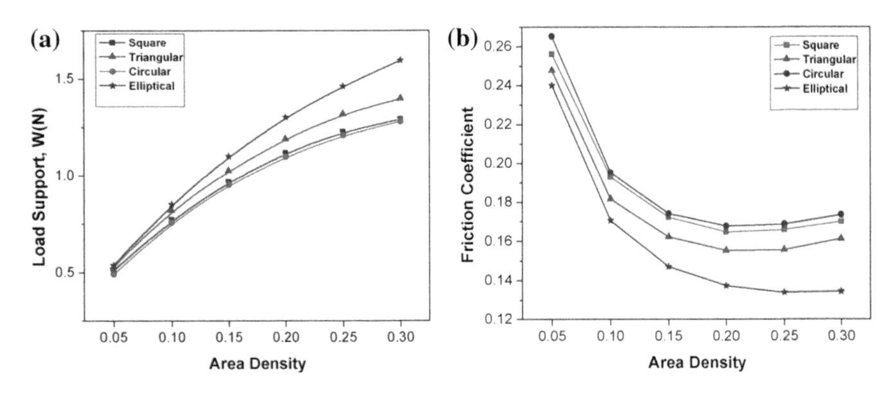

Fig. 4 Influence of area density on tribological performance parameters ($h_t = 10$; $U = 4$ m/s)

minimizes the friction coefficient. In general, the optimum area density is found to be in between 0.2 and 0.25, and in particular 0.25 for elliptical shape and 0.2 for all other texture shapes. Furthermore, at an area density of 0.2, the percentage reduction in friction coefficient by elliptical textures with respect to other shapes, i.e., circular, square and triangular, is 18.1%, 17.2% and 11.4%, respectively. From the results, it can be concluded that at higher area densities, the texture shape has a more prominent effect on load support and friction coefficient.

3.3 Textured Ring Surface Under Mixed Lubrication ($h_0/\sigma < 4$)

The impact of area density on the performance parameters under mixed lubrication conditions is shown in Fig. 5 by considering various texture shapes. Among all texture shapes, the load support is more for elliptical shape while the circular shape exhibit minimum at all area densities (see Fig. 5a). However, at lower area densities the effect of texture shape on load support is limited as compared to higher area densities. Because, as shown in Fig. 5b, the percentage of hydrodynamic load support is higher at lower area density. Furthermore, of all the shapes, elliptical textures are superior in forming the strong hydrodynamic effect. The impact of area density on friction coefficient is shown in Fig. 5c. It is observed that initially the friction coefficient decreases and then slowly increases with the area density. This indicates that there exists an optimal value for area density that minimizes the friction coefficient. The optimum area density value is found at 0.15 for elliptical shape and 0.1 for all other shapes. Furthermore, the elliptical textures are found to be superior in reducing the friction coefficient when compared with other shapes. It is seen that at an area density of 0.1, there is a reduction of friction coefficient by 16.7, 15.1 and 7.6% with the elliptical textures, as compared to other texture shapes, i.e., circular, square and triangular, respectively.

4 Conclusions

From the results obtained, the conclusions can be summarized as follows:

- All textured surfaces with an area density of 0.3 show the maximum load support under pure hydrodynamic lubrication and mixed lubrication.
- Under pure hydrodynamic lubrication, the friction reduction is found to be maximum at a texture area density of 0.25 for elliptical shape and 0.2 for all other shapes while in the case of mixed lubrication, area density of 0.15 for elliptical shape and 0.1 for all other shapes show the maximum friction reduction.

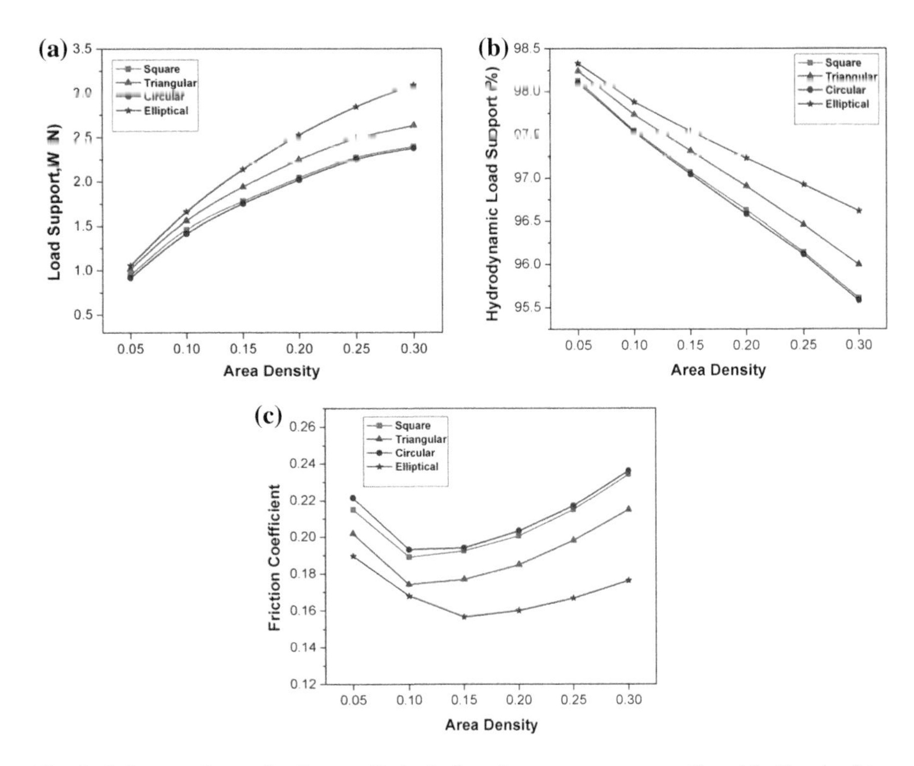

Fig. 5 Influence of area density on tribological performance parameters (h_t = 10; U = 4 m/s)

- Under both hydrodynamic and mixed lubrication, texture shape has a significant effect on the tribological performance of PR-CL contact at higher values of a texture area density.
- Among all the texture shapes, elliptical textures are better in generating higher load support and lower friction coefficient under pure hydrodynamic as well as mixed lubrication conditions.

The present study has a limitation that it has been assumed a constant sliding velocity and a minimum film thickness. The varying effect of these parameters on the tribological performance can be considered as the future work.

References

1. Holmberg K, Andersson P, Erdemir A (2012) Global energy consumption due to friction in passenger cars. Tribol Int 47:221–234
2. BŠhre D, Schmitt C, Moos U (2012) Analysis of the differences between force control and feed control strategies during the honing of bores. Proc CIRP 1:377–381

3. Ma C, Duan Y, Yu B, Sun J, Tu Q (2017) The comprehensive effect of surface texture and roughness under hydrodynamic and mixed lubrication conditions. Proc Inst Mech Eng Part J J Eng Tribol 231(10):1307–1319
4. Etsion I, Sher E (2009) Improving fuel efficiency with laser surface textured piston rings. Tribol Int 42(4):542–547
5. Wan Yi, Xiong Dang-Sheng (2008) The effect of laser surface texturing on the frictional performance of face seal. J Mater Process Technol 197(1–3):96–100
6. Gachot C, Rosenkranz A, Hsu SM, Costa HL (2017) A critical assessment of surface texturing for friction and wear improvement. Wear 372:21–41
7. Gadeschi GB, Backhaus K, Knoll G (2012) Numerical analysis of laser-textured piston-rings in the hydrodynamic lubrication regime. J Tribol 134(4):041702
8. Kligerman Y, Shinkarenko A (2015) Analysis of friction in surface textured components of the reciprocating mechanism. Proc Inst Mech Eng Part J J Eng Tribol 229(4):336–349
9. Patir N, Cheng HS (1978) An average flow model for determining effects of three-dimensional roughness on partial hydrodynamic lubrication. J Lubr Technol 100 (1):12–17
10. Patir N, Cheng HS (1978) Application of average flow model to lubrication between rough sliding surfaces. ASME Lubr Div
11. Meng F, Wang QJ, Hua D, Liu S (2010) A simple method to calculate the contact factor used in the average flow model. J Tribol 132(2):024505
12. Greenwood JA, Tripp JH (1970) The contact of two nominally flat rough surfaces. Proc Inst Mech Eng 185(1):625–633
13. Venkateswara Babu P, Syed I, Satish Ben B (2018) Influence of positive texturing on friction and wear properties of piston ring-cylinder liner tribo pair under lubricated conditions. Ind Lubr Tribol. https://doi.org/10.1108/ILT-07-2017-0203
14. Gu C, Meng X, Xie Y, Li P (2016) A study on the tribological behavior of surface texturing on the nonflat piston ring under mixed lubrication. Proc Inst Mech Eng Part J J Eng Tribol 230(4):452–471

Short-Term PV Power Forecasting for Renewable Energy Using Hybrid Spider Optimization-Based Convolutional Neural Network

Debom Ghosh

Abstract In this paper, we concentrate to optimize the power demand using forecasting of the solar radiation power, temperature and wind speed. We propose the hybrid technique (CNN-SSO), i.e. convolutional neural network (CNN) and short-term power forecasting model, and it is combined with the social spider optimization (SSO). The SSO algorithm is used to optimize the design constraints in order to inevitably choose the suitable widespread constraint cost of the PV power estimation. The results from simulation work have shown the estimation of the CNN-SSO method that successfully selected by the appropriate operating mode to achieve streamlining of the general vitality effectiveness of the framework utilizing every external parameter. The simulation results demonstrate the viability of CNN-SSO technique in standings of computational feasibility, accuracy and increased robustness.

Keywords Load forecasting · Neural network · Cuckoo search · Levy flight · Hybrid neural network

1 Introduction

In today's world, when energy demands are drastically increasing, load forecasting goes a long way in predicting the energy needs of the consumers and planning the entire power transmission and distribution efficiently [1, 2]. There are essentially three types of load forecasting possible—long-term load forecasting (spanning across the electric requirement for a year), medium-term forecasting (spanning across a week to a year) and short-term forecasting on a daily basis. Short-term forecasting involves establishing a predetermined relationship between the power output of the energy system and certain independent variables influencing it like temperature and time [3]. This is achieved by establishing information about the

D. Ghosh (✉)
Department of Electronics and Electrical Engineering, KIIT University, Bhubaneswar, India

BBVL. Deepak et al. (eds.), *Innovative Product Design and Intelligent Manufacturing Systems*, Lecture Notes in Mechanical Engineering,
https://doi.org/10.1007/978-981-15-2696-1_78

states of these individual parameters affecting the load output of the system and inferring it for a future span of 1 day or a week. Conversely, it becomes difficult to predict the load capacity in these cases when we do not have knowledge of the independent parameters influencing the power output in a system like a temperature, wind speed or the PV output of the area. In this paper, a hybrid CNN-SSO technique is proposed for load forecasting in the predictive control strategy. The main objective of the proposed CNN-SSO technique is to compute the demand level of maximum load in micro-grid and control load using optimal capacity configurations. In Sect. 2, the related works and the problem methodology have been stated in details, and a system model of the proposed CNN-SSO technique is given in Sect. 3. In Sect. 4, the detailed working function of the proposed CNN-SSO technique is described with proper mathematical models. The results and performance analysis are discussed in Sect. 5. The conclusions are given in Sect. 6.

2 Related Works

Hong et al. [4] have proposed a chaotic back-propagation (CBP) NNA grounded with the quality of the Chebyshev map. Pan et al. [5] have proposed a half and half determining methodology that consolidates ensemble empirical mode decomposition (EEMD), chaotic self-adaptive flower pollination algorithm (CSFPA) and BPNN. Li et al. [6] have presented an information-driven linear clustering (DLC) technique to explain the long-term framework load gauging issue brought about load variance in some urban communities. Genteel et al. [7] have proposed an artificial neural network (NNs) model with variational mode decomposition (VMD) for a momentary breeze speed anticipating. Wang et al. [8] have proposed a probabilistic load forecasting (PLF) strategy to use existing point load conjectures by demonstrating the restrictive figure remaining. The strategy conducts point gauging utilizing the authentic burden information and related elements to get the point conjecture. Eibl et al. [9] have examined load anticipating an issue with whether energy providers can profit by totalled shrewd metering information which is gained in a protection cordial manner. Luo et al. [10] have presented the Vanilla benchmark model in GEFCom2012 for extremely short-term load gauging VSTLF. Zhang et al. [11] have proposed an inventive strategy on probabilistic burden estimating utilizing both info vulnerability and yield variety. It turned out this model performs superior to anything normally utilized benchmark models. McPherron et al. [12] have proposed a changed nonparametric technique for building dependable forecast interims (PIs). Rafidi et al. [13] have exhibited a half and half strategy for probabilistic power load estimating, including summed up generalized extreme learning machine (GELM) for preparing an improved wavelet neural network (IWNN), wavelet before processing and bootstrapping.

3 Problem Methodology and System Model

3.1 Problem Methodology

Zang et al. [14] have a mixture technique dependent on a profound CNN that is presented for present moment PV control gauging. In the proposed strategy, distinctive recurrence parts are first extracted from the recorded time arrangement of PV control through vibrational mode disintegration (VMD) and mapped against time. In the previous couple of decades, a few models have been created to figure electric burden all the more precisely. Those researches have evaluated the problems too great extent but still, modification is going on to achieve efficient throughput. The main contribution of the proposed CNN-SSO technique is as follows:

- It is used to optimize the power demand using forecasting of solar radiation, temperature and wind speed.
- The hybrid technique (CNN-SSO), i.e. CNN STP forecasting model, and it is combined with the social spider optimization (SSO).
- The SSO algorithm is used to optimize the design constraints to consequently choose the fitting spread parameter esteem for the CNN control estimating model.
- The outcomes from the recreation work have exhibited the estimation of the SSO-CNN approach effectively chosen the fitting working mode to accomplish streamlining of the general productivity of the framework utilizing every accessible asset. The re-enactment results demonstrate the adequacy of SSO-CNN procedure as far as computational feasibility, accuracy and increased robustness.

3.2 System Model of the Proposed CNN-SSO Technique

In CNN-SSO technique, the variables such as solar power, wind power, inverter power, diesel power and water consumption are used as inputs; the sessions are 7, 15 and 30 days; per every hour load is forecasted; NN-type network and training are used.

The system model of the proposed CNN-SSO technique is showed in Fig. 1.

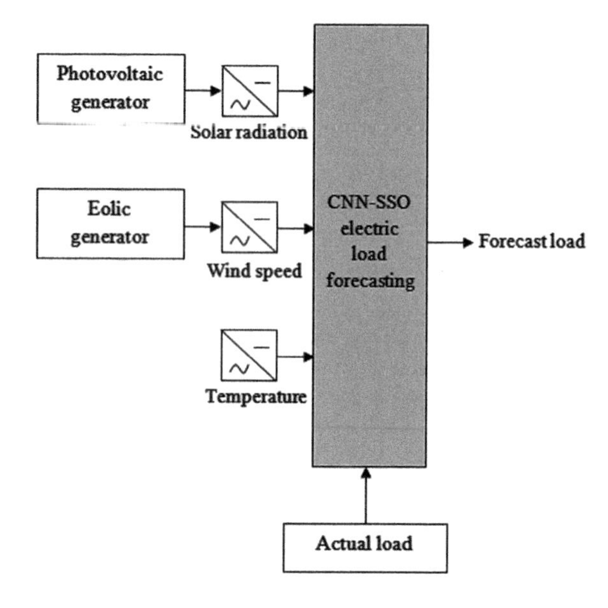

Fig. 1 System model of the proposed CNN-SSO

4 CNN-SSO Technique for Renewable Energy

CNN-SSO limits the operational expenses while providing the water and burden requests, thinking about 48 h ahead estimating of the climate conditions, water utilization and electrical burden. The inputs of CNN-SSO are predicted maximum and minimum solar power $\left(\max_{s_p}, \min_{s_p}\right)$, wind power $\left(W_p\right)$, initial battery charge $\left(C_i\right)$, battery bank voltage $\left(V_i\right)$, current $\left(I_i\right)$, diesel power $\left(D_p\right)$, water consumption $\left(W_c\right)$, water tank level $\left(L_{wt}\right)$, and measured load $\left(P_l\right)$. From this, the objective function is formulated as follows:

$$F(x) = \delta_t \sum_{t=t_0}^{T} \left(C_{S_p}(t) + C_{W_p}(t) + C_{D_p}(t)\right) + C_{L_{wt}} \sum_{t=t_0}^{T} C_{W_c}(t) + C_{P_l} \delta_t \sum_{t=t_0}^{T} P_l(t) \quad (1)$$

where $C_{S_p}(t)$, $C_{W_p}(t)$, $C_{D_p}(t)$ represent the cost function of solar, wind and diesel power at a discrete timestep δ_t; $C_{L_{wt}}$ is the cost of un-served water; $C_{W_c}(t)$ is the water consumption level; and C_{P_l}, $P_l(t)$ are the cost of un-served energy load and total load. The rule base comprises of the fuzzy for that the guidelines as follows:

$$\text{If } x \text{ is } A \text{ and } y \text{ is } B \text{ than } z \text{ is } f(x, y)$$

where An and B are the fuzzy sets in the predecessors and $z = f(x, y)$ is a fresh capacity in the subsequent. For a two principles fuzzy derivation framework, the two rules might be expressed as follows:

$$\text{Rule 1}: \text{If } S_p \text{ is rule 1, } W_p \text{ is rule 1}, D_p \text{ is rule 1, and } W_c \text{ is rule 1 than}$$
$$f_1 = m_1 S_p + n_1 W_p + o_1 D_p + r_1 W_c$$

$$\text{Rule 2}: \text{If } S_p \text{ is rule 2, } W_p \text{ is rule 2}, D_p \text{ is rule 2, and } W_c \text{ is rule 2 than}$$
$$f_2 = m_2 S_p + n_2 W_p + o_2 D_p + r_2 W_c$$

The last yield is the weighted normal of each standard's yield. The relating proportional CNN-SSO is given in Fig. 2.

Each point x_i in the inhabitants is calculated as :

$$x_i^t \leftarrow n_t + \delta(m_t - n_t); \quad t = 1, 2, \ldots, n \tag{2}$$

where δ is thought to be consistently disseminated somewhere in the range of 0 and 1 and means the upper and lower limitations of the beginning stage. The best and most noticeably awful capacity esteems are a figure as pursues:

$$f_b = \min\{f(x_i); \quad i = 1, 2, \ldots, n\} \tag{3}$$

$$f_w = \max\{f(x_i); \quad i = 1, 2, \ldots n\} \tag{4}$$

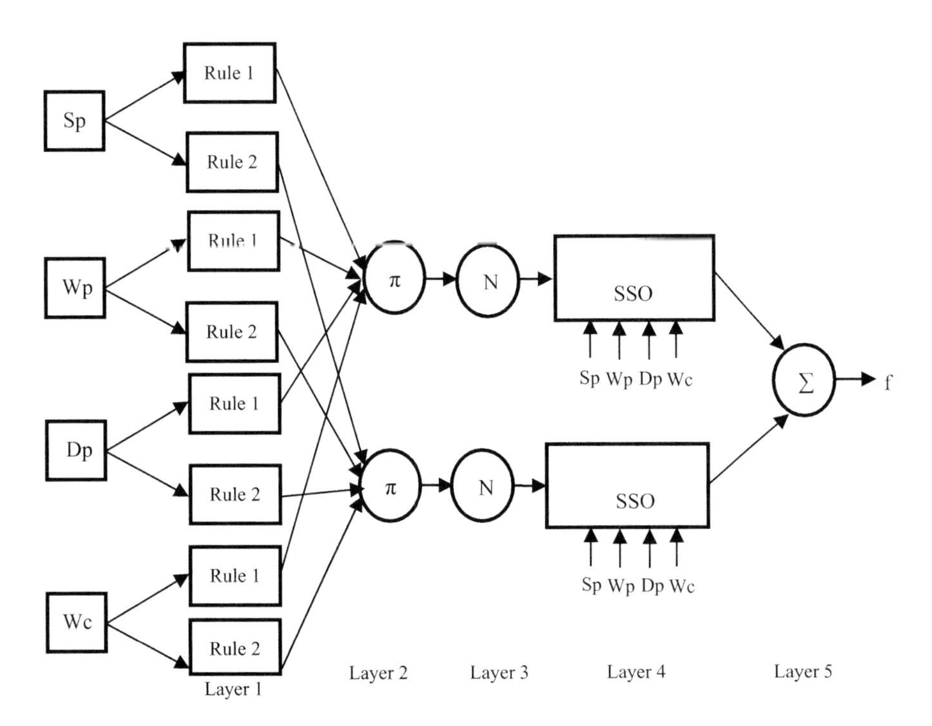

Fig. 2 Equivalent structure of CNN-SSO

A static visual incentive for all the populace relies upon the bound requirements of the issue which will characterize by fashioner.

$$V = \max_{t\in\{1,\ldots n\}} (m_t - n_t) \tag{5}$$

At the point when V is vacant methods, the point of populace moves haphazardly and generally selects point as an irregular way which is superior to x_i. The swarming, pursuing and seeking practices can be viewed as nearby practices. At the point when V is not filled, the calculation actuates the seeking conduct and haphazardly chooses a point inside the visual degree, i.e. a list is arbitrarily chosen and the point x_i is stimulated in the direction of it if the circumstance $f(x_r) < f(x_i)$ holds. The way of direction is denoted below:

$$\text{direction}_i = x_r - x_i \tag{6}$$

The development towards a specific point, i.e. along a specific course, is completed a part by segment and considers the permitted development towards the upper and lower bounds in the set. The recently produced trail focuses are selected for next dimensions which rely upon the condition as,

$$x_i = \begin{cases} t_i; & \text{iff}(t_r) < f(x_i) \\ x_i; & \text{otherwise} \end{cases} \tag{7}$$

After advance, the time changing limitations of every vehicle hub register claim quality (V_s) as follows:

$$V_s = x_1 + x_2 + \cdots \tag{8}$$

5 Simulation Result

This segment contains the trial assessment of the heap anticipating strategy that depicted in detail. The information is used to prepare and approve the model spoken by the suspicion of Sunday = 0, Monday = 1, Tuesday = 2, Wednesday = 3, Thursday = 4, Friday = 5 and Saturday = 6. It has been exhibited that this change improves the execution of the CNN-SSO. On account of the day of the week and the month, every factor entered as sins and cosines that improve the forecast.

Table 1 presents the prediction errors which are MAPE at 1-h, 1-day and 2-day ahead for electric burden, utilizing the moving skyline procedure and preparing each 7, 15 and 30 days. For each case, three models are utilized and prepared with 30, 60 and 90 days. Tables portray the mistakes scores of the diverse models amid the test.

Table 1 Performance analysis of mean MAPE (%)

Predictor (h)	Data for training (days)	No. of training frequency (days)					
		7		15		30	
		CNN-SSO	CNN-VMD	CNN-SSO	CNN-VMD	CNN-SSO	CNN-VMD
1	30	15.9003	14.122	15.6955	14.698	16.1343	14.927
	60	15.5700	14.107	15.4568	14.654	15.5245	14.541
	90	14.1630	14.137	14.1987	14.023	14.2569	14.124
6	30	15.5260	14.789	15.2391	14.001	15.6270	14.078
	60	15.2763	14.203	15.2345	14.0025	15.2777	14.005
	90	13.9682	14.001	13.9801	14.142	14.0975	13.789
12	30	15.5171	13.784	15.6394	13.994	16.2135	13.546
	60	15.4031	13.457	15.4151	13.854	15.4399	13.487
	90	14.4017	13.278	14.3877	13.472	14.5323	13.247

The CNN-VMD technique is assessed on all segments of the information collection for correlation. The best found the middle value of results is acquired from the preparation and test set. It demonstrates the best score with a normal RMSE of 0.0812, a normal MAE of 0.0412 and a normal total deviation of 0.127. The correlation is made through the RMSE records, for 60 min, 24 h and 48-hour expectation skylines. The parameter Ts is exchanged to the following dimension of burden estimating. This was made utilizing moving skyline techniques, making the heap forecast for the following *N* hours in each progression, utilizing genuine information. As watched, distinctive zones are giving distinctive forecast by utilizing diverse expectation models that coordinate best by their nearby elements and show changing exactness levels, which implies the inalienable decent variety among the expectations given by various sub-networks.

6 Conclusion

We have proposed a hybrid CNN-SSO technique for optimal electrical load forecasting techniques to overcome the complexity and accuracy problems in a micro-grid. The proposed CNN-SSO technique is utilized the SSO algorithm to compute the optimal load and also denotes as forecasted load. The SSO algorithm is utilized to compute the highest load mandate of the micro-grid to finalize the forecasted load with the ideal size conformations of manageable masses for individually preparation test set. The computed masses in the micro-grid are exact by the way of instruction necessities and manageable tons utilized; the working performance of micro-grid might be heightened by condensed application stashes. The simulation results proved the effectiveness of the proposed CNN-SSO technique over existing state-of-the-art techniques.

References

1. Deng Y, Ren X (2003) Optimal capacitor switching with fuzzy load model for radial distribution systems. IEE Proc Gener Transm Distrib 150(2):190
2. Huang Shyh-Jier, Shih Kuang-Rong (2003) Short-term load forecasting via ARMA model identification including non-gaussian process considerations. IEEE Trans Power Syst 18(2): 673–679
3. Baczyski D, Parol M (2004) Influence of artificial neural network structure on the quality of short-term electric energy consumption forecast. IEE Proc Gener Transm Distrib 151(2):241
4. Hong Z, Li X, Chen B (2014) Adaptive control based particle swarm optimization and chebyshev neural network for chaotic systems. J Comput 9(6)
5. Pan L, Feng X, Sang F, Li L, Leng M, Chen X (2017) An improved back propagation neural network based on complexity decomposition technology and modified flower pollination optimization for short-term load forecasting. Neural Comput Appl
6. Li Y, Han D, Yan Z (2017) Long-term system load forecasting based on data-driven linear clustering method. J Mod Power Syst Clean Energy 6(2):306–316

7. General M, Yuxian Z, Aoqi H (2018) Performance comparison of ANNs model with VMD for short-term wind speed forecasting. IET Renew Power Gener 12(12):1424–1430
8. Wang Yi, Chen Q, Zhang N, Wang Y (2018) Conditional residual modeling for probabilistic load forecasting. IEEE Transactions Power Systems 33(6):7327–7330
9. Eibl G, Bao K, Grassal P, Bernau D, Schmeck H (2018) The influence of differential privacy on short term electric load forecasting. Energy Inform 1(1)
10. Luo J, Hong T, Yue M (2018) Real-time anomaly detection for very short-term load forecasting. J Mod Power Syst Clean Energy 6(2):235–243
11. Zhang W, Quan H, Srinivasan D (2018) An improved quantile regression neural network for probabilistic load forecasting. IEEE Trans Smart Grid:1–1
12. McPherron R, Siscoe G (2004) Probabilistic forecasting of geomagnetic indices using solar wind air mass analysis. Space Weather 2(1)
13. Rafiei M, Niknam T, Aghaei J, Shafie-Khah M, Catalao J (2018) Probabilistic load forecasting using an improved wavelet neural network trained by generalized extreme learning machine. IEEE Trans Smart Grid 9(6):6961–6971
14. Zang H, Cheng L, Ding T, Cheung KW, Liang Z, Wei Z, Sun G (2018) A hybrid method for short-term photovoltaic power forecasting based on deep convolutional neural network. IET Gener Transm Distrib 12(20):4557–4567
15. Chen C, Duan S, Cai T, Liu B, Hu G (2011) Smart energy management system for optimal microgrid economic operation. IET Renew Power Gener 5(3):258

Virtual Manufacturing

Contingency Management of a Power System Using Rapid Contingency Management Technique and Harmony Search Algorithm

B. Sravan Kumar, R. Uma Maheswari, B. Sateesh, B. Venkateswara Rao and G. V. Nagesh Kumar

Abstract Optimal power flow (OPF) is an ideal method of optimally utilizing power system resources. Its effect further enhances the presence of FACTS devices. Performing OPF in combination with a FACTS device may also be helpful for the improvement of power system stability during outage conditions. In the present work, a combined index-based strategy for the optimal placement of Thyristor-Controlled Series Compensator (TCSC) and optimal tuning of generators using the harmony search algorithm is proposed for improving the system stability. The projected technique is verified and implemented on IEEE 30 bus system. The system is tested at both normal and contingency conditions. The contingency analysis is done using a new method, namely rapid contingency ranking technique (RCRT). The TCSC has been placed on the basis of an index which is a combination of line utilization factor (LUF) and fast voltage stability index (FVSI). A multi-objective function has been chosen for tuning the generators. The multi-dimensional function includes deviation in voltage, cost of power generation, and loss of transmission line. The outcomes of the proposed method are also compared to a method, i.e., genetic algorithm.

Keywords Optimal reallocation · Rapid contingency ranking technique · TCSC · Harmony search algorithm · Voltage stability

B. Sravan Kumar
GITAM University, Visakhapatnam, India

R. Uma Maheswari · B. Sateesh
Vignan's Institute of Information Technology, Visakhapatnam, India

B. Venkateswara Rao (✉)
V R Siddhartha Engineering College, Vijayawada, India

G. V. Nagesh Kumar
JNTUA College of Engineering, Pulivendula, India

BBVL. Deepak et al. (eds.), *Innovative Product Design and Intelligent Manufacturing Systems*, Lecture Notes in Mechanical Engineering,
https://doi.org/10.1007/978-981-15-2696-1_79

1 Introduction

Due to the increase in the competition in the electrical industry, optimum usage of the available power has become obligatory. On the contrary, with the rise of the power stream, transmission lines will continuously face the problem of congestion as power is carried at the highest transmission limit and at times higher. Continuous overloading of transmission lines will face the risk regarding safety, dependability, and immovability of the power systems. OPF is a way of optimizing an objective function in the presence of operational constraints by the method of nonlinear programming. Many methods have also been developed to resolve the OPF problem [1]. Metaheuristic methods are one of the most recent methods used for the OPF problem.

Chang [2] implemented a multi-dimensional particle swarm optimization method (PSOM) to fix SVC into progressing the transmission system loading margin (LM) to an extent so as to reduce the network expansion cost. Nam and Van [3] recommended the best possible position of SVC in the power market in which the voltage stability is increased by PV curve besides a high increase in social welfare by locational marginal price (LMP). Shaheen et al. [4] have used computational intelligence method, namely DE is adapted to find the maximal place and volume of UPFC based on the performance index is used to get $N - 1$ contingency condition. Mishra and Gundavarapu [5] planned a position for IPFC through an index, an amalgamation of real power performance index (RPPI), and line stability index in order to manage contingency. It is also found that the use of line voltage stability index is a good option for the measurement of voltage stability for series devices. Optimal reallocation of generators is necessary for the optimal utilization of the available power system resources. The advantages of the method are further improvised by the placement of FACTS devices. Series FACTS devices are most suitable for enhancing transmission capabilities. The FACTS device should be correctly placed in the system in order to enhance its effectiveness [6, 7]. Harmony search algorithm [8] was introduced in the year 2001 where the algorithm mentioned above was also used in different fields like economic load dispatch(Al-Betar et al. [9], Kumar and Dhanasekaran [10]), automatic generation control, mitigation of blackout, (Balachennaiah et al. [11]), and the AGC optimization of multi-unit power systems (Kumar Shiva and Mukherjee [12]).

This paper is analyzed using the OPF method to enhance system stability in normal and contingency conditions. Harmony search algorithm is used for the OPF in the presence of TCSC. The contingency analysis has been performed using one of the latest methods, namely rapid contingency ranking technique.

2 Contingency Analysis

As the foremost task is to decide the best site for placement of TCSC in energy system network to minimize the losses and control power flow through the lines. The following performance index is selected. Line contingency analysis consists of the removal of a line from the network and analyzing the stability of the system. The above type of analysis is a typical case of $n - 1$ contingency analysis. Various methods are available for analyzing a system in contingency condition. Most of the methods available in the literature are either very complicated or cumbersome. The method used for contingency analysis in this study is known as rapid contingency ranking technique suggested by Mishra and Gundavarapu [13]. The authors, in their study, have shown that all contingencies do not cause severity in the system and hence may be discarded from the contingency analysis study. The method comprises of analyzing only a few select lines for contingency. On the basis of this method, lines connected to an only slack bus, generator buses, load bus with the maximum number of lines are selected for the analysis of line contingency.

3 Combinatory Index Proposed

Employing LUF and FVSI indices, combinatory index is formed and can be calculated using Eq. (1).

$$\mathrm{CI} = W_{11} * I_1 + W_{12} * I_2 \tag{1}$$

where W_{11} and W_{12} are weighting factors.

I_1 is an index used for resolving the congestion of the transmission lines, called LUF. LUF is found using Eq. (2)

$$I_1 = \frac{\mathrm{MVA}_{ij}}{\mathrm{MVA}_{ij}^{\max}} \tag{2}$$

MVA_{ij} (max) Maximum MVA rating of the line between bus i and bus j.
MVA_{ij} Actual MVA rating of the line between bus i and bus j.

LUF gives an estimate of the percentage of a line being utilized.

FVSI gives voltage stability for the prescribed bus under any loading conditions. It is defined as follows: I_2 is the FVSI and can be estimated from Eq. (3)

$$I_2 = \frac{4Z^2Q_j}{V_i^2X} \tag{3}$$

where Z is the impedance of the line, X is the line reactance, V_i is the voltage at the sending end. Q_j is the reactive power at the receiving end. Both LUF and FVSI have a stable region when the value of the index is less than 1. A multi-dimensional function is considered for the purpose. The multi-dimensional function consists of voltage deviation, vigorous MW production cost, and transmission line loss.

4 Results and Discussion

4.1 OPF for Normal Condition

The value of CI determined is presented in Table 1. From the obtained values, it is visible that line number 9–10 has extreme CI value of 0.1858 per unit. Consequently, line 9–10 is the severe line of the system.

Diverse permutation of weights of the objective function is utilized, and Table 2 refers to the objective function values. It is noticed that $w1 = 0.7$, $w2 = 0.15$, $w3 = 0.15$ provide the least objective function and are considered for analysis.

Active power productions of the system and at individual generators for different cases are compared in Table 3. It is keen that OPF accompanied by TCSC is more effectual compared to without TCSC. Thus, this device is highly efficient in the optimization of the generators. Parameters of TCSC are given in Table 4.

Table 1 Severe line in IEEE 30 bus system

Rank	Line connected		CI	Rank	Line connected		CI
	Bus from	Bus to			Bus from	Bus to	
1	9	10	0.1858	17	10	17	0.0379
2	3	4	0.1725	18	10	22	0.0379
3	4	6	0.1616	19	27	29	0.0355
4	4	12	0.1489	20	25	27	0.0344
5	6	7	0.139	21	12	14	0.03
6	28	27	0.108	22	19	20	0.0274
7	10	21	0.0913	23	12	16	0.0221
8	6	10	0.0815	24	23	24	0.0202
9	12	15	0.0568	25	15	18	0.0181
10	6	28	0.0548	26	16	17	0.0174
11	6	9	0.0545	27	29	30	0.0174
12	10	20	0.0526	28	24	25	0.0127
13	12	13	0.05	29	15	23	0.0118
14	22	24	0.0443	30	21	23	0.0113
15	27	30	0.0434	31	18	19	0.0095
16	25	26	0.0417	32	14	15	0.0053

Table 2 Weights versus objective function

S. No.	Weights			*f*1	S. No.	Weights			*f*1
	*w*1	*w*2	*w*3			*w*1	*w*2	*w*3	
1	0.7	0.15	0.15	192.3	4	0.25	0.6	0.15	773.56
2	0.55	0.3	0.15	379.52	5	0.1	0.75	0.15	958.9
3	0.4	0.45	0.15	567	6	0.3	0.4	0.3	509.2

*w*1, *w*2, and *w*3 are weighted for objective function *f*1

Table 3 Analysis of OPF solution for 30 bus system by HS-OPF without and with TCSC

Parameter		HS-OPF without TCSC	GA-OPF without TCSC	HS-OPF with TCSC	GA-OPF with TCSC
Real power generation (MW)	PG1	135.556	126.656	92.771	131.378
	PG2	32.6893	27.3374	32.689	12.9481
	PG5	29.415	27.3348	29.415	23.945
	PG8	42.8081	21.3279	42.808	19.1834
	PG11	40.5583	84.8224	40.558	96.9945
	PG13	10	3.9926	50	5.0686
Total MW generation		291.027	291.471	289.34	289.518
Total active MW loss		7.6274	8.0713	5.9474	6.1185
Total MVAR loss		19.38	35.35	8.26	25.59
Voltage deviation (p.u.)		1.9507	2.5013	0.4122	0.4227
Total active power generation cost ($/h)		1360.7	1366.9	1254.3	1259.9
Objective function value		209.73	211	192.37	193.336

Table 4 TCSC parameters under normal condition

Method	TCSC location (from bus no to bus no)	TCSC parameters
NR with TCSC	9–10	$X = 0.1$, PTCSC = 0.2712, QTCSC = 0.1690
HS algorithm-based optimal power flow	9–10	$X = 0.02$, PTCSC = 0.4314, QTCSC = 0.1786

4.2 *OPF for Contingency Condition*

The projected method is implemented on a 30 bus system of IEEE, and a line contingency is considered to test the efficiency of the method in adverse conditions.

Contingency analysis by traditional method is performed for the IEEE 30 bus system, and the details of the indices after every contingency are mentioned in Table 5. CI gives an estimate of the overall stress on the lines as a result of various contingencies. In Table 6, the contingency analysis by RCRT technique is

Table 5 Severe lines for various line outages in descending order of CI by the traditional method

Line Outage		Severe line		LUF value	Severe line		FVSI	Severe line		CI
FB	TB	FB	TB		FB	TB		FB	TB	
9	10	3	4	0.3289	6	10	0.4292	4	12	0.2992
4	12	4	6	0.4236	9	10	0.1964	9	10	0.2569
28	27	3	4	0.3208	4	12	0.207	9	10	0.2326
4	6	4	12	0.2365	4	12	0.2202	4	12	0.2283
6	10	3	4	0.3055	4	12	0.168	9	10	0.2102
3	4	9	10	0.2397	4	12	0.1784	9	10	0.2047
12	15	4	6	0.3107	9	10	0.1492	9	10	0.1988
25	27	3	4	0.304	9	10	0.1658	9	10	0.1966
6	28	3	4	0.3105	4	12	0.164	9	10	0.1925
12	16	3	4	0.3025	4	12	0.1446	9	10	0.1922
15	18	3	4	0.303	4	12	0.1488	9	10	0.1896
12	14	3	4	0.3034	4	12	0.1503	9	10	0.1885
16	17	3	4	0.3024	4	12	0.1546	9	10	0.1877
24	25	3	4	0.3027	9	10	0.3589	9	10	0.1876
18	19	3	4	0.3026	4	12	0.152	9	10	0.187
27	30	3	4	0.3052	4	12	0.1548	9	10	0.1869
6	7	3	4	0.2666	4	12	0.144	9	10	0.1867
27	29	3	4	0.3046	4	12	0.1541	9	10	0.1866
14	15	3	4	0.3027	4	12	0.1521	9	10	0.1861
29	30	3	4	0.3033	4	12	0.153	9	10	0.1861
10	21	3	4	0.3089	4	12	0.1995	4	12	0.1853
23	24	4	6	0.2492	4	12	0.1495	9	10	0.1849
21	23	3	4	0.3028	4	12	0.1591	9	10	0.1833
6	9	3	4	0.3057	9	10	0.1596	9	10	0.1832
19	20	3	4	0.3044	4	12	0.1706	9	10	0.1818
10	22	3	4	0.3035	4	12	0.1534	9	10	0.1818
22	24	3	4	0.3035	4	12	0.1534	9	10	0.1818
10	20	3	4	0.3055	4	12	0.1764	9	10	0.1813
10	17	3	4	0.3044	4	12	0.1934	4	12	0.1783

performed. The most severe line detected by this technique is line 4–12 for line 9–10 contingency, which is the same as that obtained by the traditional method. It is observed that the number of lines required to be analyzed by this method is much less in comparison with the traditional method.

The MW production of the system and at individual generators for different cases is compared in Table 7. It is keen that OPF accompanied by TCSC is more effectual compared to without TCSC. It can be concluded that this device is

Table 6 Contingency analysis by RCRT technique

Line outage		Severe line		LUF value	Severe line		FVSI	Severe line		CSI
FB	TB	FB	TB		FB	TB		FB	TB	
9	10	3	4	0.328	6	10	0.429	4	12	0.2992
28	27	3	4	0.320	4	12	0.207	9	10	0.2326
4	6	4	12	0.236	4	12	0.220	4	12	0.2283
6	10	3	4	0.305	4	12	0.168	9	10	0.2102
6	28	3	4	0.310	4	12	0.164	9	10	0.1925
6	7	3	4	0.266	4	12	0.144	9	10	0.1867
10	21	3	4	0.308	4	12	0.199	4	12	0.1853
6	9	3	4	0.305	9	10	0.159	9	10	0.1832
10	22	3	4	0.303	4	12	0.153	9	10	0.1818
10	20	3	4	0.305	4	12	0.176	9	10	0.1813
10	17	3	4	0.304	4	12	0.193	4	12	0.1783

Table 7 Comparison of active power loss, cost, and voltage deviation with and without the contingency of TCSC placed at 4–12

System condition	Parameter		HS-OPF without TCSC	GA-OPF without TCSC	HS-OPF with TCSC	GA-OPF with TCSC
Without contingency	Real power generation (MW)	PG1	135.5	126.6	92.77	131.3
		PG2	32.68	27.33	32.68	12.94
		PG5	29.41	27.33	29.41	23.94
		PG8	42.80	21.32	42.80	19.18
		PG11	40.55	84.82	40.55	96.99
		PG13	10	3.992	50	5.068
	Total real power generation (MW)		291.0	291.4	289.3	289.5
	Real losses (MW)		7.627	8.071	5.947	6.118
	Reactive loss (MVAR)		19.38	35.35	8.26	25.59
	Total cost ($/h)		1360.	1366.	1254.	1259.
	Vol. deviation (p.u.)		1.950	2.501	0.412	0.422
With 9–10 line contingency	Active power generation (MW)	PG1	138.6	130.2	134.9	167.0
		PG2	32.68	27.33	32.68	45.70
		PG5	29.41	27.33	29.41	28.19
		PG8	42.80	21.32	42.80	10.19
		PG11	40.55	84.82	40.55	34.51
		PG13	10	3.992	10	6.671
	Total MW generation		293.5	295.0	290.4	292.2
	Real losses (MW)		10.13	11.63	7.007	8.899
	Reactive loss (MVAR)		45.43	79.76	20.84	27.84
	Total cost ($/h)		1370.	1380	1358.	1367.
	V.D (p.u.)		3.787	4.642	0.900	0.908

Table 8 TCSC parameters under a contingency

Method	TCSC location (from bus no- to bus no)	TCSC parameters
NR with TCSC	4–12	$X = 0.1$, PTCSC = 0.3854, QTCSC = 0.0124
HS algorithm-based optimal power flow	4–12	$X = 0.02$, PTCSC = 0.4842, QTCSC = 0.0124

extremely efficient in the optimization of the generators. Parameters of TCSC are shown in Table 8. The voltage deviation of the system improves greatly when harmony search with the presence of TCSC.

5 Conclusion

Proper planning may help the engineers to protect the system in case of severe contingencies. In this paper,

- The RCRT technique for contingency analysis has been verified and compared with the traditional method.
- In order to overcome the volatility issues of the power systems due to line outages and reduction of losses, TCSC is placed and sized.
- OPF in the presence of TCSC is found to be a very effective method of reducing the severity of the power system.
- Although TCSC is basically a series device, the voltage profile of the system improves considerably.

References

1. Momoh JA, El-Hawary M, Adapa R (1999) A review of selected optimal Power flow literature to 1993. II. Newton, linear programming and interior point methods. IEEE Trans Power Syst 14(1):105–111
2. Chang YC (2012) Multi-objective optimal SVC installation for power system loading margin improvement. IEEE Trans Power Syst 27(2):984–992
3. Nam TP, Viet La Van Ut DT (2014) Optimal placement of SVC in power market. In: IEEE 9th conference on industrial electronics and applications (ICIEA), Hangzhou, China, pp 1713–1718
4. Shaheen HI, Rashed GI, Cheng SJ Application and comparison of computational intelligence techniques for optimal location and parameter setting of UPFC. Eng Appl Artif Intell 23 (2):203–216
5. Mishra A, Gundavarapu VNK (2016) Contingency management of power system with interline power flow controller using real power performance index and line stability index. Ain Shams Eng J 7(1):209–222

6. Rao BV, Kumar GVN (2014) Firefly algorithm based optimal power flow with static VAR compensator for improvement of power system security under network contingency. Int Electr Eng J IEEJ 5(12):1639–1648
7. Rao BV, Kumar GVN (2015) A comparative study of BAT and firefly algorithms for optimal placement and sizing of static VAR compensator for enhancement of voltage stability. Int J Energy Optim Eng 4:68–84
8. Geem ZW, Kim JH, Loganathan GV (2001) A new heuristic optimization algorithm: harmony search. Simulation 76(2):60–68
9. Al-Betar MA, Awadallah MA, Khader AT, Bolaji AL (2016) Tournament-based harmony search algorithm for non-convex economic load dispatch problem. Appl Soft Comput 47:449–459
10. Kumar N, Dhanasekaran R (2014) Optimal power flow with FACTS controller using hybrid PSO. Arab J Sci Eng 39(4):3137–3146
11. Balachennaiah P, Suryakalavathi M, Nagendra P (2016) Optimizing real power loss and voltage stability limit of a large transmission network using the firefly algorithm. Eng Sci Technol Int J 19(2):800–810
12. Kumar Shiva C, Mukherjee V (2016) A novel quasi-oppositional harmony search algorithm for AGC optimization of three-area multi-unit Power system after deregulation. Eng Sci Technol Int J 19(1):395–420
13. Mishra A, Gundavarapu VNK (2016) Line utilization factor-based optimal allocation of IPFC and sizing using firefly algorithm for congestion management. IET Gener Transm Distrib 10 (1):115–122

Numerical Study of Warm Incremental Forming Limits of AZ31B Magnesium Alloy

Rohit Kumar Sharma and Shahul Hamid Khan

Abstract The main objective of this paper is to study the warm incremental forming of the magnesium alloy to achieve the forming limits. The influence of the different forming temperatures on the von Mises stresses, sheet thickness and plasticity of the material is a major concern. In this paper, a single-point incremental forming (SPIF), a tool containing hemispherical end, was used to describe the formability of AZ31B magnesium alloy sheet. The results obtained indicate that the temperature has a significant effect on the von Mises stresses which decrease obviously with the increase of temperature. For the simulation purpose, ABAQUS software is used. The hemispherical tool is assumed to be an analytical rigid body.

Keywords Warm incremental forming · Finite element method · Forming limits

1 Introduction

Single-point negative incremental forming is an emerging new technology which uses the tool having a hemispherical shape to get the desired 3D model. The tool used is fixed at spindle which is rotating at certain rpm controlled by a computed numerically control machine (CNC) to get desired geometry. The idea of choosing AZ31B magnesium alloy is because of its high strength-to-weight ratio. The magnesium is having high specific strength and low density. It has poor formability at room temperature but acquired excellent formability when the temperature is near around 250 °C. The warm incremental forming technique of magnesium alloy sheet forming is pulling the attention of scholars toward it. Ji and Park [1] made a study on warm incremental forming of AZ31 magnesium alloy and found as the temperature of

R. K. Sharma (✉) · S. H. Khan
The Mechanical Department, Indian Institute of Information Technology, Design and Manufacturing, Kancheepuram, Chennai, Tamil Nadu 600127, India
e-mail: mds17m002@iiitdm.ac.in

S. H. Khan
e-mail: bshahul@iiitdm.ac.in

BBVL. Deepak et al. (eds.), *Innovative Product Design and Intelligent Manufacturing Systems*, Lecture Notes in Mechanical Engineering,
https://doi.org/10.1007/978-981-15-2696-1_80

sheet metal increases, the formability of sheet metal also increases. Ambrogio et al. [2] carried out experimentation on the warm negative incremental forming of AZ31B alloy and found that magnesium shows high forming ability when the temperature of sheet metal is around 250 °C. Many studies have been carried out on increasing the formability of magnesium alloys using the warm incremental forming technique. This technique of warm incremental forming is also utilized for material like titanium alloy which is difficult to deform at room temperature. Zhang [3] said about the lubrication and usage of AZ31 sheet for warm negative incremental forming. For this simulation, AZ31B magnesium sheet metal of 1 mm thickness was used. Tool material taken is to be analytically rigid having hemispherical head at the bottom with 5 mm radius, vertical step down of 0.3 mm/revolution, feed rate of 1000 mm/min, forming angle of 45°, and sheet metal temperatures 100, 150, 250, and 250 °C were taken. For lubrication purpose, different lubricants were used to get low friction between the tool head and sheet metal. Two lubricating methods were used—one is by applying K2Ti4O9 and another one is by applying pulsed anodic oxidation to the surface of the sheet to reduce friction and to obtain the good surface finish and to reduce friction. Ambrogio et al. [4] takes three lightweight alloys, commonly used in the aviation and aerospace industries, were developed to produce a local heating by providing a continuous current. Ambrogio et al. [5] investigated the various process parameter to analyse the material formability using experimental setup and stastical analysis.

2 Methodology

2.1 Part Model

The incremental forming mainly contains four parts, such as sheet metal, hemispherical tool, tool holder, and computer numerical control machine. The standard dynamic explicit model is created with two parts—one is the sheet of 170 × 170 mm and other is the hemispherical deforming tool of radius 5 mm. The thickness used for sheet metal is of 1 mm (Fig. 1).

2.2 Material Property

In this, material properties of the magnesium alloy such as AZ31B are defined. The material properties such as modulus of elasticity, plastic strain, coefficient of thermal expansion, and Poisson ratio are also defined. The plasticity of the material is obtained according to the graph plotted during the tensile test of magnesium alloys in a universal testing machine. The strain values to the corresponding values of stress are obtained and fed into the ABAQUS software. The modulus of elasticity of the material magnesium AZ31B alloy is with a change in temperature. The ductility of the material improves after heating. The graph after tensile is plotted and observed

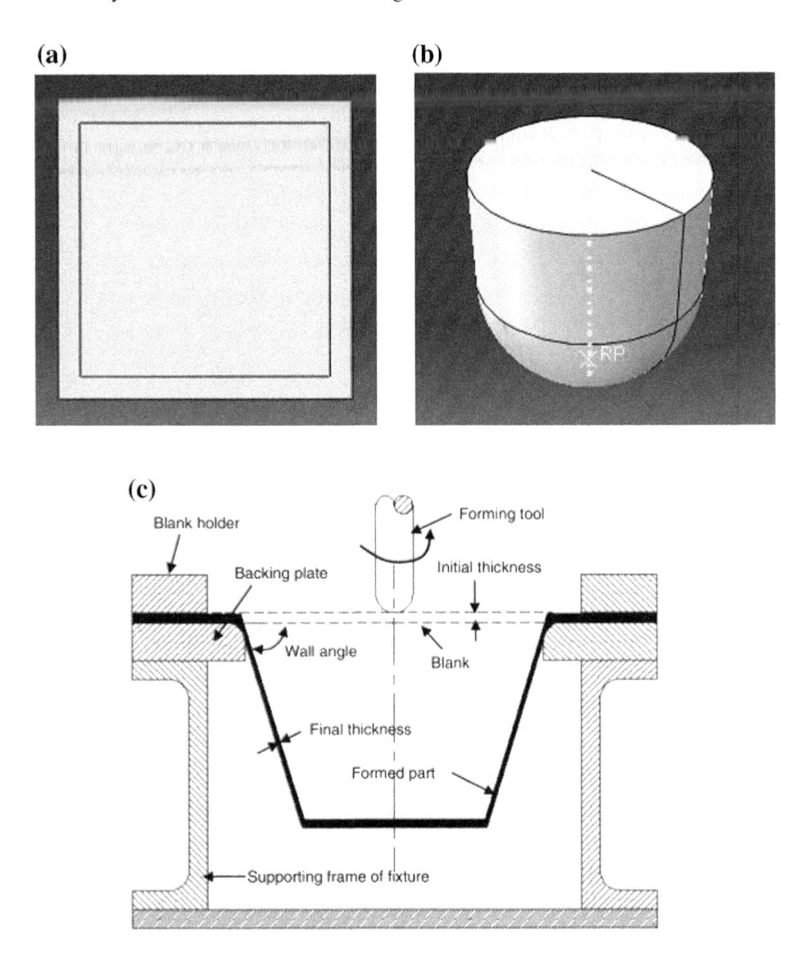

Fig. 1 **a** Magnesium sheet, **b** hemispherical tool, and **c** schematic of incremental forming

that the material property is changed with an increase in temperature. The sheet shows excellent results when the temperature is near around 250 °C. The material property also plays an important role in warm incremental forming (Fig. 2; Table 1).

2.3 Finite Element Method

In this paper, we use ABAQUS to obtain a theoretical analysis of warm incremental forming. For this numerical simulation, the sheet metal used of AZ31B magnesium alloy and tool took as an analytical rigid tool. Initially, the sheet was held by with the help of a blank holder. The sheet metal is heated till it achieved the required temperature, and then, the tool will move according to the defined tool path and deformed the sheet metal by incrementally of given step size.

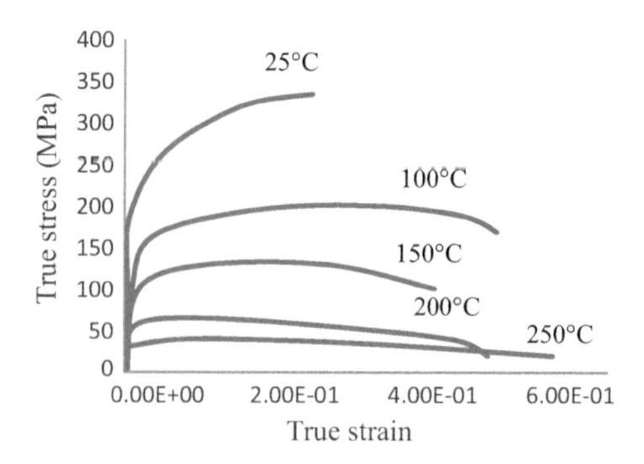

Fig. 2 Stress versus strain for AZ31B magnesium alloy

Table 1 Mechanical property of a material

Alloys	Density (g/cm^3)	E (GPa)	G (GPa)	Poisson's ratio	Thermal conductivity (W/mK)	Thermal expansion (μm/m°C)
AZ31B	1.77	45.0	17	0.33	96	26
AZ61A	1.80	44.8	17	0.34	70	26
AZ91A	1.81	44.8	17	0.35	72.7	26

2.4 *The Meshing of Sheet Metal*

The simulation contains three steps, namely initial, heating, and deform. In the initial, there is an interaction between tool and sheet, and further steps are for deformation. The S4R meshing takes less simulation time as compared to other types of meshing used like S3R, etc. During the numerical simulation of the warm incremental process, the S3R meshing takes too much of simulation time, but when the S4R meshing is used, there is a reduction in a simulation time of the process. Also it is observed that deformation is good while using S4R meshing and because of that it is giving accurate results (Table 2).

2.5 *Boundary Condition*

The incremental forming contains mainly four parts, such as sheet metal, tool, sheet blank holder, and supporting frame for the fixture. The sheet is divided into two parts—one is deformable and the other is like blank holder. First, we made one reference point on the tool, and then, we assigned the displacement along x, y, and z, and now the sheet is clamped by restricting all its degree of freedom at the edge of the sheet (Fig. 3).

Table 2 Element used in meshing

Element type	Elements	Nodes
Four-node quadrilateral	1612	1684

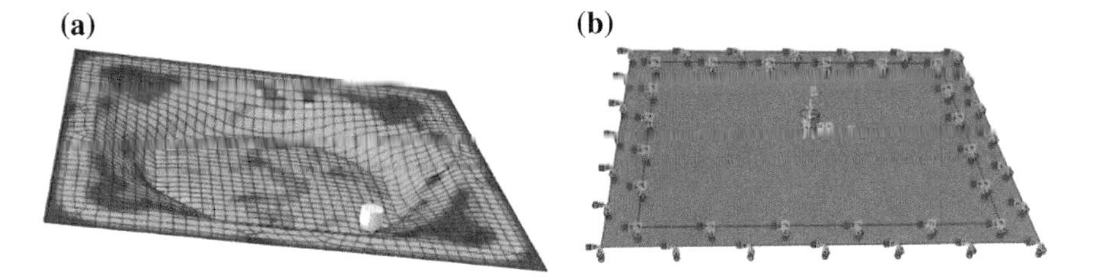

Fig. 3 **a** Profile map of an inverted cone and **b** boundary conditions

2.6 Heating of Sheet Metal

The initial temperature of the sheet is taken as 20 °C. In temperature–displacement coupled dynamic model, the temperature boundary condition is applied at 100, 150, and 250 °C. The sheet is heated in a very short time interval by Joule effect and considered the uniform temperature in the whole sheet. Once the sheet is heated up to decent temperature, then the dynamic explicit model is used to deform the sheet into desired 3D geometry. For the analysis purpose, the 36 elements shown in the figure are taken into consideration. The stress analysis and effect of temperate on this element will be studied. The entire sheet is divided into two half—the middle elements are taken into consideration. The main focus is to obtain the stress analysis of these 36 elements. Also, analysed the effect of the temperature on the plasticity and thickness distribution of these 36 elements (Fig. 4).

2.7 Objective

The objective of this paper is to analyze the effect of temperature on the formability of several magnesium alloys. The magnesium alloys AZ61A, AZ31B, and AZ91 are mainly used in automotive, aerospace, and structural applications. Because of having good mechanical and other physical properties such as relative low density

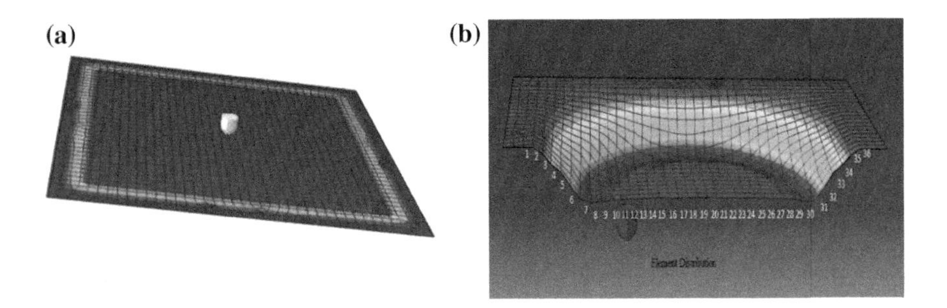

Fig. 4 **a** Heating of sheet metal and **b** element taken for consideration

and high strength-to-weight ratio, magnesium alloy is mainly used in automotive, aerospace, etc. Therefore, these materials have been considered for this study. It has poor deformability of room-temperature ductility but a good forming property when the temperature is around 250 °C. Therefore, the processing technique of warm forming is mostly adopted in magnesium alloy sheet forming at present.

3 Results

3.1 Average Von Mises

The results of the numerical analysis of magnesium alloy AZ31B are successfully obtained. The average von Mises stresses are calculated for different temperature, and the average von Mises stresses for 36 elements are obtained at different temperature. The above graph shows the variation of stresses with respect to the increase in temperature. It shows that stresses are decreasing as the temperature increases. The analysis is done for 36 elements which were taken into consideration. The average von Mises stresses for all three temperature are evaluated which is clearly shown in Fig. 5.

3.2 Variation of the Thickness of the Sheet

The thickness variation of the sheet metal for the 36 elements at different temperature is obtained successfully, and it can be easily concluded from the figure that variation of the thickness of AZ31B alloy sheet metal parts at specified temperatures is almost same and there is no such variation of sheet thickness with respect to temperature (Fig. 6).

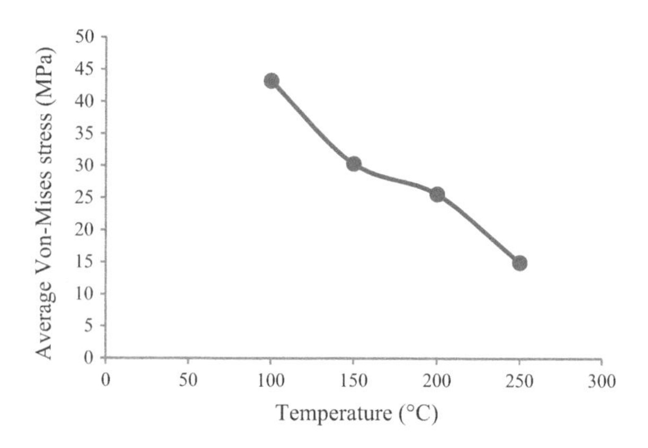

Fig. 5 Average von Mises temperature at different forming temperature

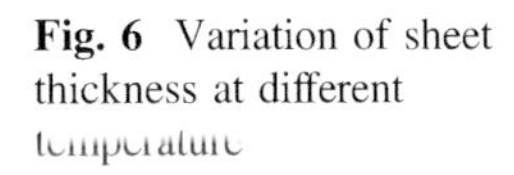

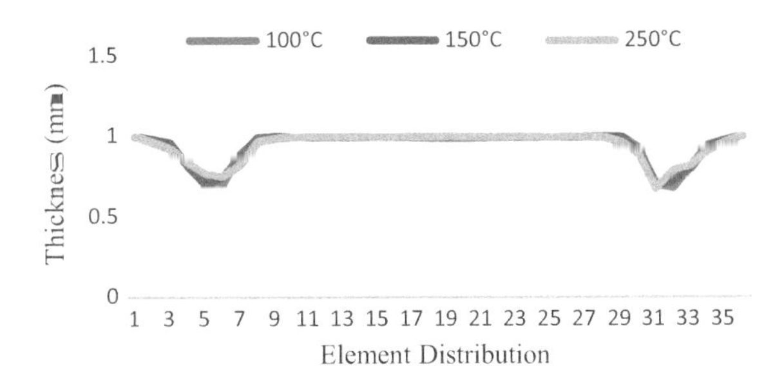

Fig. 6 Variation of sheet thickness at different temperature

3.3 *Equivalent Plastic Strain*

Magnesium alloy has relatively low plasticity and poor formability at room temperature. But as the temperature increases, non-basic slip systems will be activated along with dynamic recrystallization that causes ductility transition of the forming property of magnesium alloy and makes the plasticity of magnesium alloy stronger. As the temperature increases, the yield stress of the magnesium alloy decreases which led to an increase in the plasticity of the material (Fig. 7).

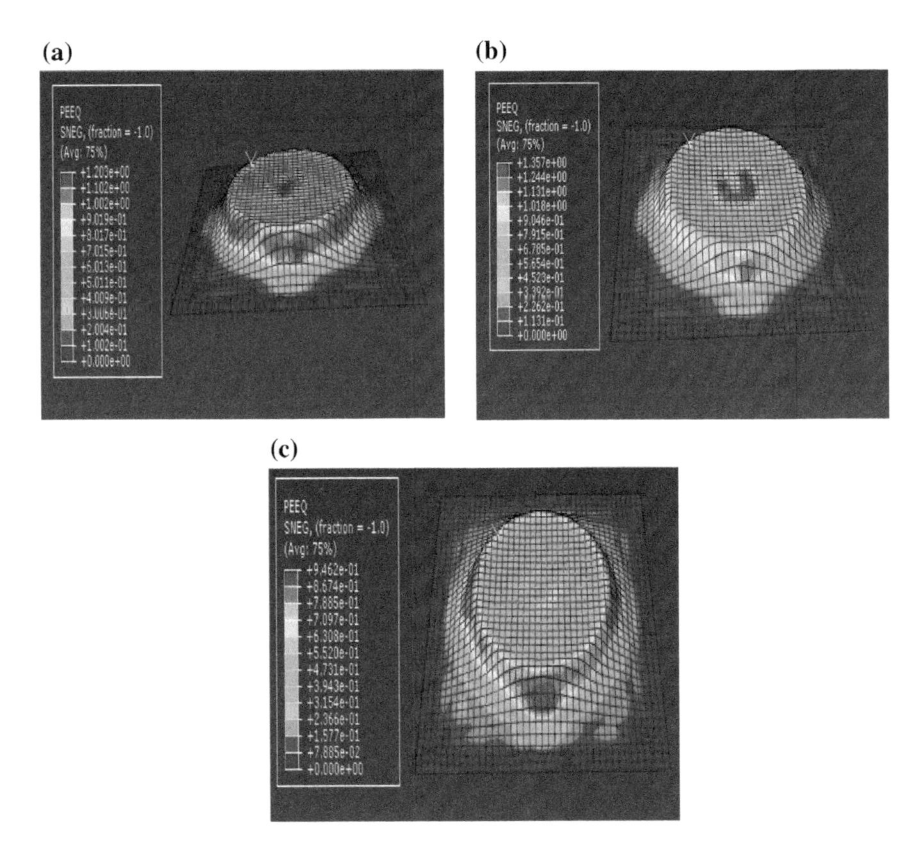

Fig. 7 Equivalent plastic strain at different temperature **a** 100 °C, **b** 150 °C, and **c** 250 °C

4 Conclusions

From the numerical simulation, it is observed that temperature has a remarkable impact on von Mises stress, which decreases as the increase of temperature which further increases the forming limits of the AZ31B magnesium alloy.

For a variety of sheet thickness, it can be known from the figure that thickness variation of AZ31B alloy sheet parts at different specified temperatures is almost the same.

So it is concluded that the temperature has not made any impact on the sheet thickness. For the equivalent plastic strain, yet as the temperature increases, non-basic slip systems can be activated along with dynamic recrystallization that causes ductility transition of the forming property of magnesium alloy and makes the plasticity of magnesium alloy stronger which further led to increase in formability of the magnesium alloy.

References

1. Ji YH, Park JJ (2008) Formability of magnesium AZ31 sheet in the incremental forming at warm temperature. J Mater Process Technol 201(1):354–358
2. Ambrogio G, Bruschi S, Ghiotti A (2009) Formability of AZ31 magnesium alloy in warm incremental forming process. Int J Mater Form 2(1):5–8
3. Zhang GX (2015) Development of heating equipment for the heat-incremental forming of magnesium alloy sheets. Manuf Technol Mach Tool 8:158–160
4. Ambrogio G, Filice L, Gagliardi F (2012) Formability of lightweight alloys by hot incremental sheet forming 34:501–508
5. Ambrogio G, Filice L, Manco G (2008) Warm incremental forming of magnesium alloy AZ31. CIRP Ann-Manuf Technol 57:257–260

Implicit Knowledge-Oriented New Product Development Based on Online Review

Huiliang Zhao, Zhenghong Liu and Jian Lyu

Abstract As one of the most important stages of product design, the early stage of new product development is knowledge-intensive creative work, whose essence is the evolution of knowledge. There is a lot of complex tacit knowledge in the early stage of new product development. Therefore, organizing and applying this knowledge is the key to the success of product conceptual design and even the whole product design. It is also the embodiment of the user-centered design concept. This paper presents a method of implicit knowledge-oriented new product development based on online review. The method of user requirement acquisition and product feature characterization based on online review is studied. This method can provide accurate user requirement analysis for the early stage of new product development, providing reference and support for product positioning.

Keywords Implicit knowledge · New product development · Online review

1 Introduction

With the increasing diversification and complexity of product demand, the shortening of the product life cycle, and the acceleration of product innovation, enterprises are urgently required to respond to market demand with the fastest speed, the lowest cost, the shortest time, and the best quality in the development of new products. Modern product design is user-centered design, focusing on user perception, experience, emotions, preferences, and other tacit knowledge. The process

H. Zhao (✉)
School of Fine Arts, Guizhou Minzu University, Guiyang 550025, China
e-mail: fightingzhl@163.com

Z. Liu
School of Mechanical Engineering, Guiyang University, Guiyang 550005, China

J. Lyu
Key Laboratory of Advanced Manufacturing Technology, Ministry of Education, Guizhou University, Guiyang 550025, China

BBVL. Deepak et al. (eds.), *Innovative Product Design and Intelligent Manufacturing Systems*, Lecture Notes in Mechanical Engineering,
https://doi.org/10.1007/978-981-15-2696-1_81

of new product development is also the knowledge management process of new product development. User requirements treated as knowledge in this paper play an important role in the product design process and perform the function of connecting the designer and user. Therefore, user requirement acquisition is the premise of product design [1]. In general, user requirements can be divided into explicit requirements and implicit requirements [2]. The explicit user requirements can be easily illustrated and acquired by the designer. However, because implicit user requirements consist of fuzzy intentions, the designer cannot acquire the requirements directly. In addition, the users themselves cannot describe their requirements in a crystal-clear fashion. Normally, implicit requirements reflect user potential requirements [3], and they affect users' perceptions and satisfaction with the product. Therefore, this paper presents a method of implicit knowledge-oriented new product development based on online review.

2 Related Studies

Michael Polanyi, a British philosopher and physicists, first proposed the concept of implicit knowledge in 1958. According to his point of view, implicit knowledge refers to the knowledge that exists in the individual's mind in a specific environment and is difficult to standardize and communicate. This knowledge is mainly generated by the accumulation of people's knowledge and experience of things and is the key content of knowledge innovation. Researchers have processed explicit requirements by constructing user requirement acquisition models with different methods [4, 5]. Osgood proposed semantic differential (SD) [6] which divides users' perception of product semantics (shape, color, etc.) into different intervals through opposite adjective pairs and reflects them on Likert scale. Then, the user data obtained are processed by mathematical statistics method, and its regularity is analyzed. Now, the SD method has become the basic theory of emotional design, user-centered design, perceptual design, and other research.

A plethora of useful information about product online reviews contains can provide user requirements and preferences to the designers [7]. Online review research was systematically reviewed in the previous studies. Studies revealed that the effects of quantitative features of online reviews and the effects of product's attributes on consumers' opinions about products or consumers' perceptions and behaviors have received the most attention from scholars [8]. Several studies performed from the perspective of affective experience examine the effects of online reviews. Wang et al. examined the relationship between preference scores and affective vocabularies based on user reviews on an online hotel information site [9]. Decker and Trusov developed a methodology for estimating overall product preferences based on online review data by improving the conjoint analysis used to understand how people evaluate various attributes of a product or service [10].

Studies have focused on online review extraction and clustering, using helpful online reviews to identify user requirements to support product design and optimization [11]. These studies show that online reviews can be used to understand users' requirements.

3 Research Methods and Implementation Procedures

This study proposed an implicit knowledge-oriented new product development based on an online review, shown in Fig. 1. The approach integrates the knowledge engineering approach and text mining techniques. Following the procedure of new product development mentioned in [12], phase 1 is to choose the design domain considering target product and user. Phase 2 is to analyze the product features and user requirement and span the semantic space of implicit knowledge (i.e., the collection of users' implicit requirement words which are used to describe the feelings for the product as design knowledge) and property space (i.e., the collection of product features) of the target product. Phase 3 aims to build the relationship of semantic space and property space (i.e., users' Kansei reaction to service or product elements), which is done by (1) synthesizing the users' semantic space and product's property space, (2) testing the validity of semantic space and property space, and (3) building models for the relationship analysis, named as design mapping. Text mining techniques are applied to analyze the user and provider-generated online reviews regarding the target product. The three phases of the proposed implicit knowledge-oriented new product development based on online review are detailed as follows.

3.1 Phase 1: Design Domain

The new product development is a series of orderly, organized, and targeted design activities from analyzing users' needs to generate conceptual products. It is an evolutionary process from rough to fine, from vague to clear, and from abstract to concrete. The main tasks of this process are to identify customer needs, business plans and marketing strategies, product and process benchmarks, product and process preliminary assumptions, product reliability analysis, research, and customer input. To determine the design objectives, and to determine the design objectives, reliability and quality objectives, initial material list, initial process flow chart, product/process special characteristics list, product assurance plan, and ultimately need the support of managers. Generally, the first stage of new product development is to choose the domain of design.

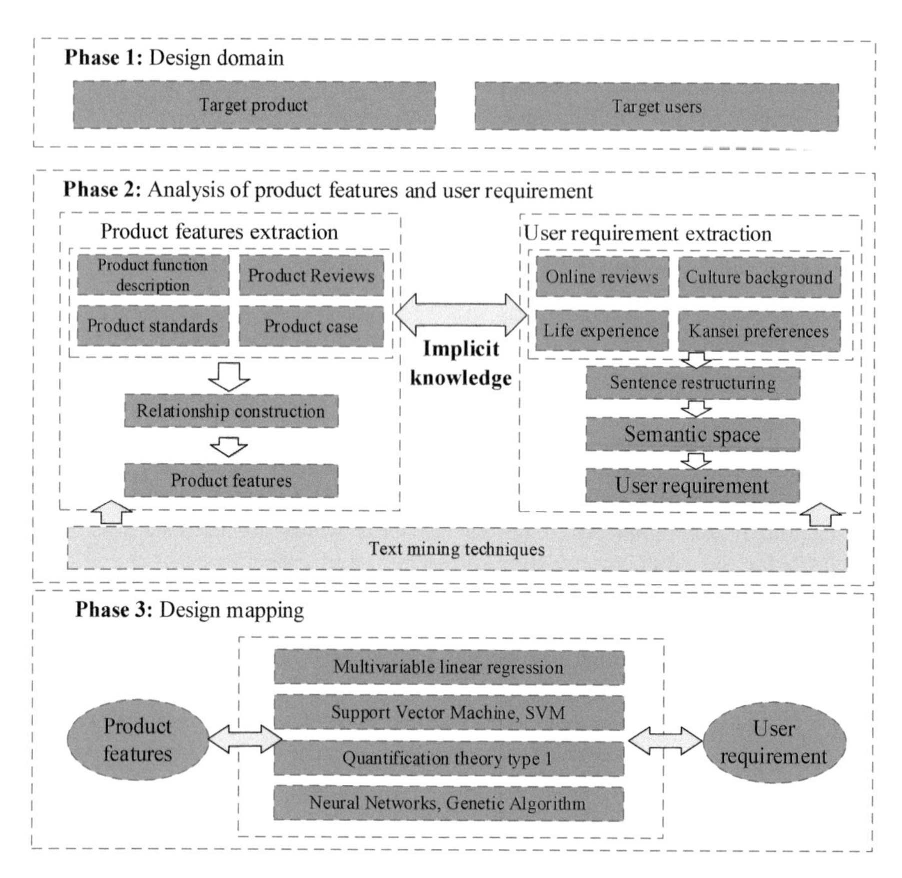

Fig. 1 Structure design for implicit knowledge-oriented new product development based on online review

3.2 Phase 2: Analysis of Product Features and User Requirements

By using text mining technology, on the basis of product function parameter description, product review, product standard, and product case base, product intention scale analysis is carried out, and relationship reconstruction property space is established to obtain product features. User needs are the user's demand for products, including the basic functional requirements of products, user's browsing, purchasing, and using of products. There is a lot of implicit knowledge about the purchasing behavior and emotional needs of users. Whether this implicit knowledge can be effectively excavated and accurately acquired is the decisive factor of whether the new product can meet the needs of users, whether it meets the market, and whether the design will succeed or fail.

3.3 Phase 3: Design Mapping

It is to build the relationship model between semantic space and property space. The importance of knowledge tools to transform user implicit requirements into explicit design elements has been highlighted; however, in view of the complex cognitive process of customer implicit response, capturing customer implicit requirements and transforming them into corresponding product design features are often challenging. Many studies have been devoted to the analysis of the relationship between customer implicit requirements and product design feature elements in order to provide support for design. In addition, the previous studies mostly used morphological analysis to deconstruct products, so as to obtain independent design features but lack of consideration of the correlation between product features. The relative position, combination, layout, size ratio, color matching, and other relationships among the design features of a complete product convey emotional information to users in a holistic form.

4 Implicit Knowledge Acquisition for the Product Features and User Requirements

The goal of modern product modeling design is not only to meet the basic functional requirements of products but also to meet the implicit needs of users in aesthetic, emotional, value recognition, experience, and other perceptual aspects. The solution of product modeling design problem is the solution of the typical ill-conditioned problem. The initial condition (the demand of modeling design), the objective requirement (the effect of modeling design), and the solution method (the rule of modeling design) of the problem are not clear. Therefore, there is no specific rule to follow in product modeling design problem but only according to the knowledge of users' needs and designers. Knowledge, experience, inspiration, skills and other implicit knowledge need to be solved. According to the research content of this paper, in order to effectively acquire implicit knowledge, implicit knowledge sources are divided into three categories: user needs and perception, designers' design experience, and product design value.

For mass consumer products purchased through shopping Web sites, people have left a lot of subjective feelings in the process of consumption, that is, commodity reviews. This massive textual information implies people's emotional attitudes toward products or events. It has the characteristics of complex content sources, diverse forms, and serious fragmentation, namely implicit knowledge. Aiming at this kind of product, this paper proposes to mine and acquire them based on emotional analysis technology.

4.1 Implicit Knowledge Acquisition for User Needs

Sentiment analysis can also be called opinion mining or comment mining. It mainly extracts and analyzes the subjective comments of users, so as to identify the opinions, positions, and attitudes that users or commentators want to express. It uses data mining and text mining technology to automatically extract the effective information of text based on vocabulary, sentence, and grammar, deeply analyze text semantics, and identify the emotional tendency of text. The implicit knowledge acquisition method based on the emotional analysis in this paper includes the following five processes (see Fig. 2). For access to comment data on the Internet, it is the most direct way to obtain from the internal database of relevant Web sites. However, due to the restrictions of confidentiality and technology of relevant companies, Web crawler technology is generally used to obtain Web page data.

4.2 Online Reviews Based on Product Feature Mining

Product features refer to the product attributes that users pay attention to in the comments. There are significant differences between commodity characteristics and implicit ones. Explicit features refer to nouns or noun phrases that directly describe the attributes of commodities, while implicit features need to be understood according to context and semantics. The commodity features extracted in this paper refer to the dominant features.

The more users pay attention to the characteristics of commodities, the higher the probability and frequency of commodities. Therefore, firstly, the frequent pattern mining method based on association rule algorithm is used to extract commodity features, and then, the omitted feature objects are selected to improve the recall rate. Secondly, the initial feature sets are screened and filtered to improve the accuracy rate. The detailed process of mining product features is shown in Fig. 3.

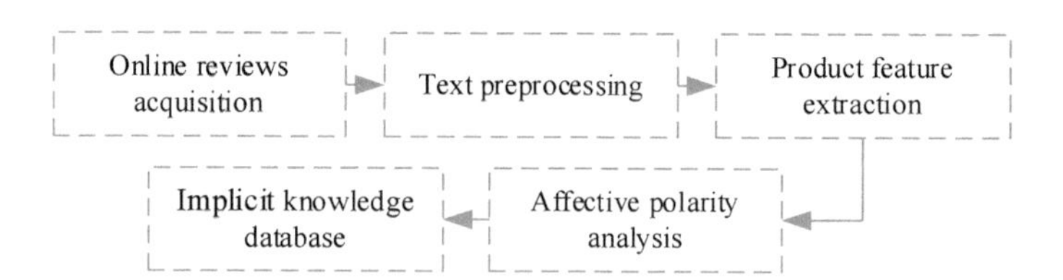

Fig. 2 Flowchart of implicit knowledge acquisition method based on the emotional analysis

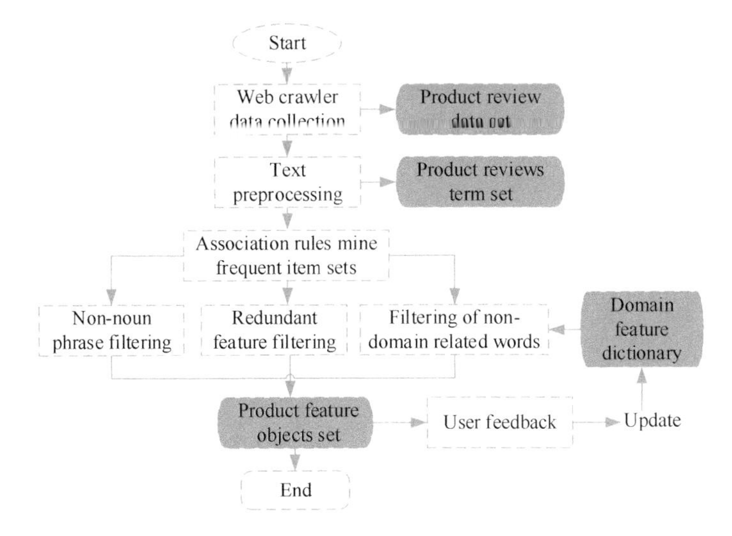

Fig. 3 Product feature mining process

5 Conclusion

Online reviews, as users' personal experiences or feelings for related products or services, have become an important source of knowledge to obtain users' needs. The uneven quality of reviews seriously interferes with the accuracy and credibility of demand mining. Identifying useful online reviews helps to accurately analyze users' explicit and implicit needs. This paper proposes an implicit knowledge-oriented new product development based on online review. This method integrates the knowledge engineering and text mining techniques, studies the design domain considering target product and user, analyzing the product features and user requirements, span the semantic space of implicit knowledge, and property space of target product based on online review. This method can provide reference support for accurate user demand analysis and product positioning in the early stage of new product development. But, the effectiveness of the technique is doubtful until it is implemented to certain products, and it will be done in the future work.

Acknowledgements Project supported by the Natural Science Foundation of the Guizhou Higher Education Institutions of China (Grant No. [2018]152, No. [2017]239). Project supported by the Humanity and Social Science Foundation of the Guizhou Higher Education Institutions of China (Grant No. 2018qn46). Technology projects of Guizhou province (LH [2016]7467, [2017]1046, [2017]2016, [2018]1049, [2016]12, YJSCXJH (2018) 088).

References

1. Guo Q, Xue C, Yu M, Shen Z (2019) A new user implicit requirements process method oriented to product design. J Comput Inf Sci Eng 19:011010
2. Münte TF, Brack M, Grootheer O, Wieringa BM, Matzke M, Johannes S (1997) Event-related brain potentials to unfamiliar faces in explicit and implicit memory tasks. Neurosci Res 28:223–233
3. Yongtai L, Zhengying L (2006) Analysis on demand and definition of implicit demand. Nankai Bus Rev 3:22–27
4. Polanyi M (2012) Personal knowledge. Routledge, Abingdon
5. Borgianni Y, Rotini F (2015) Towards the fine-tuning of a predictive Kano model for supporting product and service design. Total Qual Manage Bus Excellence 26:263–283
6. Osgood CE (2010) Semantic differential technique in the comparative study of cultures. Am Anthropol 66:171–200
7. Jian J, Ping J, Rui G (2016) Identifying comparative customer requirements from product online reviews for competitor analysis. Eng Appl Artif Intell 49:61–73
8. Mauri AG, Minazzi R (2013) Web reviews influence on expectations and purchasing intentions of hotel potential customers. Int J Hospitality Manage 34:99–107
9. Wang H, Yue L, Zhai C Latent aspect rating analysis on review text data: a rating regression approach. ACM SIGKDD Int Conf Knowl Disc Data Min
10. Decker R, Trusov M (2010) Estimating aggregate consumer preferences from online product reviews. Int J Res Mark 27:0–307
11. Zhang H, Rao H, Feng J (2018) Product innovation based on online review data mining: a case study of Huawei phones. Electron Commer Res 18:3–22
12. Hsiao YH, Chen MC, Liao WC (2017) Logistics service design for cross-border e-commerce using Kansei engineering with text-mining-based online content analysis. Telematics Inform 34:S0736585316303136

Mechanical Characterization and Microstructural Study of Carbon Steel Welded Joint Made Under SMAW and GMAW Processes

Pradipta Kumar Rout and Pankaj C. Jena

Abstract With the increasing demand for safety, emission reduction, production efficiency, cost-effectiveness and quality of the product, it becomes necessary to think about most suitable welding process among existing techniques for the better output to weld a specific kind of job. Increasing demand to weld different high alloys and dissimilar materials becomes a challenge for the fabrication industry. In fusion welding alone, several techniques are there. SMAW, GMAW, GTAW, SAW, Oxy-Acetylene welding and Resistance welding are coming under this. All these techniques have a unique identity. Hence, it is important to find their character with application suitability. In this paper, attempt has been made to characterize two different most popular welding processes, i.e., SMAW and GMAW. Using these processes, the mechanical characterization and microstructure grain distribution of carbon steel welded joint has been studied. Both the results are compared to understand their suitability and industrial applications.

Keywords SMAW · GMAW · Welded joint · Heat-affected zone · Microstructure · Comparison

1 Introduction

Shielded metal arc welding (SMAW) process has been used in all fabrication industries for many decades. More pollution due to the burning of flux is a major threat to the environment as well as to the operator. The operator has to change electrode rod frequently for a long run welding work. Thus, the process consumes more times and lowers the efficiency of production. More welding skill is required for the process to control the welding speed and arc length in manual metal arc

P. K. Rout
Kalinga Institute of Industrial Technology, Deemed to be University, Bhubaneswar, Odisha 751024, India

P. C. Jena (✉)
Veer Surendra Sai University of Technology, Burla, Odisha 768018, India

BBVL. Deepak et al. (eds.), *Innovative Product Design and Intelligent Manufacturing Systems*, Lecture Notes in Mechanical Engineering,
https://doi.org/10.1007/978-981-15-2696-1_82

welding (MMAW) process, which affects the welding quality significantly. In the other hand, another familiar welding process to the industry is gas metal arc welding (GMAW) process. GMAW process needs additional set up such as gas cylinder, wire spool, wire feed mechanism as compare to the SMAW process. Again to move the GMAW equipment is a little bit tough because of the high-pressure cylinder. But high deposition rate, lesser spatter, low hydrogen deposits and low-cost consumables make this process more suitable. GMAW can weld almost all metals in all position. In the SMAW process, welding current has a direct relation with electrode diameter, flux type and thickness of base metal. Current (I) remains stable during a particular pass, but voltage varies as it is a function of the arc length [1]. The heat generated by the arc in the arc welding process is

$$\mathbf{W} = V \times A \times T \quad (1)$$

where **W** in joule, A in ampere and T in second. Hence, more or less heat affects directly to the quality of the weld. Another aspect of welding is to know in details about the chemical composition base metals and their weld ability properties. Steels with carbon content less than 0.3% are reasonably easy to weld, while steels with over 0.5% carbon are difficult. Carbon concentration on weld vicinity affects largely to weld mate in terms of hardness. Other elements in steel like manganese, molybdenum, chromium, nickel, silicon and vanadium have a lesser effect than carbon. Carbon equivalent is directly proportional to hardenability. Higher carbon equivalent metal is more susceptible to hydrogen cracking. To overcome such problem, low hydrogen filler metals should be used, and preheat to base metal is essential. Preheat temperature is determined according to the carbon equivalent of the base metal. The equation of the International Institute of Welding (IIW) is widely used to calculate the carbon equivalent (CE) of low carbon steel is given in Eq. (2).

$$(\mathrm{CE}) = \left[\%\mathrm{C} + \frac{\%\mathrm{Mn}}{6} + \frac{(\%\mathrm{Ni} + \%\mathrm{Cu})}{5} + \frac{(\%\mathrm{Cr} + \%\mathrm{Mo} + \%\mathrm{V})}{15}\right] \quad (2)$$

In SMA welding process, current, voltage and welding speed are inversely proportional to the depth of penetration [2]. If current and arc voltage increase, then amount deposition increases. But if travel speed increases, weld deposition decreases substantially. Suitable parameters to weld mild steel in multi-pass welding are current (I) 90 A, voltage (V) 24 V and welding speed (S) 40 mm/min [3]. The properties and microstructure of HAZ depend on the rate of heat input and rate of cooling. The microstructure of HAZ is completely different than the base metal. Large grains of ferrite, Widmanstattern ferrite and colonies of perlite with a maximum hardness, compared to weld zone and base metal zone, are found at HAZ [4]. The maximum hardness is found at the weld zone and base metal because of low heat input [2, 3]. Toughness is another ability of a material to withstand the applied load. In low heat input, the value is more than high heat input [2]. Impact value is more in high heat input as compare to low heat input [5]. The material welded by low heat input is more ductile than high heat input welded joint material [5]. In GMA welding process, voltage, current, wire speed rate and gas flow rate are the different functions to achieve a better weld. Thus, it is necessary to optimize all these necessary parameters. Many authors had already worked in this direction. To achieve maximum transverse and longitudinal yield strength, optimum values are given in Table 1 [6]. Another analysis states that root gap plays an important role, which affects the tensile strength of the welded joints with a maximum contribution of 38% followed by welding current 32% and arc voltage with 14% by GMAW process [7]. Arc voltage is another significant factor, which affects the hardness value of weld. Hardness contributes 62% followed by root gap 28%, and welding current has a minimum effect of 6% to weldment.

Again this investigation reveals that the microstructure of weld metal has ferrite and perlite formation after cooling and has no martensite formation even at HAZ. Thus, the mild steel weldment has very lower hardness [7]. Increase in current slightly increases the hardness of mild steel, so ultimate tensile strength increases significantly. Further increase in voltage increases penetration [8]. If the arc voltage is less than 24 V, then the weldment is more prone to porosity, and if the voltage is greater than 32 V, then porosity, spatter and undercuts are observed in the GMAW process. Further, if the shielding gas flow rate is less than 10 lpm blow holes and porosities are measure problems and if it is higher than 18 lpm, then the gas entrapment is observed. Welding speed less than 160 mm/min has shallower penetration and wider weld bead. On the other hand, if the weld speeds are more than 220 mm/min that leads to incomplete fusion and low material deposition rate. Wire feed rate of less than 4.5 m/min causes less penetration and incomplete fusion. Wire feed rate more than 9.5 min/min causes bad-shaped weld bead and spatter. Optimized parameters to weld 6 mm thickness IS 2062 mild steel plate are mentioned in Table 2 [9].

Table 1 Maximum transverse and longitudinal yield strength, optimum values are given [3]

Yield test	Yield strength (MPa)	Voltage (V)	Current (A)	Wire speed (m/min)	Gas flow rate (l/min)
Transverse	374.571	22.5	190	2.4	12
Longitudinal	398.907	22.5	210	2.4	12

Table 2 Optimized parameters to weld 6 mm thickness IS 2062 mild steel plate

Wire feed rate	4.5 m/min
Voltage	32 V
Gas flow rate	10 lpm
Travel speed	160 mm/min
Bead width	7.53 mm
Bead height	0.891 mm
Depth of penetration	4.02 mm

2 Experimental Study

Twelve pairs of IS: 2062 mild steel plates of 200 mm × 50 mm × 6 mm dimensions are taken as the base metal for the experiment. Plates are machined by the milling machine and made free from rust and dirt. Chemical composition of base metal is checked by spectroscope is given in Table 3. Chipping hammer, tong, leather gloves, leather apron and welding screen are made ready for the welding. Then, 'V' grooves of included angle 60° are prepared by a hand grinder as shown in Fig. 1a. Then, samples were wrapped with thick clean white paper to avoid oxidation and from other contamination. E6013 and ER7066 filler material are chosen for SMAW and GMAW, respectively.

Back plates are prepared for the welding works to avoid any possible distortion. First EWAC makes model ZuperI SMAW machine with DCRP is selected for welding. Electrode specification E6013 of 2.25 mm dia. is used for the root run and 3.15 mm dia. for subsequent runs. Electrodes were baked for 30 min at 120 °C and kept aside to the welding booth. Back plates are tacked maintaining root gap of 2 mm. Root run is completed with current 90 A, arc voltage 24 V. Then, another hot pass made with current 110 A. Slag of each pass is cleaned thoroughly by a chipping hammer, and welded plates are allowed to be cooled in room temperature. Then, EWAC makes Zuper ARC 400 GMAW machine is used for welding with copper-coated filler wire of dia 1.2 mm. The machine, which is used for welding, has auto configuration. As current increases, then wire feed rate increases. The current was set at 170 A. 100% CO_2 shielding gas is used to ensure good penetration. Arc voltage is found during welding is 32 V. Further, the welded plates are cooled in room temperature. A few spatters are seen in shielded metal arc welded plates (Fig. 1b), while no spatters found in gas metal arc welded plates (Fig. 1c). DP Test was carried out for all welds. A surface discontinuity was detected in SMA welded plate and that is not included for further study and testing. Then, magnetic particle test is carried out, and no flaws detected in rest five pair welded plates.

Further samples are prepared from the welded plates to check their mechanical properties such as tensile, bending, impact and hardness as per ASTM standard. Microstructures of different zones are studied. Samples (Test specimens) shown in Fig. 2 an are cut from the weld as per the ASTM E8M standard. Tinius Olsen makes of capacity 50KN Universal Testing Machine is used for tensile test. The test is performed at a strain rate of 0.5 mm per minute, and the elongation in gauge length is measured by the strain gauge. During the experiment, the yield stress, UTS values are observed. The stress beyond UTS necking is seen and finally narrowed down and got fracture (Fig. 2a, b). SMAW specimen failed at weld area, whereas GMAW specimen failed at the base metal area. Results of both specimens are given in (Table 4).

Table 3 Chemical composition of base metal

C%	Mn%	Si%	P%	S%	Cu	Ni	Cr	V	Mo	CE
0.227	1.478	0.495	0.038	0.045	0.022	0.015	0.025	0.006	–	0.48

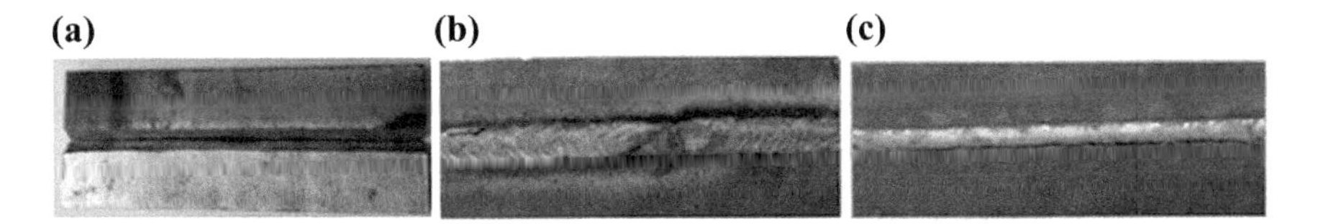

Fig. 1 **a** Base metal with V-groove before welding, **b** SMAW welded plate, **c** GMAW welded plate

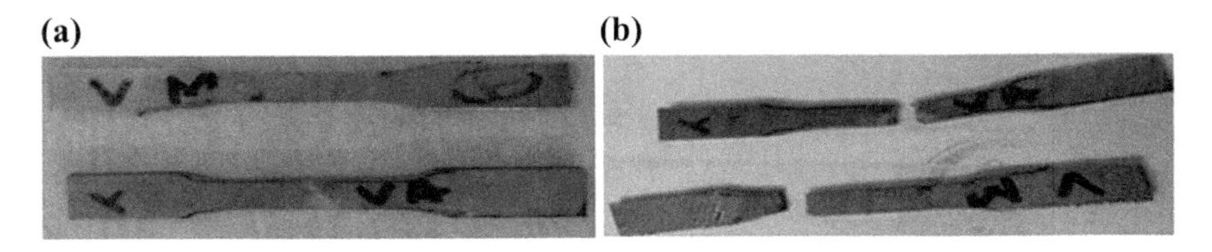

Fig. 2 Tensile test specimens **a** before testing, **b** after testing

Table 4 Comparisons of welded metals: yield strength, ultimate tensile strength and % of elongation

Welding process	Yield strength (MPa)	UTS (MPa)	% of elongation
SMAW	232	365	20
GMAW	276	418	22

Samples are prepared as per ASTM E290-14 standard method. Same UTM machine is used for bend test using bending fixture. Samples for root bend test were kept one by one between two rollers fixed in a distance of 50 mm apart, and the central mandrel acted on the center of the weld with a strain rate of 0.5 mm per minute. A LVDT facing the backside of the specimen measures the defection exhibited by that weld under load and the strain rate. The load versus deflection plot was obtained and kept for further discussion. Bending test and their results are shown in Fig. 3a–c.

The impact test is associated with the toughness of the material, where the potential energy of pendulum converted to kinetic energy and breaks the specimen. The notch of the specimen plays a critical role in determining the energy required to break the material. ‘V’ notches of 60° angles with 2 mm depth were cut at the middle portion of the samples for Charpy test, and results are given in Table 5.

Specimens are cut at the middle of the joints in transverse direction and polished. Brinell hardness is conducted for both welded specimens as per ASTM standard A370-14, and hardness measurements are taken on the weld zone, heat-affected zone and on parent metal. Though it is difficult to identify the HAZ, specimens are etched with a nital solution and observed through an optical microscope, and HAZ areas are marked carefully (Table 6).

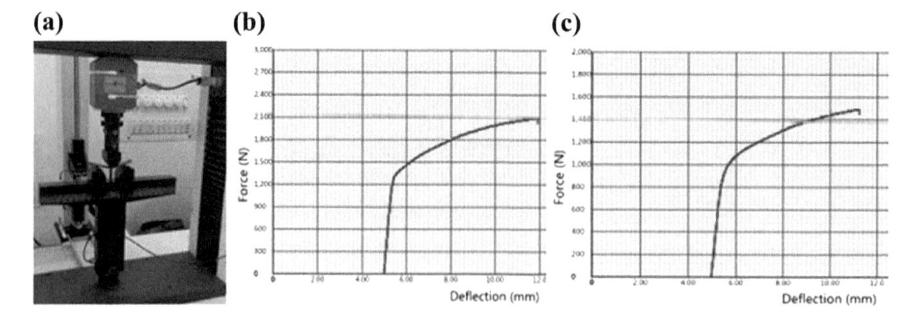

Fig. 3 **a** Three-point bending test, **b** deflectometer reading for GMAW, **c** deflectometer reading for SMAW

Table 5 Comparisons of welded metals: potential energy using Charpy test

Specimen	Energy (J)
SMAW	34
GMAW	52

Table 6 Hardness value using Brinell hardness test

Specimen	Base metal (BHN)	HAZ (BHN)	Weld zone (BHN)
SMAW	220	235	266
GMAW	233	239	280

2.1 Metallographic Study

Samples are cut from the weld on identified locations, rough grinded by a surface grinder and polished through different grades of SiC emery papers, i.e,220, 320, 400, 600, 800, 1000, 1200, 1500 and followed by diamond polishing of 3–4 microns sizes and with applying diamond aerosol as a lubricating medium. The mirror-like polished surface further etched with the nital solution (a solution of 98% HNO_3 and 2% methyl alcohol). The etched surfaces are analyzed on the optical microscope in 100 X and 200 X magnifications in order to identify the phases and the microstructural features. Microstructure of weld zone in SMAW process (Fig. 4e) reveals the presence of partly dendrite structure showing the elongated appearance of phases of ferrite and perlite. But microstructure of weld zone in GMAW (Fig. 4f) shows equiaxed grain size with the appearance of perlite and ferrite phase. HAZ of SMAW specimen (Fig. 4c) shows coarse grains than base metal and its weld zone, whereas GMAW specimen (Fig. 4d) has fine grains than the SMAW heat-affected zone. Now, it is clear from Fig. 4a–e SMAW processed weldment has bigger grain in size than GMAW processed weldment at weld zone and HAZ. Smaller or fine grain size has more yield strength as compare to coarse grains metal as the yield strength of the material is a function of its grain size [10] (Table 7).

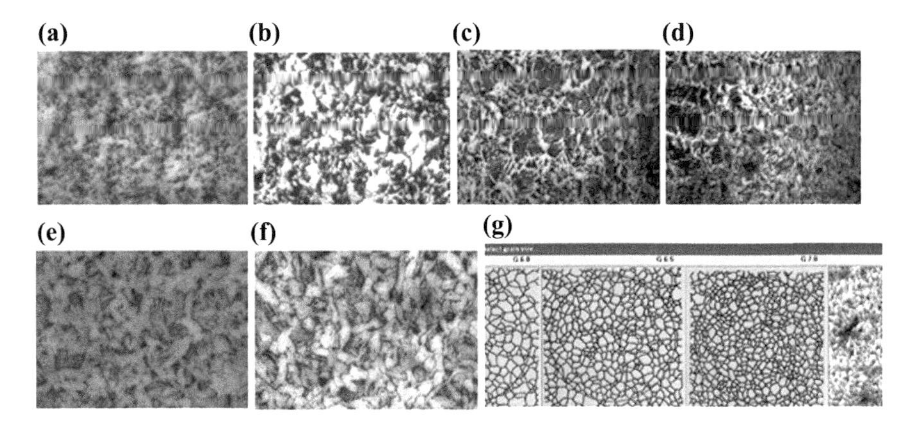

Fig. 4 Microstructure of **a** SMAW parent metal, **b** GMAW parent metal, **c** SMAW at HAZ, **d** GMAW at HAZ, **e** SMAW at weld zone, **f** GMAW at weld zone, **g** grain size of weld zone by SMAW and GMAW, respectively

Table 7 Comparisons of grain size at weld zone

Specimen	Base metal grain number (G)	Weld zone grain number (G)	HAZ grain number (G)
SMAW	7.5	5.5	4.5
GMAW		7.0	5.5

3 Results and Discussion

IS: 2062 plates were welded with optimized parameters by SMAW and GMAW, and experimental observations were recorded through different tests. NDT methods such as visual inspection die penetration test and the magnetic test were carried out. In NDT inspections, both processes found suitable without any appreciable discontinuities though one plate welded by SMAW process was rejected due to the presence of porosity. Spatters are seen on SMAW welded plates, whereas no spatters seen on GMAW welded plates. Thus, the later post welding works reduce significantly. Bead width and reinforcement of GMAW weld mate are found less than SMAW weld mate as the welding speed of the GMAW process is faster than the SMAW process. Production efficiency of the SMAW process is reasonably less than GMAW process. The results of tensile tests are shown in Table 4. It is clearly indicated that yield strength, ultimate tensile strength and % of elongation of GMAW welds have greater value than SMAW welds. As seen in Fig. 2b, figure failure occurred at welding area in SMA welded, whereas failure found at base metal in GMA welded specimen. The toughness characteristics of both welds under impact loading by using a Charpy test found greater in GMAW in comparison with SMAW under similar conditions such as pendulum size, weight and speed. Result of bending test exhibits good resisting force in both welded specimens with respect to their deflections. But GMAW specimen withstands greater bending force than SMAW specimen under similar conditions such as strain rate and velocity. Brinell hardness tests are carried out on both specimens. Greater hardness shows at weld zone as compare to another zone in both processes. Microstructures of all three zones in both welded specimens were compared. The area of HAZ of SMAW weld specimen found around 20% larger than the GMAW weldment which is amounting lesser strength in case of SMAW specimen. The microstructure of weld zones of GMAW weld specimen has perlite and ferrite phases showing the equiaxed distribution in comparison with the structural matrix present in the weld zone of SMAW weld. Apart from this, there is a remarkable difference in grain size in both welded specimens which are shown in Fig. 4a–f which utterly encounters the difference in terms of strength, hardness and bending. As weld zone is an auto cooled zone, so the grain size for both welds is smaller in size than another zone. More fine grains are seen at GMAW specimen, and thus, the hardness is more than SMAW specimen. There is no big difference in grain size at both HAZ. SMAW grains size at HAZ is a little bit more in size than the GMAW because of its slow welding speed and more heat input.

4 Conclusion

Mechanical properties (tensile, three-point bending, potential energy) of weldment by both SMAW process and GMAW process are examined using ASTM standard and tabulated. Microstructural study of base metal, HAZ and weld zone is studied, and the following conclusions are drawn.

- GMAW welds exhibit better yield strength because of their fine grain and faster welding speed.
- GMAW seems more appropriate technique because of less HAZ area, reasonable hardness value, greater toughness, etc.
- GMAW technique has better production efficiency and operational reliability.
- GMAW process is more echo-friendly as it produces lesser fumes and toxic gases with the absence of flux than SMAW process.

References

1. Cary HB, Scott CH (1979) Modern welding technology: 166–169
2. Asibeluo IS, Emifoniye E (2015) Effect of Arc welding current on the mechanical properties of A36 carbon steel weld joints. SSRG Int J Mech Eng SSRGIJME 2
3. Ranjan R (2014) Parametric optimization of shielded metal arc welding processes by using factorial design approach. Int J Sci Res Publ 4(9):01–04

4. Boumerzoug Z, Derfouf C, Baudin T (2010) Effect of welding on microstructure and mechanical properties of an industrial low carbon steel. Engineering 2(07):502
5. VijayeshRathi H (2015) Analyzing the effect of parameters on SMAW process
6. Hooda A, Dhingra A, Sharma S (2012) Optimization of MIG welding process parameters to predict maximum yield strength in AISI 1040. Int J Mech Eng Robot Res 1(3)
7. Kumar R, Kumar S (2014) Study of mechanical properties in mild steel using metal inert gas welding. Int J Res Eng Technol 3(4):751–756
8. Yadav PK, Abbas M, Patel S (2014) Analysis of heat affected zone of mild steel specimen developed due to MIG welding. Int J Mech Eng Robot Res 3(3):399
9. Radhakrishnan VM (2005) Welding technology and design. New Age Publishers
10. Lakhtin Y Engineering physical metallurgy. Mir Publisher, Muscow

Brain Tumor Segmentation Using Chi-Square Fuzzy C-Mean Clustering

G. Anand Kumar and P. V. Sridevi

Abstract Accurate brain tumor segmentation is an interesting and challenging task of magnetic resonance imaging (MRI) in the field of medical image processing. For this purpose, we propose a chi-square fuzzy c-mean-based segmentation via clustering to segment the abnormal tissues from the normal region. Initially, based on improved threshold and center-symmetric LBP, the preprocessing is performed to extract the region of interest. Then, we compare the preprocessing output and original MRI image using Bhattacharya similarity metrics to obtain the region of interest from the imaging technology. Finally, chi square distance-based fuzzy c-mean (CS-FCM) segmentation is performed to cluster the region according to the feature based on the region of interest (ROI), including entropy, contrast, and mean for necrosis, edema, and enhanced tumor regions. BRATS 2015 dataset is used to evaluate the performance in terms of Jaccard matching, specificity, positive predictive value (PPV), and dice similarity coefficient (DSC). The existing approaches are not efficient and predictive, whereas our proposed method performs better in clustering the tumor into three regions (necrosis, edema, and enhanced tumor) based on the region of interest.

Keywords Image clustering · Brain tumor segmentation · Fuzzy algorithm · Threshold · Magnetic resonance imaging

1 Introduction

Tumor is defined as the abnormal, uncontrolled growth of tissue in the brain that affects the human health condition. Brain tumors can be detected and diagnosed with the help of the MRI brain images obtained by using the medical image technologies. Brain tumor could be curable when it is detected in the earlier stage

G. Anand Kumar (✉) · P. V. Sridevi
Department of Electronics and Communication Engineering, Andhra University College of Engineering (Autonomous), Visakhapatnam, India
e-mail: ganand@gvpce.ac.in

BBVL. Deepak et al. (eds.), *Innovative Product Design and Intelligent Manufacturing Systems*, Lecture Notes in Mechanical Engineering,
https://doi.org/10.1007/978-981-15-2696-1_83

else this is a serious threat to human life. Medical MRI imaging [1] is used by the physicians to find abnormalities in the human body, to diagnose and provide earlier treatment to them. Accurate segmentation of brain tumors has a great impact on computer-aided diagnosis and treatment planning. The most frequently used imaging technique is MRI, to segment and predict the tumor region which divides an MRI image into the component object or region. In the traditional approaches, physician, radiologist, and neurologist detect the abnormalities and diseases from the manually obtained information from the images. Nowadays, to improve the accuracy for detecting the brain diseases and abnormality, numerous automatic approaches have been developed. Normally, brain tumor image contains a large amount of information, so manual segmentation is a time-consuming and complex process. In order to overcome these complications, automatic segmentation methods were introduced to detect brain tumor such as region growing [2], thresholding [3], artificial neural network [4], and clustering [5], etc. But still, segmentation of brain tumor is still a challenging problem in image processing and analysis. The structure of the brain is complicated, so it is difficult to determine the accurate segmentation of necrosis, edema, and enhanced tumor.

Several tissues present in the brain consist of three normal tissue [6] regions namely gray matter (GM), white matter (WM), and Cerebrospinal Fluid (CSF), which is significant to analysis and treatment for diseases such as multiple sclerosis, Alzheimer's disease, and epilepsy. These three regions are identified by the segmentation of the brain image by utilizing the gray level distribution of pixels. The main goal of brain tumor segmentation [7] is to identify the extensive and location of the tumor region [8], such as edema, active tumorous tissue, and necrotic tissue. Mean-shift algorithm [9] was used to detect the brain tumor in MRI image. The most widely used automatic segmentation technique in bioinformatics application [10] is clustering.

Nowadays, clustering-based image segmentation on pixels is used in imaging technique, which organizes a given database into a group. The significant role of clustering in MRI image is generally used to detect the brain diseases and abnormalities, to monitor, diagnose, and treat disease. Several clustering techniques are used in the existing work to detect the abnormalities such as fuzzy k-mean clustering, adaptive fuzzy k-mean clustering, modified k-mean clustering, and fuzzy c-mean clustering. The goal of these clustering is to detect the abnormalities-based algorithm to minimize the objective function based on certain criteria.

2 Related Work

Zhao et al. [11] introduce a novel method for brain tumor segmentation, which integrates both conditional random fields and fully convolutional neural networks to produce the result with spatial consistency and good appearance. In this, train these network using image slices and 2D image patches, which consists of three steps for segmentation. First, image patches could be used to train fully convolutional neural networks. Then image slices were used to train the conditional random fields. At last, these two networks were fine-tuned using image slices from coronal, sagittal view, and axial view, respectively.

Ding et al. [12] presented a multimodal image segmentation based-stacked de-noising auto-encoders (SDAE), which extricate different tumor tissues from normal tissue from various modality images. BRATS dataset adopted bottom-hat, and top-hat transformation was used for preprocessing the patches to enhance the image contrast. Finally, the overshoot area could be eliminated using the threshold method on the segmented image. As a consequence, the performance of the SDAE method was degraded because the texture feature was not used in the modality image.

Shivhare et al. [13] introduced a fully automatic segmentation and detection-based parameter-free clustering. This technique consists of two main modules: parameter-free clustering and morphological operation. T1c modality of the image was taken as an input and cluster based on k-mean algorithm. Then, in the morphological process, clustered input was applied to dilation and hole filling operation to obtain the segmented tumor region. In this, performance evaluation can be achieved by using the BRATS 2015 challenge dataset.

Kumar and Sridevi [14] introduce a novel T-Spline intensity inhomogeneity correction and 3D deep learning for automatic brain MR tumor segmentation. It comprises four steps. Initially, preprocessing is applied which diminishes the bias field distortion, intensity variations, and noises. Secondly, feature extraction is performed by extended gray level co-occurrence matrix (GLCM) to extract the texture patches for segmentation. Then, automatic segmentation was adopted to divide the abnormal tissue from the raw data. In this, the performance metrics such as Jaccard index, specificity, PPV, dice similarity coefficient, sensitivity, and accuracy are evaluated by standard BRATS 2015 dataset for segmentation.

Bal et al. [15] introduced a fuzzy-possibilistic c-means (FPCM), which integrate both fuzzy C mean and possibilistic C mean were accurately segment the brain tumor in MRI images. Once the skull stripping is performed, FPCM was applied over the brain tissue and tumor region in the brain. Finally, the shape-based feature was used to segment the brain tumor from the original image.

Shen et al. [16] introduced three fully convolutional networks to segment the brain tumor using BRATS 2015 dataset. They follow three major steps, i.e., preprocessing, segmentation, and post-processing. In this, mean filter, median filter, and Gaussian filter were used to preprocess the multimodality MR images. Here, BRATS 2015 dataset is used to evaluate the performance matrices such as dice and sensitivity.

3 Center-Symmetric Local Binary Pattern and Chi-Square Fuzzy C-Mean Clustering

For segmentation via clustering, we introduce a novel-center-symmetric LBP and chi-square fuzzy c-mean in imaging technology. The proposed method process is classified into three stages as preprocessing, feature extraction, and segmentation. Initially, (a) preprocessing is performed by splitting the MRI image into similar size blocks, and tumor region is detected by means of improved threshold method, and then, (b) ROI extraction is performed based on center-symmetric LBP method. Finally, (c) chi-square FCM clustering method is used to segment the image based on edema, necrosis, and enhancing tumor regions.

3.1 Preprocessing

Initially, we read the brain tumor image from medical imaging technology. Here, we divide an image into similarly sized blocks, where m represents the number of block in an image and the pixel intensity values, where p is the number of pixel in an image. After partitioning, the tumorous block is segmented by means of the improved thresholding function.

Threshold-based segmentation: Threshold method is the simplest and important in the image segmentation process, which is very sensitive to noise. This method is based on the pixel with identical intensity value to detect the specified region. Then, the output of this approach obtains the binary image which is based on the given grayscale image. There are two types of thresholding methods normally used for segmentation, namely global and local. In global threshold we have some limitations when segmenting the object and thresholding that image, contrast is low and threshold selection is difficult. But, we use a local threshold to determine the specified threshold value based on the intensity value of the image. Determination of tumor block: Tumor from the block of the image is determined by using the threshold segmentation. The neighboring block has the same feature as the tumor block is extracted from the input MRI image. For this, the common LBP-based feature extraction method is used to extract the feature based on the adjacent block. Here, the local spatial relationship between the center pixel and neighboring pixels is given by this operator. In the local binary pattern, the image is divided into small same size cells. Each cell has eight neighboring pixels, and the comparison can be made by using the center pixel and the neighboring pixel. But, in CS-LBP, the pixel value is compared with the symmetrically opposing pixel with respect to the center pixel.

3.2 ROI Extraction

In this, the tumor block is acquired by using the CS-LBP, and the original image is compared to form the region of interest (ROI) as illustrated in Fig. 1. Next, we calculate the similarity between the original image and the tumorous block using Bhattacharya similarity metric, and it is determined as,

$$B(h(O), h(T)) = \sum h(O) * h(T) \tag{1}$$

where $h(O)$ be the original image and $h(T)$ be the tumorous block.

ROI clustering based on edema, necrosis, and enhancing tumor: Feature extraction transforms the input data into the more manageable features, which contains the information related to size, color, and texture. The feature can be extricated from the region of interest using numerous methods such as texture features, color features, spatial features, and transform features. In the MRI brain images, texture-based feature is the important characteristic in the identification of ROI. This proposed approach utilizes an enhanced GLCM technique to extricate the feature from the texture patches for every sequence. For segmentation purpose, GLCM feature technique extracts the particular feature and forms a matrix. This particular feature is used for segmentation of tissue structure. After ROI extraction, clustering is performed by CS-FCM method to cluster the MRI images into edema, enhanced tumor, and necrosis regions.

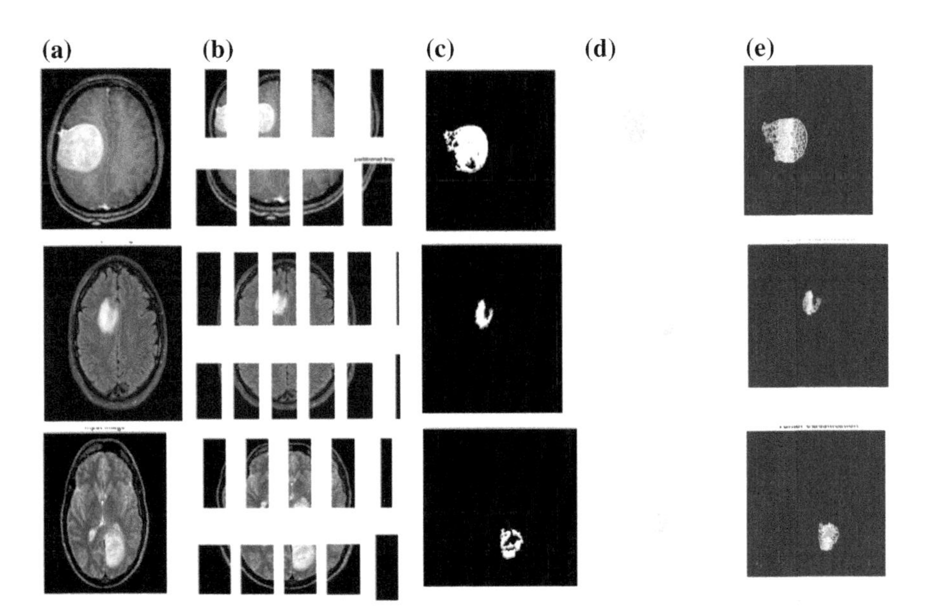

Fig. 1 Output result obtained with the three collected input MRI images obtained from the BRATS 2015 dataset **a** input image, **b** partitioned image, **c** detected tumor, **d** ROI extraction, and **e** classified tumor

3.3 Chi-Square-Based Fuzzy C-Mean Segmentation

FCM algorithm partitions the finite collection of data point-based pixel value into c fuzzy cluster with respect to the given criteria. Let $x = x_1, x_2, \ldots x_n$ be the set of data point from the image pixel to be clustered into c fuzzy cluster. Various fuzzy clustering methods have been developed to segment the image, and most of them are based on distance criteria. In this work, we introduce a chi-square FCM algorithm for the segmentation through clustering in the MRI image. Here, the execution time of chi-square distance is better than the Euclidean distance measurement, and the membership function of fuzzy logic value ranges from 0 to 1. There are no sudden changes which occur between nonmembership and membership function. The main goal is to minimize the objective function (J_{cf}) of CS-FCM algorithm. All the features are extracted from the image data to get the matrix occurrence. Some of the vital features for segmentation have been listed as follows:

Entropy: Necrosis region is segmented based on entropy feature. This provides the amount of information related to the original images required for image compression and segmentation, which calculate the region of interest (ROI) and dissimilarity in ROI and measure randomness. The accurate prediction of necrosis region is obtained by applying the entropy feature.

The entropy (C_E) is calculated as follows:

$$C_E = -\sum_{i=1}^{\text{max row}} \sum_{j=1}^{\text{max col}} R_I(i,j) \log R_I(i,j) \tag{2}$$

Contrast: The edema region is segmented based on a contrast feature. Contrast is a measure of pixel intensity and its neighbor over the image. Edema region is darker than the surrounding region; hence, we choose contrast feature for edema. It is defined as,

$$C_n = \sum_{i=1}^{\text{max row}} \sum_{j=1}^{\text{max col}} (i-j)^2 \sum_{i=1}^{\text{max row}} \sum_{j=1}^{\text{max col}} R_I(i,j) \tag{3}$$

Mean: It measures the average of the intensity value to determine the enhanced tumor region. This is calculated as follows;

$$\text{mean} = \sum_{i=1}^{\text{max row}} \sum_{j=1}^{\text{max col}} (i,j) \times R_{I_{i,j}} \tag{4}$$

This approaches to segment the tumor based on chi-square FCM algorithm with high accuracy and enhances the tumor segmentation in MRI images.

Table 1 Comparison between our proposed methods with other state-of-the-art method using BRATS 2015 dataset

Methods	DSC	PPV	Sensitivity (TPR)	Jaccard index
FCNN + CRF [17]	0.80	0.82	0.81	–
SDAE [18]	0.90	–	–	0.83
Parameter-free clustering [19]	0.75	–	–	–
3D deep learning [11]	0.89	0.92	0.94	0.80
CFCN + CRF [13]	0.86	–	0.91	–
Proposed	0.93	0.91	0.92	0.91

4 Experimental Result

The experiment can be performed on the basis of the image provided by the BRATS 2015 dataset on a computing server with Intel E5-2620 CPUs and multiple Tesla K80 GPU's. MATLAB R2017a is used to simulate this experiment. Here, we use MRI brain images for every patient under different modality, namely T1, T1c, T2, and FLAIR from BRATS benchmark dataset.

This dataset comprises 274 MRI images for training and 110 for testing. Analysis can be carried out using the testing data. Here, we partition the entire dataset into two groups, i.e., high-grade glioma images and low-grade glioma images. And then, the tumor is segmented into three parts, namely necrotic core, edema, and enhanced tumor. Performance metrics can be calculated using Jaccard matching, specificity, PPV, and dice similarity coefficient.

Table 1 shows the comparison of our proposed method with the existing methods. BRATS 2015 evaluation metrics for tumor segmentation using FCM clustering for necrotic core, edema, and enhanced tumor are Jaccard matching, specificity, PPV, and dice similarity coefficient.

Figure 1 shows the output result of the proposed technique obtained by the three sample input MRI images. Initially, partitioning is performed, and then, the tumor region is detected from the input MRI image by means of CS-LBP method. Finally, chi-square-based FCM clustering method is used to segment the region into edema, necrosis core, and enhanced tumor regions.

5 Conclusion

In this work, we introduced a novel clustering-based segmentation of MRI image. In the first step, preprocessing is applied to extricate ROI-based adaptive thresholding with CS-LBP feature. Once we extract the ROI, we use chi-square distance-based FCM to cluster the image with respect to edema, necrosis, and enhanced tumor region using the feature. Segmentation via chi-square distance-based FCM can segment the brain tumor predictively than the other existing methods. Experimentation can be carried out using BRATS 2015 dataset. Performance metrics for this proposed method achieve better in terms of Jaccard matching, specificity, PPV, and dice similarity coefficient.

References

1. Kaya IE, Pehlivanlı AÇ, Sekizkardeş EG, Ibrikci T (2017) PCA based clustering for brain tumor segmentation of T1w MRI images. Comput Methods Programs Biomed 1(140):19–28
2. Khaloo A, Lattanzi D (2017) Robust normal estimation and region growing segmentation of infrastructure 3D point cloud models. Adv Eng Inform 1(34):1–6
3. Pare S, Bhandari AK, Kumar A, Singh GK (2018) A new technique for multilevel color image thresholding based on modified fuzzy entropy and Lévy flight firefly algorithm. Comput Electr Eng 1(70):476–495
4. Raith S, Vogel EP, Anees N, Keul C, Güth JF, Edelhoff D, Fischer H (2017) Artificial neural networks as a powerful numerical tool to classify specific features of a tooth based on 3D scan data. Comput Biol Med 1(80):65–76
5. Song JH, Cong W, Li J (2017) A fuzzy C-means clustering algorithm for image segmentation using nonlinear weighted local information. J Inf Hiding Multimedia Signal Process 8(9):1
6. Zanaty EA (2012) Determination of gray matter (GM) and white matter (WM) volume in brain magnetic resonance images (MRI). Int J Comput Appl 45(3):16–22
7. Havaei M, Davy A, Warde-Farley D, Biard A, Courville A, Bengio Y, Pal C, Jodoin PM, Larochelle H (2017) Brain tumor segmentation with deep neural networks. Med Image Anal 1(35):18–31
8. Pezoulas VC, Zervakis M, Pologiorgi I, Seferlis S, Tsalikis GM, Zarifis G, Giakos GC (2017) A tissue classification approach for brain tumor segmentation using MRI. In: Imaging systems and techniques (IST), 2017 IEEE international conference on IEEE, pp 1–6, 18 Oct 2017
9. Salve MV, Salve MA, Jondhale MK (2017) Brain tumor segmentation using MS algorithm. Brain
10. Mirzaei G, Adeli H (2018) Segmentation and clustering in brain MRI imaging. Rev Neurosci
11. Zhao X, Wu Y, Song G, Li Z, Zhang Y, Fan Y (2018) A deep learning model integrating FCNNs and CRFs for brain tumor segmentation. Med Image Anal 1(43):98–111
12. Ding Y, Dong R, Lan T, Li X, Shen G, Chen H, Qin Z (2018) Multi-modal brain Tumor image segmentation based on SDAE. Int J Imaging Syst Technol 28(1):38–47
13. Shivhare SN, Sharma S, Singh N (2019) An efficient brain tumor detection and segmentation in MRI using parameter-free clustering. In: Machine intelligence and signal analysis 2019. Springer, Singapore, pp 485–495
14. Kumar GA, Sridevi PV (2018) 3D deep learning for automatic brain MR tumor segmentation with T-spline intensity inhomogeneity correction. Autom Control Comput Sci 52(5):439–450
15. Bal A, Banerjee M, Sharma P, Maitra M (2018) Brain tumor segmentation on MR image using k-means and fuzzy-possibilistic clustering. In: 2018 2nd international conference on electronics, materials engineering & nano-technology (IEMENTech), IEEE, pp 1–8, 4 May 2018
16. Shen G, Ding Y, Lan T, Chen H, Qin Z (2018) Brain tumor segmentation using concurrent fully convolutional networks and conditional random fields. In: Proceedings of the 3rd international conference on multimedia and image processing, ACM, pp 24–30, 16 Mar 2018
17. Ganesh M, Naresh M, Arvind C (2017) Mri brain image segmentation using enhanced adaptive fuzzy k-means algorithm. Intell Autom Soft Comput 23(2):325–330

18. Kumar R, Mathai KJ (2017) Brain tumor segmentation by modified k-mean with morphological operations. Brain 6(8)
19. Mane DS, Gite BB (2017) Brain tumor segmentation using fuzzy c-means and k-means clustering and its area calculation and disease prediction using Naive-Bayes algorithm. Brain 6(11)

Assessment of Lean Manufacturing Using Data Envelopment Analysis (DEA) on Value-Stream Maps

L. N. Pattanaik and Ch. Koteswarapavan

Abstract Lean manufacturing is considered to be a holistic approach to improve manufacturing practice by reducing various types of wastes. This area of research attracts the attention of researchers from academia and industries in recent times. This research paper presents a leanness assessment methodology using value-stream mapping, a prominent lean tool. Slack-based measure, a method of the data envelopment analysis, is used here to assess the leanness. A case study of a manufacturing system along with its value-stream map is considered to illustrate the application of the methodology. The value-added and non-value-added activities are also considered during the lean assessment. The leanness measure quantitatively expresses the extent of leanness in the firm and the scopes for further improvement. By utilizing lean score, the improvements in lean activities can be initiated. The methodology for leanness assessment can be useful in industrial practice for achieving the goal of manufacturing excellence.

Keywords Lean manufacturing · Lean assessment · Data envelopment analysis · Value-stream mapping · Slack-based measure

1 Introduction

After the second World war, Japan's automotive industry faced a severe crisis of low productivity, low demand and inability to cater product variety. To overcome this, the Toyota production system (TPS) was developed with an objective of

L. N. Pattanaik · Ch.Koteswarapavan (✉)
Department of Production Engineering, Birla Institute of Technology,
Mesra, Ranchi 835215, India
e-mail: lnpattanaik@bitmesra.ac.in

BBVL. Deepak et al. (eds.), *Innovative Product Design and Intelligent Manufacturing Systems*, Lecture Notes in Mechanical Engineering,
https://doi.org/10.1007/978-981-15-2696-1_84

removing waste, conserving the resources and producing a large variety. Lean manufacturing concept was conceived at Toyota production system to achieve these objectives. There are eight kinds of wastes according to the TPS.

• Over production	• Over processing
• Waiting time	• Defective products
• Transportation waste	• Non-utilized talent
• Inventory	• Waste of movement

There are several lean tools which can be applied to reduce these wastes in a system. Some of them are 5S, JIT, SMED, Poka-yoke, Heijunka, Kanban, Ishikawa and value-stream mapping. Value-stream mapping (VSM) is a lean tool which clearly indicates the material flow and the information flows in the system. The time at each work station and the number of products undergoing the process at the work station give the processing time and production lead time which can be clearly identified by drawing a VSM of a system.

In the present twenty-first century, every organization is trying to transform itself to become more competitive to survive and excel. To achieve that, they have to be responsive to the short product life cycle and be able to mass customization and optimum utilization of limited resources. Further, they should have a quality assurance system and an efficient supply chain which help in the delivery of the defect-free products in the required time. This compels an organization to transform into lean in all the functions. This is applicable for the manufacturing sector, services or hybrid organizations. Assessing leanness level of industries is challenging owing to the involvement of subjective aspects. However, it is required to judge the level of leanness in order to improve upon it. The survey method does not give the exact leanness level of the company as the score depends upon the perception and opinion of individuals. In this paper, a slack-based measure (SBM) of data envelopment analysis (DEA) model is applied on a value-stream map to calculate the lean score of typical manufacturing industry. The cost, time and product value are the three important factors which are taken into consideration for DEA. The value-added inputs are separated from actual values to form the ideal decision-making units (IDMUs). Leanness frontier is determined as the standard for actual decision-making unit's (ADMU's) leanness score.

The SBM fractional model by Tone [1] is used here to measure the leanness. The score obtained from this model varies between 0 and 1, and this score continuously decreases depending upon the slacks. The leanness score is essential for any industry in order to know the quantity of waste present in the industry. Cost and agility approaches will have different leanness score. This data envelopment-based leanness measure is not limited to the manufacturing industry, but it can be applied to any other industries.

2 Literature Review

Rother and Shook [2] made the lean path easier by developing a new method called value-stream mapping (VSM). They developed some standard symbols for the construction of the VSM. The VSM clearly shows the production flow in a system from raw material stage to the hands of the customer. It became more useful for the manufactures as it visualizes more than one process level, and it also helps in identifying the waste in the system. Most importantly, it is the only tool which displays the connection between the material flow and information flow. The current state map can be improved by performing required changes to achieve the future state map as proposed in VSM.

Wan [3] discussed about the assessment of lean manufacturing and identified the leanness target by the help of agility. As most of the lean manufacturing index is based on the survey, in this research, a quantitative cost–time analysis leanness measure using the decision-making units (DMUs) for lean assessment and Excel©-based data envelopment analysis (DEA) leanness solver was developed for ease of solving the models. An integrated, lean and agile performance index was proposed.

Charnes et al. [4] were the first ones to introduce a mathematical model for DEA. Based on the improvement direction, they developed two types of models, and for other situations, different DEA models were developed. The slack-based measure by Charnes et al. [4] was improvised by Tone [1].

Tone [1] developed the slack-based measure (SBM) for efficiency measurement of a system by the decision-making units. He has clearly explained and compared the relationship of the slack-based measure with the Charnes and Cooper model. The Charnes–Cooper–Rhodes (CCR) model does not consider the slacks in the system, whereas the SBM considers the slacks in the system. The SBM satisfies the properties of efficiency continuously decreasing with respect to the slacks, and it is unit invariant also.

Hjalmarsson and Olsson [5] presented a case study on the regular activities of a Swedish logistics company. Value-stream mapping is performed in this case study to reduce the processing time. Data envelopment analysis is applied for calculating the leanness score of the company. Cost, time and value are the main parameters taken into consideration during the calculation of the lean score. They improved the current state map by adding some changes to it, so that it can become leaner. They developed one current state map and two future state maps and finally made the system achieve 100% leanness. This research work practically proved that extended leanness measure of Wan [3] helps to increase the leanness of the system.

Wan and Chen [6] presented the data envelopment analysis fractional method which is proposed by Charnes–Cooper–Rhodes is utilized for measuring the leanness score. A software-based program solver is developed for solving the linear program. A hypothetical case of batch production leanness score is measured. The time and cost changes for improvement in the leanness score are explained in a tabular form. This proposed leanness measure gives features like integrated leanness index which discusses three important dimensions and also scopes variability for the applications and the trade-off for the competitive strategies.

Rahani and Ashraf [7] applied a value-stream map on a Malaysian-based automotive industry and used the Gemba technique for removing the hidden waste in the industry. The non-value-added time in the industry was identified and modified the current state map for the improvements in the system and reduced the waste.

Forno et al. [8] in a literature review presented an exhaustive survey of lean manufacturing papers related to VSM. They divided the papers into eleven categories for analyzing the problems faced during the implementation of the VSM. The major problems faced while implementing the VSM were of three categories; process, product and people.

Korakot et al. [9] implemented the value-stream mapping technique on a rope manufacturing industry. The present state map is drawn, and the areas of improvement are considered. The future state map is created by the assistance of the enhancements considered from the present state map. The results obtained shows that material handling time and production lead time are improved by the help of lean tools.

Chandandeep [10] presented a specific methodology in a flowchart form for implementation of the value-stream mapping for a small company. The time was considered as an important factor. The values of the takt time, lead time, cycle time and change over time were reduced after constructing the future map.

Paola et al. [11] presented a systematic review which is carried out on the lean measurement methods used in the manufacturing industries. Thirty-one techniques are distinguished and investigated depending on similar measurements. The principle qualities and weakness of each methodology are featured. A definition for the leanness is proposed, and research gaps from the systematic review are mentioned for future research.

Adwait et al. [12] constructed a value-stream map for a plastic bag manufacturing unit. The improvements required in the unit are identified from the current state map and applied in the future state value-stream map. The production lead time, takt time and processing time are improved compared to the current state. The production of the bags also increased after implementing the technique.

The literature review indicates that more research is needed in quantitative lean assessment using value-stream maps. Based on the leanness score, implementing the improvement strategies on a current value-stream for the construction of a future state map is still unexplored in the manufacturing field. As VSM can be applied to various systems, this method for measuring leanness score can be applied to other systems also.

3 Methodology

The value-stream mapping (VSM) which is a popular graphical tool in lean manufacturing was developed by Rother and Shook [2]. The value-stream designing for a manufacturing system is important for becoming lean, as it is a graphical tool which is easy to understand, and it maps the process of a system. The current state VSM can be improved as it gives information about the material flow and information flow of the system. Lean experts mostly utilize self-evaluation tools for designing their system status. In any case, surveys are mostly subjective and the default lean pointers of a poll may not fit all systems. A coordinated and quantitative measure for total leanness of a VSM is discussed in this paper.

3.1 Data Envelopment Analysis (DEA)

The DEA mathematical model concept is developed by Charnes et al. [4]. DEA is a methodology used for the measurement of performance. A standard for every data set is identified after investigating the inputs and outputs. According to Wan [3], decision-making units for DEA leanness ought to be characterized as the workpiece that moves through the system. By comparing input and output ratio, one can calculate leanness with the help of DEA models. The components of a DEA model are as follows:

(i) Decision-making units (DMUs)
  - Actual decision-making units (ADMUs)
  - Ideal decision-making units (IDMUs)

(ii) Input variables

(iii) Output variables

(iv) Transformation process.

An actual decision-making unit is monitoring the production of a single work piece. The real time and expenses contributed on the work piece are the inputs for the ADMU. The value created is output variable only. To push the frontier to the ideal leanness, virtual DMUs are used by the DEA-leanness measure. The ideal decision-making units are created virtually by taking out non-value-added activities. The ideal performance of production process is represented by value-added time and cost only. The mathematical model developed by Charnes–Cooper–Rhodes (CCR) which derives the efficiency score does not consider the slack inefficiency. Due to this, the CCR model overestimates the efficiency if any slack exists.

The fractional DEA-leanness measure [3]

$$Min\ \rho_{\text{lean}} \frac{\left[1-\left(\frac{1}{2}\right)\left(\frac{S_T^-}{X_{T0}}+\frac{S_C^-}{X_{C0}}\right)\right]}{1+\frac{S_V^-}{Y_{V0}}}$$

$$\text{Subject to}\quad X_{T0}=\sum_{i=1}^{n} X_{Ti}\lambda_i + S_T^- \tag{1}$$

$$X_{C0}=\sum_{i=1}^{n} X_{Ci}\lambda_i + S_C^-$$

$$Y_{V0}=\sum_{i=1}^{n} Y_{Vi}\lambda_i - S_V^+$$

where $S_T^-, S_C^-, S_V^+, \lambda \geq 0$

ρ_{lean}: Lean Score, X_{T0}: DMU0 Input Time, X_{C0}: DMU0 Input Cost, Y_{V0}: DMU_0 Output Value, n: Number of DMUs, λ: DMUs SBM Weights, S_T^-, S_C^-, S_V^+: Slacks Related to Inputs/outputs.

3.2 *Slack-Based Measure*

Slack-based measure (SBM) improved by Tone [1] which was improvised from the CCR model to deal with excess output and input deficits of DMU. The slack-based measure considers the ADMU and IDMU values into consideration for the calculation. Slack-based measure deals with the input and output slacks directly. The score obtained from the SBM is constant and continuously decreases depending on the slacks. By considering all the inefficiencies, the SBM score ranges from 0 to 1. The modified form of the expression (1) for finding the leanness score using SBM [3] is given here.

$$\text{Leanness score} \frac{\left\{1-\left(\frac{1}{2}\right)\left[\frac{X_{\text{TA}}-X_{\text{TI}}}{X_{\text{TA}}}+\frac{X_{\text{CA}}-X_{\text{CI}}}{X_{\text{CA}}}\right]\right\}}{1+\left[\frac{X_{\text{VI}}-X_{\text{VA}}}{X_{\text{VA}}}\right]} \tag{2}$$

4 Case Study for Measuring Leanness

A case study of a manufacturing system along with its VSM is considered to illustrate the application of the lean assessment methodology.

A total of four stages are there in the VSM. The supplier supplies 1.5 tons of raw materials in a month. A total of 15 pieces were made daily or 450 pcs per month. Detailed VSM with information and material flow is shown in Fig. 1.

The production lead time and process time and the number of products delivered on a daily basis are extracted from the VSM. The remaining data like the cost of daily production, retail price, customer satisfaction, daily value-added cost of a value-added process and scrap rate are appropriately assumed. The work time available in this case is considered as 27,000 s (7.5 h). The production lead time is 0.55 day which is 247.5 min. The daily cost of production is 3000 for delivering 15 products. The daily cost of production divided by a number of products delivered gives the actual cost. The retail price of the product with customer satisfaction gives the actual value. For the ideal decision-making units' calculation, the value-added (VA) time and cost are considered while the value remains the same. Calculations of the ADMU and IDMUs are more clearly given in Table 1. After calculating the ADMUs and IDMUs, the values of time, cost and value are substituted in the (2) for measuring the leanness score of the manufacturing system.

ADMU $(X_{TA}, X_{CA}, Y_{VA}) = (247.5, 200, 45)$
IDMU $(X_{TI}, X_{CI}, Y_{VI}) = (15, 144.18, 45)$

$$\text{Leanness Score} \frac{\left\{1-\left(\frac{1}{2}\right)\left[\frac{X_{TA}-X_{TI}}{X_{TA}}+\frac{X_{CA}-X_{CI}}{X_{CA}}\right]\right\}}{1+\left[\frac{X_{VI}-X_{VA}}{X_{VA}}\right]}=0.39$$

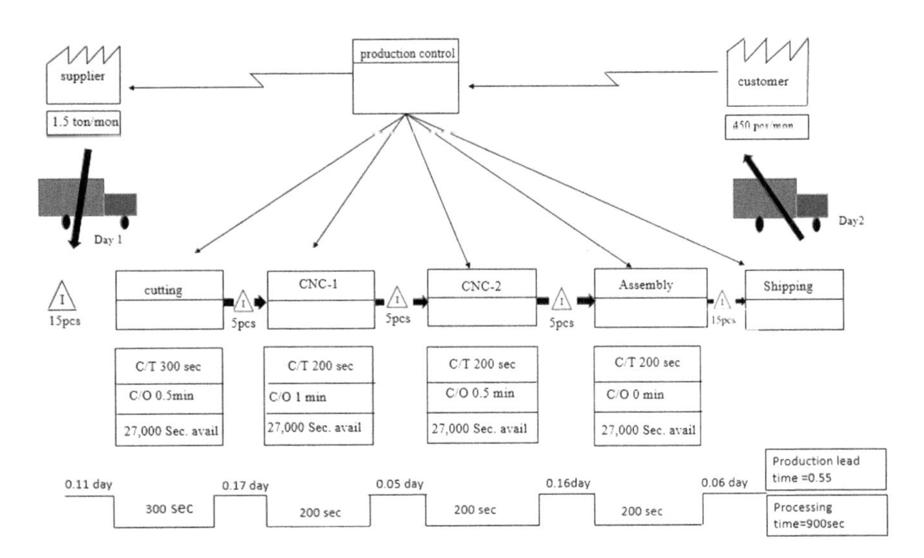

Fig. 1 Example of the four production process's value-stream map

Table 1 Calculations of leanness measure on VSM

Variable	Procedure	Example	
Time	Production lead time from VSM	0.55 days = 247.5 min	ADMU (actual decision-making units)
Cost	$\frac{\text{(Daily cost of production)}}{\text{(Daily amount of delivered products)}}$	$\frac{3000}{15} = 200$	
Value	(Retail price) × (customer satisfaction)	₹ 50 × 90% = 45	
Value-added time	Processing time from VSM	900 s = 15 min	IDMU (ideal decision-making units)
Value-added cost	$\text{Sum}\left[\frac{\text{(Daily VA cost of a VA process)}}{\text{Delivered amount} \times (1 + \text{scrap rate})}\right]$	$\frac{600}{[15 \times 1.02]} + \frac{500}{[15 \times 1.01]} + \frac{500}{[15 \times 1.03]} + \frac{600}{15 \times 1.01} = 144.18 \text{ per unit}$	
Value	(Retail price) × (customer satisfaction)	₹ 50 × 90% = 45	

Table 2 Leanness score and slack

DMU's	Lean score	S		
		T	C	V
Actual DMU (ADMU)	0.39	232	55.82	0
Ideal DMU (IDMU)	1	0	0	0

S slack; *T* time; *C* cost; *V* value

The leanness score of the given VSM is 0.39. This score indicates the leanness of the present system. It also indicates "how much leaner the present system can be improved". Improvised individual dimensions are given as slack in Table 2. The value (V) will be the same for both the ideal and actual decision-making unit as the customer satisfaction remains the same for both. The actual DMU's slack is given in Table 2 which is calculated by eliminating value-added (VA) time, value and cost from actual time, cost and value. In the case of the ideal decision-making units, the lean score is 1 and time, cost and value will be 0. As the DEA leanness continuously increases with the slacks, the increment in the non-value-added cost or time will reduce the score and the other way around. By utilizing the same procedure, multiple pairs of ADMU and IDMU can be created and different time period leanness level can be compared. The leanness score represents the volume of the waste in the system at present.

5 Conclusion

Lean manufacturing is considered as an effective approach for achieving best practice in many domains of production and service sector. However, assessment of leanness is always remaining a challenge for the practitioners and researchers. Although some feedback-based assessment of lean success exists, those methodologies are highly subjective and fail to give a correct quantitative score to rely upon. This paper presented a methodology for measuring the leanness score of an industry using the value-stream mapping and the slack-based measure. The leanness score obtained in this approach shows the volume of waste existing in the system at present. This methodology is only applied when there is only one pair of DMUs being compared. As VSM can be applied to various systems, this method for measuring leanness score can be applied to other systems also. Future state VSM with continuous improvements can be drawn until it reduces the waste and improves the lean score of the system.

References

1. Tone K (2001) A slack based measure of efficiency in DEA. Eur J Oper Res 130(3):498–509
2. Rother M, Shook J (1998) Learning to see-value stream mapping to add value and eliminate muda, 1st edn. Lean Enterprise Institute, Brookline

3. Wan H (2006) Measuring leanness of manufacturing and identifying leanness target by considering agility. Ph.D. Dissertation, Virginia Polytechnic Institute and State University, Virginia
4. Charnes A, Cooper W, Rhodes E (1978) Measuring the efficiency of decision-making units. Eur J Oper Res 2(6):429–444
5. Hjalmarsson V, Olsson L (2017) Quantifying leanness combining value stream mapping with a data envelopment analysis-based method—a case study based on swedish logistics company. In: IEEE conference on industrial engineering and engineering management, Singapore, pp 740–744
6. Wan H, Chen F (2008) A leanness measure of manufacturing systems for quantifying impacts of lean initiatives. Int J Prod Res 46(23):6567–6584
7. Rahani A, Ashraf M (2012) Production flow analysis through value stream mapping: a lean manufacturing process case study. Proc. Eng. 41:1727–1734
8. Forno D, Pwewira A, Forcellini A (2014) Value stream mapping: a study about the problems and challenges found in the literature from the past 15 years about the application of lean tools. Int. J. Adv. Manuf. Technol. 72(5–8):779–790
9. Korakot Y, Lee J, Kanjicai D (2017) Value stream mapping of rope manufacturing: a case study. Int J Manuf Eng 2017:1–11
10. Chandandeep G (2008) An initiative to implement lean manufacturing using value stream mapping in small company. Int J Manuf Technol Manage 15(3–4):404–417
11. Paola C, Filippo M, Alberti M, Schiavini D (2018) Leanness measurement methods in manufacturing organisations: a systematic review. Int J Prod Res. https://doi.org/10.1080/00207543.2018.1521016
12. Adwait D, Saily K, Giri J, Vivek K (2018) Design and evaluation of a lean manufacturing framework using value stream mapping (VSM) for a plastic bag manufacturing unit. Mater Today Proc 5:7668–7677

Reinforcement Learning for Inventory Management

Shraddha Bharti, Dony S. Kurian and V. Madhusudanan Pillai

Abstract The decision of "how much to order" at each stage of the supply chain is a major task to minimize inventory costs. Managers tend to follow particular ordering policy seeking individual benefit which hampers the overall performance of the supply chain. Major findings from the literature show that, with the advent of machine learning and artificial intelligence, the trend in this area has been heading from simple base stock policy to intelligence-based learning algorithms to gain near-optimal solution. This paper initially focuses on formulating a multi-agent four-stage serial supply chain as reinforcement learning (RL) model for ordering management problem. In the final step, RL model for a single-agent supply chain is optimized using Q-learning algorithm. The results from the simulations show that the RL model with Q-learning algorithm is found to be better than Order-Up-To policy and 1–1 policy.

Keywords Supply chain · Ordering policy · Inventory management · Reinforcement learning · Q-learning

1 Introduction

A supply chain is an integrated network of multiple agents consisting of retailers, distributors, manufacturers, and suppliers. Each agent has to make replenishment decisions on "how much to order" with the aim of minimizing long-term total supply chain inventory cost. Furthermore, the decisions made by the human agents while operating the supply chain are often biased due to the behavior of agents [1, 2].

In a decentralized supply chain, each agent has to make an independent decision based on its interacting environment [3]. The order decision hence depends on factors such as the size of the downstream demand, quantity received from the

S. Bharti · D. S. Kurian · V. M. Pillai (✉)
Department of Mechanical Engineering, National Institute of Technology Calicut, Kozhikode 673601, India
e-mail: vmp@nitc.ac.in

BBVL. Deepak et al. (eds.), *Innovative Product Design and Intelligent Manufacturing Systems*, Lecture Notes in Mechanical Engineering,
https://doi.org/10.1007/978-981-15-2696-1_85

upstream, and inventory level. In addition, the complexity of the decision process is increased if the agents are subjected to uncertain system parameters (e.g., customer demand and lead time) and bullwhip effect [4, 5]. In such complex situations, agents must take decision based on the system's state rather than following a fixed decision rule.

Various approaches have been proposed by past researchers regarding the optimality of inventory order decisions. The base stock policy is found to be optimal in a multi-agent inventory system when several assumptions are incorporated in the model [6]. When the agent faces deterministic demand with penalty cost for unfulfilled orders, the best ordering policy is "pass order" or "one for one" (1–1) policy [7]. In 1–1 policy, each agent gives order to the upstream which is equal to the order received from the downstream.

The classical example of a decentralized supply chain is the MIT's beer distribution game. Sterman [2] points out that in a beer game environment where agents act irrationally there is no known optimal policy for an agent wishing to act optimally. In his work, a formula-based method to model the agent's decision-making behavior is proposed. Like Sterman [2], Mosekilde and Larsen [8] and Strozzi et al. [9] also adopted a formula-based approach which attempts to model agent-based decision-making. But these approaches fail to determine optimal decisions. The minimum supply chain cost under beer game settings is obtained when the different agents adopt different ordering policies rather than a single ordering policy [9, 10].

Interest in harnessing the power of learning algorithms in the supply chain has increased due to the emergence of artificial intelligence. Genetic algorithm (GA), an evolutionary learning algorithm, is used by Kimbrough et al. [7] to design a multi-agent system under beer game settings. Results from their study show that the artificial agents learned via GA are able to play a beer game effectively compared to human agents.

There are mainly three categories of machine learning techniques, namely supervised learning, unsupervised learning, and reinforcement learning (RL). Labeled set of accurate training data is needed for supervised learning. Unsupervised learning is used to find a hidden pattern from the collection of unlabeled data. Unlike other forms of machine learning, in RL, there is no exact action to perform; instead, a learning agent learns by trial and error based on its state and acts based on the current state.

Many practical problems are found to be stochastic in nature, and such discrete-time stochastic processes are formulated as Markov decision process (MDP). Reinforcement learning (RL) is a technique to solve MDP [11]. In recent years, RL has been implemented to solve several problems in supply chain management. The RL framework is used to address the coordination problem of global supply chains [12]. Preliminary work on RL for ordering management has been proposed by Giannoccaro and Pontrandolfo [13]. They have employed a semi-markov average reward technique to solve inventory decision problem in a three-stage supply chain. In another study, Chaharsooghi et al. [5] have applied Q-learning method, widely used RL algorithm, to optimize inventory order decisions of four-stage supply chain. Kara and Dogan [14] have addressed ordering policies for perishable products using Q-learning and SARSA algorithm based on RL. Performance of both the algorithms proves to be better than the genetic algorithm. Recently, Oroojlooyjadid et al. [15] proposed a RL algorithm based on Deep Q Networks (DQN) and a transfer learning approach to find a near-optimal ordering policy in the beer game environment. Collectively, these studies provide evidence that by addressing the supply chain inventory management problem as RL problem, it is possible to achieve near-optimal ordering policy.

This paper aims to put forth a modified learning mechanism for an artificial agent to make inventory decisions on ordering size. In the previous works [5, 13, 15], the agent has adopted $X + Y$ rule as an ordering policy where X represents the downstream demand while Y indicates the quantity determined by the learning agent. This paper primarily centers on formulating a multi-agent four-stage supply chain as RL model for deriving near-optimal solutions. Subsequently, a Q-learning algorithm with modified ordering rule is adopted to solve a single-agent supply chain to obtain minimum inventory cost.

The remainder of this paper is organized as follows. Section 2 provides a brief description of RL and MDP. Section 3 discusses the problem formulation using RL approach. Q-learning algorithm is employed to solve the single-agent supply chain problem in Sect. 4 and the results are discussed in Sect. 5.

2 Reinforcement Learning

The fundamental concept behind RL is the interaction between the learning agent and its environment. During the learning process, an agent at time-step t selects an action a_t based on the environmental state s_t. As an outcome of its action, the agent receives a reward r_{t+1} and transits from state s_t to new state s_{t+1} (see Fig. 1). In the long term, the objective of any RL agent is to maximize the cumulative reward by learning what to do and how to map situations [16].

There should be Markovian property associated with every state of the RL model, and even if the states are non-Markov yet it is appropriate to approximate it as a Markov state [14]. Markov decision process (MDP) is a RL satisfying the Markov property. MDP, in brief, is a sequential decision-making model which

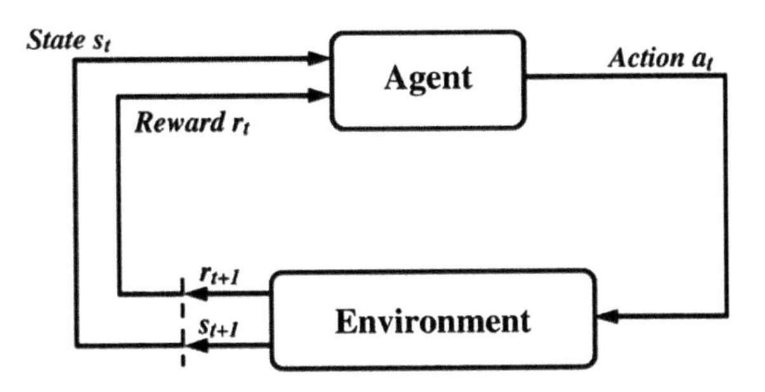

Fig. 1 Agent and environment interaction in RL [16]

consists of decision periods, system states, available actions, rewards or costs, and transition probabilities. The information provided by the system state assists the decision-maker to choose an action at each decision epoch and gets the corresponding reward. Combined previous states and actions or present state determines the action to be chosen in the present state, and this is termed as a decision rule. A decision rule forms policy, and a reward is acquired by implementing the policy. The aim is to maximize the reward sequence by choosing a suitable policy [17].

3 Problem Description and Formulation

A serial four-stage supply chain as shown in Fig. 2 is considered for the study. It consists of one retailer (i = 1), one wholesaler (i = 2), one distributor (i = 3), and one factory (i = 4). The retailer agent faces the stochastic demand ($D(t)$) from its customer. It is assumed that the factory has an unlimited production capacity and whatever quantity demanded by the factory is released ($R(t)$) after production. Each agent i of the supply chain tries to place independent decisions on ordering size ($O_i(t)$) to its upstream agent, and shipping orders ($S_i(t)$) placed by its downstream agent over a series of time period t = 1, 2, 3 …, T. The parameters of the RL including reward function, state variable, and agent's policy are described in the next subsection.

3.1 Reward Function

At time-step t, the agent observes the state and takes an action. Agent receives a reward as a feedback corresponding to the action taken. The reward is the payoff for taking the right decision. The aim of the inventory ordering management problem is to minimize the cost; therefore, the reward function is defined as a loss function and it is estimated as in Eq. (1).

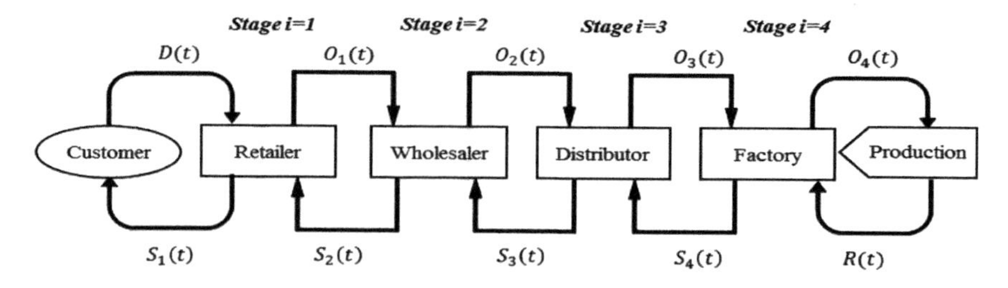

Fig. 2 Order and shipment flow in a supply chain

$$r(t+1) = \sum_{i=1}^{4} \left[C_i^h \times I_i^+(t) + C_i^s \times I_i^-(t) \right] \tag{1}$$

The reward function $r(t+1)$ is a function of the holding cost and lost sales cost occurring at period t. In the Q-learning algorithm, at every epoch, the value of Q has to be updated. For this purpose, reward value (R) is required at each step as shown in Fig. 3. Reward value (R) depends upon the reward function ($r(t+1)$). If the value of the current period reward function is less than or equal to the value of the previous period reward function, then R equals to +1 else R takes −1. In the above equation, C_i^h is the unit holding cost and C_i^s is the unit lost sales cost. $I_i^+(t)$ represents inventory at the end of the period t, and $I_i^-(t)$ is the lost sales quantity.

3.2 State Variable

State variable defines the state of the system and provides appropriate information to the decision-maker. In this study, inventory position ($\mathrm{IP}_i(t)$) at time-step t is considered as a state variable of the agent i. Inventory position at time-step t is the sum of end-period inventory at t and on-order inventory at t. State variable vector $s(t)$ at time-step t is represented as follows:

$$s(t) = [\mathrm{IP}_1(t), \mathrm{IP}_2(t), \mathrm{IP}_3(t), \mathrm{IP}_4(t)] \tag{2}$$

1-Set the initial conditions
Iteration = 0, t = 0, Q(s,a) = 0
2-While iteration ≤ Maximum_iteration
Set the inventory position of retailer:
IP = 2
While t < weeks
(a) Select an action vector according to ε-greedy policy with probability of exploration
(b) Observe next state (s′)
(c) Calculate immediate reward (r(t+1)) and corresponding R
(d) Update Q(s,a) using:
$Q(s,a) \leftarrow Q(s,a) + \alpha\,[R + \max_a Q(s',a') - Q(s,a)]$
(e) s ← s′
(f) t ← t′
t = 0
Iteration = Iteration + 1

Fig. 3 Q-learning algorithm

3.3 Ordering Policy

A modified $X + Y$ ordering policy is proposed in this study in contrast to ordering rule discussed in the literature [5, 15]. According to this modified policy, X indicates the quantity determined via Order-Up-To (OUT) decision rule and Y denotes the agent's policy which takes the values positive, negative, or zero implying that the agent can order more, less, or equal to X. OUT inventory policy is a periodic review type inventory system where OUT level is the maximum (target) inventory level. In OUT policy, order quantity (X) is determined by taking the difference between OUT level and inventory position only if the OUT level is greater than inventory position; otherwise, X is equal to zero [18]. At every time-step t, agent orders $X + Y$ quantity to his upstream member.

The modified $X + Y$ ordering rule helps to limit the state space under consideration; that is, inventory position variation can be limited to a finite number which in turn helps the agent to learn faster.

3.4 Agent's Policy

The agent's policy $Y_s(t)$ for the state s is given as:

$$Y_s(t) = [Y_{1s}(t), Y_{2s}(t), Y_{3s}(t), Y_{4s}(t)] \cdot \tag{3}$$

where $Y_{is}(t)$ represents the value of Y determined by the agent i for the state s at time-step t.

4 Q-Learning Algorithm for Single-Agent Supply Chain

Most widely used algorithm for solving RL model is Q-learning algorithm. This is because of the model-free nature of the algorithm and it does not require complete knowledge of the system [16].

Q-learning is a temporal difference method and it learns from experience [16]. After a certain number of episodes (iterations), the algorithm finds the best state-action pair values ($Q(s, a)$) which are called Q-values. Q-values are stored in Q-table and get updated through iterations. The rows of the table represent states (s) and columns represent actions (a). The table elements are initialized as zero and then get updated as algorithm proceeds. At the end of the learning, a table with learned Q-values is obtained. In addition, the best course of action is selected for each state based on the Q value. In this paper, Q-learning method is proposed to solve the RL ordering model for a single-agent supply chain and the algorithm is described in Fig. 3.

The retailer faces stochastic demand from the customer which is randomly generated from a uniform distribution between [1, 4]. The generated sequence of demand distribution is [1–4]. Lead time considered is zero and the period of operation is from 1 to 25 weeks. The inventory position is the sum of beginning inventory and the quantity ordered in the previous period but not yet received (on-order quantity). Beginning inventory at the start of every iteration for the agent is initialized by 2 based on mean demand and review period. The algorithm is run for 10^5 iterations with a ε-greedy policy where ε is the probability of exploration which indicates that the agent performs random action with probability ε and performs greedy action (exploitation) with $(1 - \varepsilon)$ probability. In every iteration, the supply chain operates for 25 weeks. In this model, the probability of exploration is reduced linearly with increase in iteration number. Probability of exploration is taken as 98% in the first episode and 10% in the last episode. In each specific episode also, it is getting reduced from 1st week to 25th week linearly to 2% as reported in the literature [5]. Learning rate (α) can take values from 0 to 1 where the value of one indicates that the agent tries to learn everything, while the value of zero means the agent learn nothing. Learning rate (α) of 0.3 is found to be appropriate on trying out different values ranging from 0.1 to 0.7.

Results from simulation indicate that the proposed ordering policy facilitates in limiting the state space. Thus, for the given demand distribution with OUT level equal to three and for the agent's action $Y = [-1, 0, 1]$, the range of inventory position lies between [0, 3]; that is, the states [0, 1, 2, 3] are suitable for this case. Unit holding cost and unit lost sales cost are considered to be 5 and 10, respectively. After learning, Q-table gives the optimal policy when greedy action is taken for each state.

5 Experimental Result and Validation

The entire programming for solving the single-agent supply chain ordering decision problem is carried out in Python 3.5 language. The aim of this study is to minimize the total inventory cost (TC) for 25 weeks which is the sum of holding cost and lost sales cost. The convergence of the total cost is obtained by solving the RL model using Q-learning algorithm. The convergence obtained after the training can be shown by plotting a graph between the number of iterations and its corresponding cost obtained and is depicted in Fig. 4. For the same supply chain, order management is simulated under 1–1 policy and OUT policy using Microsoft Excel. The performance of the three order management policies is compared. The comparison has been carried out using the total inventory cost for different policies and is shown in Table 1.

The average cost of 25 weeks for last 100 episodes obtained from RL model is better than the cost obtained from the other two policies. The result obtained from the RL model outperforms OUT policy by 5% and 1–1 policy by 20.83%. For instance, a graph is plotted indicating cumulative cost obtained from three policies for 25 weeks and is shown in Fig. 5.

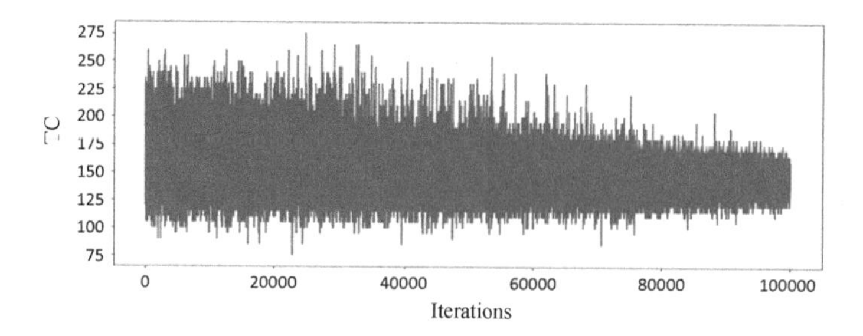

Fig. 4 Convergence plot (number of iterations versus total cost)

Table 1 Total inventory cost comparison

Policy	1–1 policy	OUT policy	RL model
Total inventory cost (TC)	180	150	142.5

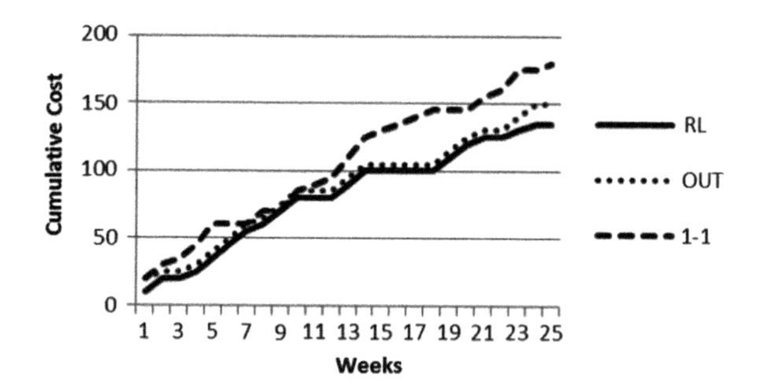

Fig. 5 Cumulative cost versus weeks

6 Conclusion

This paper is focused on supply chain ordering decision problem. Initially, a four-stage serial supply chain is formulated as a RL model for inventory management. As a next step, the RL model for a single-agent supply chain where the agent faces stochastic demand from the customer is solved using Q-learning. This study particularly has made use of a modified ordering rule which is much more effective than the ordering policy employed in the literature. Results from the study show that the RL model is found to be efficient in deciding order size. The total inventory cost obtained using the RL model is lesser than the supply chains simulated for 1–1 policy and OUT policy. The achieved results through RL model are promising, and there is a good scope of solving inventory decision problems using RL approaches in a multi-agent supply chain.

References

1. Lee HL, Padmanabhan V, Whang S (1997) Information distortion in a supply chain: the bullwhip effect. Manag Sci 43(4):546–558
2. Sterman JD (1989) Modeling managerial behavior: misperceptions of feedback in a dynamic decision making experiment. Manag Sci 35(3):321–339
3. Claus C, Boutilier C (1998) The dynamics of reinforcement learning in cooperative multiagent systems. In: Proceedings of the fifteenth national conference on artificial intelligence. AAAI, Madison, Wisconsin, pp 746–752
4. Forester JW (1961) Industrial dynamics, 1st edn. MIT Press; Wiley, New York
5. Chaharsooghi SK, Heydari J, Zegordi SH (2008) A reinforcement learning model for supply chain ordering management: an application to the beer game. Decis Support Syst 45(4):949–959
6. Clark AJ, Scarf H (1960) Optimal policies for a multi-echelon inventory problem. Manag Sci 6(4):475–490
7. Kimbrough SO, Wu DJ, Zhong F (2002) Computers play the beer game: can artificial agents manage supply chains? Decis Support Syst 33(3):323–333
8. Mosekilde E, Larsen ER (1986) Deterministic chaos in the beer production-distribution model. Syst Dyn Rev 4(1–2):131–147
9. Strozzi F, Bosch J, Zaldivar JM (2007) Beer game order policy optimization under changing customer demand. Decis Support Syst 42(4):2153–2163
10. Edali M, Yasarcan H (2016) Results of a beer game experiment: should a manager always behave according to the book? Complexity 21(S1):190–199
11. Gosavi A (2009) Reinforcement learning: a tutorial survey and recent advances. INFORMS J Comput 21(2):178–192
12. Pontrandolfo P, Gosavi A, Okogbaa OG, Das TK (2002) Global supply chain management: a reinforcement learning approach. Int J Prod Res 40(6):1299–1317
13. Giannoccaro I, Pontrandolfo P (2002) Inventory management in supply chains: a reinforcement learning approach. Int J Prod Econ 78(2):153–161
14. Kara A, Dogan I (2017) Reinforcement learning approaches for specifying ordering policies of perishable inventory systems. Expert Syst Appl 91:150
15. Oroojlooyjadid A, Nazari M, Snyder L, Takáč M (2017) A deep Q-network for the beer game: a reinforcement learning algorithm to solve inventory optimization problems. arXiv preprint arXiv:1708.05924 [cs. LG]
16. Sutton RS, Barto AG (1998) Reinforcement learning: an introduction, 1st edn. MIT Press, Cambridge
17. Puterman ML (1994) Markov decision processes: Discrete stochastic dynamic programming. Wiley, New York
18. Daniel JSR, Rajendran C (2005) A simulation-based genetic algorithm for inventory optimization in a serial supply chain. Int Trans Oper Res 12(1):101–127

Quality Improvement in Organic Food Supply Chain Using Blockchain Technology

G. Balakrishna Reddy and K. Ratna Kumar

Abstract Agriculture sector is the backbone of the Indian economy which employs more than 50% of Indian citizenry. But in the recent past, this sector is entangled with many problems like food adulteration, price hijacking, lack of transparency in the supply chain, improper communication medium, etc. The objective of the present work is to make use of advanced technologies that are available to improve the sustainability of food supply chain. In this endeavor, AHP analysis is used for studying the different technologies that can be used to improve the effectiveness of the food supply chain. From the study carried out, it is found that Blockchain technology is superior to others. Further, using Blockchain a network is developed which will ensure fair-trading and circular economy.

Keywords Organic farming · AHP analysis · Blockchain technology · Smart contract · Fair-trade

1 Introduction

"One cannot think well, love well, and sleep well if one has not dined well." Day by day the food quality has been degrading in Indian markets mainly due to food adulteration, improper standardization, and chemical-based farming. Few decades before, after independence, many times India faced drought–food deficiency due to lack of agricultural skills, irrigation problems, higher population growth rate, climate change, economic recession, etc. To solve these problems, in 1965, the Indian government adopted advanced technologies and chemical-based farming for acquiring self-sufficiency in the field of agriculture. This revolution is named as 'Green Revolution.' Simultaneously, the 'White Revolution' is launched in 1970 (NDDB, India). Later the Indian government took numerous steps in the favor of agricultural growth and farmers, so the situations get improved.

G. Balakrishna Reddy · K. Ratna Kumar (✉)
National Institute of Technology Calicut, Kozhikode 673601, India
e-mail: ratna_kumar@nitc.ac.in

BBVL. Deepak et al. (eds.), *Innovative Product Design and Intelligent Manufacturing Systems*, Lecture Notes in Mechanical Engineering,
https://doi.org/10.1007/978-981-15-2696-1_86

After a few years of the Green Revolution and White Revolution, it had been noticed that synthetic-based farming badly impacts on living beings and the environment [1, 3, 6, 14]. Same things have been noticed in the field of milk production. Due to the adoption of harmful vaccines and genetic modification procedures, milk and meat quality also degraded up to a large extent [22]. After both these revolutions, farmers lose their self-dependency and started depending on cooperative companies for fertilizers, fungicides, insecticides, seeds, hormones, vaccination, etc. Gradually, companies started to increase the price of these products which adversely affected the economic growth of the farmers. This is one of the main reasons behind the increase in farmers' suicide rate (Senses India 1950–2018).

Another strong reason for food quality degradation is food adulteration. Recently, it has been found out that 68.7% of milk and milk products that have been selling in Indian markets are not as per the standards set by the government [19]. The recent survey conducted by the Ministry of Science and Technology, Government of India, found that chemicals like detergent, urea, starch, glucose, caustic soda, refined oil, white paint, etc. in around 90% milk products in Indian markets. This issue can lead to very large number of Indians to suffer from chronic disease like Cancer in coming years. Similarly, adulteration has been taking place in meat production also [17]. Another reason for food quality degradation is improper standardization on packed-food products. Often in the newspaper, it is observed that some food items which are banned in many countries are widely available in Indian markets. In this situation, organic farming and proper supply chain management are the only way of revival. This system is cost-effective, eco-friendly and richer in nutrients than chemical-based farming [3, 18]. The only disadvantage in organic farming is its production rate which is 10–20% less than chemical-based farming [20].

Generally, it is seen that customers face difficulty in identifying the toxicity present in food item due to micro-level adulteration. So, sensory, physiochemical, chromatographic, and spectroscopic instruments are used for identifying toxic particles in food items [10]. In a pork supply chain, the belly meat is used to test whether toxins are present in meat or not [15]. In recent times, strategic tools are available to identify the fraud in these types of chains which can be used at retailer's level itself [22]. But they are tedious and time taking job. So, every time it is not possible to test and identify the quality in front of the customers.

In the other side, in the organic food supply chain, Marques Vieira et al. [13] developed a supply chain, which will be guided by the retailers who are responsible for building the customers' faith on the products. But this method is overfitting the derived objectives of this study. After two years, this method became obsolete by geo-tracking technology [4]. Further, in China, this technology is improved by adopting methods like hologram stickering, secure packaging, and quick response code [8]. Still, the problem of adulteration is not completely solved. TongKe [21] implemented cloud computing technology in the agricultural supply chain, but it has some real-world difficulties in implementing like data is insecure and expensive for data storage.

In the other side, most of the Indian farmers are struggling with financial troubles because of dependency on the costly agro-chemicals and agro-machineries. Even though the farmers can adopt organic-based farming, customers lack trust in organic products available at retail stores. Another major problem of our Indian farmers is the strong bargaining power of middlemen in the food supply chain. No doubt that various technologies have introduced through the government to solve farmers' problems along with food products' quality improvement but none of those techniques or technologies have brought a remarkable change in farmers' life. Dania [5] find out the most important ten human behaviors required to build a successful supply chain system. Presently, not a single paper has been published which finds out the suitable technology corresponding to major factors which are required for effective food supply chain management.

2 Objectives

After the literature review, the first objective is to select a robust technology by AHP analysis for effective food supply chain management. The second objective of this study is to build the network with the help of selected technology in such a way that fair-trading and circular economy can be accomplished. By achieving this objective, following benefits can be achieved: farmers will get a good price, avoid middlemen in the supply chain, increase transparency, punish the guilty, support organic food supply chain, enhance customer satisfaction level, avoid blame culture, etc. The third objective of this study is to facilitate channels separately so that similar groups of peers can share information among them, increase the unity among them, and can access the data easily.

3 Methodology

3.1 AHP Analysis

After literature review, the following factors or behaviors identified which are required to accomplish through the selected technology for an effective food supply chain management. The factors are: supply chain cost—F_1 (initial cost and running cost), traceability—F_2 (for trace product, process, genetic change, input material, pest and disease, other required values, etc.), information sharing and data management—F_3, scope for employment generation—F_4 (the ability of the technology for offer jobs), flexibility—F_5, delivery time—F_6 (how much this technology helps to minimize the time), and adaptability—F_7 (by the organization and customer) [5, 9, 15, 21]. Similarly, most highlighted supply chain technologies used in food supply chains are: conventional supply chain—T_1 (without any advanced technologies), social media—T_2 [2], radio technologies with RFID—T_3, software (like ERP)—T_4 [12], IoT with RFID—T_5 [11], and Blockchain technology—T_6 [14, 16].

All the above-mentioned technologies have their own advantages and disadvantages. So, AHP analysis is used to select the best technology for the food supply chain. For this analysis, the values are collected from professors and research scholars of the relevant field from our college.

The geometric mean of the collected data is considered for further study. The results of this study are shown in Table 1 which depicts that the Blockchain technology has the highest potential to be implemented in the food supply chain.

3.2 Common Types of Food Supply Chains and Its Drawbacks

The different structures of the existing food supply chains are shown in below figures. Figure 1 shows a common type of food supply chain. In this supply chain, farmers are restricted to sell limited quantities of food products through mandies where they get good revenue. Rest products are sold through market and ENAM where they bitterly bargained by middlemen and traders, respectively. The dotted lines in the figure represent the intermediate process, which are shown in Fig. 2. Another problem exists in this supply chain, i.e., farmers do not get revenue instantly because of the long cash flow structure (Fig. 2). Due to the lack of transparency, there is no customers' trust and there is no opportunity for organic business.

Figure 3 shows the structure of an organic food supply chain. This structure is certified by the government. These types of organizations are spread all over the country. These organizations have their own shops in their locality. Demand of the customers at the stores is met either by their own cluster or importing from other certified organizations. There are mainly two problems which arise in this structure. Firstly, it is difficult for customers to trust the quality of the products. Secondly, price-hijacking takes place due to the outsourcing. Due to these two reasons, in these types of chains, sales volume of organic products is marginal (Fig. 4).

3.3 Proposed Model with Blockchain Technology

To solve all the above-mentioned problems, the following model of the food supply chain is proposed (Fig. 5). This model is supported by Blockchain technology. This model is basically developed for small- and medium-scale farmers who are doing business locally. In the case of large-scale producers, the business activities can be carried out by installing several similar types of networks in parallel. This type of

Table 1 Result of AHP analysis

Supply chain technologies	F_1 (21.15%)	F_2 (24.17%)	F_3 (7.26%)	F_4 (13.58%)	F_5 (4.82%)	F_6 (6.92%)	F_7 (12.08%)	Overall importance
(T_1)	1.47	1.07	0.56	2.91	0.16	0.49	3.34	10.00
(T_2)	2.00	2.69	0.84	2.55	0.46	1.86	0.67	11.06
(T_3)	1.62	3.23	1.12	2.30	1.09	3.04	2.81	15.20
(T_4)	4.31	2.69	1.96	1.45	0.64	2.41	2.00	15.46
(T_5)	7.35	5.37	1.40	0.48	0.81	3.15	2.26	20.81
(T_6)	4.41	9.69	1.40	3.88	1.67	5.97	1.00	28.03

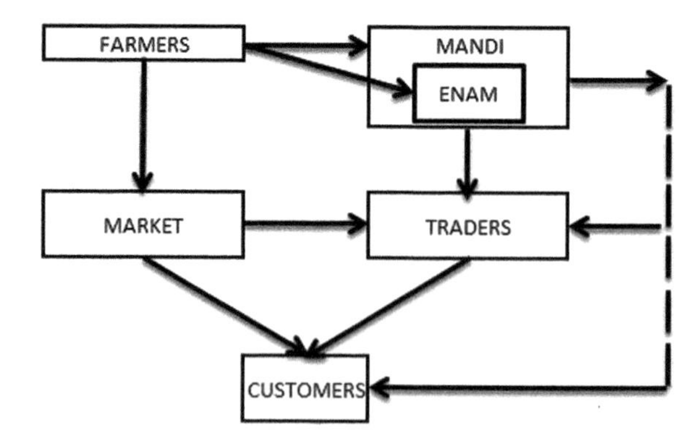

Fig. 1 General food supply chain in India

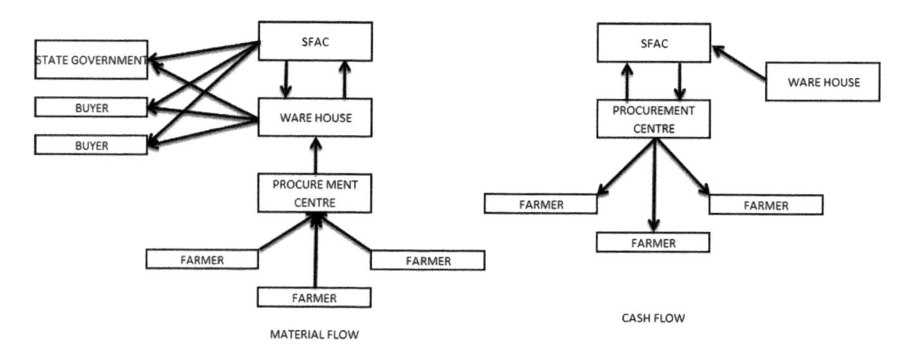

Fig. 2 Food supply chain of mandi system

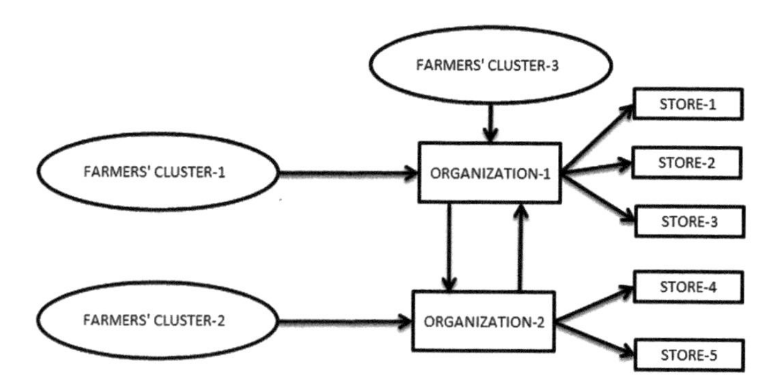

Fig. 3 Organic food supply chain in Kozhikode

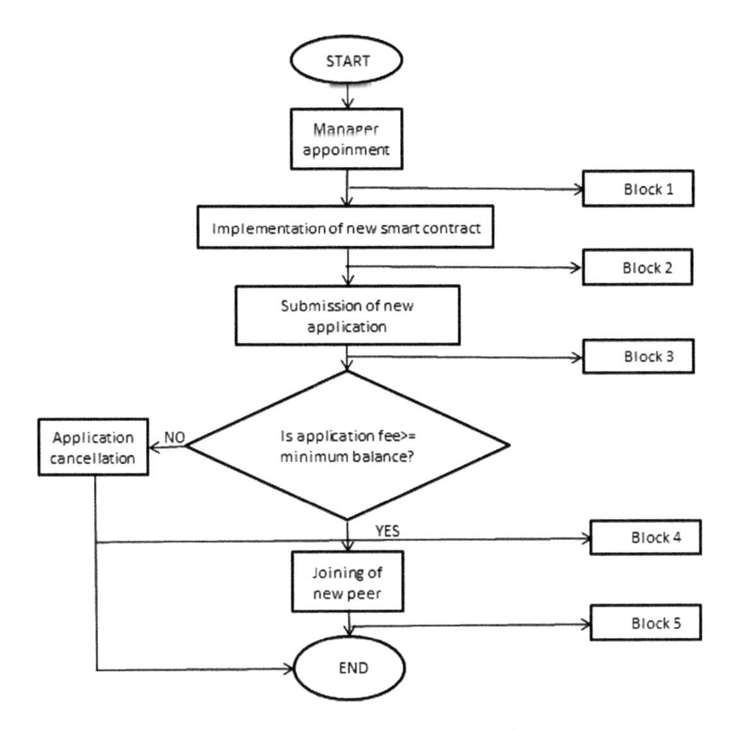

Fig. 4 Flowchart for adding and cancelation of peers in network

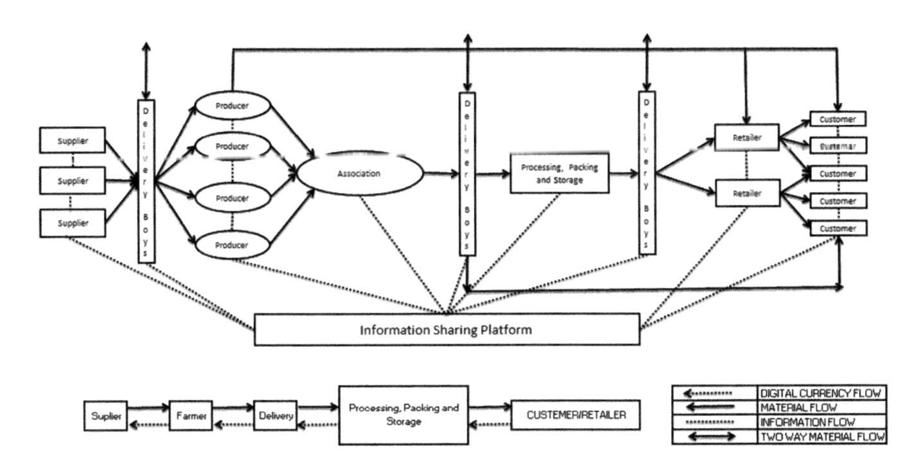

Fig. 5 Proposed model frame work

supply chain is suitable for both perishable (vegetables, milk, meat, fish, fruits, etc.) and durable items (grains, spices, etc.). Here, the product can flow from farmers to the customer in different ways based on the type of product, demand, quality, availability, etc. When the demand is very less, e.g., spices, retailer plays a better role in order to decrease the cost of the product. In the case of stock out, outsourcing can be done from nearer parallel networks. All these processes will be automatically guided by the information platform, and at the end of every event, a block will generate and added in the chain.

Initially, after production, the farmer or producer will announce their availability of products along with the quantity, location, and approximate quality. Simultaneously, the customers will mention their demand. If the availability is matching with the demand, then the delivery boy will go to the farmer and check the product with the pocket molecular sensor (IoT). Then, the data will broadcast to the information platform and the sent data will be categorized to find out the quality of the product. In this study, qualities are categorized as Q_1, Q_2, and Q_3 (numerically derived by ENAM). If the farmers' information is correct or matching with the customer's demand, then the product will be accepted by the delivery boy. Then according to the type of product, it will be delivered to processing centers, retailers, customers, or to the nearer networks. In every stage, the product will be tested which will maintain the quality of the product till the end of delivery. When the ownership of the product transfers instantaneously, some crypto-currencies will flow in the opposite direction. In this way, the information, product, and currencies flow in the entire network. If anyone found guilty in network, then they will be penalized as coded in smart contract.

Here, the Etherium platform is used for building the public Blockchain and to implement the smart contract. For this, VS code is used as a code editor. The steps involved in smart contract deployment are: Employ the manager, Start peer registration, Allow for set and get data, Change ownership on products, Generate crypto-currency, Update accounts of each peer, Quality determination from input data, Guide delivery boy, Penalize the guilty, Profit Distribution among peers, Build required number of channels, and Create block after every event in network.

4 Discussion

Following points can be achieved by implementing the proposed framework in order to maintain fair-trading and circular economy in the network:

1. For less electricity consumption, zero-target is set for hashing. Hashing will be done automatically (no need of miner) and there will be no artificial demand for mining the coin.
2. Due to the transparency in the network, adulteration and another malfunctioning can be controlled which will lead to the minimization of food-related diseases and deaths in our country.

3. Farmers can get fair price for their contribution. There is no chance of price hijacking and getting cheated by the middlemen. This will decrease the number of middlemen in the system and increase the farming jobs in the organic field.
4. Creating separate channels for farmers will facilitate in developing the skills and unity among them.

5 Conclusion

Farming is the way of life, not money-making process. This is what the Indians belief. According to ancient Indian scriptures, farming is one of the major factors to maintain a balance in the ecosystem. It is our duty to pay gratitude toward the farming and farmers. Currently, in India, 54.6% of people are directly or indirectly involved in the farming process and every year it is getting decreased by 3.6% more as compared to 2001. Among these farmers, around 76% are not satisfied with their job (Down to Earth 2019). To support the farmers and to improve the food quality in the supply chain, Blockchain technology can play a major role.

This developed model is very theoretical in nature and more research is required before field implementation. It is also required to improve the numerical data set by ENAM to define the quality of organic products also not related to develop. Transportation model for minimizing the transportation cost and delivery time is not considered in this study.

References

1. Abhilash PC, Singh N (2009) Pesticide use and application: an Indian scenario. J Hazard Mater 165(1–3):1–12
2. Aral S, Dellarocas C, Godes D (2013) Introduction to the special issue—social media and business transformation: a framework for research. Inf Syst Res 24(1):3–13
3. Brandt K, Molgaard JP (2001) Organic agriculture: does it enhance or reduce the nutritional value of plant foods? J Sci Food Agric 81(9):924–931
4. Corpes A (2015) Food fraud and allergen management. Persp Public Health 135(4):172
5. Dania WAP, Xing K, Amer Y (2018) Collaboration behavioural factors for sustainable agri-food supply chains: a systematic review. J Clean Prod 186:851–864
6. De Boer IJ (2003) Environmental impact assessment of conventional and organic milk production. Livestock Prod Sci 80(1–2):69–77
7. Down To Earth. India's deepening farm crisis: 76% farmers wants to give-up farming shows study. Available at: https://www.downtoearth.org.in/news/indias-deepening-farm-crisis-76-farmers-want-to-give-up-farming-shows-study-43728
8. El Benni N, Stolz H, Home R, Kendall H, Kuznesof S, Clark B, Zhong Q (2019) Product attributes and consumer attitudes affecting the preferences for infant milk formula in China. Food Qual Prefer 71:25–33

9. Flores-Munguia ME, Bermudez-Almada MC, Vázquez-Moreno L (2000) A research note: detection of adulteration in processed traditional meat products. J Muscle Foods 11(4): 319–325
10. Hong E, Lee SY, Jeong JY, Park JM, Kim BH, Kwon K, Chun HS (2017) Modern analytical methods for the detection of food fraud and adulteration by food category. J Sci Food Agric 97(12):3877–3896
11. Kelle P, Akbulut A (2005) The role of ERP tools in supply chain information sharing, cooperation, and cost optimization. Int J Prod Econ 93:41–52
12. Lin Q, Wang H, Pei X, Wang J (2019) Food safety traceability system based on blockchain and EPCIS. IEEE Access 7:20698–20707
13. Marques Vieira L, Dutra De Barcellos M, Hoppe A, Bitencourt da Silva S (2013) An analysis of value in an organic food supply chain. Br Food J 115(10):1454–1472
14. Mäder P, Fliessbach A, Dubois D, Gunst L, Fried P, Niggli U (2002) Soil fertility, biodiversity in organic farming. Science 296(5573):1694–1697
15. Oliveira G, Alewijn M, Boerrigter-Eenling R, van Ruth S (2015) Compositional signatures of conventional, free range, and organic pork meat using fingerprint techniques. Foods 4(3): 359–375
16. Opara LU (2003) Traceability in agriculture and food supply chain: a review of basic concepts, technological implications, and future prospects. J Food Agric Environ 1:101–106
17. Sari K (2010) Exploring the impacts of radio frequency identification (RFID) technology on supply chain performance. Eur J Oper Res 207(1):174–183
18. Schader C, Lampkin N, Christie M, Nemecek T, Gaillard G, Stolze M (2013) Evaluation of cost-effectiveness of organic farming support as an agri-environmental measure. Land Use Policy 31:196–208
19. The Economic Times. Available at: https://economictimes.indiatimes.com/industry/cons-products/food/68-milk-milk-products-in-india-not-as-per-fssai-standard-official/articleshow/65689621.cms?from=mdr. Accessed on 15/01/2019
20. The Guardian. Can we feed 10 billion people on organic farming alone? Available at: https://www.theguardian.com/sustainable-business/2016/aug/14/organic-farming-agriculture-world-hunger
21. TongKe F (2013) Smart agriculture based on cloud computing and IOT. J Converg Inf Technol 8(2)
22. Woodford MH, Dudley JP (2002) Bioweapons, biodiversity, and ecocide: potential effects of biological weapons on biological diversity: bioweapon disease outbreaks could cause the extinction of endangered wildlife species, and the extirpation of indigenous cultures. Bioscience 52(7):583–592

A New Heuristic for Solving Open Vehicle Routing Problem with Capacity Constraints

Bapi Raju Vangipurapu, Rambabu Govada and Narayana Rao Kandukuri

Abstract In this paper, a new heuristic is developed to solve an open vehicle routing problem (OVRP). This heuristic considers minimizing the number of vehicles as the primary objective and minimizing the total distance travelled as a secondary objective. The new method proposed uses a modified sweep algorithm that produces a solution with the least number of vehicles, in a relatively short amount of time. This objective is achieved by loading the vehicles nearly to their full capacity by skipping some of the customers if necessary. The new heuristic is tested on standard test instances found in the literature. The results are then compared with the eleven other methods found in the literature. This new heuristic is found to be quite effective compared to other methods if both the speed of computation and closeness to the best value are considered important. The complexity of the algorithm is only $O(n)$, and hence the algorithm produces the results in a relatively short time. The output from this method can be further improved by using metaheuristics like genetic algorithm.

Keywords OVRP · Minimum number of vehicles · Heuristic · Skipping customer

B. R. Vangipurapu (✉)
Department of Mechanical Engineering, V. R. Siddhartha Engineering College, Vijayawada, Andhra Pradesh 520007, India

R. Govada
Department of Mechanical Engineering, Andhra University, Visakhapatnam, Andhra Pradesh 530003, India

N. R. Kandukuri
Department of Mechanical Engineering, Government Polytechnic College, Vijayawada, Andhra Pradesh, India

BBVL. Deepak et al. (eds.), *Innovative Product Design and Intelligent Manufacturing Systems*, Lecture Notes in Mechanical Engineering,
https://doi.org/10.1007/978-981-15-2696-1_87

1 Introduction

Transportation is one of the important costs of logistics. Capacitated vehicle routing problem (CVRP) is one of the interesting optimization problems. Open vehicle routing problem (OVRP) is a special type of vehicle routing problem in which the vehicle is not required to return to the depot. Minimizing the number of vehicles and minimizing the total distance are two main objectives which are considered by many researchers in OVRP. Many times, it is seen hiring or maintaining a new vehicle is costlier than using existing vehicles for longer distances. Hence, most methods use hierarchical objectives, which consider minimizing the number of vehicles used as the primary objective and minimizing the total distance as the secondary objective. Most of the time, these two objectives are conflicting meaning when the number of vehicles is less the distance travelled is more and vice versa. Hence, an attempt should be made to first get a solution satisfying the primary objective of minimizing the number of vehicles and later this initial solution can be improvised to reduce the total distance travelled without increasing the number of vehicles.

2 Literature Survey

Vehicle routing problem (VRP) was first proposed by Dantzig and Ramser [1] and has been proved to be an NP-hard problem. Since OVRP is a special case of VRP where vehicles do not return to the depot, it is also an NP-hard problem. Exact methods of solving OVRP are not practical as the computational time increases exponentially with the larger number of customers. Hence, heuristics and meta-heuristics are used to get the solution. Sariklis and Powell [2] proposed a two-phase heuristic which first assigns customers to cluster and then build a Hamiltonian path for each cluster, and Tarantilis et al. [3] described a population-based heuristic, while Tarantilis et al. [4, 5] presented threshold accepting metaheuristics, and Brandão [6], Fu et al. [7, 8] proposed tabu search heuristics. Pisinger and Ropke [9] presented an adaptive large neighbourhood search heuristic which follows a destruct-and-repair paradigm. Li et al. [10] described a record-to-record travel heuristic. Bapi Raju et al. [11] developed a heuristic for VRP problem which minimizes the number of vehicles even when the tightness ratio (total demand/total capacity) is high.

In this paper, a new heuristic, based upon sweep algorithm, is developed to solve OVRP problems. The output from this heuristic always has the least number of vehicles even when the tightness (total demand/total capacity) is equal to 1. This is achieved by loading the vehicles nearly to their full capacity whenever the tightness is near 1. Note that this algorithm uses at most 4 different ways of grouping the customers where n is the number of customers. Hence, the complexity increases only linearly with an increase in the number of customers. The complexity of the algorithm is only $O(n)$. Hence, the computational time is only minimal even when the number of customers and the number of vehicles are large. The new modified sweep is based on two new propositions. The first proposition is that sweeping should start from a node whose angular distance from the next consecutive node with respect to depot is very large. This is explained using Fig. 1. As can be seen from Fig. 1, it makes sense to start sweeping from customer 1 in the anticlockwise direction or from customer 12 in the clockwise direction. This prevents customer 1 and customer 12 being in the same vehicle and subsequently increasing the travelled distance. In other words, this would ensure routes are densely packed with customers and would reduce the total distance travelled.

The second proposition is that vehicles can be loaded nearly to their full capacity if some of the customers can be skipped during sweeping. This would result in less number of vehicles. This is explained using Figs. 2 and 3. In these figures, capacity is mentioned in the brackets for each customer. The capacity of each vehicle is assumed to be 100. In Fig. 2, normal sweeping is done from customer 1 in the anticlockwise direction and a new vehicle is formed whenever the total demand exceeds the capacity of the vehicle. This results in a total of four vehicles. Figure 3 corresponds to the modified sweep. Here, after adding customers 5, 6, 7 to vehicle 2 customers 8, 9, 10, 11 are skipped (since they result in violating capacity constraint) and customer 12 is added. This would result in only three vehicles. Depending upon the tightness of the vehicles, two separate algorithms are developed, one using the normal sweep and the other using the modified sweep. For a given problem, first Algorithm 1 which uses normal sweep is used to obtain a solution. If this does not result in the least number of vehicles, then Algorithm 2 is used.

A modified travelling salesman problem is used which considers the distance from the last vehicle to the depot to be zero (because this is an OVRP).

Rest of the document is organized as follows. The methodology is explained in the third section. Next section presents the results of the computational experiments that are conducted on test data and the discussion on the results obtained. Finally, in the last section, the conclusion of this work is summarized.

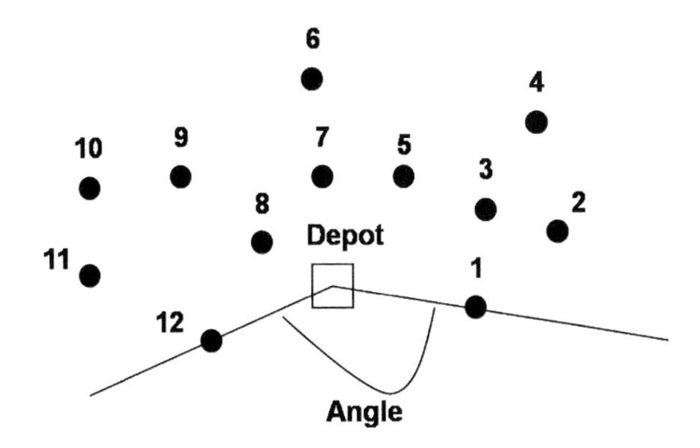

Fig. 1 Angle between successive customers

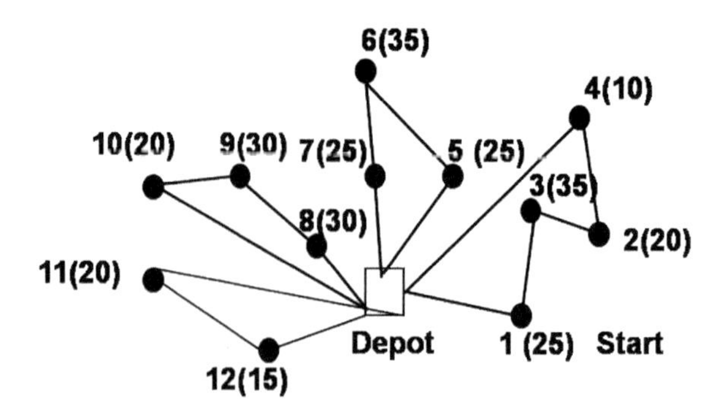

Fig. 2 Solution using a normal sweep (vehicle capacity 100)

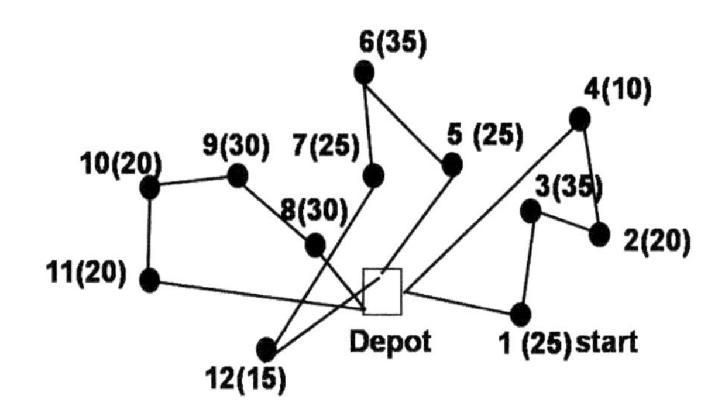

Fig. 3 Solution using a modified sweep (vehicle capacity 100)

3 Methodology

In order to solve a given OVRP problem, a minimum number of vehicles, required to transport, is calculated by using the total demand and total capacity of the vehicles. Then, customers are assigned to vehicles using any one of the two algorithms. The first algorithm (Algorithm 1) is based on the normal sweep algorithm, and it does not skip any customers. The second algorithm (Algorithm 2) skips some of the customers so as to load the vehicle to its maximum capacity. This may increase the distance but decreases the number of vehicles required. A solution is tried with Algorithm 1, and if this does not result in the least number of vehicles Algorithms 2 (which can handle problems with tightness close to 1) is used. After the customers are grouped, the individual routes of each vehicle are formed by modified travelling salesman problem to get the final solution.

The flow chart for the above process is shown in Fig. 4.

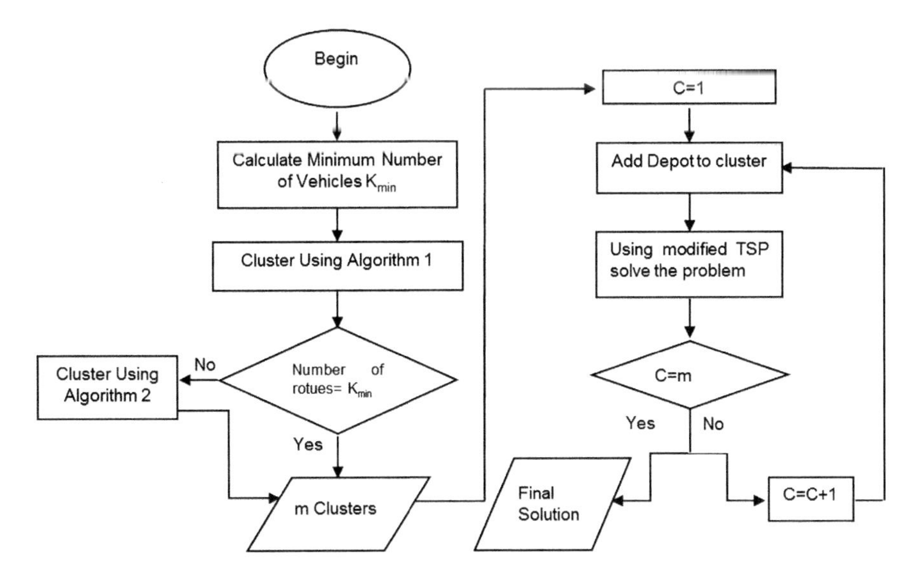

Fig. 4 Flow chart showing the methodology

3.1 Algorithm 1

1. Calculate the minimum number of vehicles $K_{\min}$ using the formula. $K_{\min}$ = ((Total Demand)/(Capacity of each vehicle)) rounded up to the nearest integer.
2. Locate the depot as the centre. Compute the polar coordinates of each customer with respect to the depot. Sort all customers with respect to polar angle. Calculate the angular distances between the successive nodes with respect to depot as shown in Fig. 1. Identify the two successive nodes which form the maximum angular distance. Let these nodes be N1 and N2.
3. Starting from customer N1, sweeping is done by increasing polar angle in the clockwise direction. Clusters are formed using the standard sweep. The formation of the route for each cluster is done by using modified TSP, and the distance travelled for each route is calculated. The total distance (dist 1) for the solution is obtained by summing up the solutions for each route. Starting from customer N2, sweeping is done by increasing polar angle in the anticlockwise direction. The total distance (dist 2) for the solution is obtained by using the same procedure as above.
4. The best solution from the above two (i.e., solution corresponding to the minimum of dist 1 and dist 2) is selected as the final solution.

3.2 *Algorithm 2*

1. Calculation of K_{min} and identification of the two successive nodes between which the angular distance is maximum is done using the method which is explained in Algorithm 1. Let these nodes be N1 and N2.
2. Starting from customer N1, sweeping is done by increasing polar angle in the clockwise direction. Assigning of customers is continued until constraints are violated. If the current vehicle is not having minimum specified per cent of capacity (example 95%), swap the last customer in the current route with a nearest unrouted customer who would meet the specified capacity level. If no customer is found, then the swapping is tried with the last but one customer. This process is repeated until a suitable customer is found who would meet the specified capacity level. The minimum specified per cent of capacity is dependent on the tightness ratio. This would result in the loading of the vehicle to its maximum capacity. A new vehicle is started after this. This process is repeated until all customers are covered. If this solution does not result in the least number of vehicles, this solution is ignored. Otherwise, the solution for each route is calculated by using the modified travelling salesman problem (TSP) and the distance travelled for each route is calculated. The total distance (dist 1) for the solution is obtained by summing up the solutions for each route.
3. Starting from customer N2, sweeping is done by increasing polar angle in the anticlockwise direction. A second solution is obtained by using a similar step as above to get total distance (dist 2).
4. If both the solutions result in the least number of vehicles, then the solution corresponding to the least distance is the best solution. Otherwise, the solution with the least number of vehicles is the best solution.

4 Computational Experiments

Seven test instances from Christofides et al. [12] work have been selected. The new heuristic is run on the test instances, and the results are given in Table 1. The results from other papers mentioned in the literature survey are also presented. The new algorithms are implemented on MATLAB. The experiments have been done on a PC (Intel Core i3-3470 CPU at 3.20 GHz CPU, 4 GB RAM) with Windows 7 OS.

The total distance travelled and the time taken for all the test instances is calculated for each of the methods and is presented in Table 1. The solutions from methods 4, 5, 6, 7, 8 yield a higher number of vehicles than the minimum required. Hence, total distances for the solutions, corresponding to these methods, are not calculated. They are removed from further comparison. Method 1 is fastest taking about 7 s, but the solution is much inferior. The total distance travelled is approximately 39% more than the best (least distance). Methods 2, 9, 10, 11 provide reasonably good solutions (distance travelled is not more than 2% than the

Table 1 Comparison of results from various methods

Problem	n	$K_{\min}$	Distance	Time (s)	Distance	Time (s)	Distance	Time (s)	Distance	Time (s)
			1-CFRS [2]		2-TSAK [6]		3-TSAN [6]		4-BR [3]	
C1	50	5	488.204	0.22	416.1	88.8	438.2	1.7	(6)412.96	7.2
C2	75	10	795.334	0.16	574.5	167.5	584.7	4.9	(11)564.06	25.8
C3	100	8	815.042	0.94	641.6	325.3	643.4	12.3	(9)641.77	28.8
C4	150	12	1034.139	0.88	740.8	870.2	767.4	33.2	735.47	75
C5	199	16	1349.709	2.2	953.4	1415	1010.9	116.9	(17)877.13	226
C11	120	7	828.254	1.54	683.4	696	713.3	15.7	(10)679.38	29.4
C12	100	10	882.265	0.76	535.1	233.6	543.2	7.8	534.24	14.4
Total			6192.947	6.7	4544.9	3796.4	4701.1	192.5	–	406.6
			5-BATA [4]		6-LBTA [5]		7-TSF [7]		8-TSR [8]	
C1	50	5	(6)412.96	38.62	(6)412.96	28.75	408.5	170	413.3	65
C2	75	10	(11)564.06	68.89	(11)564.06	61.21	587.8	202	570.6	197.8
C3	100	8	642.42	56.54	(9)639.57	53.78	644.3	720	617	367.6
C4	150	12	736.89	81.69	733.68	84.13	734.5	1610	741.1	1094
C5	199	16	879.37	98.13	(17)870.26	96.47	(17)878.0	2061	(17)886.6	1279.9
C11	120	7	(9)679.60	37.67	(10)678.54	25.36	753.8	736	716.5	38.9
C12	100	10	534.24	84.54	534.24	64.59	549.9	413	534.8	30.9
Total			–	466.08	–	414.29		5912		3124.1
			9-ALNS 25 K [9]		10-ALNS 50 K [9]		11-ORTR(10)		(New method developed by the author)	
C1	50	5	416.06	120	416.06	230	416.06	6.2	436.83032	4.221
C2	75	10	567.14	360	567.14	530	567.14	31.3	666.96628	5.276

(continued)

Table 1 (continued)

			9-ALNS 25 K [9]		10-ALNS 50 K [9]		11-ORTR(10)		(New method developed by the author)	
C3	100	8	641.76	850	641.76	1280	639.74	39.5	674.21606	9.521
C4	150	12	733.13	1790	733.13	2790	733.13	128.6	811.3622	14.19
C5	199	16	897.93	1240	896.08	2370	924.96	380.6	955.37176	17.91
C11	120	7	682.12	730	682.12	1410	682.54	121.6	854.52124	14.28
C12	100	10	534.24	800	534.24	1180	534.24	32.9	604.88708	8.348
Total			4472.38	5890	4470.53	9790	4497.81	740.7	5004.155	73.746

best), but the time taken by these methods is very large (not less than 740 s). Method 3 provides a solution which is approximately 5% higher than the best, but the time taken is 28 times more than the best (which is 7 s by Method 1).

The newly developed heuristic provides the solution which is only 12% higher distance than the best, and the time taken is nearly 11% higher than the best. Hence, this algorithm can be used to get an approximate solution somewhat quickly. Further, this solution can be used as an initial solution for other methods for improving the solution.

5 Conclusion

A new heuristic for solving OVRP is developed. The main objective of this heuristic is to minimize the number of vehicles. The secondary objective is to minimize the total distance travelled. This heuristic provides a reasonably good solution using a short amount of processing time. The output from this heuristic is compared with eleven other methods from the literature. The comparison of the outputs shows that the new heuristic can be quite useful if both the parameters, the speed of computation and closeness to the best value, are important. The output from this heuristic can be further improved by metaheuristics if necessary. This heuristic can be used to get a rough idea about the final solution of any OVRP problem within a short interval of time.

References

1. Dantzig GB, Ramser JH (1959) The truck dispatching problem. Manag Sci Informs 6(1):80–91
2. Sariklis D, Powell S (2000) A heuristic method for the open vehicle routing problem. J Oper Res Soc 51:564–573
3. Tarantilis CD, Diakoulaki D, Kiranoudis CT (2004) Combination of geographical information system and efficient routing algorithms for real life distribution operations. Eur J Oper Res 152:437–453
4. Tarantilis CD, Ioannou G, Kiranoudis CT, Prastacos GP (2004) A threshold accepting approach to the open vehicle routing problem. RAIRO Oper Res 38:345–360
5. Tarantilis CD, Ioannou G, Kiranoudis CT, Prastacos GP (2005) Solving the open vehicle routing problem via a single parameter metaheuristic algorithm. J Oper Res Soc 56:588–596
6. Brandão J (2004) A tabu search algorithm for the open vehicle routing problem. Eur J Oper Res 157:552–564
7. Fu Z, Eglese R, Li LYO (2005) A new tabu search heuristic for the open vehicle routing problem. J Oper Res Soc 56:267–274
8. Fu Z, Eglese R, Li LYO (2006) Corrigendum: a new tabu search heuristic for the open vehicle routing problem. J Oper Res Soc 57:1018
9. Pisinger D, Ropke S (2007) A general heuristic for vehicle routing problems. Comp Oper Res 34:2403–2435

10. Li F, Golden B, Wasil E (2007) The open vehicle routing problem: algorithms, large-scale test problems, and computational results. Comput Oper Res 34(10):2918–2930
11. Bapi Raju V, Rambabu G, Narayana Rao K (2019) A heuristic for solving CVRP. ICRDME (in press)
12. Christofides N, Mingozzi A, Toth P (1979) The vehicle routing problem. In: Christofides N, Mingozzi A, Toth P, Sandi C (eds) Combinatorial optimization. Wiley, Chichester, pp 313–338

Barriers in Sharing of Agricultural Information in Odisha (India): A Critical Study of Small-Scale Farmers

Suchismita Satapathy and Debesh Mishra

Abstract Based on extensive review of the literature, the barriers of information sharing in agricultural sectors of Odisha in India were identified in this study. The responses were obtained from 64 farmers using questionnaires on the barriers of information in the Likert scale (1 = strongly disagree, 2 = disagree, 3 = neither disagree nor agree, 4 = agree, 5 = strongly agree). Then, the factor analysis was done to find out the most significant barriers of information. Subsequently, the BWM ranking was performed to rank the most significant barriers of information.

Keywords Farmers · Agriculture · Information · Barriers · BWM · Odisha

1 Introduction

The paddy growing small-scale farmers need information for doing betterment in agriculture. The community needs to set their knowledge accordingly in order to get information [15]. This indicates that for socio-economic development, the information plays a vital role; as for attaining better livelihoods, people get empowered to make conversant choices. By assessing, using, and understanding, only the realization of the value of information can be made. For planning and the execution of programs, flow of information is reported to be essential [28]. Matovelo [19] has argued that the looking for evaluation and utilization of information can be better continued into the ways of life of target group, and subsequently, to become self-sustaining, the information-seeking practice needs to be internalized. Mchombu [21] has found positive impacts of information services for the improvement of agricultural practices in Tanzania. For rural development, one of the most valuable resources was reported to be information [5, 22, 23]. Information not only assists

S. Satapathy (✉) · D. Mishra
SME, KIIT Deemed to be University, Bhubaneswar, Odisha, India

BBVL. Deepak et al. (eds.), *Innovative Product Design and Intelligent Manufacturing Systems*, Lecture Notes in Mechanical Engineering,
https://doi.org/10.1007/978-981-15-2696-1_88

the small-scale farmers to make informed decisions but also helps to take proper actions. Burton [4] has reported that most of the people in underdeveloped areas are unaware of their lacking of information that is available to solve their problems. Between the information flow and development in agricultural sectors, there exists a positive relationship [14]. The flow of information in agricultural sectors plays a key role for small-scale agricultural improvement, in linking increased production to markets, improvement in food security, rural livelihoods, and national economy. The realization of agricultural productivity will occur only when the farmers and the market information are linked together [17].

The objective of the study is the identification of the barriers to access the agricultural information experienced by small-scale farmers in Odisha (India), such that it will help in designing better information systems to meet the needs of farmers. The growth of paddy farming by small-scale farmers in Odisha (India) has been stagnating in the recent years. Among the factors that lead to this stagnation is the lack of knowledge of acquiring information by the farmers. Lack of agricultural information could be gender specific, such that certain gender could be richer in information than the others. In developing countries, the rural women are normally overloaded, and even though education programs are accessible, they get no time to learn, read, or acquire information [30]. The lower literacy and financial incomes of women than the corresponding males make them unable to read or purchase pesticide information [24]. Poor communication facilities and information infrastructure are stated as the physical barriers to access information [7, 17]. When using printed information materials, illiteracy was stated as a key barrier for information accessing [5, 20]. Furthermore, radio and television are good sources of information, but its uses are limited as these are costly, expensive use of batteries, improper timing of the programmes, poor quality messages, and also, there is a lack of electrification in rural areas [6, 13]. Similarly, access to Internet and use of online information systems are reported of excessive costs [26]. Other barriers to the flow of information are as follows: language barriers of small-scale farmers in developing countries [28, 34], less number of agricultural extension workers [3, 11, 29], attitude and practices related to gender [18], distance of information centre, domestic responsibilities and cultural inhibitions of women [10, 25], lacking of agricultural libraries in the vicinity of farmers [2], ignorance of information sources, extension agents with insufficient knowledge [1, 6, 9, 27], insufficient financial resources, and incomplete support from agricultural actors [16].

Siyao [35] has revealed that the barriers in accessing information of the agricultural sectors in Tanzania have led to the stagnation of sugar cane growth. The lacking in the means and facilities of information was reported as the barriers to access agricultural information. Moreover, information transfer is dependent on the level of trust and interaction in the information system, which is a result of regular interactions among the network actors to learn more about each other [8]. Phiri et al. [31] have considered the rural smallholder farmers in Mzimba North in Malawi, to investigate their information needs and barriers. Crop husbandry was found as the major information need of rural smallholder farmers, i.e. 149 (77.6%), and majority of farmers, i.e. 180 (94.8%) were aware of information sources. The study also found that personal experiences were the predominant information sources by rural smallholder farmers 185 (96%), and the major challenge faced was lack of mobility 147 (76.6%).

For the most part, in a basic decision-making process, an option from an arrangement of choices are recognized and chosen in light of the preferences of the choice maker(s). As much of the time, a few criteria are associated with this procedure, so these problems are famously known as multicriteria decision-making (MCDM) problems. The criteria included are esteemed contrastingly by various chiefs. In spite of the fact, few MCDM techniques have been proposed to discover the estimations of the criteria and the choices in light of their inclinations/ preferences. The best worst method (BWM) was suggested as one of the most recently developed methods based on comparisons. Additionally, this method required less information with more reliable comparisons [32].

1.1 Best Worth Method

The "best worst method (BWM)" has been reported for giving more consistent results as compared to AHP and is used for computing attribute weights with smaller number of pair-wise comparisons [29]. This method has been applied successfully by Rezaei et al. [31], Salimi, and Rezaei [32]. For instance, both the best worst method and VIKOR methodology have been used to evaluate the service quality of airline industry [21].

The steps involved in BWM method were as summarized below:

Step 1. Identifying a set of decision criteria:
Based on the literature review and experts opinion used to make a decision, the n criteria $\{C_1, C_2, \ldots, C_n\}$ were identified.
Step 2. Determining the most important and the least important criterion, such as "the best and the worst criteria"
Step 3. Determining the preference of the best criterion over all other criteria by using a number between 1 and 9. The obtained best to others vector would be:

$$A_B = \{a_{B1}, a_{B2}, \ldots, a_{Bn}\},$$

where a_{Bj} denoted the preference of the best criterion B over criterion j, and $a_{BB} = 1$.
Step 4. Determining the preference of all criterion over the worst criteria by using a number between 1 and 9. The obtained others to worst vector would be:

$$A_W = \{a_{1W}, a_{2W}, \ldots, a_{nW}\}^{\mathrm{T}},$$

where a_{jW} denotes the preference of the criterion j over the worst criterion W, and $a_{WW} = 1$.

Step 5. Estimating the optimal weights as $\{w_1^*, w_2^*, \ldots, w_n^*\}$.

The primary goal in this step was the optimal weights determination of the criterion to minimize the maximum absolute differences $\{|w_B - a_{Bj}w_j|, |w_j - a_{jw}w_w|\}$ for all j. Subsequently, the following minimax model can be generated:

$$\text{Minimax}\{|w_B - a_{Bj}w_j|, |w_j - a_{jw}w_w|\}$$

Subjected to,

$$\sum_j w_j = 1 \qquad (1)$$
$$w_j \geq 0, \quad \text{for all } j$$

Model (1) can be converted to the following linear model as:

$$\text{Min } \xi^*$$

Subjected to,

$$|w_B - a_{Bj}w_j| \leq \xi^*, \text{for all } j \; |w_j - a_{jw}w_w| \leq \xi^*, \text{for all } j \qquad (2)$$
$$\sum_j w_j w_j \geq 0, \quad \text{for all } j$$

The optimized weights $\{w_1^*, w_2^*, \ldots, w_n^*\}$ and the optimal value ξ^* were obtained by solving the above model (2). According to Rezaei [30], the consistency (Ksi*) of comparisons was also required to be evaluated, and for consistency, a value closer to 0 was necessary.

2 Research Methodology

Initially, based on the extensive review of the literature, the barriers of information sharing in agricultural sectors of Odisha in India were identified. Then, questionnaire was formed to find out the extent of influence of those barriers of information in agriculture, and these questionnaires were distributed to 75 farmers of Odisha to get their responses in the Likert scale (1 = strongly disagree, 2 = disagree, 3 = neither disagree nor agree, 4 = agree, 5 = strongly agree). Based on the total responses obtained from 64 farmers, the factor analysis [12] was done using Minitab 2017 version to find out the most significant barriers of information. Then, by considering eight numbers of experts' opinion, the BWM ranking was done for the most significant barriers of information using "BWM-Solver.xlsx software" by Rezaei [33].

3 Results and Discussion

From the extensive review of the literature, the following 16 barriers of information sharing in agricultural sectors of Odisha in India were obtained such as poor communication facilities, poor information infrastructure, illiteracy, excessive costs, language barriers, less number of agricultural extension workers, attitude and practices related to gender, distance of information centre, domestic responsibilities and cultural inhibitions of women, lack of agricultural libraries in the vicinity of farmers, ignorance of information sources, extension agents with insufficient knowledge, insufficient financial resources, incomplete support from agricultural actors, lack of trust, and lack of mobility, respectively. Factor analysis of these 16 barriers of information by using "Minitab 2017 version" was done as illustrated in Table 1, to obtain the most significant barriers of information with factor loading values of more than or equal to 0.5. It was found that only 9 barriers of information were found as the most significant barriers of information in agriculture. It was seen that four variables represented the dominant factor 1, three variables represented the dominant factor 2, and one variable represented the dominant factor 3, respectively (Table 1). Variables or barriers of information under factor 1 were illiteracy, attitude and practices related to gender, domestic responsibilities and cultural inhibitions of women, and lacking of agricultural libraries in the vicinity of farmers. Variables or barriers of information under factor 2 were poor information infrastructure, language barriers, and distance of information centre. Similarly, the variable or barrier of information under factor 3 was poor communication facilities.

The obtained nine most significant barriers of information in the agricultural sectors of Odisha were further considered as the criteria for the BWM ranking as follows: poor communication facilities (A), poor information infrastructure (B),

Table 1 Factor analysis of the barriers of information ($N = 64$)

Variables	Factor 1	Factor 2	Factor 3	Communality
Poor communication facilities			0.613	0.746
Poor information infrastructure		0.535		0.518
Illiteracy	0.845			0.741
Language barriers		0.700		0.808
Attitude and practices related to gender	0.932			0.888
Distance of information centre		0.635		0.621
Domestic responsibilities and cultural inhibitions of women	0.765			0.596
Lacking of agricultural libraries in the vicinity of farmers	0.542			0.814
Insufficient financial resources			0.658	0.735
Variance	4.9581	3.0595	2.5991	10.6167
% Var.	0.177	0.109	0.093	0.379

N = Total number of farmers

illiteracy (C), language barriers (D), attitude and practices related to gender (E), distance of information centre (F), domestic responsibilities and cultural inhibitions of women (G), lacking of agricultural libraries in the vicinity of farmers (H), and insufficient financial resources (I).

3.1 BWM Ranking of Significant Barriers of Information in Agriculture

The best and worst criteria were selected among all the nine criterion by experts, and the preference of best criteria over all other criterion was determined by experts on a scale of 1–9. Similarly, the preference of other criterion to the worst criteria was determined on a scale of 1–9. After getting preference rating for all the nine criterion, optimized weights of all the criterion as well as consistency value were obtained using "BWM-Solver.xlsx" software which uses Eq. (2). The obtained optimized weights of all the criterion and consistency value were $\{w_1^* = 0.08058842$, $w_2^* = 0.13431404$, $w_3^* = 0.06715702$, $w_4^* = 0.29101375$, $w_5^* = 0.05756316$, $w_6^* = 0.04477135$, $w_7^* = 0.10073553$, $w_8^* = 0.02238567$, and $w_9^* = 0.20147106\}$, and Ksi* = 0.11192837, respectively, as shown in Fig. 1. Then, based on the corresponding optimized weight values of different criteria, the graph was plotted (Fig. 2) taking criteria on *X*-axis and optimized weight values on *Y*-axis, respectively. It was clear from Fig. 2 that language barriers ranked the first, whereas lacking of agricultural libraries in the vicinity of farmers ranked the last as the barriers of information in the agricultural sectors of Odisha (India).

Criteria Number = 9	Criterion 1	Criterion 2	Criterion 3	Criterion 4	Criterion 5	Criterion 6	Criterion 7	Criterion 8	Criterion 9
Names of Criteria	A	B	C	D	E	F	G	H	I

Select the Best	G
Select the Worst	C

Best to Others	A	B	C	D	E	F	G	H	I
G	5	3	6	1	7	9	4	8	2

Others to the Worst	C
A	2
B	5
C	6
D	9
E	3
F	7
G	8
H	1
I	4

Weights	A	B	C	D	E	F	G	H	I
	0.08058842	0.13431404	0.06715702	0.29101375	0.05756316	0.04477135	0.10073553	0.02238567	0.20147106

Ksi*	0.11192837

Fig. 1 BWM ranking of barriers of information in agricultural sectors of Odisha (India)

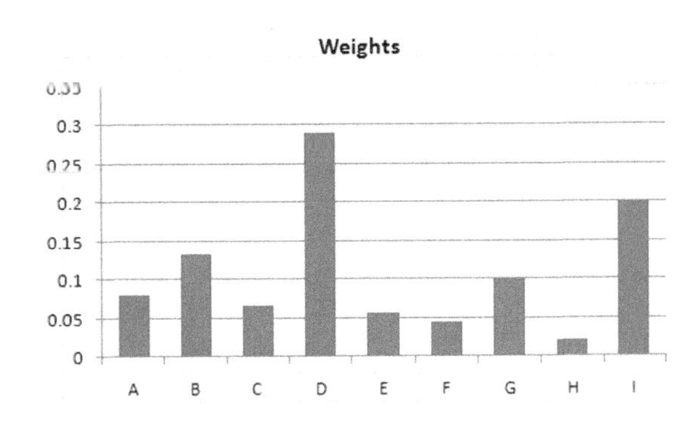

Fig. 2 Weights distribution of barriers of information in agricultural sectors of Odisha (India)

4 Conclusion

From the extensive review of the literature, sixteen barriers of information sharing in agricultural sectors of Odisha in India were obtained such as poor communication facilities, poor information infrastructure, illiteracy, excessive costs, language barriers, less number of agricultural extension workers, attitude and practices related to gender, distance of information centre, domestic responsibilities and cultural inhibitions of women, lacking of agricultural libraries in the vicinity of farmers, ignorance of information sources, extension agents with insufficient knowledge, insufficient financial resources, incomplete support from agricultural actors, lack of trust, and lack of mobility, respectively. Then, nine most significant barriers of information obtained after factor analysis were further considered as the criteria for the BWM ranking were as follows: poor communication facilities, poor information infrastructure, illiteracy, language barriers, attitude and practices related to gender, distance of information centre, domestic responsibilities and cultural inhibitions of women, lacking of agricultural libraries in the vicinity of farmers, and insufficient financial resources, respectively. By using the BWM method, it was observed that language barriers ranked the first, whereas lacking of agricultural libraries in the vicinity of farmers ranked the last.

Acknowledgements We would like to express our sincere thanks to all the farmers who participated in this survey.

Author contributions Debesh Mishra collected, analysed, and interpreted all the data related to this study, and Dr. S. Satapathy performed a major contribution in writing the manuscript. Both the authors read and approved the final manuscript.

Funding There was no funding for carrying out this research.

Compliance with ethical standards

Conflict of interest The authors declare of having no conflict of interest.

Ethical approval All studies were conducted in accordance with principles for human experimentation as defined in the declaration of Helsinki and its later amendments or comparable ethical standards.

Informed consent Informed consent after explaining the nature of investigation was obtained from each participant in this study.

References

1. Abubakar SC (2007) Approaches, strategies and challenges for information management and usage in fostering sustainable agricultural growth and rural economic development in Sub-Saharan Africa. In: Conference paper at the IFLA fifth session of the committee on development information (CODI-V), 29 April–04 May 2007, Addis Ababa, Ethiopia
2. Aina LO, Dulle FW (1999) Meeting the information needs of small-scale dairy farmers in Tanzania. Int Assoc Agric Libr Docum Q Bull 44(3/4):173–176
3. Aina O (2006) Information provision to farmers in Africa: the library-extension services linkage. In: World library and information congress: 72nd IFLA general conference and council 20–24 Aug, Seoul
4. Burton S (2002) Development at any cost: ICTs and people's participation in South Africa. Communication 28(2):43–53
5. Carter I (1999) Locally generated printed materials in agriculture: experience from Uganda and Ghana. Education Research Report No. 31. London: Department for International Development (DfID)
6. Dutta R (2009) Information needs and information-seeking behaviour in developing countries: a review of the research. Int Inf Libr Rev 41(1):44–51
7. Ellen D (2003) Telecenters and the provision of community based access to electronic information in everyday life in UK. Inf Res 8(2):146. http://informationr.net/ir/8-2/paper146.html. Assessed 23 Sept 2018
8. Hilary RS, Sseguya H, Kibwika P (2017) Information quality, sharing and usage in farmer organizations: the case of rice value chains in Bugiri and Luwero Districts, Uganda. Cogent Food Agric 3:1350089
9. Ikoja-Odongo R, Ocholla DN (2003) Information needs and information seeking behaviour of artisan fisher folk of Uganda. Libr Inf Sci Res 25(1):89–105
10. Ikoja-Odongo JR (2008) Strengthening women farmers' information networks to contribute to the millennium development goals. http://www.referenceglobal.com/doi/abs/10.1555/9783598441257.161. Assessed 23 Sept 2018
11. Isinika AC, Mdoe NSY (2001) Improving farm management skills for poverty alleviation: the case of Njombe District. REPOA Research Report No. 01.1. UDSM: Mkuki na Nyota
12. Kaiser HF (1960) The application of electronic computers to factor analysis. Educ Psyc Measur 20:141–151
13. Kalusopa T (2005) The challenges of utilizing information communication technologies (ICTs) for the small-scale farmers in Zambia. Libr High Technol 23(3):414–424
14. Manda P (2002) Review of the state of research methodology in African librarianship. Univ Dar es Salaam Libr J 4:1–2
15. Marchionini G (1995) Information seeking in electronic environments. Cambridge University Press, Cambridge
16. Masambuka-Kanchewa F (2013) Agricultural information perceptions and behaviours of smallholder farmers in the central region of Malawi (Master's Thesis). In: Available from ProQuest dissertations and thesis, UMI No. 12 information development XX(X) 1544423. Available at http://search.proquest.com/docview/1437202152. Accessed 26 Nov 2018
17. Masuki KF, Kamugisha R, Mowo JG, Tanui J, Tukahirwa J, Mogoi J, Adera EO (2010) Role of mobile phones in improving communication and information delivery for agricultural development: lessons from South Western Uganda. ICT and development-research voices from Africa. In: International federation for information processing (IFIP) technical commission 9-relationship between computers and society. workshop at Makerere University, Uganda
18. Materu-Behtsa M (2004) Information support for gender and development issues. In: Njau A, Mruma T (eds) Gender and development in tanzania: past, present and future, Dar es Salaam, Women Research and Documentation Project

19. Matovelo DS (2008) Enhancing farmers' access to and use of agricultural information for empowerment and improved livelihood. Unpublished Doctoral Dissertation, EJISDC-2012, 51(6):1–19. University of Dar es Salaam, Tanzania
20. Mbozi P (2002) Ground up: facilitating networking and sharing in sub-Saharan Africa. LEISA 18(2):13–15
21. Mchombu KJ (2003) Information dissemination for development: an impact study. Inf Dev 19(3):111–126
22. Meyer HWJ (2003) Information use in rural development. New Rev Inf Behav Res 4(1): 109–125
23. Morrow K, Nielsen F, Wettasinha C (2002) Changing information flows. LEISA 18(2):4–5
24. Naidoo S, London L, Burdorf A, Naidoo R, Kromhout H (2008) Agricultural activities, pesticides use and occupational hazards among women working in small-scale farming in Northern KwaZulu-Natal, South Africa. Int J Occup Environ Health 14(3):218–224
25. Nath V (2001) Empowerment and governance through information and communication technologies: women's perspective. Int Inf Libr Rev 33:317–339
26. Nicholas D (1996) Assessing information needs: tools and techniques. ASLIB, London
27. Odini S (2014) Access to and use of agricultural information by small scale women farmers in support of efforts to attain food security in Vihiga County, Kenya. J Emerg Trends Econ Manag Sci 5(2):80–86
28. Olorunda O, Oyelude A (2003) Professional women's information needs in developing countries: ICT as a catalyst. In: IFLA Women's issues, paper presented at IFLA women's issues section satellite meeting, Berlin, Germany
29. Ozawa VN (1995) Information needs of small scale farmers in Africa: the Nigerian example. Quarterly bulletin of the international association of agricultural information specialists, IAALD/CABI, University of Agriculture Makurdi, Nigeria 40(1). http://www.worldbank.org/html/cgiar/newsletter/june97/9nigeria.html. Assessed 26 Sept 2018
30. Park C (2007) Information needs of women in developing countries: meeting the information needs of rural and informal economy women in developing countries. http://www.scribd.com/doc/92460/Women-in-Developing-Countries. Assessed 26 Sept 2018
31. Phiri A, Chipeta GT, Chawinga WD (2015) Information needs and barriers of rural smallholder farmers in developing countries: a case study of rural smallholder farmers in Malawi. Inf Dev 1–14:421. https://doi.org/10.1177/0266666918755222
32. Rezaei J (2015) Best-worst multi-criteria decision-making method. Omega 53:49–57
33. Rezaei J (2016) Best-worst multi-criteria decision-making method: some properties and a linear model. Omega 64:126–130
34. Rwazo AJM (2007) Readers forum: small-scale farmers need information to reduce pesticides risks. Southern Africa Pesticides Newsl 2(1):1–2. http://web.uct.ac.za/depts/oehru/dox/vol2issue1.pdf. Assessed 23 Sept 2018
35. Siyao PO (2012) Barriers in Accessing Agricultural Information in Tanzania with a Gender Perspective: The Case Study of Small-Scale Sugar Cane Growers in Kilombero District. Elect J Inf Syst Dev Count 51(6):1–19

Development Inventory Model of Price-Dependent Perishable Products in Sustainable Environment

Bapi Raju Vangipurapu, Rambabu Govada and Narayana Rao Kandukuri

Abstract In this paper, an inventory model, which maximizes the profit of the vendor, in a sustainable environment is developed. It is assumed that the vendor supplies price-sensitive perishable goods. This model determines the optimal order quantity and the optimal price of the items. First, a model is developed without taking sustainability into consideration. This model is then used as a basis for the second model which takes sustainability into account. The second model incorporates carbon tax, imposed on freight transportation, into the base model. It is found that, in the second model, the optimal order quantity decreases and the optimal price of the item increases. A sensitivity analysis is also carried out on the second model by varying the carbon tax. It is found that the profit decreases more or less linearly with the tax rate in the second model.

Keywords Perishable items · Price-dependent · Sustainability · Carbon print

1 Introduction

There is growing pressure from governments to reduce emissions. Agriculture is a significant driver of global warming. Livestock agriculture produces around half of all man-made emissions, and it is growing continuously. For example, in India, the share of livestock in agricultural output grew steadily over the past two decades [1].

B. R. Vangipurapu (✉)
Department of Mechanical Engineering, V. R. Siddhartha Engineering College, Vijayawada 520007, India

R. Govada
Department of Mechanical Engineering, Andhra University, Visakhapatnam, Andhra Pradesh 530003, India

N. R. Kandukuri
Department of Mechanical Engineering, Government Polytechnic College, Vijayawada, Andhra Pradesh, India

BBVL. Deepak et al. (eds.), *Innovative Product Design and Intelligent Manufacturing Systems*, Lecture Notes in Mechanical Engineering,
https://doi.org/10.1007/978-981-15-2696-1_89

Apart from production, transportation also adds to greenhouse gas production. Freight traffic has contributed nearly 3% to global CO emissions as per Márquez-Ramos [2]. Techniques to improve sustainability include better operational procedures like reverse logistics, route optimization, organic farming, process re-engineering, use of renewable energy, optimization of various dairy plant operations, shifting to electric vehicles from diesel vehicles, technology-driven processes, multimodal freight transportation, carbon emission monitoring, sustainable performance measurement, etc [3]. However, all these methods would take a long time for implementation. An immediate solution would be to reduce the consumption and transportation of livestock products through government regulations. Most of these products generating high green gas emissions are perishable in nature, e.g., beef, lamb, etc. Hence, inventory management of these sorts of perishable goods is impacted by these regulations. It would be interesting to see how vendors, selling the price-sensitive perishable goods, would react to this situation. Hence, some study is made in this direction.

Literature Survey: Ghare and Schrader [4] were the first to develop an EOQ model for an item with exponential decay and constant demand. Nahmias [5] reviewed the relevant literature for determining suitable ordering policies for both fixed life perishable inventory and inventory subject to continuous exponential decay. Raafat [6] reviewed deteriorating inventory systems until 1991. Goyal and Giri [7] reviewed deteriorating inventory systems until 2001. Bakker et al. [8] reviewed deteriorating inventory systems from 2001 to 2012. Janssen et al. [9] reviewed deteriorating inventory models from 2012 to 2015. Sana et al. [10] have derived a production–inventory model of a deteriorating item, with linear time-varying demand. In this model, they have determined the optimal number of production cycles. Mukhopadhyay et al. [11] developed joint pricing and ordering policy for a deteriorating inventory. Alfares and Ghaithan [12] have studied an inventory model for a deteriorating product where the demand is a random variable with a known probability distribution. Ellerman et al. [13] discussed how emissions are capped, traded, and priced in the European Union's Emission Trading Scheme (EU ETS).

Integrating sustainability, while delivering perishable goods, would require innovative and reliable solutions. There are very few papers which deal with the management of perishable products under carbon constraints. Glover et al. [14] discussed various sustainable practices across the dairy supply chain. The sustainable operational practices in this category include the changing of order policies, production policies, etc. Gautam and Khanna [15] developed a production inventory model which considers carbon emission costs, apart from other costs. Hua et al. [16] found the optimum profit of perishable inventory system with freshness-dependent demand under two different carbon constraints which are price-based and cap-based. In this case, the author assumed that the carbon emissions come from shipping, holding, and deteriorating of perishable products. These emissions are translated into cost which is added to the already existing costs like holding cost and deterioration cost. Then, the profit/unit time is maximized considering all these costs. The only decision factor, in this case, is the order quantity (order cycle). As per the knowledge of the authors, no study has been done in the area of price-dependent perishable inventory system under carbon constraints. Hence, a study is made in this direction. Here, decision factors are order quantity and price of the perishable goods.

In this paper, two inventory models are developed. The first model maximizes the profit of a vendor selling perishable goods without considering the carbon constraints. The second model shows how profit is impacted when a carbon tax is introduced into the first model. Both the models determine the optimum values of price and order quantity for profit maximization. Later, a sensitivity analysis is carried out to see how the optimum output values vary with a change in carbon tax.

2 Assumptions and Notations

2.1 Assumptions

- Lead time is zero.
- Shortages not permitted.
- Deterioration rate is constant.
- Demand is primarily dependent on price only. Other factors which impact the demand remain constant.
- The deteriorated items are removed from the system and are not replaced. This involves some cost.
- Only a single item is considered.
- The carbon footprint is generated mainly from production and transportation. Some portion of the carbon print, generated from production and freight transportation, contains a fixed portion which is independent of the quantity transported, while the other one contains a variable portion which depends upon the quantity of the products shipped [16].
- Carbon print generated from all other sources like deterioration, holding of inventory, etc., is negligible.

2.2 Notations

- $I(t)$ Inventory level of the retailer as a function of time
- Q Order quantity (decision variable)
- h Holding cost
- O Ordering cost
- C Purchasing cost
- dt Deterioration cost
- S Retail selling price (decision variable)
- α Deterioration rate of the finished material

- t Time
- D Demand per unit time which is a function of selling price only
- $F'(D)$ The first derivative of selling price as a function of demand
- $f''(D)$ The second derivative of price as a function of demand
- T Time
- Tp Total profit/unit time
- ef Fixed carbon content for single order
- ev Variable carbon content (depends on quantity ordered)
- ct Carbon tax
- Optimum values of any variable are represented by a '*' superscript.

3 Development of Mathematical Models

3.1 *Profit Maximization Without Sustainability (Model 1)*

First, a mathematical model without considering any sustainability is considered. In order to maximize the profit/unit time of the vendor, optimal order quantity and selling price of the price-sensitive products are determined.

Inventory decreases due to demand and deterioration. Demand is a function of only selling price (independent of time). Rate of decrease in inventory due to demand is constant. However, decrease in inventory due to deterioration is dependent on time. Therefore, the rate of change of inventory w.r.t. time is given by

$$\frac{\mathrm{d}I}{\mathrm{d}t} = \alpha \times I(t) - D(s) \tag{1}$$

Using the boundary condition $I(t) = 0$ and solving

$$I(t) = \frac{D(s)}{\alpha} \mathrm{e}^{\alpha(T-t)} \tag{2}$$

Substituting $t = 0$ in (2), order quantity Q is obtained

$$Q = \frac{D(s)}{\alpha} \left(\mathrm{e}^{(\alpha T)} - 1 \right) \tag{3}$$

Total profit/unit time under these situations is calculated by Ata Allah Taleizadeh et al. [17] and is given by the following equation.

$$\begin{aligned}\mathrm{Tp} = S \times D(s) - \frac{O}{T} - \frac{h}{\alpha}\left[\frac{D(s)}{\alpha \times T}\left(\mathrm{e}^{(\alpha T)} - 1\right) - D(s)\right] \\ (C + \mathrm{d}t)\left[\frac{D(s)}{\alpha \times T}\left(\mathrm{e}^{(\alpha T)} - 1\right)\right] - \mathrm{d}t \times D(s)\end{aligned} \tag{4}$$

Ai et al. [18] and Shabani et al. [19] have shown that if αT is small $\mathrm{e}^{\alpha T}$ can be approximated by the first three terms of Taylor's expansion. Typical deterioration rate, α, of perishable products is of the order of 0.03%, and cycle time T is also small (since the products are perishable, they cannot have a long cycle time). Hence, $\mathrm{e}^{\alpha T}$ can be approximated by the first three terms of Taylor's expansion. After substituting the approximate value of $e^{\alpha T}$ in Eq. (4) and simplifying.

$$\mathrm{Tp} = S \times D(S) - (C \times D(S)) - (K \times D(S) \times T) - \left(\frac{O}{T}\right) \tag{5}$$

where K is a constant given by

$$K = \frac{h}{2} + C \times \frac{\alpha}{2} + \mathrm{d}t \times \frac{\alpha}{2} \tag{6}$$

Since demand D is a function of selling price S alone, S can be represented as a function of D. In other words, D is made independent variable and S is treated as a dependent variable. Let the relation between S and D be given by the following equation

$$S = f(D) \tag{7}$$

Substituting the values of S from Eq. (7) into Eq. (5), the following equation is obtained.

$$\mathrm{Tp} = f(D) \times D - (C \times D) - (K \times D \times T) - \left(\frac{O}{T}\right) \tag{8}$$

Differentiating Eq. (8) partially w.r.t. T and equating it to zero, we get

$$T = \sqrt{\frac{O}{K \times D}} \tag{9}$$

The above equation gives the values of T at which Tp is stationary w.r.t. T. Substituting the value of T into Eq. (8), partially differentiating w.r.t. D, and equating it with zero, we get

$$\text{Tp}' = f'(D) \times D + f(D) - C - \sqrt{\frac{O}{K \times D}} = 0 \tag{10}$$

This is an equation containing a single variable D so we can find out the roots of the equation by standard mathematical methods to obtain D^*. These roots will give stationary points w.r.t. to both the variables D and T and hence optimum profit.

Necessary conditions D^* to be optimum

1. The second derivative of the profit function should be negative (this ensures that this is the maxima of the profit function).
2. Total profit should be positive at the maximum point (this ensures that the solution is feasible).

Solution algorithm:

1. Solve Eq. (10) for all the roots which satisfy the necessary conditions mentioned above.
2. Select the global maximum from the above solutions. This will give optimum demand rate D^*.
3. Using the value of D^*, the value of T^* is obtained from Eq. (9).
4. The value of decision variable S^* is obtained by using Eq. (7).
5. Using the value of D^* and T^*, the other decision variable Q^* is obtained from Eq. (3).
6. Optimum profit/unit time Tp* is calculated using Eq. (8).

3.2 *Profit Maximization with Carbon Tax (Model 2)*

Next, consider the case where a carbon tax is imposed on every unit of carbon print produced. So, there is an extra cost carbon tax incurred and hence profit function needs to be modified. The additional input to this model is the carbon tax, per unit carbon print generated, which is stipulated by the government. Here, we assume some portion of the carbon print is fixed for a given order and some portion of the carbon print generated is dependent on the quantity of the product.

Total carbon print/unit time is given by

$$\text{Total carbon print/unit time} = \frac{\text{ef} + Q \times \text{ev}}{T} \tag{11}$$

Taking the value of the Q from Eq. (3), approximating the exponential term by Taylor's series as before, and simplifying, we get

$$\text{Total carbon print/unit time} = \frac{K1}{T} + K2 \times D + K3 \times D \times T \quad (12)$$

where $K1$, $K2$, $K3$ are constants defined by

$$K1 = \text{ef} \times \text{ct}, K2 = \text{ev} \times \text{ct}, K3 = \text{ev} \times \text{ct} \times \frac{\alpha \times T}{2} \quad (13)$$

Now, total profit/unit time is obtained by subtracting carbon tax/unit time from the gross profit/unit time.

$$Tp = f(D) \times D - (C \times D) - (K \times D \times T) - \left(\frac{O}{\mathrm{T}}\right) - \frac{K1}{\mathrm{T}} - (K2 \times D) - (K3 \times D \times T) \quad (14)$$

Differentiating Eq. (14) partially w.r.t. T, equating it to zero, and simplifying, we get the value of T as

$$T = \sqrt{\frac{O + K1}{(K + K3) \times D}} \quad (15)$$

Substituting the value of T into Eq. (14), partially differentiating w.r.t. D, and equating it with zero, we get

$$\mathrm{Tp}' = f'(D) \times D + f(D) - C - \sqrt{\frac{(O + K1) \times (K + K2)}{D}} \quad (16)$$

This is an equation containing a single variable D. The optimum values are obtained by the procedure which is explained in the above section for model 1.

4 Illustrative Examples

4.1 *Profit Maximization Without Sustainability (Model 1)*

Assume that demand and price are linearly related and are given by equation $D = a - (b \times S)$ where a and b are constants. The following data is assumed $a = 2100$, $b = 13$, $h = 3$, $O = \text{Rs. } 5000$, $C = \text{Rs. } 110$, $\mathrm{d}t = \text{Rs. } 15$, and $\alpha = 0.02$. A program written in MATLAB software is used to solve the problem.

4.2 *Profit Maximization with Carbon Tax (Model 2)*

In addition to the data present above, the following data is assumed: ef = 20 units of carbon print and ev = 0.2 units of carbon print/item. Let carbon tax, cut, be equal to Rs. 3/unit of carbon print. Remaining input data is taken from model 1. A program written in MATLAB software is used to solve the problem.

5 Results and Discussion

The results obtained from both models are summarized in Table 1. It is observed that in a sustainable environment, profit per unit time is always less. In the carbon tax environment, the profit/unit time decreases by nearly 4.5%.

6 Sensitivity Analysis

A sensitivity analysis is carried for the model. The carbon tax is varied from 0 to 5, and the resulting optimum values are plotted in a graph as a function of a carbon tax. This is depicted in Fig. 1.

As the carbon tax rate increases, the optimum quantity to be ordered reduces and the optimum price increases. This is to be expected since, as the carbon tax increases, the model tries to reduce the carbon prints by reducing the quantity to be shipped and compensating the loss partially by increasing the price.

Next, the variations of 'optimum profit per unit time' and 'generated carbon print per unit time,' with respect to the carbon tax rate, are shown in Fig. 2. It may also be noted that as the carbon tax decreases the optimum profit moves toward the optimum profit in model 1. This is because, as the carbon tax decreases, the restrictions from carbon prints would gradually decrease.

Table 1 Comparison of outputs from two different models

	Model 1: no regulation	Model 2: carbon tax regulation (tax at the rate of Rs. 3/unit carbon print generated)
D^*	290.262782	285.671410
T^*	2.502782	2.535183
S^*	Rs. 139.210555	Rs. 139.563738
Q^*	744.953412	742.904051
Tp^*	Rs. 4483.182869	Rs. 4281.559968
Carbon print*	–	66.496512 units

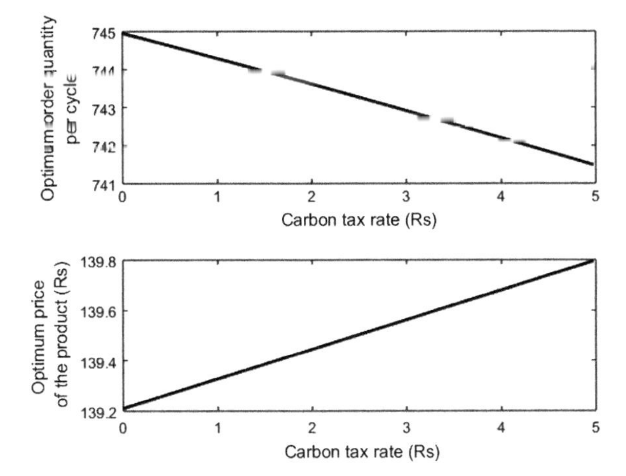

Fig. 1 Variation of the price and order quantity as a function of the carbon tax rate

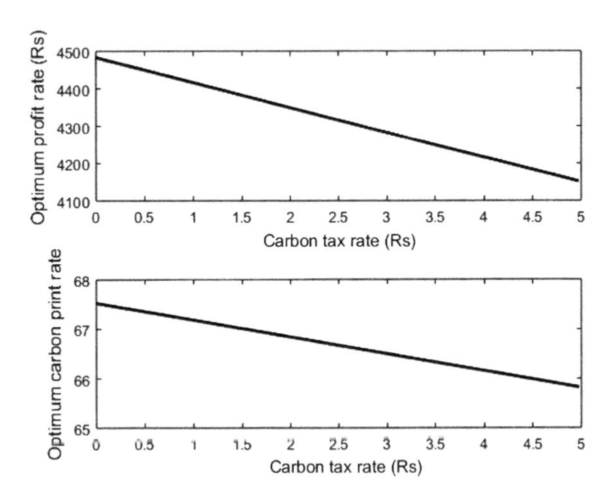

Fig. 2 Variation of the optimum profit and carbon print generated as a function of the carbon tax rate

7 Conclusion

Two different models are developed for maximizing profit of the retailer, who is selling perishable goods, like fruits, meat, etc. First, a model is developed, without taking sustainability into consideration. This model is then taken as a basis for developing the second model with carbon tax environment. In general, the models under a sustainable environment try to reduce the order quantity in order to meet the government regulations resulting in a reduction of the profit. But the loss in profit is partially compensated by the price increase. These sorts of models can be applied by vendors selling price-dependent perishable goods and operating under carbon tax regulations. Limitations of this model are that this is applicable to only those perishable products which are highly priced sensitive. Future scope of this model is as follows:

- The model can be expanded to take trading of emissions also into consideration.
- The model can be expanded to include combined transshipment of multiple perishable items.
- This model can be expanded to take the other modes of carbon emissions into considerations (e.g., carbon emissions generated from storage).
- The model can be updated so as to consider other forms of transportation, which emit less amount of greenhouse gases, like transportation by electric vehicles.

References

1. Leitch H, Ahuja V, Jabbar M India's livestock sector: demand growth, food security and public investment–issues and options
2. Márquez-Ramos L (2015) The relationship between trade and sustainable transport: a quantitative assessment with indicators of the importance of environmental performance and agglomeration externalities. Ecol Ind 52:170–183
3. Mathivathanan D, Kannan D, Haq AN (2018) Sustainable supply chain management practices in Indian automotive industry: a multi-stakeholder view. Resour Conserv Recycl 128:284–305
4. Ghare PM, Schrader GH (1963) A model for exponentially decaying inventory system. Int J Prod Res 21:449–460
5. Nahmias S (1982) Perishable inventory theory: a review. Oper Res 30(4):680–708
6. Raafat F (1991) Survey of literature on continuously deteriorating inventory models. J Oper Res Soc 42(1):27–37
7. Goyal SK, Giri BC (2001) Recent trends in modeling of deteriorating inventory. Eur J Oper Res 134(1):1–16
8. Bakker M, Riezebos J, Teunter RH (2012) Review of inventory systems with deterioration since 2001. Eur J Oper Res 221(2):275–284
9. Janssen L, Claus T, Sauer J (2016) Literature review of deteriorating inventory models by key topics from 2012 to 2015. Int J Prod Econ 182:86–112
10. Sana S, Goyal SK, Chaudhuri KS (2004) A production–inventory model for a deteriorating item with trended demand and shortages. Eur J Oper Res 157(2):357–371
11. Mukhopadhyay S, Mukherjee RN, Chaudhuri KS (2004) Joint pricing and ordering policy for a deteriorating inventory. Comput Ind Eng 47(4):339–349
12. Alfares HK, Ghaithan AM (2016) Inventory and pricing model with price-dependent demand, time-varying holding cost, and quantity discounts. Comput Ind Eng 94:170–177
13. Ellerman AD, Convery FJ, De Perthuis C (2010) Pricing carbon: the European Union emissions trading scheme. Cambridge University Press
14. Glover JL, Champion D, Daniels KJ, Dainty AJD (2014) An institutional theory perspective on sustainable practices across the dairy supply chain. Int J Prod Econ 152:102–111
15. Gautam P, Khanna A (2018) An imperfect production inventory model with setup cost reduction and carbon emission for an integrated supply chain. Uncertain Supply Chain Manag 6(3):271–286
16. Hua GW, Cheng TCE, Zhang Y, Zhang JL, Wang SY (2016) Carbon-constrained perishable inventory management with freshness-dependent demand. Int J Simul Modell (IJSIMM) 15:3

17. Taleizadeh AA, Noori-daryan M, Cárdenas-Barrón LE (2015) Joint optimization of price, replenishment frequency, and replenishment cycle and production rate in vendor managed inventory system with deteriorating items. Int J Prod Econ 159:285–295
18. Ai XY, Zhang JL, Wang L (2017) Optimal joint replenishment policy for multiple non-instantaneous deteriorating items. Int J Prod Res 55(16):4625–4642
19. Shabani S, Mirzazadeh A, Sharifi E (2014) An inventory model with fuzzy deterioration and fully backlogged shortage under inflation. SOP Trans Appl Mathematics 1(2):161–171

Robotics, Mechatronics, Automation

Optimum Navigation of Four-Wheeled Ground Robot in Stationary and Non-stationary Environments Using Wind-Driven Optimization Algorithm

Nilotpala Bej, Anish Pandey, Abhishek K. Kashyap and Dayal R. Parhi

Abstract In this article, the atmospheric motion-based inspired wind-driven optimization (WDO) algorithm is implemented to minimize the traveling path length of a four-wheeled ground robot (FWGR) in different stationary and non-stationary environmental conditions. This optimization algorithm works on the principle of atmospheric motion of very small air particles, which revolves over the multi-dimensional search area. In the present study, WDO algorithm is employed to search a minimal or near-minimal steering angle for the (FWGR); this steering angle minimizes the path length during motion, orientation, and collision avoidance. The objective function for the WDO algorithm has been created for two reasons: for obstacle avoidance and traveling path optimization in the environments from the source point to the endpoint. Simulation results demonstrate that the FWGR covers a shorter path length using WDO algorithm as compared to the path length obtained by the FWGR using particle swarm optimization (PSO) algorithm and genetic algorithm (GA).

Keywords Wind-driven optimization algorithm · Four-wheeled ground robot · Minimal · Steering angle · Objective function

1 Introduction

Motion planning and collision avoidance are the most crucial problems in the field of robotics, which are being solved by different authors in the last two–three decades. The primary motive of motion planning is to explore a minimal or near-minimal path from the source to the end with collision avoidance ability.

N. Bej · A. Pandey (✉)
School of Mechanical Engineering, KIIT Deemed to be University, Patia, Bhubaneswar 751024, India
e-mail: anish06353@gmail.com

A. K. Kashyap · D. R. Parhi
Department of Mechanical Engineering, NIT Rourkela, Sundergarh 769008, India

BBVL. Deepak et al. (eds.), *Innovative Product Design and Intelligent Manufacturing Systems*, Lecture Notes in Mechanical Engineering,
https://doi.org/10.1007/978-981-15-2696-1_90

Met heuristic optimization algorithms such as firefly algorithm [1], ant colony optimization algorithm [2], simulated annealing algorithm [3], particle swarm optimization algorithm [4], genetic algorithm [5], multi-objective particle swarm optimization [8], and other developed bio-inspired optimization methods have been adopted by the various authors to solve the path minimization problems of the wheeled ground robot in stationary and non-stationary environment conditions. Other computational intelligence or deterministic methods such as fuzzy [3], ANFIS [9], neural network [10], and its hybrid methods are also applied for FWGR motion planning and collision avoidance in different conditions. In the present work, a new-type atmospheric motion-based bio-inspired optimization algorithm called WDO is implemented for FWGR motion planning and obstacle avoidance in stationary and non-stationary conditions. WDO algorithm works based on the principle of atmospheric motion of very small air molecules, which revolves over the multi-dimensional search area. This algorithm searches the best air particles (in terms of the best pressure value and its location) from the group of air particles. This optimization algorithm has successfully implemented in the different fields of engineering applications [6, 7]. Due to its broad area of application and performance, therefore, the authors have chosen this algorithm and tried to solve the path minimization problem of the FWGR. According to the literature survey, this is the first research work to apply the WDO algorithm for motion planning of FWGR in different environments. In order to illustrate the effectiveness of this new optimization technique, it is tested in various graphical user interface (GUI) simulation platforms and compared it with previous developed bio-inspired algorithms like particle swarm optimization [4], genetic optimization algorithm [5] and found good agreement in context of traveling path length between source to endpoint.

2 Path Minimization by Applying WDO Algorithm

WDO algorithm is motivated by the atmosphere of the earth, where the particles of wind try to normalize the horizontal imbalance in the wind pressure. It is a bio-inspired algorithm, which works based on the principle of the atmospheric flow of air [7]. WDO is analogous to similar existing bio-inspired optimization methods, in which swarm-based iterative method has been applied for solving the multi-objective local and global optimization problems. The algorithm is started from Newton's second law of motion. According to this law, the net applied force on a wind particle causes it to speed up by an acceleration a in the same way as the net applied force:

$$m \times a = \sum F_n \tag{1}$$

where m denotes the mass of air for very small air particles and F_n denotes summation of force magnitudes, which acts on the wind particles. To relate the wind pressure to the wind particles mass and its temperature, the ideal gas equation has been applied:

$$P = mRT \tag{2}$$

where R represents the gas constant, T denotes the temperature, P denotes the air pressure. Four important forces are put in the Eq. (1), which supports the wind to flow in a definite direction in the desired speed or that deflect it from its previous path. The most important force causing the wind to flow is the pressure gradient force F_{GF} listed in Eq. (3). Next force is the frictional force F_{FF} given in Eq. (4) that is acting opposite to the pressure gradient force. In the multi-dimensional atmosphere, the F_{GR} gravitational force in Eq. (5) is a perpendicular force acted toward the surface of the earth. The F_{CF} (Coriolis force) in Eq. (6) has arisen due to the orientation of the earth, which turns the way of air from one place to another.

$$F_{GF} = -\delta P \times \delta V \tag{3}$$

$$F_{FF} = -m \times \alpha \times u \tag{4}$$

$$F_{GR} = m \times \delta V \times g \tag{5}$$

$$F_{CF} = -2 \times \Omega \times u \tag{6}$$

where the δP represents the change in pressure, the infinite volume of air has been denoted by δV, Ω denotes the orientation of the earth, g represents the acceleration of gravitational, α shows the coefficient of friction, and u denotes the speed of the air. The summation of all forces (F_{GF}, F_{FF}, F_{GR}, and F_{CF}) stated in Eqs. (3–6) will be put on the right-hand side of the Eq. (1), which shows:

$$m \times \frac{\delta u}{\delta t} = (m \times \delta V \times g) + (-\delta P \times \delta V) + (-m \times \alpha \times u) + (-2 \times \Omega \times u) \tag{7}$$

In Eq. (1), the acceleration can be modified as $a = \delta u/\delta t$ with the interval of time $\delta t = 1$ is supposed to simplify the equation. For very little, dimensionless wind particles, the volume is written as $\delta V = 1$ that modifies the Eq. (7) in the following form:

$$m \times \delta u = (m \times g) + (-\delta P) + (-m \times \alpha \times u) + (-2 \times \Omega \times u) \tag{8}$$

Substituting the Eq. (2) in Eq. (8), the mass m will be replaced in the context of the pressure P, with gas law constant R, and temperature T:

$$u_{new} = (1 - \alpha) \times u_{cur} - g \times x_{cur} + \left(RT \left| \frac{1}{i} - 1 \right| (x_{opt} - x_{cur}) \right) + \left(\frac{c \times u_{cur}^{other\,dim}}{i} \right) \tag{9}$$

In Eq. 9, i provides the ranking between all wind particles, u_{new} specifies the speed in the second iteration, the speed in present iteration has been denoted as u_{cur}, x_{cur} denotes the present position of the wind particle, x_{opt} represents the optimum position of the air particle, $u_{\text{cur}}^{\text{other dim}}$ represents the speed influence from another randomly selected dimension of the same wind particle, and rest of the coefficients are listed into a single term $c = -2 \times \Omega \times RT$. Equation (9) provides the final term of the speed update applied in WDO. The following function modifies the location of the wind particle:

$$x_{\text{new}} = x_{\text{cur}} + (u_{\text{new}} \cdot \delta t) \tag{10}$$

where x_{new} denotes the new location of the wind particle in the second iteration. If the modified speed u_{new} crosses the specified top speed ($u_{\text{max}} = 0.3$) in any dimension, then the speed in that dimension is bounded according to the given criteria:

$$u_{\text{new}}^{*} = \begin{cases} u_{\max}\ \textit{if}\ u_{\text{new}} > u_{\max} \\ -u_{\max}\ \textit{if}\ u_{\text{new}} < -u_{\max} \end{cases} \tag{11}$$

where the path of movement is bounded but the magnitude is constrained and will not be more than $|u_{\max}|$ in any dimension and u_{new}^{*} denotes the modified speed, which is limited to the top speed.

Here, the fitness function (steering angle) has been minimized through the WDO algorithm. The algorithm collects the obstacle distances (front, right, and left) from the group of sensors to search the minimum steering angle of the FWGR by taking the fitness function. The parameters C_1, C_2, and C_3 are optimized through WDO algorithm to acquire the near-minimal steering angle for FWGR navigation using the objective function. This optimized steering angle helps the FWGR to obtain the shortest possible path in the given environment condition. The fitness function in Eq. (12) calculates the minimal or near-minimal steering angle for an FWGR:

$$\text{S.A.} = C_1 * \text{F.F.O.D.} + C_2 * \text{R.F.O.D.} + C_3 * \text{L.F.O.D.} \tag{12}$$

where

$25 \leq \text{F.F.O.D.} \leq 150$	**(Similar to sensor reading from 25 to 120 cm)**
$25 \leq \text{R.F.O.D.} \leq 150$	**(Similar to sensor reading from 25 to 120 cm)**
$25 \leq \text{L.F.O.D.} \leq 150$	**(Similar to sensor reading from 25 to 120 cm)**
F.F.O.D.	Front Forward Obstacle Distance,
R.F.O.D.	Right Forward Obstacle Distance,
L.F.O.D.	Left Forward Obstacle Distance,
S.A.	Steering Angle.

The obstacle distances (F.F.O.D., R.F.O.D., and L.F.O.D.) are received from the various attached sensors like LIDAR sensor, SONAR sensor, and other sensors. The sensing range of sharp infrared range sensor is between 25 and 150 cm, and the ultrasonic range finder sensor is 2 cm to 4 m.

3 Computer Simulation Results

In this section, the successful simulation snapshots of the FWGR navigation between the various moving and non-moving obstacles in the environments have been presented. The FWGR has a differentially steer control drive, and its motion and orientation are controlled by independent four center shaft DC geared motors, which provides the necessary torque to all four driving wheels. The orientation of the FWGR is achieved by the differential steered control (motor speed) of all wheels. Four separate 12 V DC motors are attached to a dual DC motor driver (L298), and each driver's direction and velocity control pins are connected to the Arduino MEGA microcontroller to drive each motor to facilitate turn left and right, backward and forward movements. The voltage control signal regulates the motor velocity. The mobile robot uses two kinds of sensors such as two SONAR sensors and one LIDAR sensor.

The computer simulations are done in the GUI platform of MATLAB software on the Lenovo 2.40 GHz processor machine. The developed flowchart of FWGR navigation based on WDO method is given in Fig. 1. The WDO algorithm has run with the different population size of 20, 30, 40 air molecules and three number of dimensions for a maximum of 500 epochs (iterations). Other necessary parameters used for WDO, GA, and PSO algorithms are listed in Tables 1, 2, and 3, respectively. Simulation results in Figs. 2, 3, 4, 5, and 6 demonstrate the motion and orientations of FWGR between the different size of stationary and non-stationary obstacles in the platforms. In the simulation, the locations of the source point and endpoint have been predefined for FWGR. In contrast, the positions of all the hurdles in the scenario have been unknown for the FWGR. The obstacle distance data collected from the equipped sensors are the inputs of the WDO, which gives the necessary near-minimal steering angle control command as an output for navigation. We have set a threshold distance between the FWGR and the obstacles, and if the FWGR detects the obstacles within the threshold, then the proposed algorithm will be activated and controls the steering angle of the FWGR. To check the efficiency of this adopted algorithm, it is applied in unknown simulation platforms and compared it with other optimization techniques like GA, PSO, and the results are found good agreement in the context of traveling path distance and time covered to reach the endpoint, which is shown in Figs. 7 and 8 and listed in Tables 4 and 5. The average deviation path length between WDO versus GA is 0.7%, and WDO versus PSO is 1.640%.

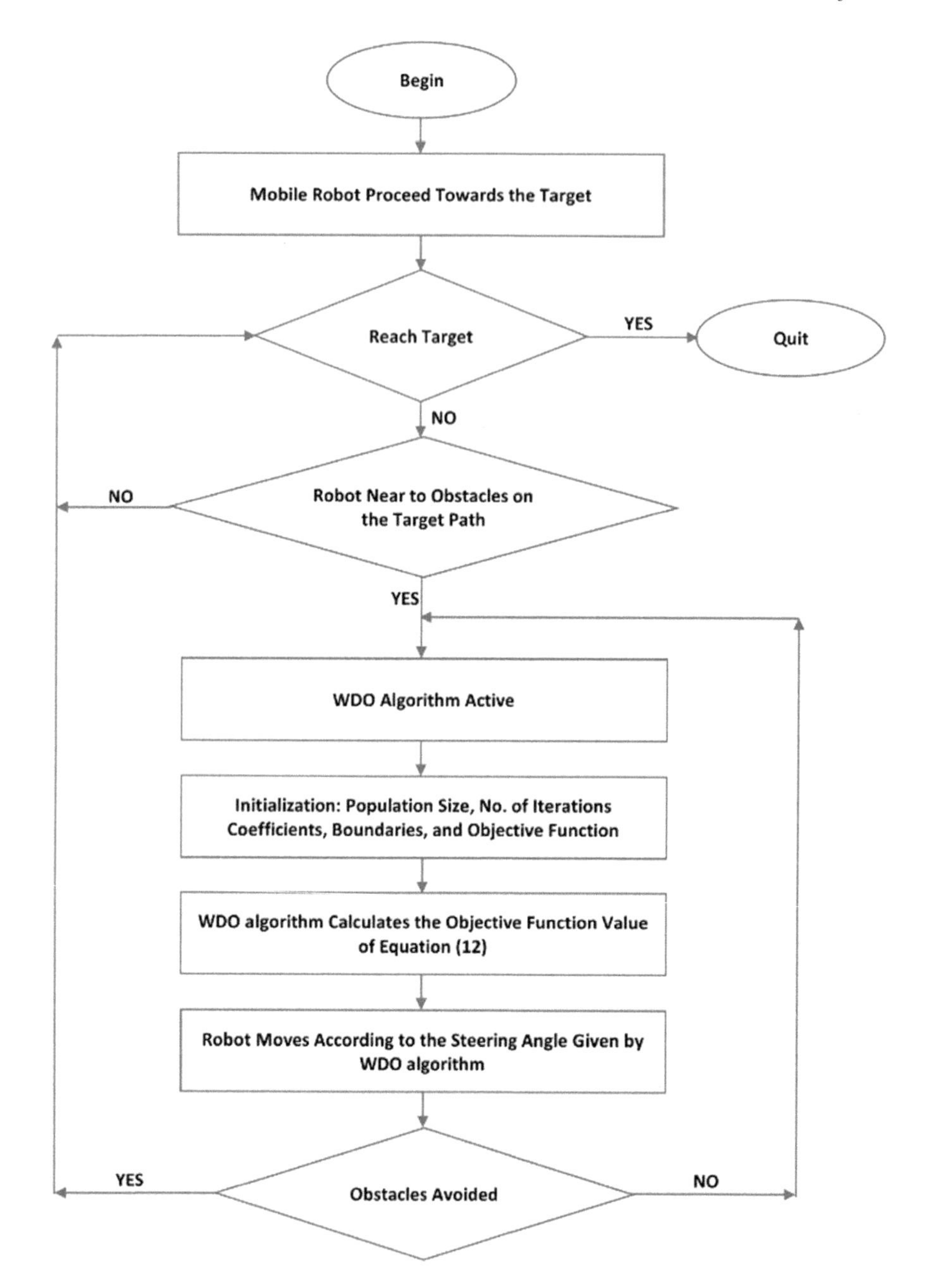

Fig. 1 Flowchart of FWGR navigation based on WDO method

Table 1 Parameters selected in WDO

S. No.	Parameters	Values
1	Population size	20
2	The dimension of the problem	3
3	Maximum number of iterations	500
4	RT coefficient	3
5	Gravitational constant	0.2
6	Friction coefficient	0.4
7	Coriolis effect	0.4
8	Speed limit	0.3

Table 2 Parameters selected in GA

S. No.	Parameters	Values
1	Population size	20
2	Selection function	Stochastic uniform
3	Elite count	1.8
4	Crossover fraction	0.7
5	Mutation function	Constraint dependent
6	Crossover function	Scattered
7	Number of generation	100

Table 3 Parameters selected in the PSO algorithm

S. No.	Parameters	Values
1	Swarm size	150
2	Maximum number of iterations	500
3	Social acceleration factors	2
4	Cognitive acceleration factors	2

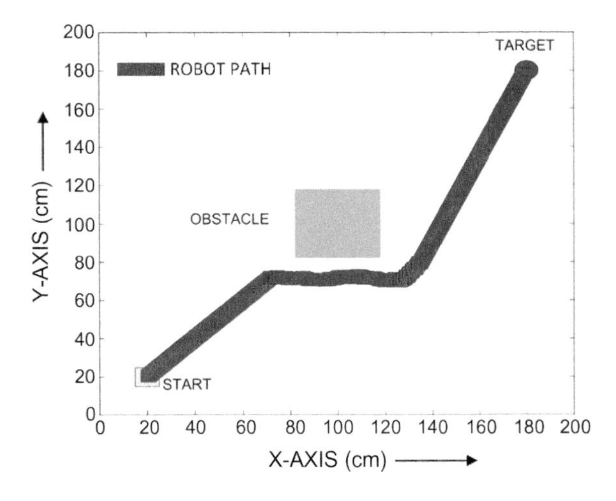

Fig. 2 Navigation of FWGR by taking WDO algorithm

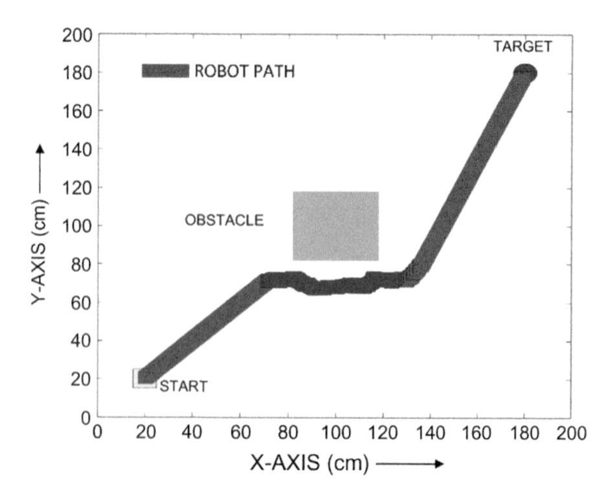

Fig. 3 Navigation of FWGR by taking GA algorithm

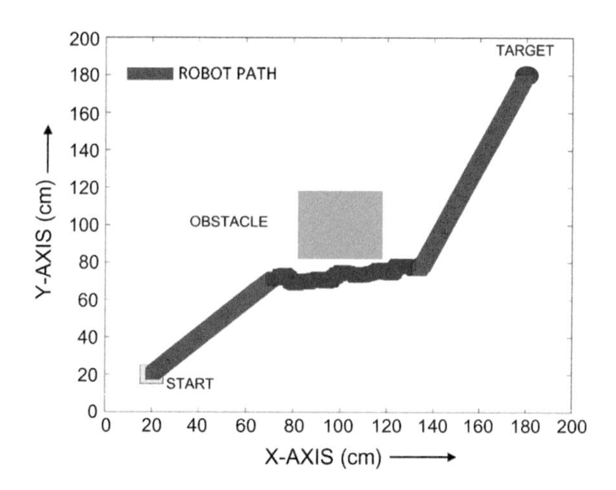

Fig. 4 Navigation of FWGR by taking PSO algorithm

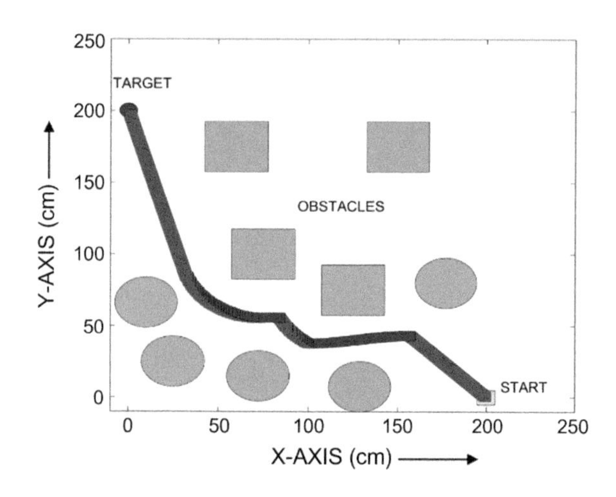

Fig. 5 Navigation of FWGR applying WDO in a complicated environment

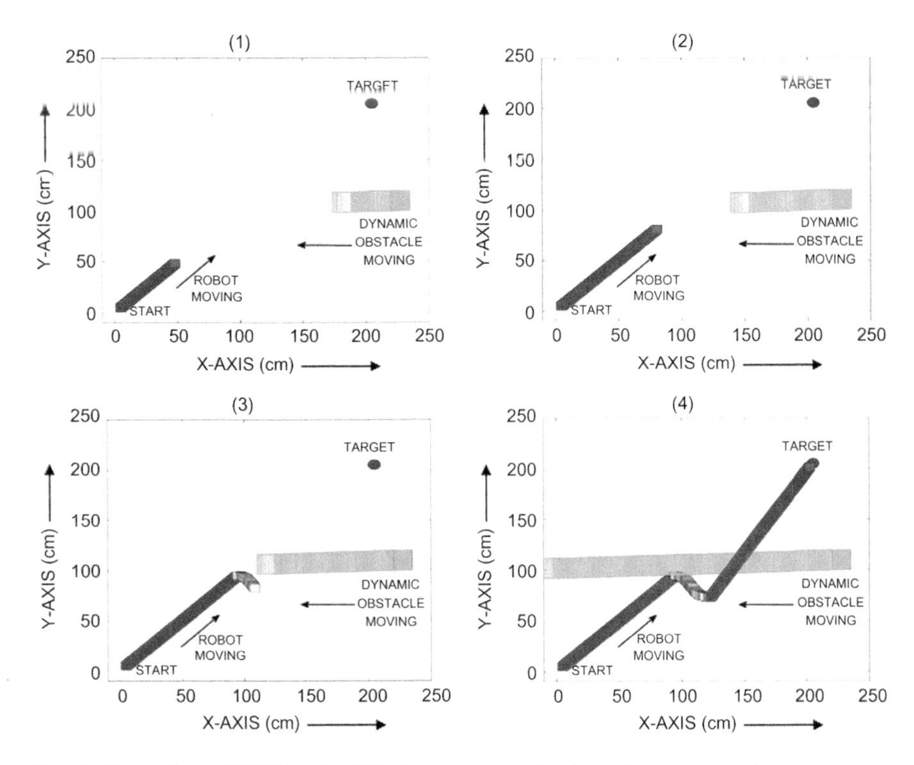

Fig. 6 Navigation of FWGR using WDO technique in the dynamic environment

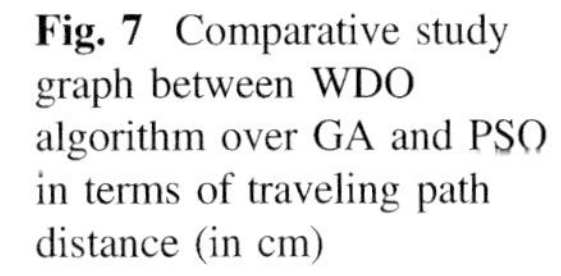

Fig. 7 Comparative study graph between WDO algorithm over GA and PSO in terms of traveling path distance (in cm)

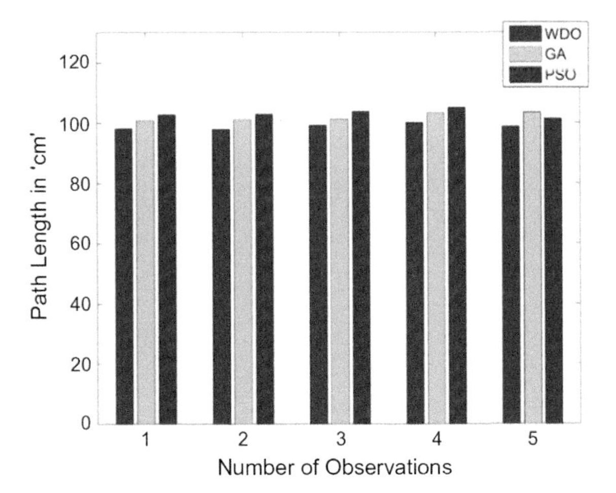

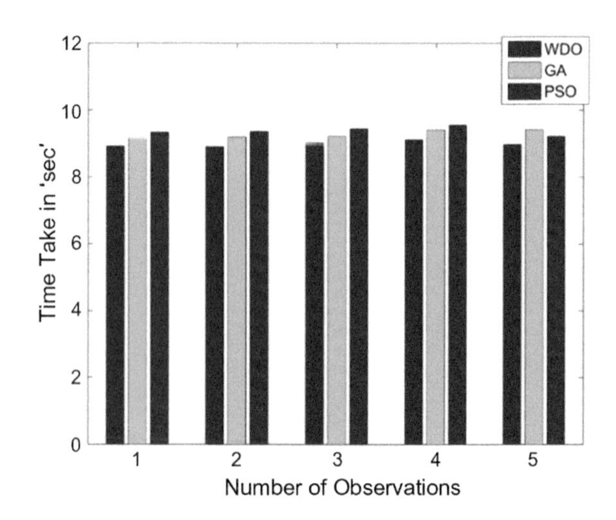

Fig. 8 Comparative study graph between WDO algorithm over GA and PSO in terms of time covered to reach the goal (in second)

Table 4 Comparative study of WDO algorithm over GA and PSO in terms of traveling path distance (in cm)

Traveling path length (cm) of FWGR using WDO in Fig. 2	Traveling path length (cm) of FWGR using GA in Fig. 3	Traveling path length (cm) of FWGR using PSO in Fig. 4
98	100	103
97	**101**	**102**
99	101	103
100	103	105
98	104	101

Note Bold value indicates the minimum path length

Table 5 Comparative study of WDO algorithm over GA and PSO in terms of time covered to reach the goal (in second)

Time covered to reach the goal FWGR using WDO (s) in Fig. 2	Time covered to reach the target by FWGR using GA (s) in Fig. 3	Time covered to reach the target by FWGR using PSO (s) in Fig. 4
8.92	9.16	9.33
8.9	**9.19**	**9.35**
9.01	9.21	9.44
9.11	9.4	9.55
8.97	9.41	9.21

Note Bold value indicates the minimum time to reach the target

4 Conclusion

In order to illustrate the efficiency and effectiveness of this WDO technique, it is applied to the different simulation conditions. Comparisons have been done with previous techniques like particle swarm optimization [4], adaptive genetic algorithm [5] and are found to agree in the context of traveling path length. This WDO algorithm searches the minimum path from a source point to the endpoint for FWGR in given environmental conditions. The simulation result shows the proposed optimization algorithm is feasible and practical for FWGR motion planning and collision avoidance in different stationary and non-stationary conditions. In the future, this technique can be combined with other bio-inspired optimization techniques to improve the navigational behavior of the FWGR.

References

1. Patel BK, Pandey A, Jagadeesh A, Parhi DR (2018) Path planning in the uncertain environment by using firefly algorithm. Def Technol 14(6):691–701
2. Chen X, Kong Y, Fang X, Wu Q (2013) A fast two–stage ACO algorithm for robotic path planning. Neural Comput Appl 22(2):313–319
3. Pandey A, Parhi DR (2016) Autonomous mobile robot navigation in the cluttered environment using hybrid Takagi-Sugeno fuzzy model and simulated annealing algorithm controller. World J Eng 13(5):431–440
4. Deepak BBVL, Parhi DR, Raju BMVA (2014) Advance particle swarm optimization-based navigational controller for a mobile robot. Arabian J Sci Eng 39(8):6477–6487
5. Jianguo W, Yilong Z, Linlin X (2010) Adaptive genetic algorithm enhancements for path planning of mobile robots. In: International conference on measuring technology and mechatronics automation. IEEE, Changsha City, China, pp 416–419
6. Bhandari AK, Singh VK, Kumar A, Singh GK (2014) Cuckoo search algorithm and wind-driven optimization based study of satellite image segmentation for multilevel thresholding using Kapur's entropy. Expert Syst Appl 41(7):3538–3560
7. Bayraktar Z, Turpin JP, Werner DH (2011) Nature-Inspired optimization of high-impedance metasurfaces with ultrasmall interwoven unit cells. Antenn Wirel Propag Lett 10:1563–1566
8. Mac TT, Copot C, Tran DT, De Keyser R (2017) A hierarchical global path planning approach for mobile robots based on multi-objective particle swarm optimization. Appl Soft Comput 59:68–76
9. Pandey A, Kumar S, Pandey KK, Parhi DR (2016) Mobile robot navigation in unknown static environments using ANFIS controller. Persp Sci 8:421–423
10. Singh NH, Thongam K (2019) Neural network-based approaches for mobile robot navigation in static and moving obstacles environments. Intel Serv Robot 12(1):55–67

Analysis on Inverse Kinematics of Redundant Robots

G. Bhavani, K. Harish Kumar, K. S. Raghuram and Hari Shankar Bendu

Abstract The objective of the present work is to finalize a numerical solution that operates on the inverse kinematic mechanism of redundant robots leading to a robust method. After considering the consequences of all numerical ways of solving the inverse kinematics problem with their limitations and difficulties, it aimed to receive the best one of them and find a final effective solution. Now, the results obtained till now are implemented to the task space trajectory planning and redundancy resolution.

Keywords Inverse kinematic mechanism · Redundant robots · Limitations · Difficulties

1 Introduction

The forward kinematic mechanism illustrates the relationship between the joints of the robot manipulator. This kinematic mechanism also enumerates the position and orientation of the tool or end-effectors [1]. The forward kinematic mechanism is to figure out the position and orientation of the end-effectors, with the known values for the joint variables of the robot. The joint variables considered for the present work are the angles between the links in revolute and/or rotational joints [2] and also taken into consideration of the link extension in the case of prismatic or sliding joints. The present work is carried out for all joints that have only a single degree of freedom [3]. The ball and socket joint having two degrees of freedom or a spherical wrist having three degrees of freedom can always have a succession of single degree of freedom joints with links of length zero in between.

The forward kinematic mechanism has a unique solution, and the joints are provided with the desired transformation. In inverse kinematics, one solution

G. Bhavani · K. Harish Kumar · K. S. Raghuram (✉) · H. S. Bendu
Department of Mechanical Engineering, Vignan's Institute of Information Technology, Visakhapatnam, India

BBVL. Deepak et al. (eds.), *Innovative Product Design and Intelligent Manufacturing Systems*, Lecture Notes in Mechanical Engineering,
https://doi.org/10.1007/978-981-15-2696-1_91

cannot be achieved [4]. These problems are over-constrained. At the same time, there are some instances where more than a single solution exists. It is to be chosen for the best solution for the inverse kinematic method, and the solver's performance is considered for the realistic the solution [5–8].

2 Denavit–Hartenberg Representation

Figure 1 gives the serial open-chain manipulator. It is a systematic approach of expressing the forward kinematics using Denavit–Hartenberg convention. This mechanism is defining the relative position and orientation works for two consecutive links. This mechanism also determines a transformation matrix between them by performing the number of operations [9–11].

3 Effect of Varying Damping Constant (Λ) on the Algorithm

On increasing the value of λ, the convergence rate of the algorithm slows down. Hence, computational time increases. This implies that number of iterations also increases. Now, we will consider a particular point in the workspace and observe the behavior of the algorithm for different damping constants. Suppose the target point is to be reached at point (2, 4).

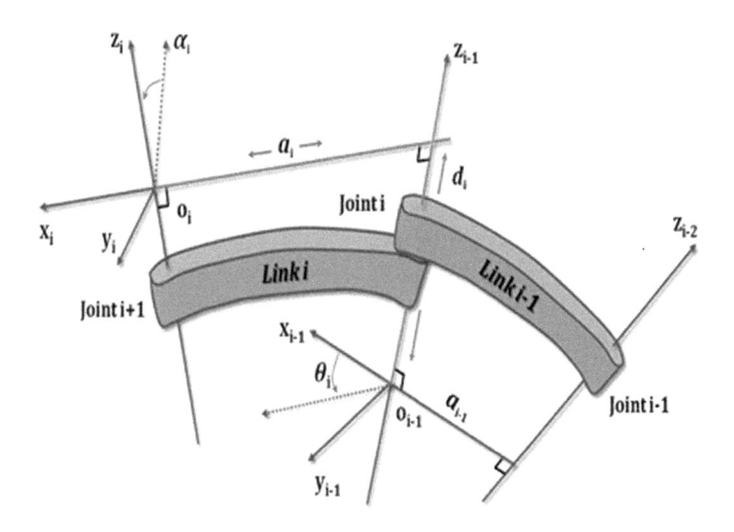

Fig. 1 Graphical interpretation of the Denavit–Hartenberg parameters

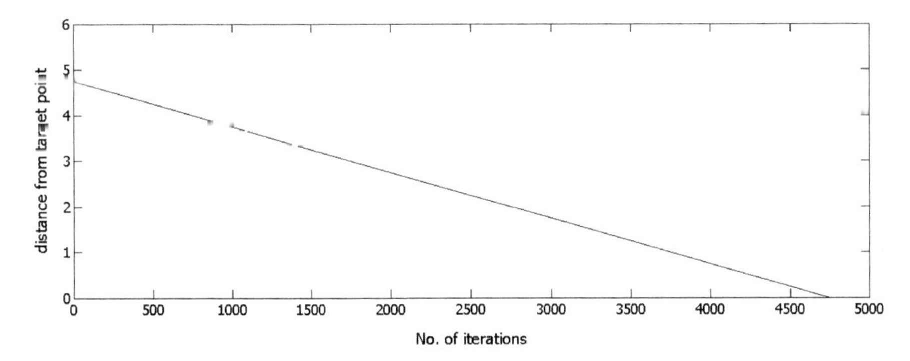

Fig. 2 Rate of convergence of inverse kinematic algorithm using DLS method for $\lambda = 0$

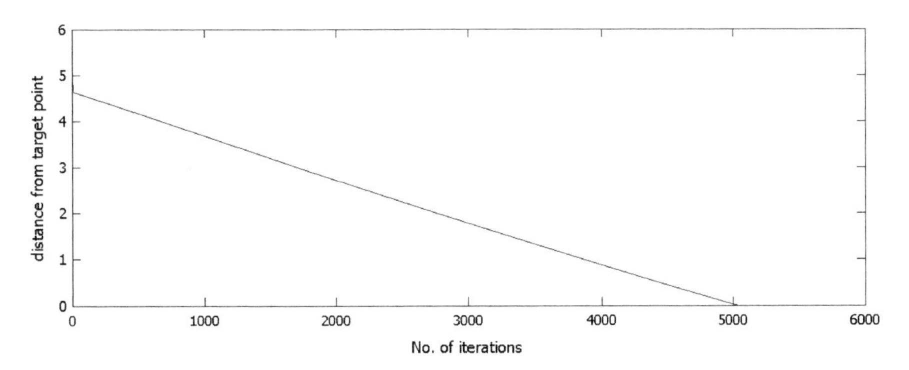

Fig. 3 Rate of convergence of inverse kinematic algorithm using DLS method for $\lambda = 1$

When damping constant $\lambda = 0$, the algorithm converges after 4725 iterations as shown in Fig. 2.

When damping constant $\lambda = 1$, the algorithm converges after 5050 iterations as shown in Fig. 3.

When damping constant $\lambda = 5$, the algorithm converges after 15,900 iterations as shown in Fig. 4.

When damping constant $\lambda = 10$, the algorithm converges after 49,100 iterations as shown in Fig. 5.

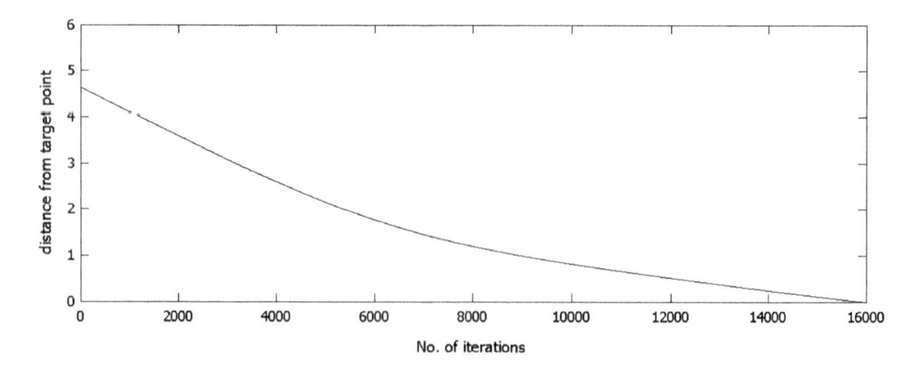

Fig. 4 Rate of convergence of inverse kinematic algorithm using DLS method for $\lambda = 5$

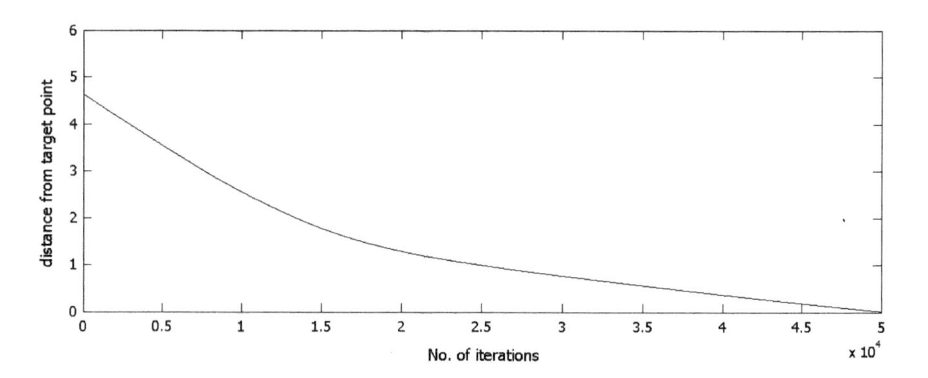

Fig. 5 Rate of convergence of inverse kinematic algorithm using DLS method for $\lambda = 10$

4 The Simulated Results

After simulating the results, graphs were generated where the graph between a number of iterations, error e, and damping constant is represented as shown in Fig. 6.

Blue-colored region indicates the zero values. As we move away from the blue-colored region toward the red-colored region over the mesh generated, values in the corresponding axis go on increasing.

It is clear from Fig. 7 that a higher value of damping constant increases the number of iterations required to reach end-effector to a target point in the inverse solution algorithm. This means the convergence rate becomes slow. Hence, computational time increases.

In Fig. 8, the end-effector of manipulator reaches the target point in different numbers of iterations for different values of damping constant when the algorithm is started from the same initial guess values of joint angles.

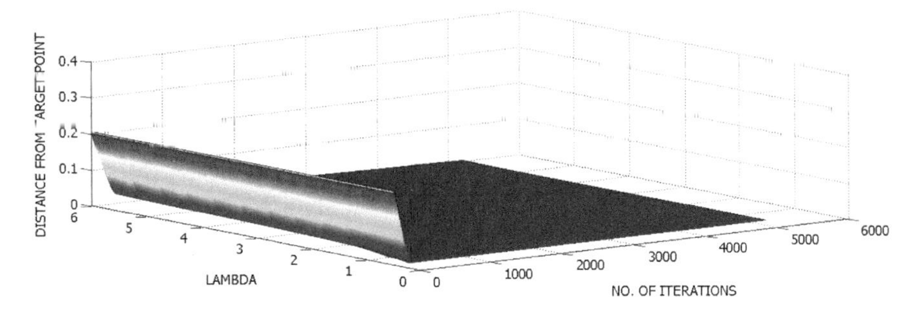

Fig. 6 Graph between damping constant (λ), distance from target point (error 'e'), and number of iterations in damped least square method

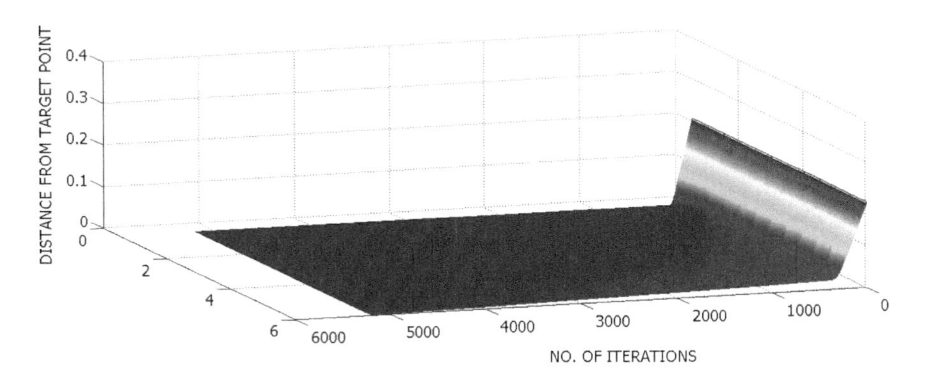

Fig. 7 3-D plot of convergence rate with varying damping constant

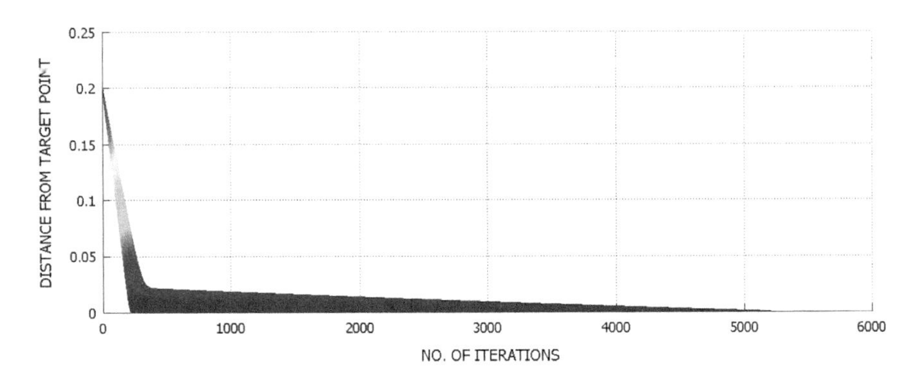

Fig. 8 Distance of end-effector from target point with number of iterations in DLS algorithm

The trajectory specified in Cartesian coordinates may force the robot to run into itself as shown in Fig. 9a. The trajectory may require a sudden change in the joint angles as shown in Fig. 9b.

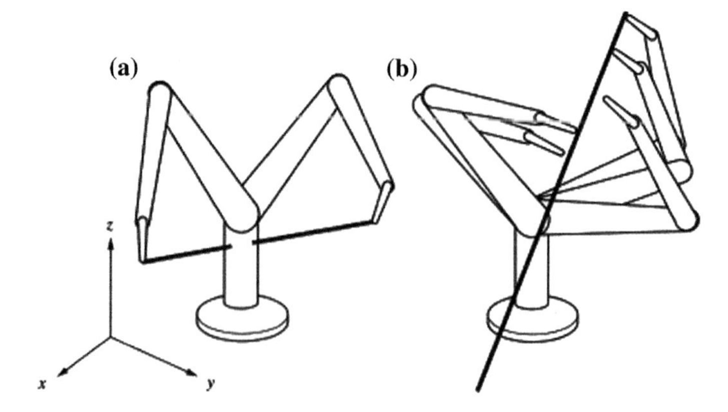

Fig. 9 Cartesian space trajectory resulting in poor configurations

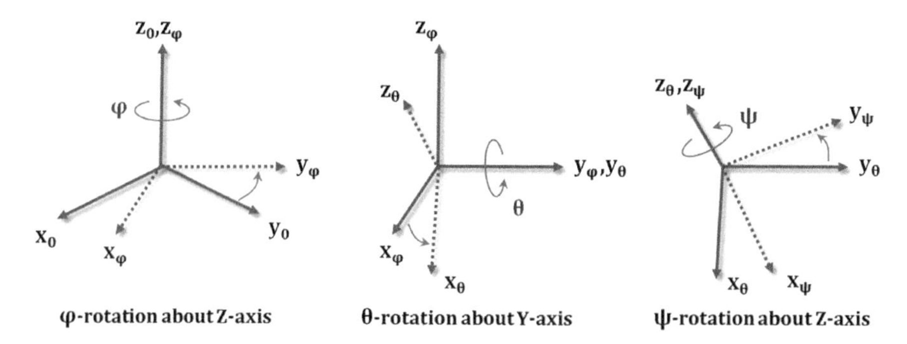

Fig. 10 Illustration of Euler angle representation of arbitrary rotations

Now considering the orientation of the end-effector in task space trajectory path. In theory, a rotation matrix can be composed of an infinite number of arbitrary independent rotations, but for practical applications, only three independent rotations are needed. Three well-recognized methods of specifying an arbitrary rotation matrix using only three independent quantities are the Euler angle representation, axis–angle representation, and the roll-pitch-yaw representation. We used the Euler angle convention as illustrated in Fig. 10.

5 Conclusion

This thesis has presented various numerical methods for solving inverse kinematics of kinematically redundant manipulators. Among these methods, damped least square solutions are most accurate and reliable. This method gives correct results within the workspace of the redundant manipulator and shows its stability even in the singular configurations and near singular configurations. Optimization is done successfully by introducing manipulability as a secondary function to the primary solution. Other optimization parameters can also be used as a secondary function to optimize them.

A methodology is proposed for trajectory planning in task space, and then, joint space trajectory is formed from the task space trajectory which uses inverse kinematic technique illustrated in the report. Interval arithmetic is introduced in the last chapter and successively solved inverse kinematics for two-link planar manipulator using this method.

6 Future Work

The interval analysis is potentially capable of computing all solutions of the inverse kinematic problem using interval arithmetic. Therefore, interval **arithmetic is an** important and suitable tool to solve inverse kinematics and redundancy resolution of the redundant manipulator. Hence, interval arithmetic is yet to be implemented in this project work in the future for the optimization of the proposed inverse kinematic technique.

References

1. Baillieul J (1985) Kinematic programming alternatives for redundant manipulators. Proc 1985 IEEE Int Conf Robot Autom 2:722–728
2. Hammond FL III: Task-specific morphological design optimization of kinematically redundant manipulators. M.S., Mechanical Engineering, University of Pennsylvania
3. Chan TF, Dubey RV (1995) A weighted least-norm solution based scheme for avoiding joint limits for redundant joint manipulators. IEEE Trans Robot Autom 11:286–292
4. Chiaverini S (1997) Singularity-robust task-priority redundancy resolution for real-time kinematic control of robot manipulators. Robot Autom IEEE Trans, pp 398–410
5. Chiaverini S, Egeland O, Kanestrom RK (1991) Achieving user-defined accuracy with damped least-squares inverse kinematics. In: Advanced robotics robots in unstructured environments, ICAR, pp 672–677
6. Chiaverini S, Siciliano B, Egeland O (1994) Review of the damped least-squares inverse kinematics with experiments on an industrial robot manipulator. Contr Syst Technol IEEE Trans, pp 123–134
7. Falco P, Natale C (2011) On: the stability of closed-loop inverse kinematics algorithms for redundant robots. Robot IEEE Trans, pp 1–5

8. en.wikipedia.org/wiki/surjective_function
9. Klein CA, Chu-Jenq C, Ahmed S (1995) A new formulation of the extended jacobian method and its use in mapping algorithmic singularities for kinematically redundant manipulators. Robot Autom IEEE Trans 50–55
10. Lee J (1997) A study on the manipulability measures for robot manipulators. In: Proceedings of the 1997 IEEE/RSJ international conference on intelligent robots and systems, IROS'97, vol 3, pp 1458–1465
11. AdriaColome: Smooth inverse kinematic algorithms for serial redundant robots (2011)

Placement and Sizing of Distributed Generation Units for Improvement of Voltage Profile and Congestion Management Using Particle Swarm Optimization

Manikonda Lavanya and Gummadi Srinivasa Rao

Abstract In this paper, the methodology used for optimal placement and best sizing of distributed generation units is particle swarm optimization algorithm. The objectives include improvement of voltage profile using voltage profile improvement index and congestion management using locational marginal price approach. In order to reduce the congestion, the difference of the locational marginal price between various buses is reduced. The IEEE 14 bus system is presented to represent the usefulness of the particle swarm optimization with locational marginal price-based approach as an objective function in relieving congestion and improving voltage profile. In this paper, voltage criteria-based approach is used to improve the voltage profile of the system and locational marginal price-based approach is used to reduce the congestion by using particle swarm optimization.

Keywords Congestion management · Distributed generation · Locational marginal price · Particle swarm optimization · Voltage profile improvement index

1 Introduction

Distributed generation (DG) is a renewable energy source, and these are normally placed near to the load centers. Integration of power system with the DG results in voltage profile (VP) improvement, loss reduction, total harmonic distortion (THD) and power quality problems [1–3]. Many references determined the area and size of DG units in basic radial systems. These services depend on the optimal placement and sizing (OPS) of DG units. The output results will be sparse, if the DG is installed in non-optimal place with non-optimal size. The studies additionally reveal that the greatest advantages from DG can be acquired just if proper DG arranging is performed [4]. A hybrid technique, a combination of artificial neural

M. Lavanya (✉) · G. S. Rao
V R Siddhartha Engineering College, Vijayawada, India

BBVL. Deepak et al. (eds.), *Innovative Product Design and Intelligent Manufacturing Systems*, Lecture Notes in Mechanical Engineering,
https://doi.org/10.1007/978-981-15-2696-1_92

networks (ANN) and PSO, is initiated for OPS of DG [5]. Whenever the power system is under unhealthy conditions which require a minimum generation to supply the demand, DG units can provide that minimum generation for recovery of distribution systems [6]. For a proper decision about OPS of DG units, independent system operator (ISO) will consider the aspects of VP, congestion and operating costs. LMP-based approach is taken for the allocation of DG units to reduce the congestion and to maximize social welfare [7]. In a highly congested area, DG units may have a great priority where LMPs are higher than elsewhere. In such a situation, the placement of DG should be optimum for the maximum benefit of DG. Many approaches are projected so far to deal with the feasibility of DGs in the power system [8–10].

The technique for optimal DG placement to decrease the real power loss (RPL) in the distribution side system using GA is initiated [11]. This manuscript presents a particle swarm optimization (PSO) algorithm to improve the VP and decrease the system congestion. Section 2 presents the problem formulation of the taken problem, Sect. 3 presents the methodology to solve the problem and Sect. 4 presents the simulation outcome.

2 Problem Formulation

In optimization problems, evaluating the proper objective function to achieve the desired result is important. Problem formulation defines and formulates the criteria to improve the VP and eliminate the congestion.

2.1 Voltage Criteria

To discover the OPS of DG units for VP improvement, the subsequent voltage-based criteria can be taken [12]. The voltage profile $\mathrm{VP_j}$ of the jth bus of the system is determined as

$$\mathrm{VP}_j = \frac{\left(V_j - V_{\mathrm{minimum}}\right)\left(V_{\mathrm{maximum}} - V_j\right)}{\left(V_{\mathrm{nominal}} - V_{\mathrm{minimum}}\right)\left(V_{\mathrm{maximum}} - V_{\mathrm{nominal}}\right)} \tag{1}$$

where V_{minimum}, V_{maximum} and V_{nominal} are the minimum, maximum and nominal voltage values respectively and V_j is the voltage at that node. According to the above equation, the VP at that bus is maximum when V_j is equal to the nominal voltage value, and the VP at that bus will be negative when V_j is more than the maximum or less than the minimum voltage, and also, VP_j will be zero if V_j is equal to the minimum or maximum values. The complete network voltage profile index $\mathrm{VP_{avg}}$ of the system is specified

$$\mathrm{VP_{avg}} = \frac{1}{m}\sum_{j=1}^{m}\mathrm{VP}_j \tag{2}$$

where m is the number of buses in the power system. The VP improvement index (VPII) for the system is defined as

$$\mathrm{VPII} = \frac{\mathrm{VP_{avg\ with\ DG}}}{\mathrm{VP_{avg\ without\ DG}}} \tag{3}$$

$$\mathrm{OBJ1} = \max(\mathrm{VPII}) \tag{4}$$

2.2 *Congestion Criteria*

The congestion in the system occurs when the transmission lines are overloaded and/or having the large differences in the LMP values. The difference in LMP values appears when lines are constrained. The congestion of the system is reduced by reducing the LMP differences between different buses. LMP is the rate of delivering the next increase in electric energy (load) to a particular node. If the line flow limits are not considered in the optimization problem, LMPs will be the same for all buses.

LMP can be calculated [13] as

$$\mathrm{LMP} = \lambda + \lambda_{L,i} + \lambda_{C,i} \tag{5}$$

$$\lambda_{L,i} = \lambda\frac{\partial P_L}{\partial P_i},\ \lambda_{C,i} = \sum_{k\in K}\mu_k \mathrm{GSF}_{ik} \tag{6}$$

where λ is the cost of energy component at the reference bus, and it is the same for all the buses, $\lambda_{L,i}$, $\lambda_{C,i}$ are the loss component and congestion component. $\lambda_{L,i}$, $\lambda_{C,i}$ are different for every bus. So, the price on every bus depends on its site in system, congestion and losses. P_L, P_i are the system losses and power injection at bus i, respectively. μ_k, GSF_{ik} and K are the constraint cost of line k, generation shift factor for bus i on line k and set of congested lines, respectively.

LMP values are calculated using the interior point method, and the objective function to calculate LMP is the minimization of social cost. The objective function to reduce the congestion is defined to limit the general production expenses for the value of real power in the system. This incorporates real power generation from generators.

$$C = \sum_{i=1}^{n_g} C_i P_{gi} \tag{7}$$

$$\text{OBJ2} = \min(C) \tag{8}$$

where n_g is the number of generators and P_{gi} is the real power output of the generator at ith bus and C_i is the real power cost function of ith generator.

2.3 Constraints

The above objectives subjected to the constraints are given below. The power balance equations which are equality constraints are given below,

$$P_{Gi} - P_{Li} = \text{Re}\left[V_i \sum_{k=1}^{N} Y^*_{\text{bus}_{ik}} V^*_k\right] \tag{9}$$

$$Q_{Gi} - Q_{Li} = \text{Im}\left[V_i \sum_{k=1}^{N} Y^*_{\text{bus}_{ik}} V^*_k\right] \tag{10}$$

Equations (11), (12), (13) and (14) are voltage, active power, reactive power and line flow limits, respectively.

$$V_{\min} \leq V_i \leq V_{\max} \tag{11}$$

$$P_{G\min i} \leq P_{Gi} \leq P_{G\max i} \tag{12}$$

$$Q_{G\min i} \leq Q_{Gi} \leq Q_{G\max i} \tag{13}$$

$$S_{ij} \leq S_{ij}^{\max}, S_{ji} \leq S_{ji}^{\max} \tag{14}$$

3 Particle Swarm Optimization

PSO is one of the intelligent methods, and it was initially presented by Dr. Eberhart and Dr. Kennedy in 1995 which requires less memory and computational time. Because of its simplicity, it has found its applications to solve a wide range of engineering problems quickly. PSO is effortless in coding implementation and is less sensitive to the character of the objective function compared to other heuristic optimization techniques.

It is inspired by performance of birds rushing or fish tutoring. Every arrangement is considered as a bird, called particle. Every one of the particles has a fitness value, and it can be determined using an objective function. Each particle preserves their individual best execution. They likewise know the best execution of their gathering. They modify their velocity considering their best execution and furthermore considering about the best execution of the best particle. With a swarm consisting of m number of particles in search space, the representation of the position and velocity vectors are

$$S_i = [x_1^i, x_2^i, \ldots, x_n^i] \quad i = 1, \ldots, m \tag{15}$$

$$V = [v_1, v_2, \ldots, v_n] \tag{16}$$

The updated position for particle i [14] is

$$S_i^{(t+1)} = S_i^{(t)} + v_i^{(t+1)} \tag{17}$$

The updated velocity vector for particle i is

$$v_i^{(t+1)} = wv_i^{(t)} + c_1 r_1 (p_{\text{besti}}^{(t)} - S_i^{(t)}) + c_2 r_2 (g_{\text{besti}}^{(t)} - S_i^{(t)}) \tag{18}$$

The flow chart of the PSO algorithm is presented in Fig. 1 where C_1, C_2 are the learning factors and ω is the inertia weight. r_1, r_2 are the random numbers, and the range of these variables varies from [0, 1]. $p_{\text{besti}}^{(t)}$, $g_{\text{besti}}^{(t)}$ are the local and global best particles, respectively. $S_i^{(k)}$, $v_i^{(k)}$ are the previous particles position and velocity, respectively.

4 Simulation Results

The proposed PSO algorithm is applied to the IEEE 14 bus system for OPS of DG.

4.1 Voltage Criteria

VP of IEEE 14 bus system before integration of DG and after integration of DG is shown in Table 1. Simulation is carried out for OPS of DG using MATLAB software. Optimal place (bus) to employ the DG is 10th bus, and the optimal size is 9 MW. The convergence characteristics of VPII are shown in Fig. 2.

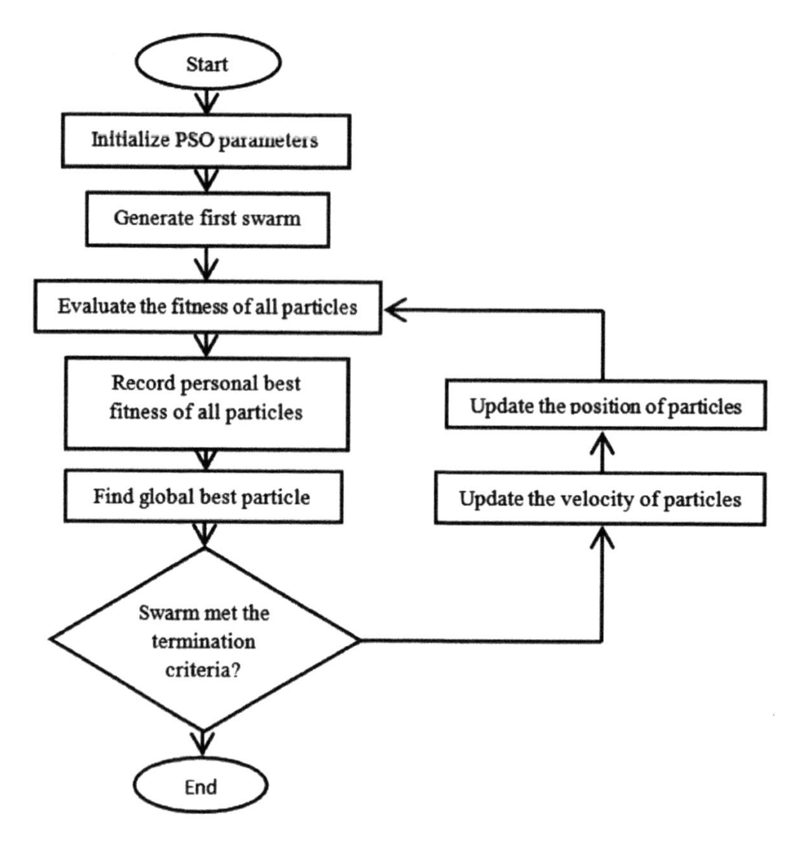

Fig. 1 Flowchart of PSO

4.2 *Congestion Criteria*

Cost coefficients of IEEE 14 bus system are shown in Table 2. Table 3 shows the generator data under normal condition of the system and under the congested system.

Congestion is created by making the line flow of line 7–9 from 27 to 20 MW. The total demand of system is 259.3 MW. DG is added to the congested system to reduce congestion.

LMP values of IEEE 14 bus system are calculated using the MATLAB software. LMP can be calculated using ACOPF method and DCOPF method, but DCOPF is advantageous over ACOPF. In this paper, LMP is calculated using DCOPF method. Table 4 shows the LMP of the system before congestion and after congestion.

DG is taken as to inject real power. Generally, the bus with higher LMP is the first priority to place the DG. Optimal place (bus) to employ the DG is 14 and the optimal size is 19.95 MW. LMP of each bus became to 3.51 ($/MWh) at each bus

Table 1 Voltage profile of IEEE 14 bus system

Bus number	Voltage profile (without DG)	Voltage profile (with DG)
1	1.060	1.060
2	1.035	1.035
3	1.000	1.000
4	0.985	0.989
5	0.990	0.993
6	1.000	1.000
7	0.980	0.991
8	0.980	1.000
9	0.967	0.979
10	0.964	0.974
11	0.978	0.983
12	0.980	0.983
13	0.976	0.977
14	0.951	0.957

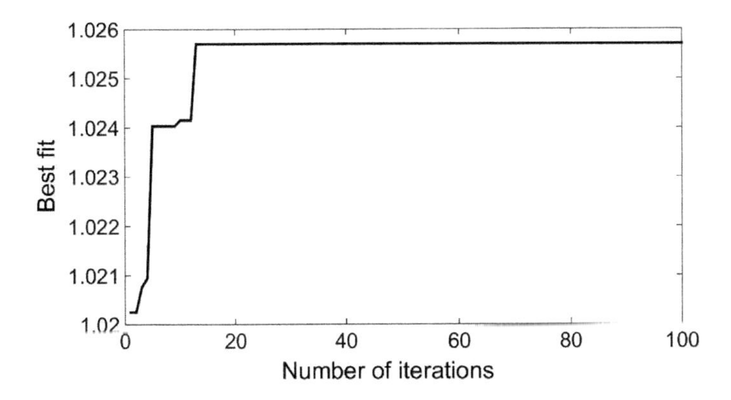

Fig. 2 Convergence characteristics of VPII

Table 2 Coefficients of IEEE 14 bus system

Generator	a	B	c	P_{min}	P_{max}
G_1	0.005	2.45	105	10	160
G_2	0.005	3.51	44.1	20	80
G_3	0.005	3.89	40.6	20	50

after placement of DG in a system at the optimal place. The differences in LMP values became zero. DG cost characteristics are not considered for now.

The convergence characteristics of LMP-based criteria are shown in Fig. 3, and the generator data after the placement of DG is shown in Table 5.

Table 3 Power generation of generators

Generator data	Before congestion (MW)	After congestion (MW)
G_1	160	160
G_2	79.3	53.2
G_3	20	46

Table 4 LMP of IEEE 14 bus system

Bus number	LMP before congestion ($/MWh)	LMP after congestion ($/MWh)
1	3.51	3.5285
2	3.51	3.5237
3	3.51	3.51
4	3.51	3.4982
5	3.51	3.5466
6	3.51	3.89
7	3.51	2.9193
8	3.51	2.9193
9	3.51	4.2801
10	3.51	4.2108
11	3.51	4.0532
12	3.51	3.9208
13	3.51	3.9449
14	3.51	4.1336

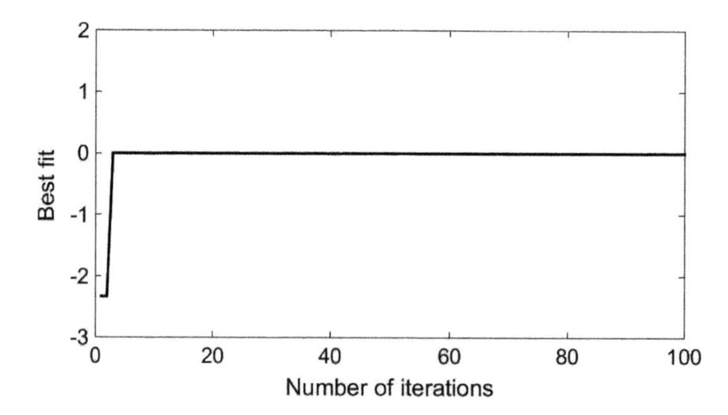

Fig. 3 Convergence characteristics of LMP-based criteria

Table 5 Power generation of IEEE 14 bus system after placement of DG

Generator data	Generation (MW)
G_1	160
G_2	59.34
G_3	20

5 Conclusion

In this paper, OPS of DG units to improve VP using voltage profile improvement index (VPII) is carried out using PSO. VP is improved after the addition of DG to the IEEE 14 bus system. LMP-based criteria are used to limit the congestion on power systems. OPS of DG units to reduce the congestion is achieved by using PSO. Simulation results showed that the VP is improved after the optimal placement of DG, and congestion is reduced by minimizing the LMP differences between various buses. This work can be further carried by including the DG cost characteristics and different types of DG technologies.

References

1. Le AD, Kashem MA, Negnevitsky M, Ledwich G (2005) Maximising voltage support in distribution systems by distributed generation. In: IEEE region 10 conference TENCON 2005, Melbourne, pp 1–6
2. Le DT, Kashem MA (2007) Optimal distributed generation parameters for reducing losses with economic consideration. In: IEEE power engineering society general meeting 2007, Tampa, pp 1–8
3. Bhargava V (2018) Voltage stability enhancement of primary distribution system by optimal DG placement. In: Malik H, Srivastava S, Sood Y (eds) Applications of artificial intelligence techniques in engineering. Advances in intelligent systems and computing 2016, vol 697, pp 65–78. Springer, Singapore
4. Srinivasa Rao G (2013) Voltage profile improvement of distribution system using distributed generating units. Int J Electr Comput Eng 3(3):337–343
5. Srinivasa Rao G (2013) Optimal location of DG for maintaining distribution system stability: a hybrid technique. Int J Electr Power Energy Convers 4(4):387–403
6. Nabavi SMH (2011) Placement and sizing of distributed generation units for congestion management and improvement of voltage profile using particle swarm optimization. In: IEEE PES innovative smart grid technologies 2011, pp 1–6. IEEE, Australia
7. Mithulananthan DGN (2007) Optimal DG placement in deregulated electricity market. Electr Power Syst Res 77(12):1627–1636
8. Celli G, Pilo F (2011) Optimal distributed generation allocation in MV distribution networks. In: 22nd IEEE PES international conference on power industry computer applications PICA 2001, pp 81–86. Sydney, Australia
9. Celi G (2002) Penetration level assessment of Distributed generation by means of genetic algorithms. In: IEEE proceedings of power system conference 2002, Clemson, SC

10. Kashyap M, Mittal A (2017) Optimal placement of distributed generation using genetic algorithm approach. In: Nath V, Mandal J (eds) Proceeding of the second international conference on microelectronics, computing & communication systems 2017, vol 476, pp 587–597. Springer, Singapore
11. Mithulananthan N (2004) Distributed generator placement technique in power system distribution system using genetic algorithm to reduce losses. Thammasat Int J Sci Technol 9:56–62
12. Iyer H, Ray S (2005) Voltage profile improvement with distributed generation. In: IEEE power engineering society general meeting 2005, vol 3, pp 2977–2984
13. Momoh JA, Boswell D (2008) Locational marginal pricing for real and reactive power. In: IEEE power and energy society general meeting-conversion and delivery of electrical energy in the 21st century, Pittsburgh, PA, 2008, pp 1–6
14. Rao S (2009) Engineering optimization, 4th edn. Wiley, Hoboken, NJ (2009)

Analysis for Material Selection of Robot Soft Finger Used for Power Grasping

Chiranjibi Champatiray, G. B. Mahanta, S. K. Pattanayak and R. N. Mahapatra

Abstract The geometric relationship between contact parameters such as deformation, contact area and touch angle of robot soft finger is developed. The suggested nonlinear cylindrical soft finger deforms on the application of normal load and contact surface and touches angle grow correspondingly. The force relationship between contact parameters and geometrical data is derived. The developed theoretical model enables to determine the total contact force at the contact surface manipulation. The theoretical model is validated by conducting experiments with the artificial finger of hyperelastic material (Silicone Ecoflex 00-30). The contact width is measured by conducting compression testing from 10 N to 100 N with an increment of 10 N on silicone finger to leave a vivid print on recording paper, and contact area is calculated from those data. The importance of coefficient of friction cannot be overlooked. The value of coefficient of friction is also calculated from inclined test result. The developed model and soft finger can be used for tackling real-life problems related to object manipulations.

Keywords Contact model · Power grasping · Hyperelastic · Soft finger

1 Introduction

From various literature, it has been seen that robot hand of rigid fingers used for handling objects, and they may get damaged throughout the time of operation, i.e., grasping/manipulation. The effect of the deformation of the soft finger and/or object is a very common problem in the development of a robotic finger.

C. Champatiray (✉) · R. N. Mahapatra
NIT Meghalaya, Shillong, Meghalaya, India
e-mail: Chiranjibi@nitm.ac.in

G. B. Mahanta
NIT Rourkela, Rourkela, Odisha, India

S. K. Pattanayak
NIT Silchar, Silchar, Assam, India

BBVL. Deepak et al. (eds.), *Innovative Product Design and Intelligent Manufacturing Systems*, Lecture Notes in Mechanical Engineering,
https://doi.org/10.1007/978-981-15-2696-1_93

The fundamental of soft manipulation is highly needed for soft manipulations. Shimoga et al. [1] discussed about the three main problems associated with multi-fingered hand, and the problems are (a) impact force, attenuation arising during grasping of rigid object, thereby affecting the functioning of fingertip sensors, (b) conformability associated with grasping of objects having uneven surfaces, (c) repetitive strains, dissipation induced into the fingers during manipulation task. Had these three factors been taken care of, weak functioning of fingertip sensors, shorter life of finger skeletal structure and jerky manipulation would have occurred, respectively. To overcome these three problems, six materials, viz. rubber, plastic, sponge, a fine powder and gel, are carefully chosen. Their ability was compared experimentally, resulting in sponge being the most suitable and plastic being the least for the application. Moreover, the gel was a good compromise over sponge. The results helped in recommending soft finger for robotic hand, especially soft-tipped fingers made out of carefully chosen materials. According to Fukaya et al. [2], the humanoid hand is able to grasp and holds objects with the fingers and the palm by adapting the grip to the shape of the object, through self-adjustment functionality. Elango et al. [3] investigated on soft materials for the manipulations of fragile robots. They reviewed on the properties of human skin and mentioned that human skin has complex mechanical characteristics and so suggest some soft materials having close resembled to human skin like natural rubber, synthetic rubber, elastomeric, polymer composite and nano-particulate polymer composite. Park et al. [4] deduced a contact model for robot fingertip and validated the model through experiments. Shao et al. [5] used artificial fingertip with viscoelastic core for virtual model development to replicate the softness and the shape of human fingerprints. Different artificial fingertips were analyzed and compared, and the results showed that silicon fingertip was showing different properties than the real fingertip. Derler et al. [6] investigated variations of lipid content on the surface of the skin. The coefficient of friction of human skin was compared with the reference value i.e. 0.27 to 0.71. The same results were obtained for different loads and were compared to different polyurethane and silicon materials. Tiezzi et al. [7] investigated viscoelastic contact and friction characteristics with pressure distribution at interfaces. He also studied the effect of relaxation nature on robotic manipulations. Haibin et al. [8] presented a grasping force model for a soft robotic gripper with variable stiffness. Pawlaczyk et al. [9] analyzed human skin behavior over the ages. The main value of skin deformation before the bursting was found to be 75% for the newborn but 60% for the elder people. It was concluded that the skin was getting thinner, less flexible and stiffer with the age which in turn reduced its protective capabilities against mechanical injuries. Controzzi et al. [10] developed an artificial fingertip to get a similar response as that of a human fingertip for the improvement of stability while grasping with robotic hands. The response of the prototypes was the same as that of actual skin.

The idea to use elastomer as a robot soft finger comes from medical science, i.e., from the literature survey, it has been observed that silicone is very popular in the field of medical science as silicone is used for the prosthetic purpose. Therefore, the silicone rubber can also be used as a robot finger. To prove the assertion, an experiment is conducted with Silicone Ecoflex 00-30.

2 Geometric Analysis of a Soft Finger

From Figs. 1 and 2 show that a soft finger of the cylindrical shape of radius r, length l in contact with a rigid object making an angle α. The contact surface area is assumed to be rectangular.

Figure 1 is for hemispherical soft finger and development of soft finger (cylindrical shape as shown in the above figure) and contact mechanics, and it can be modeled as follows;

The horizontal component (C_x) can be written as

$$C_x = (r - d_0) \times \tan\alpha \quad (\text{From } \Delta\text{ODC}) \tag{1}$$

Equation of a circle is

$$x^2 + y^2 = r^2 \tag{2}$$

Combining both the above equations in the y-direction

$$C_y = \sqrt{r^2 - (r - d_0)^2 \times \tan^2\alpha} \tag{3}$$

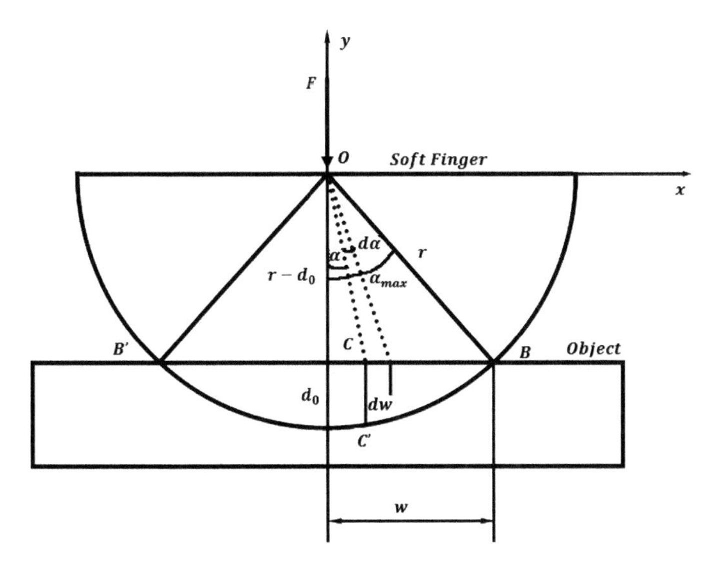

Fig. 1 Model of a cylindrical soft finger pressed against a rigid plane

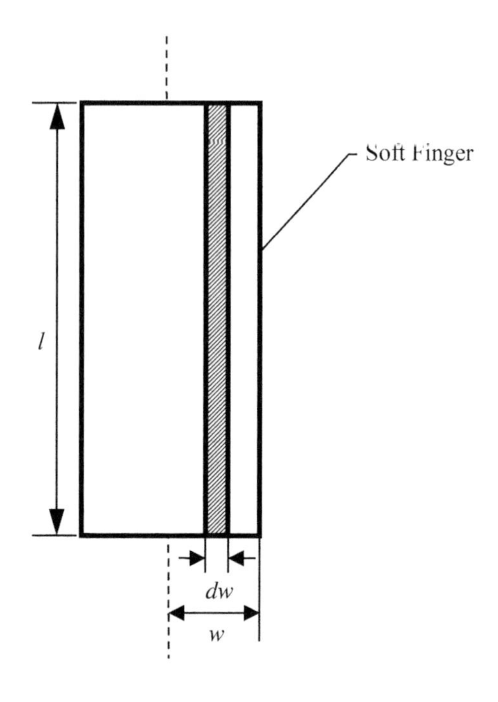

Fig. 2 Cylindrical shaped soft finger

$\overline{CC'}$ in y-direction,

$$\overline{CC'} = C_y - (r - d_0) \tag{4}$$

$\overline{CC'}$ is the soft finger's deformation at surface of the contact point. Now, a force is applied over the finger, and it deforms to d_0 and makes touch angle maximum, so it can be written as,

$$\text{Touch angle,} \quad \alpha_{\max} = \tan^{-1}\sqrt{\frac{r^2 - (r - d_0)^2}{(r - d_0)^2}} \tag{5}$$

$$\text{Half contact width,} \quad w = (r - d_0)\tan\alpha_{\max} \tag{6}$$

$\mathrm{d}w$ = Small change in the contact width (referring the above figure) and $\mathrm{d}\alpha$ = Small change in touch angle

$$\mathrm{d}w = \frac{r - d_0}{\cos^2\alpha}\mathrm{d}\alpha \tag{7}$$

Increased contact surface because of the deformation ds is given by,

$$\mathrm{d}s = \mathrm{d}wl.$$

3 Modeling of Soft Finger

We have,

$$\text{Stress,} \quad \sigma = \frac{F}{S}$$

where F = Contact force and S = Area of contact

$$\text{Pressure,} \quad p = \frac{F}{S}$$

The decrement in the finger radius along Y-axis is observed when a compressive force acts on the soft particle, but at the same time, the increment in contact width is also observed to define the change in the radius in the compressive strain,

$$\text{Compression strain} \quad e = \frac{\Delta r}{r}$$

$$\text{where} \quad \Delta r = d_0$$

Again,

$$E = \frac{\sigma}{e} = \frac{F/S}{\Delta r/r} \tag{8}$$

And

$$\text{Pressure,} \quad P = \frac{F}{S}$$

$$P = E\frac{\Delta r}{r} \tag{9}$$

After performing the integration of soft fingers force distribution,

$$f_c = \int_{0}^{\alpha_{\max}} P\mathrm{d}s$$

$$f_c = \int_0^{\alpha_{\max}} E\frac{\Delta r}{r} \mathrm{d}wl \tag{10}$$

By substitution and reduction

$$f_c = \mathrm{El} \times \frac{d_0(r - d_0)}{r}\sqrt{\frac{r^2 - (r - d_0)^2}{(r - d_0)^2}} \tag{11}$$

Again, contact width = 2 × half contact width.

$$w = \frac{2f_c r}{\mathrm{El}d_0} \tag{12}$$

And the contact area $= w \times l$

$$A = \frac{2f_c r}{Ed_0} \tag{13}$$

4 Preparation of Soft Finger Sample

The mold size was first decided by measuring the finger size of a group of people aged between 24 and 30 years old. The average diameter of finger measured was 17 ± 0.5. So, from the anthropomorphic study of a human finger, the size of mold was decided to have 17 mm diameter and 100 mm length (assuming cylindrical in shape and size). The mold was designed by means of 123D Design. The soft file of desired shape and size of the mold was converted into .stl file so that the 3D printer can read it. The SD card, where the .stl file was saved, fed into the 3D printing machine (Adroitec RxP2200i 3D printer) for the operation where filament (Peptide Nucleic Acid) was fed into the nozzle to generate required shape and size of the object (cylindrical with 18 mm diameter and 100 mm height). Then, the mold was prepared by the 3D printer as shown in Fig. 3, and then, the final product (see Fig. 4) was obtained by pouring Silicone Ecoflex into the mold by maintaining proper steps.

Fig. 3 Mold/die

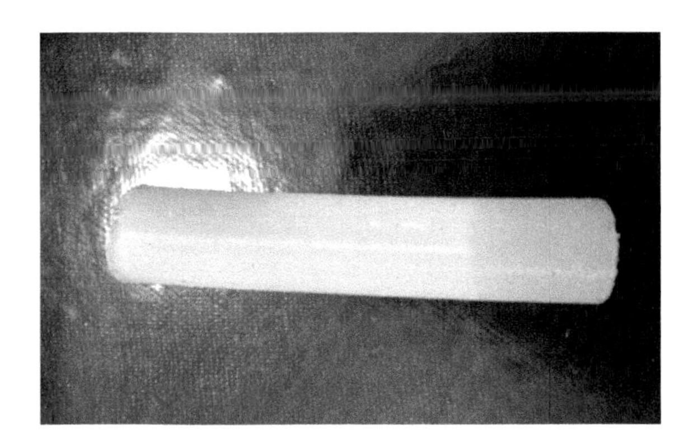

Fig. 4 Final product (soft finger)

5 Determination of Contact Width and Friction Coefficient

The Silicone Ecoflex 00-30 sample was placed over the mild steel plate, and the force was applied to the flat surface of the finger ranging from 0 N to 100 N with the incremental of 10 N to get the deformation and contact width of the finger. The deformation of the soft finger against mild steel (soft-to-hard Contact) was directly measured from the machine for each load. In order to record the contact width that the finger produced during compression, the hydraulic oil (petroleum Jelly/vaseline) of very high density was applied over the finger before testing. A recording paper (tracing paper) was kept between the silicone finger and the target plate. It was found that paper thickness had no effects on the contact area. The requirement of the ideal soft finger is that it should cover the entire contact area when the normal load is applied. With the change of loading conditions, the contact area will not vary. When the load was applied over the finger, it contacts the target surface, deformed and left a lucid print over the recording paper. The load-deformation measurement was repeated three times to ensure the repeatability of the experiment, and only the average value was taken. Then, the image was scanned by scan-jet G3010, and the scanned file was measured using Motic image plus version 2. The contact width (w) and deformation (d_0) of soft finger for various normal loads were obtained from recorded fingerprint, andthe values of the material of elasticity were calculated by using the values of contact width and deformation. A sample was set up on the surface over a sine bar along with the slip gauge set up as shown in Fig. 5.

$$\mu = \tan \emptyset = \frac{h}{l} \tag{14}$$

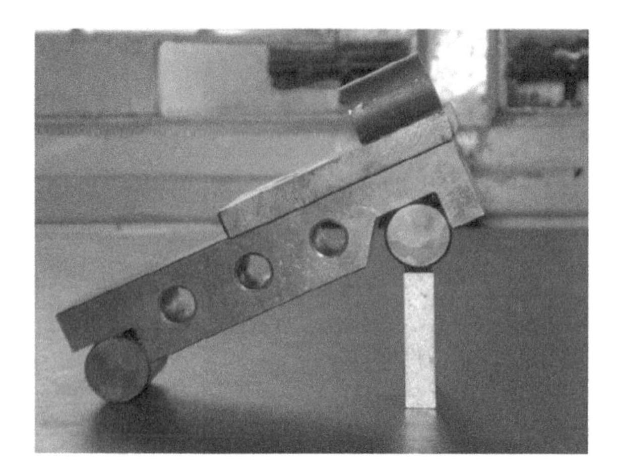

Fig. 5 Inclined testing using sine bar

where

h height of slip gauge
l length of the sine bar
θ angle of inclination.

6 Results and Discussion

To obtain the mentioned objective, the specimen was tested (compression test) to get the following data and graphs in different loading conditions. The following Fig. 6 shows the value compressive strength of the silicone specimen.

It has been observed that the compressive strength of the artificial finger made up of silicone is 2.51 MPa with a strain of 0.78.

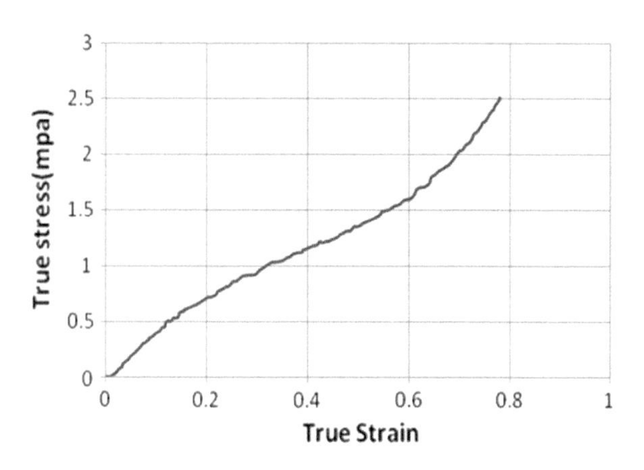

Fig. 6 True stress versus true strain for silicone

Table 1 Values of COF

Target materials	Silicone (Ecoflex 00-30) (μ)
Glass	0.612
Steel	0.918
Rubber	0.736

It can be observed from Fig. 6 that the compressive strength of artificial finger made up of silicone is 2.51 MPa with a strain of 0.78. It was also noticed that the growth of distortion on the silicone finger is more on the lower load, but this load gets much lesser than the load, and for silicone, the material of elasticity it is 93 to 98 N/mm^2 where the value of deformation is from 0.327 to 0.919 mm. Many researchers have measured the friction of human skin and come up with COF value of human skin which lies between 1.0 (under special condition) and 0.5 (Cutkosky and Wright 1986). By applying the equation number 11, COF has been found as shown in Table 1.

It is clearly seen from Table 1, that the COF is varying from material to material, and in dry and clean condition silicone is having higher a coefficient of friction when slide against mild steel but less if moved against glass and rubber. COF does not depend upon mass, force, weight, but depends upon nature of surfaces, i.e., roughness, textures, etc.

7 Conclusion

For the power grasping, a soft finger of cylindrical shape is developed and a force function is established by taking into account the contact surface's force distribution. The constitutive relationship relates a touch angle with contact width and deformation. The experiment is carried out with a soft finger made of silicone and load versus deformation was plotted. The contact width is measured from the compression testing supplied by INSTRON. The variation of the material property was calculated. The model described in this work can also be applied to hemi-cylindrical fingertip and can validate the same. Hence, it can be said that silicone is a suitable material for making robot soft finger, and the proposed soft finger can be used in making multi-fingered robot hand for real-life applications like prostheses, soft tools for surgery, diagnosis and drug delivery, wearable and assistive devices, artificial organs and tissue-mimicking active simulators for training and biomechanical studies. The use of soft finger may be difficult for industrial application where very heavy or big job needs to handle.

References

1. Shimoga KB, Goldenberg AA (1992) Soft materials for robotic fingers. In: Proceedings 1992 IEEE international conference on robotics and automation. IEEE, pp 1300–1305
2. Fukaya N, Toyama S, Asfour T, Dillmann R (2000) Design of the TUAT/Karlsruhe humanoid hand. In: Proceedings of the international conference on intelligent robots and systems, vol 3, pp 1754–1759, Japan
3. Elango N, Faudzi AAM (2015) A review article: investigations on soft materials for soft robot manipulations. Int J Adv Manuf Technol 80(5–8):1027–1037
4. Park K-H, Kim B-H, Hirai S (2003) Development of a soft fingertip and its modeling based on force distribution. IEEE Int Conf Robot Autom 3:3169–3174
5. Shao F, Childs TH, Henson B (2009) Developing an artificial fingertip with human friction properties. Tribol Int 42(11–12):1575–1581
6. Derler S, Schrade U, Gerhardt IC (2007) Tribology of human skin and mechanical skin equivalents in contact with textiles. Wear 263:1112–1116
7. Tiezzi P, Kao I (2007) Modeling of viscoelastic contacts and evolution of limit surface for robotic contact interface. IEEE Trans Robot 23(2):206–217
8. Haibin Y, Cheng K, Junfeng L, Guilin Y (2018) Modeling of grasping force for a soft robotic gripper with variable stiffness. Mech Mach Theory 128:254–274
9. Pawlaczyk M, Lelonkiewicz M, Wieczorowski M (2013) Age-dependent biomechanical properties of the skin. Adv Dermatol Allergol (Postępy Dermatologii i Alergologii) 30(5):302
10. Controzzi M, D'Alonzo M, Peccia C, Oddo CM, Carrozza MC, Cipriani C (2014) Bioinspired fingertip for anthropomorphic robotic hands. Appl Bion Biomech 11(1–2):25–38

Multivariate Statistical Process Monitoring Strategy for a Steel Making Shop

Anupam Das

Abstract Monitoring of a manufacturing process ensures production of consistently good quality end production. In this paper, an attempt has been made to develop a monitoring strategy for a serial multistage manufacturing facility based on multi-block partial least squares regression, a multivariate regression technique. The developed monitoring strategy has been applied to a medium scale steel making shop. The monitoring strategy thus developed was employed for detection as well as for diagnosis of the faults responsible for the poor quality end product. The results obtained were found to be in sync with actual conditions.

Keywords Hotelling T^2 statistic · Monitoring chart · Multi-block partial least squares regression · Process monitoring · Steel making shop

1 Introduction

Process monitoring refers to surveillance of the concerned process or facility for the production of good quality end product. Monitoring of the process is also necessary to ensure lesser breakdown, smooth flow of material and for greater safety of the personnel and the equipment involved. The process monitoring strategies can be broadly categorized as model-based and data-based strategies. The data-based strategies particularly those based on statistical methods [1, 2] have gained widespread acceptance because of the ease with which they can be developed and implemented.

The process monitoring strategies based on statistical methods are commonly known statistical quality control (SQC) [3, 4]. The emergence of SQC can be traced back to the early part of the twentieth century with the advent of the control charts by W. A. Shewhart also known as Shewhart control charts. The 1990s witnessed the emergence of multivariate statistical quality control (MSQC) [5, 6] which is

A. Das (✉)
National Institute of Technology Patna, Patna 800005, Bihar, India
e-mail: anupam.das@nitp.ac.in

BBVL. Deepak et al. (eds.), *Innovative Product Design and Intelligent Manufacturing Systems*, Lecture Notes in Mechanical Engineering,
https://doi.org/10.1007/978-981-15-2696-1_94

based on simultaneous monitoring of multiple variables. For large and complex processes, there may be multiple variables which need to be monitored simultaneously and for such cases MSQC techniques play a key role. The most commonly employed MSQC techniques include Hotelling's T^2 charts, techniques or strategies based on principal component analysis (PCA) [7] and partial least squares regression (PLSR) [8, 9]. The T^2 charts are the multivariate variant of the commonly employed X-bar chart. PCA is a multivariate dimension reduction technique which has found widespread application in the area of social sciences [10]. PLSR on the other hand is a multivariate regression technique, and it varies from PCA on the ground that it works with two data sets, the predictor variables and the response variables.

In this paper, an attempt has been made to devise a monitoring strategy for a serial multistage manufacturing facility based on a multivariate regression method, namely multi-block partial least squares regression (MBPLSR) [11, 12]. MBPLSR is a variant of the ordinary PLSR. In MBPLSR, the predictor variables are segregated into multiples blocks based on their similarity, where the similarity may refer to variables belonging to a particular manufacturing stage or variables belonging to a particular operation region. MBPLSR are best suited for highly correlated data which is normally the case with large and complex multistage manufacturing facility.

The monitoring thus devised has been applied to a medium scale steel making shop (SMS) [13]. The SMS considered here can be thought as a serial multistage manufacturing facility. A detailed description of the SMS under consideration has been provided in the succeeding section. The monitoring strategy thus devised was used for detection of the faults and their subsequent diagnosis.

2 Case Study—A Steel Making Shop

The demonstration of the developed strategy was carried out on a medium scale SMS located in Eastern India. The SMS under consideration is engaged in the production of steel billets and blooms. A schematic diagram of the SMS is depicted in Fig. 1. As evident from the schematic diagram, the SMS is composed of three major components: the Electric Arc Furnace (EAF), the Ladle Refining Furnace (LRF) and the Continuous Casting Machine (CCM). The EAF is used for the melting of the charge which is composed of direct reduced iron (DRI) and scraps. The refining of the molten charge also occurs at the EAF. At the LRF, the alloying additives are added. Here further heating and refining of the molten charge take place. No further change in the chemical composition of the molten charge takes place once it exists in the LRF. Finally, the molten charge is fed into the CCM where it is converted into billets and blooms of the requisite dimension.

As shown in Fig. 1, the end product of the SMS considered is billets of a particular grade. The grade is determined by the chemical compositions of the billet produced that are termed here the quality characteristics of the end product (billets).

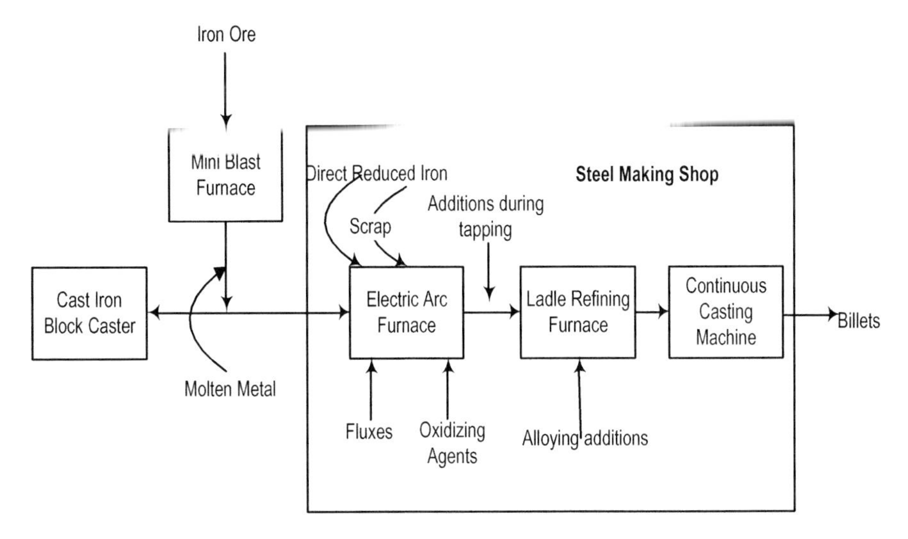

Fig. 1 Process flow of the SMS

Seven qualities are considered as important and included in the proposed methodology. However, the mechanical and dimensional properties of the billet are not considered. Therefore, the EAF and the LRF are considered as the two stages of the SMS as the chemical compositions of the molten steel are changed in these two stages. The CCM is not considered as in the CCM no changes take place in the chemical composition of the molten steel. The EAF and the LRF stage comprise 16 and 8 process and feedstock characteristics, respectively.

3 Methodology

The work involves the development of a monitoring strategy for a SMS which is a representation of a multistage manufacturing facility with the EAF, the LRF and the CCM forming the individual stages. The prime task involved is the building of a process representation or model of the process to be monitored. For building the model for the proposed monitoring strategy, MBPLSR and a multi-block variant of the PLSR were employed.

The steps for building the monitoring strategy can be segregated into three broad phases: (i) planning for model building and data collection phase, (ii) model building phase and (iii) fault detection and diagnosis phase. The planning for model building and data collection phase involves the identification of the relevant process, feedstock and quality characteristics and the collection of the in-control measurements of the chosen process, feedstock and quality characteristics. In the model building phase, construction of nominal models based on the in-control observations or measurements of the process, feedstock and quality characteristics is carried out. Prior to the building of a nominal model, the data is checked for the presence of any outlaying values. The fault detection and diagnosis phase encompass activities such as collection of new observations, determination of the control limits of the monitoring charts, charting of appropriate statistic(s) in the control charts and determination of the root cause(s) in the event of the detection of an out-of-control observation resembling an upset or abnormal condition. In the MBPLSR model that has been proposed, the predictor variables (process and feedstock characteristics) are grouped into several significant blocks with each block representing the process and the feedstock characteristics of a particular manufacturing stage of the multistage manufacturing facility.

The MBPLSR modeling primarily involves three broad steps. The first step is the estimation of predictor variables block scores. As pointed out earlier, each predictor variable block contains the process and the feedstock characteristics of a particular manufacturing stage of the concerned multistage manufacturing facility. The predictor variables block scores ($T^{(1)}$ and $T^{(2)}$) are estimated which are the linear combinations of the process and feedstock characteristics of the concerned stage. The second step of the MBPLSR modeling involves the estimation of the super scores $\left(T(s)\right)$ from individual block scores. In the third step, the super scores are used for the estimation of the response variable scores which are then used for estimation of the response variables and vice versa. The response variables are the quality characteristics of the billets. Faults represented by the out-of-control observations are detected by charting of an appropriate monitoring statistic. The control limit which acts as a threshold for detection of a fault is first determined. The signaling of a fault is accomplished when the value of the monitoring statistic crosses the threshold or the control limit. Table 1 lists the monitoring statistics and the control limit [11].

Where $T(s)_g$ = super score vector of the gth observation, $T_g^{(\mathrm{EAF})}$ = predictor variables block 1 (EAF process and feedstock characteristics) scores, $T_g^{(\mathrm{LRF})}$ = predictor variables block 2 (LRF process and feedstock characteristics) scores, q = number of MBPLSR components extracted, n = number of in-control observations with which the nominal MBPLSR models were constructed, k = no. of predictor variables (process and feedstock characteristics of all the stages combined), $F_{q,k-q,\alpha}$ = F-statistic with degrees of freedom q and $(k - q)$ and α level of significance, Λ_s = covariance matrix of the predictor variables (EAF and LRF process and feedstock characteristics combined) super scores in the form given below.

Table 1 Monitoring statistics and control limit

T_s^2 (Super T^2 statistic)	$T(s)_g^T \Lambda_s^{-1} T(s)_g$
T_{EAF}^2 (*Block* 1 T^2 *statistic*) (EAF *stage*)	$T_g^{(\mathrm{EAF})T} \Lambda_{\mathrm{EAF}}^{-1} T_g^{(\mathrm{EAF})}$
T_{LRF}^2 (*Block* 2 T^2 *statistic*) (LRF *stage*)	$T_g^{(\mathrm{LRF})T} \Lambda_{\mathrm{LRF}}^{-1} T_g^{(\mathrm{LRF})}$
Control limit	$q(n^2 - 1)/n(n - q)F_{q,k-q,\alpha}$

$$\Lambda_s = \begin{pmatrix} t(s)_{1,1} & \cdots & 0 \\ \vdots & \ddots & \vdots \\ 0 & \cdots & t(s)_{q,q} \end{pmatrix}$$ is a diagonal matrix since the super scores are orthogonal to each other

Similarly, Λ_{EAF} = covariance matrix of the EAF process and feedstock characteristics scores

$$\Lambda_{\text{EAF}} = \begin{pmatrix} t_{1,1}^{(\text{EAF})} & \cdots & 0 \\ \vdots & \ddots & \vdots \\ 0 & \cdots & t_{q,q}^{(\text{EAF})} \end{pmatrix}$$ is a diagonal matrix since the EAF process and feedstock characteristics scores are orthogonal to each other, and Λ_{LRF} = covariance matrix of the LRF process and feedstock characteristics scores

$$\Lambda_{\text{LRF}} = \begin{pmatrix} t_{1,1}^{(\text{LRF})} & \cdots & 0 \\ \vdots & \ddots & \vdots \\ 0 & \cdots & t_{q,q}^{(\text{LRF})} \end{pmatrix}$$ is a diagonal matrix since the LRF process and feedstock characteristics scores are orthogonal to each other.

Monitoring of the new observations is carried out by charting T_s^2, T_{EAF}^2 and T_{LRF}^2 statistics in separate monitoring charts. Upon the detection of an out-of-control observation, the causes responsible for it are determined by finding the contribution of the individual stages as well as the process and feedstock characteristics within the stages. Fault diagnostic statistics proposed by [11] were employed for diagnosis of faults, i.e., determination of the percent contribution of the stages and the characteristics within the stages for a particular out-of-control observation.

4 Analysis and Results

The SMS under consideration specializes in the production of plain as well as alloy carbon steel. Approximately, 110 grades of steel billets are being produced by the SMS. In this study, a particular grade of low carbon plain steel is selected for the analysis. The heat-wise measurements of the process, feedstock and quality characteristics at each stage are recorded in the logbook. In parlance to the steel industries, a heat corresponds to a batch of billets, blooms or slabs produced from a single melting and refining operation at various stages. Overall, 316 observations pertaining to the process, feedstock and quality characteristics were collected. Of the 316 observations, the first 173 were considered for the building of the nominal model, and the rest 143 observations, termed as the new observations, are considered for fault detection and its subsequent diagnosis. The 173 observations

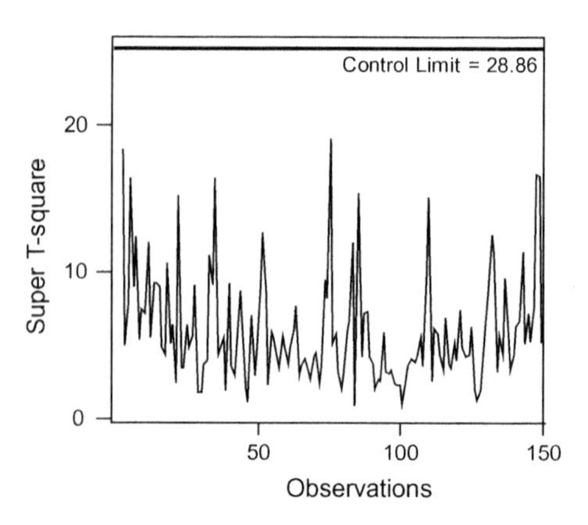

Fig. 2 Super T^2 chart for the in-control observations

selected for building of the nominal model were pretreated for the presence of outliers.

The data pre-treatment revealed a total 23 observations as outliers which were subsequently rejected. Hence, the nominal MBPLSR model was constructed with 150 in-control observations. A total of seven components were extracted. MATLAB computing software was used for writing the requisite code of model building. Two monitoring charts, namely the super T^2 chart and block T^2 charts are used for the detection of the out-of-control observations. The control limits for both set of monitoring charts are estimated from the 150 pretreated in-control observations. The super T^2 chart for the 150 in-control observations is depicted in Fig. 2.

The values of the monitoring statistics, the super T^2 statistic and block T^2 statistic are estimated for each of the new observations and are plotted against their respective control limits in the super and block T^2 monitoring charts. A fault is said to have occurred when the value of the super T^2 statistics for any particular observation crosses control limit of the super T^2 monitoring chart. The super scores and block scores for the new observations are determined with the aid NIPALS algorithm. The super T^2 chart for monitoring of the new observations detected a total of five observations as out-of-control observations. The next involves the diagnosis of the detected out-of-control observations. The diagnosis of the faults involves the determination of the contribution of the manufacturing stages and the process and feedstock characteristics within the stages to a detected fault represented by the out-of-control observations. Table 2 lists the percent contribution of the stages (EAF and LRF) to the out-of-control observations. As evident from Table VII, the percent contribution of the LRF stage to the out-of-control observations is more than the EAF stage for majority of the cases albeit for out-of-control observation number 114 where the percent contribution of the EAF stage is more.

Similarly, the contributions of the characteristics within the stages were also determined. For the EAF stage, the characteristic *EAF Lime* has been the chief contributor for out-of-control observation numbers 64 and 114 with percent

Table 2 Stage-wise percent contribution for out-of-control observations

Observation no.	Percent contribution of EAF stage	Percent contribution of LRF stage
55	22.78	77.22
59	27.49	72.51
60	5.18	94.82
64	3.28	96.72
114	51.74	48.26

contribution of 57.13% and 28.69%, respectively. The other major contributor includes *Hot Metal Carbon percentage* for observation number 114 with percent contribution of 37.01%.

5 Conclusion

In this paper, an attempt has been made to devise a monitoring strategy for a serial multistage manufacturing facility with the aid of a multivariate regression-based approach. A two-stage SMS has been considered to represent the multistage manufacturing facility. The multivariate regression technique that has been employed is MBPLSR. The notable advantage of MBPLSR lies in its ability to model highly correlated data which is often the case associated with a large and complex manufacturing facility. Another notable feature of MBPLSR is that the predictor variables can be segregated into blocks based on their similarity which subsequently aids a more elaborate diagnosis of the faults by identifying the contribution of the individual manufacturing stages to detected faults. The developed monitoring strategy performed the task of detection of faults and their subsequent diagnosis with satisfactory results.

References

1. Ganguly A, Patel SK (2015) Computer-aided design of X and R charts using teaching-learning-based optimization algorithm. Int J Prod Qual Manage 16(3):325–346
2. Doshi JA, Desai DA (2016) Statistical process control an approach for continuous quality improvement in automotive SMEs-Indian case study. Int J Prod Qual Manage 19:387–407
3. Franco BC, Celano G, Castagliola P, Costa AFB (2014) Economic design of Shewhart control charts for monitoring auto correlated data with skip sampling strategies. Int J Prod Econ 151:121–130
4. Reinikainen S, Hoskuldsson A (2007) Multivariate statistical analysis of a multi-step industrial processes. Anal Chim Acta 595(1/2):248–256
5. Kourti T (2006) The process analytical technology initiative and multivariate process analysis, monitoring and control. Anal Bioanal Chem 384(5):1043–1048

6. Han SW, Zhong H (2014) A comparison of MCUSUM-based and MEWMA-based spatiotemporal surveillance under non-homogeneous populations. Qual Reliab Eng Int 31 (8):1449–1472
7. Li G, Hu Y (2018) Improved sensor fault detection, diagnosis and estimation for screw chillers using density-based clustering and principal component analysis. Energy Build 173 (1):502–515
8. Botre C, Mansouri M, Karim MN, Nounou H, Nounou M (2017) Multiscale PLS based GLRT for fault detection of chemical processes. J Loss Prev Process Ind 46:143–153
9. Moreira SA, Sarraguça J, Saraiva DF, Carvalho R, Lopes JA (2015) Optimization of NIR spectroscopy based PLSR models for critical properties of vegetable oils used in biodiesel production. Fuel 150:697–704
10. Doble P, Sandercock M, Du Pasquier E, Petocz P, Roux C, Dawson M (2003) Classification of premium and regular gasoline by gas chromatography/mass spectrometry, principal component analysis and artificial neural networks. Forensic Sci Int 132(1):26–39
11. Choi SW, Lee I (2005) Multiblock PLS-based localized process diagnosis. J Process Control 15(3):295–306
12. Kourti T, Nomikos P, MacGregor JF (1995) Analysis, monitoring and fault diagnosis of batch processes using multiblock and multiway PLS. J Process Control 5(4):277–284
13. Tupkary RH, Tupkary VR (1998) An introduction to modern steel making, 6th edn. Khanna Publishers, New Delhi

Mathematical Modeling and Comparative Study of 12-DoF Biped Robot Using Screw Theory and Denavit–Hartenberg Convention

K. K. Rohith, Navaneeth Varma, A. P. Sudheer and M. L. Joy

Abstract The studies on legged robots have gained huge attention in recent years because of the agility that makes them applicable to a variety of environments. Biped robots are two-legged robots which replicate human anthropomorphism. Locomotion of biped robots is highly nonlinear, and modeling such a biped system demands some critical assumptions in foot rotation, ground contact, foot, and hip trajectories, etc. The gait pattern of biped walking can be expressed as a combination of single support phases (SSP) and double support phases (DSP). Kinematic and dynamic analysis is required for both configurations so that a realistic behavior can be modeled. There are various methods used for modeling of biped robots and similar open-chain robotic systems. The scope of this paper is to compare the kinematic models of a 12 DoF biped robot using Screw theory framework and Denavit–Hartenberg (D-H) convention. Further, the dynamic modeling is carried out using recursive Newton–Euler method and Lagrangian-Euler method for torque variations for single support phase. Joint angle, velocity, acceleration, and torque variations are analyzed during dynamic walking.

Keywords Biped robot · Screw theory · Denavit–Hartenberg convention · Recursive Newton–Euler formulation

1 Introduction

Recently, the studies on the development of legged robots have gained great attention due to the huge potential capabilities in risky environments. Human locomotion is inherently nonlinear, and hence, gait generation of a bipedal robot is a challenging task due to the limitations in computation time and energy utilization.

K. K. Rohith · N. Varma · A. P. Sudheer (✉) · M. L. Joy
Department of Mechanical Engineering, National Institute of Technology Calicut, Kozhikode 673601, India
e-mail: apsudheer@nitc.ac.in

BBVL. Deepak et al. (eds.), *Innovative Product Design and Intelligent Manufacturing Systems*, Lecture Notes in Mechanical Engineering,
https://doi.org/10.1007/978-981-15-2696-1_95

Human-like walking can be generated on bipedal robots by choosing appropriate kinematic and dynamic modeling methods with some assumptions.

Geometry-based kinematics is the most basic and simple method used for modeling of robots. The kinematics of a five-link biped robot geometrically on the sagittal plane is solved [1]. Joint variations in frontal and transverse planes are excluded in this model. Dynamic modeling of biped is carried out using the Lagrangian method. Ground reaction forces are measured for human walking. Position variation of trunk center of mass (CoM) and joint angle variations of all the links of the biped in the sagittal plane are plotted for the analysis. Denavit–Hartenberg (D-H) convention is widely used in kinematic modeling due to the minimum number of parameter requirement. Modeling of 8 degrees of freedom (DoF) biped in D-H framework is carried out in [2]. Though the method is easy for understanding and usage, the possibility of convergence is questionable as there are chances for multiple solutions and internal singularities. A comparative study of the kinematics using D-H and Screw theory for a robot manipulator is presented in [3]. The properties of both methods are discussed in detail and compared. Possibility of numerical solution for inverse kinematics in Screw theory is also mentioned in this paper.

The inverse kinematics in Screw theory framework is firstly introduced [4]. The position level inverse kinematics is calculated using invariant points on the screw axes for a 6R manipulator. The fundamental rule is that any point in space can be moved to another location by rotation about an arbitrary axis followed by translation with respect to the same axis. This method is also named as Paden–Kahan subroutine which is mostly used in modeling of bipeds based on Screw theory. Lie groups and Screws are used in [5] for modeling of RH0 humanoid. Inverse kinematics is solved in this work using closed-form solutions with Paden–Kahan subroutine. As an extension of this work, the same model is used in [6] for the upgraded humanoid RH1. Though the availability of the literature of kinematic models based on Screw theory is less, most of them employ Paden–Kahan subroutine for inverse kinematics. Screw theory and Paden–Kahan subroutine combination are applied for solving inverse kinematics of bionic arms connected to the upper body of a humanoid [7]. The same methodology is employed for kinematic modeling of a general 6R manipulator in [8]. Another approach for kinematic modeling of humanoid is presented based on Screw theory in [9, 10]. The whole body differential kinematics is solved using the Davies method and virtual chains taking hip joint frame as a floating reference. The humanoid is considered as a group of five serial chains working collaboratively to achieve humanoid tasks.

Lagrangian approach for solving the biped dynamics is applied in [11] that include toe rotation in the gait. Later, an upper body humanoid with wheeled base is developed and modeled using Screws and Lie groups [12]. Multiple kinematic chains are formed from base to arms and base to head for serial kinematic solutions. The model is then linked to Lagrangian dynamics for deriving the equations of motion. Analogous to the above, [13–16] have used Lagrangian dynamics in biped modeling. On the contrary, the suitability of recursive Newton–Euler dynamics over Lagrangian dynamics is proved for a six-link planar biped robot in both single support phases (SSP) and double support phases (DSP) [17]. Authors also stated that the former method is suitable for complex serial robotic configurations.

In most of the literature, biped robots are modeled in the sagittal plane, and the kinematic models are developed either using Screw theory or D-H convention. Suitability of these methodologies for kinematic modeling of biped robots with higher DoF should be analyzed for avoiding the complexity. In this paper, the Screw theory-based approach and D-H convention are used for kinematic modeling of a 12 DoF biped. A numerical iterative method is used for an inverse solution instead of the classic approaches. The kinematic model is generated in SSP, and the solutions are compared for both the approaches on the basis of computational efficiency and accuracy. The dynamic equations of motion are developed using recursive Newton–Euler method and Lagrangian method for joint torque evaluation.

2 Kinematic Modeling of Biped Robot

2.1 Kinematic Modeling Based on Screw Theory

A 12 DoF biped robot is considered in this paper for the complete analysis. There are three DoF at the hip, one at knee and two at the ankle for each leg. The link dimensions are decided based on anthropomorphic data. The screws assigned to each joint are passing through the corresponding axis of rotation. Figure 1 shows the allocation of screws to the biped mechanism along with the link dimensions. S1, S2, S3, etc., correspond to the screw axes assigned for each DoF in the exact order. The base frame $\{b\}$ is allocated to right leg ankle, and the end effector frame is attached to left leg ankle denoted as $\{s\}$.

The kinematic equation is developed using product of exponential (PoE) formula and Rodrigues equation for exponential mapping of screws. θ_i corresponds to the joint variable at ith joint. The exponential mapping of screw axis can be expressed as Eq. (1).

$$e^{[S]\theta} = \begin{bmatrix} e^{[\hat{\omega}]\theta} & \left(I\theta + (1-\cos\theta)[\hat{\omega}] + (\theta - \sin\theta)[\hat{\omega}]^2\right)v \\ 0 & 1 \end{bmatrix} \quad (1)$$

where $R(\hat{\omega}, \theta) = e^{[\hat{\omega}]\theta} = I + \sin\theta[\hat{\omega}] + (1-\cos\theta)[\hat{\omega}]^2$ and is the skew-symmetric form of angular velocity vector expressed as $[\hat{\omega}] = \begin{bmatrix} 0 & -\omega_3 & \omega_2 \\ \omega_3 & 0 & -\omega_1 \\ -\omega_2 & \omega_1 & 0 \end{bmatrix}$.

The final transformation matrix from base frame to end effector frame is represented as Eq. (2).

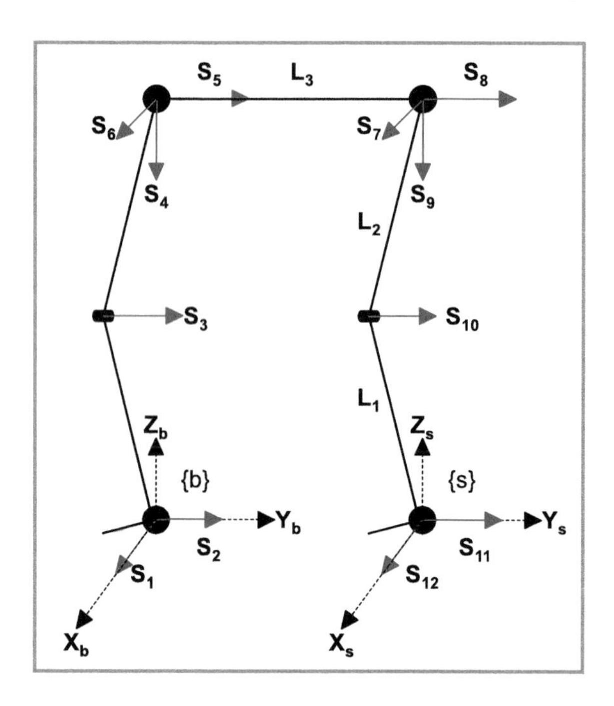

Fig. 1 Kinematic model with screw axis

$$T(\theta) = e^{[S_1]\theta_1} . e^{[S_2]\theta_2} . e^{[S_3]\theta_3} . e^{[S_4]\theta_4} . e^{[S_5]\theta_5} . e^{[S_6]\theta_6} . e^{[S_7]\theta_7} . e^{[S_8]\theta_8} . e^{[S_9]\theta_9} . e^{[S_{10}]\theta_{10}} . e^{[S_{11}]\theta_{11}} . e^{[S_{12}]\theta_{12}} . M \quad (2)$$

where the initial configuration, $M = \begin{bmatrix} 1 & 0 & 0 & 0 \\ 0 & 1 & 0 & L_3 \\ 0 & 0 & 1 & 0 \\ 0 & 0 & 0 & 1 \end{bmatrix}$.

2.2 *Kinematic Modeling Based on Denavit–Hartenberg (D-H) Method*

Kinematic modeling of 12 DoF biped robot is done with D-H method by assigning joint frames as shown in Fig. 4. The base frame $\{0\}$ is assigned to the ankle joint of the biped, and the ground frame $\{g\}$ is fixed at the ground. The link lengths L_1, L_2, L_3, and L_4 are as shown in Fig. 2. Frames are assigned from base (right leg angle joint) to the end effector (left leg foot). These frame assignments are clearly indicated in Fig. 2. All the angles are measured positive for counter-clockwise direction about the rotational axis. D-H parameter table corresponding to the frame assignment is shown in Table 1. The homogeneous transformation matrix T_i^{i-1} describes

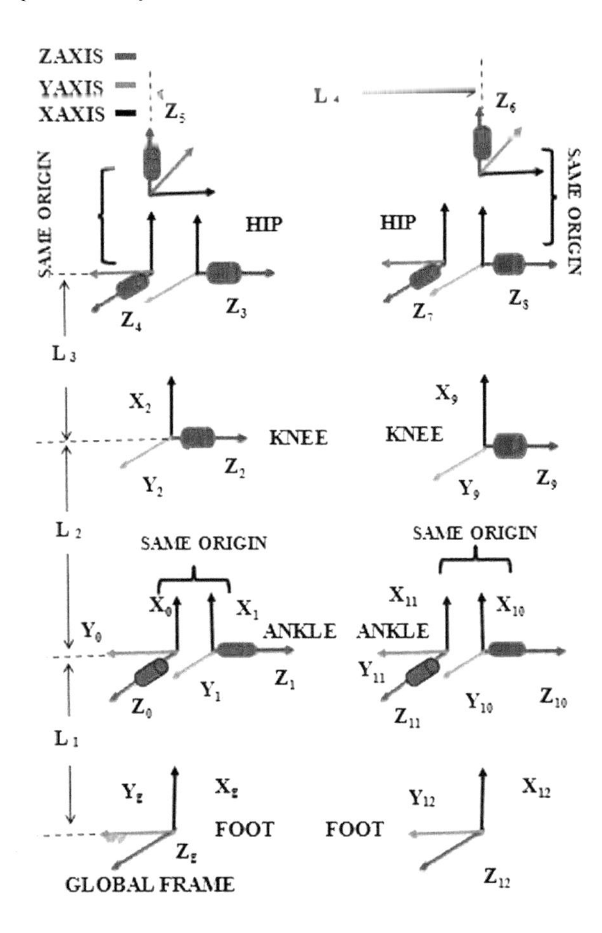

Fig. 2 Frame assignments of the biped model

Table 1 D-H parameters of biped

Link	θ_i	d_i	a_i	α_i
g	0	0	L_1	0
1	θ_1	0	0	90
2	θ_2	0	L_2	0
3	θ_3	0	L_3	0
4	θ_4	0	0	–
5	$\theta_5 - 90$	0	0	−90
6	θ_6	0	L_4	0
7	θ_7	0	0	90
8	$\theta_8 + 90$	0	0	90
9	θ_9	0	$-L_3$	0
10	θ_{10}	0	$-L_2$	0
11	θ_{11}	0	0	−90
12	θ_{12}	0	$-L_1$	0

the position, and orientation of frame $\{i\}$ relative to frame $\{i-1\}$ is given in Eq. (3).

$$T_i^{i-1} = \begin{bmatrix} C\theta_i & -S\theta_i C\propto_i & S\theta_i S\propto_i & a_i C\theta_i \\ S\theta_i & C\theta_i C\propto_i & -C\theta_i S\propto_i & a_i S\theta_i \\ 0 & S\propto_i & C\propto_i & d_i \\ 0 & 0 & 0 & 1 \end{bmatrix} \quad (3)$$

The pose of the foot from the global fixed frame is obtained by pre-multiplying individual transformation matrices. The final transformation matrix corresponds to the pose of the end effector (left foot) with respect to the inertial frame $\{g\}$ that is given by Eq. (4).

$$T_{12}^{g} = T_0^g * T_1^0 * T_2^1 * T_3^2 * T_4^3 * T_5^4 * T_6^5 * T_7^6 * T_8^7 * T_9^8 * T_{10}^9 * T_{11}^{10} * T_{12}^{11} \quad (4)$$

3 Inverse Kinematics

Inverse kinematic solutions of the biped robot in both Screw-based model and D-H model are determined using an iterative method, namely Levenberg–Marquardt algorithm (LMA) in MATLAB. The transformation matrices and the required pose can be written in Screw theory and D-H convention, respectively, as in Eq. (5).

$$T(\theta) = \begin{bmatrix} r11 & r12 & r13 & r14 \\ r21 & r22 & r23 & r24 \\ r31 & r32 & r33 & r34 \\ 0 & 0 & 0 & 1 \end{bmatrix} \quad \text{and} \quad T_{12}^{g} = \begin{bmatrix} r11 & r12 & r13 & r14 \\ r21 & r22 & r23 & r24 \\ r31 & r32 & r33 & r34 \\ 0 & 0 & 0 & 1 \end{bmatrix} \quad (5)$$

3.1 Recursive Newton–Euler Dynamics

The dynamic equations of motion for biped robot are derived using recursive N-E method. The velocities and accelerations are determined from the forward iteration, and torque and force are determined from backward iteration in the N-E formulation [18]. For joints with more than one DoF, imaginary links with negligibly small masses are assumed between the corresponding joints. The joint friction loads and external disturbances are not included in this model. The torque equations are developed in the empirical form shown below.

$$M(\theta)\ddot{\theta} + C\left(\theta, \dot{\theta}\right)\dot{\theta} + N(\theta) = \tau \quad (6)$$

3.2 Lagrangian Dynamics

The biped dynamics is modeled using Lagrangian method for verifying the obtained joint torque plots for dynamic walking obtained in the previous section. The Lagrange formulation is an energy-based method to obtain the dynamic model of robots [19]. Lagrangian function L is defined as the difference between the total kinetic energy and potential energy of the entire mechanism. The joint variables of 12 DoF biped robot are expressed as q_i and joint velocities as $\dot{q}_i$. The generalized Lagrange equation of motion for the evaluating ith joint torque (τ_i) is given by,

$$\frac{\mathrm{d}}{\mathrm{d}t}\left(\frac{\partial L}{\partial \dot{q}_i}\right) - \frac{\partial L}{\partial q_i} = \tau_i \quad \text{for} \quad i = 1, 2, 3, \ldots, n \tag{7}$$

Equation of motion for evaluating ith joint torque (τ_i) is given as in Eq. (8).

$$\tau_i = \sum_{j=1}^{n} M_{ij}(q)\ddot{q}_j + \sum_{j=1}^{n}\sum_{k=1}^{n} h_{ijk}\dot{q}_j\dot{q}_k + G_i \tag{8}$$

4 Results and Discussion

4.1 Comparison of Kinematic Models

The forward kinematic equations using Screw theory are formulated in MATLAB. The cycloidal trajectory is chosen as the foot trajectory for walking on a flat surface. The target poses on the cycloid are used with a kinematic model for inverse kinematic solutions using a numerical method. The Levenberg–Marquardt algorithm is employed for determining the corresponding joint variables. The total time taken for a single step is assumed as 2 s. Figure 3 shows continuous walking of a bipedal robot. The inverse kinematic solutions are determined for the continuous walking steps. Left and right feet are moving through a series of cycloids. Instantaneous positions on the cycloid trajectory are selected for inverse kinematic solutions and plotting of dynamic walking.

The forward transformation matrices in the D-H method are formulated in MATLAB. Cycloid path used in Screw-based model is chosen here as the foot trajectory for walking on a flat surface. The target poses on the cycloid are used with the final transformation matrix for iterative inverse kinematic solutions. Levenberg–Marquardt algorithm is employed here also to get optimized results. Instantaneous positions are used for a single step and then generated the continuous gait as shown in Fig. 4.

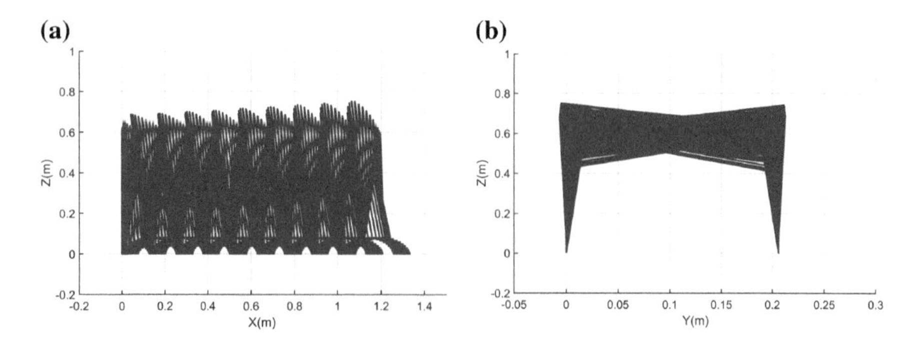

Fig. 3 Biped plot for continuous walking **a** side view **b** front view

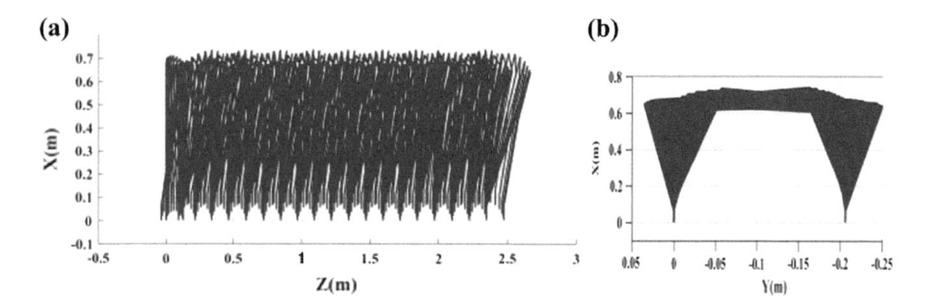

Fig. 4 Biped plot for continuous walking in D-H framework

When both views of Figs. 3 and 4 are compared, there is much deviation of the hip in the frontal plane in D-H modeling. Hence, Screw theory-based gait is better compared to D-H-based gait and ensures more human-like locomotion.

Joint velocities and accelerations are plotted for a single step in both Screw theory and D-H convention. The variation of angular velocity and acceleration is shown in Figs. 5 and 6. The change in velocity and acceleration is not gradual in D-H-based analysis. Sudden change in acceleration can create jerks on the robot while executing the gait. This change in velocity and acceleration is due to the hip trajectory variations in the frontal plane.

4.2 *Joint Torques for Dynamic Walking*

Joint torques plotted for both dynamic models are compared. The peak torque values from recursive N-E method based on Screw theory are found to be agreeing with the torque values from the Lagrangian model based on D-H convention. The

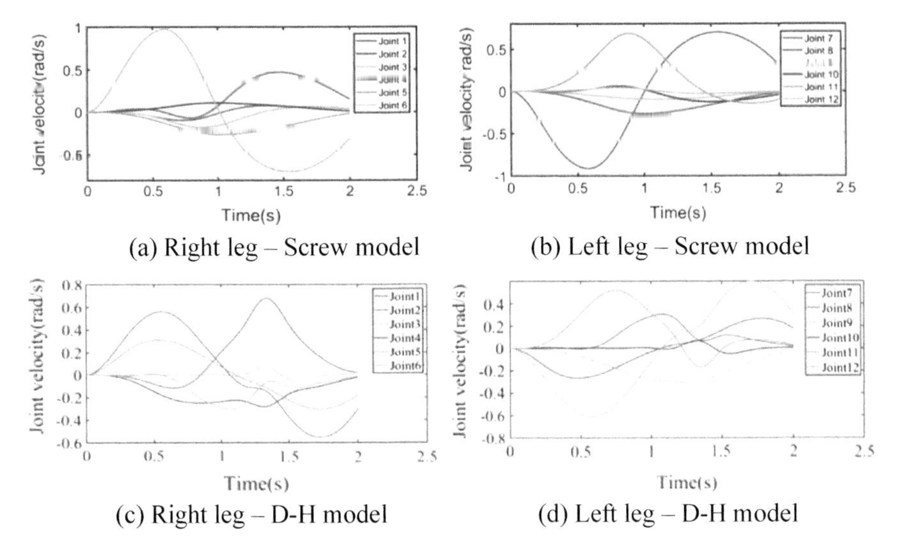

(a) Right leg – Screw model (b) Left leg – Screw model

(c) Right leg – D-H model (d) Left leg – D-H model

Fig. 5 Joint velocity

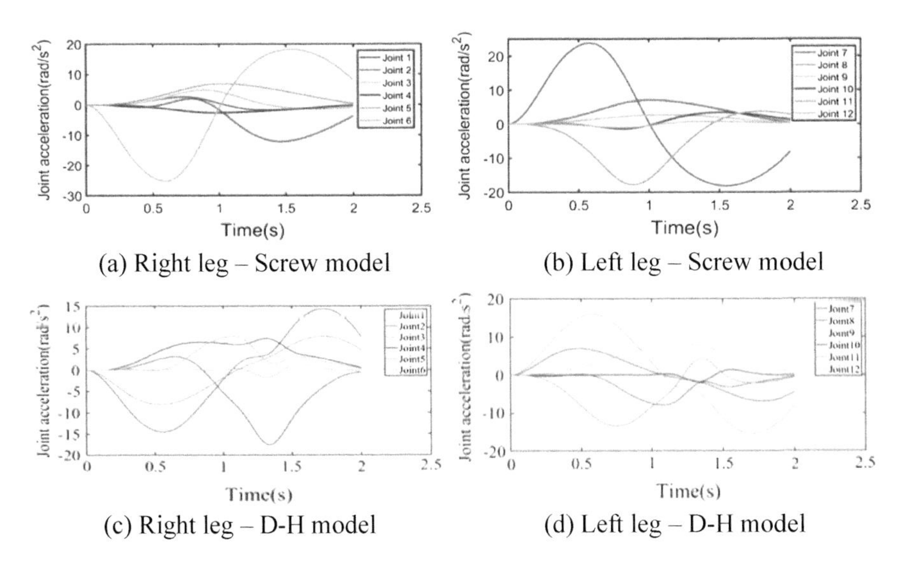

(a) Right leg – Screw model (b) Left leg – Screw model

(c) Right leg – D-H model (d) Left leg – D-H model

Fig. 6 Joint acceleration

direction change in these torque plots is due to the change in the direction of base frame axes. Torque plots for all the joints are displayed in Fig. 7. Torque plot gives the theoretical joint torque (τ_{th}) values for any given operation. A safety factor is added with the theoretical value as shown in Eq. (9). Using the actual torque τ_{actual},

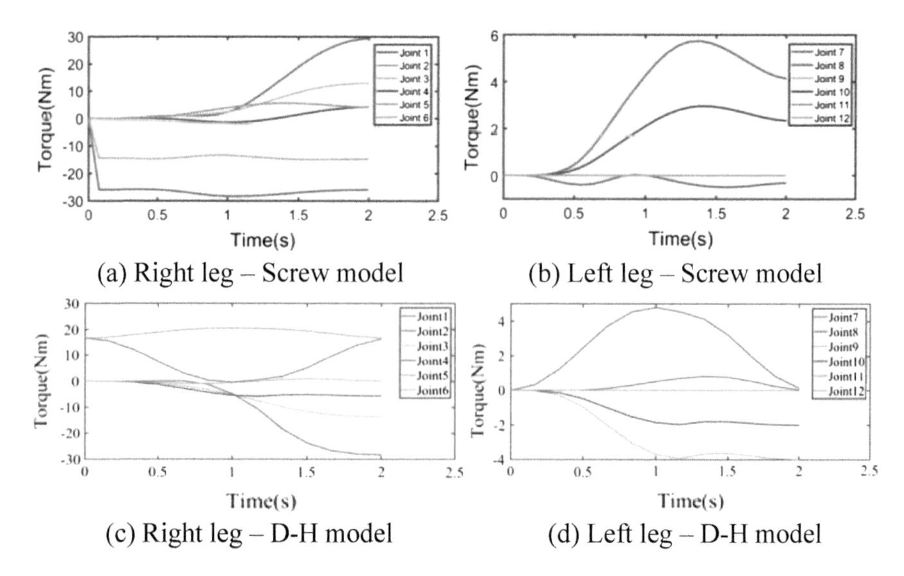

Fig. 7 Joint torque variation

actuator selection can be done from the datasheet based on the speed, voltage, and ampere rating.

$$\tau_{\text{act}} = \tau_{\text{th}} * \text{safety factor} \tag{9}$$

5 Conclusion and Outlook

The mathematical modeling of 12 DoF biped robot has been completed using Screw theory and D-H method. The numerical inverse method is used for the inverse kinematics solution in both cases. The accuracy and computational efficiency are superior while using screws due to the absence of local coordinate frames and internal singularities. Further, the dynamic model has been generated using recursive Newton–Euler method, and the obtained torque values are verified using Lagrangian dynamics. The joint torques are compared for dynamic walking on a cycloid trajectory using both methods. Joint friction and external disturbance loads are not included in this model. This work can be extended in several ways such as stability analysis for various modes of motion like climbing, jumping, running, stepping over, etc. Different controller designs can be implemented and compared; intelligent gaits can be generated for various applications, and human interactions can also be experimented.

References

1. Luxman R, Zielinska T (2017) Robot motion synthesis using ground reaction forces pattern; analysis of walking posture. Int J Adv Rob Syst 14(4)
2. Sarkar A, Dutta A (2015) 8-DoF biped robot with compliant-links. Rob Autonom Syst 63 (1):57–67
3. Rocha CR, Tonetto CP, Dias A (2011) A comparison between the Denavit–Hartenberg and the screw-based methods used in kinematic modeling of robot manipulators. Rob Comput-Integr Manuf 27(4):723–728
4. Paden B, Sastry S (1988) Optimal kinematic design of 6R manipulators. Int J Rob Res 7 (2):43–61
5. Pardos JM, Balaguer C (2005) RHO humanoid robot bipedal locomotion and navigation using Lie groups and geometric algorithms. In: IEEE/RSJ international conference on intelligent robots and systems. IEEE, Alta, pp 3081–3086
6. Arbulu M, Balaguer C (2007) Real-time gait planning for Rh-1 humanoid robot, using Local Axis Gait algorithm. In: 7th IEEE-RAS international conference on humanoid robots. IEEE, Pittsburgh, pp 563–568
7. Man C, Fan X, Li C, Zhao Z (2007) Kinematics analysis based on Screw theory of a humanoid robot. J China Univ Mining Technol 17(1):49–52
8. Wu A, Shi W, Li Y, Wu M, Guan Y, Zhang J, Wei H (2015) Formal kinematic analysis of a general 6R manipulator using the Screw theory. Math Prob Eng 1(7)
9. Toscano GS, Simas H, Castelan EB (2014) Screw-based modeling of a humanoid biped robot. In: Anais do XX Congresso Brasileiro de Automática. Belo Horizonte, pp 944–951
10. Toscano GS, Simas H, Castelan EB, Martins D (2018) A new kinetostatic model for humanoid robots using Screw theory. Robotica 36(4):570–587
11. Furusho J, Sano A (1991) Development of biped robot. Adv Psychol 78(1):277–303
12. Wang J, Li Y (2009) Dynamic modeling of a mobile humanoid robot. In: IEEE international conference on robotics and biomimetics. IEEE, Bangkok, pp 639–644
13. Novaes C, Silva P, Rouchon P (2014) Trajectory control of a bipedal walking robot with inertial disc. In: IFAC proceedings volumes, Vol 47, Issue 3. Elsevier, Cape Town, pp 4843–4848
14. Sadedel M, Yousefi-Koma A, Khadiv M, Iranmanesh F (2018) Heel-strike and toe-off motions optimization for humanoid robots equipped with active toe joints. Robotica 36 (6):925–944
15. Kanehiro F, Hirukawa H, Kajita S (2004) Open HRP: open architecture humanoid robotics platform. Int J Rob Res 23(2):155–165
16. Hernández-Santos C, Rodriguez-Leal E, Soto R, Gordillo JL (2012) Kinematics and dynamics of a new 16 DOF humanoid biped robot with active toe joint. Int J Adv Rob Syst 9(5):190
17. Al-Shuka HFN, Corves B, Zhu WH (2014) Dynamic modeling of biped robot using Lagrangian and recursive Newton-Euler formulations. Int J Comput Appl 101(3):1–8
18. Lynch KM, Frank CP (2017) Modern robotics mechanics, planning, and control, 1st edn. Cambridge University Press, USA
19. Mittal RK, Nagrath IJ (2017) Robotics and control, 1st edn. McGraw Hill Education, India

Kinematic and Static Structural Analysis of a Humanoid with a Wheeled Mobile Base

Manoj Kumar Mallick and A. P. Sudheer

Abstract Humanoid robots are facing issues such as self-collision, low speed, instability, and complex control system, when putting to work in different working environments. Humanoid robots with legs are generally not dynamically stable during high-speed motions. One of the solutions for avoiding this challenging issue is to replace the legs with wheelbase. This paper mainly deals with the development of a humanoid having a wheeled mobile base for indoor and outdoor applications. The upper body of the humanoid robot is designed with 3-DOF in the torso, 2-DOF in the neck, dual arm with 5-DOF in each arm, and mobile base with four wheels and 3-DOF. Forward and inverse kinematic modeling is done for the torso and arms. Workspace analysis is performed and the maximum work volume is 2.678 m^3. Static structural analysis is carried out for dimension synthesis.

Keywords Humanoid robot · Wheeled mobile base · Kinematic modeling · Workspace analysis · Static structural analysis

1 Introduction

Presently, humanoid robots are used both in industrial and service sectors. Humanoid robots have made an impact in many other applications such as space exploration, rehabilitation, and healthcare. Robots can be used in hazardous environments like nuclear plants or radioactive areas which have life-threatening issues for humans.

However, humanoids with legs have issues such as instability, low speed, self-collision, and complex control system. Humanoid with wheelbase solves the above-specified issues along with providing the flexibility of a robotic manipulator and workspace of the mobile robot. This robot can be used to do exhaustive and

M. K. Mallick · A. P. Sudheer (✉)
Department of Mechanical Engineering, National Institute of Technology Calicut, Kozhikode 673601, India
e-mail: apsudheer@nitc.ac.in

BBVL. Deepak et al. (eds.), *Innovative Product Design and Intelligent Manufacturing Systems*, Lecture Notes in Mechanical Engineering,
https://doi.org/10.1007/978-981-15-2696-1_96

repetitive works such as pick and place job, delivery task, and domestic services with better efficiency and accuracy when compared to humans [1]. However, when the robot is designed to interact with the humans, the robot also should have human-like structure along with anthropomorphic movements resulting in the use of higher degrees of freedom.

Most of the humanoid robots developed and have human-like upper body design. The variation can be observed in the design for locomotion. Humanoids can take two forms of locomotion in the form of being Bi-pedal (two legs) as ASIMO [2] by Honda, or wheeled like Rollin Justin [3] and Robonaut 2 by NASA [4] for space exploration. Humanoid robots with the wheeled mobile base offer faster mobility as compared to legged humanoid robots and also it decreases the complexities in case of Bi-pedal humanoid. Rollin' Justin [3] has 3-DOF in the torso consisting of only pitching and yawing motion. 2-DOF in the pitching generates redundant DOF. Justin [5], the table-mounted humanoid upper body system is transformed into Rollin' Justin by developing a four-wheeled mobile platform. Extendable wheels are used for the platform instead of statically stable mobile platforms. McGinn et al. [6] proposed a wheeled and legged humanoid with limited DOF in the upper body. The robot is able to adjust its aspect ratio in the sagittal plane. Lack of flexible motion in the upper body may lead to instability of the robot. Combination of DC geared motors and pneumatic artificial muscles are used for the actuation that results in a complex control system.

Keneko et al. [7] developed the first humanoid robot HRP-2 that have human-size and that can lie down and stand up. The joints are designed according to the human movable motion to make the robot able to perform tasks of humans. Mizuuchi et al. [8] developed a musculoskeletal humanoid using muscle driven endoskeleton which gives the robot a high degree of flexibility. The complete body has 91 DOF including the flexible spine. The control design becomes very complex due to a large number of sensors and mechatronic components. Rojas et al. [9] developed a wheeled humanoid robot "Melo," to study the stability of a single inverted body on a mobile platform. The top of the body is attached with 6-DOF manipulators. The study focuses on shifting the center of mass of system by formulating fundamental motion and gesture compliance control to avoid tipping over. Deepak and Parhi [10] have developed a wheeled mobile manipulator using an artificial immune system. The mobile platform and the manipulator are designed as such that both can be interchanged for locomotion and manipulation for various tasks. The autonomous mobile platform has the criteria of the human immune system that corresponds to different environmental situations and obstacles in the path.

The robot system requires the joint variables in the Cartesian coordinates system from which the end-effector position and orientation is determined by the inverse kinematic solution of the robot. Many approaches are used to compute the inverse kinematic solutions such as iterative approach, analytic approach, geometric approach, and Jacobian approach.

The geometric method is commonly used to obtain inverse kinematic solutions [11]. This approach becomes difficult when the robot mechanism has more degrees of freedom and cannot be generalized for all the robot mechanisms. Ali et al. [12] proposed a closed-form inverse kinematic joint solution for humanoid robots by decoupling the mechanism into positioning and orientation mechanism and solving the inverse kinematics by an analytical approach. Here, the inverse kinematic solution is obtained directly by considering only the position matrix by neglecting the orientation matrix which may not give the proper inverse kinematic solution of the end-effector. Although the geometric method or closed-form solution are commonly used to obtain inverse kinematic solutions, but cannot be efficiently used for robots with high DOF. Here, the robot upper body has 15 DOF. Above methods cannot be used because of the complexity and computational difficulty. Levenberg–Marquardt algorithm is one of the efficient iterative methods suitable for higher degrees of freedom.

The novelty of the designed model is the rolling motion that is incorporated in the robot for exact human-like motion with the simplest configuration and control systems. The humanoids developed in recent years are more focused on yawing and pitching motion instead of rolling motion. The advantage of this rolling motion is that the robot will be able to mimic human motions while picking objects lying on the lateral side of the torso. Rolling motion in the waist helps to pick up objects in the lateral side by actuating only one DOF. These results in less time consumption in comparison to the actuation of the pitch and yaw motion for picking the same object. This paper presents a wheeled humanoid robot to overcome the issues faced in a legged humanoid robot. The upper body of the humanoid has human-like anthropomorphic movements with roll, pitch, and yaw motion in the waist. The neck of the humanoid has a pitch and yaw motions. Kinematic modeling of the humanoid robot along with workspace analysis is presented. Static structural analysis of the parts used in fabrication is carried out using ANSYS software. The forward kinematics is done using DH representation and the inverse kinematic solution is obtained by the iterative approach. This iterative method is more consistent and provides a better solution when the degrees of freedom are more and the mechanism cannot be decoupled. This approach can be generalized to model any other humanoid robots.

2 CAD Model of the Wheeled Base Humanoid

The humanoid robot comprises of two parts, the upper part consists of the torso, neck, and dual arm. The lower part is the mobile base with four wheels. The upper part of the humanoid robot is designed by considering the body ratios of a 5 feet human [13].

The upper part of the humanoid has 3-DOF in the waist, 2-DOF in the neck, 5-DOF in each of the arms. The mobile base has four wheels with 3 DOF. The left and right wheels are provided with actuation, whereas the front and the back wheels are free to rotate about its own axis. The front and back wheels are castor wheels, whereas the left and right wheels are conventional wheels. The wheelbase is

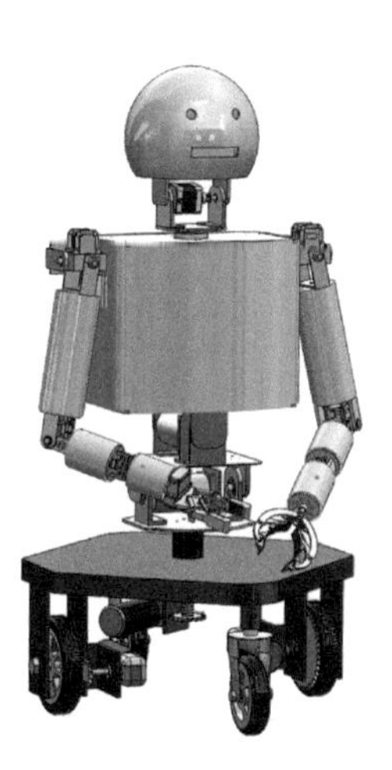

Fig. 1 3D model of the humanoid robot

designed to work in flat terrains for both indoor and outdoor applications. The CAD model of the designed mobile base is shown in Fig. 1.

The height of the upper body from the base to the top of the head is 71.9 cm, whereas the total height of the humanoid robot from the ground including the wheelbase is 100.9 cm. The body of the humanoid is made by aluminum alloy (Al-6061) and the wheelbase is made up of mild steel. The mass of the upper body is 10.15 kg and the wheelbase is 22 kg. The mass properties are obtained from SolidWorks after assigning the material property of each component.

3 Kinematic Modeling of the Upper Body Humanoid

3.1 Forward Kinematic Solution for the Upper Body Humanoid

The forward kinematics equation is used to solve the position and orientation of the end-effector when the joint angles are known. This can be determined by using the coordinate frames from the robot geometry, which are given by the DH parameters [14]. The link coordinate frames are assigned along the positive direction of rotation of the joints as shown in Fig. 2.

From Fig. 2, frame 3 is the base frame for both the arms and acts as a static joint during the modeling of the arms. The upper body has 5-DOF with 3-DOF in the waist and 2-DOF in the neck. Frame 3 acts as a revolute joint during the modeling of the upper body. The length of the links represented in Fig. 2 is tabulated in Table 1.

The pose of the end-effector is obtained by multiplying the matrices together to obtain the spatial displacement of the 5th coordinate frame with respect to the base frame. Similarly, the spatial displacement of the 10th and 16th coordinate frame

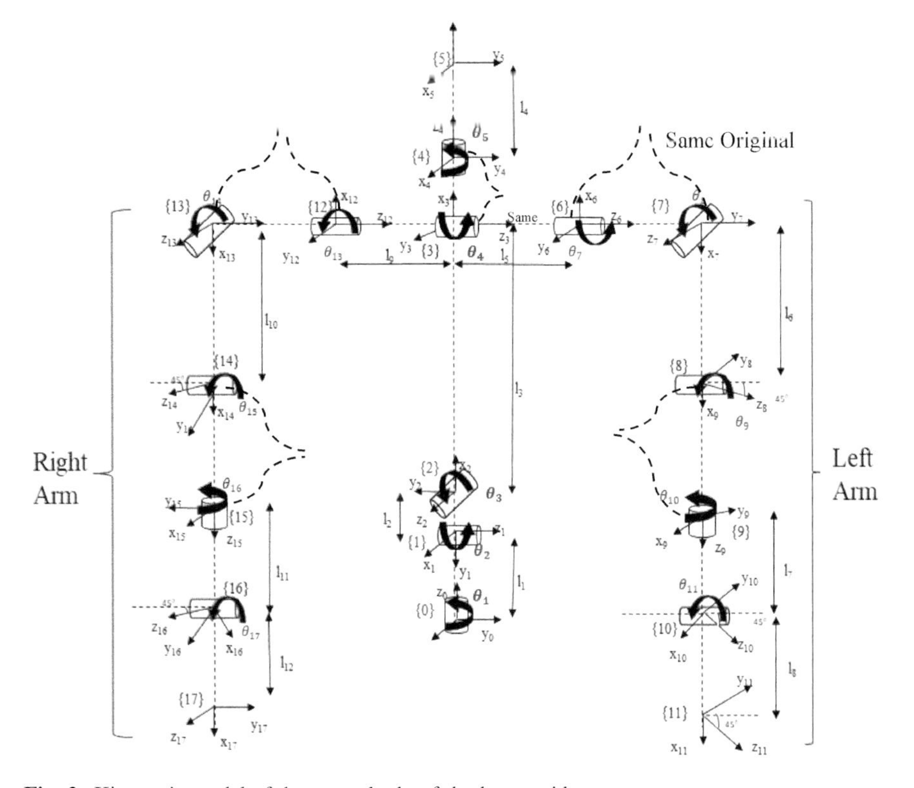

Fig. 2 Kinematic model of the upper body of the humanoid

Table 1 Link lengths of the upper body humanoid

Length of torso links		Link length of arms	
Link	Length (mm)	Link	Length (mm)
l_1	47	l_5, l_9	193.5
l_2	96	l_6, l_{10}	279
l_3	306	l_7, l_{11}	219
l_4	61	l_8, l_{12}	162
l_5	209		

with respect to the 3rd coordinate frame is obtained. DH parameters for the torso left arm and right arm are shown in Tables 2, 3, and 4, respectively.

Final transformation matrix of the torso with respect to the base as shown in Fig. 1 is given by Eq. 1,

$$T_0^5 = T_0^1\ T_1^2\ T_2^3\ T_3^4\ T_4^5 \tag{1}$$

Table 2 DH parameters for the torso and neck

Coord. Frame i	θ_{i+1}	d_{i+1}	a_{i+1}	α_{i+1}
1	θ_1	l_1	0	−90
2	$\theta_2 - 90$	0	l_2	−90
3	θ_3	0	l_3	90
4	$\theta_4 + 90$	0	0	90
5	θ_5	0	0	0

Table 3 DH parameters for the left arm

Coord. Frame i	θ_{i+1}	d_{i+1}	a_{i+1}	α_{i+1}
3	0	l_5	0	0
6	$\theta_7 + 180$	0	0	90
7	θ_8	0	l_6	-45
8	$\theta_9 - 90$	0	0	-90
9	θ_{10}	l_7	0	90
10	$\theta_{11} + 90$	0	l_8	0

Table 4 DH parameters of the right arm

Coord. Frame i	θ_{i+1}	d_{i+1}	a_{i+1}	α_{i+1}
3	0	l_9	0	0
12	$\theta_{13} + 180$	0	0	90
13	θ_{14}	0	l_{10}	45
14	$\theta_{15} + 90$	0	0	90
15	θ_{16}	l_{11}	0	−90
16	$\theta_{17} + 90$	0	l_{12}	0

Final transformation matrix of the left arm with respect to the neck/reference frame as shown in Fig. 1 is written as Eq. 2,

$$T_{10}^{3} = T_{6}^{3}\, T_{7}^{6}\, T_{8}^{7}\, T_{9}^{8}\, T_{10}^{9}\, T_{11}^{10} \tag{2}$$

Final transformation matrix of the right arm with respect to the neck/reference frame as shown in Fig. 1 is written as Eq. 3,

$$T_{17}^{3} = T_{12}^{3}\, T_{13}^{12}\, T_{14}^{13}\, T_{15}^{14}\, T_{16}^{15}\, T_{17}^{16} \tag{3}$$

The final transformation matrix of the end-effector of the gripper with respect to the base frame is given by Eq. 4,

$$T_{17}^{0} = T_{3}^{0}\, T_{17}^{3} \quad \text{and} \quad T_{10}^{0} = T_{3}^{0}\, T_{10}^{3} \tag{4}$$

This forward kinematic equation is used for getting inverse kinematic solutions by an iterative method.

3.2 *Inverse Kinematic Solution of the Upper Body*

The inverse kinematic solution is used to obtain a different joint solution for the desired position and orientation of the end-effector. Iterative Levenberg–Marquardt algorithm (LMA) is used to obtain the inverse kinematic solution by solving the non-linear equations. The LMA is a curve fitting method that adaptively varies the parameter updates between gradient descent method and the Gauss–Newton method [15]. In this method, the set of equations are represented as "$f(x) = 0$," where x represents all the joint variables, $x = [x_1, x_2, x_3, \ldots, x_{24}]$. The objective function of this algorithm is to minimize the RHS and approach to zero.

4 Workspace Analysis

It is necessary to analyze the shape and volume of the workspace for enhancing the working capabilities of the humanoid robot. The torso of the humanoid robot can undergo yaw, pitch, and roll motions. The neck of the humanoid can undergo yaw and pitch motion. Workspace is plotted using the forward kinematic equations obtained from the DH model explained in the previous section. The workspace analysis of the various links of the humanoid robot is done using MATLAB, in which the joint parameters are varied as described in Table 5. The joint parameters are varied according to an average human being. The algorithm used to plot the workspace is as follows:

Step 1: The end-effector position of each arm is obtained by the forward kinematics equation in terms of the joint variables.

Step 2: Consider the position terms in the obtained pose matrix, i.e., p_x, p_y, and p_z. The joint variables are varied (as shown in Table 5) using "for" loop and on the substitution of the angles, the position of the end-effector is marked in three-dimensional space.

Step 3: The cluster of points obtained at the end represents the workspace of the upper body of the humanoid robot as shown in Fig. 3.

After obtaining all the points, the work volume is calculated using the "Convhull" tool in MATLAB. The work volume of the upper body with the dual arm is computed as 2.8567 m^3.

Table 5 Workspace parameters [7]

Torso	Arms
$-40° < \theta_1 < 40°$	$-180° < \theta_{13}, \theta_7 < 50°$
$-30° < \theta_2 < 45°$	$-180° < \theta_{14}, \theta_8 < 0°$
$-50° < \theta_3 < 50°$	$-145° < \theta_{15}, \theta_9 < 0°$
$-50° < \theta_4 < 60°$	$-90° < \theta_{16}, \theta_{10} < 90°$
$-70° < \theta_5 < 70°$	$-70° < \theta_{17}, \theta_{11} < 90°$

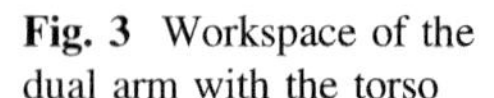

Fig. 3 Workspace of the dual arm with the torso

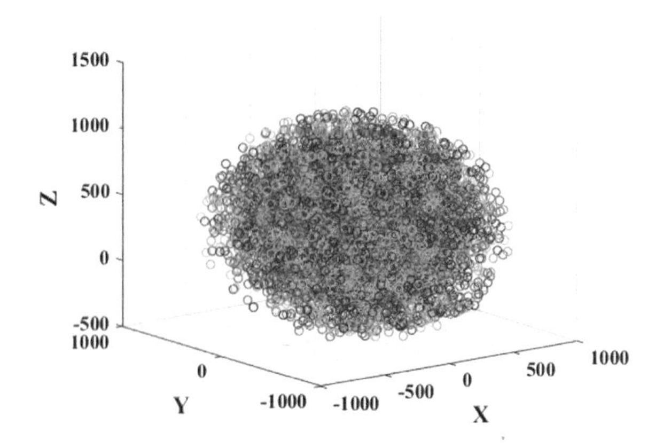

5 Static Structural Analysis

The static structural analysis is done using ANSYS workbench 18.2. It is done to compute the effects of static loading conditions on the structure of the upper body humanoid robot. The analysis is done to compute the maximum stress and total deformation at extreme loading conditions on the robotic mechanism. The simulation results are used to determine the weaker parts in the mechanism and changes can be made accordingly in the dimensions to withstand the induced force and moment in the body. The optimum dimensions for every component are obtained on specifying the maximum and minimum limits possible.

Assumptions considered for static structural analysis are (a) the material is assumed to have linear elastic behavior, (b) the forces are assumed to be statically applied, i.e., forces do not vary with respect to time, and (c) Inertial effects such as mass, and damping are not included.

The material assigned to the part is aluminum alloy that is readily available in the material library of ANSYS. The analysis is done for a payload capacity of 1 kg in the gripper of the left arm in a horizontal position. The obtained force reaction and the induced moment at the fixed joint are applied to the next consecutive component to compute the stress and deformation in the next component. Similarly, the simulation for each component is carried out and corresponding stress and deformation is obtained. ANSYS result of some of the critical components like a gripper, shoulder joint, and chest that are more susceptible to failure is shown in Figs. 4 and 5, respectively. The figures represent the von Mises stress and the total deformation of the component.

The maximum yield strength of aluminum alloy (Al-6061) is 110 MPa. The induced stress in the gripper, shoulder joint, and chest are 34.761 MPa, 23.923 MPa, and 15.008 MPa, whereas the deformation is obtained as 0.14196 mm, 0.10629 mm, and 0.25175 mm, respectively. The total deformation is also taken into consideration for the proper positioning of the end-effector. The optimization is done for minimization of the stress and deformation.

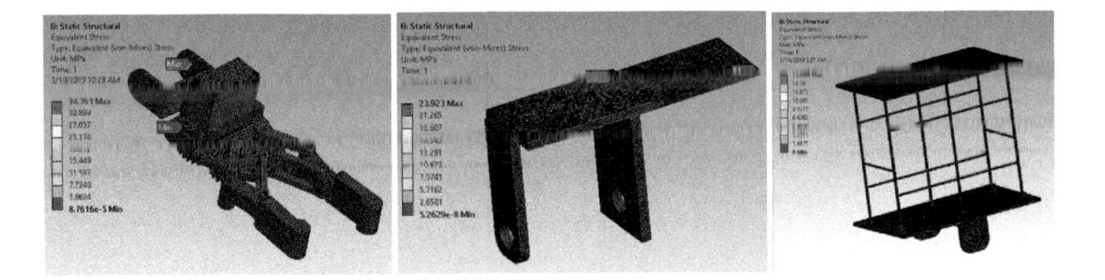

Fig. 4 ANSYS results showing the induced stress in the gripper, shoulder joint, and chest

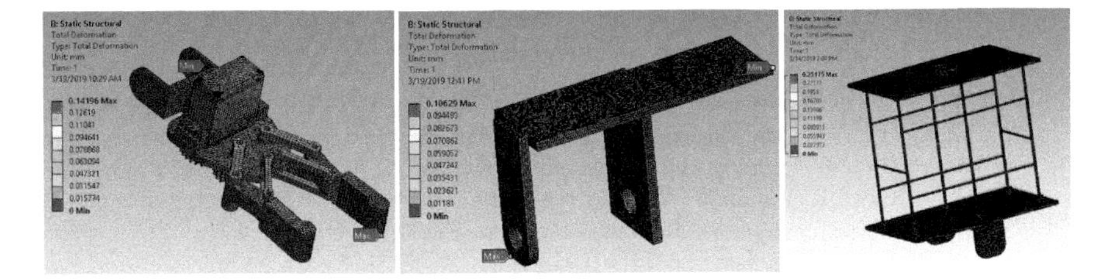

Fig. 5 ANSYS results showing deformation in the gripper, shoulder joint, and chest

6 Results and Discussions

This paper presented the design and modeling of an upper body humanoid robot with a wheelbase. Kinematic, workspace, and static structural analysis results are given in Sects. 3, 4 and 5, respectively. The workspace analysis of the complete body is carried out in MATLAB that shows the reachability of the designed model. The work volume obtained is found to be 2.856 m^3. The static structural analysis of the robot structure is done to determine the optimum link dimensions to withstand the loads and stresses acting on each component. Static structural analysis is carried out with a payload capacity of 1 kg in the arms. The induced stress in each component is lesser than the yield strength of aluminum alloy (yield strength = 110 MPa) by considering a factor of safety of three. Thus, the designed model is safe for fabrication.

7 Conclusion

In this paper, an upper body of the humanoid robot with wheelbase is proposed which is capable of attaining human-like motion. The proposed design has five revolute joints in the torso and five revolute joints in each arm. The limitation of the designed model is the working environment. The wheelbase used presently limits the usage of the robot only in flat terrains which can be eliminated by using hybrid or legged wheels. The future scope of the present work can be attributed toward integrating vision system, implementing artificial intelligence based on the applications that include human face recognition, object recognition, speech recognition, etc. Further, coordination of the robotic arms results in the development of collaborative robots that can be used in various industrial and manufacturing sectors.

References

1. Yamashiro M, Xie Z, Yamaguchi H, Ming A, Shimojo M (2009) Home service by a mobile manipulator system. In: IEEE international conference on automation and logistics. IEEE, China, pp 295–300
2. Sakagami Y, Watanabe R, Aoyama C, Matsunaga S, Higaki N, Fujimura K (2002) The Intelligent ASIMO: system overview and integration. In: IEEE/RSJ international conference on intelligent robots and system, vol 3. IEEE, New York, pp 2478–2483
3. Fuchs M, Borst C, Giordano PR, Baumann A, Kraemer E, Langwald J, Gruber R, Seitz N, Plank G, Kunze K, Burger R (2009) Rollin'Justin-design considerations and realization of a mobile platform for a humanoid upper body. In: IEEE international conference on robotics and automation. IEEE, Japan, pp 4131–4137
4. Diftler MA, Mehling JS, Abdallah ME, Radford NA, Bridgwater LB, Sanders AM, Askew RS, Linn DM, Yamokoski JD, Permenter FA, Hargrave BK (2011) Robonaut 2-the first humanoid robot in space. In: IEEE international conference on robotics and automation. IEEE, China, pp 2178–2183
5. Borst C, Ott C, Wimbock T, Brunner B, Zacharias F, Bäuml B, Hillenbrand U, Haddadin S, Albu-Schäffer A, Hirzinger G (2007) A humanoid upper body system for two-handed manipulation. In: Proceedings on ieee international conference on robotics and automation. IEEE, Italy, pp 2766–2767
6. McGinn C, Cullinan MF, Otubela M, Kelly K (2019) Design of a terrain adaptive wheeled robot for human-orientated environments. Autonom Rob 43(1):63–78
7. Kaneko K, Kanehiro F, Kajita S, Hirukawa H, Kawasaki T, Hirata M, Akachi K, Isozumi T (2004) Humanoid Robot HRP-2. In: Proceedings IEEE international conference on robotics and automation. IEEE, USA, pp 1083–1090
8. Mizuuchi I, Yoshikai T, Sodeyama Y, Nakanishi Y, Miyadera A, Yamamoto T, Niemela T, Hayashi M, Urata J, Namiki Y, Nishino T (2006) Development of musculoskeletal humanoid Kotaro. In: Proceedings 2006 IEEE international conference on robotics and automation. IEEE, Florida, pp 82–87
9. Rojas S, Shen H, Griffiths H, Li N, Zhang L (2017) Motion and gesture compliance control for high performance of a wheeled humanoid robot. In: ASME 2017 international mechanical engineering congress and exposition. American Society of Mechanical Engineers, United States, pp V04AT05A058–V04AT05A058
10. Deepak BBVL, Parhi DR (2016) Control of an automated mobile manipulator using artificial immune system. J Exp Theor Artif Intell 28(1–2):417–439
11. Zannatha JI, Limon RC (2009) Forward and inverse kinematics for a small-sized humanoid robot. In: 2009 international conference on electrical, communications, and computers. IEEE, Mexico, pp 111–118
12. Ali MA, Park HA, Lee CG (2010) Closed-form inverse kinematic joint solution for humanoid robots. In: 2010 IEEE/RSJ international conference on intelligent robots and systems. IEEE, Taiwan, pp 704–709

13. Winter DA (2009) Biomechanics and motor control of human movement, 4th edn. Wiley, New York
14. Denavit J, Hartenberg RS (1955) A kinematic notation for lower-pair mechanisms based on matrices. Trans ASME J Appl Mech 23:215–221
15. Gavin H (2019) The Levenberg-Marquardt method for nonlinear least squares curve-fitting problems. Department of Civil and Environmental Engineering, Duke University

Modelling and Analysis of Seeding Robot for Row Crops

Cino Mathew Jose, A. P. Sudheer and M. D. Narayanan

Abstract Agriculture in India relies on light and heavy machinery managed manually by farmers. In such cases, the output is affected by the availability of manual labour. This brings about a scope of automation to ease the efforts taken by farmers. Currently, available agriculture robots are very expensive and big sized. This makes it difficult for small/medium farmers to introduce automation into their fields. This paper mainly focuses on modelling of a farming robot that is compact and suitable for small-scale farming. An innovative precision seeding mechanism is designed based on vacuum pressure. A dual prismatic–revolute (2PR) robotic manipulator with a soil drill as the end effector guides the seed into its position after digging the soil to appropriate depth. The seeder is designed to seed at desired inter-crop distances in multiple rows at a single pass. The 2PR robotic manipulator is supported on a wheeled base to enable mobility through rough terrains.

Keywords Precision agriculture · Agricultural robotics · Denavit–Hartenberg convention · Wheeled mobile robot

1 Introduction

The extent of mechanization for seeding and planting is only 29% in India [1]. The major portion of seeding is done using manual labour. The lack of labour force and wage rate escalation has resulted in a significant rise in the cost of cultivation in India. The imported machinery are not aligned with the Indian conditions. Hence, there is a need to develop agriculture robots for indigenous conditions. In recent times, agriculture robotics has been a major topic for research and development elsewhere in the world and is at initial stages in India [2, 3].

C. M. Jose · A. P. Sudheer (✉) · M. D. Narayanan
Department of Mechanical Engineering, National Institute of Technology Calicut, Kozhikode 673601, India
e-mail: apsudheer@nitc.ac.in

BBVL. Deepak et al. (eds.), *Innovative Product Design and Intelligent Manufacturing Systems*, Lecture Notes in Mechanical Engineering,
https://doi.org/10.1007/978-981-15-2696-1_97

Approaches for precision farming in small-farm agriculture proposed by Shibusawa [4] assert the need for a variable rate technology to adjust the agricultural inputs. Haibo et al. [5] designed a robot with four-wheel drive, the drive system using a servo motor and the steering system using a stepper motor. Using independent steering motors imparted extra cost on the robot. An agricultural robotic platform with four-wheel steering for weed detection designed by Bak et al. [6] had the excursion from the path by two-wheel steering significantly larger than four-wheel steering. The steering motor had limited response whilst turning sharp turns. Kim et al. [7] developed a tracked robotic platform on a paddy field using sensor fusion. The tracked robot compacted the paddy fields affecting the growth of rice.

The optimum vacuum pressure for seeder was determined by Karayel et al. [8], and a mathematical model was developed. Minfeg et al. [9] optimized a seed metering device with fluted rollers, which was applicable only for wheat. Iacomi and Popescu [10] developed a precision seeder for granulated seeds using a solenoid. Based on the literature, vacuum seeders were developed either for a particular seed or granulated seeds.

Muir et al. [11] proposed a general approach to derive the wheeled mobile robot (WMR) kinematics. Using the Sheth–Uicker convention, transformations were obtained relating wheel to a robot body, and these were differentiated to relate wheel to body velocities. A 3D kinematics of WMRs on uneven terrain was proposed by Tarokh et al. [12] and presented a 3D version of Muir and Neuman's transformation-based approach. From the literature survey, the development of agricultural robots for seeding in India is scarce. The agricultural robots for Indian conditions are rare to find in the literature. The cost of precision seeding mechanisms is very high. A precision vacuum seeder for ungranulated seeds, seeds of varied size and shape was lacking in the literature. Hence, modelling of a low-cost, precision seeding mechanism for a variety of seeds mounted on a wheeled mobile base is presented in this paper.

The seeder works with the help of a vacuum pump capable of intaking the seeds from the seed bin with its suction heads. The 2PR soil drill manipulator makes the soil ready for seed deposition. The seeding mechanism is mounted on a skid-steered four-wheeled mobile base.

2 Modelling of Agricultural Seeding Robot

The seeder robot is a four-wheeled robotic platform mounted with seeding implements as shown in Fig. 1. The robot wheelbase is powered with four DC motors. The robot is skid-steered. The use of skid-steered steering eases remote operations as well as the controller design. The seeding implement is mounted on the robotic platform. The storage bin stores the seed to be sowed which is attached to the body frame. The seed flows into the suction bin, and it is sucked by the tip of vacuum seeder. The required vacuum pressure is provided by the vacuum pump. The

Fig. 1 Solidworks model of the agricultural seeding robot

Fig. 2 Vacuum seeder mechanism for seeding

vacuum seeder has six tips arranged around a cylinder. The seed is then removed from the suction tip using a brush attached to a minimotor and transfers the seed to the soil in a flexible tube as shown in Fig. 2. The seed drill manipulator makes the soil ready for deposition. It has two prismatic joints to position the seed and a revolute joint to drill the soil. The end effector has a seed drill to drill the soil to appropriate depth. The 2PR manipulator positions the seed using a flexible tube along multiple rows in a single pass. The flaps attached to the mechanism cover the soil once the seed is deposited. The wheeled mobile robot moves to a new location, and then seeds can be placed at another set of rows.

The agriculture seeding robot is designed for small-scale farms. Depending upon the seed spacing requirement, the 2PR mechanism positions the seed. Table 1 mentions the specifications of the agriculture seeding robot.

Table 1 Specification of agricultural seeding robot

Wheelbase	950 mm
Track: Front/rear	600 mm
Length × Width × Height	1230 mm × 765 mm × 710 mm
Ground clearance	350 mm
Wheel diameter	280 mm
Drive	AWD
Total weight (without any attachments)	26.5 kg
Body frame	Aluminium 6063
Steering	Skid-steered steering
Top speed	0.5 m/s

2.1 Kinematic Modelling

Kinematic modelling of wheeled mobile base. Kinematic modelling of the wheeled mobile base is done using Muir and Neuman's transform-based approach [11]. It is assumed that the wheeled mobile base moves on flat terrain. Four numbers of conventional wheels are used in the mobile platform which is having three degrees of freedom (DOF). It allows translation in the 2D plane and rotation about the vertical axis through the centre of mass. The assignment of a coordinate system to the wheeled mobile base is as shown in Fig. 3, and the coordinate nomenclature is given in Table 2.

Position transformation matrix of robot coordinates with respect to ground is given in Eq. (1). Here ${}^{A}T_{B}$ represents constant transformation matrices and ${}^{A}\phi_{B}$ represents variable transformation matrices.

$$^{\overline{R}}\Pi_R = {}^{F}T_{\overline{R}}^{-1}\,{}^{F}T_{\overline{Ci}}\,{}^{\overline{Ci}}\phi_{Ci}\,{}^{Si}T_{Ci}^{-1}\,{}^{Hi}\phi_{Si}^{-1}\,{}^{R}T_{Hi}^{-1} \tag{1}$$

Velocity transformation matrix of robot coordinates with respect to ground is given in Eq. (2).

$$^{\overline{R}}\dot{\Pi}_R = {}^{F}T_{\overline{R}}^{-1}\,{}^{F}T_{\overline{Ci}}\,{}^{\overline{Ci}}\dot{\phi}_{Ci}\,{}^{Si}T_{Ci}^{-1}\,{}^{Hi}\phi_{Si}^{-1}\,{}^{R}T_{Hi}^{-1} + {}^{F}T_{\overline{R}}^{-1}\,{}^{F}T_{\overline{Ci}}\,{}^{\overline{Ci}}\phi_{Ci}\,{}^{Si}T_{Ci}^{-1}\,{}^{Hi}\dot{\phi}_{Si}^{-1}\,{}^{R}T_{Hi}^{-1} \tag{2}$$

For a conventional steered wheel, the relation between wheel velocity and robot velocity is given in Eq. (3).

$$^{\overline{R}}\dot{P}_R = J_i\,\dot{q}_i \tag{3}$$

where wheel Jacobian matrix is $J_i = \begin{bmatrix} -R_i \sin^{R}\theta_{C_i} & {}^{R}d_{C_iy} & -{}^{R}d_{H_iy} \\ -R_i \cos^{R}\theta_{C_i} & -{}^{R}d_{C_ix} & {}^{R}d_{H_ix} \\ 0 & 1 & -1 \end{bmatrix}$

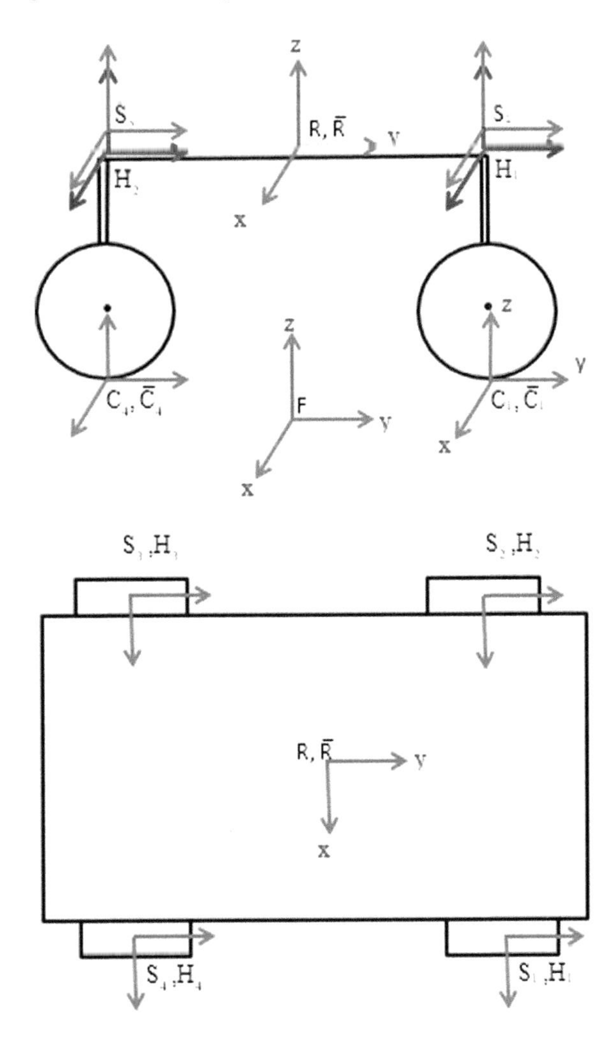

Fig. 3 Assignment of coordinate system to the wheeled mobile base

Table 2 Denavit–Hartenberg parameters of 2PR mechanism

R	Robot coordinate system
F	Floor coordinate system
H_i	Hip coordinate system
S_i	Steering coordinate system
C_i	Contact point coordinate system
$\overline{R}$	Instantaneously coincident robot coordinate system
$\overline{C_i}$	Instantaneously coincident contact point coordinate system

Wheel velocity is given as $\dot{q}_i = \begin{bmatrix} \omega_{wix} \\ \omega_{wiz} \\ \omega_{siz} \end{bmatrix}$

$$^{R}\theta_{Hi} + {}^{Hi}\theta_{Si} + {}^{Si}\theta_{Ci} = {}^{R}\theta_{Ci}$$

$$^{Si}d_{Cix}\cos({}^{R}\theta_{Hi} + {}^{Hi}\theta_{Si}) - {}^{Si}d_{Ciy}\sin({}^{R}\theta_{Hi} + {}^{Hi}\theta_{Si}) + {}^{R}d_{Hix} = {}^{R}d_{Cix}$$

$$^{Si}d_{Cix}\sin({}^{R}\theta_{Hi} + {}^{Hi}\theta_{Si}) + {}^{Si}d_{Ciy}\cos({}^{R}\theta_{Hi} + {}^{Hi}\theta_{Si}) + {}^{R}d_{Hiy} = {}^{R}d_{Ciy}$$

The composite robot equation for the four-wheeled mobile robot is given in Eq. (4).

$$\begin{bmatrix} I_1 \\ I_2 \\ I_3 \\ I_4 \end{bmatrix} \dot{p} = \begin{bmatrix} J_1 & 0 & 0 & 0 \\ 0 & J_2 & 0 & 0 \\ 0 & 0 & J_3 & 0 \\ 0 & 0 & 0 & J_4 \end{bmatrix} \dot{q} \tag{4}$$

Kinematic modelling of seed drill manipulator. The seed drill manipulator has two prismatic joints (2P) and one revolute joint (R). The seed drill indents the soil to the required depth depending upon the seed used and farm conditions. The kinematic analysis is done using the Denavit–Hartenberg method.

The frame assignment and kinematic modelling diagram of seed drill manipulator are as shown in Fig. 4. Denavit–Hartenberg parameters of the 2PR mechanism are given in Table 3 (Fig. 5).

The final transformation matrix from the base frame to the end-effector frame is equal to the desired pose matrix which is given in Eq. (5).

$$T_3^0 = \begin{bmatrix} C\theta & -S\theta & 0 & 0 \\ 0 & 0 & 1 & a_1 + a_3 + d_2 \\ -S\theta & -C\theta & 0 & d_1 \\ 0 & 0 & 0 & 1 \end{bmatrix} = \begin{bmatrix} n_x & o_x & a_x & p_x \\ n_y & o_y & a_y & p_y \\ n_z & o_z & a_z & p_z \\ 0 & 0 & 0 & 1 \end{bmatrix} \tag{5}$$

The inverse kinematic relationships for the desired pose matrix are given in Eqs. (6), (7) and (8).

$$d_2 = p_y - a_1 - a_3 \tag{6}$$

$$d_1 = p_x \tag{7}$$

$$\theta = \operatorname{atan2}(-n_x, o_x) \tag{8}$$

Fig. 4 Frame assignment of seed drill 2PR manipulator

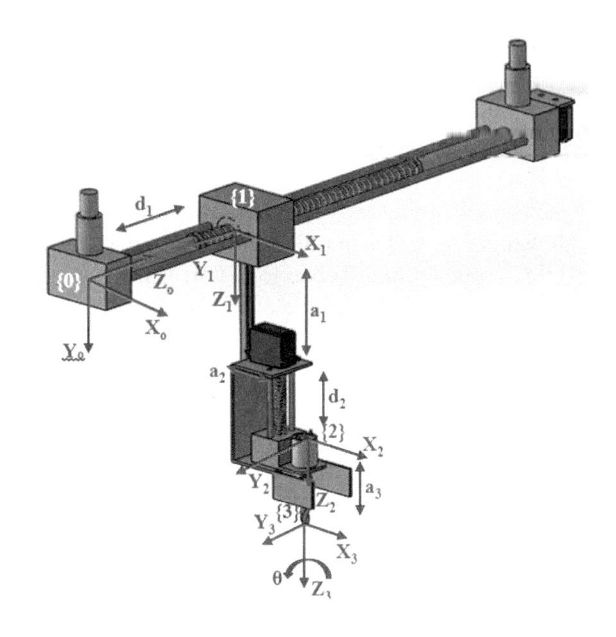

Table 3 Coordinate nomenclature

Link	θ_i	d_i	a_i	α_i
1	0	d_1	0	−90°
2	0	$d_2 + a_1$	a_2	0
3	θ	a_3	0	0

Fig. 5 Kinematic diagram and mass distribution of 2PR manipulator

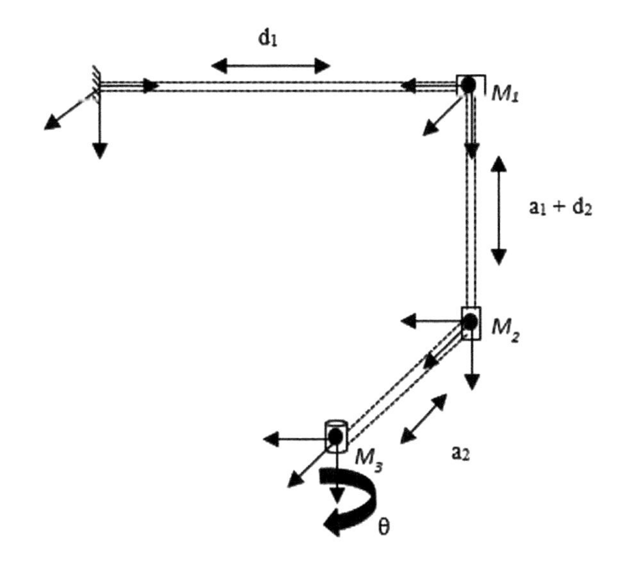

2.2 Dynamic Analysis

Dynamic analysis of wheeled mobile base using multibody dynamics software. The multibody dynamics (MBD) software, MSC Adams, is used for the dynamic analysis of the wheeled mobile base. A model is made, and the inertial and geometries properties of components are added from the actual solid model as shown in Fig. 6. The torque is plotted for a wheeled mobile base, with a velocity of 0.5 m/s along a straight line for a time period of 2 s. The maximum torque obtained from Fig. 7 is 14.5 Nm. Selection of the drive motor for wheeled mobile base is based on maximum torque value.

Dynamic analysis of soil drill manipulator using Lagrangian–Euler method. The total kinetic and potential energy of link 1, link 2 and link 3 of the 2PR mechanism of soil drill is given in Eqs. (9) and (10).

$$K = \frac{1}{2}M_1\dot{d}_1^2 + \frac{1}{2}M_2\left(\dot{d}_1^2 + \dot{d}_2^2\right) + \frac{1}{2}I\dot{\theta}^2 \quad (9)$$

$$P = M_1gh + M_2g(a_1 + d_2) + M_3g(a_1 + d_2) \quad (10)$$

where M_1 (3 kg), M_2 (1.2 kg) and M_3 (0.4 kg) are the lumped masses of link 1 (L1), link 2 (L2) and link 3 (L3) in Fig. 5; a_1, a_2 and a_3 are the constant distances as in Fig. 4. Based on Lagrangian–Euler formulation, the dynamic model is obtained from Lagrangian as Eqs. (11), (12) and (13)

$$M_1\ddot{d}_1 + M_2\ddot{d}_1 = F_1 \quad (11)$$

$$M_2\ddot{d}_2 + M_2g = F_2 \quad (12)$$

Fig. 6 Dynamic analysis of farm robot using MSC Adams

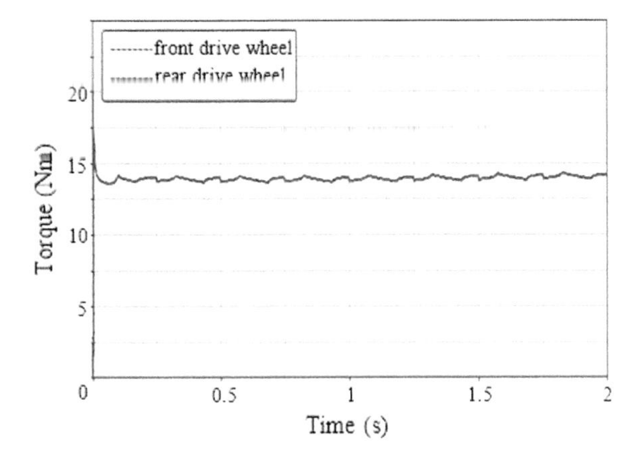

Fig. 7 Torque variation of drive wheel

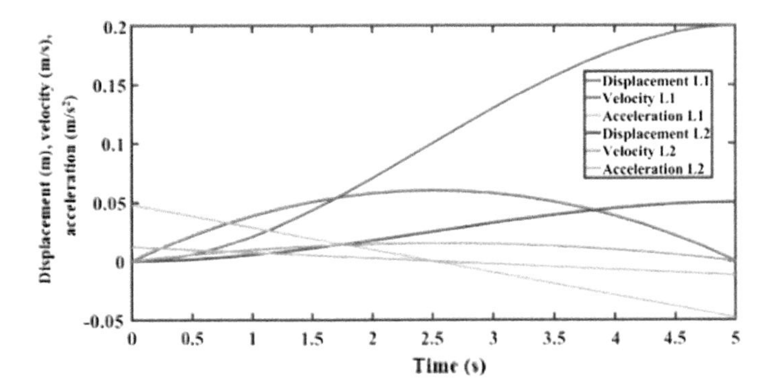

Fig. 8 Displacement, velocity and acceleration plot of link 1 and link2

$$I\ddot{\theta} = \tau_1 \tag{13}$$

Given a cubic trajectory for joint displacements, the displacement, velocity and acceleration plot for the main lead screw is as shown in Fig. 8. The force acting on the link 1 is obtained using the L-E formulation. The torque obtained by designing link 1 and link 2 as power screws is shown in Fig. 9. The maximum torque obtained for link 1 is 1.03 Nm, link 2 is 0.82 Nm and link 3 is 0.4 Nm. The obtained maximum torque value decides the actuator selection for lead screws.

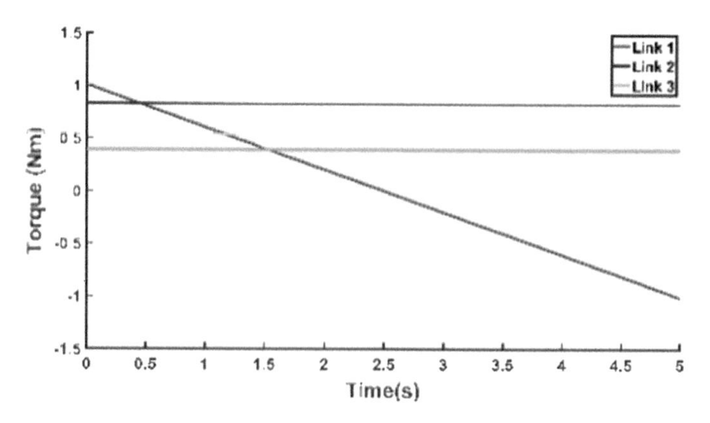

Fig. 9 Joint torques for seed drill

3 Static Structural Analysis of Body Frame

The static structural analysis of the body frame under three different conditions is considered. The first condition is when the robot is travelling in a straight line. The body frame is fixed at the wheel frame pivot point, and the external load is applied. The body frame is made of aluminium 6063 with a tensile strength of 195 MPa.

The maximum stress obtained is 34.95 MPa, and deformation is 0.49 mm as shown in Figs. 10 and 11. The factor of safety with respect to the maximum tensile strength is 5.57. The second extreme condition is when the body moves along an inclined plane with 30° slope. Here, the maximum stress obtained is 35.68 MPa, deformation is 0.47 mm and the factor of safety is 5.46. The third condition is when the body frame moves over an obstacle. During this time, only two wheels touch the ground. The maximum stress obtained is 37.06 MPa, and deformation is 1.89 mm. In the third extreme condition, a safety factor of 5.1 is obtained.

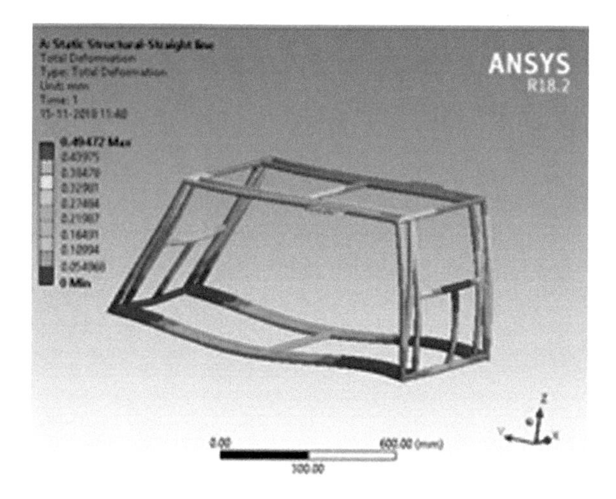

Fig. 10 Total deformation of body frame

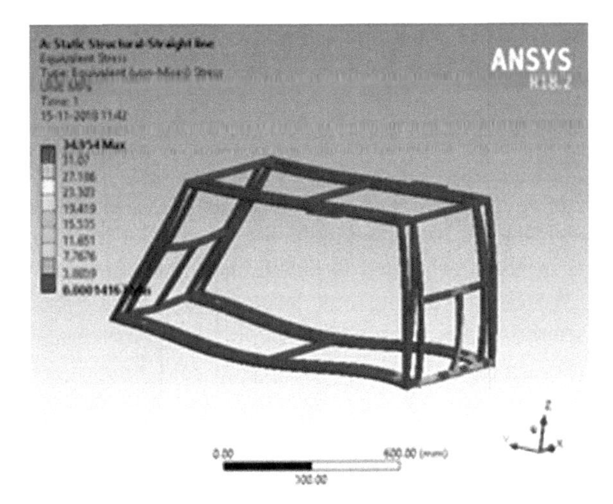

Fig. 11 Equivalent stress on body frame

4 Selection of Vacuum Pump

A vacuum generator is used to determine the vacuum pressure required for the seeder to pick up seeds from the seed bin. A compressor is attached to the vacuum generator. The vacuum pressure is varied depending upon the compressor pressure. The seed used is groundnut. The compressor pressure is varied from 0 to 6 bar. The ability of the seed suction tip to pick up the seed is noted down. It is found that the pressure required is between 48 and 63 kPa.

5 Results and Discussion

In this paper, an agricultural seeding robot is proposed which can seed along multiple rows at required depth in a single pass. The seeding robot comprises a vacuum seeder along with a 2PR manipulator which is mounted on a four-wheeled robotic base. The kinematic modelling of seed drill is done using Denavit–Hartenberg method and that of the wheeled mobile base is done using Muir and Neuman's transform-based approach. The kinematic modelling has been explained in Sect. 2.1. Static structural analysis of body frame is done using ANSYS to ensure a safe design. The frame dimensions are finalized based on this analysis. The dynamic analysis of the wheeled mobile base is done using MSC Adams. The maximum torque acting on the wheel is 14.5 Nm. Lagrangian–Euler formulation is used to determine the joint torques of the seed drill. Suitable actuators are chosen based on the maximum torque obtained from the dynamic modelling.

6 Conclusion and Scope for Future Work

Kinematic and static structural analysis is performed for the dimensional synthesis, and the torque plot is obtained using dynamic analysis. Actuator selection is done based on the maximum torque and a safety factor. Suitable controller selection and energy optimization can be done based on the dynamic equations formulated. Vision sensors to guide the robot, artificial intelligence, seed monitoring, IoT, etc. can be incorporated to enable the autonomous working of the agricultural robot. Weeding, watering and fertilization implements can be attached to make it a multi-purpose agricultural robot.

References

1. Labour in Indian agriculture: a growing challenge, FICCI Agriculture report 2015. http://www.ficci.in. Last accessed 03 March 2019
2. Kushwaha H, Sinha J, Khura T, Kushwaha D, Ekka U, Purushottam M, Singh N (2016) Status and scope of robotics in agriculture. In: International conference on emerging technologies in agricultural and food engineering, Kharagpur, India
3. Yang H, Zhang L (2014) Research on the development of agricultural mechanical automation in mechanical engineering. Appl Mech Mater 454(2014):23–26
4. Shibusawa S (2003) Precision farming approaches for small scale farms. IFAC ProcV 34 (11):22–27
5. Haibo L, Shuliang D, Zunmin L, Chuijie Y (2015) Study and experiment on a wheat precision seeding robot. J Rob 2015:1–9
6. Bak T, Jakobsen H (2004) Agricultural robotic platform with four wheel steering for weed detection. Biosys Eng 87(2):125–136
7. Kim G, Kim S, Hong Y, Han K, Lee S (2012) A robot platform for unmanned weeding in a paddy field using sensor fusion. In: 2012 IEEE international conference on automation science and engineering (CASE). IEEE, Seoul, pp 904–907
8. Karayel D, Barut Z, Özmerzi A (2004) Mathematical modelling of vacuum pressure on a precision seeder. Biosys Eng 87(4):437–444
9. Minfeng J, Yongqian D, Hongfeng Y, Haitao L, Yizhuo J, Xiuqing F (2018) Optimal structure design and performance tests of seed metering device with fluted rollers for precision wheat seeding machine. In: IFAC-papers on line, pp 509–514
10. Iacomi C, Popescu O (2015) A new concept for seed precision planting. In: Agriculture and agricultural science procedia, pp 38–43
11. Muir P, Neuman C (1986) Kinematic modelling of wheeled mobile robots. Technical Report CMURI-TR-86–12
12. Tarokh M, McDermott G (2015) Kinematics modelling and analyses of articulated rovers. IEEE Trans Rob 21(4):539–553

Bandwidth Enhancement in MEMS-Based Energy Harvester for Cochlear Implants

Ayesha Akhtar, Neela Chattoraj and Sudip Kundu

Abstract In this paper, a MEMS piezoelectric energy harvester is designed to convert vibrational energy in the range of 150–230 Hz into electric energy using piezoelectric effect for cochlear implants. The simulation is done in COMSOL Multiphysics. The comparison of different MEMS structures has been done with the same piezoelectric material, ZnO. The thickness of the piezoelectric material is kept constant in all the three structures which are equal to 2 μm. The single cantilever beam structure with a silicon anchor is designed which consist of four layers, namely silicon substrate, electrodes layer, and a piezoelectric layer. A sinusoidal acceleration of $0.1g$ is applied to three structures which are preferred for the proposed structure. The objective of this paper is to get lower resonant frequency, high output voltage, and larger bandwidth. The performance analysis is carried by considering the different designs of cantilever structures on the same substrate.

Keywords MEMS · Piezoelectric energy harvester · Cochlear implants · ZnO

1 Introduction

The microelectromechanical system (MEMS) has guided the world in a wide range of applications for miniaturization. In today's era, energy harvesting is becoming a prominent solution for replacing the usage of conventional batteries. Energy harvesting systems can harvest energy from ambient vibrations present in the environment and convert it to useful electrical energy. A large number of applications of energy harvesting system include wireless sensor nodes, health monitoring devices [1], and implantable devices like a pacemaker, cochlear implants, implanted sensors [1, 2]. Because of the limited lifespan, conventional batteries need to be recharged

A. Akhtar (✉) · N. Chattoraj · S. Kundu
Department of Electronics & Communication Engineering, Birla Institute of Technology, Mesra, Ranchi, Jharkhand, India
e-mail: nchattoraj@bitmesra.ac.in

BBVL. Deepak et al. (eds.), *Innovative Product Design and Intelligent Manufacturing Systems*, Lecture Notes in Mechanical Engineering,
https://doi.org/10.1007/978-981-15-2696-1_98

or replaced frequently. This leads to a problem in the biomedical field as well as for patients in the long run. Generally, there are three mechanisms for converting vibration to electricity, i.e. electromagnetic, electrostatics, and piezoelectric [3]. Amongst the three, piezoelectric technique is more suitable because of the following reasons: (i) no external source needed (ii) have a better electromechanical coupling, (iii) easy fabrication and many more [3].

The auditory frequency of human being lies in the range of 20–2000 Hz. The frequency of voice for a male is nearly about 125 Hz, whilst for female it is around 225 Hz [4]. A cochlear implant is widely used to treat an intense hearing problem called the sensorineural deafness in which the eardrum is operational while cochlea is almost damaged. Thus, the total process of hearing is changed by the electronic system. In cochlear implants, the harvester has to be placed on the eardrum which vibrates as the sound waves enter the ear. If the vibrational frequency near the eardrum matches with the resonant frequency of the harvester, it generates maximum output voltage which can be used to power the battery of the cochlear implants. Not, many researches have been done in the field of cochlear implants. Beker [4] has designed the harvester which has length 3.5 mm and the resonant frequency of 474 Hz which comes in the daily sound frequency range. The thicker piezoelectric material is used which is around 10 μm. This model gives a peak voltage output of 0.9 V at an acceleration of value 0.1g. Alrashdan [5] has designed a piezoelectric energy harvester of rectangular shape for cochlear implants having PZT5H as piezoelectric material with a thickness of 500 nm. The harvester is having a length of 3 mm and is designed at the resonant frequency of 589 Hz and giving a voltage of 4×10^{-15} $V_{\text{peak to peak}}$. Chaudhari et al. [6] have designed a tapered perforated piezoelectric energy harvester for cochlear implants with bio-compatible zinc oxide as a piezoelectric material. The thickness of the piezoelectric material is 2 μm. The voltage obtained is 0.75 V at 126 Hz at an acceleration of 0.1g which is higher than the voltage of basic rectangular structure of the same area.

The rectangular structure for vibrational-based energy harvester is generally preferred because the average strain generated in rectangular structure is highest for a given input and thus more power output [7]. Various modifications have been done in the design of energy harvester for different applications. As energy harvester is implanted in the human body, the selection of material should be bio-compatible. Generally, lead zirconate titanate (PZT) is used as a piezoelectric material in MEMS energy harvester. As PZT contains lead, so it is harmful to human health as well as for the environment.

2 Governing Equations and Selection of Material

The resonant frequency of a rectangular cantilever beam is given by (1) [6]

$$f_n = \frac{v_n^2}{2\pi l^2}\sqrt{\frac{EI}{m}} \tag{1}$$

where l is the length of the cantilever, E represents modulus of elasticity, I is called the area moment of inertia, v_n and f_n represent the nth mode of the eigen value and the resonant frequency, respectively, and m is mass per unit length of the cantilever beam. The voltage generated in piezoelectric energy harvester in d_{31} mode (voltage generated is perpendicular to the applied stress) is given by (2) [2]

$$V = P \times g_{31} \times t_p = \frac{d_{31} \times P \times t_p}{\varepsilon_0 \varepsilon_r} \tag{2}$$

where $g_{31} = \frac{d_{31}}{\varepsilon_0 \varepsilon_r}$ is called the piezoelectric voltage coefficient, P is the stress applied, and t_p is the thickness of the piezoelectric layer which equals to the distance between the two electrodes. Therefore, for a given input stress, the material which has a higher value of the piezoelectric voltage coefficient gives the higher output voltage. Table 1 represents the comparison of voltage coefficients of different piezoelectric materials. ZnO and AIN are bio-compatible materials [2], and from Table 1, it is concluded that ZnO has a higher value of voltage coefficient which is desired for the proposed structure.

3 Structure Modelling

In this paper, a simple rectangular cantilever structure is designed. The rectangular structure consists of silicon as a substrate which has a density of 2329 kg/m^3 and Young's modulus of 170 GPa, zinc oxide (ZnO) as piezoelectric material with 2 μm as thickness [8] having density 5680 kg/m^3 and Young's modulus 6.83 GPa.

Table 1 Comparison between the different piezoelectric materials [3, 6]

Materials	Charge coefficient (d_{31}) (C/N)	Dielectric constants (ε_r)	Voltage coefficient (g_{31}) ($\frac{V_m}{N}$)
ZnO	11.34	12.64	0.897
AIN	3.84	10.256	0.374
GaAs	2.6	13.1	0.198
Barium titanate	191	1700	0.112
PZT-5H	741	3400	0.217

Aluminium as top and bottom electrodes is chosen which has a density of 2700 kg/m^3 and Young's modulus of 69 GPa. A silicon anchor is used in the structure. Further, the rectangular cantilever is modified by removing a small portion from the rectangular section to yield a dual cantilever. Also, the dual cantilever structure is modified to a parallel structure [9] in which the anchor is removed and top electrodes are connected with aluminium. There is an electromechanical coupling between the two beams as top electrodes are connected. The material thickness and layers compositions of all the three structures are the same. All the three structures have four proof mass which is used to reduce the resonant frequency [10, 11]. The dimension of the inner beam in dual and parallel cantilever structure is (4450 × 700 × 17) μm. The spacing between the two proof mass is equal to 50 μm. The proof mass has a density of 7500 kg/m^3 and Young's modulus of 1e8 Pa which is quite large than silicon substrate.

Table 2 Dimensions of the single cantilever structure

Material	Length (μm)	Width (μm)	Thickness (μm)
Silicon (Si)	5000	2000	14
Aluminium (Al)	5000	2000	0.5
Zinc oxide (ZnO)	5000	2000	2
Proof mass	500	500	250

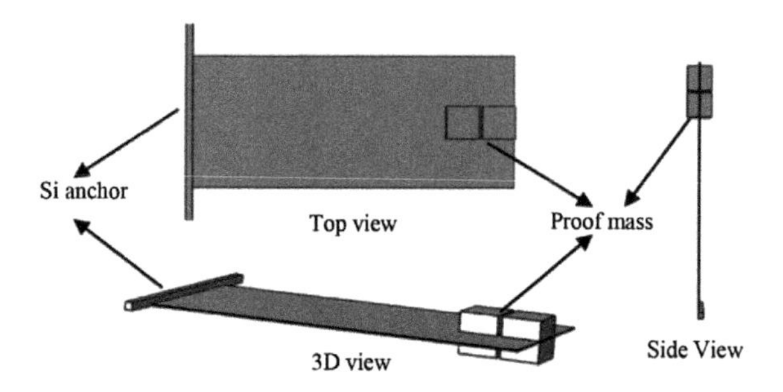

Fig. 1 Top view, side view, and 3D view of the single cantilever structure

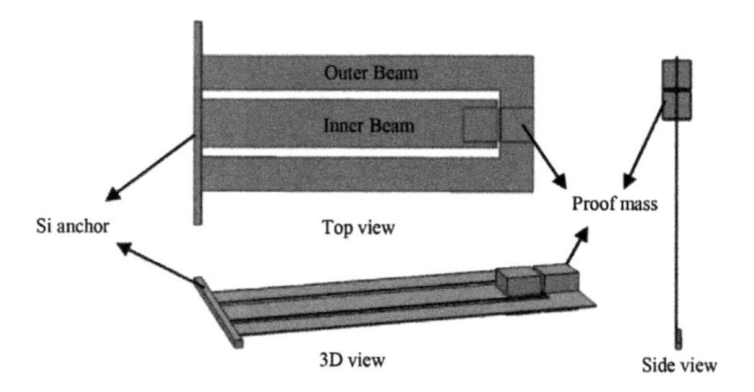

Fig. 2 Top view, side view, and 3D view of the dual cantilever structure

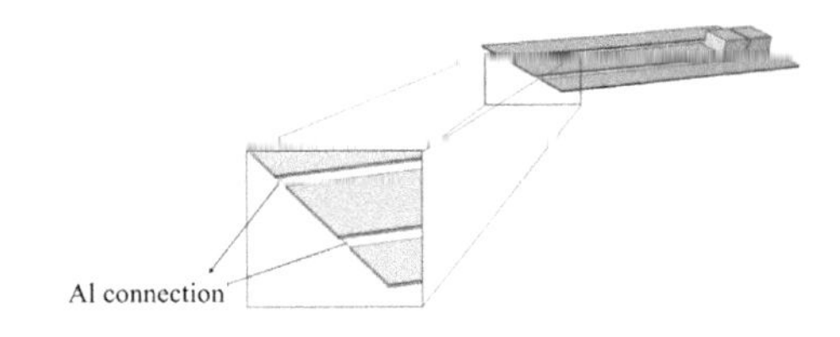

Fig. 3 3D view of the parallel cantilever structure showing the aluminium connection

4 Results and Discussion

To simulate the energy harvester due to ambient vibration present near the eardrum, assume that the system is getting vibration by an external sinusoidal body load.

An acceleration of $0.1g$ is applied to the three structures, and transient analysis is carried out on the structures in COMSOL software. The first eigen frequency of a single cantilever is 223.71 Hz. For dual cantilever structure, the first and second eigen frequencies are 200.92 and 207.71 Hz. At first eigen frequency, outer beam vibrates while the second beam vibrates at second eigen frequency. Similarly, for the parallel cantilever, outer beam vibrates at 166.44 Hz while inner beam vibrates at 175.84 Hz. For dual and parallel cantilever, the first two resonant frequencies are very close to each other which help in giving better bandwidth than the single cantilever. The single cantilever gives a peak voltage of 2.205 V at the resonance of 221 Hz as shown in Fig. 4a. At resonant frequency of 204 Hz, dual cantilever gives a peak voltage of 1.88 V as shown in Fig. 4b and a better bandwidth of 12 Hz as shown in Fig. 5. The parallel cantilever gives a peak voltage of 1.004 V at resonant frequency of 164 Hz as shown in Fig. 4c and thus giving more bandwidth than the dual cantilever as shown in Fig. 5. Table 3 represents the comparison of the resonant frequency, peak voltage, and bandwidth of all the three structures.

5 Conclusion

In this paper, zinc oxide is selected as the piezoelectric material. From Fig. 5 and Table 3, it shows that the bandwidth is enhanced in case of parallel cantilever structure by 14 Hz and peak voltage is 1.004 V. In case of single beam structure, it gives 0.3 V in the range of 217–223 Hz, while in the case of dual cantilever structure it is around 194–206 Hz. The parallel structure gives 0.3 V in the range of 161–174 Hz. In this paper, it is seen that structural modification gives lower resonant frequency and more bandwidth as ambient vibrations are random and varying in nature. Therefore, bandwidth enhancement helps to improve the performance of energy harvester used for cochlear implant application as it covers a wide range. In future, the proposed structure can be fabricated by using flow process required in microfabrication technology.

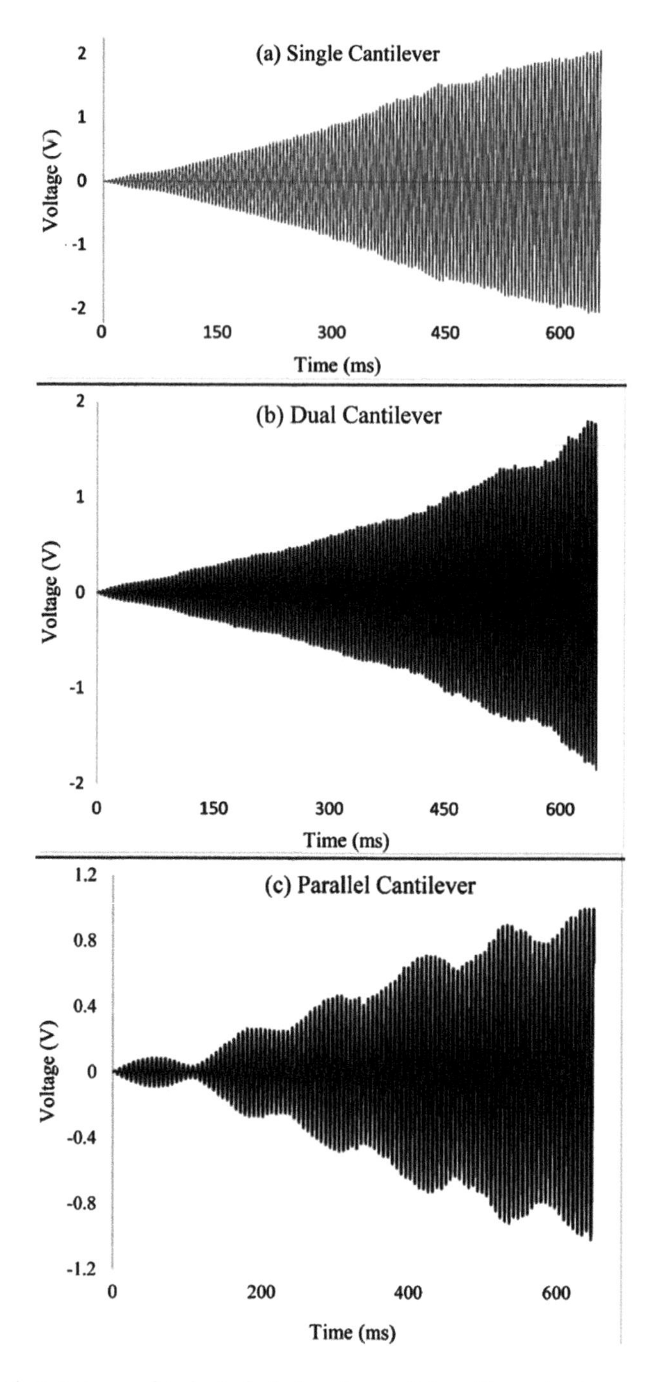

Fig. 4 Transient response of **a** single, **b** dual, and **c** parallel cantilever structure

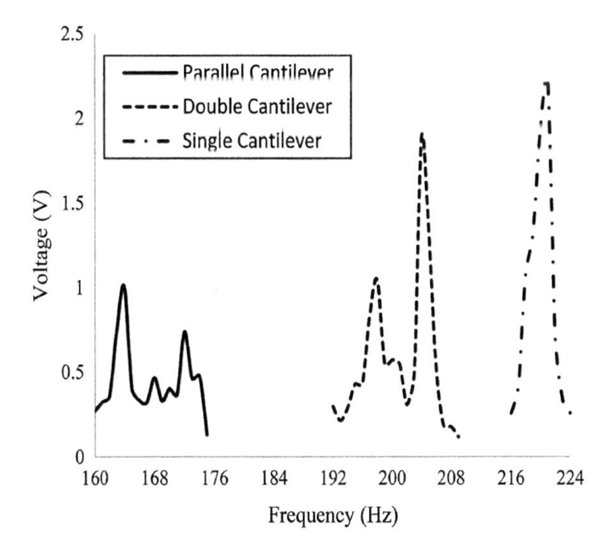

Fig. 5 Graph showing the voltage (V) versus frequency (Hz) of a single, dual, and parallel cantilever structure at an acceleration of 0.1*g*

Table 3 Comparison of resonating frequency, voltage, and bandwidth of three different structures

Structure	Resonating frequency (Hz)	Peak voltage (V)	Bandwidth (Hz)
Single cantilever	221	2.205	7
Dual cantilever	204	1.88	12
Parallel cantilever	164	1.004	14

References

1. Galchev TV, McCullagh J, Peterson RL, Najafi K (2011) Harvesting traffic-induced vibrations for structural health monitoring of bridges. J Micromech Microeng 21(10):104005
2. Kim SG, Priya S, Kanno I (2012) Piezoelectric MEMS for energy harvesting. MRS Bull 37 (11):1039–1050
3. Priya S, Inman DJ (eds) (2009) Energy harvesting technologies. Springer, New York
4. Beker L (2013) MEMS Piezoelectric Energy Harvester for Cochlear Implant Applications. M. Tech Thesis, Middle East Technical University
5. Alrashdan MH, Majlis BY, Hamzah AA, Marsi N (2013) Design and simulation of piezoelectric micro power harvester for capturing acoustic vibrations. In: IEEE regional symposium on micro and nanoelectronics, pp 383–386
6. Chaudhuri D, Kundu S, Chattoraj N (2017) Harvesting energy with zinc oxide bio-compatible piezoelectric material for powering cochlear implants. In: Innovations in power and advanced computing technologies (i-PACT), pp 1–5
7. Bindu RS, Kushal MP, Potdar M (2014) Study of piezoelectric cantilever energy harvesters. Int J Innov Res Dev, 2278-021
8. Nagayasamy N, Gandhimathination S, Veerasamy V (2013) The effect of ZnO thin film and its structural and optical properties prepared by sol-gel spin coating method. Open J Metal 3 (02):8

9. Xue H, Hu Y, Wang QM (2008) Broadband piezoelectric energy harvesting devices using multiple bimorphs with different operating frequencies. IEEE Trans Ultrason Ferroelectr Freq Control 55(9):2104–2108
10. Roundy S, Wright PK, Rabaey JM (2003) Energy scavenging for wireless sensor networks. Norwell, 45–47
11. Shen Z, Liu S, Miao J, Woh LS, Wang Z (2013) Proof mass effects on spiral electrode d 33 mode piezoelectric diaphragm-based energy harvester. In: IEEE 26th international conference on micro electro mechanical systems (MEMS), pp 821–824

Design, Analysis and Development of a Flying Wing UAV for Aerial Seeding and 3D Mapping

Movva Srilakshmi Sai, Kamlesh Kumar and Bhanu Prakash

Abstract Unmanned aerial vehicle (UAV) commercialization has increased the possibility of performing monitoring, data collection, survey mapping and aerial seeding in the modern world. The main components used in unmanned aircraft systems are ground control station (GCS) which includes mission planner software and two-way communication system (Telemetry) between the GCS and the UAV, transmitter, and receiver, autopilot on board, payload, remote video terminal, etc. This paper deals with the entire design of flying wing UAV taking into account wing and fuselage design and later implementing payload attachment for aerial seeding and mapping. Initially, UAV has been designed using dimensions and wing loading parameters calculated from basic aerodynamics. The complete UAV was designed in CATIA, and flow and structural analysis was performed in ANSYS. UAV was fabricated using balsa wood to test the prototype. Modified UAV design is further fabricated using composite for aerial seeding.

Keywords Unmanned aerial vehicle (UAV) · Flying wing · Aerial seeding · CATIA · CFD · Mapping

M. S. Sai
Department of Instrumentation and Control Engineering, Manipal Institute of Technology, MAHE, Manipal, Udupi, Karnataka 576104, India

K. Kumar (✉)
Center of Excellence in Avionics and Navigation, MAHE, Department of Aeronautical and Automobile Engineering, Manipal Institute of Technology, MAHE, Manipal, Udupi, Karnataka 576104, India
e-mail: kamlesh.kumar@manipal.edu

B. Prakash
Department of Aeronautical and Automobile Engineering, Manipal Institute of Technology, MAHE, Manipal, Udupi, Karnataka 576104, India

BBVL. Deepak et al. (eds.), *Innovative Product Design and Intelligent Manufacturing Systems*, Lecture Notes in Mechanical Engineering,
https://doi.org/10.1007/978-981-15-2696-1_99

1 Introduction

The use of unmanned aerial vehicle has become more common in the modern world because of its wide range of applications including efficient coverage of a large area in the least amount of time. There is major forest destruction by stakeholder. Day by day, the environment is degrading due to heavy deforestation and most parts of the countries are not accessible for forestation. Hence, human involvement is required with the new technology. Drone deployment is a very innovative solution for forestation without any human intervention in the field. So, we need drone/UAV facilities for seeding in areas impossible to seed with traditional methods. There are basically two types of unmanned aerial vehicle (UAV) of which one is remote controlled and the other is pre-programmed flight plans used mainly for complicated automated systems.

UAVs are mainly categorized [1] into three types based on wing design such as fixed wing design, rotary wing design and a combination of both fixed wing and rotary wing.

Fixed-wing UAVs fly by utilizing the lift generated by the forward motion [2] of aircraft or can be hand-launched by the operator by simply throwing them into the air, while larger and heavier drones require a more complex method of getting airborne, such as catapult or a runway, or being launched from a larger aircraft. It is usually categorized [3] into three types, namely flying wing, blended wing body (BWB) and delta wing; as its name suggests it is flattened and triangular in shape, respectively.

In this paper, the flying wing is preferred because of its high aerodynamic efficiency and structural efficiency making it more fuel efficient and environmental-friendly, when compared to all other fixed-wing UAVs. Implementing an aerial seeding technique to this further helps in reducing erosion, increasing the organic content of the soil and increasing the overall cash crop yield. This paper attempts to give a conceptual design of flying wing UAV through a simplified analytical method and later fabricating the same.

2 UAV Design

2.1 Planform Design

The planform design [3] plays a crucial role in the design of the wing. Planform is designed considering wing area, aspect ratio, taper ratio, one-fourth of swept chord line and wing twist angle. The schematic representation of planform design is shown in Fig. 1.

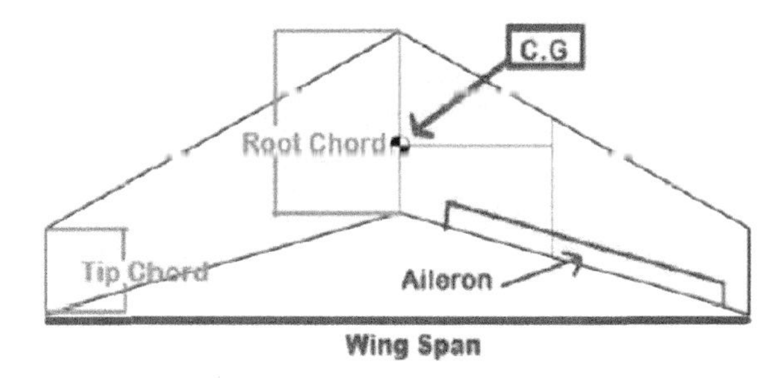

Fig. 1 Schematic representation of planform [3]

2.2 *Design Parameters*

Wing design parameters considered [4] are given in Table 1, focusing mainly on the wingspan, wing reference area, aspect ratio [5] and taper ratio.

2.3 *Selection of Airfoil*

The cross-sectional shape of the wing is used to produce a lift force called airfoil. Certain conditions are considered while selecting airfoils [6, 7] such as flying regime, space and weight requirements. Below are the criteria considered while selecting the airfoil [8]:

(1) Airfoil should be unsymmetrical and cambered.
(2) Coefficient of the moment (C_{m}) should be negative for the centre body and positive for the outer body.
(3) High lift to drag ratio (*L*/*D*).
(4) Coefficient of lift (C_{l}) should be more than 1.

SA7035 airfoil [9] is selected for the wing design, and the angle of attack for specified airfoil is considered to be close to zero degrees.

Table 1 Design parameter of the wing

Parameters	Calculated values
Wingspan (b)	2.4 m
Wing reference area (S)	0.945 m^2
Aspect ratio (AR)	7.68
Tip chord (C_{t})	0.225 m
Root chord (C_{r})	0.450 m
Sweep angle (υ)	20.94°
Selected airfoil	SA7035
Taper ratio (λ)	0.5
Mean aerodynamic centre	53 cm from C_{r} (half-wing)

The aerodynamic results are tabulated using XFLR5 software [10] and are shown in Figs. 2, 3 and 4.

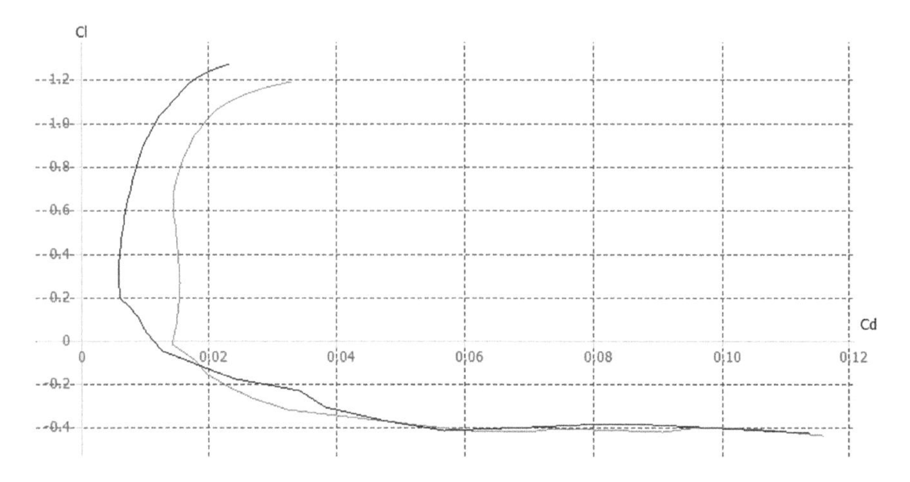

Fig. 2 C_l versus C_d plot of SA7035 airfoil

Fig. 3 C_l versus Alpha plot at different angles of attack (AoA)

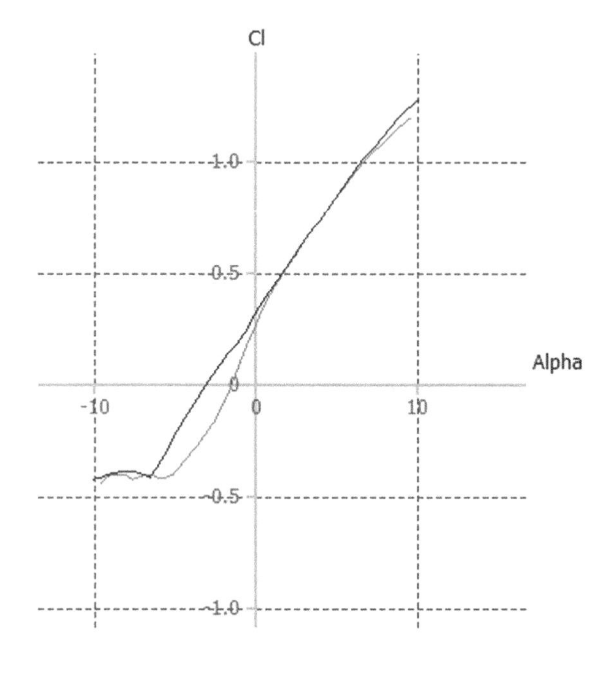

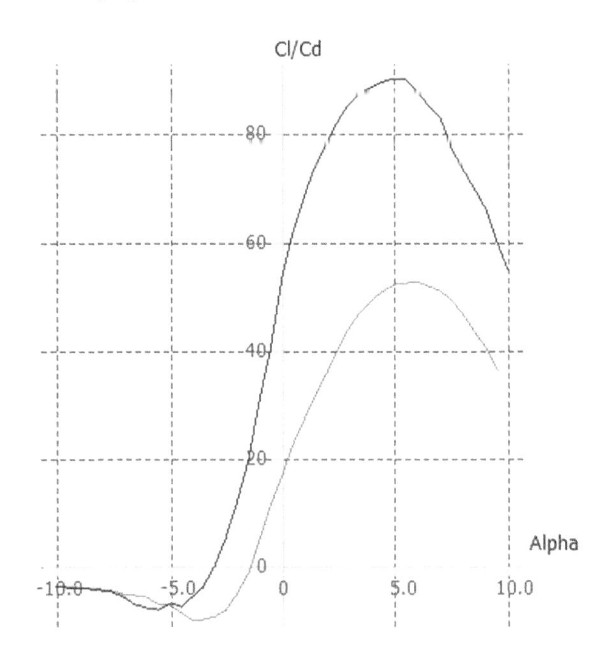

Fig. 4 C_l/C_d versus Alpha of airfoil

2.4 Modelling of UAV

Unmanned aerial vehicle model begins with planform drawing and fuselage inboard profile. The wing design [11] was approximated using a sweep of an airfoil section. Two splines were created in the airfoil section, one on the upper surface and other on the lower surfaces setting them proportional to each other. Few points were considered for smooth results. Further, twist angles, taper angles and sweep angles are specified to get accurate wing design. The entire 3D modelling [1] design was carried out in CATIA as shown in Fig. 5. Winglets are also considered here to reduce lift-induced drag [12].

Fig. 5 Conceptual design of UAV

2.5 Conceptual Analysis of UAV

After importing [13] the conceptual UAV design from CATIA V5R19, the entire model is meshed in ANSYS pre-processor. To get better results, grid independence was carried out to check the quality of the mesh. The entire model was considered along all the three axes (u, v and w), assuming along w axes is negligible.

In order to perform the flow analysis, certain conditions had to be considered. In this numerical analysis, boundary conditions [14] are considered as follows:

i. The fluid flow is inviscid.
ii. Inlet temperature considered is 288.17 K.
iii. Velocity is 20 m/s.

With the above conditions, the solver is run for specific iterations until the solution gets converged.

2.6 Avionics Selection

The motor and propeller selection was done based on the thrust required by the aircraft. The selected avionics components are shown in Fig. 6. T-motor MN3515 400 kV with a 16-inch propeller was used to meet the thrust requirements. According to the load testing data of the motor, the system would draw a maximum current of 14 A. Electronic speed controller (ESC) was used to control the speed of brushless motors mentioned above. The current rating of ESC was chosen to be higher than the maximum current drawn from the motor for the margin of safety. A pair of Turnigy A55h metal-geared servos having a torque rating of 7 kg-cm was chosen to safely handle the aerodynamic loads on the elevons (elevator + ailerons). A 6-cell lithium polymer (LiPo) battery having a capacity of 8000 mAh and discharge rating 15 °C was chosen to give enough endurance for flight test.

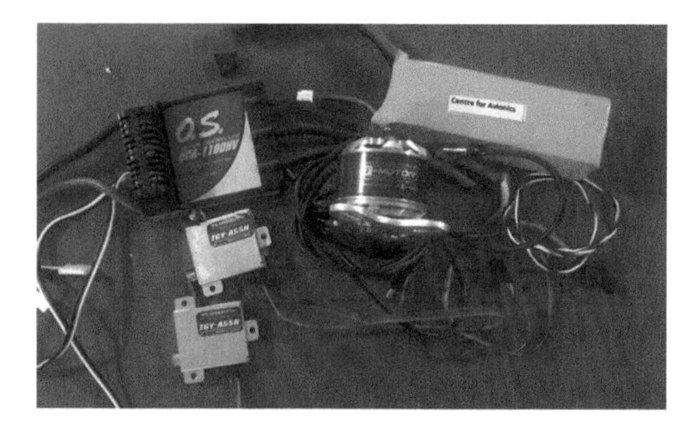

Fig. 6 Selected avionics components for UAV

2.7 Formulas

The above design parameters for the required aircraft [15] are calculated using the below formula.

$$\text{Aspect Ratio (AR)} = b^2/S \tag{1}$$

where
'b' is wingspan and 'S' is wing reference area, ($S = b * \overline{C}$)

$$\text{Mean Wing Chord}(\overline{C}) = b/\text{AR} \tag{2}$$

$$\text{Taper ratio } (\lambda) = C_t/C_r \tag{3}$$

where
C_r, C_t are root and tip chord of the wing

$$\text{Mach number } (M) = V/C \tag{4}$$

where
V, C are velocity and speed of sound

$$\text{Sweep angle } (\upsilon) = \tan^{-1}\left(\frac{C_r - C_t}{b}\right) \tag{5}$$

$$\text{Mean Aerodynamic Center (MAC)} = \frac{2}{3}C_r\left(\frac{1+\lambda+\lambda^2}{1+\lambda}\right) \tag{6}$$

$$\text{Aerodynamic Center } (X_{ac}) = C_r - \overline{C} + 0.25\overline{C} \tag{7}$$

3 Fabrication of UAV

For prototype, balsa wood and carbon boom was considered to design the wing and fuselage as internal structure. It was coated with monokote to provide skin strength. The fabricated wing is shown in Figs. 7, 8, 9 and 10. Finally, composite UAV models are fabricated for aerial seed and 3D mapping with the payload.

Balsa wood [16] is selected for fabricating UAV because of its high strength and low density making it easier to shape, sand, glue and paint. It is non-toxic, biodegradable and can easily absorb shock, heat and vibration. Balsa is relatively strong [17] in relation to its density and weight. Modulus of elasticity of balsa is 3.8 GPa, density varies from 40 to 340 kg/m^3, and modulus of rupture is 19 MPa. In this paper, wing ribs and control surfaces are fabricated using balsa wood of

Fig. 7 Wing fabrication using balsa wood

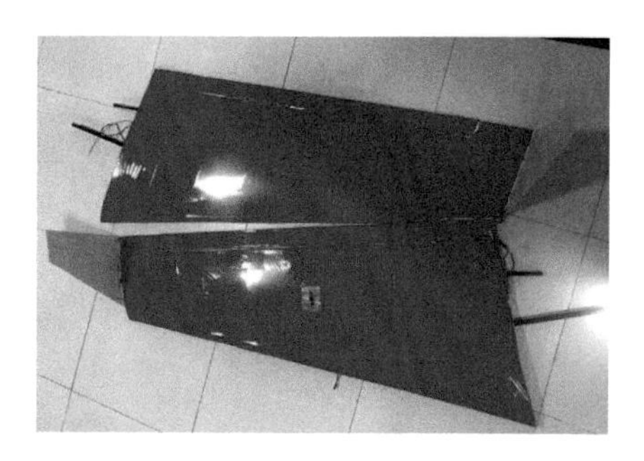

Fig. 8 Monokote wings with winglets and control surfaces of UAV

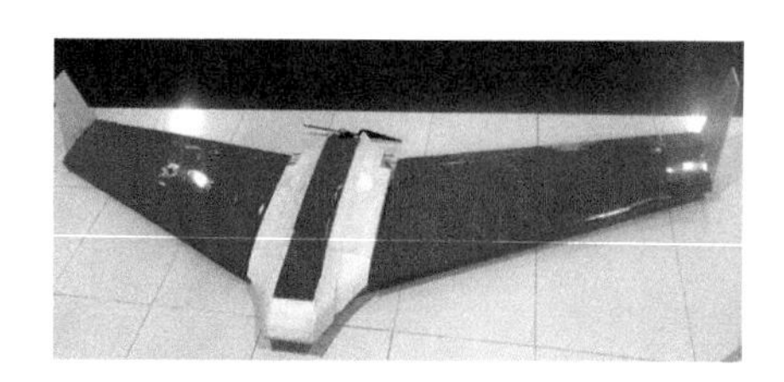

Fig. 9 Balsa wood flying wing UAV

Fig. 10 Pre-flight check [in the field]

different thicknesses and carbon fibre booms are used for connecting the fuselage and different sections of the airfoil of the wing structure. Monokote covering as shown in Fig. 8 was done to support aerodynamic shape as well as to provide stiffness. Carbon fibres are selected because of its high strength to weight ratio, good tensile strength and fatigue resistance.

4 Results and Discussion

The preliminary design flow analysis on UAV was carried out considering zero angle of attack. The entire design was carried out using the finite element method (FEM) to analyse the structural strength of UAV and CFD (ANSYS). Results of aerodynamic analysis for selected airfoil are shown in Figs. 2, 3 and 4 and complete UAV model results are shown in Figs. 11 and 12. Prototype model is tested, and it shows successful flight as shown in Fig. 13.

Fig. 11 Pressure contours on UAV

Fig. 12 Velocity distribution of UAV

Fig. 13 Flight test

5 Conclusion

This paper deals with the entire design of flying wing UAV taking into account wing and fuselage design. The complete UAV was designed in CATIA, and flow and structural analysis was performed in ANSYS. The analyzed model is fabricated using balsa wood, monokote and carbon booms to test the prototype. Aircraft instrumentation has integrated with a fabricated model for flight test.

Once the design is verified with the test flight, modified UAV design will be recommended for further fabrication using composite for aerial seeding and 3D mapping. This will be including payload, container design and seed releasing mechanism. Payload as the 4K camera will be used for 3D-mapping to support prerequisite of land surveying for aerial seeding.

Acknowledgements The project was completed with the full support of "Centre of Excellence in Avionics and Navigation" Manipal Academy of Higher Education (MAHE), Manipal.

References

1. Ashraf Y, Rahman A, Haji MT, Mahdi AS (2018) Design and fabrication of small vertical-take-off-landing unmanned aerial vehicle, vol 02023, pp 1–15
2. Yayli UC, Acdoyan AC (2017) Design optimization of a fixed wing aircraft. Int J Adv Aircraft Spacecraft Sci 4
3. Teli SN, Madan J (2014) Unmanned aerial vehicle for surveillance. Int J Sci Technol 3
4. Abdus Samad Shohan GAA, Moon FA (2016) Conceptual design, structural and flow analysis of an UAV wing. IOSR J Mech Civil Eng (IOSR-JMCE) 13
5. Prasanna Kumar GRTJ, Das S (2018) Design and stress analysis of swept back wing with Nastran and Patran. Int Res J Eng Technol (IRJET) 05
6. Ngo Khanh Hieu DAB, Vinh PQ (2016) Design of a small UAV combined between flying-wing and quadrotor with XFLR5. Int J Transp Eng Technol 2
7. Prisacariu V (2017) Performance analysis of the flying wing airfoils 18
8. Hieu NK, Loc HT (2016) Airfoil selection for fixed wing of small unmanned aerial vehicles. In: Lecture notes in electrical engineering, vol 371, pp 881–890
9. UIUC Airfoil Co-ordinate Database. https://m-selig.ae.illinois.edu/ads/coord_database.html
10. XFLR5 Airfoil Analysis Software. See (http://www.xflr5.com/xflr5.htm)
11. Khan S, Aabid A (2018) Design and fabrication of unmanned aerial vehicle for multi-mission tasks. Int J Mech Prod Eng Res Dev 8
12. Dagur R, Singh BB (2018) Design of flying wing UAV and effect of winglets on its performance. Int J Emerg Technol Adv Eng 8:414–428
13. Theja BR, Gupta DMS (2015) Design and fluid flow analysis of unmanned aerial vehicle (UAV). Int J Sci Technol 4
14. Kandwal S, Singh S (2018) Computational fluid dynamics study of fluid flow and aerodynamic forces on an airfoil. Int J Eng Technol
15. Sadrey MH (2013) Aircraft design: a systems engineering approach, p 778. Wiley, New Hampshire

16. Harasani WI (2016) Design, build and test an unmanned air vehicle. JKAU: Eng Sci 21:105–123
17. Khan MI, Salam MA, Afsar MR, Huda MN, Mahmud T (2015) Design, fabrication and performance analysis of an unmanned aerial vehicle. In: 11th international conference on mechanical engineering. ICME, vol 1754. https://doi.org/10.1063/1.4958448

Conceptual Design and Analysis of Three Jaw Robotic Gripper with Flexural Joints

Golak Bihari Mahanta, Amruta Rout, BBVL. Deepak and B. B. Biswal

Keywords Gripper · Flexural joint · Design · Analysis · Soft gripper

1 Introduction

In this modern era, robotic is an advanced field of innovation that pushes the traditional engineering boundaries toward a new era of automation. For a considerate, understanding of the difficulty of robots and their uses in everyday life needs learning of electrical, mechanical engineering, basic science, and software engineering. Some of the leading sectors where industrial robot showing their presence include pick and place operation, assembly, welding and spray-painting in automobile industry, army, home as well as in medical applications. Some of the key industries where industrial robots are effectively incorporated into the manufacturing line and play a crucial role to meet the high-quality product for the customer are the automobile, aircraft, and micro-nano fabrication industries [1]. The grippers are the devices that enable the holding of the objects that are aimed to be manipulated. Due to the more and more applications of the robots in the industry, there increased the necessity for using the grippers. Grippers could be fixed to the robot or they could also be used separately to perform the grasping and holding operations. According to the types of necessity, the end effector can be altered such as for grasping, gripper is attached to the end effector, and for welding operation, welding gun can be fixed. Industrial robots are linked with the gripper for holding the object. Industrial robotic grippers are the dedicated tool developed for execution of a particular type of job. The gripper is designed for grasping a specific size, shape, and weight of an object. Different types of robotic grippers are developed depending on the types of the object to be grasped and operated by various medium like pneumatic, electrical, magnetic, and hydraulic. It is necessary to develop the optimal gripper configuration before the production of the gripper. With advancement in technology, scientists, designers are coming up with new innovative and

G. B. Mahanta (✉) · A. Rout · BBVL.Deepak · B. B. Biswal
Industrial Design Department, NIT Rourlela, Odisha 769008, India

BBVL. Deepak et al. (eds.), *Innovative Product Design and Intelligent Manufacturing Systems*, Lecture Notes in Mechanical Engineering,
https://doi.org/10.1007/978-981-15-2696-1_100

creative ideas to design a product which probably seems to be the most appropriate and optimized product. Optimal gripper configuration of a two jaw parallel gripper is obtained using genetic algorithm [2]. They proposed the fitness function by considering the difference of the minimum and maximum forces exerted by the tip of the gripper. Evolutionary algorithm and weighted min–max approach are combined to find the optimum value of a two jaw parallel gripper [3]. Single-objective approach has been considered using accelerated PSO algorithm for finding the optimal configuration of robotic gripper which needed while developing the gripper for stable grasping [4]. Designing and developing a gripper which works effectively to perform a specific task with accuracy is still a challenging role for many designers. Different types of existed grippers are shown in Fig. 1.

The production method of the traditional rigid grippers has their own difficulties because of their heavy weight. Some of the well-known dexterous hand available which able to mimic the human hand are the KIST Hand [5], KITECH-Hand [6], and many more. Soft robotics has gain momentum in recent years because of its inherent nature like lightweight, high compliance, and less cost [7]. Though soft material robots have so many advantages, it inherent some of the disadvantages such as limitation prone to the high repeatability work, the fabrication process is more complicated compared to that of the rigid structure, may face the ruptures while working if not appropriately fabricated. Passive particle jamming is used instead of vacuum jamming [8] for stiffening the soft gripper. Tendon-driven mechanism [9] also used to actuate the soft robotic gripper to make the system less complicated.

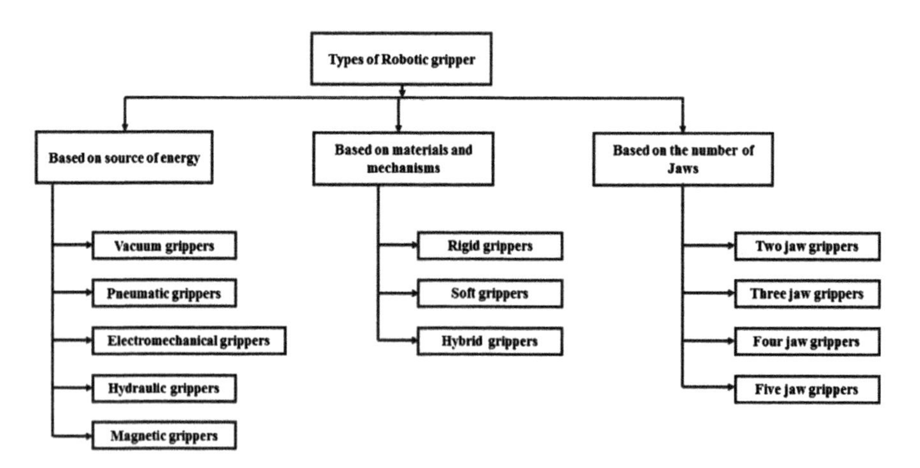

Fig. 1 Different types of existed gripper

2 Materials and Methods

The main aim of this paper is to develop a three jaw robotic gripper with flexible hinges to grasp small delicate and complex objects. Just like our human hand consists of 5 fingers out of which four fingers are arranged one after the other and the fifth finger is arranged in a different direction in order to balance the object without falling down due to the force applied by the other fingers. Also the thumb finger plays a very important role in holding any object. Taking the same concept, in the proposed method to develop the three jaw robotic gripper will be designed. The main goal of this work is to design a gripper that is cost effective, should possess light weight, adaptable, and effective for grasping objects. All the fingers of the proposed three jaw robotic gripper are actuated with the help of tendon-driven mechanism. Each finger is consisted of two section joined by a flexural materials have the capability to bend freely. The structure of the single finger along with the tendon routes are illustrated in Fig. 2. Tendon-driven mechanism is integrated into the proposed three jaw gripper to reduce the number of actuators to be used to operate the gripper, i.e., results in under-actuated mechanism. So, here, the tendon wires are taken to perform grasping which are actually treated as the soul of the product. Tendon wires are flexible but are inelastic which means they can facilitate in easy movements.

Figure 2 represents the schematic diagram of all the component of the proposed three jaw gripper connected with flexural joints.

The methodology followed for the conceptual design and analysis of the proposed three jaw gripper are as follows:

1. Design (both conceptual and technical drawings).
2. Modeling (using CATIA).
3. Analysis of a jaw for finding out the deformations in the jaws of the gripper when force is applied on them. Analysis is performed using ANSYS R15.0 workbench.

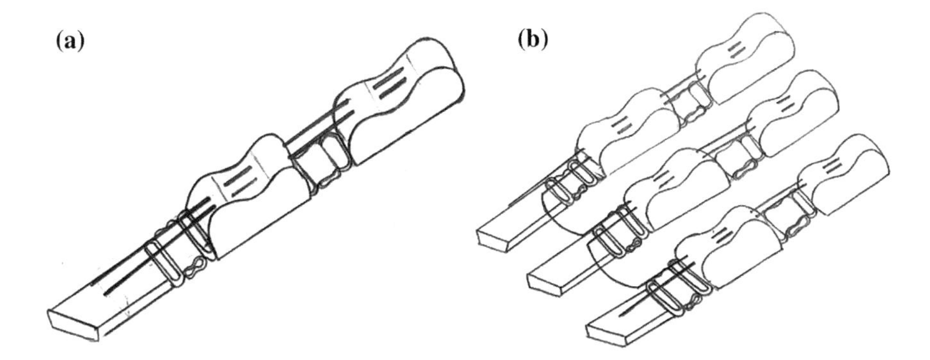

Fig. 2 **a** Finger of the three jaw soft gripper. **b** Connections of the three fingers with tendon wires

All the required dimensions of the gripper along with the assembly in front, top, and isometric view are shown in Fig. 3.

Modeling is done using CATIA V5R19 software. Firstly, all the parts are made followed by assembly of all the parts of a three jaw soft gripper. The CAD models are shown in Figs. 4 and 5. In Fig. 4, all the parts are separately shown whereas in Fig. 5, a single robotic finger and the whole assembled finger are presented. Each finger part is connected with flexural soft material. There are so many soft materials available in the market out of which silicone rubber, an elastic polymer suits best for this work. Material selection is done by dimensionless ranking in which all the factors like volume criteria, weight criteria, and cost criteria are considered.

3 Results and Discussion

The analysis of a single finger is done in order to find out the deformations of the phalanges when there arises the need for the phalanges to apply certain force on the product which is aimed to grasp. The analysis is performed using ANSYS R15.0. The force that is required to open and close the gripper is calculated using the

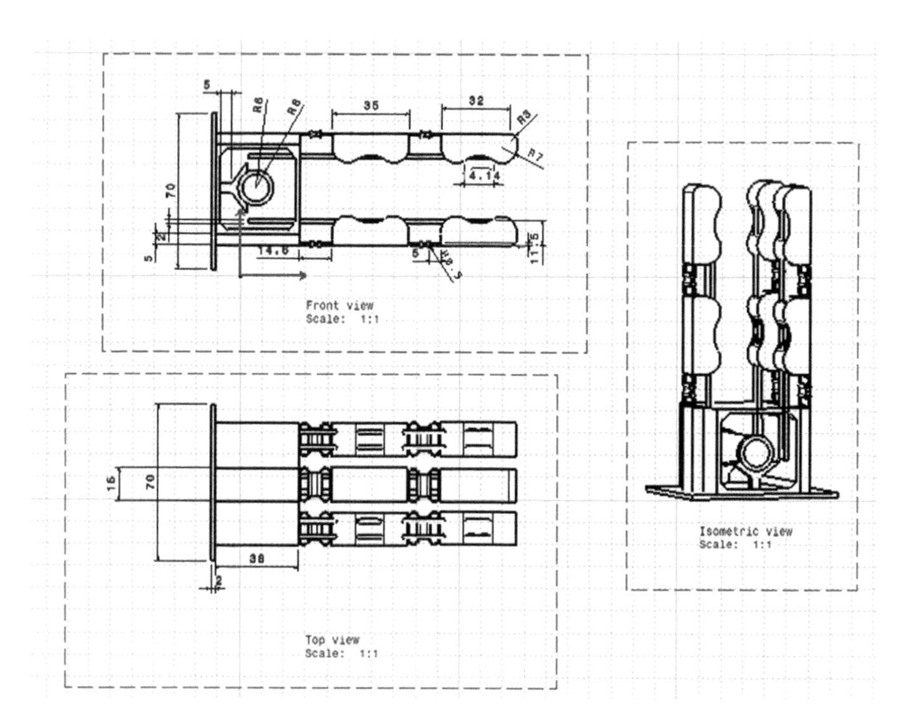

Fig. 3 Finger assembly with dimensions

following methods. Taking this force as input value and the total deformation, equivalent (Von-Mises) stress and equivalent elastic strain values are the output as a result of the input values to the finger of a three jaw soft gripper.

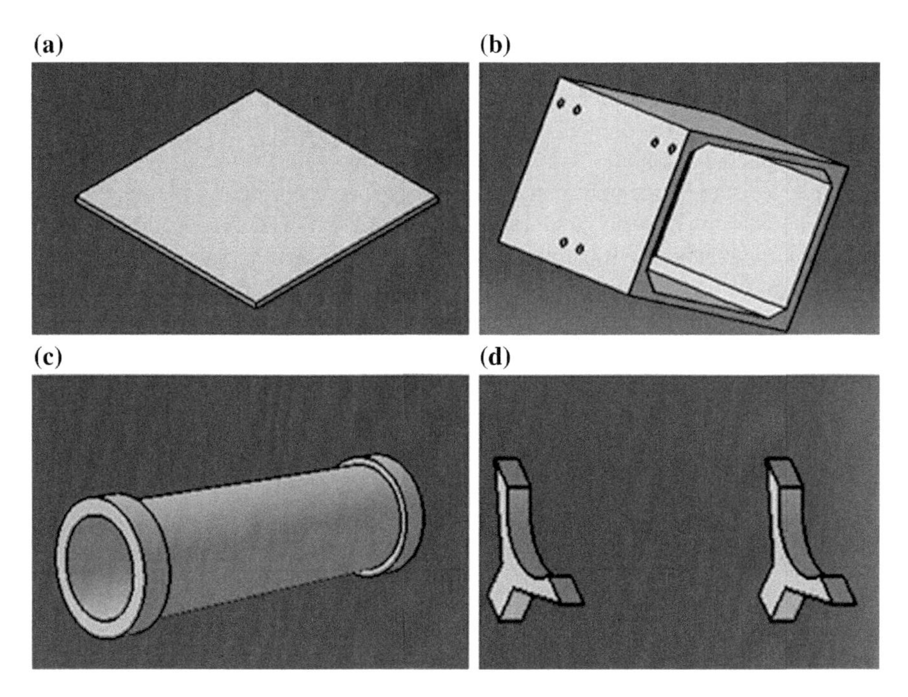

Fig. 4 **a** Base plate. **b** Supporting box. **c** Support for the spool, and **d** Spool

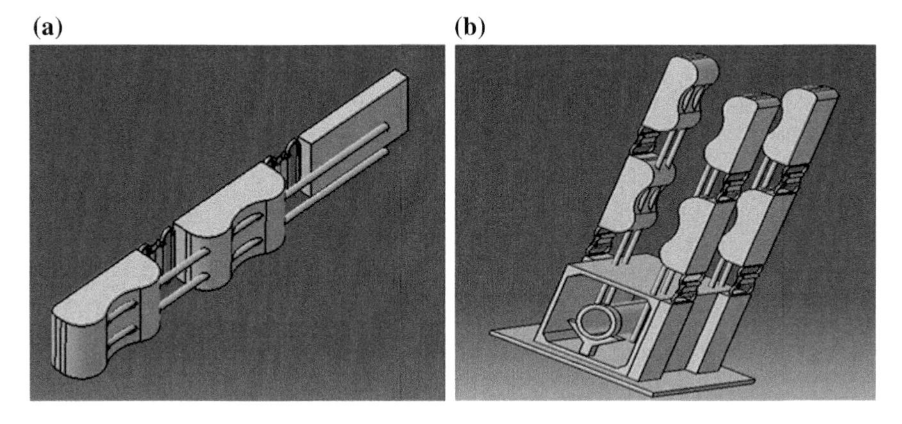

Fig. 5 **a** Finger of a three jaw soft gripper. **b** Complete assembly of three jaw soft gripper

$$\text{We know that, } F = \text{mg} \tag{1}$$

$$\text{Gripper force } F_g = \text{Part Weight} \times ((1 + 1.5) \times \text{Jaw Factor}) \tag{2}$$

There are two types of the jaw factors such as friction grip jaw factor and encompassing grip jaw factor. The friction grip jaw factor value is considered to be 4 and encompassing grip jaw factor is 1 for this study. With an object weight of 10 g, the gripping force is found to be 49 N. The boundary conditions applied by fixing the bottom part of the gripper phalanges. A force of 10 N is applied on both the phalanges and the tendon wire is given the displacement.

The boundary conditions for this study along with the results are shown in Fig. 6. The static structural analysis for the gripper considering a single finger is done in order to calculate the deformations in the jaw when application of load of 10 N on the phalanges. It is quite interesting to note that the jaws undergo friction grip when it comes to grasping which means the body will not undergo any damage while holding the object. The displacement of the jaw happens because of the U-shaped tendon wire that is fixed at certain height from the base of the phalange so as to ensure the functioning of the gripper. From Fig. 7a–d, it is clear that the maximum deformation occurs at the phalanges so that while grasping the objects, the phalange itself undergoes deformation without deforming the object which is our main concern. Optimization is performed so that a set of deformations could be known for a given range of forces without performing the task for multiple numbers of times for some specific force values.

4 Conclusion

A three jaw soft gripper was designed and analyzed that enables the user for grasping even delicate objects and is adaptive due to the shape-changing ability of the flexible material used, i.e., silicone rubber. Due to its elasticity property, the flexural material at the joint, the gripper can retain the original shape when the

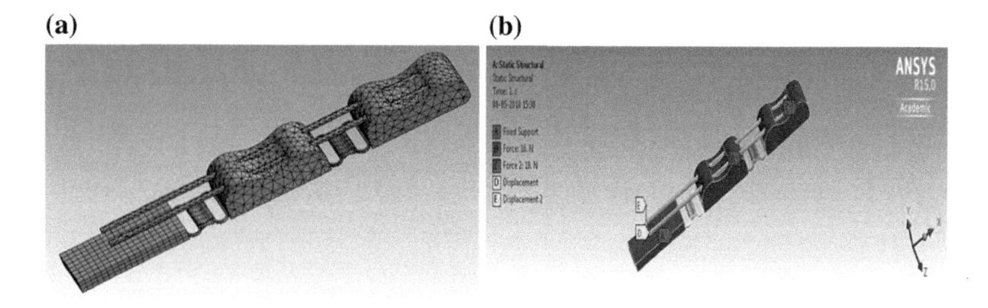

Fig. 6 **a** Fine meshing of the finger of a three jaw soft gripper. **b** Boundary conditions of the jaw

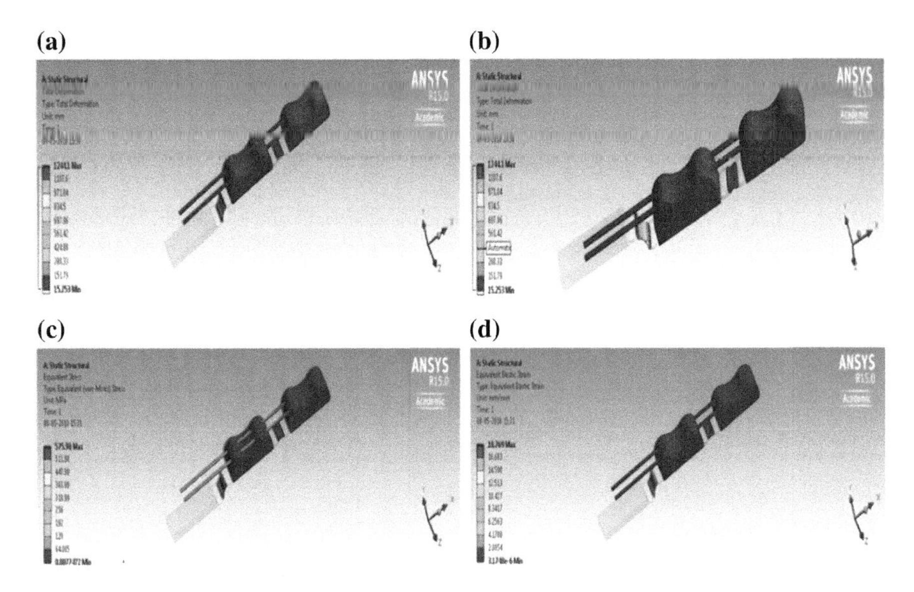

Fig. 7 **a** Total deformation in a finger in closing the gripper. **b** Total deformation in a finger in opening the gripper. **c** Equivalent stress of a finger in case of closing the gripper, and **d** Equivalent strain of a finger in closing the gripper

grasping task gets completed. The phalanges are connected by means of links so that while bending, there does not arise any problem like colliding with another phalange and hence the phalanges navigate freely without any distraction to perform the specified task. The three jaws of the gripper are driven by means of tendon wires which possess the bending property so that when the wires are made tight they could bend and retain the original structure after completing the task. Each jaw consists of wires in the U shape and hence there exists two ends and one end that is interior is connected with the other jaw's interior side of a wire. In the future, we are going to develop the prototype that will able to grasp the delicate object.

References

1. Tai K, El-Sayed A-R, Shahriari M, Biglarbegian M, Mahmud S (2016) State of the Art robotic grippers and applications. Robotics 5:11. https://doi.org/10.3390/robotics5020011
2. Osyczka A, Krenich S (2004) Some methods for multicriteria design optimization using evolutionary algorithms. J Theor Appl Mech 42:565–584
3. Krenich S (2006) Multicriteria design optimization of robot gripper mechanisms. IUTAM Symp Evol Methods Mech 207–218. https://doi.org/10.1007/1-4020-2267-0_20
4. Mahanta GB, Deepak BBVL, Biswal BB, Rout A, Bala Murali G (2018) Design optimization of robotic gripper links using accelerated particle swarm optimization technique. https://doi.org/10.1007/978-981-10-8228-3_31

5. Kim EH, Lee SW, Lee YK (2011) A dexterous robot hand with a bio-mimetic mechanism. Int J Precis Eng Manuf 12:227–235. https://doi.org/10.1007/s12541-011-0031-x
6. Lee DH, Park JH, Park SW, Baeg MH, Bae JH (2017) KITECH-hand: a highly dexterous and modularized robotic hand. IEEE/ASME Trans Mechatrons 22:876–887. https://doi.org/10.1109/TMECH.2016.2634602
7. Rus D, Tolley MT (2018) Design, fabrication and control of soft robots. Nat Rev Mater 3:101–112. https://doi.org/10.1038/nature14543
8. Li Y, Chen Y, Yang Y, Wei Y (2017) Passive particle jamming and its stiffening of soft robotic grippers. IEEE Trans Robot 33:446–455. https://doi.org/10.1109/TRO.2016.2636899
9. Hassan T, Manti M, Passetti G, D'Elia N, Cianchetti M, Laschi C (2015) Design and development of a bio-inspired, under-actuated soft gripper. Proc Annu Int Conf IEEE Eng Med Biol Soc EMBS 2015-November, 3619–3622. https://doi.org/10.1109/EMBC.2015.7319176

Path Planning of the Mobile Robot Using Fuzzified Advanced Ant Colony Optimization

Saroj Kumar, Krishna Kant Pandey, Manoj Kumar Muni and Dayal R. Parhi

Abstract Ant colony optimization (ACO) is a probabilistic optimization method. In this analysis, its application has been explored in mobile robotics for path planning. It provides multi-feedback information and robustness to the mobile robot during navigation. Due to the robustness of the advanced fuzzified ant colony optimization (FACO), the path planning task has been executed in the unstructured environment, and collision-free navigation has been achieved smoothly. For fuzzified advanced ant colony optimization (FAACO), path pheromone update scheme is divided into two categories like favorable and unfavorable path. Using these, path pheromone as the problems of conventional ACO like slow convergence has been sorted out. The advanced FACO improves the evaporation rate of pheromone to accelerate the convergence speed. Finally, the simulation results show the proposed method conquered the previous drawback.

Keywords FAACO · Path planning · Optimization · Robot

1 Introduction

Since the last decade, the development of mobile robotics science has been reached at a higher level of research, due to its autonomy and effectiveness as compared to a human. Nowadays, robots are becoming advance and it is replacing approximately every phase of human life. Among the different types of robotic development, the development that replaces the human and their work is a very challenging task. The mobile robots have many applications, but mobile robots are frequently used in a big platform such as medical, automation, manufacturing, mining, and industries. However, for the smooth operation of the robot, it is necessary to conduct path planning and navigational control task during the operational time. Therefore, in this paper, an advanced FACO technique has been proposed for the path planning of the

S. Kumar (✉) · K. K. Pandey · M. K. Muni · D. R. Parhi
National Institute of Technology, Rourkela, Odisha 769008, India

BBVL. Deepak et al. (eds.), *Innovative Product Design and Intelligent Manufacturing Systems*, Lecture Notes in Mechanical Engineering,
https://doi.org/10.1007/978-981-15-2696-1_101

mobile robot from start to goal point in an unstructured environment. Lee et al. [1] have proposed a global path planner for the mobile robot without post-processing of path. They have used heterogeneous-ants-based path planning (HAB-PP) as a global path planner. A path cross-over scheme is applied to the HAB-PP. Using HAB-PP, a new path is generated which is smoother and shorter than others. Authors [2] proposed an improved version of the ant colony methodology to facilitate the mobile robot to find their path. They have implemented a dimensional grid method for representing the space. In this paper, they show that ACO-PDG strengthens the search process and reduces the number of ant with incomplete paths. Also simulation and experimental results comparison shows, the time and distance or length opted is better than other developed methods. Wang et al. [3] proposed an improved ACO theory, which has taken Oliver 30 TSP problem as a reference problem and finds the time taken. Using this algorithm, they obtained a path that is much accurate and smooth. Xue et al. [4] show the basics of the ant colony optimization algorithm on the traveling salesman problem (TSP) to identify the shortest path and also to solve the mathematical functions of higher degree. They also simulated the result for a higher degree of mathematical function and the optimal solution found in a very quick manner. Another dynamic path planning technique has developed by Hasan et al. [5]. The algorithm is the combination of max-min ant colony optimization and D* algorithm. The tour construction probabilities have used to identify the best solution to proceed from the start point to end point inside the environment. Yan et al. [6] have presented a navigational algorithm for the mobile robot in indoor and dynamic environment toward the direction. They used particle filter-simultaneous localization and map building (PF-SLAM) algorithm which is the combination of the particle filter and FAST SLAM algorithm. The algorithm can apply in real-life application and able to detect the static as well as dynamic obstacles during path planning. Dewang et al. [7] proposed an adaptive particle swarm optimization (PSO) for mobile robot path planning in an unknown environment. They show that APSO is taking less computational time during path planning than other traditional methods. Parhi et al. [8–13] have presented the fuzzy approach toward the planning and navigation of mobile robot using different optimization and hybrid of fuzzy. The objective of this analysis is to develop a fuzzified advanced ACO technique for path planning of mobile robot in an unstructured environment. Akka and Khaber [16] have proposed an improved ACO for path planning of mobile robot. Shan [17] has proposed a modified ACO for submarine path planning.

2 Fuzzified Ant Colony Optimization (FACO)

Overview: Ant colony optimization (ACO) is a probabilistic optimization method. This method is widely used for solving data set computational problem. This algorithm is inspired by the natural phenomenon of ant's foraging. It was initially proposed by 'Macro Dorigo' [2] in the 1990s, encouraged by the communication concept between ants. Ants are using pheromone as the communication tool to

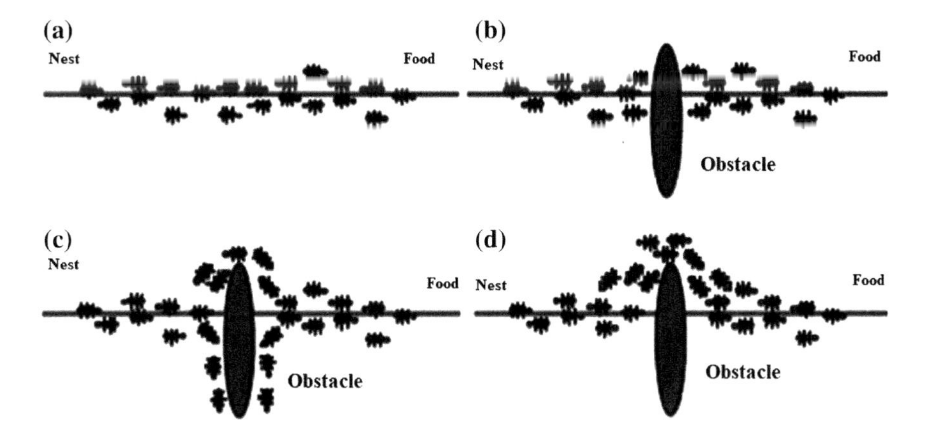

Fig. 1 Food search behavior of Ants [14]

follow the path. This theory is based on the simulation of their advanced process of ants wander for food. The movement of ants gives a very informative path; the most informative path is always a favorable path. The movement of ants from their nest is always in an arbitrary manner. As ant move, there is a tendency to release a chemical substance, is known as pheromone. This is released on land; by this ants can move forward. To move forward, ants follow the strength of pheromone in the path [14], which paths have more strength of pheromone, ants move toward that path.

The above Fig. 1 shows the food search behavior of ants. The ants visit from nest to the food places and vice versa with the help of pheromone. Now, at the time '*t*,' an ant (k) travels from its current position '*a*' to an unvisited position '*b*.' To travel from point '*a*' to point 'b,' the ant follows the path according to the probability Eq. (1) [15];

$$p_{ab}^{k} = \frac{\{\Gamma_{ab}(t)\}^{\alpha}\{\Upsilon_{ab}(t)\}^{\beta}}{\sum_{b=1}^{n}\{\Gamma_{ab}(t)\}^{\alpha}\{\Upsilon_{ab}(t)\}^{\beta}} \quad \alpha \geq 0, \quad \beta \geq 0 \tag{1}$$

where

α Is the parameter used for effect control Γ_{ab}
β Is the parameter used for effect control Υ_{ab}
Γ_{ab} Amount of pheromone (from '*a*' to '*b*')
Υ_{ab} Attractive coefficient.

A hybrid fuzzy logic technique has been introduced to optimize the effect control parameter. There are five membership functions in the input and output variable with their range. By this fuzzy technique, the effect control has been optimized to find a more accurate probability of the path in an advanced manner. Due to this the pheromone update scheme has been identified as favorable and unfavorable path and same has updated. The total of 125 fuzzy rules has been formed for this analysis (Figs. 2, 3, 4, and 5).

After choosing a path with the help of the action choice rule, an ant has to indicate its movement, and the quality of food source (solution) obtained so far to its fellow ants. This is achieved with the help of pheromone–mediated indirect communication, which creates a feedback mechanism. This helps fellow ants to

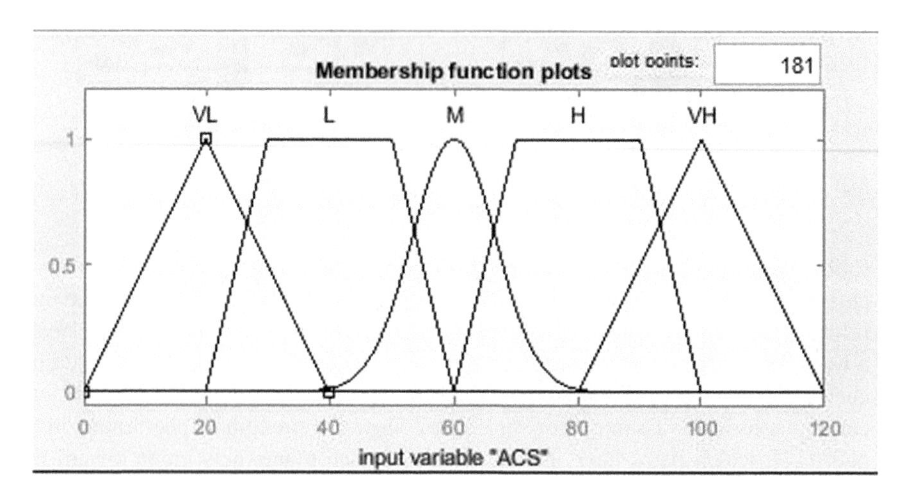

Fig. 2 Ant colony size as the input

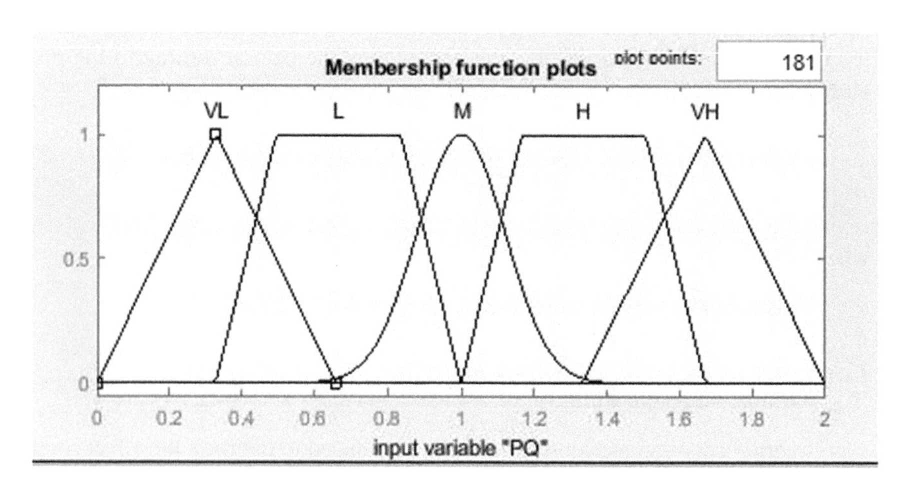

Fig. 3 Pheromone quantity as input

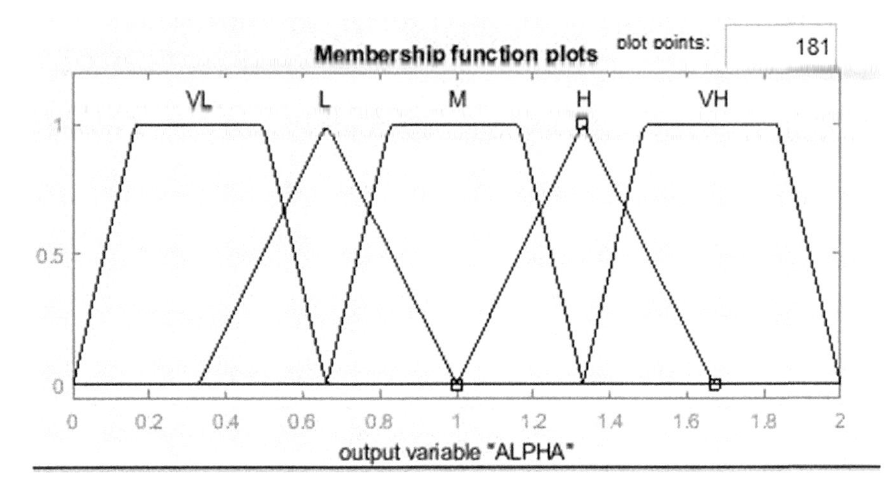

Fig. 4 Effect control parameter 'α' as output

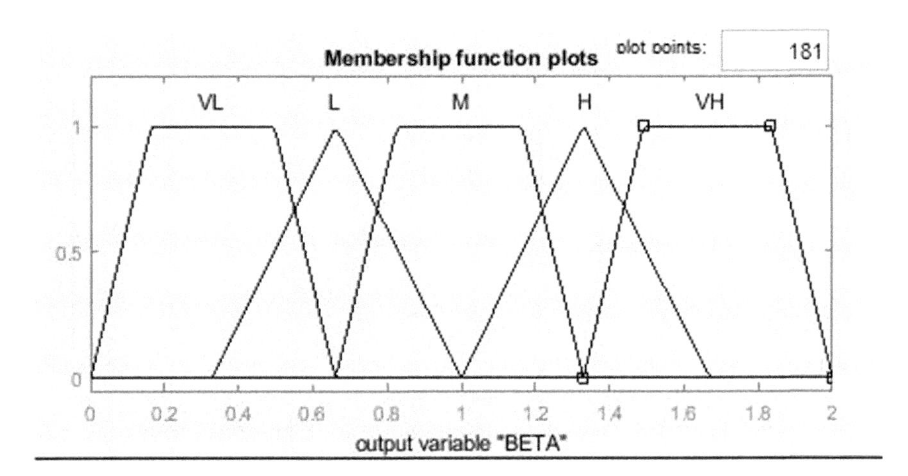

Fig. 5 Effect control parameter 'β' as output

decide whether to take the present route or select another one [7]. When the ants complete one round tour from the nest to nest, the pheromone in the path is updated as follows

$$\Gamma_{ab} = (1 - \rho) \times \Gamma_{ab} + \Delta\Gamma_{ab}^{k} \quad (2)$$

3 The Architecture of Fuzzified Advanced Ant Colony System

In this ACO system, the search algorithm and the pheromone update scheme have been advanced to find the optimal path in the quicker and accurate way. In this scheme, two new terms are introduced as favorable and unfavorable; the favorable and unfavorable depend upon the density of pheromone in the path. If the path has maximum pheromone density, the path is termed as favorable otherwise unfavorable. The advanced pheromone update scheme in ACO is;

$$\Gamma_{ab} = (1-\rho)\times\Gamma_{ab} + \sum_{k=1}^{m}\Gamma_{ab}^{k} + \Delta\Gamma_{ab}^{\text{favorable}} - \Delta\Gamma_{ab}^{\text{unfavorable}} \tag{3}$$

where

ρ The coefficient of evaporation of pheromone
$\Delta\Gamma_{ab}^{k}$ Is change in Γ_{ab}^{k}.

So, the favorable and unfavorable path pheromone updated amount ($\Delta\Gamma_{ab}^{k}$) can be calculated as,

$$\Delta\Gamma_{ab} = \sum_{k=1}^{m}\Gamma_{ab}^{k}, \text{ and} \tag{4}$$

$$\Delta\Gamma_{ab}^{k} = \begin{vmatrix} \frac{Q}{L_k}\times\eta_{\max/\min} & \text{Distance travelled by } K\text{th ant from `a' to `b'} \\ 0 & \text{otherwise} \end{vmatrix} \tag{5}$$

m is the total number of ants considered, and Q is a constant related to the quantity of the pheromone in the trail. In this update scheme, $\eta_{\max/\min}$ are used for the favorable/unfavorable paths, respectively.

The flow chart of the proposed algorithm has been presented below and using this algorithm robot finds a collision-free path (Fig. 6).

4 Result and Discussion

The main objective of this paper is to predict and generate a collision-free shortest path for the mobile robot using fuzzified advanced ant colony optimization (FAACO) method in an unstructured environment. In this analysis, an intelligent controller has been adopted for path planning. The simulation experiments are conducted in MATLAB environmental platform. Figures 7 and 8 show the

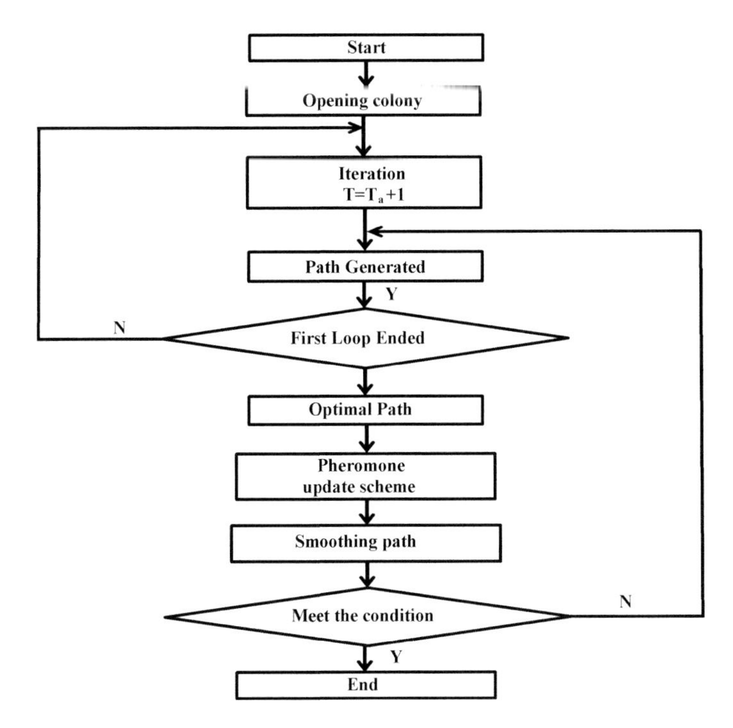

Fig. 6 Flow chart of AACO algorithm

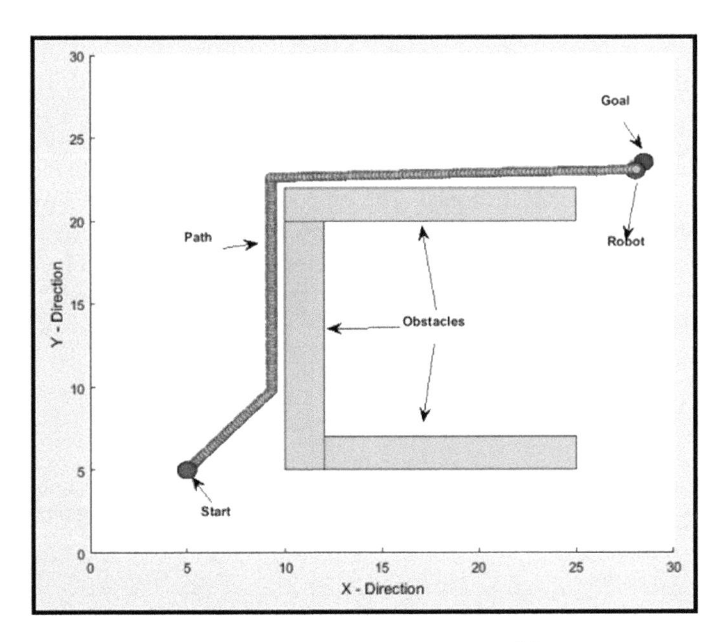

Fig. 7 Trajectory of the path in C-type environment using FAACO

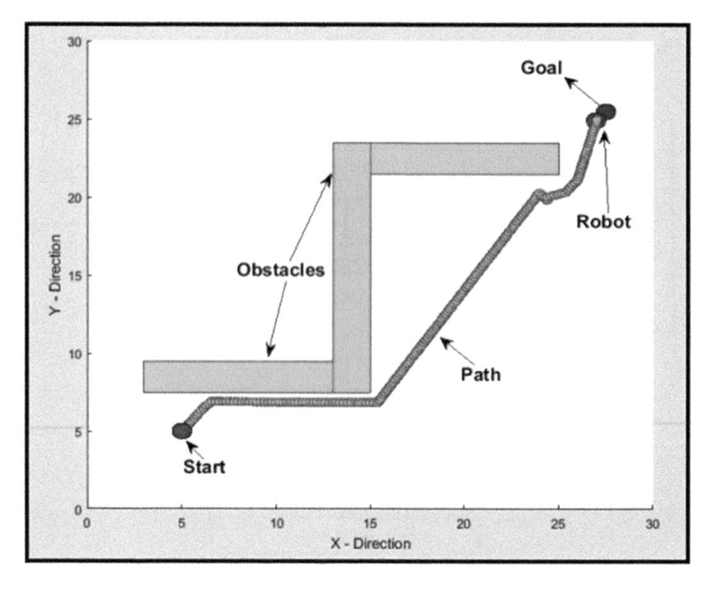

Fig. 8 Trajectory of the path in Z-type environment using FAACO

simulation (MATLAB) experiments which are conducted on 30 × 30 cm search space. The environment contains wall type obstacle, in which robot successfully achieved the goal using FAACO.

4.1 Simulation Experiment Results

To validate the efficiency and performance of the proposed method, simulation experiments have been performed. The proposed algorithm and FAACO method find the shortest path in the given unknown environment. In this process, the mobile robot continued their paths until the obstacles are not detected. Figures 7 and 8 show the smooth trajectory of the robot up to the target. Tables 1 and 2 are the data of path length and time taken by the robot to achieve the target in a different environment.

Table 1 Path length and time taken in simulation platform

Sl. No.	Search space	Travel distance (in cm)	Path planning time (in Sec)
1.	Figure 7	32.48	11.34
2.	Figure 8	34.27	10.21

Table 2 Simulation data set for different search space

No. of Exp.	Path length (cm)	Time taken (S)	Avg. deviation in path (cm)	Avg. deviation in time (S)
1	30.12	10.01	31.35	11.066
2	31.24	10.48		
3	30.52	11.24		
4	32.11	11.21		
5	32.11	12.10		
6	30.01	10.01		
7	30.21	10.01		
8	31.21	12.11		
9	32.78	11.23		
10	33.19	12.26		

5 Conclusion

This paper presents a new approach, FAACO for path planning of mobile robot in an unstructured environment. The output of the method is collision-free shortest path and the same has been verified by the simulation results. The simulation results have been recorded in terms of path length and time taken. In the simulation environment, using FAACO algorithm, the robot successfully achieved the goal from the start point. No collision has detected during path planning. Finally, results show the effectiveness of proposed FAACO technique regarding mobile robot path planning. In the future, this research may be carried out by some other environment with static and dynamic obstacles.

References

1. Lee J (2017) Heterogeneous-ants-based path planner for global path planning of mobile robot applications. Int J Control Autom Syst 15(4):1754–1769
2. Liu J, Yang J, Liu H, Tian X, Gao M (2017) An improved ant colony algorithm for robot path planning. Soft Comput 21(19):5829–5839
3. Wang T, Zhao L, Jia Y, Wang J (2018) Robot path planning based on improved ant colony algorithm. In: WRC symposium on advanced robotics and automation (WRC SARA), IEEE, pp 70–76
4. Xue XD, Xu B, Wang HL, Jiang CP (2010) The basic principle and application of ant colony optimization algorithm. In: International conference on artificial intelligence and education (ICAIE). IEEE, pp 358–360
5. Hasan AH, Mosa AM (2018) Multi-robot path planning based on max–min ant colony optimization and D* algorithms in a dynamic environment. In: 2018 international conference on advanced science and engineering (ICOASE). IEEE, pp 110–115
6. Yan Y, Wong S (2018) A navigation algorithm of the mobile robot in the indoor and dynamic environment based on the PF-SLAM algorithm. Clust Comput 1–12

7. Dewang HS, Mohanty PK, Kundu S (2018) A robust path planning for mobile robot using smart particle swarm optimization. Procedia Comput Sci 133:290–297
8. Mohanty PK, Parhi DR (2014) Navigation of autonomous mobile robot using adaptive network based fuzzy inference system. J Mech Sci Technol 28(7):2861–2868
9. Pandey A, Parhi DR (2014) MATLAB simulation for mobile robot navigation with hurdles in cluttered environment using minimum rule based fuzzy logic controller. Procedia Technol 14:28–34
10. Pandey A, Kumar S, Pandey KK, Parhi DR (2014) Mobile robot navigation in unknown static environments using ANFIS controller. Perspect Sci 8, 421–423
11. Parhi DR (2005) Navigation of mobile robots using a fuzzy logic controller. J Intell Rob Syst 42(3):253–273
12. Mohanty PK, Parhi DR (2015) A new hybrid intelligent path planner for mobile robot navigation based on adaptive neuro-fuzzy inference system. Aust J Mech Eng 13(3):195–207
13. Parhi DR, Singh MK (2010) Navigational path analysis of mobile robots using an adaptive neuro-fuzzy inference system controller in a dynamic environment. Proc Inst Mech Eng, Part C: J Mech Eng Sci 224(6):1369–1381
14. Kumar PB, Sahu C, Parhi DR (2018) A hybridized regression-adaptive ant colony optimization approach for navigation of humanoids in a cluttered environment. Appl Soft Comput 68:565–585
15. Shuyun H, Shoufeng T, Bin S, Minming T, Mingyu J (2018) Robot path planning based on improved ant colony optimization. In: 2018 international conference on robots and intelligent system (ICRIS). IEEE, pp 25–28
16. Akka K, Khaber F (2018) Mobile robot path planning using an improved ant colony optimization. Int J Adv Rob Syst 15(3):1729881418774673
17. Shan Y (2018) Study on submarine path planning based on modified ant colony optimization algorithm. In: 2018 IEEE international conference on mechatronics and automation (ICMA). IEEE, pp 288–292

Dynamics Analysis of Frictionless Spherical Joint with Flexible Socket

Dhaneshwar Prasad Sahu, Mukesh Kumar Singh, Soumitra Singh and Nohar Kumar Sahu

Abstract This paper presents the investigation of dynamic modeling and anaylsis of spherical joint with flexible socket joint model. Ball and socket are composing spherical joints which are modeled as two individual colliding components. A continuous force model is introduced for the normal contact–impact force. For the analysis of energy dissipation during contact process, Hertzian-based contact model is used. Generally, Hertzian contact model is used for the analysis of energy dissipation during the dynamic conditions of ball joint because it gives normal deformation of sphere parts of ball and socket joint. This model also describes the viscosity and shear response of the components with friction and adhesion. The pseudo-penetration that occurs between the potential contact points of the ball and the socket surface, as well as the indentation rate, plays a crucial role in the evaluation of the normal contact forces. In addition to this, different force models, such as Coulomb's law, come into the picture. A friction model is taken for the analysis of friction at the interface of friction between ball and socket joint. The normal and tangential force is evaluated and included in the dynamic model of multibody dynamic system. In modern car, different types of ball joint are used for the better performance of the spherical joint. Geometrical and material parameters are modified to understand the behaviour of the failure and heavily loaded section. The manufacturing process also plays an important role in designing of the component of spherical joint.

Keywords Spherical joint with clearance · Frictional force model · Multibody dynamic system

D. P. Sahu (✉) · M. K. Singh · S. Singh · N. K. Sahu
Guru Ghasidas Vishwavidyalaya, Bilaspur, Chhattisgarh, India

BBVL. Deepak et al. (eds.), *Innovative Product Design and Intelligent Manufacturing Systems*, Lecture Notes in Mechanical Engineering,
https://doi.org/10.1007/978-981-15-2696-1_102

1 Introduction

Ball and socket joint is used for guiding function in different automotive hub carriers. It is used to transfer motion from one component to another components. Ball and socket joint allows three degrees of freedom that is rotational motion. It removes all translation displacement. It is generally used in automobile suspension system to provide a comfortable ride Chen et al. [1]. In spherical joint, the ball is rotating inside the socket, and in between ball and socket, the lubrication is filled to reduce the friction and to increase the life of joint. The clearance exists in between the ball and socket joint. So, we need to optimize this clearance to minimize various phenomena like unwanted shake response and chaotic behavior Tian et al. [2, 3]. The basic aim and objective of this research work are to increase the bending stiffness of the joint. So, the life of ball and socket joint is increased. There are many examples around us such as automobile suspension system in the vehicles the existence of spherical joints and compliance like bushing are needed for the proper operation of the system. There are many important phenomena which is associated with clearance joint models, such as wear, non-smooth behavior, optimization and control, chaos, and uncertainty and links' flexibility, are also analysis and will be optimize Watrin et al. [4]. Accurate dynamic analysis of mechanical system with clearance joints plays a significant role in the design of spherical joints. The numerical modeling and simulation of mechanisms with clearance joints are still a challenging work for researchers and engineers. Ball joints allow the articulation of suspension system around the vertical axis and same time manage the movement of suspension system of the vehicle. There are two types of force acting in the ball and socket joint; these are element elastic force and lubricant pressure force. The above two forces are needed to analyze and optimize for better life of joint structure Sin et al. [5, 6].

1.1 Modeling of Spherical Clearance Joint

A mathematical model for spherical joint that consists of two components (A) ball (B) socket with clearance can be analyzed under the framework of multibody dynamics system and is formulized. Socket and ball are two links i and j as shown in Fig. 1. The radii of socket and ball links are as R_i, R_j, respectively. The difference between radius of socket and ball gives the clearance of spherical joint Flores et al. [7].

$$c = R_i - R_j \tag{1}$$

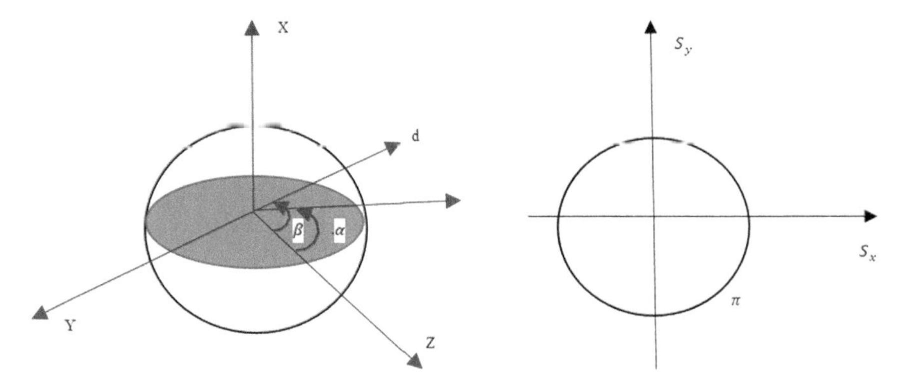

Fig. 1 Ball and socket joint with clearance in spatial mechanism

1.2 Lagrange–Euler Formulation

Lagrange–Euler and Newton–Euler equations can be used to develop the equation for motion of joint in any mechanical member. Rigid body joint velocity can be evaluated by these two formulae. The above formula has the following feathers

(A) Joint torque can also be achieved by this second-order coupled nonlinear differential equation.
(B) Inertia loading.
(C) Coupling reaction force between joints such as Coriolis and centrifugal and gravity effect.
(D) Joint velocity and acceleration.

There are two methods for the analysis of the above problem

(1) Forward dynamics problem
(2) Inverse dynamics problem.

Forward Dynamics Problem
In forward dynamics problem, the desired torque/force has been provided, the dynamics equation is used to solve for the joint acceleration which are then integrated to solve for the generalized coordinates and their velocities.

Inverse Dynamics Problem
In inverse dynamics problem it is given the generalize coordinates and their first two time derivatives, the generalized forces/Torque to be computed.

4*4 homogeneous coordinates transformation matrix T_i^{i-1} which describes the spatial relationship between the ith & $(i-1)$th link coordinate frames.

Lagrange–Euler equation for rotary joint

$$\frac{\mathrm{d}}{\mathrm{d}t}\left(\frac{\partial L}{\partial \dot{\theta}_i}\right) - \frac{\partial L}{\partial \theta_i} = \tau_i \quad (2)$$

where

L $= K - P$ Lagrangian function
K Total kinetic energy of the spherical joint
P Total potential energy of the spherical joint
θ_i Generalize coordinate of the spherical joint
$\dot{\theta}_i$ First time derivative of the generalize coordinate system
τ_i Generalize force/torque applied to the system at the joint i to derive link.

1.3 Joint Velocity of Spherical Joint

The Lagrange-Euler formulation requires knowledge of the kinetic energy of the physical system, which gives the velocity in each joint.

The velocity of a point fixed in the link i will be derived, and the effect of the motion of other joints on all points in this link will be explored Sponget al. [8].

Let r_i^i be a point fixed and at rest in a link i and expressed in homogeneous coordinates with respect to the ith link coordinate frame,

$$r_i^i = (x_i, y_i, z_i, 1)^T \quad (3)$$

Let r_i^0 be the same point r_i^i with respect to the base coordinate frame, T_i^{i-1} be the homogeneous coordinate transformation matrix which relates the spatial displacement of the ith link coordinate frame to the $(i-1)$th link coordinate frame and T_i^0 be the coordinate transformation matrix which relates the ith coordinate frame to the base coordinate frame; then r_i^0 is related to the point r_i^i by

$$r_i^0 = Tr_i^i \quad (4)$$

where

$$T_i^0 = T_1^0 T_2^1 \ldots T_i^{i-1} \quad (5)$$

If joint i is revolute, it follows Eulerian angle system which states that any orientation can be achieved by composing three elemental rotations about the axes of coordinate rotation.

(A) Rotation of φ angle about the OZ axis $(R_{z,\emptyset})$
(B) Rotation of θ angle about the axis OU $(R_{u,\theta})$

(C) Finally, rotation of ψ about OW axis ($R_{w,\psi}$)

$$T_i^{i-1} = R_{z,\phi} R_{u,\theta} R_{w,\psi} \tag{6}$$

$$T_i^{i-1} = \begin{bmatrix} C\phi & -S\phi & 0 \\ S\phi & C\phi & 0 \\ 0 & 0 & 1 \end{bmatrix} \begin{bmatrix} 1 & 0 & 0 \\ 0 & C\theta & -S\theta \\ 0 & S\theta & C\theta \end{bmatrix} \begin{bmatrix} C\psi & -S\psi & 0 \\ S\psi & C\psi & 0 \\ 0 & 0 & 1 \end{bmatrix} \tag{7}$$

$$T_i^{i-1} = \begin{bmatrix} C\psi C\theta - S\phi C\theta S\psi & -S\psi C\phi - S\phi C\theta C\psi & S\phi S\theta & 0 \\ S\phi C\psi + C\phi C\theta S\psi & -S\phi S\psi + C\phi C\theta C\psi & -C\phi S\theta & 0 \\ S\theta S\psi & S\theta C\psi & C\theta & 0 \\ 0 & 0 & 0 & 1 \end{bmatrix} \tag{8}$$

Since the point r_i^i is the at rest in link i, and assuming rigid body motion, other points as well as the point r_i^i fixed in the link i and expressed with respect to the ith coordinate frame will have zero velocity with respect the ith coordinate frame can be represented as

$$v_i^0 \equiv v_i = \frac{\mathrm{d}}{\mathrm{d}t}\left(r_i^0\right) = \frac{\mathrm{d}}{\mathrm{d}t}\left(T_i^0 r_i^i\right) \tag{9}$$

The above compact form is obtained because $\dot{r}_i^i = 0$. The partial derivative of T_i^0 with respect to θ_i is easily calculated with the help of a matrix Q_i which is for revolute joint is

$$Q_i = \begin{bmatrix} 0 & -1 & 0 & 0 \\ 1 & 0 & 0 & 0 \\ 0 & 0 & 0 & 0 \\ 0 & 0 & 0 & 0 \end{bmatrix} \tag{10}$$

It then follows that

$$\frac{\partial T_i^{i-1}}{\partial \theta_i} = Q_i T_i^{i-1} \tag{11}$$

$$\frac{\partial T_i^{i-1}}{\partial \theta_i} = \begin{bmatrix} S\phi S\theta S\psi & S\phi S\theta C\psi & S\phi C\theta & 0 \\ -C\phi S\theta S\psi & -C\phi S\theta C\phi & C\phi C\theta & 0 \\ C\theta S\psi & C\theta C\psi & -S\theta & 0 \\ 0 & 0 & 0 & 0 \end{bmatrix} \tag{12}$$

Hence, for $i = 1, 2, 3 \ldots n$

$$\frac{\partial T_i^{i-1}}{\partial \theta_i} = \begin{cases} T_1^0 T_2^1 \ldots T_{j-1}^{j-2} Q_j T_j^{j-1} \ldots T_i^{i-1} & \text{for } j \le i \\ 0 & \text{for } j > i \end{cases} \tag{13}$$

Equation 12 can be interpreted as the effect of the motion of joint j on all the points on link i. In order to simplify the notation, let us define $U_{ij} \triangleq \partial T_i^0 / \partial \theta_j$; then, Eq. 12 can be written as

$$U_{ij} = \begin{cases} T_{j-1}^0 Q_j T_i^{j-1} & \text{for } j \le i \\ 0 & \text{for } j > i \end{cases} \tag{14}$$

$$v_i = \left(\sum_{j=1}^{i} U_{ij} \dot{\theta}_j \right) r_i^i$$

For rotary joint,

$$\frac{\partial U_{11}}{\partial \theta_1} = Q_1 Q_1 T_1^0 \tag{15}$$

1.4 Kinetic Energy of the Spherical Joint

After obtaining the joint velocity of each link, we need to find the kinetic energy of link i. Let K_i be the kinetic energy of link i, $i = 1, 2, 3 \ldots n$, as expressed in the base coordinate system, and let d K_i be the kinetic energy of the particle with differential mass dm in the link i; then

$$\begin{aligned} \mathrm{d}K_i &= \frac{1}{2}\left(\dot{x}_i^2 + \dot{y}_i^2 + \dot{z}_i^2\right)\mathrm{d}m \\ &= \frac{1}{2}\operatorname{trace}\left(v_i v_i^T\right)\mathrm{d}m \end{aligned} \tag{16}$$

The matrix U_{ij} is the rate of change of the point r_i^i on link i relative to the base coordinate frame as θ_j changes. It is constant for all points on link i and independent of mass distribution of the link i. Also, $\dot{\theta}_i$ are the independent of the mass distribution of link i, So, summing all kinetic energies for both links

$$K_i = \int \mathrm{d}K_i$$

Inertia of all points on the link i can be represented as

$$I_i = \int r_i^i \left(r_i^i\right)^T \mathrm{d}m \tag{17}$$

$$I_i = \begin{bmatrix} \int x_i^2 \mathrm{d}m & \int x_i y_i \mathrm{d}m & \int x_i z_i \mathrm{d}m & \int x_i \mathrm{d}m \\ \int x_i y_i \mathrm{d}m & \int y_i^2 \mathrm{d}m & \int z_i y_i \mathrm{d}m & \int y_i \mathrm{d}m \\ \int x_i z_i \mathrm{d}m & \int z_i y_i \mathrm{d}m & \int z_i^2 \mathrm{d}m & \int z_i \mathrm{d}m \\ \int x_i \mathrm{d}m & \int y_i \mathrm{d}m & \int z_i \mathrm{d}m & \mathrm{d}m \end{bmatrix}$$

$r_i^i = (x_i, y_i, z_i, 1)^T$ as defined before. If we use inertia tensor J_{ij} which is defined as

$$J_{ij} = \int \left[\delta_{ij}\left(\sum_k x_k^2\right) - x_i x_j\right] \mathrm{d}m$$

where the indices i, j, k indicate principal axes of the ith coordinate frame and δ_{ij} is so called Kronecker delta, then J_i can be expressed in inertia tensor as.

$$I_i = \begin{bmatrix} \frac{-k_{i11}^2 + k_{i22}^2 + k_{i33}^2}{2} & k_{i12}^2 & k_{i13}^2 & \overline{x_i} \\ k_{i12}^2 & \frac{k_{i11}^2 - k_{i22}^2 + k_{i33}^2}{2} & k_{i23}^2 & \overline{y_i} \\ k_{i13}^2 & k_{i23}^2 & \frac{k_{i11}^2 + k_{i23}^2 - k_{i33}^2}{2} & \overline{z_2} \\ \overline{x_i} & \overline{y_i} & \overline{z_i} & 1 \end{bmatrix}$$

where k_{i23} is the radius of gyration of link about the yz axes and $\overline{r_i^i} = (\overline{x_i}, \overline{y_i}, \overline{z_i}, 1)^T$ is the center of mass vector of link i from the ith link coordinate frame and expressed in the ith link coordinate frame. Hence, the total kinetic energy K of the joint is

$$K = \frac{1}{2}\mathrm{trace}\left[\sum_{j=1}^{n}\sum_{k=1}^{n}\frac{\partial T_i}{\partial \theta_j} I_i \frac{\partial T_i^T}{\partial \theta_k}\dot{\theta}_j\dot{\theta}_k\right] \tag{18}$$

Since the trace of sum of matrices is the sum of individual traces, we may interchange summations and the trace operator to obtain

$$K = \frac{1}{2}\sum_{j=1}^{n}\sum_{k=1}^{n} I_{jk}(\theta)\dot{\theta}_j\dot{\theta}_k = \frac{1}{2}\dot{\theta}^T I(\theta)\dot{\theta} \tag{19}$$

where the $n \times n$ joint inertia matrix $I(\theta)$ has elements defined as

$$I_{jk}(\theta) = \sum_{i=1}^{n} \mathrm{trace}\left[\sum_{j=1}^{n}\frac{\partial T_i}{\partial \theta_j} I_i \frac{\partial T_i^T}{\partial \theta_k}\right]$$

Since $\frac{\partial T_i}{\partial \theta_j} = 0$ for j > i, we may write this more efficiently

$$I_{jk}(\theta) = \sum_{i=\max(j,k)}^{n} \text{trace}\left[\sum_{j=1}^{n} \frac{\partial T_i}{\partial \theta_j} I_i \frac{\partial T_i^T}{\partial \theta_k}\right] \tag{20}$$

1.5 Potential Energy of the Spherical Joint

Let the total potential energy of the joint be P and P_i be the potential energy of each link

$$P_i = -m_i g \bar{r}_i^0 = -m_i g \left(T_i^0 \bar{r}_i^i\right) \quad i = 1, 2, 3 \ldots n$$

where $r_i = \theta_i \times$ Radius of curvature

Total potential energy of the joint can be obtained by summing all the potential energies in each link

$$P = \sum_{i=1}^{n} P_i \tag{21}$$

where $g = \left(g_x, g_y, g_z, 0\right)$ is the gravity row vector expressed in the base coordinate system.

Note that P depends only on joint variable θ but not depends on the joint velocity $\dot{\theta}$

$$P(\theta) = -\sum_{i=1}^{n} \theta^T T_i(\theta) I_i e_4 \tag{22}$$

where e_4 is the identity matrix of 4×4 (i.e., $e_4 = (0, 0, 0, 1)^T$)

1.6 Motion Equation of the Spherical Joint

From Eqs. 19 and 22, we can get Lagrangian function $L = K - \text{P}$ given by

$$L\left(\theta, \dot{\theta}\right) = K\left(\theta, \dot{\theta}\right) - P\left(\theta, \dot{\theta}\right) = \frac{1}{2} \dot{\theta}^T I(\theta) \dot{\theta} - P(\theta) \tag{22}$$

It is the fundamental property that the kinetic energy is a quadratic function of the joint velocity vector, and potential energy is independent of $\dot{\theta}$

$$\frac{\partial L}{\partial \theta} = \frac{1}{2}\frac{\partial}{\partial \theta}(\dot{\theta}^T I(\theta)\dot{\theta} - \frac{\partial P(\theta)}{\partial \theta}$$

The dynamic equation is given by

$$I(\theta)\ddot{\theta} + \dot{I}(\theta)\dot{\theta} - 1/2\frac{\partial}{\partial \theta}\left(\dot{\theta}^T I(\theta)\dot{\theta}\right) + \frac{\partial P(\theta)}{\partial \theta} = \tau \tag{24}$$

Defining the Coriolis/centripetal vector

$$\mathrm{V}(\theta)\ddot{\theta} = \dot{I}(\theta)\dot{\theta} - 1/2\frac{\partial}{\partial \theta}\left(\dot{\theta}^T I(\theta)\dot{\theta}\right) = \dot{I}_{\dot{\theta}} - \frac{\partial K}{\partial \theta} \tag{25}$$

And the gravity vector

$$G(\theta) = \frac{\partial P(\theta)}{\partial \theta}$$

We may write the equations

$$I(\theta)\ddot{\theta} + V(\theta, \dot{\theta}) + G(\theta) = \tau \tag{26}$$

1.7 Modified Form of the Dynamics Equation

In this section, we investigate the detailed structure and properties of the dynamical equation. The dynamic equation always influences by friction and disturbances. Therefore, the generalized dynamic equation of the arm ball has been given below [17]:

$$I(\theta)\ddot{\theta} + V(\theta, \dot{\theta}) + F(\dot{\theta}) + G(\theta) + \tau_d = \tau \tag{27}$$

where friction terms $F\left(\dot{\theta}\right) = F_v\dot{\theta} + F_d$ with F_v the coefficient matrix of viscous friction and F_d a dynamics terms also added is disturbance τ_d which could represent, for instance, any inaccurately modeled dynamics.

2 Result and Discussion

By using finite element method (FEM), it is easy to identify the heavy loaded cross-sectional area, and design is required for that area. In this way, it is feasible to reach a specific conclusion on the cause and possible solution to avoid this kind of failure Ossa et al. [9]. In this analysis, the ball joint is modeled geometrically with the following boundary and loading condition: (A) a fixed contact zone in ball (B) a lateral at the end of solid circular shaft which is connected to the ball component of ball joint is shown in Fig. 2 with the value 500 N. This load has been provided to the ball joint for experimental analysis, used in suspension system of ball joint. This value of load represents the realistic estimate to the applied load [10] (Figs. 3 and 4).

The analysis and calculation are done to recognize the heavily loaded cross-sectional area, and design is done for that particular area. The contact support between the element and its cage is modeled as frictionless support because failed element did not show any wear between them [11, 12]. The element used in FE mesh was triangular element with mesh size 2 mm. The type of fracture is ductile failure because generally, material used for the ball component of spherical joint is stainless steel SAE 4135. Stainless steel is a ductile material in nature, and von Mises stress theory gives good agreement in ductile material.

The mesh had a total number of element 3354. The size of element and mesh was selected after various analyses with same load and different mesh sizes. The mesh size is selected as a triangular shape for better accuracy in results because when we use the triangular meshing, this increases per element number of node which gives smooth results. The ball joint without contact support at point C gives von Mises stress value of 76.08 MPa and ball joint element with contact support, that is frictionless support at point C, gives the value of von Mises stress value of 52.092 MPa. Both values obtained from analysis are lower than the bulk fatigue endurance stress of steel under normalize condition which is 416 MPa. There are many things that influence the stress concentration such that (i) variation in properties of materials (ii) load application (iii) discontinuities in the components. (iv) machining scratches. So, I have found it theoretically that stress concentration is high at the curvature location. So, the stress concentration factor in that area is more predominantly.

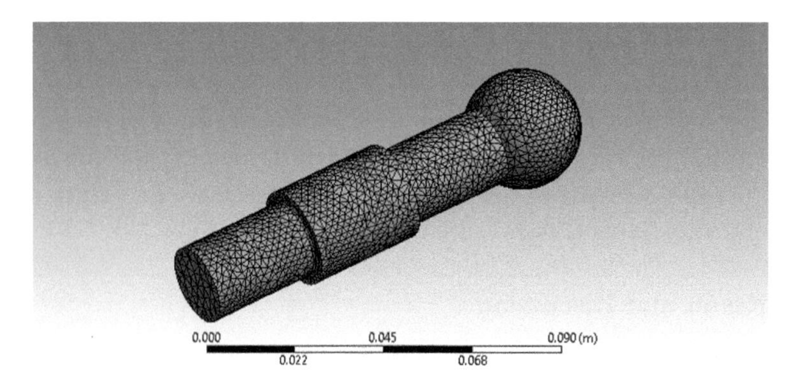

Fig. 2 Ball joint model with triangular mesh generation

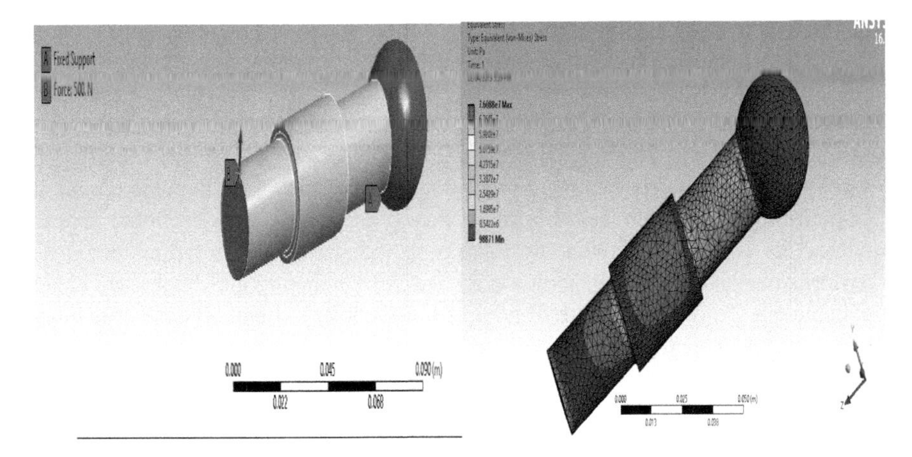

Fig. 3 Ball joint model with (**A**) fixed support (**B**) applied load 500 N and equivalent von Mises stress of ball joint without contact support at C

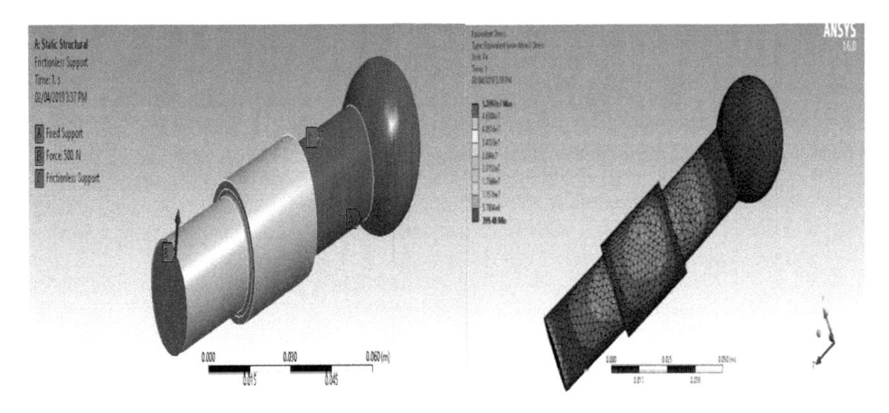

Fig. 4 Boundary condition of ball joint with contact support at C and equivalent von Mises stress with contact support at point C

3 Conclusions and Future Challenges

The analysis shows the fatigue fracture of the ball joint. Fatigue fracture is a tendency of material to fracture the material by the progressive fracture of material due to cyclic or repetitive stress of an intensity considerably below normal strength. The ball joint works on the dynamic condition, so it is experimentally proved that failure is due to fatigue fracture. This fracture initiates the ball joint element and its cage where the stress concentration is much high [13, 14]. Along with the stress concentration factor, the material structure plays an important role for reducing the fatigue endurance limit of the component on almost 40%. In addition to this, initiating cracks that grew with the application of cycling loading cause sudden failure in the ball joint. The material used for the manufacturing of ball joint is appropriate for this kind of application Wang et al. [15, 16]. During the manufacturing of ball joint, mainly two factors are influencing the design of component: (A) defective heat treatment process and (B) defective geometrical modification. Wrong selection of heat treatment process and temperature can affect the fatigue life of ball joint component. In addition to this, reduction in cross-sectional area can reduce the life of component by increasing the stress concentration factor. The performance and life of ball joint are increased by reducing the shocks and chaotic phenomena. To optimize the ball joint of a multibody system, whether it is human body or mechanical system is still a challenging task for many researchers and engineers. The future work regarding this field is to analyze the endurance strength of ball joint according to the number of stress cycle, and corresponding to that material, property may change for better life of component.

References

1. Chen Y, Sun Y, Peng B, Cao C (2016) A comparative study of joint clearance effects on dynamic behavior of planar multibody mechanical systems. Lat Am J Solids Struct 13 (15):2515–2533
2. Tian Q, Flores P, Lankarani HM (2018) A comprehensive survey of the analytical, numerical and experimental methodologies for dynamics of multibody mechanical systems with clearance or imperfect joints 122:1–57
3. Nemade KV, Tripathi VK (2013) A mathematical model to calculate contact stresses in artificial human hip joint 6(12):119–123
4. Watrin JC, Makich H, Haddag B, Nouari M, Grandjean X (2013) Analytical modelling of the ball pin and plastic socket contact in a ball joint 1–26
5. Sin BS, Lee KH (2014) Process design of a ball joint, considering caulking and pull-out strength. Sci World J 2014
6. Shinde J, Kadam S, Patil A, Pandit S (2016) Design modification and analysis of suspension ball joint using finite element. Int J Innov Res Sci Eng Technol 5(7):12797–12804
7. Flores P, Lankarani HM, Flores P, Lankarani HM (2011) Spatial rigid-multibody systems with lubricated spherical clearance joints : modeling and simulation to cite this version : HAL Id : hal-00568395
8. Spong MW, Hutchinson S, Vidyasagar M (2004) Robot dynamics and control
9. Ossa EA, Palacio CC, Paniagua MA (2011) Failure analysis of a car suspension system ball joint. Eng Fail Anal 18(5):1388–1394
10. Shinde J, Kadam S (2016) Design of suspension ball joint using FEA and experimental method. Int Res J Eng Technol 03(07):1853–1858
11. Kato K, Adachi K (2000) Wear mechanisms
12. Zmitrowicz A (2006) Wear patterns and laws of wear—a review. J Theor Appl Mech 44 (2):219–253
13. Taylor RL (2013) FEAP-a finite element analysis program. no. Dec 2013
14. Erleben K (2005) Stable, robust, and versatile multibody dynamics animation. Apr 2005

15. Wang G (2015) Dynamics analysis of spatial multibody system with spherical joint wear. 137: Apr 2015
16. Hall GW, Crandall JR, Pilkey WD, Park M (1998) Development of a dynamic multibody model to analyze human lower extremity impact response and injury. Sept 1998
17. Fu KS, Gonzalez RC (2008) Robotics control, sensing, vision, and intelligence. McGraw Hill Education (India) edn 2008

Investigation on the Effect of Different Dielectric Fluids During Powder Mixed EDM of Alloy Steel

T. Sree Lakshmi, Sagnik Sarma Choudhury, K. Gnana Sundari and B. Surekha

Abstract The machining characteristics of electric discharge machining (EDM) are dependent on various factors, namely work material, process parameters, electrode materials as well as the dielectric fluids. It is to be noted that the addition of semiconductive particles in dielectric fluid improves its properties by reducing the insulating strength of the dielectric fluid and increasing the spark gap between the tool electrode and workpiece. This mechanism results in a more stable process, higher metal removal, and good surface finish. In the present paper, an experimental investigation on the influence of various dielectric fluids namely paraffin, kerosene, distilled water, and transformer oil mixed with aluminum powder during EDM of EN-19 work material. Brass is used as the electrode material during the machining process. The effects of Al powder mixed dielectric fluid on the material removal rate (MRR), electrode wear rate (EWR), surface roughness (SR), and radial over cut (ROC) were measured at different levels of input given parameters like gap voltage (V_g), peak current (I_p), and pulse on time (T_{on}). From the experimental results, it is observed that the transformer oil is found to provide high MRR, surface finish, EWR, and ROC among all other dielectric fluids.

Keywords PMEDM · Aluminum powder · Transformer oil · Paraffin · Distilled water

---

T. S. Lakshmi (✉) · S. S. Choudhury · B. Surekha
School of Mechanical Engineering, KIIT Deemed to be University, Patia, Bhubaneswar 751024, India

K. Gnana Sundari
School of Mechanical Engineering, Karunya Deemed to be University, Coimbatore 641114, India

BBVL. Deepak et al. (eds.), *Innovative Product Design and Intelligent Manufacturing Systems*, Lecture Notes in Mechanical Engineering,
https://doi.org/10.1007/978-981-15-2696-1_103

1 Introduction

EDM is a well-established non-traditional machining process suitable for the manufacturing of engineering components that are very much difficult-to-machine by other conventional machining processes and geometrically complex features. A small amount of gap between the workpiece (anode) and the electrode (cathode) allows the erosion pulse discharge that leads to the material removal from both anode and cathode. This removal of unwanted material through melting and vaporization in the presence of the dielectric fluid. In the beginning, Russian researchers discovered dielectric fluid to replace an ordinary dielectric medium (air) to provide good insulation, quick ionization, material cooling, and removal of waste particles while machining. Thereafter, research is continuing to introduce dielectric fluids to increase the material removal rate. Later on, mineral oil was used as a dielectric. From 1960 onwards, the mineral oil industry was established and began to use mineral oil in spark erosion machines [1–3]. Zhang and Liu [4] investigated the effect of gaseous and liquid dielectrics such as air, oxygen, deionized water, and kerosene. They have also studied the geometry of craters and recast layers and also related the material removal characteristic to the discharge pulse through simulation. It was concluded that the pressure above the discharge point was affecting material removal. It is observed that the insulation and ionization rates of dielectric fluids affect the productivity, costs, and quality of machined parts. Jagdeep et al. [5] carried out the experiments with different dielectrics like kerosene, distilled water, and EDM oil with the addition of graphite powder on tungsten carbide–cobalt alloy. A critical review has been made [6] to discuss the effects of different dielectric fluids like hydrocarbons and bio-diesel-based, water and water-based additives, and gaseous-based dielectric fluid on the responses. They also discussed the effect of the addition of powders, namely titanium, silicon, graphite, copper, and aluminum oxide on the responses. A comparative study [7] has been made on electric discharge machining of Al6061 alloy with kerosene and distilled water as dielectric mediums. They found that the dielectric fluid kerosene was giving high MRR and TWR. Many other researchers [8–10] conducted an experimental study to understand the effect and suitability of dielectric media such as EDM oil, kerosene, bio-diesel, paraffin, and natural oils using the design of experiments. Despite using different plain dielectric fluids, semiconductive metallic powder has been added to the dielectric fluid to enhance machining characteristics [11]. The addition of powders to dielectric medium enhances electric conduction property by lowering resistance and resulting in a high material removal rate. Most of the research work on powder mixed EDM has been carried out with different powders like aluminum, graphite, CNT, silicon carbide, and titanium carbide. Tzeng and Lee [12] worked on powder mixed EDM to study the influence of powder particle size, concentration, particle density, thermal conductivity, and resistivity on the responses, namely effect of powders on spark gap, material removal rate, and tool wear rate. Surekha et al. [13] conducted a study to predict the effect of various metallic powders on the EDM responses such as MRR,

TWR, and surface finish on the EN-19 alloy steel with the brass electrode. In the present investigation, the machining characteristics of EN-19 alloy steel using aluminum powder mixed EDM with different dielectrics have been studied. A comparative study was carried out to analyze the effect of various dielectric mediums in terms of high MRR, EWR, and good surface finish. Most of the earlier researches suggesting the use of pure dielectric medium without addition of powder in it. So, efforts have been made in the current work to study the effects of machining characteristics when the powder is being added to various dielectric fluids.

2 Experimentation

2.1 Materials

An alloy steel EN-19 is the most commonly used material for the manufacturing of crankshafts, gears, axial shafts, and connecting rods where the strength and the shock resistance are the most desirable properties for their smooth functioning. It is known that the durability can be increased with a good surface finish. Hence, in the present work, EN-19 has been considered as the work material, and an attempt is made to increase the surface finish by the addition of powder mixed electric medium during the electric discharge machining process (Table 1).

Brass is considered as an electrode material because of its wear resistance and creates a controlled spark to remove material and easily available. Aluminum powder is considered to mix with four dielectric fluids such as kerosene, paraffin, distilled water, and transformer oil. Properties of four dielectric fluids are shown in Table 2 [10].

2.2 Machine and Methods

ELECTRONICA–SMART ZNC die sink EDM was used to carry out an experimental investigation with required modifications for powder mixed EDM (Fig. 1).

To minimize the usage of the dielectric fluid, a cuboidal container of 27.5 × 22.5 × 15 cm was utilized.

Dielectric medium (5 lt) is poured in the container until the specimen was completely sunk in it. Aluminum powder (0.5 gr/lt) was added and mixed

Table 1 Chemical composition of EN-19 alloy steel [14]

Elements	Fe	C	Si	Mn	P	S	Cr	Mo	Ni	Al
wt%	96.86	0.38	0.21	0.91	0.01	0.01	1.04	0.23	0.23	4.21

Table 2 Properties of kerosene, paraffin, distilled water, and transformer oil

Dielectrics	Density (g/cm^3)	Kinematic viscosity (CST)	Flash Point (°C)	Dielectric constant	Oxygen wt%
Kerosene	0.81	2.71	38	1.8	0
Paraffin	0.9	2.375	165	1.6	0
Distilled water	1	1.781	100	29.3	3
Transformer oil	0.89	27	38	2.1	0

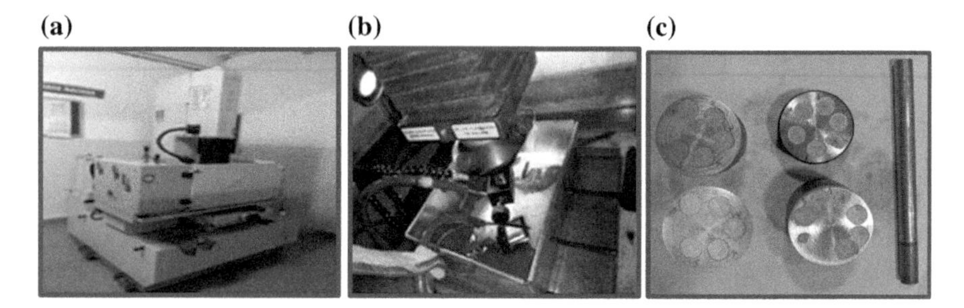

Fig. 1 **a** EDM machine. **b** Setup of PMEDM. **c** Workpieces and tool

thoroughly with the help of a mechanical stirrer at a speed of 200 rpm to attain uniform mixture with the dielectric medium. However, dielectric fluid was circulated at the machining zone through a nozzle (10" from suction) with the help of a pump to maintain the uniform concentration of the medium. At the time of powder mixed EDM, the machining gap was occupied with added aluminum particles. When a voltage is applied in range between 40–60 V is applied between the anode (specimen) and the cathode (electrode) which is separated by a short stand-off distance of 20–25 µm, an electric field shall be initiated, and the particle exactly below the electrode gets galvanized and behaves in a criss-cross fashion. This criss-cross formation helps in building between the specimen and the tool. Due to this voltage in the gap, the strength of deionization of the dielectric will be reduced and results in an early bombardment in the gap. As a consequence, a number of discharges occur below the electrode space. This results in the incrimination of the frequency of discharge leading to fast sparking which results in the speedy withdrawal of the matter from the workpiece. Simultaneously, the added powder widens and enlarges the plasma channel which leads to a reduction in the electrical density, and hence, sparks get distributed uniformly finally leading to increased uniformity on the surface of the workpiece.

2.3 Experimental Procedure

In the current study, experiments were done to study the influence of different dielectric fluids like kerosene, paraffin, distilled water, and transformer oil during the aluminum powder mixed EDM process. A small rectangular tank is located in the main working container of EDM, and machining process is conducted on EN-19 specimen with the brass electrode at the preset process variables in that machining container itself. From the study of literature, researchers have used 0.5 g/lt of dielectric for aluminum powder. For improving circulation of the powder in a dielectric medium, a small capacity pump is located at the bottom of the container. To eliminate the buildup of added particles and the removed particles in the machining zone, circulation pump with the nozzle is placed in that tank then the powder will be mixed easily in the dielectric. A suitable discharge gap is provided between the electrode and the specimen which leads to more material removal rate. In the present study, process variables are peak current (I_{p}), gap voltage (V_{g}), and pulse on time (T_{on}), whereas responses are MRR, EWR, SR, and ROC. The different parameters that were used in this study along with their set values are given in Table 2. The input is given as parameters, namely the V_{g}, I_{p}, and T_{on} are controlled to estimate the effects during EDM. The set of first five experiments was done by changing input parameters for aluminum powder mixed kerosene as a dielectric in EDM. The set of remaining experiments was done again for the same values of V_{g}, I_{p}, and T_{on} with paraffin, distilled water, and transformer oil as a dielectric medium. While conducting experiments, necessary readings are noted down for estimating output parameters (Table 3).

Calculation of material removal rate (MRR) is done as shown below:

$$\mathrm{MRR} = \frac{(M_b - M_a)}{t} \tag{1}$$

where W_a and W_b represent the mass of workpiece post and prior to machining and t represents the time taken for machining.

Electrode wear rate (EWR) is calculated as shown below

$$\mathrm{EWR} = \frac{(T_b - T_a)}{t} \tag{2}$$

where T_b and T_a represent the mass of electrode prior and post to machining and t represents time taken for machining. Surface roughness and radial over the cut (ROC) of the machined surface were measured by using Tally Surf and profile projector, respectively.

Table 3 Various levels of input parameters for various dielectric medium

Parameter/S. No	I_p (Amp)	V_g (Volt)	T_{on} (μs)	Type of dielectric
1	7	45	75	Kerosene
2	8	45	75	
3	8	50	100	
4	9	50	100	
5	9	55	100	
6	7	45	75	Paraffin
7	8	45	75	
8	8	50	100	
9	9	50	100	
10	9	55	100	
11	7	45	75	Distilled water
12	8	45	75	
13	8	50	100	
14	9	50	100	
15	9	55	100	
16	7	45	75	Transformer oil
17	8	45	75	
18	8	50	100	
19	9	50	100	
20	9	55	100	

3 Results and Discussions

Experimental studies have been done to find the influence of different dielectric fluids like kerosene, paraffin, distilled water, and transformer oil during aluminum mixed dielectric EDM on an alloy steel EN-19 workpiece. The investigation has been done by using a fixed value of the concentration of the aluminum powder in the dielectric medium.

Figure 2 indicates the comparison of the experimental results obtained for different response factors, namely TWR, MRR, SR, and ROC on the machined surfaces. From Fig. 2, it has been understood that transformer oil is giving more MRR and EWR than other dielectric fluids, and it might be due to its high dielectric strength, flash point, and resistance to thermal oxidation which leads to quick spark generation resulting in higher material and electrode material removal. At the same time, transformer oil is also giving high SR than other dielectrics as the presence of increased energy leads to the higher removal of atoms irregularly from the surface. This fact also justifies the reasoning of response ROC. For distilled water, low MRR and high EWR are observed when compared to kerosene, paraffin, and transformer oil because of the creation of oxides on the surface of workpiece and tool which does not allow the medium to ionizing quickly. It is showing good

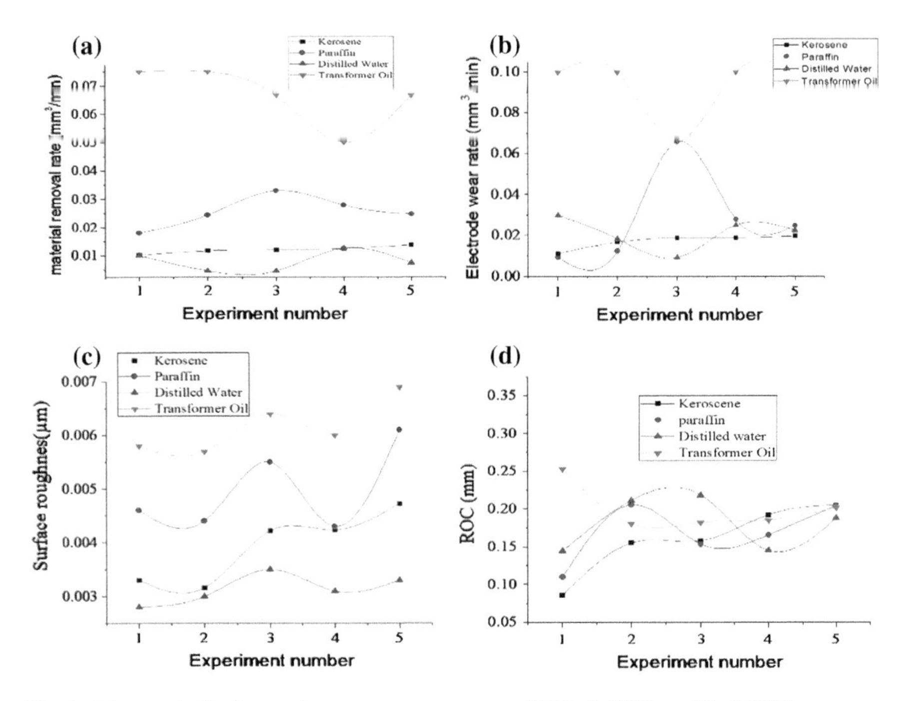

Fig. 2 Diagram indicating various response patterns. **a** MRR. **b** TWR. **c** SR. **d** ROC

surface finish and lower ROC which are indirectly proportional to the material removal rate.

Paraffin is also giving higher MRR, EWR, and very poor surface finish than the kerosene. This might be due to the formation of discontinuous large number of removed particles on the surface and resulting recast layer on it. Figure 3 shows the comparative analysis of mean response values for different dielectric mediums used in aluminum powder mixed EDM.

4 Conclusion

An attempt is made to study the effect of different dielectric fluids during the aluminum powder mixed EDM of EN-19 alloy steel with the brass electrode. It is found that the transformer oil has shown 17.75%, 6.41%, and 70.21% of higher MRR, EWR, and ROC, respectively, when compared with other dielectric mediums. It is also found that the surface roughness is seen to be almost doubled when compared with other dielectric fluids. Distilled water is not advised for powder mixed EDM as it has shown very low of MRR and EWR, even though the values of surface finish and ROC are found to be within the preferable limits. When kerosene is taken as the dielectric fluid, the obtained MRR value is higher than the MRR

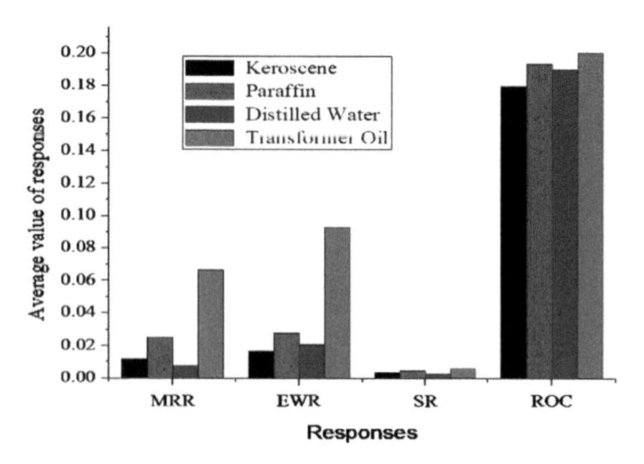

Fig. 3 Diagram of comparative analysis of mean responses for various dielectric fluids

value obtained with paraffin and distilled water but not higher than the value obtained when transformer oil is used as a dielectric fluid. The other responses such as EWR, SR, and ROC have also shown lower values than paraffin. While machining the material with distilled water, intermittent machining was observed, and as when the aluminum powder was added to the distilled water, it led to the formation of oxides which hindered the sparking. Hence, it is concluded that the transformer oil will be used if the production rate is of more important criterion, whereas kerosene is used as a dielectric when the quality of the component is of major concern.

References

1. Singh K, Gianender AKA, Ajit RY(2017) Effect of dielectric fluids used on EDM performance: a review. Int J Emerg Technol Eng Res 5(10)
2. Kumar D, Kumar K, Payal HS, Mer KKS A state-of-the-art review on dielectric fluid in electric discharge machining: uses and its effects. Int Res J Eng Technol (IRJET), 04(2395–0056)
3. Jagtap VL (2016) A review of EDM process for difficult to cut materials. Int Res J Multidiscip Stud 2:2454–8499
4. Zhang Y, Liu Y (2014) Investigation on the influence of the dielectrics on the material removal characteristics of EDM. J Mater Process Technol 214:1052–1061
5. Singh J, Sharma RK Assessing the effects of different dielectrics on environmentally conscious powder-mixed EDM of difficult-to machine material (WC-Co). Front Mech Eng 11(4):374–387
6. Singh AK, Mahajan R, Tiwari A, Kumar D, Ghadai RK (2018) Effect of dielectric on electrical discharge machining: a review. J Mater Sci Eng 377:012184
7. Niamata M, Sarfraz S, Aziza H, Jahanzaiba M, Shehabb E, Ahmada W, Hussaina S (2017) Effect of different dielectrics on material removal rate, electrode wear rate and microstructures in EDM. Procedia CIRP 60:2–7
8. Shehata F, El Mahallawy N, El Hameed MA, El Aal MA (2006) Effect of kerosene and paraffin oil as dielectrics on the electric discharge machining characteristics of Al/SIC metal matrix composite. Prod Eng Des Dev 7–9

9. Shabgard M, Seyedzavvar M, Oliaei SNB (2011) Influence of input parameters on the characteristics of the EDM process. J Mech Eng 57(9):689–696
10. Sadagopan P, MoulIprasanth B (2017) Investigation on the influence of different types of dielectrics electrical discharge machining. Int J Adv Manuf Technol 017–039
11. Kolli M, Kumar A (2015) Effect of dielectric fluid with surfactant and graphite powder on electric discharge machining of titanium alloy using Taguchi method. Eng Sci Technol Int J 18(4):524–535
12. Tzeng YF, Lee CY (2001) Effects of powder characteristics on electrodischarge machining efficiency. Int J Adv Manuf Technol 17:586–592
13. Surekha B, Sree Lakshmi T, Jena H, Samal P (2019) Response surface modeling and application of fuzzy grey relational analysis to optimize the multi-response characteristics of EN-19 machined using powder mixed EDM. Aust J Mech Eng 2204–2253
14. Diwakar NV, Bhagyanathan C, David Rathnaraj J (2014) Analysis of mechanical properties of En19 steel and En41b steel used in diesel engine camshaft. Int J Curr Eng Technol 162–167

Path Planning and Obstacle Avoidance of UAV Using Adaptive Differential Evolution

P. Nagendra Kumar, Prases K. Mohanty and Shubhasri Kundu

Abstract Presently, the path planning and obstacle avoidance of unmanned aerial vehicle (UAV) are attracting research field. A variety of techniques have been introduced by the researchers for obtaining optimal path and avoiding obstacles in the path. This paper presents the implementation of adaptive differential evolution (DE) algorithm for collision avoidance as well as obtaining the optimal path in a static environment whereas former being given more importance. Compared to classical DE algorithm, the proposed adaptive DE allows the UAV to reach the target in an optimal path while avoiding obstacles in a collective manner. The overall performance of the proposed algorithm is verified by simulation results.

Keywords Path planning · DE technique · Collision avoidance · Adaptive DE

1 Introduction

Over the years, humans are developing a variety of technology and machines to ease the work time and increase output. The evolution of robotics has been revolutionizing the world and aerial robotics has taken this to the next level by almost stepping into all the major work fields. Liu et al. [1] presented a review about developments and uses of UAV. University researchers are testing their new ideas using UAVs in robotics, navigation, and flight control theory. These are also used by the military for surveillance and rescue search operations. Path planning is one of the major issues in aerial robotics. Proper path planning helps UAV in reaching the goal in optimum time and also opts for the optimal path. It also helps in avoiding the obstacles thereby ensuring the safety of the UAV during its navigation.

P. Nagendra Kumar · P. K. Mohanty (✉)
National Institute of Technology Arunachal Pradesh, Yupia 791112, India
e-mail: prases@nitap.ac.in

S. Kundu
School of Electrical Engineering, KIIT University, Bhubaneswar 751024, India

BBVL. Deepak et al. (eds.), *Innovative Product Design and Intelligent Manufacturing Systems*, Lecture Notes in Mechanical Engineering,
https://doi.org/10.1007/978-981-15-2696-1_104

Goel et al. [2] proposed glowworm swarm optimization technique to ensure an optimal flight height of UAV for collision avoidance path planning. Mohammadreza et al. [3] presented a paper on Grey wolf optimization to find out the optimal path for UAV in an unknown dynamic environment. Nikolas et al. [4] presented a paper on evolutionary algorithm-based framework for UAV path planning. Ruan et al. [5] proposed a genetic algorithm-based path planning which uses nine different objective values that are structured with three levels of priorities for a realistic UAV model. Wang et al. [6] presented the use of multiobjective ant colony system based on Voronoi for UAV path planning. Li et al. [7] presented a paper on the use of Heuristic A* algorithm for path planning. Foo et al. [8] presented the formulation and solution of 3D path planning problem using PSO technique. Zhan et al. [9] proposed the improved A* algorithm for UAV path planning to obtain a better survival rate with minimum fuel consumption. Peng et al. [10] presented a path planning method with improved flight height using Lyapunov guidance vector field (LGVF) and collision-free path mimicking the fluid flow behavior using interfered fluid dynamical strategy (IFDS). Lu et al. [11] presented the computer vision and vision-based methods for UAV navigation which is useful in GPS denied areas. Kitamura et al. [12] used Octree and an artificial potential field method for 3D path planning. Mac et al. [13] presented a solution to obtain a collision-free path using an improved potential field method for UAV in an indoor environment. Hoy et al. [14] proposed and compared collision-free navigation of an autonomous unmanned helicopter using sliding mode and MPC approaches. Differential Evolution (DE) [15] is a promising evolutionary computational technique, used for multidimensional real-valued functions. DE is highly preferable as it can be used on optimization problems that are not continuous and non-differentiable. Storn et al. [16, 17] proposed Differential Evolution for obtaining global optimization over continuous spaces. Das et al. [18] presented a review on the DE variants and its applications over different types of problems.

It has been concluded from the above literature study that variety of techniques had been implemented for obtaining UAV path planning and as per author's knowledge, DE being a better global optimizer has not been implemented to fullest in UAV navigation planning. DE is adapted in UAV 3D path planning in our study a lot of research is carried out on the DE control parameters. Adaptive DE in our study based on varying control parameters will be presented.

1.1 Mathematical Formulation for UAV Path Planning

Figure 1 represents the basic navigation model that consists of an obstacle, UAV, start point, and goal point. The global path planning of UAV in a three-dimensional environmental model is presented in this study. The UAV is considered as the point mass starting from the start point $S(X_S, Y_S, Z_S)$. A variety of spherical obstacles O_i (where $i = 1, 2, \ldots, n$) are scattered all over the space and agent has to navigate to the target/goal point $G(X_G, Y_G, Z_G)$ avoiding the obstacles.

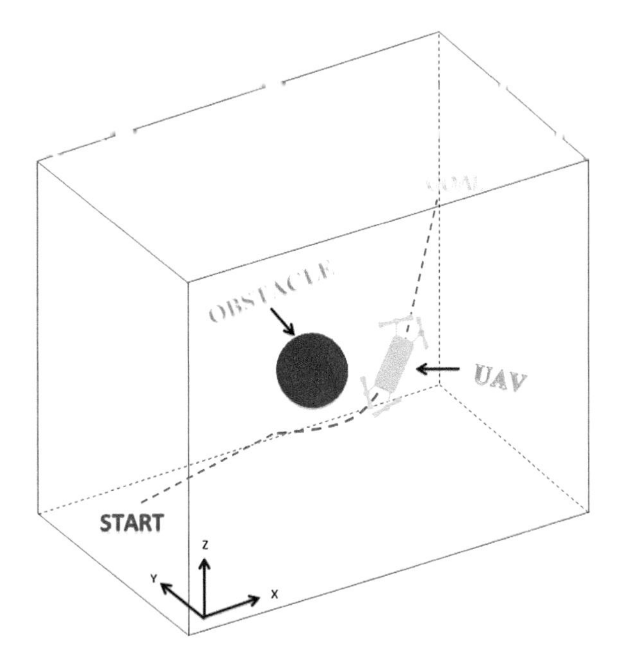

Fig. 1 Basic environment model of UAV navigation

Start point, goal point, and obstacle location are the three essential components for path planning of the agent. Entire search space is uniformly divided into the small search spaces which consist of many possible positions. Every next possible position of the agent after it starts from the start point is selected by considering the distances between start point, goal point, and nearest obstacle center each with the next possible position. These three distances can form three separate objective functions which in turn can sum up to final objective function.

Let the current position of the agent be $C_P(X_i, Y_i, Z_i)$ and the next possible position be $N_P(X_{i+1}, Y_{i+1}, Z_{i+1})$. The obstacle centers and radii for n obstacles are denoted as $O_1(X_{O1}, Y_{O1}, Z_{O1}; R_1)$, $O_2(X_{O2}, Y_{O2}, Z_{O2}; R_2), \ldots, O_n(X_{On}, Y_{On}, Z_{On}; R_n)$. The Euler distance between any two consecutive points say C_P and N_P, is given as

$$d_{CN} = \sqrt{(X_{i+1} - X_i)^2 + (Y_{i+1} - Y_i)^2 + (Z_{i+1} - Z_i)^2} \quad (1)$$

Similarly, the other two distances are given as

$$d_{NG} = \sqrt{(X_G - X_{i+1})^2 + (Y_G - Y_{i+1})^2 + (Z_G - Z_{i+1})^2} \quad (2)$$

$$d_{ON} = \sqrt{(X_{i+1} - X_{Oi})^2 + (Y_{i+1} - Y_{Oi})^2 + (Z_{i+1} - Z_{Oi})^2} \quad (3)$$

The objective function using Eqs. (1), (2), and (3) is given as

$$\text{ObjVal} = d_{CN} + d_{NG} + d_{ON} \tag{4}$$

d_{CN} is the distance between the current position and next possible position of agent.

d_{NG} is the distance between goal and next possible position of agent.

d_{ON} is the distance between the nearest obstacle and the next possible position of agent.

ObjVal is the objective function value obtained using Eq. (4).

For obtaining the optimum values, the minimum distances in Eqs. (1) and (3) with respect to all the possible positions available in the search space are calculated. For ensuring the safety of the agent, certain safe distance between the agent and the obstacle is always maintained. In other words, the boundary over each obstacle in which an agent is not permitted is initially specified. The necessary conditions are programmed in the MATLAB code which omits the selection of possible positions that are within the boundary over the spherical obstacles. Thus, out of all the rest of the available positions in the search space, the position (outside the threshold boundary) is selected which gives the minimum distance between the agent and obstacle in Eq. (2). As minimization is the primary criteria for obtaining optimum values, the objective function Eq. (4) is changed into Eq. (5) as follows

$$\text{ObjVal} = \min(d_{CN}) + \min(d_{NG}) + \min(d_{ON}) \tag{5}$$

1.2 *Differential Evolution Technique*

Differential evolution is first developed by Price and Storm [16, 17]. It optimizes a problem by updating the candidate solution iteration by iteration to yield a better solution. Unlike other techniques, it uses the vector differences between the vectors to form new solutions. It starts with randomly generating a set of solution vectors termed an initial population $\{x_p; p = 1, 2, \ldots, P\}$ with D dimensions in the uniformly distributed search space. Each population member obtains its objective value and then a target vector with minimum objective value is selected from the initial population set. In mutation operation, a random vector (x_{r1}) is chosen and it is added to the product of amplification factor (F) and vector difference of randomly chosen pair of vectors (x_{r2}, x_{r3}). It is followed by recombining the mutant vector attributes with the target vector with the crossover constant (CR) probability in crossover operation resulting in child or trail vector. This trail vector is compared with the target vector for the minimum objective value and the selected vector is used in the further iterations to yield even better results until the termination criteria are met. DE does not allow degradation as it mostly allows the improvement in the objective value of population members. The basic Eqs. (6) and (7) used in DE are as follows

$$x'_p = x_{r1} + F * (x_{r2} - x_{r3}) \quad (6)$$

$$x''_p = \begin{cases} x'_p & \text{if}(\text{rand}(0,1) < \text{CR}) \\ x_p & \text{otherwise} \end{cases} \quad (7)$$

$f(x''_p) < f(x_p)$ then x''_p is selected as the next best position.

$f(x''_p) > f(x_p)$ then x_p remains as position (objective value almost being constant) and DE is run again for next best position of the agent.

The above equations can be considered for the ith iteration out of N iterations.

x_{r1}, x_{r2}, x_{r3} are random vectors selected from the population P.

$\text{rand}(0,1)$ gives the random number in interval $(0,1)$.

x'_p is the mutated vector formed after the mutation process.

x''_p is the trail or child vector produced after the crossover process.

$f(x''_p), f(x_p)$ are the objective values obtained subjected to Eq. (5).

1.2.1 Adaptive DE

The parameter selection plays a crucial role in optimizing the problem. The major control parameters involved in DE are amplification factor (F), crossover constant (CR), and number of population (P). Selection of mutation strategies depending upon the problem is important as well. By fixing the F and CR values, we can obtain the results in classical DE technique. However, varying the F and CR values throughout the iterations resulted in better results compared to classical DE. This new method is highly appreciable and is known as adaptive technique. The population number is based on the problem dimension and the F and CR values adapt themselves during the process thereby adjusting to exploration and exploitation in the search space to find the optimum solution. The variation in F and CR values combined with the proper mutation strategies gives the scope to cover most of the search space and yield the better result. As per early research, in early iterations, high value of F and small CR value provides the diversity in the population. As iterations progress, the controlled value of F and increase in CR value is appreciable in providing promising results with faster convergence. Many researchers [19–22] had suggested a various range of values for F and CR to improve the DE. The regular DE consists of five mutation strategies and these are defined [16, 17] (for ith iteration) by

DE/rand/1/bin: $x'_p = x_{r1} + F * (x_{r2} - x_{r3})$
DE/best/1/bin: $x'_p = x_{\text{best}} + F * (x_{r1} - x_{r2})$
DE/rand-to-best/1/bin: $x'_p = x_p + F * (x_{\text{best}} - x_p) + F * (x_{r1} - x_{r2})$
DE/rand/2/bin: $x'_p = x_{r1} + F * (x_{r2} - x_{r3}) + F * (x_{r4} - x_{r5})$
DE/best/2/bin: $x'_p = x_{\text{best}} + F * (x_{r1} - x_{r2}) + F * (x_{r4} - x_{r5})$

x'_p is the mutated vector; $x_{r1}, x_{r2}, x_{r3}, x_{r4}, x_{r5}$ are the different random vectors from the population; x_{best} is the population member with the best optimum value in ith iteration and x_p is the target vector.

Brest et al. [23] suggested a self-adapting approach for control parameters after testing them on the benchmark functions. This new approach is used in the present study to find the better optimal path positions of the agent. $F_{p,i+1}, \text{CR}_{p,i+1}$ are the new values of the $(i+1)$th iteration and are calculated using Eqs. (8) and (9) whereas $F_l = 0.1$, $F_u = 0.9$. $\text{rand}_k, k = \{1, 2, 3, 4\}$, $\tau_1 = 0.1$, $\tau_2 = 0.1$.

$$F_{p,i+1} = \begin{cases} F_l + \text{rand}_1 * F_u & \text{if}(\text{rand}_2 < \tau_2) \\ F_{p,i} & \text{otherwise} \end{cases} \tag{8}$$

$$\text{CR}_{p,i+1} = \begin{cases} \text{rand}_3 & \text{if}(\text{rand}_4 < \tau_2) \\ \text{CR}_{p,i} & \text{otherwise} \end{cases} \tag{9}$$

1.3 Implementation of Adaptive DE for UAV Path Planning

Adaptive DE is supposed to initialize after detecting the presence of the obstacle. The red-colored points shown in Fig. 2 are the initial population selected in the small defined search space boundary. The member giving the least objective value as per Eq. (5) is chosen for the upgradation using the adaptive DE. The black-colored points are the new population (child/trail vectors) generated using the F and CR values as per mentioned in Eqs. (8) and (9). From Fig. 2 initial population is generated within the defined search space. A target vector is selected from that population and every other population member is compared with the target vector to find out the best position for the agent. Once the best point is chosen from the population it now undergoes mutation and crossover. Adaptive DE uses Eqs. (8) and (9) to define the F and CR values which will be used during mutation and crossover operations. The blackpoints equal in number to the initial population represent the child/trail vector after undergoing mutation and crossover operations. Now each trail vector is compared with the best point chosen initially from the population to upgrade the best position. Either of the child vector or target vector

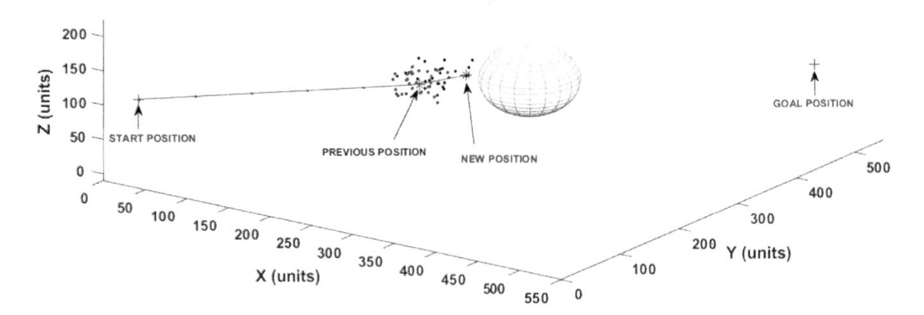

Fig. 2 Generation of population and choosing the best position using adaptive DE

remains as the new position. Now the same process is repeated for a specified number of iterations to further upgradation of the best position. During the start of the mutation process in a new iteration, adaptive DE checks the condition defined in Eqs. (8) and (9) to specify the F and CR values. The adaptive DE decides whether to update the value by upgrading the F and CR values or else use the same value as in the previous iteration. This step balances the exploration and exploitation of search space by avoiding the loss of any best points. After the completion of specified iterations, adaptive DE chooses newly produced child vector as a new position in Fig. 2 and this entire procedure is repeated until adaptive DE meets the termination criteria.

2 Simulation Results and Discussion

The simulation results are compiled through MATLAB in a PC of Windows 10 OS, Intel(R) Core(TM) i3-6006U CPU with 2.00 GHz and 4 GB RAM. Initially, the 3D environment of dimensions 550 * 550 * 235 units is created. Each unit is considered as a centimeter. UAV is considered as a point mass. Locations of start position (25 25 105) and goal position (500 500 105) are indicated by the blue-colored plus (+) sign and they are kept constant for three different conditions which are shown in Figs. 3, 4 and 5: (1) an environment with a single obstacle; (2) environment with three obstacles lined horizontally between the start and goal positions; and (3) environment with obstacles scattered all over the space. The portion of the path framed by using blue-colored points in all the conditions is obtained using classical DE whereas the green-colored points give the result of adaptive DE. The magenta-colored line joining the rest of the points gives the normal path following the shortest distance criteria.

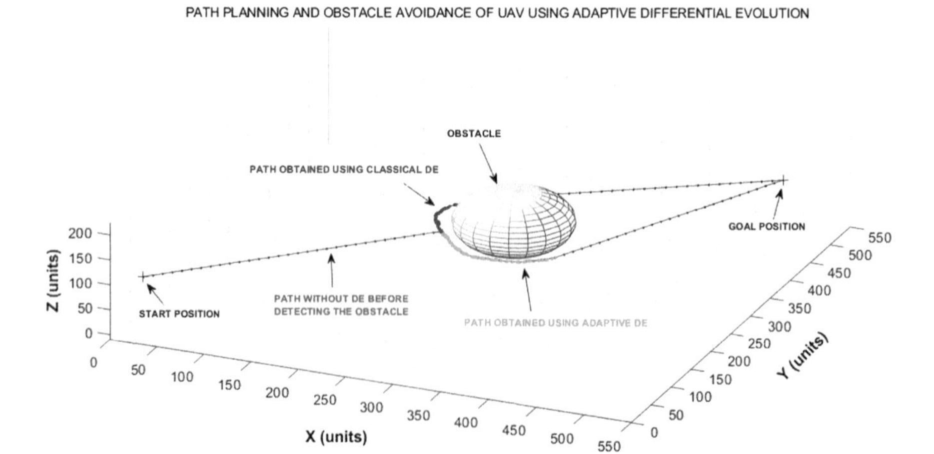

Fig. 3 Path planning of UAV in a single obstacle environment

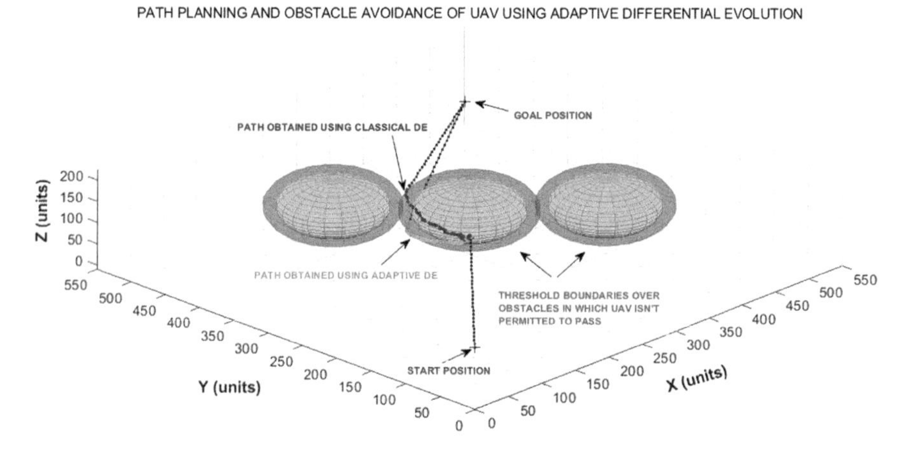

Fig. 4 Path planning of UAV in a wall-shaped obstacle environment

In Fig. 4, UAV is started from the start position following the shortest distance path toward the goal. When it sensed the presence of obstacle located at (200 400 105) whose radius is 60 units, DE algorithm gets activated to avoid the collision and find the next possible positions to frame the path until it disappears from its presence. The light red-colored sphere over the spherical obstacle defines the boundary that UAV is supposed to avoid for ensuring its safety. The 15 units of safe distance all over the obstacle surface area are considered as a boundary in which UAV did not choose its path in both classical and adaptive DE. UAV then finds out the best path using Eqs. (5)–(9) in adaptive DE till it avoids the obstacle boundary. UAV after entering into no threat zone quits the DE using the termination criteria and followed the shortest distance path and reached the goal position.

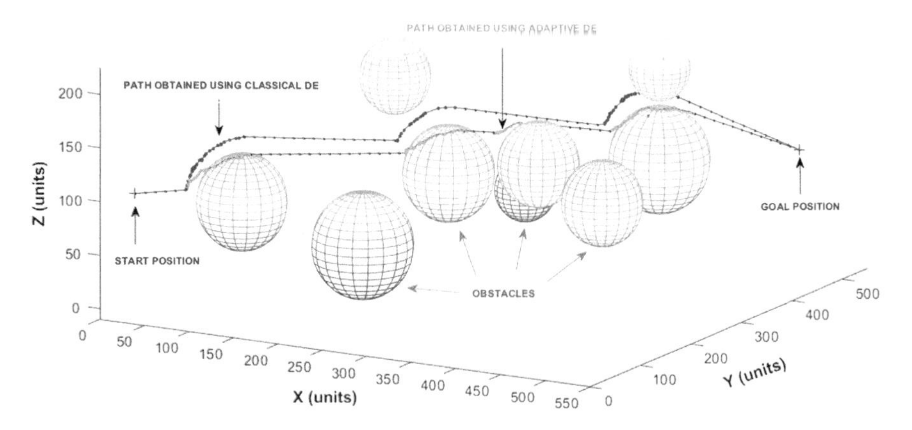

Fig. 5 Path planning of UAV in a scattered obstacle environment

Table 1 Comparison of path length, objective function values in classical and adaptive DE

Figure No.	Path length covered by UAV (units)		Min. value of the objective function at the time of DE termination (units)	
	Classical	Adaptive	Classical	Adaptive
Figure 3	913.5684	828.9704	368.6259	355.7086
Figure 4	806.9358	802.0114	372.0424	368.8777
Figure 5	821.3995	791.9668	206.5818	200.2112

The results obtained from the simulations corresponding to three figures are tabulated in Table 1 and they show that path length covered by UAV using adaptive DE gives the shortest path when compared to classical DE approach. Adaptive DE also selected the better path positions with minimum objective function value when compared to classical DE. It depicts that adaptive DE can search better optimum points beyond the classical DE search points.

3 Conclusions

The following conclusions have been drawn from the proposed study.

- The proposed adaptive DE has been successfully implemented for UAV path planning in static environments.

- The simulation results have shown that adaptive DE has given better performance in terms of path length and objective function values. The proposed algorithm has capable of generating the safe and optimal path in any type of complex environments.
- The adaptive DE has been validated with the classical DE for the same environment and it provided better results than classical DE.
- By incorporating flexibility for UAV path planning in all the three directions, the proposed work can be extended for UAV path planning in dynamic environments as well as multiple UAV path planning.

References

1. Liu P, Chen AY, Huang YN, Han JY, Lai JS, Kang SC, Wu TH, Wen MC, Tsai MH (2014) A review of rotorcraft unmanned aerial vehicle (UAV) developments and applications in civil engineering. Smart Struct Syst 13(6):1065–1094
2. Goel U, Varshney S, Jain A, Maheshwari S, Anupam S (2018) Three dimensional path planning for UAVs in a dynamic environment using Glow-worm swarm optimization. Proc Comput Sci 133:230–239
3. Mohammadreza R, Kumar M (2015) Grey wolf optimization based sense and avoid algorithm for UAV path planning in the uncertain environment using a Bayesian framework. In: International conference on unmanned aircraft systems (ICUAS), June 7–10. Arlington, VA, USA
4. Nikolos IK, Valavanis KP, Tsourveloudis NC, Kostaras AN (2016) Evolutionary algorithm based offline/online path planner for UAV navigation. IEEE Trans Syst Man Cybern B Cybern 33(6):898–912
5. Ruan D, Montero J, Lu J, Martinez L, Dhondt P, Kerre EE (2008) Multiobjective path planner for UAVs based on genetic algorithms. In: Proceedings of international conference on fuzzy logic and intelligent technologies in nuclear science, pp 997–1002. Madrid, Spain, September 21–24
6. Wang ZH, Zhang WG, Shi JP, Han Y (2008) UAV route planning using multiobjective ant colony system. In: Proceedings of IEEE conference on cybernetics and intelligent system, pp 797–800. Chengdu, China, September 21–24
7. Li X, Xie J, Cai MY, Xie M, Wang ZK (2009) Path planning for UAV based on improved heuristic A* algorithm. In: Proceedings of international conference on electronic measurement & instruments, pp 3488–3493. Beijing, August 16–19
8. Foo JL, Knutzon J, Kalivarapu V, Oliver J, Winer E (2009) Path planning of unmanned aerial vehicles using B-Splines and particle swarm optimization. J Aerosp Comput Inf Commun 6 (4):271–290
9. Zhan W, Wang W, Chen N, Chao W (2014) Efficient UAV path planning with multiconstraint in a 3D large battlefield environment. Hindawi Publishing Corporation, Mathematical Problems in Engineering, vol 2014, Article ID 597092, 12 p
10. Peng Y, Honglun W, Zikang S (2015) Real-time path planning of unmanned aerial vehicle for target tracking and obstacle avoidance in a complex dynamic environment. Aerospace science and technology, Elsevier
11. Lu Y, Xue Z, Xia GS, Zhang L (2017) A survey on vision-based UAV navigation. Geo-Spatial Information Science, Taylor & Francis
12. Kitamura Y, Tanaka T, Kishino F, Yachida M (1995) 3-D Path planning in a dynamic environment using an Octree and an artificial potential field. IEEE, New York

13. Mac TT, Copot C, Tran DT, Keyser RD (2016) Heuristic approaches in robot path planning: a survey. Rob Autonom Syst 86:13–28
14. Hoy M, Matveev AS, Garratt M, Savkin AV (2012) Collision-free navigation of an autonomous unmanned helicopter in unknown urban environments: sliding mode and MPC approaches. Robotica 30:537–550
15. https://en.wikipedia.org/wiki/Differential_evolution
16. Price K, Storn R, Lampinen J (2005) Differential evolution: a practical approach to global optimization. Springer, Berlin
17. Storn R, Price KV (1997) Differential evolution: a simple and efficient heuristic for global optimization over continuous spaces. J Global Opt 11(4):341–359
18. Das S, Suganthan PN (2011) Differential evolution: a survey of the state-of-the-art. IEEE Trans Evol Comput 15(1):4–31
19. Gamperle R, Muller SD, Koumoutsakos P (2002) A parameter study for differential evolution. In: Grmela A, Mastorakis NE (eds) Advances in intelligent systems, fuzzy systems, evolutionary computation. WSEAS Press, Interlaken, pp 293–298
20. Ronkkonen J, Kukkonen S, Price KV (2005) Real parameter optimization with differential evolution. In: IEEE Congress on evolutionary computation (CEC), vol 1, Edinburgh, 5 September 2005, pp 506–513. IEEE, New York
21. Kaelo P, Ali MM (2006) A numerical study of some modified differential evolution algorithms. Eur J Oper Res 169:1176–1184
22. Karaboga D, Akay B (2009) A comparative study of artificial bee colony algorithm. Appl Math Comput 214(12):108–132
23. Brest J, Greiner S, Boskovi B et al (2006) Self-adapting control parameters in differential evolution: a comparative study on numerical benchmark problems. IEEE Trans Evol Comput 10(6):646–657

Effect of Crack Severity on a Curved Cantilever Beam Using Differential Quadrature Element Method

Baharul Islam, Prases K. Mohanty and Dayal R. Parhi

Abstract In the present paper, investigation of a crack in a curved cantilever beam is studied. The crack in the beam is modeled using a rotational spring with stiffness defined in terms of depth of the crack. The equation of motion for the cracked cantilever beam is solved using Differential Quadrature Element Method (DQEM). The Finite Element Analysis (FEA) using ANSYS is also carried out for the cracked curved beam. The severity of the crack with its location on cross-section of the beam is studied using DQEM and FEA. Many numerical examples are demonstrated to verify the aforementioned effects. It is concluded that both the approaches are very close to each other for different crack locations and severities.

Keywords Crack · Differential quadrature element method · Curved beam · FEA

Nomenclature

$\vec{\tau}(\phi, t)$	Shear force
$\overrightarrow{F_n}(\phi, t)$	Normal force
$\vec{M}(\phi, t)$	Bending moment of curved beam
t	Time period
A_1	Cross-sectional area of curved beam
I	Moment of inertia of the curved beam
R	Mean radius of the circular arc
γ	Mass per unit volume of curved beam
E	Modulus of elasticity of the curved beam
α	Bending slope
v	Inward radial displacement
x	Tangential displacement
ϕ_t	Total angle between deformed and undeformed neutral axis

B. Islam · P. K. Mohanty (✉)
National Institute of Technology Arunachal Pradesh, Yupia 791112, India
e-mail: prases@nitap.ac.in

D. R. Parhi
National Institute of Technology Rourkela, Rourkela 769008, India

BBVL. Deepak et al. (eds.), *Innovative Product Design and Intelligent Manufacturing Systems*, Lecture Notes in Mechanical Engineering,
https://doi.org/10.1007/978-981-15-2696-1_105

K_o	Rotational spring stiffness
k	Numerical share factor of cross section
G	Modulus of rigidity

1 Introduction

The curved beam is the important basic structural component with a wide range of applications in various fields such as in C clamp, pressing machine, building, crane hook and many more. The durability of design components gets affected by various factors among which Crack is the notable one. Crack decreases the stiffness of the component which leads to its failure. Thus early detection of crack can ensure the safety of the beam.

The radial and tangential displacement of circular arcs due to in-plane vibration are studied by Takahashi [1]. Finite element method is used to solve the exact differential equation of vibrating element by Davis et al. [2]. In their study, considering shear deformation and rotational inertia they found that the thick beam gives sufficiently accurate value of frequency if shear coefficient is used as according to the straight beam theory. Issa et al. [3] derived the dynamic stiffness matrix to calculate the free vibration of curved beam. A new parameter curvature mode was studied by Pandey et al. [4] to find out the damage location in structure. They found that with increase in damage size the curvature mode shape is also increases. Kang et al. [5] used differential quadrature element method to find the nature frequency of arches having rectangular and circular cross section without any crack. In double hinged beam, Cerri et al. [6] replaced the crack by a torsional spring and divided the beam into two sections. Oz et al. [7] studied Euler-Bernoulli beam considering only bending and extension effects to calculate the natural frequency with crack at different locations and depths by using FEM. Out-of -plane vibration analysis of nonprismatic curved beam was studied by Chang-New [8] considering shear deformation effect. The DQEM method was used to solve the governing equation. Byoung et al. [9] studied horizontal curved beam. They had studied three type of beam, circular, parabolic and sinusoidal. Ebrahim et al. [10] had developed a higher order theory to study free vibration response of a debonded curved sandwich beam. They found that the curvature angle and depth of crack had a great effect in vibration response. Das et al. [11] studied the composite beam having a transverse open crack. They found that natural frequency and mode shape decrease with increasing depth of the beam. Wang et al. [12] studied frequency analysis of non-uniform beamlike structure with three dimensional sub-parametric quadrature element method. Mazahari et al. [13] in their study had replaced crack by rotational spring and FEM method was used to solve the governing equation of concrete beam. Prawin et al. [14] studied linear and non-linear behavior of cracked structure with multi-level singular spectrum analysis.

FEM is used by most of the researchers till date as per the literature study and as per author's knowledge DQEM can give us precise result compared to FEM. Curved beam made up of Aluminium alloy is widely used in aerospace, construction, electrical engineering, automobiles and marine etc. Crack detection of aluminum alloy curved beam using DQEM is presented in this study. In order to verify the accuracy of proposed method, a comparative study is presented for crack and uncrack structure. It is found that they are in good agreement.

2 Theoretical Analysis of Cracked Curved Cantilever Beam

Let us consider a curved beam having radius '*R*', depth '*d*' and width '*w*' as shown in Fig. 1. A small element of length '*ds*' which undergoes small in-plane vibration. Considering the shear force, normal force and bending moment acting on the beam, we get three equations (2.1), (2.2) and (2.3) from the equilibrium conditions [3].

$$\frac{EA_1}{R}\left(\frac{\partial^2 x}{\partial \phi^2}-\frac{\partial v}{d\phi}\right)-\frac{kA_1G}{R}\left(\frac{\partial v}{d\phi}+x-R\alpha\right)=\gamma A_1 R\frac{\partial^2 x}{\partial t^2} \tag{2.1}$$

$$\frac{kA_1G}{R}\left(\frac{\partial^2 v}{\partial \phi^2}+\frac{\partial x}{\partial \phi}-R\frac{\partial \alpha}{\partial \phi}\right)+\frac{EA_1}{R}\left(\frac{\partial x}{\partial \phi}-v\right)=\gamma A_1 R\frac{\partial^2 v}{\partial t^2} \tag{2.2}$$

$$\frac{EI}{R}\frac{\partial^2 \alpha}{\partial \phi^2}+kA_1G\left(\frac{\partial v}{d\phi}+x-R\alpha\right)=\gamma IR\frac{\partial^2 \alpha}{\partial t^2} \tag{2.3}$$

For harmonic excitation at radial frequency 'ω_n' the displacement and bending slope can be expressed as in Eqs. (2.4), (2.5) and (2.6)

$$x(\phi,t)=X(\phi)e^{i\omega_n t}, \tag{2.4}$$

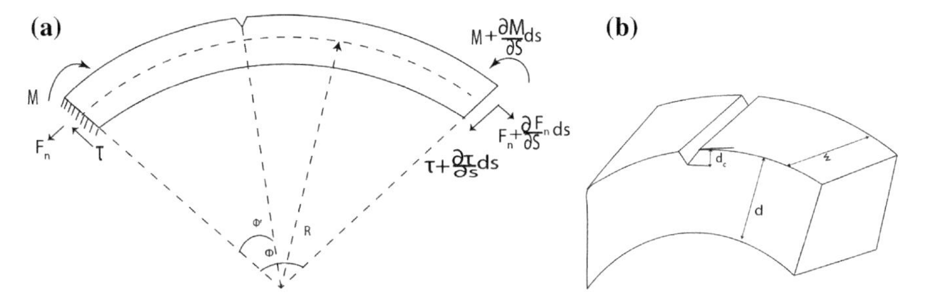

Fig. 1 **a** crack cantilever curved beam, **b** crack section of beam

$$v(\phi,t) = V(\phi)e^{i\omega_n t}, \tag{2.5}$$

$$\alpha(\phi,t) = \alpha(\phi)e^{i\omega_n t} \tag{2.6}$$

Now from Eq. (2.1)

$$\begin{aligned}
&\frac{EA_1}{R}\left[\frac{\partial^2 X(\phi)e^{i\omega_n t}}{\partial\phi^2} - \frac{\partial V(\phi)e^{i\omega_n t}}{d\phi}\right] - kA_1G\left[\frac{\partial V(\phi)e^{i\omega_n t}}{d\phi} + x - R\alpha\right] = \gamma A_1 R\frac{\partial^2 X(\phi)e^{i\omega_n t}}{\partial t^2} \\
&\Rightarrow \frac{EA_1}{R}\left[\frac{\partial^2 X(\phi)}{\partial\phi^2} - \frac{\partial V(\phi)}{d\phi}\right] - kA_1G\left[\frac{\partial V(\phi)}{d\phi} + x - R\alpha\right] = -\omega_n^2\gamma A_1 R\frac{\partial^2 X(\phi)}{\partial\phi^2}
\end{aligned} \tag{2.7}$$

From Eq. (2.2)

$$\begin{aligned}
&\frac{kA_1G}{R}\left[\frac{\partial^2 V(\phi)e^{i\omega_n t}}{\partial\phi^2} + \frac{\partial X(\phi)e^{i\omega_n t}}{\partial\phi} - R\frac{\partial X(\phi)e^{i\omega_n t}}{\partial\phi}\right] + \frac{EA_1}{R}\left[\frac{\partial X(\phi)e^{i\omega_n t}}{\partial\phi} - V(\phi)e^{i\omega_n t}\right] = \gamma A_1 R\frac{\partial^2 V(\phi)e^{i\omega_n t}}{\partial t^2} \\
&\Rightarrow \frac{kA_1G}{R}\left[\frac{\partial^2 V(\phi)}{\partial\phi^2} + \frac{\partial X(\phi)}{\partial\phi} - R\frac{\partial X(\phi)}{\partial\phi}\right] + \frac{EA_1}{R}\left[\frac{\partial X(\phi)}{\partial\phi} - V(\phi)\right] = -\omega_n^2\gamma A_1 R\frac{\partial^2 V(\phi)}{\partial\phi^2}
\end{aligned} \tag{2.8}$$

From Eq. (2.3)

$$\begin{aligned}
&\frac{EI}{R}\frac{\partial^2\alpha(\phi)e^{i\omega_n t}}{\partial\phi^2} + kA_1G\left(\frac{\partial V(\phi)e^{i\omega_n t}}{d\phi} + x - R\alpha(\phi)e^{i\omega_n t}\right) = \gamma IR\frac{\partial^2\alpha(\phi)e^{i\omega_n t}}{\partial t^2} \\
&\Rightarrow \frac{EI}{R}\frac{\partial^2\alpha(\phi)}{\partial\phi^2} + kA_1G\left(\frac{\partial V(\phi)}{d\phi} + x - R\alpha(\phi)\right) = -\omega_n^2\gamma IR\frac{\partial^2\alpha(\phi)}{\partial\phi^2}
\end{aligned} \tag{2.9}$$

Using DQEM Method:

If $V_{rj}^m(\phi)$, $X_{rj}^m(\phi)$ and $\alpha_{rj}^m(\phi)$ be the function of order 'm' then from DQEM method [8], we can write Eqs. (2.10), (2.11) and (2.12) given below

$$V_{rj}^m(\tilde{\phi}) = D_{ik}^m V_{rj}(\tilde{\phi}), \tag{2.10}$$

$$X_{rj}^m(\tilde{\phi}) = D_{ik}^m X_{rj}(\tilde{\phi}), \tag{2.11}$$

$$\alpha_{rj}^m(\tilde{\phi}) = D_{ik}^m \alpha_{rj}(\tilde{\phi}) \tag{2.12}$$

Where $\tilde{\phi}$ is dimensionless parameter

$$\tilde{\phi} = \frac{\phi}{l_\theta},$$

D_{ik}^{m} = weighting coefficient

$V_{rj}(\tilde{\phi})$, $X_{rj}(\tilde{\phi})$ and $\alpha_{rj}(\tilde{\phi})$ are the displacement vector.
By Applying DQEM method in Eqs. (2.7), (2.8) and (2.9) we get

$$\frac{kG}{R}\left[\frac{1}{l_{\phi}^{2}}D_{rj}^{2}\{V_{rj}\}\right]+\frac{kGl_{\phi}}{R^{2}\gamma}D_{ik}^{1}\{X_{rj}\}-\frac{kGl_{\phi}}{R\gamma}D_{ik}^{1}\{\alpha_{rj}\}+\frac{El_{\phi}}{R^{2}\gamma}D_{ik}^{1}\{X_{rj}\}-\frac{El_{\phi}}{R^{2}\gamma}\{V_{rj}\} = -\omega^{2}D_{rj}^{2}\{V_{rj}\} \tag{2.13}$$

$$\frac{E}{R^{2}\gamma}D_{ik}^{2}\{X_{rj}\}-\frac{El_{\phi}}{R^{2}\gamma}D_{ik}^{1}\{V_{rj}\}-\frac{kGl_{\phi}}{R^{2}\gamma}D_{ik}^{1}\{V_{rj}\}-\frac{kGl_{\phi}^{2}}{R^{2}\gamma}\{X_{rj}\}+\frac{kGl_{\phi}^{2}}{R\gamma}\{\alpha_{rj}\} = -\omega^{2}D_{rj}^{2}\{X_{rj}\} \tag{2.14}$$

$$M^{r}D_{ik}^{1}\{\alpha_{rj}\}+\frac{kA_{1}Gl_{\phi}}{R\gamma I}D_{ik}^{1}\{V_{rj}\}+\frac{kA_{1}Gl_{\phi}}{R\gamma I}\{X_{rj}\}-\frac{kA_{1}Gl_{\phi}}{R\gamma I}\{\alpha_{rj}\}=-\omega^{2}D_{rj}^{2}\{\alpha_{rj}\} \tag{2.15}$$

Let the crack occurs at an angular location 'ϕ^{r}'. Then from the two adjacent elements 'r' and '$r + 1$', we get

$$V_{N^{r}}^{r}=V_{1}^{r+1}, X_{N^{r}}^{r}=X_{1}^{r+1}, \alpha_{N^{r}}^{r}=\alpha_{1}^{r+1}$$

The bending moment at the crack location

$$M^{r}=K_{o}\left(\alpha_{N^{r}}^{r}-\alpha_{1}^{r+1}\right)$$

Where 'K_o' is the rotational spring stiffness at the crack location

$$K_{o}=\frac{(d-d_{c})^{3}Ewd^{2}}{6[(d-d_{c})^{3}-d^{3}]}$$

d The thickness of the beam
d_c Depth of crack
w Width of beam.

The bending moment, normal force and shear force continuity equation at the inter-element boundary of two adjacent element r and $r + 1$ are given in Eqs. (2.16), (2.17) and (2.18)

$$\frac{E^{r}A_{1}^{r}}{R^{r}}\left[\frac{1}{l_{\phi}}D_{ik}^{1}\{X_{rj}\}-\{V_{rj}\}\right]=\frac{E^{r+1}A_{1}^{r+1}}{R^{r+1}}\left[\frac{1}{l_{\phi}}D_{ik}^{1}\{X_{(r+1)1}-\{V_{(r+1)1}\}\right] \tag{2.16}$$

$$\frac{k^r A_1^r G^r}{R^r}\left[\frac{1}{l_\phi} D_{ik}^1\{V_{rj}\} + \{X_{rj}\} - R^r\{\alpha_{rj}\}\right] = \frac{k^{r+1} A_1^{r+1} G^{r+1}}{R^{r+1}}\left[\frac{1}{l_\phi} D_{ik}^1\{V_{(r+1)1}\} + \{X_{(r+1)1}\} - R^{r+1}\{\alpha_{(r+1)1}\}\right] \tag{2.17}$$

$$\frac{E^r I^r}{R^r}\left[\frac{1}{l_\phi} D_{ik}^1\{\alpha_{rj}\}\right] = \frac{E^{r+1} I^{r+1}}{R^{r+1}}\left[\frac{1}{l_\phi} D_{ik}^1\{\alpha_{(r+1)1}\}\right] \tag{2.18}$$

The boundary condition for cantilever curved beam

$$V_1^1 = 0, X_1^1 = 0, \alpha_1^1 = 0$$

By applying all the boundary conditions to above Eqs. (2.13)–(2.18) we get

$$[A_r][B_r] = -\omega_n^2[B_r] \tag{2.19}$$

Where

The element of $[A_r]$ consist of the coefficient of displacement vectors

$$[B_r] = [\{V_{rj}\} \quad \{X_{rj}\} \quad \{\alpha_{rj}\}]^T, r, j = 1, 2, 3, \ldots, N,$$

By solving Eq. (2.19) the natural frequency of cracked curved structure 'ω_n' is yielded.

3 Finite Element Modeling of Cracked Curved Cantilever Beam

For validation of the DQEM results, FEA simulations are carried out using ANSYS 14.5 for the cracked and uncracked beam. The natural frequency and mode shapes at different locations with varying crack severity for the beam are analyzed. The details of the specimen of the beam are given below

Opening angle of curved beam $\phi = 120°$
Width of the beam w = 14 mm
Depth of beam d = 14 mm
Density of the specimen γ = 2770 kg/m^3
Modulus of elasticity E = 71 Gpa.
Poisson's ratio = 0.33
Relative crack location (RCL) varies from 0.04 to 0.95
Relative depth crack severity (RCS) varies from 0.03 to 0.28

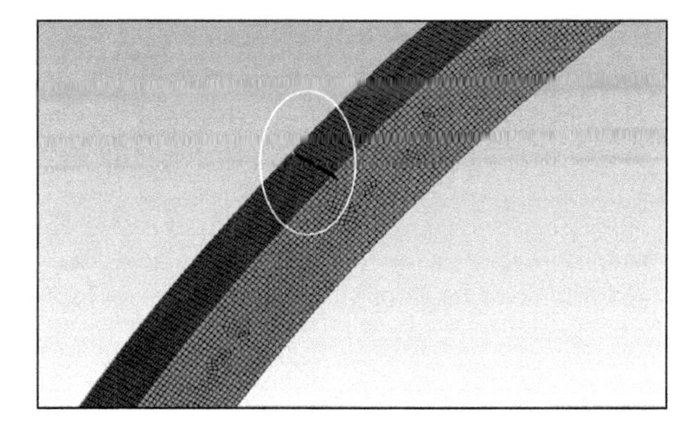

Fig. 2 Magnified view of crack

The finite element mesh modeling with a magnified view of the cracked zone is shown in Fig. 2.

Figure 3 represents the mode shape of vibration of a cracked cantilever beam using ANSYS at RCL = 0.54 with RCS = 0.14.

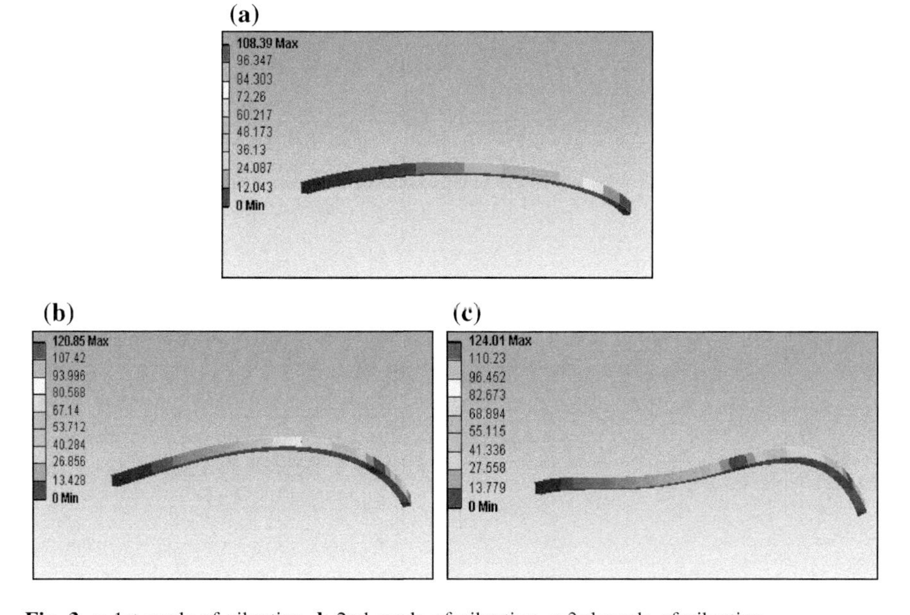

Fig. 3 **a** 1st mode of vibration, **b** 2nd mode of vibration, **c** 3rd mode of vibration

4 Results and Discussion

The natural frequencies and mode shapes of the vibrating beam in the presence of V shape crack are studied. It is noticed that if the relative crack location is increasing at constant severity there is increasing in natural frequency for three modes of vibration as shown in Figs. 4, 5 and 6. It is also distinguished that with an increase in crack severity at the constant location the natural frequency decreases for 1st, 2nd and 3rd mode shape of vibration as represent in Figs. 7 and 8. The cracked beam

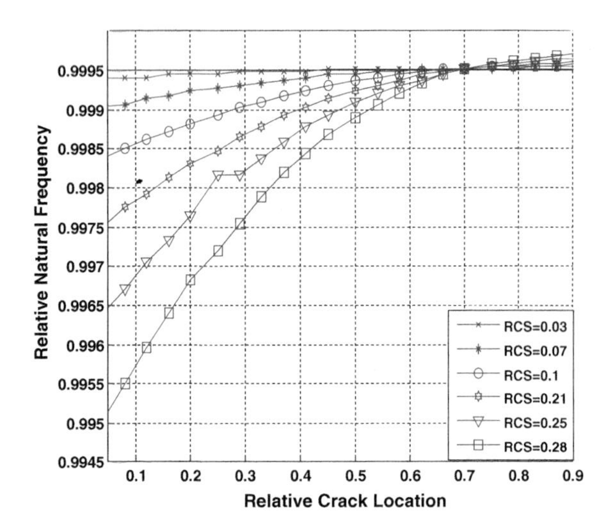

Fig. 4 1st mode shape of relative natural frequency versus relative crack location

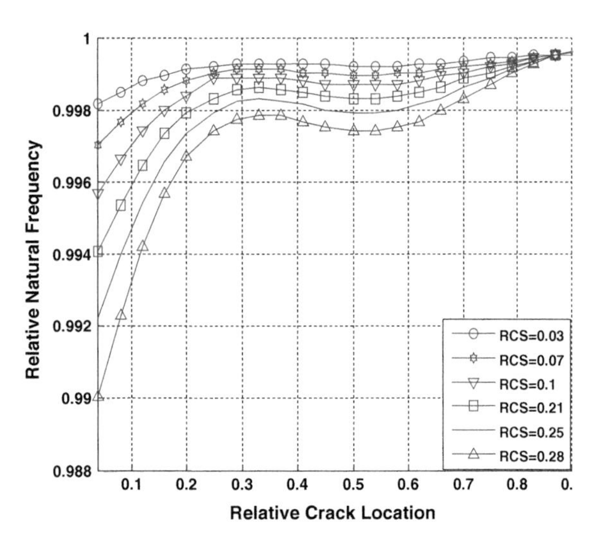

Fig. 5 2nd mode shape of relative natural frequency versus relative crack location

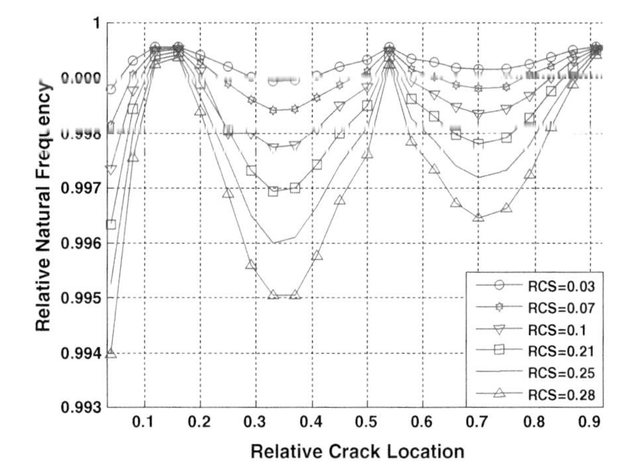

Fig. 6 3rd mode shape of relative natural frequency versus relative crack location

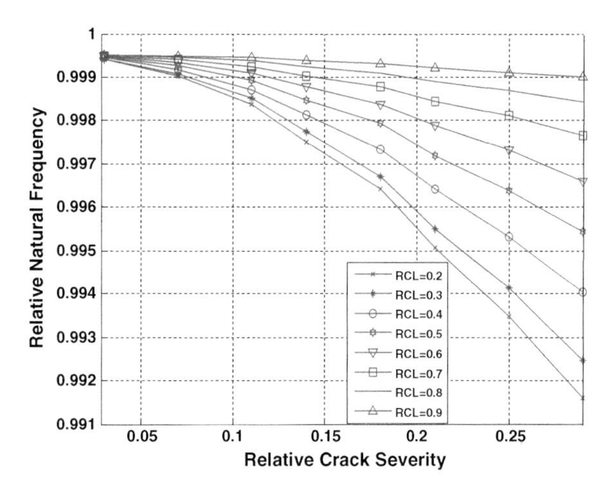

Fig. 7 The variation of relative natural frequencies with respect to relative crack severity at a fixed crack location

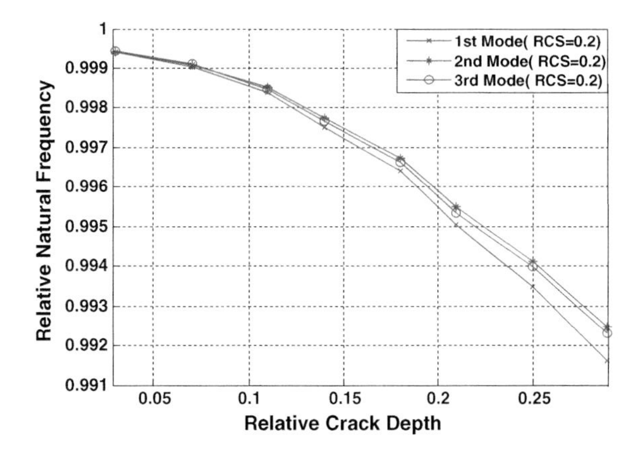

Fig. 8 The variation of relative frequencies of 1st, 2nd and 3rd mode shape with respect relative crack severity

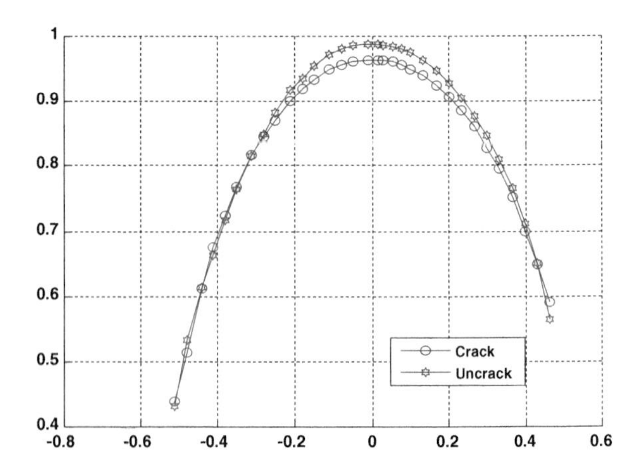

Fig. 9 1st mode shape of vibration of cracked beam

deformation with respect to natural frequencies at the relative crack location of 0.54 and relative crack depth of 0.14 for 1st, 2nd, and 3rd mode shape is plotted in Figs. 9, 10 and 11. It is found that an increase in the deformation occurs for all the three mode shapes at the crack position of the structure.

The first three modes of relative natural frequencies for different crack positions and crack severity calculated from the FEA and DQEM are presented in Table 1. The obtained results showed the average error of 2.5% from in Table 1.

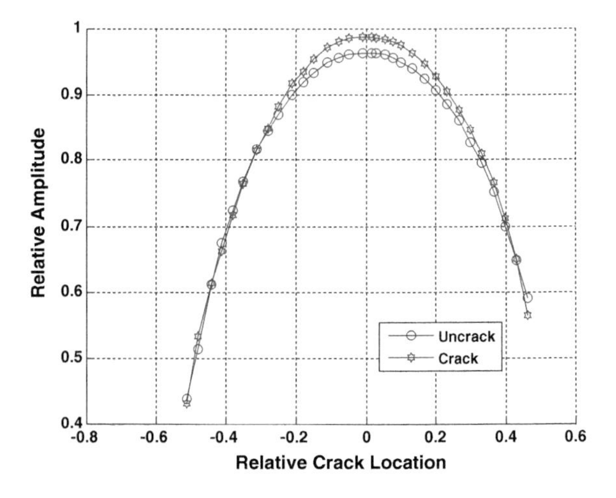

Fig. 10 2nd mode shape of vibration of cracked beam

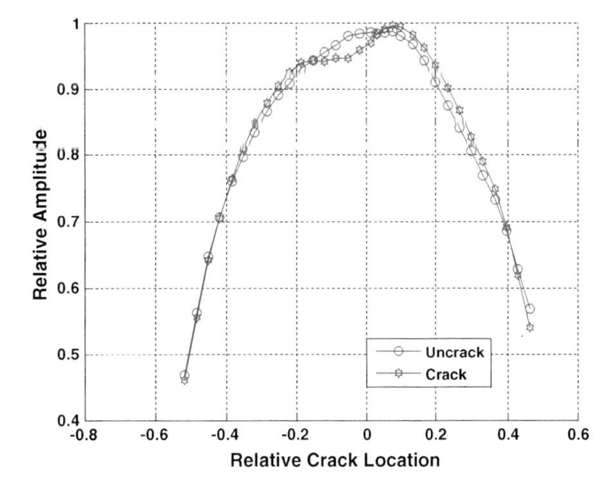

Fig. 11 3rd mode shape of vibration of cracked beam

Table 1 Comparision of natural frequencies (FEA and DQEM) of the crack curved beam at a various location with different crack severity

RCL	RCS	Relative natural frequency					
		FEA			DQEM		
0.33		1st mode	2nd mode	3rd mode	1st mode	2nd mode	3rd mode
	0.03	0.99948	0.9995	0.9995	0.9994	0.9991	0.9905
	0.1	0.9910	0.9992	0.9989	0.9942	0.9970	0.9965
	0.2	0.9978	0.9986	0.9969	0.9978	0.9986	0.9960
	0.28	0.9965	0.9978	0.9949	0.9965	0.9950	0.9901
0.41	0.03	0.9994	0.9995	0.9995	0.9964	0.9945	0.9905
	0.1	0.9992	0.9992	0.9990	0.9912	0.9912	0.9910
	0.2	0.9984	0.9984	0.9974	0.9994	0.9984	0.9994
	0.28	0.9976	0.9976	0.9957	0.9906	0.9926	0.9978
0.5	0.03	0.9995	0.9995	0.9995	0.9935	0.9955	0.9975
	0.1	0.9993	0.9991	0.9993	0.9993	0.9971	0.9943
	0.2	0.9988	0.9983	0.9985	0.9998	0.9973	0.9965
	0.28	0.9985	0.9974	0.9976	0.9925	0.9945	0.9966
0.62	0.03	0.9995	0.9995	0.9995	0.9905	0.9925	0.9935
	0.1	0.9994	0.9992	0.9992	0.9994	0.9972	0.9932
	0.2	0.9993	0.9974	0.9983	0.9903	0.9996	0.9933
	0.28	0.9992	0.9976	0.9973	0.9942	0.9956	0.9973
0.7	0.03	0.9995	0.9995	0.9995	0.9985	0.9965	0.9925
	0.1	0.9995	0.9993	0.9991	0.9910	0.9983	0.9971
	0.2	0.9995	0.9988	0.9978	0.9905	0.9928	0.9978
	0.28	0.9995	0.9983	0.9964	0.9905	0.9973	0.9904

5 Conclusion

The investigation of a cracked curved cantilever beam having a uniform cross-section has been studied in the present paper. The following observations have been summarized:

- The initiation of crack brings noticeable changes in the natural frequency and mode shape of the structure.
- When the crack severity increases the natural frequency of cracked beam decreases at a constant crack position.
- When the crack shifts position toward the extreme end of cantilever curved beam the natural frequency increases for a particular crack depth.
- The position of the crack can be forecasted from the modification of the fundamental modes between the crack and without crack beam.
- In the future, the study will be extended to analyse the shape of crack in a composite beam having a complex shape.

References

1. Takahashi S (1963) Vibration of circular arc bar in its plane. Jpn Soc Mech Eng (JSME) 6(9)
2. Davis R, Henshel RD, Warburton GB (1972) Constant curvature beam finite element for in-plane vibration. J Sound Vib 25(4):561–576
3. Issa MS, Wang TM, Hsiao BT (1987) Extensional vibration of the continuous circular curved beam with rotational inertia and shear deformation. J Sound Vib 144(2):297–308
4. Pandey AK, Biswas M, Samman MM (1991) Damage detection from changes in curvature mode shapes. J Sound Vib 145(2):321–332
5. Kang K, Bert CW, Striz AG (1995) Vibration analysis of shear deformable circular arches by the differential quadrature method. J Sound Vib 181(2):353–360
6. Cerri MN, Rutta GC (2004) Detection of localized damage in-plane circular arches by frequency data. J Sound Vib 270:39–59
7. Oz HR, Das MT (2006) In-plane vibrations of circular curved beams with a transverse open crack. Math Comput Appl 11(1):1–10
8. Chang-New C (2008) DQEM analysis of out-of-plane vibration of nonprimitive curved beam structures considering the effect of shear deformation. Adv Eng Software 39:466–672
9. Byoung KL, Kwang P, Tae EL, Hee MY (2014) Free vibration of horizontally curved beams with constant volume. KSCE J Civ Eng 18(1):199–212
10. Ebrahim S, Mojtaba S, Abdolreza O (2016) Free vibration analysis of a debonded curved sandwich beam. Euro J Mech A/Solid 57:71–84
11. Das MT, Ayse Y (2018) Experimental modal analysis of curved composite beam with transverse open crack. J Sound Vib 436:155–164
12. Wang X, Zhangxian Y (2018) Three-dimensional vibration analysis of curved and twisted beams with irregular shapes of the cross-section by sub-parametric quadrature element method. Comput Math Appl 76:1486–1499
13. Matahari H, Hossein R, Ali K (2018) Static and dynamic analysis of crack curved beams using experimental study and finite element analysis. Periodica Polytech Civil Eng 62(2):337–345
14. Prawin J, Lakshmi K, Rao AM (2019) A novel vibration based breathing crack localization technique using a single sensor measurement. Mech Syst Sign Process 122:117–138